Croser
314-7032

BIOLOGY
UNDERSTANDING LIFE

BIO
UNDERST

THIRD EDITION

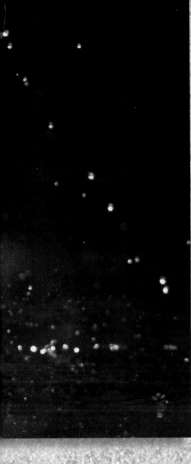

LOGY
ANDING LIFE

Sandra Alters, Ph.D.

JONES AND BARTLETT PUBLISHERS
Sudbury, Massachusetts

About the Author

Sandra Gottfried Alters brings the perspective of over 25 years of teaching and 20 years of textbook writing to *Biology: Understanding Life.* As a former Associate Professor of Biology and Education at the University of Missouri–St. Louis, and Associate Professor of Biology at Salem State College (MA), Dr. Alters's academic responsibilities included teaching the general biology course to non-majors in both large lecture halls and small classroom settings. As a science educator and biologist, she incorporates current learning theory into her teaching materials, blending biological content and educational process.

Dr. Alters received a B.A. in biology from St. Joseph College, an M.A. in biology from Wesleyan University, and a Ph.D. in science education from the University of Connecticut.

Dr. Alters has authored three textbooks in addition to *Biology: Understanding Life,* and has written over 60 chapters and features in books. She also developed the instructional design for a variety of science-related software products for students. Currently, Dr. Alters resides in Montreal, Canada, where she is concentrating on writing projects in biology, evolution, and science education.

On the Cover The basilisk lizard lives in Central America and is known for its ability to walk on water. To keep from sinking, the lizard applies a forceful plunge with each step; this motion creates little pockets of air that support the lizard for the split second before its next step. By dashing across the water, the basilisk lizard can escape from hungry predators on land.
© Stephen Dalton/Photo Researchers, Inc.

Library of Congress Cataloging-in-Publication Data
Alters, Sandra.
  Biology : understanding life / Sandra Alters. — 3rd ed.
    p. cm.
  ISBN 0-7637-0837-2 (hardcover)
  1. Biology.   2. Human biology.   I. Title.
QH302.2.A45   1999
570--dc21   99-29957
                                                      CIP

World Headquarters
Jones and Bartlett Publishers
40 Tall Pine Drive
Sudbury, MA 01776
978-443-5000
info@jbpub.com
www.jbpub.com

Jones and Bartlett Publishers Canada
2406 Nikanna Road
Mississauga, Ontario
CANADA L5C 2W6

Jones and Bartlett Publishers International
Barb House, Barb Mews
London W6 7PA
UK

Chief Executive Officer Clayton Jones
Chief Operating Officer Don Jones, Jr.
President Tom Walker
V.P., Sales and Marketing Tom Manning
V.P., Managing Editor Judith H. Hauck
V.P., College Editorial Director Brian L. McKean
V.P., Director of Interactive Technology Mike Campbell
Director of Design and Production Anne Spencer
Director of Manufacturing and Inventory Therese Bräuer
Special Projects Editor Mary Hill
Senior Marketing Manager Jennifer Jacobson
Interactive Technology Product Manager W. Scott Smith
Assistant Production Editor Jennifer Angel
Editorial/Production Assistant Anne Trafton
Text, Cover & Walkthrough Design Anne Spencer
Web Site Design Anne Spencer, Stephanie Torta, Mike DeFronzo
Art Development Kathy Naylor, Denise Thomas, Mary Hill, Jennifer Angel
Photo Research Sharon Donahue, Jennifer Angel
Text Artwork Graphic World Publishing Services, Imagineering, J/B Woolsey Associates, Precision Graphics, Rolin Graphics
Composition Graphic World Publishing Services
Production Service Kathy Smith
Cover Manufacture John Pow Company
Book Manufacture Courier Kendallville

Printed in the United States

03 02 01 00   9 8 7 6 5 4 3 2

Brief Table of Contents

Table of Contents

Part One: An Introduction

Part Two: Cells and How They Transform Energy

Part Three: Systems

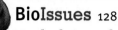

Part Four: Genetics, Reproduction, and Development

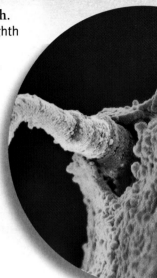
Part Five: How Species Change over Time

Part Six: An Overview of the Five Kingdoms

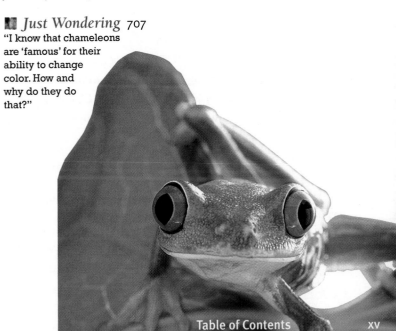

Part Seven: Interactions among Organisms and with the Environment

Biology: Understanding Life has been written and revised with one objective in mind: helping students understand and learn about biology and how it relates to their everyday lives. From discussions with my colleagues and from survey data collected for this revision, I know that you likely share this goal as an instructor of an introductory course intended for a general audience of biology non-majors.

This goal may sound simple to accomplish, yet together we face difficult challenges. Many of our students are science-shy and academically underprepared. They are also a diverse group, with a mix of learning styles and various motivations for learning.

Students' diverse needs are not the only challenge we face. Introductory courses are often taught in large lecture-hall settings to hundreds of students at once. Others of us teach both large and small classes, and must adjust our methods to best serve the education of students in both settings. New and exciting technologies are available to enhance our teaching. Classroom preparation then expands from preparing lecture notes to preparing materials that integrate various computer-based tools as well as other audiovisual technologies.

For over 25 years I have been teaching biology to a variety of students in a variety of settings. Over those years, non-majors biology has become both an increasing concern and interest of mine. In 1993, together with colleagues from the National Association of Biology Teachers (NABT), I studied issues and developed recommendations regarding college biology teaching.*

Our committee realized that the traditional survey course needed to change. The many reports we examined suggested that introductory science courses should focus on the strengths and limitations of scientific processes and should help students make connections among science, technology, and their everyday lives. The literature also emphasized that students should share ideas with one another, become active learners, and develop critical thinking skills.

My most recent teaching experiences at both the University of Missouri–St. Louis and Salem State College (MA) showed me that those foci have become even more important in today's teaching. My colleagues and I tried to move away from content-driven courses with memorization as their theme. We struggled for insight into approaches that would help us take a new, workable direction with our students. So many of us are looking for that blend of content and process not only to help students think, but to give them something to think about.

In this revision of *Biology: Understanding Life*, I identified six goals designed to meet these challenges. And I developed multiple pedagogical tools in support of these goals. Being a science educator as well as a biologist, the pedagogical tools of the textbook and supplements utilize many of the same techniques that I teach to prospective teachers of science.

* Gottfried, et al. (1993). College biology teaching: A literature review, recommendations, and a research agenda. *The American Biology Teacher, 55*, pp. 340-348.

Goal # 1: Help students understand biological concepts.

Biology: Understanding Life helps students understand biological concepts through the use of pedagogical tools based on current learning theory. Over 25% of our entering freshmen, who often make up the bulk of our introductory students, are concrete learners. Nearly 50% are in transition from being concrete learners to those capable of formal thought. Concrete learners and those in transition often have difficulty with abstract thinking and need the aid of concrete objects to understand concepts. Additionally, concrete learners grasp content better when it is presented in manageable portions.

Features that support this goal

- **Numbered, self-contained sections.** Chapters are divided into a series of one- and two-page conceptual "chunks." This format is easier for students to grasp. The illustrations and photographs within each section are placed near the associated text material so that the student easily can use the words, diagrams, and photographs together to learn new concepts. Although self-contained, each section blends with those that come before and after, to produce an integrated whole rather than unrelated pieces.

- **Organizing statements.** These statements facilitate the integration of the numbered sections of the textbook. At the top of each two-page spread, these organizing statements are depicted as labels on file folders, which contain all sections of the chapter pertaining to each statement.

- **Concept summaries and chapter summaries.** These related features help students identify key ideas. Research results tell us that students learn how to identify the key ideas of stories in elementary school, but that they often have a difficult time identifying key ideas in expository text (such as science textbooks) in their later years. Therefore, we have provided concept summaries at the end of each numbered section and chapter summaries at the end of each chapter to help students with this task. The chapter summaries follow the organization of the chapter exactly so that students can always identify the body of text from which the key idea was derived.

- **Analogies.** The effective use of analogies helps students construct meaning built on their own knowledge and past experiences. Analogies are used frequently in both the text and the *Study Guide* following the Teaching-with-Analogies Model whenever possible. This model was developed as a part of the National Reading Research Project of the University of Georgia and University of Maryland. A good example of the implementation of this model is on p. 360 in Chapter 17, Hormones. The analogy explains how hormones affect target cells (and only target cells) by comparing this interaction to the reception of radio waves by a radio. The analogy follows the model by mapping the similarities between the target concept (the reception of particular hormone "messages" by target cells) and the analog (the reception of particular radio waves by a radio).

- **Pronunciation guides.** Pronunciations for difficult words are placed within the narrative, helping students build confidence and familiarity with science terminology. These pronunciations and those for less difficult words are included in the glossary as well.

- **The use of both a systems and a phylogenetic approach.** The systems chapters begin with simplified comparative anatomy and physiology sections to help students see a "complete picture" of several groups of animals. These sections help point out similarities and differences of form and function throughout the animal kingdom. Using the same or similar animals in each systems chapter helps students build on prior learning. The systems content is cross-referenced within the text to chapters occurring later in the book that provide a basic phylogenetic approach.

- **Focus on Plants.** Information on plants is also included within many of these sections to illustrate how plant systems compare to the animal systems described in the chapter.

Organization

Organisms are cells or are made up of cells.
4.1 The cell theory

Most cells are microscopically small.
4.2 Why aren't cells larger?

All cells can be grouped into two broad categories.
4.3 Prokaryotes and eukaryotes
4.4 Eukaryotic cells: An overview
4.5 The plasma membrane
4.6 The cytoplasm and cytoskeleton
4.7 Membranous organelles
4.8 Bacterialike organelles
4.9 Cilia and flagella
4.10 Cell walls
4.11 Summing up: How bacterial, animal, and plant cells differ

Small molecules move into and out of cells by passive and active transport.
4.12 Passive transport: Molecular movement that does not require energy
4.13 Active transport: Molecular movement that requires energy

Large molecules and particles move into and out of cells by endocytosis and exocytosis.
4.14 Endocytosis
4.15 Exocytosis

---

diastole (dye-AS-tl-ee)

Focus
on Plants

Goal #2: Offer a range of visual learning aids to meet students' diverse learning styles.

Features that support this goal

- **Consistent use of color.** Similar biological structures share color in a consistent manner. The continuity of color in the art program makes it easier for students to recognize the same structures depicted in different contexts in different chapters.

- **Numbered labels.** Descriptions of processes in the text are keyed to accompanying illustrations using numbered labels. By closely relating the art to the text description, students (especially visual learners) are better able to understand complex processes in biology and "follow" how the narrative supports the art and vice-versa.

- **Hierarchical illustration labels.** Hierarchical labeling creates an easy way for students to quickly review illustrations and make note of structures of greater importance or emphasis versus those of lesser importance or emphasis.

- **Visual summary figures.** These illustrations are constructed to summarize related concepts in one location using a visual format. Put simply, they show students "the big picture." Visual summary figures are interspersed throughout the text and will help students make connections among concepts that they may miss by reading the narrative alone or by viewing separate diagrams.

- **End of chapter visual thinking questions.** These questions are direct links to in-text figures. Questions for these figures were developed around key concepts within the chapter and provide a good visual review for the student.

Goal #3: Make biology relevant to students' daily lives and foster critical thinking skills.

Features that support this goal

- **Bioissues essays and decision-making framework.** Each part of the book begins with a socially significant issue in biology that students can further research and discuss. These issues relate to the content of the part they introduce. After reading the part opener narratives, which introduce the issues and provide a baseline understanding, students work through a model for decision-making. In the process they learn how to make informed choices. The questions invite students to research the issue, take a look at its many sides, and think about the variety of resolutions possible, even those to which they may not subscribe. By working through this decision-making model, students will be better prepared to make thoughtful and informed decisions.

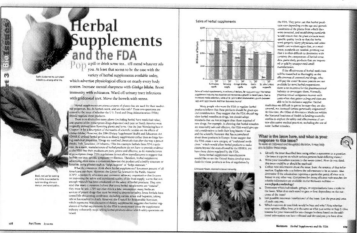

- **Just Wondering boxes.** These boxes answer questions from students across the country enrolled in introductory biology courses for non-majors. In the Third Edition, nearly 40% of the questions are new, coming from students at the University of Wisconsin–LaCrosse, North Carolina State University, Middle Tennessee State University, and Valdosta State University. We thank these students and their professors.

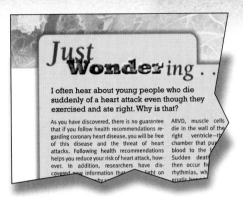

- **Health Connections sections.** The systems chapters appear early in the book and help students relate biological concepts to their own bodies. Many of these chapters contain Health Connections sections as well, which emphasize the relevance of biology to students' everyday lives.

- **Web Enhancements.** All three of the features just outlined are also expanded upon on-line with additional background material and reference sources on this book's dedicated web site **www.jbpub.com/biology**.

Goal #4: Help students gain an understanding of scientific processes and methods.

Features that support this goal

- New Chapter 1 on methodologies across the sciences. This chapter helps students understand that scientific methods are variable yet have commonalities, are based on common sense, are used in many disciplines, and are applied in everyday life.

- Research paper in Chapter 1 and Appendix. This new "science methods" chapter is unique in that it concludes with a discussion of a scientific journal article. The "anatomy" of the article is shown and discussed in Chapter 1, and an annotated reprint of the full journal article is located in Appendix E. This article helps students see first-hand how scientists communicate with one another and provides a model for students who will be developing similar reports for independent investigations they may conduct in this course.

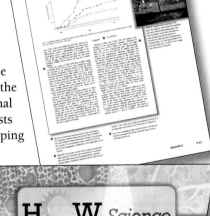

- How Science Works boxes. This boxed feature gives students insight into the applications, processes, and methods of science as well as the tools and discoveries of scientists. Students learn about such things as the groundbreaking experiments of Hershey and Chase to determine whether DNA or protein is the hereditary material, how scientists use isotopes to detect and treat disease, and how scientists' work can be biased by the context of their education, their prior work, and other factors.

- Expanding Your Understanding. Web icons placed in the margins of the text indicate connections to specific web sites that give students insight into products and processes of science today. For example, students can observe anatomically detailed, 3-dimensional representations of the human male and female bodies developed from raw data supplied by the National Library of Medicine's Visible Human Project. Students can read an article that discusses tissue engineering, published in April 1999 in *Scientific American,* one of the most respected science magazines in the world.

Goal #5: Review, test, and assess student understanding at varying levels and in various formats.

Features that support this goal

- **End-of-chapter questions** span a hierarchy of question types:
  - **Knowledge questions** center on recall of information such as terminology, facts, generalizations, or methods.
  - **Comprehension questions** go a step further and ascertain whether a student understands the information. Students show their comprehension of recalled information by paraphrasing, explaining, and summarizing.
  - **Application questions** ask students to use information in new situations.
  - **Analysis questions** require students to break down information into its component parts.
  - **Synthesis questions** ask students to put together information from different sources such as various places in the chapter, other chapters, or from their own background knowledge.
  - **Evaluation questions** direct students to make judgments.

  The cognitive level of each question is noted in parentheses at the end of the question. These labels and categories are derived from *Benjamin Bloom's Taxonomy,* a well-known guide for classifying the cognitive types of question. They will help students understand how to process information to answer a particular question. A key is included with the questions at the end of the chapter to remind students of the processes they are being asked to perform. Answers are found at the end of the textbook.

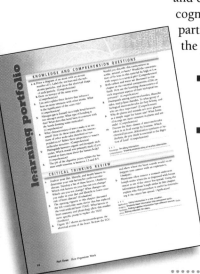

- *Study Guide, Test Bank,* **and the on-line review questions** all follow the same hierarchy of question types. There is a "fit" in question types between the text and these supplements so that students are assessed at the same cognitive levels at which they study.
- **Opportunities to build learning portfolios.** The text and supplements provide a variety of methods to assess learning so that students may build learning portfolios that reflect their academic progress. A portfolio contains examples of student work produced during the course, and may be one way in which an instructor can measure a student's academic achievement. The *Study Guide* also provides students with opportunities for portfolio building by developing concept maps and writing descriptions as prompted in the Concept Maps/Telling the Story sections of the *Study Guide.*

Goal #6: To integrate every aspect of the text and supplements program into a comprehensive teaching and learning package.

Integration of the text and supplements is crucial to the ease of use of the instructors and understanding of the students. Every effort has been made to ensure that these pieces "fit" in concept and approach to reinforce the main text. The complete package of teaching and learning materials is described below.

Instructor and Student Supplements

Jones and Bartlett Publishers offers a variety of carefully developed print and interactive supplements to assist instructors and aid students in mastering general biology. Additional information and review copies are available through your Jones and Bartlett Publisher's Representative or simply log on to **www.jbpub.com/biology** for an on-line demonstration.

For the Instructor

- **Instructor's Manual,** written by Patricia E. Mancini of Bridgewater State College, contains chapter summaries and objectives, comprehensive lecture outlines, teaching tips, suggested answers to frequently asked student questions and tips for handling common student misconceptions. The manual also contains labeled and unlabeled reproducible masters for the figure-labeling exercises.

- **Test Bank,** written by Jay M. Templin of Montgomery County Community College, Pottstown, PA, contains many helpful features to accurately assess student learning. Question types include multiple choice, true/false, fill-in, and matching. Questions are categorized using the same taxonomy in the text and *Study Guide*—by knowledge, comprehension and application. Each question is keyed to the appropriate chapter section in the text and a brief explanation of the correct response is provided. Questions are also available in a computerized test bank format and electronic text files.

- **Instructor's ToolKit** The *Instructor's ToolKit* CD-ROM contains a suite of programs and files to help you in teaching your course. All of the materials are cross-platform for Windows and Macintosh systems.

  - The **Image Bank** program offers full-color art files of more than 600 of the text's illustrations in formats appropriate for computer projection and for printing overhead transparencies or labeled and unlabeled transparency masters.
  - The **Computerized Test Bank** helps you choose an appropriate variety of questions, create multiple versions of a test, even administer and grade tests on-line.
  - A complete set of **PowerPoint Lecture Outline Slides** provides a template for building your lecture presentation.
  - **Electronic text files** of the *Instructor's Manual* and *Test Bank* make it simple for you to customize handouts for your students.

- **Color Transparency Acetates and Masters** Qualified adopters may choose either a standard or customized set of high-quality color acetates. The standard set contains more than 200 of the most frequently used illustrations from the text. Or you can select the illustrations that work best for your lectures and Jones and Bartlett will create a custom set of color acetates. We have specially formatted and upsized labels for all of the illustrations to maximize their effectiveness in the classroom. An additional set of black and white transparency masters is included in the *Instructor's Manual.*

- **Custom Video and Video Resource Library** Jones and Bartlett Publishers is pleased to offer a customized videotape designed specifically to accompany *Biology: Understanding Life,* 3/e. This 120-minute color video includes 17 segments which average 7 minutes in length. These video segments run longer than CD-ROM animations, thereby allowing for more narrative explanation. The videos also film actual structures and processes rather than recreating stylized animations. In collaboration with Films for the Humanities and Sciences, we also offer a wide variety of high quality biology videos for qualified adopters. Please contact Jones and Bartlett for a complete listing.

- **CyberClass On-line Resources** Welcome to the fastest and easiest way to add an on-line component to your course. CyberClass is a customizable, web-based teaching and learning environment. It allows instructors to post quickly and easily material specific to their course, such as a syllabus, assignments, and favorite hot links. Instructors can also administer tests on-line, receive and respond to e-mail from students in the course, and maintain a secure on-line grade book. Students can use CyberClass to access course materials, use on-line practice exams and study tools, contact classmates, and post messages to a class bulletin board.

For the Student

- **Study Guide** Written by Patricia E. Mancini of Bridgewater State College, the *Study Guide* contains chapter objectives, study tips related to the material, self-test questions to help learn the facts and vocabulary, multiple-choice questions to test understanding of the concepts and how they relate to one another, as well as critical thinking questions. The preface includes general tips for studying biology at the college level. Careful attention was given to consistency among the questions in the text, *Study Guide,* Web Site, and *Test Bank.*

- **eLearning Student CD-ROM** Upon the instructor's request, this free CD-ROM will be included in every new copy of the textbook. The CD provides a variety of activities designed to help students study for class and build their learning portfolios:

  - **Review Questions** test student's knowledge of important concepts and applications in each chapter. The review provides feedback for each correct or incorrect answer.
  - **Virtual Flash Cards** help students master the new vocabulary for each chapter.
  - **Figure-Labeling Exercises** allow students to sharpen their visual thinking by matching terms and labels for important illustrations from the text.
  - **PowerPoint Outlines** provide an overview of key concepts presented in each chapter.

- **Biology Web Site (www.jbpub.com/biology)** The Biology web site offers students an unprecedented degree of integration between their text and the on-line world through the following features:

  - *eLearning.* The on-line version contains all of the features just described in the *eLearning* CD-ROM, plus Just Wondering—a direct link to the author, Sandy Alters, for feedback on questions that arise when reading the text.
  - **BioIssues.** Each part of the book begins with a current biological issue of social significance that students can further research and discuss. This on-line component to each BioIssue helps students locate appropriate research material and reason through the issue.
  - **Health Connections.** Many of the systems chapters in the text contain a Health Connections section. Icons at the end of each of these sections point students to a wealth of additional on-line resources for important information on health and disease.
  - **Expanding Your Understanding.** Icons in the margins of every chapter of the text point to web sites that best demonstrate the process of science in action.

- **Student Art Notebook** The *Student Art Notebook* contains black-and-white versions of the same figures you choose for color acetates. Chapter, figure, and page numbers are included for easy reference to the text. With the notebook, students can concentrate on what is said in lecture and efficiently take notes. They will not need to redraw complex structures and processes during the lecture.

Acknowledgments

A textbook such as this is truly a team effort. The author is only one player in a cast of many, with the publisher being the crucial integrator in the process. I feel fortunate as an author to be a part of the Jones and Bartlett family. Jones and Bartlett is one of the few publishing companies left today that has resisted becoming part of a large corporate amalgamation. Instead, they have committed to a personalized team approach to book development and production, which results in high quality in all aspects of book publishing. I thank the management team at Jones and Bartlett for having such vision and promoting such excellence.

Each individual working on the team that produced this book is an outstanding professional. Brian McKean welcomed *Biology: Understanding Life* to Jones and Bartlett and has continued to be a consistent driving force in maintaining focus and excellence in this revision. Judy Hauck contributed her remarkable creative expertise as the Third Edition was developed. Anne Spencer is responsible for the design, which is not only visually stunning, but useful, uncluttered, and extremely helpful to students and professors alike. Mary Hill, Jennifer Angel, and Anne Trafton focused their energies on *Biology: Understanding Life* throughout production, keeping that incredibly complex process running smoothly—as impossible as that might have been at times. Many thanks to you all.

The person primarily responsible for the extraordinary excellence of the art program is Kathy Naylor of Editing By Design, Inc. Kathy worked tirelessly for over a year to develop the art program for this edition. She painstakingly analyzed illustrations for their content accuracy and fit with the text, creating a situation in which the words and the art mesh with one another incredibly well, and making this edition a superb learning tool. I cannot thank Kathy enough. Additional thanks go to Denise Thomas, Mary Hill, and Jennifer Angel, who also helped in this endeavor and aided us tremendously in creating an excellent art program.

Kathy Smith did a marvelous job with copyediting and tying up a multitude of loose ends. Sharon Donahue and Jennifer Angel located the beautiful photographs used throughout the book.

Jones and Bartlett is a company on the cutting edge of integrating technology with print materials for learning. The superb web-based materials that are a part of the *Biology: Understanding Life* program could not have been developed without Mike Campbell, Mike DeFronzo, and Scott Smith of the Interactive Technology group.

Other key members of the Jones and Bartlett team are Tom Manning, Jennifer Jacobson, and Paul Shepardson. These members of the Sales and Marketing group have helped all involved in the book's development to remain focused on the needs of our audience.

Many thanks go to authors James Quinn and Jane Sirdevan, and to the journal *Biological Conservation* (Elsevier Science Ltd.) for allowing us to annotate and reprint the research article introduced in Chapter 1 and reprinted in Appendix E. We believe that this model will be useful to students and professors alike. Thanks to Dr. Quinn and Ms. Sirdevan for reviewing Chapter 1 as well.

Rebecca Keyston of Nexia Biotechnologies shared her expertise as a cell and molecular biologist to help with the development of an illustration and related text on nuclear transfer. Pat Mancini shared her outstanding teaching skills and content knowledge to prepare the *Study Guide* and the *Instructor's Manual.* Both of these supplements are excellent and marvelously useful additions to the wealth of instructor and student support materials that accompany the text. Sincere appreciation also goes to Jay M. Templin for his preparation of the *Test Bank.*

Finally, a heartfelt thanks to all of those instructors and students who have offered their comments, suggestions, and criticisms throughout this project. Since the inception of the First Edition through the planning and development of this Third Edition, they remain a constant source of inspiration and guidance. We acknowledge the following instructors of introductory biology who contributed their time and talent to review portions of the manuscript and those who participated in the market survey conducted in preparation of this Third Edition.

Third Edition Manuscript Reviewers

Janice Ackerman, Mott Community College
Paul D. Anderson, Middlesex Community College
David H. Arnold, University of Texas–Austin
Anne M. Galbraith, University of Wisconsin–La Crosse
Richard Gross, Motlow State Community College
Greg Hampikian, Clayton College and State University
Kerry Hull, Bishop's University
Martin A. Kapper, Central Connecticut State University
Suzanne Kempke, Armstrong Atlantic State University
Robert S. Lishak, Auburn University
Amy Luttinger, Salem State College
Patricia E. Mancini, Bridgewater State College
Patricia Matthews, Grand Valley State University
David L. Pearson, Arizona State University
James S. Quinn, McMaster University
Chris C. Romero, Front Range Community College
Jane Sirdevan, McMaster University
Carol S. Steele, Metropolitan State College

Herbert H. Stewart, Florida Atlantic University
Callie Vanderbilt White, San Juan College

Third Edition Market Survey

Dawn Adams, Baylor University
David M. Armstrong, University of Colorado–Boulder
David H. Arnold, University of Texas–Austin
Gregorio Begonia, Jackson State University
Beverly Brown, Kent State University
Brad Chandler, Palo Alto College
James Courtright, Marquette University
Wayne Daugherty, San Diego State University
George Davis, Bloomsburg University
Jean DeSaix, University of North Carolina–Chapel Hill
Mary Ellen Dillon, University of Dayton
Anne Galbraith, University of Wisconsin–La Crosse
Shirley Porteous-Gafford, Fresno City College
David Goldstein, Wright State University
Janice Schnake Green, Southwest Missouri State University

William J. Grimes, University of Arizona
Blanche C. Haning, North Carolina State University
David Huffman, Southwest Texas State University
Mark Jensen, Morehead State University
Martin A. Kapper, Central Connecticut State University
Suzanne Kempke, Armstrong Atlantic State University
Peter Kish, Southwestern Oklahoma State University
Dana Krempels, University of Miami
Laura Leff, Kent State University
Sandy Ligon, University of New Mexico
Sally Frost-Mason, University of Kansas
David Muehleisen, University of Utah
John B. Pascarella, Valdosta State College
Rhoda Perozzi, Virginia Commonwealth University
Thoniot Prabhakaran, Southwest Texas State University
Les Reichelt, Dawson Community College
John Rueter, Portland State University
Michael L. Rutledge, Middle Tennessee State University
Carol Stiles, Valdosta State College
Lori Stevens, University of Vermont
Herbert H. Stewart, Florida Atlantic University
Joanne Westin, Case Western Reserve University
Albert Will, Broward Community College

Second Edition Reviewers

Brian Alters, University of Southern California
Ed Anderson, University of Arkansas
Susan Bard, Howard Community College
Clyde Bottrell, Tarrant County Junior College
John Clausz, Carroll College
Bill Cockerham, Fresno Pacific College
Carl Colson, West Virginia Wesleyan College
John Conroy, University of Winnipeg
Warren Ehrhardt, Daytona Beach Community College
Salman Elawad, Pensacola Junior College
H.W. Elmore, Marshall University
Merrill Emmett, University of Colorado
Wayland Ezell, St. Cloud State University
Rick Firenze, Broome Community College
Robert Galbraith, Crafton Hills College
Gregory Gillis, Bunker Hill Community College

Keith Morrill, South Dakota State University
Cynthia J. Moore, Washington University, Missouri
Dean Nelson, University of Wisconsin
Richard Peirce, Pasadena City College
John Phythyon, Saint Norbert College
Michael Postula, Parkland College
Kenneth Raymond, Eastern Washington University, Cheney
Steven Salaris, Maryville University
Kristi Sather-Smith, Hinds Community College
Martin Spalding, Iowa State University
George Spomer, University of Idaho
Steve Taber, University of Austin
Sam Tarsitano, Southwest Texas State University
Richard Thomas, Illinois Central College
Todd Thuma, Macon College, Georgia
Carl Thurman, University of Northern Iowa
Fred Wasserman, Boston University
Michael Wood, Del Mar College

First Edition Reviewers

William Barnes, Clarion University
Kathy Burt-Utley, University of New Orleans
James Conkey, Truckee Meadows Community College
Neil Crenshaw, Indian River Community College
Don Emmeluth, Fulton-Montgomery Community College
Robert Grammar, Belmont College
Ronald Hoham, Colgate University
Ed Joern, University of Missouri–St. Louis
Alan Karpoff, University of Louisville
Miriam Kitrell, Kingsborough Community College
John Knesel, Northeast Louisiana University
Kenneth Mace, Stephen F. Austin State University
Virginia Michelich, Dekalb College
Randy Nolan, University of Missouri–St. Louis
Michael Postula, Parkland College
June Ramsey, Pensacola Junior College
David Rayle, San Diego State University
Michael Smiles, State University of New York–Farmingdale
Carl Thurman, University of Northern Iowa
Ann Wilke, University of Missouri–St. Louis
Nancy Yurko, Prince Georges Community College

BIOLOGY
UNDERSTANDING LIFE

How Scientists Do Their Work

This mountainous road near northern California's redwood forests does not look much like a laboratory. However, in April of 1983, a new and important scientific process was being formulated in the mind of American biochemist Kary B. Mullis as he drove a winding course to his cabin in the woods. Dr. Mullis is the developer of a procedure capable of making unlimited copies of genes, a method now known as the polymerase chain reaction, or PCR. (For a description of PCR, see pp. 425-426.) As Dr. Mullis tells it, he "just ran into" the idea one night, an idea that has enabled scientists to amplify the hereditary material found in drops of blood, saliva, or semen at crime scenes as well as in microscopic bits of hereditary material in fossils.

Dr. Mullis's experience is not unique among scientists. Scientific discoveries and breakthroughs are not all conceived in the laboratory . . . and the ones that are do not necessarily result from scientists following a single method. For example, the discovery of the antibiotic penicillin occurred in 1928 by accident. While performing experiments with *Staphylococcus* bacteria, Scottish bacteriologist Sir Alexander Fleming noted that some of his bacterial cultures had become contaminated with mold. Surrounding the mold colonies were clear zones where no bacteria grew. Fleming hypothesized that the mold produced an antibacterial substance. He continued to explore his discovery. By the early 1940s, other scientists had purified penicillin, demonstrated its potency, and developed preparations that could be injected into humans to fight certain bacterial infections. Fleming's "accident" and subsequent work provided the basis for the development of antibiotics that are still used today.

CHAPTER 1

Organization

Scientific process, although variable, is a unifying theme among the sciences.

Scientists report their work in journals and at conferences.

Scientific process, although variable, is a unifying theme among the sciences.

1.1 The focus of science

Although scientists may differ in the focus of their work, the aim of all science is to better understand the *natural world*. In doing so, scientists also share scientific methods that often (but not always) include the following key features: observation, hypothesis development, and hypothesis testing. A **hypothesis** (hi-POTH-uh-sis) is a tentative explanation that guides inquiry. Scientists test their hypotheses in a variety of ways, producing results that are open to verification by others in their disciplines. Because a hallmark of science is the testable hypothesis, science does not explore supernatural explanations to understand how the natural world works.

As just noted, an important process of science is observation, but observation alone does not constitute science, (nor is observation always present in scientific endeavor, see p. 6). The predecessors of scientists— the ancient Greeks and Romans such as Plato and Aristotle— were skilled observers, but they are usually referred to as *naturalists* rather than scientists. Their "science" was called *natural history*. Until the mid-1800s, the study of life (the emerging science of biology) continued to have a natural history perspective, focusing on descriptions and comparisons of organisms and providing explanations to questions based on observation alone. At this time, inquiry in the physical sciences (chemistry and physics) often involved hypothesis testing, but many naturalists felt that the living world transcended this mode of inquiry. A critical shift in biological investigation occurred during the nineteenth century: A preoccupation with form moved to an investigation of function and process. Biologists went beyond observation, joining the chemists and physicists in putting forth hypotheses and testing them.

www.jbpub.com/biology

Scientists study the natural world using a variety of scientific methods. These methods often include observation, hypothesis formation, and hypothesis testing.

1.2 Developing hypotheses

The specific pathways of thinking that scientists follow in developing hypotheses vary greatly. In general, however, a scientist considers the information available that might help answer a question (including explanations successful in other contexts) and uses this information to develop possible explanations to answer the question. Scientists may gather information regarding their questions by searching the scientific literature and synthesizing the information they find. They may make observations of their own or find anecdotal information (individual occurrences) that relate to their question. They develop their hypotheses based on this information and their observations.

Hypothesis testing isn't just for scientists; it can be used to solve everyday problems. For example, you may observe that many of your friends grow healthy and robust houseplants, while you struggle to keep yours alive. You may be asking yourself, "Why can't I grow healthy houseplants?" The explanations that might come to mind immediately are the following: The plants are getting too much or too little water, the plants are getting too much or too little sun, or the plants need fertilizer. Your tentative explanations are preliminary alternative hypotheses.

You still need to refine your hypotheses and gather information pertinent to your question. You read magazine articles in botanical journals that discuss how to grow healthy houseplants, and determine that most of your actions in nurturing your houseplants are proper. You are giving your plants the right amount of water and sunlight. However, you realize that you are not doing one important thing: You are not fertilizing them. As you think about nutrients and plants, you also remember that you potted your plants using soil that you used to grow kitchen herbs last spring. You remember that overused soil can become depleted of nutrients. So from your research and current knowledge, you decide that the best hypothesis as to why your houseplants do not thrive is that they do not receive sufficient nutrients.

A hypothesis is a tentative explanation that guides the investigation of a question. Scientists develop hypotheses by considering information available about a question that includes information from similar situations and using this information to develop possible explanations.

1.3 Testing hypotheses

As you refine your hypothesis on plant growth, you may come up with additional questions before you test it. What should you use as a source of nutrients for your plants? How much should you give them? Will a low dose work as well as a high dose? How often should the nutrients be added to the plants' containers? Gathering additional information may help you refine your hypothesis for testing.

You remember reading advertisements stating that Brand X fertilizer causes houseplants to grow 30% taller and produce healthier, greener plants. You look up articles about fertilizing houseplants in botanical journals and find that they suggest a specific proportion of nutrients as best for certain houseplants. You note that Brand X fertilizer has these specific proportions. Some of your friends may use Brand X fertilizer and say that it works well for their plants. From these observations and testimonials, you might decide to use Brand X fertilizer to test your hypothesis that your plants are dying due to lack of nutrients.

To test your idea, you state your hypothesis in an if/then format. Your newly refined hypothesis, which is based in part on the company advertising, might be the following: *If* houseplants are given Brand X fertilizer, *then* they will grow at least 30% taller and be greener than plants with no fertilizer. What you are actually doing is generating an argument that states an expected outcome.

Testing a hypothesis involves producing data or evidence that either supports or contradicts the hypothesis. Scientists test hypotheses in many ways. This "testing" process involves the use of *deductive reasoning*. Deduction begins with a general statement (for example, Brand X fertilizer will increase growth of houseplants) and proceeds to a specific statement (Brand X fertilizer will increase growth of *my* houseplants).

While scientists are formulating hypotheses, they are also thinking about ways to test their hypotheses and what might happen as a result. Then, as scientists test their hypotheses, they compare their mental arguments for or against their hypotheses with what their testing actually shows. The ways in which a hypothesis is tested and the ways in which data and evidence are collected depend on the hypothesis itself as well as on the scientist doing the testing. For example, paleontologists often test their hypotheses about prehistoric life by gathering evidence from both living ancestors of prehistoric species and fossils, the preserved remains (or other traces) of organisms. Some population biologists test their hypotheses regarding interactions among populations using computer (mathematical) models. Biomedical researchers may test hypotheses by comparing two or more groups of subjects with regard to an intervention that occurs naturally, such as comparing the percentage of unimmunized persons contracting the flu with the percentage of immunized persons contracting the disease during the same time period.

Many scientists test hypotheses by means of controlled experiments. In a **controlled experiment**, a researcher manipulates (changes) one factor and observes how unmanipulated factors respond. All other factors are kept constant (consistent) throughout the experiment, that is, they are controlled. In analyzing the simplified hypothesis in our previous houseplant scenario (hypotheses can be much more complex), notice that the factor you are manipulating (the fertilizer) is stated after the "if" part of the hypothesis. This factor is called the **independent variable**, or manipulated variable. (The word *variable* refers to a factor that changes, or varies.) To test your hypothesis, you will need to add different amounts of Brand X fertilizer to groups of houseplants. Also notice that the factor stated after the "then" part of the hypothesis depends on the manipulated (independent) variable. This factor is called the **dependent variable**, or responding variable, and it varies in response to changes in the independent variable. In your experiment, plant growth and color are dependent variables.

The controlled variables are the factors in the experimental set-up that are not allowed to change. In this case, they include the type of soil, the amount of time the plants are exposed to the light, the temperature in which the plants are grown, the amount of water the plants are given, and the type of plant tested. Can you think of any other variables that may need to be controlled?

Dependent Variables
Plant Growth and Color

Independent Variable Amount of Fertilizer

Scientific process, although variable, is a unifying theme among the sciences.

Controlled Variables

Type of Soil

Exposure to Light

Temperature

Amount of Water

Type of Plant

As you design a controlled experiment to test your hypothesis, you should first outline how you would manipulate the independent variable. In this case, you may decide to give one group of plants no fertilizer and another group of plants the "dose" of fertilizer suggested on the fertilizer container. You also may choose to give some plants slightly more fertilizer than the suggested dose and others slightly less. If the suggested dose is 30 milliliters (ml) of properly diluted liquid fertilizer, you may decide to give a second group of plants 15 ml, a third group of plants 45 ml, and a fourth group of plants 60 ml. Each group of plants should be the same type of plant, such as all periwinkle plants of approximately the same size. You should pot them all in the same soil in the same type and size of pot, place them approximately the same distance from the same light source, and water them all with the same amount of water at the same time. These plants will receive the *treatment*—in this case, the fertilizer. They are the experimental plants. A fifth group of plants should be treated the same as the others except that it will not receive the treatment (fertilizer). This group serves as the control. The **control** is the standard against which the experimental effects on the treatment plants can be compared. Any changes in the control are the result of factors other than the treatment.

Over a preselected time (possibly based on the advertised length of time needed for the fertilizer to cause change), you should observe your plants and collect *data* regarding leaf color and plant height. (You could also choose other indicators of plant growth such as number of leaves or number of flowers per plant.) The data regarding the height of your plants are called **quantitative data** because these data are based on numerical measurements (centimeters, for example). The data regarding leaf color are called **qualitative data** because these data are descriptive and not based on numerical measurements.

At the beginning of your experiment, you should measure the height of each plant in each group and record these measurements in a data table. Find the *mean* (average) height of the plants in each group by summing the heights of the plants and dividing each sum by the number of plants. Record your data for each group. At the end of your predetermined growing period, 6 weeks for example, you measure plant height again, determine the mean heights for each group, and record these data. Your summary data table may look like this:

Mean Plant Height in Centimeters (cm)

Amount of Fertilizer	0 ml	15 ml	30 ml	45 ml	60 ml
0 weeks	145	150	148	151	149
6 weeks	166	180	181	179	184
Increase in plant height	21	30	33	28	35

Regarding the collection of qualitative data, you should determine whether the color of the leaves of each plant is light green, medium green, or dark green. Record this information in a data table as well. Find the *mode* (the value that occurs most often) for each group. Record the mode for each group in a summary data table. Do the same at the end of your predetermined growing period. Your table may look like this:

Description of Green Color of Leaves (mode)

Amount of Fertilizer	0 ml	15 ml	30 ml	45 ml	60 ml
0 weeks	medium	medium	medium	medium	medium
6 weeks	light	medium	dark	dark	dark
Change in leaf color	negative	none	positive	positive	positive

Set Up of Controlled Experiment

| Amount of Fertilizer | 15 ml | 30 ml | 45 ml | 60 ml | Control (no fertilizer) |

Do your data support your hypothesis: If I add Brand X fertilizer to the soil of my houseplants, they will grow at least 30% taller and be greener than plants with no fertilizer? Looking at the qualitative data, the plants with fertilizer were greener than the control with no fertilizer. In this experiment, 30 ml to 60 ml produced the greenest plants.

Looking at the quantitative data, can you determine if the plants were at least 30% taller than the control? First, let's compare the plants treated with 15 ml fertilizer with the control plants. The control plants grew 21 centimeters and the test plants grew 30 centimeters — 9 centimeters more than the control. Is this difference at least 30% more (30% greater than the control growth)? This question is really asking "What percent of 21 is 9?" To find out, you divide 9 by 21, which yields 0.43. Multiplying by 100 to change this number into a percentage yields 43%. In other words, the plants receiving 15 ml of Brand X fertilizer grew 43% taller than the control! Using the same reasoning: (1) the plants receiving 30 ml of Brand X fertilizer grew 57% taller, (2) the plants receiving 15 ml and 45 ml grew 33% taller, and (3) the plants receiving 60 ml grew 67% taller. In summary, these data do support your hypothesis; in fact, your test data suggest that Brand X fertilizer may help houseplants grow even greener and taller than the manufacturer suggests!

Why do you think the manufacturer of Brand X fertilizer is not making greater claims regarding this product? Your experiment certainly shows much greater than a 30% increase in growth. The answer lies in the limitations of your experiment. First, you experimented with one type of houseplant. Brand X fertilizer is probably sold for use on a variety of houseplants. Not all plants may grow as vigorously in response to the fertilizer as yours did. Second, you probably experimented on five small groups of plants. Scientists often perform experiments over and over again, in **repeated trials**, before drawing con-

clusions from their data. Using repeated trials increases the reliability of the results by reducing the effects of chance or error that may occur in a single trial. Using groups of plants for each test dose of fertilizer is considered repeated trials, but rerunning the experiment with larger numbers of plants would yield more reliable data. The Brand X fertilizer manufacturing company probably performs hundreds of experiments on a variety of houseplants before drawing conclusions from their data and generalizing to the population of houseplants grown throughout the United States (or possibly throughout the world).

What if the fertilized plants had not grown better than the control? Evidence that contradicts a hypothesis weakens the likelihood that it is accurate, and in many cases in the physical sciences, it disproves the hypothesis. However, evidence that supports a hypothesis does *not* establish that further testing will also produce supporting evidence. Unforeseen factors may affect the outcome of a future experiment. In your experiment, for example, the data collected using your plants supported your hypothesis. But data collected using other plants in someone else's home may not support your hypothesis. In summary, then, scientists cannot actually *prove* hypotheses (or anything!); they can only *support* hypotheses with evidence.

Testing a hypothesis involves producing data or evidence that either supports or does not support the hypothesis. The ways in which a hypothesis is tested and the ways in which data and evidence are collected depend on the hypothesis itself and on the scientist doing the testing. Many scientists test hypotheses by means of controlled experiments, in which researchers manipulate factors and observe other factors that change in response to the manipulated factors.

1.4 Diversity of methodology

Although some scientists find it quite useful to first make observations, then develop hypotheses, and finally to test their hypotheses by means of controlled experiments, not all scientists proceed in this manner. One reason for this diversity of process is that not all sciences are based on the accumulation of observational data. For example, a chemist cannot observe directly the chemical bonds he or she may be studying.

Another reason for diversity of scientific process is that testing hypotheses by means of controlled experiments is often difficult, if not impossible, in some scientific disciplines. For example, paleontologists cannot conduct controlled experiments to see what happened in the past. Additionally, paleontologists are not able to predict the past. Hypotheses cannot be taken back in time to see if the prediction (expectation of a particular outcome) comes true. Therefore, paleontologists often use historical explanations to "postdict" the past. Although historical explanations often differ from controlled experimental results, they are both valid procedural approaches.

Another example of a field in which the testing of hypotheses by controlled experimentation is often difficult is in biomedical research. It is impossible to control all factors when dealing with humans. When feasible, subjects are chosen randomly for both a "control" group and a "test" group in an effort to obtain groups of people that are as similar to one another as possible. In addition, scientists cannot conduct on humans controlled experiments in which a treatment might be harmful. In such cases, biomedical researchers may use methods other than controlled experiments to test their hypotheses. Such investigators often conduct observational studies in which they do not manipulate an independent variable (often called an *intervention* in biomedical research). Instead, the researchers may report observations that have biomedical importance, such as the numbers of persons receiving flu vaccinations during September of a particular year. Or, they may test a hypothesis by comparing two or more groups of subjects with regard to an intervention that occurs naturally. For example, researchers may compare the death rates from lung cancer of cigarette smokers and non-smokers. When research with human subjects involves a planned intervention (a controlled experiment), the investigator must be sure that it is ethical to conduct an experimental study to answer the research question and test the hypothesis. In other words, the researcher must be sure that participants are not subjected to undue risk and that they are informed as to the possible benefits and risks of the experiment; the researcher must also have written consent of the participants in the study. Using human subjects in research often raises many ethical questions that might not come up when other forms of life such as bacteria or plants are studied.

For this reason and others, scientists who study the workings of the human body often use tissue cultures, biochemical systems, and mathematical models. These systems are useful when a problem must be studied under simplified, well-controlled conditions and when a controlled study in humans is unethical. However, such approaches are limited in their applications, because non-living systems cannot mimic the actual workings of a living organism. Often, therefore, certain animals are used in limited numbers as experimental models. The use of animals in biomedical research has helped scientists make advances in knowledge regarding a wide array of human diseases. (See the Part 6 BioIssues, p. 588, for a discussion of the use of animals in research.)

In addition to the fact that there is no "one" scientific method, many people refer to science as "organized common sense" and suggest that the methods scientists use have no special link with science. People across disciplines and in their everyday lives engage in these generalized processes. Think about the steps that you might take to solve a problem. First, you have to identify the problem, figure out how to solve the problem and try that, attempt another approach and see if you get the same results, and then compare your results with those of others or with your previous knowledge. Are you thinking like a scientist in this case, or do scientists simply use organized, logical processes to solve their problems (and answer their questions) as you might?

The processes scientists use are diverse for many reasons, which include the following: Not all sciences are based on the accumulation of observational data, and testing hypotheses by means of controlled experiments is difficult in some fields of scientific inquiry. Many methods used by scientists are common to other disciplines.

Just Wondering . . . Real Questions Students Ask

Why is it that there's "no truth" in science?

When most people think of *truth*, they usually are referring to something that is always accurate. In science, however, hypotheses and theories may change throughout time as new data emerge. Scientists collect data and gather evidence that may support a theory, or the data may contradict a theory and it may need to be modified or even discarded. Therefore, the very nature of the scientific process is one of an openness to change. In other words, science is tentative, while most people consider truth to be absolute.

Some hypotheses and theories have not changed significantly in hundreds of years, however, and they may not change in the future. We cannot know what evidence will come to light to affect a particular theory in the future. For example, prior to the time of Copernicus (1473–1543), the widely accepted view was that the sun orbited the Earth. Copernicus published a book in which he pointed out the

logical reasons why the Earth orbited the sun. Galileo (1564–1642) later defended this theory, which still stands today after continued testing as technology advanced over the centuries. In another example, the theory of evolution by natural selection proposed by Charles Darwin in 1859 still stands today. Scientists argue about the mode and tempo of evolution. You can read about these opposing theories, punctuated equilibrium and phyletic gradualism, on page 548.

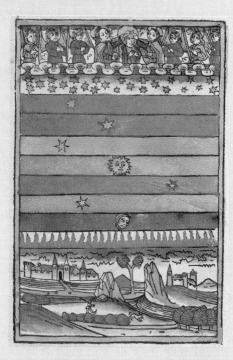

Colored woodcut from Konrad von Mengleberg's *Buch der Nature* (printed in 1481). In it we see the concept of the universe prevalent at the time: Earth at the center (shown here at the bottom), with the moon, stars, and sun surrounding it.

1.5 Facts, theories, and laws

As scientists repeat, or replicate, each other's work, their data may uphold a hypothesis again and again. As the explanations of such consistently supported and related hypotheses are woven together, "grand explanations" that account for existing data and that consistently predict new data are developed. Interwoven hypotheses upheld by overwhelming evidence are called **scientific theories**. The work of scientists is focused on generating and testing hypotheses and theories.

The word theory in a scientific context has a much stronger meaning than the everyday use of the term. A scientific theory is a synthesis of hypotheses and is a well-established group of statements that comprise a powerful concept that helps scientists make predictions about the world and explain particular phenomena. Theories are supported by such an overwhelming weight of evidence that they are accepted as scientifically valid statements. However, analyses of data from different scientific perspectives can result in opposing theories based on the same evidence. Often, continued analyses over time result in one scientific theory prevailing over others. In addition, the possibility always remains that future evidence will cause a theory to be revised, since scientific knowledge grows and changes as new data are collected, analyzed, and then synthesized with previous information.

Some theories, such as Watson and Crick's theory on the structure of the hereditary material DNA or Charles Darwin's theory on evolution, are upheld over time so consistently that some consider them to be facts. A **fact** is a proposition that is so uniformly upheld (or replicated) that little doubt exists as to its accuracy. For example, it is a fact that the Earth orbits the sun.

Scientific theories can never become **scientific laws**, because scientific laws *describe* patterns of regularity with respect to natural phenomena; they do not *explain* them as do hypotheses and theories. For example, the law of inertia states that an object in motion tends to stay in motion and an object at rest tends to stay at rest. In general, theories explain what laws describe. However, a satisfactory explanation for a law may not exist. For example, when Newton was asked to give a theory for his gravity law, he admitted he had none.

> Scientific theories are hypotheses or groups of related hypotheses that consistently account for (explain) existing data and consistently predict new data. Some theories are upheld over time so consistently that they are considered to be facts, that is, little doubt exists as to their accuracy. Scientific laws describe patterns of regularity with respect to natural phenomena.

H⚙W WORKS Science
Processes and Methods

Is there bias in science?

People usually think about the work of scientists as being objective and controlled, and envision scientists as designing experiments that eliminate any personal biases they may have. Although scientists usually strive for these goals, achieving total objectivity and eliminating all bias may be impossible. In science, as in investigation in any discipline, a person's background and previous experiences can significantly influence their work. Everything a person observes, whether a sight, sound, smell, or touch, is processed through the cognitive filter of their past. No one can make an observation that is not biased by the context of their education, their prior work, and other factors.

In addition to this inherent bias in anyone's work, scientific inquiry often begins with statements of a researcher's expectations of the outcome—yet another type of bias. These statements are hypotheses. As scientists develop their hypotheses and design studies to test them, they make subjective decisions about the relative importance of their observations and choose methods that appear appropriate based on their prior knowledge and experience. As scientists collect and analyze data, they also make decisions about how data sets relate to one another, or decide which data are useful and which are not. It is impossible to be entirely objective in making such decisions; by its very nature, science incorporates such subjective decision making.

www.jbpub.com/biology

Stephen Jay Gould, a world-renowned research scientist and Harvard professor, provides an interesting example that probes the power of subjectivity and bias in his book *The Mismeasure of Man*. He relates the story of Paul Broca, one of the many nineteenth century scientists who were engaged in craniometry, the measurement of the skull and facial bones, and thus measurement of the contents of the cranium. The nineteenth century sketch included here shows a measuring device used in this practice.

During the nineteenth century, European leaders and intellectuals viewed some races (especially their own) as being superior to

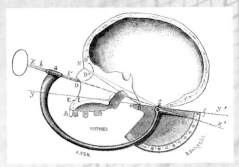

others, and believed that those races at the bottom of their ranking had inferior characteristics of anatomy and/or physiology, such as inferior intelligence. Accepting this view of a racial hierarchy, Broca, a professor of clinical surgery in Paris, believed that brain size was related to intelligence, with the more intelligent races (in his view) having larger brains, and the less intelligent races having smaller brains. When another scientist challenged his belief, Broca set out to defend himself. He hypothesized that white Europeans would have larger brains than other races, thus explaining their (perceived) higher intelligence. He meticulously gathered quantitative data on cranial capacity to test his hypothesis. He carefully designed controls for his procedures so that error would not affect his measurements. Broca concluded that the data supported his hypothesis.

A modern day re-evaluation of Broca's data suggests that while his individual measurements are reliable, Broca's conclusions are invalid. Bias influenced both the formulation of his hypothesis and the collection of his "supporting" data. As Dr. Gould points out, "Broca's cardinal bias lay in his assumption that human races could be ranked in a linear scale of mental worth." In fact, brain size is only one of many factors that affect intelligence (some of which are hard to quantify). Brain size alone does not predict intelligence with any accuracy. Furthermore, Gould suggests that this same bias influenced the type of data Broca collected, as he *subconsciously* searched for those data that would support his hypothesis. If the data clearly did not support his hypothesis, he would develop other criteria that would supersede the importance of the data that didn't "fit."

Broca's story is not just a historical curiosity. This example shows that although data may be reliable—collected meticulously and flawlessly—the interpretation of those data may not be objective if affected by factors such as prejudice and bias. Scientists today usually try to eliminate as much bias in their work as possible in a variety of ways.

To remove the subjective element from the evaluation of the effects of a new drug, for example, medical researchers often choose a double-blind methodology. In such a study, half of the participants are randomly chosen to receive the new drug. The other half of the participants receive a placebo, a pill that resembles the drug but contains no medication. Neither the researchers nor the participants know which pill is the drug (thus, double-blind); the pills are stamped with code letters by the drug company and the code is revealed only after the study ends.

In another example, the American Association for the Advancement of Science (AAAS) addressed the issue of bias in science in its publication *Benchmarks for Science Literacy*, a set of guidelines for teachers of science. AAAS recognizes that working groups of scientists tend to see things alike and may have trouble being entirely objective about their methods and findings. *Benchmarks* suggests, therefore, that scientific teams seek out possible sources of bias in the designs of their investigations and in their data analyses by checking each other's results and explanations.

Publishing the results of scientific studies—making them immediately available to scientists and laypeople of many different backgrounds and points of view—is another very important way to control for bias. If another researcher or team cannot reproduce the results of the published research, the validity of the results becomes questionable, and may eventually be refuted. If the interpretation of data is heavily biased, as in Broca's case, those in the scientific community will often perceive that bias and reject the study's conclusions. Thus, though it is not perfect, the scientific process has built-in self-monitoring systems that eventually succeed in reducing the amount of bias in research. ⚙

Scientists report their work in journals and at conferences.

1.6 The scientific journal article

In order to replicate and build on each other's work, scientists must communicate with one another. The two primary avenues of communication among scientists are scientific conferences and scientific journals. Professional organizations exist for nearly every subdiscipline of science, and most of these organizations hold meetings or conferences in which members come together to discuss the latest research findings in their particular field. Usually, scientists present papers in a lecture format, outlining their work and their results, or they may participate in poster sessions in which they discuss their work with others one-on-one, as shown below, or in a small group. Increasingly, scientists are communicating with other scientists via the internet, and journals are publishing papers on-line.

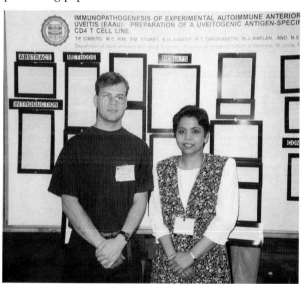

Research papers that scientists write for submission to journals in their fields follow a format that is common to scientific disciplines and usually includes these parts: title, abstract, introduction, methods, results, discussion, and references. Likewise, most student science competitions require students to report their research in this format. Science courses in which there is a laboratory component and in which professors require laboratory reports often follow this format or an adaptation of it, as well.

On the next two pages, you can examine the format of a scientific paper, which was published in 1998 in the journal *Biological Conservation*. (Please note that this journal is a British publication and therefore uses British spellings, which differ occasionally from American spellings.) This paper, written in a format that is stan-

dard for most scientific papers, is an example of scientific process and how scientists conduct and report their work. The sidenotes next to the article explain its organization. (Also note that the journal article is presented at larger size in Appendix E at the end of this book with additional annotations and comments that help explain its content.)

The article describes research conducted by James Quinn and Jane Sirdevan of McMaster University in Hamilton, Ontario (Canada). These researchers tested a variety of nesting materials (substrates) to see on which types of materials Caspian terns (a species of waterbird) were most likely to nest. They also investigated whether the survival of chicks (the birds' offspring) varied with substrate type. This information was used to prepare nesting sites for the terns on artificial islands built in an effort to relocate them. The land on which the terns were currently nesting was privately owned and likely to be developed.

Caspian terns typically nest on islands (or in this case on shore areas cut off from the mainland) where there is little or no vegetation. They nest in colonies, creating nests that are shallow depressions in the sand (or other substrate). Hamilton Harbour, the site of the research, is located on the west end of Lake Ontario in Canada, just west of Niagara Falls, New York. It is an important nesting site for colonial waterbirds in the Great Lakes in terms of species diversity and numbers of breeding birds.

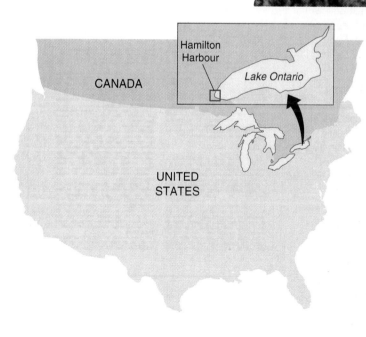

Scientists report their work in journals and at conferences.

AUTHORS OF THE STUDY Contact information is provided in a footnote at page bottom.

TITLE OF THE PAPER The title should be descriptive of the study and should have a clear focus.

ELSEVIER

Biological Conservation 85 (1998) 63–68

BIOLOGICAL CONSERVATION

Experimental measurement of nesting substrate preference in Caspian terns, *Sterna caspia*, and the successful colonisation of human constructed islands

James S. Quinn *, Jane Sirdevan

Biology Department, McMaster University, Hamilton, Ontario, Canada, L8S 4K1

Received 7 June 1997; accepted 30 September 1997

ABSTRACT An abstract is a concise summary of the research paper.

Abstract

Caspian terns, *Sterna caspia*, recently bred in Hamilton Harbour, at the western end of Lake Ontario, on private property that is likely to be developed in the next decade. To reduce this land-use conflict and to promote the current level of biodiversity of colonial nesters in the area, artificial islands were built in the winter of 1995-1996 with different areas designated for a variety of nesting waterbirds including Caspian terns. In 1994, prior to island construction, we tested three substrate types for tern nesting preferences so that an appropriate substrate could be placed on the Caspian tern designated portion of the new islands. We found a preference for sand over pea-gravel and crushed stone, and indirect evidence for a preference favouring the experimental substrates over the pre-existing substrate of hard-packed ground. Based on these results, the small area of the island designed for Caspian tern nesting was surfaced with sand and was subsequently colonised successfully. The colony established and reproduced successfully on the designated site in 1996 and grew in numbers of nesting pairs in 1997. © 1998 Elsevier Science Ltd. All rights reserved

Keywords: Habitat preference; Restoration; Colonial nesting waterbirds

INTRODUCTION In this section, the investigators present the rationale (statements of their reasons) for their study and its foundation in the literature. This section typically concludes with the purpose of the study and research hypotheses.

1. Introduction

Caspian terns, *Sterna caspia*, have probably never been abundant in North America, but they have nested in the Great Lakes since as early as 1896 (Blokpoel and Scharf, 1991). In 1986, a small colony of 48 nests was established in Hamilton Harbour (43° 16'N, 79° 46'W; Dobos et al.; 1988) at the western end of Lake Ontario. Between 1990 and 1994, the main colony was located on the site of the experiment reported here and numbers in the harbour increased from 184 to 331 pairs, representing over 16% of Lake Ontario's Caspian terns (Moore et al., 1995). This immigration to Hamilton Harbour and subsequent growth of the colony, particularly the increase to 134 pairs in 1987 (Dobos et al., 1988), coincided with the decline and desertion of a Caspian tern colony site at the Eastern Headland, a man-made area extending into Lake Ontario from the Toronto waterfront (Morris et al., 1992). The success of this thriving new colony in Hamilton Harbour is encouraging given the questionable status of the species, ranging from rare to endangered (proposed) in Ontario and the States surrounding the Great Lakes (Blokpoel and Scharf, 1991).

Development plans for an area of the Hamilton Harbour shoreline included the Caspian tern main colony site. During the winter of 1995–1996 three islands were constructed to provide nesting habitat for the six species of colonial nesters currently in the area, a major component of the remedial action plan for the rehabilitation of Hamilton Harbour (Quinn et al., 1996). The motivation for island construction was to maintain current levels of diversity of colonial nesters in the harbour and to reduce land-use conflict with the current property owners. We tested substrate preferences of Caspian terns on an experimental site on the mainland where the colony had been previously to facilitate the establishment of a colony on one of the new islands.

Little is known about the nesting substrate preferences of Caspian terns. Caspian terns generally nest in dense colonies situated in open and largely unvegetated areas (Peck and James, 1983). Although descriptions

* Corresponding Author. Tel.: (905) 525-9140 ext. 23194; fax: (905) 522-6066; email: quinn@mcmaster.ca

0006-3207/98/$19.00
PII:S0006-3207(97)00142-0

of nesting habitat are available (Peck and James, 1983; Quinn et al., 1996), experimental studies have not been reported and descriptive studies may not take habitat availability into account. For example, at sites where ring-billed gulls, *Larus delawarensis*, nest, only areas not used by gulls are available to terns as terns begin nesting after ring-billed gulls have become well established (Quinn et al., 1998). The later nesting terns are unable to take over areas that have become occupied by ring-billed gulls, a problem encountered also by common terns, *Sterna hirundo* (Morris et al., 1992). Ring-billed gulls began nesting as a colony of 17 pairs in Hamilton Harbour in 1978 (Dobos et al., 1998). Since then the colony has grown to about 40000 pairs when last counted in 1990 (Moore et al., 1995).

The Hamilton Harbour Caspian tern colony is one of only five on the lower Great Lakes. In 1994, this colony was located on a site that is slated for development in the next 4 to 9 years. Caspian terns are sensitive to human disturbance and, unlike ring-billed gulls, will not nest in close proximity to human activity. The focus of this study was to evaluate Caspian tern nesting preferences when given the opportunity to nest on one of three commercially available substrates, or on the hard-packed ground found on the colony site. The choice of materials for the experiment was based on examinations of Caspian tern nesting substrates in the Great Lakes and the Gulf of Mexico (J. Quinn, pers. obs.). Results of this experiment were used to determine substrate type placed on a designated area of recently built wildlife islands in Hamilton Harbour. The implementation resulted in the successful establishment of a colony of Caspian terns on the designated site. This colony has nested successfully for two seasons (1996 and 1997).

METHODS In the methods section, investigators describe the materials and procedures they used to conduct the study. Enough detail is included so that another researcher could repeat (replicate) the study.

2. Methods

Our study site was Pier 26 in Hamilton Harbour, at the west end of Lake Ontario. Caspian terns nested at a main site in the midst of a large colony of ring-billed gulls, estimated at about 40 000 pairs in 1990 (Moore et al., 1995), and at a separate sub-colony site about 100m to the North-East. On 8 April 1994, prior to the arrival of Caspian terns in the area, we established a 9 × 9 m frame subdivided into 3 × 3 m cells within the area of the 1993 Caspian tern colony site. Nest markers from the 1993 nesting season had been left in place and the number of clutches that had been initiated in each experimental cell was recorded (Fig. 1). We removed the old stakes and then placed the substrates within the grid, in a 3 × 3 Latin Square design, so that each substrate type was represented once in each row and once in each column, thus controlling for any effects due to position. The substrates tested were: (1) construction grade sand with a few small stones; (2) crushed stone

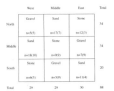

	West	Middle	East	Total
North	Gravel n=5(5)	Sand n=17(7)	Stone n=12(3)	34
Middle	Sand n=18(10)	Stone n=9(9)	Gravel n=7(9)	34
South	Stone n=6(3)	Gravel n=3(9)	Sand n=11(4)	20
Total	29	29	30	88

Fig. 1. Distribution of substrates in the experimental grid. Each grid cell was 3x3m. Numbers of clutches initiated in each cell is given for 1994 and (1993)—prior to establishment of the grid. Row and column totals are for 1994 data.

with sharp edges, approximately 1 cm in diameter; and (3) 'pea' gravel with rounded edges, approximately 1 cm in diameter. During the period of ring-billed gull nest building, prior to the arrival of terns, a field assistant visited the Caspian tern nesting area five times per day from 5 to 26 April 1994 and destroyed all ring-billed gull nests as they were started.

Upon tern clutch initiation we marked nests initially with numbered tongue depressors (wooden spatulas) placed about 15 cm from the edge of the nest scrape. Tongue depressors were replaced with 20 cm tall numbered wooden stakes on 17 May. Prior to 2 June we placed nest covers (Quinn, 1984) over most nests with eggs, particularly those near the periphery of the colony, to protect them from the ring-billed gull depredation. We typically checked nests every second day until 2 June, and every sixth day thereafter, weather permitting (Fig. 2). We placed chick shelters (Burness and Morris, 1992) near where eggs had or were about to hatch to provide shelter from harsh weather or predators during parental absences.

To examine whether offspring survival varied with substrate types we followed the fates of eggs and chicks which were banded by age 7 days post-hatching. Caspian tern nestlings generally remain in the vicinity of the nest until fledging. However, to restrict their movements when we checked the colony, we fenced the grid with 30 cm tall 6mm² mesh hardware cloth on 8 June, before chicks became mobile, and used hatching and survival to at least age 24 days post-hatching as measures of breeding success.

On several occasions we observed a ring-billed gull that flew into the colony and removed an egg from an unattended clutch. Each time, the bird flew into a particular

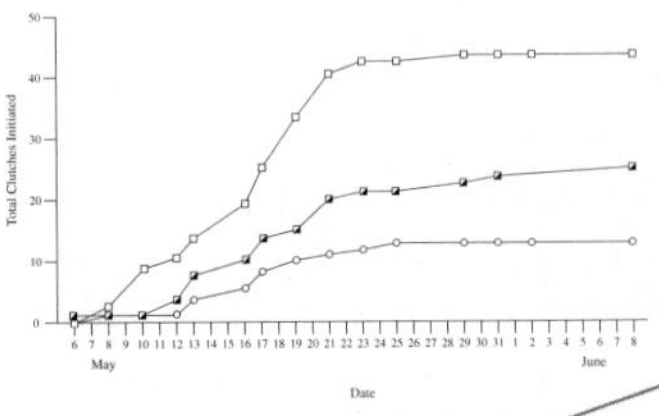

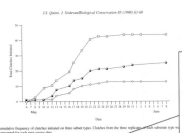

Fig. 2. Cumulative frequency of clutches initiated on three subset types. Clutches from the three replicates of each substrate type we[…] Data are presented for each nest census date.

part of the colony just outside the grid fence and ate the egg. Yolk stains on the gull's plumage and behavioural idiosyncrasies made it individually recognisable. Our observations from a blind located 13m south of the edge of the colony suggest that the same individual took at least 30 eggs. On 15 June, this bird was shot under Canadian Wildlife Service permit and egg loss virtually stopped.

During the winter of 1995–1996 three wildlife islands measuring about 100 by 20m were constructed under the direction of the Hamilton Harbour Fish and Wildlife Habitat Restoration Project, part of the Hamilton Harbour Remedial Action Plan (Quinn et al., 1996). We covered the tern designated areas with heavy gauge plastic sheeting anchored with rocks prior to the beginning of gull nesting (to prevent nesting by gulls) and removed sheeting shortly after the arrival of the terns. Nesting activities were monitored in 1996 and 1997 through twice-weekly visits to the islands which ceased when chicks became mobile.

We tested the spatial and temporal patterns of clutch initiations and reproductive parameters with the likelihood L^2 statistic (Hays, 1981). We employed a Spearman Rank Correlation (Siegal, 1988) on numbers of clutches initiated within the specific area of each of the cells between 1993 (prior to establishment of the experimental site; nest markers were left in place over winter) and 1994 (after establishment of the experimental site) to test for the possibility that pairs of terns established nests in the same location between years. To show that the frequencies of clutches initiated in the cells between 1993 and 1994 we used a multi-sample χ^2 test with 2 degrees of freedom.

RESULTS The results section of the paper presents the data collected. The data are summarized in narrative form and often have accompanying charts, graphs and/or tables.

3. Results

Between 6 May and 8 June 1994, 88 cl[…] initiated on the experimental substrates. A[…] available area of the Caspian tern colony […] than that of the experimental grid (ap[…] 280 m² vs 81 m²), 20 of the first 21 clutche[…] ated on the experimental substrates sugg[…] Caspian terns preferred the experimental sub[…] the natural hard-packed ground at the […] Additionally, despite smaller numbers of nests initiated in the main colony site in 1994 (n=177) compared with 1993 (n=242), the number initiated on the experimental substrates in 1994 exceeded that on the same area in 1993 by about a third (Fig. 1). Patterns of nest placements were independent of nest positions in 1993 as the number of nests positioned within the location of each cell of the experimental grid in 1993 and 1994 were not correlated (Spearman Rank Correlation, n_1=59, n_2=88, r_s=0·15, ns; Fig. 1). Furthermore, the nest frequencies in the cells (grouped by 1994 substrate-type) changed significantly between 1993 and 1994 (multi-sample χ^2=9·07, df=2, p<0·025).

The number and timing of clutches initiated on each substrate were analysed using pooled values of the three replicates. Row position had an effect on the number of clutches initiated, with significantly fewer clutches initiated on the South row than on the middle or North rows of the experimental substrates (χ^2=4·76, df=1, p<0·050). Because substrate preference patterns were constant for each row, replicates were pooled for analysis.

Clutch initiations were not at random with regard to substrate type (L^2=16.8, p<0·0005). We found that

Table 1
Reproductive parameters for caspian tern pairs nesting on experimental substrates in 1994

Substrate type	Sand	Gravel	Stone
Clutches (n)	46	15	27
Clutch size (mean±1SD)	2·17±0·74	2·00±0·63	2·07±0·87
Clutches hatching ≥1 egg	28	9	10
Hatching success[a]	0·61	0·60	0·37
Clutches fledging ≥1 Chick	21	6	7
Fledging Success[b]	0·75	0·67	0·70
Chicks fledged/pair[c] (mean ± 1 SD)	0·61±0·74	0·67±0·90	0·44±0·80

a Number of clutches hatching one or more eggs divided by total number of clutches initiated.
b Number of clutches fledging one or more chicks divided by total number of clutches hatching one or more eggs. Chicks were considered fledged if they survived to age 24 days.
c Number of clutches where one or more chicks survived to age 24 days divided by total number of clutches initiated.

more clutches were initiated on sand than on the other substrates (Fig. 2; L=113, p <0·0005). There were no significant differences in the tendency to nest on stone compared with gravel (J=3·5, ns). Nests initiated on sand were generally edged with a circle of pea gravel and/or crushed stone along the edge of the nest scrape. Although more pairs laid on sand, the timing clutch initiations did not differ among substrate types. The median clutch initiation date, 17 May, was the same for all three substrates.

Reproductive parameters were similar for pairs nesting on all three substrates (Table 1). Hatching success was defined as the proportion of successful clutches in which one or more eggs hatched, whereas fledging success was determined as the proportion of clutches where one or more eggs hatched and at least one chick fledged (survived to at least 24 days). Values for hatching and fledging success were highest on sand, but the differences were not statistically significant (J=21 and 0·38, p <0·05). Egg failures were categorised as: (1) failed to hatch; (2) cracked or broken; (3) disappeared. The proportion of eggs which disappeared or were cracked or

Table 2
Classification of egg failures during the 1994 breeding season

			Proportion of eggs		
Substrate	Row[*]	Total eggs	Disappeared	Cracked/broken	Failed to hatch
Sand	N	37	0.19	0.05	0.05
	M	37	0.51	0	0.03
	S	26	0.69	0.08	0
Gravel	N	10	0	0.10	0
	M	14	0.50	0	0
	S	6	0.83	0.17	0
Stone	N	28	0.39	0	0
	M	16	0.75	0.13	0
	S	12	0.92	0.08	0

* Row indicates position in experimental grid (N, north; M, middle; S, south).

broken was the greatest in the South row of the experimental grid (Table 2). Chick loss prior to fledging was classified as death or disappearance. The most common source of chick loss was death (chicks found dead; sand 77%; stone 67%; gravel 71%). While we tried to minimise the effect of our nest checks, we cannot dismiss the possibility that out infrequent nest visits affected these results. Notably, the activities of a family of red foxes (Vulpes vulpes) observed on Pier 26 in the area of the colony may have affected chick survival, however we have no direct data from 1994 to address this possibility. In 1995, fox activity was observed to cause colony desertion at night for periods of up to 42 min (J. Sirdevan, pers. obs.).

During the winter of 1995 to 1996 three wildlife islands were constructed in Hamilton Harbour (Quinn et al., 1996). The design of the islands utilised the results reported here for the construction of one 200m² knoll surfaced with sand and a small amount of 1 cm diameter pea gravel for lining of nest rims. Most of the area was covered with plastic sheeting (held down with stones) to discourage nesting by ring-billed gulls until the terns arrived. Despite intentions to use decoys to attract Caspian terns to the site in 1996 (Quinn et al., 1996) they had already begun to lay eggs on the exposed sand that remained uncovered before we were able to remove the plastic and install decoys. After uncovering the sand covered habitat terns nested over the entire sandy knoll. In total there were 226 clutches (86% of clutches initiated in Hamilton Harbour) initiated on the designated Caspian tern site in 1996. About 37 clutches (14%) were initiated on a raft (H. Blokpoel, pers. comm.) which was positioned in a confined disposal facility near the original Caspian tern colony (Lampman et al., 1996). Again in 1997 we covered most of the sandy Caspian tern habitat with plastic. The first Caspian tern nests were found on 6 May, 7 days after the plastic was removed. There were 319 clutches initiated (80% of clutches initiated in Hamilton Harbour) on the Caspian tern site with a median initiation date of 21 May. A small proportion of pairs nested on the CDF tern raft (estimated

N=15, 4%; H. Blokpoel, pers. Comm.). Sixty-three late-nesting or re-nesting pairs initiated clutches (16%) on another site on the wildlife islands with a median initiation date of 10 June. The secondary colony on an island slightly to the south, was located on a site that had been targeting common terns. The area was open and covered with pea gravel (1 cm diameter).

4. Discussion

Caspian terns demonstrated a statistical preference for nesting on sand over the other substrates. A preference for a particular substrate may be due to a variety of parameters. Substrates differed in colour, size and shape of grains. Construction grade sand is light brown in colour while crushed stone and pea gravel are shades of grey. Caspian tern eggs are light tan with dark brown spots. While they match each of the substrates quite well, they are relatively easy to see, even on sand. Chaniot (1970) noted changes in the frequency of tan-coloured relative to darker and lighter phase downy young Caspian terns in San Francisco Bay since 1943 and attributed this frequency change to natural selection favouring the colour morph which best matched the nesting substrate colour. Nesting substrate preference could reflect coloration of eggs or chicks. Nevertheless, the scrapes in sand were usually edged with crushed stone or pea gravel gathered by the birds from nearby cells with other substrates. This lining of nest edges made the nests more visible.

Sand may be marginally easier for terns to make scrapes. The nest depression is formed by resting the breast on the substrate while kicking substrate out behind. Additionally, sand, being more dense and darker in colour may carry thermal advantages.

The higher density of terns nesting on sand did not lead to diminished reproductive success as both hatching and fledging success values were comparable or slightly better than the alternative substrates. In contrast, Richards and Morris (1984) found that common terns nesting on their preferred experimental substrate, which included structures providing shelter from predators, realised significantly improved fledging success. Larger sample sizes or different environmental conditions (e.g. absence of terrestrial predators) might be required to demonstrate differences in reproductive success if they exist. Fledging success reported here was low compared with previously reported fledging success rates for Hamilton Harbour Caspian terns (Ewins et al., 1994). Our sampling techniques differed from Ewins et al. (1994). Additionally, the presence of foxes in 1994 may have resulted in temporary desertions of the colony leaving chicks exposed to ambient conditions, particularly at night when foxes are active (Southern et al., 1995). Indeed, subsequent observations in 1995 revealed

that Caspian tern adults left... greater than 45 min when... vicinity (J. Sirdevan, pers... during periods of noctu... caused ring-billed gull chi... exposure to low ambient c... 1985). All Caspian terns r... bour mainland sites in 19... eggs, probably due to the... observed on the site. Foxes... 1996 and 1997 and no... observed on the mainland at...

The first eggs in 1996... sandy knoll designed for... the plastic sheeting. Appar... to the habitat, as no deco... used to entice them. The c... of nests. The 200 m² Casp... of the human-made island... and represented a very sm... (Quinn et al., 1996). It was... (about 1m higher than surrounding substrate) on the islands. The other, not included in the earlier design plan (Quinn et al., 1996), was a ridge placed on the middle island to encourage herring gull nesting. While it is possible that topography played a role, it should be noted that the first nests were at the edge of the plastic at the same elevation as the rest of the island. The substantial increase in the number of pairs nesting on this site in 1997 is a very good sign suggesting that the site will continue to be used. The Caspian tern nesting activity observed on the middle island in 1997 began after hatching had begun on the crowded main Caspian tern site. It is likely that the high density and the feeding activities at nests with chicks made the main colony site less attractive to the late nesters. The site used on middle island was one of two surfaced with a smaller pea gravel and intended for common terns. While not the most preferred of our substrates, smaller-sized pea gravel was shown to be acceptable to Caspian terns in our experiment.

This study shows a statistical preference for nesting on sand by Caspian terns. Additionally, we found that terns nesting at greater densities on sand were not disadvantaged reproductively compared with those nesting on other substrates. We concluded that construction grade sand is a suitable substrate for attracting nesting Caspian terns and recommended its use, with a small addition of pea gravel for nest lining, in a habitat creation project. The sandy raised knoll on the Northern-most wildlife island in Hamilton Harbour (Quinn et al., 1996) has attracted most of the Caspian terns nesting in Hamilton Harbour and this colony has grown in numbers from 1996 to 1997. The application of the results of our experiment to this habitat creation project has been a great success.

Acknowledgements

The authors thank John Hall and the Hamilton Harbour Remedial Action Plan team for getting us involved in this research and for incorporating our findings and ideas into their plans. They gratefully acknowledge the reliable assistance of Cynthia Pikerik, who prevented nesting by ring-billed gulls on the study site. Robert Dawson, Sarah Hopkin, Andrea Kirkwood, Vanessa Lougheed, Brent Murray, Joe Minor, Salmon Cuso, Cynthia Pekaric, Jennifer Startek, and Carole Yauk worked hard to help construct the substrate experimental grid for peanuts (and beer). Dedicated field assistance was provided by Carrie Rongits in 1994, Cynthia Pekaric and Angelo Nicassio in 1996 and Cheryl Fink and Cindy Anderson in 1997. The Hamilton Harbour Commissioners kindly provided access to their property and permission to construct the substrate grid. The authors also thank Chip Weseloh (CWS) for providing the services of Cynthia Pekaric and the Canadian Wildlife Service for permission to carry on our studies of the colonial nesters. Finally, they are pleased to acknowledge the McMaster Eco-Research Project (MERP) for logistical and financial support. Funding for MERP was provided courtesy of the Tri-council Green Plan. Ralph D. Morris and two anonymous reviewers provided valuable comments on earlier versions of this paper.

References

Blokpoel, H., Scharf, W.C., 1991. Status and conservation of seabirds nesting in the Great Lakes of North America. ICBP Technical Publication 11, pp. 17-41.
Burness, G.P., Morris, R.D., 1992. Shelters decrease gull predation on chicks at a common tern colony. Journal of Field Ornithology 63, 186-189.
Chaniot, G.E. Jr., 1970. Notes on color variation in downy Caspian terns. Condor 72, 460-465.
Dobos, R.Z., Struger, J., Blokpoel, H., Weseloh, D.V., 1988. The status of colonial waterbirds nesting at Hamilton Harbour, Lake Ontario, 1959-1987. Ontario Birds 6, 51-60.
Ewins, P.J., Weseloh, D.V., Norstrom, R.J., Legierse, K., Auman, H.J., Ludwig, J.P., 1994. Caspian terns on the Great Lakes: organochlorine contamination, reproduction, diet and population. Canadian Wildlife Service, Occasional paper No. 85.
Hays, W.L., 1981. Statistics. 3rd ed. Holt, Rinehart, and Winston, New York.
Lampman, K.P., Taylor, M.E., Blokpoel, H., 1996. Caspian terns Sterna caspia, breed successfully on a nesting raft. Colonial Waterbirds 19, 135-138.
Moore, D.J., Blokpoel, H., Lampman, K.P., Weseloh, D.V., 1996. Status, ecology, and management of colonial waterbirds, Hamilton Harbour, Lake Ontario, 1988-1994. Technical Report Series No. 213, Canadian Wildlife Service, Ontario Region.
Morris, R.D., Blokpoel, H., Tessier, G.D., 1992. Management for the conservation of common tern Sterna hirundo colonies in the Great Lakes: two case histories. Biological Conservation 60.
Peck, G.K., James, R.D., 1983. Breeding Birds of Ontario: Nidiology and Distribution, Vol. 1: Nonpasserines. Royal Ontario Museum Life Science, Toronto.
Quinn, J.S., 1984. Egg predation reduced by nest cover... researcher activities, in a Caspian tern colony. Colonial Waterbirds 7, 419-151.
Quinn, J.S., Morris, R.D., Blokpoel, H., Weseloh, D.V., Ewins, P.J., 1996. Design and management of bird nesting habitat: conserving colonial waterbird biodiversity on artificial islands in Hamilton Harbour Ontario. Canadian Journal of Fish and Sciences 53 (Suppl.), 44-56.
Richards, M.R., Morris R.D., 1984. An experimental study of nest site selection in common terns. Journal of Field Ornithology 55, 457-466.
Siegal, S., 1988. Nonparametric Statistics for Behavioural Sciences, 2nd ed. McGraw-Hill, New York.
Southern, W.E., Patton, S.R., Southern, L.K., Hanners, L., 1985. Effects of nine years of fox predation on two species of breeding gulls. Auk 102, 827-833.

DISCUSSION The focus of this section is interpretation of the data. Scientists compare their results with their hypotheses to determine whether the data support the hypotheses. They compare findings with other research, propose explanations, note discrepancies they may have found, and draw conclusions. Sometimes, researchers then make suggestions for improvements in their procedures if they are to continue study in this area and make recommendations for further study.

ACKNOWLEDGMENTS In this section, the authors of the paper (the investigators) give credit to all who helped them conduct their study or who reviewed their paper before it was submitted to the journal. Sources of funding for the study are usually noted here as well.

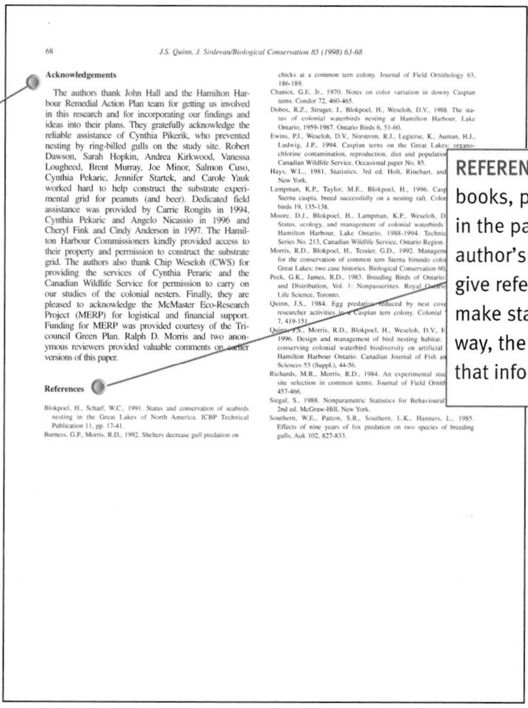

REFERENCES The reference section lists all the books, papers, and journal articles that are cited in the paper. They are listed alphabetically by author's last name. Authors of scientific papers give references whenever they cite prior work or make statements based on prior work. In this way, the reader can access the original source of that information.

Key Concepts

Scientific process, although variable, is a unifying theme among the sciences.

1.1 Scientists study the natural world using a variety of scientific methods.

1.1 The methods scientists use often, but not always, include observation, hypothesis formation, and hypothesis testing.

1.2 A hypothesis is a tentative explanation that guides inquiry.

1.2 Scientists usually formulate hypotheses by synthesizing information from various sources and considering explanations that are successful in other analogous contexts.

1.3 Testing a hypothesis involves producing data or evidence that either supports or does not support the hypothesis.

1.3 Scientists test hypotheses in a variety of ways.

1.3 Many scientists test hypotheses by means of controlled experiments, in which one factor is manipulated while others are observed that respond to the change.

1.3 The factor that is manipulated in a controlled experiment is called the independent variable; those that respond are called the dependent variables.

1.3 Controlled variables are factors kept constant in the experimental setup.

1.3 Scientists often perform experiments over and over again, in repeated trials, before drawing conclusions from their data.

1.4 The processes scientists use are diverse for many reasons, which include the following: Not all sciences are based on the accumulation of observational data, and testing hypotheses by means of controlled experiments is difficult in some fields of scientific inquiry.

1.4 Many methods used by scientists are common to other disciplines.

1.4 Scientific theories are hypotheses or groups of related hypotheses that consistently account for (explain) existing data and consistently predict new data.

1.5 Some theories are upheld over time so consistently that they are considered to be facts; that is, little doubt exists as to their accuracy.

1.5 Scientific laws describe patterns of regularity with respect to natural phenomena.

Scientists report their work in journals and at conferences.

1.6 The two primary avenues of communication among scientists are scientific conferences and scientific journals.

1.6 Research papers describing scientific studies usually include these parts: title, abstract, introduction, methods, results, discussion, and references.

Key Terms

control *4*
controlled experiment *3*
dependent variable *3*
fact *7*

hypothesis (hi-POTH-uh-sis) *2*
independent variable *3*
qualitative data *4*
quantitative data *4*

repeated trials *5*
scientific laws *7*
scientific theory *7*

KNOWLEDGE AND COMPREHENSION QUESTIONS

1. List three key features of scientific inquiry. Are these features always present? (*Knowledge*)

2. Define the term *hypothesis*. The text states that scientists cannot prove hypotheses. Explain this statement. (*Comprehension*)

3. What is a scientific theory? What is the difference between a scientific theory and a scientific law? (*Knowledge*)

4. In the context of a controlled experiment, define the term *variable*. What is the difference between an independent variable and a dependent variable? (*Comprehension*)

5. Define the terms *quantitative data* and *qualitative data*. (*Knowledge*)

6. Do all scientists perform controlled experiments as part of their work? Why or why not? (*Comprehension*)

7. Why do scientists conduct repeated trials within an experiment? What purpose do repeated trials serve? (*Comprehension*)

8. In a scientific paper, what is an abstract and what is its purpose? (*Knowledge*)

9. List seven components that are most often present in a scientific research paper. (*Knowledge*)

KNOWLEDGE AND COMPREHENSION QUESTIONS

10. Your friend is "hooked" on telephoning for psychic readings. Are psychic readings considered scientific? Why or why not? (*Comprehension*)

K E Y

Knowledge: Recalling information.

Comprehension: Showing understanding of recalled information.

CRITICAL THINKING REVIEW

1. You've been hearing about how regular exercise (say, walking for 1 hour, 4 days a week) could help people lose weight. You decide to do a series of controlled experiments to test this idea and persuade 50 students at your school to participate. State your hypothesis and identify the independent and dependent variables. (*Application*)

2. What is a control? How would you set up the control in the preceding experiment? (*Application*)

3. Enzymes help chemical reactions take place in living systems. These biological molecules are present in a variety of grocery items. Susan noticed that her contact lens cleaner listed "enzymes" in the ingredient list. She wondered whether the cleaner contained protease, which degrades protein. She decided to test this by determining whether the cleaner degraded milk protein. Using an if/then format, state a hypothesis for Susan's investigation. (*Application*)

4. Deductive reasoning is often used in both scientific and criminal investigations. In this context, how is science sometimes like crime investigation? Give specific examples. (*Synthesis*)

5. State two examples in which the study of biology could help you be a more enlightened citizen. (*Synthesis*)

K E Y

Application: Using information in a new situation.

Synthesis: Putting together information from different sources.

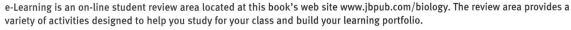

e-Learning

Location: http://www.jbpub.com/biology

e-Learning is an on-line student review area located at this book's web site www.jbpub.com/biology. The review area provides a variety of activities designed to help you study for your class and build your learning portfolio.

Review Questions The review questions test your knowledge of the important concepts and applications in each chapter. The review provides feedback for each correct or incorrect answer. This is an excellent test preparation tool.

Figure-Labeling Exercises Sharpen your visual thinking by matching terms and labels for important illustrations.

Flash Cards Studying biology requires learning new terms. Virtual flash cards help you master the new vocabulary for each chapter.

Just Wondering As you read and study from this text, you may find that you have unanswered questions. Through this site you can ask the author, Sandy Alters, your "just wondering" questions.

 Do you prefer the speed and reliability of a CD-ROM? All of the features contained on the eLearning portion of the web site are also available on the Student CD-ROM.

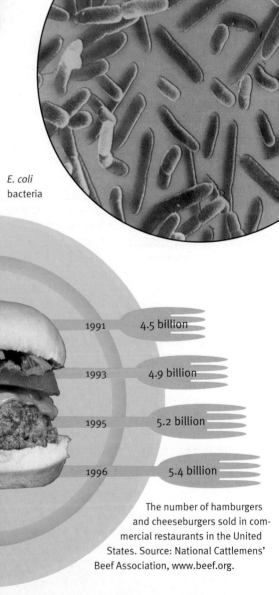

Irradiation

and the Safety of the Food Supply

On any given day, approximately 10 million Americans consume restaurant hamburgers. Our taste for burgers is increasing (see chart below) despite several incidents of contamination by *E. coli* in undercooked meats and other raw foods. These outbreaks have spurred debate about how to protect consumers from dangerous food microorganisms without creating new safety hazards.

Illness or death from *E. coli* infections had been rare until 1986, when two elderly women in Washington state succumbed to *E. coli* 0157:H7 after consuming tainted ground beef at a fast-food restaurant. Their deaths prompted Washington officials to change the cooking requirement on restaurant hamburgers and to order that *E. coli* infection be a reportable disease in that state.

In 1993, another outbreak of 0157:H7 occurred. This time it sickened 600 people and killed four children who had eaten contaminated fast-food hamburgers. Although these are already high numbers, the outbreak could have been worse. Health officials attribute the quick recognition of the 1993 outbreak, the identification of the source of the outbreak, and the ability to warn citizens so that they could receive early treatment if infected, to the new reporting requirements implemented after the 1986 outbreak.

Since the 1993 outbreak there have been other incidents. In 1996, unpasteurized apple juice contaminated with *E. coli* was linked to the illness of 13 young people in the Seattle area. In that same year in Japan over 6000 children were sickened by *E. coli* found in beef in their school lunches. In 1997, a Nebraska beef production facility recalled 25 million pounds of hamburger and shut down production because of contamination by *E. coli* bacteria.

E. coli bacteria

The culprit in the 1993 epidemic: contaminated hamburgers

E. coli is only one of many organisms that contaminate the food supply. Salmonella bacteria are a common contaminant of poultry. Certain insects and their larvae contaminate wheat and wheat flour. A wide range of organisms cause not only foodborne illness but also the spoilage of food. What can be done? Inspection by touch, sight, and smell is one way to help ensure the safety of the food we eat. Proper storage and handling is another. Killing bacteria and pathogens is a third.

One method of killing organisms in food is irradiation: that is, treating food with radiation. Radiation is energy emitted by the unstable nuclei of certain atoms in the form of rays or waves (see Chapter 3.) Food is irradiated within its packaging by exposing it to either gamma (γ) rays (a form of electromagnetic radiation

Year	Number sold
1991	4.5 billion
1993	4.9 billion
1995	5.2 billion
1996	5.4 billion

The number of hamburgers and cheeseburgers sold in commercial restaurants in the United States. Source: National Cattlemens' Beef Association, www.beef.org.

Would this apple be as nutritious for you if it were irradiated?

similar to X-rays) or to high energy electron beams produced by electron accelerators. Radiation kills living organisms within the food as the energy passes through it. And just as a dentist's X-ray does not make your teeth radioactive, irradiation does not transform food into a radioactive substance.

The food irradiation process was patented in the U.S. in 1921, but was not approved for use on the first food products (wheat, wheat flour, and white potatoes) by the Food and Drug Administration (FDA) until the early 1960s. Since then, food irradiation has been a contentious issue in the United States. Approximately 40 years after its approval, irradiation remains in limited use, although the FDA has approved its use on fresh produce, herbs, spices, pork, poultry, and red meat.

So what is the debate about? Supporters point out that irradiation has been shown to be the only way to rid ground beef of *E. coli* 0157:H7 before cooking. (Cooking ground beef thoroughly also kills this pathogen.) Irradiation also kills other bacteria, as well as insects and fungi that can make people sick or cause food spoilage. Additionally, food irradiation can be used to inhibit the sprouting of vegetables and delay the ripening of fruits. Food irradiation proponents hold, therefore, that using this process would make the food supply safer, would provide a better quality of food, and would extend the "shelf life" of food.

Opponents point out that irradiation induces chemical changes in the nutrients that compose food (lipids, proteins, carbohydrates, water, and vitamins). Therefore, they argue, these changes can affect the color, odor, and texture of food, and may change the food's chemistry to create toxins that are harmful, or even cancer-causing to humans. In addition, they charge that irradiation lowers the nutritional value of food; some vitamins, for example, are affected by radiation.

Some opponents argue that widespread approval of irradiation will cause the government to ignore other food safety problems caused by poor animal health or restaurant cleanliness. They argue that even irradiated food can be recontaminated if it is handled improperly following irradiation. Finally, some people simply oppose any technologies that involve radiation, citing problems such as nuclear waste.

What is the issue here, and what is your response to this issue?

To make an informed and thoughtful decision, it may help you to follow these steps.

1. Identify the issue described here using either a statement or a question. (An issue is a point on which various persons hold differing views.)
    For Example: Should irradiation be used to kill harmful organisms in the food supply?
2. Write your immediate reaction to the issues raised. How do you think this issue could be or should be resolved?
3. Collect new information about the issue. State the sources of that information. Explain why you believe the information to be accurate. Also determine if the information expresses a particular point of view or is biased in any other way. (Guidelines for doing efficient web searches for reliable information are available on the BioIssues website—**www.jbpub.com/biology**)
    For Example: Search the library and the Web to find information on food irradiation. Do you find any additional arguments or new information that are not noted in this BioIssues feature? Try to determine the biases of the persons writing the information before you draw conclusions from the information. Also, determine the reliability (dependability, trustworthiness) of the source of the information if possible.
4. Determine which individuals, groups, or organizations have a stake in the issue. What does each stand to gain or lose depending on the outcome of the issue?
    For Example: The BioIssues article states that The American Medical Association (AMA) a proponent of food irradiation. Its mission statement—that it seeks to "promote the betterment of public health"—can help you determine the stake that the organization has in the issue. Food suppliers also have a stake in the issue. They stand to lose money if they irradiate their products and consumers do not buy them. Or they could gain if irradiation reduced the number of lawsuits resulting from foodborne illnesses.
5. List possible outcomes (resolutions) of the issue. List the pros and cons of each outcome.
    For Example: One outcome might be that some major food suppliers decide to irradiate their products. One "pro" would be that the product (ground beef, for example) would be less likely to cause foodborne illness. One "con" might be that the price of the product would be increased due to the cost of irradiation.
6. Which outcome do you think would be best, and why? Note whether your opinion differs from or is the same as what you wrote in Step 2. Give reasons for your views and for any changes in them based on the additional information you have collected and the analysis you have done.
    For Example: Two alternative outcomes could be that (1) some major food suppliers decide to irradiate their products and (2) that food irradiation remains virtually unused as a technology to kill unwanted and disease-causing pathogens in food. You may have concluded that outcome #1 is the best based on information (cite your sources) that the low doses of irradiation needed to kill insects and microbes in food will not appreciably affect the taste or color of the food. Additionally, scientific studies (again cite your sources) do not show that cancer-causing chemicals are produced within foods due to irradiation. You could discuss other reasons (economic, cultural) as well.

Biology: Understanding Life

Science or fantasy? If you're a fan of Fox Mulder and Dana Scully, you might have wondered whether *The X-Files* is based on science or simply on writers' vivid imaginations. Well, imagination has a lot to do with this television show, but science is sometimes there, often hidden among alien abductions, fanciful creatures, and mental telepathy.

One of the people who makes certain that the few bona fide scientific processes portrayed on this program are accurate and reasonable is Dr. Anne Simon, a plant virologist at the University of Massachusetts. She researches questions about how viruses infect plants and replicate within them. The micrograph at right—done in brilliant technicolor at a magnification of 9,000,000—shows a true invasion of the body snatchers: the Pea Enation Mosaic Virus (PEMV). The virus (orange particles) has infected the nucleus of a tobacco-plant cell and hijacked its DNA, instructing the DNA to produce viral protein (the pink and orange tubelike structures). PEMV infection leads to mottling of leaves, lesions, or the outright death of the plant.

Dr. Simon is only one member of an extensive community of biologists around the world who study life in various ways and from particular points of view. That is what biology is all about—the scientific study of life.

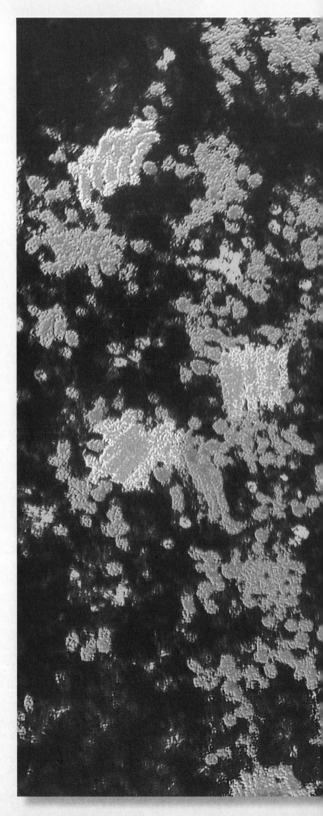

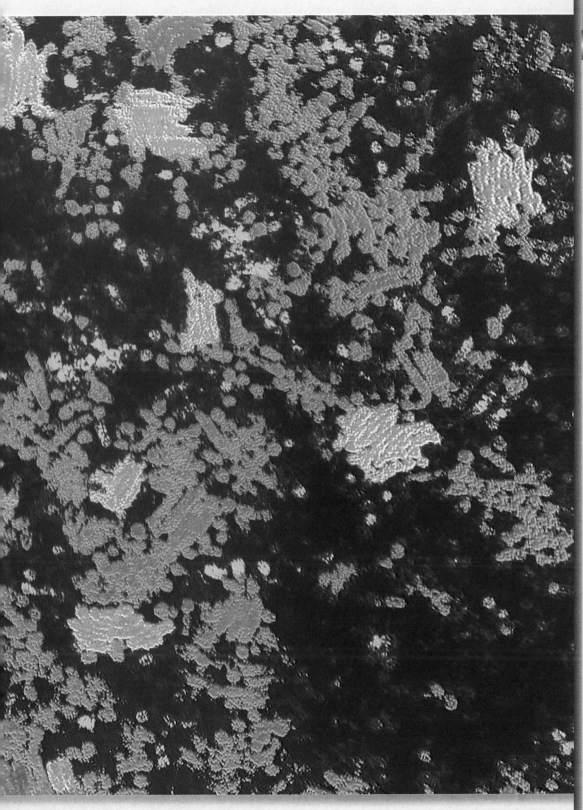

Organization

2.1 The science of biology

Biology comprises many subdisciplines, which are narrower areas of study within the broader scope of biology. For example, molecular biologists study life at the chemical level, probing the workings of the hereditary material and the molecular "chain of command" within cells. Plant virologists such as Dr. Simon are also considered molecular biologists because their work is focused on this level of life. Cell biologists study life from a slightly different perspective. They study individual cells or groups of cells, often by growing them outside of organisms in cultures; they examine cell-to-cell interactions and the effects of the environment on cells. Many cancer researchers are also cell biologists because they study this disease on the cellular level.

Dr. Anne Simon

Many biologists work at the organism level, studying animals, plants, and other multicellular organisms. Some biologists study populations—individuals of the same species occurring together at one place and at one time—and are interested in interactions among them. These population biologists study topics such as the changes in population sizes and the causes of these fluctuations. Yet other biologists work with a global orientation and are therefore interested in questions having worldwide impact. For example, some study the effects of burning billions of acres of tropical rain forest on global weather patterns and trends.

Today, the study of biology includes these focuses and many others. Biologists ask questions that probe the intricacies of life, calling on the knowledge and techniques of related fields. They answer questions that add to our knowledge base and probe areas such as the transfer of hereditary material from organism to organism, the relationship between diet and disease, and the development of new food plants. This probing advances knowledge in applied fields such as medicine, agriculture, and industry and results in the creation of products that enhance and lengthen lives. As you read *Biology: Understanding Life,* you will find out more about these and other topics that are a part of biology today. As the ancient Greeks put it, you will have a *bios logo*—a discourse on life.

> Biology comprises areas of study that focus on life at a variety of levels and from a diversity of perspectives.

2.2 The themes of life

Biology is often viewed by nonscientists as an accumulation of facts, but it is much more than that. Science is a way, or process, of understanding the world. Studying biology involves learning about problem solving, the way scientists go about their work, and worlds you may never have known existed—like the world of microbes. Studying biology involves learning how to protect the planet and how to take care of your body. But most of all, when you study biology you are learning about yourself—your evolutionary roots and your connectedness to all things living on this Earth.

As a result of observing living things, their environments, and the interactions between them, biologists have posed questions, put forth hypotheses based on those observations, and tested their predictions of the living world. Rising out of this abundance of tested predictions are the themes, or accepted explanations, that permeate the science of biology. Although the specific details of these themes may be updated or changed as biologists modify their hypotheses, the themes themselves transcend time, describing the characteristics of life and helping to answer the question, "What distinguishes the living from the nonliving?" However, as you read the following descriptions of these themes (they are also listed in Table 2.1), you may realize that nonliving things possess certain of these characteristics too. Taken together, however, these ideas embody life; and that is where we begin—with the idea that the whole is equal to *more* than the sum of its parts.

Table 2.1

The Characteristics of Life

- Cellular composition and hierarchical organization (cells → tissues → organs → organ systems)
- Interaction with one another and the environment
- Emergent properties (the characteristics of an organism result from its organization, not simply from the sum of its parts)
- DNA as hereditary material, directing structure and function, which is passed on through reproduction
- Transformation of energy and maintenance of a steady state
- Diversity in type with unity in patterns among organisms
- Form that fits function
- Evolution of species

> From their study of life, biologists have developed an understanding of characteristic qualities and traits that, when taken together, distinguish living from nonliving things.

2.3   Cellular and hierarchical organization

First, what constitutes the "parts" of living things? You have probably known the answer to this question since middle school—possibly before: All living things are composed of cells. Cells make up the intricate fanlike pattern in the cross section of a plant stem shown at the left.

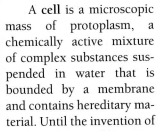

A **cell** is a microscopic mass of protoplasm, a chemically active mixture of complex substances suspended in water that is bounded by a membrane and contains hereditary material. Until the invention of the microscope around 1600, naturalists could not see or study this invisible level of organization of living things. It was not until the mid-1800s when advances in the technology of the microscope allowed botanist Matthias Schleiden and zoologist Theodor Schwann to determine (through repeated lines of inquiry) that the unit of structure of all living things is the cell. Shortly thereafter, the German medical microscopist Rudolf Virchow argued that all cells could arise only from preexisting cells. The cells, he wrote, are "the last constant link in the great chain of mutually subordinated formations that form tissues, organs, systems, the individual. Below them is nothing but change."

Virchow's statement refers to the hierarchy, or levels, of organization seen in all living things. Living things, or **organisms**, are either *multicellular* (composed of many cells, like the plant above) or *unicellular* (composed of a single cell). Single-celled organisms that work and live together as a team are called *colonial organisms* (like the *Volvox* shown at the left). Whether unicellular, multicellular, or colonial, every organism has the cell as its simplest level of structure and function.

Smaller units, such as atoms and molecules, make up cells but are not living units. The hereditary material, or DNA, present in almost every cell of an organism's body is composed of atoms, for example. Although this huge molecule is critical to life and is passed on from parent to offspring, it is still not a living unit by itself.

As shown in **Figure 2.1** on the following page, cells in multicellular organisms are organized to form the structures and perform the functions of the organism. The next more complex, or more inclusive, level of organization is the tissue. **Tissues** are groups of similar cells that work together to perform a function. Grouped together, various tissues form a structural and functional unit called an **organ**. An **organ system** is a group of organs that function together to carry out the principal activities of the organism—its most complex level of organization.

Interactions take place within an organism among its levels of organization. For example, when a chameleon sees an insect land on a nearby plant, a complex series of events occur within the chameleon. First, the nerve *cells* embedded in the back wall of its eyes (*organs*) conduct impulses to its brain (an *organ*). The vision center within the brain (*nervous tissue*) interprets these impulses, resulting in the chameleon seeing the insect. Impulses speed to other brain centers to coordinate a rapid response in the form of nerve impulses to the muscles in its tongue (*organs*). The muscles respond at incredible speed, allowing the tongue to extend to its prey. Had it been a better day for the insect, its nervous system would had reacted more quickly in perceiving and responding to the threat.

Every organism has the cell as its simplest level of structure and function. Cells work together in hierarchical fashion, forming the tissues, organs, and organ systems of the organism.

Life has a variety of characteristics.

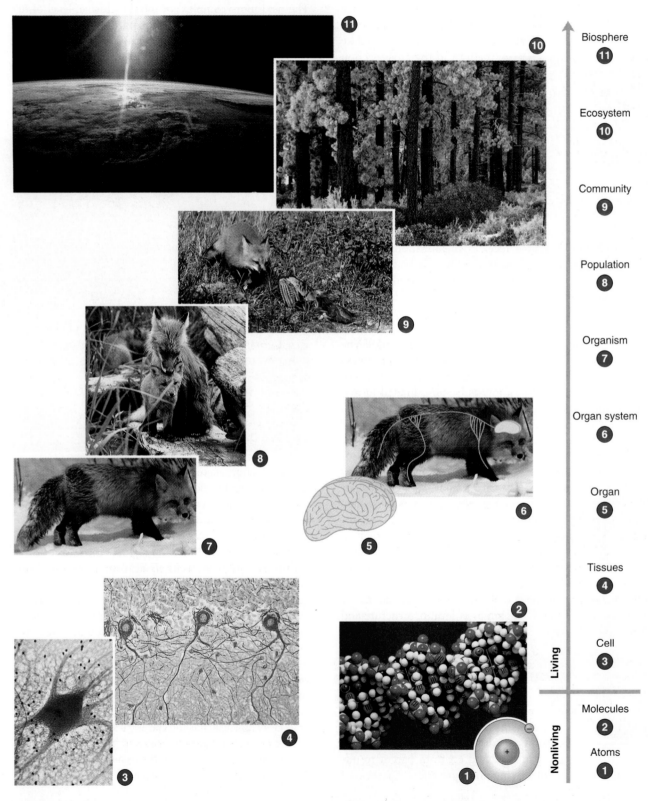

Visual Summary **Figure 2.1** **The levels of biological organization.** The progression of organization from the "lowest" level of structure and function, the atom, to the "highest" level of structure and function, the biosphere.

Biosphere ⑪

Ecosystem ⑩

Community ⑨

Population ⑧

Organism ⑦

Organ system ⑥

Organ ⑤

Tissues ④

Cell ③

Living

Molecules ②

Atoms ①

Nonliving

2.4 Interaction

Interactions occur not only among the levels of organization within organisms but also between organisms and their external environments. Biologists usually classify the interactions between organisms and their environments into the following hierarchy of levels of organization: populations, communities, ecosystems, and the biosphere. These levels of organization build on the levels of organization of individual organisms and are depicted in Figure 2.1.

A **population** consists of the individuals of a given species that occur together at one place and at one time. To continue the example, the foxes living in a small forest make up a population. The foxes interact with one another in a variety of ways. Sometimes they compete for the same limited resources and for mates. Conversely, individuals within animal populations may also work cooperatively for common purposes. Foxes often hunt together, for example, working with one another to overtake and trap prey.

Populations of different species that interact with one another make up a **community** of organisms. A forest community may be made up of populations of bacteria, fungi, earthworms, plant-eating and animal-eating insects, mice, deer, salamanders, frogs, foxes, snakes, hawks, trees, and grasses. These populations within the forest community compete with one another for resources, as do organisms within populations. The foxes, snakes, and hawks, for example, compete with one another to capture the mice for food. Other types of interactions may also exist within the community, such as mutualistic relationships in which two different species live in a close association for the benefit of both. For example, fungi and plants are often interdependent on one another. Some fungi live near the roots of many plants, such as trees in a forest. These fungi (seen as white in the photo below) envelop the roots and send billions of minute cell extensions into the soil. These microscopic "fingers" of fungus absorb water and nutrients better than the plants' roots could alone, and they pass the substances to the plant. In turn, the fungus uses certain products that the plant makes by photosynthesis.

An **ecosystem** is a community of plants, animals, and microorganisms that interact with one another and their environments and that are interdependent. The forest ecosystem includes all the organisms previously discussed as well as nonliving components of the environment such as air and water, which contribute substances needed for the ecosystem to function.

The **biosphere** is the part of the Earth where biological activity exists. Most people refer to the biosphere as simply "the environment." Within this global environment, living things interact with each other and with nonliving resources in various ways. Organisms other than humans use only *renewable* (replaceable) *resources*. When they die, decomposers return the nutrients held within the dead organisms to the soil and air. Humans, however, use many *nonrenewable resources* and fill the land with wastes. Scientists are testing and refining alternative renewable energy sources such as solar, wind, water, and geothermal energy to help curb the use of nonrenewable energy sources such as coal, oil, and natural gas. The most severe crisis of natural resource destruction is occurring today in the tropical rain forests. Species are dying out as their rain forest habitats, or homes, are lost. In addition, humans are overpopulating the land and polluting the water and the air. The destiny of future generations of all living things depends on our solving the problems humans have created and our learning to live on this planet without harming it.

> Interactions occur among the levels of organization within organisms, and between organisms and their external environments. The levels of interaction between organisms and the environment are those of populations, communities, ecosystems, and the biosphere.

2.5 Emergent properties

As you were reading about organisms and how they interact with one another, you may have gained insight into the earlier statement, "the whole is equal to *more* than the sum of its parts." It is an important concept in biology. Cells are much more than the atoms and molecules that compose them. Tissues are much more than the cells that make up their structure, and so on. Otherwise, all organisms made out of the same components would be the same. Certainly, you are much different from a polar bear sunning on a chunk of ice or from a salmon swimming upstream to spawn. In other words, the workings of a cell, tissue, organ, system, or organism cannot be predicted simply by knowing which components make up its structure.

To illustrate further, you might decide to build a cabinet to house your CD player and CDs. If your neighbor came over and looked at the materials you bought to build the cabinet—lumber, nails, hinges, and wood stain—he or she would not be able to tell what you were building and how it would function. Only when you organize the materials—build the cabinet—will your neighbor be able to see the properties of your CD cabinet. Additionally, the characteristics of some of the components would change after becoming a part of the finished cabinet. Stain, for example, no longer retains its liquid properties after being used on wood. It permeates the wood and is forever different. The wood, too, is different, with its grain enhanced and shaded by the stain. Pieces of wood, unable to function as doors by themselves, now take on the properties of doors when attached with hinges to the body of the cabinet. These new characteristics have emerged at this higher level of organization of the wood, nails, hinges, and stain. So, too, do unique properties of living things emerge at each level of their organization. These emergent properties are unable to be predicted simply by knowledge of an organism's component parts.

An organism is more than just the parts of which it is composed. Unique properties emerge from interactions within an organism at its various levels of organization and from interactions with its environment.

2.6 Reproduction

Living things all have biological information that directs their structure and function, and ultimately, therefore, their emergent properties. This biological information is the hereditary material, or *genes*. Genes are made up of molecules of **DNA**, or **deoxyribonucleic acid** (de-ok-see-RYE-boh-new-KLAY-ick). In these molecules of DNA exist the "code of life," instructions that are translated into a working organism. These instructions take the form of molecular subunits of the DNA molecule called *nucleotides*. Using only four different nucleotides, DNA codes for all the structural and functional components of an organism. DNA, its structure, and its functioning are described in detail in Chapter 18. (A model of this complex molecule is shown at left.)

The secret to DNA's ability to carry information regarding the variety of structural and functional components of all living things lies in its code. The four nucleotides of DNA are used like code letters to produce code words, each composed of three letters. These code words are then placed in a particular sequence to produce code sentences. When translated in a living organism, the code sentences result in the formation of a specific molecule designed to dō a specific job. Thus, the same code letters and code words produce the diversity of life by producing code sentences unique to each of the many species of organisms of the living world.

DNA is passed from organism to organism during reproduction, as organisms give rise to others of their kind. Some organisms reproduce sexually and some asexually, while others reproduce using both methods. In sexual reproduction, two parents give rise to offspring; in asexual reproduction, a single parent gives rise to offspring. In either case, genes—the units of heredity—are passed from one generation to the next.

Living things reproduce and pass on biological information to their offspring.

2.7 Transformation of energy

Organisms need energy to do the work of living. This work involves many processes, such as movement, cell repair, reproduction, and growth. It also involves maintaining a stable internal environment in spite of a differing external environment. Energy drives the chemical reactions that underlie all these activities.

What is the source of the energy that fuels life processes and helps organisms maintain an inner equilibrium? Ultimately, it comes from the sun. For example, certain organisms within our forest ecosystem such as the trees and grasses make their own food by capturing energy from the sun in a process called *photosynthesis* (fote-oh-SIN-thuh-sis). These organisms are called **producers**. During photosynthesis, producers (primarily green plants) convert the energy in sunlight into chemical energy by locking it within the bonds of the food molecules they synthesize. Organisms that cannot make their own food, such as the insects, mice, deer, salamanders, frogs, foxes, snakes, and hawks in the forest, are called **consumers**. They feed on the producers and on each other, passing along energy that was once captured from the sun (Figure 2.2). Both the producers and the consumers release the stored energy in food by breaking down its molecules bit by bit. As food molecules are broken down, much of the energy that is released is used to do work, but some of it is lost as heat and is therefore unusable. As a result, organisms need a continual input of energy to fuel the chains of chemical reactions that move, store, and free energy needed to perform the activities of life. **Decomposers**, such as many types of bacteria and fungi, break down the organic molecules of dead organisms, serving as the last link in the flow of energy through an ecosystem and contributing to the recycling of nutrients within the environment.

> Organisms use energy from the sun and the energy locked up in the molecules of other living things to fuel cell work. Energy flows through an ecosystem, from the sun to producers. Consumers eat producers and other consumers, transforming the energy within them. The decomposers, the last link in the chain of flow through an ecosystem, unlock the energy in dead organisms.

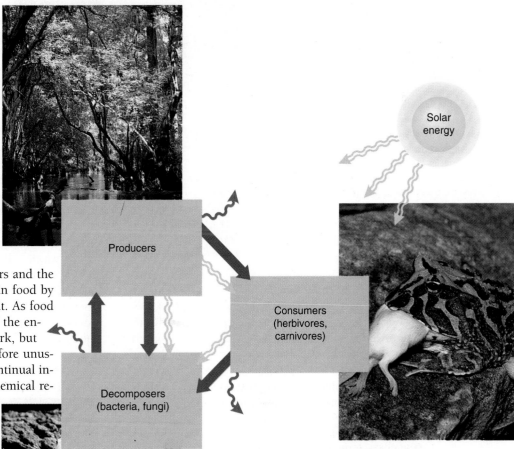

Nutrients

Energy

Heat

Figure 2.2 Flow of energy through an ecosystem. Producers transform the sun's energy (yellow wavy lines) to chemical energy by means of photosynthesis. Nutrients (blue arrows) are then transferred from producer to consumer and from consumer to consumer. Decomposers such as fungi break down the organic molecules of dead organisms, making these nutrients available for reuse. Some energy is lost as heat (red wavy lines); therefore, organisms need a constant input of energy to perform the activities of life.

2.8 Diversity and unity

The diversity, or variety, of living things is astounding. Biologists estimate that from 5 million to 30 million *different* species exist on Earth. As you can see from this range of numbers, scientists are unsure how many species exist; they have discovered, described, and catalogued probably fewer than half of them. Unfortunately, many species are becoming extinct, or dying out, as their habitats are destroyed, and before scientists can even study and classify them.

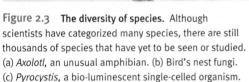

Figure 2.3 **The diversity of species.** Although scientists have categorized many species, there are still thousands of species that have yet to be seen or studied. (a) *Axolotl,* an unusual amphibian. (b) Bird's nest fungi. (c) *Pyrocystis,* a bio-luminescent single-celled organism.

A **species** (SPEE-shees *or* SPEE-sees) is a population of organisms that interbreed freely in their natural settings and do not interbreed with other populations. Put simply, members of one species are reproductively isolated from members of other species. Organisms that do not reproduce sexually such as bacteria and certain plants, animals, protists, and fungi are designated as species by means of their morphological characteristics (those relating to their form and structure) and biochemical characteristics (those relating to their chemical composition and processes).

Only a sampling of the array of the species humans have seen and categorized is shown in **Figure 2.3**. Although these organisms are very different from one another, each has characteristics that are common to all species. (These shared characteristics, however, may or may not be visible to the naked eye.) The other themes of life describe these characteristics and describe how and why organisms are diverse in their types but unified in their patterns.

> Although there is incredible diversity among the types of organisms that exist, each organism has characteristics that are common among all organisms.

2.9 Form that fits function

Would it make sense to try to turn a screw with a hammer or eat soup with a knife? Of course not—tools and kitchen utensils are structured in specific ways to do specific jobs. Just as the shape and structure of a screwdriver or a spoon fits its function, the structures of living things fit their functions. Biologists sum up this idea with the phrase "form fits function." By analyzing form, inferences can be made regarding function. Conversely, knowing function gives insight into form.

For example, there are a variety of types of bird feathers (**Figure 2.4**). Consider the form of a feather that would serve to insulate a bird against the cold. First, determine the characteristics of a good insulator from your everyday experience. Are you warmer in cold weather when you wear one thick sweater, or are you warmer when you wear many thinner layers that together may have the same thickness as the sweater? The answer is that many layers provide better insulation against the cold because they trap air between the layers. The heat radiating from your body then warms this trapped air. In essence, you create a blanket of warm air around your body. A down feather (shown at right) serves this function well.

> The structures of living things fit their functions.

(a)

(b)

Figure 2.4 **Different types of feathers: Form fits function.** (a) The brilliantly colored feathers on the head of this cockatoo are display feathers. (b) Quite different in form are the feathers used for flight.

2.10 Change over time

All organisms have common characteristics because they are related to one another. Just as you have a history and a family tree, so does the Earth's family of organisms. Yet all organisms have differences because as they changed over time, they diverged from one another. The Earth itself has changed from its beginnings some 4.6 billion years ago.

Scientists know very little about what "incubator Earth" was like nearly 4 billion years ago, but they do agree that it was a harsh environment different from that of today. Under these conditions, many scientists hypothesize that the elements and simple compounds of the primitive atmosphere reacted with one another, forming complex molecules. In a way that scientists can only hypothesize, biochemical change took place over time and resulted in the appearance of single-celled organisms approximately 3.5 billion years ago. The remains of these early cells (and of any organisms) preserved in rocks are called *fossils*. Fossils provide scientists with a record of the history of life and document the changes in living things that have taken place over billions of years (Figure 2.5). By 1.4 to 1.2 billion years ago, the fossil record documents the existence of cells more complex than the first cells, and by 500 million years ago, an abundance of multicellular organisms—with members of groups similar to those that exist today—had appeared.

The fossil record is only one piece of evidence suggesting that living things have changed over time. Using this and other types of evidence, Charles Darwin, a nineteenth century English naturalist, developed what was then a hypothesis of organismal change over time or, as he put it, "descent with modification." He termed his hypothesis **evolution**. Darwin also proposed a mechanism by which evolution took place. Since Darwin's time, his hypothesis has been consistently supported by an overwhelming amount of scientific data and has therefore become a well-accepted theory. This theory embodies the ideas that organisms alive today are descendants of organisms that lived long ago and that organisms have changed and diverged from one another over billions of years. Scientists still ponder, examine, and develop hypotheses regarding details of the mechanisms of evolution, but they agree that evolution has and is taking place.

How does evolution take place? A key idea is that some of the individuals within a population of organisms possess inheritable characteristics that favor their survival. These individuals are more likely to live to reproductive age than are individuals not possessing the favorable characteristics. These reproductively advantageous traits (called adaptive traits or adaptations) are passed on from surviving individuals to their offspring. Over time, the individuals carrying these traits will increase in numbers within the population, and the nature of the population as a whole will gradually change. This process of survival of the most reproductively fit organisms is called *natural selection.*

Scientists disagree regarding the pace of evolutionary change. Darwin suggested that new species develop slowly and gradually as an entire species changes over time. In the early 1970s, two American scientists, Niles Eldridge and Stephen Jay Gould proposed that a species might remain unchanged over long periods of time, punctuated with relatively rapid periods of change. Chapters 24 and 25 discuss these ideas and the major concepts of evolution in more detail.

> Organisms have changed and diverged from one another over billions of years. Organisms alive today are descendants of organisms that lived long ago.

Figure 2.5 **Fossil links between birds and dinosaurs.** In June, 1998, a team of scientists from the United States, China, and Canada announced fossil finds of turkey-sized dinosaurs with birdlike feathers. The photo above shows one of the "finds"—an impression in rock of *Caudipteryx*, which means "tail feathers." These display feathers are shown in the fossil photo; *Caudipteryx* probably fanned them out much like a peacock trying to win a mate. Also birdlike are *Caudipteryx*'s teeth and gastroliths, stones that the animal ate to grind food as in the gizzards of modern-day birds. Its arms and fingers are like those of a dinosaur.

2.11 The kingdoms of life

As in any family tree, some organisms are more closely related than others are. The species of the world are no exception. In the family tree of life, organisms having a common ancestor in the not-too-distant past are closely related. Organisms having a common ancestor farther down the family tree are distantly related.

Modern **taxonomy** (tack-SAHN-uh-me), the classification of the diverse array of species, categorizes organisms based on their common ancestry (their evolutionary history). Therefore, organisms that are close relatives and that resemble one another in a variety of ways are placed in common groupings reflecting their similarities and closeness; organisms not closely related are placed in groupings separate from one another that reflect their differences.

Taxonomists group all living things into broad categories called **kingdoms**. However, not all taxonomists agree on a single way to classify organisms; therefore, many taxonomic (classification) schemes exist.

The taxonomic scheme used in this book recognizes five kingdoms of life: Bacteria, Protista (pro-TISS-tah), Plantae (PLAN-teye), Fungi (FUN-jeye), and Animalia (ah-nih-MAHL-ee-ah) **(Figure 2.6)**. This scheme has been widely accepted for about 30 years; the five kingdoms are described shortly.

Recently, the three-domain system has gained some support, primarily among microbiologists. This system places all of life into three domains: Archaebacteria, Eubacteria, and Eukaryota. The bacteria, which make up one kingdom in the five-kingdom system, comprise two domains: the archaebacteria and the eubacteria. All the rest of life is classified in one domain: the Eukaryota.

In the three-domain system, organisms are primarily classified based on sequences of molecules in particular genetic (hereditary) material called ribosomal RNA (see p. 392). In the five-kingdom system, organisms are classified based on a variety of characteristics such as genetic similarities, developmental patterns, morphologies (body forms), and behaviors. At this time, most biologists use the five-kingdom system because it takes into account a wide variety of organismal features.

Kingdom Bacteria

Bacteria differ from the other four kingdoms of life in that their cells lack membrane-bounded intracellular structures called *organelles*. (They do contain ribosomes, but ribosomes are not bounded by membranes.) In addition, bacteria do not have a membrane that sur-

rounds the hereditary material of the cell, forming a nucleus. Instead, their genetic material is located within a particular region within the cell called a *nucleoid*. Such unicellular organisms, having no membrane-bounded organelles and nuclei, are called **prokaryotes** (pro-KARE-ee-oats). This word means before (pro-) the nucleus (-karyon). The organisms of the other four kingdoms are made up of cells, each containing membrane-bounded organelles and a membrane-bounded nucleus. Such cells are called **eukaryotes** (you-KARE-ee-oats). This word means true (*eu-*) nucleus (*-karyon*). This distinction among organisms as being either prokaryotic or eukaryotic is denoted as a *superkingdom*. Superkingdom Prokarya contains only the prokaryotes (bacteria). Superkingdom Eukarya contains only the eukaryotes (all other organisms—protists, fungi, animals, and plants).

Finding bacteria in the environment is an easy task—they live almost everywhere. If you took a moist cotton swab and drew it across any surface (except one that had been sterilized), you would collect an array of bacteria that is visible under a microscope. Although you may be most familiar with their role in causing disease, bacteria play key roles in the wide range of habitats in which they live.

The Kingdom Bacteria is divided into two Subkingdoms: *Archaebacteria* (or "ancient bacteria") and *Eubacteria* (or "true" bacteria). These two groups of bacteria are primarily distinguished from one another by their ribosomal RNA, as noted previously. Other differences include chemical variations within their cell walls, within their cell membranes, and in certain of their physiological processes. Many taxonomists consider these differences to be so great that they place the archaebacteria in a separate, sixth kingdom or use the three-domain system of classification.

Kingdom Protista

Ancestors of the bacteria gave rise to organisms in the Kingdom Protista. The protist kingdom consists partially of single-celled eukaryotes, whose cells are much more complex and much larger than the bacteria. You may be familiar with some common single-celled protists such as the blob-like ameba or the oval-shaped paramecium. The protist kingdom also includes the algae, which have both multicellular and unicellular forms. Algae are classified as protists because they seem to be more closely related to this group than to the plants, animals, or fungi.

By collecting a small amount of pond water and looking at it under a microscope, you can see some of the organisms in the protist kingdom. The remaining three kingdoms originated from the protists.

www.jbpub.com/biology

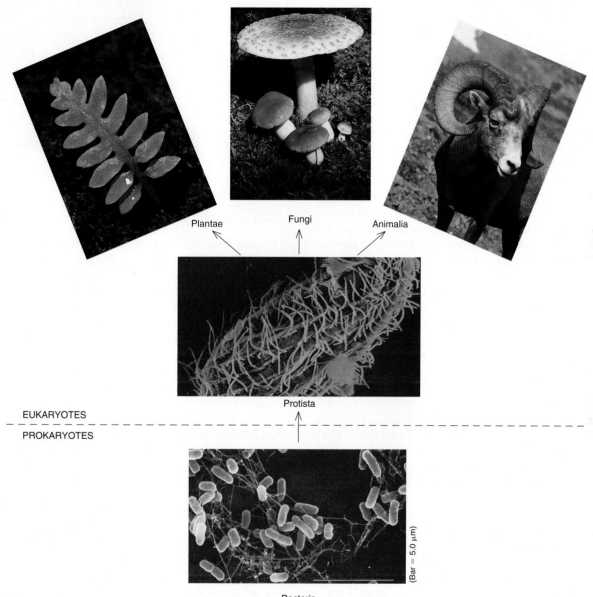

Plantae   Fungi   Animalia

Protista

EUKARYOTES
PROKARYOTES

(Bar = 5.0 μm)

Bacteria

Kingdom Fungi

Fungi are decomposers (as are many species of bacteria) and survive by breaking down substances and absorbing the breakdown products. If you've ever seen mold growing on stale bread or on rotting fruit, or mushrooms growing in the forest, you are familiar with some members of the fungi kingdom. Most fungi are multicellular; a few, such as the yeasts, are unicellular.

Most fungi are **saprophytes**; that is, they live on organisms that are no longer alive, such as fallen trees and leaves, and dead animals. Some fungi, however, are **parasites**—they live on living organisms. Certain human diseases and conditions such as athlete's foot and ringworm are caused by fungi, as are plant diseases such as potato blight.

Kingdom Plantae

The plants are multicellular eukaryotic organisms that make their own food. Such organisms are called **autotrophs**. (Autotrophs are producers; see p. 23.) Plants are photosynthetic

Just Wondering . . . *Real Questions Students Ask*

I usually find scientific words hard to understand. Why are they always so long and hard to pronounce?

The scientific names of organisms are usually descriptive Latin or Greek names. In this way, scientists worldwide have a common language to refer to the myriad species that have been described. In addition, these languages are "dead" languages; they are not used conversationally today. Therefore, the meanings of the descriptors do not change with the times as do certain words in the English language. Latin and Greek are also often the languages from which the names of structures, processes, and diseases are derived. Students not familiar with these languages often find names and terms based on these languages difficult to learn. Studying the meanings of prefixes and suffixes and parts of words that are used in many contexts is often helpful. In *Biology: Understanding Life*, the derivation of many unusual and difficult words is explained to help students understand the meanings behind the names.

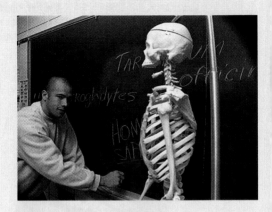

autotrophs; that is, they make their own food by the process of photosynthesis, in which they use carbon dioxide from the air, light energy from the sun, and water. Chapter 6 describes this process. Plants are usually distinguished from the algae in that plants live on land, while algae live in the water. Some botanists disagree with this distinction because aquatic plants exist. However, the few aquatic plants that exist evolved on land and then became adapted to life in the water. Examples of plants are mosses, ferns, pine trees, and flowering plants.

Kingdom Animalia

Animals are multicellular eukaryotic organisms that cannot make their own food. They are **heterotrophs**; that is, they obtain food by eating and digesting other organisms. (Heterotrophs are consumers or decomposers; see p. 23.)

The animal kingdom is an extraordinarily diverse group of organisms, which can be thought of as two types of groups: those without a backbone (the invertebrates) and those with a backbone (the vertebrates). Vertebrae are the series of bones that stacked one upon the other comprise the vertebral column, or backbone. Examples of invertebrates are worms, sea stars, spiders, crabs, and insects. Examples of vertebrates are dogs, cats, fish, birds, frogs, snakes, and, of course, humans.

Modern taxonomy, the classification of the diverse array of species, categorizes organisms based on their common ancestry. The largest groupings of organisms in the taxonomic scheme used in this book are kingdoms: Bacteria, Protista, Plantae, Fungi, and Animalia. A taxonomic scheme gaining some interest groups all of life into three domains.

2.12 Subdivisions of the kingdoms

The kingdoms of organisms are subdivided into groupings that reflect an increasing closeness in evolutionary history. Many types of evidence, including organisms' similarities in their existing behavioral (in the case of animals) and biochemical characteristics determine this closeness.

The kingdoms are each subdivided into **phyla** (singular, **phylum** [FYE-lum]). The animal kingdom, for example, has approximately 19 phyla depending on the particular classification system that is referenced.*

The top box in **Figure 2.7** shows some representatives of the animal kingdom ❶. One of the phyla within this kingdom is the phylum Chordata: animals with a backbone.† The next box down ❷ pictures the chordates from among the animals in the top box. Notice that the sea star and the butterfly are not shown. These

*The classification system used in this textbook is based on Margulis, L. and Schwartz, K.V. (1998). *Five Kingdoms: An Illustrated Guide to the Phyla of Life on Earth,* 3rd ed. New York: W.H. Freeman & Co. (See Appendix B.)

†This definition of the term *chordate* (phylum Chordata) is used here for simplicity. Please see Chapter 31 and the glossary for a more complete and precise definition.

two organisms, although both animals, are not chordates. All the other animals originally shown are included. These animals (the chordates) are more closely related to one another than to the sea star or butterfly.

The next subcategory is **class ③**. The class Mammalia is one subgroup of the phylum Chordata. The snake and the fish are not included in this group because they are not mammals. Mammals are characterized by having skin with hair and by nourishing their young with milk secreted by mammary glands (see photo above). These are characteristics that fish and snakes do not have.

The next subcategory is **order ④**. One order of class Mammalia is the order Rodentia, which includes most gnawing mammals. Humans are not gnawing animals and are not included in this subgroup. The last three subgroupings, in order of increasing relatedness, are **family ⑤**, **genus ⑥**, and *species* **⑦**. Of all the organisms shown, which one is most closely related to squirrels?

The last two categories in the hierarchy of classification, genus and species, provide the scientific name of an organism. The "first name" or genus name is the same for all organisms in the same genus. For example, cabbage and rutabaga are closely related and are in the same genus. The "first name" (genus) for both of them is *Brassica*. The "second name" is the species within the genus. Their species names are different: cabbage is *Brassica oleracea* and rutabaga is *Brassica rapus*. This system is known as **binomial nomenclature**—literally, "two-name naming." Scientifically, humans are *Homo sapiens,* meaning "wise man." House cats are *Felis domesticus,* meaning "domesticated cat." Notice that the genus name is capitalized and the species name begins with a lowercase letter. Both are italicized. What is the scientific name for the squirrel in our example?

The kingdoms of organisms are subdivided into groupings that reflect an increasing closeness in evolutionary history. These groupings are: phylum (or division), class, order, family, genus, and species.

① KINGDOM

Animalia
- multicellular
- eukaryotic
- heterotrophs

② PHYLUM

Chordata
- animals with a backbone

③ CLASS

Mammalia
- skin with hair
- offspring are nourished with milk from mammary glands

④ ORDER

Rodentia
- most gnawing mammals

⑤ FAMILY

Sciuridae
- tree squirrels
- ground squirrels
- woodchucks
- chipmunks

⑥ GENUS

Tamiasciurus
- tree squirrels

⑦ SPECIES

Tamiasciurus hudsonicus

Red squirrel

Figure 2.7 **Classification scheme for a squirrel.** The classification moves from the large taxonomic category, kingdom, to the small subcategory, species. Fewer and fewer organisms are included in each subcategory as the organisms are separated into groups reflecting their evolutionary closeness. Many types of animals are in the animal kingdom. Only tree squirrels are in the genus *Tamiasciurus*.

Key Concepts

Biology is the study of life.

2.1 Biology comprises areas of study that focus on life at a variety of levels and from a diversity of perspectives.

2.2 The themes of biology are science-based explanations regarding the structure, function, and behavior of living things as well as their interrelationships.

Life has a variety of characteristics.

2.3 Living things are composed of cells and are hierarchically organized.

2.4 Living things interact with each other and their environments.

2.5 Living things have emergent properties.

2.6 Living things reproduce and pass on biological information to their offspring.

2.7 Living things transform energy and maintain a steady internal environment.

2.8 Living things display both diversity and unity.

2.9 Living things exhibit forms that fit their functions.

2.10 Species change over time, or evolve.

Classifying life is based on common ancestry.

2.11 Modern taxonomy, the classification of the diverse array of species, categorizes organisms based on their common ancestry (their evolutionary history).

2.11 Organisms that are close relatives and resemble one another in a variety of ways are placed in common groupings reflecting their similarities and closeness.

2.11 Taxonomists group all living things into broad categories, or kingdoms. This book uses a five-kingdom system of classification: Bacteria, Protista, Plantae, Fungi, and Animalia.

2.12 The kingdoms of organisms are subdivided into groupings that reflect an increasing closeness in evolutionary history as determined by a variety of criteria.

Key Terms

autotrophs *27*
binomial nomenclature (bye-NO-mee-uhl NO-men-clay-chur) *29*
biosphere *21*
cell *19*
class *29*
community *21*
consumers *23*
decomposers *23*
deoxyribonucleic acid (de-ok-see-RYE-boh-new-KLAY-ick) *22*

ecosystem *21*
eukaryotes (you-KARE-ee-oats) *26*
evolution (ev-oh-LOO-shun) *25*
family *29*
genus (GEE-nus) *29*
heterotrophs *28*
kingdoms *26*
order *29*
organ *19*
organ system *19*
organisms *19*

parasites *27*
phyla (FYE-luh) *28*
population *21*
producers *23*
prokaryotes (pro-KARE-ee-oats) *26*
saprophytes *27*
species (SPEE-shees *or* SPEE-sees) *24*
taxonomy (tack-SAHN-uh-me) *26*
tissues *19*

KNOWLEDGE AND COMPREHENSION QUESTIONS

1. Describe three topics that biologists might study. (*Comprehension*)

2. List the eight unifying themes of biology (the characteristics of life) described in this textbook. (*Knowledge*)

3. List and describe the four levels of internal organization in multicellular organisms. (*Comprehension*)

4. Name and describe the different levels of interactions that occur between organisms and their environments. (*Comprehension*)

5. How do humans' interactions with the biosphere differ from those of all other organisms? (*Comprehension*)

6. From what ultimate source do you obtain the energy that keeps you alive? Explain how you obtain energy from that source. (*Comprehension*)

7. What is DNA? Explain its role in living things. (*Comprehension*)

8. Briefly state the theory of evolution. What are fossils and how do they relate to this theory? (*Knowledge*)

9. Name the five kingdoms of living things and give an example of an organism from each. To which kingdom do you belong? (*Comprehension*)

10. Place the following terms in their correct sequence, starting with the term that refers to the group containing the largest number of organisms: phylum, species, genus, order, family, class, kingdom. (*Comprehension*)

KEY
Knowledge: Recalling information.
Comprehension: Showing understanding of recalled information.

CRITICAL THINKING REVIEW

1. Using the information that two organisms are members of the same order, determine whether the following statements are true or false. Explain your answer.
   a. The organisms must belong to the same class.
   b. The organisms must belong to the same genus.
   c. The organisms must have the same "first name" using binomial nomenclature.
   (*Analysis*)

2. If all living things are composed of one or more cells, it follows that a computer-driven machine cannot be alive. And yet such a machine can transform energy, do work, make copies of itself, and evolve over time to better suit the challenges of its environment. Can machines be considered living? From a biological perspective, what constitutes life? (*Analysis*)

3. Choose an organism and state its common name and kingdom. Describe three characteristics it shares with an organism from another kingdom. (State the common name and kingdom of that organism.) Additionally, describe three ways in which these two organisms are different from one another. (*Application*)

4. Although you never considered yourself to have a "green thumb," you recently developed an interest in gardening and took an adult-education course on gardening offered at your local library. You've just planted vegetable seeds and young plants. Assuming that your course will help you manage your garden well, use your knowledge of the characteristics of life to predict five observations you will make over the growing season. (*Synthesis*)

5. Structurally, most viruses are made up of hereditary material encased in protein. Some biologists suggest that viruses are a bridge between the living and the nonliving world. Other scientists suggest that viruses are simply nonliving things. Evaluate both points of view regarding viruses using your personal knowledge of having had a viral infection, such as the flu, your knowledge of the characteristics of life, and the information stated in this question. (*Evaluation*)

KEY
Application: Using information in a new situation.
Analysis: Breaking down information into component parts.
Synthesis: Putting together information from different sources.
Evaluation: Making informed decisions.

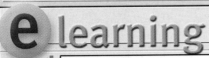

Location: http://www.jbpub.com/biology

e-Learning is an on-line student review area located at this book's web site www.jbpub.com/biology. The review area provides a variety of activities designed to help you study for your class and build your learning portfolio.

Review Questions The review questions test your knowledge of the important concepts and applications in each chapter. The review provides feedback for each correct or incorrect answer. This is an excellent test preparation tool.

Figure-Labeling Exercises Sharpen your visual thinking by matching terms and labels for important illustrations.

Flash Cards Studying biology requires learning new terms. Virtual flash cards help you master the new vocabulary for each chapter.

Just Wondering As you read and study from this text, you may find that you have unanswered questions. Through this site you can ask the author, Sandy Alters, your "just wondering" questions.

 Do you prefer the speed and reliability of a CD-ROM? All of the features contained on the eLearning portion of the web site are also available on the Student CD-ROM.

The Chemistry of Life

Water is essential for life on Earth. In fact, life as we know it could not have arisen and evolved without water, and could not go on if deprived of this amazing liquid. Water is the medium in which many species live. Fish, like the one shown here, filter not only food from water but also the oxygen they "breathe." Terrestrial animals, like the kingfisher, are no less dependent upon water. They must take in water to support their internal watery environment (the body). Water also is necessary to support their food supply, be it plant- or animal-based. What is it about water that makes it so important to life?

The clues needed to answer this question lie in the chemical properties of water. Put another way, water is important because of its characteristics, which in turn depend on its parts and how they are put together. In the photo above, you can easily observe some of these characteristics. Notice how the water sticks together to form droplets in the air. Notice, too, that its surface behaves like a thin film, caving in and bulging out in tiny waves as the airborne droplets fall on it. The story of how the parts of water relate to its characteristics and the answer to why water is important to you will soon be told—as you begin your study of the chemistry of life.

CHAPTER 3

Organization

Atoms interact by means of their electrons.

Water has properties important to life.

The primary biological molecules are carbohydrates, lipids, proteins, and nucleic acids.

3.1 The structure of atoms

Chemistry is the science of matter, the physical material that makes up everything in the universe. Matter is anything that takes up space and has a measurable amount of substance, or mass. All matter (including water) is made up of submicroscopic parts called **atoms**. Although scientists know a great deal about atoms, a simplified explanation provides a good starting point in understanding their complex structures.

Every atom is made up of particles smaller than the atom itself. These subatomic particles are of three types: *protons, neutrons,* and *electrons.* Protons and neutrons are found at the core, or nucleus, of the atom. Electrons surround the nucleus.

Protons have mass and carry a positive (+) charge. **Neutrons**, although similar to protons in mass, are neutral and carry no charge. **Electrons** have very little mass and carry a negative (−) charge. For this reason, the *atomic mass* of an atom is defined as the combined mass of all its protons and neutrons without regard to its electrons.

Elements are pure substances that are made up of a single kind of atom (atoms with the same number of protons) and cannot be separated into different substances by ordinary chemical methods. The identity of an atom is determined by its number of protons. For example, an atom containing two protons is helium, a gas you have probably seen used to blow up balloons. An atom possessing seven protons is nitrogen (see Figure 3.1b). Table 3.1 (p. 43) lists the most common elements on Earth. The number of protons in each type of element is also listed in this table and is called the *atomic number.*

The number of protons in an atom is the same as the number of electrons in that atom. Atoms are therefore electrically neutral; the positive charges of the protons are balanced by the negative charges of the electrons. The number of neutrons in an atom, however, may or may not equal the number of protons. Atoms that have the same number of protons but different numbers of neutrons are called **isotopes**.

Isotopes of an element differ in atomic mass but have similar chemical properties. For example, the three naturally occurring isotopes of hydrogen are shown at left. Each isotope has a single proton in its nucleus but has a different number of neutrons. Thus, deuterium, tritium, and hydrogen differ in their atomic masses but have the same chemical properties as one another because they all

Hydrogen (H)

Deuterium (^{2}H)   Tritium (^{3}H)

have the same number of electrons. In general, electrons determine the chemical properties of an element because atoms interact by means of their electrons (not their protons or neutrons).

The key to the chemical behavior of atoms lies not only in its number of electrons, but in their arrangement. Although scientists cannot precisely locate the position of any individual electron at a particular time, they can predict where an electron is most likely to be. This volume of space around a nucleus where an electron is most likely to be found is called the **shell**, or **energy level** of that electron.

Atoms can have many electron shells. Atoms with more than two electrons have more than one electron shell. In **Figure 3.1a**, the nucleus is shown as a small circle surrounded by spheres. These spheres represent electron shells. In Figure 3.1b, electrons are shown within the shells, which are depicted here two-dimensionally as

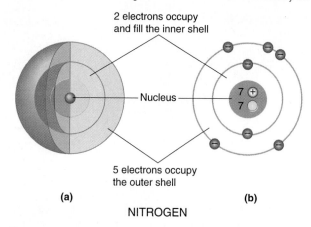

2 electrons occupy and fill the inner shell

Nucleus

7 +
7

5 electrons occupy the outer shell

(a)   (b)

NITROGEN

Figure 3.1 Two types of shell models of the same atom.

circles. Notice that the atom of nitrogen pictured in Figure 3.1b has electrons occupying two shells. The innermost shell contains two electrons. The second (outermost) shell has five electrons.

Because the energy of electrons increases as their distance from the attractive force of the nucleus increases, the various electron shells of atoms are also called *energy levels*. Electrons occupying increasingly distant shells from the nucleus have a stepwise increase in their levels of energy. The number of protons in the nucleus also influences the energy of electrons at the various energy levels, since the attractive force of the nucleus increases as its number of protons increases.

An atom is a core (nucleus) of protons and neutrons surrounded by electrons. Its number of protons identifies an atom, but its electrons largely determine its chemical properties.

H⚙W Science WORKS Applications

Using Isotopes to Detect and Treat Disease

Isotopes of an element differ from one another in their numbers of neutrons. They also differ in that some isotopes are *radioactive*. That is, they have unstable nuclei and emit three types of charged particles and rays: alpha (α) particles (high-energy helium nuclei), beta (β) particles (electronlike particles formed when neutrons are transformed into protons), and gamma (γ) rays (a form of electromagnetic radiation similar to X rays).

When used in diagnosis, isotopes are called *tracers* or *labels* because they can be used to follow the fate of certain substances in the body. The body is unable to distinguish between radioactive isotopes and stable ones of the same element. Therefore, physicians can use radioactive isotopes as substitutes for their stable counterparts.

For example, suppose your doctor wants to investigate the rate at which your thyroid gland is using iodine, a substance in your diet that your body uses to make thyroid hormone.

To determine this rate, you drink a solution of radioactive iodine, which your body will use in the same way it uses the "normal" iodine you acquire in your food. Equipment that counts the emission of particles from the nuclei of the radioactive iodine is placed over your thyroid gland and measures how fast the tracer iodine is entering your thyroid. The doctor can then see *exactly* how fast your thyroid is using iodine by simply reading the information on the detector. A visual representation of radioactivity levels is shown in the photos in Figure 3.A.

In a similar way, physicians can gather other data about your metabolism; substances such as iron, hormones, or drugs can be labeled and their pathways followed within your body. Brain tumors can be found as they concentrate certain radioactive substances within them. A low red blood cell count (anemia) can be diagnosed when red blood cells are tagged and their life spans measured.

Bones and lungs can be scanned, liver function determined, and kidneys imaged. Radioactive isotopes play roles in a variety of diagnostic tasks.

Isotopes are also effective in treating cancer. Three methods are used: (1) aiming high-energy radiation at the cancer, (2) implanting seeds or beads of isotope within the cancerous region, or (3) ingesting or injecting the isotope. High-energy radiation is often used to treat deep tumors. Methods using intense, short periods of irradiation help reduce the damage to surrounding tissues. The implantation of radioactive seeds is used to treat certain cancers such as prostate and cervical cancer. These cancers seem to respond well to a constant beam of radiation. The ingestion method is often used to treat thyroid conditions such as overactive thyroid. In this case, much larger doses of isotope are used than in diagnosis. The isotope accumulates in the gland and damages the hormone-producing cells, thereby lowering production of thyroid hormone.

Radioactivity can be detrimental to health or even fatal. When alpha, beta, and gamma rays strike normal cells, they can cause parts of their molecules to break and ionize. Radiation can also cause the formation of free radicals, which are uncharged but unstable and highly reactive atoms or compounds. Free radicals and ions can interact with other molecules in the body, causing their destruction. They may also produce chemical changes that may contribute to various disease conditions, including cancer. However, when used properly, radiation can be an important medical tool, enhancing health by assisting physicians in diagnosing and treating disease. ⚙

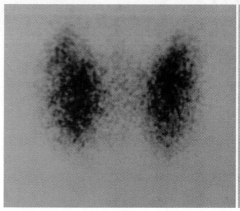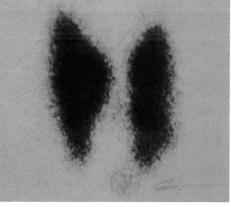

Figure 3.A **Thyroid scans after consumption of radioactive sodium iodide.** At left is a normal thyroid. At right is an enlarged thyroid.

3.2 Molecules and compounds

The atoms of most elements interact with one another. If the interaction is a sharing of electrons (see "Covalent bonds," p. 38), the interacting atoms are called a **molecule**. Molecules are two or more atoms held together by shared electrons and can be made up of atoms of the same element or atoms of different elements. Molecules made up of atoms of different elements are called **compounds**. (Compounds are also atoms of different elements held together by electrostatic attraction [ionic bonds]; see p. 37.)

The oxygen in the air consists of molecules made up of pairs of atoms of the same element—oxygen. These pairs are represented by the chemical formula O_2. A *chemical formula* is a type of "shorthand" used to describe a molecule. The atoms are represented by symbols, such as O for oxygen. (Chemical symbols are shown in Table 3.1 on p. 43.) A subscript shows the number of these atoms present in the molecule. An example of a compound with which you are all familiar is water, H_2O. In this chemical formula, the symbol H stands for the element hydrogen. The chemical formula H_2O shows that water is made up of two atoms of hydrogen and one atom of oxygen. A space-filling model of one molecule of water is shown to the left.

Three factors influence whether an atom will interact with other atoms. In addition, these factors influence the type of interactions likely to take place. These factors are (1) the tendency of electrons to occur in pairs, (2) the tendency of atoms to balance positive and negative charges, and (3) the tendency of the outer shell, or energy level, of electrons to be full. This third factor is often called the **octet rule**.

The word octet means "eight objects" and refers to the fact that eight electrons is a stable number for an outer shell; the outer electron shell of many atoms contains a maximum of eight electrons. (The first energy level is an exception to this rule because it contains a maximum of two electrons.) The octet rule states that an atom with an unfilled outer shell has a tendency to interact with another atom or atoms in ways that will complete this outer shell. Although the octet rule does not apply to all atoms, it does apply to all biologically important ones—those involved in the structure, energy needs, and information systems of living

things. These atoms and the molecules they make up are discussed later in this chapter.

Atoms of elements that have equal numbers of protons and electrons and have full outer-electron energy levels are the only ones that exist as single atoms. Atoms with these characteristics are called **noble gases**, or *inert gases,* because they do not react readily with other elements. Many stories explain why they are called noble, but they all center around the concept of nobility—those who have everything (in this case a full outer shell), need nothing, and interact little with others. Most of the noble gases—helium, neon, argon, krypton, xenon, and radon—are rare. In addition, they are relatively unreactive and unimportant in living systems. Helium and radon are probably the best known and the least rare noble gases. Radon, in fact, has gained more recognition in recent years. Formed in rock or soil particles from the radioactive decay of radium, radon gas can seep through cracks in basement walls and remain trapped in homes that are not well ventilated. Prolonged exposure to radioactive radon in levels greater than those normally found in the atmosphere is thought to lead to lung cancer. **Figure 3.2** shows the areas of the United States in which radon gas is most prevalent.

Most atoms interact with one another. They may form molecules, which are two or more atoms held together by shared electrons. If molecules are made up of different elements, they are called compounds. Various factors influence whether and how an atom will interact with other atoms.

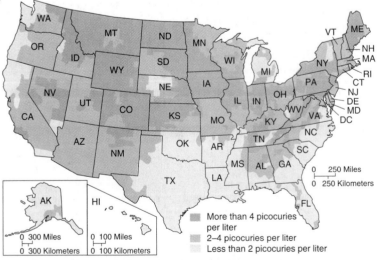

Figure 3.2   Areas of the United States in which radon gas is most prevalent. Note: A picocurie is a unit used to describe the rate of disintegration of the nucleus of a radioactive element.

3.3 Chemical bonds

An atom having an incomplete outer shell can satisfy the octet rule in one of three ways:

1. It can gain electrons from another atom.
2. It can lose electrons to another atom.
3. It can share one or more electron pairs with another atom.

Such interactions among atoms result in **chemical bonds**, forces that hold atoms together. If the force is caused by the attraction of oppositely charged particles formed by the gain or loss of electrons, the bond is called **ionic** (eye-ON-ick). If the force is caused by the electrical attraction created by atoms sharing electrons, the bond is called **covalent** (ko-VAY-lent). Other weaker kinds of bonds also occur.

Ions and Ionic Bonds

Electrons stay in their shells because they are attracted to the positive charge of the nucleus. However, electrons far from the nucleus are not held as tightly as electrons closer to the nucleus. In addition, they have more energy and they interact with other atoms more easily than close, tightly held, energy-poor electrons. Therefore, atoms typically interact with other atoms by means of the electrons in their outermost (highest) energy levels, or shells. The types of interactions that occur tend to result in atoms with completed outer shells, thus satisfying the octet rule.

An atom with a nearly completed outer shell tends to "take" enough electrons to complete its outer shell

from an atom that has only one or two electrons out of the eight needed for completion of its outer shell. Once these outer electrons are gone from the atom giving up electrons, the next shell in becomes its new, complete outer shell. As a result of this interaction, the number of protons no longer equals the number of electrons in either atom. The atom taking on electrons acquires a negative charge. The atom giving up electrons acquires a positive charge. They are no longer called atoms because they are not electrically neutral anymore. Such charged particles are called **ions** (EYE-onz).

Figure 3.3a illustrates the formation of ions with a specific example of electron "give and take." Notice in the upper left portion of the illustration that sodium (Na) is an element with 11 protons and 11 electrons. It is a soft, silver-white metal that occurs in nature as a part of ionic compounds. One familiar compound is sodium chloride, or table salt. Of sodium's 11 electrons, 2 are in its innermost energy level (full with 2 electrons), 8 are at the next level, and 1 is at the outer energy level. Because of this distribution of electrons, its outer energy level is not full, and therefore the octet rule is not satisfied.

Chlorine (Cl) is an element with 17 protons and 17 electrons, as shown in the upper right portion of the illustration. In its molecular form (Cl_2), it is a greenish-yellow gas that is poisonous and irritating to the nose and throat. In the ionic compound sodium chloride, however, it does not have these characteristics. Of chlorine's 17 electrons, 2 are at its innermost energy level, 8 at the next energy level, and 7 at the outer energy level. Chlorine, like sodium, has an outer energy level that is not full. Sodium and chlorine atoms can interact with one another in a way that results in both having full outer energy levels.

When placed together, the metal sodium and the gas chlorine react explosively. The single electrons in the outer energy levels of the sodium atoms are lost to the chlorine, as shown in the lower portion of Figure 3.3a. The result is the production of Na^+ and Cl^- ions. These ions come together as their opposite charges attract one another. This type of attraction is called *electrostatic attraction* and results in ionic bonding of the sodium and chloride ions. As these ions are drawn to one another, they form geometrically perfect crystals of salt as shown in Figure 3.3b.

The transfer of electrons between atoms is an important chemical event—one type of chemical reaction. In

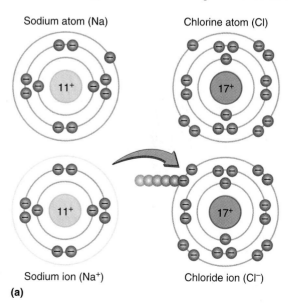

(a)

Sodium atom (Na)   Chlorine atom (Cl)

Sodium ion (Na⁺)   Chloride ion (Cl⁻)

Na⁺
Cl⁻

(b)

Figure 3.3   **The formation of an ionic bond.**

fact, this type of chemical interaction has a special vocabulary. When an atom loses (gives up) an electron, it is *oxidized*. The process by which this occurs is called an **oxidation** (ok-si-DAY-shun) meaning "to combine with oxygen." The name reflects that in biological systems, oxygen, which strongly attracts electrons, is the most frequent electron acceptor. Therefore, atoms that give up electrons to oxygen are "acted upon" by oxygen, or oxidized. For example, when iron combines with oxygen in the presence of moisture, it becomes oxidized. The product of this oxidation is commonly known as rust.

Conversely, when an atom gains an electron, it becomes *reduced*. The process is called a **reduction**. This name reflects that the addition of an electron reduces the charge by one. For example, if a molecule had a charge of +2, the addition of an electron (−1) would reduce the molecule's charge to +1.

Oxidation and reduction always take place together because every electron that is lost by one atom (oxidation) is gained by some other atom (reduction). Together they are called **redox reactions**. In a redox reaction, the charge of the oxidized atom is increased, and the charge of the reduced atom is lowered.

Covalent Bonds

Covalent bonds form when two atoms share electrons. Hydrogen is a simple example of an atom that usually shares electrons with other atoms. As you can see in the figure below, a hydrogen atom has a single electron and an unfilled outer electron shell. A filled outer shell at this energy level requires only two electrons. When hydrogen atoms are close enough to one another, they form pairs, with each of their single electrons moving around the two nuclei. These paired atoms of hydrogen are called *diatomic molecules* (die-uh-TOM-ick) and are represented by

the molecular formula H_2. The figure below left depicts this sharing as two types of shell models and a structural formula (middle).

As a result of sharing electrons, the diatomic hydrogen gas molecule is electrically balanced because it now contains two protons and two electrons. In addition, each hydrogen atom has two electrons in its outer shell, completing this shell. This relationship also results in the pairing of two free electrons. Thus, by sharing their electrons, the two hydrogen atoms form a stable molecule.

Covalent bonds can be very strong, that is, difficult to break. *Double bonds,* those bonds in which two pairs of electrons are shared, are stronger than *single bonds* in which only one pair of electrons is shared. As you might expect, *triple bonds,* those bonds in which three pairs of electrons are shared, are the strongest of these three types of covalent bonds. In chemical formulas that show the structure of covalently bonded molecules, single bonds are represented by a single line between two bonded atoms, double bonds by two lines, and triple bonds by three lines. For example, the structural formula of hydrogen gas is H—H, oxygen gas is O=O, and nitrogen gas is N≡N.

An atom can also form covalent bonds with more than one other atom. Carbon (C), for example, contains 6 electrons: 2 in the inner shell and 4 in the outer shell. To satisfy the octet rule, it must gain 4 additional electrons by sharing its 4 outer-shell electrons with another atom or atoms, forming 4 covalent bonds. Because there are many ways that 4 covalent bonds may form, carbon atoms are able to participate in many different kinds of molecules in living systems, such as proteins and carbohydrates.

The strength of a covalent bond refers to the amount of energy needed to make or break that bond. The energy that goes into making the bond is held within the bond and is released when the bond is broken. Therefore, covalent bonds are actually a storage place for energy as well as a type of chemical "glue" that holds molecules together. Living things store and use energy by means of making and breaking covalent bonds, thereby using molecules as a type of energy currency.

An ionic bond is an attraction between ions of opposite charge. These bonds form when atoms gain and lose electrons to one another. The loss of an electron by an atom is an oxidation. The gain of an electron is a reduction. Together, these paired electron transfers between atoms are called redox reactions. A covalent bond is the sharing of one or more pairs of electrons.

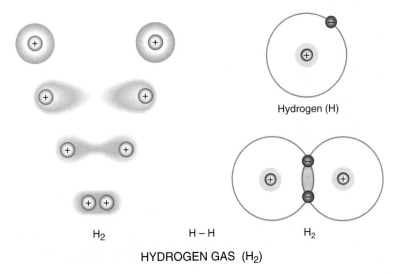

Hydrogen (H)

H_2 H – H H_2

HYDROGEN GAS (H_2)

3.4 Polarity

One covalently bonded molecule that plays a major role in living systems is water—H_2O. In fact, water is the most abundant molecule in your body, making up about two-thirds of your body weight. Although it seems to be a simple molecule, water has many surprising properties. For example, of all the common molecules on Earth, only water exists as a liquid at the Earth's surface. When life on Earth was beginning, this liquid provided a medium in which other molecules could move around and interact. Life evolved as a result of these interactions. And life, as it evolved, maintained these ties to water (Figure 3.4). Three-fourths of the Earth's surface is covered by water. Where water is plentiful, such as in

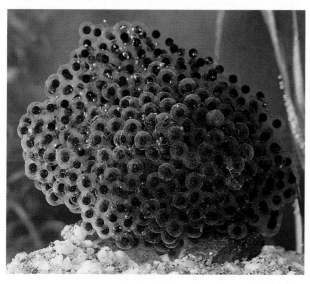

Figure 3.4 **Water is the cradle of life.** This mass of frog eggs is attached to a rock in the watery environment of a stream bottom.

the tropical rain forests, the land abounds with life. Where water is scarce, such as in the desert, the land seems almost lifeless except after a rainstorm. No plant or animal can grow and reproduce without some amount of water.

The chemistry of life, then, is water chemistry. Water has a simple molecular structure: one oxygen atom bonded by single covalent bonds to two hydrogen atoms. The resulting molecule satisfies the octet rule, and its positive and negative charges are balanced. Because the oxygen atom in each water molecule contains eight protons and each hydrogen atom contains only one proton, the electron pair shared in each covalent bond is more strongly attracted to the oxygen nucleus than to either of the hydrogen nuclei. Although the electrons surround both the oxygen and hydrogen nuclei, the negatively

charged electrons are far more likely to be found near the oxygen nucleus at a given moment than near one of the hydrogen nuclei. Because of this situation, the oxygen end of the water molecule has a partial negative charge. The hydrogen end has a partial positive charge. (The shell model and space-filling model both show this.) Molecules such as water that have opposite partial charges at different ends of the molecule are called **polar molecules**. Water is one of the most polar molecules known.

SHELL MODEL

SPACE-FILLING MODEL

The polarity of water contributes to its ability to attract other molecules and form special types of chemical bonds with them. These weak electrical attractions between molecules are called **hydrogen bonds**, and have approximately 5% to 10% the strength of covalent bonds. In fact, water forms hydrogen bonds with other water molecules because the hydrogen atoms of some water molecules are attracted to the oxygen atoms of other water molecules. (These bonds are shown as dotted lines between water molecules in the illustration of the molecular structure of water and ice in the Just Wondering box.) The ability of water molecules to form weak bonds among themselves and with other molecules is the reason for much of the organization and chemistry of living things. Although hydrogen bonds are weak, these bonds are constantly made and broken. (Each lasts only 1/100,000,000,000 of a second!) The cumulative effect of very large numbers of hydrogen bonds is responsible for the many important physical properties of water. *Surface tension*—the ability of water molecules to "stick together" at the water's surface—is one such property. It is the reason that water molecules can bear the weight of organisms such as the water strider shown to the right.

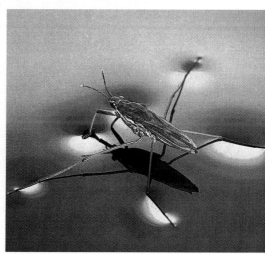

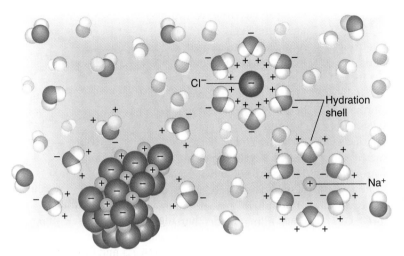

Figure 3.5   How salt dissolves in water.

Water Is a Powerful Solvent

Water molecules gather closely around any particle that exhibits an electrical charge, such as ions and polar molecules. For example, sodium chloride (table salt) is made up of the positively charged sodium (Na^+) and negatively charged chloride (Cl^-) ions. These ions are attracted to one another and cluster in a regular pattern, forming crystals. When you put salt in water, some ions break away from the crystals because the positive ends of some

water molecules are attracted to the Cl^- ions, while the negative ends of other water molecules are attracted to the Na^+ ions. These attractions are shown in the right-hand portion of Figure 3.5. These attractions are stronger than the attraction between the ions that keeps the crystal together. Therefore, the ions are pulled from their positions in the crystal as is shown on the left. Water molecules then surround each ion, forming a *hydration shell,* which keeps the ions apart. The salt is said to be *dissolved.*

Similarly, hydration shells form around all polar molecules and ions. Compounds that dissolve in water this way are said to be **soluble** (SOL-you-ble) in water. Chemical interactions readily take place in water because so many kinds of compounds are water soluble and therefore move among water molecules as separate molecules or ions.

Water Organizes Nonpolar Molecules

Remember the old saying that "oil and water don't mix?" This statement is true because oil is a nonpolar molecule and cannot form hydrogen bonds with water. Instead, the water molecules form hydrogen bonds with each other, causing the water to exclude the nonpolar molecules. It is almost as if nonpolar molecules move away from contact with the water. For this reason, nonpolar molecules are referred to as **hydrophobic** (HI-dro-FO-bik). The word hydrophobic comes from Greek words meaning "water" (*hydros*) and "fearing" (*phobos*). This tendency for nonpolar molecules to band together in a water solution is called *hydrophobic bonding* (see left). Hydrophobic forces determine the three-dimensional shapes of many biological molecules, which are usually surrounded by water within organisms.

Much of the biologically important behavior of water results because its oxygen atom attracts electrons more strongly than its hydrogen atoms do. As a result, water molecules each have electron-rich and electron-poor regions, giving them positive and negative poles.

Crude petroleum from an oil spill floats on the surface of the ocean because it is hydrophobic and less dense (lighter) than water.

3.5 Ionization and pH

The covalent bonds of water molecules sometimes break spontaneously. When this happens, one hydrogen atom nucleus (a proton) dissociates from the rest of the water molecule, leaving behind its electron. Because its positive charge is no longer balanced by an electron, it is a positively charged hydrogen ion, H^+. The remaining part of the water molecule now has an extra electron. It is therefore a negatively charged hydroxyl ion, OH^-. This process of spontaneous ion formation is called **ionization**:

$$H_2O \rightarrow OH^- + H^+$$

Only very few water molecules are ionized at a single instant in time. Scientists calculate that the fraction of water molecules dissociated (ionized) at any given time in pure water is 0.0000001. This tiny number can be written another way by using exponential notation. This is done by counting the number of places to the right of the decimal point. Because there are seven places, this number is written as 10^{-7}. The minus sign means that the number is less than 1.

To indicate the concentration of H^+ ions in a solution, scientists have devised a scale based on the slight degree of spontaneous ionization of water. This scale is called the **pH scale**. (The letters pH stand for the power of the hydrogen ion $[H^+]$). The pH values of this scale generally range from 0 to 14. The pH of a solution is determined by taking the negative value of the exponent of its hydrogen ion concentration. For example, pure water has a hydrogen ion concentration of 10^{-7} and therefore a pH of 7. When water ionizes, hydroxyl ions (OH^-) are produced in a concentration equal to the concentration of hydrogen ions. In pure water, the concentrations of H^+ and OH^- ions are equal to 10^{-7}. Because these ions join spontaneously, water, at pH 7, is neutral.

Any substance that dissociates to form H^+ ions when it is dissolved in water is called an **acid**. The more hydrogen ions an acid produces, the stronger an acid it is. Although an acid produces a higher concentration of H^+ ions than pure water (0.00001 as opposed to 0.0000001, for example), its pH is lower. Using the above numbers to illustrate: The first number (0.00001 or 10^{-5}) represents the hydrogen ion concentration of an acid. The negative value of its exponent results in a pH of 5. The second number (0.0000001 or 10^{-7}) represents the hydrogen ion concentration of water. The negative value of its exponent results in a pH of 7. Each one-unit decrease in pH, however, does not correspond to a one-fold increase in acidity. In fact, depending on the acid and how completely it ionizes, a change of one unit may correspond to as much as a tenfold increase in acidity.

Figure 3.6 shows many common acids and their pH values. The pH of champagne, for example, is about 4. This low pH is due to the dissolved carbonic acid that causes champagne to bubble. Some bodies of water are acidic, such as peat bogs (pH 4 to 5). The hydrochloric acid (HCl) of your stomach ionizes completely to H^+ and Cl^-. It forms a strong acid with a pH of 2 to 3. Stronger acids are rarely found in living systems.

Any substance that combines with H^+ ions, as OH^- ions do, is said to be a **base**. Any increase in the concentration of a base lowers the H^+ ion concentration. Bases therefore have pH values higher than water's neutral value of 7. For example, the environment of your small intestine is kept at a basic pH of between 7.5 and 8.5. Strong bases such as sodium hydroxide (NaOH) have pH values of 12 or more. As with acids, a change of 1 in the pH value of a base may reflect up to a tenfold change in pH, depending on the base.

---

pH refers to the relative concentration of H^+ ions in a solution. Low pH values indicate high concentrations of H^+ ions (acids), and high pH values indicate low concentrations.

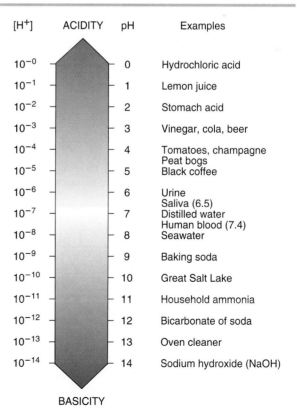

$[H^+]$	ACIDITY	pH	Examples
10^{-0}		0	Hydrochloric acid
10^{-1}		1	Lemon juice
10^{-2}		2	Stomach acid
10^{-3}		3	Vinegar, cola, beer
10^{-4}		4	Tomatoes, champagne
			Peat bogs
10^{-5}		5	Black coffee
10^{-6}		6	Urine
			Saliva (6.5)
10^{-7}		7	Distilled water
			Human blood (7.4)
10^{-8}		8	Seawater
10^{-9}		9	Baking soda
10^{-10}		10	Great Salt Lake
10^{-11}		11	Household ammonia
10^{-12}		12	Bicarbonate of soda
10^{-13}		13	Oven cleaner
10^{-14}		14	Sodium hydroxide (NaOH)

BASICITY

Figure 3.6 **The pH scale.**

*J*ust **Wonder**ing . . . *Real Questions Students Ask*

If water (a liquid) is made up of two gases, hydrogen and oxygen, then why isn't water a gas?

You have made an interesting observation and questioned one of water's many fascinating properties: At room temperature, both hydrogen (H_2) and oxygen (O_2) are gases, but water is not. Additionally, other substances whose molecules are similar in size to those of water, such as ammonia and methane, *are* gases at room temperature. Why not water?

The answer to your question and the underlying cause of many of water's properties such as its surface tension and its expansion upon freezing all have to do with the way in which the water molecule is constructed. Water is a polar molecule, as described on p. 39. Because of this polarity, its positively charged portions are attracted to the negatively charged portions of other water molecules and vice-versa. These attractive forces are called *hydrogen bonds*, and are described in more detail on page 39. Hydrogen bonding results in each water molecule being bonded to others, with each of those molecules bonded to others, and so forth, forming a extensive bonding network among water molecules at room temperature. These water molecules, therefore, do not move independently as in a gas, but move together, forming a liquid.

At room temperature, water molecules are in constant motion, so their hydrogen bonds are constantly broken and remade. However, at other temperatures, the properties of water change because temperature affects hydrogen bonding. To illustrate: Above 100°C (212°F) the hydrogen bonds weaken and water molecules begin to break free from one another, forming steam, or water vapor. However, the bonds within the water molecules do not break, so steam is made up of water, not hydrogen and oxygen gases—as shown in the illustration at top right. Below 0°C (32°F) the water molecules are no longer moving vigorously enough to break and remake hydrogen bonds so the water molecules become locked into a crystalline lattice with the hydrogen bonds holding each water molecule apart at bond's length as shown in the illustration, bottom right. Therefore, the water expands and becomes the solid we call ice. At the surface of ice, however, hydrogen bonds continue to be broken and remade (as shown by the dotted lines in the ice diagram), causing ice to be slippery, which allows us to skate, slide, and fall on this amazing compound.

Steam (gas)

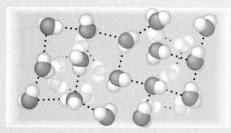
Water (liquid)

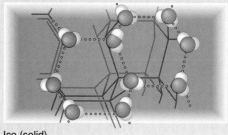
Ice (solid)

3.6 Biological molecules are organic compounds

With the exception of water, most molecules that are formed by living organisms and make up their structures contain the element carbon. This fascinating element lends itself to being the basis of living material because of its ability to form four bonds with other atoms and therefore interact with other molecules in myriad ways. Silicon, an atom with similar bonding properties, is abundant in the Earth's crust but not in living things. However, silicon is used by some organisms in building outer skeletons or shells. An example of a group of such organisms is the diatoms (shown at right), one of the most abundant protists on this planet.

The carbon-containing molecules that make up living things are called **organic compounds**. Organic compounds are often very large and are usually held together by covalent bonds. The study of organic molecules and their interactions is called *organic chemistry*.

The study of molecules and substances not containing carbon is called *inorganic chemistry*. Inorganic compounds are often quite small and are usually held together by ionic bonds. A few carbon-containing, covalently bonded inorganic molecules exist (such as carbon dioxide and carbonic acid), but they are considered

100X

www.jbpub.com/biology

inorganic molecules because they do not make up the structure of living things.

Although carbon is the underlying component of biological molecules, 10 other elements are common in living organisms. These elements are listed in **Table 3.1** and are compared according to their frequency in the Earth's crust. Notice that the great majority of atoms in living things (96.3%, in fact) are oxygen, carbon, hydrogen, or nitrogen.

It is often helpful to think of an organic molecule as a carbon-based core with other special parts attached. Each of these special parts is really a group of atoms

called a **functional group** and has definite chemical properties. A hydroxyl group (−OH), for example, is a functional group. The most important functional groups are illustrated in **Figure 3.7.** These groups are important because most chemical reactions that occur within organisms involve the transfer of a functional group from one molecule to another. Other frequent chemical reactions involve the breaking of carbon-carbon bonds.

Some of the molecules of living things are simple organic molecules, often having only one functional group. Other far larger molecules, called *macromolecules,*

Table 3.1

The Most Common Elements on Earth and Their Distribution in the Human Body

Element	Symbol	Atomic Number	Percent of Human Body By Weight	Approximate Percent of Earth's Crust By Weight	Importance or Function
Oxygen	O	8	65.0	46.6	Necessary for cellular respiration, component of water
Carbon	C	6	18.5	0.03	Backbone of organic molecules
Hydrogen	H	1	9.5	0.14	Electron carrier, component of water and most organic molecules
Nitrogen	N	7	3.3	Trace	Component of all proteins and nucleic acids
Calcium	Ca	20	1.5	3.6	Component of bones and teeth, trigger for muscle contraction
Phosphorus	P	15	1.0	0.07	Backbone of nucleic acids, important in energy transfer
Potassium	K	19	0.4	2.6	Principal positive ion in cells, important in nerve function
Sulfur	S	16	0.3	0.03	Component of most proteins
Chlorine	Cl	17	0.2	0.01	Principal negative ion bathing cells
Sodium	Na	11	0.2	2.8	Principal positive ion bathing cells, important in nerve function
Magnesium	Mg	12	0.1	2.1	Critical component of many energy-transferring enzymes
Iron	Fe	26	Trace	5.0	Critical component of hemoglobin in the blood
Copper	Cu	29	Trace	0.01	Key component of many enzymes
Zinc	Zn	30	Trace	Trace	Key component of some enzymes
Molybdenum	Mo	42	Trace	Trace	Key component of many enzymes
Iodine	I	53	Trace	Trace	Component of thyroid hormone
Silicon	Si	14	Trace	27.7	—
Aluminum	Al	13	Trace	6.5	—
Manganese	Mn	25	Trace	0.1	—
Fluorine	F	9	Trace	0.07	—
Vanadium	V	23	Trace	0.01	—
Chromium	Cr	24	Trace	0.01	—
Boron	B	5	Trace	Trace	—
Cobalt	Co	27	Trace	Trace	—
Selenium	Se	34	Trace	Trace	—
Tin	Sn	50	Trace	Trace	—

The most common elements in the human body are highlighted in blue and are listed in decreasing amounts based on percentages.

The primary biological molecules are carbohydrates, lipids, proteins, and nucleic acids.

often have many functional groups and contain thousands of atoms. There are four major groups of biologically important macromolecules: complex carbohydrates, lipids, proteins, and nucleic acids. The four major classes of biologically important macromolecules are presented in Table 3.2.

Although these classes of macromolecules are each composed of different building blocks, the process by which their building blocks are put together is the same. All except the lipids are polymers, macromolecules that are built by forming covalent bonds between similar building blocks, or monomers, to form long chains. **Dehydration synthesis** is a process by which monomers (and the building blocks of lipids) are put together. Dur-

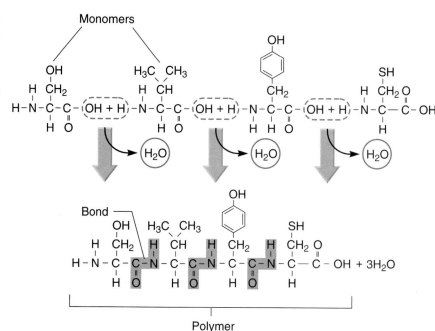

Figure 3.8 Dehydration synthesis.

ing dehydration synthesis, one molecule of water is removed (dehydration) from each two monomers that are joined (synthesis). One monomer loses its hydroxyl group ($-OH$), and the other loses an atom of hydrogen (H). Having lost electrons they were sharing in covalent bonding, both monomers bond covalently with one another as shown in Figure 3.8. (Also see Figures 3.10, 3.13, and 3.15.) The process of dehydration synthesis uses energy, which is stored in the bond that is made, and takes place with the help of special molecules called *enzymes* (see Chapter 5).

Polymers are disassembled in an opposite process called **hydrolysis** (see Figure 3.15). During hydrolysis, the bonds are broken between monomers with the addition of water (and in the presence of enzymes). In fact, the term hydrolysis literally means "to break apart" (*lysis*) "by means of water" (*hydro*). The hydroxyl group of a water molecule bonds to one monomer, and the hydrogen atom bonds to its neighbor. The energy held in the bond is released.

Functional Group	Structural Formula	Models
Hydroxyl	$-OH$	
Carbonyl	$\begin{array}{c} -C- \\ \parallel \\ O \end{array}$	
Carboxyl	$-C\begin{array}{c} O \\ \diagup\diagdown \\ OH \end{array}$	
Amino	$-N\begin{array}{c} \diagup H \\ \diagdown H \end{array}$	
Sulfhydryl	$-SH$	
Phosphate	$\begin{array}{c} H \\ \mid \\ -O-P-OH \\ \parallel \\ O \end{array}$	

Figure 3.7 **The chemical building blocks of life, functional groups.** Six of the most important functional groups involved in chemical reactions.

The carbon-containing molecules that make up living things are organic compounds. Organic compounds are often very large and are usually held together by covalent bonds. The organic molecules of living things, which are proteins, lipids, carbohydrates, and nucleic acids, are molecules built by dehydration synthesis and broken down by hydrolysis.

Table 3.2

Biologically Important Macromolecules

Macromolecule	Subunit	Function	Example
Carbohydrates			
Starch, glycogen	Glucose	Stores energy	Potatoes
Cellulose	Glucose	Makes up cell walls in plants	Paper
Chitin	Modified glucose	Makes up the exterior skeleton in some animals	Crab shells

Lipids			
Fats (triglycerides)	Glycerol + three fatty acids	Store energy	Butter
Phospholipids	Glycerol + two fatty acids + phosphate	Make up cell membranes	All membranes
Steroids	Four carbon rings	Act as chemical messengers	Cholesterol, estrogen

Triglyceride

Proteins			
Globular	Amino acids	Help chemical reactions take place	Hemoglobin
Structural	Amino acids	Make up tissues that support body structures and provide movement	Muscle

Nucleic Acids			
DNA	Nucleotides	Helps code hereditary information	Chromosomes
RNA	Nucleotides	Helps decode hereditary information	Messenger

3.7 Carbohydrates

Carbohydrates are molecules that contain carbon, hydrogen, and oxygen, with the concentration of hydrogen and oxygen atoms in a 2:1 ratio. Abundant energy is locked in their many carbon-hydrogen bonds. Plants, algae, and some bacteria produce carbohydrates by the process of photosynthesis. Most organisms use carbohydrates as an important fuel, breaking these bonds and releasing energy to sustain life.

Among the least complex of the carbohydrates are the simple sugars or **monosaccharides** (MON-oh-SACK-uh-rides). This word comes from two Greek words meaning "single" (*monos*) and "sweet" (*saccharon*) and reflects the fact that monosaccharides are individual sugar molecules. Some of these sweet-tasting sugars have as few as three carbon atoms. The monosaccharides that play a central role in energy storage, however, have six. The primary energy-storage molecule used by living things is glucose ($C_6H_{12}O_6$), a six-carbon sugar with seven energy-storing carbon-hydrogen bonds. Notice in **Figure 3.9** that glucose, like other sugars, exists as a straight chain or as a ring of atoms.

Glucose is not the only sugar with the formula $C_6H_{12}O_6$. Other monosaccharides having this same formula are fructose and galactose. Because these molecules have the same molecular formula as glucose but are put together slightly differently, they are called *isomers*, or alternative forms, of glucose. Your taste buds can tell the difference: Fructose is much sweeter than glucose.

Two monosaccharides linked together form a **disaccharide** (dye-SACK-uh-ride). Many organisms, such as plants, link monosaccharides together to form disaccharides, which are less readily broken down while they are

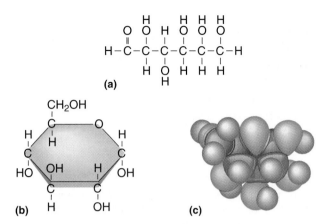

Figure 3.9 **Structure of a glucose molecule.** (a) The structural formula of glucose in its linear form and (b) as a ring structure. (c) Space-filling model of glucose. (Hydrogen, blue; Oxygen, red; Carbon, black).

transported within the organism. Sucrose (table sugar) is a disaccharide formed by linking a molecule of glucose to a molecule of fructose by dehydration synthesis as shown in **Figure 3.10.** It is the common transport form of sugar in plants. Lactose, or milk sugar (glucose + galactose), is a disaccharide produced by many mammals to feed their young.

Not only do organisms unlock, use, and transport the energy within carbohydrate molecules, they store this energy. To do this, however, organisms must convert soluble sugars such as glucose to an insoluble form to be stored. Sugars are made insoluble by joining together into long polymers called **polysaccharides**. Plants store energy in polysaccharides called **starches**. The starch amylose, for example, is made up of hundreds of glucose molecules linked together in long, unbranched chains. Most plant starch is a branched

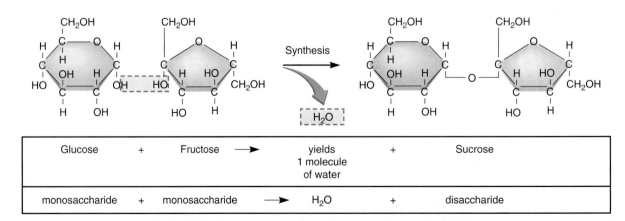

Figure 3.10 **Disaccharides are formed from monosaccharides by dehydration synthesis.**

version of amylose called *amylopectin* (AM-ih-low-PECK-tin). Animals store glucose in highly branched polysaccharides called **glycogen** (GLYE-ko-jen) (Figure 3.11).

The chief component of plant cell walls is a polysaccharide called *cellulose.* Cellulose is chemically similar to amylose but is bonded in a way that most organisms cannot digest (Figure 3.12). For this reason, cellulose works well as a biological structural material and occurs widely in this role in plants. The structural material in insects, many fungi, and certain other organisms is a modified form of cellulose called *chitin* (KITE-n). Chitin is a tough, resistant surface material that is also relatively indigestible (see Figure 16.1).

Sugars are among the most important energy-storage molecules in living things. Starch and glycogen, both types of polysaccharides, are storage forms of glucose. Plants store sugar as starch, whereas animals store sugar as glycogen.

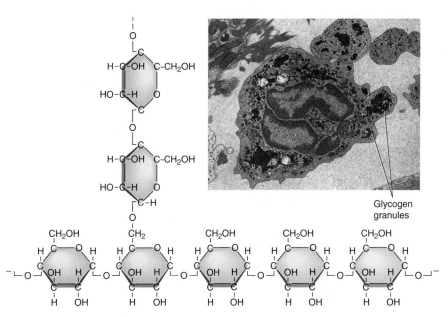

Figure 3.11 Glycogen, a starch. Glycogen storage granules can be seen in the cytoplasm of many types of cells, such as liver cells, muscle cells, and certain types of white blood cells. The electron micrograph shows a neutrophil, a type of white blood cell that is abundant in the body and that phagocytizes foreign material.

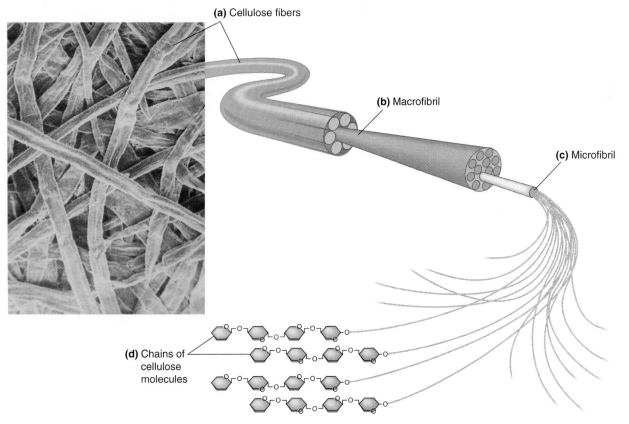

Figure 3.12 Structure of cellulose. (a) Cellulose fibers from a ponderosa pine. (b) Macrofibrils compose each fiber. (c) Each macrofibril is composed of bundles of microfibrils. (d) Microfibrils, in turn, are composed of bundles of cellulose chains. Cellulose fibers can be very strong; this is one reason why wood is such a good building material.

3.8 Fats and lipids

When organisms store glucose molecules for long periods, they usually store them as **fats** rather than as carbohydrates. Fats are large molecules made up of carbon, hydrogen, and oxygen, as are the carbohydrates, but their hydrogen-to-oxygen ratio is higher than 2:1. For this reason, fats contain more energy-storing carbon-hydrogen bonds than carbohydrates. In addition, fats are nonpolar, insoluble molecules, so they work well as storage molecules.

Fats are only one kind of **lipid**. Lipids include a wide variety of molecules, all of which are soluble in oil but insoluble in water. This insolubility results because almost all the bonds in lipids are nonpolar carbon-carbon or carbon-hydrogen bonds (see p. 40). Three important categories of lipids are (1) oils, fats, and waxes; (2) phospholipids; and (3) steroids.

Lipids are composite molecules; that is, they are made up of more than one component. Oils and fats are built from two different kinds of subunits:

1. *Glycerol:* Glycerol is a three-carbon molecule with each carbon bearing a hydroxyl (−OH) group. The three carbons form the backbone of the fat molecule.
2. *Fatty acids:* Fatty acids have long *hydrocarbon* chains (chains consisting only of carbon and hydrogen atoms) ending in a carboxyl (−COOH) group. Three fatty acids are attached to each glycerol backbone (Figure 3.13). Because there are three fatty acids, the resulting fat molecule is called a **triglyceride:**

$$
\begin{array}{l}
\text{H} \\
| \\
\text{H}-\text{C}-\text{O}-\text{Fatty acid} \\
\text{H}-\text{C}-\text{O}-\text{Fatty acid} \\
\text{H}-\text{C}-\text{O}-\text{Fatty acid} \\
| \\
\text{H}
\end{array}
$$

Most dietary fat is in the form of triglycerides, an abundant type of lipid.

The difference between fats and oils has to do with the number of double bonds in their fatty acids. As Figure 3.14 shows, a fatty acid with only single bonds between its carbon atoms can hold more hydrogen atoms than a fatty acid with double bonds between its carbon atoms. A fatty acid that carries as many hydrogen atoms as possible, such as the fatty acid in Figure 3.14a, is saturated. Fats composed of fatty acids with double bonds are unsaturated because the double bonds replace some of the hydrogen atoms. If a fat has more than one double bond, it is polyunsaturated. Polyunsaturated fats (Figure 3.14b) have low melting points and are therefore liquid fats, or oils.

The fatty acids of most plant triglycerides such as vegetable oils are unsaturated. (Exceptions are the tropical oils.) Animal fats, in contrast, are often saturated and occur as hard fats. Human diets with large amounts of saturated fats may contribute to clogged arteries and raise the risk of developing diseases of the circulatory system.

Waxes, which are used by land plants and some animals as a waterproofing material, differ from fats and oils by having a chemical backbone slightly different from

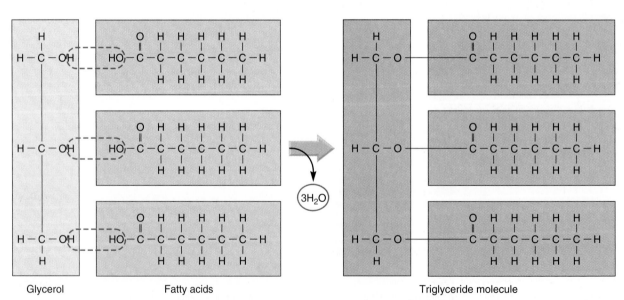

Glycerol   Fatty acids   Triglyceride molecule

Figure 3.13   Structure of a triglyceride. Triglycerides are composite molecules, made up of three fatty acid molecules bonded to a single glycerol molecule. This bonding takes place by dehydration synthesis.

glycerol. Phospholipids also differ from oils in that one of their fatty acids is replaced by a phosphate group attached to a nitrogen-containing group. Phospholipids play a key role in the structure of cell membranes (see Chapter 4). Membranes often contain steroids, a lipid having a structure very different from oils. Steroids are composed of four carbon rings. Most of your cell membranes contain the steroid cholesterol. Male and female sex hormones (discussed in Chapter 22) are also steroids.

Fats are important energy-storing molecules. They are made up of three fatty acid chains attached to a glycerol backbone.

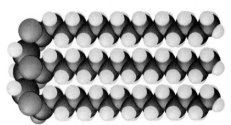

(a) Palmitic acid (saturated)

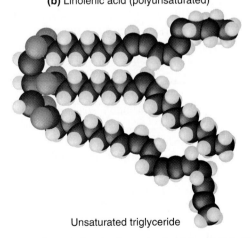

Saturated triglyceride

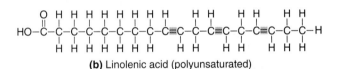

(b) Linolenic acid (polyunsaturated)

Unsaturated triglyceride

Figure 3.14 **Saturated and polyunsaturated fats.** (a) Palmitic acid, a fatty acid with only single bonds between its carbon atoms, has a maximum of hydrogen atoms and is a saturated fat. (b) Linolenic acid, with three double bonds and thus fewer than the maximum number of hydrogen atoms bonded to the carbon chain, is a polyunsaturated fatty acid.

3.9 Proteins

Proteins are the third major group of macromolecules that make up the bodies of organisms. Proteins play diverse roles in living things. Perhaps the most important proteins are **enzymes**, proteins capable of speeding up specific chemical reactions. Other short proteins called **peptides** are used as chemical messengers within your brain and throughout your body. Collagen, a structural protein, is an important part of bones, cartilage, and tendons. Despite their varied functions, all proteins have the same basic structure: a long chain of amino acids linked end to end.

Amino acids are small molecules containing an amino group ($-NH_2$), a carboxyl group ($-COOH$), a hydrogen atom, a carbon atom, and a *side chain* that differs among amino acids. In a generalized formula for an

$$H_2N-\overset{\overset{\displaystyle R}{|}}{\underset{\underset{\displaystyle H}{|}}{C}}-COOH$$

amino acid, the side chain is shown as R. The identity and unique chemical properties of each amino acid are determined by the nature of the R group.

SOME COMMON AMINO ACIDS

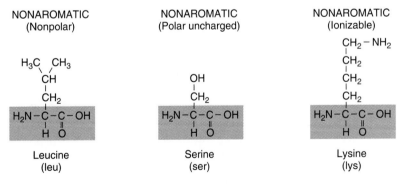

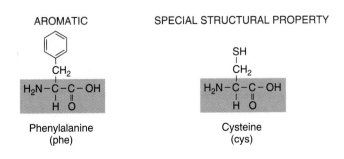

The primary biological molecules are
carbohydrates, lipids, proteins, and nucleic acids.

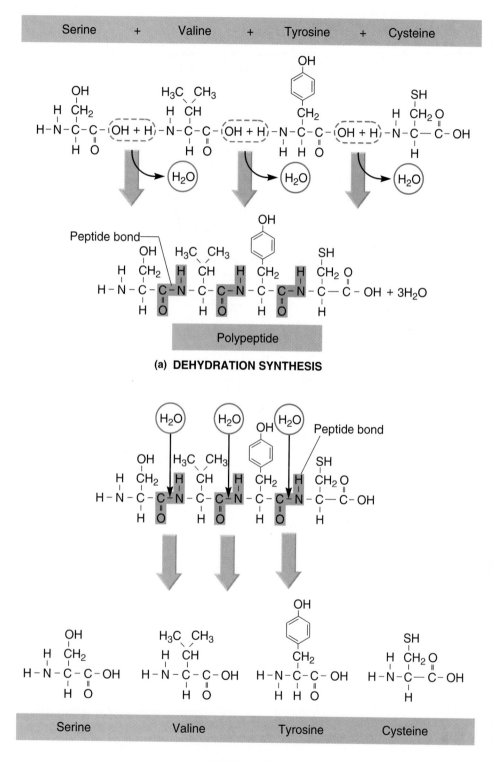

Serine + Valine + Tyrosine + Cysteine

(a) DEHYDRATION SYNTHESIS

Peptide bond

Polypeptide

(b) HYDROLYSIS

Serine   Valine   Tyrosine   Cysteine

Figure 3.15   How a polypeptide chain is formed and broken. (a) During dehydration synthesis, peptide bonds are formed between adjacent amino acids, forming a polypeptide. (b) During hydrolysis, a molecule of water is added to each peptide bond that links adjacent amino acids, breaking the bond between them. This separates the molecules into individual amino acids.

Only 20 different amino acids make up the diverse array of proteins found in living things. Each protein differs according to the amount, type, and arrangement of amino acids that make up its structure. Some common amino acids are illustrated at the bottom of page 49. Notice that each has the same chemical backbone (shown by the box) as in the generalized formula. The R groups are outside these boxes. Those amino acids with R groups that form ring structures are termed *aromatic* compounds and those without ring structures are *nonaromatic*. (This term was coined as chemists discovered that many fragrant compounds had this distinctive ring structure.) Those amino acids that are ionizable have R groups that become charged when in solution. Those with special chemical (structural) properties play important roles in forming links between protein chains or forming kinks in their shapes.

Each amino acid has a free amino group ($-NH_2$) at one end and a free carboxyl group ($-COOH$) at the other end. During dehydration synthesis, each of these groups on separate amino acids loses a molecule of water between them, forming a covalent bond that links the two amino acids (**Figure 3.15a**). This bond is called a *peptide bond*. A long chain of amino acids linked by peptide bonds is a *polypeptide*. Proteins are long, complex polypeptides. The great variability possible in the sequence of amino acids in polypeptides is perhaps the most important property of proteins, permitting tremendous diversity in their structures and functions.

The sequence of amino acids that makes up a particular polypeptide chain is termed the *primary structure* of a protein (**Figure 3.16 ❶**). This sequence determines the further levels of structure of the protein molecule resulting from bonds that form between these groups. Having the proper sequence of amino acids, then, is crucial to the functioning of a protein. If the protein does not assume its correct shape, it will not work properly or at all. Because different amino acid functional groups have different chemical properties, the shape of a protein may be altered by a single amino acid change.

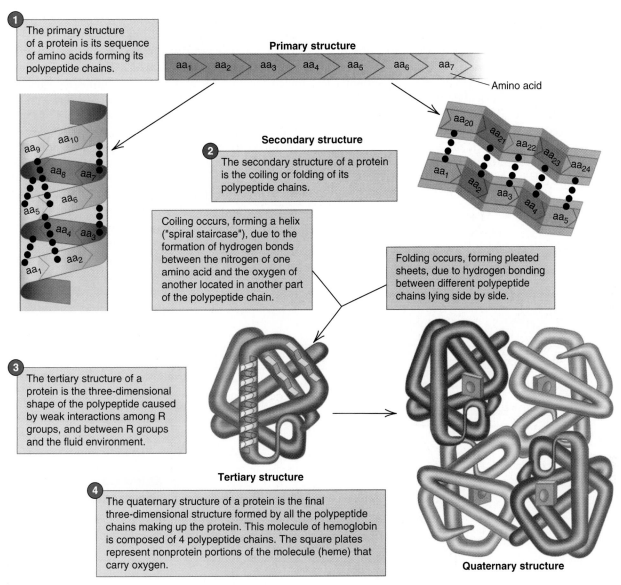

1 The primary structure of a protein is its sequence of amino acids forming its polypeptide chains.

Primary structure

aa₁ aa₂ aa₃ aa₄ aa₅ aa₆ aa₇

Amino acid

Secondary structure

2 The secondary structure of a protein is the coiling or folding of its polypeptide chains.

Coiling occurs, forming a helix ("spiral staircase"), due to the formation of hydrogen bonds between the nitrogen of one amino acid and the oxygen of another located in another part of the polypeptide chain.

Folding occurs, forming pleated sheets, due to hydrogen bonding between different polypeptide chains lying side by side.

3 The tertiary structure of a protein is the three-dimensional shape of the polypeptide caused by weak interactions among R groups, and between R groups and the fluid environment.

Tertiary structure

4 The quaternary structure of a protein is the final three-dimensional structure formed by all the polypeptide chains making up the protein. This molecule of hemoglobin is composed of 4 polypeptide chains. The square plates represent nonprotein portions of the molecule (heme) that carry oxygen.

Quaternary structure

Figure 3.16 Primary structure determines a protein's shape due to bonding along the chain.

The functional groups of the amino acids in a polypeptide chain interact with their neighbors, forming hydrogen bonds (see p. 39). In addition, portions of a protein chain with many nonpolar side chains tend to be shoved into the interior of the protein because of their hydrophobic properties. Because of these interactions, polypeptide chains tend to fold spontaneously into sheets or wrap into coils. This folded or coiled shape is called its *secondary structure* ❷. Proteins made up largely of sheets often form fibers such as keratin fibers in hair, fibrin in blood clots, and silk in spiders' webs. Proteins that have regions forming coils frequently fold into globular shapes such as the globin subunits of hemoglobin in blood.

Hydrogen bonding can also result in proteins with more complex shapes than the secondary structure. The next level of structure is called *tertiary structure* ❸. For proteins that consist of subunits (separate polypeptide chains), the way these subunits are assembled into a whole is called the *quaternary structure* ❹.

Twenty amino acids are used in various sequences and combinations to make up a wide variety of proteins. These amino acids have different properties from one another. These differences and the sequence in which the amino acids are arranged determine the shape of a protein, which is crucial to its proper functioning.

The primary biological molecules are
carbohydrates, lipids, proteins, and nucleic acids.

3.10   Nucleic acids

Organisms store information about the structures of
their proteins in macromolecules called **nucleic acids**.
Nucleic acids are long polymers of repeating subunits
called *nucleotides*. Each nucleotide is made up of three
smaller building blocks (Figure 3.17):

1. A five-carbon sugar
2. A phosphate group $(-PO_4^{-2})$
3. An organic, nitrogen-containing molecule called a
   base

To form the nucleic acid chain, the sugars and phosphate groups making up the nucleotides are linked; a
nitrogenous base protrudes from each sugar as shown in
Figure 3.17. The order in which the nucleotides are
linked together forms a code that ultimately specifies
the order of amino acids in a particular protein.

Organisms have two forms of nucleic acid. One
form, deoxyribonucleic acid (DNA), stores the information for making proteins. The other form, ribonucleic
acid (RNA), directs the production of proteins. Details
of the structure of DNA and the ways it interacts with
RNA are presented in Chapter 18.

Hereditary information is a code composed of a sequence of nucleotides that signifies the sequence
of amino acids in the proteins of living things.

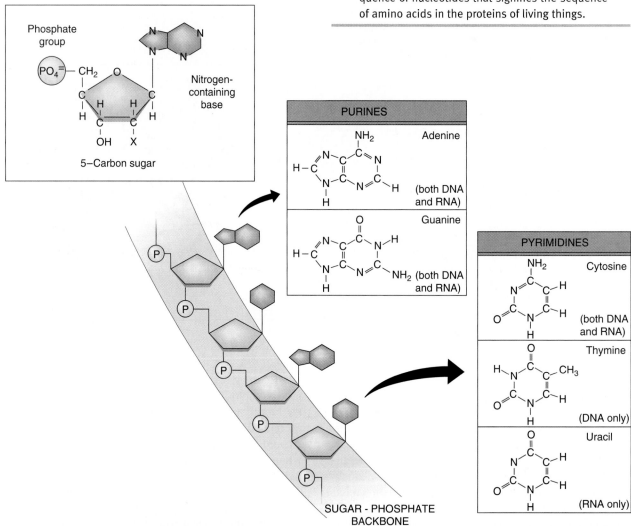

Figure 3.17   The structure of a nucleotide and the formation of nucleic acid chains. As shown in the inset, a nucleotide is composed of a
five-carbon sugar, a phosphate group, and an organic nitrogen base. The sugar and phosphate groups make up the backbone of a nucleic acid
chain, while the nitrogenous bases link the two sides of the chain. The five nitrogen bases that occur in the nucleic acids of DNA and RNA are
shown on the right.

Key Concepts

Atoms interact by means of their electrons.

3.1 Atoms make up all matter and are composed of three types of subatomic particles: protons, neutrons, and electrons.

3.1 Protons and neutrons are found at the core of the atom, and electrons surround this core.

3.2 The chemical behavior of an atom is largely determined by the distribution of its electrons, particularly the number of electrons in its outermost energy level.

3.2 Molecules are two or more atoms held together by sharing electrons.

3.2 Molecules can be made up of atoms of the same element or atoms of different elements.

3.2 Molecules made up of atoms of different elements are called compounds. Compounds are also atoms of different elements held together by electrostatic attraction.

3.3 Atoms are most stable when their outer energy levels are filled. Electrons are lost, gained, or shared until this condition is reached.

3.3 A covalent bond is the sharing of one or more pairs of electrons.

3.3 An ionic bond is an attraction between ions of opposite charge.

Water has properties important to life.

3.4 Water is a polar molecule, having positive and negative ends, and attracts other charged particles.

3.4 Water dissolves polar substances and excludes nonpolar substances.

3.4 Chemical interactions readily take place in water because so many kinds of molecules are water soluble and therefore move among water molecules as separate molecules or ions.

3.5 Water spontaneously ionizes, forming hydrogen ions (H^+) and hydroxide ions (OH^-).

The primary biological molecules are carbohydrates, lipids, proteins, and nucleic acids.

3.6 The chemistry of life is concerned with the interactions among biological molecules, or organic molecules.

3.6 Organic molecules have a carbon core with various functional groups attached to the carbon atoms.

3.7 Most organisms use carbohydrates as a primary fuel.

3.7 The most important of the energy-storing carbohydrates is glucose, a six-carbon sugar.

3.8 Certain lipids—the fats—are used in the long-term storage of energy.

3.8 Phospholipids are a major component of cell membranes.

3.9 Proteins are linear polymers of amino acids.

3.9 Because the 20 amino acids that occur in proteins have side chains with differing chemical properties, the shape and therefore the functioning of a protein is critically affected by its particular sequence of amino acids.

3.10 Hereditary information is stored as a sequence of nucleotides in the nucleotide polymer DNA.

3.10 A second form of nucleic acid, RNA, directs the production of proteins.

Key Terms

acid 41
atoms 34
base 41
carbohydrates 46
chemical bonds 37
compounds 36
covalent (ko-VAY-lent) 37
dehydration synthesis 44
disaccharide (dye-SACK-uh-ride) 46
electrons 34
elements 34
enzymes 49
fats 48
functional group 43
glycogen (GLYE-ko-jen) 47

hydrogen bonds 39
hydrolysis (hi-DROL-ul-sis) 44
hydrophobic (hi-dro-FO-bick) 40
ionic (eye-ON-ick) 37
ionization (EYE-uh-ni-ZAY-shun) 41
ions (EYE-onz) 37
isotopes (EYE-suh-topes) 34
lipids 48
molecule 36
monosaccharides (MON-o-SACK-uh-rides) 46
neutrons (NOO-trons) 34
noble gases 36
nucleic acids (new-KLAY-ick) 52
octet rule (OCK-tet ROOL) 36

organic compounds 42
oxidation (OK-si-DAY-shun) 38
peptides 49
pH scale 41
polar molecules 39
polysaccharides (POL-ee-SACK-uh-rides) 46
proteins 49
protons 34
redox reaction 38
reduction 38
shell (energy level) 34
soluble (SOL-you-ble) 40
starches 46
triglyceride (try-GLISS-er-ide) 48

learning portfolio

KNOWLEDGE AND COMPREHENSION QUESTIONS

1. a. Draw a diagram of an atom with an atomic number of 1. Label the nucleus and the subatomic particles, and show the electrical shape of each particle. (*Comprehension*)
   b. Draw an isotope of the same atom. (*Comprehension*)

2. List and explain three factors that influence how an atom interacts with other atoms. What is the significance of the octet rule? (*Comprehension*)

3. Nitrogen gas is formed via a triple bond between two nitrogen atoms. What type of bonding is this? What does this bond have in common with the bonds forming water molecules? (*Comprehension*)

4. What characteristics of water make it so unusual? How do these traits affect the interactions of other molecules dissolved or suspended in it? Relate these properties to water's molecular structure. (*Comprehension*)

5. Distinguish between organic and inorganic molecules. Which would you primarily study if you wanted to learn more about the human body? (*Comprehension*)

6. The pH of the digestive juices within the human small intestine is between 7.5 and 8.5. Would you describe this environment as acidic, neutral, or basic? Would the concentration of H^+ ions in this material be high or low with respect to a neutral pH? (*Comprehension*)

7. Both carbon and water are discussed in this chapter as the essential components of life on Earth. How are their bonding characteristics significant aspects of their biological importance? (*Comprehension*)

8. Distinguish among monosaccharides, disaccharides, and polysaccharides. To what group of biological macromolecules do they belong, and why are they important? (*Comprehension*)

9. Why do plants and animals not store glucose for future use? What forms of storage are most common in plants and animals? (*Comprehension*)

10. Discuss the three classes of macromolecules taken in as food energy by humans. Which chemical process, dehydration synthesis or hydrolysis, do you think is essential in the digestion of food? (*Comprehension*)

KEY
Knowledge: Recalling information.
Comprehension: Showing understanding of recalled information.

CRITICAL THINKING REVIEW

1. Based on what you learned in this chapter about the interactions of oil and water, suggest at least one reason why it is necessary to clean up an oil spill in the ocean. (*Synthesis*)

2. While doing your grocery shopping, you find this label on a product:

Nutritional Information Per Serving	
Calories 270	Calories from fat 45
Total fat	5 grams
Saturated fat	1.5 grams
Total carbohydrates	48 grams
Sugar	4 grams
Protein	8 grams

What macromolecules are present in this food? Would this food be part of a heart-healthy diet? Defend your answer with evidence. (*Synthesis*)

3. Name two functions of proteins in living systems. How might an error in the DNA of an organism affect protein function? (*Application*)

4. Sarah decided that she needed to lose weight, and planned to exercise daily and eat only 1,500 calories per day. She was shocked to learn that one gram of fat contained more than twice as many calories as a gram of carbohydrate or protein. Based on what you know about the structures of these molecules, propose a reason to explain Sarah's finding. (*Synthesis*)

5. John went to a health food store to buy vitamins. While there, the store clerk explained that vitamins derived from "natural" sources were better than synthetic (laboratory-made) vitamins. John noted that the "natural" vitamins were also more expensive. If you were with John, what could you tell him about the differences between these two types of vitamins based on your reading of this chapter? (*Application*)

KEY
Application: Using information in a new situation.
Synthesis: Putting together information from different sources.

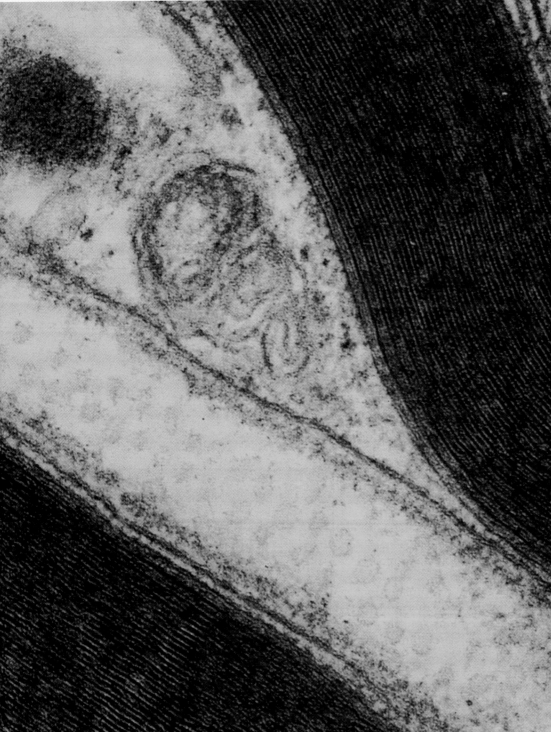

Organization

www.jbpub.com/biology

4.1 The cell theory

Cells are much more than the empty "boxes" seen by the British microscopist Robert Hooke in the mid-1600s. Scientists today know that cells are the smallest unit of life that can exist independently. Cells can take in nutrients, break them down to release energy, and get rid of wastes. They can reproduce, react to stimuli, and maintain an internal environment different from their surroundings. In multicellular organisms such as humans, cells work together to maintain life. In single-celled organisms such as the paramecium in **Figure 4.1a**, each cell survives independently. This chapter will help you become familiar with the structure of cells and how they work. By studying their structure and function, you will also see their tremendous diversity and complexity (Figure 4.1) and will begin to understand the cellular level of organization of the human body.

In 1839, botanist Matthias Schleiden and zoologist Theodor Schwann formulated the theory that all living things are made up of cells. In other words, they realized that cells make up the structure of such things as houseplants and people—but not of rocks or soil. It took another 50 years and the work of Rudolf Virchow for scientists to understand another basic concept about cells: Living cells can only be produced by other living cells. As Virchow put it, "all cells from cells." Together, these concepts are called the **cell theory**.

The cell theory, which is today recognized as fact, is a profound statement regarding the nature of living things. It includes three basic principles:

1. All living things are made up of one or more cells.
2. The smallest *living* unit of structure and function of all organisms is the cell.
3. All cells arise from preexisting cells.

When these statements were formulated in the mid-1800s, scientists discarded the idea of *spontaneous generation*: that living things could arise from the nonliving. The theory of spontaneous generation had suggested that frogs could be born of the mud in a pond and that rotting meat could spawn the larvae of flies. After the development of the cell theory, scientists recognized that life arose directly from the growth and division of single cells. Today, scientists think that life on Earth represents a continuous line of descent from the first cells that evolved on Earth.

All organisms on Earth are cells or are made up of groups of cells. In addition, all organisms are descendants of the first cells.

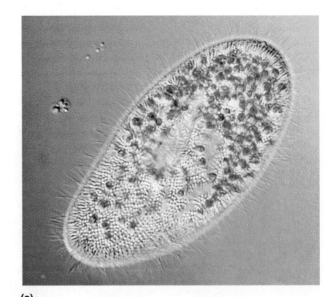

(a)

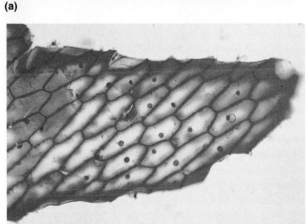

(b)

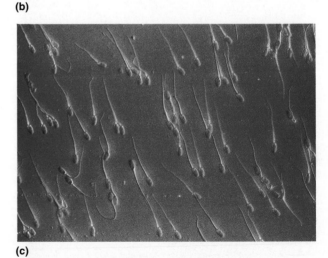

(c)

Figure 4.1 **Cells diverse in shape, structure, and function.** (a) *Paramecium*, a single-celled protozoan. (b) Onion cells. (c) Sperm cells.

Most cells are microscopically small.

4.2 Why aren't cells larger?

Most animal cells are extremely small, ranging in diameter from about 10 to 30 micrometers (μm). There are 1000 μm in 1 millimeter (mm)—the width of a paper clip's wire. So, as you might expect, most cells are invisible to the naked eye without the aid of a microscope. Your red blood cells, for example, are so small that it would take a row of about 2500 of them to span the diameter of a dime. Only a few kinds of cells are large. Individual cells of the marine alga *Acetabularia*, for example, are up to 5 centimeters long **(Figure 4.2)**. If you had eggs for breakfast, you were eating single cells! Other kinds of cells are long and thin, such as nerve cells that run from your spinal cord to your toes or fingers. However, very few cells are as large as a hen's egg or as long as a nerve cell.

To understand why cells are so small, you must first realize that most cells are constantly working, doing such jobs as breaking down molecules for energy, producing substances that cells need, and getting rid of wastes. Each cell must move substances in and out across its boundary—the cell membrane—quickly enough to meet its needs. Therefore, the amount of membranous surface area a cell has in relationship to the volume it encloses is crucial to its survival.

To illustrate, suppose that a cell is a cube like the one pictured in **Figure 4.3a,** with sides 2 centimeters (cm) in length. The surface area of each side of the cell is 2 cm × 2 cm, or 4 cm². The total surface area of the six sides is 6 × 4 cm², or 24 cm². Now let's compare surface area to the volume of this cell. Its volume is calculated by multiplying its length × width × height: 2 cm × 2 cm × 2 cm, or 8 cm³. Our hypothetical cell, then, has 24 cm² of

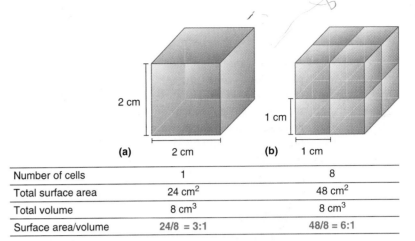

	(a) 2 cm	(b) 1 cm
Number of cells	1	8
Total surface area	24 cm²	48 cm²
Total volume	8 cm³	8 cm³
Surface area/volume	24/8 = 3:1	48/8 = 6:1

Figure 4.3 **Cells maintain a large surface-to-volume ratio.** The smaller the cell, the larger its surface area-to-volume ratio. Multicellular organisms are made up of many microscopic cells. This organization increases the total surface area of the cell membranes within living things.

surface area for 8 cm³ of volume, or 3 cm² of surface area for each cm³ of volume—a 3:1 ratio.

Suppose, however, that most cells were never able to grow this large. Suppose that as their sides grew to 1 cm, the cells were stimulated to divide. Figure 4.3b shows eight cells having sides of 1 cm each. These eight cells comprise the same volume as the single cell 2 cm on a side. Yet how does the surface area compare?

The surface area of each smaller cell is 1 cm × 1 cm × 6 sides, or 6 cm². The total surface area of the 8 cells is 48 cm². These eight smaller cells, then, have 48 cm² of surface area for 8 cm³ of volume, or 6 cm² of surface area for each cm³ of volume—a 6:1 ratio. The smaller cells have twice as much surface area for the same amount of volume as the larger cell!

In general, then, the smaller the cell, the larger its surface area-to-volume ratio. Cells that are too large have a surface area-to-volume ratio that is too small to move substances in and out (across their membranes) fast enough to meet their needs.

Another reason that cells generally do not grow large has to do with their controlling centers, or nuclei. Scientists think that a nucleus could not control all the workings of an active cell if it grew too large. In fact, some large complex cells, such as the unicellular paramecia, have two nuclei. Large cells with only one nucleus, such as unfertilized egg cells, are usually inactive cells.

Figure 4.2 **The marine green alga *Acetabularia*.** *Acetabularia* is a large, single-celled organism with clearly differentiated parts, such as the stalks and elaborate "hats" visible here. Each cell is a different individual several centimeters tall.

Most cells are microscopically small, resulting in a surface-area-to-volume ratio sufficient to move substances across their cell membranes quickly enough to meet their needs.

4.3 Prokaryotes and eukaryotes

All cells can be grouped into two broad categories: **prokaryotic cells** (pro-KARE-ee-OT-ick) and **eukaryotic cells** (yoo-KARE-ee-OT-ick). Cells are placed into one of these categories based on their type of structure. Prokaryotic cells, or prokaryotes, have a simpler structure than eukaryotes and were the first type of cell to exist as life arose on Earth billions of years ago. Eukaryotic cells evolved from these simpler cells.

You know the prokaryotes as bacteria. The cells of organisms within the other four kingdoms of life—plants, animals, protists, and fungi—are eukaryotes. Although the structure of cells in their group may vary among species, prokaryotes and eukaryotes each have distinctive features common to their group. In addition, because members of both groups are living cells, they also have some of the same features. Almost all cells have the following four characteristics:

1. A surrounding membrane
2. A thick fluid enclosed by this membrane that, along with the other cell contents, is called *protoplasm*

3. Organelles, or "little organs," located within the protoplasm (of eukaryotes) that carry out certain cellular functions
4. A control center called a *nucleus* (in eukaryotes) or *nucleoid* (in prokaryotes) that contains the hereditary material DNA

The structure of bacterial cells is discussed in Chapter 26, but Table 4.2 located on p. 77 will help you clarify and classify the differences among three cell types: plant cells and animal cells (both eukaryotes) and bacterial cells (prokaryotes). Also, **Figure 4.4** (a bacterial cell), **Figure 4.5** (an animal cell), and **Figure 4.6** (a plant cell) will help you visualize prokaryotic and eukaryotic cell structures.

Cells are either prokaryotic or eukaryotic based on their structure. Prokaryotic cells are bacteria and are simpler cells than eukaryotic cells, which make up the other four kingdoms of life. Both types of cells are surrounded by a membrane that encloses the protoplasm, DNA, and other cell structures.

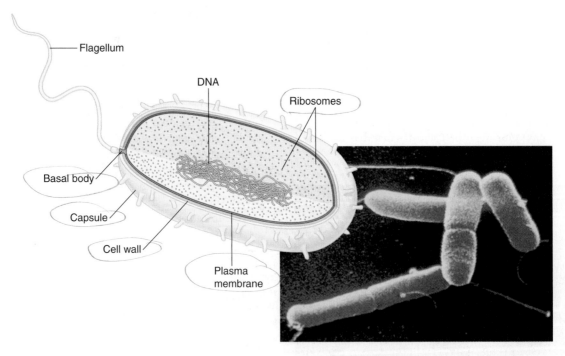

Figure 4.4 A bacterial cell. Bacterial cells are prokaryotic and have a simpler structure than eukaryotic cells. Prokaryotic cells have no membrane-bounded organelles.

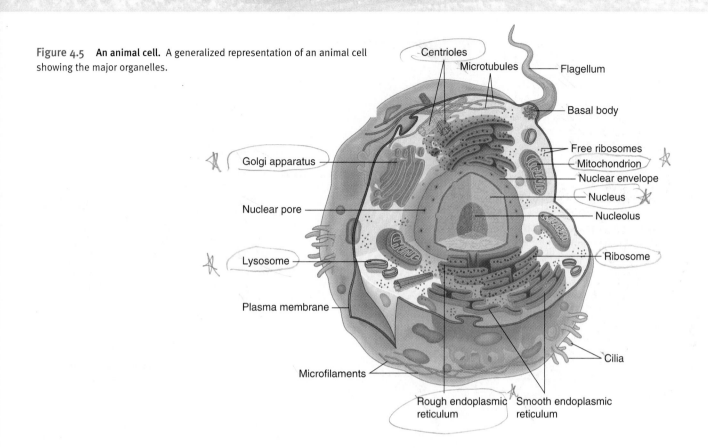

Figure 4.5 **An animal cell.** A generalized representation of an animal cell showing the major organelles.

Centrioles
Microtubules
Flagellum
Basal body
Free ribosomes
Mitochondrion
Nuclear envelope
Nucleus
Nucleolus
Ribosome
Cilia
Smooth endoplasmic reticulum
Rough endoplasmic reticulum
Microfilaments
Plasma membrane
Lysosome
Nuclear pore
Golgi apparatus

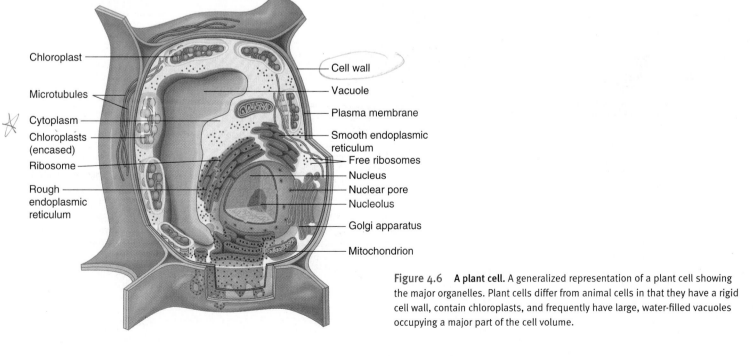

Chloroplast
Microtubules
Cytoplasm
Chloroplasts (encased)
Ribosome
Rough endoplasmic reticulum

Cell wall
Vacuole
Plasma membrane
Smooth endoplasmic reticulum
Free ribosomes
Nucleus
Nuclear pore
Nucleolus
Golgi apparatus
Mitochondrion

Figure 4.6 **A plant cell.** A generalized representation of a plant cell showing the major organelles. Plant cells differ from animal cells in that they have a rigid cell wall, contain chloroplasts, and frequently have large, water-filled vacuoles occupying a major part of the cell volume.

HOW Science WORKS

Exploring the Microscopic World

The Tools of Scientists

A world invisible to the naked eye had its roots of discovery in the late 1600s with the work of two pioneers in the field of microscopy: Anton van Leeuwenhoek and Marcello Malpighi. Microscopy is the use of a microscope—an optical instrument consisting of a lens or a combination of lenses for magnifying things that are too small to see clearly or at all. Leeuwenhoek's and Malpighi's inventions allowed them to observe such things as plant and animal tissues, blood cells, and sperm; little escaped their observant and technologically aided eyes.

Microscopes improved somewhat over the next two hundred years, but it was not until the beginning of the 20th century that this technology began to advance in sophistication at a fast pace, resulting in the array of light microscopes and electron microscopes that scientists routinely use today. Photographs prepared from three types of microscopes—the compound light microscope, the transmission electron microscope, and the scanning electron microscope—are the ones most often used in this book.

The compound light microscope (Figure 4.A) is an instrument that uses two lenses (therefore, compound) to magnify an object. A mirror focuses light from the room up to the eye from beneath the specimen, or a lamp is used for illumination (therefore, light). The compound light microscope visualizes eukaryotic cells (such as plant and animal cells) and prokaryotic cells (bacteria) and can magnify them up to 1500 times. Special stains are often used to visualize particular structures, as is shown in Figure 4.B. Brightfield microscopy, the technique used to visualize this stained specimen, passes light directly through the specimen. Unstained organisms can be seen quite well using a technique called phase-contrast microscopy. This technique allows the organism to be viewed while still alive (staining kills cells) and uses direct and indirect lighting to intensify the variations in density within the cell as shown in Figure 4.C. A variety of other techniques can be used with the compound microscopes to view organisms, but these two are most commonly used.

The electron microscope does not use light to visualize specimens, but instead uses a fine beam of electrons transmitted to a speci-

www.ibpub.com/biology
WWW

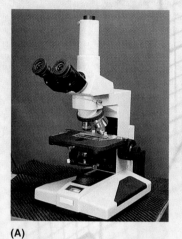
(A)

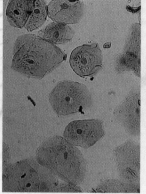
(B)

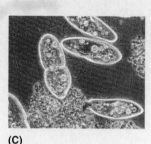
(C)

(D)

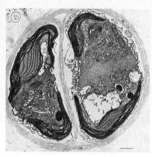
(E)

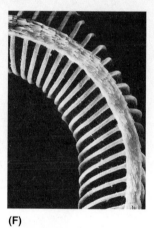
(F)

men in a vacuum. The transmitted electrons, after partial absorption by the object, are focused by magnets to form the image of the specimen. Electrons have a shorter wavelength than visible light; this difference gives the electron microscope a greater resolving power than the compound light microscope. Resolving power is the ability to distinguish two points as being separate from one another. Therefore, specimens can be magnified more highly with the electron microscope and still transmit a clear image. As with light microscopy, a variety of techniques can be used to prepare specimens. In addition, there are two types of electron microscopes: the transmission electron microscope (TEM) and the scanning electron microscope (SEM). Figure 4.D is a scanning electron microscope.

The transmission electron microscope is generally used to visualize slices of cells that have been specially prepared. The images are therefore of the interiors of cells, as is shown in Figure 4.E. In this photograph (called an electron micrograph), you can see cell organelles. The cells shown here are algae.

The scanning electron microscope is usually used to visualize surfaces of specimens that have been glued or taped to a metal slide and then covered with a microscopically thin layer of a metal, usually gold or platinum. The metal gives off secondary electrons when excited by a beam of electrons scanned across its surface. These secondary electrons are collected on a screen, which results in an image of the surface of the specimen. Figure 4.F is a scanning electron micrograph of the antenna of a moth.

4.4 Eukaryotic cells: An overview

Although eukaryotic cells are quite a diverse group, they share a basic architecture. They are all bounded by a membrane that encloses a semifluid material crisscrossed with a supporting framework of protein. All eukaryotic cells also possess many organelles (Table 4.1). These organelles are of two general kinds: (1) membranes or organelles derived from membranes and (2) bacterialike organelles.

Most biologists agree that the bacterialike organelles in eukaryotes were derived from ancient *symbi-otic bacteria*. The word symbiosis means that two or more organisms live together in a close association. An organism that is symbiotic within another is called an *endosymbiont*. The major endosymbionts that occur in eukaryotic cells are mitochondria, which occur in all but a very few eukaryotic organisms, and chloroplasts, which occur in algae and plants.

Eukaryotic cells, although a diverse group, all possess both membrane-bounded and bacterialike organelles. The latter likely developed when ancient bacteria lived in close association with ancestors to eukaryotic cells.

Table 4.1

Eukaryotic Cell Structures and Their Functions

Structure	Description	Function
Exterior Structures		
Cell wall	Outer layer of cellulose or chitin, or absent	Protection, support
Plasma membrane	Lipid bilayer in which proteins are embedded	Regulation of what passes in and out of cell, cell-to-cell recognition
Cytoskeleton	Network of protein filaments, fibers, and tubules	Structural support, cell movement
Flagella (cilia)	Cellular extensions with 9 + 2 arrangement of pairs of microtubules	Motility or moving fluids over surfaces
Interior Structures and Organelles		
Endoplasmic reticulum	Network of internal membranes	Formation of compartments and vesicles; modification and transport of proteins; synthesis of carbohydrates and lipids
Ribosomes	Small, complex assemblies of protein and RNA, often bound to ER	Sites of protein synthesis
Nucleus	Spherical structure bounded by a double membrane, site of chromosomes	Control center of cell
Chromosomes	Long threads of DNA associated with protein	Sites of hereditary information
Nucleolus	Site within nucleus of rRNA synthesis	Synthesis and assembly of ribosomes
Golgi apparatus	Stacks of flattened vesicles	Packaging of proteins for export from cell
Lysosomes	Membranous sacs containing digestive enzymes	Digestion of various molecules
Mitochondria	Bacterialike elements with inner membrane highly folded	"Power plant" of the cell
Chloroplasts	Bacterialike elements with inner membrane forming sacs containing chlorophyll, found in plant cells and algae	Site of photosynthesis in plant cells and algae

4.5 The plasma membrane

Every cell is bounded by a **plasma membrane** (Figure 4.7), so named because it encloses the *protoplasm*—the semi-fluid cell contents, including the cell organelles and nucleus (control center) of the cell. The electron micrograph in Figure 4.7a shows that the plasma membrane is a double-layered structure. The plasma membrane has two main components: *phospholipids* (shown as the blue balls with tails in Figure 4.7b) and *proteins* (shown as the purple globular structures). Lipids, as you may recall from Chapter 3, are made up of a glycerol backbone attached to three fatty acid chains. Phospholipids have only two fatty acid chains; phosphoric acid takes the place of the third (Figure 4.7c). Phosphoric acid is a phosphorylated alcohol: an alcohol molecule bonded to a phosphate group. [The "kink" in one of the fatty

acid chains means that it is unsaturated (see Chapter 3, p. 48). The straight fatty acid chain is saturated.]

Lipids are nonpolar molecules and will not dissolve in (form hydrogen bonds with) water. Phosphorylated alcohols, however, *are* polar and *do* interact with (form hydrogen bonds with) water. Therefore, phospholipids have a portion of the molecule termed **hydrophilic** (HI-droe-FIL-ick), meaning "water loving," and a portion of the molecule that is **hydrophobic** (HI-droe-FO-bic), meaning "water fearing" (Figure 4.7c).

When a collection of phospholipid molecules is placed in water, their hydrophobic fatty acid chains are pushed away by the water molecules that surround them. The water molecules move toward the molecules that form hydrogen bonds with them—the phosphoric acid. As a result of these orientations of the two parts of phospholipids to water, they form two layers of mole-

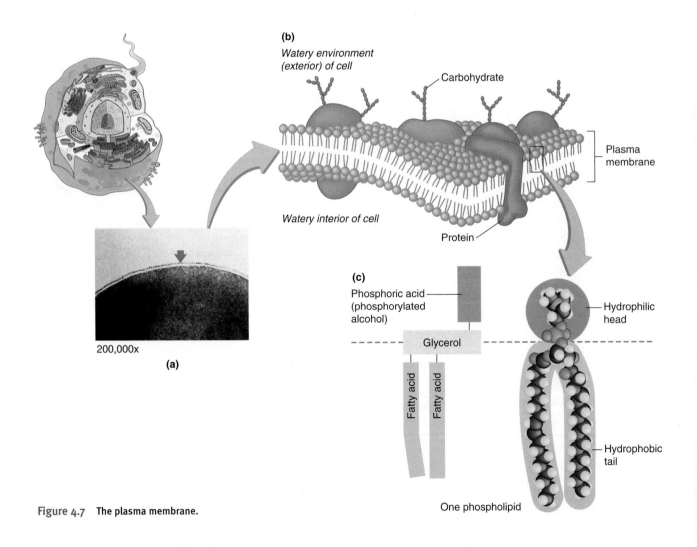

Figure 4.7 The plasma membrane.

cules, with their hydrophobic "tails" oriented inward (away from the water) and their hydrophilic "heads" oriented outward (toward the water, Figure 4.7b). This double-layered structure is called a *lipid bilayer* and is the foundation of the membranes of all living things.

Did you ever blow soap bubbles when you were a child? Take a moment to think about how the bubbles looked. Can you remember how the soap film forming the bubble seemed to swirl and move as the bubble floated in the air? Lipid bilayers are "fluid" much like the film of the soap bubbles you remember. The fluid nature of the bilayer is caused by movement of the phospholipid molecules. Although water constantly forces the phospholipid molecules to form a bilayer, the hydrogen bonds between the water molecules and the hydrophilic heads of the phospholipids are constantly breaking and re-forming. Therefore, individual molecules move about somewhat freely.

If cells were encased with pure lipid bilayers, however, they would be unable to survive because the interior of the lipid bilayer is completely nonpolar. It repels most water-soluble molecules that attempt to pass through it. In fact, as **Figure 4.8** shows, the lipid bilayer allows only a few small uncharged polar molecules, such as individual water (H_2O) and ammonia (NH_3) molecules, to pass through (Figure 4.8a). (Ammonia is a common cellular waste product formed from the breakdown of proteins.) In addition, substances that are soluble in oil, such as vitamins A, D, and E and other hydrocarbons (molecules that contain only hydrogen and carbon such as fatty acids), can also move across the lipid bilayer (Figure 4.8b). It prevents, however, any water-soluble substances such as sugars, amino acids, and proteins (Figure 4.8c) as well as charged particles (important ions) such as calcium (Ca^{2+}) and potassium (K^+) from entering or leaving the cell (Figure 4.8d). Such molecules pass through the cell membrane by means of proteins that are embedded in the lipid bilayer. Proposed in the early 1970s by Singer and Nicholson, this model of the cell membrane is widely accepted today. It is called the **fluid mosaic model**, a name that describes the fluid nature of a lipid bilayer studded with a mosaic of proteins.

The variety of proteins that penetrate the lipid layer (the purple globular structures shown in Figure 4.7b) control the interactions of the cell with its environment. Some of these proteins, called *channels*, act as doors that let specific molecules into and out of the cell. Other pro-

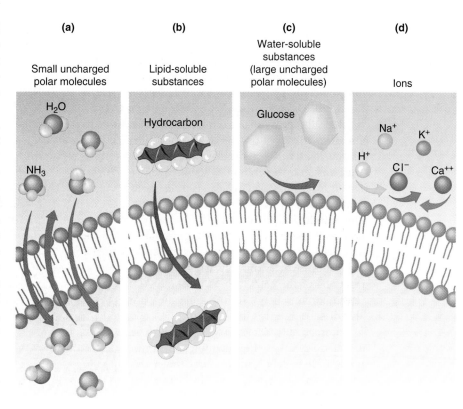

(a)
Small uncharged polar molecules

(b)
Lipid-soluble substances

(c)
Water-soluble substances (large uncharged polar molecules)

(d)
Ions

H_2O

Hydrocarbon

Glucose

NH_3

Na^+ K^+

H^+

Cl^- Ca^{++}

Figure 4.8   Some molecules can pass across the lipid bilayer; others cannot.

teins, called *receptors,* recognize certain chemicals such as hormones that signal cells to respond in particular ways. Receptor proteins cause changes within the cell when they come in contact with such chemical "messengers" or signals. A third type of protein identifies a cell as being of a particular type. These cell surface markers help cells identify one another, an ability that helps cells function correctly in a multicellular organism (such as your body).

The basic foundation of biological membranes is a lipid bilayer. Lipid bilayers form spontaneously, with the nonpolar tails of the phospholipid molecules pointed inward and the polar heads pointed outward. A lipid bilayer forms a fluid, flexible covering for a cell and keeps the watery contents of the cell on one side of the membrane and the watery environment on the other. Proteins embedded within this plasma membrane regulate interactions between the cell and its environment.

4.6   The cytoplasm and cytoskeleton

The **cytoplasm** (SYE-toe-PLAZ-um) of the cell is the viscous ("thick") fluid containing all cell organelles *except* the nucleus. (The protoplasm includes the nucleus.) The word cytoplasm literally means "living gel" (*plasm*) "of the cell" (cyto). The major components of the cytoplasm are (1) a gel-like fluid (the cytosol); (2) storage substances; (3) a network of interconnected filaments and fibers; and (4) cell organelles. The fluid part of the cytoplasm is made up of approximately 75% water and 25% proteins. The proteins, mostly enzymes and structural proteins, make the cytoplasm viscous— much like thickening gelatin. Nonprotein molecules involved in the various chemical reactions of the cell, such as ions and ATP (an important energy-storing molecule), are also dissolved in this fluid.

Storage substances vary from one type of cell to another. For example, the cytoplasm of liver cells contains large molecules of glycogen, a storage form of glucose. Fat cells contain a large lipid droplet.

The cytoplasm also contains a *cytoskeleton*—a network of filaments and fibers that do many jobs. Although the term "skeleton" implies rigidity, the cytoskeleton is not a

Figure 4.10   The cytoskeleton. (a) Micrograph of the fibers making up the cytoskeleton. A = actin filaments (microfilaments), IF = intermediate filaments, MT = microtubules. Notice that the intermediate filaments appear solid, while the microtubules have a hollow core. The actin filaments are somewhat indistinct. Bar = 0.25 μm. (b) This illustration shows a cross-section of a eukaryotic cell and the network of proteins that lies beneath the plasma membrane and helps support and shape the cell.

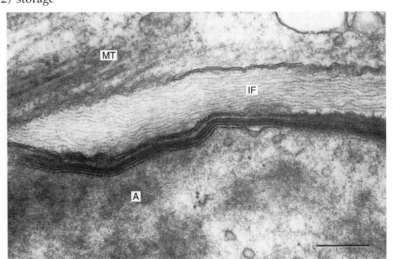

(a)

(b)

(a) Microfilament

8 nm

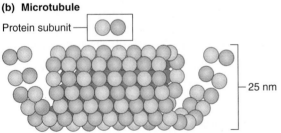

(b) Microtubule

Protein subunit

25 nm

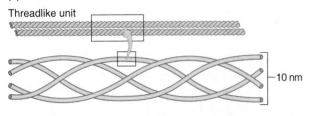

(c) Intermediate filament

Threadlike unit

10 nm

Figure 4.9   Types of fibers found in the cytoskeleton.

permanent, inflexible network; it undergoes rapid and continuous change.

The cytoskeleton is made up of three different types of fibers: *microfilaments, microtubules,* and *intermediate filaments.* The microfilaments are the thinnest—a twisted double chain of protein—and are not visible with an ordinary light microscope (**Figure 4.9a**). The microtubules, a chain of proteins wrapped in a spiral to form a tube, are the thickest members of the cytoskeleton (Figure 4.9b). As the name suggests, the intermediate filaments are an in-between size. These are thread-like protein molecules that wrap around one another to form "ropes" of protein (Figure 4.9c).

The microfilaments and intermediate filaments provide the protein network that lies beneath the plasma membrane to help support and shape the cell. This network of proteins extends into the cytoplasm and looks something like the web of proteins illustrated in **Figure 4.10**. Microtubules are also part of this "skeleton" of the cytoplasm. Together, these three types of protein fibers provide the cell with mechanical support and help anchor many of the organelles (mitochondria are an exception). They also help move substances from one part of the cell to another.

The cytoskeleton is a network of filaments and fibers within the cytoplasm that helps maintain the shape of the cell, move substances within cells, and anchor various structures in place.

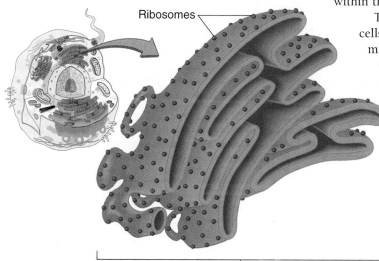

Figure 4.11   Rough endoplasmic reticulum (rough ER).

4.7   Membranous organelles

Enmeshed within the cytoskeletal fibers, the cell organelles constantly work, each contributing in a special way to the life and well-being of the cell. Most organelles of eukaryotic cells are bounded by membranes. In this way, these subcellular structures form organized compartments within the cytoplasm, an important characteristic of eukaryotic cells. By means of compartments, enzymes and their substrates are kept in particular places in cells and at optimum concentrations. Metabolic pathways can therefore run more efficiently, with their components grouped together. Additionally, toxic waste products, when they arise, can also be separated from the rest of the cell, a safe distance from enzymes and other substances that might be damaged by toxins. The compartmentalized eukaryotic cell is an efficiently running living machine.

The Endoplasmic Reticulum

The **endoplasmic reticulum** (EN-doe-PLAZ-mik ri-TIK-yuh-lum), or ER, is an extensive system of interconnected membranes that forms flattened channels and tubelike canals within the cytoplasm, almost like a cellular subway system. The channels are used to help move substances from one part of the cell to another. The name endoplasmic reticulum may sound very strange. It is, however, descriptive of its location and appearance. The word endoplasmic means "within the cytoplasm." The word reticulum comes from a Latin word meaning "a little net." That is exactly what the ER looks like—a net within the cytoplasm.

There are two types of endoplasmic reticula in cells: **rough ER** and **smooth ER**. *Rough* refers to the minute spherical structures called **ribosomes** (RYE-bow-somes) covering the surface of one type of ER. With ribosomes dotting their surfaces, rough ER membranes look like long sheets of sandpaper (**Figure 4.11**).

Ribosomes are the places where proteins are manufactured. Some ribosomes are attached to the rough ER, but others are in the cytoplasm bound to cytoskeletal fibers. The cytoplasmic ribosomes help produce proteins for use in the cell, such as those making up the structure of the cytoskeletal fibers or certain organelles. The proteins made at the ribosomes of the rough ER are most often destined to leave the cell. Cells specialized for secreting proteins, such as the pancreatic cells that manufacture the hormone insulin, contain large amounts of rough ER. After the proteins are manufactured at the ribosomes on the surface of the ER, they enter the inner space of the ER. Within this channel, the proteins may be changed by enzymes bound to the inner surface of the ER membrane. Carbohydrate molecules are often added to them.

When the proteins reach the end of their journey in the ER, they are encased in tiny membrane-bounded sacs called *vesicles*. These vesicles are formed at sections of smooth ER that are continuous with the rough ER. "Buds" of the smooth ER pinch off as the newly formed proteins reach them. These vesicles eventually fuse with the membranes of another organelle called the **Golgi apparatus** (GOL-gee).

Smooth ER has no ribosomes attached to its surfaces. Therefore, it does not have the grainy appearance of rough ER and does not manufacture proteins. Instead, smooth ER has enzymes bound to its inner surfaces that help build carbohydrates and lipids. Cells specialized for the synthesis of these molecules, such as animal cells that produce male or female sex hormones, have abundant smooth ER.

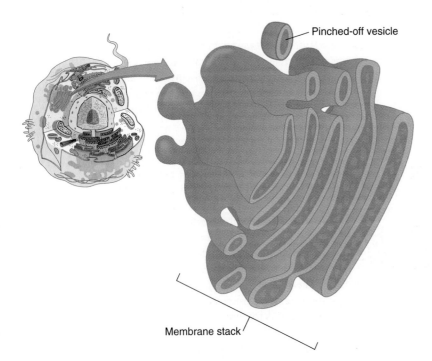

Pinched-off vesicle

Membrane stack

Figure 4.12 **A Golgi apparatus.** A Golgi apparatus is a structure that collects, modifies, packages, and distributes molecular products. When these molecular products are ready for transport, they are pinched off into vesicles that move either to the cell membrane to exit the cell or to locations within the cell.

Golgi Apparatus

First described by the physician Camillo Golgi in the last half of the 19th century, the Golgi apparatus looks like a microscopic stack of pancakes in the cytoplasm (Figure 4.12). Animal cells each contain 10 to 20 sets of these flattened membranes. Plant cells may contain several hundred because the Golgi apparatus is involved in the synthesis and maintenance of plant cell walls (a structure animal cells do not have).

The Golgi can be thought of as the packaging and shipping station of the cell. Molecules come to a Golgi apparatus in vesicles pinched off from the ER (Figure 4.13 ❶). The membranes of the vesicles fuse with the membranes of a Golgi apparatus ❷. Once inside the space formed by the Golgi membranes ❸, the molecules may be modified by the formation of new chemical bonds or by the addition of carbohydrates. For example, mucin, which is a protein with attached carbohydrates that forms a major part of the mucous secretions of the body, is put together in its final form in the Golgi. This refining of molecules occurs in stages in the Golgi, with different parts of the stack of membranes containing enzymes that do specific jobs.

When molecular products are ready for transport, they are sorted and pinched off in separate vesicles ❹. Each vesicle travels to its destination and fuses with another membrane. The vesicles containing those molecules that are to be secreted from the cell fuse with the plasma membrane ❺. In this way, the contents of the vesicle are liberated from the cell. Other vesicles fuse with the membranes of organelles such as lysosomes, delivering the new molecules to their interiors.

Lysosomes

The new molecules delivered to **lysosomes** (LYE-so-somes) are digestive enzymes—molecules that help break large molecules into smaller molecules. Lysosomes are, in fact, membrane-bounded bags of many different digestive enzymes (Figure 4.14). Several hundred of these organelles may be present in one cell alone.

Lysosomes and their digestive enzymes are extremely important to the health of a cell. They help cells function by aiding in cell renewal, constantly breaking down old cell parts as they are replaced with new cell parts. During development, lysosomes help remodel tissues, such as the reabsorption of the tadpole tail as the tadpole develops into a frog. In some cells, lysosomes also break down substances brought into the cell from the environment. For example, one job of certain white blood cells is to get rid of bacteria invading the body.

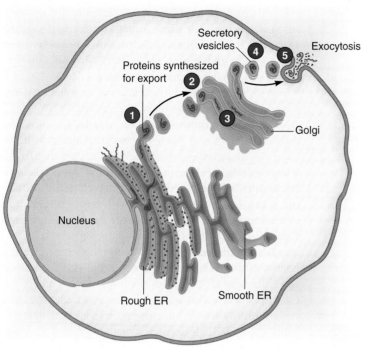

Secretory
vesicles

Exocytosis

Proteins synthesized
for export

1 **2** **3** **4** **5**

Golgi

Nucleus

Rough ER   Smooth ER

Figure 4.13   How a molecule is packaged for transport in the Golgi apparatus.

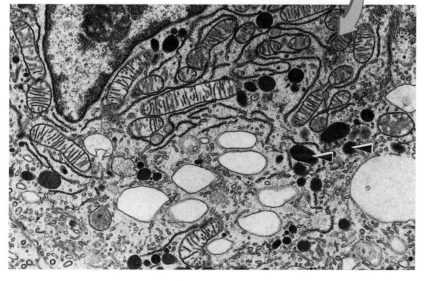

The cytoplasm of these cells flows around their prey, engulfing them in a membrane-bounded sac called a *vacuole*. A lysosome then fuses with this vacuole, and digestion of the invader begins. In the past, lysosomes were thought to be active during the aging process, breaking open and destroying whole cells. Scientists now know this idea to be false, since lysosome enzymes work only in the acid environment of the lysosome and become inactive at the relatively neutral pH of the cytoplasm.

What happens to the digestion products of lysosomes? Substances such as parts of bacteria are packaged in a vesicle, transported to the plasma membrane, and exit the cell. Other molecules, such as the breakdown products of old cell parts, may simply be released into the cytoplasm. These cellular building blocks can be recycled—used once again to build new cell parts.

Vacuoles

The word **vacuole** (VACK-yoo-ol) comes from a Latin word meaning "empty." However, these membrane-bounded storage sacs only look empty. Within them can be found such substances as water, food, and wastes. Their number, kind, and size vary in different kinds of cells.

Vacuoles are most often found in plant cells. These giant water-filled sacs play a major role in helping plant tissues stay rigid. In mature plant cells, the vacuole is often so large that it takes up most of the interior of the cell. Certain single-celled eukaryotes (protists) and some yeasts also contain vacuoles. Some fresh water protists contain contractile vacuoles. These organelles take up extra water that tends to flow into these organisms. Periodically, these vacuoles fuse with the plasma membrane and expel their contents, keeping the cell from swelling and bursting. Many protists have food vacuoles, places where food is stored and eventually digested by fusion with lysosomes.

Figure 4.14   Lysosomes. Electron micrograph of lysosomes (black arrow heads). The blackened areas within the lysosomes are digestive enzymes. These enzymes break down old cell parts and other materials brought into the cell.

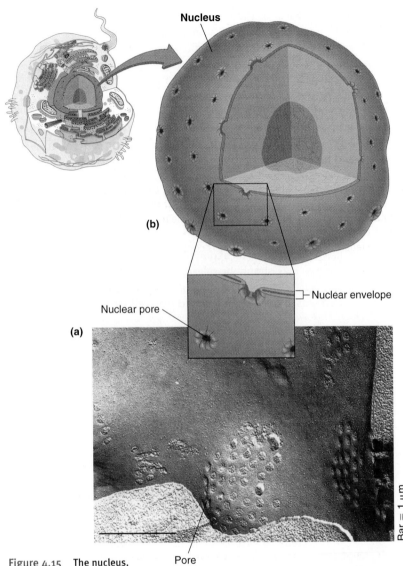

Nucleus

(b)

Nuclear envelope

Nuclear pore

(a)

Bar = 1 μm

Pore

Figure 4.15   **The nucleus.**

The Nucleus

The **nucleus**, or control center of the cell, is made up of an outer, double membrane that encloses the chromosomes and one or more nucleoli. In fact, the word eukaryote means "true nucleus" and refers to the fact that the nucleus is a closed compartment bounded by a membrane. Other than large fluid-filled vacuoles in plants, this compartment is the largest in eukaryotic cells.

The outer, double membrane of the nucleus is called the *nuclear envelope*. The inner of the two membranes actually forms the boundary of the nucleus, and the outer membrane is continuous with the ER. At various spots on its surface, the double membrane fuses to form openings called *nuclear pores*. **Figure 4.15a** shows the surface of the nuclear membrane of a cell that was frozen and then

cracked during a special type of preparation for the electron microscope. The pores look like pockmarks on the surface of the membrane. These ringlike holes are lined with proteins and serve as passageways for molecules entering and leaving the nucleus (Figure 4.15b).

Within the nucleus is the hereditary material **deoxyribonucleic acid** (de-OK-see-RYE-boh-new-KLAY-ick), or **DNA**. DNA determines whether your hair is blond or brown or whether a plant flowers in pink or white. It controls all activities of the cell. To accomplish these amazing feats, DNA performs one job: It directs the synthesis of ribonucleic acid, or RNA, which in turn directs the synthesis of proteins. (The structure, interactions, and roles of both DNA and RNA are discussed further in Chapter 18.)

DNA is bound to proteins in the nucleus, forming a complex called *chromatin*. In a cell that is not dividing, the chromatin is strung out, looking like strands of microscopic pearls. But as a cell begins to divide, the DNA coils more tightly around the proteins, condensing to form shortened, thickened structures called **chromosomes (Figure 4.16).**

The **nucleolus** (noo-KLEE-oh-lus, plural *nucleoli*) is a darkly staining region within the nucleus **(Figure 4.17).** Most cells have two or more nucleoli. Making up the bulk of the nucleolar material is a special area of DNA that directs the synthesis of **ribosomal ribonucleic acid, or rRNA.** As rRNA is made at this DNA, it forms clumps of molecules that are structural components of ribosomes. Proteins, another component of ribosomes, can also be found in the nucleolus. They are brought into the nucleus from the cytoplasm. After the ribosomes are manufactured (or partly manufactured) in the nucleolus, they move through the pores in the nuclear membrane and pass into the cytoplasm.

Figure 4.16   **Chromosomes.** As a cell begins to divide, DNA coils and supercoils, condensing to form shortened, thickened structures called chromosomes. The chromosomes in this cell have separated from one another in a way that will provide a complete set of hereditary material to each new daughter cell.

Just Wondering . . . *Real Questions Students Ask*

How do scientists know what is in the nucleus of a cell?

The story of scientists' quests to understand the structure of the nucleus began in the early 1700s. Microscopes at that time, although primitive by today's standards, were sophisticated enough to allow scientists to see the nu-

clei of plant cells and some animal cells. (The German microscope to the left is an example of microscope technology of the late 1700s.) However, it wasn't until 1833 that a nucleus was thought to be in every cell. Prior to that time, scientists held the view that some cells had no nuclei because they

could not see them. Today we realize that scientists were simply observing some cells during cell division, when the nucleus loses its membrane and the nuclear material looks much different from that of a nondividing cell.

As time went on, microscope technology improved with the invention of the high-power oil immersion lens in 1870. The microtome, which could slice individual cells into thin sec-

tions, was invented in 1866. In addition, new methods of fixing and staining cells were developed. Some dyes were produced that were highly specific for staining certain cell parts, such as the nucleus. All of these factors allowed scientists to view the nuclei of cells with much greater magnification, detail, and clarity.

As the quality of optical equipment and cell preparation techniques improved, the precision with which cells "at rest" and undergoing nuclear division could be studied increased. The nucleus no longer looked like a granule-filled compartment that split indiscriminately during cell division. Scientists could now see a nucleus filled with threads and bands of material, which they named *chromatin* after the Greek word for color, since this material readily took on color when stained. The word *chromosome,* meaning "colored body," was proposed in 1888.

During the 1800s, many scientists were also studying the principles of heredity and the nature of fertilization as well as the events of cell division. So, investigations into the function of the nuclear material paralleled the study of its structure. By 1866, Gregor Mendel published his theories of inheritance. (His work is described in Chapter 20.) In the late 1870s, scientists described the process of nuclear division and the process of nuclear re-

duction-division that occurs in sex cells. By 1884, scientists realized that fertilization in both animals and plants consists of the fusion of maternal and paternal sex cells. Taking all of this information into account, an American graduate student named Walter Sutton hypothesized that the hereditary factors were on the chromosomes. At the turn of the century, scientists began to look toward the chemical nature of this hereditary material. Then, for more than half a century, chemists joined biologists in their quest to understand the structure of nuclear material.

The rest of this fascinating story of discovery is continued later in this book. Chapter 18, "DNA, Gene Expression, and Cell Reproduction," chronicles investigations that experimentally determined that the hereditary material is located in the nucleus. It goes on to discuss the discovery of the molecular makeup of the hereditary material, culminating with the Nobel prizewinning work of two young scientists, James Watson and Francis Crick. As you can see, your question took scientists from many fields over one hundred years to answer. Today, using electron microscopes and highly sophisticated laboratory techniques, molecular biologists and others are still asking and answering questions about the nucleus and the nature of the hereditary material.

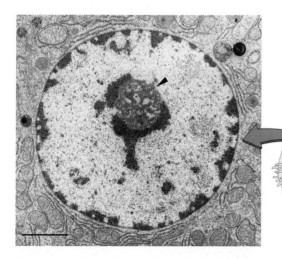

Figure 4.17 The nucleolus. Electron micrograph of a nucleolus (black arrow head) within the nucleus of the cell. The nucleolus consists of DNA, ribosomal RNA (rRNA), and ribosomal proteins. Bar = 2 μm.

Membrane-bounded eukaryotic organelles are the endoplasmic reticulum (ER), Golgi, lysosomes, vacuoles, and the nucleus. The ER is an extensive system of channellike membranes within eukaryotic cells. Rough ER makes and transports proteins destined to leave the cell. Smooth ER helps build carbohydrates and lipids. The Golgi apparatus is the packaging and delivery system of the eukaryotic cell. It collects, modifies, packages, and distributes molecules that are made at one location within the cell and used in another. Lysosomes are vesicles that contain digestive enzymes. These enzymes break down old cell parts or materials brought into the cell from the environment. Vacuoles are storage compartments for such substances as food, water, and wastes. The nucleus contains the hereditary material, or DNA. When the cell divides, the DNA condenses into compact structures called chromosomes.

4.8   Bacterialike organelles

The idea of the symbiotic origin of mitochondria and chloroplasts has had a controversial history, and a few biologists still do not accept it. The endosymbiotic hypothesis is, however, the most widely agreed-upon model at this time regarding the origin of mitochondria and chloroplasts in eukaryotic cells.

Mitochondria

The **mitochondria** (MITE-oh-KON-dree-uh) that occur in most eukaryotic cells are thought to have originated as symbiotic bacteria. According to this hypothesis, the bacteria that became mitochondria were engulfed by eukaryotic cells early in their evolutionary history. Before they acquired these bacteria, the host cells were unable to carry out chemical reactions necessary for living in an atmosphere that had increasing amounts of oxygen. The engulfed bacteria were able to carry out these reactions *using oxygen* and are considered to be the precursors to mitochondria.

Mitochondria are oval, sausage-shaped, or thread-like organelles about the size of bacteria that *have their own DNA* (which suggests that they were once free-living bacteria). They are bounded by a double membrane. The outer of the two membranes is smooth and defines the shape of the organelle. The inner membrane, however, has many folds called *cristae* that dip into

the interior of the mitochondrion. These cristae resemble the folded membranes that occur in various groups of bacteria. Notice in **Figure 4.18** that this arrangement of membranes forms two mitochondrial compartments.

The job of the mitochondria is to break down fuel molecules, releasing energy for cell work. The two most important fuels of cells are glucose and fatty acids. Some organisms, such as plants, make their fuel (glucose) using the raw materials of carbon dioxide, water, and sunlight. Other organisms, such as animals, eat food and digest it to produce glucose and fatty acids, which can then be transported to the cells. In both plants and animals, these fuels are broken down by cells, releasing energy by means of a series of oxygen-requiring reactions called *cellular respiration*. The energy released during these reactions is stored for later use in special molecules called ATP. (See Chapters 5 and 6 for a detailed description of ATP and cellular respiration.) Using chemical symbols, the reactions of cellular respiration can be summarized as follows:

$$C_6H_{12}O_6 \; + \; 6O_2 \; \longrightarrow \; 6CO_2 \; + \; 6H_2O$$

Glucose   Oxygen   Carbon dioxide   Water

Energy released

Cellular respiration begins in the cytoplasm, but most of the energy from the breakdown of glucose and fatty acids is generated in the mitochondria. The enzymes that are used in this breakdown are bound to the membranes of the mitochondrion. Within this closed compartment, the complex series of reactions needed to break apart glucose and to capture the liberated energy and store it are accomplished in an orderly and efficient way separated from the rest of the cell.

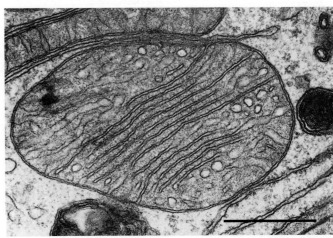

Figure 4.18   **The mitochondrion.** Bar = 0.25 μm.

Chloroplasts

Symbiotic events similar to those postulated for the origin of mitochondria also seem to have been involved in the origin of **chloroplasts**. They are thought to be derived from symbiotic photosynthetic bacteria. These energy-capturing organelles are found in the cells of plants and algae. In these organelles, the energy in sunlight is used to power the reactions that make the cellular fuel—glucose—by using molecules of carbon dioxide from the air. Together, the complex series of chemical reactions that perform these tasks is known as *photosynthesis*. Using chemical symbols, this series of reactions can be summarized as follows:

$$6CO_2 \ + \ H_2O \ \xrightarrow{\textit{Energy stored}} \ C_6H_{12}O_6 \ + \ 6O_2$$

Carbon dioxide   Water   Glucose   Oxygen

The glucose can be either broken down in the mitochondria to release its energy to ATP for immediate use or short-term storage, or stored as complex carbohydrates for later use.

Chloroplasts have a structure similar to the mitochondria. Like mitochondria, chloroplasts have their own DNA and are bounded by a double membrane. The inner membrane forms an extensive array of saclike invaginations called *thylakoids* (THIGH-leh-koidz). Stacks of thylakoids are called *grana* (**Figure 4.19**). Chlorophyll, a chemical that can absorb light energy from the sun and that allows photosynthesis to take place, is found within the thylakoids. As with the mitochondria, chloroplasts provide an orderly, closed compartment within the cell in which a series of reactions can occur.

The mitochondria apparently originated as endosymbiotic bacteria. In a complex series of reactions using oxygen, cell fuel is broken down to release the energy within mitochondria. Chloroplasts, which are located within the cells of plants and algae, apparently originated as endosymbiotic photosynthetic bacteria. Within chloroplasts, carbon dioxide, water, and light energy are used to produce the cell fuel glucose during a series of reactions called photosynthesis.

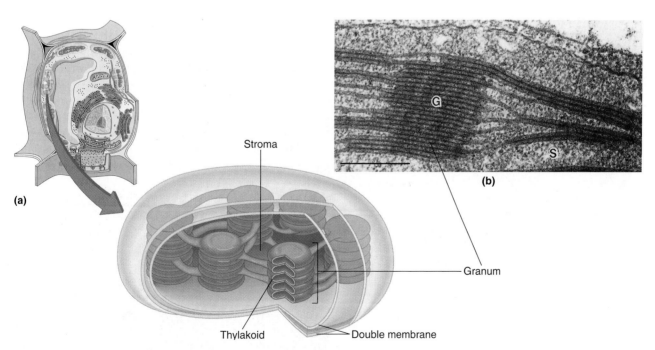

(a)

Stroma

Granum

Thylakoid

Double membrane

(b)

Figure 4.19   The chloroplast. (a) Chloroplasts have a double outer membrane and a system of interior membranes that form sacs called thylakoids. A stack of thylakoids is called a granum. (b) Electron micrograph of a chloroplast showing the grana, thylakoids, and stroma. Bar = 0.5 μm.

4.9   Cilia and flagella

Single-celled eukaryotic cells are often motile—able to move within their environments. One way in which these cells move is by means of cell extensions that look somewhat like hairs. These structures, called **cilia** (SILL-ee-uh), are short and often cover a cell. In human cells, cilia are found only on sections of certain cells and are used to move substances across their surfaces. Figure 9.6 shows the cilia that line the trachea, or windpipe. They help sweep invading particles and organisms up the trachea and away from the lungs. Some cells have whiplike extensions called **flagella** (fluh-JELL-uh). Used strictly for movement, these structures are longer than cilia, but fewer are usually present on a cell. Sperm are the only human cells that have flagella. Normal human sperm have a single flagellum, as you can see in Figure 4.1c.

Although cilia and flagella differ in length, they have the same structure: They are bundles of microtubules covered with the plasma membrane of the cell.

As you can see in **Figure 4.20,** nine pairs of microtubules surround a single, central pair. As these microtubules dip into the cell beneath the level of the plasma membrane, they connect with another structure called the *basal body*. Also composed of microtubules, a basal body serves to anchor a cilium or flagellum to the cell as shown in Figure 4.4.

Cilia and flagella in eukaryotes beat in a whiplike fashion because of microtubules sliding past each other in a manner similar to the sliding of actin and myosin filaments in muscle contraction (see Chapter 16, p. 349). The structure of the flagella and basal body in bacteria differs from eukaryotic flagellar and basal body structure. In addition, bacteria swim by rotating their flagella rather than by whipping them as eukaryotes do.

> Cilia and flagella are whiplike organelles of motility that protrude from some eukaryotic cells. Certain bacteria have flagella as well, but the structure differs from eukaryotic structure.

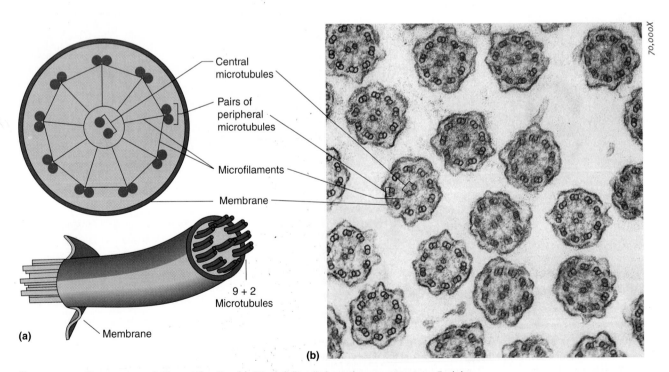

Figure 4.20   The structure of cilia and flagella. (a) Cilia and flagella have the same structure. Each has nine pairs of peripheral microtubules surrounding a central pair of microtubules. The microtubules are connected by microfilaments. (b) Micrograph showing a cross-section of cilia.

4.10 Cell walls

Plants, fungi, and many protists (especially algae) have a rigid structure called a **cell wall** that surrounds the plasma membrane, as can be seen in the plant cell below. Almost all plant cell walls are made up of cellulose, large molecules formed by the linking of glucose units,

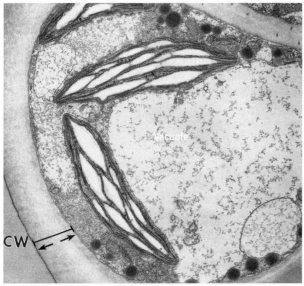

polysaccharides, and protein. The cellulose acts like a supporting mesh within the wall, with its fibers running in various directions. Many single-celled eukaryotes have a similar type of cell wall. However, other single-celled eukaryotes, such as *Paramecium* and *Euglena* species, have an outer structure called a *pellicle* that is made up of protein. Fungi have cell walls that contain chitin, the same substance that is found in the shells of organisms such as grasshoppers and lobsters.

Cell walls perform many jobs for cells. In plants and fungi, they help impart stiffness to the tissues. They also provide some protection from a drying environment. In single-celled organisms, cell walls give shape to the organisms and help protect them. Pellicles are even more specialized in their function, often serving as anchors for defense mechanisms such as organelles that discharge sticky or toxic threads.

> Plants, fungi, and many protists have rigid cell walls surrounding the cell membrane. Cell walls perform many jobs, including protection from drying and imparting shape and stiffness.

4.11 Summing up: How bacterial, animal, and plant cells differ

Table 4.2 lists the differences among bacterial, animal (including human), and plant cells. As the table shows, prokaryotes (bacterial cells) and eukaryotes (plant and animal cells) differ greatly in structure and complexity. The most distinctive structural difference between these two cell types is the extensive subdivision (compartmentalization) of the interior of eukaryotic cells by membranes. Not shown in the table are differences in physiology in prokaryotes and eukaryotes. Eukaryotic cells (which are larger than prokaryotes) generally have lower metabolic and growth rates, and divide much less often. (Prokaryotic cells are described in more detail in Chapter 26.)

Table 4.2

A Comparison of Bacterial, Animal, and Plant Cells

	Bacterium	Animal	Plant
Exterior Structures			
Cell wall	Present (protein polysaccharide)	*Absent*	Present (cellulose)
Plasma membrane	Present	Present	Present
Flagella (cilia)	Sometimes present	Sometimes present	Sperm of a few species possess flagella
Interior Structures and Organelles			
Endoplasmic reticulum	*Absent*	Usually present	Usually present
Microtubules	*Absent*	Present	Present
Centrioles	*Absent*	Present	*Absent*
Golgi apparatus	*Absent*	Present	Present
Nucleus	*Absent*	Present	Present
Mitochondria	*Absent*	Present	Present
Chloroplasts	*Absent*	*Absent*	Present
Chromosomes	A single circle of naked DNA	Multiple units, DNA associated with protein	Multiple units, DNA associated with protein
Ribosomes	Present	Present	Present
Lysosomes	*Absent*	Usually present	Present
Vacuoles	*Absent*	Absent or small	Usually a large single vacuole in mature cell

4.12 Passive transport: Molecular movement that does not require energy

Cells usually live in an environment where they are bathed in water. When you consider that bacteria live on your body and that protists live in the soil, this fact may not seem to be true. However, all cells must move substances across their membranes to survive, and water is the liquid in which their molecules are dissolved. In addition, water provides a fluid environment within which molecules can move. Therefore, water must surround cells, even in microscopic amounts that may not be readily apparent.

How do molecules get where they are going? To answer this question, you must understand that molecules cannot move *purposefully* from one spot to another. Molecules move randomly. All molecules and small particles have a constant, inherent "jiggling" motion called *Brownian movement*. Because molecules are always jiggling, they tend to bump into things, such as other molecules. A bump may push a molecule in a particular direction until it bumps into another molecule and gets pushed again.

This random motion in all directions often results in a net movement of molecules in a particular direction in response to differences in concentration, pressure, or electrical charge. These differences are referred to as a **gradient**. The term *net movement* means that although individual members of a group of molecules are traveling in different directions, the resulting movement of the group is in one direction. This movement results in a uniform distribution of molecules.

The net movement of molecules often requires no input of energy. In this type of movement, molecules travel from regions of high concentration to those of low concentration or from regions of high pressure to regions of low pressure. Because ions each carry an electrical charge, they move toward unlike charges or away from like charges (see Chapter 3, p. 37).

As molecules or ions move from regions of high concentration to regions of low concentration, they are referred to as going *down* a concentration, pressure, or electrical gradient. Some molecules move into and out of cells as they travel down gradients. Molecular movement down a gradient but across a cell membrane is called **passive transport**. There are three types of passive transport: diffusion, osmosis, and facilitated diffusion.

Diffusion

Diffusion is the net movement of molecules from a region of higher concentration to a region of lower concentration, eventually resulting in a uniform distribution of the molecules. Everyone has had experience with diffusion. Did you ever wake up to the smell of freshly brewed coffee? Did you ever spill a bottle of a liquid that had a strong odor, such as ammonia or perfume? The molecules that reached the smell receptors in your nose moved from an area of high concentration (the spill) to an area of low concentration (your nose). They moved down the concentration gradient, eventually becoming evenly spread. This process is illustrated in **Figure 4.21** with a lump of sugar.

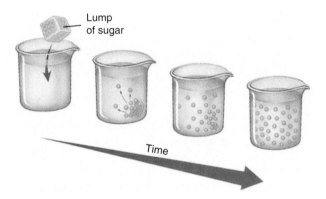

Figure 4.21 **Process of diffusion.** Sugar molecules move from an area of higher concentration (the cube) to an area of lower concentration (the water in the beaker); eventually there will be an even distribution of sugar molecules throughout the water.

Osmosis

Osmosis (os-MOE-sis) is a special form of diffusion in which water molecules move from an area of higher concentration to an area of lower concentration across a differentially permeable membrane. A *differentially permeable membrane* allows only certain types of molecules to pass through it, or permeate it, freely. Cell (plasma) membranes are differentially permeable membranes (and are also called *semipermeable* or *selectively permeable membranes*). In fact, most types of molecules that occur in cells cannot pass freely across the plasma membrane (see p. 66).

The cytoplasm of a cell consists of many different types of molecules and ions dissolved in water. The mixture of these molecules and water is called a *solution*. Water, the most common of the molecules in the mixture, is called the *solvent*. The other kinds of molecules dissolved in the water are called *solutes*.

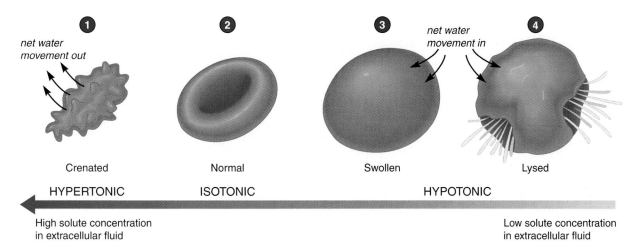

Figure 4.22 Effects of hypertonic, isotonic, and hypotonic solutions on red blood cells.

Because of diffusion, both solvent and solute molecules in a cell move from regions where the concentration of each is greater to regions where the concentration of each is less. When two regions are separated by a membrane, however, what happens depends on whether the molecule can pass freely through that membrane.

The cytoplasm of living cells contains approximately 1% dissolved solutes. If a cell is immersed in pure water, interesting things begin to happen. Because the water has a lower concentration of solutes than the cell, water molecules begin to move into the cell. The reason is simple: Water moves down the concentration gradient, or from a region of higher concentration of water molecules to a region of lower concentration of water molecules. The pure water is said to be **hypotonic** (Greek *hypo,* "under") with respect to the cytoplasm of the cell because its concentration of solutes is less than the concentration of solutes in the cell.

As you can see in **Figure 4.22,** a cell in a hypotonic solution begins to blow up like a balloon ❸. Likewise, if a cell is placed in a **hypertonic** (Greek *hyper,* "over") solution, one with a solute concentration higher than the cytoplasm of the cell, water will move out of the cell, and the cell will shrivel ❶. If a cell is placed in a solution with the same concentration of solutes as its cytoplasm, an **isotonic** (Greek *isos,* "equal") solution, water will move into and out of the cell, but no net movement will take place ❷.

Intuitively, you might think that as "new" water molecules diffuse into a cell placed in a hypotonic solution, the pressure of the cytoplasm pushing against the cell membrane would build. This is indeed what hap-

pens. At the same time, however, the water molecules that continue to diffuse into the cell also exert a pressure—**osmotic pressure.** Because the pressure of the cytoplasm within the cell opposes osmotic pressure in this case, diffusion of water molecules into the cell will not continue indefinitely. The cell will eventually reach an equilibrium—a point at which the osmotic force driving water inward is counterbalanced exactly by the pressure outward of the cytoplasm. However, the pressure within the cell may become so great that the cell bursts like a balloon ❹. The cell is said to *lyse.*

Single-celled and multicellular organisms have various mechanisms that work to keep their cells from swelling and bursting or shriveling like prunes. Single-celled organisms that live in fresh water, for example, battle a constant influx of water. Some of these organisms have one or more organelles that collect water from the cell's interior and transport it to the cell's surface. Plant cells have cell walls that support their membranes, so plant cell membranes press outward against cell walls in hypotonic solutions and pull in away from cell walls in hypertonic solutions.

Many multicellular animals, such as humans, circulate a fluid through their bodies that bathes cells in isotonic liquid. By controlling the composition of its circulating body fluids, a multicellular organism can control the solute concentration of the fluid bathing its cells, adjusting it to match that of the cells' interiors. Your blood, for example, contains a high concentration of a protein called *albumin,* which serves to elevate the solute concentration of the blood to match that of your tissues so that the movement of water molecules is regulated.

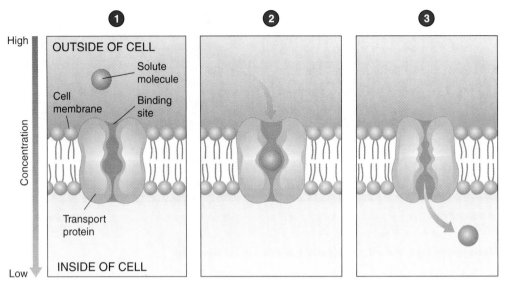

Figure 4.23 Facilitated diffusion.

Facilitated Diffusion

One of the most important properties of any cell is its ability to move substances necessary for survival into its interior and to get rid of unnecessary or harmful substances. Your cells, for example, move glucose into the cytoplasm from the bloodstream to be used for fuel. Without the ability to move glucose into your cells, you would die.

Cells can perform these feats because they have differentially permeable membranes. The cell membrane can select which molecules are to enter and leave, and in which concentrations, because most molecules must enter or leave the cell through protein doors, or channels. These doors are not open to every molecule that presents itself; only a particular molecule can pass through a given kind of door. Some of the channels in the cell membrane help certain molecules and ions enter or leave the cell and speed their movement by providing them with a passageway. These passageways are most likely transport proteins that extend from one side of the membrane to the other. After the transport protein binds with a solute molecule (Figure 4.23 **1**), its shape changes **2**, and the molecule moves across the membrane **3**. This type of transport process, in which molecules move down the concentration gradient by means of transport proteins but without an input of energy, is called **facilitated diffusion**. Facilitated diffusion helps rid the cell of certain molecules present in high

concentrations and moves molecules into the cell that are present on the outside in high concentrations.

Facilitated diffusion has two essential characteristics: (1) it is *specific*, with only certain molecules being able to traverse a given channel, and (2) it is *passive*, with the direction of net movement being determined by the relative concentrations of the transported molecule inside and outside the membrane.

Diffusion is the net movement of molecules from a region of higher concentration to a region of lower concentration, eventually resulting in a uniform distribution of the molecules. This movement is the result of random, spontaneous molecular motions. Osmosis is the diffusion of water molecules across a differentially permeable membrane. A solution with a solute concentration lower than that of another fluid is said to be hypotonic to that solution. A solution with a solute concentration higher than that of another fluid is said to be hypertonic to that solution. Solutions having equal solute concentrations are isotonic to one another. Water moves into cells placed in hypotonic solutions and out of cells placed in hypertonic solutions. Facilitated diffusion is the movement of selected molecules across the cell membrane by specific transport proteins along the concentration gradient and without an expenditure of energy.

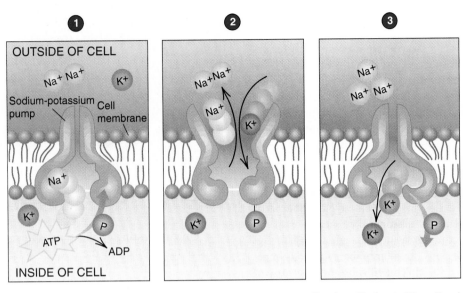

OUTSIDE OF CELL

Na⁺ Na⁺ K⁺

Sodium-potassium pump

Cell membrane

Na⁺

K⁺

P

ATP → ADP

INSIDE OF CELL

Na⁺Na⁺

Na⁺

K⁺

K⁺

P

Na⁺

Na⁺ Na⁺

K⁺

K⁺

P

Figure 4.24   The sodium-potassium pump uses energy to move sodium ions (Na⁺) out of the cell and potassium ions (K⁺) into the cell.

4.13   Active transport: Molecular movement that requires energy

Cells often move substances into or out of the cell *against* the gradients of concentration, pressure, and electrical charge. A cell uses energy to move molecules against a gradient much like you might use energy to move something against gravity. For example, if you are driving downhill, you can put your car into neutral and coast (although that is not the safest way to drive downhill). The car will continue to move without a push from the engine. As soon as you come to a hill, however, you must put the car in gear and press on the accelerator or the car will soon come to a stop. Cells, too, need to use energy to move molecules "uphill," or against a gradient. Cells also need to expend energy to move large molecules or particles into the cell that cannot move across the cell membrane.

There are many molecules that a cell takes up or eliminates against a concentration gradient. Some, such as sugars and amino acids, are molecules the cell needs to use for energy or to build new cell parts. Others are ions such as sodium and potassium that play a critical role in such functions as the conduction of nerve impulses. These many kinds of molecules enter and leave cells by way of a variety of selectively permeable transport channels. In all these cases, a cell must expend energy to transport these molecules against the concentration gradient and maintain the concentration difference.

This type of transport is called **active transport.** Active transport is one of the most important functions of any cell. Without it, the cells of your body would be unable to maintain the proper concentrations of substances they need for survival.

More than one-third of all the energy expended by a cell that is not dividing is used to actively transport sodium (Na⁺) and potassium (K⁺) ions. The type of channel by which *both* ions are transported across the cell membrane in opposite directions is a coupled channel, which has binding sites for both molecules on one membrane transport protein. This remarkable coupled channel is called the *sodium-potassium pump,* and it uses energy to move these ions across the cell membrane (**Figure 4.24 ❶**). Sodium ions are moved out of the cell to maintain a low internal concentration relative to the concentration outside the cell ❷. Conversely, potassium ions are moved into the cell to maintain a high internal concentration relative to the concentration outside the cell ❸. (Three sodium ions are moved out for every two potassium ions that are moved in by the channel.) This transport mechanism is important in most cells of the human body but is particularly important for the proper functioning of muscle and nerve cells.

Active transport is the movement of a solute across a membrane against the concentration gradient with the expenditure of chemical energy. This process requires the use of a transport protein specific to the molecule(s) being transported.

Large molecules and particles move into and out of cells by endocytosis and exocytosis.

4.14 Endocytosis

Certain types of cells transport particles, small organisms, or large molecules such as proteins into their cells. In humans, for example, white blood cells police the body fluids and ingest substances as large as invading bacteria. In nature, some single-celled organisms often eat other single-celled organisms whole. How can cells move such large substances into their interiors?

Cells such as these ingest particles or molecules that are too large to move across the membrane by a process called **endocytosis**, which literally means "into the cell." During endocytosis, a cell envelops the particle or large molecules with fingerlike extensions of the membrane-covered cytoplasm (steps ❶ and ❷ in Figure 4.25). The edges of the membrane eventually meet on the other side of the enveloped substance. Because of the fluid nature of the lipid bilayer, the membranes fuse together, forming a vesicle around the particle (step ❸ in Figure 4.25).

If the material that is brought into the cell contains an organism or some other fragment of organic matter as in Figure 4.25a, the endocytosis is called **phagocytosis** (FAG-oh-sye-TOE-sis) (Greek *phagein,* "to eat," and *cytos,* "cell"). If the material brought into the cell is liquid—contains dissolved molecules as in Figure 4.25b—the endocytosis is referred to as **pinocytosis** (Greek *pinein,* "to drink"). Pinocytosis is common among the cells of multicellular animals. Human egg cells, for example, are "nursed" by surrounding cells that secrete nutrients the maturing egg cell takes up by pinocytosis.

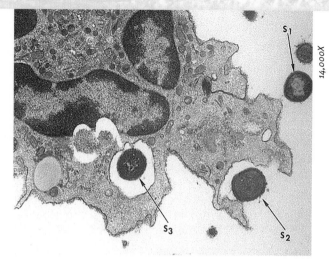

One interesting application of this ability is as a defensive mechanism in the body. Whole bacteria can be engulfed as shown in the colorized micrograph above; all three stages are shown at once for three different bacterial cells (S). Virtually all animal cells are constantly carrying out endocytosis, trapping extracellular fluid in vesicles and ingesting it. Within the cell, these vesicles fuse with lysosomes (see Figure 4.14), tiny cellular bags of digestive enzymes, to break down these large particles into molecules usable to the cell. Rates of endocytosis vary from one cell type to another but can be surprisingly large. Some types of white blood cells, for example, ingest 25% of their cell volume each hour!

Endocytosis is a process in which cells engulf large molecules or particles and bring these substances into the cell packaged within vesicles.

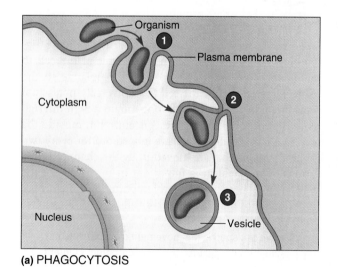

(a) PHAGOCYTOSIS

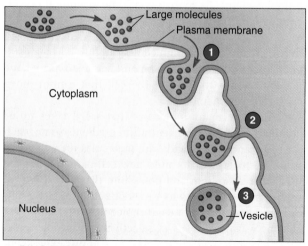

(b) PINOCYTOSIS

Figure 4.25 **(a)** Phagocytosis and **(b)** pinocytosis.

4.15 Exocytosis

The reverse of endocytosis is **exocytosis** (Figure 4.26). During exocytosis, a cell discharges material by packaging it in a vesicle ❶ and moving the vesicle to the cell surface ❷. The membrane of the vesicle fuses with the cell membrane, and the contents are expelled ❸. In plants, exocytosis is the main way that cells move the materials from the Golgi apparatus and out of the cytoplasm to construct the cell wall. In animals, many cells are specialized for secretion using the mechanism of exocytosis, including cells that produce and secrete digestive enzymes or hormones.

> Exocytosis is a process in which cells use vesicles to transport large molecules or particles through the cytoplasm to the cell membrane where they are discharged.

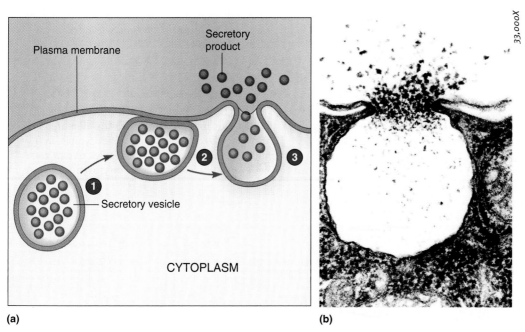

(a) **(b)**

Figure 4.26 **Exocytosis.** (a) Proteins and other molecules are secreted from cells in small pockets called vesicles, whose membranes fuse with the cell membrane. When this fusion occurs, the contents of the vesicles are released to the cell surface. (b) An electron micrograph showing contents of a vesicle being expelled from a cell.

Key Concepts

Organisms are cells or are made up of cells.

4.1 A cell is a membrane-bounded unit containing hereditary material and cytoplasm that can release energy from fuel and use that energy to grow and reproduce.

4.1 The cell theory states that all living things are made up of cells and that all cells arise from the growth and division of other cells.

Most cells are microscopically small.

4.2 Most cells divide before they grow too large, maintaining a large surface area-to-volume ratio and the ablity to move substances across their cell membranes quickly enough to meet their needs.

All cells can be grouped into two broad categories: prokaryotic or eukaryotic.

4.3 All cells can be classified as either prokaryotic or eukaryotic.

4.3 Prokaryotic cells are the bacteria and have a relatively simple structure.

4.4 Eukaryotic cells are more complex than prokaryotes and contain many organelles.

4.4 Eukaryotic cells make up the bodies of plants, animals, protists, and fungi.

4.5 All cells are bounded by a plasma membrane, which is a lipid bilayer with embedded proteins.

4.5 Lipid bilayers allow only certain molecules to pass across them.

4.5 Membrane proteins facilitate the movement of many types of molecules into and out of the cell.

4.6 The cytoplasm of the cell is the viscous fluid containing all cell organelles except the nucleus.

4.6 The cytoskeleton is a network of filaments and fibers in the cytoplasm, which undergoes continuous change and performs many jobs for the cell.

4.7 Eukaryotic cells contain membrane-bounded organelles: endoplasmic reticula, Golgi apparatus, lysosomes, vacuoles, and nucleus.

4.7 Ribosomes, which are small, spherical organelles located in the cytoplasm or on the ER, are the sites of protein synthesis.

4.8 Eukaryotic cells also contain bacterialike organelles: mitochondia and chloroplasts.

4.9 In eukaryotes, cilia and flagella are composed of microtubules and help propel cells; in some eukaryotic cells, cilia are used to move substances across surfaces.

4.10 Plants, fungi, and many protists have rigid cell walls surrounding the cell membrane; cell walls perform many jobs, including protecting the membrane from drying and imparting shape and stiffness.

4.11 Bacterial, animal, and plant cells differ in a variety of significant ways.

Small molecules move into and out of cells by passive and active transport.

4.12 Molecules move randomly with a constant jiggling motion.

4.12 Molecules and ions undergo net movement in response to differences, or gradients, in concentration, pressure, and electrical charge, resulting in an equal distribution of these molecules and ions.

4.12 Movement down a gradient requires no added energy to take place.

4.12 Molecular movement down a gradient but across a cell membrane is called passive transport.

4.12 The three types of passive transport are diffusion, osmosis, and facilitated diffusion.

4.13 When cells move substances against a gradient, energy is required; this movement is called active transport.

Large molecules and particles move into and out of cells by endocytosis and exocytosis.

4.14 The transport into the cell of particles or molecules that are too large to move across the membrane takes place by a process called endocytosis, in which the cell engulfs large molecules or particles.

4.15 The reverse of endocytosis is exocytosis.

Key Terms

active transport *81*
cell theory *60*
cell wall *77*
chloroplasts (KLOR-oh-plasts) *75*
chromosomes
    (KROW-muh-somes) *72*
cilia (SILL-ee-uh) *76*
cytoplasm (SYE-toe-PLAZ-um) *68*
deoxyribonucleic acid (DNA) (de-OK-
    see-RYE-boh-new-KLAY-ick) *72*
diffusion *78*
endocytosis (ENN-doe-sye-TOE-sis)
    82
endoplasmic reticulum (EN-doe-PLAZ-
    mik ri-TIK-yuh-lum) *69*
eukaryotic cells
    (yoo-KARE-ee-OT-ick) *62*

exocytosis (EK-so-sye-TOE-sis) *83*
facilitated diffusion *80*
flagella (fluh-JELL-uh) *76*
fluid mosaic model *67*
Golgi apparatus (GOL-gee) *70*
gradient *78*
hydrophilic (HI-droe-FIL-ick) *66*
hydrophobic (HI-droe-FO-bic) *66*
hypertonic (HI-per-TAWN-ick) *79*
hypotonic (HI-poe-TAWN-ick) *79*
isotonic (EYE-so-TAWN-ick) *79*
lysosomes (LYE-so-somes) *70*
mitochondria
    (MITE-oh-KON-dree-uh) *74*
nucleolus (noo-KLEE-oh-lus) *72*
nucleus (NOO-klee-us) *72*
osmosis (os-MOE-sis) *78*

osmotic pressure *79*
passive transport *78*
phagocytosis
    (FAG-oh-sye-TOE-sis) *82*
pinocytosis
    (PIE-no-sye-TOE-sis) *82*
plasma membrane *66*
prokaryotic cells
    (pro-KARE-ee-OT-ick) *62*
ribosomal ribonucleic acid (rRNA)
    (RYE-buh-SO-mull RYE-boh-new-
    KLAY-ick) *72*
ribosomes (RYE-buh-somes) *69*
rough ER *69*
smooth ER *69*
vacuole (VACK-yoo-ol) *71*

KNOWLEDGE AND COMPREHENSION QUESTIONS

1. What are the basic principles of the cell theory? *(Knowledge)*

2. Humans are multicellular organisms. Explain what this means, and discuss some of the tasks performed by the cells of your body. *(Comprehension)*

3. Some of the skeletal muscle cells that allow you to move your legs can be as long as 30 to 40 centimeters. These cells have many nuclei. Explain why. *(Comprehension)*

4. A human liver cell has produced a protein to be used outside of the cell. Using what you know about the endoplasmic reticulum and Golgi apparatus, trace the path of protein from production to removal from the cell. What was the role of the smooth ER in this process? *(Comprehension)*

5. What would happen inside your cells if the lysosomes stopped working? *(Comprehension)*

6. Make a table that briefly describes each of the following and lists its function(s): plasma membrane, flagella (cilia), cytoskeleton, endoplasmic reticulum, ribosomes, nucleus, chro-

mosomes, nucleolus, Golgi complex, lysosomes, mitochondria, and chloroplasts. *(Knowledge)*

7. You're examining three cells under the electron microscope. You know that one is a bacterium, one is a plant cell, and the third is an animal cell. From their structures, how can you tell which is which? *(Comprehension)*

8. Many scientists theorize that chloroplasts and mitochondria evolved from ancient bacteria living within the cells of host organisms. Does this mean that the mitochondria in your cells are endosymbionts? *(Comprehension)*

9. What characteristics do diffusion and facilitated diffusion have in common? In what ways do they differ? *(Comprehension)*

10. Draw a diagram of a cell engaging in endocytosis. Give an example of this process in the human body. *(Comprehension)*

KEY
Knowledge: Recalling information.
Comprehension: Showing understanding of recalled information.

CRITICAL THINKING REVIEW

1. Trace the paths by which a nutrient outside a cell could pass into the interior of a cell. (The nutrients are carbohydrates, lipids, proteins, vitamins, minerals, and water.) (*Application*)

2. Identify the type(s) of molecular movement involved in each of the following. Which are active and which are passive? (*Application*)
   a. Glucose molecules leave your bloodstream (where they are usually in higher concentrations) and enter your cells by attaching to specialized carrier proteins.
   b. You dissolve a spoonful of instant coffee crystals in a mug of hot water.
   c. Your cells "import" sodium ions by transporting them against the concentration gradient.

3. In question 2b, identify the solute, solvent, and solution. (*Application*)

4. When a patient is suffering from dehydration, the physician may recommend that an 0.9% saline solution be added to the patient's bloodstream rather than water. Explain the reasoning behind this recommendation. What would happen if water were used instead? (*Application*)

5. Cigarette smoking reduces the effectiveness of cilia lining the respiratory tract in humans. What is the role of these cilia and what would you hypothesize might be a result of smoking for many years? (*Synthesis*)

KEY

Application: Using information in a new situation.
Synthesis: Putting together information from different sources.

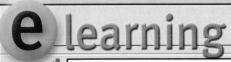 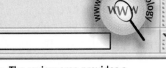

 Location: http://www.jbpub.com/biology

e-Learning is an on-line student review area located at this book's web site www.jbpub.com/biology. The review area provides a variety of activities designed to help you study for your class and build your learning portfolio.

Review Questions The review questions test your knowledge of the important concepts and applications in each chapter. The review provides feedback for each correct or incorrect answer. This is an excellent test preparation tool.

Figure-Labeling Exercises Sharpen your visual thinking by matching terms and labels for important illustrations.

Flash Cards Studying biology requires learning new terms. Virtual flash cards help you master the new vocabulary for each chapter.

Just Wondering As you read and study from this text, you may find that you have unanswered questions. Through this site you can ask the author, Sandy Alters, your "just wondering" questions.

 Do you prefer the speed and reliability of a CD-ROM? All of the features contained on the eLearning portion of the web site are also available on the Student CD-ROM.

Label the structures in both the plant cell and animal cell below by filling in the blanks.

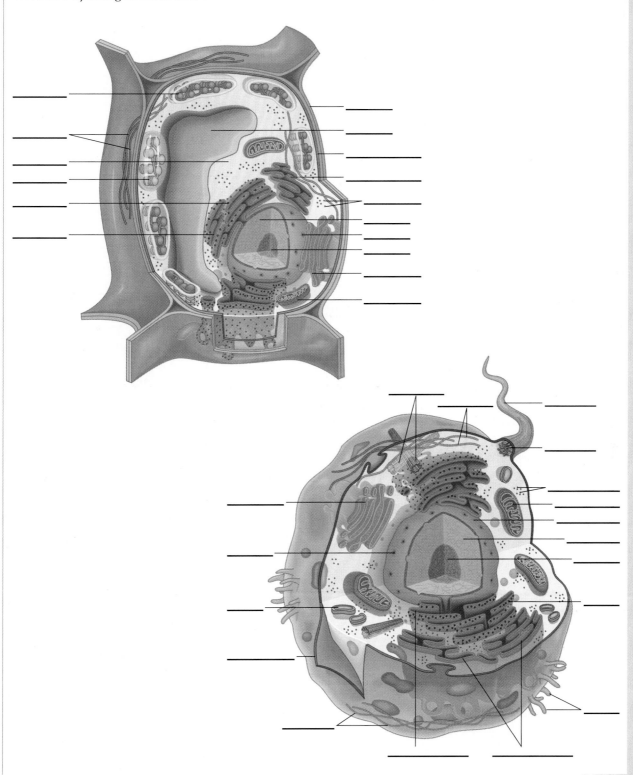

The Flow of Energy within Organisms

Did you know that your body takes constant "rollercoaster rides" that you can't even feel? At the beginning of a rollercoaster ride, the cars are pulled up a high hill by a cable. For the cable to pull the cars, it must be run by an engine that supplies energy for the cable to do its work. Energy is, in fact, the ability to do work. In the case of the rollercoaster, energy is needed to pull the cars up the hill; otherwise, they would go nowhere. After the cars reach the top and are freed of the cable, they swoop down the hill, releasing this input of energy and coasting around the rails at breakneck speed.

All the chemical reactions within your body—your metabolism—take place in a similar way. Chemical reactions are changes in molecules in which one substance is changed into another. For chemical reactions to take place, they need an input of energy to get started. This input of energy helps molecules "climb" an energy "hill" and interact with one another.

CHAPTER 5

Organization

Energy is stored and released during chemical reactions.

Enzymes regulate chemical reactions in living systems.

ATP is the primary energy carrier of living systems.

5.1 Starting chemical reactions

Figure 5.1 is a graph of the varying energy states of the molecules in a chemical reaction. Doesn't this graph look like the beginning hill of a rollercoaster? The **reactants** are the molecules entering into the chemical reaction—the rollercoaster cars. The energy needed to pull the reactants up the energy hill is called the *free energy of activation*. This energy input destabilizes the bonds of the reactants. Molecules with unstable bonds interact with one another more easily than molecules with stable bonds do. In this way, the free energy of activation allows the chemical reaction to "go." The reactants undergo a chemical change resulting in new bonding arrangements between the molecules. This chemical change may involve the breaking of bonds and the making of new bonds. The changed reactants are called **products.**

Sometimes reactants are chemically broken down to yield products. When a reactant is broken down, energy is released from the chemical bonds that were holding it together. This type of reaction is called **exergonic,** meaning "energy out." In an exergonic reaction, the products contain less energy than the reactant. The excess energy is released (see Figure 5.1). Exergonic reactions happen spontaneously; that is, they absorb free energy (energy that is useable, such as heat energy) from their surroundings and occur with no additional input of energy from another chemical reaction. Spontaneous reactions are often referred to as "downhill" reactions because, like a ball rolling down a hill, spontaneous processes occur without help other than the

input of activation energy. With the ball analogy, the activation energy can be likened to the "help" the ball received to somehow get to the top of the hill. (In living systems, there is usually a significant activation energy barrier that requires an enzyme to initiate the reaction, which will then proceed without further energy input.)

Exergonic reactions take place in your body too. In fact, a series of exergonic reactions breaks down the glucose in your cells to supply your body with energy. The heat released from these reactions helps keep your body warm. Reactions such as this, which release energy by breaking down complex molecules into simpler molecules, are called **catabolic reactions** (CAT-uh-BOL-ick).

In some chemical reactions, reactants are chemically joined to yield a product. When reactants are joined, energy is used to build the chemical bonds holding the product together. This type of reaction is called **endergonic,** meaning "energy in." In an endergonic reaction, the product contains more energy than the reactants (**Figure 5.2**).

Endergonic reactions *do not occur spontaneously;* that is, they require more free energy of activation to drive the reaction than the reactants are able to absorb from their surroundings. Such nonspontaneous reactions are often referred to as "uphill" reactions because, like a ball rolling up a hill, nonspontaneous processes can only occur with help—an input of energy in addition to activation energy. Baking a cake is an example of an endergonic reaction. In order for batter to turn into cake, it must have an input of heat energy by baking in an oven. No matter how long it sits out on the counter at room temperature, it will not rise and become firm.

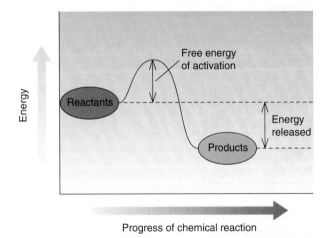

Figure 5.1 **An exergonic reaction.** In an exergonic reaction when a reactant is broken down, energy is released from its chemical bonds. The reaction product(s) contain less energy than the reactant.

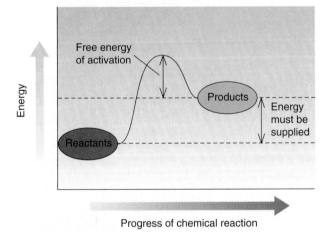

Figure 5.2 **An endergonic reaction.** In an endergonic reaction, the products of the reaction contain more energy than the reactant(s). Extra energy must be supplied for the reaction to proceed.

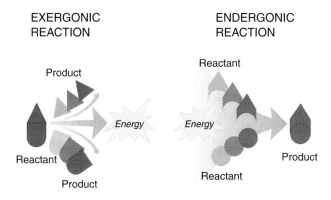

EXERGONIC REACTION

Product

Energy

Reactant

Reactant

Product

ENDERGONIC REACTION

Reactant

Energy

Product

Reactant

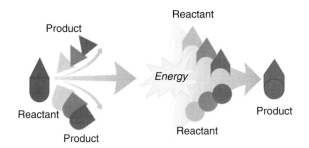

COUPLED REACTIONS

Product

Reactant

Product

Reactant

Energy

Reactant

Product

Reactant

Figure 5.3 **Coupled reactions.** In coupled reactions, the energy released in an exergonic reaction is used to drive an endergonic reaction.

Endergonic reactions play important roles in your body, such as putting molecules together to build muscles and bones. Reactions like this, which use energy to build complex molecules from simpler molecules, are called **anabolic reactions** (AN-uh-BOL-ick).

In living things, exergonic and endergonic reactions are *coupled reactions,* meaning that they occur in conjunction with one another. In this way, the energy released when molecules are split (exergonic reactions) is used to power the combining of molecules (endergonic reactions) **(Figure 5.3).**

Free energy of activation is needed to start a chemical reaction. Reactions that break apart reactants release energy. These reactions are exergonic. Reactions that bond reactants together store energy. These reactions are endergonic. In living systems, exergonic reactions are coupled with endergonic reactions, supplying the energy for endergonic reactions to take place.

5.2 Energy flow and change in living systems

Life can be viewed as a constant flow of energy that is channeled by organisms to do the work of living. As you learned at the beginning of this chapter, energy is the ability to do work, and work is many things, such as the pull of a cable on a rollercoaster car or the swift dash of a horse. It is also heat, such as the blast from an explosion or a warming fire. Energy can exist in many forms, including mechanical force, heat, sound, electricity, light, radioactivity, and magnetism. All of these forms of energy are able to create change—to do work.

Energy exists in two states. Energy not actively doing work but having the capacity to do so is called **potential energy.** A rollercoaster car perched atop that first hill, for example, possesses this type of stored energy (as does a flying squirrel in a tree). As the car rolls down the hill (or the squirrel jumps), each is actively engaged in doing work. This form of energy is called **kinetic energy,** or the energy of motion. Much of the work performed by living organisms involves changing potential energy to kinetic energy. The exergonic reactions that supply your body with energy, for example, change the potential energy stored in the food you eat to kinetic energy for cellular work.

All the changes in energy that take place in the universe, from nuclear explosions to the buzzing of bees, are governed by the two **laws of thermodynamics.** The **first law of thermodynamics** states that energy can change from one form to another and from one state to another, but it can never be lost. New energy cannot be made. The total amount of energy in the universe remains constant.

The **second law of thermodynamics** states that all objects in the universe tend to become more disordered and that the disorder in the universe is continually increasing. Stated simply, disorder is more likely than order. You can relate this to keeping your personal environment in order. Your bedroom continually becomes messy unless you make an effort to keep it neat.

Molecules, cell systems, and organisms also become increasingly disordered. It takes energy to overcome this disorder. In other words, organisms must have an input of energy so that their cell systems will continue to function, maintaining molecular "order." Molecular disorder results in death. For example, you need to eat (take in energy) or you will die (your chemical pathways will become disordered and you will no longer function). Likewise, plants need sun, or they will die.

As an organism uses energy from its energy sources (food, sunlight), it loses some of this energy as heat when energy transfers are made between molecules during chemical reactions. (This "lost" heat energy helps keep your internal body temperature at 98.6°F.) Although the total amount of energy in the universe does not change, the amount of useful energy available to do work decreases as progressively more energy is degraded to heat.

The energy lost to disorder is referred to as **entropy** (ENN-truh-pee). In fact, entropy is a measure of the disorder (or randomness) of a system. The greater the disorder, the greater the entropy. Sometimes the second law of thermodynamics is simply stated as "entropy increases." So, although energy cannot be destroyed, the universe is constantly moving toward increasing entropy. Eventually, the universe will have wound down like a forgotten clock. Scientists speculate that this will occur approximately 100 billion years from now.

Although energy cannot come into or go out of the universe, the Earth is constantly receiving "new" energy from the sun. Much of this energy heats up the oceans and continents. Some of it is captured by photosynthetic organisms such as green plants. In photosynthesis, the energy from sunlight is changed to chemical energy, combining small molecules into more complex molecules by means of endergonic reactions. This stored energy can be shifted to other molecules by forming different chemical bonds. In addition, this stored energy can be changed into kinetic energy: motion, light, electricity—and heat. Thus, energy continuously flows into and through the biological world, with new energy from the sun constantly flowing in to replace the energy that is lost as heat.

> Energy is the capacity to do work or bring about change. It can exist in many forms, such as mechanical force or heat. Energy actively doing work is called kinetic energy; stored energy is called potential energy. The first law of thermodynamics states that energy cannot be created or destroyed; it can only be changed from one form or state to another. The second law of thermodynamics states that disorder in the universe constantly increases. Energy spontaneously converts to less organized forms. Photosynthetic organisms such as green plants change energy from the sun to other forms of energy that drive life processes. This energy is never destroyed, but as it is exchanged in chemical reactions, much of it is changed to heat, a form of energy that is not useful for performing work.

Enzymes regulate chemical reactions in living systems.

5.3 Enzymes lower activation energy

The flow of energy within an organism like yourself consists of a long series of coupled reactions. Energy is moved from one molecule to the next by means of exergonic and endergonic reactions. Some of this energy is stored in the bonds of molecules that make up the structure of your body. Some is freed to do cellular work. Some is lost as heat.

These chains of reactions that move, store, and release energy are called *metabolic pathways*. Metabolic pathways accomplish jobs such as obtaining energy from the food you eat and repairing tissues that are worn out or damaged. By means of such pathways, your body and the bodies of all living things work to maintain order and avoid increasing entropy. Put simply, all the metabolic pathways in your body work to help you survive. However, for these chemical reactions to occur, they must be pushed over the hill of activation energy. In addition, they must be controlled so that they occur on schedule.

As you read earlier in this chapter, chemical reactions require activation energy to get started. This energy is needed for various reasons. In some chemical reactions, it helps break old bonds, which occurs before new bonds are formed. In others, it excites electrons, which then achieve a higher energy level or shell prior to "pairing up" in covalent bonds. In certain chemical reactions, activation energy helps molecules overcome the mutual repulsion of their many electrons, allowing them to get close enough to react. In any chemical reaction, the free energy of activation performs one or more of these jobs.

If molecules are heated, they move very quickly and bump into one another forcefully. In such a situation, the

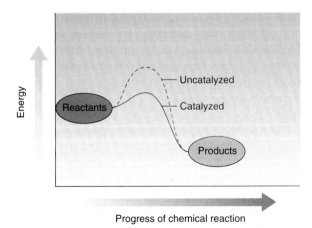

Figure 5.4 The function of enzymes. Enzymes are able to catalyze particular reactions because they lower the amount of activation energy required to initiate the reaction.

kinetic energy of the bump can provide enough energy for activation. This situation rarely occurs in living organisms because the chemical reactions of living things take place in the moderate temperatures of living cells. However, living systems contain proteins called **enzymes** (ENN-zymes) that lower (or lessen) the free energy of activation that is needed, allowing chemical reactions to take place.

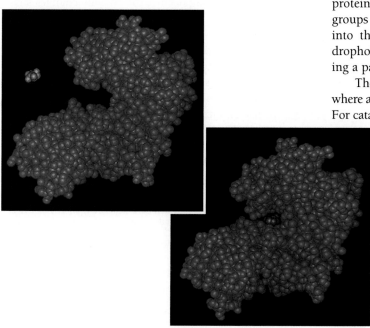

Figure 5.5 A computer simulation for the induced fit of an enzyme (hexokinase, shown in blue) and its substrate (glucose, shown in red). Note the differences in the shape of the enzyme before (top photo) and after (bottom photo) the substrate has bound.

Enzymes are proteins that act as *catalysts*. A catalyst is a substance that increases the rate of a chemical reaction but is not chemically changed by the reaction. Catalysts (and enzymes) lower the barrier of the free energy of activation (**Figure 5.4**). That is, they reduce the amount of energy needed for a reaction to occur by bringing substrates (reactants) together or by placing stress on the bonds of a single substrate, thereby making its bonds more reactive and easy to break. (**Substrates** are the substances [reactants] on which enzymes act.) Enzymes are biological catalysts and control all the chemical reactions making up the metabolic pathways in living things. Life is therefore a process regulated by enzymes.

Enzymes are biological catalysts that reduce the amount of free energy of activation needed for a chemical reaction to take place, thus speeding up the reaction.

5.4 Enzymes bind substrates at active sites

Enzymes are proteins that have one or a few grooves or furrows on their surfaces. These surface depressions are called **active sites**. These sites are a result of the tertiary structure of the protein, which is determined by the side groups (R groups) of the amino acids making up the protein chain (see Figure 3.16 **⑤**). Nonpolar side groups of amino acids, for example, tend to be shoved into the interior of the protein because of their hydrophobic (water-fearing) interactions, thereby imparting a particular shape to that portion of the protein.

The active sites are the locations on the enzyme where a reaction is catalyzed—where *catalysis* takes place. For catalysis to occur, a substrate must fit into the surface depression of an enzyme so that many of its atoms nudge up against atoms of the enzyme. The fit between a substrate and an enzyme is much like putting your hand (the substrate) into a glove (the enzyme). The shapes are complementary. The binding of a substrate causes the enzyme to adjust its shape slightly, allowing a better fit called an *induced fit* (**Figure 5.5**). Induced fit is also shown in **Figure 5.6** using the example of a glove. The glove has the shape of a hand, but takes on that shape more exactly after a hand is placed in the glove. Additionally, just as your hands will only fit into certain gloves, substrates can only fit into, or bind with, the active site of certain enzymes. Enzyme/substrate interactions are specific. Therefore, enzymes typically catalyze only one or a few different chemical reactions. Because of this specificity, each cell in your body contains from 1000 to 4000 different types of enzymes.

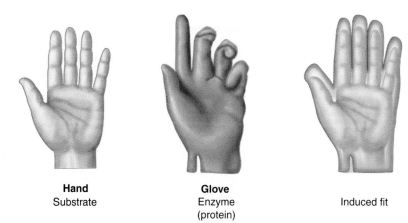

| **Hand** | **Glove** | |
| Substrate | Enzyme (protein) | Induced fit |

Figure 5.6 **The induced fit model of enzyme activity.** A substrate fits an active site of an enzyme much like a hand fits a glove.

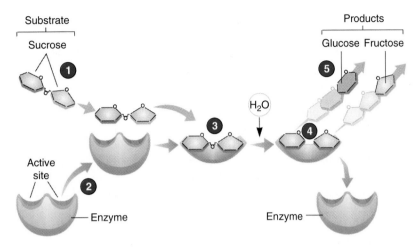

Figure 5.7   How an enzyme catalyzes an exergonic reaction.

Enzymes catalyze both endergonic and exergonic reactions. **Figure 5.7** shows how an enzyme catalyzes an exergonic reaction. In this example, a molecule of sucrose ❶ binds to the active site of an enzyme ❷. After binding takes place ❸, certain atoms within the active site of the enzyme chemically interact with the sucrose. This interaction causes a slight change in the shape of both the enzyme and the sucrose molecule. This change in shape places stress on the bonds joining the glucose and fructose subunits, reducing the amount of free energy of activation needed to be absorbed by the molecules for them to react. The bond then breaks ❹, and the products are released from the enzyme ❺. (This reaction is a hydrolysis; note the addition of water.)

Enzymes also catalyze reactions that bind two substrates together. **Figure 5.8a** shows a reaction in which two reactants bond with one another. (The energy needed for this reaction to take place is not shown.) Figure 5.8b shows that without an enzyme, this reaction takes place only when the reactants collide by chance. Figure 5.8c shows how an enzyme facilitates this reaction. Both substrates bind at the active site on the enzyme ❶. As binding occurs, the enzyme changes shape for a better fit ❷. The binding of the substrates to the enzyme distorts and weakens their chemical bonds. In addition, their interaction with the enzyme orients the substrates ❸ so that their reactive sites are in contact with one another and react ❹. The products from the reaction are released and the enzyme returns to its normal shape ❺. Moreover, enzymes do not get used up after one interaction takes place. They are only intermediaries in chemical reactions and are then released, available to catalyze yet another reaction ❻.

In summary, an enzyme acts like a dating service. It gets the substrates together so that their meeting does not occur simply by chance. Interestingly, the Chinese characterize enzymes in a similar way, calling them *tsoo mei*, or "marriage brokers."

> Enzymes work by bringing substrates together so that they react more easily and by placing stress on certain bonds, which lowers the amount of free energy of activation that must be absorbed by the substrates to react.

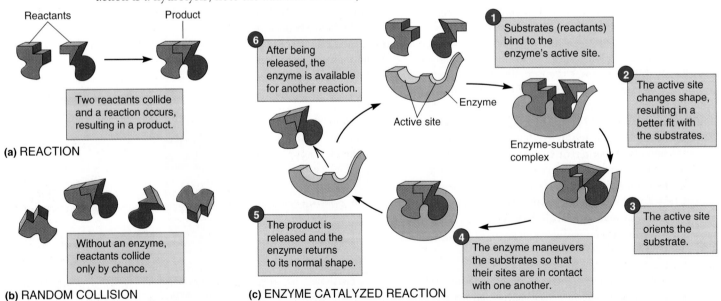

(a) REACTION

Two reactants collide and a reaction occurs, resulting in a product.

(b) RANDOM COLLISION

Without an enzyme, reactants collide only by chance.

(c) ENZYME CATALYZED REACTION

❶ Substrates (reactants) bind to the enzyme's active site.

❷ The active site changes shape, resulting in a better fit with the substrates.

❸ The active site orients the substrate.

❹ The enzyme maneuvers the substrates so that their sites are in contact with one another.

❺ The product is released and the enzyme returns to its normal shape.

❻ After being released, the enzyme is available for another reaction.

Figure 5.8   **Enzymes catalyze reactions between two substrates.** (a) A collision between two reactants can result in a reaction, producing a product. (b) Without an enzyme, reactants collide only by chance. (c) With an enzyme, a reaction is more likely to occur.

5.5 Factors that affect enzyme activity

The activity of an enzyme is affected by anything that changes its three-dimensional shape. If it loses its shape, or is denatured, an enzyme cannot bind with a substrate. Temperature is one factor that denatures protein. You can see the protein in an egg become denatured when you cook it. The heat permanently changes the shape of the egg proteins, making the egg firmer and more opaque. Two other environmental conditions that affect protein shape and therefore enzyme activity are pH and the binding of specific chemicals.

Most human enzymes function best between 36°C and 38°C, close to body temperature. Likewise, the enzymes of other living things work best at temperature ranges specific to the organism. Bacteria that live in hot springs, for example, have enzymes that function at temperatures between 74°C and 76°C (approximately 167°F) **(Figure 5.9a)**. At temperatures colder or warmer than the enzyme's optimum range, the bonds between the amino acids that determine the enzyme's shape become weak or rigid, changing the shape of the active site and causing chemical reactions to stop.

Enzymes also work best within a particular range of pH values, pH being a measure of the hydrogen ion concentration of a solution (see Chapter 3). Most enzymes, trypsin for example, work best within the range of pH 6 to 8. Some enzymes, however, function in environments with a high hydrogen ion concentration (low pH). The enzyme pepsin digests proteins in your stomach, a highly acidic environment, at pH 2 (Figure 5.9b).

The activity of an enzyme is sensitive not only to pH and temperature but also to the presence of specific chemicals that bind to the enzyme and cause changes in its shape. By means of these specific chemicals, a cell is able to turn enzymes on and off. When the binding of a chemical changes the shape of the protein and the active site so that it can catalyze a chemical reaction, the chemical is called an *activator*. When the binding causes a change in the active site that shuts off enzyme activity, the chemical is called an *inhibitor*. Enzymes usually have special activator and inhibitor binding sites, and these binding sites are different from their active sites. Common examples of enzyme inhibitors are the antibiotic penicillin and the element lead (found in such products as lead-based paints and car batteries). Penicillin kills certain disease-causing bacteria by inhibiting the main enzyme involved in the synthesis of their cell walls. Lead inhibits a wide variety of enzymes, especially in nervous tissue. It can cause poisoning in humans and result in devastating effects in the central nervous system.

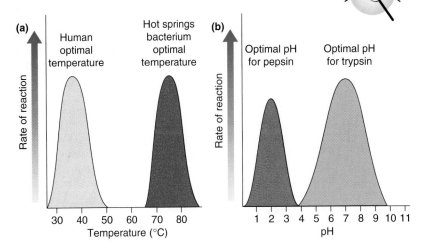

Figure 5.9 Enzymes are sensitive to the environment.

Enzyme activity within an organism is often regulated by inhibitors, using a process called **negative feedback**. In this process the enzyme catalyzing the first step in a series of chemical reactions has an inhibitor binding site to which the end product of the pathway binds. As the concentration of the end product builds up in the cell, it begins to bind to the first enzyme in the metabolic pathway, shutting off that enzyme. In this way, the end product is feeding information back to the first enzyme in the pathway, shutting the pathway down when an additional end product is not needed **(Figure 5.10)**.

Such *end-product inhibition* is a good example of the way many enzyme-catalyzed processes within cells are self-regulated. For example, the enzyme threonine deaminase catalyzes the first reaction in a series that converts the amino acid threonine to another amino acid, isoleucine. The cells of the body need both amino acids in proper concentrations to make proteins. The presence of an adequate amount of isoleucine for protein manufacture shuts off the enzyme threonine deaminase by binding to its inhibitor binding site and thereby stopping any further conversion of threonine to isoleucine.

The activity of an enzyme is affected by anything that changes its three-dimensional shape. Three environmental conditions that can affect enzyme activity in this way are pH, temperature, and the binding of specific chemicals. In living systems, the activity of enzymes is regulated by changes in enzyme shape; these changes result when activator and inhibitor molecules bind to specific enzymes.

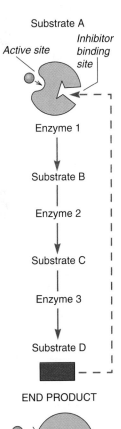

Figure 5.10
End-product inhibition. Substrate D controls the rate of its own synthesis by acting on enzyme 1, the catalyst for the first reaction in the pathway.

Just Wondering . . . *Real Questions Students Ask*

Enzymes regulate chemical reactions in living systems.

It's hard not to notice when the press reports that an athlete like baseball player Mark McGwire uses the dietary supplement creatine to help develop his physical strength. Does it work? Is it safe?

Mark McGwire of the St. Louis Cardinals became one of the big sports stories of 1998 when he broke Roger Maris's 37-year-old record of 61 home runs in a single season. McGwire, in the end, got to watch 70 home run balls leave the park from the end of his bat. His obvious athletic prowess also drew attention to his training regimen and inadvertently made "stars" of two dietary supplements he used: creatine monohydrate (or simply creatine) and androstenedione (or andro, see Chapter 17.

Creatine is a compound that is advertised to enhance the speed with which ADP is converted back to ATP in muscles (see Section 5.8). This conversion is presumably accomplished as creatine combines with a phosphate group, forming phosphocreatine, and then interacts with ADP to form ATP.

ATP supplies energy for muscle contraction but is in limited supply in the muscles. Therefore, supplementing the muscles with creatine is supposed to make more energy (more ATP) available to them more quickly. As a result, athletes taking creatine are supposed to have greater endurance and be able to perform high-intensity, short-term sports activities better than athletes not taking the supplement.

A variety of studies have tested the effects of creatine on muscle activity and the results are not conclusive. Some studies suggest that creatine supplementation may enhance an athlete's ability to maintain his or her maximal rate of power output in high-intensity, intermittent exercise such as sprinting or repeated sets of bench presses. The results of some studies suggest that this effect is likely to persist for at least a week after treatment. However, many studies show no effects at all.

Biochemical studies have also been conducted that show that supplementation has no effect on muscle ATP content or post-exercise blood lactate. (Increased blood lactate levels would show increased levels of ATP resynthesis via the glycolytic pathway and anaerobic metabolism. See Chapter 6.) These results suggest to researchers that creatine monohydrate does not promote a more efficient use of energy in the muscles.

Although short-term creatine supplementation has been found to increase body mass in males, researchers find that the initial increase is most likely due to water retention. Chronic creatine supplementation in conjunction with physical training involving resistance exercise may increase lean body mass.

Creatine monohydrate supplementation of up to 8 weeks has not been associated with any major health risks. However, the safety of more prolonged creatine supplementation has not been established. The potential side effects are unknown, but could possibly include effects to organs that process this supplement such as the liver and the kidneys. Additionally,

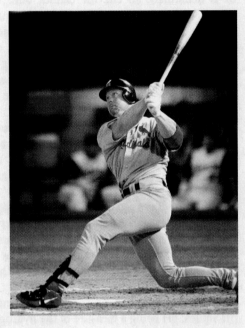

the U.S. Food and Drug Administration (FDA) does not regulate the quality and purity of creatine monohydrate, so health effects could arise from other substances present in the supplement. The short-term effects of taking creatine may include an upset stomach, intestinal gas, abdominal or stomach cramping, and water retention (bloating).

5.6 Cofactors and coenzymes

Enzymes often have additional parts in their structures that are made up of molecules other than proteins, as shown in the illustration. These additional chemical parts are called **cofactors**. Cofactors help enzymes catalyze chemical reactions. For example, many enzymes have metal ions such as zinc, iron, or copper locked into their active sites. These ions help draw electrons from substrate molecules. One of your digestive enzymes, carboxypeptidase, breaks down proteins in foods by using a zinc ion to draw electrons away from the bonds being broken in the food. Many trace elements necessary for your health, such as manganese, help enzymes in this way.

When the cofactor is an organic (carbon-containing) molecule, it is called a **coenzyme**. Many of the vitamins that your body requires, such as members of the B-vitamin complex, are used to synthesize coenzymes to maintain health. These coenzymes perform many jobs in the body, playing key roles in the reactions of cellular respiration, amino acid synthesis, and protein metabolism.

Special nonprotein molecules called cofactors help enzymes catalyze chemical reactions. Organic cofactors are called coenzymes.

Substrates — *Cofactor* — *Enzyme* — *Enzyme-substrate*

5.7 The structure of ATP

Many metabolic pathways in your body break down complex substances into simpler substances, releasing energy in the process. What happens to the energy that is released in these exergonic reactions? What if your body could only use this energy when it was released? You might run out of energy a few hours after lunch, leaving you without enough energy to eat another meal. Obviously, living things could not survive under such circumstances. They have evolved ways to capture the energy released from food after it is eaten and absorbed so that this energy can be used to do cell work in the future.

One way your body stores energy from the food you eat is by converting it to fat. In fact, many of us think we have too much of this stored energy. Fat, however, cannot serve our cells' immediate energy needs. It is used for long-term storage.

Glycogen is another molecule used by the body to store energy. Although glycogen can be more readily converted into energy than body fat can, it is still unable to meet immediate energy demands. The primary molecule used by cells to capture energy and supply it at a moment's notice is called **adenosine triphosphate** (uh-DEN-oh-SEEN try-FOS-fate), or **ATP**.

Each ATP molecule is made up of three subunits:
1. A five-carbon sugar called *ribose*
2. A double-ringed molecule called *adenine*
3. Three phosphate groups (PO_4) linked in a chain called a *triphosphate group*

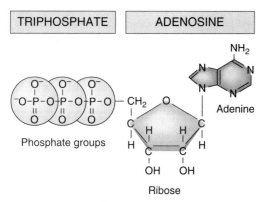

Figure 5.11 Adenosine triphosphate (ATP) is the primary energy currency of the cell.

Together, the ribose sugar and the adenine rings are called *adenosine*. The "working end" of the molecule, however, is the triphosphate group, especially the terminal (end) phosphate (**Figure 5.11**).

ATP is made up of ribose, adenine, and three phosphate groups.

5.8 How ATP stores, transfers, and releases energy

The covalent bonds that link the phosphate groups in ATP are fairly unstable and require very little free energy of activation to be broken. Because of their instability and the negative charges that repel one another in the triphosphate group, ATP has a high *group transfer potential*—its terminal phosphate group easily transfers from one molecule to another during coupled reactions.

ATP transfers chemical energy as it transfers its terminal phosphate group. The entire ATP molecule contains energy "stored" in its molecular configuration. When ATP is coupled with endergonic reactions, the bond that links its terminal phosphate group to the rest of the ATP molecule is broken by hydrolysis (see p. 44) and some of that energy is released. The energy is transferred, along with the phosphate group, to another molecule in a process called **phosphorylation** (FOS-for-ih-LAY-shun). This process is extremely important in living systems because most cellular work depends on ATP energizing other molecules by phosphorylation.

ATP is such a universally used molecule to store and transfer energy in living systems that it is referred to as the energy "currency" of the cell. The analogy goes like this: ATP saves energy (like we save money) when it captures the energy that is released from exergonic reactions. Likewise, when cells need energy to drive endergonic reactions, the cell can "spend" ATP to provide this energy. In so doing, ATP connects the metabolic steps in the metabolic machinery of living systems.

Figure 5.12 shows an example of an endergonic reaction coupled with the exergonic reaction of the hydrolysis of ATP. During this reaction, the terminal

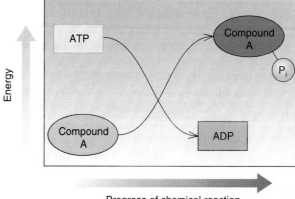

Figure 5.12 **A coupled reaction.** The energy that is released from conversion of ATP to ADP + P_i (an exergonic reaction) is used to drive the conversion of Compound A to Compound A + P_i (an endergonic reaction).

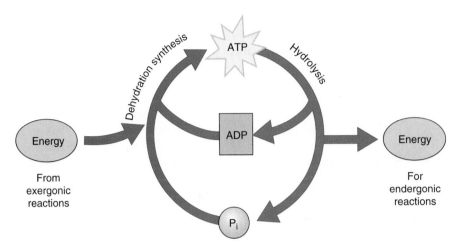

Figure 5.13 **The ATP-ADP cycle.** When ATP is used to drive the energy-requiring activities of living things, the high-energy bond that links the last phosphate group to the ATP molecule is broken, releasing energy. Cells always contain a pool of ATP, ADP, and phosphate (P$_i$), so ATP is continually being made from ADP, phosphate, and the energy released from exergonic reactions.

phosphate group of ATP is hydrolyzed and then bonded to Compound A. Free energy is transferred to Compound A with the phosphate group. That is, Compound A *gains* energy when it acquires a phosphate group from ATP. Compound A might be a muscle fiber that contracts when phosphorylated, or an intermediate product formed as a plant stores the energy from sunlight in molecules of sugar during photosynthesis. ATP loses free energy when its phosphate group is cleaved. Some of this energy is lost as heat, but most of it is transferred to Compound A. The molecule that remains is called **adenosine diphosphate**, or **ADP.**

ATP is used to fuel a variety of cell processes. It is used, for example, when glow worms, fireflies, and deep-water fishes produce light. It is used to build larger, more complex molecules such as proteins from smaller, simpler molecules such as amino acids. It also provides energy for you to move, fueling the reactions that take place in your muscles causing muscle fibers to contract. Cells also use ATP to help move substances against a gradient—it is the fuel of active transport. Finally, the energy in ATP is changed into electrical energy, primarily in nerves.

ATP is constantly being split into ADP and phosphate to drive the endergonic, energy-requiring processes of the cell by means of coupled reactions. In addition, however, ATP is continually being made from ADP, phosphate, and energy during coupled exergonic reactions. The energy for these exergonic reactions comes from the breakdown of energy molecules such as fats and carbohydrates. (These processes are described in Chapter 6.) This recycling happens quickly and is depicted in **Figure 5.13.** In fact, if you could mark every ATP molecule in your body at one instant in time and then watch them, they would be gone in a flash. Most cells maintain a particular molecule of ATP for only a few seconds before using it.

Where does the energy come from that is captured in molecules of ATP? The ultimate source of energy, of course, is energy from the sun that is captured by plants during the process of photosynthesis. Plants and the animals that eat them are food sources for other animals. Your body breaks down the food you eat into molecules such as glucose that are useable for fuel by your cells. Your cells break down these fuel molecules, releasing energy that is then captured in molecules of ATP.

ATP is called the universal energy currency of all cells because it can capture and release amounts of energy in coupled reactions that are used in many of the exergonic and endergonic reactions in living systems. ATP, ADP, and phosphate are continually being recycled within living cells as cells use group transfer to capture, carry, and release energy.

Key Concepts

Energy is stored and released during chemical reactions.

5.1 Chemical reactions involve changing one substance into another.

5.1 The substances entering a chemical reaction are reactants. The changed reactants are products.

5.1 Energy is needed to initiate a chemical reaction. This energy is the free energy of activation.

5.1 In exergonic reactions, reactants are broken down to yield products. The products have less energy than the reactants; the excess energy is released.

5.1 In endergonic reactions, reactants are chemically joined to yield a product. The product has more energy than the reactants.

5.1 In living systems, endergonic reactions are coupled with exergonic reactions to provide the energy needed to drive the endergonic process.

5.2 Energy exists as either potential (stored) energy and kinetic energy (energy actively doing work).

5.2 Energy can be changed from one form to another during chemical reactions.

5.2 Energy cannot be created or destroyed; it can only be changed from one form or state to another.

5.2 Energy is continually being lost as heat during chemical reactions. Heat is a form of energy that is not useful for performing work.

Enzymes regulate chemical reactions in living systems.

5.3 Chains of chemical reactions that move, store, carry, and release energy in living systems are called metabolic pathways.

5.3 Special proteins called enzymes regulate the reactions in metabolic pathways.

5.3 Enzymes lower the free energy of activation and speed up reactions.

5.4 Enzymes work by bringing substrates together so that they react more easily and by placing stress on certain bonds, which lowers the amount of free energy of activation that must be absorbed by the substrates to react.

5.5 The activity of an enzyme is affected by anything that changes its three-dimensional shape, such as pH, temperature, and the binding of specific chemicals.

5.5 The activity of enzymes is regulated by changes in enzyme shape; these changes result when activator and inhibitor molecules bind to specific enzymes.

5.6 Special nonprotein molecules called cofactors and coenzymes help enzymes catalyze chemical reactions.

ATP is the primary energy carrier of living systems.

5.7 Adenosine triphospate, or ATP, is made up of ribose, adenine, and three phosphate groups.

5.8 ATP transfers chemical energy as it transfers its terminal phosphate group, a process called phosphorylation.

5.8 ATP is the energy currency of the cell in that it can capture energy from an exergonic reaction, carry this energy, and then "spend" it when needed.

Key Terms

active sites *93*
adenosine diphosphate (ADP) (uh-DEN-o-seen dye-FOS-fate) *98*
adenosine triphosphate (ATP) (uh-DEN-o-seen try-FOS-fate) *97*
anabolic reactions (AN-uh-BOL-ick) *91*
catabolic reactions (CAT-uh-BOL-ick) *90*
coenzyme (KO-ENN-zyme) *96*

cofactors *96*
endergonic reaction (ENN-der-GON-ick) *90*
entropy (ENN-truh-pee) *92*
enzymes (ENN-zymes) *93*
exergonic reaction (EK-sur-GON-ick) *90*
first law of thermodynamics *91*
kinetic energy (kuh-NET-ick) *91*
laws of thermodynamics *91*

negative feedback *95*
phosphorylation (FOS-for-ih-LAY-shun) *97*
potential energy *91*
products *90*
reactants *90*
second law of thermodynamics *91*
substrates *93*

KNOWLEDGE AND COMPREHENSION QUESTIONS

1. What is the role of free energy of activation in a chemical reaction? (*Knowledge*)
2. Cells within your bone marrow are continuously producing blood cells for the body. Reactions occur that involve creating complex molecules from simple molecules. Are these processes dependent on anabolic or catabolic reactions? Bone marrow cells also receive nutrients, such as glucose, to fuel the cell. What type of reaction would you hypothesize occurs as glucose is broken down, releasing energy? (*Comprehension*)
3. State the first and second laws of thermodynamics. (*Knowledge*)
4. When asked about his messy room, a friend who's studied science simply replies, "Entropy increases." Explain what this means. (*Comprehension*)

5. What are metabolic pathways? Explain their significance. (*Comprehension*)
6. If a catalyst is not changed by a reaction, what purpose does it serve? (*Knowledge*)
7. Explain why the cells of your body require so many different enzymes (1000 to 4000 per cell). (*Comprehension*)
8. What do enzyme activators and inhibitors have in common? Which plays a key role in the process of negative feedback? (*Comprehension*)
9. What are cofactors and coenzymes? Name a biologically important example of each. (*Knowledge*)
10. ATP has been called the "energy currency of the cell." Explain what that phrase means. (*Comprehension*)

K E Y
Knowledge: Recalling information.
Comprehension: Showing understanding of recalled information.

CRITICAL THINKING REVIEW

1. Our cells exhibit a continual cycle of ATP build-up and breakdown involving exergonic and endergonic reactions. What do you think might be the biological advantage of such a simple energy exchange system? (*Application*)
2. Industrial pollution can change the pH of a pond or river to make the water more acidic. How can this affect the metabolic pathways of the plants that live in the water? (*Application*)
3. The chapter states that "life is a process regulated by enzymes." What might be the sources of these enzymes? If a particular enzyme is not available in a person's cells, what sequence of events might result to produce it? (*Synthesis*)
4. The cellular cycle of ATP = ADP + P_i + energy = ATP is occurring continuously in all of our

cells. Why is the input of fuel (food energy) needed to run this cycle? (*Application*)
5. Energy is continually entering the Earth's atmosphere from the sun. Explain the role played by photosynthetic organisms in capturing this energy and contributing to the biological flow of energy on Earth. Why is the role of plants important? And why doesn't the addition of the sun's energy to the amount of energy available on Earth contradict the first law of thermodynamics? (*Application*)

K E Y
Application: Using information in a new situation.
Synthesis: Putting together information from different sources.

Explain what is happening at each step of the enzyme catalyzed reaction shown below.

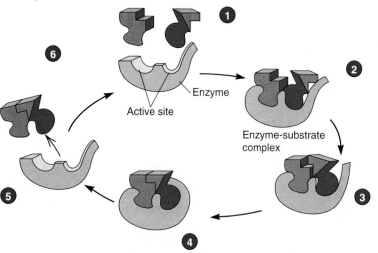

e-Learning is an on-line student review area located at this book's web site www.jbpub.com/biology. The review area provides a variety of activities designed to help you study for your class and build your learning portfolio.

Review Questions The review questions test your knowledge of the important concepts and applications in each chapter. The review provides feedback for each correct or incorrect answer. This is an excellent test preparation tool.

Figure-Labeling Exercises Sharpen your visual thinking by matching terms and labels for important illustrations.

Flash Cards Studying biology requires learning new terms. Virtual flash cards help you master the new vocabulary for each chapter.

Just Wondering As you read and study from this text, you may find that you have unanswered questions. Through this site you can ask the author, Sandy Alters, your "just wondering" questions.

Do you prefer the speed and reliability of a CD-ROM? All of the features contained on the eLearning portion of the web site are also available on the Student CD-ROM.

Gathering, Storing, and Using Energy from the Sun

Whether directly or indirectly, living organisms depend on the sun for survival. Imagine what might happen if the light of the sun were blocked from the Earth forever. You might predict that the temperature of the Earth's atmosphere would drop, and you would be correct. Obviously, you might also think that the Earth would become shrouded in darkness. And that is true, too. But more devastating consequences would occur. Life on the surface of the Earth—as we know it today—would totally and completely end.

The living organisms of Earth depend on the sun and have for billions of years. Your life depends on a process called photosynthesis, which is powered by the sun. By means of photosynthesis, energy from the sun is captured and stored in carbohydrate molecules. This food ultimately builds and fuels the bodies of all living things.

A byproduct of photosynthesis is oxygen. At right is a waterweed, *Elodea canadensis*, releasing oxygen bubbles from its leaves as it photosynthesizes underwater.

6

Organization

6.1 Autotrophs and heterotrophs

Plants, algae, and certain protists and bacteria are the members of the living world that are able to carry out photosynthesis. Organisms that produce their own food by photosynthesis, along with a few others that use chemical energy in a similar way, are called **autotrophs** (AW-toe-trofs). (Chapter 8, p. 154, discusses unusual nonphotosynthetic autotrophs.) The word autotroph literally means "self-feeder." All organisms live on the food produced by autotrophs, including the autotrophs themselves.

Photosynthetic autotrophs make their own food by harvesting light energy from the sun and changing it into stored chemical energy within food. More specifically, the energy from the sun powers the synthesis of glucose molecules, using the raw materials of carbon dioxide and water (Figure 6.1, ❶). Oxygen is a byproduct of this process ❷. Photosynthetic organisms are critical to life on Earth not only because they manufacture food for themselves as well as other organisms but also because they generate large amounts of oxygen.

The glucose produced during photosynthesis ❸ is a sugar that plants and other photosynthetic organisms use as a source of energy (from which they manufacture ATP) and as building blocks to construct larger molecules. From glucose, plants and other autotrophs manufacture cellulose (found in the cell walls of many autotrophs) and starch ❹. Starch is a storage form of sugar,

which autotrophs can convert to glucose when they need energy. Autotrophs also use inorganic nutrients such as phosphates and nitrates to build large organic molecules from glucose. (This is why plants need fertilizer, which contains phosphates and nitrates.) Some autotrophs use what might be considered unusual nutrients. For example, the diatoms (as shown in Chapter 3 p. 42, and in Figure 27.15), which are single-celled protists that are dominant photosynthesizers in the ocean, use silicates to construct their glasslike shells.

Organisms that cannot produce their own food (like the cows shown above) are called **heterotrophs** (HET-ur-oh-trofs), literally "other-feeders." Heterotrophs

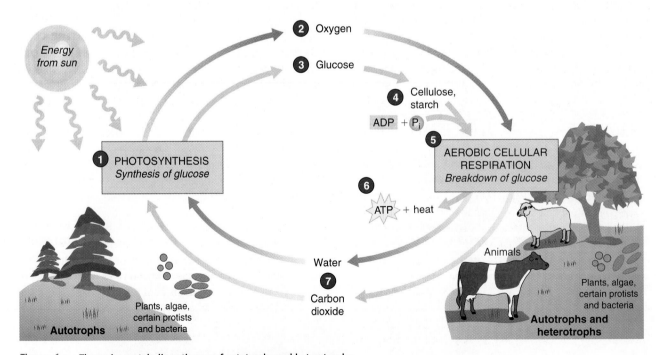

Figure 6.1 The major metabolic pathways of autotrophs and heterotrophs.

consume other organisms and, ultimately, the food produced by autotrophs. At least 95% of the species of organisms on Earth—all animals, all fungi, and most protists and bacteria—are heterotrophs. They live by feeding on the chemical energy that is fixed (incorporated into carbohydrates) by photosynthesis. Like photosynthetic autotrophs, heterotrophs use glucose as a source of energy, from which they manufacture ATP.

The energy that organisms use to make ATP, then, is actually energy from the sun, captured during photosynthesis. By eating plants, and by eating animals that eat plants (or other photosynthetic autotrophs such as algae), humans and other animals can harvest the energy from the sun that was originally captured and stored by autotrophs. Therefore, all living things—plants, bacteria, and you—share the same ultimate dependency on the sun.

The primary process by which organisms unlock the energy in food (which ultimately comes from the sun) is called **cellular respiration ❺**. During this complex series of chemical reactions, ATP is made ❻ by breaking down nutrient molecules such as glucose to release the energy stored in its bonds, capturing it in ATP. (Some energy is released as heat, see photo above right.) The term cellular refers to the fact that this process takes place in the cells of living things. The term respiration describes the process: the breakdown of fuel molecules with a resulting release of energy in which an inorganic compound (such as oxygen) serves as the terminal electron acceptor. (The terminal electron acceptor is the last compound to accept electrons in the series of chemical reactions that comprise cellular respiration.)

Aerobic cellular respiration is oxygen dependent. *All* organisms that use oxygen—including plants, which produce the oxygen that they (and we) need—break down nutrient molecules by means of *aerobic respiration*. The byproducts of aerobic respiration—carbon dioxide and water ❼—are the raw materials of photosynthesis. As Figure 6.1 shows, these two processes are intimately connected, with one fueling the other. These major metabolic pathways are the energy crossroads of cells.

Organisms that do not "breathe" oxygen, such as certain microorganisms, make ATP by either *anaerobic cellular respiration* or *fermentation*. During anaerobic respiration (an-er-OH-bick; literally, "without oxygen"), organisms (such as certain bacteria) use the pathways of aerobic respiration but use compounds such as nitrate or sulfate, rather than oxygen, as the terminal electron acceptor. Fermentation, on the other hand, consists of glycolysis (an initial series of reactions of cellular respiration) and one or two additional reactions that take

A skunk cabbage generates enough heat internally, through cellular respiration, to melt surrounding snow when it emerges in the spring.

place anaerobically. Baker's yeast, for example, can generate ATP by fermentation. During this process, the yeast produces carbon dioxide gas, an end product that causes bread dough to rise. Certain yeasts produce alcohol during fermentation and are used in the production of beers, wines, and other alcoholic beverages. Additionally, muscle cells, when they are deprived of sufficient oxygen for aerobic respiration, undergo a fermentative process that produces lactic acid.

All organisms use either cellular respiration or fermentative pathways to generate ATP, suggesting a strong evolutionary relationship among all living things. The only organisms that use neither of these two ATP-generating processes are single-celled parasites called *Chlamydia*. One species of *Chlamydia* causes certain diseases of the eyes as well as one of the most prevalent sexually transmitted diseases in the United States: nongonococcal urethritis. Another species causes diseases in animals. These organisms, among the smallest of the bacteria, can grow only within other cells, and they obtain their ATP from the cells they infect. For this reason, scientists call *Chlamydia* "energy parasites." *Chlamydia* are thought to have lost the ability to generate ATP during their evolution.

Most autotrophs manufacture their own food by converting the energy in sunlight into chemical energy by means of the process of photosynthesis. Heterotrophs cannot make their own food and ultimately depend on autotrophs—and the sun—for survival. All cells that make their own ATP do so by means of cellular respiration or fermentation.

6.2   The energy in sunlight

How do photosynthetic organisms "harvest" light energy from the sun? And how do they use this energy to create chemical bonds in food molecules? First, it is important to understand what light energy is.

As you may recall from Chapter 5, energy can exist in many forms, such as mechanical force, heat, sound, electricity, light, radioactivity, and magnetism. All these forms of energy are able to create change—to do work. Sunlight is a form of energy known as *electromagnetic energy*, also called *radiation*.

Electromagnetic energy travels as waves. The waves can be envisioned as ripples on the surface of a still pond after a pebble has been thrown into the water. Electromagnetic "ripples," however, are disturbances in electric and magnetic fields moving through space.

The array of electromagnetic waves coming from the sun vary in length. Their lengths are measured from the crest of one wave to the crest of the next and are expressed in meters or in billionths of meters (nanometers, or nm). The shortest wavelengths are gamma rays; the longest are radio waves.

The full range of electromagnetic radiation in the universe is called the *electromagnetic spectrum* (Figure 6.2). Visible light is only a small part of this spectrum, but it, too, is made up of many different wavelengths. Remember the last rainbow you saw? Its array of colors was caused by the separation of the various wavelengths of visible light as they passed through tiny droplets of water in the air.

Although electromagnetic energy travels as waves, it also behaves like individual particles—discrete "packets" of energy—which are called *photons*. The energy content of each photon is inversely proportional to the wavelength of the radiation. Put simply, short-wavelength radiation (such as gamma rays and x-rays) contains photons of a high energy level; long-wavelength radiation (such as microwaves and radio waves) contains photons of a low energy level. Within the range of visible light, then, the shorter wavelengths of violet light carry a greater "energy punch" than the longer wavelengths of red light.

Sunlight can be described both as waves, or disturbances in electric and magnetic fields, and as discrete packets of energy called photons.

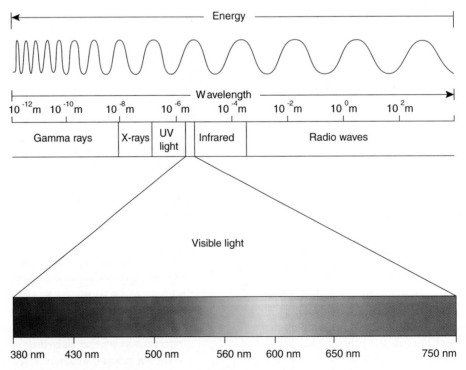

Figure 6.2   The electromagnetic spectrum. Visible light represents only a small part of the electromagnetic spectrum, that between 380 and 750 nanometers (nm).

6.3 Capturing light energy in chemical bonds

A molecule "captures" the energy of sunlight when a photon of energy boosts an electron of the molecule to a higher energy level than it already occupies. This electron is then said to be "excited." (See Chapter 3 for a description of the energy levels of electrons.) Boosting an electron to a higher energy level requires just the right amount of energy—no more and no less. For example, when you climb a ladder, you must raise your foot just so far to climb a rung, not 1 centimeter more or less. Likewise, specific atoms can absorb only certain photons of light—those that correspond to available energy levels. Therefore, a given atom or molecule has a characteristic range, or *absorption spectrum,* of photons it is capable of absorbing depending on the electron energy levels that are available in it.

Molecules that absorb some visible wavelengths and transmit (allow them to pass through) or reflect others are called **pigments**. Pigments are responsible for all the colors you see. For example, in **Figure 6.3,** red artist's paint looks red because it is manufactured using a pigment that absorbs variable amounts of different wavelengths, including red, but reflects light enriched in red wavelengths. These red wavelengths bounce off the pigment and are reflected back to your eyes. Pigments located in special cells within your eyes then absorb the photons of this red light. The resulting electron excitations ultimately generate nerve impulses that are sent to the brain. The brain interprets these impulses as red. Likewise, white objects absorb no light and reflect all the wavelengths of visible light to your eye. A black object absorbs all wavelengths, reflecting none.

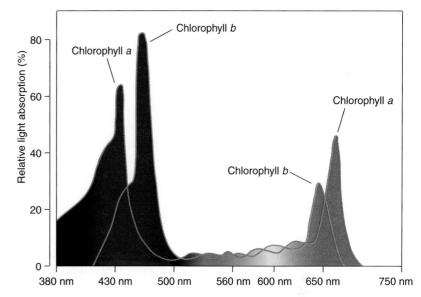

Figure 6.4 **Absorption spectra for chlorophylls.** Chlorophylls, such as chlorophyll *a* and chlorophyll *b*, absorb predominantly violet-blue and red light in two narrow bands of the spectrum.

Organisms have evolved a variety of pigments. Two groups of pigments that are important to the process of photosynthesis are the **carotenoids** (kuh-ROT-uh-noids) and the **chlorophylls**. Carotenoids absorb photons of green, blue, and violet wavelengths and reflect red, yellow, and orange. Chlorophylls, as shown in **Figure 6.4,** absorb photons of violet-blue and red wavelengths and reflect green and yellow.

Chlorophyll is used as the primary light gatherer in all plants and algae and in certain photosynthetic bacteria—the cyanobacteria. The carotenoids are important, however, because they absorb wavelengths of light that chlorophyll cannot, and then pass the energy to chlorophyll, thereby increasing the spectrum of light that can be absorbed. Chlorophyll masks the presence of the carotenoids, which become visible only when chlorophyll breaks down. This breakdown occurs in autumn (in the areas of the world that have seasons) as the days shorten and the amount of sunlight decreases. As a result, the carotenoids become visible, providing the magnificent yellows and oranges of fall foliage. The bright reds are caused by the unmasking of other non-light-gathering pigments.

> A pigment is a molecule that absorbs some wavelengths of light and transmits or reflects others. The wavelengths absorbed depend on the energy levels available in the molecule to which light-excited electrons can be boosted.

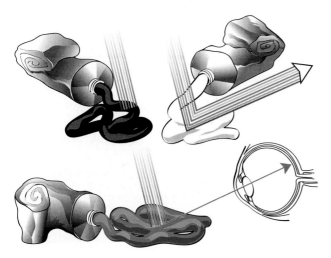

Figure 6.3 **Pigments absorb some visible wavelengths of light and transmit others.**

6.4   An overview of photosynthesis

The process of **photosynthesis** is described chemically
in the following equation:

$$6CO_2 + H_2O \xrightarrow{\text{Light energy}} C_6H_{12}O_6 + 6O_2$$

Carbon dioxide   Water   Glucose   Oxygen

This chemical equation represents the beginning and
end points in a series of chemical reactions and merely
summarizes complex events, much as you might summarize the details of a book in a single paragraph. It
shows that the net effect of photosynthesis is the
production of one molecule of glucose and six molecules of oxygen from six molecules of carbon dioxide
and six molecules of water in the presence of sunlight. (The main product of photosynthesis is actually
glyceraldehyde-3-phosphate [G3P], which can be

synthesized into glucose, sucrose, complex carbohydrates, and other organic compounds. Glucose is used
to simplify our discussion and to show the relationship
between photosynthesis and respiration.)

A more accurate summary of the chemical reactions of photosynthesis is represented by the following
equation:

$$6CO_2 + 12H_2O \xrightarrow{\text{Light energy}} C_6H_{12}O_6 + 6H_2O + 6O_2$$

Here, water appears on both sides of the equation—as a
substrate and as a product. Writing the equation in this
way points out that the water molecules on the left side
of the equation and those on the right side are *not* the
same water molecules. The six molecules of oxygen
(12 atoms) produced by photosynthesis are derived
from those in the water molecules on the left side of the
equation. (The colored boxes will help you track the

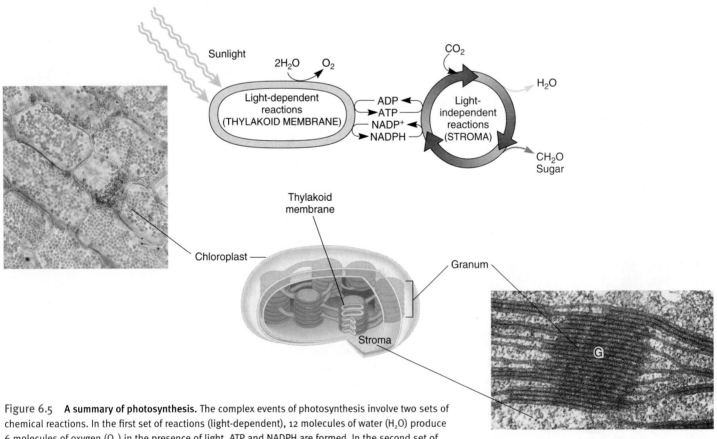

Figure 6.5   A summary of photosynthesis. The complex events of photosynthesis involve two sets of
chemical reactions. In the first set of reactions (light-dependent), 12 molecules of water (H₂O) produce
6 molecules of oxygen (O₂) in the presence of light. ATP and NADPH are formed. In the second set of
reactions (light-independent), the newly formed ATP and NADPH drive the formation of glucose (sugar)
from carbon dioxide (CO₂). These reactions take place in the chloroplasts of plant cells.

atoms and molecules in this equation.) Plants, algae, and the cyanobacteria produce oxygen in this way. In addition, the water molecules on the right side of the equation derive their oxygen atoms from the oxygen in carbon dioxide. Glucose (an organic molecule) is constructed from the carbon and oxygen in carbon dioxide (an inorganic molecule) and the hydrogen in water. During the process of photosynthesis, inorganic molecules are broken apart and their atoms are re-shuffled and put back together again to form organic molecules—molecules of living things. Photosynthesis, then, is a process during which the living and nonliving worlds meet.

Light-Dependent Reactions

The complex events of photosynthesis summarized by the previous equation involve two sets of chemical re-actions as shown in **Figure 6.5**. During the first set of reactions ❶, adenosine triphosphate (ATP) and the re-duced form of nicotinamide adenine dinucleotide phos-phate (NADPH, see next paragraph) are formed using energy captured from sunlight. These molecules store the energy harvested from the sun and are used to man-ufacture sugars in the second set of reactions. The reac-tions that produce ATP and NADPH take place only in the presence of sunlight and are therefore called the **light-dependent reactions**.

Redox Reactions and Electron Carriers

NADPH, an electron carrier, is an important molecule in both the light-dependent and light-independent reac-tions of photosynthesis. *Electron carriers* transfer elec-trons as hydrogen atoms from one molecule to another during oxidation-reduction (redox) reactions. (Chapter 3, p. 38, discusses redox reactions.) In biological sys-tems, electrons often do not travel alone from one atom to another, but instead take along a proton; together the electron and the proton are a hydrogen atom. $NADP^+$ (the oxidized form of NADPH) is only one of many electron carriers that are key players in the processes of photosynthesis and cellular respiration. The $^+$ sign indi-cates that this molecule is oxidized (has given up elec-trons). It can accept two electrons and one proton (H^+) to become reduced to NADPH. Notice that this mole-cule (NADPH) no longer bears a positive charge be-cause it has been reduced (has accepted an electron). In addition, the extra H^+ and electron (hydrogen atom) are shown by the addition of one H to NADP. The electrons that $NADP^+$ accepts contain a great deal of energy; thus, they produce molecules that carry energy in the form of energized electrons.

Light-Independent Reactions

A second series of reactions uses the ATP and NADPH formed during the light-dependent reactions to provide energy for the formation of glucose from carbon diox-ide. These reactions are called the **light-independent re-actions** ❷ because there is no direct involvement of light in the reactions.

Photosynthetic Membranes

The reactions of photosynthesis take place on *photosyn-thetic membranes*. In photosynthetic bacteria, these membranes are infoldings of the cell membrane. In plants, algae, and the photosynthetic single-celled pro-tists called *euglenoids*, all the reactions of photosynthe-sis are carried out within cell organelles known as **chloroplasts** (Figure 6.5). Chloroplasts, which were probably derived evolutionarily from photosynthetic bacteria (like the cyanobacterium shown here), have a

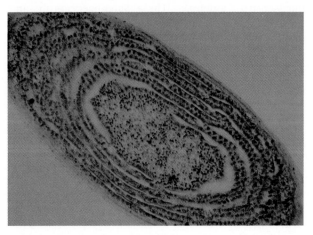

system of internal membranes that are organized into flattened sacs called **thyla-koids** (THIGH-luh-koidz). Stacks of thylakoids are referred to as **grana** (GRA-nuh). Each thylakoid is a closed compartment. Chorophyll, acces-sory pigments, and enzymes are embedded in the thylakoid membranes. The light-dependent reactions take place here. Surrounding the thylakoids is a fluid called the **stroma** (STROH-muh), which contains the enzymes of the light-independent reactions.

Photosynthesis, an energy-requiring series of reac-tions, produces carbohydrates from carbon dioxide and water. Oxygen is a byproduct of this process. Photosynthesis takes place in the chloroplasts of photosynthetic eukaryotic cells and on infoldings of the cell membrane in photosynthetic bacteria. It involves two sets of reactions: the synthesis of ATP and NADPH, and the use of ATP and NADPH to pro-duce glucose.

6.5 The light-dependent reactions: Synthesizing ATP and NADPH

The light-dependent reactions of photosynthesis perform the important job of changing the energy in sunlight into usable chemical energy: ATP and NADPH. These molecules are used in the light-independent reactions to make glucose from carbon dioxide. The first event leading to ATP and NADPH production is the capturing of a photon of light by a pigment.

How Photosystems Capture Light

In all but the most primitive bacteria, light is captured by individual networks of chlorophylls, carotenoids, and other pigment molecules. These networks, located in the thylakoid membranes of the chloroplasts, are called *photocenters*. The pigment molecules within the photocenters are arranged in such a way that they act like sensitive antennae. These antennae capture and funnel photon energy to special molecules of chlorophyll called **chlorophyll a.** The variety of pigments in the photocenters allows a plant to capture energy from a range of wavelengths of light and use it to power photosynthesis.

When light of the proper wavelength (photon) strikes any pigment molecule of the photocenter as shown in Figure 6.6, one of its electrons is boosted to a higher energy level. This pigment molecule passes the energy to a neighboring molecule within the photocenter. The energy is passed from one pigment molecule to the next, until it reaches the molecule of chlorophyll a that will participate in photosynthesis. This molecule is called *P700* or *P680*. The *P* stands for pigment. The number refers to the wavelength at which this molecule of chlorophyll a absorbs light. Both P700 and P680 absorb wavelengths in the far-red portion of the light spectrum.

A simple analogy to the form of energy transfer in the photocenter is the initial break in a game of pool. If the cue ball squarely hits the point of the triangular array of 15 pool balls, the two balls at the far corners of the triangle fly off. None of the central balls move at all. The energy is transferred through the central balls to the most distant ones. In a similar way, the pigment molecules of the photocenter channel energy to a molecule of chlorophyll a in the form of excited electrons.

Plants and algae have two different types of photosystems that play a role in photosynthesis. These photosystems are named after the order in which they evolved in photosynthetic organisms: photosystem I and photosystem II. In **photosystem I**, energy is transferred to a molecule of chlorophyll a, P700 (see Figure 6.6). In **photosystem II**, energy is transferred to a molecule of chlorophyll a called P680, whose name refers to its absorption peak of 680 nanometers. These two kinds of chlorophyll a act as reaction centers in each type of photosystem.

Noncyclic Electron Flow

Both photosystems absorb light at the same time (Figure 6.7 ❶ and ⑫). It is easiest to first describe what happens in photosystem II to explain the events of the light-dependent reactions. When four photons of light are absorbed by pigment molecules in photosystem II ❶, the energy is funneled to P680 ❷. This energy causes four electrons to be ejected from the P680 molecule ❸. The electrons are captured sequentially by an electron acceptor located within the thylakoid membrane ❹. However, an electron "hole" is now left in the P680 molecule. The P680 pulls four electrons from two molecules of water ❺, normally a hard task to accomplish. The P680 strongly attracts these electrons, however, causing each molecule of water to be split into two hydrogen ions and an oxygen atom. Oxygen atoms from two split molecules of water quickly join to form a molecule of oxygen ❻—a byproduct of photosynthesis and the gas that helps keep us alive. (This process occurs one electron at a time; only a single electron is or can be energized in a P680 [or P700] molecule. Only a single electron can be ejected, and the electron hole that is formed must be filled before another electron can be energized and ejected.)

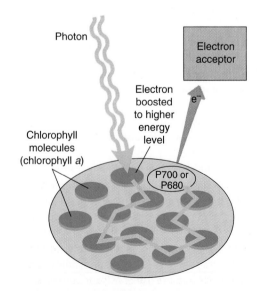

Figure 6.6 **How photocenters work.**

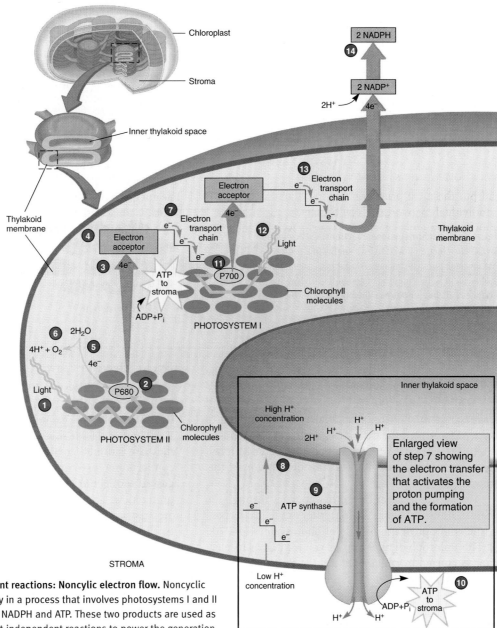

Figure 6.7 **Light-dependent reactions: Noncylic electron flow.** Noncyclic electron flow harvests energy in a process that involves photosystems I and II working together to produce NADPH and ATP. These two products are used as the energy source in the light-independent reactions to power the generation of glucose from carbon dioxide.

Meanwhile, the electron acceptor passes the four energized electrons from the P680 to the next in a series of electron acceptors. This series of electron acceptors or carrier proteins is called an *electron transport chain*, a series of electron carrier proteins such as NADP⁺ **7**. Here, electrons are passed from one carrier protein to another. As each of these carriers accepts electrons in turn, the electrons fall to lower energy levels, releasing energy in the process. The energy released from the electrons be-

ing passed along the chain pumps the protons (H^+) from the stroma through the thylakoid membrane and into the interior of the thylakoid compartment. (This "proton pumping" **8** and the ATP formation described next are shown in the box at the bottom of the illustration.) As a high concentration of protons builds up within this thylakoid space, the protons begin to diffuse back across the membrane through the membrane protein ATP synthase **9**. As the protons travel outward through these channels,

Energy from the sun is locked in carbohydrates during photosynthesis.

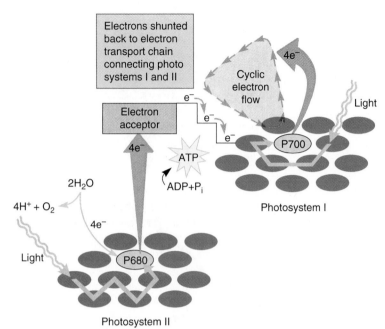

Figure 6.8 **Light-dependent reactions: Cyclic electron flow.** Cyclic electron flow harvests energy in a process that involves photosystem I working independently of photosystem II. This method of generating energy produces ATP, but it does not directly produce NADPH as does noncyclic electron flow.

enzymes couple the energy from the movement of the protons to the reaction of bonding inorganic phosphate (P_i) to ADP, forming ATP **❿**. This process in which a cell uses the potential energy in a concentration gradient to make ATP is called the theory of **chemiosmosis** (KEM-ee-oz-MOH-sis). In 1978, British biochemist Peter Mitchell was awarded the Nobel Prize for his description of this process.

When the electrons finish their journey through the electron transport chain (see **❼**), they are then accepted by the chlorophyll *a* molecule of photosystem I—P700 **⓫**. This chlorophyll molecule, as you recall, has its electrons energized by photons of light at the same time as P680 is energized **⓬**. So it, too, develops electron holes as energized electrons are ejected from their shells. The electrons of P680, now devoid of their "boost" of energy, fill these electron holes. The energized electrons of P700 have meanwhile been passed down the electron transport chain of photosystem I **⓭**. ATP molecules are not generated in this transport chain. Instead, the last acceptors are two molecules of NADP⁺, each of which picks up a proton in addition to two electrons to become reduced to NADPH **⓮**. Because the electrons it accepts are still highly energized, NADPH serves as a storage depot for this energy. This molecule is considered a source of "reducing power" for the light-independent reactions. The processes that occur in photosystems I and II make up the light-dependent reactions of eukaryotic photocenters.

Cyclic Electron Flow

Sometimes photosystem I works on its own in a modified way, independent of photosystem II **(Figure 6.8)**. In this situation, the excited electrons ejected from the P700 molecule are not accepted by NADP⁺. Instead, they are shunted to the electron transport chain that connects photosystems I and II. Here, the electrons are passed along the chain, triggering the events of ATP production. The electrons, no longer in an excited state, are then accepted back by P700 to complete a cycle. Cyclic electron flow provides eukaryotic organisms with ATP when there is no need for additional NADPH.

A photocenter is an array of pigment molecules within the thylakoid membranes of a chloroplast. It acts like a light antenna, capturing and directing photon energy toward a single molecule of chlorophyll *a* that will participate in photosynthesis. Plants and algae use a two-stage photosystem to carry out the light-dependent reactions of photosynthesis. During noncyclic electron flow, energized electrons ejected from photosystem II are passed down an electron transport chain, triggering events of ATP production. The electron hole in photosystem II is filled by electrons from the breakdown of water, a process that releases oxygen as a byproduct. Energized electrons ejected from photosystem I, along with hydrogen ions, reduce NADP⁺ to NADPH.

6.6 The light-independent reactions: Making carbohydrates

The light-dependent reactions of photosynthesis described in the preceding section use light energy to produce (1) metabolic energy in the form of ATP and (2) reducing power in the form of NADPH. But this is only half the story. Photosynthetic organisms use the ATP and NADPH produced by the light-dependent reactions to build organic molecules from atmospheric carbon dioxide, a process called **carbon fixation** or the **Calvin cycle** (after the researcher who discovered its steps). These organic molecules are food for the plant and also provide the plant with raw materials for growth. This phase of photosynthesis does not need light energy to drive its reactions, and therefore, it is referred to as the *light-independent reactions*. In plants, algae, and the euglenoids, these reactions are carried out by a series of enzymes located in the stroma of the chloroplast.

The Calvin cycle has three stages that produce one three-carbon sugar, *glyceraldehyde 3-phosphate (G3P)*, from three molecules of carbon dioxide (CO_2). Two molecules of G3P are used to manufacture one glucose molecule (and other carbohydrates) as shown at the bottom of the cycle. (Figure 6.9 depicts the Calvin cycle in simplified form; the grey balls stand for carbon atoms and will help you follow the flow of these reactions.)

In the first stage of the Calvin cycle, carbon dioxide joins a five-carbon molecule called *ribulose 1, 5 bisphosphate (RuBP)* from the "end" of the cycle ❶ in the process of carbon fixation. Thus, the inorganic molecule carbon dioxide is attached to a biological molecule, making it available for use to synthesize glucose and complex carbohydrates. For three turns of the cycle, three CO_2 molecules join with three RuBP molecules and form six three-carbon molecules (3PG) ❷. In the second stage of the cycle, these molecules are "chemically reshuffled" to produce glyceraldehyde 3-phosphate molecules ❸. Only one of these six G3P molecules is used in the manufacture of carbohydrates ❹. The other five molecules of G3P ❺ are used to resynthesize three molecules of RuBP (the third stage of the cycle, ❻). These molecules of RuBP are needed to keep the cycle going. In this way, RuBP is used as a "handle" to fix carbon dioxide, and the three-carbon molecules produced are used to build sugars for energy and storage and to reconstruct or recycle molecules of RuBP.

The three-carbon molecules formed during the first step of the Calvin cycle are also found in the process of glycolysis (see Figure 6.12). Using the ATP and NADPH from the light reactions, these molecules are bonded together and are chemically reduced by steps in the second stage of the Calvin cycle that are actually a reversal of the steps in glycolysis.

Referring back to the equation of photosynthesis, it takes six molecules of carbon dioxide to produce one molecule of glucose (see the equation on p. 108). For one molecule of G3P to be produced, the cycle must take place three times, and it must take place six times to produce one molecule of glucose.

Three main events take place during the light-independent reactions of photosynthesis: (1) carbon is fixed with the attachment of carbon dioxide to an organic molecule; (2) glucose and other carbohydrates are produced from molecules of carbon dioxide; and (3) the organic molecule that attaches to carbon dioxide is re-formed to begin the cycle again.

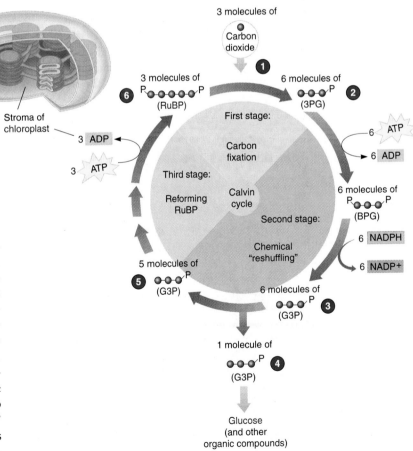

Figure 6.9 **The Calvin cycle.** For every three molecules of carbon dioxide that enter the cycle, one molecule of the three-carbon compound glyceraldehyde 3-phosphate is produced. Notice that the process requires energy, supplied as ATP and NADPH. ATP and NADPH are generated by the light-dependent reactions.

6.7 An overview of aerobic cellular respiration

The carbohydrates formed during photosynthesis contain energy from the sun, which is first captured in molecules of ATP and NADPH during the light-dependent reactions and is then incorporated into sugar molecules during the second stage of the Calvin cycle. First, ATP transfers chemical energy to 3PG as it transfers its terminal phosphate group. Then, NADPH, a source of energy-rich electrons, transfers these electrons to 1,3 BPG. The job of cellular respiration is to extract the energy locked in carbohydrates and store it in molecules of ATP that can be used to perform cell work. The process is described chemically in the following formula:

$$36ADP + 36P_i \longrightarrow 36ATP \text{ (energy)}$$

$$\underset{\text{Glucose}}{C_6H_{12}O_6} + \underset{\text{Oxygen}}{6O_2} \longrightarrow \underset{\text{Carbon dioxide}}{6CO_2} + \underset{\text{Water}}{6H_2O}$$

Although the formula is stated as if glucose were broken down to carbon dioxide and water in a single step, this is not the case. This simplified formula shows that as aerobic respiration occurs, the net effect is the breakdown

of one molecule of the substrate glucose in the presence of six molecules of oxygen to yield end products of six molecules of carbon dioxide and six molecules of water. During this process, enough energy is liberated from the glucose as it is being cleaved to power many endergonic reactions: 36 molecules of ADP are bonded to 36 atoms of inorganic phosphate—an energy yield of 36 ATPs. Therefore, the breakdown of glucose during aerobic respiration releases energy (is exergonic), which helps to produce molecules of ATP, a type of endergonic reaction (see Chapter 5).

The complex series of reactions of cellular respiration can be divided into three parts (**Figure 6.10**): **glycolysis** (glye-KOL-uh-sis), the **Krebs cycle**, and the **electron transport chain**. The word glycolysis comes from Greek words meaning "to break apart" (*lysis*) a "sugar" (*glyco*). Glycolysis takes place in the cytosol of the cell and is the first stage of extracting energy from glucose ❶. No oxygen is needed for it to take place. It is a metabolic pathway in which ATP is generated, but the total yield of ATP molecules is small—only two ATPs for each original glucose molecule. Energy-rich NADH is also produced ❷. When glycolysis is completed, the six-carbon glucose has been cleaved in half, yielding two three-carbon molecules called pyruvate, or pyruvic acid ❸. The two pyruvate molecules still contain most of the energy that was present in the one original glucose molecule.

The Krebs cycle ❹ takes place in the mitochondria and is the second stage of extracting energy from glucose. Named after the biochemist Sir Hans Krebs, this cycle of reactions begins with a two-carbon molecule ❺ that is produced by the removal of one carbon dioxide molecule (CO_2) from each pyruvate molecule formed by glycolysis. Each of these two-carbon molecules (acetyl CoA) combines with a four-carbon molecule to form a six-carbon molecule called citric acid. For this reason, the Krebs cycle is also often called the *citric acid cycle*. For every two pyruvate molecules entering the Krebs cycle as the result of the glycolytic breakdown of one glucose molecule, two more ATP molecules are made ❻ and a large number of electrons are removed from the substrates in the cycle. These electrons pass to a series of electron carriers also located in the mitochondria ❼, which form an electron transport chain ❽ similar to the electron transport chains found in the thylakoid membranes. By means of chemical reactions and processes that take place along the chain, 32 ATPs are produced.

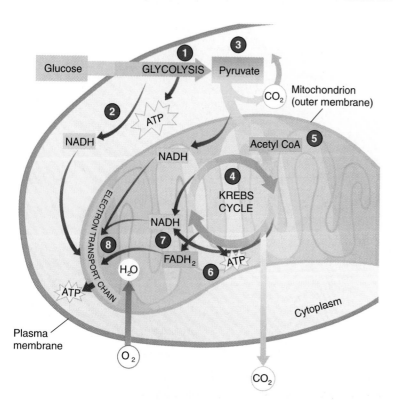

Figure 6.10 **An overview of cellular respiration.** 36 ATPs are produced for every one glucose molecule.

During aerobic cellular respiration, glucose is broken down in the presence of oxygen, yielding carbon dioxide and water and capturing energy in molecules of ATP. Three series of reactions make up the process of cellular respiration: glycolysis, the Krebs cycle, and the electron transport chain.

6.8 Preparing nutrients for aerobic respiration in the human body

Food is a complex mixture of sugars, lipids, proteins, and other molecules. The first thing that happens in food's journey toward ATP production is that digestive system enzymes break down complex molecules to simple ones. Complex sugars such as those found in vegetables, pasta, and bread, for example, are split into simple sugars such as glucose or into sugars that are changed to glucose (Figure 6.11 ❶). Proteins, which are found in foods such as meats and nuts, are broken down to amino acids ❷. Lipids, which can be found in oil-based salad dressings and ice cream, are split into fatty acids and glycerol ❸. These steps taken by the digestive system yield no usable energy, but they change a diverse array of complex molecules into a small number of simpler molecules that can be used by your cells either as fuel or as building blocks to manufacture substances for growth, repair, and maintenance.

Once carbohydrates are changed to glucose, they are completely metabolized by glycolysis, the Krebs cycle, and the electron transport chain ❹. Certain inter-mediary products formed in glycolysis may be used to build carbon compounds the body needs ❺. After the digestion of proteins to amino acids, their carbon portions are chemically modified and are then metabolized by the Krebs cycle and the electron transport chain, skipping over glycolysis ❻. The nitrogen in proteins is eliminated in the urine ❼. Often, amino acids are used to build proteins rather than being used for energy ❽. Some of the breakdown products of the fats in your diet are converted to substances that can also be metabolized by the Krebs cycle and the electron transport chain ❾ or used to build other lipids the body needs ❿.

In preparation for aerobic respiration, food is first broken down from complex to simple molecules in the digestive system and then absorbed into the bloodstream. Carbohydrates are broken down into simple sugars, lipids into fatty acids and glycerol, and proteins into amino acids. Once inside the cells, these substances are metabolized into substrates that enter the chemical pathways of aerobic respiration at some point and are completely broken down to carbon dioxide, water, and energy.

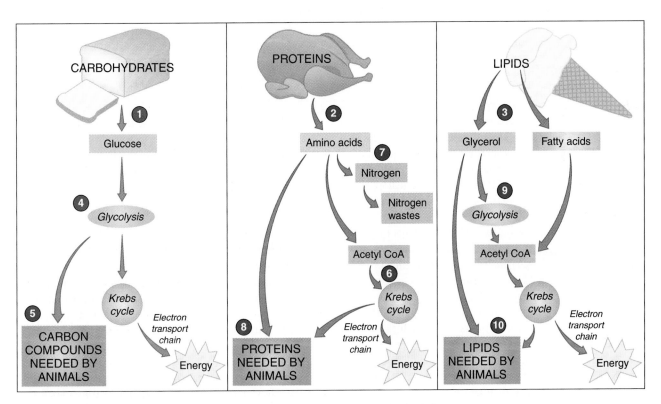

Figure 6.11 **The complex molecules in food must be broken down into simpler molecules before they can be used by cells as fuel.** Carbohydrates are broken down into simpler sugars; proteins are broken down into amino acids; and lipids are broken down into fatty acids and glycerol.

6.9 Glycolysis

Scientists think that glycolysis was one of the first metabolic processes to evolve. One reason is that this process uses no molecular oxygen and therefore occurs readily in an environment devoid of oxygen—a characteristic of the atmosphere of the primitive Earth. In addition, all of the reactions of glycolysis occur free in the cytoplasm; none are associated with any organelle or membrane structure. "Early" cells most certainly had no specialized structures such as organelles; thus, metabolic processes that evolved independently of membrane systems are probably evolutionarily older than those associated with membrane systems.

All living things (except *Chlamydia*) possess glycolytic enzymes. However, most present-day organisms are able to extract considerably more energy from glucose molecules than glycolysis does. For example, of the 36 ATP molecules you obtain from each glucose molecule that you metabolize, only 2 are obtained by glycolysis. Why, then, is glycolysis still maintained even though its energy yield is comparatively meager?

This simple question has an important answer: Evolution is a step-by-step process. Cells that were capable of glycolysis survived the early competition of life. Later changes in catabolic metabolism built on this success. Glycolysis was not discarded during the course of evolution but rather was used as the starting point for the further extraction of chemical energy.

The catabolic pathway of glycolysis is shown in **Figure 6.12.** (The gray balls represent carbon molecules.) Its reactions can be divided into four stages:

1. *Stage A—Glucose mobilization.* During reactions ❶ and ❷, glucose is changed into a compound that can be split into two molecules. During each of these two reactions, ATP transfers chemical energy to glucose as it transfers its terminal phosphate group. Before energy can be released from glucose, your body must spend two ATP molecules to "prime" the glucose "pump."

2. *Stage B—Cleavage.* During reaction ❸, the six-carbon product of glucose mobilization is split into two three-carbon molecules.

3. *Stage C—Oxidation.* During reaction ❹, two H^+ and two electrons are removed from each three-carbon molecule. NAD^+ accepts one hydrogen ion and two electrons, forming two molecules of NADH for each original glucose molecule. (The second hydrogen atom appears as a free proton [H^+].) As this chemical change takes place, an inorganic phosphate molecule is bonded to each three-carbon molecule, resulting in the three-

carbon molecule (BPG). (NAD^+ is a major electron acceptor [carrier] in cellular respiration.)

4. *Stage D—ATP generation.* During reactions ❺ through ❼, the two phosphate groups on each three-carbon molecule (BPG) are removed by enzymes and bonded to two ADP molecules, producing four ATP molecules for each original glucose molecule and two 3PG molecules. A molecule of water is removed from each 3PG molecule, producing two PEP molecules. The end product of these reactions after the removal of two phosphate groups is two three-carbon molecules called pyruvate.

Because each glucose molecule is split into *two* three-carbon molecules, the overall net reaction sequence yields two ATP molecules, as well as two molecules of pyruvate:

$$-2 \text{ ATP Stage A} + 2\ (2 \text{ ATP}) \text{ Stage D} = +2 \text{ ATP}$$

Although this is not a great amount of energy (only 2% of the energy in the glucose molecule), glycolysis does generate ATP. During the first anaerobic, or airless, stages of life on Earth, this reaction sequence was the only way for living things to extract energy from food molecules.

The three most significant changes that take place during glycolysis are as follows:

1. Glucose is converted to pyruvate.
2. ADP + P_i is converted to ATP.
3. NAD^+ is converted to NADH.

These three products can be formed continually as long as the substrates used to produce them are available. The glucose is supplied by the food you eat. ADP and P_i continually become available as ATP is broken down to do cellular work. But cells contain only a small amount of NAD^+, and the supply is quickly depleted unless NADH passes along its hydrogen atom to another electron carrier. In this way, NADH is oxidized to form NAD^+ once again. Cells recycle NADH in one of two ways:

1. *By means of the electron transport chain as part of cellular respiration.* NADH passes its proton and two electrons to a molecule that shuttles them into the mitochondria, ultimately passing them to an electron carrier in the electron transport chain. Oxygen, an excellent electron acceptor, forms the last link in the chain in aerobic respiration. A molecule other than oxygen (such as nitrate or sulfate) forms the last link in the chain in anaerobic respiration. In aerobic respiration (a process on which we are focusing) each atom of oxygen

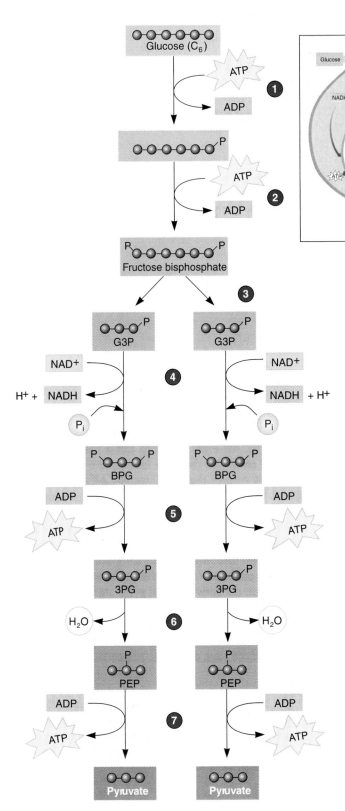

Figure 6.12 Glycolysis. Glycolysis produces a net yield of two ATP molecules for each original glucose molecule, as well as two molecules of NADH and two molecules of pyruvate.

accepts two protons and two electrons to form one molecule of water. In addition, three molecules of ATP are formed during various chemical processes that take place (see Figure 6.15).

2. *By means of fermentation.* Certain bacteria and yeasts produce an organic molecule from pyruvate that will accept the hydrogen atom from NADH and thus re-form NAD^+. Fermentation is described in more detail later in this chapter.

In all aerobic organisms—like you—the oxidation of glucose continues with the further oxidation of the product of glycolysis, which is pyruvate. In eukaryotic organisms, aerobic metabolism takes place in the mitochondria. Glycolysis, however, takes place in the cytoplasm outside the mitochondria. Therefore, the cell must move the electrons from the two NADH molecules produced during glycolysis into the mitochondrion. (The mitochondrial membrane is impermeable to the NADH.) These electrons are carried across the mitochondrial membrane by carrier molecules, and, as a result, they enter the electron transport chain in a slightly different place than the electrons from the other NADH molecules. Only two ATPs (instead of the usual three) are produced from each NADH. The net result is 36 ATPs for each molecule of glucose. Pyruvate, which is in high concentration in the cytoplasm, simply diffuses into the mitochondrion.

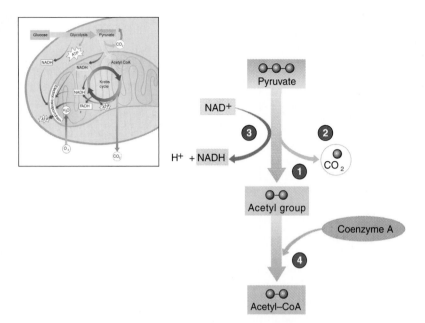

Figure 6.13   How pyruvate enters the Krebs cycle.

Before entering the reactions of the Krebs cycle, each pyruvate molecule is cleaved into a two-carbon molecule called an *acetyl group* (Figure 6.13 **1**). The left-over carbon atom from each pyruvate is split off as carbon dioxide gas (CO_2) **2**. In addition, one H^+ and two electrons reduce NAD^+ to NADH **3**. In the course of these reactions, the two-carbon acetyl fragment removed from pyruvate is added to a carrier molecule called *coenzyme A,* or *CoA,* forming a compound called *acetyl-CoA* **4**. These reactions take place in the matrix of the mitochondrion as shown in the inset figure.

The reactions that link glycolysis and the Krebs cycle produce a molecule of NADH, which is later used to produce ATP molecules. Of greater importance, however, is the acetyl-CoA, the energy-rich first substrate in the Krebs cycle. Acetyl-CoA is produced not only during the breakdown of glucose but also during the metabolic breakdown of proteins, fats, and other lipids. This molecule is the point at which many of the catabolic processes of eukaryotic cells converge (see Figure 6.11).

The sequence of reactions of glycolysis generates a small amount of ATP by reshuffling the bonds of glucose molecules. Glycolysis is a very inefficient process, capturing only about 2% of the available chemical energy of glucose. Pyruvate, the end product of glycolysis, is converted to acetyl-CoA, which is the first substrate in the Krebs cycle.

6.10   The Krebs cycle

The Krebs cycle, which oxidizes acetyl-CoA, consists of eight enzyme-mediated reactions. These reactions are shown in **Figure 6.14.** The cycle has two stages:

1. *Stage A—Preparation reactions.* These two reactions set the scene. In the first reaction, acetyl-CoA joins a four-carbon molecule from the end of the cycle to form the six-carbon molecule citric acid **1**. In the next reaction **2**, chemical groups are rearranged.

2. *Stage B—Energy extraction.* Four of the remaining six reactions are oxidations in which hydrogen ions and electrons are removed from the intermediate compounds in the cycle to form three NADH molecules for each acetyl-CoA, six for each original glucose molecule. Reaction **3** is an oxidation that produces NADH and is followed by the release of carbon dioxide. During reaction **4** oxidation occurs again, producing NADH and causing the release of a second carbon dioxide molecule. During the cycle **5**, a molecule called *GTP* is produced, which is quickly used to produce an ATP; two ATP are generated for each original molecule of glucose. In addition, one molecule of $FADH_2$ is formed **6**, two for each original glucose.

Together, the eight reactions make up a cycle that begins and ends with the same four-carbon molecule **7**, **8**. At every turn of the cycle, acetyl-CoA enters and is oxidized to CO_2 and H_2O, and the hydrogen ions and electrons are donated to electron carriers.

In the process of aerobic respiration, the glucose molecule has been consumed entirely. Its six carbons were first split into three-carbon units during glycolysis. One of the carbons of each three-carbon unit was then lost as CO_2 in the conversion of pyruvate to acetyl-CoA, and the other two were lost during the oxidations of the Krebs cycle. Part of the glucose molecule's energy and its electrons, which are preserved in four ATP molecules and the reduced state of 12 electron carriers, are all that is left.

The breakdown of the two pyruvate molecules produced from one glucose molecule during glycolysis generates 2 ATP molecules, 10 molecules of NADH, and 2 molecules of $FADH_2$. NADH and $FADH_2$ can be used to generate ATP.

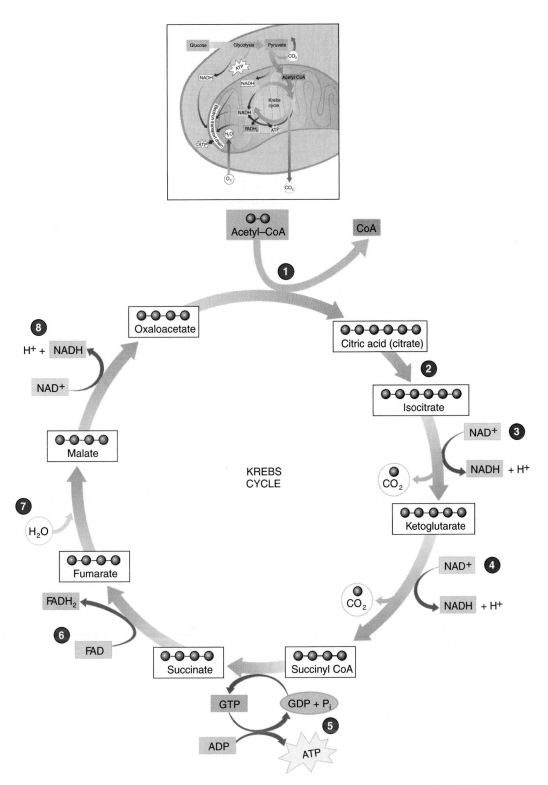

Figure 6.14 **The Krebs cycle.** The Krebs cycle produces three molecules of NADH, one molecule of FADH$_2$, and one molecule of ATP for each acetyl-CoA molecule that enters the cycle.

6.11 The electron transport chain

Embedded within the inner mitochondrial membrane are many series of electron carrier proteins. Each series is known as an electron transport chain (Figure 6.15). Here, electrons are passed from one carrier protein to another (in the same way they were in the electron transport chains of photosynthesis). Interestingly, poisons such as certain pesticides, cyanide, and carbon monoxide work by interfering with chemical reactions that occur in the chain.

The NADH molecules formed during glycolysis and the subsequent oxidation of pyruvate carry their electrons to this membrane. As you may recall, each NADH contains a pair of electrons and a proton gained when NADH was formed from NAD^+. The $FADH_2$ molecules are already attached to this membrane, each containing two protons and two electrons gained when $FADH_2$ was formed from FAD.

As each of the carriers in the electron transport chain accepts electrons in turn, the electrons fall to lower energy levels, releasing energy in the process. The energy released from the electrons being passed along the chain pumps the protons (H^+) from the inner mitochondrial compartment (matrix) through the inner mitochondrial membrane proteins and into the

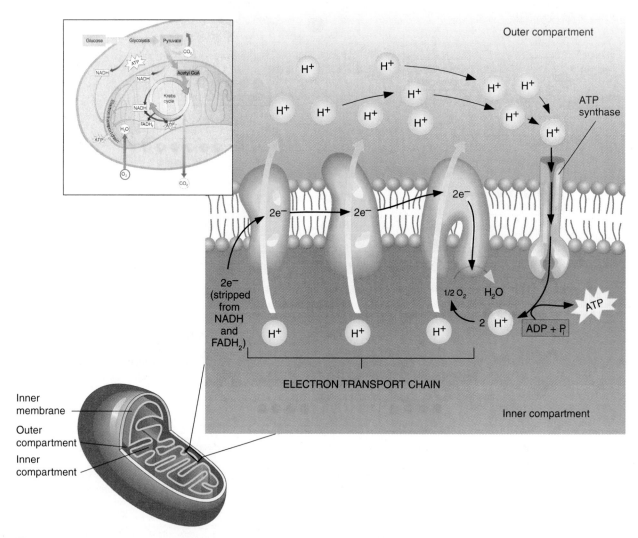

Figure 6.15 **The electron transport chain.** The NADH and $FADH_2$ formed during the Krebs cycle and glycolysis pass their electrons to electron receptors located in the inner membrane of the mitochondrion.

outer mitochondrial compartment (Figure 6.15). As a high concentration of protons builds up in the outer compartment, the protons begin to diffuse back across the membrane through the membrane protein *ATP synthase*. As the protons travel inward through these channels, proteins associated with the channels couple the energy from the movement of the protons to the reaction of bonding P_i to ADP, forming ATP (chemiosmosis). Oxygen is the final acceptor for both the electrons and protons. Water is therefore an end product of cellular respiration, with one molecule of water produced for each NADH or $FADH_2$.

Because three of the electron carriers in the chain act as "proton pumps," most NADH molecules produced during cellular respiration ultimately cause the production of *three* ATP molecules. Each $FADH_2$, which activates two of the pumps, leads to the production of *two* ATP molecules. However, eukaryotes carry out glycolysis in the cytoplasm and the Krebs cycle within the mitochondria. This separation of the two processes within the cell requires transporting the electrons of the NADH created during glycolysis across the mitochondrial membrane. In fact, only the electrons of NADH are shuttled across the membrane, not the entire molecule. The electron transfers that are made as these electrons are shuttled into the mitochondria result in the formation of only two ATP, not three as is usually the case. Thus, each glycolytic NADH produces only two ATP molecules instead of three in the final total. Figure 6.16 describes the total energy yield from the aerobic metabolism of one molecule of glucose. Figure 6.17 shows the relationships between photosynthesis and cellular respiration, adding detail to the generalized overview of these two processes shown in Figure 6.1.

The electrons harvested from glucose and transported to the mitochondrial membrane by NADH drive protons out across the inner membrane. The return of the protons through the membrane protein ATP synthase results in the generation of ATP.

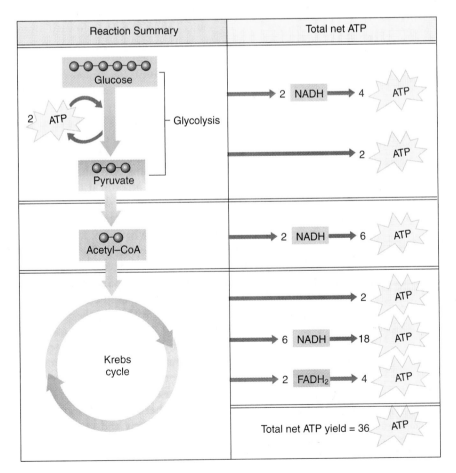

Figure 6.16 **A generalized overview of the energy extracted from the oxidation of glucose.** A net total of 2 ATP are produced during glycolysis. Two NADPH molecules are also produced, which each yield only 2 (rather than 3) molecules of ATP after their electrons are shuttled from the cytoplasm to the inner mitochondrial membrane. Two NADH, yielding 6 ATP, are produced as pyruvate is oxidized to acetyl-CoA. In the Krebs cycle, 2 ATP are produced directly; 6 NADH are produced, yielding 18 ATP; and 2 $FADH_2$ are produced, yielding 4 ATP.

NOTE: We have used whole numbers throughout this chapter to represent ATP yield. Although ATP yield varies from organism to organism, generally:
• ATP yield per NADH arising from glycolysis is 1.5–2.5
• ATP yield per NADH arising from Krebs cycle is closer to 2.5
• ATP yield per $FADH_2$ arising from Krebs cycle is closer to 1.5
Using these values, total ATP yield varies from 30–32.

Certain anaerobic organisms release
energy from carbohydrates during
fermentation.

6.12 Fermentation

Aerobic metabolism cannot take place in the absence of oxygen because oxygen is missing from the electron transport chain. Not only will the reactions of the electron transport chain come to a halt, but the Krebs cycle reactions will not take place because NADH molecules will not be recycled to NAD$^+$ molecules, the needed electron acceptors in the cycle. In such a situation, some cells (such as certain bacteria living in your large intes-

tine) have the enzymes for anaerobic respiration and are able to use a terminal electron acceptor other than oxygen. Some cells, such as certain bacteria and yeasts, have the enzymes to use glycolysis to produce ATPs. However, they must regenerate NAD$^+$ from NADH for use in the glycolytic pathway. During aerobic and anaerobic respiration, cells accomplish this task by means of the reactions in the electron transport chain. During **fermentation**, bacteria and yeasts produce an organic molecule from pyruvate that will accept the hydrogen atom from NADH and thus re-form NAD$^+$. These reactions, including the glycolytic pathway, comprise fermentation. The end products of fermentation depend on the organic molecule that is produced from pyruvate.

Bacteria and yeasts carry out many different sorts of fermentations. In fact, more than a dozen fermentative processes have evolved among bacteria, each process using a different organic molecule as the hydrogen acceptor. Often, the resulting reduced compound is an acid. In some organisms, such as yeasts, an organic molecule called *acetaldehyde* is formed from pyruvate and then reduced, producing an alcohol (**Figure 6.18a**). This particular type of fermentation is of great interest to people because it is the source of the ethyl alcohol in wine and beer. However, ethyl alcohol is an undesirable end product for yeast because it becomes toxic to the yeast when it reaches high levels. That is why natural wine contains only about 12% alcohol—12% is the amount it takes to kill the yeast fermenting the sugars.

Muscle cells undergo a similar fermentative process when they produce ATP without sufficient oxygen (Figure 6.18b). During strenuous exercise, muscle cells break down large amounts of glucose to produce ATP. However, because muscle cells are packed with mitochondria, the pace of glucose breakdown far outstrips the blood's ability to deliver oxygen for aerobic respiration. Therefore, the muscle cells switch from aerobic respiration to fermentation. Muscle cells do not change pyruvate to another organic molecule to be reduced as the bacteria and yeasts do, but instead directly reduce pyruvate to lactic

Visual Summary   Figure 6.17   **The metabolic machine.**

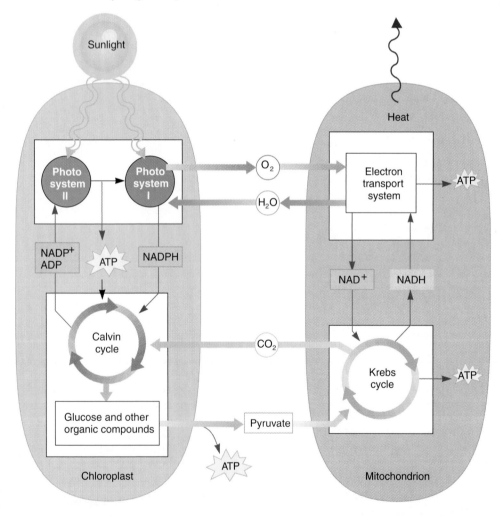

The diagram shows how the processes of photosynthesis and aerobic respiration are connected. The products of photosynthesis (glucose and other organic compounds) are converted to pyruvate and used by the nonphotosynthesizing parts of the plant to obtain energy. Leaf cells (the primary photosynthesizing parts of the plant) use the products of photosynthesis to obtain energy in the dark (when the light-dependent reactions are not taking place).

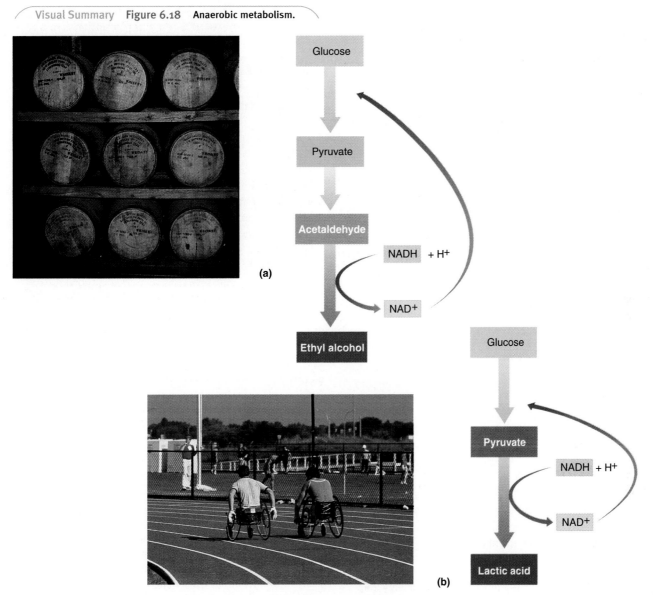

(a)

(b)

(a) Yeasts form acetaldehyde from pyruvate. This molecule accepts the hydrogen atom from NADH, re-forming NAD$^+$ to be used in glycolysis. The result is the ethyl alcohol of wine, beer, and other alcoholic beverages. (b) When animal cells do not have sufficient oxygen, pyruvate accepts the hydrogen atom from NADH to reform NAD$^+$. This converts pyruvate into lactic acid, which builds up in the muscles until oxygen becomes available. This is what happens when you exercise strenuously. The ache in your muscles as you run is the result of a buildup of lactic acid.

acid. This acid builds up in the muscle and tends to produce a sensation of muscle fatigue. Gradually, however, the lactic acid is carried by the blood to the liver, where it is broken down aerobically. This need for oxygen to break down the lactic acid produces what is termed an *oxygen debt* and is the reason a person continues to pant after completing strenuous exercise.

In fermentations, which are anaerobic processes, pyruvate molecules or organic compounds produced from pyruvate are reduced, accepting the electrons of NADH generated in the glycolytic breakdown of glucose. These last reactions of the fermentative process produce the NAD$^+$ needed for the anaerobic breakdown of glucose to continue.

Just Wondering . . . *Real Questions Students Ask*

How come the weeds and crabgrass in my lawn grow during the hot summer while the rest of my lawn dies?

The answer to your question lies in the fact that not all plants undergo photosynthesis in the same way. Kentucky bluegrass and similar grasses use C_3 photosynthesis, the type of photosynthesis described in this chapter. It is termed C_3 because molecules of CO_2 are used to form a three-carbon compound, the first stable product of photosynthesis produced in the Calvin cycle (see Figure 6.9). The weeds in your lawn and crabgrass, which thrive during a hot, dry summer, use a C_4 photosynthetic pathway.

During C_4 photosynthesis, CO_2 is incorporated into a four-carbon molecule *before* it enters the Calvin cycle. The four-carbon molecule is then decarboxylated (carbon is dropped off) to form a three-carbon molecule, which then enters into the normal Calvin cycle pathway. Although C_4 photosynthesis seems like a cumbersome way to generate three-carbon molecules, it is actually an advantage to plants growing in hot, dry, sunny regions. Why? C_4 plants use water more efficiently and use less energy than C_3 plants in high temperatures because of their unique leaf anatomy and their ability to fix CO_2 by mechanisms and under conditions that C_3 plants cannot. In essence, C_4 plants are able to grow and thrive while C_3 plants wither and die in dry heat and intense sunlight. But not only pesky weeds have this adaptation. Corn and sugarcane, two extremely important crops, are also C_4 plants.

Key Concepts

Virtually all living things depend on the sun.

6.1 Organisms capable of making their own food by photosynthesis—plants, algae, and certain protists and bacteria—are called photosynthetic autotrophs.

6.1 Photosynthesis is a process whereby energy from the sun is captured by living organisms and used to produce molecules of food.

6.1 Organisms not capable of making their own food, called heterotrophs, depend on the food produced by autotrophs.

6.1 The processes by which organisms unlock energy in food are called cellular respiration and fermentation.

6.1 The byproducts of aerobic respiration—carbon dioxide and water—are the substrates of photosynthesis.

6.1 Conversely, the byproduct of photosynthesis—oxygen—and its product of carbohydrates are the substrates of aerobic respiration.

6.2 The energy from the sun is electromagnetic radiation and behaves as both waves and discrete packets of energy called photons.

6.3 Special molecules called pigments absorb some wavelengths of light and reflect others.

6.3 Two important types of pigments that play a role in photosynthesis are the chlorophylls and the carotenoids.

6.3 Chlorophyll and carotenoid pigments are organized into photocenters and serve to capture light energy and change it into chemical energy useful in photosynthesis.

Energy from the sun is locked in carbohydrates during photosynthesis.

6.4 During photosynthesis, atmospheric carbon dioxide is used to produce molecules of sugar.

6.5 During the light-dependent reactions, ATP and NADPH molecules are produced and oxygen is given off as a byproduct.

Key Concepts, continued

6.6 During the light-independent reactions, ATP and NADPH are used to power the synthesis of glucose and other carbohydrates.

Organisms release energy from carbohydrates during cellular respiration.

6.7 Cellular respiration is composed of three series of chemical reactions: glycolysis, the Krebs cycle, and the electron transport chain.

6.7 During aerobic respiration, glucose is broken down in the presence of oxygen, capturing energy in molecules of ATP; the byproducts of this process are carbon dioxide and water.

6.8 In the human body, most carbohydrates are changed to glucose and completely metabolized by aerobic respiration; parts of protein and lipid molecules are also metabolized by components of this metabolic pathway.

6.9 During glycolysis, glucose is split in half to pyruvate, and ATP and NADH are formed.

6.9 Before entering the Krebs cycle, pyruvate is converted to acetyl-CoA.

6.10 The Krebs cycle oxidizes acetyl-CoA, producing ATP and NADH.

6.11 The NADH and $FADH_2$ molecules formed during the breakdown of glucose are themselves oxidized via the electron transport chain; these oxidations power the chemiosmotic production of ATP.

6.11 Aerobic respiration yields 36 ATP molecules from glycolysis, the Krebs cycle, and the electron transport chain.

Certain anaerobic organisms release energy from carbohydrates during fermentation.

6.12 Organisms that metabolize nutrient molecules anaerobically by the process of fermentation use pyruvate or a molecule derived from pyruvate to accept the electrons produced during the glycolytic breakdown of glucose.

6.12 Fermentation produces end products that are frequently acids or alcohols.

Key Terms

autotrophs (AW-toe-trofs) 104
Calvin cycle 113
carbon fixation 113
carotenoids (kuh-ROT-uh-noids) 107
chemiosmosis
   (KEM-ee-oz-MOH-sis) 112
cellular respiration 105
chlorophyll *a* 110
chlorophylls (KLOR-uh-fils) 107

chloroplasts (KLOR-oh-plasts) 109
electron transport chain 114
fermentation 122
glycolysis (glye-KOL-uh-sis) 114
grana (GRA-nuh) 109
heterotrophs (HET-ur-oh-trofs) 104
Krebs cycle 114
light-dependent reactions 109

light-independent reactions 109
photosynthesis
   (foe-toe-SIN-thuh-sis) 108
photosystem I 110
photosystem II 110
pigments 107
stroma (STROH-muh) 109
thylakoids (THIGH-luh-koidz) 109

KNOWLEDGE AND COMPREHENSION QUESTIONS

1. Compare autotrophs and heterotrophs. Which are you? (*Knowledge*)

2. You hold a prism in front of a window and the incoming sunlight forms a rainbow on the floor. Explain what happened. (*Comprehension*)

3. What are the products of each of the two series of reactions involved in photosynthesis? What is meant by the terms *light-dependent reactions* and *light-independent reactions*? (*Knowledge*)

4. Define *photocenter* and *chlorophyll a*, and explain their significance. (*Comprehension*)

5. Some scientists think that a large nuclear explosion could blanket the Earth with a thick haze of atmospheric dust lasting for years. Explain how this would affect life on Earth. (*Comprehension*)

6. Distinguish between cellular respiration and fermentation. What purpose do they serve? (*Comprehension*)

7. You stop at a fast-food restaurant to eat a burger and fries. Explain how your body converts these foods into usable energy. (*Comprehension*)

8. Summarize the main events of glycolysis. (*Comprehension*)

9. Create a table that summarizes the output of aerobic metabolism for glycolysis, the oxidation of pyruvate, and the Krebs cycle. (*Knowledge*)

10. We have discussed before how the chemistry of life is integrally involved with the nature of water. What is the link between water and the electron transport chain? (*Comprehension*)

KEY
Knowledge: Recalling information.
Comprehension: Showing understanding of recalled information.

CRITICAL THINKING REVIEW

1. Imagine you place a potted plant in your classroom window after weighing it carefully. Pot and all, it weighs just 3 pounds. Watering it properly, you let the plant grow for a year and then reweigh it. Pot and all, it now weighs 4-1/2 pounds. Where precisely did the 1-1/2 pounds of extra plant mass come from? (*Synthesis*)

2. The roots of plants are not exposed to sunlight—they are under the ground. How do they manufacture ATP? (*Application*)

3. Study the absorption spectrum for chlorophyll in Figure 6.4. Are all wavelengths equally effective for photosynthesis? Support your answer using evidence gleaned from Figure 6.4. (*Analysis*)

4. Is the following statement true or false? The process of glucose breakdown during cellular respiration has both endergonic and exergonic elements. Support your answer with evidence. (*Analysis*)

5. Chapter 4 briefly described the endosymbiont theory, which suggests that mitochondria originated as bacteria that were engulfed by preeukaryotic cells. What have you learned in this chapter that would lead you to more firmly hold to that theory or to discard it as false? (*Evaluation*)

KEY
Application: Using information in a new situation.
Analysis: Breaking down information into component parts.
Synthesis: Putting together information from different sources.
Evaluation: Making informed decisions.

Fill in the diagram below with the appropriate names of the stages, products, and byproducts of aerobic respiration. List the events that occur outside the mitochondrion and then those that occur inside this organelle.

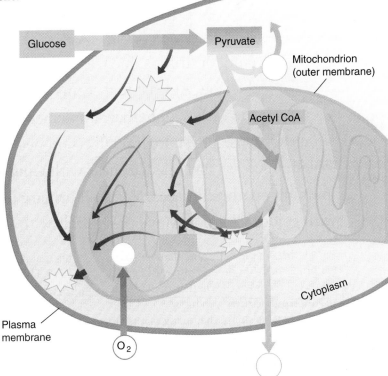

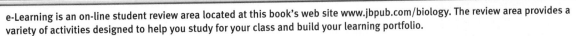

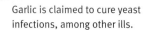
Herbal Supplements and the FDA

Pop a pill or drink some tea . . . it'll mend whatever ails you. At least that seems to be the case with the variety of herbal supplements available today, which advertise physiological effects on nearly every body system. Increase mental sharpness with *Ginkgo biloba*. Boost immunity with echinacea. Ward off urinary tract infections with goldenseal root. Move the bowels with senna.

Garlic is claimed to cure yeast infections, among other ills.

Herbal supplements are plants or parts of plants that are used for their medicinal properties. But do herbals work, and are they safe? These two questions are central to the issue of whether the U.S. Food and Drug Administration (FDA) should regulate these products.

There is no doubt that many plants (including herbs) have medicinal value; one-quarter of the prescription drugs dispensed each year in North America were originally derived from flowering plants and ferns. (See the Just Wondering box in Chapter 14 for a description of the results of scientific studies on the effects of *Ginkgo biloba*.) However, the 1994 Dietary Supplement Health and Education Act (DSHEA) defined herbal products as dietary supplements rather than as drugs having medicinal properties. (Drugs are nonfood chemicals that alter the way a person thinks, feels, functions, or behaves.) This law exempts herbals from FDA regulation. As a result, manufacturers of herbal products do not have to provide evidence to the FDA that their products are pure, potent and effective, as do manufacturers of prescription drugs. FDA regulations do not allow manufacturers to suggest that herbals can treat specific symptoms or illnesses. Therefore, herbal supplement advertising often states a connection between the product and a bodily structure or function, using phrases such as "helps support" or "helps maintain."

What do consumers think about herbal supplements? Seventeen percent of all Americans use them. However, the Center for Science in the Public Interest (CSPI), a nonprofit education and consumer advocacy organization that focuses on improving the safety and nutritional quality of the food supply, warns that not enough research has been conducted on the safety of herbal products. They contend that many consumers believe that since herbal supplements are "natural", they must be safe. CSPI says that this is a false assumption; many herbs are sources of potent drugs that must be tested to determine safety. Some herbals have caused life-threatening conditions, including cardiac arrest and hepatitis; taking others has resulted in death. However, the Council for Responsible Nutrition, which represents manufacturers of dietary supplements, suggests that further regulation of herbal supplements by the FDA is unnecessary since the supplement industry voluntarily stops selling herbal products about which safety questions are raised.

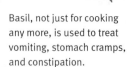

Basil, not just for cooking any more, is used to treat vomiting, stomach cramps, and constipation.

Sales of herbal supplements

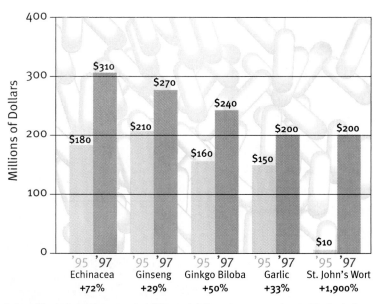

Sales of herbal supplements, in millions of dollars, for 1995 and 1997. The herbal supplement industry has experienced tremendous growth in recent years. Due to increased media attention, sales of St. John's Wort skyrocketed 1,900% between 1995 and 1997. Source: *Nutrition Business Journal*

Many people who want the FDA to regulate herbal products believe that these products should be given special consideration. They suggest that the FDA should regulate herbal remedies as drugs, but should adopt standards that are less stringent than those required of new drugs. For example, in showing that herbal remedies are reasonably safe and effective, the FDA would give special consideration to both their long history of use and the scientific literature that has accumulated about these products in Europe. Some suggest that the FDA create a new category of "traditional medicines," which would allow herbal products to make claims beyond the ones allowed by the DSHEA, yet have these claims regulated by the FDA.

Some herbal supplement manufacturers would like to see the United States develop standards for these products in lieu of regulation by

Echinacea flower, can it boost immunity?

the FDA. They point out that herbal products vary depending on the age and growth conditions of the plants from which they were extracted, and establishing standards would ensure that the plant extracts meet specific quality levels so that the herbs work properly. Many physicians and other health care workers agree that, at a minimum, standards are needed, pointing out that it is often difficult to determine with certainty the composition of herbal remedies, particularly products that are imported or sold by unsupervised small manufacturers.

If the effectiveness of herbal medicines will be researched as thoroughly as the effectiveness of conventional drugs, who will pay the costs? Because patents are not available for most herbal preparations, there is no incentive for the pharmaceutical industry to investigate them. Normally, pharmaceutical companies recover such costs when they patent a drug and then are able to be its exclusive supplier. Herbal medicines are difficult to patent because they are discovered, not invented (unless genetically engineered). At this time, the Office of Alternative Medicine within the National Institutes of Health is funding scientific studies to explore the safety and effectiveness of certain alternative medical practices, including the use of some herbal remedies.

What is the issue here, and what is your response to this issue?

To make an informed and thoughtful decision, it may help you to follow these steps.

1. Identify the issue described here using either a statement or a question. (An issue is a point on which various persons hold differing views.)
2. Write your immediate reaction to the issues raised. How do you think this issue could be or should be resolved?
3. Collect new information about the issue. State the sources of that information. Explain why you believe the information to be accurate. Also determine if the information expresses a particular point of view or is biased in any other way. (Guidelines for doing efficient web searches for reliable information are available on the BioIssues website — www.jbpub.com/biology)
4. Determine which individuals, groups, or organizations have a stake in the issue. What does each stand to gain or lose depending on the outcome of the issue?
5. List possible outcomes (resolutions) of the issue. List the pros and cons of each outcome.
6. Which outcome do you think would be best and why? Note whether your opinion differs from or is the same as what you wrote in Step 2. Give reasons for your views and for any changes in them based on the additional information you have collected and the analysis you have done.

Patterns and Levels of Organization

Cells, tissues, organs, and organ systems—the body is like a carefully honed "machine." Each level of organization in the cyclist's body plays as important a role during this race as his high-tech bike. In the legs, each muscle contracts and relaxes in synergy with the others to pedal the bicycle. In the chest, the heart contracts rhythmically, pumping blood. The vessels of the cardiovascular system channel the rushing blood from the heart to the muscles as well as to the lungs where the waste gas carbon dioxide is exchanged with oxygen. Making up the lungs are several hundred million microscopic air sacs, each with walls only one cell thick, which serve as the site for this gas transfer. In the brain, nerve cells communicate with one another as the cyclist plans his winning strategy, while his breathing deepens and his speed increases. This sophisticated integration of the various levels of organization not only helps the cyclist win the race but also helps all multicellular animals survive.

Organization

Multicellular organisms exhibit a hierarchy of organization.

Cells are organized into four basic tissue types.

Tissues are organized into organs, and organs into organ systems.

Multicellular organisms exhibit a hierarchy of organization.

7.1 Levels of organization: From cells to organ systems

The bodies of all multicellular animals are made up of many different types of cells. The human body, for example, contains more than 100 different kinds of cells! These cells are not distributed randomly, but are organized to form the structures and perform the functions of the human body, just as workers in a factory may be organized to manufacture a product.

The cells of humans and other multicellular animals (Figure 7.1) are organized into **tissues**. Tissues are groups of similar cells that work together to perform a function. Traditionally, tissues are divided into four basic types based on their function: **epithelial**, **connective**, **muscle**, and **nervous**. For example, the cells making up the walls of the air sacs of your lungs are a type of epithelial tissue. Epithelial tissue covers body surfaces and lines its cavities. All four tissue types are described later in this chapter.

Two or more tissues grouped together to form a structural and functional unit are called **organs**. Your heart is an organ. It contains cardiac muscle tissue wrapped in connective tissue and "wired" with nerves. All of these tissues work together to pump blood through your body. Other examples of organs are the stomach, skin, liver, and eyes.

An **organ system** is a group of organs that function together to carry out the principal activities of the body. For example, the digestive system is composed of individual organs concerned with the breaking up of food (teeth), the passage of food to the stomach (esophagus), the storage and partial digestion of food (stomach), the digestion and absorption of food and the absorption of water (intestine), and the expulsion of solid waste (rectum). The human body contains 11 principal organ systems (Table 7.1, p. 149), which are discussed in-depth in later chapters.

The simplest of multicellular animals, such as sponges (see Chapter 30), have no organs and, therefore, no organ systems. Sponges have a variety of kinds of cells making up their bodies, however (Figure 7.2). Collar cells, for example, line the inner surface of sponges, generating water currents within the sponges. The current produced within the sponge takes water in through pores and then pushes it out of the large opening; the sponge captures small food particles as the water flows through its body. Many scientists refer to this layer of collar cells as well as the outer layer of flattened cells as epithelial tissue. Others

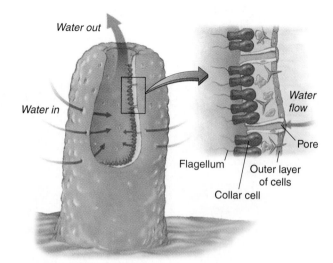

Figure 7.2 **The sponge (phylum Porifera).** Diagram of a sponge showing epitheliumlike layers.

prefer to describe these layers only as "epitheliumlike" because they (particularly the collar cells) are different from the epithelial tissues of most other animals. Still others contend that sponges have no tissue at all because each specialized cell functions on its own.

With the exception of the "controversial" sponge, most simple multicellular animals have epithelial tissue and connective tissue. Some simple multicellular animals have tissues in addition to these as well. For example, the jellyfish *Aurelia* has muscle cells that lie below its epidermis. Longitudinal muscle fibers are found in the tentacles. It even has a stomach, some nervous tissue, and reproductive organs. The flatworms are more complex still. The cross-eyed-looking flatworm called *Dugesia* has a brain and nervous system, excretory system, reproductive system, and digestive system. It has muscles and sense organs (the pigment cups that look like eyes, for example). More complex animals have more and diverse tissue types. Each animal phylum (see Chapters 30 and 31) has the same basic tissue types that build their organs and organ systems. Although organs and organ systems differ widely among animal phyla, each phylum has recognizable patterns of organization that unite its members.

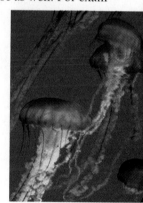

Organism

↑

Organ Systems

↑

Organs

↑

Tissues

↑

Cells

Figure 7.1 **Levels of organization.** The different levels show a hierarchical arrangement from simple to complex.

20X

7.2 Patterns of organization: Symmetry

The word symmetry refers to the distribution of the parts of an object or living thing. All animals (except the sponges) exhibit either radial symmetry or bilateral symmetry in their body plans (Figure 7.3). Sponges are asymmetrical, or without symmetry.

Radial symmetry means that body parts emerge, or radiate, from a central axis (Figure 7.3a). The jellyfish, corals, sea anemones, and their relatives are radially symmetrical. These organisms have an oral side (side containing the mouth) and an aboral side (literally, side without a mouth), but they have neither a dorsal (back) surface nor a ventral (belly) surface as does the fish in Figure 7.3b.

Animals exhibiting radial symmetry are aquatic organisms. Many are sessile; that is, they are anchored in one place. Their radially symmetrical body allows them to interact with the watery environment in all directions. Other organisms such as sea stars are also radially symmetrical, but are not grouped with the jellyfish and corals because their embryonic development and internal anatomy suggest that they are related to organisms having bilateral symmetry. In addition, sea star larvae are bilaterally symmetrical.

Bilateral symmetry means that the right side of an object or an organism is a mirror image of the left side (Figure 7.3b). Animals with bilateral symmetry have a dorsal and a ventral surface and a cephalic (head) end and a caudal (tail) end. All phyla of animals other than the sponges, jellyfish, and echinoderms exhibit bilateral symmetry.

All animals, with the exception of sponges, exhibit either bilateral (mirror-image) symmetry or radial symmetry.

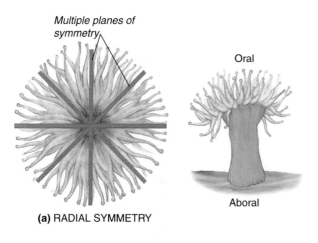

(a) RADIAL SYMMETRY

Multiple planes of symmetry

Oral

Aboral

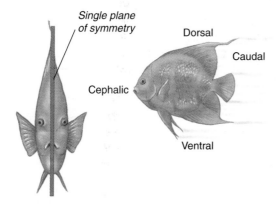

Single plane of symmetry

Cephalic

Dorsal

Caudal

Ventral

(b) BILATERAL SYMMETRY

Figure 7.3 **Patterns in symmetry.** (a) In radial symmetry, body parts are arranged regularly around a central axis. Any one of the multiple planes divides the animal into mirror-image halves. (b) In bilateral symmetry, a single plane divides the animal into mirror-image halves.

Focus on Plants

Patterns and Levels of Organization in Plants
Plants fall into one of two groups with respect to patterns and levels of organization: vascular plants or nonvascular plants. More than 95% of all living plant species are vascular plants and are familiar to you: the flowering plants, trees, bushes, grasses, and shrubs. Vascular plants are those having specialized tissues to transport fluids and have the plant organs roots, stems, and leaves (as shown in the top photo). Only certain genera of nonvascular plants have specialized vascularlike tissue. None have true roots, stems, or leaves. As a consequence, nonvascular plants usually grow close to the ground, such as the mosses shown in the bottom photo. You can learn more about patterns and levels of organization in plants in Chapter 29.

7.3 Patterns of organization: Body cavity structure

Single-celled organisms (such as amebas), small multicellular organisms (such as certain algae and some flatworms), and larger multicellular organisms having only a few layers of cells between themselves and the environment (such as sponges) are able to move substances easily to all the cells of their bodies. However, animals with large, complex bodies (such as earthworms, insects, frogs, fish, and humans) must have body plans that provide efficient means of moving substances within and between tissues—or they would not survive.

The type of body organization of most bilaterally symmetrical organisms that accomplishes this task is the *coelomate* body plan. A **coelom** (SEE-lum) is a fluid-filled body cavity lined with connective tissue. It may function as a simple circulatory system in some organisms, transporting materials from one part of the body to another. Organs are located within this body cavity, somewhat protected from injury and compression just as a fetus is protected as it develops within a fluid-filled sac within the uterus. Humans are coelomates. For example, you have an abdominal cavity lined with a thin, nearly transparent sheet of connective tissue (peritoneum). Suspended by thin sheets of connective tissue arising from the peritoneum, the stomach and intestines hang in this coelom.

Roundworms are *pseudocoelomates*. These organisms (along with tiny aquatic organisms called rotifers) have a so-called false coelom: a fluid-filled cavity that houses the organs, but it is not lined completely by connective tissue. In addition, the organs are not suspended within the cavity by thin sheets of connective tissue.

The *acoelomates* (the flatworms) have no body cavity. Since their organs are embedded within the other tissues of the body, they are compressed as the animal moves. These compressions aid the movement of substances throughout the body.

The three patterns of body cavity structure are shown in **Figure 7.4**.

In animals, body plans are of three types: coelomate (having a fluid-filled body cavity lined with connective tissue), pseudocoelomate (having a "false coelom"—a fluid-filled body cavity not lined completely with connective tissue), and acoelomate (having no body cavity).

7.4 Organization of the human body

The human body has the same general body architecture that all vertebrates (animals with a backbone) have (**Figure 7.5**), including a coelom. (However, not all organisms with a coelom are vertebrates.) This book uses the human as the vertebrate example of how living things work, so that you can gain insight into the animal world while learning more about yourself.

The human body plan includes a long tube that travels from one end of the body to the other, from mouth to anus, which is discussed in Chapter 8. This tube is suspended within the coelom. In humans the coelom is divided into two main parts: (1) the thoracic cavity, which contains the heart and lungs, and (2) the abdominal cavity, which contains organs such as the stomach, intestines, and liver. The diaphragm forms the floor of the

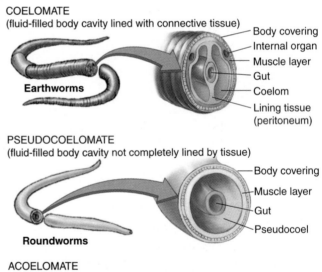

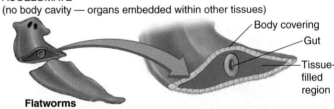

Figure 7.4 **Three plans for construction of animal bodies.** Animals can be coelomate (having a coelom), pseudocelomate (having a false coelom), and acoelomate (having no coelom or body cavity). Along with the earthworm, additional examples of coelomates include other bilaterally symmetrical organisms, such as insects, frogs, fish, and humans.

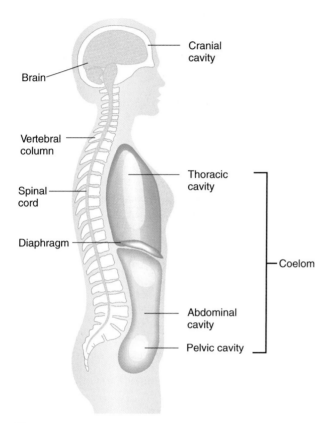

Figure 7.5 **Architecture of the human body.** Humans, like all vertebrates, have a spinal cord and brain enclosed in the vertebral column and skull. In mammals, a muscular diaphragm divides the coelom into the thoracic cavity and the abdominal cavity.

thoracic cavity and the roof of the abdominal cavity. The lower portion of the abdominal cavity is often referred to as the pelvic cavity. The body is supported by an internal scaffold, or skeleton, made up of bones that grow as the body grows. A bony skull surrounds the brain, which is located in a cavity separate from the coelom, called the cranial cavity. In addition, a column of bones called the vertebral column forms the backbone, or spine, of the human body. These bones surround a nerve cord, the spinal cord that relays messages between the brain and other parts of the body.

The human body plan (similar to all vertebrates) includes a long tube that travels from mouth to anus, suspended within the coelom. The coelom is divided into the thoracic cavity and the abdominal cavity.

7.5 Epithelial tissue

Epithelial cells guard and protect your body; these cells are really the doors to the inner you. Put simply, epithelial cells cover and line the surfaces of the body, both internal and external. Remember, you can think of your body as a tube, with your skin forming the outside of the tube and your digestive system forming the inside of the tube. Therefore, the surfaces of your inner tube, as well as your skin, are made up of epithelial cells. Because epithelial cells cover the surfaces of the body, they determine which substances enter and which do not.

All epithelial cells, collectively called the epithelium, are broadly similar in form and function. The epithelial layers of the body function in six different ways:

1. *Protection.* They protect the tissues beneath them from drying out, sustaining chemical and mechanical injury, and being invaded by microorganisms.
2. *Absorption.* Because epithelium encases all of the body's surfaces, every substance that enters or leaves the body must cross this protective barrier.
3. *Sensation.* Many sensory nerves end close to epithelial cell layers. The epithelium therefore provides a surface through which sensations are perceived.
4. *Secretion.* Certain epithelial cells (glands) are specialized to produce and discharge substances. Glands may be single cells, such as the mucus-secreting cells lining the intestine. Most glands, however, are multicellular structures, such as the thyroid and pituitary glands. Even your sweat glands are made up of many cells. Although many glands lie deep within your body and do not now cover or line a surface, they originally formed as infoldings of epithelial cell layers, during embryological growth and development.
5. *Excretion.* Specialized epithelial cells in the kidney excrete waste products during the formation of urine. Also, the epithelium forming the air sacs of the lungs excretes the waste gas carbon dioxide.
6. *Surface transport.* Some epithelial cells have hairlike projections called cilia. These cilia beat in unison, causing a wavelike movement in the thin film of mucus that bathes the cells' surfaces, sweeping particles along in the process. The ciliated cells lining your respiratory passageways keep foreign particles from entering the lungs. Cigarette smoking damages these cilia, resulting in dangerous accumulations of mucus in the throat and lungs.

Structurally, epithelial tissues share some common characteristics. Epithelial tissues are usually only one or a few cells thick (with the exception of the skin), are packed together and stacked very tightly, and have very few blood vessels running through them. The circulation of nutrients, gases, and wastes in epithelial tissue occurs by diffusion from the capillaries of neighboring tissue. In addition, although the chemical reactions of many types of epithelial cells take place at a very slow rate, many have amazing regenerative powers. Your skin cells, for example, are continually being replaced throughout your lifetime. Your liver, a gland having epithelial tissue on its absorptive and secretory surfaces, can readily regenerate substantial portions of tissue if parts are surgically removed or are damaged by certain diseases.

There are three main shapes of epithelial cells: *squamous, cuboidal,* and *columnar* (Figure 7.6). *Squamous cells* (SKWAY muss) ❶ are thin and flat. They are found in places such as the air sacs of the lungs, the lining of blood vessels, and the skin. *Cuboidal cells* ❷ have complex shapes, but look like cubes when the tissue is cut at right angles to the surface. These cells are found lining tubules in the kidney and the ducts of glands. *Columnar cells* ❸ look like tiny columns (as their name suggests) when they are viewed from the side. Much of the digestive tract is lined with columnar epithelium.

The various types of epithelial cells may also be arranged in various ways. Epithelial tissue that is only one cell thick is referred to as *simple epithelium*. It is usually found in areas where substances diffuse through the tissue and where substances are secreted, excreted, or absorbed. The cells lining your blood vessels are simple squamous epithelium, for example, as are the cells forming the air sacs of your lungs, whereas much of your digestive tract is lined with simple columnar epithelium.

Stratified epithelium is made up of two or more layers. These layers are usually protective. The surface of your skin is *stratified squamous epithelium* ❹. *Pseudo-stratified epithelium* ❺ only looks as though it is layered. In reality, this tissue is made up of only one layer of cells—some tall, some short. Some of your airways are lined with pseudostratified epithelium. *Transitional epithelium* ❻ is tissue that can stretch. The cells of transitional epithelium in the bladder, for example, are cubelike when the bladder has little or no urine. However, when the bladder fills with urine, the cells become thin and flat as the bladder expands like a water balloon to hold the urine. Transitional epithelium is also found in the ureters of the urinary tract, which bring urine from the kidneys to the bladder.

> Epithelial cells cover and line the internal and external surfaces of the body and compose the glands. These cells are of three shapes and are arranged in ways that best suit their functions.

Visual Summary **Figure 7.6** **Epithelial tissues.**

Simple squamous epithelium ❶
Lining of blood vessels, air sacs of lungs, kidney tubules, and lining of body cavities

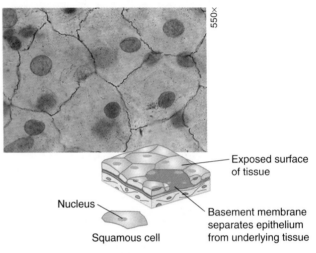

550×

Exposed surface of tissue

Nucleus

Basement membrane separates epithelium from underlying tissue

Squamous cell

DESCRIPTION Simplest of all epithelial tissues, consists of a single layer of thin, flat cells
FUNCTION Diffusion, filtration, and passage of materials where little protection is needed

Cuboidal epithelium ❷
Kidney tubules, glands and their ducts, terminal bronchioles of lungs, and surface of ovaries and retina

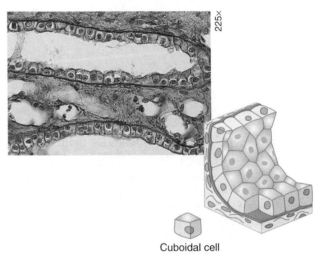

225×

Cuboidal cell

DESCRIPTION Single layer of cube-shaped cells; some have microscopic microvilli, some have cilia that protrude from their surfaces
FUNCTION Secretion, absorption, and movement of substances

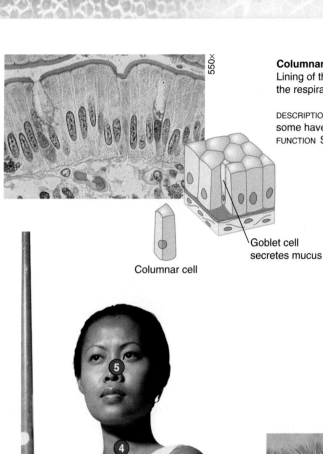

Columnar epithelium ③
Lining of the digestive tract and upper part of the respiratory tracts, auditory and uterine tubes

DESCRIPTION Single layer of tall, narrow cells; some have microvilli or cilia
FUNCTION Secretion of mucus

Goblet cell secretes mucus

Columnar cell

Stratified squamous epithelium ④
Skin, mouth, and throat lining; vaginal lining; anal lining; and cornea

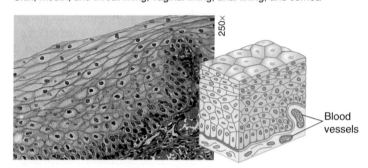

Blood vessels

DESCRIPTION Consists of several layers of cells: the lower layers are columnar and active, the upper layers are flattened at the surface
FUNCTION Protection, hard outer layer being continuously removed by friction and replaced from below

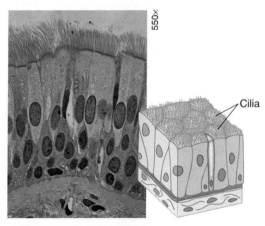

Pseudostratified epithelium ⑤
Nasal cavity and sinuses, ducts of some glands, and some ducts of the male reproductive system

DESCRIPTION Single layer of cells similar to columnar epithelium, except of varying heights: some reach the surface and others do not; some have cilia, and some may have microvilli
FUNCTION Protection and secretion of mucus

Cilia

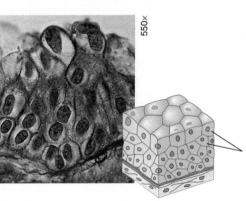

Transitional epithelium ⑥
Lining of urinary bladder and ureters

DESCRIPTION Consists of several layers of columnar cells beneath layers of surface cells; cells are flattened when tissue is stretched
FUNCTION Accommodation of fluid fluctuations in an organ or tube by stretching easily

"Unstretched" epithelium

Cells flatten when stretched

7.6 Connective tissue

The cells of connective tissue provide the body with structural building blocks and potent defenses. In addition, connective tissue joins the other tissues of the body. Connective tissue performs an assortment of important jobs for the body. The varied composition of the different types of connective tissue reflects this diversity. However, connective tissue is generally made up of cells that are usually spaced well apart from one another and are embedded in a nonliving substance called a *matrix*. In fact, connective tissue is made up of a great deal more matrix than cells. This matrix varies in consistency among the different types of connective tissue—from a fluid to a gel to crystals.

The different types of connective tissues and cells are categorized in many different ways. One way to group them is by function: defensive, structural, and isolating connective tissue. *Defensive connective tissue cells,* which protect the body from attack, float in a matrix of blood plasma or reside in certain tissues. They roam the circulatory system, hunting invading bacteria and foreign substances. An example of this type of cell is the lymphocyte, a special type of white blood cell (Figure 7.7). *Structural connective tissue cells,* such as bone and cartilage cells, stay in one place, secreting protein fibers into the empty spaces between them. These proteins provide structural connective tissue with a fibrous matrix, giving it the strength it needs to support the body and provide connections among tissues. *Isolating connective tissue cells* act as storehouses, accumulating specific substances such as fat (in adipose cells) and hemoglobin (in red blood cells).

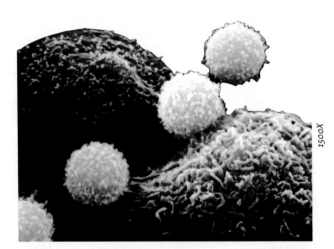

1500X

Figure 7.7 Lymphocytes and macrophages. The lymphocytes are small and spherical; the macrophages are larger and more irregular in form. Both types of cells play key roles in the body's defense against disease.

Cells That Defend

The three principal defensive cell types are lymphocytes (see Figure 7.9), macrophages (MACK-row-fayge-ez), and mast cells. All of these cells are dispersed throughout the body either in the blood or among other tissues.

Lymphocytes (7, in Figure 7.9) are a type of white blood cell that circulates in the blood or resides in the organs, vessels, and nodes of the lymphatic system. Lymphocytes and the lymphatic system itself play complex, key roles in the body's defense against infection. Your body has an amazing trillion or so lymphocytes ready to attack foreign cells or viruses that enter the body, or to produce antibodies that can act against specific substances. (The way lymphocytes and antibodies function in the body's defense against disease is described in Chapter 11.)

Macrophages (8, in Figure 7.9) are abundant in the bloodstream and also in the fibrous mesh of many tissues, such as the lungs, spleen, and lymph nodes. They develop, or differentiate, from white blood cells called *monocytes*. Usually macrophages move about freely, but sometimes they stay in one place, attached to fibers. These cells may be thought of as the janitors of the body, cleaning up cellular debris and invading bacteria by a process known as *phagocytosis* (FAH-guh-sigh-TOE-sis)— an engulfing and digesting of particles.

Mast cells (9, in Figure 7.9) produce substances that are involved in the body's inflammatory response to physical injury or trauma. One important substance produced by mast cells is histamine (the granules colored red in the photo of the mast cell shown at right). This chemical causes blood vessels to dilate, or widen. As more blood then flows through the vessels, it brings added oxygen and nutrients and dilutes any toxins, or poisons. The increased blood flow also aids the movement of defensive leukocytes (white blood cells) coming to the area. Mast cells, although important in the inflammatory response, also play a role in allergic reactions (see Chapter 11).

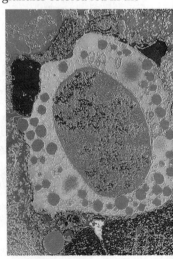

Cells and Tissues That Shape and Bind

The three principal types of cells found in structural connective tissue are fibroblasts, cartilage cells (chondrocytes), and bone cells (osteocytes). These cells produce substances that cause the tissues of which they are a part to have distinctive characteristics.

Fibroblasts

Of all the connective tissue cells, fibroblasts are the most numerous. They are flat, irregular, branching cells; one fibroblast is pictured below. Fibroblasts secrete fibers into the matrix between them. These fibers are of three basic types: collagen, reticular, and elastic.

Both the collagen fibers and the reticular fibers are made up of the protein collagen, the most abundant protein in the human body (Figure 7.8). Collagen fibers are strong and wavy. These properties allow the connective tissues that they compose to be somewhat flexible and stretchy, without the fibers themselves stretching. *Dense fibrous connective tissue* (❿, Figure 7.9) is primarily made up of collagen fibers. In Figure 7.9, you can see bundles of collagen fibers with widely spaced rows of fibroblasts. This type of connective tissue is very strong and is found as tendons connecting muscles to bones; it composes the lower layer of the skin and makes strong attachments between organs.

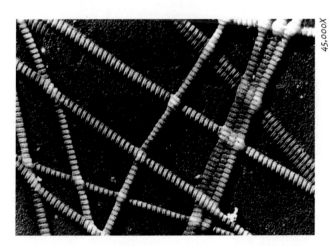

700X

Figure 7.8 Collagen fibers. Each fiber is composed of many collagen strands and is very strong.

45,000X

Reticulin is a fine branching fiber that forms the framework of many glands such as the spleen and the lymph nodes. It also makes up the junctions between many tissues. The tissue formed by fibroblasts and reticulin alone is called *reticular connective tissue* (⓫, in Figure 7.9). Elastic fibers, as the name suggests, act much like rubber bands. They are not made of collagen, but of a protein called elastin, a "stretchy" protein. *Elastic connective tissue* (⓬, in Figure 7.9)

is made up of branching elastic fibers with fibroblasts interspersed throughout. This type of tissue is found in structures that must expand and then return to their original shape, such as the lungs and large arteries.

Loose connective tissue (⓭, in Figure 7.9) contains various connective tissue cells and fibers within a semifluid matrix. Fibroblasts and macrophages are the most common cells in loose connective tissue. This tissue also contains loosely packed elastic and collagen fibers; it is therefore a somewhat strong, but very flexible tissue. Loose connective tissue is distributed widely throughout the body and is found wrapping nerves, blood vessels, and tissues; filling spaces between body parts; and attaching the skin to the layers beneath it. If you have ever skinned chicken before cooking, for example, you have seen the loose connective tissue that binds the skin to the muscle beneath.

Cartilage Cells

Chondrocytes are the cells that produce cartilage, a specialized connective tissue that is hard and strong. Cartilage is found in many places such as the ends of long bones, the airways of the respiratory system, and the spaces between the vertebrae. Cartilage is made up of cells that secrete a matrix consisting of a semisolid gel and fibers. The fibers are laid down along the lines of stress in long parallel arrays (groups or arrangements). The result of this process is a firm and flexible tissue that does not stretch.

As the cartilage cells secrete the matrix, they wall themselves off from it and eventually come to lie in tiny chambers called lacunae. In Figure 7.9, you can see that although there are three different types of cartilage—hyaline cartilage (HI-uh-lynn), elastic cartilage, and fibrocartilage—they all have cartilage cells within lacunae. Their differences lie in the matrix.

In its matrix, *hyaline cartilage* (⓮, in Figure 7.9) has very fine collagen fibers that are almost impossible to see under the light microscope. During your development before birth, most of your skeleton was composed of hyaline cartilage. As an adult, you have hyaline cartilage on the ends of your long bones, cushioning the places where these bones meet. Hyaline cartilage also forms C-shaped rings around the windpipe, keeping this airway propped open, and makes up parts of your ribs and nose. *Elastic cartilage* (⓯, in Figure 7.9), as the name suggests, has elastic fibers embedded in its matrix. It is found where support with flexibility is needed, such as in the external ear. *Fibrocartilage* (⓰, in Figure 7.9) has collagen fibers embedded in its matrix. It is therefore a very tough substance and is used in places of the body where shock absorbers are needed. It is found, for example, as discs between the vertebrae and in the knee joint.

www.jbpub.com/biology

Cells are organized into four basic tissue types.

Visual Summary Figure 7.9 Connective tissues.

STRUCTURAL CONNECTIVE TISSUE

Dense fibrous connective tissue ⑩
Tendons, ligaments, and attachments between organs and dermis of the skin

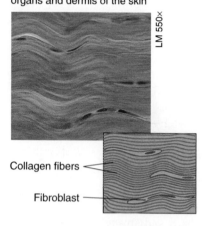

LM 550×

Collagen fibers

Fibroblast

Reticular connective tissue ⑪
Liver, lymph nodes, spleen, and bone marrow

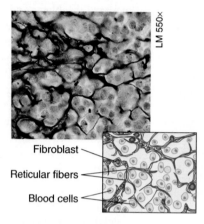

LM 550×

Fibroblast

Reticular fibers

Blood cells

Elastic connective tissue ⑫
Lung tissue, arteries

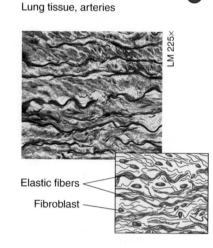

LM 225×

Elastic fibers

Fibroblast

Loose connective tissue ⑬
Packing between glands, muscles, and nerves; attachments between skin and underlying tissue

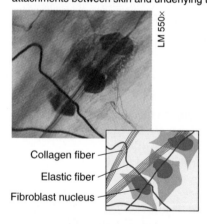

LM 550×

Collagen fiber

Elastic fiber

Fibroblast nucleus

Hyaline cartilage ⑭
Ends of long bones, joints, respiratory tubes, costal cartilage of ribs, nasal cartilage, and embryonic skeleton

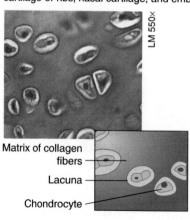

LM 550×

Matrix of collagen fibers

Lacuna

Chondrocyte

Elastic cartilage ⑮
Auditory tube, external ear, epiglottis

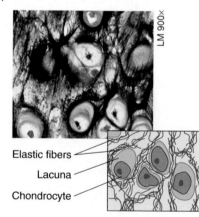

LM 900×

Elastic fibers

Lacuna

Chondrocyte

Fibrocartilage ⑯
Connection between pubic bones, intervertebral disks

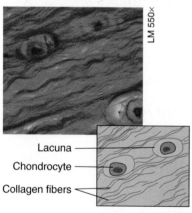

LM 550×

Lacuna

Chondrocyte

Collagen fibers

Bones ⑰

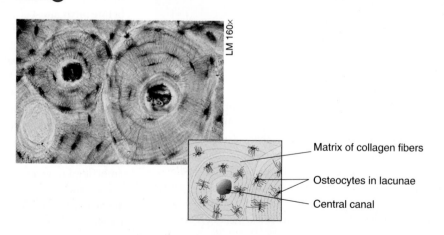

LM 160×

Matrix of collagen fibers

Osteocytes in lacunae

Central canal

ISOLATING CONNECTIVE TISSUE

Adipose tissue 18
Under the skin, insulation of organs such as the heart, kidneys, and breasts

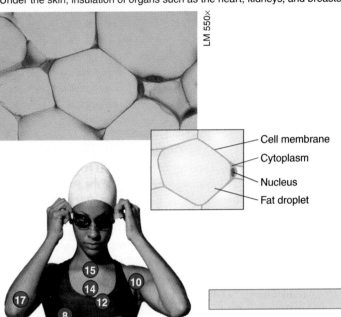

LM 550×

- Cell membrane
- Cytoplasm
- Nucleus
- Fat droplet

Blood 19
Blood vessels, heart

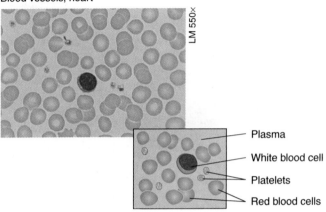

LM 550×

- Plasma
- White blood cell
- Platelets
- Red blood cells

DEFENSIVE CONNECTIVE TISSUE CELLS

Lymphocytes 7
Circulate in the blood or reside in organs, vessels, and nodes of the lymphatic system

SEM 3500×

Macrophages 8
Located throughout the body in connective tissue, especially in the lungs, spleen, and lymph nodes

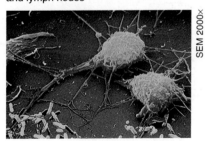

SEM 2000×

Mast cells 9
Located throughout the body in connective tissue, especially in the lungs, spleen, and lymph nodes

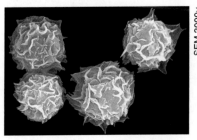

SEM 3000×

Bone Cells

Osteoblasts are the cells that produce *bone* ⓱. (Osteocytes are mature bone cells.) As in cartilage, these cells are isolated in lacunae. They lay down a matrix of collagen fibers that becomes coated with small, needle-shaped crystals of calcium salts. The calcium salts make the bone rigid, whereas the fibers keep the bone from being brittle. Bone makes up the adult skeleton, which supports the body and protects many of the organs. Bone is also a storehouse for calcium.

The bones of your skeleton have two types of internal structure: compact bone and spongy bone. *Compact bone* is denser than spongy bone and gives the bone the strength to withstand mechanical stress. It forms the outer surface of all bones. In compact bone, the cells lay down matrix in thin concentric rings, forming tubes of bone around narrow channels or canals (see Figure 7.9). These canals run parallel to the length of the bone, are interconnected, and contain blood vessels and nerves. The blood vessels provide a lifeline to the bone-forming cells, and the nerves control the diameter of the blood vessels and thus the flow through them.

Spongy bone fills the ends of long bones and the interiors of flat bones and irregular bones. It lines the interiors of long bones (Figure 7.10). Spongy bone is composed of an open lattice of bone that supports the bone just as beams support a building. It also helps keep bones somewhat lightweight. The spaces within the latticework of bone are filled with red bone marrow, the substance that produces the body's blood cells.

Cells and Tissues That Isolate

The third general class of connective tissue is composed of cells that specialize in accumulating and transporting particular molecules. Isolating connective tissues include the fat cells of adipose tissue, as well as pigment-containing cells.

As you can see in Figure 7.9, groups of fat cells as seen under a microscope bear a striking resemblance to chicken wire. The "wire" is really the cytoplasm and nucleus of each cell pushed against the cell membrane by a large fat droplet. The fat takes up much of the cell and is released when the body needs it for fuel. Although many people think that they have more fat cells than they need for fuel emergencies, fat serves other purposes too. *Adipose tissue* (⓲, in Figure 7.9) helps shape and pad the body and insulates against heat loss.

Possibly the most important "isolating" cells are red blood cells, one of the solids that float in the fluid connective tissue called *blood* (⓳, in Figure 7.9). Blood cells

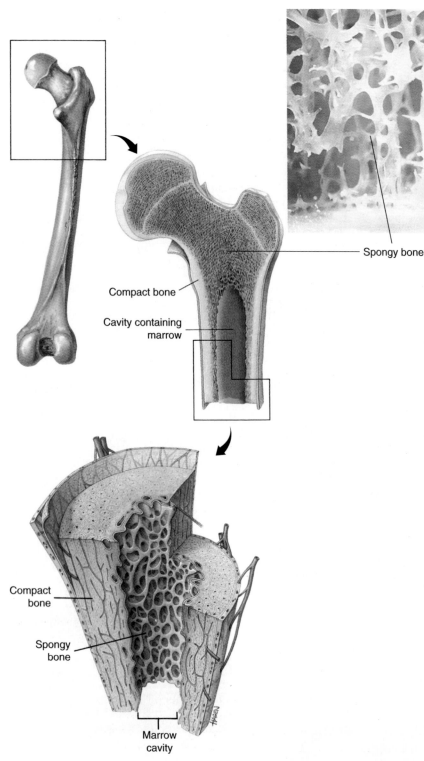

Spongy bone

Compact bone

Cavity containing marrow

Compact bone

Spongy bone

Marrow cavity

Figure 7.10 **Structure of a long bone.** A long bone, such as the bone located in your upper leg, is composed of spongy bone and compact bone. Spongy bone (photo inset) is composed of a delicate latticework and adds support which, along with the hollowed-out shaft of compact bone, keeps the bone somewhat lightweight. Compact bone adds strength.

are classified according to their appearance, either as *erythrocytes* (red blood cells) or *leukocytes* (white blood cells). Some types of white blood cells were described earlier in this chapter; they include the macrophages and lymphocytes that defend the body. The role of red blood cells is very different. Red blood cells act as mobile transport units, picking up and delivering the gases carbon dioxide and oxygen. Cell fragments called *platelets* are also present in the blood. These cell pieces play an important role in the clotting of blood.

Red blood cells are the most common of the blood cells. There are about 5 million in every milliliter of blood. During their maturation in the red bone marrow,

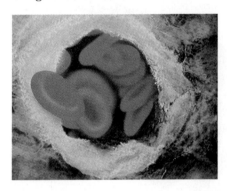

they lose their nuclei and mitochondria, and their endoplasmic reticula dissolve. As a result of these processes, red blood cells are relatively inactive metabolically, but they still perform the life-sustaining job of carrying oxygen to your tissues. This oxygen (as well as some carbon dioxide) is carried by the iron-containing pigment hemoglobin, which imparts the color to red blood cells. Hemoglobin is produced within the red bone marrow as the red blood cells are formed. An amazing 300 million molecules of hemoglobin become isolated within each red blood cell.

Blood cells float in a fluid intercellular matrix, or plasma. This fluid is both the banquet table and the refuse heap of your body because practically every substance used and discarded by cells is found in the plasma. These substances include the sugars, lipids, and amino acids that are the fuel of the body, as well as the products of metabolism such as the waste gas carbon dioxide. The plasma also contains minerals such as calcium used to form bone; proteins such as fibrinogen that help the blood to clot; albumin, which gives the blood its viscosity; and antibodies produced by lymphocytes. Every substance secreted or discarded by cells is also present in the plasma.

Connective tissue and its cells provide a framework for the body, join its tissues, help defend it against foreign invaders, and act as storage sites for specific substances.

7.7 Muscle tissue

Muscle cells are the workhorses of your body. The distinguishing characteristic of muscle cells (also called muscle fibers) is the abundance of special thick and thin filaments made of protein. These filaments are highly organized to form strands called *myofibrils*. Each muscle cell is packed with many thousands of these myofibril strands. Their arrangement is shown in skeletal muscle tissue in **Figure 7.11**. The myofibrils shorten when the filaments slide past each other, causing the muscle to contract. **Figure 7.12** shows the three different kinds of muscle cells of the human body: smooth muscle, skeletal muscle, and cardiac muscle.

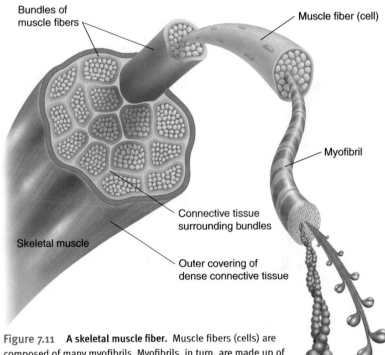

Figure 7.11 **A skeletal muscle fiber.** Muscle fibers (cells) are composed of many myofibrils. Myofibrils, in turn, are made up of microfilaments, which are responsible for muscle contraction. The arrangement of these structures is shown here in a skeletal muscle.

Smooth Muscle

Smooth muscle cells are long, with bulging middles, tapered ends, and a single nucleus. The cells are organized into sheets, forming smooth muscle tissue (**20**, in Figure 7.12). This tissue contracts involuntarily—you cannot consciously control it. Because it is found in the organs, or viscera, smooth muscle tissue is also often called visceral muscle tissue.

Some smooth muscle contracts when a nerve or hormone stimulates it. Examples of smooth muscles that contract in this way are the muscles that line your blood vessels and those that make up the iris of your eye.

Cells are organized into four basic tissue types.

Smooth muscle 20
Walls of hollow organs, pupil of eye, skin (attached to hair), and glands

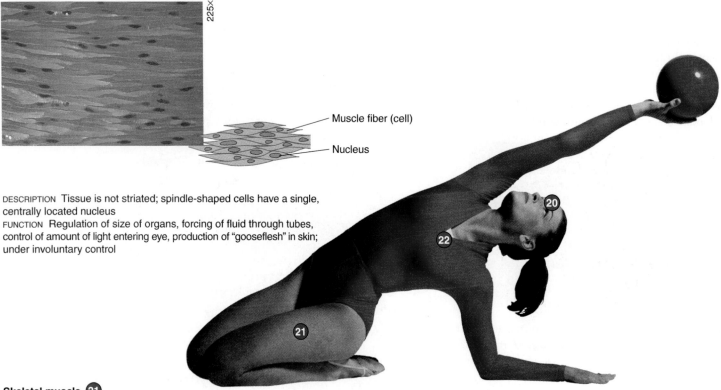

225×

Muscle fiber (cell)

Nucleus

DESCRIPTION Tissue is not striated; spindle-shaped cells have a single, centrally located nucleus
FUNCTION Regulation of size of organs, forcing of fluid through tubes, control of amount of light entering eye, production of "gooseflesh" in skin; under involuntary control

Skeletal muscle 21
Attachment to bone

225×

Muscle fiber (cell)

Nucleus

DESCRIPTION Tissue is striated; cells are large, long, and cylindrical with several nuclei
FUNCTION Movement of the body, under voluntary control

Cardiac muscle 22
Heart

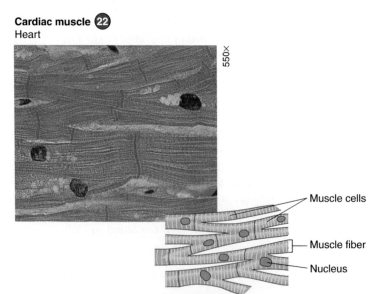

550×

Muscle cells

Muscle fiber

Nucleus

DESCRIPTION Tissue is striated; cells are cylindrical and branching with a single centrally located nucleus
FUNCTION Pumping of blood, under involuntary control

However, nerves do not reach each muscle cell. In many cases, impulses may be able to pass directly from one smooth muscle cell to another, resulting in a self-propagating wave of contraction throughout a layer of the muscle. In other smooth muscle tissue, such as that found in the wall of the intestines, individual cells might contract spontaneously when they are stretched, leading to a slow, steady squeezing of the tissue.

Skeletal Muscle

Skeletal muscles (㉑, in Figure 7.12) are attached to your bones and allow you to move your body. These muscles are called voluntary muscles because you have conscious control over their action. They are also called striated muscles because the tissue has "stripes"—microscopically visible bands, or striations. These striations result from the organization of thick and thin microfilaments within the myofibrils and the alignment of the myofibrils with one another. The myofibrils are organized so that all within a single cell contract at the same time when the muscle cell is stimulated by a nerve.

Striated muscle cells are extremely long. A single muscle cell, or *fiber*, may run the entire length of a muscle. Each fiber has many nuclei that are pushed to the edge of the cell and lie just under the cell membrane. Bundles of these muscle cells are wrapped with connective tissue and joined with other bundles to form the muscle itself.

Cardiac Muscle

The heart, which acts as a pump for the circulatory system, is composed of striated muscle fibers. These fibers are arranged differently from their organization in skeletal muscle. Instead of very long cells running the length of the muscle, *cardiac muscle* (㉒, in Figure 7.12) is composed of chains of single cells. These chains of cells are organized into fibers that branch and interconnect, forming a latticework. This lattice structure is critical to how heart muscle functions, and it allows an entire portion of the heart to contract at one time.

Muscle cells contain filaments that are capable of contraction. Smooth muscle contracts involuntarily and is located in the walls of certain internal structures such as blood vessels and the stomach. Skeletal muscle is connected to bones and allows you to move your body. Cardiac muscle makes up the heart, acting as a pump for the circulatory system.

7.8 Nervous tissue

The fourth major class of tissue in humans is nervous tissue. It is made up of two kinds of cells: (1) *neurons*, which transmit nerve impulses, and (2) *supporting cells*, which surround neurons, nourishing, protecting, insulating, and holding them in place.

Neurons are cells specialized to conduct an electrochemical "current" (Figure 7.13). The cell body of a neuron contains the nucleus of the cell. Two different types of projections extend from the cell body. One set of projections, the *dendrites*, act as antennae for the reception of nerve impulses. A single long projection called an *axon*, which may give out branches that usually split off at right angles, is usually covered with insulation. Axons often make distant connections. When axons or dendrites are long, they are referred to as *nerve fibers*. Some nerve fibers are so long, in fact, that they can extend from your spinal cord all the way to your fingers or toes. Single neurons over a meter in length are common.

The *nerves*, which appear as fine white threads when they are viewed with the naked eye, are actually com-

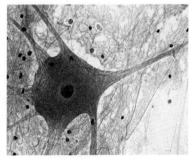

150X

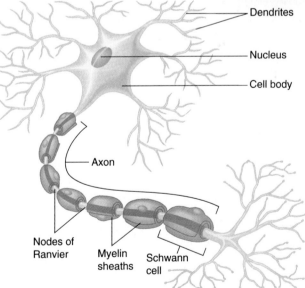

Dendrites

Nucleus

Cell body

Axon

Nodes of Ranvier

Myelin sheaths

Schwann cell

Figure 7.13   A human neuron. This diagram illustrates the generalized structure of a vertebrate neuron. The photo inset shows a vertebrate neuron.

Just Wondering . . . *Real Questions Students Ask*

> Tissues are organized into organs, and organs into organ systems.

How come some animals can regenerate certain tissues and body parts and we can't?

The idea of humans regenerating limbs is, of course, science fiction. But, as you suggested, many kinds of animals can regenerate parts. Sponges, for example (see Section 30.2), are capable of extensively regenerating and replacing lost parts. Sometimes their regeneration results in a new organism and is considered a form of asexual reproduction. This type of asexual reproduction also occurs in echinoderms such as sea stars (see Section 30.9). When one arm is removed from a sea star, the animal will grow a new arm; however, if the cut arm has a piece of the central disk attached, it, too, will grow into a new individual. Vertebrates show some capacity for tissue regeneration. The best examples are the amphibians; in fact, salamanders have been extensively studied to try to answer just the question you raised.

Developmental biologists have been intrigued by the regeneration question for years. They have found that the regenerated limbs of salamanders arise from epidermal tissue that surrounds a mass of rapidly growing cells capable of differentiating into connective tissue, bone, cartilage, vascular tissue, and lymphatic vessels. These cells apparently arise from the stump and eventually grow and develop into a new limb.

But what controls the growth and differentiation of these cells? And what controls their patterning—the ability of the proper type of cell to develop in the appropriate place, forming a fully functional and anatomically correct limb? Understanding the mechanisms by which these processes occur is central to comprehending the process of regeneration. Scientists are a long way from having complete answers to these questions. Researchers know that hormones play a role in influencing the normal course of growth during regeneration. They are also studying the role of one class of regulatory genes called homeotic genes (see p. 508), which scientists know control patterns of development. However, the exact role of these genes in pattern formation remains unclear. Research reported in 1998 also suggests that key immune system cells called T cells (p. 227) play a role in blocking regeneration in vertebrates similar to the way in which they block the overproliferation of cells in cancer.

Humans are capable of regenerating many of their tissues except mature nerve cells (when their cell bodies are damaged) and muscle cells. Injuries outside of the brain and spinal cord do not involve the cell bodies of nerve cells; thus some regeneration takes place if a wound is not too serious. Most serious injuries of the human body (and some not so serious) result in the development of scar tissue. This tissue is formed by fibroblasts (see p. 139), which divide rapidly in the injured area and secrete large quantities of collagen. This substance "fills in" the area of lost cells . . . certainly not a process nearly as complex or useful as that of salamander limb regeneration!

www.jbpub.com/biology

---

posed of clusters of axons and dendrites. Like a telephone cable, they include large numbers of independent communication channels—bundles of hundreds of axons and dendrites, each connecting a different nerve cell with a different muscle fiber or sensory receptor. In addition, the nerve contains numerous supporting cells bunched around the nerve fibers. In the brain and spinal cord, which together make up the central nervous system, these supporting cells are called glial (GLEE-uhl) cells. The supporting cells associated with nerve fibers of all other nerve cells, which make up the peripheral nervous system, are called Schwann cells. Nervous tissue is described more fully in Chapter 13.

> Neurons are cells that are specialized to conduct electrical signals. Nerve tissue is made up of these cells and supporting cells.

7.9 Organs and organ systems

The four major classes of tissues that we have discussed in this chapter (see Figures 7.6, 7.9, 7.12 and 7.13) are the building blocks of the human body. These tissues form the *organs* of the body. Each organ contains several different types of tissue coordinated to form the structure of the organ and to perform its function. A muscle, for example, is composed of muscle cells that together make up the muscle tissue. Bundles of this tissue are wrapped in connective tissue and wired with nervous tissue. Muscles can help you walk, pump your blood, and digest your food. Different combinations of tissues are found in different organs that perform different functions.

An *organ system* is a group of organs that function together to carry out the principal activities of the body. Figure 7.14 shows the 11 major organ systems of the human body. See also Table 7.1.

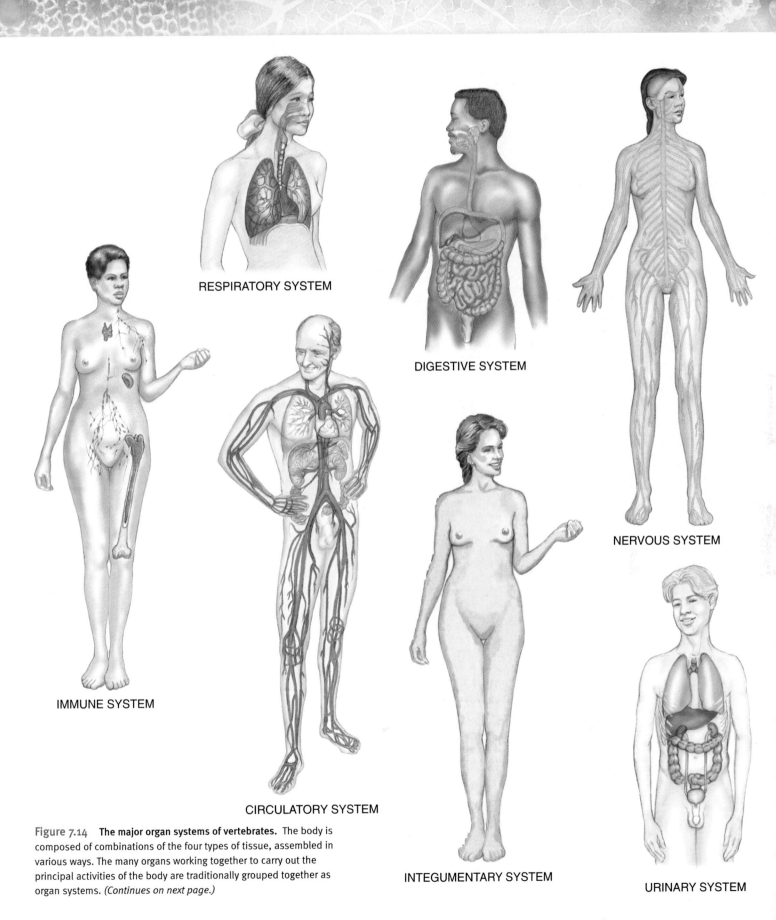

RESPIRATORY SYSTEM

DIGESTIVE SYSTEM

NERVOUS SYSTEM

IMMUNE SYSTEM

CIRCULATORY SYSTEM

INTEGUMENTARY SYSTEM

URINARY SYSTEM

Figure 7.14 **The major organ systems of vertebrates.** The body is composed of combinations of the four types of tissue, assembled in various ways. The many organs working together to carry out the principal activities of the body are traditionally grouped together as organ systems. *(Continues on next page.)*

Chapter 7 *Patterns and Levels of Organization* 147

Tissues are organized into organs, and organs into organ systems.

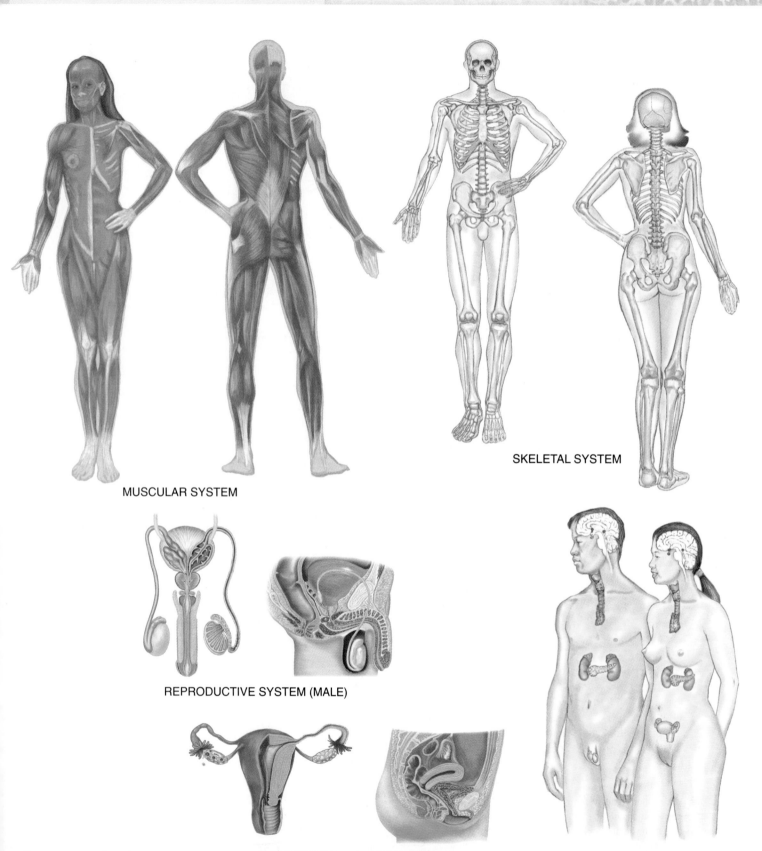

MUSCULAR SYSTEM

SKELETAL SYSTEM

REPRODUCTIVE SYSTEM (MALE)

REPRODUCTIVE SYSTEM (FEMALE)

ENDOCRINE SYSTEM

Figure 7.14, continued

Table 7.1

The Major Human Organ Systems

System	Functions	Components
Digestive	Breaks down food and absorbs breakdown products	Mouth, esophagus, stomach, intestines, liver, and pancreas
Respiratory	Supplies blood with oxygen and rids it of carbon dioxide	Trachea, lungs, and other air passageways
Circulatory	Brings nutrients and oxygen to cells and removes waste products	Heart, blood vessels, blood, lymph, and lymph structures
Immune	Helps defend the body against infection and disease	Lymphocytes, macrophages, and antibodies
Urinary	Removes wastes from the bloodstream	Kidney, bladder, and associated ducts
Nervous	Receives and helps body respond to stimuli	Nerves, sense organs, brain, and spinal cord
Skeletal	Protects the body and provides support for locomotion and movement	Bones, cartilage, and ligaments
Muscular	Produces body movement	Skeletal, cardiac, and smooth muscles
Integumentary	Covers and protects the body	Skin, hair, nails, oil and sweat glands
Endocrine	Coordinates and regulates body processes and functions	Pituitary, adrenal, thyroid, and other ductless glands
Reproductive	Produces sex cells and carries out other reproductive functions	Testes, ovaries, and other associated reproductive structures

The skeletal system supports and protects your body. It is moved by the large, voluntary muscles of the muscular system. Other muscles in this system help move internal fluids throughout your body. The nervous system regulates most of the organ systems. It can sense conditions in both your internal and external environments and help your body respond to this environmental information. The organs of your endocrine system secrete chemicals called hormones that also regulate body processes and functions. The circulatory system is the transportation system of the body. It brings nutrients and oxygen to your cells and removes the waste products of metabolism. Along with the immune and integumentary (skin) systems, it also helps defend the body against infection and disease. The respiratory system works hand in hand with the circulatory system, supplying the blood with oxygen and ridding it of the waste gas carbon dioxide. The food you eat is broken down by the digestive system and is absorbed through the intestinal walls into the bloodstream. Solid wastes are also eliminated from the body by this organ system. Liquid wastes are eliminated by the urinary system after it collects waste materials and excess water from the bloodstream. Ensuring the perpetuation of the human species, the reproductive system produces gametes, or sex cells, that can join in the process of fertilization to produce the first cell of a new individual.

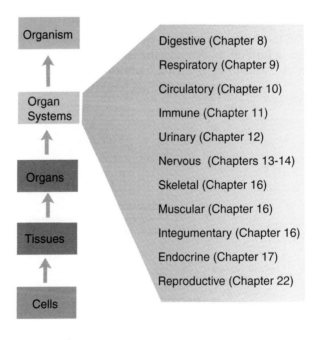

The four major classes of tissues form body parts called organs that perform particular functions. Groups of organs form organ systems. The human body has 11 major organ systems.

7.10 The organism: Coordinating it all

As you can see from their descriptions, the organ systems interact with one another to keep the organism—you—alive and well. This state of "wellness" is called **homeostasis**. Put another way, homeostasis is the maintenance of a stable internal environment despite what may be a very different external environment. To maintain this internal equilibrium, all your molecules, cells, tissues, organs, and organ systems must work together.

Your body maintains a steady state by means of feedback systems, or **feedback loops**. Feedback loops are mechanisms by which information regarding the status of a physiological situation or system is fed back to the system so that appropriate adjustments can be made. The thermostat in your house or apartment works by means of a feedback loop and is a good analogy to use to understand feedback loops in your body. If you set the temperature in your house at 68° F, for example, the furnace will run until that temperature is reached. A sensor in the thermostat monitors the temperature in the room, and this information triggers the electrical stimulation of the furnace. As soon as the air temperature near the thermostat reaches 68° F, the thermostat sends an electrical signal to the furnace to turn off. In this way, the temperature fluctuates within a small range. Such a feedback loop is called a **negative feedback loop** because the change that takes place is negative—the furnace turns off—to counter the rise in temperature. In other words, the response of the regulating mechanism (the thermostat turning off the furnace) is opposite to the output (the heat of the furnace).

Most regulatory mechanisms of your body work by means of negative feedback loops. The secretion of hormones is regulated in this way (see Chapter 17). Sufficient levels of specific hormones in the blood trigger mechanisms that result in the shutdown of their secretion. Likewise, your body temperature is maintained by means of negative feedback (Figure 7.15), much as the temperature of your home is maintained. When your body temperature rises too high ❶, your thermostat (a special portion of the brain called the *hypothalamus*) senses this rise and counters it. Messages race along your nerve fibers ❷ to your sweat glands, triggering the release of sweat, which evaporates and cools the body ❸. Blood vessels near the surface of the skin dilate, or widen, bringing blood near the

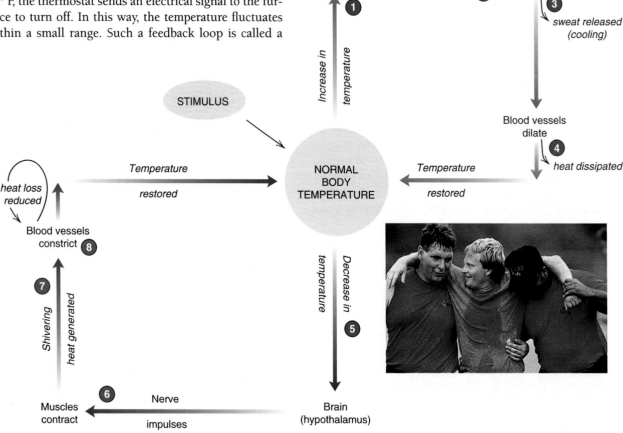

Figure 7.15 **Negative feedback loops.**

surface, where the heat is dissipated ❹ until a normal body temperature is achieved. Likewise, when your body temperature falls ❺, the hypothalamus triggers tiny muscles under the skin to contract ❻, causing a shiver that generates heat ❼. Blood vessels near the surface of the skin constrict ❽, lessening the flow of blood there and the amount of heat that radiates from the body until a normal body temperature is restored.

In all negative feedback loops, the change that takes place is opposite to whatever triggers the change. In some cases the change may be an increase rather than a decrease. When blood glucose levels fall, the body responds by releasing glucose into the bloodstream from the liver, thereby raising blood glucose levels.

Very few body mechanisms are regulated by means of **positive feedback loops**. In positive feedback loops the response of the regulating mechanism is positive with respect to the output—it moves in the same direction. You can see the problems inherent with this type of regulatory mechanism—a situation continues to intensify rather than moving in an opposite direction. For example, in a positive feedback loop, your furnace would be stimulated to stay on when the temperature in the house reached a particular setting. Positive feedback loops most

often disrupt the steady state and can even lead to death. In circulatory shock, for example, a severe loss of blood can result in such a decreased blood volume that the body cannot compensate—homeostasis cannot be maintained. The blood pressure drops and the blood flow is reduced so much that the heart and brain do not receive enough oxygen. As the heart weakens, the blood flow decreases more, weakening the heart even further. As the body becomes damaged from the loss of oxygen and nutrients, the shock worsens and can result in death. There are, however, some examples of and good uses for positive feedback in the body, including oxytocin release during childbirth, which stimulates contraction of the uterus.

The molecules, cells, tissues, organs, and organ systems that make up an organism must all work together to maintain homeostasis—a stable internal environment. Homeostasis is maintained primarily by means of negative feedback loops, regulatory mechanisms that reverse the direction of change of output systems when they reach certain levels. For example, a high internal body temperature triggers mechanisms that lower the temperature.

Key Concepts

Multicellular organisms exhibit a hierarchy of organization.

7.1 Multicellular organisms are organized hierarchically, from the lowest level of organization, the cell, to the highest level, the organ system.

7.1 Each animal phylum has certain broad structural features that occur within the context of its body plan.

7.3 All animals except the sponges exhibit either radial symmetry or bilateral symmetry.

7.4 Most bilaterally symmetrical organisms have a coelomate body plan.

Cells are organized into four basic tissue types.

7.5 Epithelium covers and lines the surfaces of the body and composes the glands. These cells are of three shapes and are arranged in ways that best suit their functions.

7.6 Connective tissue and its cells provide a framework for the body, join its tissues, help defend it from foreign invaders, and act as storage sites for specific substances.

7.7 There are three kinds of muscle tissue: smooth muscle, which is found in internal organs; skeletal muscle, which moves the body parts; and cardiac muscle, which makes up the heart—the pump of the circulatory system.

7.8 Nerve cells are specialized to conduct electrical signals. There are two kinds of nerve cells: neurons and supporting cells.

Tissues are organized into organs, and organs into organ systems.

7.9 Organs are body structures composed of several different tissues grouped into a structural and functional unit.

7.9 An organ system is a group of organs that function together to carry out the principal activities of the body.

7.10 The human body is organized so that its parts form an integrated whole. These parts work together at each level of organization to maintain homeostasis—a stable internal environment within the human organism.

Key Terms

bilateral symmetry *133*
coelom (SEE-lum) *134*
connective tissue *132*
epithelial tissue
    (ep-uh-THEE-lee-uhl) *132*
feedback loops *150*

homeostasis
    (HOE-mee-oh-STAY-sis) *150*
muscle *132*
negative feedback loop *150*
nervous tissue *132*

organ system *132*
organs *132*
positive feedback loop *150*
radial symmetry *133*
tissues *132*

KNOWLEDGE AND COMPREHENSION QUESTIONS

1. Distinguish among cells, tissues, organs, and organ systems. Give an example of each. (*Comprehension*)

2. What are the four basic types of tissue in the human body? Give an example of each. (*Knowledge*)

3. Draw and label the three shapes of epithelial cells. Give an example of where each is found in your body. (*Comprehension*)

4. What do lymphocytes, erythrocytes, and fibroblasts have in common? How do they differ? (*Comprehension*)

5. Red blood cells provide the life-sustaining role of carrying oxygen to your body cells. For what cellular process is this oxygen necessary? (*Knowledge*)

6. Everybody hates fat cells. Do they serve a purpose and if so, what is it? (*Knowledge*)

7. Which type of muscle tissue is under voluntary control and may be specifically strengthened by such exercises as lifting weights and doing push-ups? Which type is not under voluntary control and may be strengthened by aerobic exercises such as running and cross-country skiing? (*Comprehension*)

8. Compare the body plans of an acoelomate, a pseudocoelomate, and a coelomate. Give an example of each. (*Comprehension*)

9. Create a table that lists the major function(s) and at least two components of each of the principal organ systems in your body. (*Knowledge*)

10. What is homeostasis? How is it achieved? (*Comprehension*)

KEY
Knowledge: Recalling information.
Comprehension: Showing understanding of recalled information.

CRITICAL THINKING REVIEW

1. Our bones generally have reached their full growth and mature length by the age of 20 years. How then is the femur, a leg bone, of a 30-year-old man able to heal fully when broken in an accident? (*Synthesis*)

2. Your cat scratches your finger, and the skin surrounding the scratch turns reddish and feels warm. Explain why. (*Application*)

3. The text states that connective tissues are generally made up of cells spaced well apart in a nonliving matrix. Using your knowledge of cells and tissues, suggest what the functions of a matrix might be. (*Synthesis*)

4. Certain tissues in your body do not repair themselves if damaged (for example, nervous tissue within the brain and spinal cord), whereas other tissues such as skin rapidly repair themselves. Based on what you learned in this chapter, suggest a reason why nerve tissue in your central nervous system does not repair itself as readily as skin. (*Synthesis*)

5. Suggest how the coelom of your body acts as a transportation system. (*Synthesis*)

KEY
Application: Using information in a new situation.
Synthesis: Putting together information from different sources.

e learning

Location: http://www.jbpub.com/biology

e-Learning is an on-line student review area located at this book's web site www.jbpub.com/biology. The review area provides a variety of activities designed to help you study for your class and build your learning portfolio.

Review Questions The review questions test your knowledge of the important concepts and applications in each chapter. The review provides feedback for each correct or incorrect answer. This is an excellent test preparation tool.

Figure-Labeling Exercises Sharpen your visual thinking by matching terms and labels for important illustrations.

Flash Cards Studying biology requires learning new terms. Virtual flash cards help you master the new vocabulary for each chapter.

Just Wondering As you read and study from this text, you may find that you have unanswered questions. Through this site you can ask the author, Sandy Alters, your "just wondering" questions.

Do you prefer the speed and reliability of a CD-ROM? All of the features contained on the eLearning portion of the web site are also available on the Student CD-ROM.

Digestion

Unlike a giant panda, red-tipped tubeworms do not need to eat at all. How can this be? Tubeworms (shown at left) live deep in the ocean, beyond even the depths where sunlight penetrates. They use energy provided by bacteria that live within their bodies. These bacteria use a process called chemosynthesis to extract energy from the hydrogen sulfide that emerges from deep-sea vents where the tubeworms live. Deep-sea vents are places where hot, sulfur-rich water spews from cracks in the ocean floor due to volcanic activity beneath.

Among animals, tubeworms are the exception to the rule, because they do not need a digestive system. **Digestion** is a process in which food particles are broken down into small molecules that can be absorbed by the body. **Autotrophs,** or self-feeders, make food within their bodies so digestion is unnecessary. Plants, algae, some protists, and some bacteria, along with certain sea creatures like the red-tipped tubeworms, are autotrophs.

Heterotrophs, like the giant panda shown at right, meet their need for energy by consuming and digesting food. A panda has to eat a lot more than most heterotrophs; it feeds for up to 16 hours and consumes from 20 to 50 pounds (about a quarter of its body weight) each day.

Organization

8.1 Intracellular digestion

Heterotrophs digest their food either within their cells (intracellularly), outside of their cells (extracellularly), or both (Figure 8.1). Intracellular digestion usually takes place in single-celled protists and the sponges. These organisms survive on food particles that are small enough to directly enter the cells, but large enough to need further breakdown, readying them for cellular respiration. During intracellular digestion, food particles or liquids are taken into a cell by phagocytosis or pinocytosis (see p. 82). The food is enclosed within a membrane-bounded sac called a food vacuole, which merges with a lysosome, or digestive vacuole. As the two vacuoles fuse, enzymes come into contact with the food, breaking it down into molecules useable by the organism.

Figure 8.1 shows the process of intracellular digestion in the paramecium, a single-celled protist. The beating cilia of the paramecium sweep food into its gullet ❶. From the gullet, food passes into a developing food vacuole. Lysosomes fuse with the food vacuole and enzymes and hydrochloric acid aid in digestion. Wastes are expelled from the anal pore ❷.

Although sponges are multicellular organisms, they process food in a similar way. Water containing food particles passes through the sponge's body by means of its pores ❸. Collar cells, located in the interior of the vase-shaped sponge, trap food particles as their flagella generate currents in the water. Food trapped in the collars is ingested by phagocytosis and digested by merging with lysosomes ❹.

Intracellular digestion also occurs in conjunction with extracellular digestion in some members of the *Cnidaria* (hydra, jellyfish, corals, and sea anemones), *Ctenophores* (comb jellies), and *Turbellarians* (free-living [nonparasitic] flatworms). When these organisms take in large pieces of food, digestion begins in the digestive cavity as usual ❺. However, some food may remain only partially digested. These particles are taken into the cells of the digestive cavity, where they are broken down completely ❻.

> During intracellular digestion, food is taken into a cell, enclosed within a food vacuole, and broken down by enzymes found in digestive vacuoles. Intracellular digestion usually takes place in single-celled protists and the sponges.

Visual Summary Figure 8.1 Patterns in digestion.

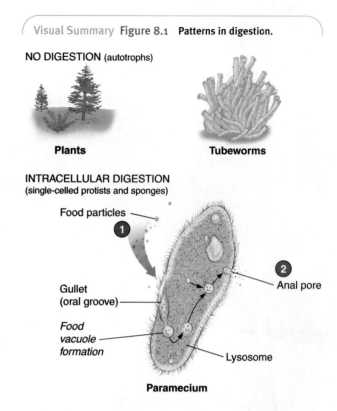

NO DIGESTION (autotrophs)

Plants   **Tubeworms**

INTRACELLULAR DIGESTION
(single-celled protists and sponges)

Food particles

❶

Gullet (oral groove)

❷ Anal pore

Food vacuole formation

Lysosome

Paramecium

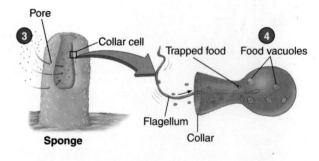

Pore

❸

Collar cell

Trapped food   Food vacuoles ❹

Flagellum

Collar

Sponge

INTRACELLULAR AND EXTRACELLULAR DIGESTION
(hydra, jellyfish, corals, sea anemones, comb jellies, flatworms)

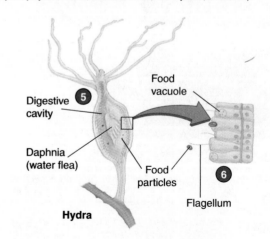

Food vacuole

Digestive cavity ❺

Daphnia (water flea)

Food particles

❻

Flagellum

Hydra

EXTRACELLULAR DIGESTION
No Digestive System (fungi and some bacteria)

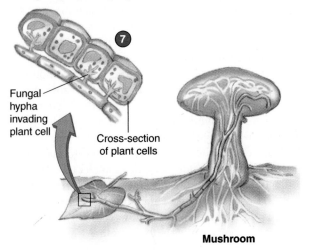

Fungal hypha invading plant cell

Cross-section of plant cells

Mushroom

EXTRACELLULAR DIGESTION
Digestive System (most animals)

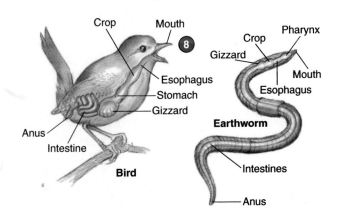

Crop

Mouth

Esophagus

Stomach

Gizzard

Anus

Intestine

Bird

Crop

Pharynx

Gizzard

Mouth

Esophagus

Earthworm

Intestines

Anus

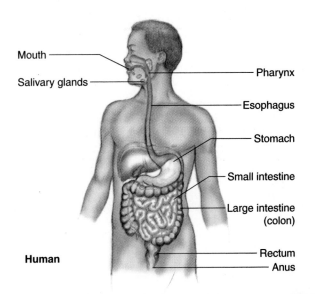

Mouth

Salivary glands

Pharynx

Esophagus

Stomach

Small intestine

Large intestine (colon)

Rectum

Anus

Human

8.2 Extracellular digestion

Extracellular digestion can take place with or without a **digestive system**, a series of organs specialized for breaking down food and ridding the body of wastes. In fungi and bacteria, extracellular digestion takes place without a digestive system. Fungi secrete enzymes onto their food source (usually a dead organism such as a tree or other type of organic matter such as a slice of bread). The enzymes digest the food and then the fungi absorb the breakdown products ❼. Some fungi derive nutrition from living sources, such as humans. Athlete's foot is an itchy, painful condition caused by a parasitic fungus.

Bacteria often take in food directly by phagocytosis or pinocytosis. However, if food particles are too large to ingest this way, some bacteria produce enzymes that digest food outside of the cell much like the fungi.

In animals, extracellular digestion usually takes place in a digestive system. Enzymes are secreted in various parts of the system, digesting the food to small molecules that can be absorbed by the body ❽. (The food you digest is really *outside* of your cells and tissues.) Some organisms have digestive systems that provide mechanical digestion of food as well as chemical. Your teeth, for example, mechanically break down food as does the gizzard of a bird. The bird's gizzard usually contains bits of sand or gravel that physically break food apart as it is churned in this second stomach. In addition to digesting and absorbing food (and temporarily storing food in a crop, for example), digestive systems also excrete indigestible material, or wastes.

Digestive systems vary depending on the complexity of the animal and its diet. Animals that have relatively simple anatomies, such as hydra, jellyfish, corals, sea anemones, and flatworms, have only one opening to the digestive system. Food is taken in through this opening, where it then enters a digestive cavity. Food is digested and absorbed there, and the wastes are expelled through the same opening. These animals continue the process of digestion intracellularly. More complex animals, such as earthworms, insects, birds, cows, and humans, have digestive systems with two openings—a mouth and an anus. Food passes in one direction through a series of organs that facilitate digestion and absorption of food. The processing of food and waste can take place uninterrupted, or, as happens with birds, food may be stored temporarily (in the crop) and digested later.

During extracellular digestion, some simple organisms such as bacteria and fungi secrete enzymes onto their food source. In animals, extracellular digestion usually takes place in a digestive system.

8.3 Carbohydrates, lipids, and proteins

www.ibpub.com/biology

There are six classes of **nutrients**: carbohydrates, lipids (fats and oils), proteins, vitamins, minerals, and water. These are the substances, in fact, that make up most of your body and the bodies of all living things. Using you as an example (assuming that you are a proper weight for your height), your body is made up of about 60% water and about 20% lipids. The other nearly 20% is mostly protein, carbohydrate, combinations of these two substances, and two major minerals found in your bones: calcium and phosphorus. Other minerals and vitamins make up less than 1% of your body.

Composition of the human body

Water

Lipids

Protein, carbohydrate, calcium, phosphorus, and other minerals and vitamins

As you may recall from Chapter 3, carbohydrates, lipids, and proteins are organic (carbon-containing) compounds and are used by heterotrophs as a source of energy, or kilocalories. These three organic compounds, therefore, are often referred to as the energy nutrients. Proteins, however, are not a preferred energy source; they are most often broken down to amino acids, and these molecules are used to build new proteins. Lipids and carbohydrates are used as building blocks for growth, repair, and maintenance also, but this use is secondary to their energy use.

Autotrophs primarily use carbohydrates that they manufacture during photosynthesis or chemosynthesis for their energy needs. They synthesize proteins, lipids, vitamins, and nucleic acids by using these carbohydrates along with inorganic nutrients such as nitrogen and the minerals potassium, calcium, magnesium, and phosphorus. When you "feed" your plants, for example, you are giving them the inorganic nutrients needed for the synthesis of these molecules.

Carbohydrates, lipids, and proteins are the energy nutrients, although dietary proteins are used most frequently for building proteins the body needs.

8.4 Vitamins, minerals, and water

Organisms do not obtain energy from vitamins, minerals, or water. Vitamins are organic molecules that help unlock the energy in carbohydrates, lipids, and proteins as well as perform functions such as helping to form red blood cells. Although required only in small amounts, each of the 13 different vitamins plays a vital role in the human body (Table 8.1). Many, such as the B vitamins, are used for the synthesis of coenzymes, small molecules that help enzymes work. For example, niacin and riboflavin, two of the B vitamins, are used to make the coenzymes nicotinamide adenine dinucleotide (NAD) and flavin adenine dinucleotide (FAD), which function in cellular respiration (see Chapter 6). Vitamin C, vitamin E, and beta carotene (a yellow pigment in vegetables that the body converts to vitamin A) are antioxidants. Antioxidants have been shown to lower the risk of heart disease as well as the risk of breast, lung, stomach, colon, uterine, and cervical cancers. (The risk of lung cancer in smokers is raised, however, because of interactions of the antioxidants with oxygen and cigarette smoke.)

Minerals are inorganic substances and are transported around the body as ions dissolved in the blood and other body fluids. Your body uses a variety of minerals that perform a variety of functions (see Table 8.1). Calcium, for example, does many jobs, including making up a part of the structure of your bones and teeth, helping your blood to clot, and helping your muscles contract. Sodium plays a key role in regulating the fluid balance within your body. Magnesium is an important player in the process of releasing energy from carbohydrates, lipids, and proteins.

Water provides the medium in which all the body's reactions take place. Along with dissolved substances and (sometimes) suspended solids, it bathes your cells, makes up much of the cytosol within your cells, and travels through your arteries and veins. It even lubricates your joints and cushions organs such as the brain and spinal cord.

Vitamins, minerals, and water help body processes take place. Some minerals are also incorporated into body structures.

Table 8.1

Major Vitamins and Minerals

Vitamins and Minerals	Major Function	Rich Food Sources	RDA[1] (mg)	Signs and Symptoms of Deficiency	Toxicity
Vitamins					
A	Formation of visual pigments, maintenance of epithelial cells	Green or yellow fruits and vegetables, milk products, liver	0.8–1.0	Night blindness, dry skin, growth failure	Headache, nausea, birth defects
Thiamin	Coenzyme in CO_2 removal during cellular respiration	Pork, whole grains, seeds, and nuts	1.1–1.5	Mental confusion, loss of muscular coordination	None from food
Riboflavin	Part of coenzymes FAD and FMN	Liver, leafy greens, dairy products, eggs, meats	1.3–1.7	Dry skin, cracked lips	None from food
Niacin	Part of coenzymes NAD[1] and NADI[3]	Wheat bran, tuna, chicken, beef, enriched breads and cereals	15–19	Skin problems, diarrhea, depression, death	Skin flushing
Pantothenic acid	Part of coenzyme A, energy metabolism	Widespread in foods	4–7[2]	Rare	None from food
B₆	Protein metabolism	Animal proteins, spinach, broccoli, bananas	1.6–2.0	Anemia, headaches, convulsions	Numbness, paralysis
B₁₂	Coenzyme for amino acids and nucleic acid metabolism	Animal foods	0.002	Pernicious anemia	None from food
Biotin	Coenzyme in carbohydrate and fat metabolism, fat synthesis	Liver, peanuts, cheese, egg yolk	0.03–0.10[2]	Rare	Rare
Folate	Nucleic acid and amino acid synthesis	Green leafy vegetables, orange juice, liver	0.18–0.20	Anemia, embryonic neural tube defects	None from food
C	Collagen synthesis antioxidant	Citrus fruits, broccoli, greens	60	Scurvy, poor wound healing bruises	Diarrhea, kidney stones
D	Absorption of calcium and phosphorus, bone formation	Vitamin D fortified milk, fish	0.005–0.010	Rickets (bone deformities)	Calcium deposits in soft tissues, growth failure
E	Antioxidant	Vegetable oils	8–10	Rare	Muscle weakness, interference with vitamin K metabolism
K	Synthesis of blood clotting substances	Green leafy vegetables, liver	0.06–0.08	Hemorrhage	Anemia
Major Minerals					
Calcium (Ca)	Component of bone and teeth, blood clotting, nerve transmission, muscle action	Dairy products, canned fish	800–1200	Osteoporosis	Kidney stones
Phosphorus (P)	Component of bone and teeth, energy transfer (ATP), component of nucleic acid	Dairy products, meat, soft drinks	800–1200	None	Bone loss if calcium intake low

[1] Recommended dietary allowances (1989) values are for adults.
[2] Estimated minimum requirements for adults (no RDA established).
[3] Estimated safe and adequate daily dietary intake for adults (no RDA established).

Continued on the following page.

Table 8.1

Major Vitamins and Minerals—cont'd

Vitamins and Minerals	Major Function	Rich Food Sources	RDA[1] (mg)	Signs and Symptoms of Deficiency	Toxicity
Magnesium (Mg)	Bone formation, muscle and nerve function	Wheat bran, green vegetables, nuts	280–350	Weakness	Rare
Sodium (Na)	Osmotic pressure, nerve transmission	Salt, seafood, processed food	500[1]	Nausea, vomiting, muscle cramps	Possible hypertension
Chlorine (Cl)	HCl synthesis, osmotic pressure, nerve transmission	Salt, processed food	700[3]	Rare	Possible hypertension
Potassium (K)	Osmotic pressure, nerve transmission	Fruits and vegetables	2000[3]	Heart irregularities, muscle cramps	Slowed heart rate
Iron (Fe)	Hemoglobin synthesis, oxygen transport	Liver, meat, enriched breads and cereals	10–15	Anemia	Constipation, death from overdose of children's iron supplements
Zinc (Zn)	Component of many enzymes, including those involved in growth, sexual development, and immune function	Seafood, liver, meats, whole grains	12–15	Skin rash, poor growth, hair loss	Diarrhea, depressed immune function
Iodine (I)	Thyroid hormone production	Iodized salt, seafood	0.15	Goiter	Interference with thyroid function
Fluorine (F)	Strengthener of teeth	Fluoridated water, tea	1.5–4.0	Increased risk of dental caries	Stained teeth during development
Copper (Cu)	Iron metabolism, component of many enzymes	Liver, cocoa, whole grains	1.5–3.0[2]	Anemia	Rare, vomiting

[1]Recommended dietary allowances (1989) values are for adults.
[2]Estimated minimum requirement for adults (no RDA established).
[3]Estimated safe and adequate daily dietary intake for adults (no RDA established).

Table 8.2

Digestive Enzymes

Source	Enzyme	Substrate	Digestion Product
Salivary gland	Amylase	Starch, glycogen	Disaccharides
Stomach	Pepsin	Proteins	Short polypeptides
Small intestine	Peptidases	Short peptides	Amino acids
	Lactase, Maltase, Sucrase	Disaccharides	Glucose, monosaccharides
Pancreas	Pancreatic amylase	Starch, glycogen	Disaccharides
	Trypsin, Chymotrypsin, Carboxypeptidase	Proteins	Polypeptides
	Lipase	Triglycerides	Fatty acids and glycerol

8.5 An overview of human digestion

To obtain nutrients and energy from the food you ingest, your body must first digest food. In the human body, digestion is carried out mechanically in the oral cavity as food is crushed by the teeth and in the stomach as food is churned by the stomach's muscular walls. Digestion is carried out chemically in three ways:

1. by hydrochloric acid (HCl), which denatures, or unfolds, protein molecules and disrupts the protein glue that holds cells together;
2. by bile salts, which emulsify, or separate, large lipid droplets into much smaller lipid droplets; and
3. by a variety of highly specific enzymes that help cleave certain chemical bonds (Table 8.2).

As you read in Chapter 5, enzymes are proteins that speed up the rate of chemical reactions in living things. In fact, without enzymes, you would die before needed chemical reactions took place!

The enzymes that help digest the energy nutrients—proteins, carbohydrates, and lipids—are of three basic types:

1. **Proteases** (PRO-tee-ACE-es), which first break down proteins to smaller polypeptides and then to amino acids
2. **Amylases** (AM-uh-lace-es), which break down starches and glycogen to sugars. (Both starch and glycogen are storage forms of polysaccharides, complex sugars.)
3. **Lipases** (LYE-pays-es), which break down the triglycerides in lipids to fatty acids and glycerol. (Most dietary fat is in the form of triglycerides, an abundant type of lipid.)

Proteins, carbohydrates, and lipids, along with their breakdown products, are described in Chapter 3.

During digestion, proteins are unfolded by hydrochloric acid and are then cleaved into peptides and individual amino acids by protease enzymes. Starch and glycogen are digested to sugars by amylases. Triglycerides are digested to fatty acids and glycerol by lipases.

8.6 Where it all begins: The mouth

Does your mouth ever water when you think about your favorite food? Does the smell of some of your favorite foods also evoke this reaction? Do you know why this reaction occurs?

The water in your mouth is really a secretion from a set of glands called the **salivary glands**. The locations and names of each of these glands are shown in **Figure 8.2**. Their secretion, saliva, is a solution that consists primarily of water, mucus, and the digestive enzyme **salivary amylase**. Other substances can be found in smaller amounts, such as antibodies and a bacteria-killing enzyme.

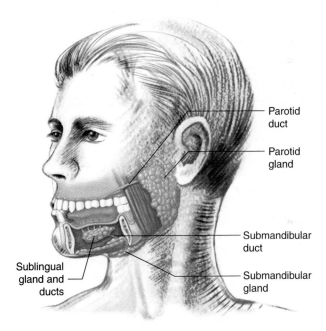

Parotid duct

Parotid gland

Submandibular duct

Submandibular gland

Sublingual gland and ducts

Figure 8.2 **The salivary glands.**

Salivary amylase breaks down starch into molecules of the disaccharide maltose. During this reaction, the starch is broken down with the addition of water—a process called hydrolysis.

$$\text{Starch} + \text{Water} \xrightarrow{\text{Salivary amylase}} \text{Maltose}$$

Enzymes that break down substances by hydrolysis are called *hydrolyzing enzymes*. Along with playing this role in digestion, the saliva also moistens and lubricates food so that it is swallowed easily and does not scratch the throat.

Secretion by the salivary glands is controlled by the nervous system. Although the nervous system works to maintain a constant secretion of saliva in the mouth, it speeds up the secretion when it is stimulated by the presence (or sometimes the sight, smell, or thought) of food. When food is in the mouth, nerve endings called chemoreceptors, which are sensitive to the presence of certain chemicals, send a signal to the brain, which responds by stimulating the salivary glands. Have you ever sucked on a slice of lemon? Then you know that the most potent stimuli are acid solutions; lemon juice can increase the rate of salivation eightfold.

Mechanical digestion begins when the teeth tear food apart into tiny pieces. In this way, the surface area of the food is increased, allowing the digestive enzymes to mix with the food and break it down more quickly and completely. The teeth of humans, as well as those of other organisms, are specialized in different ways. This specialization depends on the type of food an organism eats and how it obtains its food. Humans are omnivores. An omnivore eats both plant and animal foods. As a result, human teeth are structurally intermediate between the pointed, cutting teeth characteristic of carnivores, or meat eaters, and the flat, grinding teeth characteristic of herbivores, or plant eaters. In fact, the teeth of humans are like carnivores in the front and herbivores in the back.

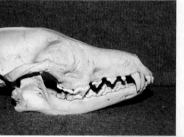

Carnivore skull (coyote)

Herbivore skull (mule deer)

The four front teeth in the upper and lower jaws of humans are incisors (Figure 8.3a). These teeth are sharp and chisel shaped and are used for biting. On each side of the incisors are pointed teeth called canines, which are used in tearing food. Behind each canine, on each side of the mouth and along both top and bottom jaws, are two premolars and three molars, all of which have flattened ridged surfaces for grinding and crushing food. In early childhood, however, humans do not have these 32 adult teeth but only 20 "baby teeth." These first teeth are lost during childhood and are replaced by the 32 adult teeth.

Each tooth is alive and is rooted in the bones of the upper and lower jaw. The gums cover this bone; the portion of the tooth protruding above the gumline is called the crown (Figure 8.3b). The portion of the tooth that extends into the bone is called the root. Inside lies a central, nourishing pulp that contains nerves, blood vessels, and connective tissue. The nerves and blood

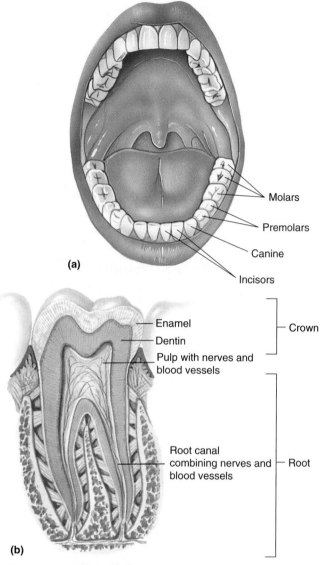

Figure 8.3 Teeth in humans. (a) The four front teeth are incisors, used for biting. Canines are used for tearing food. The premolars and molars are used for grinding and crushing. (b) Although you may forget until you are in the dentist's chair, each tooth in your mouth is alive and contains nerves and blood vessels.

vessels pass out of the tooth through holes at the root. The chewing surface of the tooth is made up of enamel—the hardest substance in the body. It is layered over the softer dentin that forms the body of the tooth.

Teeth shred and grind the plant and animal material that humans ingest as food. Saliva secreted into the mouth moistens the food, which aids its journey into the rest of the digestive system and begins its enzymatic digestion.

8.7   The journey to the stomach

As food is chewed and moistened, the tongue forms it into a ball-like mass called a bolus and pushes it into the upper part of the **pharynx** (FAIR-inks), or throat. As this happens, the soft palate raises up, sealing off the nasal cavity and preventing any food from entering this chamber ❶. The soft palate is the tissue at the back of the

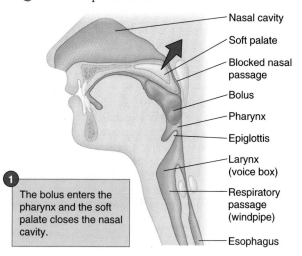

Nasal cavity
Soft palate
Blocked nasal passage
Bolus
Pharynx
Epiglottis
Larynx (voice box)
Respiratory passage (windpipe)
Esophagus

❶ The bolus enters the pharynx and the soft palate closes the nasal cavity.

roof of the mouth. The pressure of the food in the pharynx stimulates nerves in its walls that begin the swallowing reflex, an involuntary action. As part of this reflex action, the voice box, or **larynx** (LAIR-inks), raises up to meet the **epiglottis** (ep-ih-GLOT-iss), a flap of tissue that folds back over the opening to the larynx ❷. With

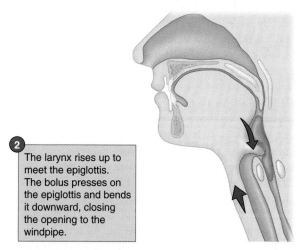

❷ The larynx rises up to meet the epiglottis. The bolus presses on the epiglottis and bends it downward, closing the opening to the windpipe.

this action the epiglottis acts much like a trapdoor, closing over the **glottis**, the opening to the larynx and trachea (your windpipe), so that food will not go down the wrong way. If you place your hand over your larynx (Adam's apple), you can feel it move up when you swallow.

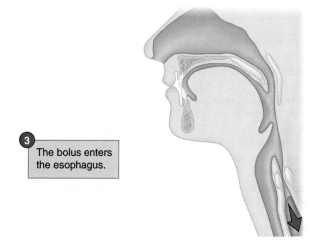

❸ The bolus enters the esophagus.

After passing into the pharynx and then bypassing the windpipe, the food enters the **esophagus** (ih-SOF-uh-gus) ❸, a food tube that connects the pharynx to the stomach. The esophagus, which is about 25 centimeters long (a bit less than a foot), pierces the diaphragm before it connects to the stomach. The diaphragm is a sheetlike muscle that forms the floor of the chest cavity; it is a muscle of breathing (see Chapter 9). The opening in the diaphragm that allows the esophagus to pass is called the esophageal hiatus.

The esophagus ends at the door to the stomach. This door is called the cardiac opening and is ringed by a circular muscle called the lower esophageal sphincter. When the ring of muscle contracts, or tightens, it closes the cardiac opening (Figure 8.4a). When it relaxes, or

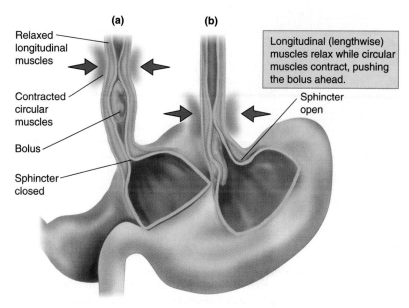

(a)   (b)

Relaxed longitudinal muscles

Contracted circular muscles

Bolus

Sphincter closed

Longitudinal (lengthwise) muscles relax while circular muscles contract, pushing the bolus ahead.

Sphincter open

Figure 8.4   **Peristalsis**.

Just Wondering . . . Real Questions Students Ask

Human digestion is mechanical and chemical, and is an extracellular process.

I get heartburn a lot. What causes it?

Heartburn results when the acid contents of the stomach splash back into the esophagus at the cardiac opening. The acid burns the esophagus—but it feels like your heart is on fire! You're not alone in your pain: 10% of adults report experiencing heartburn about once a week, and more than 30% experience it once a month.

www.jbpub.com/biology

Occasional heartburn is not a serious condition. However, frequent heartburn can irritate the delicate lining of the esophagus. In severe cases, stomach acid injures the esophagus so much that it bleeds. If your heartburn is severe and frequent, you should consult a physician.

There are a few things you can do to avoid or relieve heartburn. If you smoke, understand that cigarette smoke relaxes the lower esophageal sphincter, the muscle that closes the

cardiac opening. Therefore, quitting smoking will reduce your heartburn or get rid of it altogether. In addition, try not to overeat, which stuffs your stomach too full of food and promotes reflux (movement of stomach contents into the esophagus). Avoid eating close to bedtime, because lying down after eating with food in your stomach also promotes reflux. Try elevating the head end of your bed about six inches. Tight clothing around the middle causes some people trouble, as does obesity. Both conditions put extra pressure on the abdominal organs and may force food into the esophagus. Last, avoid the following foods and medications if you can because they all relax the lower esophageal sphincter: alcohol, chocolate, fats, peppermints, birth control pills, antihistamines, antispasmodics, some heart medications, and some asthma medications. As a last resort, many persons find relief by taking over-the-counter liquid antacids.

---

loosens, it opens this stomach door (Figure 8.4b). Although it is in the abdominal cavity, the cardiac opening is located very close to the heart, which lies just above it in the thoracic cavity.

The esophagus does not take part in digestion, but instead acts like an escalator, moving food down toward the stomach. The circular muscles in its walls produce successive waves of contractions called **peristalsis** (PEAR-ih-STALL-sis) (Figure 8.4). The movement of food to the stomach is therefore not dependent on gravity, so even astronauts in zero gravity (see photo at left)—or you standing on your head—can swallow without difficulty.

When the tongue pushes food into the pharynx, nerves stimulate the swallowing reflex. The food then enters the esophagus, or food tube, and is moved to the stomach by rhythmic muscular contractions called peristalsis.

8.8 Preliminary digestion: The stomach

The lower esophageal sphincter relaxes when food reaches it, allowing food to enter the **stomach**. The stomach is a muscular sac in which the food is collected and partially digested by hydrochloric acid and proteases (Figure 8.5). The stomach then "feeds" this food, little by little, to the primary organ of digestion, the small intestine. The stomach and intestinal tract have the same basic structural plan (Figure 8.6). Their interiors are lined with a layer of tissue called the mucosa, which consists of epithelial cells, blood and lymph vessels, and a thin layer of muscle. The mucosa covers a deeper, thicker layer of connective tissue, the submucosa, which is rich in blood vessels and nerves. Surrounding the connective tissue are layers of smooth muscle tissue—three in the stomach and two in the intestines. An envelope of tough connective tissue called serosa (seh-ROH-sa) serves as the outer covering of the digestive tract (as well as the other abdominal organs). Thin sheets of connective tissue called mesentery (MEH-zen-TAR-ee) are attached to the serosa along most of the intestinal tract, holding it in place and serving as a highway for blood vessels and nerves.

The inner surface of the stomach is dotted with **gastric glands** that are part of the epithelium that dips deeply into the mucosa. Two different kinds of cells in these glands secrete a gastric juice made up of hydrochloric acid and the protein pepsinogen. After the secretion of both, the acid chemically interacts with the pepsinogen, converting it to the protein-hydrolyzing enzyme **pepsin**. The hydrochloric acid also softens the connective tissue in foods; denatures, or unfolds, large protein molecules; and kills most bacteria that may have been ingested with the food. The pepsin digests only proteins, breaking them down into short polypeptides. Starches and lipids are not digested in the stomach. Other epithelial cells are specialized for the secretion of mucus. This mucus, produced in large quantities, lubricates the stomach wall and protects the stomach from digesting itself.

The stomach controls the production of gastric juice by means of a digestive hormone called **gastrin**. Hormones are regulating chemicals that are made at one place in the body and work in another. Gastrin is produced by endocrine (hormone-secreting) cells that are scattered throughout the epithelium of the stomach. Some stomachs greatly overproduce gastrin, however, which results in excessive acid production. The causes of this overproduction include such factors as heredity, smoking, and infection with the bacterium *Helicobacter pylori*. This extra acid may attack the walls of the first portion of the small intestine, or duodenum, burning holes through the wall. These holes are called duodenal ulcers (see the How Science Works box in Chapter 26). Because the contents of the small intestine are not normally acidic, this organ is much less able to withstand the disruptive actions of stomach acid than the wall of the stomach is. For this reason, more than 90% of all ulcers are duodenal, although other ulcers sometimes occur in the stomach.

Food stays in the stomach for approximately 2 to 6 hours. During this time, the contractions of the muscular wall of the stomach churn the food, mixing it with the gastric juice and mucus. By the time the food is ready to leave the stomach as a substance called chyme (kime), it has the consistency of pea soup. The gate to the small intestine, the pyloric sphincter, opens to allow just a bit of the chyme to pass. When the acid in this chyme is neutralized and the food is digested, the pyloric sphincter is signaled by the nervous system to open again, allowing the next bit of chyme to pass.

In the stomach, concentrated acid breaks up connective tissue and unfolds proteins. These proteins are digested by pepsin into short polypeptides. Starches and lipids are not digested in the stomach.

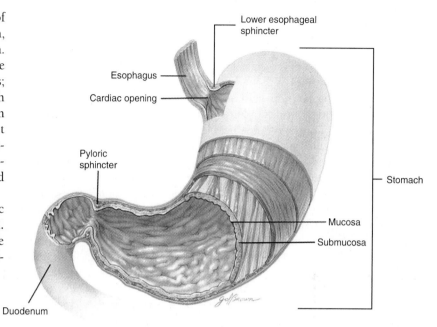

Figure 8.5   **The stomach.**

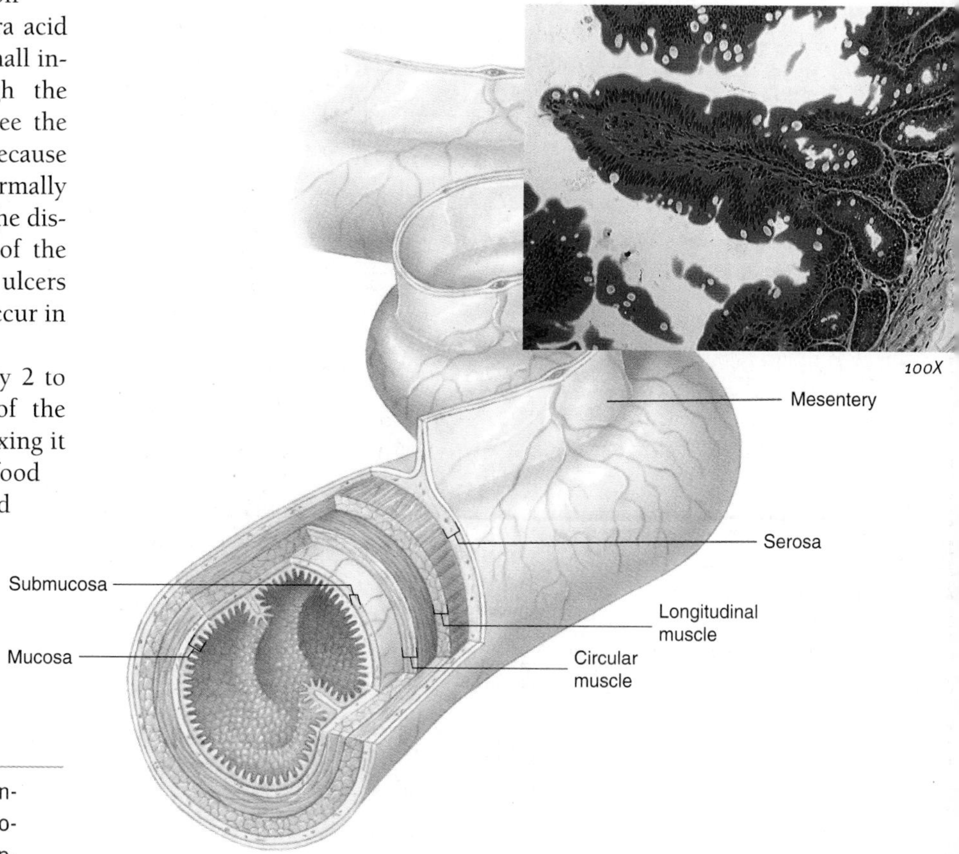

Figure 8.6   **Digestive tract.** The inset photo shows the mucosa (note fingerlike projections) with its underlying submucosa. The deeper circular muscle can be seen in the lower right corner of the photo.

8.9 The role of the small intestine and accessory organs

Most of the digestion of food takes place in the small intestine. Within this organ, starches and glycogen are broken down to sugars, proteins to amino acids, and triglycerides to fatty acids and glycerol. These products of digestion then pass through the cells of the intestinal mucosa and diffuse into the blood in underlying blood vessels. Fatty acids and glycerol move into the lymph in neighboring lymphatic vessels.

All of this activity takes place in the **small intestine**, the tubelike portion of the digestive tract that begins at the pyloric sphincter and ends at its T-shaped junction with the large intestine. The small intestine is approximately 6 meters long—long enough to stretch from the ground to the top of a two-story building. The initial portion of the small intestine (the first 25 centimeters [8 inches]) is called the **duodenum** (DOO-oh-DEE-num *or* doo-ODD-un-um). The other portions of the small intestine, the **jejunum** (ji-JOO-num) and the **ileum** (ILL-ee-um), are highly specialized (along with the duodenum) to aid in the absorption of the products of digestion by the blood and lymph.

Some of the enzymes necessary for digestion are secreted by the salivary glands, epithelial cells of the stomach, and epithelial cells of the duodenum. The **pancreas**, a long gland that lies beneath the stomach and is surrounded on one side by the curve of the duodenum (Figure 8.7), secretes the others. A tiny duct runs from the pancreas to the small intestine and serves as the passageway for the pancreatic juice. As you can see from Table 8.2, this secretion of the pancreas includes a number of digestive enzymes.

The **liver** is another organ that works with the duodenum to digest food. This organ, which weighs over 3 pounds, lies just under the diaphragm (see Figure 8.7). It is one of the most complex organs of the body and performs more than 500 functions. Although the liver produces no digestive enzymes, it does help in the digestion of lipids by secreting a collection of molecules called **bile**. One of the many components of bile is bile pigments, breakdown products of the hemoglobin from old, worn-out red blood cells. These pigments give bile its greenish color. The bile also contains bile salts, which are substances that act much like detergents, breaking lipids up into minute droplets of triglycerides, similar to droplets of cream suspended in milk. The liver manufactures bile salts from cholesterol. Excess bile is stored and concentrated in the **gallbladder** on the underside of the liver. A bile duct brings bile from the liver and gallbladder to the small intestine.

The liver and pancreas, although not organs of the digestive system, help digestion take place. The pancreas secretes a number of digestive enzymes. The liver produces bile as one of its numerous and diverse functions. One component of bile, bile salts, aids in lipid digestion.

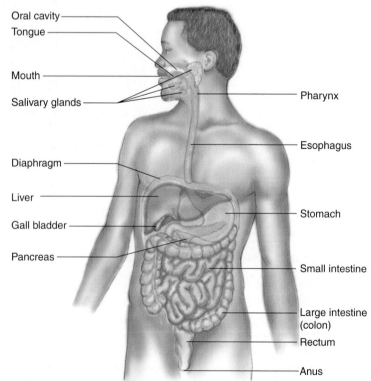

Oral cavity
Tongue
Mouth
Salivary glands
Diaphragm
Liver
Gall bladder
Pancreas

Pharynx
Esophagus
Stomach
Small intestine
Large intestine (colon)
Rectum
Anus

Figure 8.7 **The human digestive system.**

8.10 Digestion in the small intestine

Some of the food that enters the small intestine has already been partially digested. Salivary amylase has broken some of the bonds in the starches and glycogen, producing the disaccharide maltose. However, much starch and glycogen remain undigested. In the small intestine, **pancreatic amylase** breaks down these carbohydrates to maltose. Maltose, sucrose (table sugar), and lactose (milk sugar) are digested to the monosaccharides glucose, fructose, and galactose by enzymes called **disaccharidases** that are produced by specialized epithelial cells of the small intestine. Some people do not produce the enzyme to digest lactose, however, and are therefore unable to digest milk. These persons cannot drink milk without experiencing cramps and, in some cases, diarrhea. This condition is known as lactose intolerance.

Some of the proteins in the food entering the small intestine have also been partially digested. The hydrochloric acid of the stomach has unfolded these proteins, and pepsin has cleaved some of them to shorter polypeptides. Three other enzymes produced by the pancreas complete the digestion of proteins: **trypsin** (TRIP-sin), **chymotrypsin** (KYE-moe-TRIP-sin), and **carboxypeptidase** (kar-BOK-see-PEP-ti-dace). These enzymes work as a team with **peptidases** produced by cells in the intestinal epithelium, breaking down polypeptides into shorter chains and then to amino acids. The pancreatic enzymes are secreted in an inactive form and become active in the presence of a particular enzyme secreted by cells in the intestinal epithelium. This way, they do not digest their way down the pancreatic duct to the small intestine.

Virtually all lipid digestion takes place in the small intestine. Because they are insoluble in water, lipids tend to enter the small intestine as globules. Before these globules can be digested, they are emulsified, or separated into small droplets, by bile salts. Emulsification increases the surface area available for **pancreatic lipase** to act on the individual triglyceride molecules, which make up the lipids. Fatty acids and glycerol are the digestion products of triglycerides.

As you can see, the digestion of food involves so many players that the digestive team could use a manager. In fact, the key players in digestion—the liver, gallbladder, pancreas, stomach, and small intestine—have more than one manager. These managers of digestion are hormones (Figure 8.8). Earlier in this chapter you read how the hormone gastrin controls the release of hydrochloric acid in the stomach ❶. Two other hormones, **secretin** and **cholecystokinin** (KOL-uh-SIS-tuh-KINE-un) or CCK, control digestion in the small intestine.

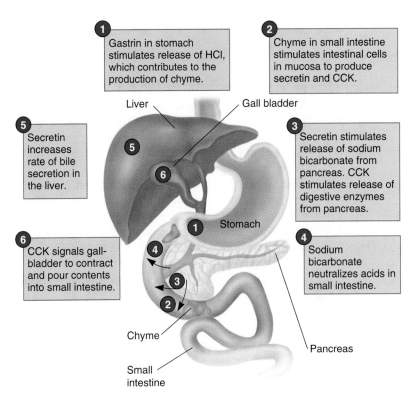

1 Gastrin in stomach stimulates release of HCl, which contributes to the production of chyme.

2 Chyme in small intestine stimulates intestinal cells in mucosa to produce secretin and CCK.

5 Secretin increases rate of bile secretion in the liver.

3 Secretin stimulates release of sodium bicarbonate from pancreas. CCK stimulates release of digestive enzymes from pancreas.

6 CCK signals gallbladder to contract and pour contents into small intestine.

4 Sodium bicarbonate neutralizes acids in small intestine.

Liver   Gall bladder

Stomach

Chyme

Small intestine

Pancreas

Figure 8.8 Digestive hormones manage many digestive enzymes.

When chyme enters the small intestine, the acid in it stimulates cells in the intestinal mucosa to produce secretin ❷. This hormone does two things: first, it stimulates the release of an alkaline fluid called sodium bicarbonate from the pancreas ❸. This solution neutralizes the acid in the chyme so that it will not damage the wall of the small intestine and produces the proper pH in which the pancreatic and intestinal enzymes will work ❹. In addition, secretin increases the rate of bile secretion in the liver ❺.

The presence of fatty acids and partially digested proteins in the chyme stimulates the mucosa to produce CCK ❷. This enzyme is a stimulus to the pancreas to release its digestive enzymes ❸. It also signals the gallbladder to contract and pour its contents into the small intestine ❻.

Digestion is completed in the duodenum. Undigested starch is digested to disaccharides and then to monosaccharides. Undigested polypeptides are digested to shorter peptides and then to amino acids. Triglycerides are made soluble and are then digested to fatty acids and glycerol. Hormones orchestrate these digestive processes.

8.11 Absorption in the small intestine

The amount of material passing through the small intestine is startlingly large. An average person consumes about 800 grams (1.75 pounds) of solid food and 1.2 liters (3 quarts) of fluid each day. To this amount is added 7 liters of fluid secreted by the salivary glands, stomach, pancreas, liver, and the small intestine itself. Of the 800 grams of solid food and 8.2 liters (8.7 quarts!) of liquid that enter the digestive tract, only about 50 grams of solid and 100 milliliters of liquid leave the intestinal tract as waste, or feces. The small intestine absorbs the 750 grams of nutrients per day and most of the water (over 8.5 quarts); the large intestine absorbs much of the remaining water.

The internal surface area of the small intestine is tremendously large, which aids in the absorption of nutrients. How does a large surface area exist in such a relatively small space? First, the mucosa and submucosa of the small intestine are thrown into folds; they do not have a smooth inner surface like a garden hose. This folded surface, in turn, is covered by fine, fingerlike projections of the epithelium. These projections are called **villi** (singular *villus*) and are so small that it takes a microscope to see them. In addition, the epithelial cells of the villi are covered on their exposed surfaces by cytoplasmic projections called microvilli **(Figure 8.9)**. The in-

foldings, the villi, and the microvilli provide the small intestine with a surface area of about 300 square meters, or 2700 square feet, an area greater than the floor space in many homes.

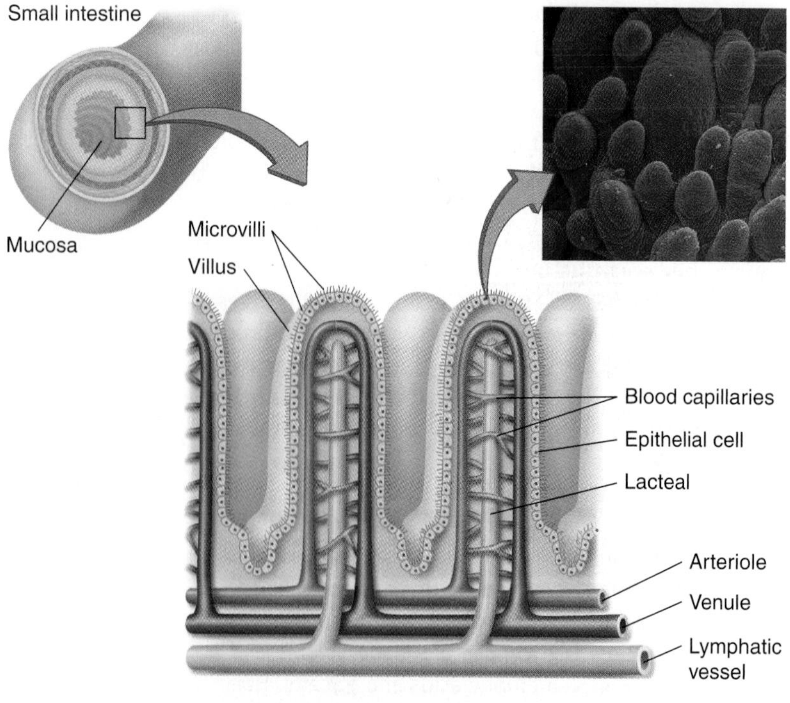

Within each villus is a network of capillaries and a lymphatic vessel called a lacteal. Each monosaccharide and amino acid must catch a ride on a carrier molecule to get into an epithelial cell of a villus. Energy is needed to ferry many of these molecules across cell membranes—a process called active transport (see Chapter 4). Others are taken up by facilitated diffusion. Once the monosaccharides and amino acids are in the epithelial cells, however, they accumulate and eventually move by facilitated diffusion through the base of the cell and into the blood. When in the bloodstream, they are quickly swept away to the liver for processing and storage. When blood levels of glucose are sufficient to supply all your cells with this fuel of cellular respiration (see Chapter 6), the liver stores glucose as glycogen. When more glucose is needed, such as between meals, the liver readily converts the glycogen back to glucose. In this way, the liver is your metabolic bank, accepting deposits and withdrawals in the currency of glucose molecules.

The absorption of lipids takes place somewhat differently. After triglycerides are broken down into fatty acids and glycerol, they become surrounded by bile salts. Packaged in this way, they move to the cell membranes of the villi. As you may recall from Chapter 4, one of the main ingredients in cell membranes is lipid. Therefore, when the fatty acids and glycerol from digestion come into contact with these cell membranes, they discard their shell of bile salts and easily move across the membrane and into the cell. Short-chained fatty acids are absorbed directly into the bloodstream. Longer-chained fatty acids are reassembled into triglycerides by the endoplasmic reticulum and are then encased in protein. After this processing, they pass out of the epithelial cells and into the lacteal. These protein-coated triglycerides are then transported in the lymphatic fluid through a system of vessels that drains the lymph into the blood at the left subclavian vein, a major blood vessel at the base of the neck.

Figure 8.9 **Structure of villi.** The microvilli are seen in the electron micrograph.

The internal surface area of the small intestine is increased by the presence of inner folds and projections, which results in the efficient absorption of nutrients and water. Monosaccharides and amino acids are absorbed into the intestinal epithelium and then diffuse into the bloodstream. Epithelial cells re-form fatty acids and glycerol to triglycerides and shuttle them to the blood by means of the lymphatic system. Water is absorbed by osmosis.

8.12 Concentration of solids: The large intestine

The large intestine, or **colon**, is much shorter than the small intestine—only about a meter and a half, or 5 feet, long. It is wide, however, having a diameter slightly less than the width of your hand. In contrast, the small intestine has a diameter only slightly larger than the width of two of your fingers. The small intestine joins the large intestine about 7 centimeters up from its end, creating a blind pouch called the cecum (SEE-kum) at the beginning of the large intestine (Figure 8.10). Hanging from this pouch is the appendix, a structure that serves no digestive function in humans, but does contain lymphatic tissue. An infection of the appendix is called appendicitis and can be quite serious and painful.

In plant-eating vertebrates other than cattle, sheep, and goats, the cecum serves as a place where the cellulose of plant cell walls is digested by the activity of intestinal bacteria and other microorganisms. Humans cannot digest cellulose, so it becomes a digestive waste. This waste, however, is important to the regular movement of the feces through the large intestine. Also called dietary fiber, undigested plant material provides bulk against which the muscles of the large intestine can push.

The junction of the small and large intestines is in the lower right side of the abdomen. From there, the large intestine goes up the right side of the abdomen to the liver. It then turns left, crossing the abdominal cavity just under the diaphragm. On the left side of the abdominal cavity, it turns downward, ending at a short portion of the colon called the **rectum**. The rectum terminates at the **anus**, the opening for the elimination of the feces.

No digestion takes place within the large intestine. Its role is to absorb sodium ions and water, to eliminate wastes, and to provide a home for friendly bacteria. These bacteria help keep out disease-causing microbes and produce certain vitamins, especially vitamin K.

Waste materials move slowly along the smooth interior of the large intestine as water and sodium are slowly reabsorbed. (Water is absorbed by osmosis, and sodium by diffusion and active transport; see Chapter 4.) As they move along, the wastes become more compacted. If the wastes move too slowly through the colon, too much water may be reabsorbed, leading to a difficulty in elimination called constipation. Conversely, if the wastes move too quickly, as happens with certain intestinal illnesses, not enough water may be removed, resulting in diarrhea. Eventually, the solids within the colon pass into the rectum as a result of the peristaltic contractions of the muscles encasing the large intestine. From the rectum, the solid material passes out of the anus through two anal sphincters. The first of these is composed of smooth muscle. It opens involuntarily in response to a pressure-generated nerve signal from the rectum. The second sphincter, in contrast, is composed of skeletal muscle. It is subject to voluntary control from the brain, thus permitting a conscious decision to delay defecation, or the elimination of waste.

The large intestine serves primarily to reabsorb water and sodium from digestive wastes and eliminate the remainder, or feces.

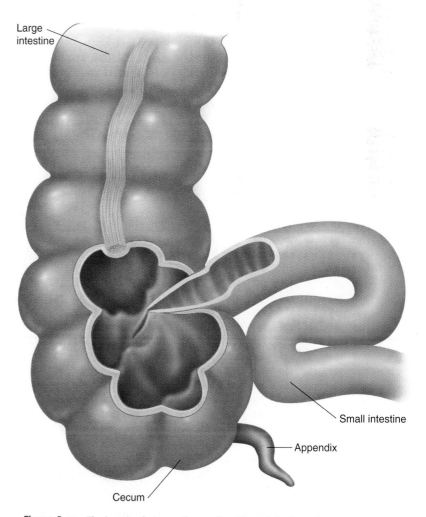

Large intestine

Small intestine

Appendix

Cecum

Figure 8.10 The junction between the small and large intestines. A one-way valve prevents waste material in the large intestine from re-entering the small intestine, yet allows waste from the small intestine to enter the colon.

For the latest health information, please visit www.jbpub.com/biology.

8.13 The nation's waistline

The digestion of food yields no usable energy, but changes a diverse array of complex molecules into a small number of simpler molecules. These molecules are absorbed by the body and are used by your cells as fuel for cellular respiration. During cellular respiration, adenosine triphosphate (ATP) molecules are produced—the energy currency of your body. (That is how you derive energy from the food you eat—by the process of cellular respiration.) As you read in Chapter 6, glucose is completely broken down by the processes of glycolysis, the Krebs cycle, and the electron transport chain to yield molecules of ATP. In addition, the carbon portions of amino acids and the glycerol backbones of triglycerides are converted to substances that can also be metabolized by glycolysis and the Krebs cycle. Fatty acids are broken apart two carbon units at a time, attached to acetyl CoA, and metabolized by the Krebs cycle (see Figure 6.14).

Any intake of food in excess of that required to maintain the blood sugar (glucose) level and the glycogen reserve in the liver results in one of two consequences. Either the excess glucose is metabolized by the muscles and other cells of the body or it is converted to

lipid and stored within adipose tissue (fat cells). Only when all the body's energy needs have been met—including the energy needed to run chemical reactions, move muscles, and digest food—will kilocalories be stored as fat. Think of your body as a giant scorecard, keeping track of kilocalories eaten and kilocalories used. If you eat more than you use, you will gain weight.

The National Heart, Lung, and Blood Institute (NHLBI), a division of the National Institutes of Health (NIH), released the first federal guidelines on the identification, evaluation, and treatment of overweight and obesity in adults on June 17, 1998. The report states that overweight and obesity affect 97 million American adults—55% of the population. Overweight and obese people are at increased risk of illness from chronic high blood pressure, blood lipid disorders, type 2 diabetes, heart disease, stroke, gallbladder disease, osteoarthritis, certain sleep disorders and respiratory problems, and certain cancers.

Are you overweight or obese? The NHLBI report uses the concept of the Body Mass Index (BMI) to determine overweight and obesity. The BMI is the weight of a person divided by his or her height and is a measure of fatness of both males and females. You can determine your BMI by using **Table 8.3**.

Table 8.3

Body Mass Index

Height	Body Weight (pounds)																									
	19	20	21	22	23	24	25	26	27	28	29	30	31	32	33	34	35	36	37	38	39	40	41	42	43	44
4'10"	91	96	100	105	110	115	119	124	129	134	138	143	148	153	158	162	167	172	177	181	186	191	196	201	205	210
4'11"	94	99	104	109	114	119	124	128	133	138	143	148	153	158	163	168	173	178	183	188	193	198	203	208	212	217
5'	97	102	107	112	118	123	128	133	138	143	148	153	158	163	168	174	179	184	189	194	199	204	209	215	220	225
5'1"	100	106	111	116	122	127	132	137	143	148	153	158	164	169	174	180	185	190	195	201	206	211	217	222	227	232
5'2"	104	109	115	120	126	131	136	142	147	153	158	164	169	175	180	186	191	196	202	207	213	218	224	229	235	240
5'3"	107	113	118	124	130	135	141	146	152	158	163	169	175	180	186	191	197	203	208	214	220	225	231	237	242	248
5'4"	110	116	122	128	134	140	145	151	157	163	169	174	180	186	192	197	204	209	215	221	227	232	238	244	250	256
5'5"	114	120	126	132	138	144	150	156	162	168	174	180	186	192	198	204	210	216	222	228	234	240	246	252	258	264
5'6"	118	124	130	136	142	148	155	161	167	173	179	186	192	198	204	210	216	223	229	235	241	247	253	260	266	272
5'7"	121	127	134	140	146	153	159	166	172	178	185	191	198	204	211	217	223	230	236	242	249	255	261	268	274	280
5'8"	125	131	138	144	151	158	164	171	177	184	190	197	203	210	216	223	230	236	243	249	256	262	269	276	282	289
5'9"	128	135	142	149	155	162	169	176	182	189	196	203	209	216	223	230	236	243	250	257	263	270	277	284	291	297
5'10"	132	139	146	153	160	167	174	181	188	195	202	209	216	222	229	236	243	250	257	264	271	278	285	292	299	306
5'11"	136	143	150	157	165	172	179	186	193	200	208	215	222	229	236	243	250	257	265	272	279	286	293	301	308	315
6'	140	147	154	162	169	177	184	191	199	206	213	221	228	235	242	250	258	265	272	279	287	294	302	309	316	324
6'1"	144	151	159	166	174	182	189	197	204	212	219	227	235	242	250	257	265	272	280	288	295	302	310	318	325	333
6'2"	148	155	163	171	179	186	194	202	210	218	225	233	241	249	256	264	272	280	287	295	303	311	319	326	334	342
6'3"	152	160	168	176	184	192	200	208	216	224	232	240	248	256	264	272	279	287	295	303	311	319	327	335	343	351
6'4"	156	164	172	180	189	197	205	213	221	230	238	246	254	263	271	279	287	295	304	312	320	328	336	344	353	361

Source: The National Heart, Lung, and Blood Institute.
To use the table, find the appropriate height in the left-hand column. Move across to a given weight. The number at the top of the column is the BMI at that height and weight. Pounds have been rounded off.

A BMI of 25 to 29.9 is considered overweight and one 30 or above is considered obese. Additionally, a waist circumference of over 40 inches in males and over 35 inches in females signifies increased risk in those who have a BMI of 25 to 34.9. The report notes that the most successful strategies for weight loss include reducing the number of calories consumed, increasing physical activity, and engaging in behavior therapy designed to improve eating and exercise habits.

> When the body's metabolic needs have been met, any excess calories taken in are stored as fat. Overweight and obesity affect 55% of the adult population in the United States. Overweight and obesity are risk factors for many medical conditions and illnesses.

Proteins

Fats

Carbohydrates

Figure 8.11 Representative proteins, fats, and carbohydrates.
The foods shown in each of the photographs are representative dietary proteins, fats, and carbohydrates. Notice that eggs and cheese are shown in both the protein group and the fat group. However, eggs and cheese yield more calories from fat than from protein.

8.14 Eating for good nutrition and health

As the digestive process breaks down food into molecules usable in cellular respiration, it also unlocks the vitamins and minerals from these foods. Vitamins fall into two general categories: water soluble and fat soluble (see Table 8.1). The water-soluble vitamins enter the cells of the intestinal mucosa and move into the bloodstream. The fat-soluble vitamins are carried across the membranes of the intestinal cells associated with the fatty acids and glycerols. They are also transported to the bloodstream by means of the lymphatic system. Minerals are absorbed into the bloodstream as ions. Most people who eat a sufficient amount and variety of foods get the vitamins they need in the food they eat.

Proteins, lipids, and carbohydrates provide more than energy for the diet—they provide the raw materials to build the substances the body needs. In fact, proteins are used primarily as building blocks and not as a source of energy. Of the 20 different amino acids that make up proteins, humans can manufacture only 12. Therefore, 8 of the amino acids, called **essential amino acids**, must be obtained by humans from proteins in the food they eat (**Figure 8.11**).

Protein foods that contain the essential amino acids in amounts proportional to the body's need for them are called high-quality, or complete, proteins. Unfortunately, many high-quality protein foods such as meat, cheese, and eggs are high in animal fat and cholesterol as well. A high percentage of animal fat and cholesterol in the diet has been shown to result in weight gain and to increase the risk of heart attack because of the clogging of blood vessels with cholesterol deposits. To reduce fat intake, yet get the amino acids you need, you should combine any legume (dried peas, beans, peanuts, or soy-based food) with any grain, nut, or seed. Alternatively, you should combine any grain, legume, nut, or seed with small amounts of milk, cheese, yogurt, eggs, red meat, fish, or poultry.

So how much protein, fat, and carbohydrate should you have in your diet? The U.S. Senate Select Committee on Nutrition and Human Needs recommends that 12% of your daily intake of kilocalories come from proteins, 30% or less come from fats, and 58% from carbohydrates (Figure 8.11).

> Eating a variety of foods daily in which a large portion of the diet contains fruits, vegetables, and whole grains will help you take in the nutrients you need for good health.

The Evolution of Food Guides

Today, more Americans than ever before are concerned about nutrition, and keeping up with the latest nutrition research can be a daunting task. So how can people be sure that their diets are nutritious without scanning the newspaper or the Web for daily updates? A practical approach is to use a food guide, an easy-to-use chart developed by nutritionists to help people monitor the quality of their diets.

A food guide groups foods of similar nutrient composition together. Guides also recommend how much of each food should be eaten to meet nutrient needs. For example, the guide presently in use in the United States, the food pyramid, groups fruits together because they contribute a major portion of vitamins A, C, and folate to the diet. This guide recommends eating two to four servings of fruit daily. However, not all foods in each grouping are equally nutritious, so nutritionists recommend eating a variety of foods from each group.

Americans have used a variety of food guides over the past six decades. In 1943, in an effort to promote nutrition education, the U.S. Department of Agriculture (USDA) introduced the basic seven food guide. This plan remained in use until 1956, when it was replaced by the basic four guide. The basic seven guide divided foods into seven groups: (1) green and yellow vegetables; (2) oranges, tomatoes, and raw salads; (3) potatoes; (4) milk and cheese; (5) meat, poultry, fish, and eggs; (6) bread, flour, and cereals; and (7) butter and margarine. The basic four guide combined the basic seven's three fruit and vegetable groupings into one group and eliminated the butter and margarine group. The basic four plan was not revised for more than 20 years, and it formed the basis for most nutrition education during that time.

By the late 1970s, critics charged that the basic four guide did not reflect current scientific findings about the role of such nutrients as fats in the development of heart disease and some cancers. Responding to these concerns, the USDA presented the *Hassle-Free Guide to a Better Diet* in 1979. This guide, sometimes referred to as the "basic five,"

added a fifth food group (fats, sweets, and alcoholic beverages) and recommended limiting intake of foods rich in these substances. However, critics still were not satisfied that these recommendations went far enough.

Again, the USDA responded with a revised food guide. *The Food Wheel: A Daily Pattern for Food Choices* recommended eating more fiber-rich fruits, vegetables, and whole grains and limiting dairy products and meats (foods that are high in fat) in the diet. By this time, the development and use of food guides had become a political and economic issue in

providing a visual model of the optimal diet. Grouped together at the base of the pyramid are cereals and other grain products. These foods have the highest number of recommended servings. Fruits and vegetables, grouped separately, form the next tier. Fewer servings of these foods are needed for a healthy diet, so the tier is smaller. Meats, meat alternatives, and dairy products form the next, still smaller tier, because they include dietary fat. At the top are high-sugar or fat foods and alcoholic beverages, which offer few vitamins and minerals in relation to calories.

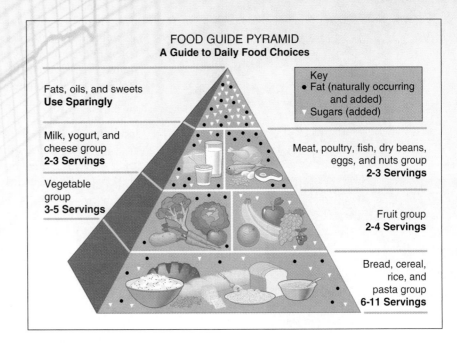

addition to one of health. Meat and dairy producers, for example, were not pleased with the latest food guide that recommended the limitation of their products in the American diet.

The latest food guide was introduced in 1992. Called the *Food Pyramid*, this guide continues to reflect health concerns. In some ways it is a return to the basic seven concept with more food groups, but it also adds recent recommendations on healthy eating by

Like the food guides of the past, the Food Pyramid is not perfect. Combination foods such as casseroles and ethnic foods such as Asian or Mexican dishes may still be difficult to classify without checking recipes. However, the Food Pyramid is still one of the easiest ways to compare your diet with the most current nutritional guidelines, even as nutritional information is modified by research and the Food Pyramid undergoes its inevitable evolution. ⚙

Key Concepts

Heterotrophs digest food inside or outside their cells.

8.1 Intracellular digestion usually takes place in single-celled protists and the sponges.

8.2 In bacteria and fungi, extracellular digestion takes place without a digestive system.

8.3 In animals, extracellular digestion usually takes place in a digestive system.

Food provides both energy and building materials.

8.3 Carbohydrates, lipids (fats and oils), and proteins provide energy and building blocks.

8.4 Vitamins, minerals, and water help various chemical reactions and bodily processes take place.

Human digestion is mechanical and chemical, and is an extracellular process.

8.5 Food is digested mechanically in the mouth and stomach, and chemically by means of enzymes in the mouth, stomach, and small intestine.

8.6 Teeth shred and grind food in the mouth, while the digestion of carbohydrates begins there with the action of salivary amylase.

8.7 When food is swallowed, it moves along the esophagus to the stomach.

8.8 The stomach juices contain a concentrated acid that unfolds proteins and in which the protein-digesting enzyme pepsin is active.

8.9 Most of the digestion of food takes place in the small intestine with the help of the liver and pancreas, accessory organs to digestion.

8.10 In the small intestine, a variety of enzymes, many synthesized in the pancreas, act to complete digestion; two hormones help orchestrate this process.

8.11 The products of digestion are absorbed across the walls of the small intestine, which possess numerous villi that create a large surface area.

8.11 During absorption, amino acids and monosaccharides pass into the bloodstream. Fatty acids and glycerols pass into the lymphatic system first and are then transported to the bloodstream.

8.12 The large intestine functions principally to absorb water from, and to eliminate, digestive waste.

☀ Health Connections

8.13 Overweight and obesity affect 55% of the adult population in the United States.

8.13 Overweight and obese persons are at higher risk than others for certain diseases such as type 2 diabetes, heart disease, and stroke.

8.14 The U.S. Senate Select Committee on Nutrition and Human Needs recommends that 12% of the daily intake of kilocalories come from proteins, 30% or less from fats, and 58% from carbohydrates.

Key Terms

amylases (AM-uh-lace-es) *161*

anus *169*

autotrophs (AW-toe-trofs) *154*

bile *166*

carboxypeptidase (kar-BOK-see-PEP-ti-dace) *167*

cholecystokinin (CCK) (KOL-uh-SIS-tuh KINE-un) *167*

chymotrypsin (kye-moe-TRIP-sin) *167*

colon *169*

digestion *154*

digestive system *157*

disaccharidases (dye-SACK-uh-rye-DAYS-is) *167*

duodenum (DOO-oh-DEE-num *or* doo ODD-un-um) *166*

epiglottis (ep-ih-GLOT-iss) *163*

esophagus (ih-SOF-uh-gus) *163*

essential amino acids *171*

gallbladder *166*

gastric glands *164*

gastrin *165*

glottis (GLOT-iss) *163*

heterotrophs (HEH-tur-oh-trofs) *156*

ileum (ILL-ee-um) *166*

jejunum (ji-JOO-num) *166*

larynx (LAIR-inks) *163*

lipases (LYE-pays-es) *161*

liver *166*

nutrients *158*

pancreas (PANG-kree-us *or* PAN-kree-us) *166*

pancreatic amylase (pang-kree AT-ick *or* pan-kree-AT-ick AM-uh-lace) *167*

pancreatic lipase (pang-kree-AT-ick LYE-pace *or* LIP-ace) *167*

pepsin *165*

peptidases (pep-teh-DACE-is) *167*

peristalsis (pear-ih-STALL-sis) *164*

pharynx (FAIR-inks) *163*

proteases (PRO-tee-ACE-es) *161*

rectum *169*

salivary amylase (SAL-i-VER-ee AM-uh-lace) *161*

salivary glands (SAL-ih-VER-ee GLANDZ) *161*

secretin (sih-KRE-tin) *167*

small intestine *166*

stomach *164*

trypsin (TRIP-sin) *167*

villi (VIL-eye) *168*

KNOWLEDGE AND COMPREHENSION QUESTIONS

1. Name the six classes of nutrients, and explain why you need each one in your diet. (*Comprehension*)

2. Describe the digestive fates of proteins, lipids, and carbohydrates. What three types of enzymes are involved? (*Comprehension*)

3. Where does the process of human digestion begin? (*Knowledge*)

4. You bite off a mouthful from a crunchy apple and chew it thoroughly. Describe the role played by each type of tooth in your mouth. (*Comprehension*)

5. Is the muscle activity of peristalsis under voluntary control or is it an involuntary process? Does digestion occur in the esophagus as peristalsis is occurring? (*Knowledge*)

6. What important tasks does the hydrochloric acid in the human stomach accomplish? What protects the stomach wall from potential damage by this chemical product? (*Knowledge*)

7. What is the duodenum? Summarize the digestive activities that take place there. (*Comprehension*)

8. Identify and give a function for each of the following: gastrin, secretin, and cholecystokinin. What do they have in common? (*Comprehension*)

9. Discuss the features of the small intestine that increase its internal surface area. Why is this increase important? (*Comprehension*)

10. Summarize how your body produces ATP from glucose, amino acids, triglycerides, and fatty acids. (*Comprehension*)

KEY

Knowledge: Recalling information.
Comprehension: Showing understanding of recalled information.

CRITICAL THINKING REVIEW

1. Many people of other countries are vegetarians. This has become more common in the United States in recent years also. Discuss how a vegetarian could create a balanced diet while eliminating meat and animal products. (*Synthesis*)

2. While on a hike, a student finds the dry skull of a small mammal. How could the student determine whether that animal ate only plant products? (*Analysis*)

3. Why do you think that the liver and pancreas are not called organs of the digestive system although they serve important accessory roles in digestion? (*Application*)

4. For years, baking soda (sodium bicarbonate) has been a popular home remedy for indigestion and heartburn. Explain why. (*Synthesis*)

5. Keep a record of everything you eat and drink for 3 days. Then (this could be painful!) analyze your diet. Compare it with the recommendations in this chapter. How could you improve your nutritional intake? (*Evaluation*)

KEY

Application: Using information in a new situation.
Analysis: Breaking down information into component parts.
Synthesis: Putting together information from different sources.

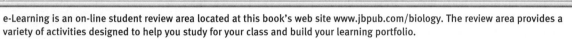

e-learning

Location: http://www.jbpub.com/biology

e-Learning is an on-line student review area located at this book's web site www.jbpub.com/biology. The review area provides a variety of activities designed to help you study for your class and build your learning portfolio.

Review Questions The review questions test your knowledge of the important concepts and applications in each chapter. The review provides feedback for each correct or incorrect answer. This is an excellent test preparation tool.

Figure-Labeling Exercises Sharpen your visual thinking by matching terms and labels for important illustrations.

Flash Cards Studying biology requires learning new terms. Virtual flash cards help you master the new vocabulary for each chapter.

Just Wondering As you read and study from this text, you may find that you have unanswered questions. Through this site you can ask the author, Sandy Alters, your "just wondering" questions.

Do you prefer the speed and reliability of a CD-ROM? All of the features contained on the eLearning portion of the web site are also available on the Student CD-ROM.

Respiration

Awkward in movement out of the water, these northern elephant seals are master divers. They often spend hours under the water. In fact, taking a mile-deep dive and staying down for about two hours is a normal occurrence for these seals and the other master divers of the ocean, the sperm whales.

The diving abilities of these animals are far superior to those of humans, who can only dive to depths of slightly over 100 meters, or about 300 feet, while holding their breath. (The world record is 133 meters.) Both humans and seals breathe in much the same way. So what confers this diving advantage upon the northern elephant seal? Put simply—myoglobin. Found in the muscles, myoglobin is an oxygen-carrying protein that is similar to hemoglobin, the oxygen-carrying protein of the blood. Myoglobin stores the oxygen needed for the muscles to burn fuel. Human muscle cells have small amounts of myoglobin. In contrast, elephant seals (and the other adept divers) have high concentrations of myoglobin in their muscles. Because their muscles can store so much oxygen, seals, whales, sea turtles, and penguins can stay submerged for long periods without coming up for a fresh gasp. In addition, many of their body processes slow down, allowing their muscles to have a large share of the oxygen supply. Amazingly, the heart rate of elephant seals can fall to a low of 3 beats per minute during a dive!

Organization

Respiration takes place over thin, large, moist surfaces.

Human respiration uses special organs and relies on pressure changes within the thoracic cavity.

Health
Connections

Respiration takes place over thin, large, moist surfaces.

9.1 What is respiration?

Although the diving abilities of elephant seals may be amazing and unusual in the animal world, their using oxygen is not. Seals, whales, penguins, and humans are only a few of the myriad organisms that **respire**; they take up oxygen from the environment and release the waste gas carbon dioxide. Oxygen is used within cells during the breakdown of fuel, primarily glucose. The chemical process by which cells break down fuel molecules using oxygen, producing carbon dioxide, and releasing energy is called **aerobic cellular respiration**. (This process is described in Chapter 6.)

Respiring organisms obtain oxygen from their environments—either the air or water. They also rid themselves of the waste gas carbon dioxide, expelling it to the environment. In order for an organism to exchange oxygen and carbon dioxide with the environment it must have a thin, moist, and relatively large respiratory surface over which these gases can move. Body fluids move past the inside of this surface, and air or water moves past the outer, environmental side of the surface.

Diffusion is the mechanism by which oxygen and carbon dioxide move across a respiratory surface. During this process, molecules of oxygen and carbon dioxide each move from areas of high concentration and pressure to areas of low concentration and pressure. (Because the pressure of each individual gas makes up part of the total, it is called partial pressure.) They are said to move down a concentration and pressure gradient. (This process is described in detail in Chapter 4, pp. 78–80.) The oxygen supply within a respiring organism is continually being depleted because it is used during cellular respiration, so oxygen diffuses from the environment (the area of higher concentration) and into the organism (the area of lower concentration). Likewise, carbon dioxide continually increases in concentration within a respiring organism because it is a waste product of cellular respiration. Therefore, carbon dioxide diffuses from the organism and into the environment.

Respiration is the uptake of oxygen from the environment and the release of the waste gas carbon dioxide.

9.2 Respiration over cell and skin surfaces

In organisms that are single cells, such as bacteria and single-celled protists, the cell membrane serves as the respiratory surface (Figure 9.1). Oxygen and carbon dioxide simply diffuse across this thin cell layer. These organisms, which live in aquatic or moist environments, are small and their cell membranes provide adequate surface area over which enough oxygen and carbon dioxide can diffuse for their survival. (See p. 61 for a description of surface-to-volume ratio.)

The outer layer of cells of certain multicelled animals (some sponges, corals, and jellyfish; many flatworms, tapeworms, and roundworms; and a few earthworms) also act as respiratory surfaces. All these organisms live in watery environments with the exception of the roundworms and earthworms, which live in soil. These soil-dwelling organisms must keep their outermost cells (the cuticle) moist or respiration will cease.

Some of these organisms are small, such as the flatworms. Others, although much larger, have great surface-to-volume ratios because of their shapes. Corals, for example, are usually branching, fanlike, or fringed, each shape providing a large surface area over which respiration can take place. Earthworms, roundworms, and tapeworms have large surface areas created by their elongated shapes. Sponges, vase-shaped masses of cells embedded in a jelly-like matrix, also have large surface areas. Additionally, pores dot the surface of these organisms, providing channels through which water can pass. The beating of whip-like extensions on cells within the sponge creates currents of oxygen-laden water that flow from the outside to the inside of the organism, exiting the sponge's open top.

Some large, multicellular organisms can breathe through their skin. The most familiar are the amphibians, such as salamanders and frogs, but certain fishes, snakes, turtles, and lizards rely on the skin as a respiratory organ to one degree or another. These animals have thin, moist skins with an abundant supply of blood vessels just beneath the skin surface. Additionally, most of these animals have lungs or gills that also provide a surface for gas exchange, supplementing their "skin breathing." (Only certain small salamanders survive by skin breathing alone and have neither lungs nor gills.)

Figure 9.1 reviews the basic respiratory plans of single-celled organisms, simple multicelled organisms, and complex multicelled animals, which are described next.

Organisms that breathe through their skin or epidermal layer are either single-celled, multicellular with a large surface-to-volume ratio, or large with thin, moist skins supplied with abundant blood vessels.

RESPIRATION OVER CELL AND SKIN SURFACES

Cell membrane
(single-celled organisms)

Outer layer of cells
(certain multicelled animals)

Skin
(amphibians and certain fishes, snakes, and turtles)

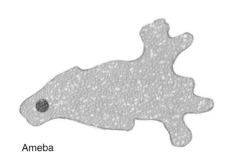

Ameba

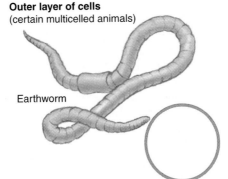

Earthworm

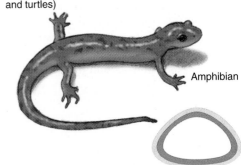

Amphibian

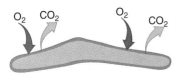

O_2 CO_2 O_2 CO_2

O_2 CO_2 O_2 CO_2

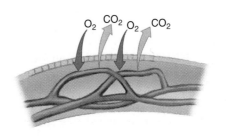

RESPIRATION USING SPECIAL ORGANS

Tracheal system
(most spiders and insects)

Gills
(most large multicellular aquatic animals)

Lungs
(most large, multicellular land-dwelling animals, including most animals that also respire through the skin)

Grizzly bear

Salmon

Grasshopper

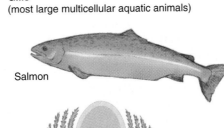

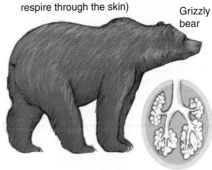

O_2

CO_2

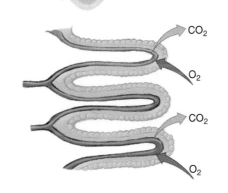

CO_2

O_2

CO_2

O_2

CO_2

O_2

O_2

CO_2

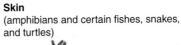

9.3   Respiration using special organs

Many animals are large and their bodies have small surface-to-volume ratios. They need special organs for oxygen uptake and carbon dioxide release. These specialized organs of respiration are shaped in ways that provide the large surface area necessary for respiration. Although there are many types of respiratory systems, three primary mechanisms are the tracheal systems of spiders and insects, the gills of fishes and other aquatic animals, and the lungs of reptiles and mammals.

Tracheal Systems

A **trachea** (TRAY-kee-uh) is an air-filled tube. (This word comes from a Latin word meaning "windpipe" and is also used to denote the windpipe in mammalian respiratory systems, such as in humans.) A **tracheal system** is a branching network of microscopic air tubes. Some spiders and most insects have tracheal systems (see Figure 9.1), and a variety of types of these systems exist. The tracheal systems of these two groups appear to have evolved independently from one another, and are a good example of convergent evolution (see p. 544).

In general, the tracheal system is open to the environment by means of small holes in the cuticle (exoskeleton) of the insect or spider. These openings are called **spiracles**, a term which comes from a Latin word meaning "to breathe." In many tracheal systems, spiracles are muscular openings that can open and close under the direction of the nervous system, triggered by carbon dioxide levels. Spiracles open and close at different times from one another, regulating the flow of air through the system.

In insects, tracheas originate at the spiracles and usually interconnect to form a pair of longitudinal trunks (**Figure 9.2a**) that extend the length of the thorax (midsection) and abdomen (last section) of the organism. Figure 9.2b shows how tracheas branch into narrower and narrower passageways called tracheoles. The tracheoles end in microscopic branches that are lined with moist epithelium and extend to all parts of the body. Some extend into individual cells. Oxygen and carbon dioxide exchange takes place at the moist ends of the tracheoles.

In small insects and those that use little energy, gases move through the tracheal system by diffusion. In large insects and ones that move about and fly, diffusion is not enough. The body movements of large, active insects compress parts of the tracheal system, thereby changing its volume (and consequently the pressure) within the tracheas. These movements and the resultant volume and pressure changes help pump air through the system.

Many aquatic insects, whose ancestors lived on land, breathe air via a tracheal system. One example with which you may be familiar is mosquito larvae. The photo of the larva shows it sticking the tip of its abdomen out of the water while keeping the rest of its body submerged. This positioning enables air to enter the spiracles located at the tip of abdomen, while the organism continues its development underwater.

Gills

Gills are organs by which large aquatic animals respire (with the exception of those mentioned previously that use cell or skin surfaces, and the aquatic mammals such as whales, seals, and dolphins). Crustaceans (such as lobsters and crabs), most molluscs (such as clams), aquatic worms, fishes, and some amphibians have gills.

Just as tracheas (and lungs) are invaginations (a turning inward) of the body surface, gills are evaginations of the body surface (a turning outward), which means they are external structures. In addition, gills often have protective coverings because they are feathery and delicate, and are therefore easily damaged. Bony fishes, for example, have a protective flap covering the gills called an operculum.

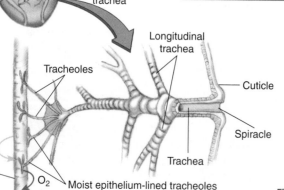

> Sets of spiracles along the thorax and abdomen open into tracheas, which branch into tracheoles where gas exchange takes place.

(a)

Spiracle

Longitudinal trachea

Longitudinal trachea

Tracheoles

Cuticle

Spiracle

Trachea

CO_2

Muscle

O_2

Moist epithelium-lined tracheoles

(b)

Figure 9.2   **The tracheal system of an ant.** (a) Longitudinal trachea. (b) Branching of trachea.

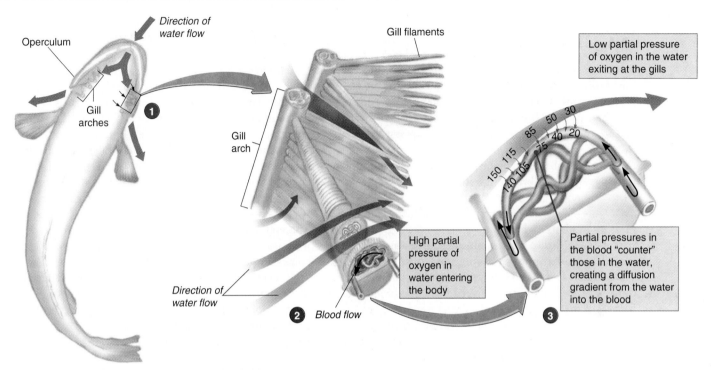

Figure 9.3 Countercurrent flow in gills. During countercurrent flow, the partial pressure of oxygen in the water is always higher than the partial pressure of oxygen in the blood. Therefore, oxygen diffuses from the water and into the blood as they flow past one another.

Gills can be distributed throughout an organism, as in the case of aquatic worms, or can be in a particular location, as in bony fishes. Figure 9.3 is an illustration of gills in a bony fish.

Just as humans inhale and exhale to ventilate (pass air through) the lungs, so do gill breathers ventilate the gills. Figure 9.3 diagrams the way in which fishes pass water over the gills. Water enters the fish's mouth, passes through slits in the throat (pharynx), flows over the gills, and exits at the back of the gill cover (operculum) **①**. Notice in the enlarged view of the gill that the water flows in a direction opposite to the flow of blood **②**. This movement is called countercurrent flow—one fluid flows counter to (against) the flow of the other. Countercurrent flow enhances the diffusion of oxygen into the blood of the fish and the diffusion of carbon dioxide into the water **③**.

Lungs

Lungs are organs by which most amphibians, aquatic mammals (such as whales, seals, and dolphins), and large land-dwelling animals respire (with the exception of the insects, spiders, roundworms, and earthworms mentioned previously). Lungs are invaginations of the body surface, but are not located throughout the body as are tracheal systems. Lungs are confined to one part of the organism. Therefore, "lung-breathers," like "gill-breathers," have a circulatory system to distribute the oxygen to other parts of the body.

Lungs are made up of microscopic air sacs that are covered with blood vessels and coated with a thin film of moisture. It is at the air sacs that oxygen enters the blood and carbon dioxide leaves. Narrow passageways connect the lungs to the outside, their thinness minimizing the loss of moisture to the environment.

As organisms evolved on land, variations in lung development occurred. Snails, for example, lack the gills of their aquatic mollusc relatives: clams, octopus, and squid. Instead, their mantle cavity acts like a lung. Most spiders have structures called book lungs, which are stacks of spaced membranes within their abdomens, connected to the outside by a tracheal system. Other organisms, such as most amphibians, reptiles, and birds, have lungs that are closer in their anatomy and physiology to human lungs (and those of other mammals), but are ventilated differently. Frogs, for example, gulp air and pump it into their simple lungs. Birds have air sacs in addition to lungs, which help ventilate the lungs.

Large animals with small surface-to-volume ratios use specialized organs for respiration. These organs provide the large, moist surface area necessary for respiration. Three types of specialized respiratory organs are the tracheal systems of spiders and insects, the gills of fishes and other aquatic animals, and the lungs of reptiles and mammals.

Human respiration uses special organs and relies on pressure changes within the thoracic cavity.

9.4 The general plan

In humans (and other mammals), air moves into the lungs through the nose, mouth, and airways of the **respiratory system**. The movement of air into and out of the lungs is called **breathing**. A breath consists of taking air into the lungs, or **inspiration**, and expelling air from the lungs, or **expiration**. Try to breathe normally, count the number of breaths you take in 15 seconds, and multiply by 4. Is your breathing rate within the average of 14 to 20 breaths per minute?

Figure 9.4 shows the general plan of the human respiratory system. As blood moves around the body and past its cells within microscopic blood vessels, carbon dioxide diffuses from the tissue fluid into the blood. *Tissue fluid* (also called interstitial fluid or intercellular fluid) is a waterlike fluid derived from the blood that bathes all the cells of the body. Oxygen moves from the blood into the tissue fluid. This process is called **internal respiration**, the exchange of oxygen and carbon dioxide between the blood and the tissue fluid. **External respiration** is the exchange of gases at the lungs.

In humans (and other mammals), air moves into the lungs through the airways of the respiratory system. At the lungs, oxygen diffuses into the bloodstream, which carries it to the cells. There, oxygen moves into the tissue fluid. The waste gas, carbon dioxide, diffuses into the blood from the tissue fluid and is carried to the lungs, where it is expelled.

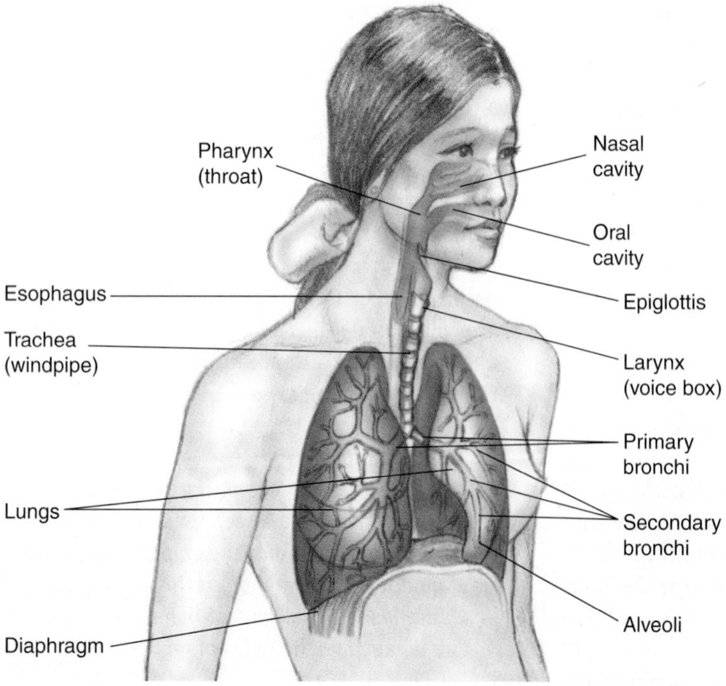

9.5 The journey to the lungs

The Nasal Cavities

As you breathe in, air first enters your body through the nostrils. The nostrils are lined with hairs that filter out dust and other particles from the air. The air is warmed and moistened as it swirls around in the **nasal cavities**. These cavities, located above your oral cavity and behind your nose, are bordered by projections of bone covered with epithelial tissue. This tissue stays moist with mucus secreted by its many mucous glands. This sticky fluid helps trap dirt and dust that you breathe in. The epithelium is also covered with tiny, hairlike projections called *cilia* (see Figure 9.6). The word cilia comes from a Latin word meaning "eyelashes." These "cell eyelashes" beat in unison, creating a current in the mucus that carries the trapped particles toward the back of the nasal cavity. From here, the mucus drips into the throat and is swallowed—at a rate of over a pint per day!

The Pharynx

After passing through the nasal cavities, air enters the **pharynx** (FAIR inks), or throat. The pharynx extends from behind the nasal cavities to the openings of the esophagus and larynx. The esophagus, as you may recall from Chapter 8, is the food tube, a passageway for food to the stomach. The **larynx** (LAIR-inks), or voice box, lies at the beginning of the trachea. The trachea is the air passageway that runs down the neck in front of the esophagus and brings air to the lungs.

The Larynx

The larynx is a cartilaginous box shaped somewhat like a triangle. Stretched across the upper end of the larynx are the **vocal cords (Figure 9.5)**. The vocal cords are two folds of elastic tissue covered with a mucous membrane; thus they are also called vocal folds. Muscles within the larynx pull on its cartilaginous walls, regulating the tension on the vocal cords. When the vocal cords are stretched tightly, the space between them—the **glottis** (GLOT-iss)—is closed. In this way the vocal cords provide a "backup"

Figure 9.4 **The human respiratory system.**

Pharynx (throat)
Nasal cavity
Oral cavity
Esophagus
Epiglottis
Trachea (windpipe)
Larynx (voice box)
Primary bronchi
Lungs
Secondary bronchi
Diaphragm
Alveoli

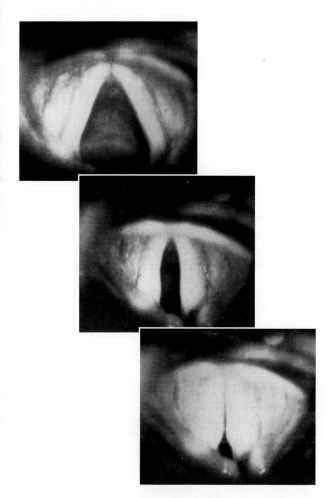

Figure 9.5 **Movement of the vocal cords (folds).** The space between the two folds of tissue is the glottis. In this series of photographs, the vocal cords are shown in the process of closing off the glottis.

for the epiglottis that flaps over the glottis during swallowing. Both structures work to prevent food and drink from going down the wrong way (see Section 8.7). The vocal cords also produce the sounds you make as air rushes by and causes them to vibrate. You can illustrate this principle by stretching a rubber band between your fingers. Have someone repeatedly pluck the band. At the same time, increase and then decrease the stretch on the band. What happens? Your vocal cords work in much the same way to produce a variety of pitches of sound. You also have a mouth with lips and a tongue to form the sounds into words. Additionally, the size and shape of the passageways in your chest, head, nose, throat, and mouth affect the quality of the sound. Your lungs add a power supply and volume control to your personal musical instrument—your voice.

The Trachea

Put your hand on your larynx, or Adam's apple, and then picture about 4 or 5 inches of garden hose attached to its bottom end. A garden hose is about the diameter of your trachea, or windpipe, which extends downward from your larynx toward your lungs. The trachea has thin walls, similar to the thickness of those in the hose. Garden hoses are reinforced with materials such as rubber or vinyl to keep them from collapsing; your trachea is reinforced with rings of cartilage. The cartilage wraps around the trachea only part way, forming C shapes that begin and end where the windpipe lies next to the esophagus. Press gently on your windpipe just below your Adam's apple, and you can feel some of these cartilaginous rings.

The inner walls of the trachea are lined with ciliated epithelium **(Figure 9.6).** Certain cells in the epithelium secrete mucus. Together, the cilia and the mucus provide your windpipe with an "up escalator" for any particles or microbes you may have inhaled. This escalator brings substances up to the pharynx, where they are swallowed and eliminated through the digestive tract. In the trachea of a cigarette smoker, however, action of the cilia is impaired, causing mucus to build up in the airway. The result is that the tars in cigarettes are not caught and expelled with the action of cilia. They move easily into the lungs and settle there. A chronic cough, often called smoker's cough, is triggered by accumulations of mucus below the larynx.

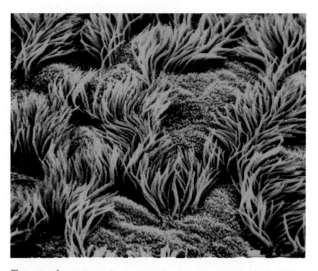

Figure 9.6 **Colorized scanning electron micrograph of cilia.** Cilia such as these line the trachea and sweep trapped particles out of the respiratory tract.

The Bronchi and Their Branches

If you could look down your trachea, as in the upper photograph in **Figure 9.7,** the two black holes you would see would be your **primary bronchi** (BRON-keye). These airways are structured much like the trachea, but are smaller in diameter. One bronchus goes to each lung, branching into three right and two left **secondary bronchi** serving the three right and two left lobes of the lungs. The heart is nestled into the left side of the lungs, taking up some of the space that a third left lobe might occupy.

The secondary bronchi divide into smaller and smaller branches, looking much like an upside-down tree until they end in thousands of passageways called **respiratory bronchioles** (BRON-kee-olz). These airways have a diameter less than that of a pencil lead. Their walls have clusters of tiny pouches that, along with the respiratory bronchioles, are the sites of gas exchange. These pouches, or air sacs, are called **alveoli** (al-VEE-uh-lye).

The Alveoli: Where Gas Exchange Takes Place

The alveoli provide a perfect place for carbon dioxide and oxygen to diffuse between the air in the lungs and the blood. Membranes made up of a single layer of epithelial cells bound these clusters of microscopic air sacs, and a network of capillaries tightly clasps each alveolar sac. The capillary walls are also only one cell thick and press against the alveolar epithelium. These two adjacent membranes provide the thinnest possible barrier between the blood in the capillaries and the air in the alveoli.

Visual Summary **Figure 9.7** **Gas exchange at the lungs.**
A section of a lung is enlarged to show the exchange of the respiratory gases, oxygen (O_2) and carbon dioxide (CO_2), between the blood and alveoli.

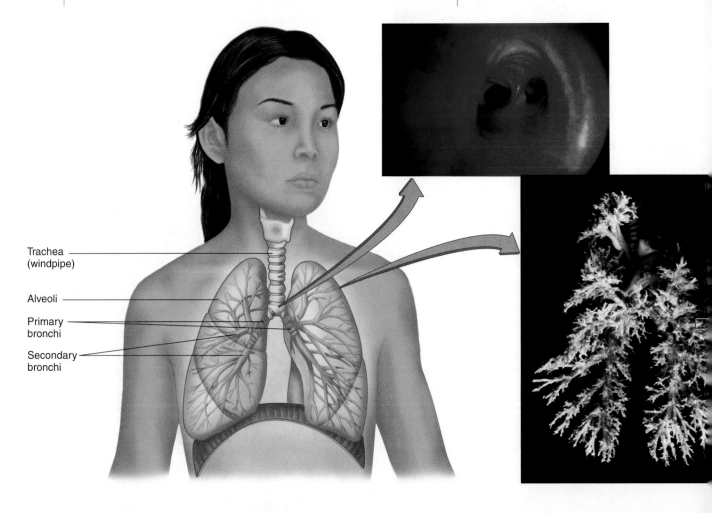

Trachea (windpipe)

Alveoli

Primary bronchi

Secondary bronchi

The alveoli also provide another important component of efficient gas exchange: a large surface area. In fact, if the epithelial membrane of all your alveoli were spread out flat, it would cover a tennis court! The capillaries cover this enormous surface, creating patterns much like tightly woven spider webs, providing nearly a continuous sheet of blood over the alveolar surface.

In the magnified view of the bronchioles and alveoli in Figure 9.7, the blue vessels denote arteries that carry deoxygenated blood from the right side of the heart to the lungs ❶. After the blood picks up oxygen at the lungs and gets rid of carbon dioxide ❷, it travels to the heart via veins (shown in red ❸), and gets a push out to the rest of the body.

Large white blood cells called macrophages are also found in the alveoli. These cells work to remove any particles or microbes that have escaped the other defenses of the airways. The macrophages transport the invaders to the bronchioles or to the lymphatic system.

Sometimes the job is too big, however, and particles remain in the lungs. In smokers, the action of the macrophages is impaired, making the lungs more susceptible to disease and injury.

During human breathing, air enters the body through the nostrils, where it is warmed and moistened. Air passes from the nasal cavities into the throat, or pharynx, and then passes over the vocal cords that are stretched across the larynx. From this voice box, air moves into the trachea, or windpipe, on its way to the lungs. The trachea branches into two bronchi that supply each lung. The bronchi divide into smaller and smaller branches ending in respiratory bronchioles having outpouchings called alveoli, the sites of gas exchange. The alveoli provide a thin, enormous surface area over which gas exchange can take place.

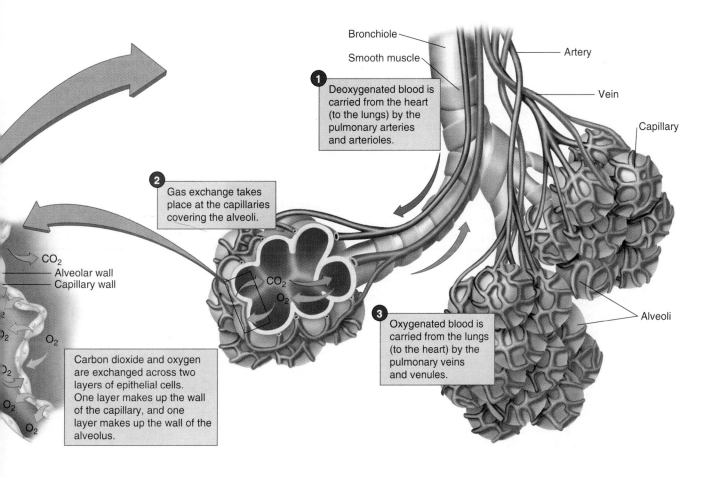

Bronchiole

Smooth muscle

Artery

Vein

Capillary

❶ Deoxygenated blood is carried from the heart (to the lungs) by the pulmonary arteries and arterioles.

❷ Gas exchange takes place at the capillaries covering the alveoli.

CO_2

Alveolar wall

Capillary wall

CO_2

O_2

O_2

O_2

O_2

O_2

Carbon dioxide and oxygen are exchanged across two layers of epithelial cells. One layer makes up the wall of the capillary, and one layer makes up the wall of the alveolus.

❸ Oxygenated blood is carried from the lungs (to the heart) by the pulmonary veins and venules.

Alveoli

9.6 The mechanics of breathing

Air moves into and out of your lungs as the volume of your thoracic cavity is made larger and smaller by the action of certain muscles. The **thoracic cavity** (thu-RASS-ick), or chest cavity, is within the trunk of your body above your diaphragm and below your neck. The **diaphragm** (DYE-uh-fram) is a sheet of muscle that forms the horizontal partition between the thoracic cavity and the abdominal cavity. Various blood vessels and the esophagus penetrate it as they traverse these two body cavities. The position of the diaphragm is shown in **Figure 9.8** (also see Figure 7.5).

Other muscles of breathing assist the diaphragm. These muscles extend from rib to rib—from the lower border of each rib to the upper border of the rib below—and are called **intercostal muscles**. The word intercostal literally means "between" (*inter*) "the ribs" (*costal*). You have two sets of intercostals: the internal intercostals and the external intercostals. The *internal intercostals* are those that lie closer to the interior of the body (as their name suggests) and that have fibers extending obliquely downward and backward (from front to back). The *external intercostals* extend from back to front, having fibers that are directed downward and forward.

Inspiration

When you are breathing quietly—not exerting yourself physically—your diaphragm and external intercostals alone are responsible for the change in the size of your thoracic cavity. This change in size results in the movement of air into and out of your lungs. During inspiration the dome-shaped diaphragm contracts, flattening somewhat and thereby lowering the floor of the thoracic cavity. The external intercostals contract, raising the rib cage ❶. Notice in Figure 9.8 how these two actions increase the size of the thoracic cavity.

The interior walls of the thoracic cavity are lined with a thin, delicate, sheetlike membrane called the *pleura*. The pleura folds back on itself to cover each lung. These two parts of the membrane are close to one another, separated only by a thin film of fluid. As the volume of the thoracic cavity increases during inspiration, the lungs also expand, held to the wall of the thoracic cavity by cohesion of the water molecules between the two membranes. Put simply, the lungs stick to the thoracic wall and move with it.

As the lungs expand in volume, the air pressure within the lungs decreases because there are fewer air molecules per unit of volume. As a result, air from the environment outside the body is pulled into the lungs, equalizing the pressure inside and outside the thoracic cavity. By means of this process, you breathe in 13,638 liters (more than 3000 gallons) of air every day!

Expiration

The lungs contain special nerves called *stretch receptors*. When the lungs are stretched to their normal inspiratory capacity, these receptors send a message to a respiratory center in the brain. This respiratory center is located in parts of the brainstem called the medulla and the pons (see Chapter 14). The respiratory center stops sending "contract" messages to the muscles of breathing, which causes them to relax—a passive process in contrast to the active process of inspiration. As the diaphragm relaxes, it assumes its domelike shape, reducing the volume of the thoracic cavity. Likewise, as the external intercostals relax, the rib cage drops (Figure 9.8 ❷), reducing the volume of the thoracic cavity further. The volume of the lungs, in turn, decreases, aided by the recoil action of the lungs' elastic tissue. The reduced volume of the lungs results in an increase in the air pressure within them. Air is forced out of the lungs, equalizing the pressure outside and inside the thoracic cavity. The cycle of inspiration and expiration is shown in **Figure 9.9**.

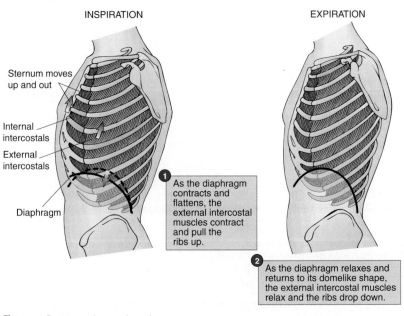

INSPIRATION EXPIRATION

Sternum moves up and out

Internal intercostals

External intercostals

Diaphragm

❶ As the diaphragm contracts and flattens, the external intercostal muscles contract and pull the ribs up.

❷ As the diaphragm relaxes and returns to its domelike shape, the external intercostal muscles relax and the ribs drop down.

Figure 9.8 **How a human breathes.**

The brainstem also contains areas sensitive to changes in the levels of carbon dioxide in the blood that control the rate of the cycle of inspiration and expiration. Hydrogen ions (H^+) that are produced from a series of chemical reactions involving carbon dioxide also play an extremely important role in influencing the activity of the respiratory center (see Internal Respiration, p. 191). Thus, when you have a high level of carbon dioxide in your blood, the level of hydrogen ions in your blood increases (blood pH decreases), and your breathing rate is "stepped up."

Deep Breathing

When you breathe deeply, such as during physical exercise, the internal intercostals as well as other muscles in the chest and abdomen help out the muscles of respiration. During inspiration, certain muscles attached to the sternum, or breastbone—the bone in the center of your chest to which your ribs are attached—pull up on it. In addition, other muscles pull up on the first two ribs. This action increases the volume of the chest cavity more than during quiet breathing, so more air flows into the lungs. By deep breathing and increasing the rate of breathing, world champion runners have been shown to increase their air intake fifteenfold.

Expiration, a passive process during quiet breathing, becomes an active process during deep breathing. The internal intercostals pull down on the rib cage. Abdominal muscles also contract, pulling down on the lower ribs and compressing the abdominal organs, which results in a push up on the diaphragm. These actions also decrease the volume of the thoracic cavity and cause more air to be expelled than during quiet breathing.

Lung Volumes

How much air do you move into and out of your lungs during inspiration and expiration? You can find out by performing this simple procedure: Fill a large jar with water and invert it in a pan of water. Be careful not to let any of the water seep out of the jar while you are turning it upside down. Mark the level of the water in the jar with tape or a wax pencil. Take a piece of rubber tubing or garden hose about a foot long and slip one end up into the inverted jar. Put the other end into your mouth and exhale normally. The air you breathe out will displace the water. Mark the new level of the water. Remove the jar from the water and fill it to this line with water. Now measure the volume of the water between the two lines with a measuring cup or a graduated cylinder. Its volume equals the volume of air you breathed out. The average adult male breathes out 500 milliliters

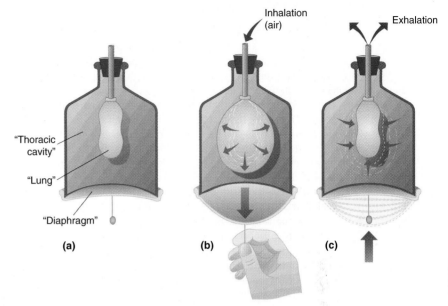

Figure 9.9 **A simple experiment that shows how you breathe.** In the jar is a balloon (a). When the diaphragm is pulled down, as shown in (b), the balloon expands. When it is relaxed (c), the balloon contracts. In the same way, air is taken into your lungs when your diaphragm contracts and flattens, increasing the volume of your chest cavity. When your diaphragm relaxes and resumes its dome shape, the volume of your chest cavity decreases and air is expelled.

of air during quiet breathing, or slightly more than 1 pint (Figure 9.10). This volume of air—the amount inspired or expired with each breath—is called the **tidal volume**.

Of the 500 milliliters of air you normally breathe in, only about 350 milliliters reach the alveoli. The other 150 milliliters is either on its way into or out of the

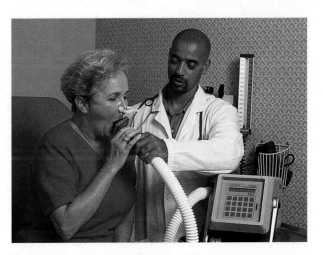

Figure 9.10 **Measuring lung volume.** This woman is having her lung volume measured in a hospital.

H⚙W Science
W⚙RKS
Discoveries

Asthma—New Findings about an "Old" Ailment

Asthma is a chronic inflammatory disorder of the airways that affects about 5 million children and adolescents, and about 10 million adults in the United States. Chronic means that a disease persists for a long time. In fact, people with asthma usually have the disease for their lifetimes, although some people affected with asthma in childhood "outgrow" it.

New drugs and preventive measures lessen the severity of asthma attacks and allow sufferers to lead normal lives. In addition, scientists have recently discovered that some persons with hard-to-treat asthma may have underlying bacterial or viral infections of the lungs. This new discovery may lead to new treatments for these sufferers.

www.jbpub.com/biology

The airways (bronchi and bronchioles) of a person with asthma are much more sensitive to certain stimuli than are the airways of other people. Triggers that are known to cause asthma attacks include allergens such as pollen, dust, tobacco smoke, dust

500X

mites (shown in photo), animal dander, or certain foods; physical factors such as coughing, sneezing, rigorous exercise, or cold temperatures; certain viral or bacterial infections; and chemical irritants such as cigarette smoke. In an asthma attack or episode, the lining of the airways swells, the bronchial muscles contract, and mucus production increases. This combination of swelling, increased mucus production, and muscle contraction narrows the airways. The person with asthma then has trouble bringing air into the lungs and also has trouble getting air out. Forcing air through the narrowed airways can produce a whistling or wheezing sound. In fact, mild wheezing in young, nonasthmatic children may be a sign that they will later develop asthma. Such children, if they also have a tendency to have allergic reactions, are more likely than their peers to develop asthma by age 11.

It was once thought that asthma was a psychosomatic illness, with no physical cause. Some physicians believed that the symptoms were all in a person's mind or that the person could somehow control the attacks. Others believed that asthma was a bad habit, like a temper tantrum, that showed a person's extreme need for attention. Today, scientists know that there is a genetic link to asthma, and that asthma is more likely to be passed to children by the mother than the father. The mechanisms involved in the higher rate of maternal transfer, such as prenatal influences, are unknown.

Immune system cells known as gamma delta T cells appear to drive the reactions that result in the asthma sufferer's hypersensitive response to allergens and other stimuli. Recently, medical researchers discovered that inner city children also have higher rates of asthma than children who live in suburban or rural environments. As of yet, scientists are unsure what factors in the city environment are to blame.

Asthma is treated by teaching the asthma patient to recognize the triggers that bring on attacks and to prevent attacks before they can occur. Asthma patients are encouraged to measure their lungs' vital capacity by exhaling into a device called a peak flow meter. The measurements shown by this device can help the physician and the patient develop an individualized treatment plan.

Treatment can involve several types of drugs. Anti-inflammatory agents lessen the swelling of the airways and are taken to prevent an attack. Bronchodilators that are inhaled are taken during an attack—these fast-acting drugs work to open constricted airways. However, bronchodilators must be used sparingly. Recent studies have shown that overuse of bronchodilator inhalers can lead to a worsening of asthma and even to death in a few cases. The inhaler itself does not cause the death. Instead, the asthma patient comes to rely on the fast-acting nature of the drug and does not notice that his or her asthma might be worsening. A severe attack of asthma can take such a person by surprise, leading to serious complications and death. That is why asthma patients must participate as much as possible in monitoring and treating their condition. This participation can lessen the severity of the disease, giving asthma patients some control over the episodes and the opportunity to lead a normal life.

The recent discovery that the lungs of some young asthma patients are chronically infected with the virus that causes bronchitis suggests that antiviral therapy might be a useful treatment for some children. Additionally, in December of 1996, researchers revealed that an upper respiratory virus or the bacterium *Chlamydia pneumoniae* is sometimes present in adult-onset asthma. Treatment for these pathogens cured or improved the asthma of over half the infected adults. In 1998, researchers reported that obesity was also a risk factor in the development of adult-onset asthma.

lungs, occupying space in the nose, pharynx, larynx, trachea, and bronchial tree. This space is called dead air space because it serves no useful purpose in gas exchange. Some of this air, in fact, will never reach the lungs. And some air, called residual air, remains in the lungs—even during deep breathing.

Inspiration occurs when the volume of the thoracic cavity is increased and the resulting negative pressure causes air to be pulled into the lungs. Expiration occurs when the volume of the thoracic cavity is decreased and the resulting positive pressure forces air out of the lungs.

9.7 Gas transport and exchange

The transport of gases throughout the body is assisted by the circulatory system. Without the help of a "highway" of blood, scientists estimate that it would take a molecule of oxygen 3 years to diffuse from your lung to your toe! Humans could not survive if gas transport were this slow.

Although the pathway of blood throughout the body will be described in detail in the next chapter, it is helpful to understand the basic routing of blood when discussing gas transport. Notice in **Figure 9.11** that the upper right chamber of the heart collects incoming blood from the upper and lower body (on the left as you face the illustration). This blood has given up much of its supply of oxygen to the tissues, so it is oxygen poor.

Along its route, however, it collected the waste product of cellular respiration, carbon dioxide. Deoxygenated blood such as this (more carbon dioxide than oxygen) is shown as blue in the diagram. The deoxygenated blood passes from the upper to the lower right chamber of the heart, where it is pumped to the lungs. Some of this blood goes to the right lung, and some goes to the left lung.

At the lungs, carbon dioxide within the blood of the capillaries surrounding the alveoli and the oxygen in the air of the alveoli are exchanged. External respiration, the exchange of gases at the lungs, works by the process of diffusion. It converts deoxygenated blood to oxygenated blood (more oxygen than carbon dioxide). This blood is shown as red in Figure 9.11. The oxygenated blood flows from the lungs to the left side of the

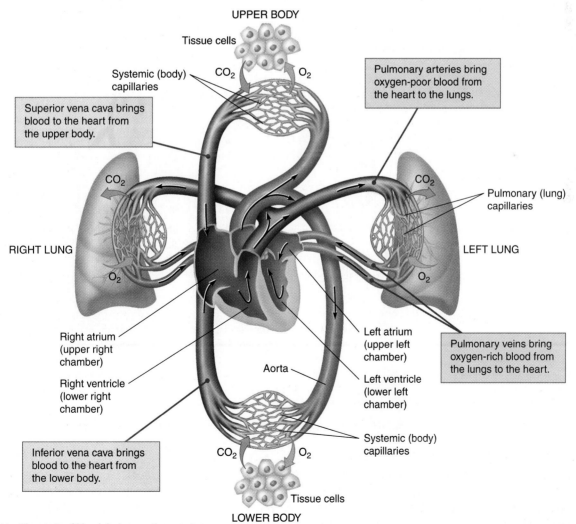

Figure 9.11 **The route of blood during gas transport.**

heart, where the lower left chamber pumps it out to other parts of the body.

External Respiration

Air is made up of many different kinds of molecules, such as oxygen, nitrogen, and carbon dioxide. Each of these gases exerts a pressure that depends on the number of molecules of the specific gas present per unit of volume. Molecules of a gas in a liquid (such as blood) also exert a pressure. Differences in pressures of a particular gas in two adjoining locations within a living system produce a *pressure gradient*. Molecules of a gas such as oxygen or carbon dioxide tend to move from an area of higher pressure to an area of lower pressure, or down the pressure gradient. Each gas moves according to its own pressure gradient, unaffected by the pressure gradients of other gases with which it might be mixed. This movement is a type of diffusion.

As deoxygenated blood (blue in Figure 9.11) arrives at the lungs, the pressure of the carbon dioxide in the blood is greater than the pressure of carbon dioxide

in the air within the alveoli. Therefore, carbon dioxide diffuses out of the blood and into the alveoli. Likewise, the pressure of the oxygen in the air within the alveoli is greater than the pressure of the small amount of oxygen in the blood. Therefore, oxygen diffuses out of the alveoli and into the blood.

Most of the carbon dioxide (69%) in the blood travels bound to water molecules in the fluid portion of the blood (**Figure 9.12**). The carbon dioxide and water combine chemically to form molecules of carbonic acid (H_2CO_3), but quickly break apart, or dissociate, into bicarbonate ions (HCO_3^-) and hydrogen ions (H^+).

$$CO_2 + H_2O \rightarrow H_2CO_3 \rightarrow HCO_3^- + H^+$$

One fourth (25%) of the carbon dioxide is carried in the red blood cells, bound to the oxygen-carrying molecule **hemoglobin (Hb)**. It is carried, however, by a different portion of the molecule than oxygen is. A small amount (6%) of the carbon dioxide is simply dissolved in the blood. At the alveoli, the dissolved carbon dioxide first moves out of the blood. This decrease in the carbon

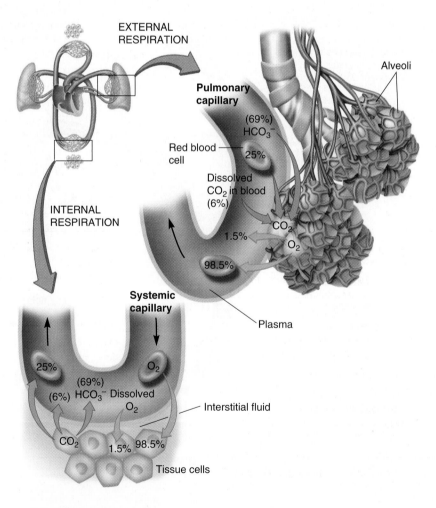

Figure 9.12 **How gases are transported during external and internal respiration.**

dioxide concentration triggers a reversal of the chemical reactions just described:

$$H^+ + HCO_3^- \rightarrow H_2CO_3 \rightarrow H_2O + CO_2$$

Carbon dioxide also dissociates from hemoglobin, and the freed carbon dioxide molecules diffuse into the alveoli. As oxygen diffuses from the alveoli, very little of it (1.5%) dissolves in the fluid portion of the blood (plasma). Instead, it combines with hemoglobin within the red blood cells. When the pressure gradient is high, as in the alveoli, hemoglobin binds with large amounts of oxygen. When oxygenated, hemoglobin turns bright red, which makes blood look red. You will notice, however, that the blood vessels on the underside of your wrists look blue. These blood vessels are veins carrying deoxygenated blood. Deoxygenated blood is dark red, but appears blue through layers of skin.

At high altitudes, such as in mountainous regions, the air is thinner and the pressure of the oxygen molecules within the air is lower than at sea level. Therefore, less of a pressure gradient is created at the alveoli. As a consequence, less oxygen diffuses into the blood, and can cause shortness of breath, nausea, and dizziness in those unaccustomed to the altitude. This condition is referred to as *high-altitude sickness*. Mountain climbers, athletes working out or playing at high altitudes (such as Mile High Stadium in Denver, Colorado), and tourists visiting an area of high altitude (such as Mexico City, Mexico) need to slowly work up to their normal levels of activity to give their bodies time to adjust to the lower oxygen pressure. The body adjusts by increasing the number of red blood cells and thus, the amount of hemoglobin.

Internal Respiration

Oxygenated blood is pumped out to the body by the left ventricle, or lower left chamber, of the heart (see Figure 9.11). It travels through the aorta to other large arteries that soon branch into smaller and smaller arteries. Eventually the blood vessels become so small that only one red blood cell at a time can pass. These vessels are the capillaries and are the sites of internal respiration.

Internal respiration is the exchange of oxygen and carbon dioxide between the blood within capillaries at the tissues and the tissue fluid that bathes the cells. As this exchange takes place, oxygenated blood becomes progressively deoxygenated. In essence, the processes of internal respiration are the reverse of the processes of external respiration. As the blood begins its journey from the lungs around the body, it is oxygen-rich. The oxygen molecules in the blood exert a higher pressure than those in the cells because the cells (in the process of cellular respiration) are continuously using oxygen. Consequently, oxygen molecules begin to dissociate from the hemoglobin, diffuse into the tissue fluid, and from there diffuse into the cells. Conversely, the levels of carbon dioxide are higher in the cells than in the blood because cellular respiration continuously produces carbon dioxide. Consequently, carbon dioxide diffuses from the cells into the tissue fluid and then into the blood.

As gas exchange at the capillaries continues, an interesting thing happens. The oxygen supply of the blood decreases, thereby decreasing the pressure of oxygen in the blood. Diffusion of oxygen into the tissue fluid does not slow down, however, because the pH at the capillaries lowers as carbon dioxide diffuses into the blood. Remember that carbon dioxide is carried in the blood primarily as bicarbonate ions. As these ions are formed, hydrogen ions are also produced. The buildup of these hydrogen ions makes the blood increasingly acidic as more carbon dioxide diffuses in. (See Chapter 3 for a discussion of the relationship between hydrogen ions and pH.) This acidic environment helps split more oxygen from hemoglobin, thereby enhancing oxygen's diffusion into the tissue fluid. (Figure 9.12 shows the processes of internal and external respiration.) Increases in temperature also help split oxygen from hemoglobin. During exercise or any type of exertion, your body needs more oxygen delivered to its cells. However, active cells also produce more carbon dioxide and heat, thereby helping to split oxygen from hemoglobin and meet their own needs. This is one example of how your body works to maintain homeostasis, a state of internal equilibrium.

The circulatory system aids in respiration by transporting gases throughout the body. During external respiration, carbon dioxide diffuses from the blood in capillaries surrounding the alveoli to air in the alveoli. Likewise, oxygen diffuses from the alveolar air to the blood in the capillaries surrounding the alveoli. Carbon dioxide is carried in the fluid portion of the blood primarily as bicarbonate ions; oxygen is carried within the red blood cells by the oxygen-carrying molecule hemoglobin. During internal respiration, carbon dioxide diffuses from the tissue fluid surrounding the body cells to the blood. Likewise, oxygen diffuses from the blood within capillaries to the tissue fluid. As carbon dioxide diffuses into the blood, the resultant increase in hydrogen ions increases the dissociation of oxygen from hemoglobin.

For the latest health information, please visit www.jbpub.com/biology.

9.8 Choking: A common respiratory emergency

Have you ever been eating with someone who started to choke? Did you know what to do? Choking is caused when food or a foreign object becomes lodged in the windpipe. When you are with someone who is choking, first notice whether the person can talk, breathe, or cough. If so, stay with the person until coughing clears the airway. Do not try to slap the person on the back. The slapping may only cause the food to become more deeply lodged in the windpipe.

If a person cannot talk or cough and appears not to be breathing, administer several short, quick abdominal thrusts. This technique is called *abdominal thrusts* or the *Heimlich maneuver*, after Dr. Henry Heimlich, who

developed the procedure. First, stand behind the choking victim as shown in the photo at left. Put your arms around the person, placing one hand over the fist of the other positioned just below the breastbone (shown above right). Then give a series of quick, sharp, upward and inward thrusts. These thrusts push in on the diaphragm and the thoracic cavity, suddenly decreasing its volume. This sudden decrease creates a surge in air pressure below the obstruction, which usually projects it forcefully from the windpipe.

If a person cannot talk or cough and appears not to be breathing, administer several short, quick, upward and inward thrusts just beneath the breastbone to dislodge any obstruction.

9.9 Chronic obstructive pulmonary disease

The term **chronic obstructive pulmonary disease (COPD)** is used to refer to disorders that block the airways and impair breathing. COPD affects 25 million people in the United States alone and is responsible for at least 50,000 deaths per year. Two disorders commonly included in COPD are **chronic bronchitis** and **emphysema**.

Chronic bronchitis is an inflammation of the bronchi and bronchioles that lasts for at least 3 months each year for 2 consecutive years with no accompanying disease as a cause. The primary cause is cigarette smoking. Air pollution and occupational exposure to industrial dust are much less frequent causes. Cigarette smoking paralyzes ciliated epithelial cells so that they can no longer effectively remove incoming particles and microbes. It also causes increased mucus production by cells lining the trachea. A continued buildup of mucus provides food for bacteria, and infection can result. The mucus also plugs up the respiratory "plumbing."

Normally, bronchioles widen, or dilate, during inspiration; they narrow, or constrict, during expiration. If mucus is plugging various bronchioles, some air may therefore be able to get to the alveoli beyond the plugged bronchioles, but may not be able to get out. Coughing spells produce pressure within these continuously inflated alveoli that ruptures their walls, decreasing the surface area over which gas exchange takes place. The lungs lose their elasticity and the ability to recoil during exhalation, staying filled with air, a disorder aptly called *emphysema*, meaning "full of air." A person with emphysema has to work voluntarily to exhale.

People with COPD find that they have more respiratory infections than they did before the disorder and that these infections last longer. They have a morning cough or may cough all day. They may also tire easily and become short of breath with minimal physical exertion. Some people with COPD feel as though they cannot breathe at times. As the disorder progresses, some find it difficult to do a day's work or accomplish the daily activities of living. Periods of breathlessness increase. Some persons suffer bouts of respiratory failure and must be hospitalized. COPD is serious and deadly—but in most cases, it is avoidable with wise health choices.

Cigarette smoking is the primary cause of chronic obstructive pulmonary disease, a group of disorders that block the airways and impair breathing.

Just Wondering . . . Real Questions Students Ask

I'm a smoker thinking about quitting. Does my smoking really affect the health of my children?

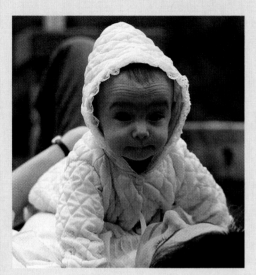

Cigarette smoking damages the lungs of children in more ways than people realize, and the damage can start in the womb. Recent reports from the Harvard Six Cities Study, a 20-year investigation into the effects of air pollution and lung hazards, conclude that smoking during pregnancy can detrimentally affect a child's lung function for the rest of the child's life! Lung function refers to the vital capacity of the lungs—the maximum volume of air that can be pushed out of the lungs during one exhalation after a single, deep inhalation. The reason for the reduced lung function in these children is unclear. However, many investigators hypothesize that it is due to their developing smaller lungs that produce less elastic tissue, a protein that helps the lungs recoil during exhalation. There is also a positive correlation between mothers who smoke and the incidence of asthma in their children. In addition, questions have been raised about the incidence of maternal smoking and hearing defects in their children, as well as the occurrence of sudden infant death syndrome (SIDS).

Researchers conducting other parts of the Six Cities Study showed that children exposed to tobacco smoke before the age of 6 had a 2 to 3 percent decrease in lung function, a decrease that normally does not occur until the late 20s or early 30s. Children growing up in households with smokers also develop more lower respiratory conditions such as shortness of breath, wheezing, coughing, and bronchitis. Statistics from the American Heart Association state that secondhand smoke causes up to 300,000 lower respiratory tract infections such as pneumonia and bronchitis in children younger than the age of 1 1/2. In 5% of these cases, the infections are so severe that the child must be hospitalized. The choice to quit smoking will not only benefit your health, but the health of your children as well.

Key Concepts

Respiration takes place over thin, large, moist surfaces.

9.1 Respiration is the uptake of oxygen and the release of carbon dioxide by an organism.

9.2 Respiration occurs across the cell membranes of single-celled organisms and the outer layer of cells of certain multicelled animals.

9.2 Respiration occurs across the skin of amphibians and certain fishes, snakes, and turtles.

9.3 Respiration occurs in the tracheal systems of most spiders and insects, the gills of most large multicellular aquatic animals, and the lungs of most large land-dwelling animals.

Human respiration uses special organs and relies on pressure changes within the thoracic cavity.

9.4 Human respiration includes the processes of cellular respiration, internal respiration, external respiration, and breathing.

9.5 Air passes through many respiratory structures on its way to the lungs: the nasal cavities; the pharynx, or throat; the larynx, or voice box; the trachea, or windpipe; two bronchi; bronchioles; and microscopic air sacs called alveoli.

9.5 The alveoli provide a thin, large, moist, surface area over which gas exchange takes place.

9.6 Inspiration occurs when the volume of the thoracic cavity is increased and the resulting negative pressure causes air to be pulled into the lungs.

9.6 Expiration occurs when the volume of the thoracic cavity is decreased and the resulting positive pressure forces air out of the lungs.

9.7 The circulatory system aids in respiration by transporting gases throughout the body.

9.7 Oxygen in the air and carbon dioxide in the blood are exchanged at the alveoli. This exchange is called external respiration.

9.7 At the capillaries in the body tissues, oxygen in the blood diffuses into the cells and carbon dioxide diffuses out of the cells and into the blood. This exchange is called internal respiration.

Health Connections

9.8 To assist a person who is choking and cannot talk or breathe, administer several short, quick, abdominal thrusts with your arms around the victim and your fists just below the breastbone.

9.9 Chronic obstructive pulmonary disease (COPD) refers to disorders that impair movement of the air in the respiratory system.

9.9 Chronic bronchitis and emphysema are two common COPD disorders that are most frequently caused by cigarette smoking.

Key Terms

aerobic cellular respiration *178*
alveoli (al-VEE-uh-lye) *184*
breathing *182*
chronic bronchitis (bron-KYE-tis) *192*
chronic obstructive pulmonary
   disease (COPD) *192*
diaphragm (DYE-uh-fram) *186*
emphysema (em-fi-SEE-muh) *192*
expiration *182*
external respiration *182*
gills *180*
glottis (GLOT iss) *183*

hemoglobin (Hb) (HEE-muh-glow-bin)
   190
inspiration *182*
intercostal muscles (inn-ter-KOS-tul)
   186
internal respiration *182*
larynx (LAIR-inks) *182*
lungs *181*
nasal cavities *182*
pharynx (FAIR-inks) *182*
primary bronchi (BRON-keye) *184*
respiration (respire) *178*

respiratory bronchioles
   (BRON-kee-olz) *184*
respiratory system *182*
secondary bronchi (BRON-keye) *184*
spiracles *180*
thoracic cavity (THU-rass-ick) *186*
tidal volume *187*
trachea (TRAY-kee-uh) *180*
tracheal system *180*
vocal cords *182*

KNOWLEDGE AND COMPREHENSION QUESTIONS

1. Describe the ways in which organisms without special organs of respiration can exchange oxygen and carbon dioxide with the environment. (*Comprehension*)

2. Distinguish among respiration, cellular respiration, internal respiration, and external respiration. (*Comprehension*)

3. Your biology instructor poses a problem in class, and you suggest a solution. Explain how you produced the necessary sounds. (*Comprehension*)

4. Explain how differences in air pressure help you to breathe. (*Comprehension*)

5. Explain what happens during gas exchange at the alveoli. What gases are exchanged, and what forces "drive" this exchange? (*Comprehension*)

6. A normal hemoglobin level is an essential circulatory function. What does the hemoglobin carry? (*Knowledge*)

7. Deoxygenated blood appears blue while in the veins. Why then, when you cut a vein and look at your injury, does your blood appear red as it leaves the wound? (*Comprehension*)

8. Explain the process of gas exchange at the capillaries. What gases are exchanged, and why does an exchange of gases occur? (*Comprehension*)

9. What symptoms indicate that a person is choking and that it would be essential to apply the Heimlich maneuver to the person? (*Knowledge*)

10. While on vacation in the mountains, you find that you feel lightheaded and short of breath. Explain why. (*Comprehension*)

K E Y
Knowledge: Recalling information.
Comprehension: Showing understanding of recalled information.

CRITICAL THINKING REVIEW

1. During the winter in cold regions, frogs often hibernate in the mud in the bottom of ponds. How do they get enough oxygen to survive? (*Application*)

2. Hiccups occur when the diaphragm begins to contract spasmodically, causing a sudden inhalation. What do you hypothesize causes the sound effect that results? (*Synthesis*)

3. How does the circulatory system assist the process of respiration? Why is this help important? (*Application*)

4. Why do you think that people with COPD and emphysema tire easily? (*Synthesis*)

5. Study Figure 9.3b regarding countercurrent exchange. Describe in detail how this process works to aid the diffusion of oxygen into the blood of organisms that use gills for respiration. (*Synthesis*)

K E Y
Application: Using information in a new situation.
Synthesis: Putting together information from different sources.

1. Blue usually denotes oxygen-poor blood that is carried in veins. In the illustration, what type of vessel carries oxygen-poor blood to the lungs?

2. Which gas is being "exchanged" with oxygen at the lungs?

3. What type of vessel carries oxygen-rich blood?

4. To which organ of the body is blood being returned after picking up oxygen at the lungs?

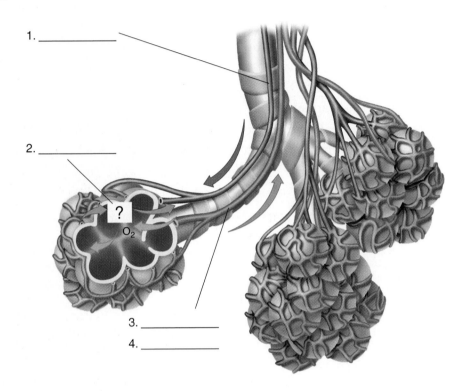

1. _____

2. _____

3. _____

4. _____

e learning

Location: http://www.jbpub.com/biology

e-Learning is an on-line student review area located at this book's web site www.jbpub.com/biology. The review area provides a variety of activities designed to help you study for your class and build your learning portfolio.

Review Questions The review questions test your knowledge of the important concepts and applications in each chapter. The review provides feedback for each correct or incorrect answer. This is an excellent test preparation tool.

Figure-Labeling Exercises Sharpen your visual thinking by matching terms and labels for important illustrations.

Flash Cards Studying biology requires learning new terms. Virtual flash cards help you master the new vocabulary for each chapter.

Just Wondering As you read and study from this text, you may find that you have unanswered questions. Through this site you can ask the author, Sandy Alters, your "just wondering" questions.

 Do you prefer the speed and reliability of a CD-ROM? All of the features contained on the eLearning portion of the web site are also available on the Student CD-ROM.

Circulation

The heart is an amazing living pump. It moves about 5 liters (5.3 quarts) of blood through each of its two sides per minute. The heart beats, on average, 72 times per minute, or over 100,000 times per day. So if you are about 20 years old, your heart has beat about 750 million times!

Impossible decades ago, open-heart surgery is now commonplace. The physicians in the photo are performing one of many open-heart surgical procedures—an aorta replacement. The aorta, the largest artery in the human body, is the vessel that receives the rush of blood pumped by the strongest chamber of the heart—the left ventricle. Branches from the aorta bring blood to the head as well as to the arms, legs, and torso.

Physicians have not only constructed replacement parts for the vessels and valves of the heart, they have designed replacements for the heart itself. The Jarvik-7 was the first Total Artificial Heart (TAH), developed in 1982. Since then, the Jarvik-7 has been implanted in approximately 180 patients, but due to problems ranging from infection to strokes induced by blood clots, the TAH is no longer approved for use. A new, smaller artificial heart has taken its place—the Jarvik 2000, also called a Left Ventricular Assist Device (LVAD). Unlike the Jarvik-7, the LVAD is not meant to permanently replace a diseased heart, but serves as a substitute heart while a patient waits for a heart transplant. The LVAD is about the size of a man's fist and sits in the upper part of a patient's abdomen. A tube shunts blood from the left ventricle of the heart to the pump, and another tube sends blood from the pump to the aorta. The pump is connected to a battery pack and computerized control system worn outside the body. Although "high tech" and amazing, the LVAD is still not as amazing as the body's own remarkable pump—the heart.

CHAPTER
10

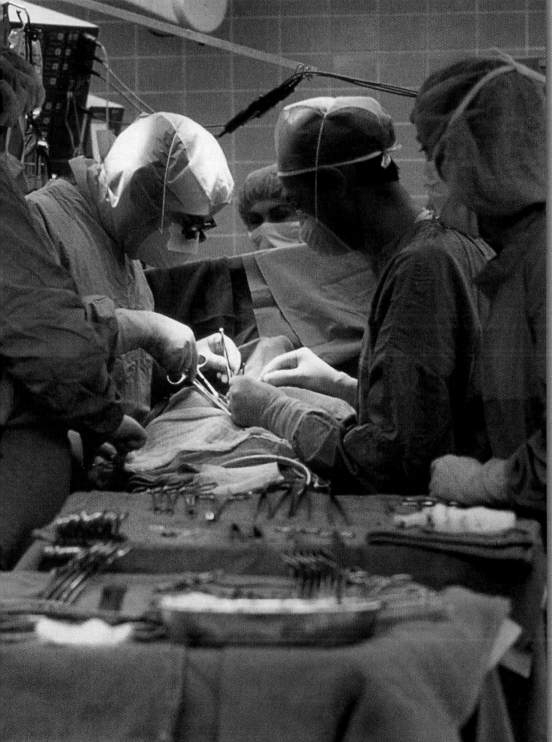

Organization

10.1 Transport over cell surfaces

Not all organisms have circulatory systems with hearts as humans do. In fact, not all organisms have circulatory systems. A **circulatory system** is a transport system that uses a fluid to move substances such as nutrients, wastes, and gases throughout an organism.

Organisms devoid of circulatory systems live in moist or watery environments. They are unicellular, are made up of only a few cells, or are structured so that all their cells are close enough to the external environment for the direct exchange of substances such as nutrients, wastes, and gases.

Single-celled organisms such as bacteria, yeasts, and single-celled protists (such as the paramecium shown in **Figure 10.1**) do not have circulatory systems. Gases and nutrients simply diffuse into the cell or are moved into the cell by means of active transport or phagocytosis (see Chapter 4). Within the cell, nutrients, wastes, and gases diffuse or are actively transported from one part of the cell to another.

Even fungi, although multicellular, have no circulatory system; their structure allows the transport of substances without the aid of a circulating fluid. Most fungi live in moist places and are composed of nearly microscopic filaments called hyphae. They are able to absorb nutrients, gases, and water directly from the soil, air, and decaying organisms. Many species have no cell partitions, which aids the movement of dissolved substances within the organism.

Some small animals have a body architecture that allows substances to move directly between their cells and the outside environment. As described in Chapter 8, aquatic animals that have relatively simple anatomies, such as hydra, jellyfish, corals, sea anemones, and flatworms, have digestive systems with a single opening. Food is taken in through this opening, and it then enters a digestive cavity. The digestive cavity, more properly called the *gastrovascular cavity*, is a dead-end sac that functions not only in digestion (*gastro-*), but also in circulation (*-vascular*). Figure 10.1 shows the gastrovascular cavity of the hydra. In the hydra, this cavity extends into the tentacles of the animal, allowing the complete interior of the hydra to be bathed in fluid, as is the exterior of the animal. No cells are far from its watery environment.

Single-celled or small organisms with simple structures that live in aquatic or moist environments generally do not have circulatory systems. Substances move into their cells by means of diffusion, active transport, or phagocytosis.

10.2 Transport with open circulatory systems

Larger animals and those that live on land have circulatory systems. Even most land plants have their own version of a circulatory system (see "A Focus on Plants" on the facing page). Terrestrial (land) animals have a wet internal environment although their external environment is dry. Their cells are bathed in *tissue fluid*, a waterlike substance. The circulatory system helps maintain the composition of this fluid.

All circulatory systems have a liquid that moves, or circulates within an organism to transport substances. Some animals have an **open circulatory system**, in which the circulating fluid bathes internal organs directly. It may or may not have a pump (heart) connected to open-ended vessels.

Open circulatory systems are found in most arthropods such as spiders, lobsters, crabs, and insects; certain molluscs such as clams and snails (but not squid and oc-  topus); and tunicates, which are saclike marine organisms (see Chapter 31 and the photo to the left). In an open circulatory system, a fluid composed of blood mixed with tissue fluid bathes the internal tissues and organs of an animal directly. It oozes through spaces or cavities that surround the organs. This mixture is usually referred to as hemolymph.

The term *hemolymph* refers to tissue fluid, which is also called *lymph*, and to hemoglobin or hemocyanin, two oxygen-carrying pigments of blood. These pigments are the reason that different organisms have bloods of different colors. Hemoglobin is a deep-red pigment containing iron. When combined with oxygen, hemoglobin becomes bright red. Hemocyanin has a structure similar to hemoglobin but the heme group contains copper instead of iron. When combined with oxygen, hemocyanin becomes blue. Hemoglobin or hemocyanin is the respiratory pigment for most molluscs and arthropods. Some organisms such as the tunicates and the insects have no oxygen-carrying pigment in the blood, so their blood is colorless.

The fluid in an open circulatory system moves sluggishly, pushed by the movements of the animal or by a simple heart connected to open-ended vessels. The circulatory system of the tunicate consists of a heart with a large vessel at each end. The heart pushes blood first in one direction and then the other. This unusual reversal of blood flow is unique to the tunicates. Figure 10.1 shows the open circulatory system of the grasshopper. Insects such as the grasshopper are highly active, so oxygen is not transported by means of this system—they would die before oxygen got to all the cells of the body. Instead, oxygen is transported by means of a tracheal system (see Chapter 9), which is a set of tubes that connect to the outside and provide the transport mechanism for this essential gas.

> Open circulatory systems are found in most arthropods, certain molluscs, and tunicates. In an open circulatory system, a fluid composed of blood mixed with tissue fluid bathes the internal tissues and organs of an animal directly; it may have a simple heart connected to open-ended vessels.

10.3 Transport with closed circulatory systems

Other organisms (such as you) have a **closed circulatory system**, in which blood is enclosed in vessels. Closed circulatory systems are found in cephalopod molluscs (that is, molluscs with a head such as squid and the octopus shown below), annelids (earthworms and leeches), echinoderms (sea stars, sea urchins, sand dollars, sea cucumbers), and vertebrates (all animals with a

backbone such as fish, amphibians, reptiles, birds, rodents, seals, elephants, dogs, and humans). In a closed circulatory system, the blood is located within blood vessels as it circulates throughout the body. This fluid is pumped either by a heart or by contractions of specialized blood vessels. The human circulatory system will be discussed in-depth later in this chapter.

One of the simplest examples of a closed circulatory system is that of the earthworm, shown in Figure 10.1. One major vessel runs along the dorsal (top) side of the worm and carries blood anteriorly (toward the front). The other major vessel runs along the ventral (bottom) side of the worm and carries blood posteriorly (toward the back). Capillaries, which are microscopic blood vessels, connect the dorsal and ventral vessels through a network of capillary beds that bring blood to the tissues. The dorsal vessel is the pump of the system, pushing the blood forward by its waves of contractions. Assisting the dorsal "heart" are five pairs of blood vessels that connect the dorsal and ventral vessels. These blood vessels pulsate, aiding the movement of the blood around the worm's body.

> In a closed circulatory system, blood is enclosed in vessels and is pumped either by a heart or by contractions of specialized blood vessels. Closed circulatory systems are found in cephalopod molluscs, annelids, echinoderms, and vertebrates.

Focus
on Plants

Most plants (more than 95% of all living species) have a system of specialized tissues to transport substances to and from their various parts. Plants with these tissues are called vascular plants and these specialized tissues are collectively called vascular tissue. Vascular tissue forms the "circulatory system" of a plant, conducting water and dissolved inorganic nutrients up the plant and the products of photosynthesis throughout the plant. Chapter 29 describes vascular plants in more detail.

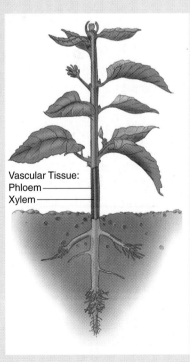

Vascular Tissue:
Phloem———
Xylem———

Visual Summary Figure 10.1 Patterns of circulation.

NO CIRCULATORY SYSTEM
(single-celled organisms, fungi, and small animals having a gastrovascular cavity)

Paramecium

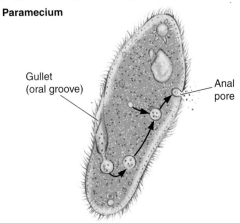

Gullet (oral groove)

Anal pore

Hydra

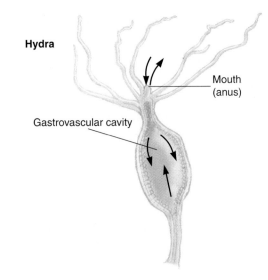

Mouth (anus)

Gastrovascular cavity

OPEN CIRCULATORY SYSTEM
(most arthropods, such as spiders, lobsters, crabs, and insects; certain molluscs, such as clams and snails; tunicates)

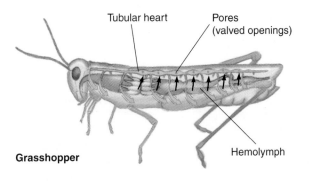

Tubular heart

Pores (valved openings)

Hemolymph

Grasshopper

Tunicate

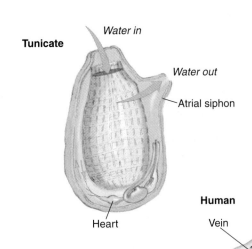

Water in

Water out

Atrial siphon

Heart

Human

Vein

Artery

Heart

CLOSED CIRCULATORY SYSTEM
(cephalopod molluscs such as squid and octopus; annelids such as earthworms and leeches; echinoderms such as starfish and sea urchins; vertebrates [all animals with a backbone])

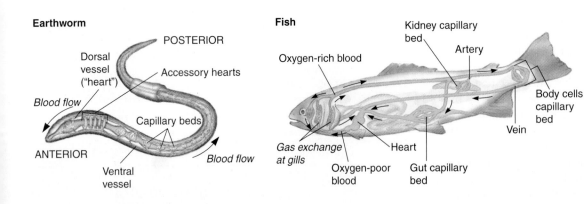

Earthworm

POSTERIOR

Dorsal vessel ("heart")

Accessory hearts

Blood flow

Capillary beds

ANTERIOR

Blood flow

Ventral vessel

Fish

Kidney capillary bed

Artery

Oxygen-rich blood

Body cells capillary bed

Vein

Gas exchange at gills

Heart

Oxygen-poor blood

Gut capillary bed

NO CIRCULATORY SYSTEM

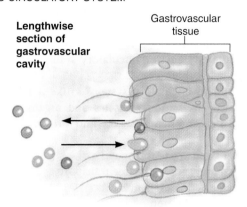

Lengthwise section of gastrovascular cavity

Gastrovascular tissue

Organisms with no circulatory systems live in moist environments and are structured in ways in which molecules move directly across their cell surfaces.

Molecules are transported across the membrane surface.

OPEN CIRCULATORY SYSTEM

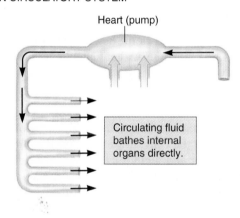

Heart (pump)

Circulating fluid bathes internal organs directly.

In organisms with open circulatory systems, a simple heart pumps blood (hemolymph) through open-ended vessels. The blood oozes through body spaces, as shown by the arrows. The blood reenters the heart through valved openings, which prevent the backflow of blood.

CLOSED CIRCULATORY SYSTEM

Earthworm

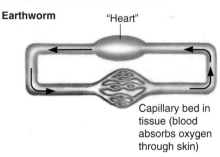

"Heart"

Capillary bed in tissue (blood absorbs oxygen through skin)

In organisms with closed circulatory systems, the blood is pumped by a heart and remains enclosed in vessels throughout its journey.

Fish

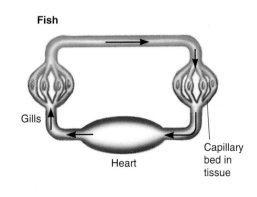

Gills

Heart

Capillary bed in tissue

Human

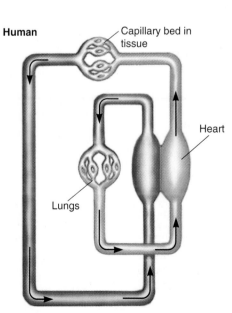

Capillary bed in tissue

Heart

Lungs

10.4   Functions of the circulatory system

The human circulatory system (Figure 10.2) is a closed system and is an excellent example of vertebrate circulation. It is made up of three components: (1) the heart, a muscular pump; (2) the blood vessels, a network of tubelike vessels that permeate the body; and (3) the blood, which circulates within these vessels. The heart and blood vessels alone are known as the **cardiovascular system**. The prefix *cardio* refers to the heart, and the suffix *vascular* refers to blood vessels. However, the terms cardiovascular system and circulatory system are often used interchangeably.

The human circulatory system is like a roadway that connects the various muscles and organs of the body with one another. It serves four principal functions:

1. Nutrient and waste transport
2. Oxygen and carbon dioxide transport
3. Temperature maintenance
4. Hormone circulation

In addition, proteins and ions in the fluid portion of the blood regulate the movement of water between the blood and the tissues. Specialized cells are also found in the blood that defend the body against invading microorganisms and other foreign substances. Many biologists argue that the functions of the circulatory system are truly functions of the blood alone; however, the blood cannot perform these functions without circulating throughout the body within vessels, pushed by a powerful pump.

Nutrient and Waste Transport

The nutrient molecules that fuel cell metabolism are transported to the cells of the body by the circulatory system. Sugars and amino acids pass into the bloodstream at the small intestine, diffusing into a fine net of blood vessels below the mucosa (see Figure 8.9). Most fatty acids are absorbed by the epithelial cells lining the small intestine. After being resynthesized into fats and packaged for transport, these fats pass into lymphatic vessels and are transported to the blood by the lymphatic system (see Section 10.10). The blood takes all of these digestive products to the liver for processing. There, some molecules are converted to glucose, which is released into the bloodstream. Others, such as essential amino acids and vitamins, pass through the liver unchanged. Still others, such as excess energy molecules and excess amino acids, are used for the synthesis of glycogen and body fat and are stored for later use.

From the liver, the blood carries glucose and other energy molecules to all the body's cells. In addition, the blood also brings to the cells molecules such as amino acids, which are used as building blocks to produce other substances. The cells, in turn, release the waste products of metabolism into the bloodstream. The blood carries most of these wastes to a cleansing organ, the kidney, which captures and concentrates them for excretion in the urine (see Chapter 12). The cleansed blood then passes back to the heart.

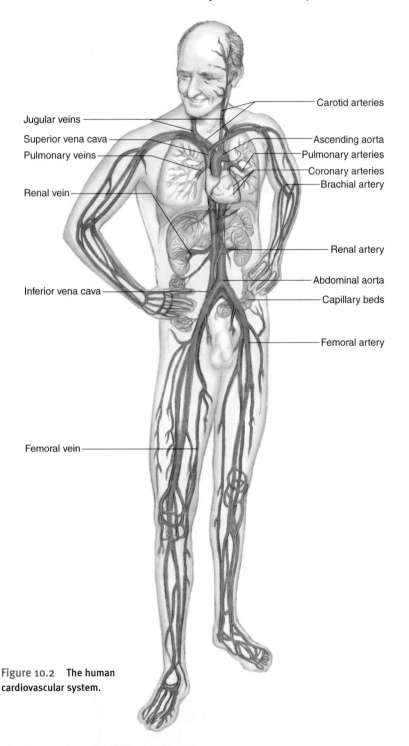

Jugular veins

Superior vena cava

Pulmonary veins

Renal vein

Inferior vena cava

Femoral vein

Carotid arteries

Ascending aorta

Pulmonary arteries

Coronary arteries

Brachial artery

Renal artery

Abdominal aorta

Capillary beds

Femoral artery

Figure 10.2   The human cardiovascular system.

Oxygen and Carbon Dioxide Transport

The cells of the body carry out cellular respiration and need oxygen for this series of reactions to take place. Oxygen is transported to the cells of the body by the circulatory system. Within the lungs, oxygen molecules diffuse into the circulating blood through the walls of capillaries, which are very fine blood vessels (see Figure 9-12). This oxygen passes into red blood cells suspended in the liquid portion of the blood. From the lungs, the blood carries its cargo of oxygen to all cells of the body. At the same time, it picks up a waste product of cellular respiration, carbon dioxide. The blood then returns to the lungs, where the carbon dioxide is released and a fresh supply of oxygen is captured.

Temperature Maintenance

As you read in Chapter 5, energy is continuously being lost as heat during the chemical reactions that take place in the cells of your body. The blood distributes this heat, helping to maintain your body temperature. As the blood circulates, it passes through delicate, microscopic networks of blood vessels that lie under your skin. The blood passing through these vessels gives up heat because the environment is usually cooler than the body's temperature of 36°C (98.6°F). The body works to balance the amount of heat produced and the amount of heat lost to maintain a stable internal temperature.

To maintain this balance, a regulatory center in the brain, the hypothalamus, acts like your own personal thermostat, constantly monitoring body temperature and stimulating regulatory processes. If your body temperature drops, for example, signals from this center cause surface blood vessels to narrow, or constrict. Constriction of these blood vessels limits blood flow to the surface of the skin and lessens heat loss. Conversely, if your body temperature rises, signals from this center cause surface blood vessels to widen, or dilate. Dilation increases blood flow to the surface of the skin and increases heat loss (**Figure 10.3**).

Hormone Circulation

Nerve signals and hormones coordinate the chemical reactions and other activities of the body. Hormones are chemical messengers; they are produced in one place in the body and affect cells or tissues in another. The circulatory system is the highway within which hormones travel throughout the body, from their site of production to the target tissues that are capable of responding to them.

The circulatory system transports oxygen and nutrients to cells, transports carbon dioxide and metabolic wastes away from cells, helps maintain a stable internal temperature, and carries hormones throughout the body. Substances within the blood regulate the movement of water between the blood and the tissues, and specialized blood cells defend the body against disease.

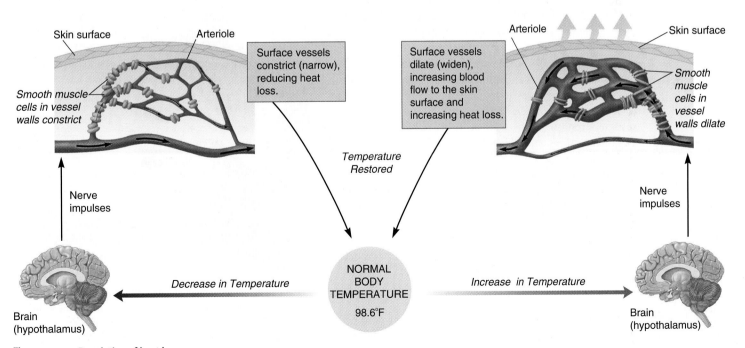

Figure 10.3 Regulation of heat loss.

10.5 The blood vessels

As the blood is pushed along by the heart to begin its journey throughout the body, it leaves the heart within vessels known as **arteries** (see Figure 10.2). From the arteries, the blood passes into a network of **arterioles**, or small arteries. From these, it eventually is forced through the **capillaries**, a fine latticework of very narrow tubes, which get their name from the Latin word *capillus*, meaning "a hair." As blood passes through these capillaries, gases are exchanged between the blood and the tissues, nutrients are delivered to the tissues, and wastes are picked up from the tissues. After its journey through the capillaries, the blood passes into a third kind of vessel, the **venules** (VAYN-yooles), or small **veins**. A network of venules and larger veins collects the circulated blood and carries it back to the heart.

Arteries and Arterioles

Arteries are vessels that carry blood away from the heart. The adult human heart pumps about 100 milliliters of blood—a little over 3 ounces—into the aorta with each beat. On average, your heart beats about 75 times per minute, pumping more than 7.5 liters, or nearly 8 quarts, of blood into your arteries each minute of your life. The vessels leading out from the heart expand slightly in diameter during a heartbeat and then recoil before the next heartbeat, responding to the pressure of blood surging into them.

The walls of the arteries are made up of three layers of tissue **(Figure 10.4a)** with a hollow core called the *lumen*. *Endothelial cells* (the inner epithelium of blood vessels and the heart) line arteries and are in contact with the blood **①**. Surrounding these cells is a thick layer of elastic fibers and smooth muscle **②**. The elastic tissue allows the artery to expand and recoil in response to the pulses of blood. The steady contraction of the muscle strengthens the wall of the vessel against overexpansion. This muscle and elastic layer is encased within an envelope of protective connective tissue **③**.

Arterioles are smaller in diameter than arteries. The walls of the largest arterioles are similar in structure to artery walls. Their muscles tighten or relax in response to messages from nerves and hormones. As arterioles near the capillaries, their diameter decreases until they consist of nothing but a layer of endothelium wrapped with a few scattered smooth muscle cells.

Capillaries

Capillaries are microscopic blood vessels that connect arterioles with venules. They have a simple structure

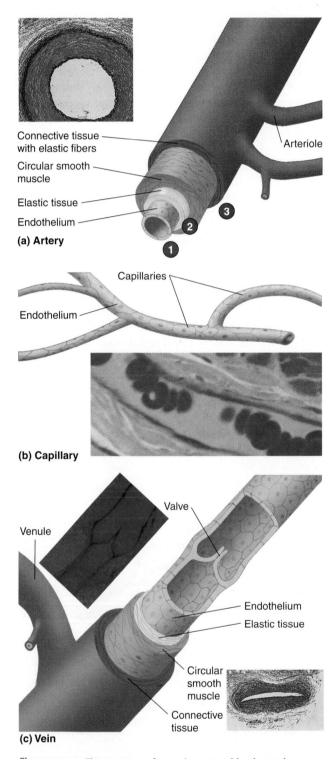

(a) Artery

Connective tissue with elastic fibers
Circular smooth muscle
Elastic tissue
Endothelium
Arteriole

(b) Capillary

Capillaries
Endothelium

(c) Vein

Venule
Valve
Endothelium
Elastic tissue
Circular smooth muscle
Connective tissue

Figure 10.4 **The structure of some important blood vessels.** Blood leaves the heart through arteries and returns to the heart through veins. The capillaries are the tubes through which the blood is forced in the tissues.

(Figures 10.4b) and are little more than tubes with walls one cell thick and with a length that would barely stretch across the head of a pin. The internal diameter of the capillaries is about the same as that of red blood cells, causing these cells to squeeze through the capillaries single file. The closeness between the walls of the capillaries and the membranes of the red blood cells facilitates the diffusion of gases, nutrients, and wastes between them—a swap of oxygen and nutrients for carbon dioxide and other metabolic waste products.

Your entire body is permeated with a fine mesh of capillaries, networks that amount to several thousand kilometers in overall length. In fact, if all the capillaries in your body were laid end to end, they would extend across the United States! These networks of capillaries are called *capillary beds*. In a capillary bed, some of the capillaries connect arterioles and venules directly (shown at right) and are called *thoroughfare channels*. From these channels, loops of true capillaries—those not on the direct flow route from arterioles to venules—leave and return. Almost all exchanges between the blood and the cells of the body occur through these loops. A ring of muscle (a precapillary sphincter) guards the entry to each loop, and when contracted, blocks flow through the capillary. Restricting blood flow in capillary beds near the surface of the skin and constricting surface arterioles provide powerful means by which the body can limit heat loss. The body can also cut down on the flow within a capillary bed when heavy flow is not needed—to your muscles, for example, when you are resting. Likewise, the body can increase the flow within a capillary bed when the need increases—to your small intestine, for example, after a meal. Interestingly, you do not have enough blood to fill all your capillary beds if they were all open at the same time. If such a situation occurred, you would faint because of lack of sufficient blood to the brain.

Veins and Venules

Venules are small veins that collect blood from the capillary beds and bring it to larger veins that carry it back to the heart. The force of the heartbeat is greatly diminished by the time the blood reaches the veins, so these vessels do not have to accommodate pulsing pressures as arteries do. Therefore, the walls of veins, although similar in structure to those of the arteries, have much thinner layers of muscle and elastic fiber. Lacking much of this supportive tissue, the walls of veins may collapse when they are empty (although this situation would never occur in the living body). Empty arteries, on the other hand, stay open like tiny pipes. In addition, the lumen of veins is larger than that of arteries. This difference in size is related to the lower pressure of blood flowing within veins back toward the heart—large vessels present less resistance to the flow than smaller vessels do.

The pathway of blood back to the heart from much of the body is an uphill struggle. The pressure pushing the blood upward in the veins of your legs, for example, approximately equals the force of gravity pulling it down. As skeletal muscles contract, they press on veins and help move blood along. In addition, veins have one-way valves that help blood move back toward the heart by preventing its backflow (Figure 10.4c). If some of the valves in a vein are weak, however, gravity can force blood back through these valves, overloading a portion of a vein and pushing its walls outward. These "stretched out" veins, called varicose veins, are often seen in the legs at the surface of the skin.

Thoroughfare channel

Venule

Capillaries

Arteriole

Precapillary sphincters

Arteries and veins are the major vessels of the circulatory system and have walls composed of three layers: (1) an innermost, thin layer of endothelial tissue that is in contact with the blood; (2) a layer of elastic fibers and smooth muscle; and (3) a layer of protective connective tissue. The walls of arteries have more elastic tissue and smooth muscle than veins, which helps them accommodate pulses of blood pumped from the heart. Capillaries are microscopic blood vessels that have walls only one cell thick. The exchange of substances between the cells and the blood takes place at the capillaries.

www.jbpub.com/biology

10.6　The heart: A double pump

The pump of the human circulatory system is the **heart**, but it is two pumps in one. **Figure 10.5** is a diagram of a frontal section (a cut that results in "front" [ventral] and "back" [dorsal] pieces) through the heart and shows the organization of this double pump. The left side (one pump) has two connected chambers, as does the right side (the other pump). The two sides, or pumps, of the heart are not directly connected with one another.

Circulatory Pathways

The journey of blood around the body starts with the entry of oxygenated blood into the heart from the lungs. Oxygenated blood from the lungs enters the left side of the heart, emptying directly into the upper left chamber of the heart, the **left atrium ❶**, through large vessels called the **pulmonary veins**. These veins are unusual in that they carry oxygenated blood; other veins, because they carry blood back to the heart from the body tissues, carry deoxygenated blood. The word pulmonary

Visual Summary　Figure 10.5　**Blood flow through the human heart.**

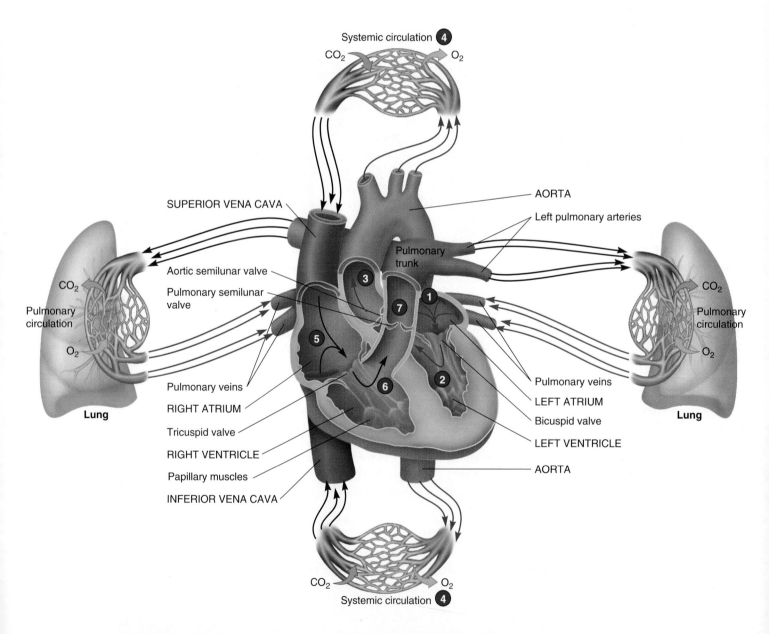

refs to the lungs. The circulation of blood to and from the lungs is therefore called **pulmonary circulation** (Figure 10.5).

From the left atrium of the heart, blood flows through a one-way valve, the *bicuspid valve*, into the lower, adjoining chamber, the **left ventricle ❷**. Most of this flow (roughly 70%) occurs while the heart is relaxed. The atrium then contracts, filling the remaining 30% of the ventricle with its blood. After a slight delay, the ventricle contracts. The walls of the ventricle are far more muscular than are those of the atrium (note the relative thickness in Figure 10.5), and as a result, this contraction is much stronger. It forces most of the blood out of the ventricle in a single strong pulse. The blood is prevented from going back into the atrium by the *bicuspid valve,* whose flaps are pushed shut after the ventricle fills from the atrium. Strong fibers are attached to the edges of the valve and prevent the flaps from moving too far when closing. If the flaps moved too far, they would project out into the atrium and allow backflow.

Prevented from reentering the atrium, the blood takes the only other way out of the contracting left ventricle: an opening that leads into the largest artery in the body—the **aorta ❸**. The aorta is closed off from the left ventricle by a one-way valve, the *aortic semilunar valve* (shown below). It is oriented to permit the flow of the blood out of the ventricle. As the blood is pushed forcefully out of the left ventricle to make its trip around the body, it rushes into the aorta, causing the elastic, muscular walls of this artery to bulge slightly outward. Quickly, however, the aorta walls recoil, as a stretched

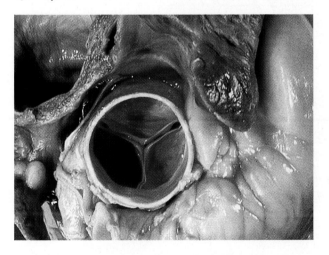

rubber band does when released. This recoil action pushes on the blood and some of it is pushed backward against the valve. The valve is constructed in such a way

that it snaps shut in response to this backflow. The other one-way valves found in the heart and blood vessels are constructed in a similar manner, preventing the backflow of blood.

Many arteries branch from the aorta, carrying oxygen-rich blood to all parts of the body. The pathway of blood vessels to the body regions and organs other than the lungs is called the **systemic circulation ❹** with the aorta being the first and largest vessel in the circuit. The first arteries to branch off the aorta are the coronary arteries, which carry freshly oxygenated blood to the heart itself; the muscles of the heart do not obtain their supply of blood from directly within the heart. From the arch of the aorta, the *carotid arteries* branch off and bring blood to networks of vessels in the neck and head. The *subclavian arteries* bring blood to the shoulders and arms. The aorta then descends down the trunk of the body, with arteries branching off to supply various organs such as the kidneys, liver, and intestines. The aorta divides into two major vessels at the lower back, one traveling to each leg.

The blood that flows into the arterial system eventually returns to the heart after delivering its supply of oxygen to the cells of the body and picking up the waste gas carbon dioxide. This exchange takes place at the capillaries. As it returns, blood passes through a series of veins, eventually entering the right side of the heart. Two large veins collect blood from the systemic circulation. The **superior vena cava** (VEE-nuh-KAY-vuh *or* VAY-nuh-KAH-vuh) drains the upper body, and the **inferior vena cava** drains the lower body. These veins dump deoxygenated blood into the right atrium ❺.

The right side of the heart is similar in organization to the left side. However, the muscular walls of the right ventricle are not as thick as those of the left ventricle. Blood passes from the right atrium into the right ventricle ❻ through a one-way valve, the *tricuspid valve*. It passes out of the contracting right ventricle through a second valve, the *pulmonary semilunar valve*, into a single **pulmonary artery** (trunk) ❼, which subsequently branches into arteries that carry deoxygenated blood to the lungs. The blood then returns from the lungs to the left side of the heart, replenished with oxygen and cleared of much of its load of carbon dioxide. (The human circulatory system as a whole is outlined in Figure 10.2.)

How the Heart Contracts

The contraction of the heart depends on a small cluster of specialized cardiac muscle cells that is embedded in the upper wall of the right atrium (Figure 10.6). This cluster of cells, called the **sinoatrial (SA) node** (sye-no-AY-tree-uhl), automatically and rhythmically sends out impulses that initiate each heartbeat. The SA node is therefore nicknamed the "pacemaker" of the heart.

The impulse initiated by the SA node causes the left and right atria to contract simultaneously and also excites a bundle of cardiac muscle cells located at the base of the atria. These cells are known as the **atrioventricular (AV) node** (AY-tree-oh-ven-TRIK-yuh-lur). The AV node conducts the impulse to a strand of specialized muscle in the **septum**, the tissue that separates the two sides of the heart. This strand of impulse-conducting muscle, known as the *atrioventricular (AV) bundle*, has branches that divide to the right and left. On reaching the apex (lower tip) of the heart, each branch further divides into conducting fibers called *Purkinje fibers* (per-KIN-gee), which initiate the almost simultaneous contraction of all the cells of the right and left ventricles.

> The circulation of the blood to and from the lungs is called pulmonary circulation. The circulation of the blood through the parts of the body other than the lungs is called systemic circulation. The heart contracts when an excitatory impulse is initiated in its SA node, which causes the atria to contract and the impulse to be passed along to other specialized groups of cardiac cells: the AV node, AV bundle, bundle branches, and Purkinje fibers. The passage of this excitatory impulse causes the almost simultaneous contraction of all cells of the right and left ventricles.

10.7 Monitoring the heart's performance

The heartbeat is really a series of events that occurs in a predictable order. A physician can gain information about the health of the heart and events occurring during the heartbeat by making several different kinds of observations. The simplest is to listen to the heart at work by placing a stethoscope on the skin over the heart.

The first sound heard, a low-pitched *lubb*, is caused by the turbulence in blood flow created by the closing of the bicuspid and tricuspid valves at the start of ventricular contraction. A little later, a higher-pitched *dupp* can be heard, signaling the closing of the pulmonary and aortic semilunar valves at the end of ventricular contraction. If the valves are not closing fully or if they open too narrowly, a slight backflow of blood occurs within the heart. This backflow can be heard as a sloshing sound; this condition is known as a heart murmur.

A second way to examine the events of the heartbeat is to monitor the blood pressure, the force exerted

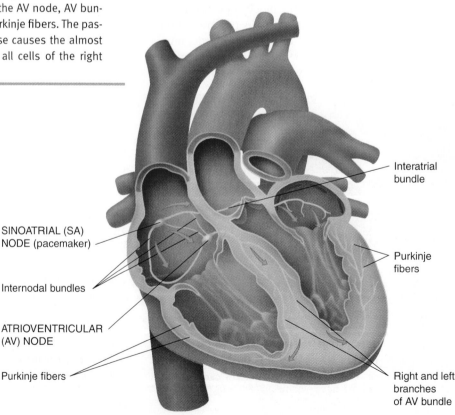

Figure 10.6 The conduction system of the heart. Four structures comprise the conduction system of the heart—sinoatrial (SA) node, atrioventricular (AV) node, AV bundle, and Purkinje fibers. These specialized groups of cardiac muscle cells initiate an electrical impulse throughout the heart. The impulse begins in the SA node and spreads to the AV node. The AV node then initiates a signal that is conducted through the ventricles by way of the AV bundle and Purkinje fibers.

Interatrial bundle

SINOATRIAL (SA) NODE (pacemaker)

Internodal bundles

ATRIOVENTRICULAR (AV) NODE

Purkinje fibers

Purkinje fibers

Right and left branches of AV bundle

by the blood on blood vessel walls. During the first part of the heartbeat, the atria are filling and contracting. At this time, the pressure in the arteries leading from the left side of the heart out to the tissues of the body decreases slightly because they are not receiving blood from the left ventricle. This period of relaxation (with respect to the ventricles) is referred to as ventricular **diastole** (dye-AS-tl-ee). The force the blood exerts on the blood vessels at this time is the diastolic pressure. During the contraction of the ventricles, a pulse of blood is forced into the systemic arterial system by the left ventricle, immediately raising the blood pressure within these vessels. This period of contraction (with respect to the ventricles) ends with the closing of the aortic semilunar valve and is referred to as ventricular **systole** (SIS-tl-ee). The force the blood exerts on the blood vessels at this time is the systolic pressure.

Blood pressure is measured in units called millimeters (mm) of mercury (Hg). These units refer to the height to which a column of mercury is raised in a tube by the force of the heart beating. On the basis of this system of measurement, normal blood pressure values are 70 to 90 (mm of Hg) diastolic and 110 to 130 systolic. Blood pressure is expressed as the systolic pressure over the diastolic pressure, such as 110 over 70. When the inner walls of the arteries accumulate fats, as they do in the condition known as **atherosclerosis** (ATH-uh-ROW-skluh-ROW-sis), the diameters of the passageways are narrowed. Such narrowing is one cause of elevated systolic and diastolic blood pressures.

A third way to monitor the events of a heartbeat is to measure the electrical changes that take place as the heart's chambers both contract and relax. Because so much of the human body is made up of water, it conducts electrical currents rather well. Therefore, as the impulses initiated at the SA node pass throughout the heart as an electrical current, this current passes in a wave throughout the body. Although the magnitude of this electrical pulse is small, it can be detected with sensors placed on the skin. A recording made of these impulses (**Figure 10.7**) is called an **electrocardiogram** (ECG).

Three successive electrical pulses are recorded in a normal heartbeat. The first pulse occurs just prior to atrial contraction; this electrical event is called the *P wave* on the ECG (Figure 10.7). There is a much stronger pulse (the *QRS complex*) 2/10 of a second later, which shows the stimulus for the contraction of the ventricles and electrical changes that precede the relaxation of the atria. Finally, a third pulse (the *T wave*) occurs, which is caused by the electrical changes that precede the relaxation of the ventricles.

The heartbeat is a series of events that occurs in a predictable order. A physician can gather data about the health of the heart by monitoring the heartbeat in various ways: by listening to the sounds the heart makes, by monitoring the blood pressure, and by measuring the electrical changes that take place as it contracts and relaxes.

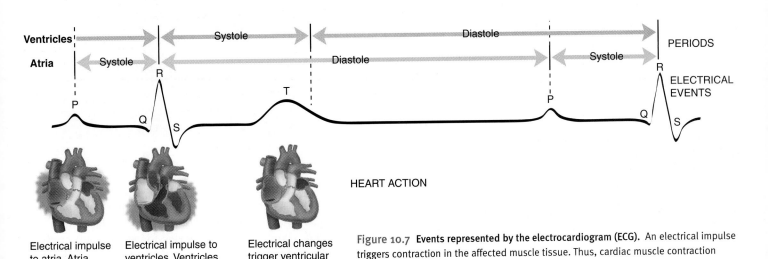

Electrical impulse to atria. Atria contract. Ventricles relax.

Electrical impulse to ventricles. Ventricles contract. Atria relax.

Electrical changes trigger ventricular relaxation.

HEART ACTION

Figure 10.7 **Events represented by the electrocardiogram (ECG).** An electrical impulse triggers contraction in the affected muscle tissue. Thus, cardiac muscle contraction occurs after this electrical triggering.

Human blood is a transport fluid.

10.8 Blood plasma

Blood **plasma** is a straw-colored liquid made up of water and dissolved substances. The dissolved substances can be grouped into three categories:

1. *Nutrients, hormones, respiratory gases, and wastes.* Dissolved within the plasma are substances that are used or produced by the metabolism of cells. These substances include glucose, lipoproteins (a soluble form of lipid), amino acids, vitamins, hormones, and the respiratory gases.

2. *Salts and ions.* Plasma is a dilute salt solution. Chemically, the word salt is applied to any substance composed of positively and negatively charged ions (see Chapter 3). In water, salts dissociate into their component ions. The chief plasma ions are sodium (Na^+), chloride (Cl^-), and bicarbonate (HCO^{3-}). In addition, there are trace amounts of other ions, such as calcium (Ca^{2+}), magnesium (Mg^{2+}), zinc (Zn^{2+}), and potassium (K^+). In living systems, these ions are called *electrolytes.*

   Electrolytes serve three general functions in the body. First, many are essential for the various physiological processes such as the proper functioning of muscles and nerves, and the proper formation of bone. Second, they play a role in the movement of water—osmosis—between various compartments within the body. Third, they help maintain the acid-base (pH) balance required for normal cellular activities.

3. *Proteins.* Blood plasma is approximately 90% water but contains a concentration of proteins that helps balance osmotic pressure between the cells and the blood. For this reason, water is not osmotically sucked out of the blood. The types of proteins in blood vary. Some of these proteins are antibodies and other proteins that are active in the immune system, while others consist of fibrinogen and prothrombin, key players in blood clotting. The majority of circulating protein in blood, however, is called serum albumin, which acts as the primary osmotic counterforce.

Blood plasma contains nutrients, hormones, respiratory gases, wastes, and a variety of ions and salts. It also contains high concentrations of the protein serum albumin, which functions to keep the blood plasma in osmotic equilibrium with the cells of the body.

10.9 Types of blood cells

Although blood is liquid, 45% of its volume is actually occupied by cells and pieces of cells, collectively called **formed elements**. There are three principal types of formed elements in the blood: erythrocytes (red blood cells), leukocytes (white blood cells), and platelets (**Figure 10.8**).

Erythrocytes

In only one teaspoonful of your blood, there are about 25 billion **erythrocytes** (ih-RITH-row-sites), or red blood cells. Each erythrocyte is a flat disk with a central depression (see below) something like a doughnut with a hole that does not go all the way through. Almost the entire interior of each cell is packed with the oxygen-carrying molecule hemoglobin. One estimate, in fact, is that each erythrocyte can carry 280 million molecules of hemoglobin!

Mature erythrocytes do not have nuclei or the ability to manufacture proteins. Red blood cells are therefore unable to repair themselves and consequently have a rather short life span. Erythrocytes live only about 4 months. Parts of old, worn-out erythrocytes are broken down for recycling by macrophages in the spleen, liver, and bone marrow. New erythrocytes are constantly being synthesized and released into the blood by cells within the soft interior marrow of bones at the amazing rate of 2 million per second!

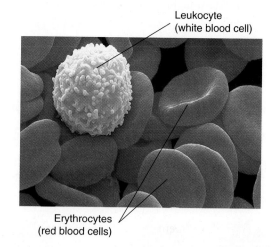

Leukocyte
(white blood cell)

Erythrocytes
(red blood cells)

Leukocytes

Less than 1% of the cells in human blood are **leukocytes** (LOO-ko-sites), or white blood cells. In fact, there are only about 1 or 2 leukocytes for every 1000 erythrocytes. Leukocytes are larger than red blood cells; they contain no hemoglobin, have nuclei, and are essentially colorless. There are several kinds of leukocytes,

	BLOOD CELL TYPE	DESCRIPTION	FUNCTION	LIFE SPAN
RED BLOOD CELLS	Erythrocyte 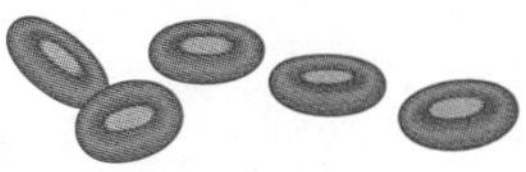	Flat disk with a central depression, no nucleus, contain hemoglobin.	Transport oxygen (O_2) and carbon dioxide (CO_2).	About 120 days.
WHITE BLOOD CELLS (LEUKOCYTES) — **Granulocytes**	Neutrophil 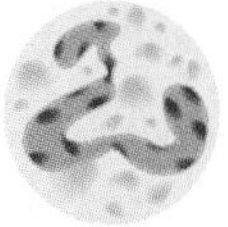	Spherical; with many lobed nucleus, no hemoglobin, pink-purple **cytoplasmic granules**.	Cellular defense-phagocytosis of small microorganisms.	Hours to 3 days.
	Eosinophil 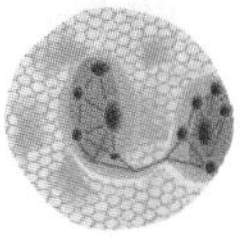	Spherical; two-lobed nucleus, no hemoglobin, orange-red staining **cytoplasmic granules**.	Cellular defense- phagocytosis of large microorganisms such as parasitic worms, releases anti-inflammatory substances in allergic reactions.	8 to 12 days.
	Basophil 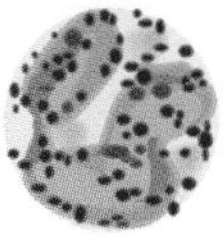	Spherical; generally two-lobed nucleus, no hemoglobin large purple staining **cytoplasmic granules**.	Inflammatory response - contain granules that rupture and release chemicals enhancing inflammatory response.	Hours to 3 days.
Agranulocytes	Monocyte 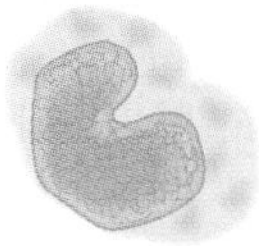	Spherical; **single nucleus** shaped like kidney bean, no cytoplasmic granules, cytoplasm often blue in color.	Converted to macrophage which are large cells that entrap microorganisms and other foreign matter.	Days to months.
	B-lymphocyte 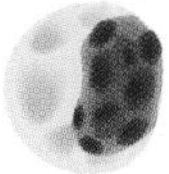	Spherical; round **singular nucleus**, no cytoplasmic granules.	Immune system response and regulation, antibody production sometimes cause allergic response.	Days to years.
	T-lymphocyte 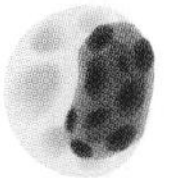	Spherical; round **singular nucleus**, no cytoplasmic granules.	Immune system response and regulation; cellular immune response.	Days to years.
PLATELETS	Platelets	Irregularly shaped fragments, very small pink staining granules.	Control blood clotting or coagulation.	7 to 10 days.

and each has a different function. All functions, however, are related to the defense of the body against invading microorganisms and other foreign substances (see Chapter 11).

There are two major groups of leukocytes: **granulocytes** (GRAN-yuh-low-sites) and **agranulocytes**. Granulocytes are circulating leukocytes and get their name from the tiny granules in their cytoplasm. In addition, they have lobed nuclei (see Figure 10.8). The granulocytes are

classified into three groups by their staining properties. About 50% to 70% of them are **neutrophils**, cells that migrate to the site of an injury and stick to the interior walls of the blood vessels. They then form projections that enable them to push their way into the infected tissues, where they engulf, or *phagocytize*, microorganisms and other foreign particles. The term phagocytosis comes from a Latin word meaning "cell eating." **Basophils** (BAY-soh-filz), a second kind of granulocyte, contain granules that rupture and release chemicals that enhance the body's response to injury or infection. They play a role in causing allergic responses. The third kind of granulocyte, the **eosinophils** (EE-oh-SIN-oh-filz), is also believed to be involved in allergic reactions. In addition, they act against certain parasitic worms.

Agranulocytes, the second major group of leukocytes, have no cytoplasmic granules, nor are their nuclei lobed. One group of agranulocytes, the **monocytes**, circulate as the granulocytes do. Monocytes are attracted to the sites of injury or infection, where they are converted into **macrophages** (MAK-row-FAY-djus), enlarged, amebalike cells that entrap microorganisms and particles of foreign matter by phagocytosis. They usually arrive after the neutrophils and clean up any bacteria and dead cells. **Lymphocytes**, the other type of agranulocyte, recognize and react to substances that are foreign to the body, sometimes producing a protective immunity to disease. Occasionally, however, these cells produce an inflammation or an allergic response (see Chapter 11).

Platelets

Certain large cells within the bone marrow called *megakaryocytes* regularly pinch off bits of their cytoplasm. These cell fragments, called **platelets**, enter the bloodstream and play an important role in controlling blood clotting, or coagulation. The clotting of blood is a complicated process initiated by damage to blood vessels or tissues.

When an injury occurs, platelets clump at the damaged area, temporarily blocking blood loss. The damaged tissues and platelets release a complex of substances called *thromboplastin*. The release of this complex begins a cascade of events, one dependent on the other. A simplified explanation is as follows: thromboplastin interacts with calcium ions, vitamin K, and other clotting factors to form an enzyme called *prothrombin activator*. Prothrombin activator brings about the conversion of prothrombin to thrombin. Thrombin converts fibrinogen, a soluble protein, to fibrin, an insoluble, threadlike protein. Fibrin threads, along with trapped red blood cells (Figure 10.9), form the clot—a plug at the damaged area so that blood cannot escape.

> The three principal types of formed elements in the blood are erythrocytes, leukocytes, and platelets. Erythrocytes, or red blood cells, carry oxygen; leukocytes, or white blood cells, defend the body against invading microorganisms and other foreign substances; and platelets play an important role in blood clotting.

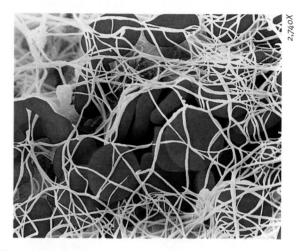

Figure 10.9 **Formation of fibrin threads.** Eventually, many red blood cells will become caught in this net of insoluble protein and a clot will form.

10.10 Lymph

Although the blood proteins and electrolytes help maintain an osmotic balance between the blood and the tissues, the blood loses more fluid to the tissues than it reabsorbs from them. Of the total volume of fluid that moves from the blood into and around the tissues, about 90% reenters the cardiovascular system. Where does the other 10% go? The answer is to the body's one-way, passive circulatory system, the **lymphatic system** (lim-FAT-ik) (Figure 10.10). This system counteracts the effects of net fluid loss from the blood.

By osmosis and diffusion, blind-ended lymphatic capillaries (see Figure 10.10) fill with tissue fluid (including small proteins that have diffused out of the blood). From these capillaries, the tissue fluid—now called **lymph** (LIMF)—flows through a series of progressively larger vessels to two large lymphatic vessels, which structurally resemble veins. These vessels drain into veins near the base of the neck. Although the lymphatic system has no heart to pump lymph through its vessels, the fluid is squeezed through them by the movements of the body's muscles. The lymphatic vessels contain a series of one-way valves (see Figure 10.10) that permits movement only in the direction of the neck.

Small, ovoid, spongy structures called **lymph nodes** are located in various places of the body along the routes of the lymphatic vessels. They are clustered in areas such as the groin, armpits, and neck, filtering the lymph as it passes through. Some lymphocytes, the cells that activate the immune response, reside in the lymph nodes. (These cells are discussed in more detail in Chapter 11.) In addition, the lymphatic system has two organs: the spleen and the thymus. The *spleen* stores an emergency blood supply and also contains white blood cells. Specific types of white blood cells in the spleen destroy old red blood cells, filter microorganisms out of the blood as it passes through, and initiate an immune response against the foreign microbes. The *thymus* plays an important role in the maturation of certain lymphocytes called *T cells*, which are an essential part of the immune system.

Approximately 10% of the fluid that moves out of the blood and into the cells and cell spaces at the capillaries does not return to the blood directly. It is collected and then returned to the blood by a system of one-way, blind-ended vessels called the lymphatic system.

10.11 Cardiovascular disease

Over 40% of all deaths in the United States are due to cardiovascular disease. More than 58 million people in this country—about one person in five—have some form of cardiovascular disease.

Arteriosclerosis (ar-TEER-ee-oh-skluh-ROW-sis) is a thickening and hardening of the walls of the arteries. Blood flow through such arteries is restricted. These arteries lack the ability to dilate and therefore have difficulty accommodating the volume of blood pumped out by the heart. The narrowing of these vessels forces the heart to work harder.

Atherosclerosis is a form of arteriosclerosis in which masses of cholesterol and other lipids build up within the walls of large and medium-sized arteries (Figure 10.11a). These masses are referred to as **plaques** (PLAKS). The accumulation of plaques impairs the arteries' proper functioning. When this condition is severe, the arteries can no longer dilate and constrict properly and the blood moves through them with difficulty. The accumulation of cholesterol is thought to be the prime contributor to atherosclerosis, and diets low in cholesterol as well as cholesterol-lowering medications are now prescribed to help prevent this condition. Atherosclerosis contributes to both heart attacks and strokes.

For the latest health information, please visit www.jbpub.com/biology.

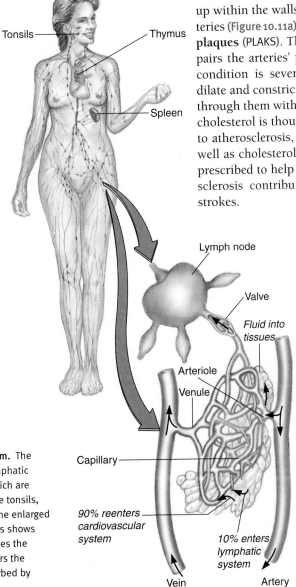

Figure 10.10 The lymphatic system. The lymphatic system consists of the lymphatic vessels and ducts, lymph nodes (which are distributed throughout the body), the tonsils, the thymus gland, and the spleen. The enlarged diagram of a lymph node and vessels shows the path of the excess fluid that leaves the arteriole end of a capillary bed, enters the adjacent tissue spaces, and is absorbed by lymphatic capillaries.

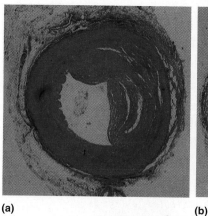
(a)

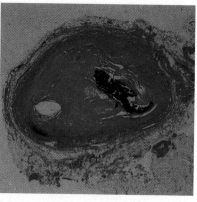
(b)

30X

Figure 10.11 Atherosclerosis. (a) The coronary artery exhibits severe atherosclerosis—much of the passage of blood is blocked by buildup of cholesterol and other lipids on the interior walls of the artery, forming masses or plaques. (b) The coronary artery is almost completely blocked. A blood clot (dark red) blocks blood flow on the right side of the artery.

Table 10.1

Risk Factors for Cardiovascular Disease

- Cigarette smoking
- Heavy alcohol use
- Physical inactivity
- Obesity (20% or more above the desirable weight for your height)
- Total serum cholesterol over 200 mg/dl
- High levels of "bad" (LDL) cholesterol (> 130 mg/dl)
- Family history (having a male first-degree relative who had a heart attack before age 55 or a female first-degree relatives who had a heart attack before age 65)
- Chronic high blood pressure (hypertension) (160/90 is significant hypertension)
- Gender (women are at lower risk than men until menopause)
- Age (risk increases with age)
- Elevated blood glucose levels present in diabetes mellitus
- Anxiety disorders

Table 10.2

How to Reduce Your Risk of Cardiovascular Disease

- Get regular medical checkups
- Do not smoke cigarettes
- Restrict alcohol consumption to no more than 1 drink (1 oz. ethanol) per day
- Exercise regularly
- Maintain an appropriate weight for your height
- Eat a low-fat diet (less than 30% of calories from fat per day)
- Eat foods rich in soluble fiber such as fruits, beans, and oats
- Maintain salt intake at less than one-half teaspoon per day if over 45 years
- Take estrogen replacement after menopause if advisable
- Manage diabetes mellitus properly
- Reduce your stress level

Heart attacks, the most common cause of death in the United States, result from an insufficient supply of blood reaching an area of heart muscle. A blood clot, or *thrombus*, may form in one of the vessels that supplies the heart with blood, thereby blocking its blood supply (see Figure 10.11b) and causing a heart attack. A floating blood clot, or *embolus*, may travel to the heart from another location. Additionally, if a vessel supplying the heart with blood is blocked sufficiently by fatty deposits, especially cholesterol and triglycerides, blood flow to the heart may be reduced or stopped altogether, even without the complication of a blood clot.

The signs of a heart attack are a pressure, squeezing, or pain in the center of the chest. This pain may extend to the neck, shoulder, or arms and is often accompanied by lightheadedness, nausea, vomiting, sweating, and shortness of breath. Any person experiencing any of these signs should be rushed to a hospital emergency room immediately or an ambulance should be called. Swift action can save the life of a heart attack victim. If a person reaches the hospital emergency room alive, he or she has a 95% chance of surviving their attack.

Angina pectoris, which literally means "chest pain," occurs for reasons similar to those causing heart attacks, but angina pain is not as severe as heart attack pain. In angina, a reduced blood flow to the heart muscle weakens the cells, but does not kill them.

The amount of heart damage associated with a small heart attack may be relatively slight and thus difficult to detect. It is important that such damage be detected, however, so that the overall condition of the heart can be evaluated properly. Electrocardiograms are very useful for this purpose because they reveal abnormalities in the timing of heart contractions—abnormalities that are associated with the presence of damaged heart tissue. Damage to the atrioventricular (AV) node, for example, may delay as well as reduce the second, ventricular pulse. Unusual conduction routes may lead to continuous disorganized contractions called *fibrillations*. In many fatal heart attacks, ventricular fibrillation is the immediate cause of death.

Many factors contribute to heart disease and heart attacks. **Table 10.1** lists risk factors for cardiovascular disease and **Table 10.2** lists ways to reduce your risk of developing this disease.

I often hear about young people who die suddenly of a heart attack even though they exercised and ate right. Why is that?

As you have discovered, there is no guarantee that if you follow health recommendations regarding coronary heart disease, you will be free of this disease and the threat of heart attacks. Following health recommendations helps you reduce your risk of heart attack, however. In addition, researchers have discovered new information that sheds light on possible reasons why some people die of heart attacks despite their heart-healthy lifestyles.

Cardiovascular disease is the number one killer of Americans, so assessing your risk level and doing something about it is important. The major risk factors of this disease are physical inactivity, elevated serum cholesterol (a fat that often sticks to artery walls), cigarette smoking, obesity, chronically high blood pressure, and a family history of heart and blood vessel disease. The more of these risk factors you have, the more likely you are to have a heart attack or stroke.

So why does someone die from a heart attack when the chances are low according to the risk factors noted here? In some cases, persons are discovered (during autopsy perhaps) to have congenital heart abnormalities—problems of heart structure or function with which they were born and of which they were unaware. Suddenly, under some physical or emotional time of stress, their heart gives out.

One inherited heart disorder that can cause sudden death is arrhythmogenic right ventricular dysplasia, or ARVD. In persons with ARVD, muscle cells die in the wall of the right

ventricle—the chamber that pumps blood to the lungs. Sudden death can then occur from arrhythmias, which are erratic heart beats, in this weakened heart. Even if a person with ARVD does not die suddenly, his or her heart eventually weakens so much that it cannot deliver sufficient oxygen to the organs. When the heart begins to fail in this way, death occurs within a few years. By the year 2001, medical researchers expect to have a genetic test for this disease, which is caused by a mutation on chromosome 3.

Other information has also come to light that may help us understand many more of these seemingly paradoxical cases. As scientists studied how arteries become clogged with fat, they came to understand that the blood levels of two cholesterol-carrying molecules, high-density lipoproteins (HDL) and low-density lipoproteins (LDL), were possibly more important than the blood level of total cholesterol. LDL, the so-called bad cholesterol, carries cholesterol to the cells (including those lining the blood vessels). HDL, the so-called good cholesterol, carries cholesterol away from the cells and facilitates its removal from the body. In fact, heart disease risk factors tend to correlate in significant ways with LDL and HDL levels. For example, cigarette smokers tend to have lower HDL levels than do nonsmokers.

Recently, a cholesterol-carrying lipoprotein called lipoprotein(a) has been shown to be an important factor in the development of heart and blood vessel disease as well.

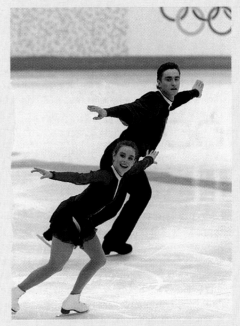

Heart failure ended the professional and personal partnership of Sergei Grinkov and Ekaterina Gordeeva. The couple had won gold medals for Russia in pairs figure skating in both the 1988 and 1994 Olympics. Grinkov died suddenly during a practice session in November 1995—he was just 28 years old and in the prime of his life.

Lipoprotein(a) has been found in high concentrations in the blood of persons with a low risk of this disease. Researchers are currently studying lipoprotein (a) to understand its role in cardiovascular disease. With this understanding may come modifications or additions to the recommendations of "heart healthy" lifestyles.

Strokes ("brain attacks") are caused by an interference with the blood supply to the brain and often occur when a blood vessel bursts in the brain. A stroke may also be caused by a thrombus, embolus, and/or fatty deposits blocking blood flow as in heart attacks. The effects of strokes depend on the severity of the damage and the specific location of the stroke. Some of the signs of a stroke are: weakness, numbness, or paralysis on one side of the body; visual dimming, particularly in one eye; difficulty speaking; sudden, severe headache; and sudden dizziness or falling. As with heart attack victims, those experiencing stroke symptoms should seek med-

ical attention immediately to help reduce brain damage and increase the possibility of survival.

Following heart-healthy behaviors throughout your life, even when you are young, will help you avoid heart disease and its debilitating effects when you are older. Scientists now know that fatty deposits can begin to develop in the arteries of children. Heart disease is not just a disease of older persons—it is a life-long process affected by diet and lifestyle. Establishing heart-healthy eating patterns and other behaviors as early in life as possible will help you maintain a healthy cardiovascular system throughout your life.

H⚙W WⓄRKS Science Applications

Creating Fat Replacements and Substitutes for "Heart Healthy" Foods

Cheeseburgers, french fries, ice cream... most Americans love fatty foods. Why? Because fats are flavorful, have a pleasant texture and "mouth feel," and make us feel full. One problem with the fats in our diets is that they provide 9 calories per gram—more than twice the amount of energy in a gram of carbohydrate or protein. Therefore, eating fats can easily result in the consumption of too many calories and bulges around our midsections.

A more serious problem with high fat consumption is that too much fat in the diet contributes to the formation in the arteries of plaque—masses of fats and other substances that block arteries and impair their proper functioning. This buildup most often occurs in the arteries that supply the heart and brain with blood, and it can result in coronary heart disease, heart attacks, and strokes.

The American Heart Association currently recommends that less than 30% of the total calories in the diet of a healthy person of normal weight come from the various kinds of fats. Of that 30%, less than 10% should come from saturated (generally "animal") fats. Currently, 34% of the typical American diet consists of calories from fat. Clearly, Americans need to find ways to reduce their fat consumption.

Food scientists have found ways to make many of our favorite, fat-laden foods "heart healthy" by using fat replacements or fat substitutes. Fat replacements are created by taking normal food ingredients, such as proteins from eggs and milk, water, or various carbohydrates, and processing and combining them to produce an ingredient that has some of the properties of fats without their calories or artery-clogging characteristics. Polydextrose, a starch-based fat replacement, is currently used in frozen desserts, puddings, and cake frostings. Maltodextrin, another starch-based product, is used in salad dressings and margarines.

Fat substitutes differ from fat replacements in that they are a single, unique ingredient and usually have even fewer calories than

fat replacements. One fat substitute is marketed under the brand name *Simplesse*. This fat substitute is used in the manufacture of such foods as ice cream, salad dressings, and cheese products. *Simplesse* has a fatlike consistency and texture because it is composed of proteins from either milk or egg whites that have been heated and blended to create nearly microscopic spheres of protein. These mistlike particles mimic the "mouth feel" of real fat globules—that smooth, creamy texture we enjoy in a chocolate bar or a scoop of ice cream. *Simplesse* has one drawback. It does not retain its fatlike consistency and taste when it is heated. Olestra, one of the newest fat substitutes, can withstand heat.

Olestra is a sucrose polyester: a sugar molecule to which six, seven, or eight fatty acid chains are attached. Triglycerides, or natural fats, are also composed of fatty acids; thus Olestra comes close to mimicking their molecular structure. Because fat-digesting enzymes do not break down Olestra, it bypasses the body's digestive and absorptive mechanisms (and has no calories). This ability, combined with Olestra's textural resemblance to

natural fat, makes it a very convincing and effective fat substitute, which can be used in a variety of products from oils to ice cream. Currently, Olestra has U.S. Food and Drug Administration (FDA) approval for use in savory snacks (potato chips, tortilla chips, cheese puffs, and snack crackers). Potato chips made with Olestra are currently available, but the warning label on the package notes that Olestra may cause abdominal cramping and loose stools. It also inhibits the absorption of fat-soluble vitamins and other nutrients, so vitamins A, D, E, and K are added to the chips.

Some nutritionists argue that the best way to lower fat consumption is to develop a taste for foods that are naturally low in fat rather than searching for substitutes for fatty foods. However, fat substitutes offer the opportunity to eat a once-forbidden treat occasionally without increasing fat consumption above healthy levels, as well as to modify the level of fat in the diet using replacements or substitutes in moderation. Fat substitutes are also useful for those on physician-prescribed no-fat diets.

Key Concepts

Circulatory (transport) systems, when present, are either open or closed.

10.1 Single-celled organisms do not have circulatory systems. Gases and nutrients move into the cell directly from the environment.

10.2 Organisms such as insects, spiders, clams, and snails have open circulatory systems.

10.2 In an open circulatory system, blood mixed with tissue fluid bathes the internal tissues; blood is not enclosed in vessels.

10.3 Closed circulatory systems are found in cephalopod molluscs, annelids, echinoderms, and vertebrates.

10.3 In a closed circulatory system, the blood is located within blood vessels as it is pumped throughout the body.

Human circulation is a closed transport system.

10.4 The circulatory system is made up of four components: the heart, the blood vessels, the blood, and the lymphatic vessels.

10.4 The circulatory system transports nutrients, wastes, respiratory gases, and hormones and plays an important role in temperature maintenance of the body.

10.5 Arteries and veins are the major vessels of the circulatory system.

10.5 Capillaries are microscopic blood vessels through which the exchange of substances between the cells and the blood takes place.

10.6 The heart is a double pump, pushing both pulmonary (lung) circulation and systemic (general body) circulation.

10.6 The impulse is passed along to other specialized groups of cardiac cells, causing the almost simultaneous contraction of all cells of the right and left ventricles.

10.7 A physician can gather data about the health of the heart by monitoring its contractions.

Human blood is a transport fluid.

10.8 The liquid portion, or plasma, of the circulating blood contains the proteins and ions that are necessary to maintain the blood's osmotic equilibrium with the surrounding tissues.

10.9 The formed elements of the blood are the red blood cells (erythrocytes), white blood cells (leukocytes), and platelets.

10.9 The red blood cells transport oxygen, the white blood cells defend the body against disease, and the platelets are essential to the process of blood clotting.

10.10 The lymphatic system gathers fluid from the body that has been lost from the circulatory system by diffusion and returns it via a system of lymphatic capillaries, lymphatic vessels, and two large lymphatic ducts to veins in the lower part of the neck.

Health Connections

10.11 Cardiovascular diseases, diseases of the heart and blood vessels, are the leading cause of death in the United States.

10.11 Heart attacks, strokes, and atherosclerosis are all serious cardiovascular diseases.

Key Terms

agranulocytes (ay-GRAN-yuh-low-sites) *211*

aorta (ay-ORT-uh) *207*

arteries (ART-uh-rees) *204*

arterioles (are-TEER-ee-oles) *204*

arteriosclerosis (ar-TEER-ee-oh-skluh-ROW-sis) *213*

atherosclerosis (ATH-uh-ROW-skluh-ROW-sis) *209*

atrioventricular (AV) node (AY-tree-oh-ven-TRIK-yuh-lur NODE) *208*

basophils (BAY-soh-filz) *211*

capillaries (KAP-uh-LARE-ees) *204*

cardiovascular system (KAR-dee-oh-VAS-kyuh-lur) *202*

circulatory system *198*

closed circulatory system *199*

diastole (dye-AS-tl-ee) *209*

electrocardiogram (ECG) (ih-LEK-trow-KARD-ee-uh-GRAM) *209*

eosinophils (EE-oh-SIN-oh-filz) *212*

erythrocytes (ih-RITH-row-sites) *210*

formed elements *210*

granulocytes (GRAN-yuh-low-sites) *211*

heart *206*

inferior vena cava *207*

left atrium (AY-tree-um) *206*

left ventricle (VEN-truh-kul) *207*

leukocytes (LOO-ko-sites) *210*

lymph (LIMF) *212*

lymph nodes (LIMF NODES) *212*

lymphatic system (lim-FAT-ik) *212*

lymphocytes (LIM-foh-sites) *212*

macrophages (MAK-row-FAY-djus) *212*

monocytes (MON-oh-sites) *212*

neutrophils (NOO-truh-filz) *211*

open circulatory system *198*

plaques (PLAKS) *213*

plasma (PLAZ-muh) *210*

platelets *212*

pulmonary artery *207*

pulmonary circulation (PUHL-muh-NARE-ee sur-kyuh-LAY-shun) *207*

pulmonary veins (PUHL-muh-NARE-ee VAYNZ) *206*

septum *208*

sinoatrial (SA) node (sye-no-AY-tree uhl NODE) *207*

superior vena cava (VEE-nuh KAY-vuh or VAY-nuh KAH-vuh) *207*

systemic circulation *207*

systole (SIS-tl-ee) *209*

veins (VAYNS) *204*

venules (VAYN-yooles) *204*

KNOWLEDGE AND COMPREHENSION QUESTIONS

1. List three differences between an open circulatory system and a closed circulatory system. (*Knowledge*)

2. What are the principal functions of the human circulatory system? (*Knowledge*)

3. You have just finished a great lunch of a chicken sandwich and a tossed salad with oil and vinegar dressing. Discuss what happens to the molecules of the food after they pass into your bloodstream. (*Comprehension*)

4. What significant change occurs in the blood from the time it enters the capillaries until it enters the venules? (*Comprehension*)

5. Mystery novels sometimes describe people who "turn pale with fear." Describe the circulatory reason for this. (*Comprehension*)

6. Compare the structure of arteries, capillaries, and veins. Relate any differences to their respective functions. (*Comprehension*)

7. The text states that the heart is actually two pumps in one; explain this duality. State a reason, backed up by information from this chapter, as to why these two sides do not make contact. (*Comprehension*)

8. Why are valves necessary in heart function? (*Comprehension*)

9. What are the SA node and the AV node? Explain their significance. (*Comprehension*)

10. List the names of all the formed elements of the blood and describe their functions. What is the function of blood plasma? (*Knowledge*)

K E Y
Knowledge: Recalling information.
Comprehension: Showing understanding of recalled information.

CRITICAL THINKING REVIEW

1. Analyze your diet, lifestyle, and family history to determine your risk factors for cardiovascular disease. Develop a list of these factors. (*Analysis*)

2. Which risk factors on your list can you change? Which can you not change? What changes can you make in your diet and lifestyle to lower your risk of heart disease? (*Evaluation*)

3. The opening vignette to this chapter discussed the left ventricular assist device that has replaced the total artificial heart (TAH). What factors in heart anatomy and physiology allow this smaller, more specific pump to replace the TAH? (*Synthesis*)

4. Figure 10.7 shows an electrocardiogram, the electrical events of the heart. Redraw the ECG and show where the heart sounds would occur. Support your answer with an explanation. (*Analysis*)

5. Physicians often remove a woman's underarm lymph nodes when she is diagnosed with breast cancer to see if the cancer has spread. Using the information about lymph nodes in this chapter, explain why this procedure is useful in determining if the cancer has spread. (*Application*)

K E Y
Application: Using information in a new situation.
Analysis: Breaking down information into component parts.
Synthesis: Putting together information from different sources.
Evaluation: Making informed decisions.

e-learning

Location: http://www.jbpub.com/biology

e-Learning is an on-line student review area located at this book's web site www.jbpub.com/biology. The review area provides a variety of activities designed to help you study for your class and build your learning portfolio.

Review Questions The review questions test your knowledge of the important concepts and applications in each chapter. The review provides feedback for each correct or incorrect answer. This is an excellent test preparation tool.

Figure-Labeling Exercises Sharpen your visual thinking by matching terms and labels for important illustrations.

Flash Cards Studying biology requires learning new terms. Virtual flash cards help you master the new vocabulary for each chapter.

Just Wondering As you read and study from this text, you may find that you have unanswered questions. Through this site you can ask the author, Sandy Alters, your "just wondering" questions.

 Do you prefer the speed and reliability of a CD-ROM? All of the features contained on the eLearning portion of the web site are also available on the Student CD-ROM.

Defense against Disease

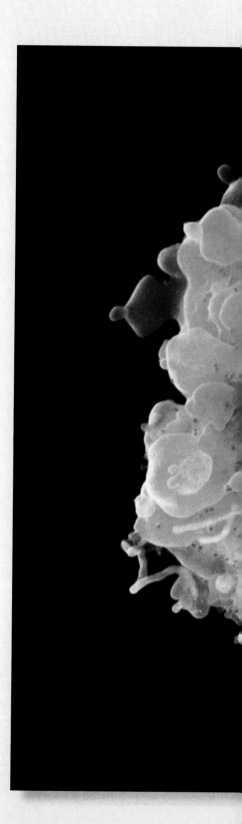

Every day you fight for your life. Some of your attackers swarm over your skin. Some enter your mouth, nose, eyes—any place where they may be able to gain a foothold in your tissues and spread. The battles between you and these invaders are silent and deadly. The fighting is one on one; chemical warfare is commonplace.

It is your nonspecific and specific immune responses that fight and win these thousands of battles. Yet the war within you rages on continually. Each day, thousands of people lose their inner wars, succumbing to invasions of the body by viruses, bacteria, fungi, or protists. Persons whose immune systems are impaired, such as those infected with the human immunodeficiency virus (HIV), have the hardest time winning battles against infectious diseases. HIV attacks the immune system itself (the body structures that perform specific immune responses), lowering its defenses. Therefore, those infected with HIV and who are not helped with the medication "cocktails" available today contract many other diseases quite easily.

Looking at the photo, you can see the human immunodeficiency virus doing some of its dirty work, attacking one of the most important cells of the immune system—a helper T cell. Helper T cells identify foreign invaders and stimulate the production of other cells to fight an infection. Although thousands of times larger than HIV particles (colored blue in this photograph), the helper T cell is inactivated and sometimes killed by this virus. The body is then an easy target for other invaders to enter—unnoticed—and win the war between life and death.

11

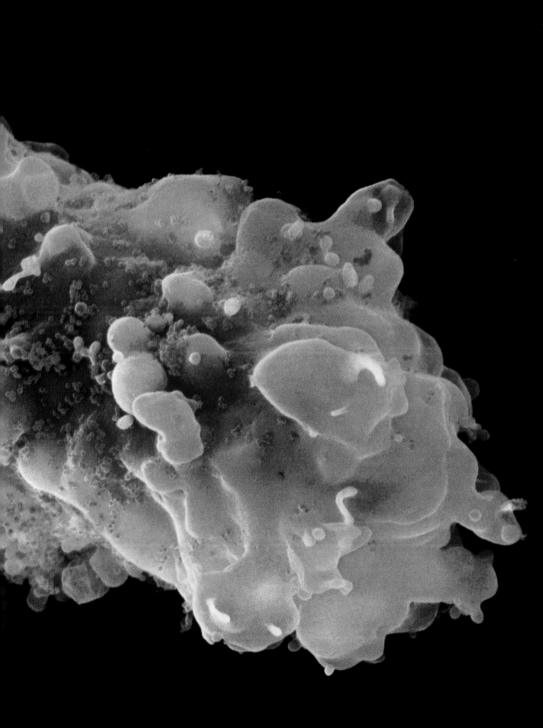

Organization

Almost all living things have some elements of immunity.

Human immunity is both nonspecific and specific.

Health
Connections

11.1 An overview of human immunity

Your body works in many ways to keep you healthy and free of infection. An *infection* results when microorganisms or viruses enter the tissues, multiply, and cause damage. **Immunity** (ih-MYOON-ih-tee) is protection from disease, particularly infectious disease. You have two types of immunity: nonspecific and specific.

Nonspecific immunity works to keep out *any* foreign invader, just as the walls and roof of your home are barriers that protect you from any of the elements of the environment—rain, hail, wind, the sun's rays, insects, or the neighbor's dog, for example. Nonspecific immunity acts against any invader *in the same way;* analogously, your roof does not change the manner in which it protects you from rain, snow, or anything else. It is simply a wooden barrier with a waterproof covering. Another characteristic of nonspecific immunity is that it does not require previous exposure to an invader to be effective. Again, your roof does not have to have been rained upon in order for it to acquire the properties of being effective against rain.

Nonspecific immunity is also called *innate immunity* because all "normal" individuals are born with these nonspecific defenses. These defenses consist of mechanical and chemical barriers, and cells that attack invaders in general. They will be discussed shortly in more detail (see Section 11.3).

The other set of defenses against disease in the human body is **specific immunity**. This specific immune response works against *each particular type* of microbe that may invade your body. A specific defense in your home, for example, may be a certain chemical you use to rid your kitchen of ants in the summer.

Specific immunity is also called *adaptive immunity* because it does not become active against a particular invader until the individual is first exposed to it. In other words, specific immunity is activated during a first infection, and adapts to protecting you the next time you contact that same pathogen (disease-causing microbe). Put simply, the specific immune response has a memory. Although it works to protect you on first contact, the response is delayed; upon a subsequent contact with that same pathogen, specific immunity acts more quickly and is therefore more effective in thwarting disease. Specific immunity consists of cellular and molecular responses to particular foreign invaders, which will be discussed shortly (see Sections 11.4–11.6).

> The defenses of the body are nonspecific, acting against any foreign invaders, and specific, acting against particular invaders.

11.2 Patterns of resistance in nonhuman organisms

Certain elements of immunity are found in almost all living things. However, not all living things have an immune system. This complex system of body structures that can recognize specific pathogens is found only in the vertebrates (see Figure 11.5). Although there are differences among vertebrate immune systems, their mechanisms of specific immunity are remarkably similar. This chapter provides a detailed description of the human immune system and uses it as a model to explain how the vertebrate system works.

Invertebrates, on the other hand, as well as all other organisms on Earth, have only nonspecific immunity. They have neither lymphatic systems nor immune systems (see Sections 10.10 and 11.4). Their cells do not cooperate during defensive reactions as do the immune cells of vertebrates. In addition, they cannot recognize foreign molecules with specificity as can vertebrate immune systems. However, their nonspecific defense systems work quite well for them as defenses against infection.

One of the nonspecific defenses found in animals and the protozoa is **phagocytosis** (fag-oh-si-TOE-sis). During this process, a host cell surrounds an invader, engulfing it **(Figure 11.1)**. After being enclosed within the cell, the pathogen is digested. This same process occurs

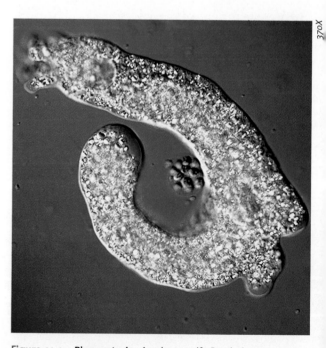

370X

Figure 11.1 **Phagocytosis.** Ameba engulfs *Pandorina morum.*

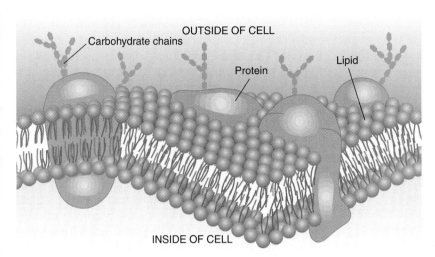

OUTSIDE OF CELL

Carbohydrate chains

Protein

Lipid

INSIDE OF CELL

Figure 11.2   The outer surface of a cell is imbedded with proteins, carbohydrates, and lipids. The glycoproteins (carbohydrate-protein complexes) often serve as highly specific cell surface markers that identify particular cell types and also identify the cell as "self."

in your body when white blood cells attack, engulf, and digest bacteria. This phagocytic process was observed as far back as 1882, when Elie Metchnikoff, the "father" of cellular immunology, observed phagocytes surrounding a rose thorn inserted into the transparent larva of a sea star.

Key to the process of phagocytosis (and to immunity) is the ability of invertebrates (or any organism) to distinguish self from nonself. Humans distinguish self from nonself by means of **antigens**, which are complex molecules (usually proteins or carbohydrates) present on the surface of foreign substances that are different from host molecules (**Figure 11.2**). Other organisms recognize nonself in much the same way. Protozoa, for example, recognize other species of protozoa as different because they have different enzymes. Sponges, as well as other invertebrates and the vertebrates, will reject grafts of nonself tissues.

Certain elements of immunity are found in almost all living things; however, only the vertebrates have specific immunity. Invertebrates, protists, and plants have only nonspecific immunity. Phagocytosis is one important nonspecific defense in invertebrates and protozoans, and is also a mechanism used extensively in vertebrates.

11.3   Nonspecific immunity

The human body has a wide variety of nonspecific defenses, as shown in **Table 11.1**. Together, they are considered the first line of defense against disease.

Mechanical barriers keep foreign substances and disease-causing microbes from entering the body. Unbroken skin is an effective mechanical barrier to any foreign substance. Most microorganisms cannot penetrate the skin of a healthy individual. The mucous membranes of the body, such as those lining your mouth, nose, throat, eyelids, urinary tract, and genital tract, also defend the body against invaders. These membranes are sticky and trap microbes and foreign particles much as flypaper traps flies.

Some of the mucous membranes of the body also have other physical or chemical aids for combating infection.

Focus on Plants

Plants do not have immune systems, but they exhibit immunity against disease. Plants recognize pathogens, or non-self molecules (a key to immunity), and they have a variety of cellular mechanisms that defend against these foreign invaders.

At the site of an infection, plants respond with the death of plant cells in the immediate area, the production of chemicals to kill the invader, the deposition of a substance to wall off the area, and the strengthening of cell walls. (Notice how the tree in the photo "sealed off" its injured tissues.) Interestingly, these responses are much like the inflammatory response in humans. These plant responses take place between minutes and days after infection. Additionally, certain types of pathogens induce defense responses in parts of a host plant far from the site of infection. These defenses take from 1 to 2 weeks to occur, but last for weeks to months. They enhance the resistance of the plant against secondary infections by a broad range of pathogens.

For example, the membranes covering the eyes and eyelids are constantly washed in tears, a fluid that contains a chemical called lysozyme, which is deadly to most bacteria. The ciliated mucous membranes of the upper respiratory system have a comparatively thick coating of mucus that not only traps invaders, but efficiently transports them from these respiratory passageways to the upper part of the throat, where they are swallowed.

Table 11.1

The Nonspecific Defenses of the Human Body

- mechanical barriers: skin and mucous membranes
- lysozyme in tears and saliva
- chemical barriers: acid environment of the stomach, and urinary and genital tracts
- normal microbiota
- inflammatory response
- phagocytic cells
- natural killer cells
- complement system
- interferon

In the digestive system, lysozyme in the saliva and the acid environment of the stomach kill microbes possibly eaten with food or taken in with liquids. The environments of the urinary and genital tracts are somewhat acidic too—an unfavorable situation for many foreign bacteria. (Figure 11.3 illustrates some of these nonspecific defense mechanisms of the body.) Foreign bacteria are also kept from invading the body by bacteria that are normal inhabitants there (called *normal microbiota*). These normal microbiota colonize such areas as the mouth, throat, colon, vagina, and skin. Their presence makes it difficult for other organisms to flourish, since the normal microbiota compete for the substances bacteria need to live.

If the skin is injured, many types of microorganisms will grow and multiply at the injured site. Additionally, pathogens may survive the other mechanical and chemical barriers of the body. An inflammation or infection may result.

An *inflammation* is a nonspecific response of the body to damage by microbes, chemicals, or physical injuries. Characterized by redness, pain, heat, and swelling, an inflammation consists of a series of events that removes the cause of the irritation, repairs the damage that was done, and protects the body from further invasion and infection. During this nonspecific response, blood vessels dilate, or widen, bringing an increased supply of blood to the injured area (Figure 11.4 ❶). This increased

flow brings defensive substances to the site of injury as fluid and phagocytic white blood cells pass out of the capillaries ❷. Lymphatic drainage removes dissolved poisonous substances that may accumulate there ❸. Blood clots wall off the area, preventing the spread of microbes or other injurious substances to other parts of the body. Phagocytes, both neutrophils and monocytes, migrate to the area and ingest microbes and other foreign substances ❹. Nutrients stored in the body are also released to the area to support these defensive cells.

Specialized white blood cells called *natural killer cells (NK cells)* are also part of the body's repertoire of nonspecific defenses. NK cells react to cell surface changes that occur on cancer cells or virally infected cells, and attack them, breaking them apart. Rather than ingesting infected or diseased cells as macrophages do, NK cells secrete proteins that create holes in the membranes of the cells they attack. In addition, NK cells secrete toxins that poison these cells. The cells under attack break apart and die.

The **complement** system is a nonspecific defense that acts both alone and in concert with the specific immune response. In fact, its name comes from the fact that

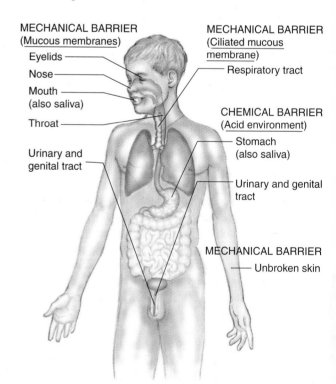

MECHANICAL BARRIER
(Mucous membranes)
- Eyelids
- Nose
- Mouth (also saliva)
- Throat

Urinary and genital tract

MECHANICAL BARRIER
(Ciliated mucous membrane)
- Respiratory tract

CHEMICAL BARRIER
(Acid environment)
- Stomach (also saliva)
- Urinary and genital tract

MECHANICAL BARRIER
- Unbroken skin

Figure 11.3 **Some nonspecific defenses of the human body.** These nonspecific defenses are the body's first line of defense against foreign invaders. Also included in the body's arsenal of nonspecific defenses are phagocytes, natural killer cells, the complement system, the inflammatory response, and interferon.

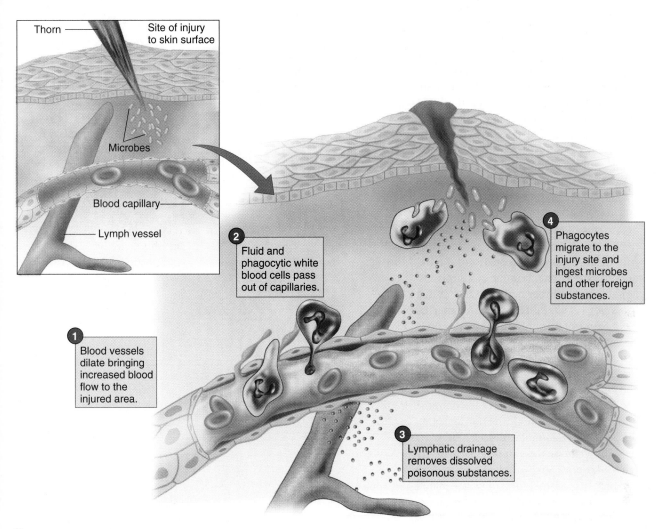

Thorn — Site of injury to skin surface

Microbes

Blood capillary

Lymph vessel

1 Blood vessels dilate bringing increased blood flow to the injured area.

2 Fluid and phagocytic white blood cells pass out of capillaries.

3 Lymphatic drainage removes dissolved poisonous substances.

4 Phagocytes migrate to the injury site and ingest microbes and other foreign substances.

Figure 11.4   **The inflammatory process.**

these proteins help (complement) the specific response. Complement proteins are produced in the liver and circulate in the blood in an inactive form. The presence of foreign antigens activates these proteins. The ultimate result is that the activated proteins form channels, or holes, in the cell membranes of the invading pathogen. The cell contents leak out and the organism dies.

Interferon is another nonspecific defense that helps fight infection by viruses. When a virus infects a body cell, it induces the cell to make interferon, a family of proteins. The cell secretes these proteins, which circulate in the bloodstream and travel to both nearby and distant cells. Interferon binds with receptors on the cell membranes of healthy cells, triggering the cell to produce virus-blocking enzymes.

Preparations of interferon are also used as medications to treat certain types of cancers, such as leukemia (cancer of the blood), Kaposi's sarcoma (a skin cancer prevalent in AIDS patients), and melanoma (a particularly deadly form of skin cancer). The mechanism of action of interferon in combating certain cancers is not entirely understood, but it seems to prevent the growth of tumor cells.

This initial defense by phagocytes, natural killer cells, and interferon (depending on the type of pathogen) peaks within a day or two of the infection. This response "buys time" for the activation of the specific immune response.

The nonspecific defenses of the body include the skin and mucous membranes, chemicals that kill bacteria, the inflammatory process, natural killer cells, the complement system, and interferon.

11.4   Specific immunity

If many pathogens invade the body at one time, or if pathogens are extremely virulent, that is, if they are good at establishing an infection and damaging the body, the nonspecific defenses may not be able to stop an infection. However, while the nonspecific immune response is at work, another line of defense is being summoned: the specific immune response. In a healthy person, this specific defense system staves off most infections and usually overcomes those that take hold.

The body structures that perform specific defense responses constitute the **immune system**. The immune system is different from the other body systems in that it cannot be identified by interconnected organs. In fact, the immune system is not a system of organs, nor does it have a single controlling organ. The immune system is, instead, made up of cells residing in organs and tissues that are scattered throughout the body. However, the cells of the immune system have a common function: reacting to specific foreign molecules.

The cells of the immune system are white blood cells, or leukocytes (see Chapter 10). Two types of white blood cells are involved in the immune response: phagocytes and **lymphocytes** (LIM-foe-sites). Two classes of lymphocytes play roles in specific resistance: *T cells* and *B cells*. These cells arise in the bone marrow, circulate in the blood and lymph, and reside in the lymph nodes, spleen, and thymus **(Figure 11.5)**. Although not bound together, these white blood cells exchange information and act in concert as a functional, integrated system—your "army" of 200 billion defenders—that is called into action when antigens invade your body.

The body structures that perform specific defense responses constitute the immune system. The cells of the immune system are particular white blood cells: phagocytes and lymphocytes. The lymphocytes are either T cells or B cells.

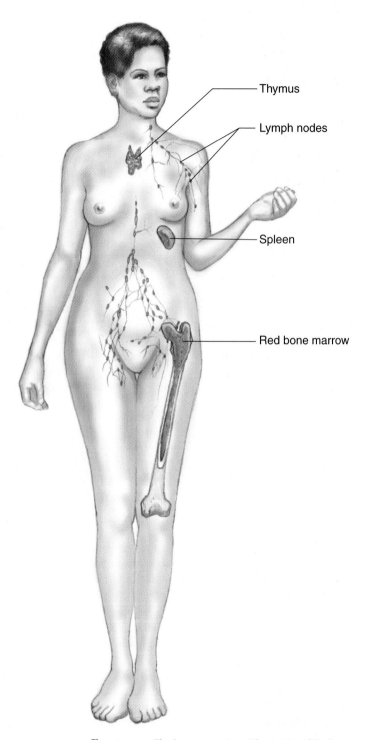

Figure 11.5   **The immune system.** The organs of the immune system are the red bone marrow and the thymus gland (in which disease-fighting cells [lymphocytes] arise and/or develop) and the spleen and lymph nodes (in which mature lymphocytes accumulate and function).

Thymus

Lymph nodes

Spleen

Red bone marrow

11.5　How specific immunity works

The specific immune response works in several ways to defend the body against foreign invaders. Some of the cells of the immune system react immediately to invasion by foreign antigens. Other cells work to protect your body against future attack. The specific immune response is a complex yet coordinated effort, and it involves several types of protection.

Sounding the Alarm

Roving armies of phagocytic white blood cells constantly patrol the body. These white blood cells are the neutrophils and the macrophages. Both play a role in nonspecific immunity. Macrophages also play a role in the specific immune response.

Large, irregularly shaped cells, the macrophages act as the body's scavengers. Macrophages phagocytize anything that is not normal, including cell debris, dust particles in the lungs, and invading microbes. When the body is not under attack, only a small number of macrophages circulate in the body's bloodstream and lymphatic system. In response to infection, precursors of macrophages called *monocytes* develop into mature macrophages in large numbers.

When macrophages encounter foreign microbes in the body, they attack (**Figure 11.6 ①**) and engulf them and then display parts of these microbes on their surfaces ②. Helper T cells, which initiate the specific immune response, recognize the macrophages that have foreign antigens displayed on their surfaces and bind to them ③. This binding stimulates the macrophage to secrete a hormonelike protein called a *monokine* (from the word "monocyte," the precursor of the macrophage) ④. The monokine (interleukin-1) allows communication between cells of the immune system, triggers the maturation of monocytes into macrophages (thereby increasing their numbers), and signals the brain to raise the body temperature, producing a fever. The higher temperature aids the immune response and inhibits the growth of invading microorganisms. The monokine also stimulates the helper T cell to produce other hormonelike proteins called *lymphokines* ⑤, which trigger the specific immune response by stimulating both B cells (see p. 229) and T cells to multiply.

The Two Branches of the Specific Immune Response

T cells, or **T lymphocytes**, develop in the bone marrow, but then migrate

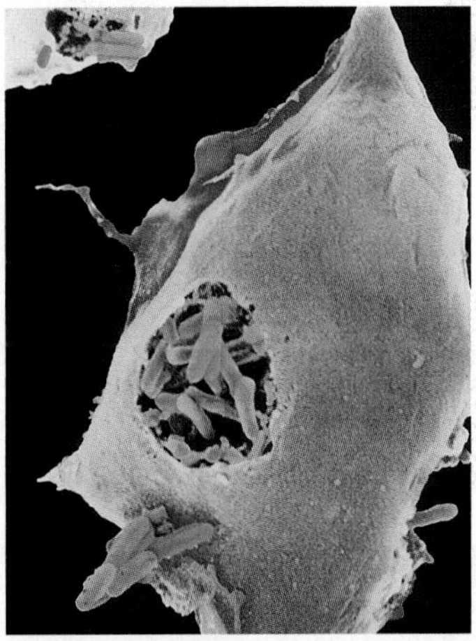

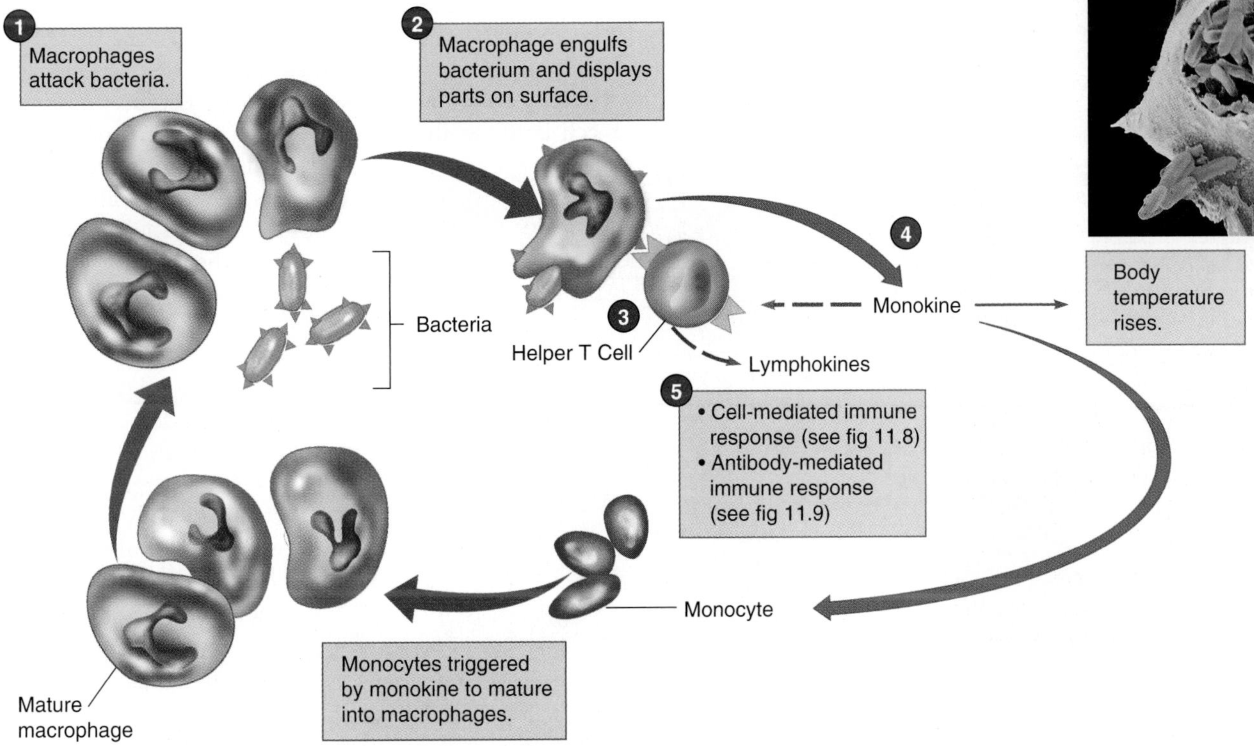

① Macrophages attack bacteria.

② Macrophage engulfs bacterium and displays parts on surface.

Bacteria

Helper T Cell

③

④ Monokine

Body temperature rises.

Lymphokines

⑤ • Cell-mediated immune response (see fig 11.8)
• Antibody-mediated immune response (see fig 11.9)

Monocyte

Monocytes triggered by monokine to mature into macrophages.

Mature macrophage

Figure 11.6　**Activation of the specific immune response.** The colorized micrograph shows a macrophage (yellow) phagocytizing tuberculosis-causing bacteria, *Mycobacterium tuberculosis* (green).

through the circulatory system to the thymus. The thymus, a small gland located in the upper chest and extending upward toward the neck (see Figure 11.5), is the place where T (thymus) cells mature. This organ is large and active in children; it dwindles in size throughout childhood until, by puberty, it is quite small. In the thymus, T cells develop the ability to identify invading bacteria and viruses by the foreign molecules (antigens) on their surfaces. Tens of millions of different T cells are produced, each specializing in recognizing one particular foreign antigen. No invader can escape being recognized by at least a few T cells. There are five principal kinds of T cells:

1. *Helper T cells*, which initiate the specific immune response
2. *Cytotoxic T cells*, which break apart cells that have been infected by viruses and break apart foreign cells such as incompatible organ transplants
3. *Inducer T cells*, which oversee the development of T cells in the thymus
4. *Suppressor T cells*, which limit the specific immune response
5. *Memory T cells*, which respond quickly and vigorously if the same foreign antigen reappears

When they are triggered by the monokine interleukin-1 and the presentation of antigens, the helper T cells stimulate both branches of the specific immune response: the *cell-mediated immune response* and the *antibody-mediated immune response*.

The Cell-Mediated Immune Response

The activation of helper T cells by monokine and the binding of antigen to these activated helper T cells unleash a chain of events. One of these events is the **cell-mediated immune response**. The primary response of this arm of specific immunity is that cytotoxic T cells ("cell-poisoning" cells) recognize and destroy infected body cells (Figure 11.7).

When a helper T cell has been activated, it produces chemical substances collectively called *lymphokines* (Figure 11.8 ❶), which help the specific immune response. One type of lymphokine attracts macrophages to the site of infection ❷, and another inhibits their migration away from it. Another of the lymphokines stimulates T cells that are bound to foreign antigens to undergo cell division many times. This cell division produces enormous quantities of T cells capable of recognizing the antigens specific to the invader ❸. Each type of activated T cell does a specific job.

The activated inducer T cells trigger the maturation of immature lymphocytes in the thymus into mature T cells. Helper T cells ❹ continue to stimulate the specific immune response. Activated cytotoxic T cells ❺ kill the body's own cells that have been infected with fungi, some viruses, and bacteria that produce slowly developing diseases such as tuberculosis. In addition, they act to kill cells that have become cancerous. Cytotoxic T cells bind to these infected or abnormal cells by means of molecules on their surfaces that specifically fit antigens, much as a key fits a lock. Because the entire cell binds to the abnormal cells (by means of specific cell-surface proteins), this response is called *cell-mediated*. The cytotoxic T cells then secrete a chemical that breaks apart the foreign cell (see Figure 11.7).

Unfortunately for patients undergoing organ transplantation, cytotoxic T cells also recognize foreign body cells in a similar way. Because of this, cytotoxic T cells attack transplanted tissue, leading to the rejection of transplanted organs. Although tissue donors are matched closely with recipients to minimize rejection, trans-

Figure 11.7 Cytotoxic T cell: A killer cell. The cell-mediated response involves cytotoxic T cells binding to antigens on the surfaces of infected cells and secreting a chemical that breaks the infected cells apart. The photo on the left shows a cytotoxic T cell (the smaller cell) binding with a tumor cell (the larger cell); at right, the cytoplasm of the tumor cell is leaking from holes in its membrane, which were created by the T cell.

Cytotoxic T cell

Cytotoxic T cell

plant patients must take immunosuppressive drugs that selectively subdue cell-mediated immunity while leaving antibody-mediated immunity active. At this time, scientists are researching the use of custom-built antibodies that would block specific facets of the rejection process so that donors and recipients would not have to be close matches.

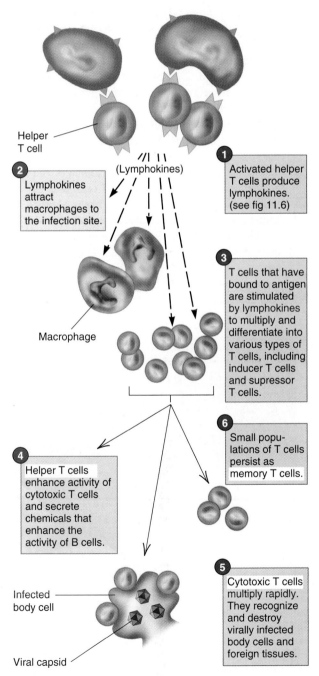

Helper T cell

2 Lymphokines attract macrophages to the infection site.

(Lymphokines)

1 Activated helper T cells produce lymphokines. (see fig 11.6)

Macrophage

3 T cells that have bound to antigen are stimulated by lymphokines to multiply and differentiate into various types of T cells, including inducer T cells and supressor T cells.

6 Small populations of T cells persist as memory T cells.

4 Helper T cells enhance activity of cytotoxic T cells and secrete chemicals that enhance the activity of B cells.

Infected body cell

Viral capsid

5 Cytotoxic T cells multiply rapidly. They recognize and destroy virally infected body cells and foreign tissues.

Figure 11.8   The cell-mediated immune response.

After the cytotoxic cells and macrophages do their jobs, the cell-mediated immune response begins to shut down. The cells in charge of shutdown are the suppressor T cells. The number of suppressor T cells slowly begins to rise after activation by the helper T cells. However, it takes about 1 to 2 weeks for their numbers to increase to a point where they are able to suppress the cytotoxic T cell response.

After suppression, or shutdown, a population of T cells persists, probably for the life of the individual. Referred to as *memory T cells* **6**, these cells provide an accelerated and larger response to any later encounter with the same antigens.

The Antibody-Mediated Immune Response

When helper T cells are stimulated to respond to foreign antigens, they activate the cell-mediated immune response as described and activate a second, more long-range defense called the **antibody-mediated immune response**. Depending on the types of antigens present, the helper T cells may stimulate either or both of these branches of the specific immune response.

The key players in antibody-mediated immunity are lymphocytes called **B cells**, or **B lymphocytes**. The B cells are named after a digestive organ in birds called the *bursa of Fabricius* in which these lymphocytes were first discovered. However, B cells mature in the bone marrow of humans, which may be a convenient way for you to remember these cells. The antibody response is sometimes called the *humoral response*, which refers to the fact that B cells secrete antigen-specific chemicals into the bloodstream—one of the body fluids called "humours" long ago.

On their surfaces, B cells each have about 100,000 copies of a protein receptor that binds to antigens. Because different B cells bear different protein receptors, each recognizes a different, specific antigen. At the onset of a bacterial infection, for example, the receptors of one or more B cells bind to bacterial antigens (**Figure 11.9 1**). The B cell processes and displays bacterial antigens on its surface **2**. These antigen-bound B cells are detected by helper T cells, which have also bound to macrophages displaying bacterial antigens on their surfaces **3**. The helper T cells then bind to the antigen-B cell complex **4**. After this "double" binding, the helper T cells release lymphokines that trigger cell division in the B cell **5**.

After about 5 days and numerous cell divisions, a large clone of cells is produced from each B cell that was stimulated to divide. A clone is a group of identical cells that arise by repeated mitotic divisions from one original

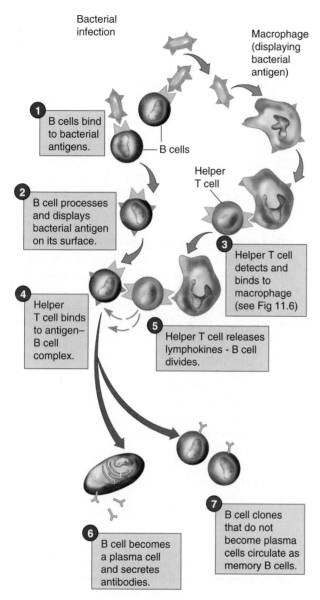

Figure 11.9   The antibody-mediated immune response.

1 B cells bind to bacterial antigens.

B cells

Macrophage (displaying bacterial antigen)

Helper T cell

2 B cell processes and displays bacterial antigen on its surface.

3 Helper T cell detects and binds to macrophage (see Fig 11.6)

4 Helper T cell binds to antigen–B cell complex.

5 Helper T cell releases lymphokines - B cell divides.

6 B cell becomes a plasma cell and secretes antibodies.

7 B cell clones that do not become plasma cells circulate as memory B cells.

activity of macrophages is enhanced by the binding of both antibodies and complement to the antigen.

The B cell clones that did not become plasma cells live on as circulating lymphocytes called *memory B cells* (Figure 11.9 **7**). These cells provide an accelerated response to any later encounter with the stimulating antigen. This is why immune individuals are able to mount a prompt defense against infection. As in the case of the cell-mediated immune response, the antibody response is shut down after several weeks by suppressor T cells.

Figure 11.11 shows an overview of both nonspecific and specific immunity.

T cells carry out the cell-mediated immune response, during which cytotoxic T cells recognize and destroy body cells infected with certain bacteria, viruses, and fungi. In addition, they destroy transplanted cells and cancer cells. Helper T cells initiate the response, attracting macrophages and activating cytotoxic T cells, inducer T cells, and, as the infection subsides, suppressor T cells. B cells carry out the antibody-mediated immune response in which B cells recognize foreign antigens and, if activated by helper T cells, produce large quantities of antibody molecules directed against the antigen. The antibodies bind to the antigens they encounter and mark them for destruction.

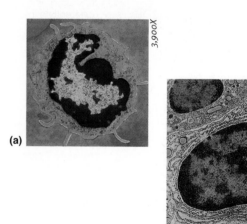

3,900X

6,300X

(a)

(b)

Figure 11.10   Comparison of a B cell and a plasma cell.
(a) Colorized micrograph of a B cell. These cells are key players in antibody-mediated immunity. (b) Colorized micrograph of a plasma cell. Plasma cells are formed from B cells and secrete antibodies. The large amount of endoplasmic reticula (ER) on which antibodies are made can be seen in the cytoplasm as green lines. The B cell shown in (a) does not have large quantities of ER.

cell. (See Chapter 18 for a discussion of mitosis.) After these clones are formed, most of the B cells stop dividing, but begin producing and secreting copies of the receptor protein that responded to the antigen **6**. These receptor proteins are called **antibodies**, or *immunoglobulins*. The secreting B cells are called *plasma cells*. After B cells become plasma cells, they live for only a few days, but secrete a great deal of antibody during that time (**Figure 11.10**). In fact, one plasma cell will typically secrete more than 2000 antibodies per second!

Antibodies do not destroy a virus or bacterium directly, but rather mark it for destruction by either the complement system or macrophages. The phagocytic

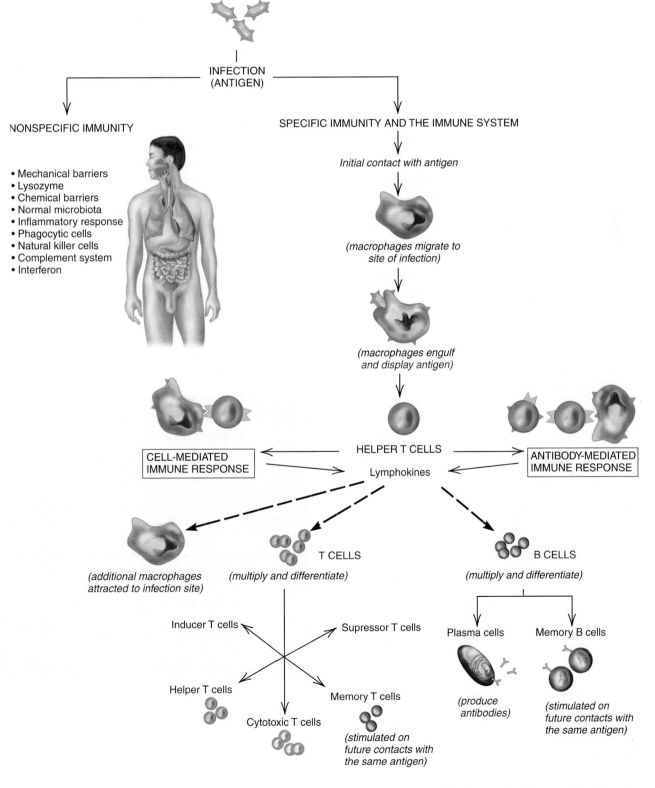

Visual Summary Figure 11.11 An overview of nonspecific immunity and specific immunity of the human body. Notice the central role of the helper T cells in both the antibody-mediated and cell-mediated immune responses.

INFECTION
(ANTIGEN)

NONSPECIFIC IMMUNITY

- Mechanical barriers
- Lysozyme
- Chemical barriers
- Normal microbiota
- Inflammatory response
- Phagocytic cells
- Natural killer cells
- Complement system
- Interferon

SPECIFIC IMMUNITY AND THE IMMUNE SYSTEM

Initial contact with antigen

(macrophages migrate to site of infection)

(macrophages engulf and display antigen)

HELPER T CELLS

CELL-MEDIATED IMMUNE RESPONSE

Lymphokines

ANTIBODY-MEDIATED IMMUNE RESPONSE

(additional macrophages attracted to infection site)

T CELLS
(multiply and differentiate)

B CELLS
(multiply and differentiate)

Inducer T cells

Supressor T cells

Plasma cells

Memory B cells

Helper T cells

Cytotoxic T cells

Memory T cells
(stimulated on future contacts with the same antigen)

(produce antibodies)

(stimulated on future contacts with the same antigen)

11.6 How immune receptors recognize antigens

The cell surface receptors of lymphocytes (T cells and B cells) can recognize specific antigens with great precision. Even single amino acid differences between proteins can often be discriminated, with a receptor recognizing one form and not the other. This high degree of precision is a necessary property of specific immunity because without it, the identification of foreign antigens would not be possible in many cases. The differences between self and foreign (nonself) molecules can be very subtle.

A typical antibody, or immunoglobulin molecule, consists of four polypeptide chains. A polypeptide is a chain of amino acids linked end-to-end by peptide bonds (see Chapter 3). There are two identical short strands, called *light chains,* and two identical long strands, called *heavy chains.* The four chains are held together by disulfide (-S-S-) bonds, forming a Y-shaped molecule (**Figures 11.12 and 11.13**). The two "arms" of the Y determine which antigen will bind to the antibody. Antibodies recognize, or lock onto, antigens by means of special binding sites in their arms. These binding sites are made up of specific sequences of amino acids that determine the shape of the site. The specificity of the antibody molecule for an antigen depends on this shape. An antigen fits into the binding site on the antibody like a hand into a glove. Changes in the amino acid sequence of an antibody can alter the shape of this region and, by doing so, change the antigen that can bind to that antibody, just as changing the size of a glove will alter which hand can fit into it.

The stem of the Y determines what role the antibody plays in the specific immune response. An antibody can have one of five different stems and therefore be a member of one of the five classes of antibodies. For example, the antibodies produced in the first week of an infection are primarily class M antibodies and are called IgM (immunoglobulin M). After the first week, IgG antibodies are primarily produced.

An antibody molecule recognizes a specific antigen because it possesses binding sites into which an antigen can fit, much as a key fits into a lock. Changes in the amino acid sequence at the binding sites alter the antibody's shape and thus change the identity of the antigen that is able to fit into them.

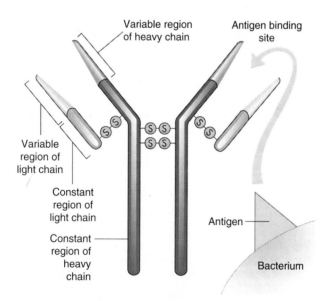

Figure 11.12   **The structure of an antibody molecule.** Each antibody molecule is composed of two identical light chains and two identical heavy chains. Disulfide bonds hold the chains together. The antibody binds the antigen to which it is specific at the antigen binding site (the variable region of the molecule).

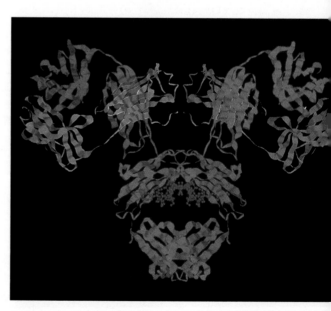

Figure 11.13   **A computer-generated image of an antibody molecule.**

11.7 Immunization: Protection against infection

A vaccine is made up of disease-causing microbes or toxins (poisons) that have been killed or changed in some way so as not to produce the disease. Injection with these antigens causes an antibody (B cell) response, with the production of memory cells. A booster shot induces these memory cells to differentiate into antibody-producing cells and more memory cells. So, **vaccination** causes your body to build up antibodies against a particular disease without getting the disease. This type of specific immunity is called **active immunity**.

Scientists can now use the techniques of genetic engineering to produce vaccines. Instead of growing cultures of an agent of infection such as a bacterium or a virus, scientists can, in the laboratory, produce large quantities of antigenic proteins from disease-causing microbes. These antigens can then be used as vaccines.

Vaccination to produce active immunity works well for many diseases, but it is a somewhat slow process. After the injection of the antigen, it takes weeks for the body to develop sufficient antibodies and memory cells to combat the disease. However, another type of specific immunity, **passive immunity**, can be used when protection is needed quickly. One example is the use of *antitoxin* when a person has been bitten by a snake.

When a poisonous snake bites a person or other animal, it injects a toxin (poison) into the body. This toxin circulates in the bloodstream. Certain toxins damage the nerve cells they reach and may result in death. An antitoxin is a preparation of antibodies specific for a toxin. These antibodies are injected into the victim and they bind with the toxin, preventing the toxin from causing damage. Although effective, this type of "borrowed" specific immunity lasts for only a short time. The body soon uses or eliminates these antibodies, so they will not be available if another snakebite occurs.

Newborns have a type of passive immunity borrowed from their mothers. Antibodies that pass from mother to fetus through the placenta provide babies with a short-term specific immunity that subsides by the time the baby has produced its own system of antibodies. In addition, if a mother nurses her baby, antibodies in the mother's milk provide the baby with some protection.

> A vaccine is made up of killed or altered disease-causing microbes or toxins, or laboratory-made antigenic protein. Vaccination confers active specific immunity because your body builds up antibodies. Passive specific immunity is conferred by receiving antitoxins or antibodies.

11.8 Defeat of the immune system: AIDS

For the latest health information, please visit www.jbpub.com/biology.

Helper T cells are the key to the entire specific immune response because they initiate the proliferation of both T cells and B cells. Without helper T cells, the immune system is unable to mount a response to foreign antigens. The **human immunodeficiency virus (HIV)** was discovered to be a cause of human disease in the early 1980s and is deadly because it mounts a direct attack on T lymphocytes, mainly targeting the helper T cells (also called CD4 cells). The HIV particles kill helper T cells by entering them, using the cells' own machinery to produce more virus particles, and then bursting open the cells to release the new viruses (see Chapter 26). These newly formed viruses then infect other helper T cells.

Infection with HIV does not mean that a person has **acquired immunodeficiency syndrome (AIDS)**. During the first years of the disease process, the HIV-infected individual is usually asymptomatic, which means that no symptoms are apparent. This phase of HIV disease usually lasts for about 7 to 10 years. Recently, many HIV-infected individuals have been extending the asymptomatic phase of their disease with drug "cocktails," most notably with drugs called protease inhibitors. These drug combinations appear to work well in holding HIV to low levels in the bloodstream and retarding its debilitating effects on the immune system. However, there is no cure for HIV infection at this time and it usually progresses to AIDS.

During the symptomatic phase of HIV infection (not yet AIDS), the helper T cell count has usually declined from the normal level of 800 to 1200 helper T cells/mm³ to about 500 helper T cells/mm³. Signs and symptoms of the symptomatic phase of HIV infection include chronic diarrhea, minor infections of the mouth, fever, night sweats, headache, and fatigue. (Signs are observable, measurable features of an illness such as a fever. Symptoms are subjective complaints of an ill person, such as reporting a headache.)

A person with HIV infection is usually diagnosed as having AIDS when the helper T cell count falls below 200 cells/mm³ and certain diseases typical of this stage are present. (AIDS is a syndrome, a set of signs and symptoms occurring together.) Such diseases include *Pneumocystis carinii* pneumonia (a fungal infection of the lungs) and Kaposi's sarcoma (a type of cancer). In a normal individual, helper T cells make up 60% to 80% of circulating T cells; in patients with AIDS, helper T cells often become too rare to detect. Because the specific immune response cannot be initiated, any one of a variety of infections proves fatal. This is the primary reason that AIDS is a particularly devastating disease.

HIV is *not* highly infectious; that is, it is not transmitted from person to person by casual contact. It is transmitted only by the direct transfer of body fluids, typically in semen or vaginal fluid during sexual activity, in blood during transfusions, or by contaminated hypodermic needles. It also crosses the placenta to infect babies within the womb of an infected mother.

To *eliminate* your risk of becoming infected with HIV, do not have sex with HIV-infected individuals and abstain from using drugs. Not only do contaminated needles transmit HIV from an infected person to another, persons using illegal drugs often engage in risky sexual practices when under the influence of these substances. These precautions will not eliminate the risk of infection for persons in certain professions, such as health care workers and police officers, who may become infected by HIV-contaminated blood during the course of their work.

To help *reduce* your risk of becoming infected with HIV, remember these key points. Use a new latex condom during each act of sexual intercourse, never share needles and syringes, avoid having sex while under the influence of drugs (including alcohol), reduce your number of sexual partners, avoid having sex with persons who practice risky behaviors with regard to HIV infection, and avoid having sex with persons you do not know well. Asymptomatic individuals may not know they are infected, so being sure that a partner has not engaged in risky behavior while you have known them does not guarantee that they are not infected.

In June of 1998, the U.S. Food and Drug Administration (FDA) approved the start of a 4-year HIV vaccine trial, which includes 7500 volunteers from across the United States. The volunteers were injected with a vaccine called AIDSVAX, which consists of a viral protein called *gp120*. Although the vaccine induces antibody formation in 100% of the people that are inoculated, it may not protect against contracting the AIDS virus because the virus mutates (changes) so readily. The volunteers are persons at high risk of becoming infected with HIV, such as injection drug users and those whose sexual partners are HIV positive.

Figure **11.14** shows the decline of AIDS-related deaths in recent years. In 1997, HIV infection fell from the eighth to the fourteenth leading cause of death in the United States. This decline is thought to be the result of increasingly effective drug therapies. Although the incidence of AIDS has dropped as well, the decline in new cases is not as dramatic as the decline in deaths.

The AIDS virus weakens the ability of the immune system to mount a defense against infection because it attacks and destroys helper T cells.

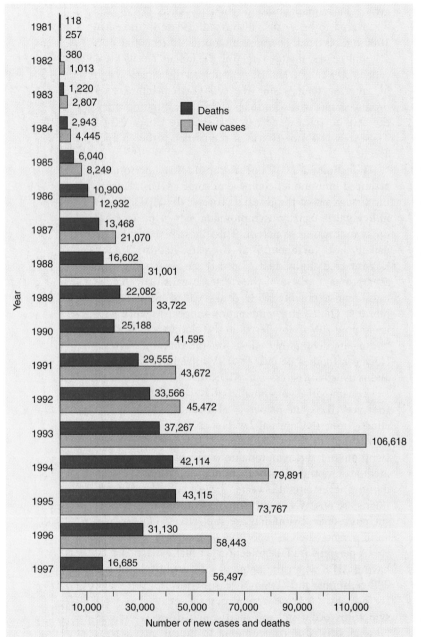

Figure **11.14** **Reported cases and known deaths from AIDS, United States, 1981–1997.** The extreme rise in reported cases in 1993 is the result of a change in the definition of AIDS, which broadened the list of conditions reportable as AIDS and included measures of immune system function.
Source: National Center for Health Statistics, National Vital Statistics System.

Just Wondering . . . *Real Questions Students Ask*

Can I get AIDS from an insect bite?

No. Mosquitos and other insects do not transmit HIV infection. The primary way in which scientists know that this does not happen is by looking at the statistics for HIV. If mosquitos spread HIV, they would expect to see more HIV infection in areas where mosquitos live. Although HIV does occur in areas such as Florida and Texas where mosquitos live, researchers tend to see higher concentrations of HIV in places such as San Francisco and New York, which do not have a mosquito problem.

Another way in which scientists know that HIV infection is not transmitted by mosquitoes is that they would expect to see HIV more equally distributed among men, women, and children if this route of transmission existed. Insects do not discriminate according to a person's gender or age; they just want blood. In fact, if mosquitos transmitted HIV, it would be likely that more children and women would be infected than men because mosquitos tend to attack women and children more often. In the United States, there are still many more men infected than women or children.

It is important to remember that mosquitos would have to inject blood into people to create a risk of HIV infection. Mosquitos don't inject blood when they bite, they inject saliva. Since HIV does not get into the saliva of mosquitos, they do not inject HIV.

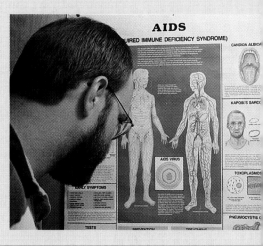

11.9 Allergy

Although the human immune system provides effective protection against viruses, bacteria, parasites, and other microorganisms in healthy people, sometimes it does its job too well, mounting a major defense against a harmless antigen. Such specific immune responses are called **allergic reactions**. Hay fever, the sensitivity that many people exhibit to proteins released from plant pollen, is a familiar example of an allergy. **Figure 11.15** shows the ragweed plant and its pollen, a common cause of allergic reaction. Many people are also sensitive to proteins released from the feces of a minute house-dust mite. This microscopic arthropod lives in the house dust present on mattresses and pillows and eats the dead epithelial tissue that everyone sheds from their skin in large quantities daily. Many people sensitive to feather pillows are actually allergic to the feces of mites that are residents of the feathers.

What makes an allergic reaction uncomfortable and sometimes dangerous is involvement of class E (IgE) antibodies. The binding of antigens to these antibodies initiates an inflammatory response; powerful chemicals (such as histamines) cause the dilation of blood vessels and a host of other physiological changes. Sneezing, runny nose, and fever often result. In some instances, allergic reactions can be far more dangerous, resulting in anaphylactic shock, a severe and life-threatening response of the immune system. The chemicals that dilate blood vessels are released suddenly into the bloodstream, causing many vessels to widen abruptly. Their widening causes the blood pressure to fall. In addition, the muscles of the trachea, or windpipe, may contract, making breathing difficult.

Not all antigens are *allergens*, initiators of strong immune responses. Nettle pollen, for example, is as abundant in the air as ragweed pollen, but few people are allergic to it. Nor do all people develop allergies; the sensitivity seems to run in families. Allergies require both a particular kind of antigen and a high level of class E antibodies. The combination of an appropriate antigen on the one hand and inappropriately high levels of IgE antibodies on the other hand produces the allergic response.

> An allergic reaction is a specific immune response to a harmless antigen. During an allergic reaction, the binding of antigens to IgE antibodies initiates a response in which chemicals (such as histamines) cause the dilation of blood vessels and a host of other physiological changes.

650X

Figure 11.15 **Ragweed plant with an inset of a scanning electron micrograph of ragweed pollen.**

Key Concepts

Almost all living things have some elements of immunity.

11.1 The defenses of the body are nonspecific, acting against any foreign invaders, and specific, acting against particular invaders.

11.2 Only the vertebrates have specific (acquired) immunity.

11.2 Invertebrates, protists, and plants have only nonspecific immunity.

11.2 Phagocytosis is one important nonspecific defense in invertebrates and protozoans, although this mechanism is used extensively in vertebrates.

Human immunity is both nonspecific and specific.

11.3 The nonspecific defenses, those acting against any foreign invader, include the skin and mucous membranes, acidic secretions of the body, chemicals and cells that kill bacteria, and the inflammatory process.

11.4 The body structures that perform specific defense responses constitute the immune system.

11.5 Two classes of lymphocytes play roles in specific immunity: T cells and B cells, which are capable of recognizing foreign substances by means of receptor proteins.

11.5 When stimulated by macrophages and antigens, helper T cells can activate the two branches of the specific immune response: the cell-mediated immune response (T cell response) and the antibody-mediated immune response (B cell response).

11.5 In the cell-mediated immune response, cytotoxic T cells attack infected body cells.

11.5 In the antibody-mediated immune response, B cells are converted to plasma cells that secrete proteins called antibodies.

11.5 Antibodies specifically bind circulating antigen and mark cells or viruses bearing antigens for destruction.

11.6 The high level of specificity of an antibody for a particular antigen is caused by the three-dimensional shape of the ends of the arms, or antigen-binding sites, of the molecule.

 Health *Connections*

11.7 Vaccination, which is an injection with disease-causing microbes or toxins that have been killed or altered in some way so as to be harmless, causes the body to build up antibodies against a particular disease.

11.7 Scientists can now use laboratory-made antigenic proteins as vaccines.

11.8 The AIDS virus is deadly because it primarily attacks helper T cells, the key to the entire specific immune response.

11.8 Because helper T cells are no longer present to initiate the specific immune response, any one of a variety of otherwise commonplace infections proves fatal.

11.9 An allergic reaction is a specific immune response against a harmless antigen.

11.9 Allergic reactions are produced by the combination of a particular kind of antigen and a high level of IgE antibody.

Key Terms

active immunity *223*
acquired immunodeficiency syndrome (AIDS) *233*
allergic reactions *235*
antibodies *230*
antibody-mediated immune response *229*
antigens (ANT-ih-jens) *223*

B cells or B lymphocytes (LIM-foe-sites) *229*
cell-mediated immune response *228*
complement *224*
human immunodeficiency virus, or HIV (IM-yoo-no-de-FISH-un-see VYE-rus) *233*
immune system *226*

immunity (ih-MYOON-ih-tee) *222*
lymphocytes (LIM-foe-sites) *226*
nonspecific immunity *222*
passive immunity *233*
phagocytosis (fag-oh-sigh-TOE-sis) *222*
specific immunity *222*
T cells or T lymphocytes *227*
vaccination (VAK-suh-NAY-shun) *233*

KNOWLEDGE AND COMPREHENSION QUESTIONS

1. Name your body's nonspecific defenses, and summarize how each protects you against infection or injury. *(Comprehension)*

2. How is your immune response involved if you get a particle of foreign matter, such as dust, in the mucous membranes of your eyelid? *(Comprehension)*

3. Earlier chapters described how a cell membrane may have proteins or carbohydrates on its surface to act as "cell markers." How do you think these would be interpreted if this cell were placed in the tissues of another organism? *(Comprehension)*

4. What are the five principal types of T cells? Summarize the function of each. *(Comprehension)*

5. Summarize the events of the cell-mediated immune response. What cells are involved? *(Comprehension)*

6. Summarize the events of the antibody-mediated immune response. What cells are involved? *(Comprehension)*

7. How is an antibody molecule able to discriminate between millions of cells in a tissue and bind to only one specific antigen? *(Comprehension)*

8. A nurse vaccinates a child against the tetanus bacterium. Later, the child receives a booster shot. Describe how these two vaccinations affect the child's specific immune response during subsequent exposures to tetanus. *(Comprehension)*

9. Which virus is responsible for AIDS? Explain why this virus is so deadly. *(Comprehension)*

10. Many people—including you, perhaps—are allergic to something. Explain what an allergic reaction is and why it occurs. What causes its uncomfortable symptoms? *(Comprehension)*

KEY
Knowledge: Recalling information.
Comprehension: Showing understanding of recalled information.

CRITICAL THINKING REVIEW

1. Why do you think it is important that there are phagocytes constantly circulating in the bloodstream and in body tissues? *(Application)*

2. You had measles as a child. Now your younger brother has measles, but you don't catch it from him. Explain why *(Application)*

3. Pediatricians often encourage mothers to breastfeed infants because maternal antibodies circulate in milk and provide the baby with some protection against infection. What sort of immune response is involved here? Why is this immunity only temporary? *(Application)*

4. Recent studies have shown that exhaustive exercise (such as jogging 60 miles per week) suppresses the immune system, while moderate exercise (such as jogging or walking 20 miles per week) stimulates the immune system.

Evaluate your weekly level of activity and determine whether you are regularly suppressing your immune function or enhancing it. Taking your level of activity into account, what recommendations would you make for yourself to maintain the highest level of resistance to disease? *(Evaluation)*

5. If you have allergies, you've probably taken over-the-counter medications containing antihistimines. Using your understanding of allergic reactions, what do you think is the physiological role of antihistamines? Why do they make you feel better? *(Application)*

KEY
Application: Using information in a new situation.
Evaluation: Making informed decisions.

1. The nonspecific defenses of the body are often called the body's "first line of defense." Explain why this is so.
2. Helper T cells are central to immune system function. Using this illustration to guide your thinking, explain why these cells, more than any other, are critical to the body's specific resistance to infection. Relate the importance of helper T cells to the progression of disease in an HIV-infected individual.
3. Using this illustration to guide your thinking, explain how the immune reaction is triggered.
4. List the jobs of each of the defense systems illustrated here: (a) nonspecific resistance, (b) the cell-mediated immune response, and (c) the antibody-mediated immune response.

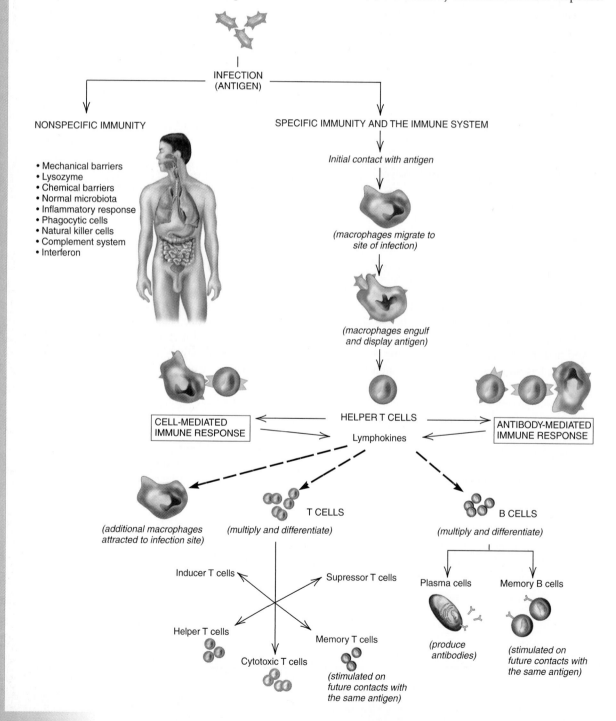

INFECTION
(ANTIGEN)

NONSPECIFIC IMMUNITY

- Mechanical barriers
- Lysozyme
- Chemical barriers
- Normal microbiota
- Inflammatory response
- Phagocytic cells
- Natural killer cells
- Complement system
- Interferon

SPECIFIC IMMUNITY AND THE IMMUNE SYSTEM

Initial contact with antigen

(macrophages migrate to site of infection)

(macrophages engulf and display antigen)

CELL-MEDIATED IMMUNE RESPONSE

HELPER T CELLS

Lymphokines

ANTIBODY-MEDIATED IMMUNE RESPONSE

(additional macrophages attracted to infection site)

T CELLS
(multiply and differentiate)

B CELLS
(multiply and differentiate)

Inducer T cells

Supressor T cells

Plasma cells

Memory B cells

Helper T cells

Memory T cells

Cytotoxic T cells

(produce antibodies)

(stimulated on future contacts with the same antigen)

(stimulated on future contacts with the same antigen)

e learning

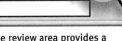

Location: http://www.jbpub.com/biology

e-Learning is an on-line student review area located at this book's web site www.jbpub.com/biology. The review area provides a variety of activities designed to help you study for your class and build your learning portfolio.

Review Questions The review questions test your knowledge of the important concepts and applications in each chapter. The review provides feedback for each correct or incorrect answer. This is an excellent test preparation tool.

Figure-Labeling Exercises Sharpen your visual thinking by matching terms and labels for important illustrations.

Flash Cards Studying biology requires learning new terms. Virtual flash cards help you master the new vocabulary for each chapter.

Just Wondering As you read and study from this text, you may find that you have unanswered questions. Through this site you can ask the author, Sandy Alters, your "just wondering" questions.

Do you prefer the speed and reliability of a CD-ROM? All of the features contained on the eLearning portion of the web site are also available on the Student CD-ROM.

Excretion

The story of excretion is, to a large extent, a story about water and salt. And that story explains why the sea turtle below appears to be crying, and why the butterflies are visiting the freshwater turtle at right.

In animal bodies, water is contained within compartments—any space bounded by a membrane. Cells themselves are compartments, as are structures such as the interiors of the blood vessels and the heart. Water moves among the compartments within an organism by osmosis, moving from areas of high concentration of water to areas of lower concentration of water. The concentration of salts—compounds that form ions in water—affects the concentration of water in the various body compartments and therefore water's movement. To regulate the water in the various body compartments, animals must therefore regulate the amount of salt as well—a balancing act of sorts.

Animals have evolved various mechanisms that deal with their water and salt problem. The production of salt water "tears" is only one mechanism. Because the sea turtle lives in a salty environment, water tends to leave its less salty body. To compensate, the turtle drinks a lot of seawater, but gets rid of sea salt via excretory organs located in the corners of its eyes. So, shedding salty tears is one way the sea turtle accomplishes its balancing act and maintains a steady internal environment.

While excretion ends the salt/water balance story for one animal, it may begin the story for another kind of animal. Butterflies primarily feed on the nectar of flowers, a rich energy source packed with sugars. Still, like other animals, butterflies need salts and minerals, too. They sip the moisture from the eyes and nose of the side-necked turtle at right. These secretions are not as salty as the tears of the sea turtle, but saltier than the water in their Amazon River habitat.

CHAPTER 12

Organization

Excretion removes metabolic wastes and balances water and salt.

Human excretion relies primarily on the lungs and the kidneys.

Health
Connections

Excretion removes metabolic wastes and balances water and salt.

12.1 Functions of excretion

Although the story of **excretion** is, to a large extent, a story about water and salt balance, excretion also regulates the concentrations of other substances in the body and removes metabolic wastes (end products). Table 12.1 lists the functions of excretion.

Table 12.1

Functions of Excretion

- Maintaining proper concentrations of dissolved substances throughout an organism's cells and fluids
- Maintaining a proper concentration of water throughout an organism's cells and fluids
- Removing metabolic wastes (such as carbon dioxide) from an organism
- Removing foreign substances (such as drugs) and/or their breakdown products from an organism

Do not confuse metabolic wastes removed by excretion with digestive wastes removed during **elimination**. Digestive wastes arise from the process of digestion, not from cellular metabolism, which comprises all the chemical reactions (such as cellular respiration) that take place within the cells of an organism. Digestive wastes never pass *into* an organism's cells or blood; they remain outside as they pass along a digestive tract or a gastrovascular cavity. In unicellular (single-celled) organisms, however, food is really digested outside the organism. Digestion occurs after food has been trapped within a vacuole, a sac whose contents are separated from the cell by a membrane. The wastes remain in the vacuole and leave the cell by exocytosis (see Figure 8.1). Metabolic wastes, on the other hand, are produced within the cell itself, within the cytoplasm or cellular oganelles.

Metabolic wastes are removed during excretion and digestive wastes during elimination.

Focus on Plants

Plants have no excretory organs, so how do they regulate the concentration of water in their cells? How do they rid themselves of oxygen, the end product of photosynthesis? Plants take in water via their roots and lose water as water vapor, principally from openings in their leaves called stomata. (See transpiration, p. 669.) Oxygen passes from the plant in this way also, providing most of the oxygen on Earth. Plant roots take

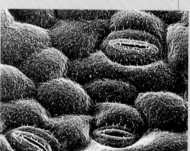

Stomata on a rose leaf

up minerals by active transport, which is the way plants concentrate the minerals they need. These minerals do not leave the plant, but are used for plant growth and metabolism.

12.2 Patterns of excretion

No Excretory Organs

In simple single-celled organisms such as bacteria, which have no excretory organs, the processes of diffusion, osmosis, and active transport get the job done. During diffusion and osmosis, small molecules and water move across the semipermeable membrane of the unicellular organism, *down* their respective concentration gradients. Some substances that are kept in a higher or lower concentration inside than outside the cell are moved across the membrane by active transport. This process uses carrier proteins within the membrane and requires energy to move molecules and ions against the concentration gradient. Chapter 4 provides an in-depth explanation of these processes.

Other living things that have no excretory organs are plants (see a Focus on Plants), fungi, the plantlike and funguslike protists (such as algae and mildew), the cnidarians (such as hydra, jellyfish, and corals), and the echinoderms (such as sea stars and sea urchins). Diffusion, osmosis, and active transport are mechanisms at work in all these organisms to help regulate their balance of water, salts, and other substances.

Contractile Vacuoles

More complex unicellular organisms such as protozoans and simple multicellular organisms such as sponges use organelles called **contractile vacuoles** for excretion (Figure 12.1). Freshwater organisms (including all fresh water protozoans and sponges) are always hypertonic to the surrounding water. That is, their body fluids contain more dissolved substances than fresh water. (The fresh water is hypotonic to the organism, see p. 79.) Therefore, water flows from a higher concentration of water (outside the organism) to a lower concentration of water (inside the organism). If the water that moves into the organism were not excreted, the organism would burst. The contractile vacuoles in protozoans and sponges collect this excess water that is channeled to them by canals. The vacuoles gradually increase in size. When the vacuoles reach a critical size they contract, squeezing out their contents through tiny pores that open in the membrane. This method works just like bailing out the water in a boat that has a small hole in its hull; as quickly as the water flows in, you take a bucket and bail it out. So, too, contractile vacuoles in freshwater protozoans and sponges keep up with the inflow of water.

Some marine protozoans, like their freshwater counterparts, have contractile vacuoles. However, the contractile vacuoles are hypothesized to serve other excretory functions because a marine protozoan does not

NO EXCRETORY ORGANS
(simple single-celled organisms including bacteria, fungi, plants, plantlike and fungilike protists [algae and mildew], cnidarians [hydra, jellyfish, and corals], and echinoderms [sea stars and sea urchin])

Process at work: diffusion, osmosis, and active transport

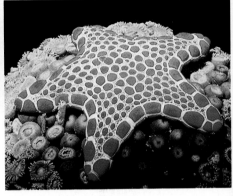

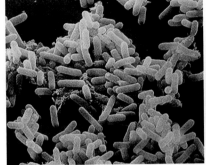

CONTRACTILE VACUOLES
(complex single-celled organisms [protozoans] and simple multicellular organisms [sponges])

Paramecium

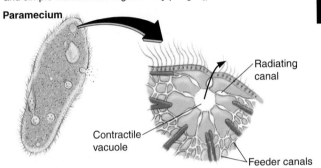

Radiating canal

Contractile vacuole

Feeder canals

NEPHRIDIA
(platyhelminthes [flatworms and tapeworms] and annelids [earthworms and leeches])

Flatworm

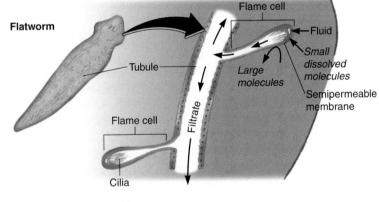

Flame cell

Fluid

Tubule

Large molecules

Small dissolved molecules

Semipermeable membrane

Flame cell

Filtrate

Cilia

MALPIGHIAN TUBULES
(insects and spiders)

Grasshopper

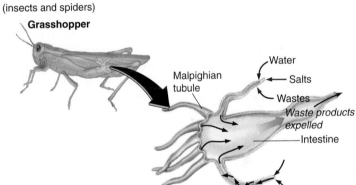

Malpighian tubule

Water

Salts

Wastes

Waste products expelled

Intestine

ANTENNAL GLAND
(crustaceans [lobster and crayfish])

Crayfish

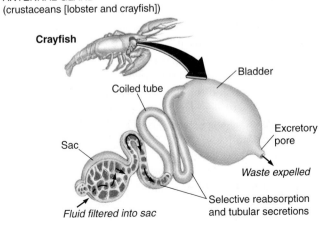

Bladder

Coiled tube

Excretory pore

Sac

Waste expelled

Fluid filtered into sac

Selective reabsorption and tubular secretions

KIDNEYS
(all vertebrates)

Human

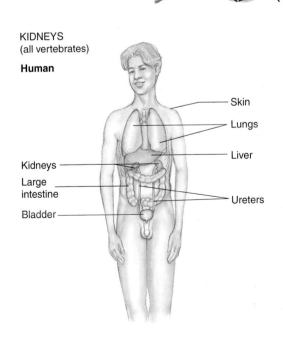

Skin

Lungs

Liver

Kidneys

Large intestine

Ureters

Bladder

take on water. It is hypotonic to its saltwater environment, meaning that its cellular contents have a lower concentration of dissolved substances than the salt water surrounding the cell so water would flow out of the protozoan rather than into it.

Nephridia

Platyhelminthes (such as flatworms and tapeworms) and annelids (such as earthworms and leeches) have excretory organs called **nephridia** (neh-FRID-ee-uh). The most simple type of nephridia (*protonephridia*) are blind-ended excretory tubes that open to the outside of the organism through a hole, or pore. A good example of protonephridia is in the excretory system of the planarian, or flatworm. Body fluids of the flatworm pass into this network of fine tubules that runs along the length of the worm. Specialized blind-ended, bulblike cells, called flame cells, are located along these tubules (see Figure 12.1). As cilia within them beat (looking like a flickering flame), they move the water and dissolved substances from the surrounding fluids of the flatworm into the tubules and out excretory pores.

Although questions remain regarding how the nephridia of flatworms work, research indicates that they exhibit one of the basic processes of excretion: *filtration*. During filtration, the body fluid of an organism is forced through a semipermeable membrane, which allows the fluid and small, dissolved molecules such as salts, sugars, and amino acids to pass, but holds back large molecules such as proteins ❶. In the case of the flatworm, the body fluid of the worm is forced through the semipermeable membrane of the flame cell. The fluid and small molecules (which are now called the filtrate) that pass through this membrane become the excreted material ❷. However, the composition of the filtrate may be changed somewhat before it is excreted.

The nephridia of earthworms are more complex than those of flatworms and open into the body cavity (coelom) at one end and to the outside on the other. In these *metanephridia*, some substances are reabsorbed by active transport. Such s*elective reabsorption* is another of the basic processes of excretion. By means of this process, certain substances are taken from the filtrate and returned to the body, which helps maintain just the right balance of dissolved substances.

Molluscs have complex metanephridial systems called kidneys (although different from vertebrate kidneys). The molluscs secrete wastes into the nephridial tubes by active transport. This process, called *tubular secretion,* is yet another basic process of excretion. These excretory processes of filtration, selective reabsorption, and tubular secretion are described again in more detail on p. 248 with the description of the kidney.

Antennal Gland

If you've ever eaten lobster or crayfish, you may remember the green material located in the head. This is the excretory organ of crustaceans, and is called the **antennal gland** (because its excretory pore opens near the base of the antennae) or *green gland* (because of its color). These paired glands each consist of a sac, a long coiled tube, and a bladder.

In crustaceans, nitrogenous wastes (nitrogen-containing wastes from protein metabolism) are removed at the gills. The green gland, on the other hand, is primarily involved with the excretion of water and salts. This is how it works: Fluid from the blood is filtered into the sac. The composition of this filtrate is adjusted by selective reabsorption and tubular secretion as it moves along the coiled tube. The bladder then stores the waste fluid (urine) until it is expelled through the pore.

Malpighian Tubules

Insects and certain other arthropods such as spiders have excretory organs called **Malpighian tubules** (mal-PIG-ee-en, named after the 17th century Italian anatomist who discovered them, Marcello Malphighi.) The blind ends of these tubes lie within the blood-filled body cavity and empty into the intestine (see Figure 12.1.) Salts and wastes move into the tubules by either diffusion or active transport. The composition of this material is adjusted by selective reabsorption and tubular secretion as it moves along the tubule. Insects and spiders that live on fresh vegetation take in large amounts of water with their food and excrete a liquid urine that is mixed with digestive wastes from the intestinal tract. Insects and spiders that live on dry food excrete very little water; their waste products are expelled in the form of a semidry paste.

Kidneys

Vertebrates (and molluscs) have excretory organs called **kidneys**. Vertebrate kidneys are made up of microscopic units called **nephrons** that carry out filtration, selective reabsorption, and tubular secretion to produce a liquid waste called *urine*. The kidney is quite a versatile excretory organ, meeting the needs of marine, freshwater, and terrestrial animals.

> Types of excretory organs are contractile vacuoles, nephridia, antennal glands, Malpighian tubules, and kidneys.

Human excretion relies primarily on the lungs and kidneys.

12.3 What substances does the body excrete?

Humans (as well as other vertebrates) excrete a wide variety of substances. Some of these products, bile pigments and the salts of certain minerals, are removed during elimination, the ridding of digestive wastes.

Bile pigments are produced by the liver from the breakdown of old, worn-out red blood cells. These pigment molecules are combined with other substances to form bile, which is carried by a duct from the liver, to the gallbladder (bile's storage pouch), and then to the small intestine. You may recall from Chapter 8 that bile plays a role in the emulsification of fat globules during the digestion of food. The bile pigments leave the body mixed with the digestive wastes and are the cause of the characteristic brown color of the feces (digestive excrement). Because of its role in the excretion of the bile pigments, the liver is considered an organ of excretion. The large intestine is also considered an organ of excretion because cells lining its walls excrete the salts of some minerals such as calcium and iron (Figure 12.2).

Bile pigments excreted by the liver and the minerals excreted by the large intestine make up only a tiny fraction of the excretory products of the human body. The major products of excretion are water, salts (primarily sodium chloride [NaCl]—table salt), carbon dioxide, and nitrogen-containing molecules. Carbon dioxide (CO_2) is a gas produced during cellular respiration and is excreted primarily by the lungs (see Chapter 9). The lungs also excrete water in the form of water vapor—which is the reason you can see your breath on a cold day as the water vapor condenses to droplets. A small amount of salt and often a large amount of water also leave the body by means of glands in the skin (see Chapter 16).

Nitrogen-containing molecules, or **nitrogenous wastes**, are produced from the breakdown of proteins and nucleic acids (DNA and RNA). As discussed in Chapter 8, humans need to take in protein as a source of amino acids. These amino acids are used to construct bodybuilding proteins. The liver breaks down any extra amino acids you may eat—those not needed for building new body parts or repairing old ones. In the liver, enzymes break down these amino acids by removing their amino groups ($-NH_2$), a process called *deamination*. The molecules that result can then be used to supply the body with energy. The amino group is a leftover that cannot be used.

In the liver, amino groups ($-NH_2$) are chemically converted to ammonia (NH_3). Ammonia is quite toxic, or poisonous, to all cells, and the body must get rid of it quickly, not allowing its concentration to increase. A bit

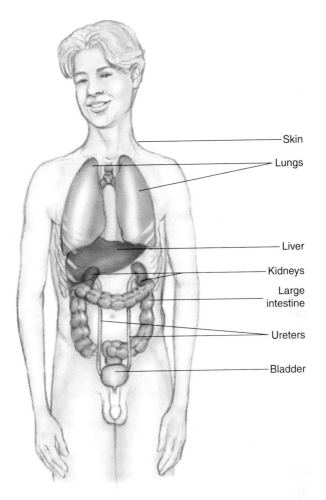

Figure 12.2 Organs that carry out excretion in humans. In addition to the kidneys, the skin, lungs, liver, and large intestine all play a role in the excretion of metabolic wastes from the body.

of chemical reshuffling takes care of this problem, as the ammonia combines with carbon dioxide to form **urea**, the primary excretion product from the deamination of amino acids. In fact, approximately 90% of all nitrogenous wastes are eliminated from the human body in the form of urea. Two other nitrogenous wastes found in small amounts in the urine are uric acid and **creatinine** (kree-AT-uh-neen). **Uric acid** is formed from the breakdown of nucleic acids found in the cells of the food you eat and from the metabolic turnover of your nucleic acids and adenosine triphosphate (ATP). Creatinine is derived primarily from a nitrogen-containing molecule called *creatine* (KREE-ah-teen) found in muscle cells.

The primary metabolic waste products of the human body are carbon dioxide, water, salts, and nitrogenous (nitrogen-containing) molecules. Most nitrogenous wastes are excreted as urea.

12.4   The organs of excretion

The skin, liver, and large intestine play a minor role in the excretion of metabolic wastes from the body. The primary organs of excretion are the lungs (see Chapter 9) and the *kidneys* (see Figure 12.2). As mentioned previously, the lungs excrete carbon dioxide and water vapor. The kidneys excrete the ions of salts, such as Na^+ (sodium), K^+ (potassium), Cl^- (chloride), Mg^{2+} (magnesium), and Ca^{2+} (calcium). In addition, they excrete the nitrogenous wastes urea, creatinine, and uric acid, along with small amounts of other substances that may vary depending on diet and general health. These substances, together with water, form the excretion product called **urine**.

The kidneys produce urine by first filtering out most of the molecules dissolved in the blood, then selectively reabsorbing useful components, and finally secreting a few other waste products into this remaining filtrate, which becomes urine. As the kidneys process the blood in this way, they regulate its chemical composition and water content and, in turn, the chemical and fluid environment of the body. For example, although almost no amino acids are removed from the blood by the kidneys, almost half its urea is removed and then excreted. In addition, the kidneys maintain the concentrations of all ions within narrow boundaries. This strict maintenance of specific levels of ion concentrations keeps the blood's pH at a constant value, maintains the proper ion balances for nerve conduction and muscle contraction, and affects the amount of water reabsorbed into the bloodstream from the filtrate.

The primary organs of excretion are the lungs and the kidneys. The kidney is a regulatory as well as an excretory organ.

12.5   An overview of how the kidney works

Your body has two kidneys, each about the size of a small fist, located in the lower back region and partially protected by the lower ribs (see Figure 12.2 and Figure 12.3). Put simply, each is a living filtration plant that balances the concentrations of water and salts (ions) in the blood and, at the same time, excretes wastes. Specifically, kidneys work in the following way: As blood flows through the millions of microscopic filtering systems, or nephrons, within each kidney, the fluid portion of the blood is forced through capillary membranes into the nephron tubules (see Figure 12.5). The formed elements of the blood (red blood cells, white blood cells, and platelets), along with most proteins, stay within the blood. Much of the fluid portion of the blood, including ions and molecules other than large proteins, passes through the membrane. The fluid that passes through the membrane is called the *filtrate*. Most of the water, as well as selected ions and molecules, is then reabsorbed from the filtrate back into the blood as the filtrate passes through the rest of the nephron.

The fluid that passes through the membrane filters of each nephron contains many molecules that are of value to the body, such as glucose, amino acids, and various salts and ions. Humans (and all vertebrates) have evolved a means of selectively reabsorbing these valuable molecules without absorbing the waste molecules that are also dissolved in the filtered waste fluid, or urine. Such selective reabsorption gives the vertebrates great flexibility because the membranes of different groups of animals have evolved specific transport channels that reabsorb a variety of different molecules. This flexibility is a key factor underlying the ability of different vertebrates to function in many diverse environments. They can reabsorb small molecules that are especially valuable in their particular habitats and not absorb wastes. In all vertebrates, the kidney carries out the processes of filtration, reabsorption, and excretion. It can function with modifications in fresh water, in the sea, and on land (see Section 12.8).

The human kidney maintains the proper balance of water in the body, retains substances the body needs, and eliminates metabolic wastes by first filtering most substances out of the blood and then reabsorbing what is needed.

Figure 12.3   The human kidney. Each kidney is about 4 inches long, 2½ inches wide, and 1 inch thick. They are located on each side of the vertebral column just above the waistline and are surrounded by a cushioning mass of fat.

12.6 The anatomy of the kidney

The kidneys look like what their name implies: two, gigantic, reddish-brown kidney beans (see **Figure 12.4a**). Substances enter and leave the kidneys through blood vessels that pierce the kidney near the center of its concave border. If you were to slice a kidney vertically as shown in Figure 12.4a, dividing it into front (ventral) and back (dorsal) portions, you would see an open area called the *renal pelvis* near the concave border. (The word renal means "kidney.") *Collecting ducts* empty into this space, carrying the urine produced by the nephrons. From here, urine exits the kidney via a tube called a **ureter** (YER-ih-ter).

Covered with a fibrous connective tissue capsule and cushioned with fat, the kidney is made up of outer, reddish tissue called the *cortex* and inner, reddish-brown tissue called the *medulla*. Lined up within the medulla are triangles of tissue called the *renal pyramids*. The cor-

tex and pyramids are made up of approximately 1 million nephrons (Figure 12.4b), their collecting ducts, and the blood vessels that surround them (Figure 12.4c).

As well as showing the cortex, medulla, and renal pyramids, the lengthwise section of the kidney illustrates the general layout of the major blood vessels within a kidney. Blood enters each kidney via the right and left *renal arteries*. Branches split off these major arteries, travel up the sides of the pyramids, and meet at the interface of the cortex and medulla. Smaller branches extend into the cortex, giving rise to the arterioles that enter each individual nephron. The total volume of blood in your body passes through this network of blood vessels every 5 minutes!

The kidney is made up of two main types of tissue: an outer cortex and an inner medulla. These tissues are composed of microscopic filtering units called nephrons, their collecting ducts, and blood vessels.

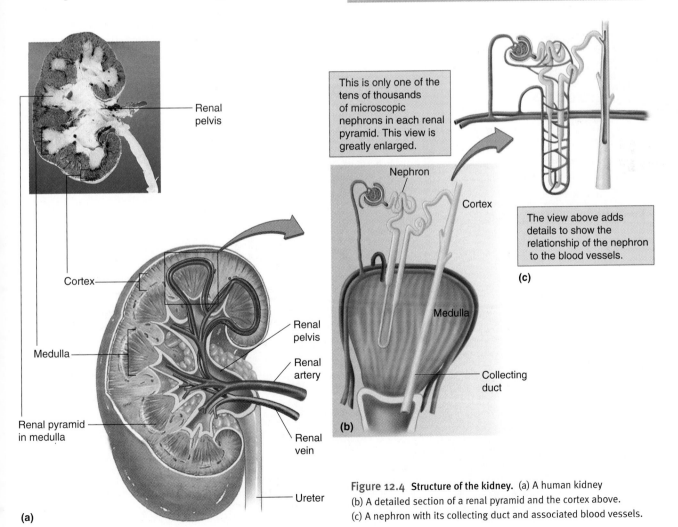

This is only one of the tens of thousands of microscopic nephrons in each renal pyramid. This view is greatly enlarged.

The view above adds details to show the relationship of the nephron to the blood vessels.

(c)

Renal pelvis

Cortex

Medulla

Renal pyramid in medulla

Renal pelvis

Renal artery

Renal vein

Ureter

(a)

Nephron

Cortex

Medulla

Collecting duct

(b)

Figure 12.4 Structure of the kidney. (a) A human kidney (b) A detailed section of a renal pyramid and the cortex above. (c) A nephron with its collecting duct and associated blood vessels.

Human excretion relies primarily on the lungs and kidneys.

12.7 The workhorse of the kidney: The nephron

Each nephron (**Figure 12.5**) is a tubule that is structured in a special way to accomplish three tasks:

1. *Filtration.* During filtration, the blood is passed through membranes that separate blood cells and proteins from much of the water and molecules dissolved in the blood. Together, the water and dissolved substances are referred to as the *filtrate.*

2. *Selective reabsorption.* During reabsorption, desirable ions and metabolites and most of the water from blood plasma are recaptured from the filtrate, leaving nitrogenous wastes, excess water, and excess salts behind for later elimination. The filtrate is now called *urine.*

3. *Tubular secretion.* During secretion, the kidney *adds* materials such as potassium and hydrogen ions, ammonia, and potentially harmful drugs to the filtrate from the blood. These secretions rid the body of certain materials and control the blood pH.

Filtration

At the front end of each nephron tube is a filtration apparatus called the *glomerular capsule.* The capsule is shaped much like a caved-in tennis ball or basketball and surrounds a tuft of capillaries called the *glomerulus.* The glomerular capillaries branch off an entering arteriole termed the *afferent arteriole* (named after the Latin verb *affero,* which means "going toward"). The walls of the glomerular capillaries, along with specialized cells of the capsule, act as filtration devices. The pressure of the incoming blood forces the fluid within the blood through the capillary walls and through spaces between capsular cells that surround the capillaries. Capillary walls are made up of a single layer of cells having differentially permeable membranes (see Figure 10.4). These walls do not allow molecules such as large proteins and the formed elements of the blood to pass through. Water and smaller molecules such as glucose, ions, and nitrogenous wastes pass through easily.

This filtrate passes into the tubule of the nephron at the glomerular capsule. From there, it passes into a coiled portion of the nephron called the *proximal convoluted tubule.* This name describes the coiled (convo-

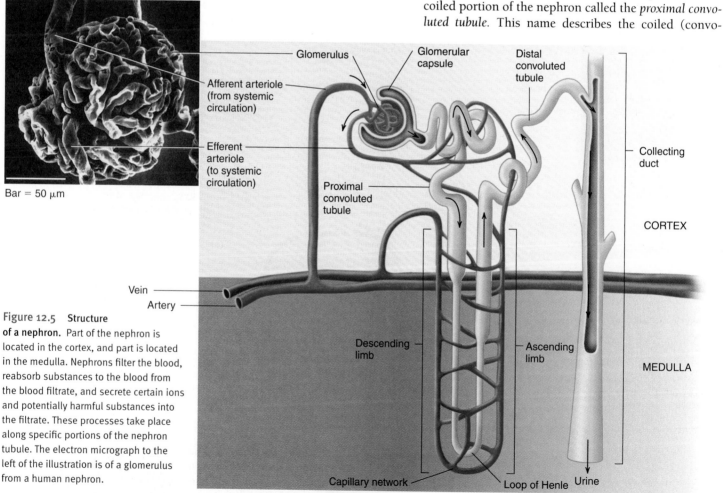

Bar = 50 μm

Figure 12.5 **Structure of a nephron.** Part of the nephron is located in the cortex, and part is located in the medulla. Nephrons filter the blood, reabsorb substances to the blood from the blood filtrate, and secrete certain ions and potentially harmful substances into the filtrate. These processes take place along specific portions of the nephron tubule. The electron micrograph to the left of the illustration is of a glomerulus from a human nephron.

luted) tubule that is closest to (proximal to) the glomerular capsule. Both these structures lie in the cortex of the kidney. As the filtrate passes through the proximal tubule, reabsorption begins.

Selective Reabsorption

Epithelial cells throughout the length of the tubule carry out reabsorption. During this highly selective process, specific amounts of certain substances are reabsorbed depending on the body's needs at the time. Substances are reabsorbed by both active and passive transport mechanisms (see Chapter 4). In the proximal convoluted tubule, glucose and small proteins are put back into the blood by active transport. These substances move out of the tubule and into the blood within surrounding capillaries. (The insets show the process being described. For ease of viewing the nephron, surrounding capillaries are not shown.)

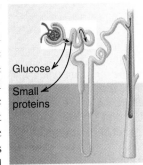

The filtrate moves from the proximal convoluted tubule to the descending limb of the loop of Henle. The *loop of Henle* is the part of the tubule that extends into the medulla of the kidney. The *descending limb* of the loop of Henle dips downward and the *ascending limb* extends in the opposite direction, back up into the cortex. The cells making up the walls of the descending limb do not permit salt and urea to pass out of the tubule; their membranes are impermeable to salt and urea. However, the walls of the descending limb are freely permeable to water. Water moves out of the tubule by osmosis, passing into the surrounding tissue fluid and then into the blood in surrounding capillaries. As water passes out of the descending limb, it leaves behind a more concentrated filtrate.

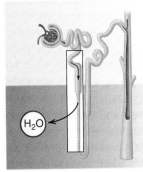

At the turn of the loop, the walls of the tubule become permeable to salt, but much less permeable to water. Salt passes out of the tubule and into the surrounding tissue fluid by diffusion. This movement

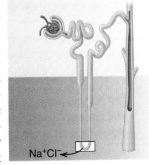

of salt produces a high concentration of salt in the tissue fluid surrounding the bottom of the loop.

In the ascending limb, the walls of the tubule contain active transport channels that pump even more salt out of the filtrate within the kidney tubule. This active removal of salt from the ascending limb causes water to diffuse outward from the filtrate just above the ascending limb at the *distal convoluted tubule*. The walls of this portion of the kidney tubule are permeable to water, which is not the case in the ascending limb. Left behind in the filtrate are some water and the urea that initially passed through the glomerulus as nitrogenous waste; eventually the urea concentration becomes very high in the tubule.

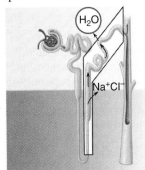

Finally, the filtrate empties into a collecting duct that passes back into the medulla. The collecting ducts of all the nephrons bring the urine to the renal pelvis, an open area in which the urine collects before it flows out of the kidney. The lower part of the collecting duct is permeable to urea. During this final passage, some of the concentrated urea in the filtrate diffuses into the surrounding tissue fluid, which has a lower urea concentration than the filtrate. A high urea concentration in the tissue fluid surrounding the loop of Henle results. This high concentration of urea helps produce (along with sodium) the osmotic gradient that caused water to move out of the filtrate as it passed down the descending limb. Water also passes out of the filtrate as it moves down the collecting duct. This movement occurs as a result of the osmotic gradient created by both the movement of urea out of the collecting duct and the movement of salt out of the ascending limb. The water is then collected by blood vessels in the kidney, which carry it into the systemic circulation.

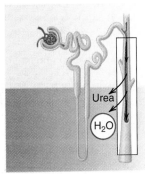

In summary, the kidney achieves a high degree of water reabsorption by using the salts and urea in the glomerular filtrate to increase the osmotic concentration of the tissue fluid surrounding the nephron. This high concentration of salts and urea creates an osmotic gradient that pulls water from the filtrate out into the surrounding tissue fluid. It is collected there by blood vessels impermeable to the high urea concentration, but permeable to water.

Tubular Secretion

A major function of the kidney is the elimination of a variety of potentially harmful substances that you may eat, drink, or inhale or that may be produced during

Just Wondering . . . Real Questions Students Ask

I play sports and we take random drug tests. Are these tests accurate? I worry they'll accuse me of something I didn't do.

In 1986, the National Collegiate Athletic Association (NCAA) instituted mandatory drug screening of student athletes. Therefore, if you attend an NCAA school and are involved in sports, you are subject to the NCAA regulations prohibiting drug use. In addition, many colleges and universities have their own drug use and drug testing policies.

In addition to its use in collegiate sports, drug testing is conducted in professional and Olympic sports as well as in the workplace. It was first used in the United States by the Department of Defense in the 1960s and early 1970s to screen servicemen and women returning from the war in Vietnam. In the 1980s, drug testing methods became more reliable, and private companies, especially those concerned about public safety (such as some in the transportation industry), began testing employees. Many laws were passed in the late 1980s that established drug-free workplaces within the federal government and random

drug-screening programs in transportation industries such as aviation, rail, mass transit, and trucking.

Major improvements have come with the introduction of tests that use recombinant DNA techniques to amplify trace chemicals in urine (see Chapter 19). The enzyme-multiplied immunoassay technique (EMIT) has become the standard approach for screening large numbers of urine samples. EMIT can be used in a robot tester that employs a light sensor to read urine samples and prints out a value for each of five or six drugs present; it can process 18,000 samples an hour. The detection period depends on the type and dose of the drug. EMIT is more than 98% accurate, with error biased toward nondetection.

Because EMIT responds to a broad range of opiate and amphetamine compounds, it sometimes produces a positive result for harmless prescription drugs such as ibuprofen and decongestants, as well as for foods such as herbal

teas and poppy seeds. For this reason, samples that show a positive result on EMIT are then retested by more cumbersome but 100% accurate procedures (gas chromatography–mass spectrometry). For example, eating poppy seeds may trigger a positive signal on EMIT, but the confirmation test tells that a key heroin breakdown product (6-0-acetylmorphine) is not present. So don't worry; the tests are highly reliable and you shouldn't be wrongly accused.

---

metabolism. The human kidney has evolved the ability to detoxify the blood through the process of tubular secretion. During this process, the cells making up the walls of the distal convoluted tubule take substances from the blood within surrounding capillaries and from within the surrounding tissue fluid and put them into the filtrate (urine) within the kidney tubule (**Figure 12.6**). Ammonia (NH_3) from the deamination of

amino acids that have not been converted to urea is removed from the blood by tubular secretion. Certain prescription drugs such as penicillin are also removed by this process. In addition, the body rids itself of harmful drugs such as marijuana, cocaine, heroin, and morphine by tubular secretion. These drugs are processed by the liver, and their breakdown products are secreted into the filtrate within the kidney tubules. Specific tests for the breakdown products of these drugs can be performed on urine samples to determine whether a person has particular drugs in his or her body.

Tubular secretion can also be thought of as a fine-tuning mechanism. By removing specific amounts of hydrogen ions (H^+) from the blood and secreting them into the filtrate, the kidney can keep the pH of the blood at a constant level (7.35 to 7.45). (See Chapter 3 for an explanation of pH.) Likewise, the potassium ion (K^+) concentration of the blood is fine tuned by tubular secretion. The proper concentration of potassium ions is important to the proper functioning of muscles, including the heart.

Visual Summary Figure 12.6 Tubular secretion and a summary of the process of selective reabsorption.

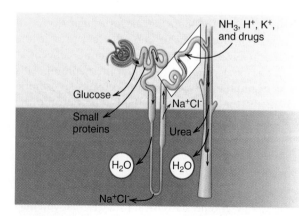

The kidneys work by removing most of the water and dissolved substances from the blood and then replacing what is needed by the body at that time.

12.8   A summary of urine formation

The roughly 2 million nephrons that form the bulk of the two human kidneys receive a flow of approximately 2000 liters of blood per day—enough to fill about 20 bathtubs! The nephron first filters the blood, removing most of its water and all but its largest molecules and its cells. The nephron then selectively reabsorbs substances back into the blood. This process is driven by two factors: (1) the varying permeability of the membranes of the cells lining the kidney tubule and (2) the development of a high osmostic gradient surrounding the loop of Henle.

Because of the varying permeability of kidney tubule cells, substances move from the filtrate back into the blood by diffusion and active transport mechanisms at specific places along the length of the tubule. Each substance will be reabsorbed only to a particular threshold level, however, with the rest remaining in the filtrate. For example, glucose is reabsorbed at the proximal convoluted tubule to a threshold level of 150 milligrams per 100 milliliters of blood. Any glucose above this threshold will be excreted in the urine. The amount of glucose in the blood of a healthy person does not normally exceed this limit. However, persons with diabetes mellitus excrete glucose because they fail to produce a hormone called *insulin* that promotes glucose uptake by the cells and the glucose level in the bloodstream rises. Urea, on the other hand, has a very low threshold that is reached quickly. Therefore most urea stays in the filtrate (except the urea that diffuses from the collecting duct, adding to the osmotic gradient surrounding the loop of Henle).

The high osmotic gradient surrounding the loop of Henle causes most of the water filtered from the blood to be reabsorbed, or conserved, in the descending limb of the loop. Water conservation is, in fact, the job of this loop.

For example, freshwater fish have no loop of Henle. Their bodies take on water

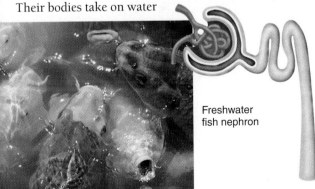

Freshwater
fish nephron

by osmosis; therefore, they rid their bodies of excess water rather than conserve it. Freshwater fish drink little water and produce large amounts of urine. These mechanisms work to maintain their body fluids hypertonic to their freshwater surroundings. That is, their body fluids contain more dissolved substances and have a lower concentration of water than the water in which they live. (Marine fish, on the other hand, lose water by osmosis to their environment. To compensate, they drink seawater and secrete the excess salt from their gills.)

Only birds and mammals have a loop of Henle and are able to produce urine that is more concentrated than their bodily fluids. Among these animals, the longer the loop, the greater the ability to concentrate the urine and conserve water. Animals such as desert rodents that

Desert
mammal
nephron

have highly concentrated urine have exceptionally long loops of Henle.

The nephron also controls blood pH, fine tunes the concentrations of certain ions and molecules, and removes potentially harmful drugs from the blood by the process of tubular secretion. In these ways, the kidney contributes to the chemical and water balance of the body, thus functioning as a major organ of internal equilibrium, or homeostasis. The resulting fluid is urine, a waste that is excreted from the body.

Urine is formed in the following way: the nephron first filters the blood, removing most of its water and all but its largest molecules and its cells. Substances move from this filtrate back into the blood by diffusion and active transport mechanisms at specific places along the length of the nephron tubule, with each substance reabsorbed only to a particular threshold level.

12.9 The urinary system

The kidneys are only one part of the **urinary system**—a set of interconnected organs (Figure 12.7) that not only remove wastes, excess water, and excess ions from the blood but store this fluid, or urine, until it can be expelled from the body. The urine exits each nephron by means of the collecting duct. From there, it flows into the renal pelvis (see Figure 12.4). This area narrows into a tube called the *ureter* that leaves the kidney on its concave border, near where the renal artery and vein enter and exit. By means of muscular contractions (peristalsis), these muscular tubes—one from each kidney—bring the urine to a storage bag called the **urinary bladder**.

The urinary bladder is a hollow muscular organ that, when empty, looks much like a deflated balloon. As the bladder fills with urine, it assumes a pear shape. On average, the bladder can hold 700 to 800 milliliters of urine, or almost a quart. However, when less than half this amount is in the bladder, special nerve endings in the walls of the bladder, called *stretch receptors,* send a message to the brain that results in the desire to urinate.

The urinary bladder empties into a tube called the **urethra**. This tube leads from the underside of the bladder to the outside of the body. The urethra in men plays a role in the reproductive system, carrying semen to the outside of the body during ejaculation. However, the urinary and reproductive systems have no connection in women. In addition, the urethra in women is much shorter than the urethra in men and is closer to the rectum and its associated bacterial population. For these reasons, women contract urinary tract (bladder) infections more easily than do men.

The urinary system is made up of the kidneys, ureters, urinary bladder, and urethra. These interconnected organs remove wastes, excess water, and excess ions from the blood. Additionally, the urinary system stores this fluid, or urine, until it can be expelled from the body.

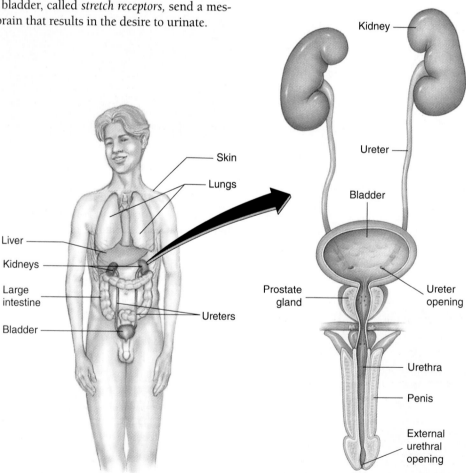

Figure 12.7 The human urinary system. The urinary system in males and females includes the kidneys, ureters, urinary bladder, and urethra. This diagram shows details of the male urinary system.

12.10 The kidney and homeostasis

The kidney is an excellent example of an organ whose principal function is homeostasis, the maintenance of constant physiological conditions within the body. It is concerned with both water balance and ion balance. To maintain homeostasis, however, the urinary system depends in part on the endocrine (hormone) system.

A *hormone* is a chemical messenger sent by a gland to other cells of the body. Although there are several hormones that influence kidney function, among the most important are ADH and aldosterone. The hormone that regulates the rate at which water is lost or retained by the body is **antidiuretic hormone (ADH)**. It is secreted by the pituitary gland at the base of the brain. Its effect on the excretion of water is easy to remember if you know that a diuretic is a substance that helps remove water from the body. Because ADH is an *antidiuretic*, it works to conserve, or retain, water.

ADH works by controlling the permeability of the distal convoluted tubules and collecting ducts to water **(Figure 12.8, top)**. (The walls of the distal tubules are impermeable to water when ADH is absent.) When ADH levels increase, the permeability to water of the collecting ducts and distal tubules increases. Water therefore moves out of the ducts and distal tubules by osmosis ❶ and back into the blood within surrounding capillaries. When ADH levels decrease, the permeability of the collecting ducts and distal tubules to water decreases. Therefore, less water is reabsorbed from the urine (filtrate). Instead, it remains in the ducts and distal tubules ❷ and is excreted from the body. Alcohol inhibits the release of ADH, resulting in decreased reabsorption of water and therefore increased urination, which can cause dehydration.

Another hormone regulates the level of sodium ions (Na^+) and potassium ions (K^+) in the blood. This hormone is called **aldosterone** (al-DOS-ter-own) (Figure 12.8, bottom). When aldosterone levels increase, the kidney tubule cells increase their reabsorption of sodium ions from the filtrate ❸ and decrease their reabsorption of potassium ions ❹. Put simply, aldosterone promotes the retention of sodium and the excretion of potassium. In addition, because the concentration of sodium ions in the blood affects the reabsorption of water, the increase in sodium ions in the blood causes water to move by osmosis from the filtrate into the blood ❺.

The principal function of the kidney is homeostasis. Two hormones help the kidney maintain homeostasis: ADH regulates the amount of water reabsorbed at the collecting duct, aldosterone promotes conservation of sodium and therefore water and promotes the excretion of potassium.

ADH LEVEL	EFFECT ON KIDNEY
Increased ADH Levels	Collecting ducts and the distal convoluted tubules become permeable to water; water moves out of ducts and into blood ❶
Decreased ADH Levels	Collecting ducts become impermeable to water; water is not reabsorbed from the filtrate and is excreted ❷

ALDOSTERONE LEVEL	EFFECT ON KIDNEY
Increased Aldosterone Levels	Tubules increase reabsorption of sodium from the filtrate and decrease reabsorption of potassium; water and sodium thus move from filtrate into the blood, and excess potassium is excreted ❸ ❹ ❺
Decreased Aldosterone Levels	Tubule absorption of sodium and potassium normal; water is not reabsorbed from the filtrate and is excreted

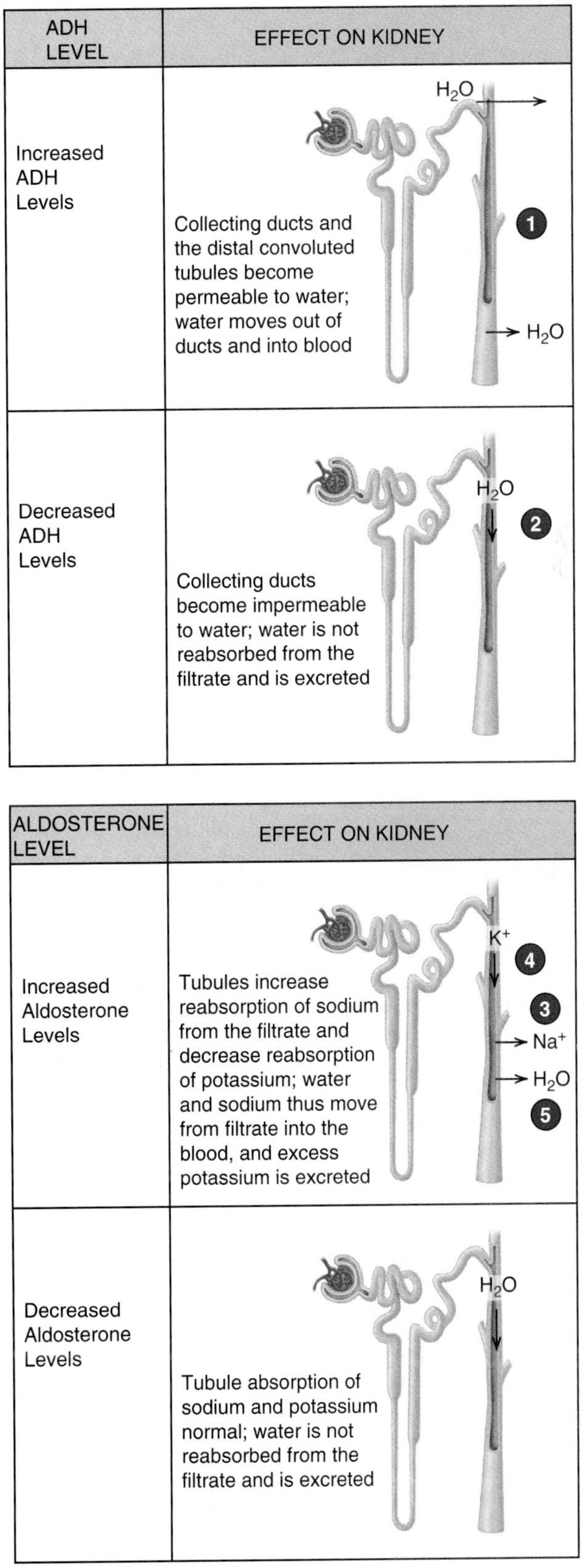

Figure 12.8 How hormones regulate kidney function.

For the latest health information, please visit www.jbpub.com/biology.

12.11 Kidney stones

During the formation of urine, salts and other wastes are dissolved in the filtrate and pass with it out of the kidney as urine. Sometimes, however, certain salts do not stay dissolved—most notably calcium salts or uric acid—but instead form crystals called **kidney stones**. Each year, more than 300,000 Americans are hospitalized because of kidney stones. They often form these stones because of diminished water intake, diets high in protein and calcium, genetic disorders, infections, and the misuse of medications.

Kidney stone attacks are excruciatingly painful; the stone moves through the kidney (some are quite jagged) and then blocks the flow of urine. Small stones—those the size of a grain of sand—pass through the ureters, bladder, and urethra within a few days. Larger stones (many are as large as a pearl and some are as large as a golf ball) require certain medical procedures to remove them. The photo at left shows a kidney stone on a fingertip.

Since 1984, a treatment for kidney stones called *shock-wave lithotripsy* (LITH-oh-TRIP-see) has been available. Before this time, surgery was the only means of relief. Surgery is still used, however, to remove stones larger than 2 centimeters in diameter or stones that are causing infection. During one form of surgery, a tube is inserted into the renal pelvis. An ultrasonic probe within the tube first bombards the stone with sonic waves, crushing it. Tiny forceps then protrude from the tube and are used to collect the pieces of the stone. In another type of surgical procedure, the kidney is cut open and forceps are used to remove the stones.

Shock-wave lithotripsy involves a procedure quite different from these two. The patient is immersed up to the neck in a tank of water. Guided by x-ray monitors, intense sound waves are directed at the stone, shattering it. (Lithotripsy is a Greek word meaning stone crushing.) Once the stone is broken apart, the pieces are excreted in the urine.

> Kidney stones are crystals of certain salts that develop in the kidney and block urine flow. They may be surgically removed or broken apart with shock-wave therapy.

12.12 Renal failure

Renal failure occurs when the filtration of the blood at the glomerulus either slows or stops. In acute renal failure, filtration stops suddenly. Acute renal failure can have many causes, such as a decreased flow of blood through the kidneys as a result of problems with the heart or blockage of a blood vessel, damage to the kidney by disease, the presence of a kidney stone blocking urine flow, or severe dehydration as occurs occasionally in marathoners and other endurance athletes.

In chronic renal failure, the filtration of the blood at the glomerulus slows gradually. This condition is usually irreversible because it is most commonly caused by injury to the glomerulus. These injuries have many causes, such as the deposit of toxins, bacterial cell walls, or molecules produced by the immune system within the glomerulus; the coagulation of the blood within the glomerulus; or the presence of a disease such as diabetes mellitus. In fact, diabetes mellitus is the primary cause of kidney failure.

If the kidneys become unable to excrete nitrogenous wastes, regulate the pH of the blood, and regulate the ion concentration of the blood because of renal failure, the individual will die unless the blood is filtered. A machine called an artificial kidney accomplishes this job (see photo below). The process of filtering blood in this way is called **dialysis**.

During *hemodialysis* (dialysis of the blood, see photo below), a patient's blood is pumped through tubes to one side of a selectively permeable membrane. On the other side of the membrane is a fluid called the dialysate. The dialysate contains the same concentration of ions as normally found in the bloodstream. Because

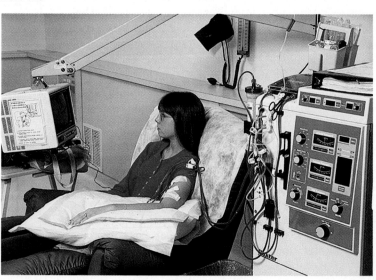

small molecules can pass across the membrane, any extra ions in the patient's blood move by diffusion into the dialysate until their concentrations on both sides of the membrane are equal. In addition, the dialysate contains no wastes, so the wastes in the patient's blood also diffuse into the dialysate.

Hemodialysis is a slow process, usually taking 4 hours, three times a week. The results of recent research studies show that some patients with kidney failure can undergo hemodialysis successfully at home, while they sleep. Dialysis then takes place for 8 hours, not 4, and for 6 to 7 nights per week. Patients using this technique feel better, eat better, are more energetic, and have less disrupted lives.

Another type of dialysis is *periotoneal dialysis*. This process uses the thin lining of the abdominal cavity, the peritoneum, as the dialyzing membrane. The patient has a catheter (a thin tube) implanted in the abdominal wall in order to perform this procedure. The dialysate flows into the abdomen through the catheter from a bag outside the body. Waste materials and excess water pass from blood vessels in the peritoneum into the dialysate. After several hours, the dialysate is drained from the abdomen.

Only 10% of persons with kidney failure survive for at least 10 years on dialysis due to the stress that toxic waste products place on the body organs. Many more persons with kidney failure survive for at least 10 years with a kidney transplant: 56% with a transplant from a cadaver and 76% with a transplant from a living donor. Each year in the United States, over 12,000 patients with renal failure receive kidney transplants.

Renal failure is a reduction in the filtration rate of blood in the glomerulus. This condition has many possible causes and may be treated by dialysis or kidney transplantation.

Key Concepts

Excretion removes metabolic wastes and balances water and salt.

12.1 Excretion is a process whereby metabolic wastes, excess water, and excess salts are removed from the blood and passed out of the body.

12.1 Elimination is a process that takes place as digestive wastes leave the body during defecation.

12.2 Some organisms have no excretory organs and rely on diffusion, osmosis, and active transport to rid them of metabolic wastes and excess water.

12.2 The simplest excretory organ is the contractile vacuole and is found in protozoans and sponges.

12.2 Other types of excretory organs are nephridia, antennal glands, Malpighian tubules, and kidneys.

Human excretion relies primarily on the lungs and the kidneys.

12.3 The excretory products of the body are bile pigments, nitrogen-containing molecules (nitrogenous wastes), carbon dioxide, water, and salts.

12.4 Bile pigments are excreted by the liver and passed out of the body by means of the digestive system.

12.4 Most carbon dioxide is excreted by the lungs.

12.4 The kidneys excrete most of the nitrogenous wastes along with excess water and salts.

12.5 The kidneys excrete wastes as urine, which is formed within microscopic tubular units of the kidney called nephrons.

12.6 The kidney is made up of an outer region called the cortex and an inner region called the medulla.

12.6 These tissues are composed of nephrons, their collecting ducts, and blood vessels.

12.7 During the formation of urine, most of the water and other small molecules are first filtered out of the blood; this fluid is called the filtrate.

12.7 As the filtrate flows through the nephron tubule, various substances are reabsorbed because the permeability of its walls varies along its length.

12.7 By means of tubular secretion, the kidney also excretes the breakdown products of a variety of potentially harmful substances from the blood, such as marijuana, cocaine, heroin, morphine, and prescription drugs.

12.8 Most of the water removed from the blood is reabsorbed from the filtrate because of a high osmotic gradient that surrounds certain sections of the tubule.

12.8 Different groups of vertebrates can function in a wide variety of environments because of adaptations in their nephrons that allow selective reabsorption of molecules valuable to their particular habitats.

12.9 Urine leaves the kidneys by means of muscular tubes called ureters and is conveyed to a storage pouch called the urinary bladder.

12.9 A tube called the urethra brings urine to the outside.

Key Concepts, continued

12.10 Two hormones, ADH and aldosterone, help the kidneys control the balance of water and ions in the blood, thus helping it perform its primary function, the maintenance of homeostasis.

 Health
Connections

12.11 Crystals of salts that sometimes form within the kidney are called kidney stones.

12.11 Surgery and shock-wave lithotripsy are two ways in which kidney stones are removed.

12.12 When the filtration of the blood at the glomerulus is seriously impaired, the blood must be filtered by means of dialysis.

12.12 In severe cases of renal failure, a patient may require kidney transplantation.

Key Terms

aldosterone (al-DOS-ter-own) *253*
antennal gland *244*
antidiuretic hormone, or ADH (AN-ti-dye-yoo-RET-ik) *253*
bile pigments *245*
contractile vacuoles *242*
creatinine (kree-AT-uh-neen) *245*
dialysis (dye-AL-uh-sis) *254*
elimination *242*
excretion (ex-SKREE-shun) *242*

kidney stones *254*
kidneys *244*
Malpighian tubules *244*
nephridia (neh-FRID-ee-uh) *244*
nephrons (NEF-rons) *244*
nitrogenous wastes (nye-TROJ-uh-nus) *245*
renal failure *254*
urea (yoo-REE-uh) *245*
ureter (YER-ih-ter) *247*

urethra (yoo-REE-thruh) *252*
uric acid (YOO-rik) *245*
urinary bladder *252*
urinary system *252*
urine *246*

KNOWLEDGE AND COMPREHENSION QUESTIONS

1. What is excretion? Explain its importance to life. (*Comprehension*)

2. How do organisms with no excretory organs rid themselves of toxic waste materials? (*Knowledge*)

3. What are the primary metabolic waste products that you excrete? (*Knowledge*)

4. What do urea, urine, and uric acid have in common? How do they differ? (*Comprehension*)

5. Draw a diagram of a human kidney showing the location of the renal pelvis, collecting ducts, ureter, cortex, medulla, renal pyramids, and renal arteries. (*Knowledge*)

6. Summarize how your kidneys help your body conserve water. (*Comprehension*)

7. Explain why urine can reveal the use of certain drugs. (*Comprehension*)

8. Briefly summarize the process by which urine is formed. (*Comprehension*)

9. What is homeostasis? Explain how ADH and aldosterone are important in maintaining homeostasis. (*Comprehension*)

10. Describe two problems that can occur with kidney function. (*Comprehension*)

KEY
Knowledge: Recalling information.
Comprehension: Showing understanding of recalled information.

CRITICAL THINKING REVIEW

1. Where do you think the carbon dioxide used in the formation of urea comes from? Where does the remainder of excess carbon dioxide go to be excreted? (*Synthesis*)

2. Imagine you are the sole survivor of a ship that sank in the ocean, and you have managed to crawl onto a raft containing nothing but bottles of beer. After several days, you are very thirsty. Without water you may die. Should you drink the beer or the ocean water, or neither? (*Analysis*)

3. The "Just Wondering" box in this chapter discusses drug testing. Describe the process by which the breakdown products of drugs become a part of the urine. (*Application*)

4. Normally, glucose is not present in the urine. Persons with diabetes mellitus often show glucose in their urine upon testing. Why? Also, using the information in this chapter, propose a reasonable hypothesis as to why persons with poorly controlled diabetes urinate frequently. (*Analysis*)

5. Organisms with contractile vacuoles have a layer of mitochondria surrounding the vacuole. What role might you propose for the high concentration of mitochondria in this area? (*Synthesis*)

KEY
Application: Using information in a new situation.
Analysis: Breaking down information into component parts.
Synthesis: Putting together information from different sources.

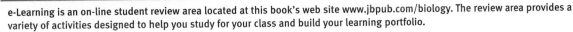

e-Learning is an on-line student review area located at this book's web site www.jbpub.com/biology. The review area provides a variety of activities designed to help you study for your class and build your learning portfolio.

Review Questions The review questions test your knowledge of the important concepts and applications in each chapter. The review provides feedback for each correct or incorrect answer. This is an excellent test preparation tool.

Figure-Labeling Exercises Sharpen your visual thinking by matching terms and labels for important illustrations.

Flash Cards Studying biology requires learning new terms. Virtual flash cards help you master the new vocabulary for each chapter.

Just Wondering As you read and study from this text, you may find that you have unanswered questions. Through this site you can ask the author, Sandy Alters, your "just wondering" questions.

 Do you prefer the speed and reliability of a CD-ROM? All of the features contained on the eLearning portion of the web site are also available on the Student CD-ROM.

Nerve Cells and How They Transmit Information

Nerves were decidedly on edge in the closing moments of the 1999 Women's World Cup Final. The Chinese and U.S. soccer teams were scoreless after 90 minutes of regulation play. Ten penalty kicks, five for each side, would decide the match. When Chinese midfielder Liu Ying walked up to face U.S. goalkeeper Briana Scurry, the match was tied 2-2. Scurry dove to her left to block Liu's penalty kick. Minutes later Brandi Chastain would fall to her knees in celebration of her tie-breaking goal, the one that gave the U.S. women the Cup at 5-4. Briana's save had set the stage for victory, but what set the stage for that all-important save?

Briana said later "I saw her body language when she was walking up to the penalty spot. She didn't look like she really wanted to be there. . . . I thought—this is the one." But what is it that translates anticipation into decision and then into action, in just a matter of seconds? It is a complex network of nerve cells spread throughout the body. In far less time than it takes 90,000 screaming spectators to rise to their feet, the nerves transmit messages from the brain to the muscles, making that famous lunge possible.

CHAPTER 13

Organization

Almost all animals have nerve cells.

The nervous system is the body's quick communication network.

13.1 Nervous transmission in invertebrates and vertebrates

With the exception of the sponges, both invertebrates (animals without a backbone) and vertebrates (animals with a backbone) have nerve cells, or **neurons** (NER-ons), and nervous systems or nerve networks. Invertebrate nerve cells have the same characteristics as vertebrate nerve cells. Additionally, all neurons look alike in that they have cell bodies containing a nucleus as well as long cell extensions (Figure 13.1).

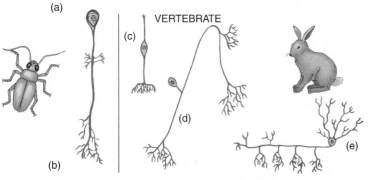

Figure 13.1   **A comparison of neurons from vertebrates and invertebrates.**
INVERTEBRATE (a) a neuron from the nerve net of a cnidarian and (b) an arthropod motor neuron. VERTEBRATE (c) a bipolar neuron from the vertebrate retina, (d) a mammalian spinal sensory neuron, and (e) a basket cell from the mammalian cerebellum.

Only the animals possess nerve cells, yet living things other than animals respond to their environments. For example, the single-celled ameba "knows" whether a particle it encounters is a grain of sand (and not its "dinner") or a bit of nutritious organic material. It makes this determination by *chemotaxis*, a kind of chemical sense. Although the ameba has no sense organs, it responds to chemical stimuli. This is the same response mechanism your white blood cells use when they encounter and phagocytize ("eat") a pathogen.

You have probably seen plants lean toward the light. This, too, is a type of response to the environment that has nothing to do with nerve cells. Plants bend toward the light due to the action of hormones. They have a variety of *hormones*, which are chemicals secreted in one part of an organism that have an effect in another part. Hormones are another method of internal communication in organisms.

> Only animals have nerve networks or nervous systems, but other organisms can respond to the environment using different mechanisms, including hormones.

13.2 Nervous and hormonal communication systems

The cells of vertebrates and most invertebrates communicate with one another in several ways, including by means of hormones and neurons. A complex communication system exists in humans, and most or all of these methods of communication occur in vertebrates as well as in many invertebrates.

In humans, a simple way in which cells communicate with one another is by direct contact: open channels between adjacent cells allow ions and small molecules to pass freely from one cell to another. This method of communication allows cells to interact with their neighbors, but is too slow and inefficient for interactions among cells that are far apart, such as between cells in different toes or in different fingers. This process can be likened to running a large corporation with only face-to-face interactions between the people who sit next to each other! It would be better, in terms of distant communication, if managers of the various departments within a corporation sent memos to their staff members, instructing them what to do.

The varied tissues within the human body communicate in a similar way by means of hormones. The hormone "memos" are produced by one of several different *endocrine glands*, secreted into the bloodstream, and carried around the body by the circulatory system. Each hormone, however, interacts only with certain target tissues, just as a memo that is sent through interoffice mail is delivered only to those persons to whom it is addressed. (Hormones are discussed in more detail in Chapter 17.)

Although hormones are an important means of communication within the human body, they do not serve all its communication needs. For example, if the message to be delivered to leg muscles is "Contract quickly, we are about to be hit by a car," hormones are too slow a message system. That is, a person in an emergency does not send a memo to the police, but uses the telephone to get help quickly. That, in effect, is what the nervous system does.

The quick message system of the body is the **nervous system**. The nervous system is made up of nerve cells, or *neurons*. Neurons are specialized cells that transmit signals throughout the body, forming networks such as the one shown on the chapter opener. Bundled in groups, the long cell extensions of neurons make up nerves. Their message signals are called **nerve impulses**. Like the dots and dashes of Morse code, all nerve impulses are the same, differing only in their frequencies, their points of origin, and their destinations.

The command center of the nervous system is the **brain**, a precisely ordered but complicated maze of interconnected neurons—a large biological computer. The brain is connected by a network of neurons to both the hormone-producing glands and the individual muscles and other tissues. This dual channel of command permits great flexibility: The signals can be slow and long-lasting (hormones), fast and short-lived (nerve signals), or any combination of the two.

Two forms of communication integrate and coordinate body functions in humans, as well as in all other vertebrate animals and most invertebrates. Hormones are chemical messengers that trigger widespread prolonged responses, often in a variety of tissues. Neurons are specialized cells that transmit rapid signals called nerve impulses, reporting information or initiating quick responses in specific tissues.

13.3 The nerve cell, or neuron

Despite the fact that individual neurons vary widely in size and shape within an individual organism and among organisms, all neurons have the same basic parts: a cell body and cell extensions called dendrites and axons. The **cell body** is the region of the neuron around the nucleus. The main organelles of the cytoplasm are in this portion of the cell. An **axon** is a single, long cell extension that often makes distant connections. It may give out branches, which usually split off at right angles, and usually has a myelin covering, a type of insulation. A **dendrite** is a cell extension that is usually much shorter than an axon, has no myelin covering (insulation), and is specialized to receive impulses from sensory cells or from axons of other neurons. Most neurons contain multiple dendrites. Some neurons have only a few cell extensions, while others are bushy with many projections.

Figure 13.2 illustrates some of the structural differences that exist among neurons in vertebrates. These differences are directly related to the varying functions of the individual nerve cells. **Sensory neurons** receive information and transmit it to the central nervous system (Figure 13.2a). Most of these neurons have one long axon bringing messages from particular receptors. (Certain specialized sensory neurons, such as some in the eye, do not fit this description and may have multiple dendrites bringing in messages.) The cell bodies of sensory neurons lie near the central nervous system.

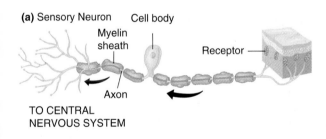

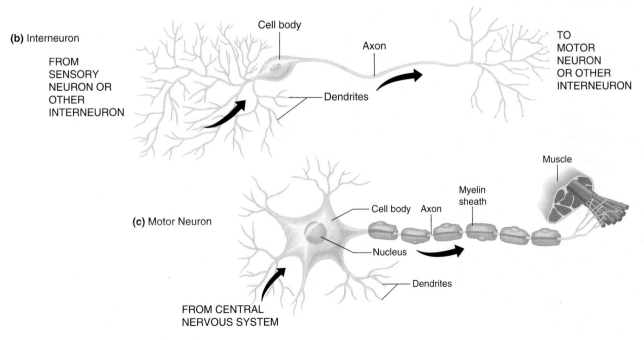

Figure 13.2 Types of vertebrate neurons.

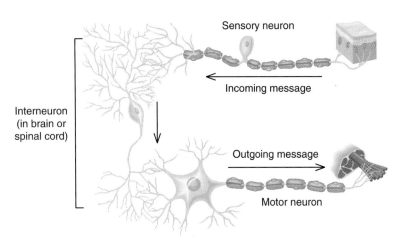

Figure 13.3 Interneurons provide a link between sensory neurons and the motor neurons.

Motor neurons transmit commands away from the central nervous system and to the muscles and glands (Figure 13.2c). Each of these neurons has one long axon, usually branched at the end, bringing messages to a muscle or a gland. The axons that control the muscular activity in the legs and feet can be more than a meter long (depending on height). Even longer axons occur in larger mammals. In a giraffe, such as the one shown at left, single axons extend from the neck to the toes in its front legs, and from the pelvis to the toes in its back legs, each spanning a distance of several meters. The cell bodies of most motor neurons lie in or near the central nervous system.

Interneurons, which are located within the brain or spinal cord, integrate incoming information with outgoing messages (Figure 13.2b and Figure 13.3). These neurons usually have a highly branched system of dendrites, which is able to receive input from many different neurons converging on a single interneuron. The axons of interneurons may not be myelinated and are usually much shorter than the axons of sensory and motor neurons that travel from the central nervous system to the muscles and glands.

The cell body of a neuron has the structures typical of a eukaryotic cell, such as a membrane-bound nucleus. Surrounding the nucleus is cytoplasm, which contains the various cell organelles. The cell body is responsible for producing substances that are necessary for the nerve cell to live. The rough endoplasmic reticulum, for example, makes proteins that are used for the growth and regeneration of the nerve cell processes, or extensions (rough ER is not shown in Figures 13.2 and 13.3). Substances such as these are able to leave the cell body and travel down the cell processes of the neuron, riding on "currents" within the cytoplasm or along "tracks" made by microtubules and microfilaments. In addition, some materials travel back to the cell body to be degraded or recycled.

Most neurons do not exist alone, but have companion cells nearby. These companion cells are called *neuroglia* (nuh-ROG-lee-uh) and provide nutritional support to neurons. Special neuroglial (nuh-ROG-lee-uhl) cells called **Schwann cells** are wrapped around many of the long cell processes of sensory and motor neurons of the peripheral nervous system. Such long cell extensions are called *nerve fibers*.

The fatty wrapping created by multiple layers of many Schwann cell membranes is a tissue called the **myelin sheath** (MY-eh-len). The myelin sheath insulates the axon; however, the Schwann cells are wrapped around the axon in such a way that uninsulated spots occur at regular intervals (Figure 13.4). These uncovered spots are called **nodes of Ranvier** (ron-VYAY). These nodes and the myelin sheath (discussed in more detail later in this chapter) create conditions that speed the nerve impulse as it is conducted along the surface of the axon. Many invertebrates, such as insects, earthworms, and crabs, have insulation covering their neurons that is quite similar to the myelination of vertebrate neurons.

A nerve cell, or neuron, is made up of a cell body that contains a nucleus and other cell organelles and two types of cellular projections: axons and dendrites. Axons are single, long cell extensions that are usually insulated and often make distant connections. Dendrites are cell extensions that are usually much shorter than axons, are not insulated, and are specialized to receive impulses from sensory cells or from axons of other neurons. Most neurons contain multiple dendrites. Structural differences exist among neurons and relate to their functional differences.

13.4 An overview of how neurons work

How do nerve cells conduct nerve impulses? The answer is not obvious from examining the amazingly complex network of nerve cells that extends throughout the body. Looking at this network and trying to understand how the nervous system operates is a little like gazing at a telephone wire and trying to understand how it transmits your voice to a receiver far away—nothing can be seen moving along the wires. The key to understanding how your telephone (and a nerve cell) transmits information is an understanding of the abstract concept of electricity.

Electricity is a form of energy—an invisible form—and you can only see, hear, or feel its effects. Electricity is invisible because it is a flow of electrons, subatomic particles so minute that they cannot be visualized even with a high-powered microscope. In a telephone line, a flow of electrons carries information from your telephone to the receiver of the person whom you called.

The internal components of the telephone change the sound energy of your voice into electrical energy, transmit it over the telephone lines, and then change it back to sound energy when it reaches its destination.

Neurons carry information in a similar way. The job of nerve cells is to transmit information from the environment to the spinal cord and brain, from one cell to another within the brain, and from the brain and spinal cord to other parts of the body. Nerve cells transmit this information in the form of electrical signals. The stimulation of specialized receptor cells, such as the rods and cones in your eyes, or specialized receptor endings of nerve cells, such as pressure receptors in your skin, causes electrical signals to be generated in these cells. Once an electrical signal is generated in a receptor, it can travel in the nervous system.

> Nerve impulses are electrical signals transmitted along the membranes of nerve cells, or neurons, within the nervous system.

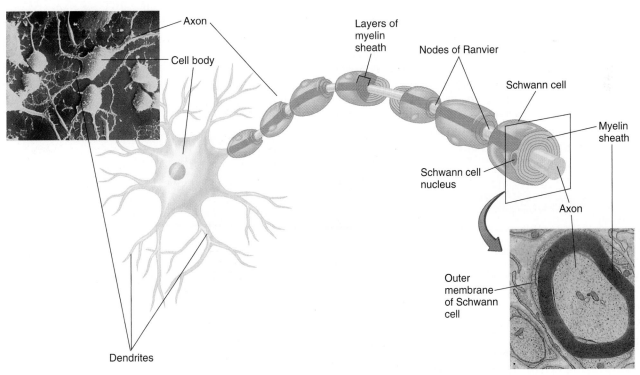

Figure 13.4 The insulation of axons. Usually, axons are encased at intervals by Schwann cells, which form the myelin sheath. The myelin sheath has many layers, visible in the electron micrograph, that insulate the axon. The nerve impulse jumps from uninsulated node to uninsulated node and so moves very rapidly along the nerve fiber.

The nervous system is the body's quick communication network.

13.5 The neuron at rest: The resting potential

How do neurons conduct electricity? The story begins with the neuron at rest—a neuron not conducting an impulse.

While at rest, a neuron is electrically charged. This electrical charge can be measured in the laboratory (by means of microelectrodes) and is approximately −70 millivolts (mV) (**Figure 13.5**, left). The negative charge means that the inside of the cell (near the membrane where the microelectrode is placed) is negatively charged relative to extracellular fluid along the outside of the membrane (where a reference electrode is placed). A millivolt is one thousandth of a volt, the unit measure of electrical potential.

The term *electrical potential* refers to the amount of potential energy created by a separation of positive and negative charges (in this case, along the inside and outside of the cell membrane of the neuron). The electrical potential of the nerve cell membrane is called the *membrane potential*. This potential energy comes from the work that is done to separate these opposite charges that are attracted to one another. For example, work is necessary to roll a boulder up a hill. Some of the actively working energy (kinetic energy) used to roll the boulder uphill is stored in the boulder (potential energy) as it sits at the top of the hill. In the same way, positive and negative charges that are separated from one another possess potential energy. The energy it takes to separate these oppositely charged particles is released if they rejoin.

The charges that are separated from one another along the nerve cell membrane are carried as ions—atoms with unequal numbers of protons and electrons (see Chapter 3). The ions that play the principal role in the development of the electrical potential along the membrane of the neuron (and that are also important to many other body processes) are sodium ions (Na+) and potassium ions (K+).

Embedded within the cell membranes of neurons are enzymes (proteins) called **sodium-potassium pumps** that actively transport these ions across the cell membrane (see Chapter 4 and Figure 13.5). These transmembrane proteins use the energy stored in molecules of adenosine triphosphate (ATP) to move potassium ions into the neuron at the same time that they move sodium ions out of the neuron ❶. However, the potassium ions can simply diffuse back out of the cell through voltage-gated channels or tunnels open to them ❷. Put simply, potassium ions actively enter the cell through one type of "door," but then exit by another.

The situation with the sodium ions is different. These

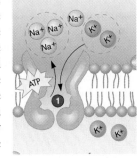

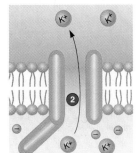

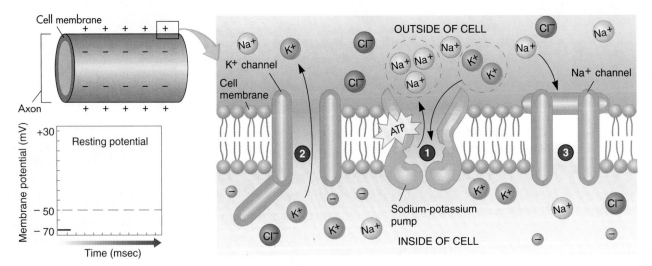

Figure 13.5 The sodium-potassium pump helps maintain the resting potential in a neuron.

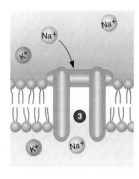

ions move back into the cell very slowly. The voltage-gated channels in the neuron membrane through which the sodium ions pass are closed at −70 mV (the resting potential) ❸. Other sodium channels (that do not work based on voltage) *are* open, however. These channels allow sodium to leak *very* slowly back into the cell, but the small amount of leakage is offset by the sodium-potassium pump. Negatively charged chloride ions (Cl⁻) are attracted to the positively charged sodium and follow the sodium ions slowly out of the cell.

The sodium-potassium pump works as a long-term mechanism that lowers the concentration of positively charged sodium ions within its interior. Other complex mechanisms work to maintain the membrane potential on a short-term basis. This variety of mechanisms establishes a chemical gradient of sodium ions, with a higher concentration of sodium ions outside the cell than inside the cell. There are also many other negative ions that stay within the cell. Proteins have a negative charge, low permeability, and are most abundant inside the cell. Therefore, an electrical gradient is also established; the inside of a nerve cell along its membrane is more negatively charged than the outside of the cell along the membrane. The membrane of the neuron is said to be *polarized*, which means there is a difference in charge on the two sides of the membrane. Resting neurons—ones not conducting an impulse—have this difference in electrical charge on either side of the membrane. The difference in charge is called the **resting potential**. This electrical potential difference across the membrane is the basis for the transmission of signals by nerves.

The action of sodium-potassium transmembrane pumps and ion-specific membrane channels produces conditions that result in the separation of positive and negative ions along the inside and outside of the nerve cell membrane. This separation of charged particles creates an electrical potential difference, or electrical charge, along the membrane of the resting neuron. This electrical potential difference is defined as the resting potential.

13.6 Conducting an impulse: The action potential

A neuron transmits a nerve impulse when it is excited by an internal or external environmental change called a *stimulus* (pl., stimuli). Examples of stimuli are pressure, chemical activity, sound, and light. Specialized receptors detect stimuli. For example, the rods and cones in the retina of your eye are sensitive to light, pressure receptors in your skin allow you to feel a hug, and certain cells in your nose allow you to smell your favorite dessert baking in the oven.

Stimuli cause a nerve impulse to be transmitted by initiating events that change the electrical potential difference of the receptor cell or nerve cell membrane at that spot. This change in electrical potential difference is called depolarization.

Appropriate stimuli open ion channels within the membrane of the receptor cell, thus depolarizing it. This depolarization is called a generator potential. A change in voltage across the nerve cell membrane opens and closes the transmembrane sodium (Na⁺) channels. You can think of these channels as being operated by electricity. What actually happens is that the sudden change in voltage changes the shape of the proteins forming the channels. As their shape changes, these channels become permeable to sodium ions and allow them to diffuse into the cell.

As the sodium channels open, a few sodium ions move rapidly into the cell, diffusing from where there are more sodium ions (outside the cell) to where there are fewer (inside the cell). This permeability for sodium ions across the cell membrane begins a depolarization of the nerve cell membrane. When a neuron is sufficiently stimulated, depolarizing the membrane to a level called the **threshold potential** (see **Figure 13.6a**), it initiates an action potential in the sensory neurons with which the receptor communicates. The **action potential** is a rapid reversal in the membrane's electrical potential across a portion of the membrane.

Amazingly, although only about 1 in 10 million ions actually moves across the membrane, the interior of the cell develops a positive charge of approximately 30 mV relative to the outside (see Figure 13.6b), a 100 mV electrical difference from the −70 mV resting potential. This electrical difference occurs because a new electrical potential develops when the permeability for sodium ions develops across the cell membrane. Although it may seem hard to believe, the depolarization of the cell membrane (the action potential) lasts for only *a few thousandths of a second* (milliseconds) because the sodium channels close quickly. They cannot reopen until after the resting potential is reestablished and another depolarization occurs, triggering them again. During this inactive state of the sodium channels, a nerve impulse cannot be conducted. This period, which is only milliseconds long, is called the *refractory period*.

When the sodium channels close, potassium ions move outward as they usually do through potassium channels. Many neurons contain voltage-sensitive potassium channels that open as the membrane depolarizes and the sodium channels shut down. It is this event—the movement of potassium ions out of the

cell—that repolarizes the membrane (see Figure 13.6c). The sodium-potassium pump (see Figure 13.5) works to maintain the resting potential on a more long-term basis. The whole process of stimulation, depolarization, and recovery in one section of a neuron happens in the blink of an eye. In fact, 500 such cycles could occur in the time it takes you to say the words "nerve impulse."

An action potential at one point on the nerve cell membrane is a stimulus to neighboring regions of the cell membrane. The change in membrane potential causes sodium channels to open, depolarizing the adjacent section of membrane. Put simply, depolarization at one site produces depolarization at the next. In this way, the initial depolarization passes outward over the membrane, spreading out in all directions from the site of stimula-

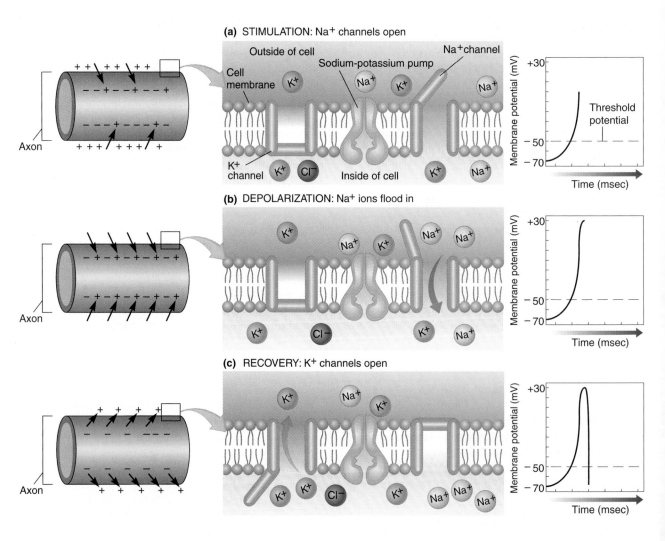

Figure 13.6 **Stages of the action potential.**

tion. Like a burning fuse, the signal is usually initiated at one end and travels in one direction, as shown in **Figure 13.7**, but it would travel out from both directions if it were lit in the middle. The self-propagating wave of depolarization that travels along the nerve cell membrane is the nerve impulse.

The nerve impulse is an *all-or-nothing response*. The amount of stimulation applied to the receptor or neuron must be sufficient to open enough sodium channels to generate an action potential. Otherwise, the cell membrane will simply return to the resting potential. For any one neuron the action potential is always the same. But any stimulus intense enough to open a sufficient number of sodium channels will depolarize the membrane to the threshold level and initiate an impulse. At this point, the neuron is said to have fired. This wave of depolarization has a constant amplitude (the height or strength of the wave). The speed at which impulses are conducted, however, varies among nerves.

A nerve impulse arises when a receptor is stimulated. This stimulation results in a rapid change in the electrical potential difference along the nerve cell membrane. This temporary disturbance, or depolarization, has electrical consequences to which nearby transmembrane proteins respond, spreading the depolarization in wavelike fashion.

13.7 Speedy neurons: Saltatory conduction

As mentioned earlier, many neurons have axons that are covered by neuroglial Schwann cells. These cells envelop many of the long cell processes of sensory and motor neurons and certain interneurons as well **(Figure 13.8)**, wrapping their cell membranes around them so many times that they form stacks of lipid-rich layers. In fact, some neurons have as many as 100 membrane layers surrounding them! This "cell wrapper" is the myelin sheath.

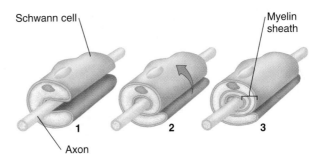

Figure 13.8 Schwann cell formation. The progressive growth of the Schwann cell membrane around the process contributes the many membrane layers characteristic of myelin sheaths.

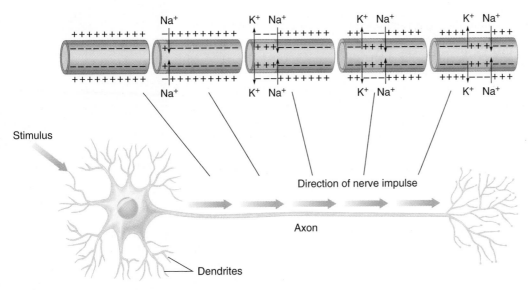

Figure 13.7 Transmission of a nerve impulse: Propagation of the action potential. Depolarization moves along a neuron in a self-propagating wave.

The nervous system is the body's quick
communication network.

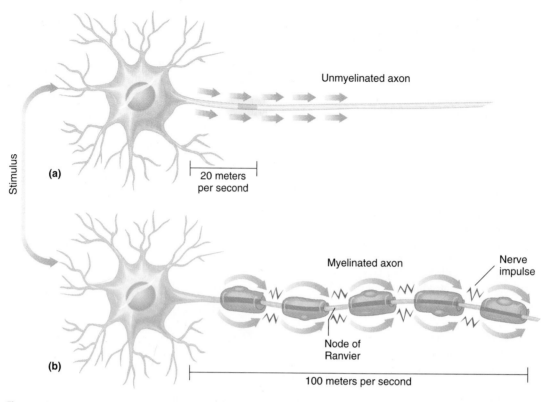

(a) Stimulus

Unmyelinated axon

20 meters
per second

(b) Stimulus

Myelinated axon

Nerve
impulse

Node of
Ranvier

100 meters per second

Figure 13.9 **Saltatory conduction.** (a) In a fiber without Schwann cells, each portion of the
membrane becomes depolarized in turn, like a row of falling dominoes. (b) In neurons of the same
diameter as in (a), the nerve impulse moves faster along a myelinated fiber because the wave of
depolarization jumps from node to node without ever depolarizing the insulated membrane
segments between nodes.

Schwann cells wrap around the length of an axon,
one after the other, with spaces separating one from the
next (**Figure 13.9**). These spaces, the nodes of Ranvier, are
critical to the propagation of the nerve impulse in
myelinated cells. The myelin sheath is an insulator; it
prevents the transport of ions across the neuron membrane beneath it. However, within the small gaps between Schwann cells, the surface of the axon is exposed
to the intercellular fluid surrounding the nerve. An action potential can be generated only at these gaps. In
fact, the pumps and channels that move ions across the
neuron membrane are concentrated at the nodes of Ranvier, enhancing ion movements at these spots.

The action potential moves along the nerve cell membrane differently in myelinated cells than in unmyelinated
cells. The wave of membrane depolarization that travels
down the axon of unmyelinated neurons is impossible in
myelinated axons due to the myelin sheath, which acts as
an electrical insulator. Instead, the action potential jumps
from one node to the next
(much as the mountain
climber to the right rap-
pels down the cliff), causing a depolarization only
at these specific points
(see Figure 13.9b). This
depolarization opens the
voltage-sensitive sodium
channels at that node, resulting in the production
of an action potential.
This very fast form of
nerve impulse conduc-
tion is known as saltatory
conduction, from the
Latin word saltare, mean-
ing "to jump."

Impulses conducted by myelinated neurons travel much faster than impulses conducted by nerve fibers of the same diameter without this insulation. In fact, impulses conducted in large-diameter myelinated neurons travel up to 270 miles per hour (120 meters per second). These myelinated neurons can transmit a signal from your toes to your brain in a fraction of a second!

Multiple sclerosis (MS), one of the leading causes of serious neurological disease in adults, affects approximately 1 out of every 800 people in the United States. This disease results in the destruction of large patches of the myelin sheath around neurons of the brain and spinal cord. Left behind are hardened scars (scleroses) in multiple places that interfere with the transmission of nerve impulses. The ensuing slowed transmission of signals in the nervous system results in a gradual loss of motor activity and an alteration in the way in which impulses are interpreted in the brain. Although the exact cause of MS is unknown, many researchers suspect that infection with certain viruses, such as the herpes virus that causes roseola in children, may be a trigger. Such a virus may lay dormant in the body, stimulating the immune system later in life to attack not only the virus, but also the nerve-insulating myelin as well.

> Not all nerve impulses propagate as a continuous wave of depolarization spreading along the neuronal membrane. Along myelinated neurons, impulses travel by jumping along the membrane, leaping over insulated portions. Impulses travel much more quickly along these myelinated neurons than along unmyelinated ones.

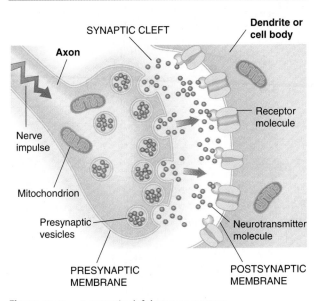

Figure 13.10 A synaptic cleft between neurons.

13.8 Transmitting information between neurons and other cells

When the nerve impulse reaches the end of an axon, it must be transmitted to another neuron or to muscle or glandular tissue. Muscles and glands are called *effectors* because they effect (or cause) responses when stimulated by nerves. This place—where a neuron communicates with another neuron or an effector cell—is called the **synapse** (SIN-aps).

Most neurons do not actually touch other neurons or cells with which they communicate. Instead, there is a minute space (*billionths* of a meter across) separating these cells called the **synaptic cleft (Figure 13.10)**. The nerve impulse must cross this gap and does so by changing an electrical signal to a chemical signal. Chemical synapses are the prevalent type of synapse in humans (and all vertebrates). (Neurons that communicate with one another by means of electrical signals *do* touch one another; they have no synaptic cleft.)

The membrane on the axon side of the synaptic cleft is called the *presynaptic membrane*. In chemical synapses, when a wave of depolarization reaches the presynaptic membrane, it stimulates a flow of calcium into the cell. The sudden rise in the cytoplasmic concentration of calcium triggers the release of organic molecules called **neurotransmitters** into the cleft (see Figure 13.10). These molecules are stored in thousands of small, membrane-bound sacs located at the tips of the axon. Each sac contains from 10,000 to 100,000 molecules of neurotransmitter. These chemicals diffuse to the other side of the gap. Once there, they combine with receptor molecules in the *postsynaptic membrane* (associated with either a dendrite or a cell body) of the target cell. When they do, they cause ion channels to open.

Chemical junctions between neurons and other neurons or effector cells have a distinct advantage over direct electrical connections—flexibility. The chemical transmitters can vary in different junctions. Just as you might take an aspirin to stop headache pain or cough syrup to subdue a cough, different neurotransmitters result in different kinds of responses. In fact, more than 60 different chemicals have been identified that act as neurotransmitters or that act to modify the activity of neurotransmitters.

Neuron-to-Muscle Cell Connections

Synapses between neurons and skeletal muscle cells are called **neuromuscular junctions** as shown in **Figure 13.11**. The neurotransmitter found at neuromuscular junctions is *acetylcholine* (uh-SEE-tl-KOH-leen). The nerve impulse travels down the axon, reaches the axon tip ❶, and stimulates a flow of calcium into the cell, which triggers the release of acetylcholine (ACh) from presynaptic vesicles ❷. Passing across the gap, the acetylcholine molecules bind to receptors in the postsynaptic (muscle cell) membrane, opening sodium channels ❸. This influx of sodium ions depolarizes the muscle cell membrane, which initiates a wave of depolarization that passes down the muscle cell ❹. This wave of depolarization releases calcium ions, which in turn trigger muscle contraction.

After an impulse has been transmitted across the synaptic cleft, the neurotransmitter must be broken down or the postsynaptic membrane will remain depolarized. The breakdown products of the neurotransmit-

ter then diffuse or are actively transported from the postsynaptic cell. In general, some or all of the breakdown products are transported back to the presynaptic cell to be reused. For example, the neurotransmitter acetylcholine is broken down to acetate and choline by an enzyme called *acetylcholinesterase* (uh-SEE-tl-koh-lin-ESS-ter-ase) ❺. Choline is transported back to the presynaptic cell, where it is used to make molecules of acetylcholine ❻. Acetylcholinesterase is one of the fastest-acting enzymes in the body, breaking down one acetylcholine molecule every 40 microseconds. The fast work of acetylcholinesterase permits as many as 1,000 impulses *per second* to be transmitted across the neuromuscular junction.

Nerve gases and the agricultural insecticide parathion work by blocking the action of acetylcholinesterase. These chemicals can cause death because they produce continuous neuromuscular transmission, which results in a continuous muscular contraction of vital muscles such as those involved in breathing and the circulation of blood. Many drugs also work by affecting synapses. For example, cocaine, local anesthetics, and some tranquilizers work by destroying the control of neurotransmitter release. (See Section 13.9.)

Neuron-to-Neuron Connections

Impulses are transmitted from one neuron to another by a variety of neurotransmitters. Some neurotransmitters depolarize the postsynaptic membrane, which results in the continuation of the nerve impulse. This type of

Visual Summary Figure 13.11 **The sequence of events in synaptic transmission at a neuromuscular junction.**

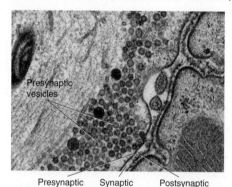

Presynaptic vesicles

Presynaptic membrane   Synaptic cleft   Postsynaptic membrane

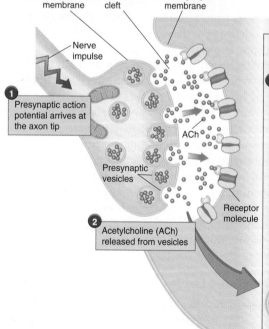

Nerve impulse

❶ Presynaptic action potential arrives at the axon tip

ACh

Presynaptic vesicles

Receptor molecule

❷ Acetylcholine (ACh) released from vesicles

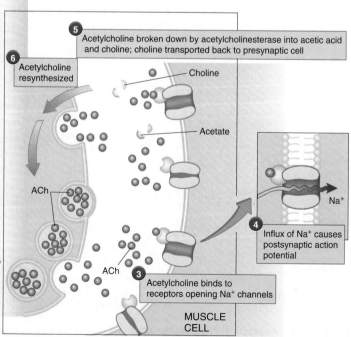

❺ Acetylcholine broken down by acetylcholinesterase into acetic acid and choline; choline transported back to presynaptic cell

❻ Acetylcholine resynthesized

Choline

Acetate

ACh

Na⁺

❹ Influx of Na⁺ causes postsynaptic action potential

ACh

❸ Acetylcholine binds to receptors opening Na⁺ channels

MUSCLE CELL

synapse is called an *excitatory synapse*. Other neurotransmitters have the reverse effect, reducing the ability of the postsynaptic membrane to depolarize. This type of synapse is called an *inhibitory synapse*. A single nerve cell can have *both* kinds of synaptic connections to other nerve cells. As you might expect, excitatory signals cancel out inhibitory signals, modifying each other's effects. The postsynaptic neuron keeps score as the impulses reach its dendrites and cell body, and it responds accordingly. For this reason, the postsynaptic neuron is called an *integrator* (Figure 13.12).

Within your body, synapses are organized into functional units with definite patterns, similar to electrical circuits. Just as your house is wired in a definite pattern to provide electricity to various appliances and light fixtures, your body is wired in a specific manner. However, the manner in which your body is wired is much more complex than that in your home, allowing you to maintain internal homeostasis as well as to adapt to the outside environment.

A synapse is a junction between an axon tip and another cell, usually including a narrow gap separating the two cells. Passage of the impulse across the gap is by an electrical current or, more likely in vertebrates, a chemical signal from the axon. At a neuromuscular junction, acetylcholine released from an axon tip depolarizes the muscle cell membrane, releasing calcium ions that trigger muscle contraction. The dendrites and cell body of a postsynaptic neuron integrate the information they receive from presynaptic neurons. The summed effect of both excitatory and inhibitory signals either facilitates depolarization or inhibits it.

13.9 How drugs affect neurotransmitter transmission

Drugs are substances that affect normal body functions. Although the specific actions of drugs vary, they all work by interfering with the normal activity of neurotransmitters. In this way, drugs affect communication between neurons or between neurons and muscles or glands.

One way in which drugs work is to decrease the amount of neurotransmitter that is released from a presynaptic neuron. Some drugs directly block neurotransmitter release. Other types of drugs work more indirectly, causing neurotransmitter molecules to leak out of their storage vesicles and then to be degraded by enzymes. Reserpine, a tranquilizer that also lowers blood pressure, is such a drug. It interferes with the storage of the neuro-

For the latest health information, please visit www.jbpub.com/biology.

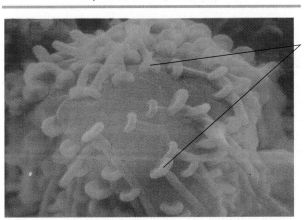

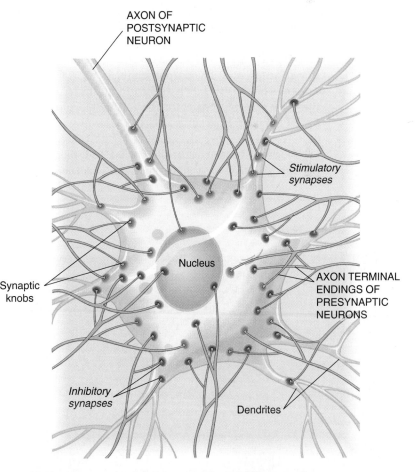

Figure 13.12 **Integration of nerve impulses takes place on the neuron cell body.** The synapses made by some axons are inhibitory, tending to counteract depolarization of the postsynaptic membrane; these synapses are indicated in *reddish-brown*. The synapses made by other axons are stimulatory, tending to depolarize the postsynaptic membrane; these synapses are indicated in *green*. The summed influences of these inputs determine whether the postsynaptic membrane will be sufficiently depolarized to initiate a nerve impulse.

Chapter 13 *Nerve Cells and How They Transmit Information*   **271**

transmitter norepinephrine (nor-EP-ih-NEF-rin), also called noradrenaline (nor-ah-DREN-ll-in). This neurotransmitter is one of two neurotransmitters active in the sympathetic nervous system, a branch of the autonomic nervous system that controls involuntary body functions (see p. 300). One of norepinephrine's functions within the sympathetic nervous system is the constriction of blood vessels, which increases blood pressure. Decreasing the amount of norepinephrine decreases the constriction of the blood vessels and lowers the blood pressure.

Some drugs increase the amount of a neurotransmitter or its effects at the synapse. These types of drugs may enhance the release of neurotransmitter molecules, inhibit the action of enzymes that degrade neurotransmitter molecules at the postsynaptic neuron, or chemically resemble the neurotransmitter and mimic its effects at the postsynaptic neuron. In all cases, the postsynaptic neuron becomes or remains stimulated by the neurotransmitter or its mimic. Amphetamines, drugs that stimulate the brain, are of this type and are described in more detail in the next section.

Some drugs that chemically resemble specific neurotransmitters act in still another way. These mimics occupy receptor sites, but do not stimulate postsynaptic neurons. They simply block the neurotransmitter molecules from the sites. Therefore, these drugs block the effects of the neurotransmitter.

Drugs are substances that affect normal body functions by interfering with the normal activity of neurotransmitters.

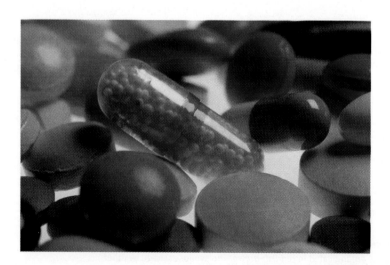

13.10 Drug addiction

Psychoactive drugs are chemical substances that affect neurotransmitter transmission in specific parts of the brain. Some psychoactive drugs are used medically to alter moods or to treat diseases or disorders. For example, the drug diazepam (die-AZ-eh-pam), commonly known as Valium, is used to control anxiety. Imipramine (ih-MIP-rah-meen) is a psychoactive drug that works as an antidepressant. Morphine is sometimes used to control pain after surgery or in terminally ill cancer patients. However, many psychoactive drugs (such as morphine) are abused. That is, they are used for nonmedical reasons, are taken in doses that may cause damage to the body, and often result in personally destructive, antisocial, and crime-related behaviors.

The chronic use (usually abuse) of psychoactive drugs results in **drug addiction**: a compulsive urge to continue using the drug, physical and/or psychological dependence on the drug, and a tendency to increase the strength (dosage) of the drug. Persons physically dependent on a drug show symptoms of this dependence when they stop taking the drug. These symptoms are called withdrawal symptoms; their effects are usually opposite to the effects caused by the drug (See Just Wondering, p. 276). For example, if a drug relieves pain, those physically addicted to it become hypersensitive to pain when they stop taking the drug. Withdrawal from drugs that cause euphoria, a feeling of intense well-being, will result in depression. Unfortunately, these feelings are often the ones addicts try to relieve by taking the drug in the first place. Symptoms of withdrawal become a new stimulus for the drug-taking response, resulting in a cycle of addiction that is hard for the addict to break.

Chronic drug abusers also tend to increase the dosage of the drug they take because they become drug tolerant. **Drug tolerance** is a decrease in the effects of the same dosage of a drug in a person who takes the drug over time. Therefore, the chronic abuser of a drug must take increasingly higher dosages of a drug for it to continue to elicit the same response. One reason behind the phenomenon of drug tolerance is that the chronic use of a drug stimulates liver enzymes to degrade, or break down, a drug with increasing swiftness. Another metabolic reason for drug tolerance is that brain cells become less responsive to a drug over time. As drug abusers increase the amount of the drug they take, dangerous side effects appear more often and become stronger.

Reverse tolerance occurs with some drugs such as marijuana and hallucinogens. With reverse tolerance, the drug abuser feels the desired effects from lesser amounts of the drug. One reason for reverse tolerance is that drugs such as marijuana are stored in fatty tissues and are later released as fat breaks down. Researchers also suspect that anticipation of the drug's effects plays a role in reverse tolerance.

Compulsive drug-seeking behavior and psychological dependence have sociological and psychological dimensions. Unfortunately, the biological factors involved in drug addiction, other than those already mentioned, are not clear. However, recent research on alcoholism (an addiction described in the next section) points to the conclusion that this addiction is a disease. Twin and adoptive studies, animal studies, and physiological studies of addicts all provide evidence supporting the hypothesis that a gene may exist that predisposes those inheriting this gene to alcohol addiction. It is hoped that future research will reveal further information and understanding regarding drug addiction that could lead to more effective treatments and preventive measures.

> The chronic use of psychoactive (mind-altering) drugs results in drug addiction, which is a compulsive urge to continue using the drug, physical and/or psychological dependence on the drug, and a tendency to increase the strength (dosage) of the drug.

13.11 Psychoactive drugs and their effects

Table 13.1 lists the major classes of psychoactive drugs. Their actions relate to the neurotransmitters to which they are chemically similar and to the parts of the brain or spinal cord that have receptors for those neurotransmitters. Just as neurons communicate with skeletal muscles by secreting acetylcholine at neuromuscular junctions, neurons in specific parts of the brain and spinal cord communicate by means of specific neurotransmitters. After reading Chapter 14, which will help you understand brain anatomy, you may choose to reread this discussion of drugs.

Depressants

Depressants are drugs that slow down the activity of the central nervous system (brain and spinal cord). The **sedative-hypnotics** are central nervous system depressants that induce sleep (sedatives) and reduce anxiety (hypnotics). This group of drugs interacts with gamma-aminobutyric acid (GABA), an inhibitory brain neurotransmitter. The interaction takes place at receptor sites found close together on postsynaptic neurons located in the amygdala (ah-MIG-dah-lah) of the limbic system and throughout the cerebral cortex. When GABA binds to these postsynaptic neurons, it slows the neurons' rate of firing. When a sedative-hypnotic drug binds to the postsynaptic neurons, it enhances GABA's inhibitory effects and results in calmness. Scientists are unsure, however,

Table 13.1

Major Classes of Psychoactive Drugs

Class	Type	Examples	Street Names
Depressants	Sedative-hypnotic	Barbiturates (Seconal, Nembutal, Phenobarbital)	Downers, dolls
		Benzodiazepines (Valium)	
		Methaqualone	Ludes, sopors
	Alcohol	Beer, Wine	
Opiates		Opium	
		Morphine	Mexican brown, black tar
		Codeine	
		Heroin	
Stimulants		Amphetamines	Uppers, speed, bennies
		Cocaine	Crack, ice
		Nicotine	
		Caffeine	
Hallucinogens		LSD	
		Marijuana (cannabis)	Pot, weed, grass
		PCP (phencyclidine)	Angel dust, trank
		Psychedelic mushrooms	

why the binding of a benzodiazepine (such as Valium) reduces anxiety whereas the binding of a barbiturate (such as Nembutal) produces sedation. However, barbiturates and benzodiazepines (BEN-zoh-die-AZ-eh-peenz) bind at separate sites near the GABA receptor site.

Sedative-hypnotics are used to treat sleep disorders, epileptic convulsions, and anxiety and are sometimes used as anesthetics for dental surgery. However, they are addictive drugs that can have dangerous side effects from either large single doses or prolonged use. Some of these side effects are the inability to think clearly, regression of the personality to childlike characteristics, emotional instability and irritability, unsteadiness when walking, and an indifference to personal hygiene. Because they suppress vital body functions, a sedative-hypnotic drug overdose can be fatal.

Another type of depressant drug is alcohol (ethanol). The amount of ethanol in an alcoholic beverage is designated by the word "proof"; the percentage of alcohol in the beverage is half that of the proof. Eighty-proof whiskey, for example, is 40% ethanol.

About 20% of the alcohol a person consumes is absorbed through the stomach. It immediately enters the bloodstream and travels all over the body. The rest passes into the small intestine and is then absorbed.

Once in the bloodstream, ethanol, like other psychoactive drugs, is able to pass through the cells that make up the walls of the blood vessels within the brain. These cells, because of the way they are joined together, screen out many substances harmful to the brain. However, lipid-soluble drugs (all psychoactive drugs) are able to pass through the cells themselves. Once in the brain, ethanol acts at the same binding site as barbiturates and has sedating effects, as do barbiturates. These close interactions may account for the potentially lethal interactions of alcohol or barbiturates and benzodiazepines. In other words, it is extremely dangerous to take benzodiazepines (such as Valium) with alcohol or a barbiturate (sleeping pill) because one enhances the activity of the other and at relatively low doses the combination can lead to death.

The effects of alcohol include increased heart rate, loss of alertness, blurred vision, and decreased coordination. Persons who use alcohol on a regular basis can proceed through the stages of addiction (use/tolerance/dependency/abuse) and display the characteristics of addicts described previously. Long-term use of alcohol can result in liver damage, ulcers, inflammation of the pancreas, nutritional disorders (due to both the behavior of the alcoholic and the metabolic changes that result from alcoholism), heart disease, and, in pregnant women, children who exhibit fetal alcohol syndrome (see p. 518).

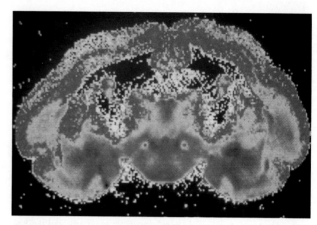

Figure 13.13 Opiate receptors in the guinea pig brain. This scan of guinea pig brain shows the concentration of opiate receptors; the highest density is indicated in red; yellow indicates moderate density; blue indicates low density; purple and white indicate very low densities.

Opiates

The **opiates** are compounds derived from the milky juice of the poppy plant *Papaver somniferum* or their synthetic (human-made) derivatives. These drugs are *narcotic analgesics*. An analgesic is a drug that stops or reduces pain without causing a person to lose consciousness. A narcotic is a drug that produces sedation and euphoria; originally it referred only to the opiates. Today, however, the term narcotic is used more loosely and refers to any addictive drug that produces narcosis, a deadened or dazed state. It is used incorrectly when referring to stimulant drugs, which are described later in this chapter.

The opiates work by mimicking naturally occurring morphinelike neurotransmitters called *endorphins* (*en*dogenous *m*orphinelike substances). Endorphins help people cope with pain and help modulate their response to emotional trauma. They bind to opiate receptors concentrated in areas of the central nervous system that include (1) the portion of the thalamus that conveys sensory input associated with deep, burning, aching pain; (2) portions of the midbrain and spinal cord involved in integrating pain information; (3) portions of the limbic system that govern the emotions; and (4) portions of the brainstem that control respiratory reflexes, pupil constriction, cough suppression, and gastric secretion and motility. This distribution of opiate receptors is shown in a guinea pig brain in **Figure 13.13**.

Some opiates have therapeutic uses. Dextromethorphan, an ingredient in some cough medicines, is a synthetic opiate that stimulates brainstem receptors that control coughing. Paregoric, a drug given to infants and small children to control diarrhea and accompanying cramps, contains a small amount of opium that acts to control pain and decrease gastric secretion and motility. Codeine is sometimes prescribed for the control of pain. Taken in an uncontrolled way, however, the opiates are highly addictive and dangerous drugs.

Stimulants

Stimulants include the *amphetamines* and *cocaine*. Stimulant drugs enhance the activity of two neurotransmitters: norepinephrine and dopamine. These neurotransmitters function in brain pathways that regulate emotions, sleep, attention, and learning. The alerting, stimulating effects produced by these drugs relate to their action in the cerebral cortex and the action of norepinephrine in the sympathetic division of the autonomic nervous system (see p. 300). The euphoria abusers feel relates to the action of these neurotransmitters in the limbic area of the brain.

Stimulant drugs act by moving into presynaptic neurotransmitter storage vesicles, which causes norepinephrine and dopamine to move into the synaptic cleft. The drugs then block the breakdown and recycling of the neurotransmitter molecules, causing an increase of the neurotransmitter in the synaptic cleft. With no recycling of neurotransmitter, however, the neurotransmitters are depleted with long-term use. When their supply of neurotransmitters is gone, drug abusers must take more and more of the stimulant to achieve their "high." In fact, the nervous system eventually becomes so depleted of neurotransmitter that drug abusers cannot get through a day without the stimulant because their nervous systems are no longer working properly.

If an abuser begins to take in large amounts of a stimulant drug such as cocaine, the drug can cause a schizophrenialike mental disorder characterized by paranoia, the hearing of voices, and irrational thought. This disorder appears to be caused by an overstimulation of dopamine receptors, as shown in Figure 13.14. Chronic cocaine and amphetamine use can also lead to long-lasting and severe physical changes in the brain. Chronic cocaine use also causes severe damage to the tissues of the nose and the lungs, heart disease, epileptic seizures, and respiratory failure.

Despite the fact that cocaine and other stimulants can be extremely dangerous drugs, they have important therapeutic uses. The drug methylphenidate (Ritalin) is used in the treatment of attention deficit disorder. The seemingly paradoxical effects of this drug (hyperactive children are slowed down) are attributed to the drug's attention-focusing effects, allowing hyperactive children to focus their attention for longer periods of time and, therefore, resulting in their not moving quickly from one activity or place to the next. Another important therapeutic use of stimulant drugs is the use of cocaine as a local anesthetic. When injected into peripheral nerves, cocaine blocks conduction on nerve fibers that transmit sensation. For this reason, synthetic cocaine derivatives are used for local anesthesia in most dental and eye surgery.

Nicotine is also a stimulant drug, although it produces much milder effects than the amphetamines and cocaine. Along with stimulating the release of norepinephrine and dopamine, nicotine also affects the nicotinic receptors of the autonomic nervous system by affecting the release of acetylcholine in the presynaptic neurons. Nicotinic receptors are those that bind and respond to nicotine as well as acetylcholine. Depending on the dosage of nicotine, it may either increase or decrease the release of this neurotransmitter.

Nicotinic receptors are found in both divisions of the autonomic nervous system. Because the divisions of the autonomic nervous system oppose one another in their activities and the effect on acetylcholine release is variable, the effects of nicotine are often variable. For example, nicotine may cause the heart rate to increase or decrease. The result is that the heart rhythm becomes somewhat irregular.

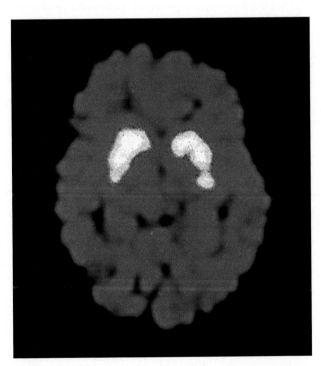

Figure 13.14 **PET scan of a human brain 15 minutes after taking cocaine.** The yellow areas indicate where the reward center (cocaine receptor area) is located.

Just Wondering . . . Real Questions Students Ask

What happens to your body when you quit using drugs?

When a person is addicted to a drug, his or her body becomes used to the drug's activity instead of the natural neurotransmitter activity the drug replaces. Under these conditions, the body actually suppresses its normal neurotransmitter function. In addition to causing these effects in the brain, drugs affect various organ systems of the body, often damaging tissues and making the organs more susceptible to disease. Therefore, when a person stops taking a drug, the body must adjust to the absence of the drug's activity and begin to use its own mechanisms once again. The bodily responses to these adjustments are referred to as withdrawal. In addition, the body begins healing the damage done to organs.

Withdrawal can be described as a rebound response. When a drug is no longer used, the neurotransmitter activity it supplied is also gone. Meanwhile, however, the body's normal neurotransmitter functions are still suppressed. In essence, the body swings from a high level of drug-related neurotransmitter activity to no activity or low levels of activity. The body reacts with symptoms such as tension, anxiety, anger, pain, and panic. The severity of these withdrawal symptoms is related to the amount of drug use and, consequently, the amount of suppression. Withdrawal effects also vary among individuals and may vary for a particular individual at different times in his or her life.

In addition to the generalized symptoms already mentioned, withdrawal may cause various other effects depending on the drug. For example, withdrawal from opiates results in certain physiological changes in the body, such as changes in blood pressure, whereas withdrawal from cocaine results in mood and behavioral changes. Withdrawal from alcohol and other depressants can be life-threatening and can result in convulsions; tremors of the hands, tongue, and eyelids; and other symptoms such as vomiting, heart palpitations, elevated blood pressure, and anxiety. One goal of drug withdrawal programs, therefore, is to stop the drug in a safe manner and allow the body's natural chemistry to return. Sometimes drug withdrawal must be managed with other drugs to achieve this goal.

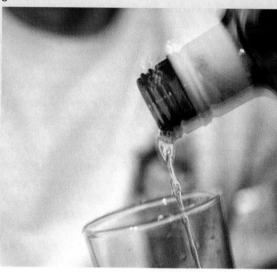

Alcohol, in a legal sense, is a "controlled" substance. Some people believe that it should be treated as a drug. The designation of "drug" carries with it more stringent control by the federal government.

The effects of nicotine on the brain and body are only one reason not to smoke cigarettes. The carbon monoxide in cigarette smoke replaces some of the oxygen in a smoker's red blood cells, thus interfering with the delivery of oxygen to the heart, brain, and other vital organs. In addition, smoking impairs the functions of the respiratory system. Cigarette smoke also contains a number of known cancer-causing agents and has been directly linked to cancers of the lungs, throat, and mouth. Cigarette smoking may also result in heart disease, stroke, high blood pressure, ulcers, earlier onset of menopause, excessive wrinkling of the skin, and sleep problems.

Hallucinogens

Hallucinogens, or psychedelic drugs, cause sensory perceptions that have no external stimuli; that is, a person hears, sees, smells, or feels things that do not exist. These drugs bear a close chemical resemblance to the neurotransmitters norepinephrine, dopamine, and serotonin (a transmitter involved with mood, anxiety, and sleep induction). However, scientists have not been able to pinpoint the exact mechanisms by which psychedelic drugs might affect the transmission of these neurotransmitters. The best hypothesis at this time is that hallucinogenic drugs act on two small nuclei in the brainstem (containing serotonin, norepinephrine, and dopamine neurons) that act as filtering stations for incoming sensory stimuli. These two nuclei are part of the reticular formation. Although small, these two nuclei give rise to axons that branch out to interact with billions of neurons in the cerebral cortex and cerebellum. Scientists think that hallucinogens may disrupt the sorting process and allow a surge of sensory data to overload the brain.

Hallucinogens can have devastating effects. For example, LSD, PCP, peyote, and mescaline can result in psychotic behavior (derangement of the personality and loss of contact with reality). These episodes can trigger chronic mental health problems or can result in suicide. Regular use of marijuana, the psychoactive plant *Cannabis*, can result in impaired eye-hand coordination, increased heart rate, panic attacks, anxiety, paranoia, depression, immune system impairment, upper respiratory system damage, and decreased levels of sex hormones. Regular use of this drug can also result in two syndromes:

1. *acute brain syndrome*, a condition marked by perceptual distortions, sleep and memory problems, disorientation with regard to time and place, and the inability to concentrate or sustain attention to important stimuli in the environment, and
2. *amotivational syndrome*, characterized by apathy, fatigue, poor judgment, loss of ambition, and diminished ability to carry out plans.

Preliminary findings from a three-year study published in 1998, conducted by Dr. Ronald Kadden and co-workers at the University of Connecticut Health Center in Farmington, suggest that marijuana is addictive (although most authorities believe that marijuana is not physically addictive). On stopping marijuana use, regular marijuana users often have physical symptoms similar to those experienced by persons withdrawing from alcohol, barbiturates, heroin, and cocaine. However, the symptoms tend to be less severe than with other abused substances.

The only hallucinogen shown to have therapeutic effects is marijuana. It has been found to relieve nausea in some patients undergoing chemotherapy treatment for cancer. It also appears to reduce pain and suffering for persons with AIDS, and reduces pressure within the eye in persons suffering from glaucoma, the leading cause of blindness in the United States. Arizona and California voters have approved laws that allow for the legal use of marijuana under a physician's supervision, but there are still federal laws denying legal medicinal use of this drug.

> The actions of psychoactive drugs relate to the neurotransmitters to which they are chemically similar and to the parts of the brain or spinal cord that have receptors for those neurotransmitters. Four major classes of psychoactive drugs are depressants, opiates, stimulants, and hallucinogens.

Key Concepts

Almost all animals have nerve cells.

13.1 With the exception of the sponges, both invertebrates and vertebrates have nerve cells and nervous systems or nerve networks.

13.1 Living things other than animals respond to their environments by mechanisms other than nerves.

The nervous system is the body's quick communication network.

13.2 The human body has two primary means of internal communication: hormones and nerve impulses.

13.2 Nerve impulses are quick and short-lived electrical signals; hormones are slower, persistent chemical signals.

13.3 Nerve cells have a cell body that contains structures typical of a cell and cellular extensions called axons and dendrites.

13.3 Axons are single, long cell extensions that often make distant connections; dendrites are cell extensions that are usually much shorter than axons, have no insulation, and are specialized to receive impulses from sensory cells or from axons of other neurons.

13.4 Nerve cells transmit information in the form of electrical signals.

13.5 A resting neuron has an electrical potential difference across its cell membrane that occurs because of the separation of positively and negatively charged ions along the inside and outside of the nerve cell membrane.

13.6 A neuron transmits an impulse when excited by an environmental change, or stimulus.

13.6 Specialized receptors detect stimuli and can initiate events that change the electrical potential difference along the nerve cell membrane, which is called depolarization.

13.6 If a neuron is depolarized to a certain threshold, voltage-sensitive membrane sodium channels open and sodium ions cross the cell membrane, further depolarizing the nerve cell membrane (action potential).

13.6 Within a few thousandths of a second following the depolarization of the nerve cell membrane, the voltage-sensitive sodium channels close and potassium ions move out of the cell, changing the electrical potential of the membrane once again (resting potential).

Key Concepts, continued

13.6 An action potential at one point on the nerve cell membrane causes depolarization of the adjacent section of membrane, which results in a "wave" of depolarization along the nerve membrane.

13.6 A self-propagating wave of depolarization is the nerve impulse.

13.7 The long cell processes of many neurons are wrapped in insulating lipids called the myelin sheath.

13.7 Spaces between insulated segments of axons help speed the nerve impulse as the action potential jumps from one space to the next.

13.8 When a nerve impulse reaches the axon tip of most vertebrate nerve cells, it causes the release of chemicals called neurotransmitters.

13.8 Neurotransmitters pass across a space called the synaptic cleft to the next cell.

13.8 In neuron-to-muscle cell connections, the neurotransmitter triggers muscle cell contraction.

13.8 In neuron-to-neuron connections, the neurotransmitter may cause either an excitatory response or an inhibitory response.

13.8 The integration of nerve impulses occurs on the cell body membranes of individual neurons, which receive both excitatory and inhibitory signals; the neuronal response depends on the mix of the signals received.

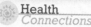 **Health Connections**

13.9 Drugs affect body functions by interfering with the normal activity of neurotransmitters.

13.10 Psychoactive drugs are chemical substances that affect neurotransmitter transmission in specific parts of the brain.

13.10 Some psychoactive drugs are used medically to alter the mood or to treat diseases or disorders.

13.10 Many psychoactive drugs are used for nonmedical reasons, are taken in doses that may cause damage to the body, and often result in personally destructive, antisocial, and crime-related behaviors.

13.10 The chronic use of psychoactive drugs results in drug addiction, which is a compulsive urge to continue using the drug, physical and/or psychological dependence on the drug, and a tendency to increase the strength (dosage) of the drug.

13.11 Major classes of psychoactive drugs are depressants (sedative-hypnotics, alcohol), opiates (opium, morphine), stimulants (amphetamines, cocaine), and hallucinogens (LSD, marijuana).

Key Terms

action potential 265
axon (AK-son) 261
brain 261
cell body 261
dendrite 261
depressant (dih-PRESS-unt) 273
drug addiction 272
drug tolerance 272
hallucinogens (huh-LOO-suh-nuh-jens) 276
interneurons 262

motor neurons 262
myelin sheath (MY-eh-len) 262
nerve impulses 260
nervous system 260
neuromuscular junctions (NOOR-oh-MUS-kyuh-ler JUNGK-shuns) 269
neurons (NER-ons) 260
neurotransmitters (NER-oh-TRANS-mit-urs) 269
nodes of Ranvier (ron-VYAY) 262
opiates (OH-pee-uts) 274

resting potential 265
Schwann cells 262
sedative-hypnotics (SED-uh-tiv hip-NOT-iks) 273
sensory neurons 261
sodium-potassium pumps 264
stimulants 275
synapse (SIN-aps) 269
synaptic cleft (sin-AP-tik KLEFT) 269
threshold potential 265

KNOWLEDGE AND COMPREHENSION QUESTIONS

1. What do neurons and hormones have in common? How are they different? (*Knowledge*)

2. Explain the term *resting potential*. What creates it, and why is it important? (*Comprehension*)

3. Explain how a nerve impulse travels through the body. (*Comprehension*)

4. What is saltatory conduction? Explain how it works and what advantages it offers. (*Comprehension*)

5. You decide to move your finger, and it moves. Explain how your nervous system communicated your decision to the muscles in your finger. (*Comprehension*)

6. In question 5, explain why your finger muscles stopped contracting when you wanted them to stop. (*Comprehension*)

7. Why are neurons of the central nervous system called *integrators*? What do they integrate? (*Comprehension*)

8. A drug may cause decreased communication between neurons or between neurons and muscles or glands. By what common action do drugs cause such varied effects? (*Knowledge*)

9. Susan takes sleeping pills regularly with no ill effects. She assumes that having a few drinks at a party is not overly dangerous either and plans to still use her regular bedtime sedative. Do you agree that this behavior is safe? Why or why not? (*Comprehension*)

10. How do cocaine and amphetamines affect neurotransmitters such as norepinephrine and dopamine? (*Knowledge*)

K E Y
Knowledge: Recalling information.
Comprehension: Showing understanding of recalled information.

CRITICAL THINKING REVIEW

1. Is the nerve impulse a chemical reaction? Explain the nature of such an impulse as you understand it. (*Synthesis*)

2. Basing your answer on drug action, how would you explain the fact that psychoactive drugs are among the most commonly abused type of drug, rather than nonpsychoactive drugs such as aspirin? (*Analysis*)

3. Can you hypothesize a link between drug tolerance and increased criminal or antisocial behavior among long-term addicts? (*Synthesis*)

4. Some people would like marijuana to be considered a legal drug in the United States for any purpose, not just for medicinal use. Evaluate this position by considering the health effects on the population if marijuana were legalized and its use increased. (*Evaluation*)

5. The nerve impulse is an all-or-nothing response, and for any one neuron the action potential is always the same. How do you explain, then, the fact that you can experience differing intensities of feeling such as a mild pain or a severe pain? (*Synthesis*)

K E Y
Analysis: Breaking down information into component parts.
Synthesis: Putting together information from different sources.
Evaluation: Making informed decisions.

For each numbered step, write a brief description of the process being shown in the spaces provided below the figure.

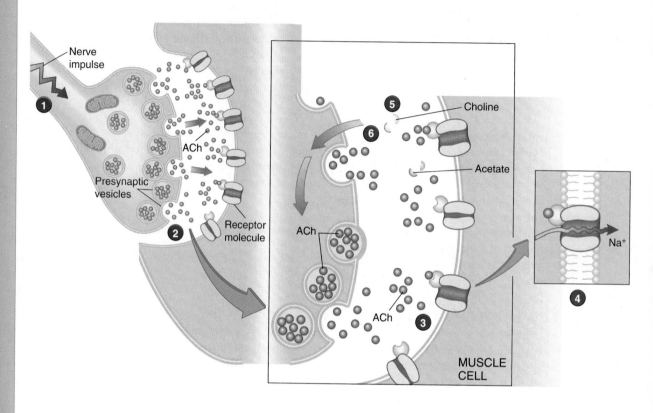

1 Description:

2 Description:

3 Description:

4 Description:

5 Description:

6 Description:

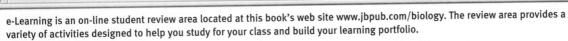

e·learning

Location: http://www.jbpub.com/biology

e-Learning is an on-line student review area located at this book's web site www.jbpub.com/biology. The review area provides a variety of activities designed to help you study for your class and build your learning portfolio.

Review Questions The review questions test your knowledge of the important concepts and applications in each chapter. The review provides feedback for each correct or incorrect answer. This is an excellent test preparation tool.

Figure-Labeling Exercises Sharpen your visual thinking by matching terms and labels for important illustrations.

Flash Cards Studying biology requires learning new terms. Virtual flash cards help you master the new vocabulary for each chapter.

Just Wondering As you read and study from this text, you may find that you have unanswered questions. Through this site you can ask the author, Sandy Alters, your "just wondering" questions.

Do you prefer the speed and reliability of a CD-ROM? All of the features contained on the eLearning portion of the web site are also available on the Student CD-ROM.

The Nervous System

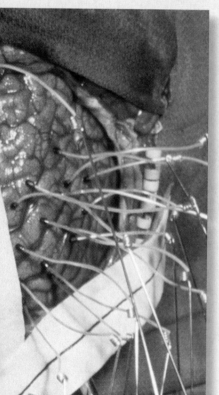

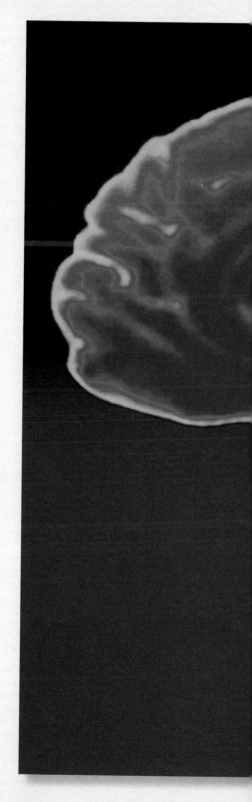

The photograph at left shows an exposed human brain undergoing an incredible procedure known as *electrocorticography.* This procedure is used to map the electrical activity of the outer, highly active layer of the brain called the *cortex*—the part of the brain that allows you to speak, smell, see, move, and remember. Sixteen electrodes are usually used in this mapping procedure, their tips fashioned into carbon balls supported by flexible wires connected to a sensing and recording device.

Epilepsy is a disorder in which patients experience recurrent episodes of sudden, high-frequency discharges of the brain's neurons. These discharges result in the patient experiencing seizures, or convulsions. During a seizure, the patient may lose consciousness, stare into space, fall to the floor, and experience involuntary muscular contractions.

Although 5% of the general population will have a seizure at some time in their lives, less than 1% will have chronic seizures classified as epilepsy. Of these persons, most will be able to manage their medical condition well with anticonvulsant drugs. However, some persons develop an epilepsy that is difficult to manage and eventually decide to have their disorder treated with surgery. For these patients, it is important that brain surgeons determine the focal area of their seizure activity so that just this part of the brain can be removed without taking tissue essential for important functions. Using various techniques to map brain function (including electrocorticography), brain surgeons are successful in about 80 percent of all epilepsy surgeries and help patients control otherwise unmanageable seizures. An MRI (magnetic resonance imaging) scan of a healthy brain is shown at right.

14

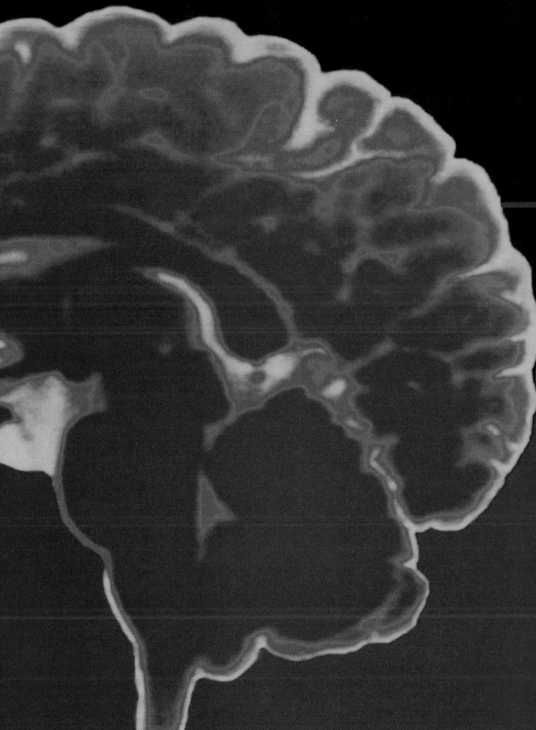

Organization

14.1 Invertebrate nervous systems

In vertebrates, the brain is only one part of the complex web of neurons known as the **nervous system**. A nervous system is a network of neurons (nerve cells) specialized for transmitting information from sensory receptors to neurons, between neurons, and from neurons to effectors, such as muscles and glands. The job of a nervous system is to gather information about an organism's internal and external environments, and then to process and respond to the information it has gathered. These nervous system responses are nerve impulses as described in Chapter 13—electrochemical commands sent out to muscles and glands, directing them to react in an appropriate way.

Although invertebrate neurons look and act very much like vertebrate neurons (see Figure 13.1), the nervous systems of invertebrates vary in their organization among invertebrate phyla. The simplest nervous system organization is the **nerve net**, found in the cnidarians (hydra, jellyfish, sea anemones).

A nerve net is simply a system of interconnecting nerve cells **(Figure 14.1)** with no central controlling area, or brain. Impulses are transmitted in all directions, resulting in a response in the stimulated region of the organism. Responses are simple, such as the withdrawing of tentacles. The stronger the stimulus, the greater the area of response, since stronger sensory impulses spread farther from the area of the stimulus than weaker impulses do.

Echinoderms (sea stars, sea urchins, sea cucumbers) are another phylum of invertebrates with no brain. The nervous system of echinoderms consists of a ring around the esophagus with five nerves radiating outward. In species with arms, such as sea stars, these nerves extend down each arm to the tube feet. Some species of echinoderms have a second nerve ring system that controls motor function.

The flatworms (*Platyhelminthes*), have a variety of nervous system plans. Some have a nerve net, while others have more complex plans. The most complex have a distinct brain and one to three pairs of nerve cords (also called trunks) that run the length of the animal. The planarian shown in Figure 14.1 has a small brain that sends nerve impulses down its single pair of nerve cords, resulting in a ladderlike nervous system.

The nematodes, or roundworms, also have simple brains connected to nerve cords. Such simple brains are aggregations of nervous tissue often referred to as *ganglia* (GANG-glee-uh), which serve as a central processing station for incoming and outgoing impulses. Ganglia are not capable of conscious thought or emotion, as is the human brain.

Annelids (earthworms and leeches) have a similar nervous system plan to the flatworms and roundworms: an anterior (head-end) brain attached to nerve cord(s). In the case of annelids, however, they have one or two nerve cords that run along the ventral ("stomach") side of the animal. These segmented worms also have segmental ganglia. That is, at each segment, the nerve cord has a swelling of nervous tissue, making this annelid nervous system a bit more complex than those of the flatworms and roundworms. Arthropods (insects, spiders, lobsters) have a similar plan: a well-defined ventral nerve cord with segmental ganglia and a prominent brain at the anterior end.

The molluscs—clams, squid, and octopus—have varying nervous system structures that correspond to the level of activity of the organism. For example, the slow-moving clam has a ring of ganglia surrounding its esophagus with nerve cords running to its major body parts. The fast-moving cephalopods such as squid (shown below) and octopus, on the other hand, have

large, complex brains. The brain of the common octopus, for example, has over 10 distinct lobes. Results of research studies also show that cephalopods are capable of memory and learning. Additionally, cephalopods have neurons specialized for rapid impulse conduction, allowing them to move with jetlike speed. The processes of these neurons are called giant fibers and connect the brain with the muscles. Giant fibers have also been important to humans; our present knowledge of how nerve impulses are generated and transmitted comes largely from studies of these cephalopod giant fibers.

The simplest nervous system organization is the nerve net, a system of interconnecting nerve cells with no central controlling area, or brain. Some phyla of invertebrates have simple brains attached to nerve cords.

NERVE NET
(all cnidarians [hydra, jellyfish,
sea anemones])

Hydra

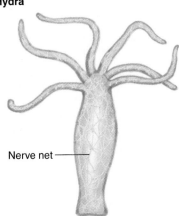

Nerve net

RINGLIKE NERVOUS SYSTEM
(all echinoderms [sea stars, sea urchins,
sea cucumbers])

Sea star

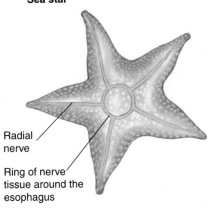

Radial
nerve

Ring of nerve
tissue around the
esophagus

LADDERLIKE NERVOUS SYSTEM
(many Platyhelminthes [flatworms])

Flatworm

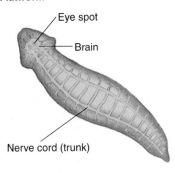

Eye spot

Brain

Nerve cord (trunk)

NERVOUS SYSTEM WITH SEGMENTAL GANGLIA,
VENTRAL NERVE CORD AND A BRAIN
(annelids [earthworms and leeches] and arthropods [insects,
spiders, and lobsters])

Earthworm

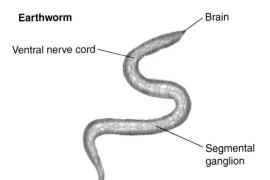

Brain

Ventral nerve cord

Segmental
ganglion

Grasshopper

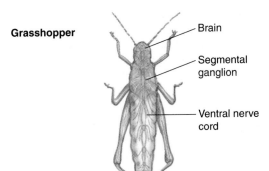

Brain

Segmental
ganglion

Ventral nerve
cord

CENTRAL (BRAIN AND SPINAL CORD) AND
PERIPHERAL NERVOUS SYSTEM
(all vertebrates)

Human

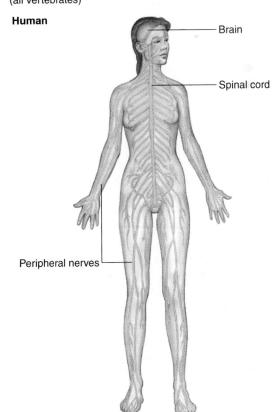

Brain

Spinal cord

Peripheral nerves

14.2 The vertebrate nervous system

Structurally, the vertebrate nervous system can be divided into two main parts (see Figure 14.1 and Figure 14.2):

1. the **central nervous system**—the site of information processing within the nervous system, which is made up of the **brain** and **spinal cord**, and

2. the **peripheral nervous system** (puh-RIF-uh-rul)—an information highway made up of nerves that bring messages to and from the brain and spinal cord. The nerves of the peripheral nervous system are made up of the long cell processes of nerve cells (nerve fibers), support cells (see Chapter 13), connective tissue, and blood vessels. The nerves of the peripheral nervous system contain the nerve fibers of two different types of neurons: **sensory neurons**, which transmit information to the central nervous system, and **motor neurons**, which transmit commands away from the central nervous system. Sensory neurons are also called *afferent neurons*, from the Latin prefix *affero* meaning "going toward." Motor neurons are also called *efferent neurons*, from the Latin prefix *effero* meaning "going away from."

One group of motor neurons controls voluntary responses, such as coordinating the movement of muscles in your legs so that you can cross the street. These responses are called voluntary because you consciously choose whether or not to do this activity. Motor neurons that control voluntary responses make up the **somatic nervous system** (soe-MAT-ik). The word somatic means "body" and refers to the fact that these neurons carry messages to your skeletal muscles—those that move the parts of your body. In addition, certain voluntary activities that may seem somewhat out of your control, such as blinking and breathing, are also directed by the somatic nervous system. Such activities are actually *reflexes*, automatic responses to stimuli that are mediated by the spinal cord or lower portions of the brain—those closest to the spinal cord.

Another group of motor neurons carries messages that control *involuntary responses*. These responses include such activities as mixing the food in your stomach with acid and enzymes after a meal and pumping adrenaline into your bloodstream when you dodge an oncoming car. Motor neurons that carry messages about involuntary activities make up the **autonomic nervous system** (awe-tuh-NOM-ik). The word autonomic comes

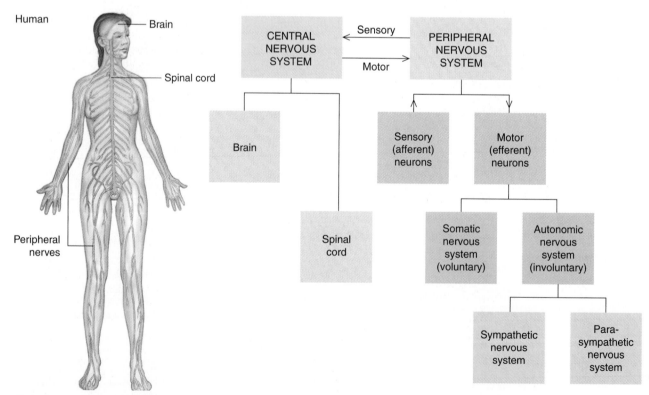

Figure 14.2 **The vertebrate nervous system.** The nervous system consists of the central nervous system (the brain and spinal cord) and the peripheral nervous system.

from Greek words meaning "self" (*auto*) "governing" (*nomos*). This portion of the nervous system, then, literally takes care of you—by itself! In general, it works to promote homeostasis (a "steady state") within your body. The autonomic nervous system accomplishes this feat by carrying opposing messages that either speed up or slow down the activities of your glands, heart muscle, and smooth muscles such as those found in the digestive, excretory, and circulatory systems. In fact, these opposing messages are carried on separate neurons, dividing the autonomic nervous system functionally into two parts: the sympathetic and parasympathetic systems. (The effect of nerve impulses from each of these systems on various organs is shown in Figure 14.14.)

> The vertebrate nervous system is made up of the central nervous system, consisting of the brain and the spinal cord, and the peripheral nervous system. Within the peripheral nervous system, sensory pathways transmit information to the central nervous system, and motor pathways transmit commands from it. One group of motor neurons makes up the somatic nervous system and controls voluntary responses. Another group of motor neurons makes up the autonomic nervous system and controls involuntary responses.

14.3 An overview of the central nervous system

In many ways, your central nervous system can be compared to the central processing unit (CPU) of your computer. Without your CPU, input (what you type on the keyboard, for example) is not processed by any of the programs in your computer. Therefore, you will have no output either on your monitor or from your printer. In a similar but much more complex way, your brain and spinal cord make sense of incoming sensory information and then produce outgoing motor impulses. This function of the central nervous system is called integration.

Each part of the central nervous system plays its own unique role in integration. Many simple nervous reactions, such as pulling your hand away from a hot stove, are integrated in the spinal cord. Other more complex nervous system reactions, such as breathing, are controlled in a lower portion of the brain called the *brainstem*. And the most complex or highest functions of the nervous system, such as thinking, remembering, and feeling, are all integrated in a portion of the brain called the *cerebrum* (seh-REE-brum).

Your brain weighs about 3 pounds and contains an amazing 100 billion (100,000,000,000) neurons, a number that does not include neuroglial (supporting) cells. Both types of cells work together to form the intricate structure of the brain. Although a complicated whole, the human brain can be described as having four main parts: the cerebrum, the cerebellum (sere-eh-BEL-um), the **diencephalon** (DIE-en-SEF-eh-lon) (thalamus and hypothalamus), and the brainstem. These four parts are shown in Figure 14.3.

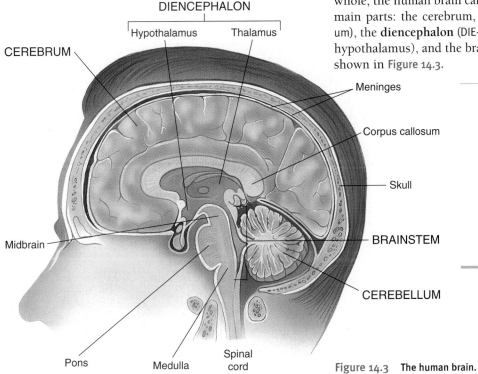

> The central nervous system makes sense of incoming sensory information and produces out- going motor impulses. Each part of the central nervous system— the cerebrum, cerebellum, diencephalon, brainstem, and spinal cord—plays its own unique role in integration.

Figure 14.3 **The human brain.**

14.4   The cerebrum and cerebellum

Cerebrum

The **cerebrum**, the dominant part of the human brain, is so large that it appears to wrap around and envelop the other three parts. In the brains of humans (and other mammals) the cerebrum is split into two halves, or hemispheres, the right and left sides of the brain. These two sides of the cerebrum are connected by a single, thick bundle of nerve fibers called the **corpus callosum** (KOR-pus kuh-LOW-sum). The corpus callosum, as shown in Figure 14.6, is a communication bridge that allows information to pass from one side of the brain to the other, so each side "knows" what the other is thinking and doing.

Most of the activity of the cerebrum takes place within the **cerebral cortex**, a thin layer of tissue that forms its outer surface. The word cortex actually means "rind" or "bark" and helps give a sense of the relative thickness of this tissue to the underlying cerebral tissue. Your cortex cap is densely packed with neuron cell bodies. Because these cell bodies and their dendrites are un-myelinated, the cerebral cortex appears gray and is referred to as *gray matter* (see Chapter 13 for a discussion of myelin). Many of these neurons send out myelinated axons; these form the central core of the cerebrum, which is mostly white matter.

Although the cerebral cortex is only a few millimeters thick, it has a tremendous surface area. It gains this surface area by lying in deep folds called *convolutions*, making the surface of the cerebrum look like a jumble of hills and valleys. These convolutions increase the surface area of the cerebral cortex 30 times. If laid out flat, it would cover an area of 5 square feet.

The cerebral cortex of each hemisphere of the brain is divided into four sections by deep grooves among its many convolutions. These sections are the **frontal lobe,**

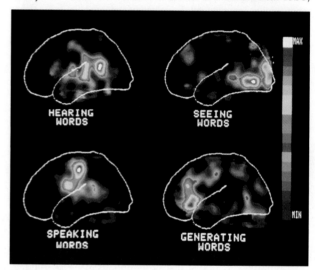

parietal lobe (puh-RYE-uh-tul), **temporal lobe**, and **occipital lobe** (ock-SIP-uh-tul) **(Figure 14.4)**. By examining the effect of injuries to particular sites on the cerebrum, scientists were first able to determine the approximate location of the various activities of each lobe of the cerebral cortex. More recently, using positron emission tomography (PET) scans, researchers have been able to determine more specifically the areas of the cerebral cortex that are used during various activities. The photo above shows PET scans of the brain of a person performing a series of intellectual tasks related to words. These scans show the increase in blood flow that occurs at the part of the brain performing the task. Red indicates the most intense activity; blue the least.

There are three major types of activities that take place within the lobes of the cerebral cortex: motor, sen-

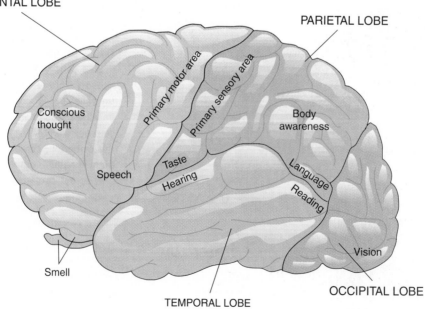

FRONTAL LOBE

PARIETAL LOBE

Conscious thought

Primary motor area

Primary sensory area

Body awareness

Speech

Taste

Hearing

Language

Reading

Vision

Smell

OCCIPITAL LOBE

TEMPORAL LOBE

Figure 14.4   **The cerebral cortex.**

sory, and association activity (Figure 14.5). The *motor area*, the part of the brain that sends messages to move your skeletal muscles, straddles the rearmost portion of the frontal lobe. The bottom portion of Figure 14.5 represents points in the left hemisphere that control motor functions. Each point on its surface is associated with the movement of a different part of the body.

Right behind the motor area, on the leading edge of the parietal lobe, lies the *sensory area*. Each point on the surface of the sensory area represents sensory receptors from a different part of the body, such as the pressure sensors of the fingertips and the taste receptors of the tongue. The top portion Figure 14.5 represents the location of sensory regions in the left hemisphere. Other sensory areas are located on other lobes. For example, the auditory (hearing) area lies within the temporal lobe; different surface regions of this area correspond to different tonal patterns and rhythms. The visual area lies on the occipital lobe, with different sites corresponding to different positions on the retina.

The remaining areas within the cerebral cortex are referred to as *association areas*. These areas are the sites of higher cognitive activities, such as planning and contemplation. The associative cortex represents a far greater portion of the total cortex in primates than it does in any other mammal and reaches its greatest extent in humans. In a mouse, for example, 95% of the surface of the cerebral cortex is occupied by motor and sensory areas. In humans, only 5% of the surface is devoted to motor and sensory functions; the remainder is associative cortex.

Although each hemisphere of the cerebrum contains motor, sensory, and association areas, these areas on each hemisphere are responsible for different associative activities. The right side of the cerebrum controls spatial relationships, musical and artistic ability, and expression of emotions. You might think of it as your "artistic" and "visual" side. The left side controls speech, writing, logical thought, and mathematical ability. It is your "logical" or "verbal" side. Injury to the left hemisphere of the cerebrum, for example, often results in the partial or total loss of speech, but a similar injury to the right side does not. Also, several speech centers control different aspects of speech. An injury to one speech center produces halting but correct speech; injury to another speech center produces fluent, grammatical, but meaningless speech; and injury to a third center destroys speech altogether. Injuries to other sites on the surface of the brain's left hemisphere result in impairment of the ability to read, write, or do arithmetic. Comparable injuries to the right hemisphere have very different effects, resulting in impairment of three-

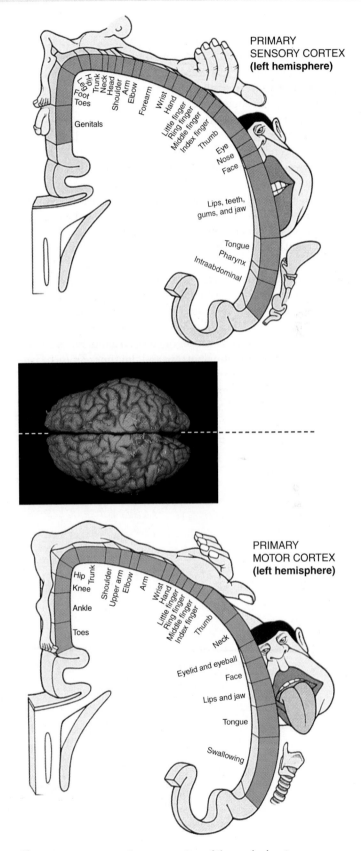

Figure 14.5   Motor and sensory regions of the cerebral cortex.

dimensional vision, musical ability, and the ability to recognize patterns and solve inductive problems. The significance of this clustering of associative activities in different areas of the brain is not clear and remains a subject of much interest.

In addition to the gray matter of the cerebral cortex, other masses of gray matter are located deep within the cerebrum. These islands of gray matter, shown in **Figure 14.6,** are often collectively referred to as the **basal nuclei.** In this context, the term "nuclei" means groups of nerve cell bodies having a similar function. (When groups of nerve cell bodies are located within the peripheral nervous system, they are called *ganglia.*) The basal nuclei are located at the base, or lowest level, of the cerebrum.

The basal nuclei are connected to one another, to the cerebral cortex, and to parts of the brain not yet described—the thalamus and the hypothalamus—by bridges of nerve fibers. These nuclei play important roles in the coordination of slow, sustained movements such as maintaining posture and the suppression of use-less patterns of movement. Injury to the basal nuclei can result in various types of uncontrolled muscular activities, such as shaking, aimless movements, sudden jerking, and muscle rigidity. Parkinson's disease is associated with the degeneration of parts of the basal nuclei and is characterized by continuous tremors (such as in the hands and arms) and a forward flexion of the trunk.

The cerebral cortex, or gray matter, covers an underlying solid white region of myelinated nerve fibers. These nerve fibers are the highways of the brain, bringing messages from one part of the brain to another. These highways, like the message highways in all parts of the nervous system, are composed of individual axons all bundled together like the strands of a telephone cable.

Within the central nervous system, bundles of nerve fibers are called *tracts.* In the peripheral nervous system, they are called *nerves.* The tracts that make up the white matter of the cerebrum run in three principal directions:

1. from a place in one hemisphere of the brain to another place in the same hemisphere,
2. from a place in one hemisphere to a corresponding place in the other hemisphere, and
3. to and from the cerebrum to other parts of the brain and spinal cord.

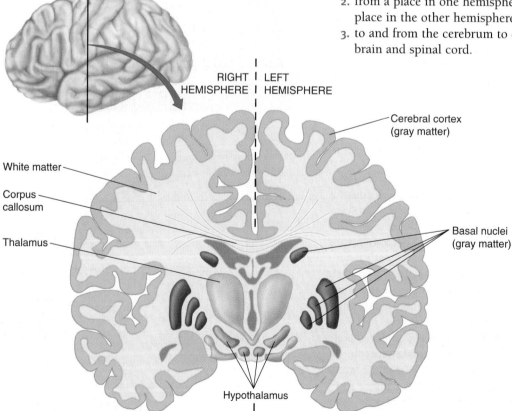

RIGHT | LEFT
HEMISPHERE | HEMISPHERE

Cerebral cortex (gray matter)

White matter

Corpus callosum

Thalamus

Basal nuclei (gray matter)

Hypothalamus

Figure 14.6   The basal nuclei. The basal nuclei, shaded in dark red, are groups of nerve cell bodies. The nuclei are connected to the thalamus and hypothalamus and play a part in controlling many subconscious behaviors.

Cerebellum

The **cerebellum** is a relatively large part of the brain, weighing slightly less than half a pound. It is located below the occipital lobes of the cerebrum, and is highlighted in the illustration below. Although its name

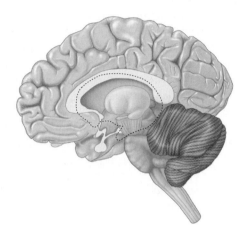

means "little cerebrum," the cerebellum does not perform cerebral functions. Instead, it coordinates subconscious movements of the skeletal muscles. Sensory nerves bring information to the cerebellum about the position of body parts relative to one another, the state of relaxation or contraction of the skeletal muscles, and the general position of the body in relation to the outside world. These data are gathered in the cerebellum and synthesized. The cerebellum then issues orders to motor neurons that result in smooth, well-coordinated muscular movements, contributing to overall muscle tone, posture, balance, and equilibrium.

The cerebral cortex is the major site of higher cognitive processes such as sense perception, thinking, learning, and memory. It makes up a thin layer on the surface of the cerebrum, the largest and most dominant part of the brain. The right side of the cerebrum controls spatial relationships, musical and artistic ability, and expression of emotions. The left side controls speech, writing, logical thought, and mathematical ability. The basal nuclei are groups of nerve cell bodies located at the base of the cerebrum that control large, subconscious movements of the skeletal muscles. The cerebral white matter, which lies beneath the cerebral cortex, consists of myelinated nerve fibers that send messages from one part of the central nervous system to another. The cerebellum coordinates subconscious movements of the skeletal muscles.

14.5 The thalamus and hypothalamus (diencephalon)

At the base of the cerebrum but not part of it, lying close to the basal nuclei, are paired oval masses of gray matter called the **thalamus** (THAL-uh-muss). The view of these nuclei in Figure 14.6 shows them connected by a bridge of gray matter and indicates how the thalamus—meaning "inner chamber"—got its name.

The thalamus acts as a relay station for most sensory information. This information comes to the thalamus from the spinal cord and certain parts of the brain. The thalamus then sends these sensory signals to appropriate areas of the cerebral cortex. In addition, the thalamus interprets certain sensory messages such as pain, temperature, and pressure.

The **hypothalamus**, located beneath the thalamus (*hypo* means "under"), controls the activities of various body organs and is highlighted along with the thalmus in the brain below. The hypothalamus works to maintain homeostasis, a steady state within the body, by

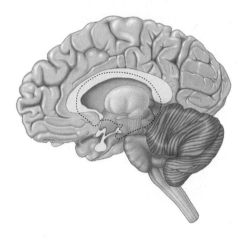

means of its various activities such as the control of body temperature and the heartbeat. It also directs the hormone secretions of the pituitary, which is located at the base of the brain.

The hypothalamus is linked by a network of neurons to the cerebral cortex. This network, together with the hypothalamus, is called the **limbic system**. The term limbic is derived from a Latin word meaning "border." As shown in **Figure 14.7**, the neurons of the limbic system form a ringlike border around the top of the brainstem. The operations of the limbic system are responsible for many of the most deep-seated drives and emotions of vertebrates, including pain, anger, sexual drive, hunger, thirst, and pleasure.

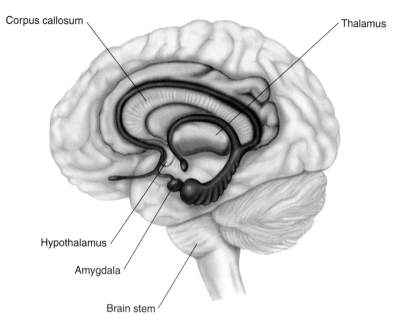

Corpus callosum

Thalamus

Hypothalamus

Amygdala

Brain stem

Figure 14.7 **The limbic system.**

www.jbpub.com/biology

Did you ever wonder why the teen years can be a highly emotional time of life? Researchers are uncovering new answers to this question, and one of these answers involves the limbic system. The *amygdala,* which deals with instinctual responses, is highly active in teens when they perform various cognitive tasks such as evaluating expressions on a series of faces. Adult brains show more activity in the frontal lobe, an area of the brain involved in rational thought. The older the teen, the more brain activity in the "rational" frontal lobe.

Dr. Deborah Yurgelun-Todd, director of neuropsychology and cognitive neuroimaging at the McLean Hospital Brain Imaging Center in Belmont, Massachusetts, and lead investigator in these brain studies, hypothesizes that these teen/adult differences show that there are ongoing developmental changes in the brain during adolescence. Additionally, she thinks that emotional turbulence in teens appears to be partly due to the active role of the amygdala during these years.

The thalamus and hypothalamus are masses of gray matter that lie at the base of the cerebrum. The thalamus receives sensory stimuli, interprets some of these stimuli, and sends the remaining sensory messages to appropriate locations in the cerebrum. The hypothalamus controls the activities of various body organs and the secretion of certain hormones. The limbic system, a network of neurons at the base of the brain that links the hypothalamus with the cerebral cortex, is the center of emotions, instincts, and drives.

14.6 The brainstem

If you think of the brain as being shaped somewhat like a mushroom, the cerebrum, cerebellum, thalamus, and hypothalamus would be its cap. The **brainstem** would be the mushroom's stalk (see Figure 14.7). Its 3 inches of length consists of three parts—the midbrain, pons (ponz), and medulla (mih-DULL-uh). Each part makes up about an equal length of the brainstem and contains tracts of nerve fibers that bring messages to and from the spinal cord. In addition, each portion of the brainstem contains nuclei that govern important reflex (automatic) activities of the body. Many of the cranial nerves, which enter the brain rather than the spinal cord, enter at the brainstem. These cranial nerves bring messages to and from the regulatory centers of the brainstem or use the brainstem as a relay station. These nerves can be seen in Figure 14.10, where they are discussed in more detail.

At the top of the brainstem sits the **midbrain (Figure 14.8)**. extending down from the lower portion of the thalamus and hypothalamus about an inch—approximately the distance from the tip of your thumb to its first knuckle. If you were to cut open the midbrain, you would see both white and gray matter. The white matter consists of nerve tracts that connect the upper parts of the brain (cerebrum, thalamus, and hypothalamus) with lower parts of the brain (pons and medulla). In addition, the midbrain contains nuclei that act as reflex centers for movements of the eyeballs, head, and trunk in response to sights, sounds, and various other stimuli. For example, if a plate falls off the counter behind you, you probably turn around quickly and automatically. That is your midbrain at work.

The term **pons** means "bridge," and it is actually two bridges. One bridge consists of horizontal tracts that extend to the cerebellum, connecting this part of the brain to other parts and to the spinal cord. The other bridge consists of longitudinal tracts that connect the midbrain and structures above to the medulla and spinal cord below. In addition, the gray matter of the pons contains nuclei that work with other nuclei in the medulla to help control respiration.

The **medulla**, the lowest portion of the brainstem, is continuous with the spinal cord below. Because of its location, a large portion of the medulla consists of tracts of neurons that bring messages up from the spinal cord and others that take messages down to the spinal cord. Most of these tracts cross over one another within the medulla. Therefore, sensory information from the right side of the body is perceived in the left side of the brain and vice versa. Likewise, the right side of the brain sends messages to the left side of the body and the left side of the brain controls the right side. In addition to these tracts, the medulla contains reflex centers that regulate heartbeat, control the diameter of blood vessels, and adjust the rhythm of breathing. Centers there also control less vital functions such as coughing, sneezing, and vomiting.

Throughout the entire length of the brainstem but concentrated in the medulla weaves a complex network of neurons called the *reticular formation* (ri-TIK-yuh-ler). All of the sensory systems have nerve fibers that feed into this system, which serves to "wiretap" all of the incoming and outgoing communication channels of the brain. In doing so, the reticular formation monitors information concerning the incoming stimuli and identifies important ones. The reticular formation also plays a role in consciousness, increasing the activity level of many parts of the brain when aroused and decreasing the activity level during periods of sleep. This is the part of the brain that causes a knockout during boxing. When the head is twisted sharply and suddenly during a punch to the jaw, the brainstem (and reticular formation) is twisted, resulting in unconsciousness.

Figure 14.8 shows the structures of the brain described in Sections 14.3 through 14.7, along with a list of their functions.

The brainstem, consisting of the midbrain, pons, and medulla, contains tracts of nerve fibers that bring messages to and from the spinal cord. In addition, nuclei located in various parts of the brainstem control important body reflexes such as the rhythm of the heartbeat and rate of breathing.

Visual Summary **Figure 14.8** **Major structures of the brain and their functions.**

Cerebral cortex

- Receives sensory information from skin, muscles, glands, and organs
- Sends messages to move skeletal muscles
- Integrates incoming and outgoing nerve impulses
- Performs associative activities such as thinking, learning, and remembering

Basal nuclei

- Plays a role in the coordination of slow, sustained movements
- Suppresses useless patterns of movement

Thalamus

- Relays most sensory information from the spinal cord and certain parts of the brain to the cerebral cortex
- Interprets certain sensory messages such as those of pain, temperature, and pressure

Hypothalamus

- Controls various homeostatic functions such as body temperature, respiration, and heartbeat
- Directs hormone secretions of the pituitary

Cerebellum

- Coordinates subconscious movements of skeletal muscles
- Contributes to muscle tone, posture, balance, and equilibrium

Brain stem

- Origin of many cranial nerves
- Reflex center for movements of eyeballs, head, and trunk
- Regulates heartbeat and breathing
- Plays a role in consciousness
- Transmits impulses between brain and spinal cord

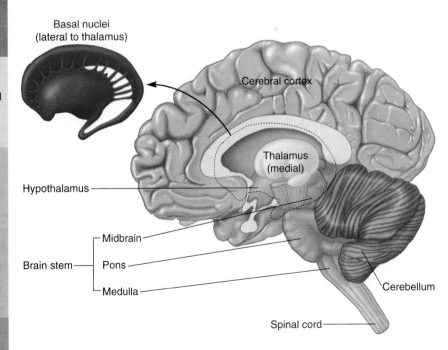

14.7   The spinal cord and meninges

www.jbpub.com/biology

The central nervous system also includes the brain's extension, the spinal cord. The spinal cord runs down the neck and back within an inner "tunnel" of the vertebral column, or spine. This tunnel is created by the stacking of vertebrae one on another, with their central spaces, or *foramina* (singular *foramen*) in alignment with one another. This bony casing protects the spinal cord from injury, just as the bones of the skull protect the brain (see Chapter 16).

The spinal cord receives information from the body by means of **spinal nerves** (Figure 14.9). It carries this information to the brain along organized tracts of myelinated nerve fibers (white matter) and similarly sends information from the brain out to the body. In addition, the gray matter of the spinal cord integrates responses to certain kinds of stimuli. These integrative pathways are called *reflex arcs* and are discussed in more detail later in the chapter.

The white matter tracts of the spinal cord, unlike the white matter of the brain, are located on the exterior of the spinal cord with gray matter in the center as shown in the cross-section in Figure 14.9. (The brain, as discussed earlier, is covered with gray matter, which surrounds the white matter hidden below.) In addition, the spinal cord has a tiny central canal that pierces its length. This tube-like space is filled with **cerebrospinal fluid** and is continuous with fluid-filled spaces, or *ventricles,* in the brain. The cerebrospinal fluid is a filtrate of the blood, formed by specialized cells in the ventricles. This fluid acts as a shock absorber, cushioning the brain and the spinal cord. In fact, your brain is actually floating in this fluid, although its volume would barely fill an average-sized glass. In addition, the cerebrospinal fluid brings nutrients, hormones, and white blood cells to different parts of the brain.

As well as being encased in the skull and vertebrae and cushioned by the cerebrospinal fluid, both the brain and spinal cord are protected by three layers of membranes called the **meninges** (muh-NIN-jeez). The outermost of these layers is called the *dura mater,* from the

medieval Latin meaning "tough mother." Surrounding the brain, the dura mater is made up of two parts: one that adheres to the inner side of the bones of the cranium and another that lies closer to the brain. Surrounding the spinal cord, however, the dura mater is only a single layer. In between it and the bones of the vertebrae is a space filled with fat, connective tissue, and blood vessels called the *epidural* (outside the dura) *space*. At the level of the lower back, the epidural space is where anesthetics are injected to numb the lower portion of the body for certain operations or childbirth.

Another space exists between the middle (arachnoid layer) and inner (pia mater) membranes of the meninges. The cerebrospinal fluid circulates within this space, called the *subarachnoid space*. The subarachnoid space is continuous with the central canal of the vertebral column and the ventricles of the brain.

> The spinal cord, the part of the central nervous system that runs down the neck and back, receives information from the body, carries this information to the brain, and sends information from the brain to the body. The brain and spinal cord are protected by a cushion of fluid called the cerebrospinal fluid, layers of membranous coverings called the meninges, and bones of the skull and vertebral column.

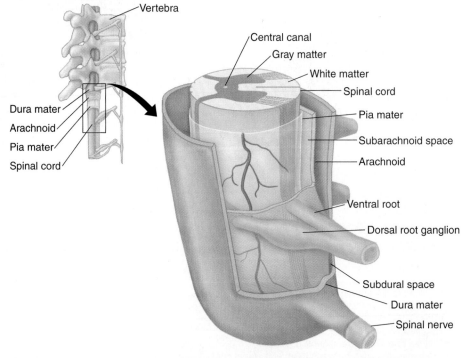

Figure 14.9   Section through the spinal cord.

H◯W Science
W◯RKS Discoveries

Solving the Mystery of Alzheimer's Disease

Alzheimer's disease was first described nearly a century ago in 1907 by Swiss psychiatrist, Dr. Alois Alzheimer. One of Dr. Alzheimer's patients, aged 51, was having substantial memory difficulties. She continued to deteriorate, forgetting words and losing her ability to reason. She died four years later. At the autopsy, the woman's brain was found to be much smaller than average for her age and gender, and the ventricles (cavities) of her brain were much larger. Her brain was also riddled with waxy-looking patches where nerve cells had once been. Today, analysis of brain tissue at autopsy is still the only conclusive way to determine if someone has Alzheimer's disease. A new diagnostic blood test created by researchers in Germany may help spot Alzheimer's early on, but the test is not yet available.

Currently affecting an estimated 2 to 3 million Americans, Alzheimer's disease is characterized by a gradual loss of memory and reasoning. Affected individuals cannot remember things that they heard or saw just a few minutes previously. They have trouble finding their way around and eventually forget how to talk, feed themselves, and even swallow. There is no cure; however, the experimental drug galantamine appears to improve memory and learning along with slowing disease progression in Alzheimer's patients. Additionally, the drug metrifonate, which belongs to the class of drugs called acetylcholinesterase (ACE) inhibitors, appears to slow mental decline and control behavior in persons with mild or moderate Alzheimer's. The drug appears to work by slowing the breakdown of the chemical messenger acetylcholine, which is found at lower than normal levels in the brains of Alzheimer's patients.

Some forms of hereditary Alzheimer's occur in persons under age 55. Recently scientists discovered forms that occur in older persons as well. However, the majority of Alzheimer's disease cases occur sporadically and are not thought to be hereditary.

Just what happens in Alzheimer's disease to cause such devastating mental deterioration? In those suffering from Alzheimer's, the synthesis of a component of nerve cell membranes called amyloid protein goes awry. Pieces of amyloid pile up in needlelike masses within the brain, punching holes in brain cells and killing them. These masses are usually referred to as *amyloid beta plaques*. (Beta refers to amyloid's structure of accordionlike sheets.)

As the population ages, Alzheimer's disease is becoming more common. By 2047, 8.6 million people in the United States are expected to have Alzheimer's. In response to this coming national health crisis, researchers are now focusing on finding ways to block the formation of amyloid fragments or prevent their attachment to the surface of brain cells to thwart the disease. Researchers at New York University Medical Center have found protein fragments called beta sheet breakers that appear to prevent amyloid plaques from forming and appear to dissolve existing plaques. Other researchers from Cornell University in Ithaca, New York, are experimenting with implanting plastic pellets that release a growth factor that may keep nerve cells from degenerating. Yet others at the University of Washington and a Washington-based biotechnology company are working with a plant extract that may be useful in disrupting the amyloid.

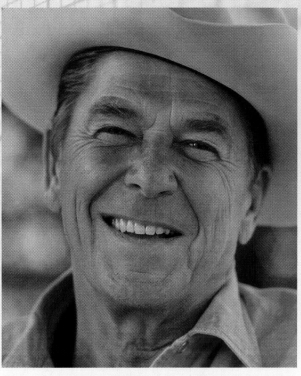

Former President Ronald Reagan disclosed in a 1994 letter to the nation that he had been diagnosed with Alzheimer's disease.

What can you do to help prevent this disease as you age? Post-menopausal hormone replacement therapy for women may prevent Alzheimer's by reducing the synthesis of amyloid fragments. Smoking cigarettes increases the risk of Alzheimer's, so to protect yourself from this risk, do not smoke. Researchers have also discovered that persons less likely to develop Alzheimer's are those who exercise regularly, have intellectually demanding occupations, have a high level of education, and stay socially active. Researchers hypothesize that the brain is like other organs of the body—it ages better when it is used.

14.8 Sensory pathways

Just as your central nervous system can be compared with the CPU of your computer, your peripheral nervous system can be compared with the cables that connect the CPU to the other pieces of hardware—the peripherals—that make up your computer. In your computer, connector cables carry information to the monitor to be visualized on the screen or to the printer to be printed. So, too, the nerve "cables" of your peripheral nervous system bring information to your muscles, instructing them to contract and relax in patterns that result in the integrated and flowing movements of your body. For example, food in your stomach (the input) sends signals along your peripheral nervous system to the medulla to be interpreted, resulting in the release of gastric juice and the churning of the muscular stomach walls (the output). In this way, the peripheral nervous system works to connect the integrative portion of your nervous system with its *receptors* (structures that detect stimuli) and *effectors* (muscles and glands that respond to that stimuli).

Seeing a sunset, hearing a symphony, experiencing pain—all these stimuli travel along sensory neurons to the central nervous system and arrive there in the same form: as nerve impulses. Put simply, every nerve impulse is identical to every other one. How, then, does the brain distinguish between pleasure and pain or sight and sound?

The information that the brain derives from sensory input is based solely on the source of the impulse and its frequency. Thus, if the auditory nerve is artificially stimulated, the central nervous system perceives the stimulation as sound. If the optic nerve is artificially stimulated in exactly the same manner and degree, the stimulation is perceived as a flash of light. Increasing the intensity of the stimulation of either receptor will result (within limits) in an increase in the frequency of nerve impulses and therefore produce a perception of a brighter light or a louder noise.

Many kinds of receptors have evolved among vertebrates, with each receptor sensitive to a different aspect of the environment. Sensory receptors are able to change specific stimuli into nerve impulses by having low thresholds for these particular types of stimuli and

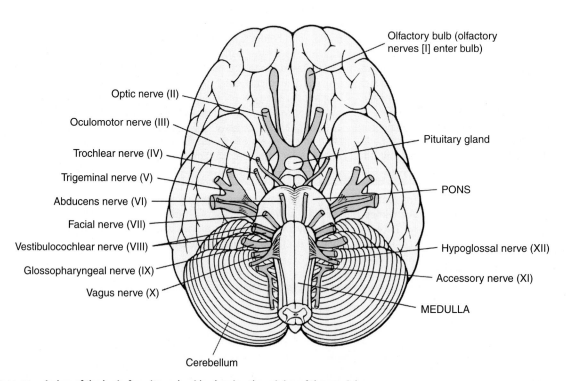

Olfactory bulb (olfactory nerves [I] enter bulb)

Optic nerve (II)

Oculomotor nerve (III)

Trochlear nerve (IV)

Trigeminal nerve (V)

Abducens nerve (VI)

Facial nerve (VII)

Vestibulocochlear nerve (VIII)

Glossopharyngeal nerve (IX)

Vagus nerve (X)

Pituitary gland

PONS

Hypoglossal nerve (XII)

Accessory nerve (XI)

MEDULLA

Cerebellum

Figure 14.10   A view of the brain from its underside showing the origins of the cranial nerves.

high thresholds for others. Receptors in the retina of your eye, for example, have a low threshold for light.

Certain cranial nerves conduct sensory information from specialized receptors to the brain. **Cranial nerves** communicate directly with the brain, without traveling through the spinal nerves and spinal cord. Their axons are bundled into groups with other sensory nerve fibers arising from the same area. The cranial nerves that conduct only sensory stimuli are those associated with sight, sound, smell, and equilibrium, such as cranial nerve I (Figure 14.10). the olfactory nerve, which brings in "smell" messages from your nose; cranial nerve II, the optic nerve, which brings in "sight" messages from your eyes; and cranial nerve VIII, which brings impulses from your ears regarding hearing and balance. Other cranial nerves called **mixed nerves** are made up of both sensory (incoming or afferent) and motor (outgoing or efferent) nerve fibers, such as cranial nerve X (the vagus nerve), which regulates the function of your heart rate, respiration rate, and digestive activities. The motor nerve fibers of cranial nerves are also axons; the cell bodies of these motor neurons are located within the brain. Cranial nerve XII, which stimulates movement in your tongue, carries outgoing (motor) nerve impulses only.

Sensory and motor nerve fibers that travel directly to and from the spinal cord make up the spinal nerves. Nerve fibers that serve the same general area of the body, whether they are sensory or motor neurons (Figure 14.11), are bundled together to form these nerves. Thus, all spinal nerves are mixed nerves. Close to the spinal cord, however, motor and sensory fibers separate from one another. The sensory (afferent) fibers are myelinated axons that extend from the source of stimulation (your fingers or toes, for example) to a swelling near their entrance to the dorsal (back) side of the spinal cord. This swelling is a ganglion that contains the nerve cell bodies of these sensory neurons. Shorter portions of the axons of these neurons extend from this ganglion to the gray matter of the spinal cord. Here, the axon ends of these sensory neurons synapse with neurons (interneurons) that play a role in integrating incoming messages with outgoing responses.

Sensory receptors change stimuli into nerve impulses. The sensory pathways of the peripheral nervous system bring messages from receptors to the brain and spinal cord.

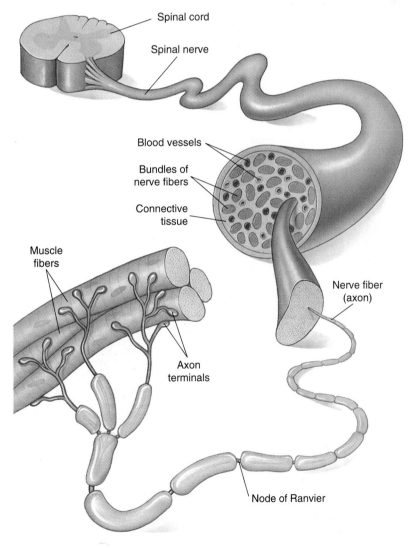

Figure 14.11 Structure of a spinal nerve.
A nerve consists of bundles of nerve fibers and various layers of connective tissue.

14.9   Integration and motor pathways

"Integrating" neurons, or **interneurons**, receive incoming messages and send appropriate outgoing messages in response. Interneurons extend from the spinal cord to the brain, make up part of the brain tissue itself, and in the case of certain reflex pathways, are located within the gray matter of the spinal cord. The pathway of a nerve impulse from stimulus to response may involve a single interneuron, as in the case of certain reflexes such as the one shown in Figure 14.12. In a few of the simplest reflexes, a sensory neuron may synapse directly with a motor neuron, with no interneuron(s) playing a role. Most pathways, however, involve many interneurons. In sequence, these interneurons receive the incoming message, direct it to other interneurons for interpretation (such as specific areas of the cerebrum), and bring the outgoing message to the appropriate motor neurons.

The short dendrites of spinal motor neurons, located in anterior portions of the gray matter of the spinal cord, synapse with interneurons. These dendrites then conduct the impulses to their cell bodies. From there, the nerve impulse sweeps along the membrane of the axon of each motor neuron to effectors: muscles or glands that produce a response. Motor pathways to skeletal muscles contain single motor neurons; that is, the distance from the spinal cord to the muscle is traversed by individual neurons. The motor pathways to smooth and cardiac muscles and to glands are made up of a series of two motor neurons.

The bundles of axons leaving the spinal cord join with the axons of sensory neurons entering the same level of the cord, forming spinal nerves. A total of 31 pairs of spinal nerves bring messages to and from specific areas of the body. Figure 14.2 shows the approximate location of the spinal nerves (those emerging from the spinal cord) and, in general, the areas of the body that they serve.

Interneurons interpret incoming sensory messages and send appropriate outgoing messages. The axons of motor neurons conduct these impulses to muscles and glands, the structures that effect a response.

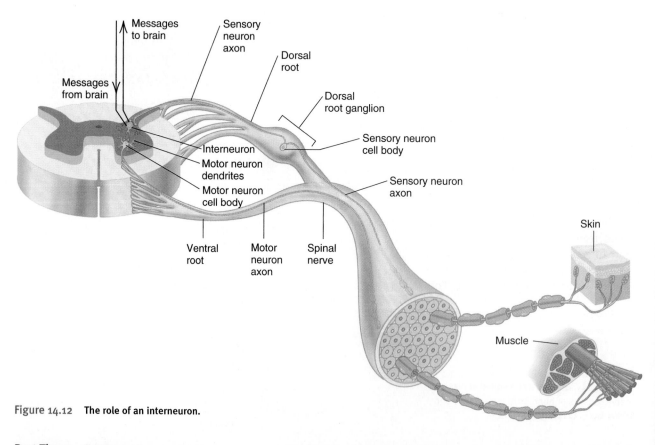

Figure 14.12   **The role of an interneuron.**

14.10 The somatic nervous system

The central nervous system directs different types of responses in different ways. Movements of the skeletal muscles are controlled by messages from the brain and spinal cord via pathways that contain single motor neurons. These motor pathways are referred to as the *somatic nervous system*. Messages along these pathways coordinate your fingers when you grasp a pencil, spin your body when you dance, and outstretch your arms when you hug a friend. These muscular movements are primarily subject to conscious control by the associative cortex of the cerebrum. However, not all movements of the skeletal muscles are conscious. Blinking your eyes, putting one foot in front of the other when you walk, and breathing can be consciously controlled, but most often take place without conscious thought.

A *reflex* is an automatic response to nerve stimulation. Very little, if any, integration (and certainly no thinking) takes place during reflex activity. The knee jerk is one of the simplest types of reflexes in the human body; a sensory neuron synapses directly with a motor neuron (Figure 14.13). This pathway—the pathway an impulse follows during reflex activity—is called a **reflex arc**.

Simple reflex arcs such as the knee jerk are called monosynaptic reflex arcs (literally, "one synapse"). A monosynaptic reflex arc works without involvement of the central nervous system. Most voluntary muscles within your body possess such monosynaptic reflex arcs, although usually in conjunction with other more complex reflex pathways. It is through these more complex paths that voluntary control is established. In the few cases in which the monosynaptic reflex arc is the only feedback loop present, its function can be clearly seen, such as in the knee jerk reaction.

If the ligament just below the kneecap is struck lightly by the edge of your hand or by a doctor's rubber mallet, the resulting sudden pull stretches the muscles of the upper leg, which are attached to the ligament. Stretch receptors in these muscles immediately send an impulse along afferent nerve fibers ❶ to the spinal cord, where these fibers synapse directly with motor neurons ❷ that extend back to upper leg muscles, stimulating them to contract and the leg to jerk upward ❸. Such reflexes play an important role in maintaining posture.

The somatic branch of the peripheral nervous system consists of motor fibers that send messages to the skeletal muscles. A reflex is an automatic response to nerve stimulation. The pathway of nervous activity in a reflex is called a reflex arc. Impulses travel along sensory neurons to the spinal cord, which sends a message out via motor neurons without involving the brain.

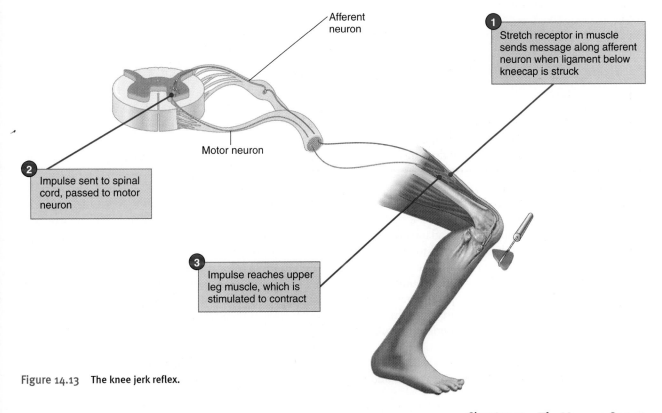

Afferent neuron

① Stretch receptor in muscle sends message along afferent neuron when ligament below kneecap is struck

Motor neuron

② Impulse sent to spinal cord, passed to motor neuron

③ Impulse reaches upper leg muscle, which is stimulated to contract

Figure 14.13   The knee jerk reflex.

Just Wondering . . . *Real Questions Students Ask*

Does *Ginkgo biloba* increase mental alertness, memory, and concentration, and if so, how?

All cultures have traditions that include the use of plants and plant products as medicinals. The American culture is no different; *Ginkgo biloba* is only one of many herbal preparations that Americans frequently buy and use. The World Health Organization (WHO) estimates that 4 billion people, or 80% of the world population, presently uses herbal medicine for some aspect of primary health care.

Although not all plant extracts have medicinal value, about 25% of all prescription drugs dispensed each year in North America were originally derived from flowering plants and ferns. (See the Just Wondering box in Chapter 29.) Prescription drugs are regulated in the United States by the Food and Drug Administration (FDA). However, herbal products are not, because they are considered dietary supplements, just like vitamins and minerals. (See the Part 3 Bio-Issues, p. 128.) Therefore, the substances sold as *Ginkgo biloba* can vary in their contents, active ingredients, quality, and purity. Substances other than *Ginkgo biloba* may be present in preparations of this supplement, which is made from the leaves of a shade tree of the same name.

As you suggest in your question, *Ginkgo biloba* is advertised to enhance memory, increase alertness and concentration, and aid

memory. However, research studies that show these effects have been primarily conducted on nonhumans (such as rats or mice) that had a deficiency of blood to the brain, or on older humans experiencing dementia. Dementia is a condition in which a person's cognitive functioning is impaired due to conditions such as strokes (which reduce blood flow to the brain), Parkinson's disease, or Alzheimer's disease (AD). Dementia most often includes memory loss, confusion, disorientation, and delusions.

The results of most studies show that particular extracts of *Ginkgo biloba* leaves (see photo) can improve the performance of cognitively impaired research subjects on learning tasks. Study results also show that persons with AD or other dementias, when treated for 3 to 6 months with extracts of *Ginkgo biloba*, have their cognitive function stabilized or show small improvements in cognitive function. Medical researchers are hopeful that *Ginkgo biloba* may be shown to be a helpful treatment for patients with AD. In 1999, the National Center for Complementary and Alternative Medicine, a branch of the National Institutes of Health (NIH), requested applications for grants to support further research studies on the effects of *Ginkgo biloba* on older individuals.

Because *Ginkgo biloba* appears to positively affect memory and cognitive functioning

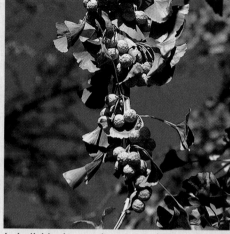

in individuals or rodent test subjects with certain conditions or disorders of the brain, it does not necessarily mean that this herb can cause persons without these impairments to function at increased levels of cognitive ability. Additionally, the way in which this herb works to improve the impairment of cognitive functioning is unclear. Scientists postulate that its function is related to its antioxidant properties. Antioxidants are compounds that can protect cell membranes by preventing or reducing the formation of free radicals. These unstable and highly reactive atoms or compounds remove electrons from other substances in a process called oxidation. Free radical formation in cell membranes produces chemical changes that may contribute to various chronic health conditions.

14.11 The autonomic nervous system

When you enter a dark movie theater, the pupils of your eyes dilate, or become larger. When you are frightened, adrenaline is pumped into your bloodstream and your heart beats faster. Neither of these reactions is governed by conscious thought—they occur automatically and involve the action of smooth muscle, cardiac muscle, or glands. The motor pathways that control such involuntary and automatic responses of the glands and nonskeletal muscles of the body are referred to as the *autonomic nervous system*. The autonomic nervous system takes your temperature, monitors your blood pressure, and sees to it that your food is properly digested. The body's internal physiological condition is thus fine-tuned within relatively narrow bounds, a regulatory process called *homeostasis*.

The autonomic nervous system is made up of two divisions, the **parasympathetic nervous system** and the

sympathetic nervous system, that act in opposition to each other, speeding up or slowing down certain bodily processes (Figure 14.14). Each of these motor pathways consists of a series of two motor neurons. In the parasympathetic system, the axons of the motor neurons leaving the spinal cord extend to ganglia located near the muscles or organs they affect. In the sympathetic system, the axons leaving the spinal cord are much shorter and extend only to ganglia located near the vertebral column. The axons of second motor neurons in each system extend from these ganglia to their targets.

The neurotransmitters used by each of these two branches of the autonomic nervous system differ at the axon ends of the second motor neuron—at the synapse with the target organ. The actions of these different neurotransmitters oppose each other. Thus, because each gland (except the inner portion of the adrenal gland), smooth muscle, and cardiac muscle is "wired" to *both*

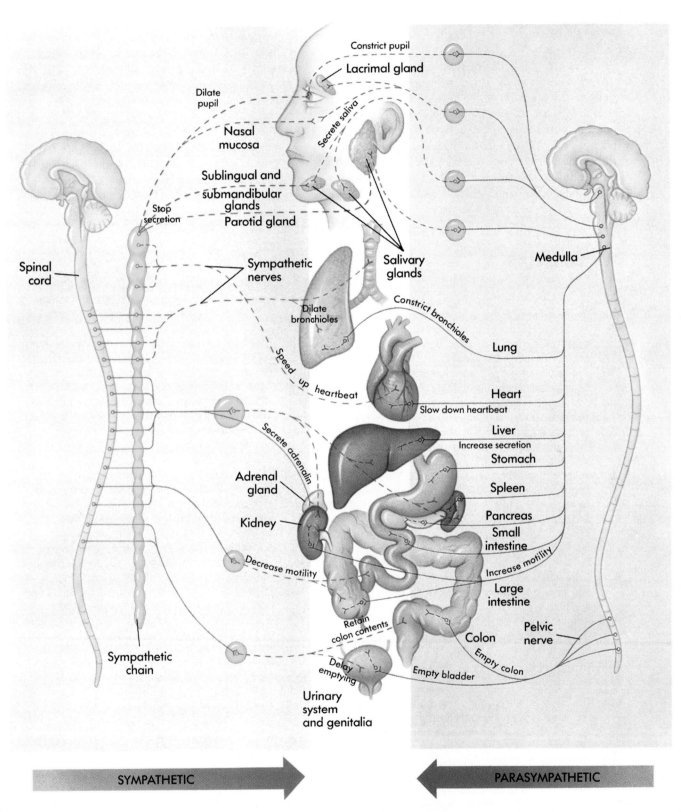

Figure 14.14 **The sympathetic and parasympathetic nervous systems.** The ganglia of sympathetic nerves are located near the spine, and the ganglia of parasympathetic nerves are located far from the spine, near the organs they affect. A nerve runs from both of the systems to every organ indicated, except the adrenal gland.

systems, an arriving signal will either stimulate or inhibit the organ.

For example, the sympathetic system speeds up the heart and slows down digestion, whereas the parasympathetic system slows down the heart and speeds up digestion. In general, the two opposing systems are organized so that the parasympathetic system stimulates the activity of normal body functions such as the churning of the stomach, the contractions of the intestine, and the secretions of the salivary glands. The sympathetic system, on the other hand, generally mobilizes the body for greater activity, as in increased respiration or a faster heartbeat. It is a key player in the "fight-or-flight" response of stress (see p. 370, Chapter 17). The "decision" to stimulate or inhibit a muscle, organ, or gland is "made" by the central nervous system.

> The autonomic nervous system is a branch of the peripheral nervous system that consists of two antagonistic sets of motor fibers, the parasympathetic and sympathetic nervous systems. Messages sent along these fibers control smooth and cardiac muscle as well as glands.

Key Concepts

Nervous systems exhibit great diversity.

14.1 A nervous system is a network of neurons specialized for transmitting information from sensory receptors to neurons, between neurons, and from neurons to effectors.

14.1 The nervous systems of invertebrates vary in their organization among invertebrate phyla.

14.1 The simplest nervous system organization is the nerve net, a system of interconnecting nerve cells with no central controlling area, or brain.

14.1 Some phyla of invertebrates have simple brains attached to nerve cords.

14.2 The vertebrate nervous system consists of the peripheral nervous system and the central nervous system.

14.2 The peripheral nervous system, made up of the nerves of the body, gathers information about the internal and external environments and brings it to the central nervous system via sensory nerve cells.

14.2 The central nervous system, made up of the brain and spinal cord, processes and responds to that information.

14.2 Responses are sent out to the body via the motor nerve cells of the peripheral nervous system.

The human central nervous system integrates sensory and motor impulses.

14.3 The cerebrum, cerebellum, diencephalon, brainstem, and spinal cord constitute the central nervous system, which processes incoming sensory information and produces outgoing motor impulses.

14.4 The cerebrum, the dominant part of the human brain, is split into two hemispheres connected by a nerve tract called the corpus callosum.

14.4 Each hemisphere is divided further by deep grooves into four lobes: the frontal, parietal, temporal, and occipital lobes.

14.4 Specific sensory areas are located in the various lobes of the cortex; the motor region is mostly a part of the frontal lobe.

14.4 Association areas connect all parts of the cerebral cortex and govern such functions as memory, reasoning, intelligence, and personality.

14.4 The basal nuclei, groups of nerve cell bodies located at the base of the cerebrum, control large, subconscious movements of the skeletal muscles.

14.4 The cerebellum is the part of the brain that coordinates unconscious movements of the skeletal muscles.

14.5 The thalamus and the hypothalamus (diencephalon), groups of nerve cell bodies also located near the base of the cerebrum, receive and interpret certain sensory stimuli and control the activities of various body organs and glands, respectively.

14.6 The brainstem consists of three parts: the midbrain, pons, and medulla.

14.6 All parts of the brainstem contain nerve fibers that bring messages to and from the spinal cord.

14.6 The brainstem contains groups of nerve cell bodies that control important body reflexes.

14.7 The spinal cord brings messages to the brain and sends messages from the brain out to the body.

14.7 The spinal cord also integrates the incoming and outgoing information of reflex arcs, with the exception of certain simple reflexes.

The human peripheral nervous system carries sensory and motor impulses.

14.8 In the peripheral nervous system, sensory receptors convert stimuli into nerve impulses.

14.8 Sensory impulses travel along the nerve fibers of sensory neurons to the brain or spinal cord.

Key Concepts, continued

14.9 At the CNS, interneurons interpret sensory information and send appropriate outgoing messages.

14.9 The axons of motor neurons conduct outgoing messages to muscles and glands, the structures that effect a response.

14.10 The motor pathways are divided into somatic pathways, which relay commands to skeletal muscles, and autonomic pathways, which stimulate the glands and other muscles of the body.

14.11 The autonomic pathways consist of nerve pairs having antagonistic neurotransmitters, one of which stimulates while the other inhibits.

14.11 In general, the parasympathetic nerves stimulate the activity of normal internal body functions and inhibit alarm responses, and the sympathetic nerves do the reverse.

Key Terms

autonomic nervous system (awe-tuh-NOM-ik) 286

basal nuclei (BAY-suhl) 290

brain 286

brainstem 292

central nervous system 286

cerebellum (SER-uh-BELL-um) 291

cerebral cortex (suh-REE-brul KOR-tecks) 288

cerebrospinal fluid (suh-REE-brow-SPY-nul) 294

cerebrum (suh-REE-brum) 288

corpus callosum (KOR-pus kuh-LOW-sum) 288

cranial nerves 297

diencephalon (DYE-un-SEF-uh-lon) 287

frontal lobe 288

hypothalamus (HYE-poe-THAL-uh-muss) 291

interneurons 298

limbic system 291

medulla (mih-DULL-uh) 292

meninges (muh-NIN-jeez) 294

midbrain 292

mixed nerves 297

motor neurons 286

nerve net 284

nervous system 284

occipital lobe (ock-SIP-uh-tul) 288

parasympathetic nervous system (PARE-uh-SIM-puh-THET-ik) 300

parietal lobe (puh-RYE-uh-tul) 288

peripheral nervous system (puh-RIF-uh-rul) 286

pons (ponz) 292

reflex arcs 289

sensory neurons 286

somatic nervous system (soe-MAT-ik) 286

spinal cord 286

spinal nerves 294

sympathetic nervous system (SIM-puh-THET-ik) 300

temporal lobe 288

thalamus (THAL-uh-muss) 291

KNOWLEDGE AND COMPREHENSION QUESTIONS

1. Name and describe the simplest type of invertebrate nervous system. *(Knowledge)*

2. Most invertebrate brains would probably more accurately be called ganglia. What is the definition of a ganglion and why is it a better descriptor than the term "brain" for these invertebrates? *(Comprehension)*

3. Distinguish between the vertebrate central nervous system and peripheral nervous system. What are the functions and components of each? *(Knowledge)*

4. Distinguish between the somatic and autonomic nervous systems. What type of response does each control? Give an example of each type. *(Knowledge)*

5. A friend accidentally steps on your foot and you quickly pull it away. Where in the nervous system do you think this reflex is integrated? *(Comprehension)*

6. A sentimental movie makes you remember your beloved grandparents as tears come to your eyes. Where in the nervous system would the connection between the movie (visual) and emotional reaction be integrated? *(Comprehension)*

7. What is the brainstem? Name its three major components and summarize its functions. *(Knowledge)*

8. You're studying in a quiet library. Suddenly someone drops an armful of books; you hear the loud crash and turn your head toward the noise. Explain how your sensory and motor pathways allowed you to detect and respond to this stimulus. *(Comprehension)*

9. In question 8, why did you perceive this stimulus as a loud noise (rather than, say, as a bright light)? *(Knowledge)*

10. The organs of your body are controlled by two branches of the autonomic nervous system, the sympathetic and parasympathetic. Each organ is reached by neurons from each system. Why *two* systems? Why use two neurons to communicate with an organ rather than one? *(Comprehension)*

KEY
Knowledge: Recalling information.
Comprehension: Showing understanding of recalled information.

CRITICAL THINKING REVIEW

1. Looking at Figure 14.5, develop a hypothesis explaining why a greater proportion of the sensory and motor area of the cerebral cortex is devoted to hands and facial structures rather than an equal distribution according to body parts. Use your own body as a model. *(Synthesis)*

2. Doctors have warned parents to avoid applying their own low-fat diet to small toddlers, because a higher fat level is essential for the body's production of myelin. Why do you think this may be so important in prenatal care and in the diet of infants and toddlers? *(Synthesis)*

3. Which would you hypothesize to be more similar in function: the limbic systems of a human being and a lion or the associative cerebral cortex of these vertebrates? Why? *(Application)*

4. If the spinal cord is severed in an injury, the closer the point of damage to the brain, the more severe the disability. For example, a break near the neck may result in paralysis from the neck down. Why do you think this occurs? *(Application)*

5. A grandmother was playing with her grandchildren when she suddenly experienced dizziness and a headache. She had numbness on the right side of her body and was unable to speak. Her daughter, who was also with her and the children, called an ambulance. After being examined by emergency room physicians, the grandmother was diagnosed as having had a mild stroke. Which side of her brain was affected? Which areas were likely affected? Explain the rationale for your answer. *(Application)*

KEY
Application: Using information in a new situation.
Synthesis: Putting together information from different sources.

The illustration below shows the pathway of incoming sensory messages and outgoing motor messages.

1. Fill in the blanks to identify the structures shown.

2. Imagine you just touched a hot stove. Describe the pathway of the incoming sensory message and the outgoing motor message.

3. What role does the spinal cord play in integrating these messages?

4. What role does the brain play?

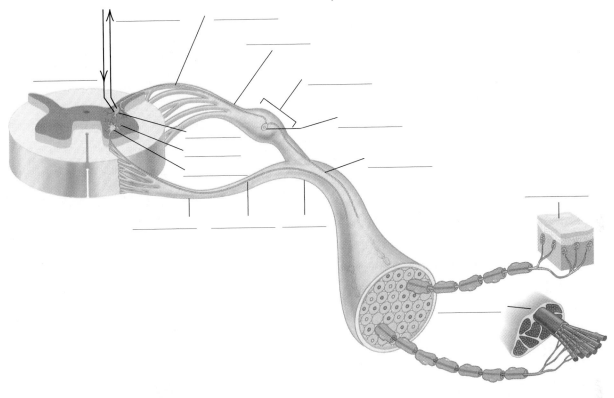

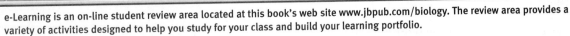

Location: http://www.jbpub.com/biology

e-Learning is an on-line student review area located at this book's web site www.jbpub.com/biology. The review area provides a variety of activities designed to help you study for your class and build your learning portfolio.

Review Questions The review questions test your knowledge of the important concepts and applications in each chapter. The review provides feedback for each correct or incorrect answer. This is an excellent test preparation tool.

Figure-Labeling Exercises Sharpen your visual thinking by matching terms and labels for important illustrations.

Flash Cards Studying biology requires learning new terms. Virtual flash cards help you master the new vocabulary for each chapter.

Just Wondering As you read and study from this text, you may find that you have unanswered questions. Through this site you can ask the author, Sandy Alters, your "just wondering" questions.

 Do you prefer the speed and reliability of a CD-ROM? All of the features contained on the eLearning portion of the web site are also available on the Student CD-ROM.

Senses

Elephants can hear it, as can homing pigeons and probably hippos, but you cannot. It's that deep rumbling you might feel (rather than hear) when loud bass sounds are played, or during thunder and earthquakes. Its name is *infrasound*, and it is low-frequency sound outside of the range of normal human hearing.

Elephants cannot only hear infrasound, they can make these low-frequency sounds to communicate with distant members of their herd, especially during mating. Using specialized elephant-sized radio collars, researchers discovered that a female elephant in estrus sends out a call that informs males for miles around of her condition. The estrus call is the same in all female elephants and is an infrasound song. The female elephant sings for only half an hour, and within a day she is surrounded by elephant bulls, with some having traveled from several miles away on the African savanna.

Infrasound is used by elephants for purposes other than mating as well. Warnings are issued by infrasound, and the whereabouts of groups of elephants are transmitted to other groups by infrasound. Observers have noted for years that entire groups of elephants will suddenly freeze, raising and spreading their ears. Researchers surmise that these elephants are listening to faint, distant calls and are remaining as silent as possible to do so. Moments later, the elephants will come out of their trance and rush off together to follow the signal. The elephants' "talk" is a factor that enhances their survival by communicating useful information and facilitating the reproductive process.

Organization

Sensory receptors detect stimuli from both the external and internal environments.

General sense receptors detect external stimuli such as touch, pressure, pain, and temperature.

Some animals can detect stimuli that humans cannot.

The special senses are taste, smell, sight, hearing, and balance.

15.1   The nature of sensory communication

Sensing the external environment is important for all animals, not just elephants. How else can animals find food and mates, and escape from environmental dangers such as predators? Animals possess an interesting array of sense organs, specialized structures that trigger nerve impulses, which ultimately provide information about the surroundings. These sense organs are described throughout this chapter.

As mentioned in Chapters 13 and 14, only animals have nerve cells. Therefore, only animals have senses. Chapter 13 described examples of ways in which organisms other than animals respond to their environments, such as the ability of the ameba to respond to chemical stimuli or the ability of plants to respond to light by means of plant hormones. This chapter gives additional examples.

We see fireworks because our eyes detect the presence of light.

Table 15.1

Types of Environmental Stimuli

Mechanical Stimuli
- Pressure
- Touch
- Motion
- Sound
- Vibration
- Gravity

Thermal Stimuli
- Heat
- Cold
- Infrared radiation

Electromagnetic Stimuli
- Visible light
- Electricity
- Magnetism

Chemical Stimuli
- Individual types of molecules

Animals can sense a variety of environmental stimuli. Their sensory receptors change environmental stimuli such as sound, light, and pressure into nerve impulses. Some receptors are composed of nervous tissue; others are not, but are capable of initiating a nerve impulse in an adjacent neuron. Many kinds of receptors have evolved among invertebrates and vertebrates, with each receptor sensitive to a different aspect of the environment (Table 15.1).

GENERALIZED SENSORY PATHWAY

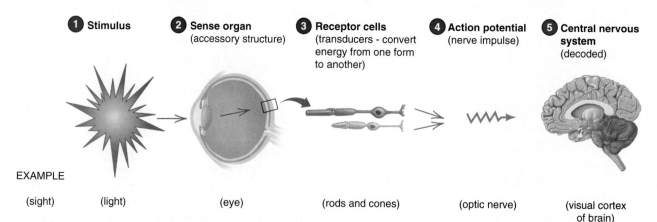

1. **Stimulus**
2. **Sense organ** (accessory structure)
3. **Receptor cells** (transducers - convert energy from one form to another)
4. **Action potential** (nerve impulse)
5. **Central nervous system** (decoded)

EXAMPLE

(sight)   (light)   (eye)   (rods and cones)   (optic nerve)   (visual cortex of brain)

Figure 15.1   Generalized sensory pathway.

For an organism to be aware of, or to sense, its internal or external environment, certain events must take place as shown in the example in Figure 15.1. First, a change in the environment (the stimulus) ❶ must occur in the presence of a sense organ ❷ and it must be of sufficient magnitude to open ion channels within the membranes of receptor cells ❸, thus depolarizing them (see Chapter 13). This depolarization is called a *generator potential*. When the generator potential reaches the threshold level, it initiates a nerve impulse (action potential ❹) in the sensory neurons with which the receptor synapses. In simple receptors (the dendrite endings of sensory neurons), the generator potential initiates an action potential along that same neuron. The impulse is conducted by nerve fibers to either the spinal cord or the brain ❺.

Impulses conducted to specific sensory areas of the cerebral cortex in vertebrates produce conscious sensations (see Chapter 14). Only the cerebral cortex can "see" a flower, "hear" a symphony, or "feel" a paper cut. Although not a part of the cortex, the thalamus can sense pain, but it is unable to distinguish its source or intensity. Impulses that end at the spinal cord or brainstem in vertebrates, or in a ganglion in invertebrates, do not produce conscious sensations. They may, however, result in reflex activity. Examples in humans are the rhythmic contraction of the muscles of breathing or turning the head toward a startling noise.

Receptors provide animals with information about:

1. The organism's internal environment
2. The organism's external environment
3. The organism's position in space

Receptors that sense the internal environment are located deep in the body within the walls of blood vessels and organs. They tell you when you are hungry, thirsty, sick, or tired. Receptors sensitive to stimuli outside the body are located at or near the body surface. They allow you to see, hear, taste, and feel various stimuli in the environment. Receptors that provide information about body position and movement are located in the muscles, tendons, joints, and inner ear. They tell you whether you are lying down or standing up and where the various parts of your body are in relationship to one another.

Sensory receptors are cells that can change environmental stimuli into nerve impulses. Specific receptors detect certain stimuli that inform the body about both its external and internal environments and its position in space.

15.2 Sensing the internal environment

Many of the neurons that monitor an organism's body functions are simply nerve endings that depolarize in response to direct physical stimulation—to temperature, to chemicals such as carbon dioxide diffusing into the nerve cell, or to a bending or stretching of the neuron cell membrane. Among the simplest of these neurons are those that report on changes in body temperature and blood chemistry. These neurons are involved in sophisticated regulatory mechanisms that maintain the body temperature and blood chemistry within a narrow range; these mechanisms are most well developed in mammals.

Temperature-sensitive neurons in the hypothalamus (highlighted in the illustration) act as your body's thermostat. These neurons constantly take your temperature by monitoring the temperature of your blood. If the temperature of the blood rises, such as when you

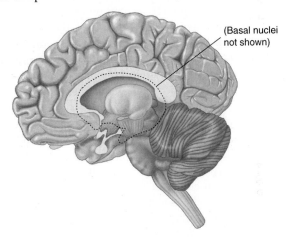

(Basal nuclei not shown)

are sick or vigorously exercising, neurons in the hypothalamus trigger your body's heat loss mechanisms. Such mechanisms include dilating, or widening, the blood vessels closest to the skin so that excess heat can be lost to the environment (see Figure 10.3). Likewise, if the temperature of the blood drops, such as when you are in a cold environment, neurons in the hypothalamus trigger your body's heat production mechanisms. Such mechanisms include shivering, a cycle of contraction and relaxation of your skeletal muscles producing body heat as a waste product of the cellular respiration that generates adenosine triphosphate (ATP) for muscle contraction.

Other receptors are sensitive to the levels of carbon dioxide and oxygen in your blood as well as its pH. These receptors are embedded within the walls of your arteries at several locations in the circulatory system. Bathed by the blood that flows through the arteries, these chemical receptors provide input to respiratory

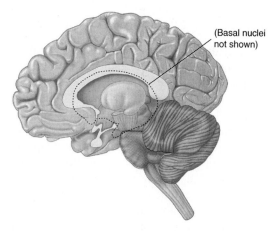

(Basal nuclei not shown)

centers in the medulla and pons (highlighted above), which use this information to regulate the rate of respiration. When carbon dioxide and pH levels in the blood rise and the oxygen level falls, the respiratory centers respond by increasing the respiration rate.

Various other receptors that sense the environment within your body have membranes whose ion channels open in response to mechanical force. Put simply, twisting, bending, or stretching these nerve endings results in depolarization of their membranes, causing these nerves to fire. These receptors differ from one another primarily in their location and in their orientation to the stimulus. Pain receptors, for example, are widely distributed throughout the body. They respond to many different types of stimuli when these stimuli reach a level that can endanger the body. Pain receptors deep within the body detect such internal environmental stresses as inadequate blood flow to an organ, excessive stretching of a structure, and spasms of muscle tissue.

Specialized mechanical receptors sensitive to pressure changes are attached to the walls of three major arteries of your body (the carotids and the aorta) and constantly monitor your blood pressure. These highly branched networks of nerve endings detect the stretching of the walls of these arteries caused by the "push" of the blood as it is pumped out of the heart. These neurons fire at a low rate at all times, sending a steady stream of impulses to the cardiac (heart) center in the medulla (shown above). A rise in blood pressure stretches the walls of these vessels, resulting in a higher frequency of nerve impulses, which inhibits the cardiac center. A drop in blood pressure lessens the stretch, resulting in a lower frequency of nerve impulses, which stimulates the cardiac center.

Receptors that sense the body's internal environment are located within the walls of blood vessels and organs.

15.3 Sensing position in space

Animals have receptors called **proprioceptors** (PRO-pree-oh-SEP-turz) that sense the position of the body in space. For example, proprioceptors tell you how much your arms and legs are bent, where they are in relationship to your body, and where your head is relative to the ground even if you cannot see.

The muscles of vertebrate animals (such as humans, fish, reptiles, and birds) have proprioceptors buried deep within them, keeping track of the degree to which the muscles are contracted. These receptors are actually specialized muscle cells called *muscle spindles*. Wrapped around each spindle is the end of a sensory neuron called a **stretch receptor (Figure 15.2)**. When a muscle is stretched, the muscle spindle gets longer and stretches the nerve ending, repeatedly stimulating it to fire. Conversely, when the muscle contracts, the tension on the fiber lessens and the stretch receptor ceases to fire.

For example, Figure 15.2 ❶ shows a student carrying books. The books are a constant load on the biceps muscle, which is balanced by a constant force. Equal impulses are sent in each direction ❷. In ❸, another student places his books in the arms of the first student to look for change for the soda machine. This sudden increase in load stretches the muscle. The muscle spindles send signals back to the spinal cord ❹. There, additional motor neurons are stimulated and send impulses back in the motor nerve ❺, causing increased contraction to balance the load ❻.

Other proprioceptors are found in vertebrate tendons, which is the connective tissue that joins muscles to bones, and in the tissue surrounding joints. These receptors are also stimulated when they are stretched and help protect muscles, tendons, and joints from excessive tension and pulling.

Gravity and motion receptors are other types of proprioceptors. Many animals are sensitive to the force of gravity and to motion; receptors for both types of stimuli help an animal maintain equilibrium, or balance. In humans, these receptors are located in the inner ear. Balance is discussed later in this chapter in Section 15.12.

Receptors called proprioceptors, located within the skeletal muscles, tendons, and inner ear, give the body information about the position of its parts relative to each other and to the pull of gravity.

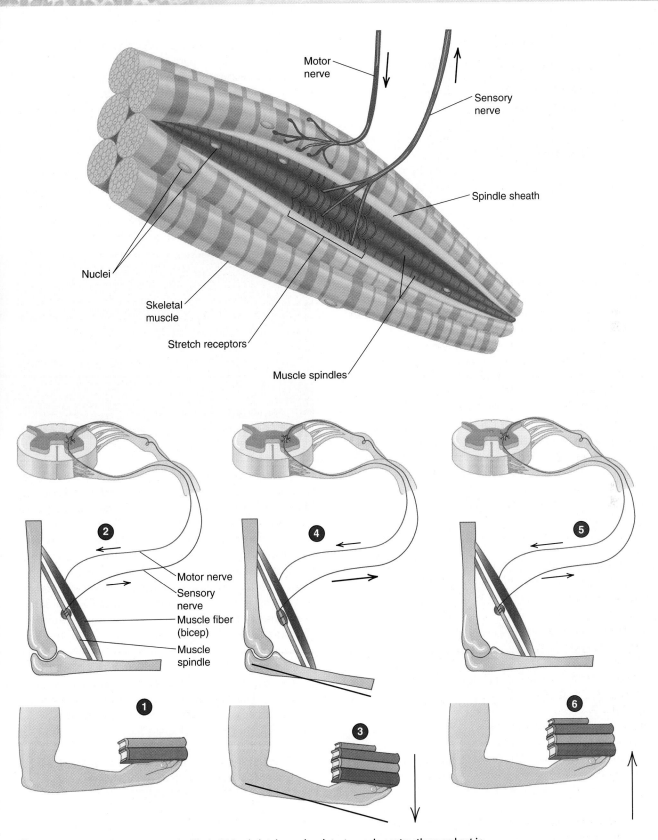

Figure 15.2 Stretch receptors embedded within skeletal muscles detect muscle contractions and act in the feedback control of muscle contraction.

15.3 Detecting mechanical energy such as touch, pressure, and pain

The body has a variety of *mechanoreceptors*, which are stimulated when they are physically deformed by mechanical energy such as pressure. **Figure 15.3** is a diagram of a cross section of skin showing these specialized nerve endings in humans.

The simplest nerve endings are free nerve endings, which are distributed throughout most of the body. In addition to detecting pain, free nerve endings detect temperature, itch, and movement. For example, free nerve endings wrap around the roots of hairs and detect any stimulus, such as the wind or an insect, that moves body hair.

Disk-shaped dendrite endings called *Merkel's disks* (MER-kelz) and egg-shaped receptors called *Meissner's corpuscles* (MYZ-nerz) are two types of touch receptors. (The word corpuscle means "little body.") Both are widely distributed in the skin, but are most numerous in the hands, feet, eyelids, tip of the tongue, lips, nipples, clitoris, and tip of the penis.

Pressure receptors called *Pacinian corpuscles* (pah-SIH-nee-an) are located deeper within the skin. Pacinian corpuscles look layered, much like an onion. The layers are made up of connective tissue with dendrites sandwiched between. These receptors are most numerous in

the nipples and external genitals of both sexes. Pacinian corpuscles are also a part of the body's internal sensing system because of their locations around joints and tendons, in muscles, and in certain organs.

Fishes and amphibians have a complex system of mechanoreceptors called a *lateral line system*. This system consists of a single row of mechanoreceptors along the sides of the body (see photo below) and others that form various patterns on the head. The lateral line system detects mechanical stimuli such as sound and movement. It also senses pressure changes (such as those produced by water currents).

Lateral lir

Invertebrates as well as vertebrates have mechanoreceptors. For example, flatworms have a variety of cells that sense pressure changes and mechanical stimuli. Many species of roundworms have external setae (SEE-tee) (thin, stiff, bristlelike structures) at various locations on the body. These are thought to be mechanical receptors. (Earthworms have setae also, but they function in locomotion.) Earthworms have nerve endings along their length that act as touch receptors and others that act as vibration receptors. Additionally, the tentacles of most cephalopods (octopus and squid) are studded with receptors sensitive to touch.

> The body has a variety of mechanoreceptors, which are stimulated when they are physically deformed by mechanical energy such as pressure. Invertebrates as well as vertebrates have mechanoreceptors. Fishes and amphibians have a complex system of mechanoreceptors along their bodies called a lateral line system, which detects stimuli such as sound, movement, and pressure changes.

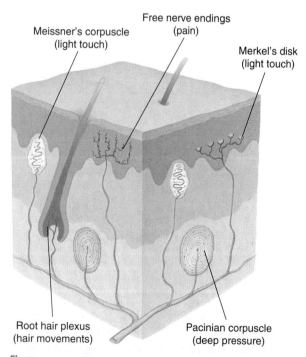

Meissner's corpuscle (light touch)

Free nerve endings (pain)

Merkel's disk (light touch)

Root hair plexus (hair movements)

Pacinian corpuscle (deep pressure)

Figure 15.3 **Receptors in the human skin.**

15.5 Detecting temperature

In humans and other vertebrates, two populations of
nerve endings buried in the skin are sensitive to changes
in temperature. The temperature of the environment is
sensed only indirectly. The temperature of the skin at
the level of these receptors is what is really sensed. The
environmental temperature, the
body's internal temperature, and
other factors that cool the body,
such as the evaporation of sweat,
all affect skin temperature.

One set of the body's ther-
moreceptors (temperature receptors) is stimulated by a lowering
of temperature (the cold receptors) and the other by a raising of
temperature (the heat receptors).
Scientists are unsure how these
receptors work. However, evidence suggests that changes in
the temperature cause changes in
the shapes of the proteins that
make up the sodium channels of
their cell membranes, leading to
the depolarization and firing of
these neurons.

In humans and other vertebrates, two populations
of nerve endings buried in the skin are sensitive to
changes in temperature.

15.6 Detecting infrared and ultraviolet radiation

Only a few animals can detect heat directly. Some
snakes, such as rattlesnakes and other pit vipers, can detect heat radiated from warm-blooded animals (infrared
radiation). Pit vipers detect infrared radiation by means

www.jbpub.com/biology

Figure 15.4 Pit organs of a rattlesnake.

Pit organ

of *pit organs*, which are located on either side of the
head between the nostrils and the eyes (Figure 15.4).
(The term pit viper refers to venomous [poisonous]
snakes and their pit organs.) Heat guides the direction
at which the rattlesnake strikes. It moves its head back
and forth; when the intensity of detected heat is equal
on both sides of the snake's head, it perceives that its
prey is straight ahead. (You may be familiar with
infrared sensors used in police and military he-
licoptors to detect humans and warm engines in
vehicles on the ground.)

While infrared radiation consists of wavelengths longer than visible light, ultraviolet
(UV) radiation consists of wavelengths shorter
than visible light. Neither UV nor infrared light
(radiation) is visible to humans. However, insects can see UV light. Many flowers emit patterns of UV light that are visual cues to insects
(see photos at left). The majority of flowering
plants depend on insects for pollination.

...ning primrose in normal light

. . . and in UV light.
Notice the pattern differences.

Some animals, such as rattlesnakes, can detect infrared radiation. Rattlesnakes and other pit vipers
determine where to strike based on the body heat
they detect. Insects can detect UV light and use
this ability to locate flowers.

15.7 Detecting electrical fields

Humans have no sensory organs to detect electric currents, but some fish, some salamanders, and the duck-billed platypus have receptors that can. *Electroreceptors* help these organisms find prey and navigate waters when the conditions are dark and/or murky.

Sharks can use their electroreceptors to detect electrical fields emitted by other fishes. Sharks have these receptors on their heads, and rays have them on their pectoral fins. Some species of fish, such as catfish, have electroreceptors in their lateral line systems.

Electric fish generate electric fields as well as detect them. These fish, such as electric eels, generate a weak electrical field around themselves by means of electric organs, which consist of modified muscle tissue. Electric fish detect objects or organisms in the environment when the objects' conductivity differs from the surrounding water and they distort the fish's electric field. The lines of flow converge on an object of higher conductivity, such as another fish, and diverge around a poorer conductor, such as a rock (Figure 15.5).

The electroreceptors of the duck-billed platypus recently have been found to be located in its bill. This egg-laying mammal finds its food in muddy streams, diving with its eyes, ears, and nose closed. Its ability to home in on weak electric fields with its bill helps it search out prey.

> Some fish, some salamanders, and the duck-billed platypus have receptors that can detect electrical fields. Electroreceptors help these organisms find prey and navigate dark waters.

15.8 Detecting magnetic fields

A wide variety of animals are affected by magnetic fields, including salmon, salamanders, turtles, hornets, honeybees, and homing pigeons. Scientists have found that most animals affected by magnetic fields have magnetic granules containing iron (magnetite) in their bodies. Certain bacteria are also affected by magnetic fields; they also contain similar magnetic particles.

There is evidence that some migratory animals (Figure 15.6) and those with a homing instinct use the Earth's magnetic lines of force to help orient themselves. Scientists have also observed that magnetic bacteria living in the ocean mud of the Northern Hemisphere consistently swim toward magnetic north. In their natural environment, these bacteria swim north and orient themselves deep in mud and away from oxygen, a toxin to them. The mechanism of a magnetic sense in animals and bacteria is still being researched; it is not understood at this time.

> There is evidence that some migratory animals and those with a homing instinct detect magnetic fields and use the Earth's magnetic lines of force to help orient themselves.

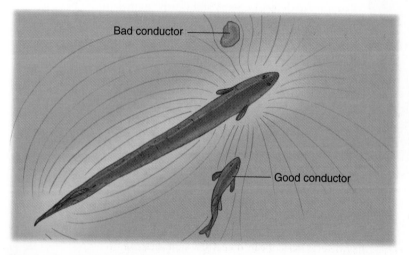

Figure 15.5 Electric field differentiation between good and bad conductors in the eel.

Figure 15.6 Loggerhead marine turtles migrate back to the same beach they left as hatchlings.

15.9 Taste and smell: Chemoreception

Chemoreception in Organisms Other than Humans

The ability to sense chemicals, or *chemoreception*, is widespread in the animal kingdom. Many flatworm species have a variety of cells that sense chemicals, such as those in potential foods. Roundworms are more complex in that they have *chemoreceptors* located on their anterior end, and some species have additional chemoreceptors on their posterior end. Earthworms have chemoreceptors distributed along the length of the body, while cephalopods (squid and octopus) have tentacles covered with chemoreceptors. Insects have their six legs studded with chemoreceptors; such receptors are also found on their mouthparts and elsewhere on the body (as can be seen dispersed over the head of the acrobat ant shown above right).

Used for finding food or mates, locating enemies, or warning an animal of danger, chemoreception is most highly developed in the vertebrates and arthropods (spiders, lobsters, and insects). However, even the single-celled protozoans respond to chemical stimuli, although they have no sense organs. Like the white blood cells that roam the human body fighting infection, protozoans detect chemicals when these stimuli reach their cell membranes. They respond with directed movements either toward or away from the stimuli. Such directed movements are called *taxes* (sing. *taxis*). A directed movement toward or away from a particular chemical is a positive or negative chemotaxis.

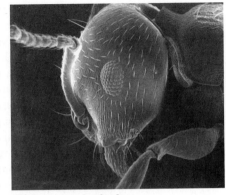

Smell and taste are the two human types of chemoreception. Smell usually means sensing chemicals in the air; taste usually means sensing chemicals in water (in solution). These terms are applied to chemoreception in other animals, too. The catfish, for example, has taste buds distributed all over its body. Imagine going for a swim and using your back or chest for tasting!

Taste

Taste receptors—**taste buds**—detect chemicals in the foods you eat. Humans detect many taste qualities, far more than the four primary tastes: salty, sweet, sour, and bitter. (A potential fifth primary taste, monosodium-glutamate-induced savory taste, or umami, has been suggested by some sensory physiologists.)

Each of the 10,000 taste buds on the upper surface of the tongue, the lips, and the throat responds in varying degrees to the four primary tastes, but generally responds to one taste more than the others. Taste buds that are most responsive to sweet tastes are grouped on the tip of the tongue (Figure 15.7). Salty and sour taste buds are along the sides of the tongue, with bitter taste buds in the back. The rich and diverse array of tastes humans perceive are composed of different combinations of impulses from just these four types of chemoreceptors. In addition, the sense of taste interacts with the sense of smell to produce the taste sensation, as you may have noticed when you had a "stuffy nose" and lost your ability to taste.

Taste buds are packed within short projections known as *papillae*. Figure 15.7 shows a drawing of a typical papilla on the tongue and points out the location of taste buds deep within the papilla. Taste buds are microscopic structures shaped like tiny onions. Each is made up of 30 to 80 receptor cells bound together by support cells. Hairlike

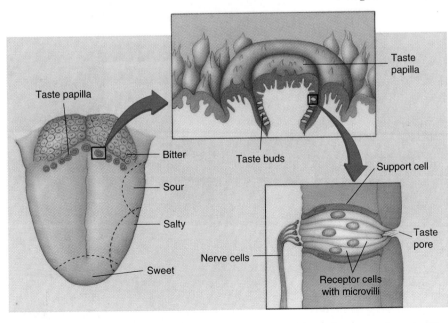

Figure 15.7 Humans have four kinds of taste buds that respond in varying degrees to each of the four primary tastes (bitter, sour, salty, and sweet).

projections of the receptor cells poke through an opening in the taste bud called a *taste pore*. The receptor cells are stimulated by the various chemicals in food as these chemicals dissolve in the saliva and come into contact with the cellular "hairs." A certain threshold of stimulation produces generator potentials within the receptor cells, firing neighboring sensory nerve fibers. These nerve fibers travel from the tongue and throat to the cerebrum, medulla, and thalamus. Some of these nerve fibers also travel to the reticular formation (see Chapter 14), the part of the brain that monitors incoming stimuli, identifying important information.

Smell

Although smell receptors are structurally the simplest of the special senses in humans, they can distinguish several thousand odors. Smell receptors are also called **olfactory receptors** (ole-FAK-tuh-ree), a name derived from the Latin words *oleo*, "a smell" and *facio*, "to make." They are *not* located (surprisingly) in your nostrils, but in a one-half-inch square of tissue in the roof of the nasal cavity—just behind the bridge of your nose (Figure 15.8).

Olfactory receptors consist of neurons whose cell bodies are embedded in the nasal epithelium, the tissue that lines the nasal cavity. Dendrites extend from these cell bodies. Cilia, microscopic projections of the dendrites, poke out of the epithelium like minute tufts of hair and are bathed by the mucus that covers this tissue. Gases in the air dissolve in the mucus and come into contact with the cilia.

Scientific evidence suggests that the olfactory receptors detect different smells because of specific binding of airborne gases with receptor proteins located within the cilia. This interaction opens ion channels within the membrane of the receptor so that a generator potential is developed, firing the neuron when it reaches a particular threshold. From here, the nerve impulse travels to the olfactory area of the cerebral cortex to be interpreted. On its way, it travels through the brain's limbic system (see Chapter 14), the area of the brain responsible for many drives and emotions. Does the odor of baking cookies please you? Does the smell of rotting garbage cause you to turn your head in disgust? That is your limbic system at work in conjunction with your cerebrum and your sense of smell.

Sensing chemicals, or chemoreception, is a widespread ability in the animal kingdom. In humans, the taste receptors, or taste buds, are microscopic chemoreceptors embedded within the papillae of the tongue, lips, and throat. They work with the sense of smell to produce the taste sensation. The smell, or olfactory, receptors are located in the nasal epithelium and detect airborne chemicals as they bind with receptor chemicals.

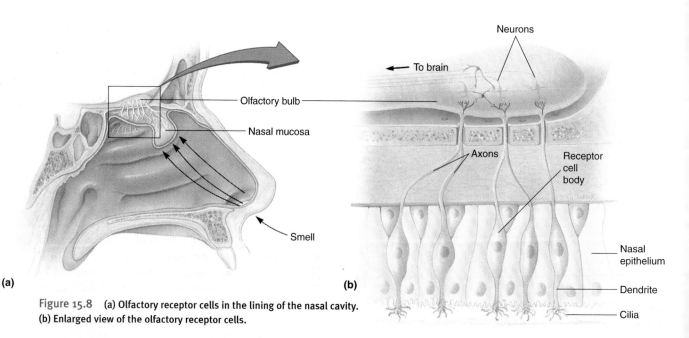

Figure 15.8 (a) Olfactory receptor cells in the lining of the nasal cavity. (b) Enlarged view of the olfactory receptor cells.

15.10 Sight: Photoreception

Photoreception is the sensing of visible light, which consists of certain wavelengths of electromagnetic energy, the repeating disturbances in electrical and magnetic fields in the atmosphere. You can visualize these disturbances as waves, much like the repeating disturbances or tiny waves caused by a stone thrown in a still pond.

Visible light is only a part of the full range of the electromagnetic waves coming from the sun; collectively, this range of waves is called the *electromagnetic spectrum* (see Figure 6.2). The waves in this spectrum vary in length. The shortest wavelengths are gamma rays and the longest are radio waves. Visible light has wavelengths in between these two extremes, but it, too, is made up of many different wavelengths. Remember the last rainbow you saw? Its array of colors was caused by the separation of the various wavelengths of visible light as they passed through tiny droplets of water in the air. The infrared radiation sensed by pit vipers (see Section 15.6) consists of wavelengths longer than visible light.

Photoreception in Organisms Other than Humans

Sensitivity to light is extremely widespread. For example, organisms other than animals, such as certain protozoans, can detect light. In fact, the red, cup-shaped, light-sensing organelle of certain flagellated protozoans, the *stigma* or *eye spot*, is one of the few true sensory

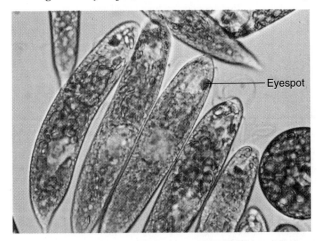

Eyespot

organelles known among protozoans. *Euglena* (shown above) is the best known in this photoreceptive group.

Although many invertebrates have "eyes," these sensory organs are not image-forming in some. For example, various species of flatworms have eyes on their anterior ends and earthworms (annelids) have photore-

ceptor cells in their epidermis along the body. However, like the protozoan eyespot, the flatworm and earthworm "eyes" are only sensitive to light and changes in its intensity. These receptors contain light-absorbing pigments that send impulses to the animal's brain when stimulated. In organisms having a *positive phototaxis,* a behavioral adaptation in which an organism moves toward the light, the animal moves until the sensations coming from the eyes are equal and strong. Many species of flying insects exhibit positive phototaxis. In organisms having a *negative phototaxis*, the animal moves until the sensations coming from the eyes are equal and weak. The common cockroach is an example of an animal that exhibits negative phototaxis.

Most cephalopods (squid and octopus) have image-forming eyes similar to the eyes of vertebrates. The traditional view of biologists is that this is an example of convergent evolution in which both groups evolved eyes independently of one another, yet their eyes are quite similar in structure. Recent work in genetics by Walter J. Gehring and his colleagues at the University of Basel in Switzerland suggest that this may not be the case. They have found evidence that the prototypic eye arose only once (not more than 40 times) in evolution. Gehring holds that subsequent convergent evolution gave rise to the image-forming eyes of vertebrates and cephalopods,

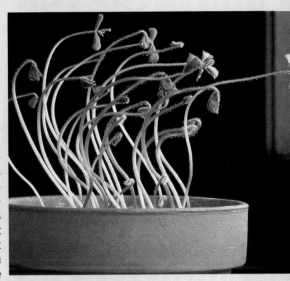

Focus
on Plants

Plants and light detection

Plants bend toward the light—this is a well-known phenomenon. However, plants have no nervous system and no sense organs. How does this happen? When a plant is exposed to light on one side only, a shadow or darker side results. Under such conditions, the plant growth hormone *auxin* redistributes, resulting in a higher concentration of this hormone on the darker side of the stem. Thus, the side of the plant away from the light grows faster, causing the plant to bend toward the light. This response of plants to light is known as *phototropism.*

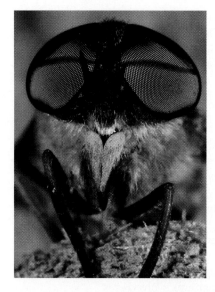

whereas the compound eyes of insects (see next paragraph) resulted from divergent evolution. (Dr. Gehring's related work in development is discussed in the opening paragraphs of Chapter 23).

Arthropods (spiders, crabs, lobsters, and insects) are another group of animals that have eyes. Some species, such as the flatworms and annelids, have only light-sensitive eyes, but many species, especially many insect species, have image-forming eyes called *compound eyes,* as shown in the photo of a horsefly to the left. Horror movies have led us all to believe that animals with compound eyes (such as flies) see thousands of identical, complete images. This is not the case. Each unit of the multifaceted compound eye "sees" only a part of the complete image. The brain puts the "pieces" together to form a single image.

Sight

Image-forming eyes are either of the multifaceted type as in insects, or the single-lens, cameralike type as in cephalopods and vertebrates. The cameralike (human)

eye works basically like this: receptors sensitive to the various wavelengths of visible light are located in the back of the eye and act somewhat like film in a camera. Light that falls on the eye is focused by a lens onto these receptors, just as the lens of a camera focuses light on film.

The structure of the human eye is shown in **Figure 15.9.** Human eyes are each about 1 inch in diameter and are covered and protected by a tough outer layer of connective tissue called the **sclera** (SKLEAR-uh). The front of the eye is transparent, allowing light to enter the eye. This portion of the eye's outer layer is called the **cornea.** Because the cornea is rounded, it not only allows light to enter the eye, but bends it as well. This bending, or refraction, of light occurs as the light waves slow down as they move from the air and pass through the tissue of the cornea. You can see this phenomenon if you put a straw in a glass of water. The parts of the straw both outside and inside the water do not appear connected. If the cornea were flat, the light waves or rays would exit the cornea traveling in a path parallel to their original path. However, the cornea is rounded; the light waves that enter above, below, or to the side of the center of the cornea are bent slightly toward one another. Therefore, the cornea is the first structure of the eye that begins to focus incoming light onto the rear of the eye.

A compartment behind the cornea is filled with a watery fluid called the **aqueous humor.** The word *hu-*

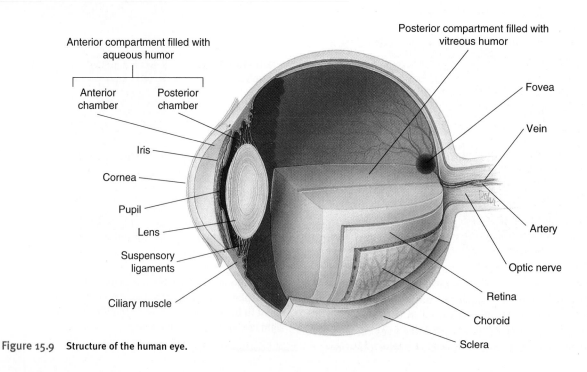

Figure 15.9 **Structure of the human eye.**

mor refers to "fluid within the body," and the word *aqueous* means "water." Fresh aqueous humor is continually produced as old fluid is drained into the bloodstream. This fluid nourishes both the cornea and the lens, because neither structure has a supply of blood vessels. In addition, the aqueous humor (along with fluid farther back in the eye called the *vitreous* [Latin for "glass"] *humor*) creates a pressure within the eyeball that maintains the eyeball's shape. It also keeps the **retina** (RET-un-nuh), the sensory portion of the eye, pressed properly against the back of the eyeball.

The **lens** lies just behind the aqueous humor and plays a major role in focusing the light that enters the eye onto the retina. It looks much like a lemon drop candy or a somewhat flattened balloon. If you were to cut a lens in half, it would look similar to an onion because it is made up of layer upon layer of protein fibers. Ligaments encircle the lens and suspend it within the eye. These ligaments are attached to a tiny circular muscle called the **ciliary muscle** (SILL-ee-err-ee) that, by contracting or relaxing, slightly changes the lens shape of the lens. The greater the curve of the lens, the more sharply it bends light rays toward one another.

In individuals with normal vision (**Figure 15.10**, top), this bending of the light results in its being focused on the retina. In focusing on distant objects, the lens is flattened ❶. In focusing on near objects, the lens is more rounded ❷. Nearsighted people, however, have an elongated eyeball ❸ or a thickened lens. In either case, the light from distant objects is focused in front of the retina. (Compare with ❶.) The near object can be brought into focus, while distant objects cannot. Nearsightedness can be corrected with a concave lens that spreads the light rays entering the eye from distant objects so that they focus on the retina ❹. Some people are farsighted, having either a shortened eyeball ❺ or a thin lens. Light from near objects is focused behind the retina (compare with ❷), whereas light from distant objects can be focused on the retina. Farsightedness can be corrected with a convex lens that converges entering light rays so that they focus on the retina ❻. (Although light rays are reflected to the eye from all points on an object, only two rays fom the top of the object and two from the bottom are shown in Figure 15.10 for simplicity.)

The **iris**, which lies between the cornea and the lens, controls the amount of light entering the eye. It works much like the diaphragm of a camera, reducing the size of the **pupil** of the eye (what you see as a black dot) through which the light passes. Together, the ciliary muscle and the iris make up two of the structures of the middle layer of the eye. The third structure is the **choroid** (KOR-oyd). As shown in Figure 15.9, the

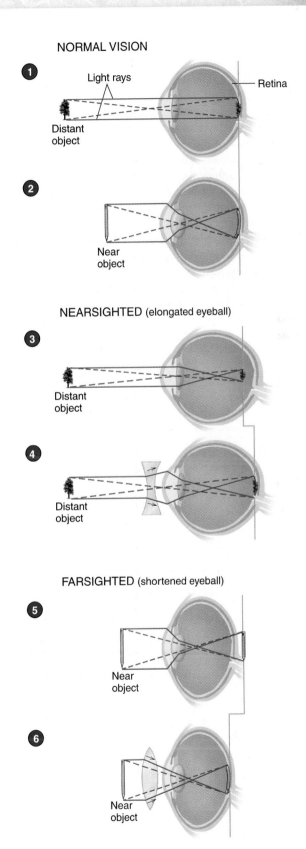

NORMAL VISION

❶ Light rays / Distant object / Retina

❷ Near object

NEARSIGHTED (elongated eyeball)

❸ Distant object

❹ Distant object

FARSIGHTED (shortened eyeball)

❺ Near object

❻ Near object

Figure 15.10 Focusing the human eye.

choroid extends toward the back of the eye as a thin, dark-brown membrane that lines the sclera. It contains vessels carrying blood that nourishes the retina and a dark pigment that absorbs light rays so that they will not be reflected within the eyeball.

The inner layer of the eye is the retina. The retina lines the back of the eye and contains the eyes' sensory receptors: the **rods** and **cones** (Figure 15.11). The tips of these cells contain the pigment *rhodopsin* that undergoes a chemical alteration when activated by light. The activated rhodopsin triggers a series of events that are responsible for *closing* sodium channels. As discussed in Chapter 13, stimuli usually lead to an opening of sodium channels, allowing sodium ions to rush into the cell, depolarizing the receptor membrane. However, in rod and cone cells, sodium channels close when they are stimulated. The result is *hyperpolarization*—the

interior of the rod becomes even more negatively charged than before. Hyperpolarization of the cell in turn activates a mechanism that *lowers* the amount of neurotransmitter released by the cell. This decrease in neurotransmitter release lessens the frequency of firing in adjacent neurons—nerve signals that eventually reach the brain and are interpreted as patterns of light and dark.

Rod cells are stimulated by low levels of light and detect only the presence of light and not its color; thus vision with rod cells alone is black, white, and shades of gray. Did you ever try to see color in extremely dim light? Your rods alone are at work, and they cannot see color. Color vision is achieved by your cone cells, which function in bright light.

There are three kinds of cone cells (although they all look the same under the electron microscope). Each

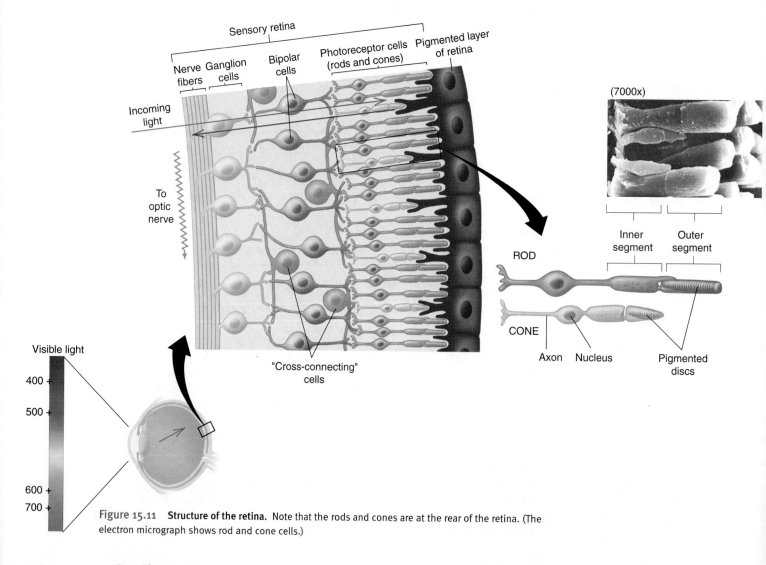

Figure 15.11   Structure of the retina. Note that the rods and cones are at the rear of the retina. (The electron micrograph shows rod and cone cells.)

type of cone cell possesses rhodopsin molecules that have slightly different shapes from one another because they have different types of opsin. These differences determine which wavelengths of light the pigment (and therefore the cone cell) will absorb. One type absorbs wavelengths of light in the 455-nanometer (blue-absorbing) range. Another type absorbs in the 530-nanometer (green-absorbing) range, and the third in the 625-nanometer (red-absorbing) range. The color you perceive depends on how strongly each group of cones is stimulated by a light source. The pigment in rod cells, on the other hand, absorbs in the 500-nanometer range. This range encompasses the three primary colors (blue, green, and red), which together make up white light.

The human retina contains about 7 million cones, most of which are located at a central region of the retina called the **fovea** (FOE-vee-uh). The lens focuses incoming light on this spot, which is the area of sharpest vision because of the high concentration of cones. There are no rods in the fovea. Some are located just outside the fovea and increase in concentration as the distance from the fovea increases. The area immediately surrounding the fovea is called the *macula lutea* (MACK-yeh-leh LOO-tea-eh) (Latin for "yellow spot") and is an area of sharp vision because of its abundance of cones. In **Figure 15.12,** you can see both the macula lutea and the blind spot, or optic disc, which is where the optic nerve leaves the eye (see next paragraph). It is devoid of photoreceptors, but the brain "fills in" the missing spot.

Each foveal cone cell makes a one-to-one connection with a special kind of neuron called a *bipolar cell* (see Figure 15.11). Each of the bipolar cells is connected in turn to a ganglion cell, whose axon is part of the **optic nerve**. The bipolar cells receive the hyperpolarization stimulus from the cone cells and transmit a depolarization stimulus to the ganglion cells. The axons of the ganglion cells transmit the impulses to the brain. The frequency of impulses transmitted by any one receptor provides information about light intensity. The pattern of firing among the different foveal axons provides a point-to-point image. The different cone cells provide information about the color of the image.

The relationship of receptors to bipolar cells to ganglion cells is one-to-one-to-one within the fovea. Put simply, each receptor (cone) cell in the fovea synapses with its own bipolar cell that in turn synapses with its own ganglion cell. Outside the fovea, however, the output of many cones and rods is channeled to one bipolar cell. Many bipolar cells, in turn, synapse with one ganglion cell. In fact, in the outer edge of the retina, more than 125 receptor cells feed stimuli to each ganglion cell in the optic nerve. In this outer, or peripheral, region,

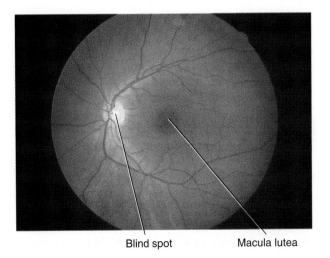

Blind spot   Macula lutea

Figure 15.12   Macula lutea and blind spot in the human eye.

many additional neurons cross-connect the ganglion cells with one another and carry out extensive processing of visual information. As a result, this portion of the retina does not transmit a point-to-point image as the fovea does, but instead transmits a processed version of the visual input that may be interpreted simply as movement. Think of the periphery of your retina as a detector, whereas your fovea is an inspector.

The positioning of your two eyes on each side of the head sends two slightly different sets of information to the brain about what you are seeing. Each set of information is slightly different because each eye views an object from a slightly different angle. This slight displacement of images gives *depth perception*—a three-dimensional quality—to your sight.

Photoreception, the sensing of visible light, is an extremely widespread characteristic among living things. In humans, light receptors are rod and cone cells located within the retina at the back of the eye. The pigment within these cells absorbs light, which causes a series of events leading to a hyperpolarization of the receptor cells and a subsequent firing of adjacent neurons. Rod cells function in dim light and detect white light only, whereas cone cells function in bright light and detect color. Light enters the pupil of the eye, and the lens focuses it on the back of the eye, on an area of the retina that is rich in cone cells. These cells, along with surrounding rod and cone cells, initiate nervous impulses that travel to the cerebrum via the optic nerve.

15.11 Hearing: Sound reception

Sound is a type of mechanical energy resulting from the vibration of an object. This vibration disturbs the air around the object, pushing the molecules in the air closer to one another, compressing them. Areas of molecules that are not compressed follow these areas of compression. If you could visualize these disturbances, they would look much like the concentric circles of waves on the surface of water that occur when a pebble is thrown into a still pond. Vibrating objects can cause sound waves in substances other than air, too, such as in water, glass, iron, or wood.

Sound Reception in Organisms Other than Humans

In invertebrates, the organs responsible for perception of vibration are very different from the vertebrate ear. Scientists have been unable to identify vibration or sound receptors in many phyla of invertebrates, but they do know that earthworms have vibration receptors located along the length of their bodies. They also know that insects perceive sounds, many of which they make themselves to attract mates.

Many animals perceive sounds humans cannot hear, such as the infrasound heard by elephants mentioned in the chapter opening paragraphs. Dogs can hear high-pitched sounds our ears cannot. Whales, dolphins, shrews, some birds, and bats can hear high-pitched sounds also—those extremely high-frequency (short wavelength) sound waves used to judge distance called *echo location* or *animal sonar*. Most well developed in bats, echo location allows an animal to gather information about the environment from hearing the "echos" of sound waves it has emitted as they bounce off living and nonliving things.

Hearing

The human ear has three parts: the outer ear, middle ear, and inner ear. The outer ear and middle ear work together to transmit sound waves to the inner ear, where sound stimuli are changed into nerve impulses (Figure 15.13).

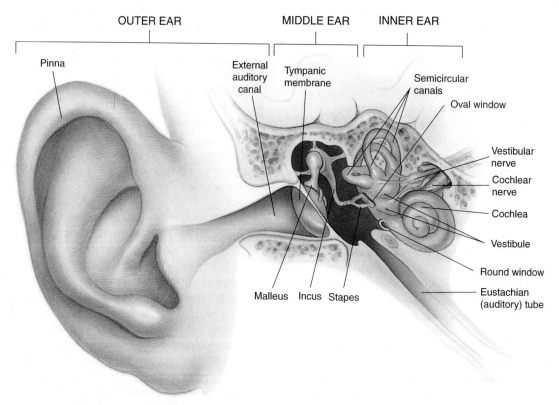

Figure 15.13 Structure of the human ear. The human ear is composed of outer, middle, and inner sections. The outer ear extends from the pinna to the tympanic membrane. The middle ear contains bones that transmit sound vibrations from the tympanic membrane to the cochlea. The cochlea contains the organ of hearing and makes up part of the inner ear. (Not to scale.)

Just Wondering . . . *Real Questions Students Ask*

How does exposure to loud noise levels contribute to a loss of hearing?

Listening to loud sounds contributes to hearing loss by damaging hair cells in the inner ear. As described in Section 15.11, the hair cells of the inner ear form part of the organ of Corti, which is the organ of hearing. These hair cells are embedded in a "roof" called the tectorial membrane, and the entire organ of Corti sits on the basilar membrane below. When the basilar membrane vibrates, the hairs move up and down, bending when they are pushed upward. Very loud sounds can cause violent vibrations of the basilar membrane, which may result in the hairs breaking or becoming permanently deformed. Such damage leads to partial hearing loss. The scanning electron micrograph shows portions of the organ of Corti with rows of hair cells. The photo on the left shows hair cells from a normal guinea pig. The photo on the right shows hair cells from a guinea pig after it experienced 24 hours of 120 decibel noise—about the level of sound at a rock concert.

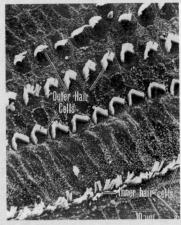
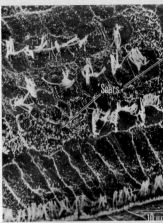

The flaps of skin on the outside of the head that are called ears are only one part of the outer ear. Each flap, or **pinna** (PIN-uh), funnels sound waves into an auditory canal. The **external auditory canal**, which is about 1 inch long, leads directly to the eardrum.

The eardrum, or **tympanic membrane**, is a thin piece of fibrous connective tissue that is stretched over the opening to the **middle ear**. On its external side, the eardrum is covered with skin. On its internal side, the eardrum is connected to one of the smallest bones in the body—the *hammer,* or **malleus** (MAL-ee-us). The malleus is connected to another tiny bone called the *anvil,* or **incus** (ING-kus). The anvil, in turn, is connected to a third bone called the *stirrup,* or **stapes** (STAY-peez). These three bones are the structures of the middle ear that pick up sound vibrations from the outer ear and transfer them to the inner ear.

Sound waves entering the outer ear beat against the tympanic membrane, causing it to vibrate like a drum. Vibrations of this membrane cause the malleus to move with a rocking motion because it is attached to the internal surface of the tympanic membrane. This rocking is transferred, in turn, to the incus and stapes. These three bones are hinged to one another in a way that produces a lever system, a mechanism that causes them to act like an amplifier and increase the force of the vibrations.

Also located in the middle ear is an opening to the **eustachian tube** (you-STAY-kee-en *or* you-STAY-shun) (a structure named after an Italian physician who lived in the 1500s). This tube is also called the auditory tube. The eustachian (auditory) tube connects the middle ear with the nasopharynx (the upper region of the throat). Its function is to equalize air pressure on both sides of the eardrum when the outside air pressure is not the same as the pressure in your middle ear. You have probably had the experience of your ears popping while you were driving up or down a mountain, taking off and landing in an airplane, or deep-sea diving; it is the result of the pressure equalization between these two sides of the eardrum.

The stirrup, the third in the series of middle ear bones, is attached to a membrane that separates the middle ear from the inner ear. This membrane covers the **oval window**, the entrance to the inner ear, the part of the ear in which hearing actually takes place. In addition, other parts of the inner ear detect motion and the effect of gravity on the body.

The oval window is the entrance to the fluid-filled **cochlea** (KOCK-lee-uh)—the part of the inner ear that contains the organ of hearing (Figure 15.14). To understand the cochlea's structure and function, imagine a tapering, blind-ended tube having two membrane-covered holes at its wider end. The upper hole is the oval window, and the lower hole is the **round window**. Also imagine that another, smaller membranous tube runs down its center, serving as a partition between the oval and round windows and bisecting the large tube into upper and lower channels, or canals. The upper channel is the *vestibular canal;* the lower channel is the *tympanic canal.* The inner tube stops short of the blind, tapered end of the outer tube, however, so the two canals connect with one another there. Imagine this tube rolled up like a jellyroll, and you have the basic structure of the cochlea. In fact, the word cochlea comes from a Latin word meaning "snail" and describes the rolled-up shape of this structure extremely well.

The inner tube of the cochlea is called the *cochlear duct,* and it contains specialized cells that are the receptors of hearing. These cells are called *hair cells* and are embedded in supporting cells. Together, the hair cells and the supporting cells are the organ of hearing, called the **organ of Corti**. The floor of the inner tube is the *basilar membrane* (BASS-ih-ler). The hairs that project from the hair cells stick up into the cochlear duct and are embedded in a covering called the *tectorial membrane* (tek-TORE-ee-ul).

How does the organ of hearing detect sound? As the stapes rocks in the oval window, it sets the fluid within the vestibular canal in motion, as shown in **Figure 15.15** ❶. As the oval window membrane rocks inward, it pushes on the fluid in the vestibular canal. This fluid moves forward in waves that pass into the tympanic canal ❷. When the waves reach the membrane-covered round window, it bulges outward ❸, compensating for the pressure increase. However, none of this movement results in the perception of sound. What you hear is the result of the following events: The waves in the fluid of the vestibular canal also push on the thin vestibular membrane ❹. This "push" transfers through the cochlear duct and causes the basilar membrane to move up and down (vibrate) ❺. The organ of Corti sits on the basilar membrane, so the hair cells move up and down when the basilar membrane vibrates. The hairs, embedded in the stiff and stationary tectorial mem-

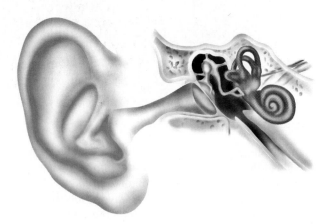

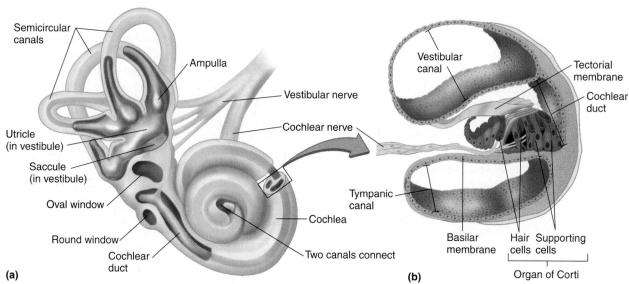

Figure 15.14 Structure of the inner ear. (a) The inner ear contains the organ of hearing (organ of Corti) and the organs that are responsible for maintaining the body's equilibrium (semicircular canals). (b) The larger view shows a section of the cochlea, which contains the organ of Corti. Hair cells in the organ of Corti detect sound and send the information to the brain via the cochlear nerve. The vestibular and cochlear nerves join to form the eighth cranial nerve.

brane, are therefore bent back and forth with the vibration **6**. The bending of the hairs are the stimuli that generate action potentials along neurons that are all part of the auditory (cochlear) nerve **7**. The brain interprets these impulses.

The brain is able to interpret pitch (the highness or lowness of a sound) because sounds of different pitches produce sound waves of different frequencies (numbers of waves per second). These differences set up differing wave patterns within the fluid of the inner ear that cause specific regions of the basilar membrane to vibrate more intensely than others. In addition, louder sound waves cause greater vibrations of the basilar membrane than can be interpreted by the brain. Repeated exposure to extremely loud noises such as gunshots, jet engines, and loud music can damage the hairs of the receptor cells and cause partial but permanent hearing loss (see Just Wondering, p. 323).

Sound is a type of mechanical energy resulting from the vibration of an object. Many animals can detect sound. Some animals can detect sounds that humans cannot. The ear is the organ of sound detection in humans. It has three parts: the outer, middle, and inner ear. The outer ear funnels sound waves in toward the eardrum, which changes these waves into mechanical energy. This energy is then transmitted to the bones of the middle ear, which increase its force. The inner ear, a complex of fluid-filled canals, contains the organ of hearing. The receptor cells of this organ change the mechanical sound energy into nerve impulses, which are transmitted to the brain, which interprets them.

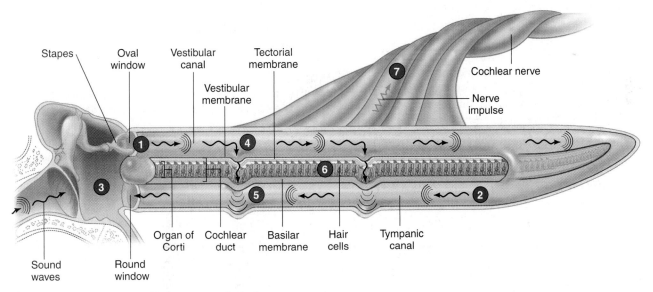

Figure 15.15   Transmission of pressure waves in the inner ear.

15.12 Balance

Sensitivity to the force of gravity is extremely widespread in the animal kingdom and allows animals to distinguish up from down. In most animals, hearing and balance are related, and the mechanisms involved in balance are strikingly similar. In invertebrates, gravity receptors are called *statocysts*. They have a similar plan to the organ of balance in vertebrates: a stonelike body (*statolith*) rests on sensory hairs. Displacement of the stone or stones relative to the force of gravity causes the statolith to stimulate sensory cells, which then send information to the central nervous system. Acceleration is sensed in a manner similar to that of vertebrates also: An enclosed fluid is set in motion when the animal moves. The fluid bends sensory cells in a direction that depends on the direction of acceleration.

In humans (our vertebrate example), the inner ear has two parts that are each structured as a fluid-filled tube within a tube like the cochlea, but their shapes are different from that of the cochlea. As shown in Figure 15.14, the cochlea makes up one side of the inner ear. The bulge in the midsection of the inner ear is called the **vestibule**, and it contains structures that sense whether you are upside down or right side up. In other words, it detects the effects of gravity on the body.

Inside this bulge are two sacs called the **utricle** (YOO-trih kul) and the **saccule** (SAK-yool) (Figure 15.16, ❶ and ❷). Both words mean "a little bag or sac." Within each of these sacs is a flat area composed of both ciliated receptor cells ❸ and nonciliated supporting cells. A layer of jellylike material is spread over the surfaces of these cells. Their long cilia and thin cell extensions are embedded within the jelly. Also embedded within the jelly are small pebbles of calcium carbonate called

otoliths (OH-teh-liths) (literally, "ear stones") ❹. When the head is moved, the otoliths pull on the jelly, which pulls on the cilia, bending them ❺. Any shift in the position of the otoliths results in different cilia being bent to specific degrees and in specific directions. These stimuli initiate generator potentials that are transmitted to neurons that make up a branch of cranial nerve VIII (see Chapter 14). The brain interprets these messages, resulting in your perception of "up" with respect to the pull of gravity.

Positioned above the saccule and utricle are three fluid-filled **semicircular canals** (see Figure 15.14). These loops are oriented at right angles to one another in three planes. At the base of each loop is a group of ciliated sensory cells that are connected to neurons. Lying above these cells is a mass of jellylike material. When the head moves, the fluid within the semicircular canals moves and pushes the jelly (and the cilia) in a direction opposite to that of the motion. (You have experienced this phenomenon when you accelerate suddenly in your car and your head is thrown backward.) This movement initiates the depolarization of the cell membrane, which triggers a nerve impulse. Because the three canals are each oriented differently, movement in any plane is sensed by at least one of them. Complex movements are analyzed in the brain as it compares the sensory input from each canal.

Sensitivity to the force of gravity is extremely widespread in the animal kingdom and allows animals to distinguish up from down. In humans, the vestibule of the inner ear detects position with respect to gravity, whereas the semicircular canals detect the direction of movement.

Figure 15.16
Balance is maintained by the semicircular canals.

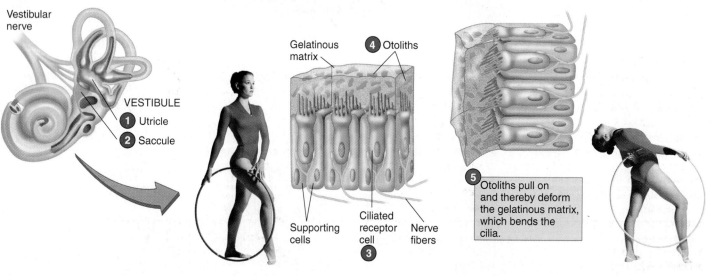

Vestibular nerve

VESTIBULE
❶ Utricle
❷ Saccule

Gelatinous matrix   ❹ Otoliths

Supporting cells

Ciliated receptor cell ❸

Nerve fibers

❺ Otoliths pull on and thereby deform the gelatinous matrix, which bends the cilia.

Key Concepts

Sensory receptors detect stimuli from both the external and internal environments.

15.1 Animals possess an interesting array of sense organs, specialized structures that trigger nerve impulses, which ultimately provide information about the surroundings.

15.2 Receptors that sense the body's internal environment are located within the walls of blood vessels and organs.

15.3 Receptors located within the skeletal muscles, tendons, and inner ear give the body information about the position of its parts relative to each other and to the pull of gravity.

General sense receptors detect external stimuli such as touch, pressure, pain, and temperature.

15.4 The body has a variety of mechanoreceptors, which are stimulated when they are physically deformed by mechanical energy such as pressure.

15.4 Fishes and amphibians have a complex system of mechanoreceptors along their bodies called a lateral line system, which detects stimuli such as sound, movement, and pressure changes.

15.4 Invertebrates as well as vertebrates have mechanoreceptors.

15.5 In humans and other vertebrates, two populations of nerve endings buried in the skin are sensitive to changes in temperature.

Some animals can detect stimuli that humans cannot.

15.6 Pit viper snakes can detect heat radiated from warm-blooded animals (infrared radiation).

15.6 Insects (especially pollinating insects) can see ultraviolet radiation, which is emitted by some flowering plants.

15.7 Humans have no sensory organs to detect electric currents, but some fish, some salamanders, and the duck-billed platypus have receptors that can.

15.8 There is evidence that some migratory animals and those with a homing instinct detect magnetic fields and use the Earth's magnetic lines of force to help orient themselves.

The special senses are taste, smell, sight, hearing, and balance.

15.9 The ability to sense chemicals, or chemoreception, is widespread in the animal kingdom.

15.9 In humans, taste receptors are located within the short projections on the tongue known as papillae.

15.9 The olfactory receptors, which are located in the nasal epithelium just behind the bridge of the nose in humans, detect various smells.

15.10 Image-forming eyes are either of the multifaceted type as in insects, or the single-lens, cameralike type as in cephalopods and vertebrates.

15.10 In humans, a lens focuses light entering the eye onto receptors located at the back of the eye; impulses are transmitted by the optic nerve to the visual cortex of the brain.

15.11 Sound waves are disturbances of molecules in the air (and in certain other substances) that result from the vibration of an object.

15.11 Scientists have been unable to identify vibration or sound receptors in many phyla of invertebrates, but they do know that earthworms and insects perceive sounds.

15.11 The ear collects sound waves and changes them into other forms of mechanical energy that stimulate the organ of hearing within the inner ear; impulses are carried from there to the brain.

15.12 Gravity is detected by the utricle and the saccule in the inner ear.

15.12 Motion is sensed by the semicircular canals, also located in the inner ear.

Key Terms

aqueous humor (AYK-wee-us HYOO-mur) *318*
choroid (KOR-oyd) *319*
ciliary muscle (SILL-ee-err-ee) *319*
cochlea (KOCK-lee-uh) *323*
cones *320*
cornea (KOR-nee-uh) *318*
eustachian tube (yoo-STAY-kee-un or yoo-STAY-shun) *323*
external auditory canal (AWD-uh-tore-ee) *323*
fovea (FOE-vee-uh) *321*
incus (ING-kus) *323*

iris (EYE-rus) *319*
lens *319*
malleus (MAL-ee-us) *323*
middle ear *323*
olfactory receptors (ole-FAK-tuh-ree) *316*
optic nerve *321*
organ of Corti (KORT-ee) *324*
oval window *323*
pinna (PIN-uh) *323*
proprioceptors (PRO-pree-oh-SEP-turz) *310*
pupil *319*

retina (RET-uh-nuh) *319*
rods *320*
round window *323*
saccule (SAK-yool) *326*
sclera (SKLEARuh) *318*
semicircular canals *326*
stapes (STAY-peez) *323*
stretch receptor *310*
taste buds *315*
tympanic membrane (tim-PAN-ik) *323*
vestibule (VES-tuh-byool) *326*
utricle (YOO-trih-kul) *326*

KNOWLEDGE AND COMPREHENSION QUESTIONS

1. Explain the events that occur at the cellular level that allow you to "sense" something. (*Comprehension*)

2. Which type of receptor would be most highly stimulated by an experience of riding on a roller coaster at an amusement park? Why? (*Comprehension*)

3. Researchers have found the hypothalamus to be involved in the sense of satiation or "feeling full" after a meal and thus integrally involved in weight control. What type of information would you hypothesize is being sent to the hypothalamus? (*Comprehension*)

4. Explain how you can smell the difference between your favorite perfume and your least favorite food. (*Comprehension*)

5. You may have noticed that certain smells are able to evoke specific memories and emotions—both good and bad. Explain why. (*Comprehension*)

6. What are taste buds? Explain their role in allowing you to taste the flavors in your favorite ice cream. What other sense (besides taste) is involved? (*Comprehension*)

7. Draw a diagram of the human eye. Label the sclera, cornea, aqueous humor, lens, iris, pupil, and fovea. (*Knowledge*)

8. Is there any difference in the manner in which you see in a well-lit environment as compared with your vision in a darkened room? What type of cell is involved in each situation? (*Comprehension*)

9. You switch on the radio to your favorite program. Explain how your ears enable you to hear the sounds. (*Comprehension*)

10. Even with your eyes closed, you can tell when your head tilts to one side. Explain how. (*Comprehension*)

KEY

Knowledge: Recalling information.

Comprehension: Showing understanding of recalled information.

CRITICAL THINKING REVIEW

1. In a very interesting series of experiments, migrating song birds were placed in a cage lined with carbon paper similar to that used in some credit card slips. The birds would peck at the wall of the cage, attempting to get out, but not at random—by far the greatest density of pecks was in the direction they were migrating when captured! When a large magnet was placed in the room, they pecked in the direction of the magnet, even though they could not see it. What sort of sensory system do you imagine the birds possess that lets them orient in this way? Why does the magnet affect the birds? Explain the rationale of your answer. (*Application*)

2. Some children are born with the inability to feel pain. Do you think that this is an advantage or a disadvantage with respect to the well-being of the person? Explain the reasons for your answer. (*Evaluation*)

3. Mr. Garcia went for an eye examination, and the eye doctor placed drops in his eyes that dilated (widened) his pupils so that he could view the retina. What structures did the drops affect? What happened within the eye so that the pupil enlarged? (*Application*)

4. Pick a sense in an animal that has been described in this chapter and suggest how this sense helps the animal survive and/or reproduce. (*Application*)

5. In their preschool years, many children experience frequent middle ear infections. Which structures of the ear are affected by such infections? How do these infections affect hearing and why? (*Application*)

KEY

Application: Using information in a new situation.

Evaluation: Making informed decisions.

Explain how the waves of pressure created in the fluid of the inner ear allow humans to hear.

Describe the process occurring at each numbered step. Where is sound being perceived?

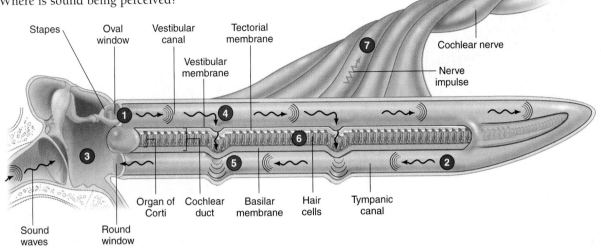

Stapes • Oval window • Vestibular canal • Tectorial membrane • Vestibular membrane • Cochlear nerve • Nerve impulse • Organ of Corti • Cochlear duct • Basilar membrane • Hair cells • Tympanic canal • Sound waves • Round window

1 Description:

2 Description:

3 Description:

4 Description:

5 Description:

6 Description:

7 Description:

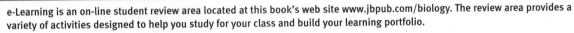

e learning

Location: http://www.jbpub.com/biology

e-Learning is an on-line student review area located at this book's web site www.jbpub.com/biology. The review area provides a variety of activities designed to help you study for your class and build your learning portfolio.

Review Questions The review questions test your knowledge of the important concepts and applications in each chapter. The review provides feedback for each correct or incorrect answer. This is an excellent test preparation tool.

Figure-Labeling Exercises Sharpen your visual thinking by matching terms and labels for important illustrations.

Flash Cards Studying biology requires learning new terms. Virtual flash cards help you master the new vocabulary for each chapter.

Just Wondering As you read and study from this text, you may find that you have unanswered questions. Through this site you can ask the author, Sandy Alters, your "just wondering" questions.

Do you prefer the speed and reliability of a CD-ROM? All of the features contained on the eLearning portion of the web site are also available on the Student CD-ROM.

Protection, Support, and Movement

A rhinoceros takes care of its skin. The skin of rhinos is very thick, and appears like plates of armor because of deep folds in the skin. Still, this skin can be penetrated by biting insects and the claws of predators. The rhino's solution: a daily mud mask. A wet mud bath cools off the rhino, helps keep the skin in good condition, and once the mud dries, provides an extra layer of protection on top of the skin.

Human skin, unlike that of the rhino, is quite thin and pliable. Yet it functions in much the same way as the rhino's "suit of armor." Even a simple puncture can compromise the skin barrier and open it to pathogens. Take, for example, the popular practice of body piercing. *Pseudomonas aeruginosa* and *Staphylococcus aureus* bacteria, which are widely distributed in nature and commonly infect wounds of all types, cause most post-piercing infections. Both these organisms can cause serious problems that can be difficult to treat successfully.

If you decide on body piercing, use a reputable piercing salon. Ask about their health and safety guidelines, such as methods for hygienically inserting jewelry and recommendations for aftercare. See a physician immediately if you experience redness, swelling, or pain in the pierced area. If you have oral piercing, visit your dentist for instructions on how to avoid infection and damage of the oral area.

16

Organization

Skin, bones, and muscles work together to help protect, support, and move animals.

Human skin, bones, and muscles perform many functions.

Health
Connections

Skin, bones, and muscles work together to help protect, support, and move animals.

16.1 The role of skin

Figure 16.1 lists examples of animals from the major animal phyla and their types of protection. The name and phylum of each representative animal is listed at the top left of each animal, while its characteristics of integument (skin) are listed at the top right.

Although one of the primary jobs of skin is being an effective barrier to infection, skin also plays a role in the body movement of humans and other animals. A more general term for the outer covering of living things is **integument** (in-TEG-you-ment). In humans (and all vertebrates), the integument consists of keratinized skin. *Keratin* (KARE-ah-tin) is a hardened, protective, fairly waterproof protein in integumentary structures such as vertebrate skin, nails, hair, horns, and feathers. Although the skins of vertebrates have certain structural differences, their integuments are all instrumental in helping to balance the gain and loss of water, which is crucial to life.

Land animals have integuments that not only protect them but conserve water. The primary component of living things is water (about 75% of body weight), and if it were to evaporate quickly through the integument, survival would be difficult if not impossible.

Vertebrate skin provides an effective (but partial) barrier to water loss in land animals. Surface differences among vertebrate skins include scales in cold-blooded (ectothermic) animals such as reptiles and mucus coverings in amphibians. (Ectothermic organisms absorb heat from their surroundings rather than generate their own body heat.) In reptiles, epidermal scales help prevent water loss as these animals bake in the sun to raise their body temperature and metabolic rate. In skin-breathers (see Section 9.2), such as the amphibians (frogs and salamanders), the naked skin is covered with mucus to help keep that respiratory surface moist yet reduce water loss.

The integument of insects is also structured to reduce the evaporation of water from the body. Insects, for example, have a hard, outer covering with a waxy surface. The wax is tremendously important in the conservation of water in these animals. In fact, if its wax covering is scratched, an insect will die from dehydration. For this reason, abrasives are mixed with stored grain to kill insect pests, both of which are later removed from the grain. The hard, celluloselike chitin cuticle (covering) of insects is more important as a skeleton than as a structure for water conservation (see Section 16.2).

Aquatic mammals and other aquatic vertebrates also have a variety of integument adaptations that help them survive in a watery environment. For example, freshwater fishes are covered by scales and mucous, which retard the movement of water *into* their bodies.

The rest of the organisms listed in Figure 16.1 live in watery environments (sponge, jellyfish, flatworm, scallop, sea star) or in the soil (earthworm and roundworm). All these organisms, with the exception of the scallop and the sea star, breathe through their body surfaces. Notice in Figure 16.1 that their integuments are adapted to this function, being simply a layer of epidermal cells. The earthworm and the roundworm, which live in the soil, must keep their integuments moist or they will be unable to respire and will lose valuable internal moisture. The multilayered but porous cuticle that covers the roundworm does not protect the worm from dehydration.

Echinoderm means "spiny-skinned." The integument of the sea star serves as a flexible, protective covering for the skeleton and internal organs. Respiration takes place at the tube feet, which cover the underside of the sea star and also play a role in movement.

Bivalve molluscs, such as scallops and clams, have a shell that functions as integument. The shell protects the animal's soft body from being eaten by most predators. (Mollusc means "soft body.") However, the sea star, a major predator of the clam, can pull on a closed clamshell for several hours, eventually opening this hard, protective integument.

In animals, the outer covering (integument) protects soft body parts beneath, provides a barrier to infection, and helps conserve water in land animals. In many aquatic animals, the integument also acts as the respiratory surface.

Focus
on Plants

Plants, like animals, have integuments.

Dermal tissue is the "skin" of a plant. It covers the plant, protecting it from water loss and injury to its internal structures.

As in animals, the most abundant type of cell found in the dermal tissue of plants is epidermal cells. These cells are often covered with a thick, waxy layer called a cuticle that protects the plant and provides an effective barrier against water loss. Plant "skin" also has outgrowths, much like hair and nails on the human body. For example, on leaves or fruits, fuzzlike outgrowths reflect sunlight, which helps control water loss as in the "Lamb's Ear" shown in the photo. Some outgrowths, such as the thorns of a rosebush, protect a plant against animal predators.

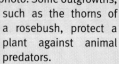

Woody shrubs and trees have protective bark in place of dermal tissue. Bark is made up of other tissue types. Chapter 29 discusses the structure and function of plants—including their dermal tissues—in more detail.

REPRESENTATIVE ANIMAL (PHYLUM)   **INTEGUMENT**

Sponge (Porifera)   Layer of epidermal cells

- Jellylike cells
- Layer of epidermal cells
- Inner cell layer (collar cells)

Jellyfish (Cnidaria)   Layer of epidermal cells

- Bell
- Layer of epidermal cells
- Jellylike layer
- Inner layer of cells

Flatworm (Platyhelminthes)   Layer of epidermal cells

- Layer of epidermal cells

Roundworm (Nematoda)   Thick cuticle covering layer of epidermal cells

- Layer of epidermal cells
- Cuticle

Earthworm (Annelida)   Thin cuticle covering layer of epidermal cells

- Layer of epidermal cells
- Cuticle

REPRESENTATIVE ANIMAL (PHYLUM)   **INTEGUMENT**

Scallop (Mollusca)   Calcium carbonate shell

- Calcium carbonate shell

Insect (Mandibulata)   Chitin cuticle with a waxy surface and underlying epidermis

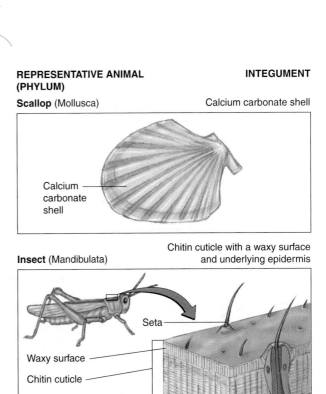

- Seta
- Waxy surface
- Chitin cuticle
- Epidermis

Sea star (Echinodermata)   Thin, prickly epidermis

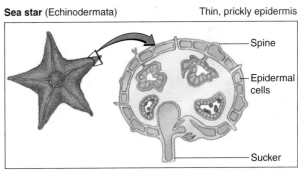

- Spine
- Epidermal cells
- Sucker

Human (Chordata)   Keratinized skin

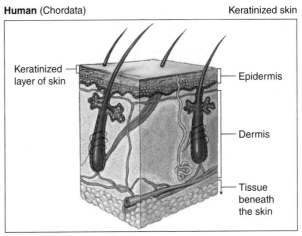

- Keratinized layer of skin
- Epidermis
- Dermis
- Tissue beneath the skin

16.2 The skeleton, muscles, and movement

Most of an animal's tissues are soft; the skeleton provides support for the animal body. On land, this support keeps the animal's body from collapsing under its own weight and becoming a formless mass. In aquatic environments, skeletal support keeps the shape of the animal's body even though water flows over it and water pressure acts on it. Muscles attach to the rigid, skeletal portion of a body; the skeleton acts as something against which muscles pull.

Look down the column labeled "skeleton" in **Table 16.1.** Aside from the sponge, which has hard, spiny pieces of tissue embedded in its gellike middle layer to support its body, three patterns of skeleton emerge: the hydroskeleton, the exoskeleton, and the endoskeleton.

A **hydroskeleton** is a support system that uses a fluid under pressure as a scaffolding for an organism or body part. For example, erection in the mammalian penis takes place this way: Blood rushes into the spongy erectile tissue and the penis becomes rigid. When the blood flows out of this tissue, the penis becomes limp.

In a true hydroskeleton, an organism has a fluid-filled, enclosed body cavity with flexible body walls. Muscles are attached to the body wall, and the body fluids are used to transmit force. Movement results when the muscles pull in on one part of the body, which compresses the fluid, causing it to push out another part of the body.

The earthworm is an excellent example of an organism with a true hydrostatic skeleton. Its body consists of enclosed segments, each housing a fluid-filled body cavity (coelom). Two sets of muscles are attached to the body wall: circular muscles and longitudinal (lengthwise) muscles. When the earthworm contracts the circular muscles in some of its segments, it elongates those segments. That is, when the segments get smaller in diameter as the circular muscles tighten, the pressure exerted on the body fluid inside each segment forces each of these segments to get longer. This process is how the earthworm pushes its front end forward.

Table 16.1

The Skeleton, Musculature, and Movement of Representative Animals

Representative animal (phylum)	Skeleton	Musculature	Movement
Sponge (Porifera)	Gelatinlike middle layer supported by hard, sharply pointed skeletal pieces	No true musculature	Sessile
Jellyfish (Cnideria)	Absent, but water in bell acts as hydroskeleton	Circular muscles in swimming bell	Propulsion: Circular muscles in bell contract, pushing water out of the bell, which moves the jellyfish forward
Flatworm (Platyhelminthes)	Hydroskeleton	Two sets of muscles underneath the epidermis: circular and longitudinal	Waves of contraction that generate thrusting and pulling forces called looping
Roundworm (Nematoda)	Hydroskeleton	Longitudinal muscles underneath the epidermis and cuticle; no circular muscles	Alternate contraction of longitudinal muscles on either side of body results in thrashing movements
Earthworm (Annelida)	Hydroskeleton	Two sets of muscles underneath the epidermis: circular and longitudinal	Alternate contraction of circular and longitudinal muscles results in peristaltic movement
Clam (Mollusca)	Shell	Muscular foot	Muscles in foot allow clam to burrow into sand or mud when shell is open
Insect (Mandibulata)	Exoskeleton (Integument is skeleton)	Primarily striated muscle that attaches to exoskeleton	Ground movement by means of limbs; flight by movement of wings
Sea star (Echinodermata)	Hydroskeleton and endoskeleton of hard calcium carbonate plates	Retractor muscles in tube feet	Movement by hydraulics; protraction and retraction of tube feet, which are connected to the water vascular system
Human (Chordata)	Bony endoskeleton	Striated muscles attached to skeleton for locomotion; smooth muscle in organs; cardiac muscle in heart	Movement by lever action of limbs

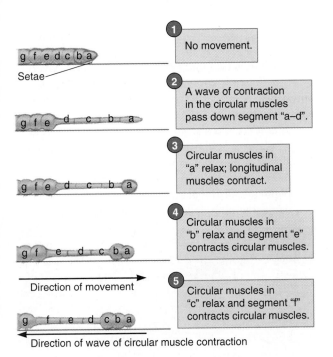

① No movement.

g f e d c b a

Setae

② A wave of contraction in the circular muscles pass down segment "a–d".

g f e d c b a

③ Circular muscles in "a" relax; longitudinal muscles contract.

g f e d c b a

④ Circular muscles in "b" relax and segment "e" contracts circular muscles.

g f e d c b a

Direction of movement →

⑤ Circular muscles in "c" relax and segment "f" contracts circular muscles.

g f e d c b a

← Direction of wave of circular muscle contraction

Figure 16.2 **Movement in an earthworm.**

Figure 16.2 shows this earthworm movement: ① through ② shows what happens to segments a, b, c, and d of an earthworm when it contracts circular muscles in these segments. The earthworm elongates in these anterior sections, and the earthworm moves forward. In ③, the earthworm relaxes its circular muscles and contracts its longitudinal muscles in segment a, shortening the segment. The fluid inside the segment forces the shortened segment to fatten. In ④ and ⑤, the contraction of circular muscles continues in segments e and f, while segments b and c shorten. This wave of circular and then longitudinal contraction that travels down the worm's body is called peristaltic locomotion, and is similar to peristalsis in humans.

Jellyfish do not have a true hydroskeleton. They use the water of their environment, rather than their own enclosed body fluids, as a type of hydroskeleton. Water is pushed out of the jellyfish as the circular muscle at the bottom of its bell contracts. As water is forced out of the bell, it pushes the animal forward, just as the air forced out of airplane jets propels the plane forward.

Flatworms are acoelomates (see Section 7.3), which means that they have no body cavity. Therefore, they have no true hydroskeleton, but their body contents do act in this fashion. Like earthworms, they do have circular and longitudinal muscles attached to their integument. By alternately contracting these sets of muscles, flatworms generate thrusting and pulling forces as their

body contents become alternately compressed and decompressed, moving them along a surface.

Roundworms are pseudocoelomates, with turgid bodies and rigid cuticles. Unlike earthworms and flatworms, they have only longitudinal muscles. To move, these worms alternately contract the longitudinal muscles on either side of the body. As the longitudinal muscles contract on one side of the body, the longitudinal muscles on the other side relax. Since the cuticle cannot expand to relieve the pressure, the nematode body forms S-shaped curves as the worm thrashes forward.

Even though the sea star has an **endoskeleton** (a rigid support system composed of hard plates), it moves by using the hydraulic action of a hydroskeleton. Put simply, water is used for the mechanics of locomotion. The sea star has a water vascular system, a unique system of water-filled canals connected to its tube feet on its underside. By a complex set of muscular actions, suction is created or released on the tube feet. The sea star crawls along by extending, attaching, and releasing the grip of its tube feet on a surface.

Arthropods (insects and spiders) have striated muscle as vertebrates do (see Section 16.11), which makes rapid movement and flight possible. (Other invertebrates have slower-acting smooth muscle.) Insects and spiders also have lightweight **exoskeletons** (rigid support systems that cover their bodies), not heavy shells. Their muscles are attached to the exoskeleton, which is paper-thin at the joints, allowing arthropods to bend their appendages at joints, using them as levers to increase force (see illustration), as do vertebrates. The arthropod exoskeleton does not grow, however, so insects and spiders periodically shed their skin, which is called molting. A new exoskeleton is present under the old one. The arthropod "puffs up" and expands this exoskeleton, and it hardens after a few hours.

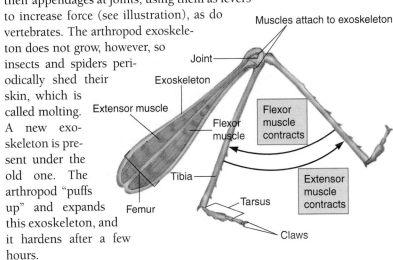

The skeleton provides support for the animal body. Muscles attach to the rigid, skeletal portion of a body; the skeleton acts as something against which muscles pull. An animal may have a hydroskeleton, an exoskeleton, or an endoskeleton.

Human skin, bones, and muscles perform many functions.

16.3 How skin, bones, and muscles work together

In humans, the skin is a soft and flexible covering, stretching to accommodate the various movements of which the human body is capable. That movement is determined largely by muscles, which are attached to bones as seen in the illustration below.

Bones not only give shape to the human body and protect its delicate inner structures but also act as levers in movement. The manner in which muscles are attached to bones, pulling on one while anchored to another, allows bones to be used as levers. The bones increase the strength of a movement, similar to the way a claw hammer increases the force exerted on a nail to

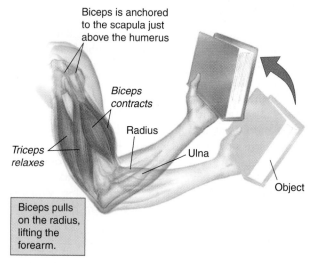

Biceps is anchored to the scapula just above the humerus

Biceps contracts

Radius

Triceps relaxes

Ulna

Object

Biceps pulls on the radius, lifting the forearm.

pull it out of a piece of wood. Rigid yet flexible, bones are able to bear a considerable amount of weight.

To supply the force needed for movement, your body uses the chemical energy of adenosine triphosphate, or ATP (see Chapter 5). Your body uses this energy to move certain filaments within your muscle cells, resulting in the shortening of those cells (see Section 16.11). When a large group of muscle cells shortens all at once, the muscle cells exert a great deal of force. For a muscle to use this force to produce movement, it must direct its force against an object. In humans, muscles pull on bones, so for example, muscle contraction results in the lifting of an arm or the bending of a finger.

> In humans, movement results from the contraction of skeletal muscles anchored to bones. These muscles use bones like levers to direct force against an object. When a body part is pulled to a new position, the skin stretches to accommodate the change.

16.4 The structure and functions of skin

Human skin is far more than simply an elastic covering of epithelial cells encasing your body's muscles, blood, and bones. Instead, it is a dynamic organ that performs many functions:

1. *Skin is a protective barrier.* It keeps out microorganisms that would otherwise infect the body. Because skin is waterproof, it keeps the fluids of the body in and other fluids out. Skin cells also contain a pigment called melanin, which absorbs potentially damaging ultraviolet radiation from the sun.

240X

Scanning electron micrograph of the surface of human skin.

2. *Skin provides a sensory surface.* Sensory nerve endings in skin act as your body's pressure gauge, telling you how gently to caress a loved one and how firmly to hold a pencil. Other sensors embedded in the skin detect pain, heat, and cold (see Figure 15.3). Skin is the body's point of contact with the outside world.
3. *Skin compensates for body movement.* Skin stretches when you reach for something and contracts quickly when you stop reaching. It also grows as you grow.
4. *Skin helps control the body's internal temperature.* When the temperature is cold, the blood vessels in the skin constrict, so less of the body's heat is lost to the surrounding air. When it is hot, these same vessels dilate, giving off heat. In addition, glands in the skin release sweat, which then absorbs body heat and evaporates, cooling the body surface.

5. *Skin manufactures Vitamin D in the presence of sunlight.* This vitamin helps the body absorb calcium from the digestive tract. Because we wear clothes and may not be in the sun for lengthy periods, our diets must be supplemented with vitamin D and milk is typically vitamin-D fortified.

Your skin is the largest organ of your body. In an adult human, 15% of the total body weight is skin. Much of the multifunctional role of skin reflects the fact that its tissues are made up of a variety of specialized cells. In fact, one typical square centimeter of human skin contains 200 nerve endings, 10 hairs with accompanying microscopic muscles, 100 sweat glands, 15 oil glands, 3 blood vessels, 12 heat receptors, 2 cold receptors, and 25 pressure-sensing receptors (**Figure 16.3**). Together with the hair and nails, human skin is called the **integumentary system**.

As mentioned previously, vertebrates (such as humans) have keratinized skin. Keratinized human skin develops as follows: As cells produced deep within the skin rise to the surface, pushed by the continual formation of new cells, they produce the strengthening protein keratin and change in shape and structure. By the time they reach the topmost layer of the skin, they are dead, flattened, and hardened, providing a surface that is resistant to abrasion and impermeable to most pathogens and noncorrosive substances.

> The skin is the largest organ of the human body and is made up of various specialized cells. It covers the surface of the body, protecting it while helping to control the body's internal temperature, providing a sensory surface, and compensating for body movement.

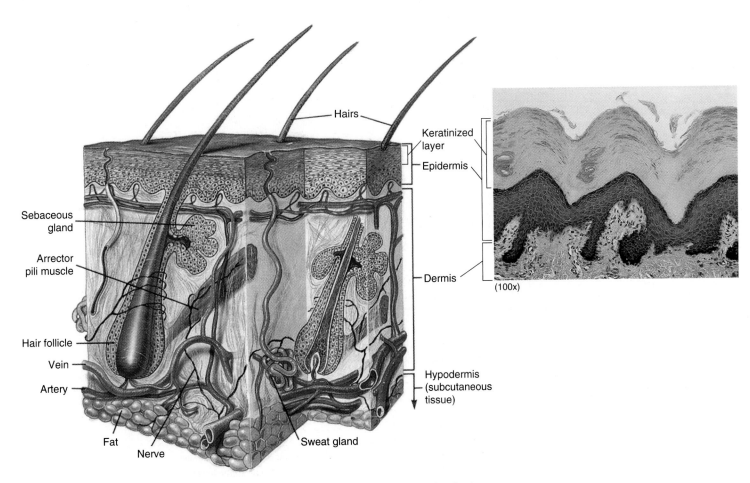

Figure 16.3 **The structure of human skin.** Human skin is composed of three layers: the epidermis, the dermis, and the subcutaneous layer. Within these layers are specialized tissues and cells that perform many specialized functions.

16.5 The structure of bone

Bone is a type of connective tissue (see Chapter 7) consisting of widely separated bone cells embedded in a matrix of collagen fibers and mineral salts in a semisolid gel. The bones' mineral salts are needle-shaped crystals of calcium phosphate and calcium carbonate. The collagen fibers (see Figure 7.8) are coated and surrounded by these mineral salts within the gel. Interestingly, the collagen and minerals give bone a structure strikingly similar to that of fiberglass, producing bones that are rigid yet flexible. You can see this interplay of flexibility and hardness by soaking a chicken bone in vinegar overnight. The acetic acid of the vinegar dissolves the mineral salts in the bone. Without the hardness of the mineral salts, the collagen fibers leave the bone so flexible that you can tie it in a knot!

Living cells are as important a component of bone as are the collagen fibers and mineral salts. Cells called *osteoblasts* form new bone. These cells secrete the collagen fibers and semisold gel in which the body later deposits mineral salts. The osteoblasts lay down bone in thin, concentric layers called *lamellae* (Figure 16.4), like layers of insulation wrapped around an old pipe. You can think of the osteoblasts as being embedded within the insulation and the open tube of the pipe as a narrow channel called the **Haversian canal**.

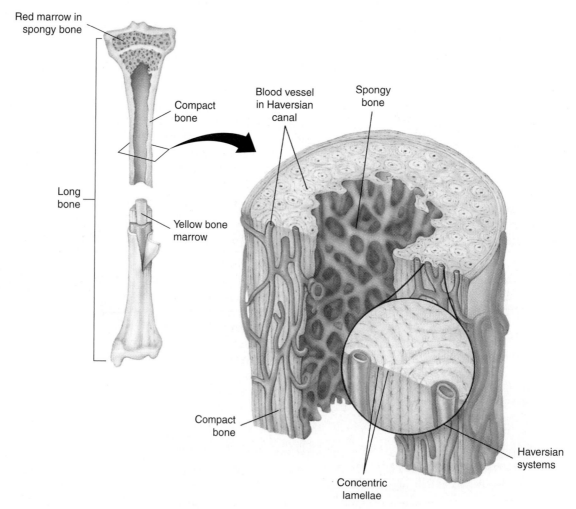

Figure 16.4 The organization of bone, shown at three levels of detail. The shaft of a long bone is dense and compact, giving the bone strength. The ends of the bone and the lining of the cavity within the long bone are spongy, with a more open lattice. Blood cells are formed within red bone marrow that fills the lattice at the ends of the bone. Inset shows a magnified Haversian system with its concentric lamellae. Yellow bone marrow (a fat storage tissue) fills the cavity in the shaft of the long bone.

Haversian canals run parallel to the length of the bone and contain blood vessels and nerves. (Haversian canals and the surrounding lamellae are together called *Haversian systems*.) The blood vessels provide a lifeline to living bone-forming cells, whereas the nerves control the diameter of the blood vessels and thus the flow through them. Nutrients and oxygen diffuse from the bloodstream into thin cellular processes of the osteoblasts, and metabolic wastes diffuse out.

The bones of the human skeleton are composed of spongy bone and compact bone. Microscopically, *compact bone* has the concentric ring structure described previously. As shown in Figure 16.4, it runs the length of long bones, such as those in your arms and legs. However, if bones were completely made up of compact bone, your body would be very heavy, and your arms and legs would be nearly impossible to move. Instead, the central core of long bones is a hollow cylinder lined with spongy bone, and this cavity is filled with a soft, fatty connective tissue called *yellow bone marrow*.

Spongy bone makes up most of the ends of long bones and most of the bone tissue of short bones (like your wrist and ankle bones), flat bones (like your ribs), and irregularly shaped bones (like some of your facial bones). Spongy bone is an open latticework of thin plates, or bars, of bone. Microscopically, its structure does not show a regular concentric ring structure like compact bone; it has a somewhat more irregular organization. The spaces within its bony latticework are filled with *red bone marrow* where the body's blood cells are formed. Surrounding the spongy bone tissue are layers of compact bone. The compact bone gives bones the strength to withstand mechanical stress, and spongy bone provides some support and a storage place for the red bone marrow, while helping to lighten bones.

Bone is a type of connective tissue made up of widely separated living cells that secrete collagen fibers into their surrounding matrix. The body deposits mineral salts on these fibers. The salts give bone hardness, whereas the collagen fibers give flexibility. Bone tissue appears under the light microscope as concentric rings surrounding a central canal containing blood vessels and nerves. The bones of the skeleton contain two kinds of tissue: spongy bone and compact bone. Compact bone runs the length of long bones and has no spaces within its structure visible to the naked eye. Spongy bone is found in the ends of long bones and within short, flat, irregularly shaped bones. It looks like interweaving bars of bone with the intervening spaces filled with red bone marrow.

16.6 The functions of the skeletal system

The 206 bones of the adult human body make up a working whole called the **skeleton**, or **skeletal system**. Together, the bones of the skeleton perform many important functions, some of which have already been mentioned. To summarize, the skeletal system:

1. provides support for the body;
2. provides for movement, with individual bones serving as points of attachment for the skeletal muscles and acting as levers against which muscles can pull;
3. protects delicate internal structures such as the brain, heart, and lungs;
4. gives rise to new red and white blood cells in its marrow, and
5. acts as a storehouse for minerals such as calcium and phosphorus.

Bone tissue is not the inert tissue it may seem to be. In fact, it is continually broken down and re-formed, its minerals being transported to other parts of the body on demand. Bone is laid down in areas exposed to physical stress, which is why weight-bearing exercise results in stronger bones. Conversely, loss of bone mass occurs when physical stress is removed, which is why people confined to bed or in outer space lose bone density. Under "normal" conditions, your body completely replaces its skeleton over a period of 7 years.

The skeleton is made up of two divisions: the axial skeleton and the appendicular skeleton (see illustration to the right). The **axial skeleton** is your axis, or central column, of bones off which the appendages (arms and legs) of your **appendicular skeleton** hang. The appendicular skeleton also includes the bones that serve to attach the appendages to the axial skeleton. These are the bones of your shoulders and hips.

Axial skeleton (blue)

Appendicular skeleton (orange)

The 206 bones of the body make up the skeletal system, which provides support for the body and attachment sites for skeletal muscles, protects delicate internal structures, stores minerals, and produces blood cells in its red marrow. It can be divided into the axial skeleton and the appendicular skeleton.

16.7 The axial skeleton

The 80 bones of the axial skeleton include the bones of the skull, vertebral column, and rib cage and are shown in Figure 16.5. The **skull** contains 8 cranial and 14 facial bones. Find the following skull bones in Figure 16.6: 1 frontal bone that forms the forehead, 2 parietal bones that form a large portion of the sides and top of the skull, 2 temporal bones that form the lower sides of the skull, and 1 occipital bone that forms the lower back portion of the skull. The irregularly shaped ethmoid and sphenoid (SFEE-noid) bones are not easy to locate in a diagram because only a tiny portion of these bones lies on the surface of the skull. The ethmoid bone sits be-

tween the eyes; only a fraction of it can be seen forming a part of each orbit, or eye socket. The rest of this bone forms part of the nasal cavity and the floor of the cranial cavity, which encloses the brain. The sphenoid bone makes up part of the back and sides of the eye sockets. Looking somewhat like wings, it extends from one side of the skull to the other, forming the middle portion of the base of the skull.

Some of the 14 facial bones are easy to see in Figure 16.6: 1 mandible, or lower jawbone; 2 maxillae (max-ILL-eye), which unite to form the upper jawbone; 2 nasal bones, which form the bridge of the nose; and 2 zygomatic (zeye-go-MAT-ik) bones, or cheekbones. Look below the part of the ethmoid bone that forms part

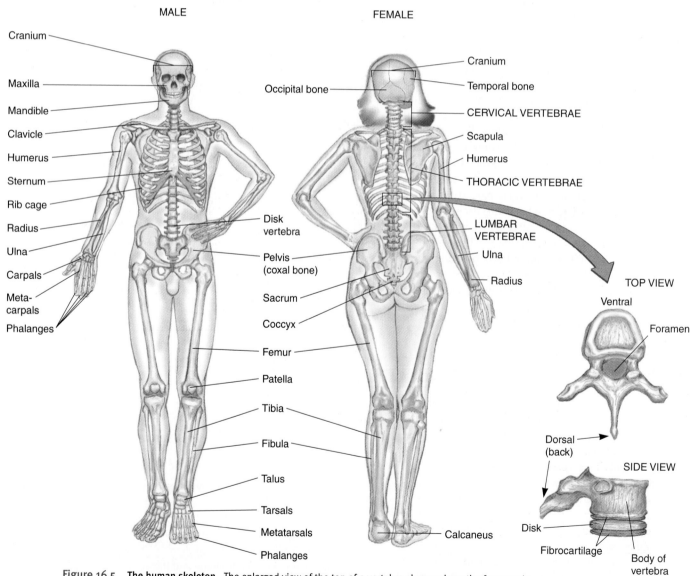

Figure 16.5   The human skeleton. The enlarged view of the top of a vertebra shows where the foramen is located. The enlarged view of the side of a vertebra shows that between each vertebra and the next, there is a disk composed of an outer layer of fibrocartilage and an inner, more elastic layer.

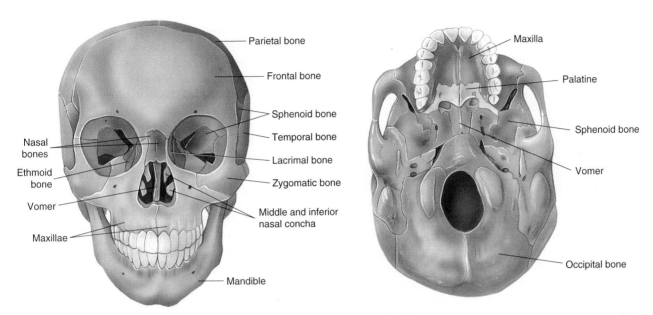

Figure 16.6 **Frontal view of the human skull.** The skull contains 8 cranial and 14 facial bones.

of the nasal septum to find the vomer. Along with the ethmoid and cartilage, the vomer (VOE-murr) helps divide the nose into right and left nasal cavities. The paired lacrimal bones are tiny and so difficult to see; they are located at the inner corners of the orbits near the nose. Each of these bones has a groove forming part of the tear ducts, canals that drain excess fluid bathing the eyes. The two L-shaped palatine (PAL-uh-tine) bones are hidden from view. They form the back portion of the roof of your mouth and extend upward to form part of the floor and sides of the nasal cavity. The two inferior nasal conchae (KONG-keye) are coiled bones and project into the nasal cavity. The structure of these bones causes the air that enters the nose to be circulated and warmed before being breathed into the lungs.

The **vertebral column** is made up of 26 individual bones called *vertebrae* (see Figure 16.5). Stacked one on top of the other, these bones act like a strong yet very flexible rod that supports the head. In addition, some of the vertebrae serve as points of attachment for the ribs. Each vertebra has a central hole, or foramen (fo-RAH-men) as seen in the enlarged view on the right of Figure 16.5. Lined up, the vertebrae form a bony canal protecting the spinal cord, which runs down much of its length.

The 7 vertebrae closest to the head are the cervical (neck) vertebrae. Next are 12 thoracic (chest) vertebrae. Following these are 5 lumbar (lower back) vertebrae. The sacrum (SAY-crum) and the coccyx (COCK-sicks), or tailbone, are the last vertebrae in the column. Positioned between the vertebrae (except the sacrum and coccyx) are disks of fibrocartilage called **intervertebral disks**. These disks act as shock absorbers, provide

the means of attachment between one vertebra and the next, and permit movement of the vertebral column.

Attached to the 12 thoracic vertebrae are 12 pairs of ribs. The ribs curve around to the front of the thoracic (chest) cavity, producing a bony cage that protects the heart and lungs. The upper 7 pairs of ribs directly connect to a flat bone that lies at the midline of the chest called the breastbone, or sternum. These 14 ribs each connect to the sternum by means of a strip of hyaline cartilage and are therefore called true ribs. The remaining five pairs of ribs, called false ribs, do not directly connect to the sternum. Instead, the cartilage strips of ribs 8 through 10 attach to each other and then to the cartilage of the seventh rib pair. Ribs 11 and 12 do not attach to the sternum but hang free, supported by muscle tissue. These ribs are therefore called floating ribs.

The remaining bone of the axial skeleton is the hyoid bone, a name that means "U-shaped." Ligaments in the neck support the hyoid, and the tongue attaches to it. (Figure 16.5 does not show the hyoid bone.) **Ligaments** (LIG-uh-munts) are bundles, or strips, of dense connective tissue that usually hold bones to bones, but in this case are supportive structures. The hyoid is the only bone in the body that does not form a joint with other bones. Because this bone is often broken when a person is strangled, the hyoid can provide important evidence in certain murder cases.

The axial skeleton includes the central axis—the skull, vertebral column, rib cage (including the sternum), and hyoid.

Human skin, bones, and muscles perform many functions.

16.8 The appendicular skeleton

The appendicular skeleton is made up of the bones of the appendages (arms and legs), the *pectoral (shoulder) girdle*, and the *pelvic (hip) girdle*. The **pectoral girdle** is made up of two pairs of bones: the clavicles (CLAH-vih-kuhlz), or collarbones, and the scapulae (SKAP-you-lie), or shoulder blades. Find these bones in Figure 16.5 and the illustration below, and then locate the edges of your own clavicles and scapulae. The bones of the pectoral girdle support and articulate with (form joints with) the arms. Together, your arms (upper extremities) contain 60 bones.

As shown in the illustration below, the bone of the upper arm is called the humerus. This bone articulates with the scapula. The other end of this long bone articulates with the two bones of the forearm: the radius and the ulna. The wrist is made up of eight short bones called carpals (KAR-pulz), lined up in two rows of four bones. These bones are held together by ligaments and articulate with the bones of the hand, or metacarpals. The five metacarpals articulate with the bones of the fingers, or phalanges (fuh-LAN-geez) (singular: phalanx [FAY-lanks]). Each finger has three phalanges; the thumb has only two phalanges.

The **pelvic girdle** is made up of two bones called coxal (COCK-suhl) bones. You know these bones as pelvic bones or hipbones. Find these bones in Figure 16.5 and the illustration below. Pelvic bones can distinguish male and female skeletons from one another. Females have a wider pelvis with a large, oval opening in the center for giving birth. The male pelvic girdle, in contrast, is narrower and has a smaller, heart-shaped opening in the center.

The bones of the pelvis support and articulate with the legs. Together, both legs, or lower extremities, contain 60 bones as the arms do. As shown in the illustration below, the bone of the upper leg, or thigh, is called the femur. This bone has a rounded, ball-shaped head that articulates with a depression, or socket, in the hipbone. The femur (FEE-mur), the longest and heaviest bone in the body, articulates at its lower end with the two bones of the lower leg: the tibia (TIB-ee-ah) and the fibula (FIB-you-lah). The tibia is the bone commonly referred to as the shinbone and can be felt at the front of the lower leg. If you have ever had shin splints, you are well aware of this bone. Shin splints are usually an inflammation of the outer covering of the tibia caused by repeated tugging by muscles and tendons. The tendons themselves may also

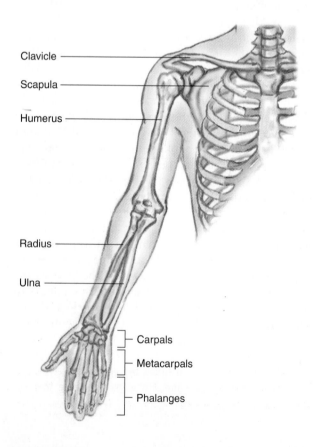

Clavicle
Scapula
Humerus
Radius
Ulna
Carpals
Metacarpals
Phalanges

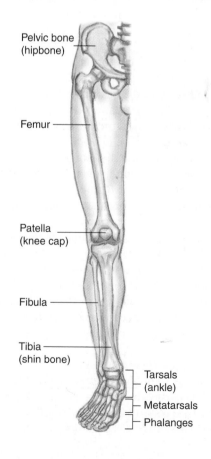

Pelvic bone (hipbone)
Femur
Patella (knee cap)
Fibula
Tibia (shin bone)
Tarsals (ankle)
Metatarsals
Phalanges

I work all day at a computer and have developed a pain in my wrist that shoots up my arm. I've heard a lot of talk about carpal tunnel syndrome, but I'm not sure what it is. Could that be my problem?

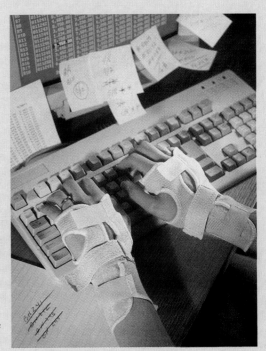

As you can see in Figure 16.5, the carpals are the wrist bones. The carpal tunnel is a narrow passageway within your wrist through which the median nerve passes on its way to your middle, ring, and little fingers. Some types of activities that use the wrists and hands repetitively, such as typing, golfing, carpentry, and certain types of factory work, can cause the tissues around the tunnel to become inflamed and swollen. When this happens, the median nerve becomes compressed, resulting in numbness and/or tingling in the hand and fingers and pain in the wrist that may shoot up or down the arm.

The results of recent studies show that conditions and illnesses such as obesity, hypothyroidism (an underactive thyroid), diabetes mellitus, rheumatoid arthritis, and lupus (a chronic connective tissue disease) may be primary causes of carpal tunnel syndrome. Although CTS is a leading cause of work-related disability among American workers, medical researchers think that physicians should examine non-workplace-related medical illness as a cause of CTS prior to investigating repetitive motion causes.

Your physician can perform tests to determine whether you have carpal tunnel syndrome. Often, the cure for this condition is as easy as resting the affected wrist and hand or wearing a splint designed to immobilize the wrist, as shown in the photo. In more serious cases, injections of steroid drugs relieve the symptoms, with surgery considered a last resort.

www.ibpub.com/biology

---

be inflamed. **Tendons** are cords of dense connective tissue that attach muscles to bones. Vigorous walking, running, or other types of exercise can sometimes result in shin splints, which can be very painful.

The patella (pah-TELL-uh), or kneecap, is a small, triangular bone that sits in front of the joint formed by the femur, tibia, and fibula. The patella is a sesamoid (SESS-uh-moid) bone, one that is formed within tendons where pressure develops.

Each ankle is made up of seven short bones called tarsals (TAR-suhlz). One of these bones forms the heel and is the largest and strongest ankle bone. The tarsals are held together by ligaments and articulate with the bones of the foot, or metatarsals. The five metatarsals articulate with the bones of the toes, or phalanges. Like the fingers, each toe has three phalanges; like the thumb, the "big toe" has only two phalanges.

> The appendicular skeleton is made up of the appendages (arms and legs) and the bones that help attach the appendages to the axial skeleton.

16.9 Joints

If you have ever broken a leg or an arm and had a cast covering a joint, you have learned firsthand just how important joints are to movement. A **joint**, or **articulation**, is a place where bones, or bones and cartilage, come together. All joints are not alike; some permit little or no movement (immovable joints), some permit limited movement (slightly movable joints), and some permit a considerable amount of movement (freely movable joints). The amount of movement allowed by a joint is a direct result of how tightly the bones are held together at that location. The enlarged views of Figure 16.7 on the next page summarize the types of movements possible at freely moveable (synovial) joints. The skeleton in the middle of the illustration provides references to slightly moveable and immoveable joints.

The bones of the skull articulate with one another in a type of immovable joint called a **suture** (Figure 16.7 ❶). The edges of each skull bone are ragged, but fit tightly with the ragged edges of adjoining bones—just

Human skin, bones, and muscles perform
many functions.

Visual Summary **Figure 16.7** **Joints.**

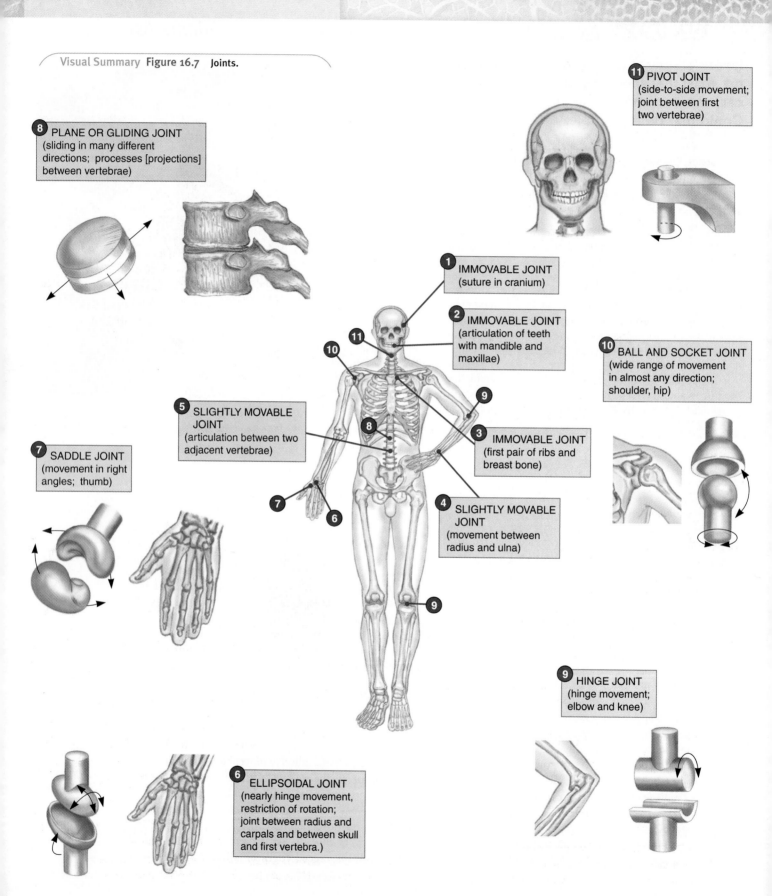

11 PIVOT JOINT
(side-to-side movement;
joint between first
two vertebrae)

8 PLANE OR GLIDING JOINT
(sliding in many different
directions; processes [projections]
between vertebrae)

1 IMMOVABLE JOINT
(suture in cranium)

2 IMMOVABLE JOINT
(articulation of teeth
with mandible and
maxillae)

10 BALL AND SOCKET JOINT
(wide range of movement
in almost any direction;
shoulder, hip)

5 SLIGHTLY MOVABLE
JOINT
(articulation between two
adjacent vertebrae)

3 IMMOVABLE JOINT
(first pair of ribs and
breast bone)

7 SADDLE JOINT
(movement in right
angles; thumb)

4 SLIGHTLY MOVABLE
JOINT
(movement between
radius and ulna)

9 HINGE JOINT
(hinge movement;
elbow and knee)

6 ELLIPSOIDAL JOINT
(nearly hinge movement,
restriction of rotation;
joint between radius and
carpals and between skull
and first vertebra.)

as puzzle pieces fit together. A layer of dense connective tissue helps keep the bones from separating (see Chapter 7).

Another example of immovable joints is the articulation of your teeth with the mandible and maxillae ❷. Here, peg-shaped roots fit into cone-shaped sockets in the jawbones. A ligament lies between each tooth and its socket, holding each tooth in place.

A third example of an immovable joint is the articulation between the first pair of ribs and the breastbone ❸. Each of these two ribs is connected to the sternum by a strip of hyaline cartilage, connections that change to bone during adult life.

As you twist your forearm to the right and left or twist your lower leg in the same manner, imagine the movement taking place between the shafts of the radius and ulna ❹ or between the tibia and fibula at their lower ends. Dense connective tissue is present at these locations, holding these bones together while permitting some flexibility. These two articulations are examples of slightly movable joints.

The articulation between one vertebra and the next is also an example of a slightly movable joint ❺. Here, a broad, flat sheet of fibrocartilage covers the top and bottom of each intervertebral disk, which is sandwiched between adjacent vertebrae, creating a somewhat flexible connection.

In slightly movable and immovable joints, there is no space between the articulating bones. The bones forming these joints are held together by dense connective tissue or cartilage. In freely movable joints, there is a space between the articulating bones. Such joints are common in your body and are called **synovial joints** (suh-NO-vee-uhl).

Figure 16.8 diagrams the components of a synovial joint. Hyaline cartilage covers the ends of the articulating bones, which are separated by a film of fluid continuous with the joint cavity. The joint cavity is the space between a connective tissue capsule (joint capsule) that surrounds the bones, and the bones themselves. This cavity is filled with fluid.

The joint capsule is composed of a double layer of tissue. The outer layer is made up of dense connective tissue that holds the bones of the joint together. Some of the connective tissue fibers are arranged in bundles forming strips of tissue, or ligaments, as shown in **Figure 16.9.** Although this outer capsule holds the bones of the joint in place, it permits a wide range of motion. The inner layer of the capsule is the synovial membrane, which secretes the fluid of the joint cavity, called synovial fluid. In fact, the synovial joint gets its name from this "egg-whitelike" fluid; the word synovial comes from

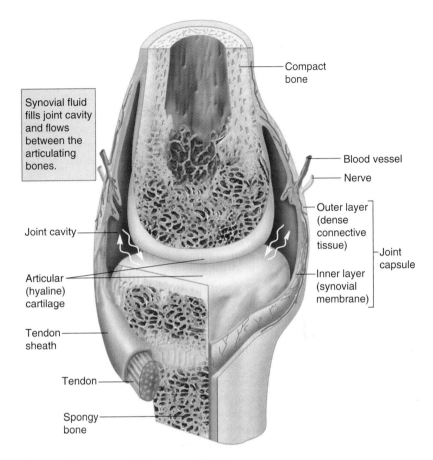

Figure 16.8 A synovial (freely movable) joint.

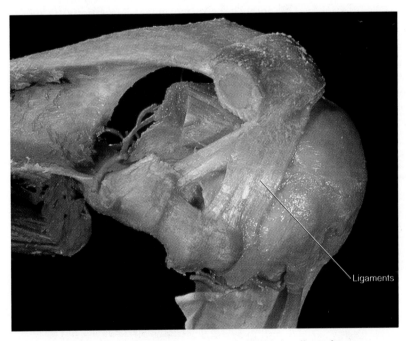

Figure 16.9 Dense connective tissue fibers are arranged in bundles to form ligaments that hold the bones of a joint together. This is a human left shoulder joint.

a Greek word and a Latin word meaning "with" *(syn)* "an egg" *(ovum)*. Synovial fluid lubricates the joint, provides nourishment to the cartilage covering the bone, and contains white blood cells that battle infection.

Although synovial joints permit a wide range of movement, this movement is limited by several factors, including the tension, or tightness, of the ligaments and muscles surrounding the joint and the structure of the articulating bones. You can see the various types of movement of synovial joints on your own body. For example, move your hand at the wrist; your radius forms an *ellipsoidal joint* (Figure 16.7 ❻) with the closest of the wrist bones, or carpals. Another joint, located in the thumb, is called a *saddle joint* ❼, so-named because one bone of the joint is shaped like a saddle and the other like the back of a horse. Processes of your vertebrae glide over one another when you move your back, as seen in the *plane* or *gliding joint* ❽. Some of your carpals and tarsals also have a gliding motion when you move your ankles and wrists.

Other joints allow angular movements, which, as the name suggests, increase or decrease the angle between two bones. When body builders flex their biceps, for example, they bring their forearms closer to their upper arms, decreasing the angle between the radius/ulna and the humerus. The elbow joint permits this movement and is called a *hinge joint* ❾. A hinge joint allows movement in one plane only, similar to how a door opens and closes by means of its hinges.

Another type of movement is rotation, the movement of a bone around an axis. You see this type of movement when a pitcher winds up and throws a baseball or when you twist your head from side to side. The type of joint the pitcher is using at the shoulder is called a *ball-and-socket joint* ❿. In this joint, the rounded head or ball of the humerus articulates with the concavity or socket formed by the ends of the clavicle and scapula. When you move your head from side to side, the type of joint you are using is a *pivot joint* ⓫, formed by the first two vertebrae, the atlas and the axis. Your skull rests on the ringlike atlas (named after the mythological Greek god who could lift the Earth) and pivots on a projection of the axis rising from below.

> A joint is a place where bones and cartilage come together. The amount of movement allowed between two bones depends on how tightly the bones are held together. Freely movable joints, those in which a space exists between articulating bones, are common in the body and include such joints as the elbow and knee.

16.10 The muscular system

Bones and joints are of no use in movement unless muscles are attached to the bones. Muscles provide the power for movement and are made up of specialized cells packed with intracellular fibers capable of shortening.

Humans have three different kinds of muscle tissue: smooth muscle, cardiac muscle, and skeletal muscle (see Figure 7.12). Each type is found in certain locations within the body and performs specific functions. Smooth muscle tissue is made up of sheets of cells (fibers) that are found in many organs, doing such jobs as mixing food in your stomach and narrowing the interior of your arteries to restrict blood flow. Cardiac muscle tissue makes up the heart—the pump of your circulatory system. Skeletal muscle tissue makes up the muscles that are attached to bones, allowing you to move your body; these muscles are the nearly 700 muscles of the muscular system.

The muscular system is shown in **Figure 16.10**. One end of a skeletal muscle is attached by connective tissue called tendons to a stationary bone; this attachment is called the *origin*. The other end is attached to a bone that will move; this attachment is called the *insertion*. The origin serves to anchor the muscle as it pulls at the

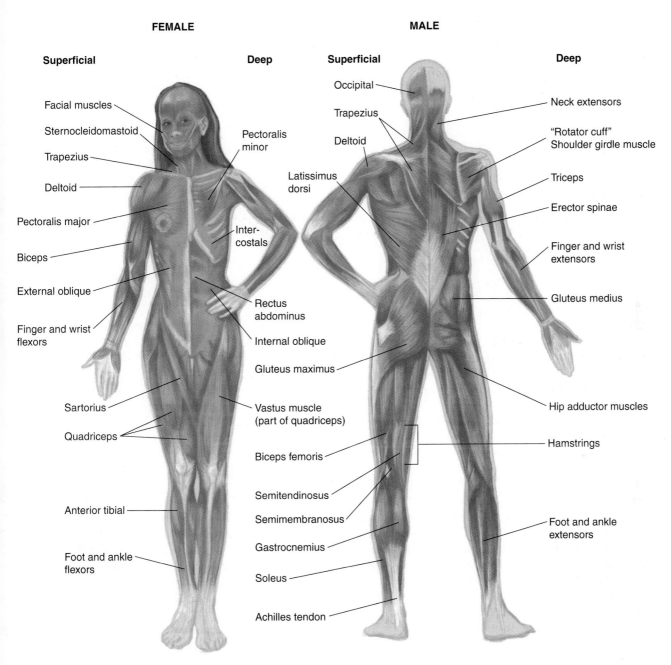

FEMALE

Superficial **Deep**

Facial muscles

Sternocleidomastoid

Trapezius

Deltoid

Pectoralis major

Biceps

External oblique

Finger and wrist flexors

Pectoralis minor

Inter-costals

Rectus abdominus

Internal oblique

Gluteus maximus

Sartorius

Quadriceps

Vastus muscle (part of quadriceps)

Biceps femoris

Semitendinosus

Semimembranosus

Anterior tibial

Gastrocnemius

Foot and ankle flexors

Soleus

Achilles tendon

MALE

Superficial **Deep**

Occipital

Trapezius

Deltoid

Latissimus dorsi

Neck extensors

"Rotator cuff" Shoulder girdle muscle

Triceps

Erector spinae

Finger and wrist extensors

Gluteus medius

Hip adductor muscles

Hamstrings

Foot and ankle extensors

Figure 16.10   The human muscular system.

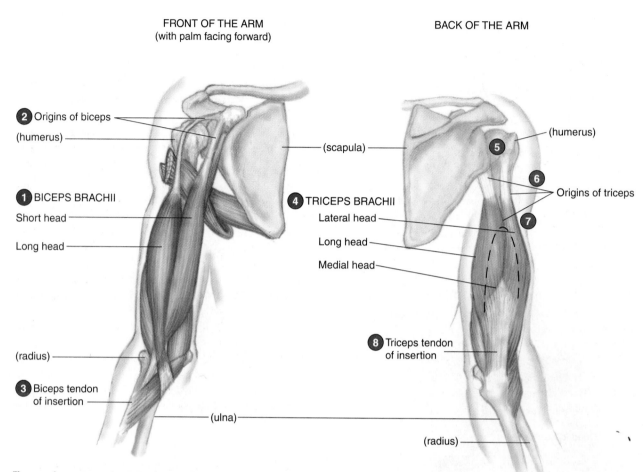

FRONT OF THE ARM
(with palm facing forward)

BACK OF THE ARM

2 Origins of biceps

(humerus)

(scapula)

(humerus)

5

6

Origins of triceps

1 BICEPS BRACHII

4 TRICEPS BRACHII

7

Short head

Lateral head

Long head

Long head

Medial head

(radius)

8 Triceps tendon
of insertion

3 Biceps tendon
of insertion

(ulna)

(radius)

Figure 16.11 **Example of the attachments of skeletal muscles to bones.**

insertion. As an example, Figure 16.11 shows the biceps and triceps muscles and their origins and insertions. The biceps brachii is located on the front of the upper arm **1**. In fact, the word brachii means "arm." This muscle has two upper ends, or heads (hence, biceps, meaning "double headed"). These two heads have origins on the edges of the scapula near the front side of the humerus **2**. Their tendons pass from the scapula over the upper end of the humerus and then blend into the fattened midsection, or belly, of the biceps. The insertion of the biceps is by means of flattened tendons into the radius **3**. When the biceps shortens, the insertion is brought closer to the origin of the muscle, an action that pulls the forearm to the upper arm.

To lower the forearm, the triceps brachii **4** goes into action. Skeletal muscles oppose each other in this way; such opposing muscle pairs are called *antagonists*. In this case, the action of the triceps opposes the action of the biceps. The triceps sits on the back of the arm

with one of its three heads originating on edges of the scapula near the back side of the humerus **5**. Another head originates along the upper half of the back side of the humerus **6** and the third along the lower half of the back side of the humerus **7**. The triceps inserts on the ulna near the elbow **8**. When the triceps shortens, its insertion is brought closer to its origin and the forearm is brought downward. The distance between the origin and insertion is measured as the length of the muscle and tendons, not as a straight line between the two points.

The nearly 700 muscles of the skeletal system are attached to bones by means of connective tissue called tendons. One end of a muscle is attached to a stationary bone (the origin) and the other to a movable bone (the insertion). The origin anchors the bone as it pulls at the insertion.

16.11 How skeletal muscles contract

Skeletal muscles are called *striated muscles* because their cells appear marked with striations, or lines, when viewed under the light microscope (see photo in Figure 16.12b). These striations are caused by an orderly arrangement of filaments called myofilaments (myo- means "muscle") within skeletal muscle cells. Special groupings of these myofilaments are the contractile units of muscle cells.

Skeletal muscle cells are extremely long cells formed by the end-to-end fusion of shorter cells during embryonic development. These long muscle cells are called *muscle fibers.* Each muscle fiber contains all the nuclei of the fused cells pushed out to the periphery of the cytoplasm (Figure 16.12). To see the relationship between muscle fibers and muscles, study Figure 16.12a, noting that a muscle is made up of bundles of muscle fibers, bound together with connective tissue, and nourished by blood vessels. As you study the diagram, you will notice that the names of certain parts of muscle cells have the prefix sarco, which comes from a Greek word meaning "flesh." Table 16.2 lists these differences in terminology.

Table 16.2

Muscle Cell Terminology

Cell Component	Muscle Cell Component
Cell membrane (or plasmalemma)	Sarcolemma
Cytoplasm	Sarcoplasm
Endoplasmic reticulum	Sarcoplasmic reticulum

Each muscle fiber is packed with *myofibrils,* which are cylindrical, organized arrangements of special thick and thin myofilaments capable of shortening the muscle fiber (see Chapter 4). The so-called thin myofilaments are made up of the protein **actin** (AK-tin). The thick myofilaments are made up of a much larger protein, **myosin** (MY-uh-sin). All the myofilaments within a muscle fiber are lined up in such a way that the cells appear to have interior bands of light and dark lines, a feature readily observable in Figure 16.12.

Figure 16.12b is a close-up view of the banding pattern in a muscle fiber as shown in an electron micrograph. This pattern is repeated throughout each muscle fiber. Below the electron micrograph, the diagram shows how the actin and myosin filaments are arranged, forming this banding pattern.

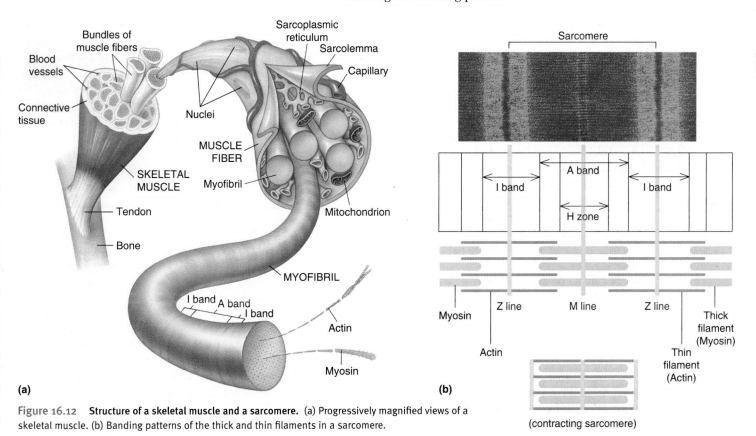

Figure 16.12 **Structure of a skeletal muscle and a sarcomere.** (a) Progressively magnified views of a skeletal muscle. (b) Banding patterns of the thick and thin filaments in a sarcomere.

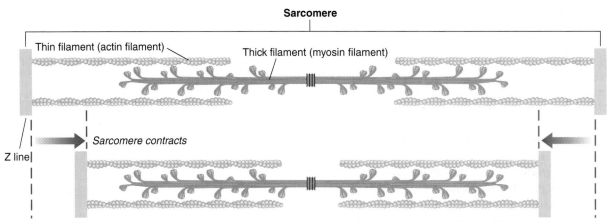

Figure 16.13 How the myofilaments move during muscle contraction.

The thin actin filaments are attached to plates of protein that appear as dark lines called *Z lines*. The actin filaments extend from the Z lines (plates) equally in two directions, perpendicular to the plates. In a resting muscle, the actin filaments that extend from two sequential Z lines are not long enough to reach each other. Instead, they are joined to one another by interdigitating myosin filaments (M line), similar to the way your fingers interlock when you fold your hands. This arrangement of protein plates and myofilaments produces the banding patterns in muscle fibers.

The I band contains the Z line and the thin actin filaments. The A band contains a portion of the thin actin filaments and the thick myosin filaments. The H zone appears as a light zone running down the center of the A band; it contains only thick filaments held in place by a series of fine threads called the *M line*. The part of the myofibrils lying between adjacent Z lines is called the **sarcomere** (SAR-koe-mere), the contractile unit of muscles.

Molecularly, each thin actin filament consists of two strings of proteins wrapped around one another, like two strands of loosely wound pearls (Figure 16.13). The result is a long, thin, helical filament. Myosin has an unusual shape: One end of the molecule is a coil of two chains that forms a very long rod, whereas the other end consists of a double-headed globular region. In electron micrographs, a myosin molecule looks like a two-headed snake or golf club. The contraction of myofilaments occurs when the heads of the myosin filaments change shape. This causes the myosin filaments to slide past the actin filaments, their globular heads "walking" step by step along the actin.

How does myofilament sliding lead to muscle contraction? Because actin myofilaments are anchored to the Z lines, the Z lines are pulled closer to one another when actin slides along the myosin, contracting the sarcomere as shown in Figure 16.13. All the sarcomeres of a myofibril contract simultaneously, shortening the myofibril. In addition, all the myofibrils of a muscle fiber usually contract at the same time. However, all the muscle fibers within a muscle do not contract simultaneously. The forcefulness, speed, and degree of a muscle contraction depend on the number of muscle fibers that contract, their positions in relation to one another, and the frequency of nerve stimuli.

The contraction of skeletal muscles is initiated by a nerve impulse, which arrives as a wave of depolarization along the nerve fiber (see Chapter 13). The end of the nerve fiber, the motor endplate, is microscopically close to the surface of the muscle fiber, forming a **neuromuscular junction.** Acetylcholine passes across the neuromuscular junction from the nerve to the muscle, initiating the process that causes the muscle to contract. This process is driven by the energy from ATP.

Although the structure and organization of fibers within skeletal, smooth, and cardiac muscle tissue differs, these three muscle types all have thin actin filaments and thick myosin filaments that slide past each other during muscular contraction. Additionally, all use ATP as their energy source.

Skeletal muscle cells are long, multinucleated cells called muscle fibers. A muscle is made up of bundles of muscle fibers. Each muscle fiber is packed with organized arrangements of myofilaments that are capable of contracting. Muscle cells contract as a result of the sliding of actin and myosin filaments past one another. Changes in the shape of the ends of the myosin molecules, which are located between adjacent actin filaments, cause the myosin molecules to move along the actin, producing contraction of the myofilament.

16.12 Exercise, bones, and joints

Exercise can help your bones, but it can also hurt your joints. Where your bones and joints are concerned, how much exercise you do is just as important as the kind you get.

Researchers have known for many years that weight-bearing exercise helps maintain bone density and helps avoid *osteoporosis,* or bone loss. Osteoporosis is a condition in which the bones gradually lose mineral density, becoming weak and brittle (Figure 16.14). Although some bone loss begins in both men and women at about age 40, postmenopausal women are at greatest risk for developing osteoporosis. At menopause, women stop producing estrogen, a hormone that stimulates bones to retain calcium. For this same reason, young women who stop menstruating due to eating disorders, and young women athletes who train so strenuously that they no longer have menses, are also at risk. Researchers have found that the bone density of the vertebrae in some female athletes in their

twenties is similar to that of women in their seventies and eighties.

To help maintain your bone density throughout life, engage regularly in moderate physical activities that force you to work against gravity, such as walking, jogging, racquet sports, and weight lifting. Also, be sure that you take in enough calcium—approximately 1200 mg to 1500 mg per day, the amount in about 4 cups of milk. Postmenopausal women at high risk for osteoporosis are often counseled by their physicians to take estrogen replacement therapy. Lastly, do not smoke; cigarette smoking tends to increase bone loss.

Although moderate exercise has a beneficial effect on bones (and even on cartilage), excessive exercising can damage the cartilage in joints. Evidence suggests that a high level of practice and training in certain sports, such as soccer, racquet sports, and track, or doing heavy manual work over a long time are risk factors for cartilage degeneration and osteoarthritis. Recent studies on animals support this connection, showing that the metabolism of the joint cartilage in horses becomes abnormal when the horses are exercised strenuously.

Osteoarthritis is a condition in which the cartilage that covers the articulating ends of bones degenerates. Therefore, when the bones move against one another, the cartilage cushion is gone and the bones move noisily and painfully. This condition usually develops because of overuse of the joints, particularly the hip and knee joints. The occurrence of osteoarthritis is highest in those athletes or workers who have irregular or sudden impacts to their limbs, apply heavy loads to their limbs, or have had previous injury to the joints most used in their work or sports activity.

The bottom line regarding bone and joint health is that exercise is good—in moderation. This idea bears out with regard to your immune system also: Moderate exercise strengthens the immune system, while overly strenuous exercise impairs immune function. The old adage that "more is better" is not true when it comes to exercise and its effects on your bones, joints, and general health.

For the latest health information, please visit www.jbpub.com/biology.

Bone and joint health is enhanced by moderate physical activity, but can be hurt by excessively strenuous, long-term physical activity.

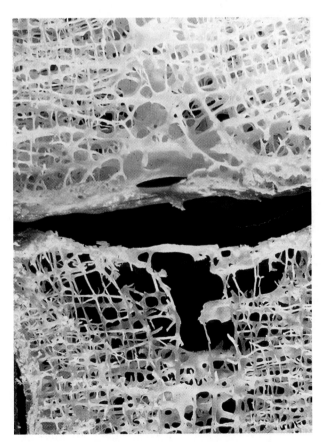

Figure 16.14 **Normal (top) and osteoporotic bone (bottom).** The osteoporotic bone is less dense than the normal bone.

Key Concepts

Skin, bones, and muscles work together to help protect, support, and move animals.

16.1 In animals, the outer covering (integument) protects soft body parts beneath, provides a barrier to infection, and helps conserve water in land animals.

16.1 In many aquatic animals, the integument also acts as the respiratory surface.

16.2 The skeleton provides support for the animal body; an animal may have a hydroskeleton, an endoskeleton, or an exoskeleton.

16.2 Muscles attach to the rigid, skeletal portion of a body; the skeleton acts as something against which muscles pull.

Human skin, bones, and muscles perform many functions.

16.3 In humans, movement results from the contraction of skeletal muscles anchored to bones, which are used like levers to direct force against an object.

16.4 Human skin provides an elastic covering for the body, serves as a protective barrier against water loss and invading microbes, provides a sensory surface, compensates for body movement, and helps control the body's internal temperature.

16.5 Bones contain living cells embedded in a matrix of collagen fibers, which provide flexibility, and mineral salts, which provide hardness.

16.5 Compact bone is found along the length of long bones, and spongy bone is found within the ends of long bones and within short, flat, and irregularly shaped bones.

16.6 The skeletal system provides support for the body and attachment sites for skeletal muscles, protects delicate internal structures, produces blood cells in its red marrow, and stores minerals.

16.7 The axial skeleton is the central column of bones and consists of the skull, vertebral column, and rib cage.

16.8 The appendicular skeleton consists of the bones of the appendages and the bones off which the appendages hang.

16.9 A joint, or articulation, is a place where bones, or bones and cartilage, come together.

16.10 There are three kinds of muscle tissue: smooth muscle tissue, which is found in many body organs; cardiac muscle tissue, which makes up the heart; and skeletal muscle tissue, which is attached to bones and move the body.

16.11 Skeletal muscle cells contract as a result of thick and thin filaments, which are packed within the muscle cells, sliding past one another.

 Health
Connections

16.12 Bone and joint health is enhanced by moderate physical activity, but can be hurt by excessively strenuous, long-term physical activity.

Key Terms

actin (AK-tin) *349*
appendicular skeleton (AP-en-DIK-you-lur) *339*
articulation (ar TIK-you-LAY-shun) *343*
axial skeleton (AK-see-uhl) *339*
ball-and-socket joint *346*
bone *338*
endoskeleton *335*
exoskeleton *335*
Haversian canal (huh-VUR-shun) *338*
hinge joint *346*

hydroskeleton *334*
integument (in-TEG-you-ment) *332*
integumentary system (in-TEG-you-MEN-tuh-ree) *337*
intervertebral disks (IN-tur-VER-tuh-brul) *341*
joint *343*
ligaments (LIG-uh-munts) *341*
myosin (MY-uh-sin) *349*
neuromuscular junction (NER-oh-MUS-kyuh-lur) *350*

pectoral girdle (PEK-tuh-rul) *342*
pelvic girdle *342*
pivot joint *346*
sarcomere (SAR-koe-mere) *350*
skeletal system *339*
skeleton *339*
skull *340*
suture (SOO-chur) *343*
synovial joints (suh-NO-vee-uhl) *345*
tendons *343*
vertebral column (VER-tuh-brul) *341*

KNOWLEDGE AND COMPREHENSION QUESTIONS

1. You raise your hand to answer a question in class. Explain the roles played by your bones and skeletal muscles in this movement. (*Comprehension*)
2. Human babies learn extensively by touching and manipulating toys and objects. What physiological role does skin serve in this process? (*Comprehension*)
3. What is bone, and what are its functions? Explain its importance. (*Knowledge*)
4. Leukemia patients sometimes receive red bone marrow transplants as treatment for their disease. From what kind of bone tissue would this marrow be obtained? (*Knowledge*)
5. What type of bone tissue makes up most of the bones of the axial skeleton? (*Comprehension*)
6. What do you hypothesize happens to excess calcium taken in by the body yet not needed at the moment? (*Comprehension*)
7. Mrs. Gorman has injured her back; her doctor tells her she has a "slipped disk." What type of "disk" is involved, and what is its function? (*Knowledge*)
8. What do hinge, pivot, and ball-and-socket joints have in common? How are they different? (*Comprehension*)
9. What connective tissue connects muscles to bones? What connects muscles to each other? (*Knowledge*)
10. What are neurosecretory cells? Give an example of an endocrine gland made up of neurosecretory cells. (*Knowledge*)

K E Y
Knowledge: Recalling information.
Comprehension: Showing understanding of recalled information.

CRITICAL THINKING REVIEW

1. From your observations, describe the changes that occur to skin and bone as people age. (*Synthesis*)
2. In recent years, there has been tremendous progress made in the development and use of metallic artificial joints and bone replacement. What activities and bone functions would be easily served by such artificial structures? Are there functions of the bone that could not be filled by such replacements? (*Analysis*)
3. The trunk of an elephant has no endoskeleton to which muscles can attach, yet the elephant is able to move its trunk quite skillfully. What type of skeleton must the elephant trunk have? Describe how you think the elephant moves its trunk. (*Synthesis*)
4. Many bacteria colonize human skin; these organisms are called the normal microbiota. Based on your knowledge of the structure of human skin, why do you think that these bacteria do not usually cause skin infections, yet can quickly cause an infection when the skin surface is broken, as in a wound? (*Application*)
5. Analyze and evaluate your patterns of exercise (or non-exercise) to determine whether you are exercising in a manner that will help strengthen your bones and joints, or in a manner that may cause damage to your bones and joints. Give reasons that support your evaluation as you describe your exercise or training routine. (*Evaluation*)

K E Y
Application: Using information in a new situation.
Analysis: Breaking down information into component parts.
Synthesis: Putting together information from different sources.
Evaluation: Making informed decisions.

Circle the correct type of joint for items 1-5.
For items 6-11, write the name of the joint in the blank,
and then describe the type of movement the joint allows.

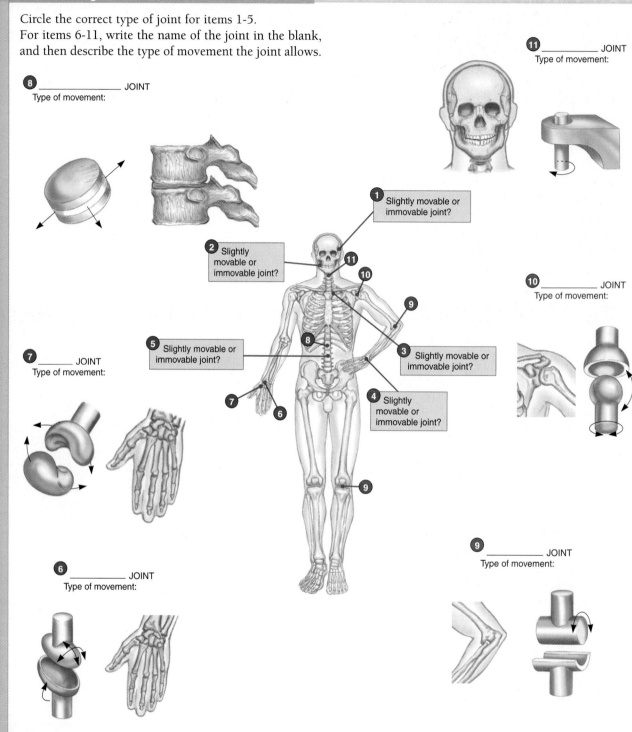

11 _____ JOINT
Type of movement:

8 _____ JOINT
Type of movement:

1 Slightly movable or immovable joint?

2 Slightly movable or immovable joint?

5 Slightly movable or immovable joint?

3 Slightly movable or immovable joint?

4 Slightly movable or immovable joint?

10 _____ JOINT
Type of movement:

7 _____ JOINT
Type of movement:

9 _____ JOINT
Type of movement:

6 _____ JOINT
Type of movement:

e learning

Location: http://www.jbpub.com/biology

e-Learning is an on-line student review area located at this book's web site www.jbpub.com/biology. The review area provides a variety of activities designed to help you study for your class and build your learning portfolio.

Review Questions The review questions test your knowledge of the important concepts and applications in each chapter. The review provides feedback for each correct or incorrect answer. This is an excellent test preparation tool.

Figure-Labeling Exercises Sharpen your visual thinking by matching terms and labels for important illustrations.

Flash Cards Studying biology requires learning new terms. Virtual flash cards help you master the new vocabulary for each chapter.

Just Wondering As you read and study from this text, you may find that you have unanswered questions. Through this site you can ask the author, Sandy Alters, your "just wondering" questions.

Do you prefer the speed and reliability of a CD-ROM? All of the features contained on the eLearning portion of the web site are also available on the Student CD-ROM.

Hormones

Anabolic steroids have been the downfall of many winners.
These controversial drugs are really synthetic hormones,
chemicals that affect the activity of specific organs or tissues.
Anabolic steroids affect the body in ways similar to the male sex
hormone testosterone and stimulate the buildup of muscles.
However, along with building a championship body, anabolic
steroids strikingly change the body's metabolism.

Female athletes on steroids experience side effects such as
shrinking breasts, a deepening voice, and an increase in body
hair. Male athletes find that their testicles shrink. Some users
also experience life-threatening kidney and liver damage.
Youngsters who take these drugs risk stunting their growth
because anabolic steroids cause bones to stop growing
prematurely. Additionally, anabolic steroids may cause
psychological effects such as "steroid rage," a state of mind in
which users attack people and things around them.

It isn't just older athletes who use these risky drugs. A 1998
study by researchers at Boston University revealed that the use
of steroids by middle-school-aged youngsters was on the rise.
Study results indicated that 2.7% of all middle school students
reported using steroids, and that young female athletes used
steroids at nearly the same rate as male athletes. Youngsters
involved in weight training or gymnastics were the ones most
likely to use these drugs.

If you are thinking about using anabolic steroids, first
consider that they can be deadly. Second, realize that if they
don't kill you, they will most likely have a negative effect on
your health. To land in the winner's circle, rely on the advice of
physicians: Practice good nutrition, proper conditioning, and
appropriate training.

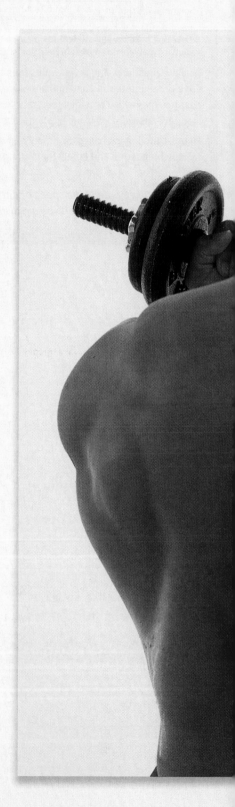

CHAPTER
17

Organization

Hormonal control is ancient and widespread.

17.1 What are hormones?

In vertebrates, **hormones** are defined as chemical messengers that are released into the bloodstream and have a specific effect on body functions or on other structures within the body. Table 17.1 below lists the central (brain) endocrine glands and their hormones in the human body. Although hormones are secreted into the bloodstream and are therefore released to *all* the tissues, hormones affect only particular tissues called target tissues. Such selective interaction takes place because hormones affect only those tissues bearing hormone-specific receptors.

In vertebrates such as humans, the hormonal system is referred to as the **endocrine system** (ENN-doe-krin). This name comes from the **endocrine glands**, which secrete hormones. Glands are individual cells or groups of cells that secrete substances. They are called "endocrine" because they release substances within (*endo-*) the body, rather than secreting substances that exit the body (such as sweat, for example). Endocrine glands

are also called *ductless glands* because they secrete hormones directly into the bloodstream rather than into ducts. Sweat glands, for example, are **exocrine glands** (EK-so-krin), routing their secretions to a specific place outside the body—the skin surface—by means of ducts.

Two types of cells make and secrete hormones: endocrine cells and specialized nerve cells called *neurosecretory cells* (NEW-row-SEE-creh-tor-ee). In addition to making and secreting hormones, neurosecretory cells conduct nerve impulses and stimulate adjacent neurons by means of neurotransmitters. As shown in Figure 17.1, both neurosecretory cells and endocrine cells secrete their hormones directly into the bloodstream. Therefore, the glands composed of either of these cell types are endocrine glands. The endocrine glands made up of neurosecretory cells are the hypothalamus, the posterior pituitary, and the adrenal medulla.

Many of the endocrine glands of the body have the sole job of secreting hormones. Such glands are the pituitary, adrenals, thyroid, parathyroids, and pineal gland. Other endocrine glands are located within organs that have other functions in addition to the synthesis and release of hormones. These organs include the hypothalamus, thymus, kidneys, digestive tract, pancreas, liver, skin, heart, ovaries, testes, and placenta.

Focus
on Plants

Hormones are abundant in the plant world.

At least five major kinds of hormones in plants direct their growth, differentiation, maturation, flowering, and many other activities. Chapter 28 describes plant hormones and their effects in more detail.

Table 17.1

Central (Brain) Endocrine Glands and their Hormones

Endocrine Gland and Hormone	Target Tissue	Principal Actions
Hypothalamus		
Releasing hormones	Anterior pituitary	Stimulate the release of hormones by the anterior pituitary
Posterior Pituitary		
Oxytocin	Uterus Mammary glands	Stimulates contraction of uterus and milk production
Antidiuretic hormone (ADH)	Kidneys	Stimulates reabsorption of water by the kidneys
Anterior Pituitary		
Follicle-stimulating hormone (FSH) (Gonadotropic hormone)	Sex organs	Stimulates ovarian follicle, spermatogenesis
Luteinizing hormone (LH) (Gonadotropic hormone)	Sex organs	Stimulates ovulation and corpus luteum formation in females
Adrenocorticotropic hormone (ACTH)	Adrenal cortex	Stimulates secretion of adrenal cortical hormones
Thyroid-stimulating hormone (TSH)	Thyroid	Stimulates secretion of thyroid hormones
Growth hormone (GH)	Cartilage and bone cells, skeletal muscle cells	Stimulates division of cartilage and bone cells, growth of muscle cells, and deposition of minerals
Prolactin	Mammary glands	Stimulates milk production
Melanocyte-stimulating hormone (MSH)	Melanocytes	Stimulates production of melanin
Pineal Gland		
Melatonin	Hypothalamus and/or reproductive organs	Possible stimulation of immune system; inhibits secretion of GnRH; other specific actions unknown

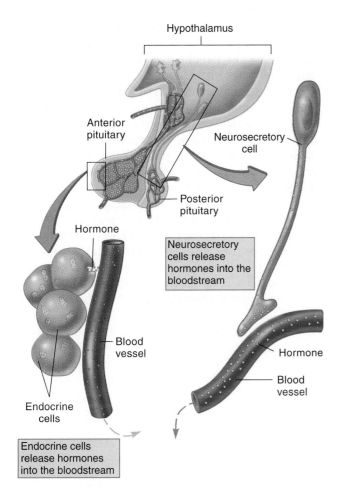

Figure 17.1 Hormones can come from neurosecretory cells or from endocrine cells.

In concert with the nervous system, the endocrine system controls body functions. In vertebrates as well as invertebrates, each of these systems performs particular control functions, yet they are well integrated. That is, one system interacts with the other. Many examples of this integration are given throughout this chapter. In general, however, the nervous system is a rapid communication system and oversees such functions as fleeing from a predator (or attacker) and keeping the heart beating regularly (Chapters 13 and 14 describe how the nervous system works). Hormonal control is a slow communication system and, as such, is involved with gradual (or slow) processes such as growth and maturation.

Hormones are chemical messengers that are secreted by cells into the blood and that affect other cells. Hormones are secreted from endocrine cells or from neurosecretory cells.

17.2 Hormones in organisms other than vertebrates

The function of chemical messengers (hormone function) has been studied in a wide variety of invertebrates such as squid, earthworms, lobsters, and insects, and in other organisms such as protozoans and plants. (See a Focus on Plants.) Researchers have discovered that these organisms have hormones, but that most of their hormones differ from vertebrate hormones in their chemical structures and modes of action. (Additionally, not all of the organisms have a bloodstream in which hormones flow.) However, a few vertebrate hormones (and cell receptor molecules) have been found in non-vertebrates. For example, adrenaline has been found in protozoans. It appears to function similarly to that of the vertebrate hormone, but its precise function is unknown. Insulin, the hormone in humans that ferries glucose molecules across the cell membrane, has been found in insects, annelids (such as worms), molluscs (such as snails, octopus, and squid), and unicellular eukaryotes (such as protozoans). Scientists deduce from these discoveries that hormones are ancient molecules. That is, hormones first appeared as simple organisms evolved millions of years ago.

Of the invertebrates, insects have been studied most extensively with regard to hormone systems. In insects, hormones play a major role in metamorphosis and growth. Metamorphosis refers to the changes that most insects go through as they develop from egg to adult. It includes such processes as molting, the shedding of skin to allow further growth (as shown in the photo above), and pupation, the development of a "cocoon" stage between larva and adult.

In bloodsucking insects such as mosquitoes, certain hormones regulate water balance as they do in vertebrates. After a blood meal, a mosquito would have a difficult time flying if it didn't quickly rid itself of all that extra water weight! This process of elimination begins even *before* it finishes its "meal."

A wide variety of organisms have hormonal control mechanisms, including invertebrates, protozoans, and plants. Hormones are ancient molecules; they first appeared as simple organisms evolved millions of years ago.

17.3 Human endocrine hormones

There are 10 major endocrine glands in the human body that make up the endocrine system (Figure 17.2). These glands are scattered throughout the body, and in some cases are parts of other glands. Together, the glands of the endocrine system produce over 50 different hormones. The messages of the endocrine hormones are varied, but can be grouped into four categories:

1. *Regulation*: Hormones control the internal environment of the body by regulating the secretion and excretion of various chemicals in the blood, such as salts and acids.
2. *Response*: Hormones help the body respond to changes in the environment and cope with physical and psychological stress.
3. *Reproduction*: Hormones control the female reproductive cycle and other reproductive processes essential to conception and birth, and control the development of sex cells, the reproductive organs, and secondary sexual characteristics (those that make men and women different) in both sexes.
4. *Growth and development*: Hormones are essential to the proper growth and development of the body from conception to adulthood.

Once molecules of a hormone are released into the bloodstream, they travel throughout the body. Although hormone molecules may pass billions of cells, specific hormones affect only target cells. Hormones recognize target cells because they bind to receptor molecules embedded within the cell membrane or located within the cytoplasm of the cell. The binding of a hormone molecule to a receptor molecule activates a chain of events in the target cell that results in the effect of the hormone being expressed.

An easy way to think about how hormones affect target cells (and only target cells) is to compare this interaction to the reception of radio waves by a radio. Radio waves (hormones) are sent through the antenna of a radio station (endocrine gland) and are transmitted in all directions (throughout the body via the bloodstream). The receiver, or radio (the target cells) must be tuned to the frequency of the radio waves (the receptor molecules) to receive the transmission. Any radio not tuned to that frequency (cells without receptor molecules for that hormone) will not receive the signal.

The glands of the endocrine system produce over 50 different hormones. Together, these hormones help regulate the internal environment of the body, help the body respond to environmental changes and stress, control reproductive processes, and direct growth and development.

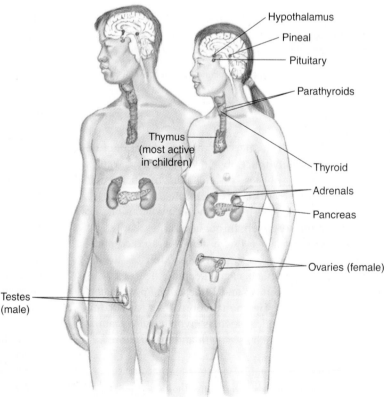

Figure 17.2 **The human endocrine system.**

17.4 How endocrine hormones signal tissues

Two major classes of endocrine hormones work within the human body: peptide hormones and steroid hormones.

Peptide Hormones

Peptide hormones are made of amino acids, but the amino acid chain length varies greatly from hormone to hormone. The smallest are actually modifications of the single amino acid tyrosine. Somewhat larger are short peptide hormones that are several amino acids in length. Polypeptide hormones have chain lengths of several dozen or more amino acids, such as the hormone insulin. Even larger are protein hormones that may have over 200 amino acids with carbohydrates attached at several positions.

Unable to pass through the cell membrane, peptide hormones bind to receptor molecules embedded in the cell membrane of target cells (Figure 17.3 ❶). The binding of a hormone to a receptor triggers an increase in the conversion of ATP to cyclic AMP, a compound referred to as the *second messenger* ❷. The second messenger activates enzymes ❸ that cause the cell to alter its functioning ❹. For example, prolactin stimulates cells of the mammary glands to produce milk. Target cells respond as enzymes go into action catalyzing reactions that produce the components of mother's milk. Other types of hormone responses in-

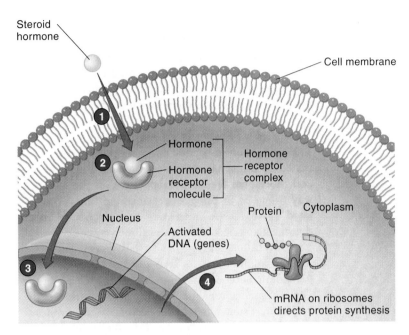

Figure 17.4 **How steroid hormones work.**

clude the secretion of substances from target cells and the closing or opening of certain protein doors within target cell membranes. Cyclic adenosine monophosphate (cAMP for short), a "cousin" of ATP (see Chapter 5), acts as a second messenger in many cells. Besides cAMP, other second messenger molecules have been discovered.

Steroid Hormones

Steroid hormones are all made from cholesterol, a lipid synthesized by the liver. You know cholesterol as that dietary devil present in certain foods such as eggs, dairy products, and beef. A characteristic of steroid hormones is their set of carbon rings.

Steroid hormones, being lipid soluble, pass freely through the lipid bilayer of the cell membrane (Figure 17.4 ❶). Once inside a cell, these hormones bind to receptor molecules located within the cytoplasm of target cells ❷. Together, the hormone-receptor complex moves into the nucleus

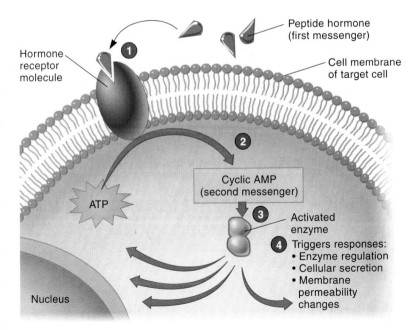

Figure 17.3 **How peptide hormones work.**

of the cell ❸, causing the cell's hereditary material, or DNA, to trigger the production of certain proteins ❹. In response to the sex hormones estrogen and testosterone, for example, the proteins produced are those involved in such processes as the development and maintenance of female and male sexual characteristics.

Hormonal Feedback Loops

The production of peptide and steroid hormones is regulated by a mechanism called a *feedback loop*. In general, hormonal feedback loops work in the following way (Figure 17.5): Endocrine glands are initially stimulated ❶ to release hormones. Stimulation of an endocrine gland occurs in one of three ways:

1. *Direct stimulation by the nervous system:* The sensation of fear, for example, can cause the autonomic nervous system to trigger the release of the hormone adrenaline from the adrenal medulla.

2. *Indirect stimulation by the nervous system by means of releasing hormones:* The hypothalamus is a specialized portion of the brain that produces and secretes releasing hormones. Some releasing hormones stimulate the release of other hormones; some prevent the release.

3. *The concentration of specific substances in the bloodstream:* The blood level of a substance such as glucose or calcium ions, for example, may signal an endocrine gland to turn on or turn off.

After an endocrine gland secretes its hormone into the bloodstream ❷, the hormone travels throughout the body via the circulatory system and interacts with target tissues ❸. The target tissues cause the desired effect to be produced ❹. This effect acts as a new stimulus to the endocrine gland ❺. Put simply, the body feeds back information to each endocrine gland after it releases its hormone. In a positive feedback loop, the information that is fed back causes the gland to produce more of its hormone. In a **negative feedback loop**, the feedback causes the gland to slow down or to stop the production of its hormone. Most hormones work by means of negative feedback loops. (Specific examples of feedback mechanisms and interactions are discussed throughout this chapter.)

Two main classes of endocrine hormones are peptide hormones and steroid hormones. Both travel within the bloodstream to all parts of the body, but affect only certain target cells. Peptide hormones bind to receptors of the cell membrane of target cells and ultimately trigger enzymes that alter cell functioning. Steroid hormones bind to receptors within the cytoplasm of target cells and ultimately cause the hereditary material of the cell to produce specific proteins. The production of these hormones is regulated by feedback loops, in which information about hormonal effects is fed back to endocrine glands.

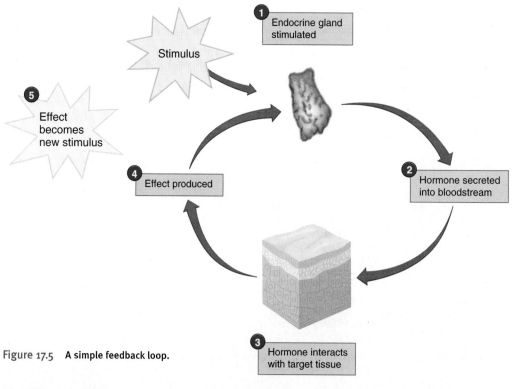

Figure 17.5   A simple feedback loop.

17.5   The pituitary and hypothalamus

The **pituitary** (puh-TOO-ih-tare-ee) is a powerful gland that secretes nine major hormones. Although it secretes so many hormones, it is amazingly tiny—slightly smaller than a marble. The pituitary gland hangs from the underside of the brain, supported and cradled within a bony depression of the sphenoid bone.

Controlling the Pituitary: The Hypothalamus

The pituitary secretes seven major hormones from its larger front portion, or lobe, the *anterior pituitary*. It secretes two from its rear lobe, the *posterior pituitary* (see Table 17.1, p. 358). The secretion of these hormones is regulated by a mass of nerve cells that lies directly above the pituitary, making up a small part of the floor of the brain. This regulatory nervous tissue, the **hypothalamus** (HYE-poe-THAL-uh-muss), is connected to the pituitary by a stalk of tissue (Figure 17.6).

The hypothalamus uses information it gathers from other parts of the brain and its own receptors to stimulate or inhibit the secretion of hormones from the pituitary. In this way, the hypothalamus acts like a production manager, receiving information about the needs of the company's customers and regulating the production of products to satisfy those needs. However, the hypothalamus manages the anterior and posterior lobes of the pituitary differently.

The hypothalamus regulates the secretion of hormones from the anterior pituitary by producing *releasing hormones*. These seven hormones flow directly from the hypothalamus to the anterior pituitary via a network of blood vessels. Five releasing hormones are stimulatory, each causing the release of one or more pituitary hormones. Two are inhibitory. One inhibits the release of growth hormone and TSH. The other inhibits the release of prolactin.

The hypothalamus also regulates the release of two hormones from the posterior pituitary. This regulation is neural, not hormonal. Neurons extend from the hypothalamus to the posterior pituitary. The hormones released from the posterior pituitary are produced in the cell bodies of the neurons, which lie in the hypothalamus. After production, the hormones travel down the axons of these neurons to the posterior pituitary. They are held there until the hypothalamic neurons signal their release into the bloodstream.

The Anterior Pituitary

The seven hormones produced by the anterior pituitary regulate a wide range of bodily functions (Figure 17.7). Four of these hormones are called **tropic hormones** (TROW-pick). The word tropic comes from a Greek word meaning "turning" and refers to the ability of tropic hormones to turn on or stimulate other endocrine glands. Of the four tropic hormones, two are **gonadotropins** (GON-ah-duh-TROP-inz).

The gonads are the male and female sex organs, the testes and the ovaries. The gonadotropins are hormones that affect these sex organs, which are considered endocrine glands because they secrete hormones. The two gonadotropins are *follicle-stimulating hormone (FSH)* and *luteinizing hormone (LH)* (LOO-tea-in-EYE-zing). In fe-

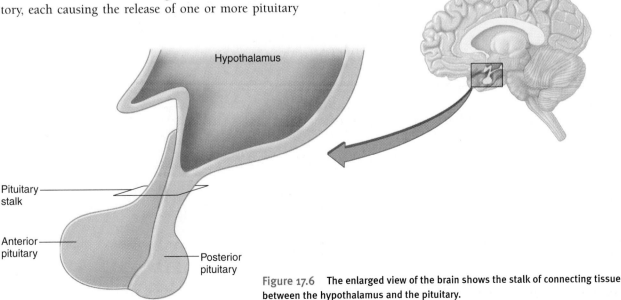

Hypothalamus

Pituitary stalk

Anterior pituitary

Posterior pituitary

Figure 17.6   The enlarged view of the brain shows the stalk of connecting tissue between the hypothalamus and the pituitary.

The hypothalamus, pituitary, and pineal are the central (brain) endocrine glands.

males, FSH targets the ovaries and triggers the maturation of one egg each month. In addition, it stimulates cells in the ovaries to secrete female sex hormones called *estrogens*. In men, FSH targets the testes and triggers the production of sperm. LH stimulates cells in the testes to produce the male sex hormone *testosterone*. In females, a surge of LH near the middle of the menstrual cycle stimulates the release of an egg. In addition, LH triggers the development of cells within the ovaries that produce another female sex hormone—*progesterone*. (See Chapter 22 for a description of the organs and processes of the reproductive system.)

The two other tropic hormones are *adrenocorticotropic hormone (ACTH)* (ah-DREE-no-kor-tih-coh-trow-pick) and *thyroid-stimulating hormone (TSH)*. ACTH triggers the adrenal cortex to produce certain steroid hormones. The adrenal glands are located on top of the kidneys (see Figure 17.2). Each of these two glands has two distinct parts: an outer cortex and an inner medulla. ACTH stimulates the adrenal cortex to produce hormones that regulate the production of glucose from noncarbohydrates such as fats and proteins. Others regulate the balance of sodium and potassium ions in the blood. Still others contribute to the development of

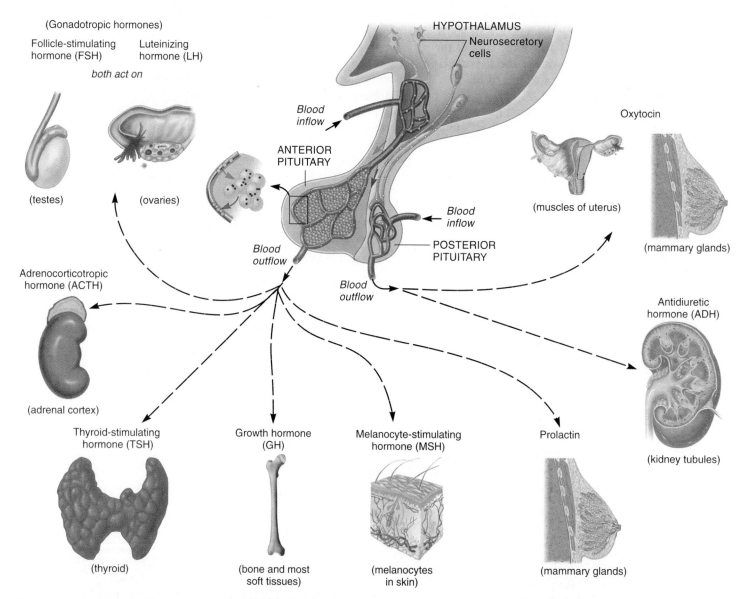

Figure 17.7 **The pituitary secretes anterior pituitary hormones (green dots) and posterior pituitary hormones (blue dots).** Releasing and inhibiting hormones secreted by the hypothalamus, which act on the anterior pituitary, are shown in orange.

Giantism is caused by the oversecretion of growth hormone.

These photographs of a woman with acromegaly, at ages 16 and 33, show the thickening of the facial bones that results from the oversecretion of growth hormones.

the male secondary sexual characteristics. TSH triggers the thyroid gland to produce two thyroid hormones. This endocrine gland is located on the front of the neck, just below the voice box (see Figure 17.2). Its hormones control normal growth and development and are essential to proper metabolism. (Further discussion of ACTH is on p. 369, and further discussion of TSH is on p. 367.)

Growth hormone (GH) (also known as somatotropin) is produced by the anterior pituitary and works with the thyroid hormones to control normal growth. GH increases the rate of growth of the skeleton by causing cartilage cells and bone cells to reproduce and lay down their intercellular matrix. In addition, GH stimulates the deposition of minerals within this matrix. GH also stimulates the skeletal muscles to grow in both size and number. In the past, children who did not produce enough GH did not grow to an average height; this condition is called hypopituitary dwarfism. However, in the past decade, scientists have been able to use the techniques of genetic engineering to insert the human GH gene into bacteria to produce human GH. Currently, this laboratory-made hormone is being used successfully for treating growth disorders caused by hyposecretion (underproduction) of GH in children. The opposite problem may also occur: During the growth years, some children produce too much GH. This hypersecretion (overproduction) can cause the long bones to grow unusually long (see photo above left) and result in a condition known as giantism. In adults, hypersecretion of GH causes the bones of the hands and face to thicken, resulting in a condition known as acromegaly (AK-row-MEG-ah-lee), as shown in the photos above right.

Prolactin is another hormone secreted by the anterior pituitary. Prolactin works with estrogen, progesterone, and other hormones to stimulate the mammary glands in the breasts to secrete milk after a woman has given birth to a child. During the menstrual cycle, milk is not produced and secreted because prolactin levels in the bloodstream are very low. Late in the menstrual cycle, however, as the levels of progesterone and estrogen fall, the pituitary is stimulated by the hypothalamus to secrete some prolactin. This rise in prolactin, although not sufficient to cause milk production, does cause the breasts of some women to feel sore before menstruation. After menstruation, estrogen levels begin to rise, and prolactin secretion is once again inhibited.

Melanocyte-stimulating hormone (MSH) acts on cells in the skin called melanocytes, which synthesize a pigment called melanin. Epidermal cells in the skin contain this pigment, which produces skin colorations from pale yellow (in combination with another pigment called carotene) to black, depending on the amount of melanin in the skin. Variations are caused by the amount of pigment the melanocytes produce; this variation is genetically determined and is an inherited characteristic. Exposure to ultraviolet (UV) light triggers increased melanin production, which is visible as a suntan in light-skinned people and as a darkening of the skin in dark-skinned individuals.

The Posterior Pituitary

The posterior lobe of the pituitary stores and releases two hormones that are produced by the hypothalamus: *antidiuretic hormone (ADH)* (AN-tih-die-yuh-RET-ick) and *oxytocin* (OK-sih-TOE-sin). ADH helps control the volume of the blood by regulating the amount of water reabsorbed by the kidneys. For example, receptors in the hypothalamus can detect a low blood volume by detecting when the solute concentration of the blood is high.

When the hypothalamus detects such a situation, it triggers its specialized neurosecretory cells to make ADH. This hormone is transported within axons to the posterior pituitary, which releases the hormone into the bloodstream. ADH binds to target cells in the collecting ducts of the nephrons of the kidneys, increasing their permeability. An increased amount of water moves out of these ducts and back into the blood, resulting in more concentrated urine. ADH also acts on the smooth muscle surrounding arterioles. As these muscles tighten, they constrict the arterioles, an action that helps raise the blood pressure despite fluid loss. Alcohol suppresses ADH release, which is why excessive drinking leads to the production of excessive quantities of urine and eventually to dehydration.

Oxytocin is another hormone of the posterior pituitary. In women, oxytocin is secreted during the birth process, triggered by a stretching of the cervix at the beginning of the birth process. Oxytocin binds to target cells of the uterus, enhancing the contractions already taking place. The mechanism of oxytocin secretion is an example of a **positive feedback loop** in which the effect produced by the hormone enhances the secretion of the hormone. For this reason, physicians use oxytocin to induce uterine contractions when labor must be brought on by external means. Oxytocin also targets muscle cells around the ducts of the mammary glands, allowing a new mother to nurse her child. The suckling of the infant triggers the production of more oxytocin, which aids in the nursing process and helps contract the uterus to its normal size.

> The pituitary is a tiny gland that hangs from the underside of the brain. The secretion of its many diverse hormones is controlled by the hypothalamus, a mass of nerve cells lying directly above it. The hypothalamus stimulates or inhibits the secretion of hormones from the anterior lobe of the pituitary by means of releasing hormones. Of these seven anterior pituitary hormones, four stimulate other endocrine glands and are called tropic hormones. The rear lobe of the pituitary, or posterior pituitary, stores and releases two hormones, ADH and oxytocin, which are produced by the hypothalamus.

17.6 The pineal

The **pineal gland** (PIN-ee-uhl) gets its name from its shape: It looks like a tiny pine cone embedded deep within the brain between the two cerebral hemispheres (see Figure 17.2). The major endocrine product of the pineal gland is *melatonin*, which is synthesized by the pineal primarily at night. Signaled by nerve impulses arising at the eyes, the pineal gland increases (nearly 10-fold) its secretion of melatonin at nighttime. Likewise, during the daytime, nervous stimulation from the eyes causes the pineal to suppress its secretion of this hormone.

One of the most widely accepted roles of melatonin is the synchronizing of the body's circadian rhythms with the light/dark cycle. Circadian rhythms (biological clocks) are body activities that change throughout the day and night or with the seasons. A daily circadian rhythm is the sleep/wake cycle. Many animals show seasonal circadian rhythms such as patterns of hibernation, migration, or mating as their melatonin production fluctuates with increasing or decreasing numbers of daylight hours. These seasonal biological patterns of behavior are discussed in more detail in Chapter 32.

Recent books and articles in the popular press promote melatonin as an anti-aging pill and a preventative therapy for heart attacks. However, there are no studies that provide evidence that melatonin has an effect on human life expectancy, nor are there human studies that support its role in preventing heart attacks. Studies do suggest, however, that low-dose melatonin supplements (fractions of a milligram) can hasten sleep and ease jet lag by helping to shift the sleep cycle. Melatonin has also been studied extensively regarding its interaction with the immune system. Results of these studies suggest that melatonin may directly or indirectly stimulate both cellular and humoral immunity. Additionally, the results of many studies suggest that melatonin inhibits the secretion of gonadotropin-releasing hormone (GnRH) from the hypothalamus in many mammals. This action, in turn, inhibits egg release in females and sperm production in males. Researchers are currently investigating whether melatonin may play a role in future contraceptives for both men and women.

> The pineal gland is located at the base of the brain. Its primary endocrine secretion is melatonin, a hormone that regulates the sleep/wake cycle in humans and certain seasonal behavior patterns in other animals.

17.7   The thyroid

Sitting like a large butterfly just below the level of the voice box, the thyroid gland can be thought of as your metabolic switch. This gland secretes hormones that determine the rate of the chemical reactions of your body's cells. Put simply, thyroid hormones determine how fast bodily processes take place.

The thyroid hormones are called amines: single, modified amino acids. They are not considered to be true peptide hormones, however, because they

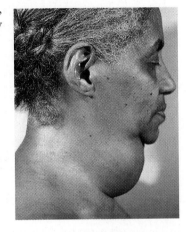

Figure 17.8   A goiter. Iodide is used in the production of the thyroid hormones; when not enough iodide is available, the thyroid cannot produce thyroid hormones, and the thyroid swells from overstimulation by the anterior pituitary.

act on the DNA of target cells as steroid hormones do. They are also unique because an inorganic ion—iodide—is part of their structures.

Your body uses iodide in the food you eat to help make the thyroid hormones. Foods such as seafood and iodized salt are good sources of dietary iodide. If the diet contains an insufficient amount of iodide, the thyroid gland enlarges. This condition is called a *hypothyroid goiter* (Figure 17.8).

The hypothalamus and the thyroid gland work together to keep the proper level of thyroid hormone circulating in the bloodstream. This level is detected by the hypothalamus (Figure 17.9 ❶). A low level of thyroid hormones stimulates the hypothalamus to secrete a releasing factor to the anterior pituitary ❷. This hormone message tells the pituitary to release more TSH ❸. The thyroid responds to stimulation by TSH, secreting more of the thyroid hormones, thereby raising their levels in the blood ❹. Shutdown occurs ❺ when thyroid hormone levels are sufficient. This mechanism of action is another example of a negative feedback loop in which the ultimate effect produced by stimulation of a gland (inhibition of TSH secretion) is opposite to the stimulus (TSH secretion).

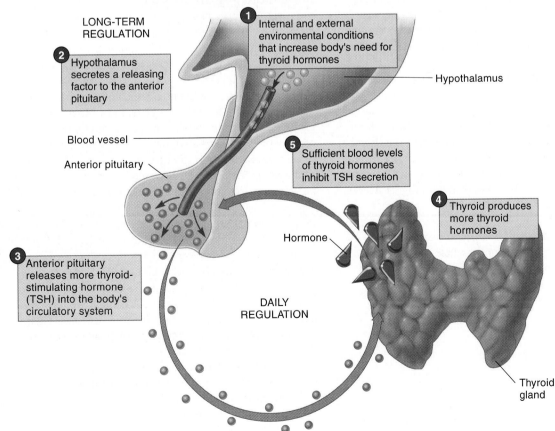

LONG-TERM REGULATION

❶ Internal and external environmental conditions that increase body's need for thyroid hormones

Hypothalamus

❷ Hypothalamus secretes a releasing factor to the anterior pituitary

Blood vessel

Anterior pituitary

❺ Sufficient blood levels of thyroid hormones inhibit TSH secretion

❹ Thyroid produces more thyroid hormones

Hormone

❸ Anterior pituitary releases more thyroid-stimulating hormone (TSH) into the body's circulatory system

DAILY REGULATION

Thyroid gland

Figure 17.9   A negative feedback loop is used in the release of thyroid hormone.

In certain disease conditions, the amount of thyroid hormones in the bloodstream cannot be regulated properly. If the thyroid produces too much of the thyroid hormones, a person may feel as though the "engine is racing," with such symptoms as a rapid heartbeat, nervousness, weight loss, and protrusion of the eyes (as shown in the photo below). This condition is called

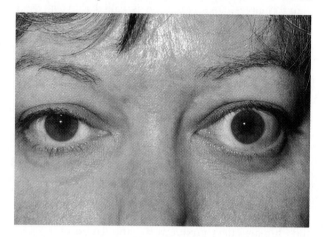

hyperthyroidism. On the other hand, if the thyroid produces too little of the thyroid hormones, a person may feel run down, with such symptoms as weight gain and slow growth of the hair and fingernails. This condition is called *hypothyroidism.* Various factors can be the underlying cause of such problems; often medication or surgery can correct the situation.

In addition to secreting the thyroid hormones, the thyroid gland secretes a hormone called *calcitonin,* or *CT.* This hormone works to balance the effect of another hormone called *parathyroid hormone,* or *PTH. PTH* and *CT* regulate the concentration of calcium in the bloodstream. (This process is described in Section 17.8) Calcium is an important structural component in bones and teeth and aids in the proper functioning of nerves and muscles.

The thyroid gland, located in the neck near the voice box, produces hormones that regulate the body's metabolism.

17.8 The parathyroids

Embedded in the posterior side of the thyroid are the **parathyroid glands**. Most people have two parathyroids on each of the two lobes of the thyroid. These glands secrete PTH, which works antagonistically to CT to help maintain the proper blood levels of various ions, primarily calcium.

PTH and CT work in the following way to keep calcium at an optimum level in the blood: If the calcium level is too low (Figure 17.10 ❶), the parathyroids are stimulated to release more PTH ❷. PTH stimulates the activity of osteoclasts, or bone-destroying cells. These cells liberate calcium from the bones and put it into the bloodstream. PTH also stimulates the kidneys to reabsorb calcium from urine that is being formed and stimulates cells in the intestines to absorb an increased amount of calcium from digested food ❸. CT acts in opposition to PTH. When the level of calcium in the blood is too high ❹, the parathyroids secrete less PTH ❺ and the thyroid secretes more CT ❻. The CT inhibits the release of calcium from bone and speeds up its absorption ❼, decreasing the levels of calcium in the blood. These interactions of PTH and CT are an example of a negative feedback loop that does not involve the hypothalamus or pituitary gland. The level of calcium in the blood directly stimulates the thyroid and parathyroid glands.

Two of the many problems related to abnormal calcium levels in the blood are kidney stones and osteoporosis. If calcium levels in the blood remain high, tiny masses of calcium may develop in the kidneys. These masses, called kidney stones (shown on a fingertip below), can partially block the flow of the urine from a kidney. If calcium levels in the blood remain low, calcium may be removed from the bones (see Figure 16.14), a disorder known as osteoporosis. Osteoporosis is most common in middle-aged and elderly women, who have stopped secreting  estrogen at menopause (see Chapter 22). Estrogen stimulates bone cells to take calcium from the blood to build bone tissue. Taking estrogen replacement therapy and engaging in regular weight-bearing exercise helps women avoid the development of osteoporosis after menopause.

The parathyroid glands, embedded in the posterior side of the thyroid, secrete a hormone that helps maintain the proper levels of various ions such as calcium in the bloodstream.

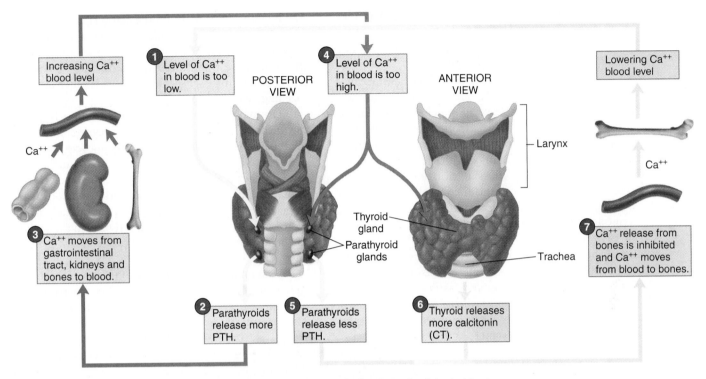

Figure 17.10   How parathyroid hormone and calcitonin work to maintain proper calcium levels in the blood.

The labels within the figure read:

- Increasing Ca++ blood level
- Ca++
- 1 Level of Ca++ in blood is too low.
- POSTERIOR VIEW
- 4 Level of Ca++ in blood is too high.
- ANTERIOR VIEW
- Lowering Ca++ blood level
- Larynx
- Ca++
- 3 Ca++ moves from gastrointestinal tract, kidneys and bones to blood.
- Thyroid gland
- Parathyroid glands
- Trachea
- 7 Ca++ release from bones is inhibited and Ca++ moves from blood to bones.
- 2 Parathyroids release more PTH.
- 5 Parathyroids release less PTH.
- 6 Thyroid releases more calcitonin (CT).

17.9   The adrenals

The two **adrenal glands** (uh-DREE-nul) are named for their position in the body: above (ad meaning "near") the kidneys (renal meaning "kidney"). Each of these triangular glands has two parts with two different functions (see illustration below). The **adrenal cortex** is the outer, yellowish portion of each adrenal gland. The word cortex comes from a Latin word meaning "bark" and is often used to refer to the outer covering of a tissue, organ, or gland. The **adrenal medulla** (muh-DULL-uh) is the inner, reddish portion of the gland and is surrounded by the cortex. Not surprisingly, the word medulla comes from a Latin word meaning "marrow" or "middle."

The Adrenal Cortex

As you may recall, the anterior pituitary gland secretes the hormone ACTH, adrenocorticotropic hormone. This hormone, as its name implies, stimulates the adrenal cortex to secrete a group of hormones known as **corticosteroids** (KORT-ik-oh-STARE-oydz). These steroid hormones act on the nucleus of target cells, triggering the cell's hereditary material to direct the manufacture of certain proteins. The two main types of corticosteroids produced by the adrenal cortex are the *mineralocorticoids* (MIN-er-al-oh-KOR-tih-koidz) and the *glucocorticoids* (GLOO-ko-KOR-tih-koidz).

The mineralocorticoids are involved in the regulation of the levels of certain ions within the body fluids. The most important of this group of hormones is *aldosterone* (al-DOS-ter-own). It affects tubules within the kidneys, stimulating them to reabsorb sodium ions and water from the urine that is being produced, putting these

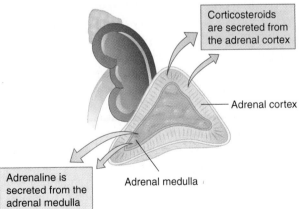

The labels within this figure read:
- Corticosteroids are secreted from the adrenal cortex
- Adrenal cortex
- Adrenaline is secreted from the adrenal medulla
- Adrenal medulla

substances back into the bloodstream. The secretion of aldosterone is triggered when the volume of the blood is too low, such as during dehydration or blood loss. Special cells in the kidneys monitor the blood pressure. When the blood pressure drops, these cells secrete an enzyme that begins a chain of reactions ending with the secretion of aldosterone. Conversely, when the blood pressure is within a normal range, the cellular detectors in the kidneys are not stimulated, the release of aldosterone is not triggered, and the kidney tubules are not stimulated to conserve sodium and water.

The glucocorticoids affect glucose metabolism, causing molecules of glucose to be manufactured in the body from noncarbohydrates such as proteins. This glucose enters the bloodstream, is transported to the cells, and is used for energy as part of the body's reaction to stress.

Almost everyone is familiar with the term stress and can give examples of stressful situations: their boss "chewing them out" in front of co-workers, their kids fighting constantly with one another, or their sustaining a physical injury. The stress reaction was first described in 1936 by Hans Selye, a researcher who has since become the acknowledged authority on stress. Dr. Selye explained how the body typically reacts to stress—any disturbance that affects the body—and called this reaction the *general adaptation syndrome*. Over a prolonged period of stress, the body reacts in three stages: (1) the alarm reaction, (2) resistance, and (3) exhaustion. Contrary to maintaining homeostasis within the body, the general adaptation syndrome works to help the body gear up to meet an emergency.

During the alarm reaction, the body goes into quick action. Imagine waking up late one morning during final exam week and realizing you've missed your biology final. Your body reacts with a quickening pulse, increased blood flow, and an increased rate of chemical reactions within your body. Why does your body react in this way? Although the adrenal cortex is involved in the stress reaction by manufacturing glucose, the adrenal medulla, the inner section of the adrenal glands, plays a central role.

The Adrenal Medulla

The adrenal medulla is different from most other endocrine tissue in that its cells are derived from cells of the peripheral nervous system (like those of the posterior pituitary). The sympathetic division of the autonomic nervous system, which controls involuntary or automatic responses, stimulates the adrenal medulla. Tropic hormones do not trigger these cells.

The principal hormone made by the adrenal medulla is *adrenaline* (also called *epinephrine*). This hormone is primarily responsible for the alarm reaction (along with a similar hormone secreted by the autonomic nervous system called *noradrenaline*). The hypothalamus is responsible for sending the alarm signal (via the autonomic nervous system) to the adrenal medulla.

The hypothalamus picks up the alarm signal as it monitors changes in the emotions via its neural connections with the emotional centers in the cerebral cortex. It can therefore sense when the body perceives an emotional stress. It can also sense physical stress, such as cold, bleeding, and poisons in the body.

The hypothalamus initially reacts to stress by readying the body for fight or flight; it first triggers the adrenal medulla to secrete adrenaline. Adrenaline and noradrenaline cause the heart rate and breathing to quicken, the rate of chemical reactions to increase, and glucose (stored in the liver) to be released into the bloodstream. In general, the actions of adrenaline and noradrenaline increase the amounts of glucose and oxygen available to the organs and tissues most used for defense: the brain, heart, and skeletal muscles.

A summary of the first stage of the stress reaction is shown in Figure 17.11. For example, when a person sees a bear, the fear of bears is detected by the hypothalamus. The hypothalamus triggers the anterior pituitary to secrete ACTH ❶, which stimulates the adrenal cortex to secrete glucocorticoids ❷. The glucocorticoids trigger the synthesis of glucose ❸. The hypothalamus also sends out a nerve signal ❹, which triggers the adrenal medulla to secrete adrenaline ❺. Adrenaline (and noradrenaline) increase the heart rate, breathing rate, and blood sugar level by releasing glucose from the liver ❻.

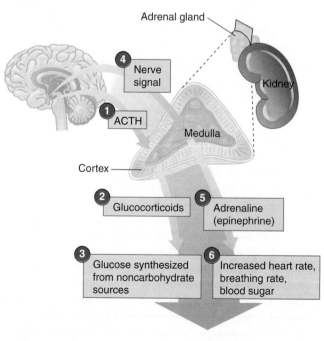

Figure 17.11   The first stage of the stress reaction.

During the second stage of the stress reaction, the resistance stage, the hypothalamus triggers continuing responses by releasing regulating factors. These factors stimulate the pituitary to release ACTH, TSH, and growth hormone (GH). TSH stimulates the thyroid to secrete thyroid hormones, which stimulate the liver to break down stored carbohydrates (glycogen) to glucose. GH also stimulates the liver to produce glucose from glycogen, providing the body with an abundant energy source. ACTH stimulates the adrenal cortex to secrete both mineralocorticoids and glucocorticoids. The mineralocorticoids cause the body to retain sodium ions and water, raising the blood pressure and providing more blood volume in the case of blood loss. The glucocorticoids also promote the production of glucose.

If a person continues to be highly stressed over a long time, the body may lose the fight and enter the third stage of the stress reaction: exhaustion. This stage is serious and can result in death. One cause of exhaustion is the loss of potassium ions, which are excreted when sodium ions are retained. This loss severely affects the body's ability to function properly. Another cause is depletion of the glucocorticoids, resulting in a sharp drop in the blood glucose level. The organs also become weak and may cease to function. To combat chronic stress, people can learn ways to psychologically handle their stress and can work toward a level of health and fitness that will help their bodies cope with the physical effects of stress.

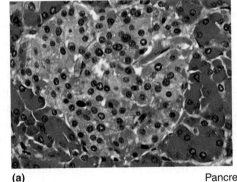

(a)

The adrenal glands, located on top of each kidney, are divided into two secretory portions: an outer cortex and an inner medulla. The hormones of the adrenal cortex, the corticosteroids, primarily regulate the level of sodium ions and consequently water in the bloodstream and stimulate the liver to produce glucose from stored carbohydrates. Adrenaline secreted from the adrenal medulla readies the body to react to stress.

17.10   The pancreas

The pancreas, located alongside the stomach, is two glands in one: an exocrine gland and an endocrine gland. As an exocrine gland, it secretes the digestive enzymes discussed in Chapter 8. As an endocrine gland, it secretes the hormones *insulin* and *glucagon*.

The endocrine portion of the pancreas consists of separated clusters of cells that lie among the exocrine cells. For this reason, these cells are called islets—the **islets of Langerhans**—and can be seen in the photo in **Figure 17.12a** as the more lightly stained cells. Separate types of cells within the islets produce insulin and glucagon. These hormones act antagonistically to one another to regulate the level of glucose in the bloodstream. Glucagon increases the blood glucose level by triggering the liver to convert stored carbohydrates (glycogen) into glucose and to convert other nutrients such as amino acids into glucose. Insulin decreases the blood glucose level by helping body cells transport glucose across their membranes. In addition, insulin acts on the liver to convert glucose into glycogen and fat for storage, as shown in Figure 17.12b.

Diabetes mellitus is a set of disorders in which a person tends to have a high level of glucose in the blood. There are many variations of the disorder and many causes. The underlying cause is the lack or partial lack of insulin or the inability of tissues to respond to insulin, which leads to increased levels of glucose in the blood. The high levels of glucose in the blood result in water moving out of the body's tissues by osmosis. The kidneys remove this excess water through increased urine production, which causes excessive thirst and dehydration.

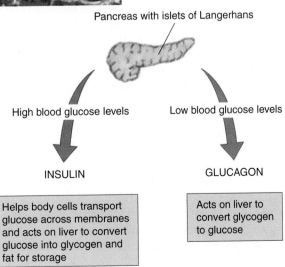

(b)

Figure 17.12
Functions of insulin and glucagon.

Just Wondering . . . Real Questions Students Ask

What is adult onset diabetes and why do people suddenly develop it?

Adult onset or maturity onset diabetes has a new name: type 2 diabetes. Another former name for this disease is non-insulin-dependent diabetes mellitus (NIDDM). Type 2 diabetes is the most common form of diabetes mellitus. Approximately 90% of all cases are of this type and it affects about 15 million Americans.

Persons with type 2 diabetes have a problem properly regulating their blood sugar level because of an underlying problem with either their manufacture of insulin or its functioning. Some persons with type 2 diabetes make insulin, but not enough to meet their needs. Others have sufficient insulin, but the cells in many parts of the body lack enough insulin receptors, so they cannot use the insulin they make. Without enough insulin, the body cannot move blood sugar (glucose) into the cells. Therefore, the level of blood glucose rises and can cause problems with the kidneys, legs and feet, eyes, heart, nerves, and

www.jbpub.com/biology

blood flow. If left untreated, these problems can lead to kidney failure, gangrene and amputation, blindness, and stroke.

Individuals with type 2 diabetes must control their blood sugar levels by following their physician's instructions regarding diet and exercise. Persons who cannot control their blood sugar level sufficiently with these measures may have to take diabetes pills or insulin injections to help prevent problems.

Type 2 diabetes only seems to develop suddenly because its symptoms are not always apparent. Many people have no symptoms, or they have such mild symptoms that they do not notice them. Older persons often ignore their symptoms, thinking they are just part of the aging process.

Common symptoms of Type 2 diabetes are an increased thirst and increased urination. Many persons also feel edgy, tired, and nauseous. Some people have an increased appetite, but they lose weight. Other symptoms of type 2 diabetes are repeated or hard-to-heal infections of the skin, gums, vagina, or

bladder; blurred vision; tingling or loss of feeling in the hands or feet; and dry, itchy skin.

Persons at highest risk for developing type 2 diabetes are those who are over 45, overweight, and have a family history of the disease. Additionally, African Americans, Hispanic Americans, Asian and Pacific Islanders, and Native Americans have a higher prevalence of type 2 diabetes and a higher lifetime incidence of complications. Although researchers do not know the cause of type 2 diabetes, they do know that a person can inherit a tendency to get type 2 diabetes. However, it usually takes another factor such as obesity to bring on the disease.

At this time, the American Diabetes Association recommends screening for persons with one or more risk factors beginning at age 45 and every 3 years thereafter. For more information on this disease, contact the branch of the American Diabetes Association in your area.

The primary types of diabetes mellitus are termed type 1 and type 2 diabetes. Insulin-dependent (type 1) diabetes was formerly known as juvenile diabetes because it strikes chiefly in childhood. Non-insulin-dependent (type 2) diabetes was known as adult onset diabetes because it strikes primarily in adulthood. This disease is described in the "Just Wondering" box.

Type 1 diabetes mellitus is a disorder in which the pancreas does not produce adequate amounts of the hormone insulin or the body becomes insensitive to it. Affecting about 5% of diabetics, type 1 diabetes is thought to be a hereditary disease in which the body's immune system attacks the pancreas, destroying its ability to produce insulin. As a result, muscles, fat, and liver cells are prevented from absorbing sugar from the blood; the sugar is excreted in urine while the under-nourished cells literally starve. The high levels of blood glucose cause the thickening of capillary and artery walls, constricting blood flow and damaging critical organs.

Diabetes is treated by adding insulin directly to the bloodstream by injection (as a protein, it would be de-

stroyed in the stomach if administered as a pill), usually twice a day. However, taking insulin injections is only part of the treatment for type 1 diabetes. Personal management plans help avoid the complications of diabetes. Such a plan usually includes nutrition and exercise recommendations, glucose monitoring instructions, and a medical plan to monitor organs and body systems that may be affected by type 1 diabetes.

Research on transplanting islets of Langerhans for persons who do not produce enough insulin, especially those with type 1 diabetes mellitus, holds much promise as a lasting treatment for this disease. Researchers are currently experimenting with ways to encapsulate islet cells collected from rats so that they will not trigger an immune response. Clinical trials of this animal-to-human transplantation are expected to begin in mid- to late 1999. Fetal cell transplantation holds promise as well (see Part 2 BioIssues).

The pancreas secretes two hormones, insulin and glucagon, that act antagonistically to one another, regulating the level of glucose in the bloodstream.

17.11   Other endocrine glands

The Thymus

The thymus gland is a small gland located in the neck a few inches below the thyroid (see Figure 17.2). In the thymus, certain immune system cells called *T lymphocytes* develop the ability to identify invading bacteria and viruses (see Chapter 11). The thymus produces a variety of hormones to promote the maturation of these cells. This gland is quite active during childhood, but is replaced by fat and connective tissue by the time a person reaches adulthood.

The Ovaries and Testes

The ovaries produce female sex cells, or eggs, and the testes produce male sex cells, or sperm. In addition, the ovaries produce the hormones estrogen and progesterone, and the testes produce testosterone. These organs become active during puberty, the time of sexual maturation. These endocrine glands and the roles of their hormones are discussed in Chapter 22.

Digestive Hormone-Secreting Cells

Certain cells in the walls of the stomach and small intestine secrete hormones that regulate the release of digestive juices by various cells. Additionally, they regulate other processes related to digestion such as the emptying of the stomach, contracting of the gallbladder, and inducing mass movements in the large intestine. **Table 17.2** reviews the peripheral (nonbrain) endocrine glands and their hormones described in Sections 17.7–17.11.

> The thymus, located in the neck, produces a variety of hormones to promote the maturation of certain immune system cells. This gland is most active in childhood. The ovaries and testes produce the sex hormones. These organs become active during puberty. Cells within the stomach and small intestine secrete digestive hormones that stimulate the secretion of digestive juices as well as regulate other digestive processes.

Table 17.2

Major Peripheral (Nonbrain) Endocrine Glands and Their Hormones

Endocrine Gland and Hormone	Target Tissue	Principal Actions
Thyroid Gland		
Thyroid hormones	General	Regulate metabolism
Calcitonin	Bone	Regulates calcium levels in the blood
Parathyroid Gland		
Parathyroid hormone (PTH)	Bone, kidney, small intestine	Regulates calcium levels in the blood
Adrenal Cortex		
Aldosterone	Kidney	Increases sodium and water reabsorption and potassium excretion
Glucocorticoids	General	Stimulate manufacture of glucose
Adrenal Medulla		
Adrenaline and noradrenaline	Heart, blood vessels, liver, fat cells	Regulate fight or flight response: increase cardiac output, blood flow to muscles and heart, conversion of glycogen to glucose
Pancreas (Islets of Langerhans)		
Insulin	Liver, skeletal muscle, fat	Decreases blood glucose levels by stimulating movement of glucose into cells
Glucagon	Liver	Increases blood glucose levels by converting glycogen to glucose
Ovary		
Estrogens	General, female reproductive organs	Stimulate development of secondary sex characteristics in females, control monthly preparation of uterus for pregnancy
Progesterone	Uterus	Completes preparation of uterus for pregnancy
	Breasts	Stimulates development
Testis		
Testosterone	General	Stimulates development of secondary sex characteristics in males and growth spurt at puberty
	Male reproductive structures	Stimulates development of sex organs, spermatogenesis

Key Concepts

Hormonal control is ancient and widespread.

17.1 Hormones are the chemical products of cells, secreted into the bloodstream and used as messengers that affect other cells within the body.

17.2 A wide variety of organisms have hormonal control mechanisms, including invertebrates, protozoans, and plants.

17.2 Hormones are ancient molecules; they first appeared as simple organisms evolved millions of years ago.

Human endocrine hormones signal target cells in two ways.

17.3 The 10 major endocrine glands make over 50 different hormones.

17.3 Together, the glands that secrete hormones are known as the endocrine system.

17.3 Hormones affect cells in one of four ways: by regulating their secretions and excretions, by helping them to respond to changes in the environment, by controlling activities related to reproductive processes, and by influencing their proper growth and development.

17.4 The two primary types of endocrine hormones are the peptide hormones and the steroid hormones.

17.4 Peptide hormones work by binding to receptor molecules embedded in the cell membranes of target cells and ultimately triggering enzymes that cause cells to alter their functioning.

17.4 Steroid hormones work by binding to receptor molecules within the cytoplasm of cells and ultimately causing the hereditary material to produce certain proteins.

17.4 The production of hormones is regulated by feedback mechanisms in the following way: an endocrine gland is initially stimulated by the nervous system or by the concentration of various substances in the bloodstream. The gland secretes a hormone that interacts with target cells, and the effect produced by the target cells acts as a new stimulus to increase or decrease the amount of hormone produced.

The hypothalamus, pituitary, and pineal are the central (brain) endocrine glands.

17.5 The pituitary gland, which hangs from the underside of the brain, secretes nine major hormones: seven from its anterior lobe and two produced by the hypothalamus that are stored in its posterior lobe.

17.5 The hypothalamus is a small mass of brain tissue lying above the pituitary; the hypothalamus controls the pituitary by means of releasing hormones.

17.5 Four of the hormones produced by the anterior pituitary control other endocrine glands; they are called tropic hormones.

17.5 In addition to tropic hormones, the anterior pituitary secretes growth hormone, which works with the thyroid hormones to control normal growth; prolactin, which works with other female sex hormones to stimulate the mammary glands to secrete milk after childbirth; and melanocyte-stimulating hormone, which affects certain pigment-producing cells of the body.

17.5 The posterior pituitary secretes two hormones produced by the hypothalamus: Antidiuretic hormone helps control the volume of the blood by regulating the amount of water reabsorbed by the kidneys and oxytocin is secreted during the birth process and enhances uterine contractions.

17.6 The primary endocrine secretion of the pineal gland, located at the base of the brain, is melatonin, a hormone that regulates the sleep/wake cycle in humans and certain seasonal behavior patterns in other animals.

Humans have a variety of peripheral (nonbrain) endocrine glands.

17.7 The thyroid gland, located in the neck near the voice box, produces hormones that regulate the body's metabolism.

17.8 The parathyroids, located on the underside of the thyroid, secrete parathyroid hormone, which works in an antagonistic manner with one of the thyroid hormones, calcitonin, and helps maintain the proper blood levels of various ions, primarily calcium.

17.9 The adrenal glands, located on top of the kidneys, have two different secretory portions: the outer cortex and the inner medulla.

17.9 The adrenal cortex secretes a group of hormones known as corticosteroids that regulate the levels of certain mineral ions, water, and glucose in the bloodstream.

17.9 The adrenal medulla secretes adrenaline, which readies the body for action during times of stress.

17.10 The pancreas secretes two hormones, insulin and glucagon, that act antagonistically to one another, regulating the level of glucose in the bloodstream.

17.11 The thymus, located in the neck and most active in childhood, produces a variety of hormones to promote the maturation of certain immune system cells.

17.11 The ovaries and testes, which become active during puberty, produce the sex hormones.

17.11 Cells in the walls of the stomach and small intestine secrete digestive hormones.

Key Terms

adrenal cortex (uh-DREE-nul KOR- tecks) *369*
adrenal glands (uh-DREE-nul) *369*
adrenal medulla (uh-DREE-nul muh-DULL-uh) *369*
corticosteroids (KORT-ik-oh-STARE-oydz) *369*
endocrine glands (ENN-doe-krin) *358*
endocrine system *358*

exocrine glands (EK-so-krin) *358*
gonadotropins (GON-ah-duh-TROP-inz) *368*
hormone *358*
hypothalamus (HYE-poe-THAL-uh-muss) *363*
islets of Langerhans (EYE-lets of LANG-ur-HANZ) *371*
negative feedback loop *362*

parathyroid glands (PARE-uh-THIGH-royd) *368*
peptide hormones *361*
pineal gland (PIN-ee-uhl) *366*
pituitary (puh-TOO-ih-tare-ee) *363*
positive feedback loop *366*
steroid hormones *361*
tropic hormones (TROW-pik) *363*

KNOWLEDGE AND COMPREHENSION QUESTIONS

1. What are hormones, and why are they important? *(Knowledge)*

2. How do the "messages" sent by the endocrine system differ from those carried by the nervous system? *(Knowledge)*

3. Why is cAMP required to act as a "second messenger" in peptide hormone-related processes? *(Comprehension)*

4. Describe how a feedback loop regulates the production of a hormone. What is the difference between a negative feedback loop and a positive feedback loop? *(Comprehension)*

5. If the fluid content of the blood is low, where would this information initially be received and processed? What hormone would be released and from which endocrine gland would it be released? *(Knowledge)*

6. Explain the term tropic hormone. Identify the four tropic hormones produced by the anterior pituitary and summarize the function(s) of each. *(Comprehension)*

7. Describe two hormones that together regulate calcium ion levels in your blood and discuss how they do it. *(Knowledge)*

8. Describe the three stages of the general adaptation syndrome. What hormones are involved? *(Knowledge)*

9. Which hormones regulate the level of glucose in your blood? Describe how they do this. Where are they produced? *(Knowledge)*

10. What are neurosecretory cells? Give an example of an endocrine gland made up of neurosecretory cells. *(Knowledge)*

KEY
Knowledge: Recalling information.
Comprehension: Showing understanding of recalled information.

1. In the past, people living in remote inland areas often suffered from goiter. Can you assess the reasons for increased goiter problems in such places? (*Synthesis*)

2. Chronic emotional stress has been linked to a greater than normal risk of various diseases. How might this link be explained? (*Synthesis*)

3. After playing soccer on a hot sunny day, you are sweating profusely and feeling very thirsty. A friend suggests getting a beer. Will the beer help your body replace the fluids it has lost? Explain. (*Application*)

4. A friend of yours read a news report that in 1998, baseball slugger Mark McGwire admitted to taking the over-the-counter hormone supplement "andro," which is banned in Olympic competition, but not in baseball. Andro (androstenedione) is a hormone that the body uses to make the male hormone testosterone. Your friend told you that he was going to purchase some andro to help increase his batting average. You told him that he would only be wasting his time and money, and that McGwire's greatness was not due to taking andro. In fact, McGwire has stopped using andro, and that hasn't hurt his performance. Your physician said that taking andro does not cause testosterone levels to rise. He also indicated that andro speeds the conversion of testosterone to estrogen. Rather than a rise in his batting average, what might your friend experience from taking andro? (*Application*)

5. Sometimes parents refer to their middle-schooled-aged children as being "containers of raging hormones." To what hormones are the parents referring? Why would parents consider these hormones to be "raging" in their children? What effects would these hormones have on adolescents? (*Synthesis*)

K E Y

Application: Using information in a new situation.

Synthesis: Putting together information from different sources.

e learning

Location: http://www.jbpub.com/biology

e-Learning is an on-line student review area located at this book's web site www.jbpub.com/biology. The review area provides a variety of activities designed to help you study for your class and build your learning portfolio.

Review Questions The review questions test your knowledge of the important concepts and applications in each chapter. The review provides feedback for each correct or incorrect answer. This is an excellent test preparation tool.

Figure-Labeling Exercises Sharpen your visual thinking by matching terms and labels for important illustrations.

Flash Cards Studying biology requires learning new terms. Virtual flash cards help you master the new vocabulary for each chapter.

Just Wondering As you read and study from this text, you may find that you have unanswered questions. Through this site you can ask the author, Sandy Alters, your "just wondering" questions.

Do you prefer the speed and reliability of a CD-ROM? All of the features contained on the eLearning portion of the web site are also available on the Student CD-ROM.

The target tissues for the anterior and posterior pituitary hormones are shown in the illustration. Name the hormone that affects each target tissue. Beneath the target tissue, state the principal actions of the hormone.

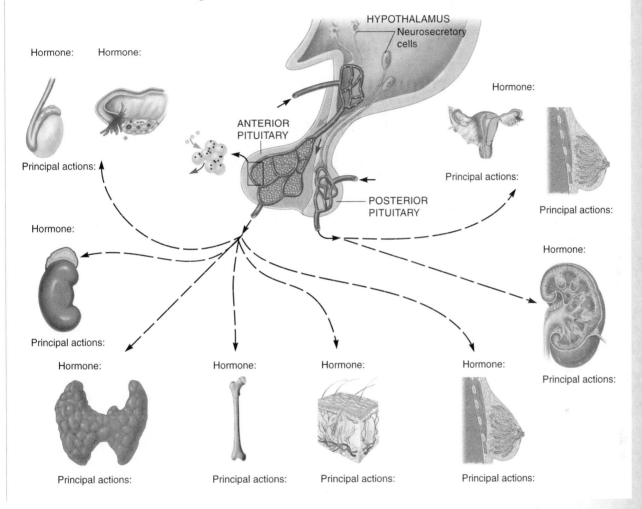

Hormone:

Hormone:

Principal actions:

Hormone:

Principal actions:

HYPOTHALAMUS
Neurosecretory cells

ANTERIOR PITUITARY

POSTERIOR PITUITARY

Hormone:

Principal actions:

Principal actions:

Hormone:

Principal actions:

Hormone:

Principal actions:

Hormone:

Principal actions:

Hormone:

Principal actions:

Hormone:

Principal actions:

Genetic Testing

"Your assistance is being requested. Collect a DNA sample by swabbing between your cheek and lower gum . . ."

It was mid-February, 1999, and the memo (with enclosed cotton swabs) had been sent to more than 100 African-American employees in the Ohio Attorney General's office. Collecting DNA samples from people belonging to various ethnic groups was part of establishing a forensic DNA lab in that state. The volunteer pool for this particular group had been exhausted, so DNA was being solicited from African-American state workers. The request was later dropped.

How would you feel if you were asked to "donate" your DNA for genetic testing and use in a database? Some people fear that genetic testing for any reason could result in discrimination in the workplace. Genetic testing includes both genetic screening and genetic monitoring. Genetic screening is a one-time test that identifies the existence of a particular disease-related gene or genes in a person's hereditary makeup. (Screening tests are now available for diseases such as muscular dystrophy, Huntington's disease, and various cancers.) Genetic monitoring involves multiple tests over time to detect whether mutations have arisen in an individual's genetic material due to exposure to chemicals, radiation, and other environmental agents that might be present in the workplace. (You will read more about DNA and genes in this part of the book.)

> "We used to think our future was in the stars. Now we know it is in our genes."
>
> James Watson, Nobel Prize Winner and developer of the Human Genome Project

In 1990, the Human Genome Project (HGP) was launched and the genetic testing debate heated up. (The HGP is a worldwide effort by groups of scientists to map the gene positions of the entire human genome [see Chapter 19, p. 430]). Then, in 1995, the debate grew hotter when scientists discovered a high level of mutation in particular genes in the gene pool of Jewish women of Central and Eastern European descent (Ashkenazi Jews). Two percent of all Jewish women (1 in 50) with Ashkenazi heritage have a specific genetic mutation on one or the other of the two "breast cancer genes," called BRCA1 and BRCA2. These mutations dramatically increase their risk of breast and ovarian cancer. The BRCA genes are tumor suppressor genes; that is, they suppress tumor growth in such organs as the breast and ovary. If either of these two genes is altered by a mutation, it will not work properly. The results can be devastating, raising a woman's lifetime risk of breast cancer from 8% to over 80%, and her lifetime risk of ovarian cancer from 2% to over 50%. Additionally, women with BRCA mutations are most likely to be stricken with cancer before they reach the age of 50, while women in the general population are most likely to be stricken after age 50.

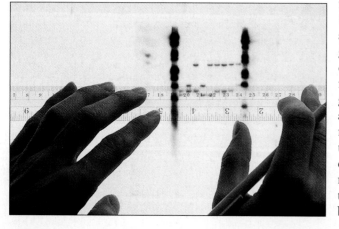

One of the uses of genetic fingerprinting is to determine paternity.

With this discovery, questions arose as to whether all Ashkenazi Jewish women should be offered genetic tests that would tell them whether they had inherited such mutations. Advocates of the genetic testing suggested that people have the right to know their genetic makeup, especially those in ethnic or family groups known to be at high risk for certain diseases. However, officials at the National Center for Human Genome Research (NCHGR) opposed the testing, suggesting that the knowledge might only bring anxiety and depression, since physicians were still unclear as to what to recommend to women with mutated BRCA genes. Should diagnosed "at-risk" women have more frequent mammograms than is usual, or would the additional radiation further increase their chances of developing breast cancer? Should at-risk women undergo surgery to remove their breasts or their ovaries, even if they might never develop breast or ovarian cancer? Would such surgery reduce their risk? Many scientists agreed that studies were first needed to determine the psychological and physiological consequences of interventions that might be undertaken, as well as the effect of interventions on lowering a woman's risk of developing the cancer she was trying to avoid.

Many women in this ethnic group raised concerns about the consequences of their being tested. Would they lose their health insurance or be unable to obtain health insurance in the future if they tested positive for a BRCA gene mutation? If they were already doing all they could to prevent breast and ovarian cancer, what good would it do to know of their increased risk? Would they be discriminated against by current or potential employers because they had a "gene defect" documented in their medical records? How would insurance companies and employers interpret their genetic testing results?

Although the merits and drawbacks of genetic testing in particular groups of healthy adults (such as Ashkenazi Jews) are widely debated in the scientific and medical communities, genetic testing is often conducted in circumstances in which the benefits of testing are clear. In fact, genetic testing is used frequently in carrier testing (to see if one or both members of a couple have deleterious recessive genes before they have children), in prenatal diagnosis, and in newborn screening (see p. 474). For example, the retinoblastoma gene was identified in 1986. Ninety-five percent of infants who carry this gene develop eye cancer by age 5. However, if newborns are tested for the retinoblastoma gene, those carrying it can be examined frequently for eye lesions and the lesions can be successfully treated. Nearly 100% of affected children survive with this treatment.

Because genetic discrimination was excluded from federal law in the Americans with Disabilities Act, some states have developed their own legislation. In March, 1992, the first genetic

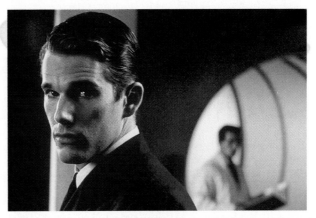

In the futuristic movie *GATTACA*, Ethan Hawke plays the role of Vincent, a young man whose deficient genes motivate him to "borrow" another man's genes so that he may participate in a space exploration program.

testing law in the United States was passed in Wisconsin. This law bars public and private employers in Wisconsin from making genetic screening a condition of employment, using information gathered from genetic screening in employment decisions, or offering employees additional pay or benefits to undergo genetic screening. Other states have since passed legislation prohibiting job and insurance discrimination on the basis of genetic test results and services, but there are no federal laws that provide such protection as of this writing. However, in 1999, several such bills were introduced in the House and Senate. Many business groups oppose such laws, saying that they just provide another mechanism by which employees can sue employers.

What is the issue here, and what is your response to this issue?

To make an informed and thoughtful decision, it may help you to follow these steps.

1. Identify the issue described here using either a statement or a question. (An issue is a point on which various persons hold differing views.)
2. Write your immediate reaction to the issues raised. How do you think this issue could be or should be resolved?
3. Collect new information about the issue. State the sources of that information. Explain why you believe the information to be accurate. Also determine if the information expresses a particular point of view or is biased in any other way. (Guidelines for doing efficient web searches for reliable information are available on the BioIssues website — **www.jbpub.com/biology**)
4. Determine which individuals, groups, or organizations have a stake in the issue. What does each stand to gain or lose depending on the outcome of the issue?
5. List possible outcomes (resolutions) of the issue. List the pros and cons of each outcome.
6. Which outcome do you think would be best and why? Note whether your opinion differs from or is the same as what you wrote in Step 2. Give reasons for your views and for any changes in them based on the additional information you have collected and the analysis you have done.

DNA, Gene Expression, and Cell Reproduction

Many first-time parents no doubt wish that their newborn child would come with a set of instructions. Indeed, each child does, but this information has little do with parenting. In fact, the instructions come from the parents themselves, hidden in the *chromosomes* each contributed to the child upon conception. Chromosomes, the brightly stained bodies seen in the photo on p. 382, contain all the information necessary for the development and maintenance of a human being. This information is a highly specialized biochemical "language," somewhat like a complex computer code that can be understood by certain other molecules in the body. For every living organism, it is the hereditary information contained within chromosomes that determines which traits will be exhibited. For a child, chromosomes determine which traits he or she will inherit from the mother and which from the father—such as whether the hair is blond or brown and whether whether the child is male or female. Most importantly, chromosomes hold the key to the mystery of what makes any one of us unique.

Organization

DNA is the hereditary material.

www.jbpub.com/biology

18.1 Chromosomes

Shown below is a *karyotype* (CARE-ee-oh-type), a particular array of chromosomes that belongs to an individual cell. In this case, we have the complete set of 46 chromosomes that are characteristic of human cells. (Each species of eukaryotic organism has its own distinctive number of chromosomes, which are linear [as in the photo]. Nearly all bacteria have a single, circular chromosome.)

Scientists develop karyotypes by specially treating and staining a cell just as it is about to divide. The cell is then photographed, and the chromosomes are either cut out of the photograph and physically arranged on a piece of paper, or, more commonly today, are manipulated on a computer. Each chromosome has a mate, so they are paired and then placed in order of decreasing size. (One purpose of karyotyping is to screen for abnormalities in the chromosomes, a topic that is discussed in Chapter 21).

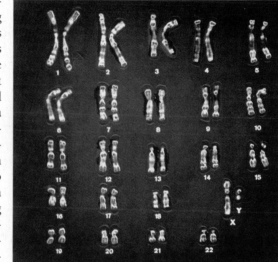

A human karyotype.

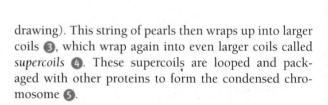

Except for gametes, each cell in the human body carries a full complement of 46 chromosomes.

Since the discovery of chromosomes, scientists have learned a great deal about their structure and function. Scientists know that the chromosomes of eukaryotes are made up of two principal components: deoxyribonucleic (de-OK-see-RYE-boh-noo-KLAY-ick) acid (DNA) and protein (primarily histone protein). The DNA contains the hereditary information, commonly called the *code of life*. The protein allows the long DNA molecule to be condensed into a more compact fiber, as shown in Figure 18.1. This complex of DNA and histone proteins is called **chromatin** (KRO-muh-tin).

DNA exists as one very long, double-stranded molecule that extends unbroken through the chromosome's entire length. DNA strands are coiled within cells, much as you might coil up your garden hose. The coiling of DNA within cells, however, is much more complicated. Figure 18.1 shows how this coiling is accomplished, providing organization, compaction, and orderliness.

First, the double strand of DNA ❶ winds like thread on spools around the histones ❷. In this form, the DNA looks similar to a string of pearls (see the micrograph of rat liver DNA next to the DNA/histone

drawing). This string of pearls then wraps up into larger coils ❸, which wrap again into even larger coils called *supercoils* ❹. These supercoils are looped and packaged with other proteins to form the condensed chromosome ❺.

$$\text{DNA} + \text{histone proteins} = \text{chromatin} \xrightarrow{\textit{coiling and supercoiling}} \text{chromosome}$$

This process of coiling and condensing of chromatin takes place just before and during cell division. (Cell division is described later in this chapter.) In fact, this process of DNA compaction is so effective that if all the DNA of your body were strung out and placed end to end, it would extend from the sun to the planet Jupiter and back again about 118 times!

Proteins and the hereditary material DNA make up the chromatin of the cell. When coiled and condensed, chromatin forms chromosomes.

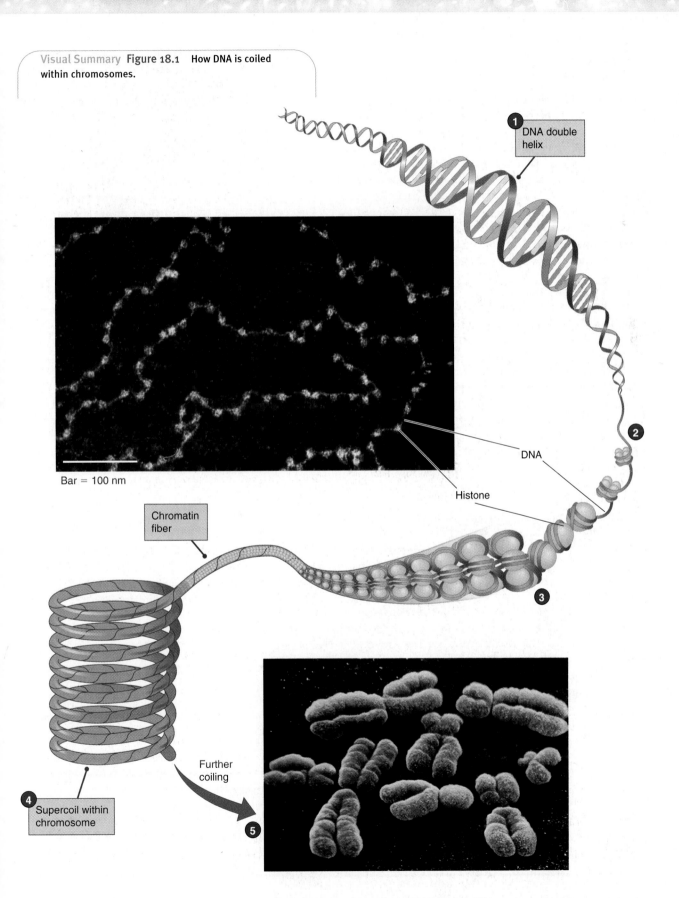

1 DNA double helix

Bar = 100 nm

DNA

Histone

Chromatin fiber

2

3

4 Supercoil within chromosome

Further coiling

5

18.2 The location of the hereditary material

Today, scientists understand (in considerable detail) the way that information in the DNA of a developing organism is translated into its eyes, arms, and brain. This understanding is the result of more than a century of work by a succession of scientific investigators.

Gregor Mendel's work, done in the late 1800s, suggested that traits are inherited as discrete packets of information. At the turn of the century, Walter Sutton suggested that these packets of information were on the chromosomes, but he had no direct evidence to support this hypothesis. In 1910, however, Thomas Hunt Morgan's experiments provided the first clear evidence upholding Sutton's theory. (The work of Mendel, Sutton, and Morgan is described in Chapter 19.) Beginning in the 1930s, experiments by biologists such as Joachim Hammerling further probed the question of where the hereditary information is stored within the cell. His experiments identified the nucleus as the likely place of the hereditary material. **Table 18.1** presents a timeline of significant events in DNA research.

Identifying the nucleus as the most likely source of hereditary information again focused attention on the chromosomes, which are located within the nucleus.

Scientists had been studying the chromosomes since the late 1800s and had suspected them to be the vehicles of inheritance. In 1869, a German scientist named Friedrich Miescher isolated the DNA contained within the chromosomes of various types of cells. He did not call it DNA (this designation would not come until the 1920s). Miescher called this chromosomal material **nucleic acid** because it seemed to be specifically associated with the cell nucleus and was slightly acidic.

In the 1920s, the American biochemist Pheobus A. Levene discovered that two types of nucleic acid were located within cells. Today, scientists understand that one type contains the hereditary message, whereas the other type helps express that message. Levene found that both were nearly identical in their structures and contained three molecular parts **(Figure 18.2a):** (1) phosphate $(-PO_4^-)$ groups, (2) five-carbon sugars, and (3) four types of nitrogen-containing bases. Although nucleic acids as a whole are acidic and tend to form hydrogen ions when in solution, the nitrogen-containing bases tend to accept hydrogen ions. For this reason, these portions of nucleic acids are called bases.

The hereditary material, DNA, is located in the cell nucleus. It is composed of phosphate groups, sugars, and four types of nitrogen-containing bases.

Table 18.1

Some Important Events Leading to the Discovery of the Genetic Code

Date	Researchers	Discoveries
1869	Friedrich Miescher	Isolated DNA (called it nucleic acid).
Late 1800s	Gregor Mendel	Discovered that traits are inherited as discrete packets of information.
Early 1900s	Walter Sutton	Hypothesized that packets of hereditary information are located on the chromosomes.
1910	Thomas Hunt Morgan	Provided the first clear evidence supporting Sutton's hypothesis that the hereditary information is located on the chromosomes.
1920s	Pheobus A. Levene	Discovered that two types of nucleic acids are located within the cell, and that both types have nearly identical structures and contain three molecular parts: phosphates, sugars, and bases.
1930s	Joachim Hammerling	Found that the nucleus was the most likely location of the hereditary information.
Late 1940s	Erwin Chargaff	Showed that the proportion of the bases varies in the nucleic acids of different types of organisms. Also noted that the proportions of particular bases are always roughly equal to the proportions of other bases.
Early 1950s	George Beadle & Edward Tatum	Developed the one-gene-one-enzyme theory, which states that the production of a given enzyme is under the control of a particular unit of heredity, or gene.
1952	Alfred Hershey & Martha Chase	Confirmed that DNA (not protein) was the hereditary material.
1953	Rosalind Franklin & Maurice Wilkins	Analyzed DNA by x-ray diffraction and concluded that it is shaped like a helix (winding staircase).
1953	James Watson & Francis Crick	Developed a model for the DNA double helix structure.
1961	Francis Crick et al.	Discovered that the DNA (genetic) code comprises sequences of three nucleotide bases.

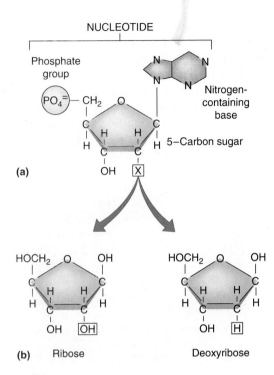

(a)

NUCLEOTIDE

Phosphate group

Nitrogen-containing base

5–Carbon sugar

(b) Ribose   Deoxyribose

Figure 18.2 **Structure of a nucleotide.** (a) A nucleotide contains three different molecular components: a phosphate group (PO_4), a nitrogen-containing base, and a five-carbon sugar. As shown in (b), the sugar can be either ribose (RNA) or deoxyribose (DNA). The difference in the structures of these sugars is a single atom of oxygen.

Figure 18.2b shows the structure of both ribose and deoxyribose sugars.

Each nucleotide making up the structure of DNA and RNA contains one of four different bases. DNA and RNA both contain the bases adenine (ADD-uh-neen) and guanine (GUAH-neen), which are double-ring compounds called **purines** (PYOOR-eens) **(Figure 18.3)**. DNA also contains the bases thymine (THEYE-mean) and cytosine (SIGH-toe-seen), which are single-ring compounds called **pyrimidines** (pih-RIM-uh-deens). RNA contains the pyrimidine cytosine as well, but contains the single-ring base uracil instead of thymine.

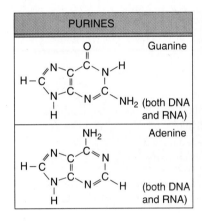

PURINES	PYRIMIDINES
Guanine (both DNA and RNA)	Cytosine (both DNA and RNA)
Adenine (both DNA and RNA)	Thymine (DNA only)
	Uracil (RNA only)

Figure 18.3 **Purines and pyrimidines.**

18.3 The nucleic acids: DNA and RNA

Levene found that nucleic acids are composed of roughly equal proportions of these three molecular parts. He concluded (correctly) from this information that these three molecular parts, bonded together, form the repeating units of structure of nucleic acids. Each unit, a five-carbon sugar bonded to a phosphate group and a nitrogen-containing base, is called a **nucleotide** (see Figure 18.2a). One of the two types of nucleic acids found in cells is composed of units that contain the five-carbon sugar ribose. It is therefore called **ribonucleic acid** or **RNA** for short. The other type of nucleic acid is composed of units containing a five-carbon sugar similar to ribose but having one less oxygen atom. This sugar is called deoxyribose, and the nucleic acid is called **deoxyribonucleic acid (DNA)**. (The name deoxyribose literally means "without-oxygen ribose.")

In the late 1940s, the experiments of Erwin Chargaff showed that the proportion of bases varies in the DNA of different types of organisms, as is shown in **Table 18.2.** This evidence suggested to Chargaff that DNA has the ability to be used as a molecular code, with the base composition of DNA varying as its code varies from organism to organism.

Along with the variations among the DNA molecules of different organisms, Chargaff also noted an important similarity: The amount of adenine present in DNA molecules is always roughly equal to the amount of thymine, and the amount of guanine is always roughly equal to the amount of cytosine (A = T and G = C). Notice the percentages of A and T, and of G and C in each organism listed in Table 18.2. Notice also that DNA molecules always have an equal proportion of purines (A + G) and pyrimidines (C + T).

Table 18.2

Chargaff's Analysis of DNA Nucleotide Base Compositions

Organism	Base Composition (Mole Percent)			
	A	T	G	C
Escherichia coli	26.0	23.9	24.9	25.2
Streptococcus pneumoniae	29.8	31.6	20.5	18.0
Mycobacterium tuberculosis	15.1	14.6	34.9	35.4
Yeast	31.3	32.9	18.7	17.1
Sea Urchin	32.8	32.1	17.7	18.4
Herring	27.8	27.5	22.2	22.6
Rat	28.6	28.4	21.4	21.5
Human	30.9	29.4	19.9	19.8

There are two types of nucleic acids located within cells: deoxyribonucleic acid (DNA) and ribonucleic acid (RNA). Both types of nucleic acid are composed of units called nucleotides, which are made up of three molecular parts: a sugar, a phosphate group, and a base.

18.4 The structure of DNA

In 1952, Alfred Hershey and Martha Chase performed now-famous experiments showing that DNA is the hereditary material, not protein. They reasoned that if DNA carried hereditary information, then bacteria infected with viruses would contain viral DNA inside their cells, which would direct the formation of new virus particles. If, on the other hand, the protein carried the hereditary message, then virally infected bacteria would contain viral proteins within their cells. The experiments of Hershey and Chase showed clearly that viral DNA, not viral protein, was inside the infected bacteria. (See the How Science Works box for details of their experiments.)

The next puzzle to solve was how the nucleotides of DNA were put together. The significance of the regularities in the proportion of pyrimidines and the proportion of purines in DNA pointed out by Chargaff became clear through the results of two later experiments.

Two British biophysicists, Rosalind Franklin and Maurice Wilkins, carried out x-ray diffraction analysis of fibers of DNA. In this process, the DNA molecule is bombarded with an x-ray beam. When individual rays encounter atoms, each ray's path is bent or diffracted; the pattern created by these diffractions can be captured on photographic film. With careful analysis of the diffraction pattern, it is possible to develop a three-dimensional image of the molecule causing the particular pattern of diffractions of the x-rays. The diffraction patterns Franklin and Wilkins obtained (Figure 18.4) suggested that the DNA molecule was a helical coil. Put simply, this complex molecule was shaped like a spring.

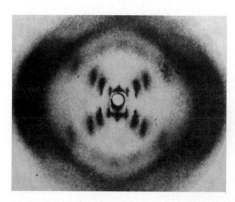

Figure 18.4 **Evidence for the helical structure of DNA.** This x-ray diffraction photograph of crystals of DNA was made in 1953 by Rosalind Franklin in the laboratory of Maurice Wilkins.

Learning informally of Franklin and Wilkins' results before they were published in 1953, James Watson and Francis Crick, two young scientists at Cambridge University in England, quickly worked out a likely structure of the DNA molecule. They built models of the nucleotides, assembled them into molecular structures, and then tested each to see whether its structure fit with what they knew from Chargaff's and Franklin and Wilkins' work. They finally hit on the idea that the molecule might be a double helix (two springs twisted together) in which the bases of the two strands pointed inward toward one another (Figure 18.5). Pairing a purine (which is large) with a pyrimidine (which is small) resulted in a helical "ladder" with "rungs" of uniform length. In fact, always pairing adenine with thymine and cytosine with guanine yielded a molecule in which A = T and C = G, a molecule consistent with Chargaff's observations. In Watson and Crick's model, the sugar and phosphate units linked in an alternating fashion to form the sides of this twisted ladder (the springs in the earlier analogy). For their groundbreaking theory, Watson and Crick shared the Nobel Prize in 1962.

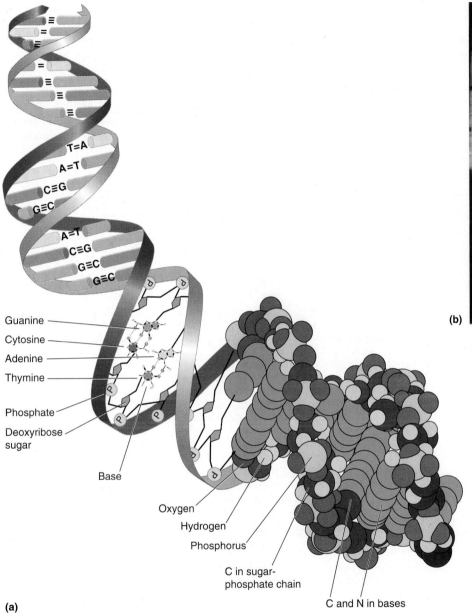

Guanine
Cytosine
Adenine
Thymine

Phosphate
Deoxyribose
sugar

Base

Oxygen
Hydrogen
Phosphorus

C in sugar-
phosphate chain

C and N in bases

(a)

(b)

Figure 18.5 **Structure of DNA.** (a) A simplified view of a DNA molecule is shown in the upper portion of the model, with the tubelike double helix representing the sugar-phosphate "uprights" of the DNA "ladder." The bases are denoted by the letters A (adenine), C (cytosine), T (thymine), and G (guanine). A G-C base pair is bonded by three hydrogen bonds, and an A-T base pair is bonded by two hydrogen bonds. The most stable base pairing occurs when adenine pairs with thymine and guanine pairs with cytosine. In the bottom portion of the DNA model, each atom is depicted by a sphere. (b) Watson and Crick with their first model of the DNA double helix.

Chargaff determined that although the proportion of the bases in DNA varied among organisms, the amount of adenine always equaled that of thymine and the amount of guanine always equaled that of cytosine. This information, coupled with the work of Rosalind Franklin, Maurice Wilkins, James Watson, and Francis Crick, led to the conclusion that the DNA molecule is a double-stranded helix. In this spiral molecular ladder, the bases of each strand of DNA form rungs of uniform length, and alternating sugar-phosphate units form the ladder sides.

18.5 How DNA replicates

The Watson-Crick model suggested that the basis for copying the genetic information is the complementarity of its bases. One side of the DNA ladder has a base sequence that results in a particular code that the body understands. This sequence of bases then determines the sequence of bases making up the other side of the ladder. If the sequence on one side were ATTGCAT, for example, the sequence of the other side would have to be TAACGTA—its complementary image.

Scientists have since learned how DNA copies (replicates) its strands to make more DNA. Before a cell divides, the bonds between the complementary bases break in short sections of the double-stranded DNA molecules, and the complementary strands separate from one another. (The bonds between the bases are shown in Figure 18.5.) This process is commonly re-

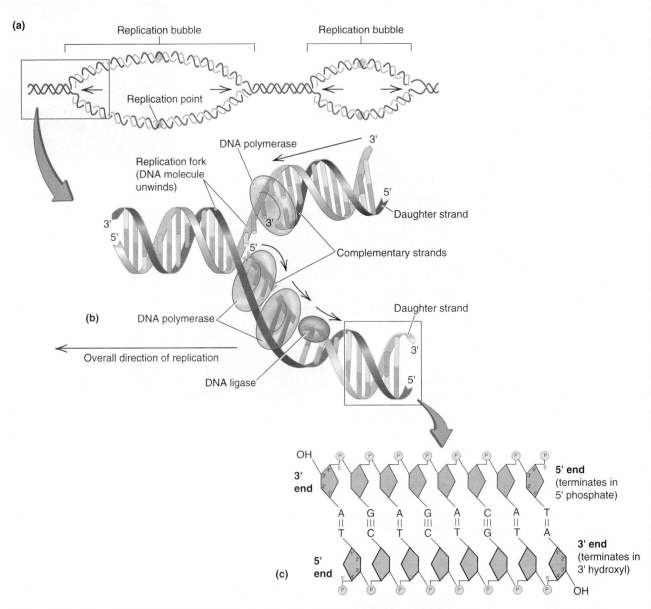

Figure 18.6 DNA replication. (a) Replication begins at various places along the DNA strand forming replication bubbles. (b) The double-stranded DNA molecule separates at its bases, forming a split, or fork. Each separated strand acts as a template while DNA polymerase and DNA ligase bond free nucleotides during the synthesis of a new complementary strand (c). Each DNA strand has a 3′ and a 5′ end.

Just Wondering . . . Real Questions Students Ask

Does an individual's DNA makeup change with age, or can the same DNA be used to identify someone even if they have become much older since the time they gave a DNA sample?

Although there are some changes that may occur in the DNA with age, DNA does not change enough to have noticeable differences when it is analyzed. Your DNA at age 5 will look virtually the same as your DNA at age 85.

Some changes do occur in DNA over time, however, as the environment takes its toll on our hereditary material. High-energy radiation (such as X-rays), low-energy radiation (such as UV light), and chemicals (such as LSD, marijuana, and pesticides) all can do damage to chromosomes. Cancer, for example, is one possible manifestation of gene mutation. The aging process also changes the DNA. Scientists only recently discovered that the ends of our chromosomes—regions called telomeres—shorten with age. However, none of these types of changes would visibly change your DNA.

Scientists analyze a person's DNA by means of restriction fragment length polymorphism analysis. Nicknamed RFLP (RIF-lip) analysis, this procedure uses methods to cut the DNA into fragments using restriction enzymes (see p. 417), and then separate and visualize these fragments using electrophoresis (see p. 419). These techniques are described in Chapter 19. The resulting banding patterns, such as those shown in the photograph, create what scientists call a DNA fingerprint. This fingerprint is unique to every individual just as the fingerprints on your hands are unique. For this reason, DNA fingerprinting is increasingly being used in criminal cases to either acquit or convict individuals of crimes. This is because

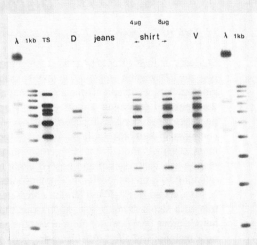

A stabbing victim's DNA profile is shown in column V; the defendant's is shown in column D. The banding patterns of the blood taken from the defendant's shirt and jeans matches that of the victim, not the defendant.

DNA fingerprinting can identify an individual with virtual certainty—even if the DNA sample and the individual have aged.

---

ferred to as *unwinding* or *unzipping*. Replication begins at various places in the DNA strand, and proceeds in both directions from each replication point, creating replication bubbles as shown in **Figure 18.6a.**

Figure 18.6b shows one end of a replication bubble, which is called a *replication fork*. At these Y-shaped forks, each separated strand serves as a template for the synthesis of a new complementary strand. In a process directed by the enzyme DNA polymerase, free nucleotide units (that are present along with the DNA in the nucleus) link to complementary bases on each of the DNA strands. However, the sugar-phosphate backbones of DNA run in opposite directions on each strand as shown in Figure 18.6c. Therefore, the process of replication differs somewhat on each strand.

Each strand of DNA has a 3′ (3-prime) end and a 5′ (5-prime) end. These names refer to positions of particular carbon atoms in the sugar molecules of the backbone. More specifically, note that in Figure 18.6c, the 5′ end of each DNA strand terminates with a phosphate group, which is attached to the carbon occupying the 5′ position. The 3′ end of each DNA strand terminates with a hydroxyl group, which is attached to the carbon occupying the 3′ position.

The attachment of nucleotides to each unzipped DNA strand proceeds from the 3′ end to the 5′ end of the template. On one strand, therefore, replication proceeds *toward* the replication fork. The daughter strand is synthesized continuously. On the other strand, however, replication proceeds *away from* the replication fork. The daughter strand is synthesized in pieces. On this strand, an enzyme called *DNA ligase* assists bonding of the backbone between adjacent pieces.

Thus, the process of DNA replication begins with one double-stranded DNA molecule and ends with two double-stranded DNA molecules. Each double strand contains one strand from the parent molecule plus a new complementary strand assembled from free nucleotides. Each of the new double-stranded molecules is identical to the other and is also identical to the original parent molecule.

DNA begins replication by unwinding at intervals along its helix before cell division. In a process initiated and coordinated by enzymes, free nucleotides bond to the exposed bases, producing two new DNA strands, each identical to one another and to the original double strand from which they were replicated.

Chapter 18 *DNA, Gene Expression, and Cell Reproduction*

H⚙W WORKS Science Discoveries

Is DNA or Protein the Genetic Material?

Students studying biology today sometimes take the information presented in textbooks for granted. For instance, it seems a given that all the genetic information about you—your appearance, how your cells function, your sex—is contained in your DNA. As little as 50 years ago, many scientists thought that the genetic information was contained in and passed along to offspring by proteins located in chromosomes. In 1952, Alfred Hershey and Martha Chase performed a series of important experiments that helped prove DNA (and not protein) was the genetic material. Building on an increasing body of experimental evidence that pointed to DNA, Hershey and Chase used viruses to carry out a clear either/or test. They showed that an infection by a virus involved an injection of the virus DNA, not protein, into a cell.

Hershey and Chase reasoned that if DNA were indeed the genetic carrier, then bacteria infected with a virus would contain DNA from the virus inside their cells. They reached this conclusion because they knew that when viruses infect cells, they somehow direct the cell to manufacture more virus particles. On the other hand, if protein were the material of which genes are made, then virus-infected bacteria would contain viral proteins within their cells.

The challenge for Hershey and Chase was to find a way to look inside an infected cell and see what parts of the infecting virus could be found. Hershey and Chase chose to work with a bacteriophage, a type of virus that infects bacteria. Like all viruses, bacteriophages have a nucleic acid core surrounded by a protein coat. Bacterial viruses also have "legs" attached to the protein coat that allow them to "dock" with a bacterium.

A key part of Hershey and Chase's experiments was the way they distinguished between DNA and protein—they used radioactive "name tags." Proteins contain sulfur, but DNA does not, so radioactive sulfur served as a label for protein. DNA contains phosphorus, but protein does not, so radioactive phosphorus served as a label for DNA. Hershey and Chase performed two experiments (Figure 18.A). In experiment 1, they labeled the protein coat of the bacteriophage with ^{35}S, an isotope of sulfur. (See Chapter 3 for a discussion of isotopes.) Then they mixed the labeled bacteriophage with *E. coli* bacteria (which are found in abundance in your digestive tract) and allowed the viruses to infect the bacteria. To examine the bacteria after infection, Hershey and Chase had to separate the protein coats that did not enter the bacteria from the infected bacteria. They knew from the previous experiments of other researchers that the virus was not engulfed and taken into the host bacterium during infection, but that its protein coat remained outside, separated from the bacterial cell, where it could be seen with an electron microscope.

Hershey and Chase separated the viral protein coats by blending the virus/bacteria mixture and then centrifuging it (spinning it rapidly). This pulled the heavier bacteria to the bottom of a test tube, leaving the lighter protein coats in the solution on top, which is called the supernatant. Hershey and Chase found all the ^{35}S-labeled protein in the supernatant, not in the bacteria. Thus, the bacteriophage protein did not enter the bacteria during infection.

In experiment 2, Hershey and Chase labeled the bacteriophage DNA with an isotope of phosphorus, ^{32}P, and performed the same mixing, blending, and centrifugation as they did in experiment 1. The ^{32}P was found in the bottom of the test tube, with the bacteria. Clearly, ^{32}P-labeled DNA had entered the bacteria.

These experiments went a long way toward convincing the scientific community that DNA, not protein, was the genetic material. Within the year, Watson and Crick's discovery of DNA's structure showed how easily DNA could fulfill its suggested genetic role. ⚙

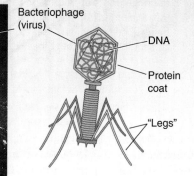

Bacteriophage (virus)
DNA
Protein coat
"Legs"

Figure 18.A

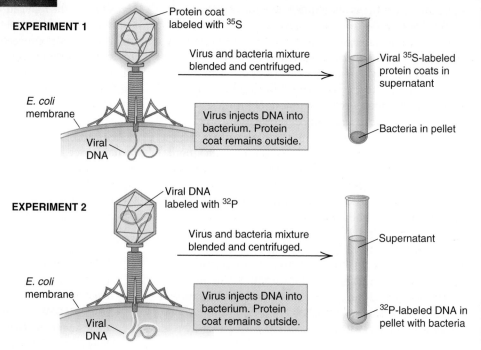

EXPERIMENT 1

Protein coat labeled with ^{35}S

E. coli membrane

Viral DNA

Virus and bacteria mixture blended and centrifuged.

Virus injects DNA into bacterium. Protein coat remains outside.

Viral ^{35}S-labeled protein coats in supernatant

Bacteria in pellet

EXPERIMENT 2

Viral DNA labeled with ^{32}P

E. coli membrane

Viral DNA

Virus and bacteria mixture blended and centrifuged.

Virus injects DNA into bacterium. Protein coat remains outside.

Supernatant

^{32}P-labeled DNA in pellet with bacteria

Units of DNA called genes direct the synthesis of polypeptides.

18.6 Genes: Units of hereditary information

Before the discovery of the structure of DNA by Watson and Crick, scientists were working to determine how DNA directs the growth and development of an organism. They asked the question, "What is the hereditary message that DNA carries, and how is that message expressed?" Experiments conducted in the late 1940s and early 1950s by two geneticists, George Beadle and Edward Tatum, answered this question. These researchers, whose work officially marked the beginning of molecular genetics, studied the biochemical characteristics of the expression of genes. They worked with the common red bread mold *Neurospora* (ner-ROS-per-ah).

At the same time Beadle and Tatum were doing their work, various biochemists were also studying the manufacture and breakdown of organic molecules within cells. They determined that cells build and degrade molecules by sequences of steps in which each step is catalyzed by a specific enzyme. These enzymatically controlled sequences of steps are called biochemical pathways.

Beadle and Tatum studied mold cultures that they had exposed to x-rays. The x-rays induced mutations, or changes, in the DNA of the organism, resulting in a variety of mutants, each unable to manufacture certain amino acids. (Amino acids, as noted in Chapter 3, are the building blocks of proteins.) Using known information about the biochemical pathways of *Neurospora*, they hypothesized that specific enzymes must be involved in the manufacture of these amino acids and that some or all of the enzymes in each biochemical pathway must not be doing their jobs.

To test their hypothesis, Beadle and Tatum chose mutants unable to synthesize the amino acid arginine. They supplied each mutant with various compounds intermediate in the arginine pathway and observed whether the mutant was then able to synthesize arginine. Using this method, Beadle and Tatum were able to infer the presence of defective enzymes in this biochemical pathway.

From these studies, Beadle and Tatum proposed the **one-gene-one-enzyme theory**, which states that the production of a given enzyme is under the control of a specific gene (Figure 18.7). If the gene mutates, the enzyme will not be synthesized properly or will not be made at all. Therefore, the reaction it catalyzes will not take place, and the product of the reaction will not be produced.

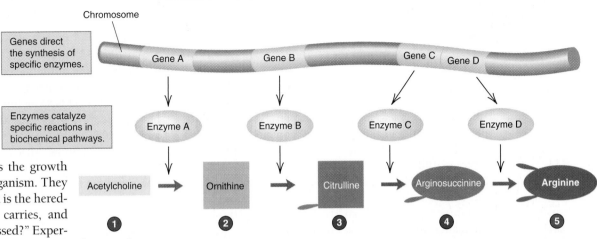

Figure 18.7 **The one-gene-one-enzyme theory.** This illustration shows the biochemical pathway of the synthesis of the amino acid arginine. Each gene directs the synthesis of one enzyme that catalyzes a specific step in this pathway.

For example, suppose the product of a biochemical pathway is arginine (pictured as an oval with two tails), as shown in Figure 18.7. In order to produce that product, acetylcholine (pictured as the rectangle ❶) is converted to ornithine. You can think of this change as enzyme A changing the rectangle to a square ❷. Next, enzyme B adds a tail to the square, forming citruline ❸. Enzyme C changes the square to an oval ❹, and enzyme D adds another tail to the oval ❺. This two-tailed oval represents arginine. Notice how blocking one step would prevent that immediate change from occurring, as well as all the other changes "down the line."

Today scientists have refined the one-gene-one-enzyme hypothesis to say that the production of a given polypeptide (a *portion* of an enzyme or other protein) is under the control of a single gene. Additional biochemical evidence has upheld this conclusion. For their groundbreaking work, Beadle and Tatum were awarded the Nobel Prize in Physiology and Medicine in 1958.

These experiments and other related ones have helped clarify the concept of "a unit of heredity." A unit of heredity—a **gene**—is a sequence of nucleotides that codes for (scientists say encodes) the amino acid sequence of an enzyme or *other protein*. Although most genes code for a string of amino acids (polypeptides), there are also genes devoted to the production of special forms of RNA, which play important roles in protein synthesis.

> DNA carries a code that directs the synthesis of polypeptides—pieces of enzymes and other proteins that orchestrate and regulate the growth, development, and daily functioning of cells. Each hereditary unit, or gene, codes for one polypeptide.

18.7   Gene expression

An Overview

Polypeptides are not made in the nucleus where DNA is located. Instead, they are made in the cytoplasm at the **ribosomes** (RYE-buh-somes) (see Chapter 4). These complex polypeptide-making factories contain more than 50 different proteins in their structure. Along with these proteins, ribosomes are also made up of RNA—a hint that RNA molecules play an important role in polypeptide synthesis. In fact, as shown in Figure 18.8, cells contain three types of RNA, and each plays a special role in the manufacture of polypeptides.

One type of RNA, called **messenger RNA (mRNA)**, brings information from the DNA (within the nucleus) to the ribosomes (in the cytoplasm) to direct which polypeptide is assembled. A second type of RNA, called **transfer RNA (tRNA)**, is found in the cytoplasm. During polypeptide synthesis, tRNA molecules transport amino acids (used to build the polypeptide) to the ribosomes. In addition, tRNA molecules position each amino acid at the correct place on the elongating polypeptide chain. The third type of RNA found in ribosomes is called **ribosomal RNA (rRNA)**.

Transcribing the DNA Message to RNA

The first step in the process of polypeptide synthesis and gene expression is the copying of the gene into a strand of messenger RNA, a process called **transcription**.

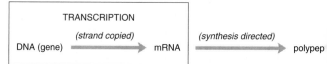

Transcription begins when the enzyme RNA polymerase (pol-IH-meh-race) binds to a sequence of nucleotides that makes up a gene on a single DNA strand (Figure 18.9, ❶). This sequence of nucleotides is located at the beginning of the gene that is being expressed. You can think of such a sequence of nucleotides as an enzyme code that says, "Start here." Scientists call this place on a DNA strand a *promoter site*.

Next, the RNA polymerase causes the DNA base-pair bonds to break, freeing the DNA strand that is to be transcribed from its partner DNA strand ❷. RNA nucleotides bind with the now-exposed DNA bases in a sequence complementary to that of the DNA ❸. For example, RNA nucleotides having the base adenine pair with DNA nucleotides having the base thymine. Likewise, RNA nucleotides having the base uracil pair with DNA nucleotides having the base adenine.

As the RNA polymerase moves along the DNA strand encountering each DNA nucleotide in turn, it adds the corresponding complementary RNA nucleotide to the growing single strand of mRNA. When the enzyme arrives at a special stop sequence (frequently a loop in the DNA) located at the end of the gene, it disengages from the DNA and releases the newly assembled mRNA strand ❹. A processing step then removes segments of RNA from the original transcript that are not used in polypeptide synthesis, as shown in Figure 18.10. These extra sequences of nucleotides that intervene between the polypeptide-specifying portions of the gene are

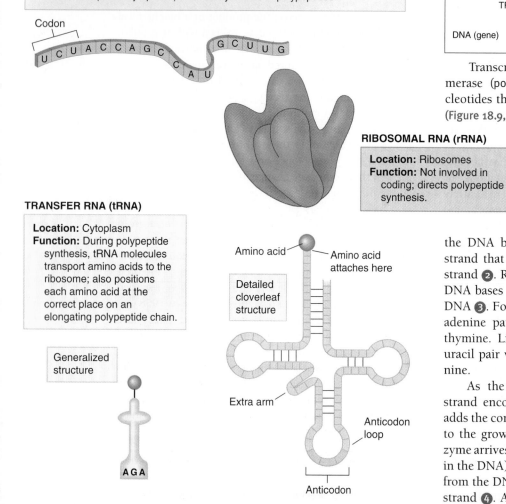

MESSENGER RNA (mRNA)

Location: Made in the nucleus; moves through nuclear pores to the cytoplasm.
Function: Brings information from the DNA (within the nucleus) to the ribosomes (in the cytoplasm) to direct synthesis of polypeptides.

Codon

U C U A C C A G C C A U G C U U G

RIBOSOMAL RNA (rRNA)

Location: Ribosomes
Function: Not involved in coding; directs polypeptide synthesis.

TRANSFER RNA (tRNA)

Location: Cytoplasm
Function: During polypeptide synthesis, tRNA molecules transport amino acids to the ribosome; also positions each amino acid at the correct place on an elongating polypeptide chain.

Generalized structure

AGA

Amino acid

Amino acid attaches here

Detailed cloverleaf structure

Extra arm

Anticodon loop

Anticodon

Figure 18.8   **The three types of RNA with their locations and functions.** The diagrammatic representations of the shapes of these molecules are used to depict each type of RNA throughout this chapter.

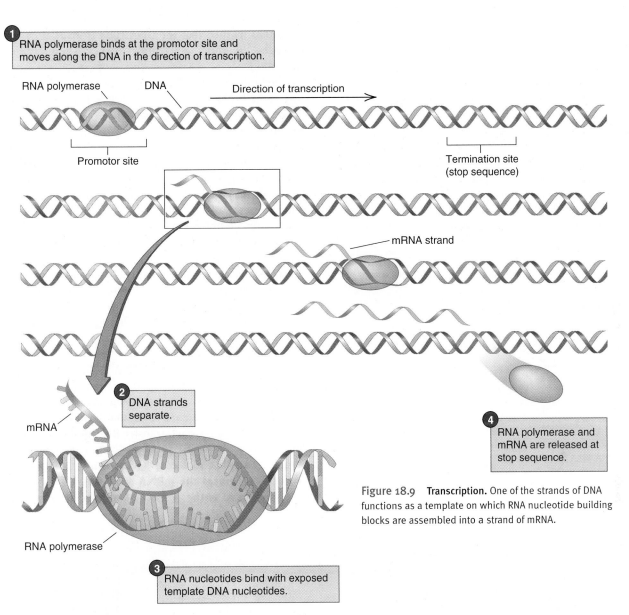

1 RNA polymerase binds at the promotor site and moves along the DNA in the direction of transcription.

RNA polymerase   DNA   Direction of transcription

Promotor site

Termination site (stop sequence)

mRNA strand

2 DNA strands separate.

mRNA

4 RNA polymerase and mRNA are released at stop sequence.

RNA polymerase

3 RNA nucleotides bind with exposed template DNA nucleotides.

Figure 18.9   Transcription. One of the strands of DNA functions as a template on which RNA nucleotide building blocks are assembled into a strand of mRNA.

called *introns* (they *intrude* into the gene but are not expressed). The remaining segments of the gene—the nucleotide sequences that encode the amino acid sequence of the polypeptide—are called *exons* (those portions of a gene that will be *expressed*). Cutting out introns and splicing together exons results in the final mRNA strand. After cutting and splicing occurs, the mRNA strands leave the nucleus through the nuclear pores and travel to the ribosomes in the cytoplasm of the cell.

Translating the Transcribed DNA Message into a Polypeptide

In the second step of gene expression, the mRNA—using its *copied* DNA code—directs the synthesis of a polypeptide, as shown in the lower portion of Fig-

ure 18.10. You can think of mRNA as copied computerized instructions or a template for an assembly line. After the instructions are copied from the master plans during transcription, they are used to build products. The products are polypeptides, and the process of making polypeptides is called **translation**.

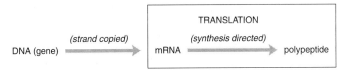

TRANSLATION

DNA (gene)   *(strand copied)*   mRNA   *(synthesis directed)*   polypeptide

What exactly is the DNA code that must be translated in living things to produce polypeptides? In 1961, as a result of experiments led by Francis Crick (one of the researchers who worked out the structure of DNA),

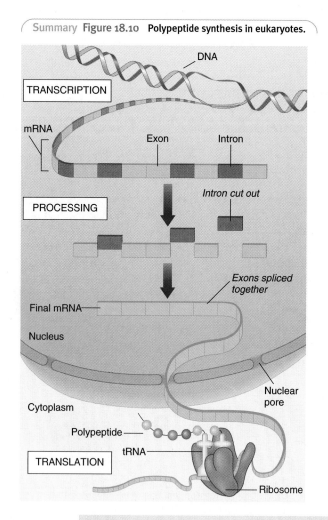

Summary Figure 18.10 **Polypeptide synthesis in eukaryotes.**

DNA

TRANSCRIPTION

mRNA

Exon Intron

Intron cut out

PROCESSING

Exons spliced together

Final mRNA

Nucleus

Nuclear pore

Cytoplasm

Polypeptide

tRNA

TRANSLATION

Ribosome

scientists learned that the DNA code is made up of sequences of three nucleotide bases. These triplet-nucleotide code words, as they appear on the transcribed mRNA, are called **codons** (KOH-dons). Researchers soon broke the code and learned which codons stand for which amino acids, the building blocks of polypeptides.

Table 18.3 shows the mRNA codons and the amino acids for which they code. The table is structured so that every triplet-base sequence of the four mRNA bases is listed. As you can see from the table, using a three-base sequence of four different bases produces 64 possible codon combinations ($4 \times 4 \times 4$, or 4^3). Of these 64 combinations, 61 code for amino acids and 3 act as "stop" signals for the process of translation. With only 20 amino acids used in polypeptide production, extra codons exist, which provide alternative code words for many of the amino acids. Additionally, the codon AUG can code for the amino acid methionine or can serve as a "start" signal.

The molecules that recognize these mRNA codons are molecules of tRNA. Each tRNA molecule has an **anticodon** loop—a portion of the molecule with a sequence of three base pairs complementary to a specific mRNA codon (see Figure 18.8). On the opposite end of the anticodon loop, each transfer RNA molecule carries an amino acid specific to its anticodon sequence. For example, a tRNA molecule having the anticodon AGA (which recognizes the mRNA sequence, UCU) carries the amino acid serine. A special family of enzymes links

Table 18.3

The Genetic Code

		Second Nucleotide in Codon			
		U	**C**	**A**	**G**	
U		UUU Phe F *Phenylalanine*	UCU Ser S *Serine*	UAU Tyr Y *Tyrosine*	UGU Cys C *Cysteine*	U
		UUC Phe F *Phenylalanine*	UCC Ser S *Serine*	UAC Tyr Y *Tyrosine*	UGC Cys C *Cysteine*	C
		UUA Leu L *Leucine*	UCA Ser S *Serine*	UAA Stop codon	UGA Stop codon	A
		UUG Leu L *Leucine*	UCG Ser S *Serine*	UAG Stop codon	UGG Trp W *Tryptophan*	G
C		CUU Leu L *Leucine*	CCU Pro P *Proline*	CAU His H *Histidine*	CGU Arg R *Arginine*	U
		CUC Leu L *Leucine*	CCC Pro P *Proline*	CAC His H *Histidine*	CGC Arg R *Arginine*	C
		CUA Leu L *Leucine*	CCA Pro P *Proline*	CAA Gln Q *Glutamine*	CGA Arg R *Arginine*	A
		CUG Leu L *Leucine*	CCG Pro P *Proline*	CAG Gln Q *Glutamine*	CGG Arg R *Arginine*	G
A		AUU Ile I *Isoleucine*	ACU Thr T *Threonine*	AAU Asn N *Asparagine*	AGU Ser S *Serine*	U
		AUC Ile I *Isoleucine*	ACC Thr T *Threonine*	AAC Asn N *Asparagine*	AGC Ser S *Serine*	C
		AUA Ile I *Isoleucine*	ACA Thr T *Threonine*	AAA Lys K *Lysine*	AGA Arg R *Arginine*	A
		AUG Met M *Methionine*	ACG Thr T *Threonine*	AAG Lys K *Lysine*	AGG Arg R *Arginine*	G
G		GUU Val V *Valine*	GCU Ala A *Alanine*	GAU Asp D *Aspartic acid*	GGU Gly G *Glycine*	U
		GUC Val V *Valine*	GCC Ala A *Alanine*	GAC Asp D *Aspartic acid*	GGC Gly G *Glycine*	C
		GUA Val V *Valine*	GCA Ala A *Alanine*	GAA Glu E *Glutamic acid*	GGA Gly G *Glycine*	A
		GUG Val V *Valine*	GCG Ala A *Alanine*	GAG Glu E *Glutamic acid*	GGG Gly G *Glycine*	G

(First nucleotide in codon (5' end) — left side; Third Nucleotide in Codon (3' end) — right side)

Codon Three-letter and single-letter abbreviations

Instructions for reading this table: The first nucleotide in the codon identifies the row of the amino acid. The second nucleotide in the codon identifies the co of the amino acid. (This process identifies a box of 4 amino acids.) The third nucleotide in the codon identifies the specific amino acid in the box of 4.

the amino acids to the tRNA molecules. Using the information in Table 18.3 and your understanding of the relationship between codons and anticodons, can you determine which amino acid a tRNA molecule with the anticodon GGG carries?

During the process of translation, the genetic code (the sequence of codons in mRNA) is deciphered. First, certain initiation events take place. The start codon of the mRNA (transcribed from DNA in the nucleus) binds to a small ribosomal subunit. A special tRNA molecule detects the start codon (AUG) and binds to it by means of its anticodon (UAC) (Figure 18.11, top). (This particular tRNA usually carries the amino acid methionine [met]). Then, a large ribosomal subunit binds to the small subunit, resulting in a functional ribosome. This functional ribosome has an A (aminoacyl) site and a P (peptidyl) site. The tRNA already present occupies the P site of the ribosome.

Next, the elongation of the polypeptide begins. *Elongation* consists of three steps that take place over and over until the entire mRNA is "read." In Step 1, the codon in the A site is "read." That means that a tRNA with an anticodon complementary to the next mRNA codon "in line" binds to the mRNA at the A site. This tRNA carries the next amino acid that will be added to the chain. In Step 2, a peptide bond is formed between the two adjacent amino acids. In Step 3, the tRNA at the P site breaks away from its amino acid and the mRNA, leaving the ribosome. The ribosome moves so that the tRNA originally located at the A site is shifted (with its attached growing amino acid chain) to the P site. One by one, as tRNA molecules bind to codons at the A site on the mRNA, amino acids are lined up in an order determined by the sequence of codons that passed through the ribosome.

When a stop codon is encountered (see Table 18.3), no tRNA exists to bind to it. Instead, it is recognized by special release

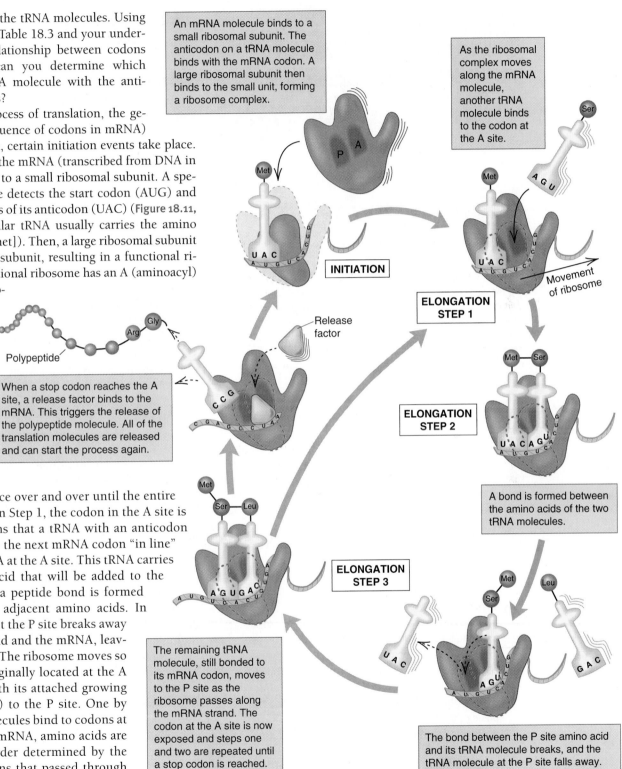

An mRNA molecule binds to a small ribosomal subunit. The anticodon on a tRNA molecule binds with the mRNA codon. A large ribosomal subunit then binds to the small unit, forming a ribosome complex.

INITIATION

As the ribosomal complex moves along the mRNA molecule, another tRNA molecule binds to the codon at the A site.

ELONGATION STEP 1

Movement of ribosome

Release factor

When a stop codon reaches the A site, a release factor binds to the mRNA. This triggers the release of the polypeptide molecule. All of the translation molecules are released and can start the process again.

Polypeptide

ELONGATION STEP 2

A bond is formed between the amino acids of the two tRNA molecules.

ELONGATION STEP 3

The remaining tRNA molecule, still bonded to its mRNA codon, moves to the P site as the ribosome passes along the mRNA strand. The codon at the A site is now exposed and steps one and two are repeated until a stop codon is reached.

The bond between the P site amino acid and its tRNA molecule breaks, and the tRNA molecule at the P site falls away.

Figure 18.11 Translation and the termination of protein synthesis. A polypeptide is formed after Steps 1–3 have been repeated many times. The small ribosomal subunit is shown as a transparent structure in all but the initiation step so that the process of translation can be viewed more easily.

(a)

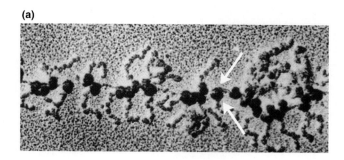

(b)

Figure 18.12 **Protein synthesis in a fly.** (a) An electron micrograph of ribosomes in the fly *Chironomus tentans*. These ribosomes are reading along an mRNA molecule from left to right, assembling polypeptides that dangle behind them like tails. Visible are the two subunits *(arrows)* of each ribosome translating the mRNA. (b) An illustration depicting the same process described above. The mRNA molecule is shown in blue, the ribosome subunits are shown in two different shades of red, and the polypeptides are shown in shades of green.

factors, proteins that bring about the release of the newly made polypeptide from the ribosome.

Figure 18.12 shows that more than one ribosome at a time reads and translates the mRNA message to synthesize a polypeptide. Here, a group of many ribosomes (called a *polyribosome*) is reading along an mRNA molecule taken from a cell of a fly. Although the process is shown in a fly, it occurs in a similar fashion in all organisms. The polypeptides being made can be seen dangling behind each ribosome.

In the first step of polypeptide synthesis, mRNA is constructed from free RNA nucleotides by the use of a single strand of DNA as a template. After noncoding sequences are removed from the mRNA strand, the mRNA leaves the nucleus and travels to the ribosomes, where polypeptide synthesis takes place. Ribosomes bind to sites at one end of an mRNA strand and then move down the strand, exposing the codons one by one. At each step of a ribosome's progress, it exposes a codon to binding by a tRNA molecule having a three-base sequence complementary to the exposed mRNA codon. The amino acid carried by that particular tRNA molecule is then added to the end of the elongating polypeptide chain. A stop codon triggers the release of the polypeptide from the translation molecules.

18.8 Regulation of eukaryotic gene expression

Cells not only know how to make particular polypeptides but also when to make them. For example, during your fetal development, specific enzymes played crucial roles at certain times, directing the series of biochemical reactions that resulted in your growth. As this growth and development took place, your genes were transcribed in a specific order, each gene for a specified time. Likewise, the cells in your adult body that produce digestive enzymes know when to manufacture these enzymes. Your red blood cells—not other types of cells—synthesize hemoglobin to carry oxygen to all parts of your body. How are these specific genes activated?

Molecular geneticists are now beginning to understand these mechanisms in eukaryotic cells. Interestingly, cells have gene switches. These switches are really proteins that interact with specific nucleotide sequences called **regulatory sites**. Regulatory sites control the transcription of genes. Control can be positive and result in turning a gene on, or it can be negative and result in turning a gene off. In eukaryotes, control is generally positive because eukaryotic genes are usually inactive, so genes must be turned on. In one type of gene switch, steroid hormones, along with protein receptor molecules from the cytoplasm, activate transcription by binding to regulatory sites on the DNA. Scientists are probing the details of this regulatory process. (See Section 17.4 and the opener in Chapter 23 for more information on gene regulation and gene switches.)

Eukaryotic cells also regulate gene expression by the number of identical genes they transcribe at one time. This use of multiple copies of genes is a way to control the amount of a particular protein that is produced at a certain time. Some genes have related, but not identical, nucleotide sequences. Transcribing these genes at the same time produces related products. Such multiple copies of genes in eukaryotes, called *gene families*, are derived from a common ancestral gene and are a reflection of the evolutionary process.

Organisms control the expression of their genes by selectively inhibiting the transcription of some genes and facilitating the transcription of others. Regulatory proteins control the transcription process.

---

mitosis), the supercoils of DNA, which are normally diffuse or strung out within the cell nucleus, begin the long process of *condensation*. During this process, the complex of DNA and proteins coil into more tightly compacted bodies that become visible as chromosomes during mitosis. In addition, during the G_2 phase, the cell begins to assemble the "machinery" (microtubules that make up the spindle) that it will later use to move and divide the chromosomes. In animal cells, the centrioles replicate. These organelles are surrounded by microtubule-organizing centers that play a key role in the mitotic process; most plants and fungi lack centrioles.

Mitosis is a continuous sequence of events that occurs just after interphase and that results in the division of the chromosomes duplicated during its S phase. To more easily understand this process, scientists divide its events into the following phases: prophase, metaphase, anaphase, and telophase (Figure 18.16).

Interphase is the portion of the cell cycle in which the cell grows and carries out normal life functions. During this time, the cell also produces an exact copy of the hereditary material DNA as it prepares for cell division.

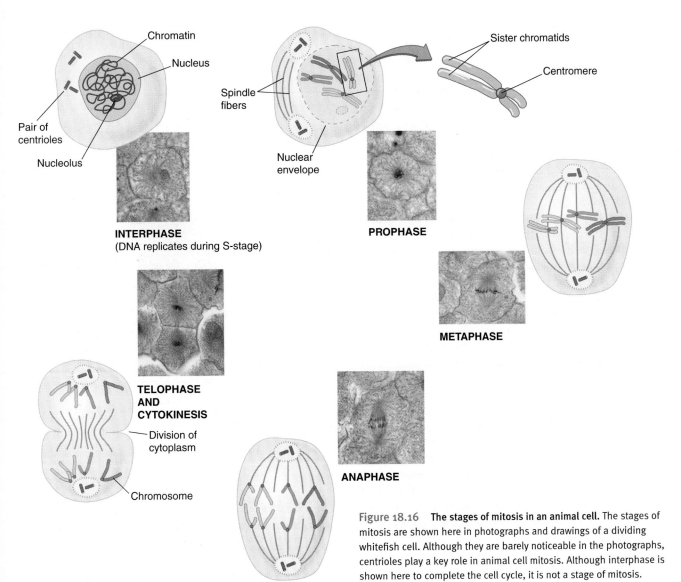

Figure 18.16   **The stages of mitosis in an animal cell.** The stages of mitosis are shown here in photographs and drawings of a dividing whitefish cell. Although they are barely noticeable in the photographs, centrioles play a key role in animal cell mitosis. Although interphase is shown here to complete the cell cycle, it is not a stage of mitosis.

18.11 Mitosis

Prophase

The first stage of mitosis, **prophase** (see Figure 18.16, p. 399), begins when the chromosomes have condensed to the point where they become visible under a light microscope. As prophase continues, the chromosomes continue to shorten and thicken, looking much bulkier at the end of prophase than at the beginning. The nucleolus, which was previously visible, disappears. This disappearance is due to the nucleolus being unable to make ribosomal RNA (rRNA) when the part of the chromosome bearing the rRNA genes is condensed. It is rRNA that makes up most of the substance of the nucleolus (see Chapter 3).

While the chromosomes are condensing, another series of equally important events is also occurring: Special microtubules (thin, tubelike, protein structures [see Chapter 3]) called the *spindle fibers* are being assembled. In animal cells, these spindle fibers extend from a pair of related microtubular structures called *centrioles* (see Figure 18.16). Although the centrioles were once thought to play a role in forming the spindle, recent evidence suggests that this is not the case. Instead, the microtubules of the spindle appear to form from granules surrounding the centrioles called the *microtubule-organizing center (MTOC)*. Interestingly, plant cells do not contain centrioles, but spindle fibers form in plant cells.

In early prophase, the centrioles of animal cells begin to move away from one another. By the end of prophase, each member of the pair has moved to an opposite end, or pole, of the cell. As the spindle fibers form, the nuclear envelope breaks down, forming small vesicles that disperse in the cytosol. Without the nucleus in the way, the spindle fibers form a bridge between the centrioles, spanning the distance from one pole to the other. Some spindle fibers extending from each pole attach to their side of the centromere of each sister chromatid (CROW-mah-tid). By this time, each chromatid has developed a *kinetochore* (kih-NET-ih-core) at its centromere. A kinetochore is the structure to which several pole-to-centromere microtubules will attach. The effect is to attach one sister chromatid to one pole and the other sister chromatid to the other pole. During later stages of mitosis, the sister chromatids separate and move to opposite poles of the cell, pulled by the spindle fibers. The proper attachment of each spindle fiber to the kinetochore is therefore critical to the process of mitosis.

When the centrioles reach the poles of the cell in animal mitosis, they radiate an array of microtubules outward (toward the cell membrane) as well as inward (toward the chromosomes). This arrangement of microtubules is called an *aster*. The function of the aster is not well understood. Evidence suggests that the aster probably acts as a support during the movement of the sister chromatids.

Metaphase

The second phase of mitosis, **metaphase**, begins when the chromatid pairs line up in the center of the cell. In reality, the chromosomes are not in a line, but form a circle. They only appear to form a line when viewed two dimensionally with a light microscope as in Figure 18.16. **Figure 18.17** is a diagram of a three-dimensional view of an animal cell at metaphase. As you can see, the chromosomes form a circle perpendicular to the direction of the spindle fibers. Positioned by the micro-

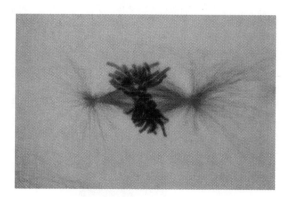

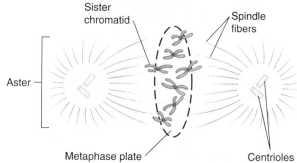

Figure 18.17   The metaphase plate. In metaphase, the chromosomes form an array around the spindle midpoint.

tubules attached at their centromeres, all the chromosomes are equidistantly arranged between the two poles at the "equator" of the cell. The region of this circular arrangement, called the *metaphase plate,* is not a physical structure but indicates approximately where the future axis of cell division will be.

Anaphase

At the beginning of **anaphase**, the sister chromatids separate at the centromere, freeing them from their attachment to each other. Before this split, the chromatids are tugged in two directions at once by opposing microtubules—somewhat like a cellular tug-of-war. With the separation of the chromatids (now called chromosomes), they move rapidly toward opposite poles of the cell, each pulled at its kinetochore by attached, shortening microtubules.

Telophase

The separation of the sister chromatids in anaphase equally divides the hereditary material that replicated just before mitosis. Therefore, each of the two new cells that are forming receives a complete, identical copy of the chromosomes. This partitioning of the genetic material is the essence of the process of mitosis.

The events of this last phase of mitosis, **telophase**, ready the cell for *cytokinesis,* or division of the cytoplasm. The spindle fibers are chemically disassembled and therefore disappear. The nuclear envelope reforms around each set of what were sister chromatids, now chromosomes. These chromosomes begin to uncoil, returning to their normally strung out, more diffuse state. The DNA of the nucleolus begins making ribosomal RNA once again; this rRNA is visible, resulting in the reappearance of the nucleolus. In summary, the events of telophase are very much like the reverse order of the events of prophase.

Prophase is the stage of mitosis characterized by the appearance of visible chromosomes. By the end of prophase, spindle fibers that radiate from opposite poles of the cell attach to each kinetochore at the centromere. During metaphase, chromosomes align equidistantly from the two poles of the cell. During anaphase, sister chromatids separate and go to opposite poles of the cell. In telophase, the mitotic apparatus assembled during prophase is disassembled, the nuclear envelope is reestablished, and the normal use of the genes present in the chromosomes is reinitiated.

18.12   Cytokinesis

At the end of telophase, mitosis is complete. The eukaryotic cell has divided its duplicated hereditary material into two nuclei that are positioned at opposite ends of the cell. While this process has been going on, the cytoplasmic organelles, such as the mitochondria, have been somewhat equally distributed to each side of the cytoplasm. These organelles replicate throughout interphase.

At this point, the process of cell division is still not complete. The division of the cytoplasm—that portion of the cell outside the nucleus—begins during telophase. The stage of the cell cycle at which cell division actually occurs is called **cytokinesis**. Cytokinesis generally involves the division of the cell into approximately equal halves. The separation of the mother cell into two separate daughter cells usually occurs shortly after mitosis is complete.

Cytokinesis in Animal Cells

In human cells and in the cells of other eukaryotes that lack cell walls, cytokinesis occurs by a pinching of the cell in two. A belt of microfilaments that encircles the cell at the metaphase plate accomplishes this pinching. These microfilaments contract, forming a *cleavage furrow* around the circumference of the cell (Figure 18.18). As contraction proceeds, the furrow deepens until the opposing edges of the membrane make contact with one

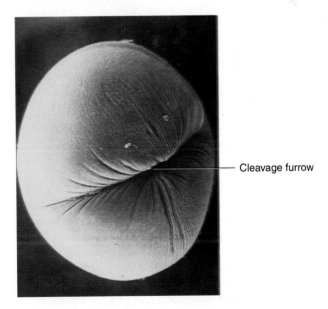

— Cleavage furrow

Figure 18.18   Cytokinesis in an animal cell. A cleavage furrow is forming around this dividing sea urchin egg.

another. Then the membranes fuse, separating the one cell into new cells.

Cytokinesis in Plant Cells

In plants and some algae, a rigid wall surrounds the cell. Therefore, cytokinesis involves the laying down of a new cell wall between the two daughter cells rather than a pinching in of the cytoplasm. Plants manufacture new sections of membrane and wall from tiny vesicles most likely derived from the Golgi complex. These membrane "parts" accumulate at the metaphase plate and

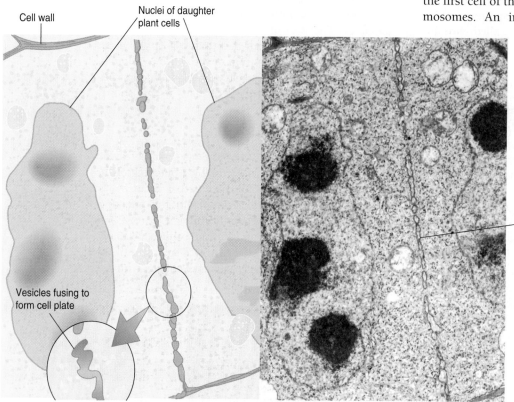

Cell wall

Nuclei of daughter plant cells

Vesicles fusing to form cell plate

Cell plate

Figure 18.19 **Cytokinesis in plant cells.** In this photograph and companion drawing, a cell plate is forming between daughter nuclei. The cell plate is composed of pieces of cell membrane (vesicles) that fuse together.

fuse, beginning the formation of a partition called the *cell plate.* A new cell wall forms between the two membranes of the cell plate. **Figure 18.19** shows cytokinesis in a plant cell and the formation of a cell plate.

> Cytokinesis is the physical division of the cytoplasm of a eukaryotic cell into two daughter cells. In animal cells, division occurs by the "pinching" of the mother cell into two daughter cells; in plant cells, a new cell wall is laid down between the two daughter cells.

18.13 Meiosis

Most animals, plants, algae, fungi, and certain protists reproduce sexually. In sexual reproduction, gametes of opposite sexes or mating types (sometimes just termed + cells and − cells) unite in the process of fertilization, producing the first cells of new individuals (see Figure 18.13).

Humans have 46 chromosomes in all of their body cells—the diploid amount. If the two cells that joined in fertilization each contained 46 chromosomes, however, the first cell of the future offspring would have 92 chromosomes. An individual born after 10 generations would have more than 47,000 chromosomes! Such a continuing addition of chromosomes to each new individual is obviously an unworkable situation. Even early investigators realized that there must be some mechanism during the course of gamete formation to reduce the number of chromosomes. They reasoned that if sex cells were formed with half the number of chromosomes characteristic of the cells of that species, then the fusion of these cells during fertilization would produce cells of new individuals with the proper number of chromosomes. Investigators soon observed this special type of cell division and named it *meiosis,* from a Greek word meaning "less."

During meiosis, the diploid number of chromosomes is reduced by half, forming haploid cells. For this reason, meiosis is often called *reduction-division.* Four daughter cells result from two divisions of one original parent cell, although in human females, only one of the four cells is functional (see Chapter 22). In animals, meiosis occurs in the cells that produce gametes.

Although meiosis is a continuous process, scientists divide it into stages as they do the process of mitosis. The two forms of cell division have much in common. They are both special forms of nuclear division (although these processes are often referred to as forms of cell division for convenience). Meiosis consists of two sets of divisions called **meiosis I and meiosis II.** Each set is divided into prophase, metaphase, anaphase, and telophase, just as in mitosis. In meiosis, however, prophase I is much more complicated.

Meiosis is preceded by an interphase that is similar to the interphase of mitosis. During interphase, the chromosomes are replicated, resulting in each chromosome consisting of two genetically identical sister chromatids held together at the constricted centromere. The centrioles also replicate.

Meiosis I

In prophase I, the individual chromosomes condense as their DNA coils more tightly, thus becoming visible under a light microscope. Because this DNA replicated before meiosis, each chromosome consists of two sister chromatids joined at the centromere.

Chromosomes, as you can see in the chapter opener photo, come in pairs. Each pair is formed during fertilization as an egg and sperm fuse. Each of these gametes contributes one chromosome to each pair of chromosomes. Thus, you received half your chromosomes (one set of genetic information) from your father and half (a second set of genetic information) from your mother. Likewise, all organisms produced by sexual reproduction receive half their chromosomes from one parent organism and half from the other. These pairs of chromosomes are called **homologous chromosomes** (hoe-MOL-uh-gus), or **homologues** (HOME-uh-logs).

Each homologous chromosome contains genes that code for the same inherited traits, such as eye color and hair color. During prophase I, homologous chromosomes line up side by side in a process called **synapsis** (sih-NAP-sis). Because each chromosome is made up of two sister chromatids, the paired homologous chromosomes together have four chromatids. These synapsed homologues are therefore called *bivalents* (meaning "too strong"), or *tetrads* (meaning "groups of four").

The process of synapsis begins a complex series of events called **crossing over**. During crossing over, homologous *nonsister* chromatids actually cross over one another. These crossed-over pieces break away from the chromatids to which they are attached and reattach to the nonsister chromatid—literally exchanging segments (Figure 18.20). Once crossing over is complete, the nuclear envelope dissolves and the homologues begin to move apart. These four chromatids cannot separate from one another, however, because (1) the sister chromatids are held together at their centromeres and (2) the paired homologues are held together at the points where crossing over occurred. As the chromosomes move apart somewhat, the points of crossing over can be seen (under a light microscope) as X-shaped structures called **chiasmata** (keye-AZ-muh-tuh; sing. chiasma keye-AZ-muh) (see photo in Figure 18.20). Crossing over is a significant event in meiosis because it produces new

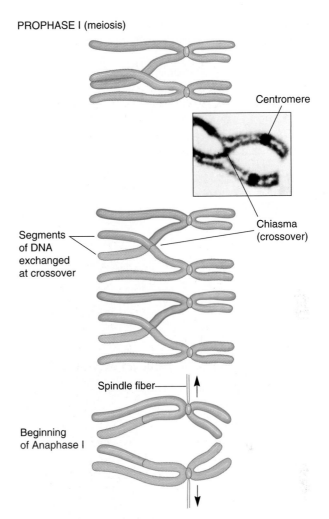

PROPHASE I (meiosis)

Centromere

Segments of DNA exchanged at crossover

Chiasma (crossover)

Spindle fiber

Beginning of Anaphase I

Figure 18.20 Crossing over. Replicated homologous chromosomes line up side by side (synapsis) during prophase I. Crossing over occurs when the paired chromosomes exchange segments at such locations. Once crossing over is complete, chromosomes move apart slightly, forming X-shaped structures called chiasmata (photograph). Crossing over produces new combinations of alleles.

combinations of alleles. This process provides one way in which offspring have a genetic makeup different from either parent.

In metaphase I, the next phase of meiosis, the microtubules have formed a spindle, just as in mitosis. A crucial difference exists between this metaphase and that of mitosis: The chromosomes line up with their homologues double file in meiosis (Figure 18.21), not single-file as in mitosis. For each pair of homologues the orientation on the metaphase plate is random; which homologue is oriented toward which pole is a matter of chance.

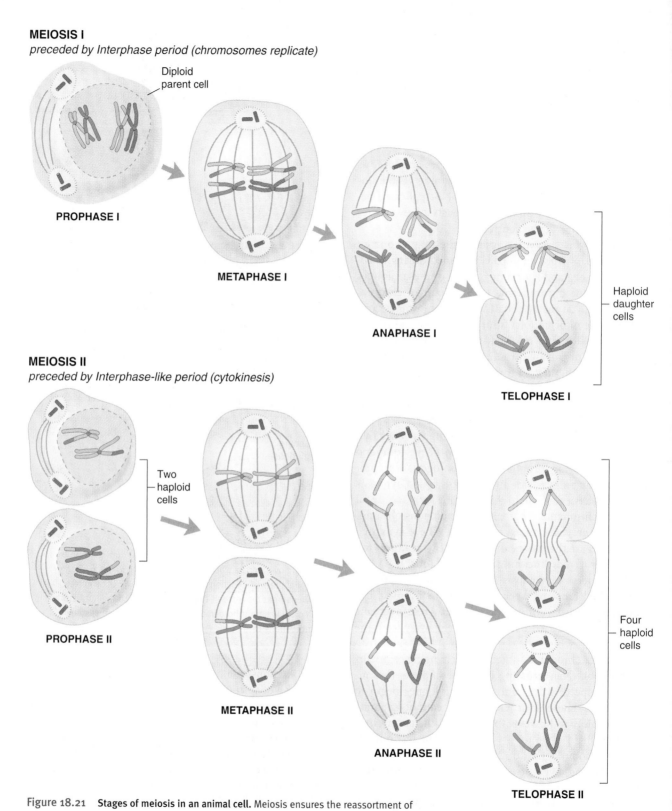

MEIOSIS I
preceded by Interphase period (chromosomes replicate)

Diploid parent cell

PROPHASE I

METAPHASE I

ANAPHASE I

TELOPHASE I

Haploid daughter cells

MEIOSIS II
preceded by Interphase-like period (cytokinesis)

Two haploid cells

PROPHASE II

METAPHASE II

ANAPHASE II

TELOPHASE II

Four haploid cells

Figure 18.21 **Stages of meiosis in an animal cell.** Meiosis ensures the reassortment of genetic material. It is preceded by interphase, during which the chromosomes are replicated. An interphaselike period of variable length often occurs between meiosis I and meiosis II. During this time there is no replication of chromosomes (DNA).

After spindle attachment to the kinetochore of each homologous pair is complete, the homologues begin to move toward opposite poles of the cell. As this movement occurs, the homologues pull apart at their crossover points. By the end of anaphase I, each pole has one member of each chromosome pair. Each chromosome still has two sister chromatids, but these chromatids are no longer perfectly identical, however, due to crossing over. During the last stage of meiosis I (telophase I), the two groups of chromosomes gather together at their respective poles, forming two chromosome clusters. A nuclear membrane forms around each group of chromosomes.

An interphaselike period of variable length often occurs between meiosis I and meiosis II. *During this time, there is no replication of DNA.* Cytokinesis occurs and the cells separate completely.

Meiosis II

At the end of anaphase I, each pole of the original cell has a haploid complement of chromosomes. That is, each pole has *half* the normal amount of chromosomes—only one member of each homologous pair. Each chromosome, however, still has two sister chromatids. Meiosis II separates these sister chromatids. Because of crossing over in the first phase of meiosis, these sister chromatids are *no longer identical* to one another.

The two haploid cells formed during meiosis I divide during meiosis II. The nuclear envelopes disappear, and the spindle fibers form. The chromosomes line up (metaphase II), their sister chromatids separate (anaphase II), and they move to the opposite poles of each cell. At this point in telophase II, the nucleoli reorganize and nuclear envelopes form around each set of chromosomes.

Meiosis II results in the production of four daughter cells, each with a haploid number of chromosomes. The cells that contain these haploid nuclei function as gametes for sexual reproduction in most animals and as spores in plants. A comparison of mitotic and meiotic cell division is shown in Figure 18.22.

The Importance of Meiotic Recombination

The reassortment of genetic material that occurs during meiosis generates variability in the hereditary material of the offspring. To understand why this is true, remember that most organisms have more than one chromosome. Humans, for example, have 23 different pairs of homologous chromosomes, one of each pair from the father and one of each pair from the mother. Each human gamete receives one of the two copies of each of the 23 different chromosomes, but which copy of a particular chromosome it receives is random. For example, the copy of chromosome 14 that a particular human gamete receives has no influence on which copy of chromosome 5 that it will receive. Each of the 23 pairs of chromosomes goes through meiosis independently of all the others, so there are 2^{23} (more than 8 million) different possibilities for the kinds of gametes that can be produced, and no two of them are alike. In addition, crossing over adds even more variability to the random assortment of chromosomes. The subsequent union of two gametes thus creates a unique individual, a new combination of 23 chromosomes that probably has never occurred before and probably will never occur again. When you study evolution in Chapters 24 and 25, you will see that variability within species is essential to the process of evolution by natural selection.

Meiosis is a process of nuclear division in which the number of chromosomes in cells is halved, forming gametes in most animals and spores in plants. Early in prophase I of meiosis, homologous chromosomes pair up in a process called synapsis. Synapsis initiates the process of crossing over. During this process, homologous chromosomes exchange segments of DNA, which produces new combinations of alleles. The first meiotic division is traditionally divided into four stages plus interphase:

1. *Prophase I.* Homologous chromosomes pair and exchange pieces of genetic material.
2. *Metaphase I.* Homologous chromosomes align on a central plane.
3. *Anaphase I.* Homologous chromosomes move toward opposite poles. Chromatids do not separate.
4. *Telophase I.* Individual chromosomes gather together at the two poles and the nuclear membranes re-form.
5. *Interphase.* The haploid cells separate completely.

During meiosis I, two haploid daughter cells are produced from one original diploid parent cell. During meiosis II, four haploid daughter cells are produced from the two haploid daughter cells of meiosis I.

MITOSIS

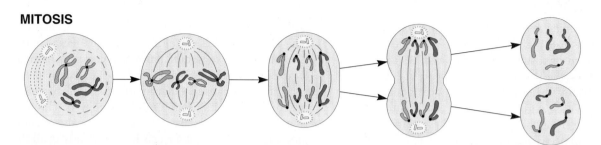

Prophase
Chromosomes condense.
The nuclear membrane disintegrates.

Metaphase
Spindle fibers attach to centromeres. Chromosomes align along metaphase plate.

Anaphase
Sister chromatids, each now called a chromosome, separate and move to opposite poles.

Telophase
Chromosomes arrive at each pole, and new nuclear membranes form. Cytoplasmic division begins.

Cell division complete. Each cell receives chromosomes that are identical to those in the original nucleus before DNA replication during interphase.

MEIOSIS I

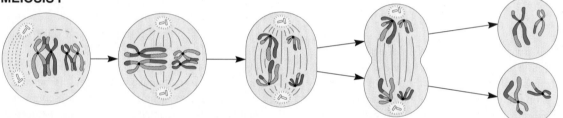

Prophase I
Homologous chromosomes condense and pair. Crossing-over occurs. The nuclear membrane disintegrates.

Metaphase I
Spindle fibers attach to centromeres. Homologous pairs align along metaphase plate.

Anaphase I
Homologous pairs of chromosomes separate and move to opposite poles.

Telophase I
One set of paired chromosomes arrives at each pole, and cytoplasmic division begins.

Each cell receives exchanged chromosomal material from homologous chromosomes.

MEIOSIS II

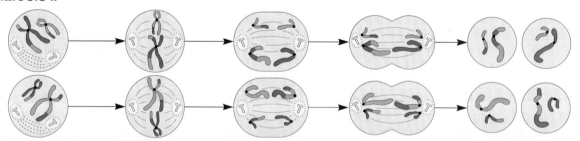

Prophase II
Chromosomes recondense.
A new spindle forms.

Metaphase II
Spindle fibers attach to centromeres. Chromosomes align along spindle.

Anaphase II
Sister chromatids, each now called a chromosome, separate and move to opposite poles.

Telophase II
Chromosomes arrive at each pole, and cytoplasmic division begins.

Cell division complete. Each cell receives half the original number of chromosomes in the original nucleus before DNA replication during interphase.

Figure 18.22 **A comparison between mitosis and meiosis.** Meiosis involves two nuclear divisions with no DNA replication between them. Meiosis therefore produces four daughter cells, each with half the original amount of DNA. Mitosis produces two identical daughter cells, each with the same number of chromosomes as the original mother cell.

Key Concepts

DNA is the hereditary material.

18.1 Proteins and DNA make up the chromatin of the cell.

18.1 When coiled and condensed, chromatin forms chromosomes.

18.2 The hereditary material is located in the nucleus of a cell.

18.3 There are two types of nucleic acids located within cells: DNA and RNA.

18.3 Both DNA and RNA are made up of nucleotides, which each contain a sugar, a phosphate group, and a base.

18.4 DNA is a double-stranded, helical molecular "ladder," with rungs of uniform length and alternating sugar-phosphate units forming the ladder uprights.

18.5 After DNA unwinds, splitting occurs at the bonds between its bases; free nucleotides then bond to the exposed bases, producing two new DNA strands.

Units of DNA called genes direct the synthesis of polypeptides.

18.6 The units of heredity are genes, lengths of DNA that each code for a polypeptide.

18.7 mRNA is made in the nucleus using DNA as a template; it then travels to the ribosomes.

18.7 Polypeptide synthesis takes place at the ribosomes using messenger RNA (mRNA).

18.7 Triplet base sequences of mRNA (codons) direct the synthesis of an amino acid chain (polypeptide) with the help of transfer RNA (tRNA).

18.8 Regulatory proteins control the expression of genes by selectively inhibiting the transcription of some and facilitating the transcription of others.

Hereditary material is passed on to new cells by mitosis or meiosis.

18.9 Mitosis, cell division that occurs as a part of growth and reproduction, produces two identical cells from an original parent cell.

18.9 Meiosis, cell division that occurs as a part of reproduction, produces four cells from one parent cell, each with half the full complement of genetic material.

18.10 Interphase is the portion of the cell cycle in which the cell grows and carries out normal life functions.

18.11 Mitosis is a continuous sequence of events that occurs just after interphase and that results in the division of duplicated chromosomes.

18.12 Cytokinesis is the physical division of the cytoplasm of a eukaryotic cell into two daughter cells.

18.13 Meiosis is a process of nuclear division in which the number of chromosomes in cells is halved, forming gametes in most animals and spores in plants.

18.13 The reassortment of DNA segments that occurs during meiosis generates variability in the hereditary material of offspring.

18.13 Variability within species is essential to the process of evolution by natural selection.

Key Terms

anaphase (ANN-uh-faze) *401*
anticodon *394*
asexual reproduction *398*
cell cycle *398*
chiasmata (keye-AZ-muh-tuh) *403*
chromatin (KRO-muh-tin) *382*
codons *394*
crossing over *403*
cytokinesis
  (sye-toe-kuh-NEE-sis) *401*
deoxyribonucleic acid (DNA) (de-OK-
  see-RYE-boh-noo-KLAY-ick) *385*
diploid *397*
fertilization *397*
gametes (GAM-eets) *397*
gametophyte
  (geh-MEE-toe-fite) *397*

gene *391*
haploid (HAP-loyd) *397*
homologous chromosomes (homologues) (hoe-MOL-uh-gus KRO-
  muh-somes/HOME-uh-logs) *403*
interphase (IN-tur-faze) *398*
life cycle *397*
meiosis I and meiosis II
  (my-OH-sis) *402*
meiosis (my-OH-sis) *397*
messenger RNA (mRNA) *392*
metaphase (MET-uh-faze) *400*
mitosis (my-TOE-sis
  or mih-TOE-sis) *397*
nucleic acid (noo-KLAY-ick) *384*
nucleotide (NOO-klee-o-tide) *385*
one-gene–one-enzyme theory *391*

prophase (PRO-faze) *400*
purines (PYOOR-eens) *385*
pyrimidines (pih-RIM-uh-deens
  or pye-RIM-uh-deens) *385*
regulatory sites *396*
ribonucleic acid (RNA)
  (RYE-boh-noo-KLAY-ick) *385*
ribosomes (RYE-buh-somes) *392*
ribosomal RNA (rRNA)
  (rye-buh-SO-mull) *392*
sexual reproduction *397*
sporophyte (SPOR-ih-fite) *397*
synapsis (sih-NAP-sis) *403*
telophase (TEL-uh-faze) *401*
transcription (trans-KRIP-shun) *392*
transfer RNA (tRNA) *392*
translation (trans-LAY-shun) *393*

KNOWLEDGE AND COMPREHENSION QUESTIONS

1. Explain how a unique characteristic of yours, such as your hair color, is a result of specific chemical instructions. Where do these instructions originate? (*Comprehension*)

2. What are the two types of nucleic acids found in your cells? Describe the structures of each. (*Knowledge*)

3. Explain the term *double helix*, and describe its structure. (*Knowledge*)

4. What characteristics of DNA did Chargaff's experiments reveal? Why was this significant? (*Knowledge*)

5. Distinguish among rRNA, tRNA, and mRNA. What does each abbreviation stand for, and what are the respective functions of each? (*Knowledge*)

6. Is mitosis or meiosis responsible for the tremendous genotypic and phenotypic variation among humans? Explain. (*Comprehension*)

7. Place these steps in the correct sequence and label each stage. What process is being described? (*Comprehension*)
   a. The chromosomes line up in the center of the cell.
   b. The nuclear envelope forms and the chromosomes uncoil.
   c. The chromosomes grow shorter and thicker, and spindle fibers form.
   d. The sister chromatids separate and move to opposite poles of the cell.

8. Place these steps in the correct sequence and label each stage. What process is being described? (*Comprehension*)
   a. Homologous chromosomes move toward opposite poles of the cell; chromatids do not separate.
   b. Chromosomes gather together at the two poles of the cell, and the nuclear membranes re-form.
   c. Homologous chromosomes pair and exchange segments.
   d. Homologous chromosomes align on a central plane.
   e. The haploid cells separate completely.

9. Compare the cells that result from mitosis with those that result from meiosis. How are they different? (*Comprehension*)

10. Explain the importance of the genetic recombination that occurs during meiosis. (*Comprehension*)

KEY
Knowledge: Recalling information.
Comprehension: Showing understanding of recalled information.

CRITICAL THINKING REVIEW

1. What do you hypothesize might occur if there is an error made in base arrangement during DNA transcription? Suggest a possible sequence of subsequent events. (*Application*)

2. There is often a tremendous amount of repetition in the genetic code for enzyme production. Why do you suppose such a backup system is necessary to an organism? (*Application*)

3. Using Table 18.3, state the order of amino acids when the sequence of bases in DNA is the following: GATTACAGATTACACCTAGCTATC. Describe how you arrived at your answer. (*Analysis*)

4. If a mutation occurred in the above DNA sequence in which one molecule of guanine was deleted, which would cause the greatest change in the polypeptide encoded by this length of DNA—the first guanine molecule or the last in the sequence? Why? (*Analysis*)

5. The meiotic division of an egg cell produced the following daughter cells: 2 haploid cells, one cell with no genetic material, and one haploid cell with still-attached sister chromatids. What could have gone wrong in the meiotic process to produce these four cells? (*Analysis*)

KEY
Application: Using information in a new situation.
Analysis: Breaking down information into component parts.

The generalized life cycles for plants and animals are shown below. Briefly describe the main differences between the life cycle of a plant and of an animal. In animals, what types of cells are the result of mitosis? In plants? If the results are different, are the processes the same? Explain your answer.

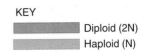

KEY

Diploid (2N)
Haploid (N)

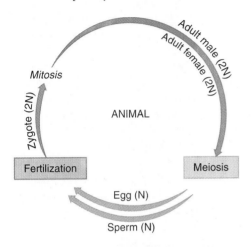

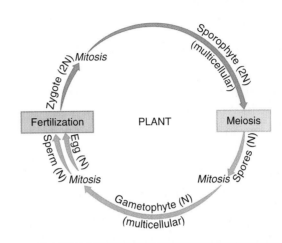

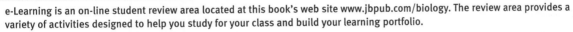

e·learning

www.jbpub.com/biology

Location: http://www.jbpub.com/biology

e-Learning is an on-line student review area located at this book's web site www.jbpub.com/biology. The review area provides a variety of activities designed to help you study for your class and build your learning portfolio.

Review Questions The review questions test your knowledge of the important concepts and applications in each chapter. The review provides feedback for each correct or incorrect answer. This is an excellent test preparation tool.

Figure-Labeling Exercises Sharpen your visual thinking by matching terms and labels for important illustrations.

Flash Cards Studying biology requires learning new terms. Virtual flash cards help you master the new vocabulary for each chapter.

Just Wondering As you read and study from this text, you may find that you have unanswered questions. Through this site you can ask the author, Sandy Alters, your "just wondering" questions.

Do you prefer the speed and reliability of a CD-ROM? All of the features contained on the eLearning portion of the web site are also available on the Student CD-ROM.

Biotechnology and Genetic Engineering

The business of manufacturing has taken on a new look. It is a look far different from the stereotyped notion of the assembly line in a noisy factory. This laboratory technician is putting bacteria to work to create products for human use. The bacteria are growing in the small, table-top fermentors lining the laboratory bench. Working in an environmentally controlled room, the technician is engaged in a scale-up experiment—figuring out how to make a procedure work on a large scale that has been perfected on a small scale. These fermentors are only an intermediate step. Eventually, the bacteria used in this process will be grown in huge vats (photos of which appear later in this chapter).

To turn bacteria into living "factories," scientists isolate individual genes (portions of the hereditary material) and transfer them from one kind of organism, a human, for example, to another such as a bacterium. The bacteria are then propagated in vast quantities, producing the substance they are now genetically programmed to manufacture.

This newly found ability to isolate individual genes and transfer them from one kind of organism to another has revolutionized scientists' ability, not only to create new products for therapeutic use, but also to improve the characteristics of plants and animals. Gene manipulation offers enormous potential for use in agriculture and medicine of the future, and it has already produced many important applications. Genetic engineering is also an important tool that helps scientists learn about gene structure, function, and regulation. Thus, these powerful techniques are used not only to produce new products but also to gain a better understanding of the molecular basis of life.

CHAPTER

19

Organization

The use of biotechnology began about 10,000 years ago.

Genes can be transferred from organism to organism.

Genetic engineering has a variety of applications.

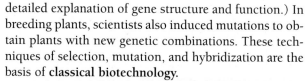

19.1 Classical biotechnology

Biotechnology is the use of scientific and engineering principles to manipulate organisms, producing one or more of the following:

1. Organisms with specific biochemical, morphological, and/or growth characteristics
2. Organisms that produce useful products
3. Information about an organism or tissue that would otherwise not be known

Although biotechnology is a relatively new word, classical biotechnology is not new. Molecular biotechnology is the "new" biotechnology.

As far back as 10,000 years ago, humans selected plants and animals with specific characteristics to propagate from the wild. In a simple way, this process is considered biotechnological: Humans were producing organisms with specific characteristics by selection. For more than 8000 years, bacteria and yeasts have been used to produce products such as beer, vinegar, yogurt, and cheese, although the processes involved were not understood at the time. Loaves of yeast breads have even been found in Egyptian pyramids built 6000 years ago. And more than 3000 years ago, the Chinese and the Central American Indians used products produced by molds and fungi to treat infections. Today we know that certain molds and bacteria are natural sources of bacteria-fighting antibiotics.

Even before the late 1800s, when Gregor Mendel and others gathered evidence regarding the nature of the variability among organisms, plant and animal breeders selectively bred organisms to develop hybrids having certain desired characteristics. (*Hybrids* are the offspring produced by crossing two genetically dissimilar varieties of a species.) For example, the cauliflower, broccoli, and cabbage shown here have all been bred from one species of wild mustard. By selecting certain characteristics of each plant they wished to cultivate, breeders were able to develop plants that are quite different from one another.

At the turn of the century, scientists began developing hybrids using scientific principles brought to light by Mendel: the selection and recombination of hereditary factors now known as genes. (See Chapter 18 for a detailed explanation of gene structure and function.) In breeding plants, scientists also induced mutations to obtain plants with new genetic combinations. These techniques of selection, mutation, and hybridization are the basis of **classical biotechnology.**

Using carefully planned and controlled breeding programs, scientists found that they could produce hybrids that were stronger or improved in certain ways from their parents, a feature termed *hybrid vigor.* During the 1930s, a worldwide effort was initiated to increase food production in developing countries. By developing hybrid varieties of food crops using the techniques of selection and recombination, agricultural researchers began a *green revolution,* dramatically increasing the yields of food crops. For example, hybrid corn developed in the United States at that time helped double crop yields. New dwarf varieties of wheat introduced to farmers in Mexico in 1960 resulted in a 300% increase in wheat yields there.

At about the same time that Mendel was conducting genetic research with pea plants in the mid-1800s, strides were also being made in classical biotechnology in the field of microbiology. For example, processes of fermentation that had been used without being fully understood for thousands of years were slowly explained during the 1800s. In 1837–1838, researchers concluded that yeasts are alive. Almost 30 years later, Louis Pasteur confirmed that certain bacteria and yeasts formed molecules such as acetic acid, lactic acid, butyric acid, alcohol, and carbon dioxide. He determined why and how wine often turned to vinegar, and he developed a process to kill the microorganisms causing this fermentative change. This process, termed *pasteurization,* is still used to help preserve wine and other foods such as milk.

Knowledge of fermentative processes was also put to work during World War I. Because of blockades during the war, the Germans were unable to obtain the vegetable oils from which they extracted the glycerol necessary for manufacturing explosives. As a result, the Germans devised a biological way to produce glycerol using the fermentative abilities of yeast. The British, likewise, developed biological methods to produce acetone and butanol using the bacterium *Clostridium acetobutylicum.* (Acetone was used in manufacturing ammunition and butanol in producing artificial rubber.)

As strides were being made in the use of microorganisms to synthesize products for humans and in the understanding of those processes, the use of classical biotechnology to fight disease progressed rapidly. In the late 1800s, Robert Koch developed the germ theory of disease, which explains that microbes cause infection. He discovered both the bacterium that causes anthrax, a disease of cattle that sometimes affects humans, and the bacterium that causes tuberculosis. During this time, Koch also developed *pure culture* methods. Using these methods, scientists could isolate and work with a specific organism, free from contamination by other organisms.

Microorganisms were also used to produce various disease-fighting products such as vaccines. Although Edward Jenner did not understand the role of microorganisms in disease, he is credited with developing the first successful vaccine against smallpox in 1796. Without understanding the microbiology and immunology behind his work, Jenner injected volunteers with the relatively harmless *Vaccinia* virus, which conferred resistance against the deadly smallpox virus. Almost 100 years later, having a greater understanding than Jenner had of the disease process, Louis Pasteur and others developed the first cholera, diphtheria, and tetanus vaccines. By the early 1900s, mammalian cell culture techniques had been developed. Scientists were then able to replicate viruses within these cultures (Figure 19.1), harvest and attenuate (weaken) them, and use them in vaccine preparations. Thus, microbes were themselves being used to fight human disease.

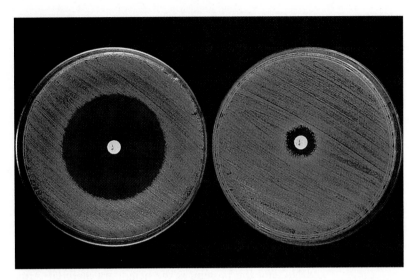

Figure 19.2 **Penicillin resistance** *(right)* **and sensitivity** *(left)* **in culture of bacteria** *Staphylococcus aureus.* Bacteria were spread on the nutrient surface of each of these two petri plates. A disk containing the antibiotic penicillin was then placed in the center of each plate. The plates were incubated, and the bacteria grew. The bacteria in the plate on the right carry a gene that confers resistance to penicillin. Therefore, these bacteria grew quite close to the disk. However, the bacteria growing in the plate to the left do not carry this resistance gene. The antibiotic inhibited the bacteria from growing in a large area surrounding the disk, which is called a zone of inhibition.

Figure 19.1 **Growing viruses in eggs.** The shells have been broken off the tops of the raw eggs, and laboratory technicians are inoculating the eggs with viruses, which will be used to develop influenza vaccine.

An important step in the development of disease-fighting products occurred quite by accident in 1927, when Alexander Fleming discovered antibiotics. He noticed that a mold, which had contaminated his bacterial cultures, inhibited the growth of surrounding bacteria (Figure 19.2). That mold was *Penicillium notatum,* which naturally produces the antibacterial product we now call penicillin. By 1940, the first purified preparations of penicillin were available. This antibiotic played a major role in preventing the death of soldiers from infection during World War II. An intensive search for new antibiotics began as pharmaceutical companies began screening bacteria and molds for their antibiotic potential.

For thousands of years, even before they had any understanding of genetics, humans have selected plants and animals with specific characteristics to propagate from the wild, and they have selectively bred organisms to develop individuals having certain desired characteristics. In addition, without knowledge of microbial biochemistry, humans have developed various products by using microbes.

19.2　The birth of molecular biotechnology

During the 1950s, a revolution took place that set the stage for the birth of molecular biotechnology: the discovery of the molecular structure of the hereditary material deoxyribonucleic acid (DNA). This revolution involved more than 50 years of work by a succession of scientists building on the work of Gregor Mendel (see Table 18.1). Mendel's work suggested that traits are inherited as discrete packets of information. As the years progressed, experimental data showed that this information is stored within the nucleus of the cell in the chromosomes (see "Just Wondering," Chapter 4). Scientists were then able to determine that the chromosomes are composed of DNA and protein. In 1952, Alfred Hershey and Martha Chase established that the DNA, not the protein, *is* the hereditary material. After Rosalind Franklin and Maurice Wilkins demonstrated that DNA molecules are helically shaped, James Watson and Francis Crick deduced the structure of the DNA molecule, for which they were awarded the Nobel Prize. (These scientific discoveries are described in more detail in Chapter 18.)

Locating and characterizing the hereditary material were the first steps in what might be termed a *gene revolution*. Now, genes and genetic diversity could be studied at the molecular level. Scientists then began to manipulate genes in bacteria, intervening in and directing their natural methods of genetic recombination.

> The advent of molecular biotechnology, which involves the manipulation of the genes themselves, began with the discovery of the structure of the genetic material DNA in the 1950s.

Model of DNA.

19.3　Natural gene transfer among bacteria

Certain bacteria can naturally transfer genetic material from one cell to another. Genetic material is exchanged between bacteria by three methods: transformation, transduction, and conjugation.

In **transformation**, pieces of DNA from a donor bacterial cell (**Figure 19.3** ❶) that has lysed (ruptured) is released into the surrounding medium ❷. Under certain conditions, some bacterial cells can become recipient cells and take up DNA from their surroundings and incorporate it into their chromosome ❸, thereby becoming transformed, or changed. (The **genome** (GEE-nome) of an organism is its total complement of genetic material.) Not all bacteria can be transformed; the ability to take up DNA fragments is an inherited characteristic. Genetic engineers, however, have found a way to induce certain bacteria to take up DNA by cultivating them in the presence of specific chemicals.

During **transduction**, DNA from a donor bacterium is transferred to a recipient bacterium by a virus. For this transfer to occur, the virus must first inject its DNA ❹ and combine its genetic material with that of the bacterium it has infected ❺. The virus particles containing host bacterial DNA burst from the host cell and infect a new cell ❻. The DNA from another bacterium now becomes incorporated with the DNA of the new bacterial cell ❼. Only certain lysogenic viruses (viruses that infect a cell but do not immediately replicate) or damaged viruses are capable of incorporating bacterial genes with their genome. Additionally, not all bacteria are capable of being infected by a virus carrying bacterial genes. However, this process does occur in certain viral-bacterial interactions and results in genes from one bacterium being transferred to another bacterium.

During **conjugation** (con-juh-GAY-shun), a donor and a recipient bacterium make contact ❽, and the DNA from the donor is transferred to the recipient cell ❾. This transfer takes place in bacteria having extra chromosomal pieces of DNA called **plasmids** (PLAZ-mids). Plasmids replicate independently of the main chromosome and make up only about 0.04% of the DNA in many bacteria. **Figure 19.4** shows how small a plasmid is relative to the size of the chromosomal DNA.

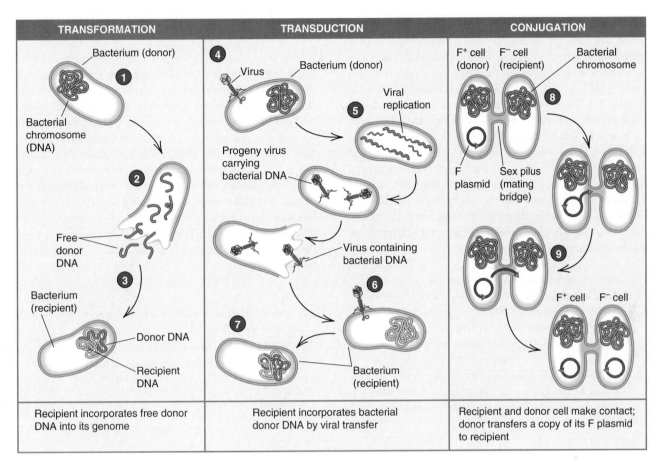

TRANSFORMATION	TRANSDUCTION	CONJUGATION

TRANSFORMATION

Bacterium (donor) ❶

Bacterial chromosome (DNA)

❷

Free donor DNA

❸

Bacterium (recipient)

Donor DNA

Recipient DNA

Recipient incorporates free donor DNA into its genome

TRANSDUCTION

❹ Virus — Bacterium (donor)

❺ Viral replication

Progeny virus carrying bacterial DNA

Virus containing bacterial DNA

❻

❼

Bacterium (recipient)

Recipient incorporates bacterial donor DNA by viral transfer

CONJUGATION

F⁺ cell (donor) — F⁻ cell (recipient) — Bacterial chromosome

❽

F plasmid — Sex pilus (mating bridge)

❾

F⁺ cell — F⁻ cell

Recipient and donor cell make contact; donor transfers a copy of its F plasmid to recipient

Figure 19.3 **Natural gene transfer in bacteria: Transformation, transduction, and conjugation.**

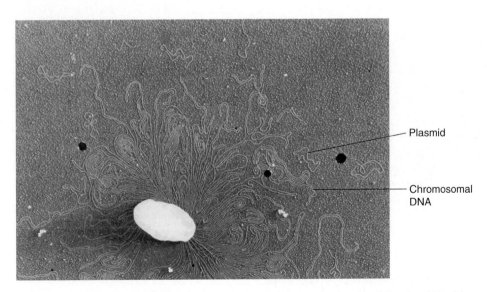

Plasmid

Chromosomal DNA

Figure 19.4 **Bacterium with DNA extruded and plasmid visible.** Notice how small the plasmid is with respect to the total chromosomal DNA.

Some plasmids have several special genes that promote the transfer of the plasmid to other cells. These genes are referred to as a fertility factor, and such plasmids are called *F plasmids*. Fertility genes code for proteins that form a tubelike mating bridge called an *F pilus* (or *sex pilus*) on the surface of the bacterial cell (see Figure 19.3 and **Figure 19.5**). When the F pilus of one cell (F⁺) makes contact with the surface of another cell that lacks F pili (F⁻), which means it does not contain the F plasmid, the replication and then transfer of the plasmid occurs (see Figure 19.3). If the F plasmid has been incorporated into the larger bacterial chromosome, as sometimes happens, the cell may copy and transfer the entire newly formed bacterial chromosome to the recipient cell.

Certain bacteria can transfer genetic material from one cell to another. In one method called transformation, free pieces of DNA move from a donor cell to a recipient cell. During transduction, a second method of natural gene transfer, pieces of DNA from a donor cell are transferred to a recipient cell by a virus. During conjugation, a third method, a donor and a recipient cell make contact, and DNA from the donor is transferred to the recipient cell.

19.4 Human-engineered gene transfer using eukaryotes

Although scientists were able to manipulate the natural mechanisms of bacterial (prokaryotic) gene transfer in the 1950s and 1960s, it was not until the mid-1970s that scientists were able to combine DNA from different species of organisms and manipulate genes in eukaryotes. These techniques of molecular biology that involve the manipulation of genes themselves, and not just the organism, are called **genetic engineering**, or **recombinant DNA technology**. In the 1980s, molecular biotechnology became a major area of growth in business. The first commercial application of recombinant DNA technology was the production of proteins for therapeutic use.

The first techniques of molecular biotechnology, the artificial manipulation of the genes of organisms, were developed in the 1970s. These techniques differ from the techniques of classical biotechnology in that the genes, not just the organisms, are used and manipulated.

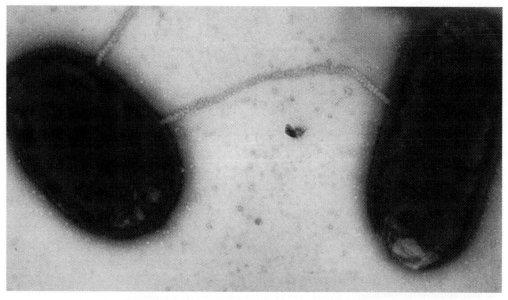

Figure 19.5   Genetic recombination between bacteria. In this colorized electron micrograph, the mating *E. coli* bacteria are seen as purple, and the F pilus (sex pilus) joining them is green.

Genetic engineering has a variety of applications.

19.5 The production of proteins for therapeutic use

The genetic engineer uses a variety of complex technological methods and laboratory equipment to study and manipulate genes. As scientists learn more about the process of genetic engineering, they develop new methods and equipment to advance their studies. The field of DNA technology is constantly expanding and changing. The following explains some of the basic methods employed today in the context of current applications.

One of the first applications of recombinant DNA technology was the insertion of human genes into bacteria and the manipulation of these organisms to produce human proteins for therapeutic use. The first two genetically engineered proteins were developed in 1979 and became available to the public after the completion of clinical trials and with the approval of the Food and Drug Administration a few years later. These proteins were human insulin, used in the treatment of diabetes mellitus, and human growth hormone, given to children who do not produce enough of this hormone for normal growth. Genetically engineered bacteria grown in large numbers, such as shown in **Figure 19.6,** produce substantial quantities of these hormones, which can be purified and used to treat patients.

Finding the Gene of Interest

The first step in producing human proteins using the recombinant DNA technology may be the hardest of all: the identification and characterization of the desired gene. In other words, before scientists can insert a gene from one organism into another, they must find the gene (such as the human insulin gene) among all the genes in the genome. There are an estimated 100,000 genes in the human genome, so finding one gene among them is not an easy task. In addition to finding a specific gene, genetic engineers need copies of the gene to insert into microbes. These microbes are usually bacteria or yeast, and they will produce the protein desired.

Scientists use various methods to find genes of interest. The methods a researcher uses in any particular situation depend on the gene itself, the amount of information known about the gene, and the research goals. One method of finding a gene is **shotgun cloning.** This is a somewhat long and complex process, and it is used when researchers know very little about the gene they are trying to find. In brief, shotgun cloning

Figure 19.6 **Large fermenters are used for the commercial production of genetically engineered bacterial products.** Large tanks such as these hold hundreds (and some hold thousands) of gallons of bacterial culture.

involves cutting the DNA of the entire genome into pieces, isolating and purifying the genomic DNA, and inserting these pieces or fragments into bacteria or yeast. These transformed organisms then reproduce and make copies, or **clones,** of the DNA fragments. The term shotgun means that no one gene is targeted for cloning—all the genes are cloned. The result is a complete **gene library**: a collection of clones of DNA fragments, which together represent the entire genome of an organism. The library is then screened to find the desired gene. Each step of this process is described in more detail in later sections.

Cutting DNA into Gene Fragments

The first step in cutting the genomic DNA into pieces is to extract the DNA from human cells that have been grown in tissue culture. The freed DNA is then clipped into pieces at the molecular level using restriction enzymes. **Restriction enzymes** act like chemical scissors, cutting long, intact DNA strands into smaller pieces of various sizes, which are more easily studied than the intact DNA. Restriction enzymes recognize certain nucleotide (base) sequences in DNA molecules and break the bonds between the nucleotides. The base sequence at which a particular restriction enzyme makes a cut is called a **restriction site**. As shown in Figure 19.7, some enzymes produce "sticky ends" when the DNA is cut. In nature, restriction enzymes are produced by bacteria to protect themselves from bacteriophage (viral) invasion. They protect the bacteria by cutting viral DNA into pieces before the virus can infect the cell. Discovered in the late 1960s, these naturally occurring bacterial enzymes have now become one of the many research tools of the genetic engineer.

Hundreds of restriction enzymes exist and are now available for use by scientists. Different restriction enzymes identify different restriction sites, allowing scientists to cleave the DNA in a variety of ways. Carefully using these enzymes, scientists can produce a range of sizes of fragmented DNA, called **restriction fragments**.

Cloning Genes

The next step in shotgun cloning is to insert the restriction fragments, each of which may contain one or more genes, into bacteria or yeast cells using either plasmids or viruses. Plasmids can be extracted from bacteria or yeasts, induced to incorporate restriction fragments into their genomes, and reinserted into the cell. Scientists can also incorporate restriction fragments into viral genomes and use the viruses to infect bacteria or yeasts. Plasmids and viruses used to insert restriction fragments of foreign DNA into cells are called **cloning vectors**.

To incorporate restriction fragments into cloning vectors (such as plasmids, for example), the genetic engineer must first cut the DNA of both fragments using the same restriction enzyme. This results in cut pieces of DNA that will bond with one another for the following reasons. Nucleotides of double-stranded DNA are complementary to one another and are "read" in opposite directions. Therefore, a restriction enzyme recognizes a base sequence (such as GGATCC) in different places on each strand. It then cuts the strand at a specific point in the sequence. In Figure 19.7, you can see the result: The cut piece of DNA has ends in which one strand is longer than the other (in most cases). These ends are called *sticky ends* because their bases are complementary. They will bond with another piece of DNA cut with the same restriction enzyme (Figure 19.8).

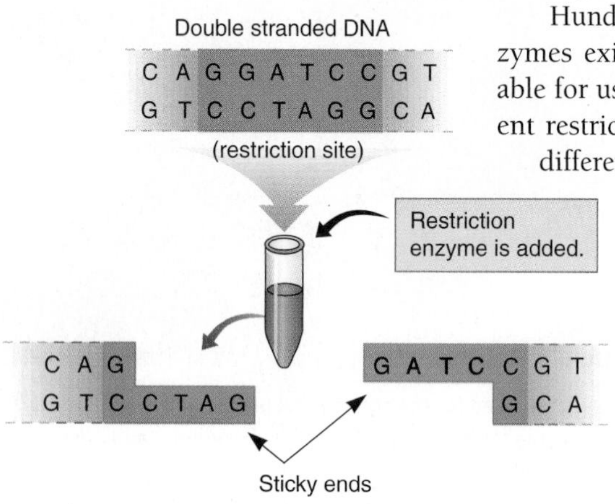

Figure 19.7 **Restriction enzymes act like chemical scissors.** Restriction enzymes recognize short sequences of nucleotides in DNA and cut them at a certain point within the sequence. The restriction enzyme in this illustration recognizes the base (nucleotide) sequence GGATCC and cuts the sequence between the two guanine (G) bases. Because each strand of DNA in double-stranded DNA is "read" in a direction opposite to the other, the cut piece of DNA has ends in which one strand is longer than the other. These ends are called sticky ends.

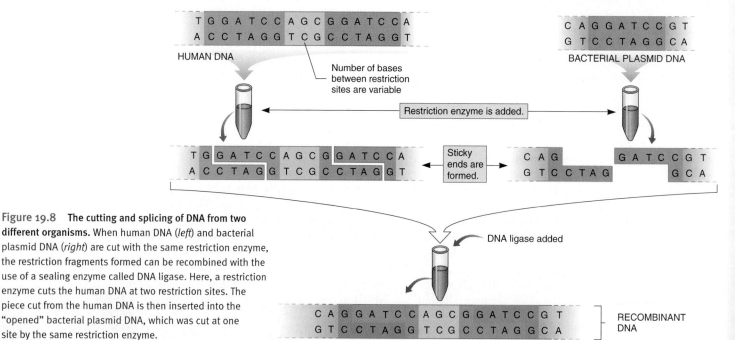

Figure 19.8 **The cutting and splicing of DNA from two different organisms.** When human DNA (*left*) and bacterial plasmid DNA (*right*) are cut with the same restriction enzyme, the restriction fragments formed can be recombined with the use of a sealing enzyme called DNA ligase. Here, a restriction enzyme cuts the human DNA at two restriction sites. The piece cut from the human DNA is then inserted into the "opened" bacterial plasmid DNA, which was cut at one site by the same restriction enzyme.

Just Wondering . . . *Real Questions Students Ask*

What is electrophoresis and how does it aid in prosecuting or defending a criminal?

If you were a genetic engineer, much of your laboratory work would involve separating, analyzing, and purifying proteins and other macromolecules such as DNA restriction fragments. One standard laboratory technique widely used by molecular biologists is *gel electrophoresis*—a method of separating charged molecules by drawing them through a filtering gel material using an electrical field. Although widely used today, this technique was developed more than 100 years ago, but has been refined and adapted to help answer various questions in new contexts over time. It is a technology that has never lost its usefulness. In fact, like the microscope, the applications of gel electrophoresis have increased as time has passed. Today, gel electrophoresis helps accomplish such procedures as analyzing DNA fragments from crime scenes and comparing the DNA of organisms to determine their evolutionary relatedness, tasks never thought of in the nineteenth century.

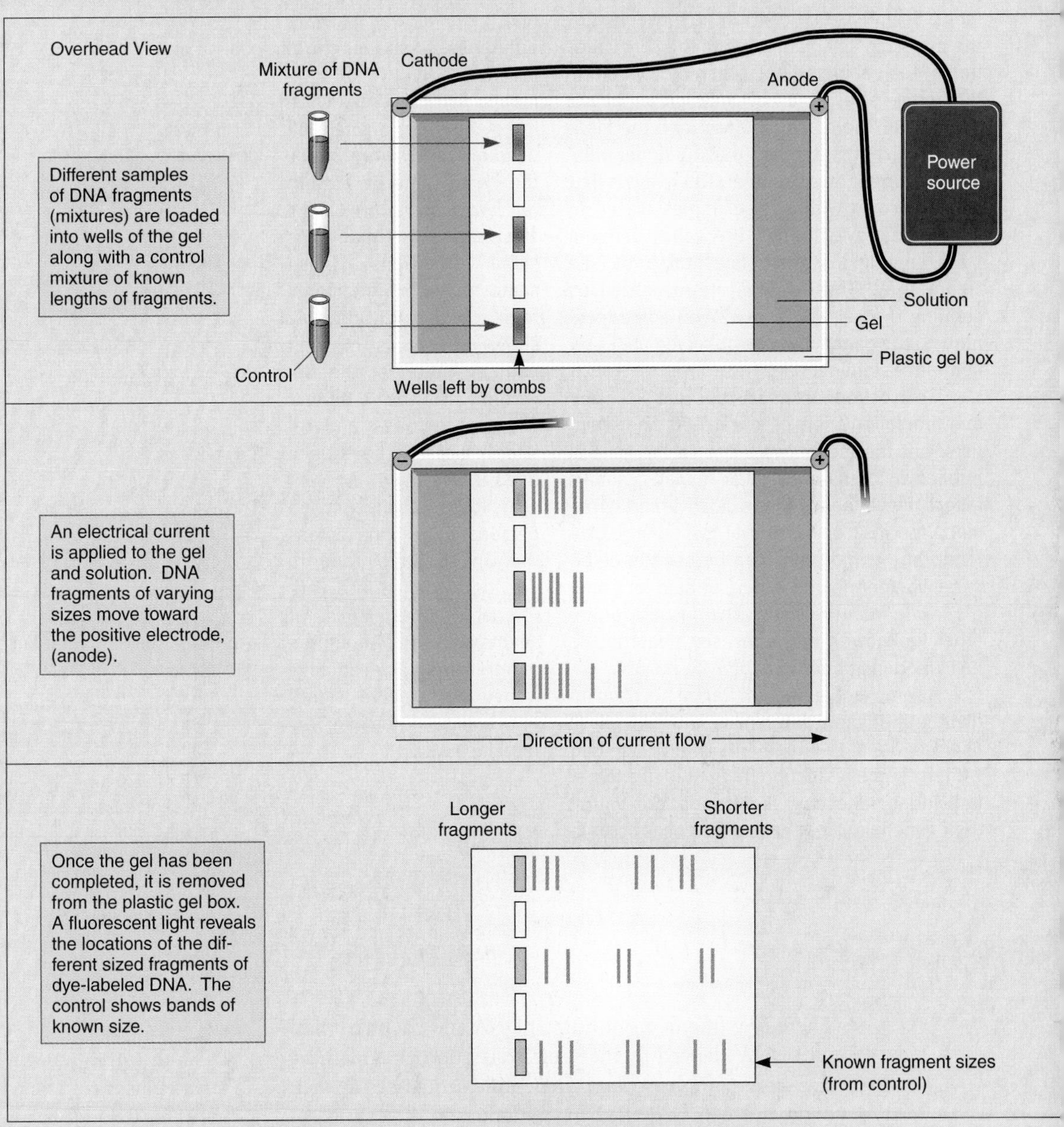

The word electrophoresis refers to a phenomenon in which molecules with net charges migrate, or move, in an electric field. Electrodes are placed at either end of a thin slab of a firm, gellike material to create an electric field. The gel is formed by pouring the hot liquid gel medium into a glass or plastic container. Notches called wells are formed with "combs" at one end of the gel as it hardens. The samples to be electrophoresed are pipetted into the wells once the gel has cooled, the combs have been removed, and a solution (through which the current will run) has been poured over the gel (Figure 19.A). (A pipette is a mechanical measuring device that looks somewhat like a pointed glass straw.) A negatively charged electrode is placed at the end of the gel containing the wells, and a positively charged electrode is placed at the opposite

Overhead View

Mixture of DNA fragments

Cathode

Anode

Power source

Different samples of DNA fragments (mixtures) are loaded into wells of the gel along with a control mixture of known lengths of fragments.

Solution

Gel

Plastic gel box

Control

Wells left by combs

An electrical current is applied to the gel and solution. DNA fragments of varying sizes move toward the positive electrode, (anode).

Direction of current flow

Longer fragments

Shorter fragments

Once the gel has been completed, it is removed from the plastic gel box. A fluorescent light reveals the locations of the different sized fragments of dye-labeled DNA. The control shows bands of known size.

Known fragment sizes (from control)

Figure 19.A

end. When an electric current is applied, the molecules move from the negative end to the positive end of the gel but at different rates depending on their net charges and their sizes.

Macromolecules often have portions of their structures that are positively or negatively charged. These charges do not usually balance each other exactly; the remaining charge is the net charge. The more negatively charged a particle is, the more quickly it will move toward the positive pole and vice versa. Another factor that affects the movement of molecules through the gel is the size and shape of the molecules. The gel acts like a strainer, having porelike spaces through which the molecules move. Molecules that are small move through the pores easily and quickly, whereas larger molecules move more slowly as they "squeeze" through the pores (Figure 19.B on next page). In other words, the distance the band of DNA moves is

(continued next page)

(continued from previous page)

proportional to the length of the DNA fragment. Scientists can control the movement of the molecules by manipulating the concentration of the gel ingredients, thereby controlling the pore size, and by manipulating the strength of The electric current and the manner in which it is applied (pulsing or not pulsing the current, for example). The result is that the molecules within the samples placed in each well move through the gel at differing rates, creating a series of lines, or bands (see Figure 19.A). These bands are visualized by staining the gel with a dye. With some techniques, the bands must be observed under UV light after staining.

Gel electrophoresis has become an important tool for a variety of researchers and scientists, but the most visible use to the general public is in forensics. Because human genomes reflect the variability that exists within the human species, any material containing DNA found at a crime scene can be used to determine whether blood, semen, or hair, for example, came from the victim, the suspect, or another (unknown) individual (see photo in the Just Wondering box in Chapter 18, p. 389).

A sample of restriction fragments of an individual will show differences in the length of one or more of the fragments when compared with the same restriction fragments of another individual, reflecting slight variations within their genetic codes. These differences occur in various locations in the human genome and result in differences in the length of a variety of restriction fragments. Particular restriction fragments from a single individual will always produce the same banding pattern (or *DNA fingerprint*) on gel electrophoresis under the same conditions. These same restriction fragments from another individual will show differences in their banding patterns; the fingerprint will be different. These techniques and related processes have been used increasingly in recent years to free falsely accused persons in prisons across the United States and to provide evidence in a variety of murder trials. It seems amazing that a nineteenth-century technology is helping to solve twentieth-century crimes in a very high-tech way.

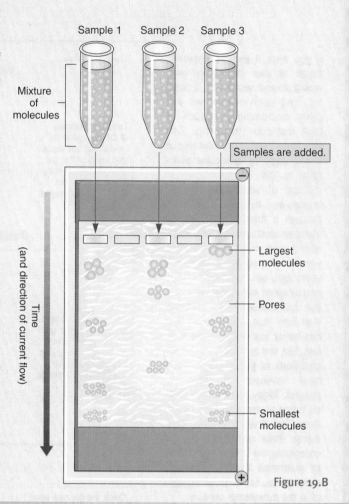

Figure 19.B

Therefore, fragments of human DNA and bacterial plasmid DNA cleaved by the same restriction enzyme have the same complementary nucleotide sequences at their ends and can be joined to one another. A sealing enzyme called a **ligase** (LYE-gase) helps re-form the bonds.

The next step is to insert these "hybrid" plasmids into bacteria (or yeasts) by transformation. When these plasmids are mixed with specially prepared bacteria under the proper conditions needed for transformation, the bacteria will take up the plasmids from the mixture. This procedure of genetically engineering bacteria is shown in Figure 19.9.

Inserting fragments of foreign DNA into bacterial or yeast cells using plasmids is a technique that researchers employ frequently. Occasionally, certain viruses are used to insert DNA fragments into bacteria, plant cells, or animal cells by introducing the DNA into the genome of a virus instead of into a bacterial plasmid. The virus then carries the DNA fragment into a cell that it is capable of infecting and begins replicating.

Once the restriction fragments have been inserted into the bacteria or yeasts, the genetic engineer then cultures the organisms in growth media. This results in large populations of the organisms that contain the inserted DNA. Both bacteria and yeasts reproduce asexually—one cell splitting into two, two into four, and so forth. If these cells are then grown on a semisolid culture medium in a small, flat dish (called a Petri dish), they grow into discrete "piles" of cells called colonies, which are visible to the naked eye as shown in Figure 19.9. All the cells in the colony are genetically iden-

Figure 19.9 **Cloning a human gene in a bacterium.** Both human and bacterial (plasmid) DNA are cut using the same restriction enzyme. The human DNA is inserted into plasmids, and bacteria take up the plasmids by transformation. The bacteria are then cultured and plated on selective media, as shown in the photo, to identify colonies made up of recombinant cells.

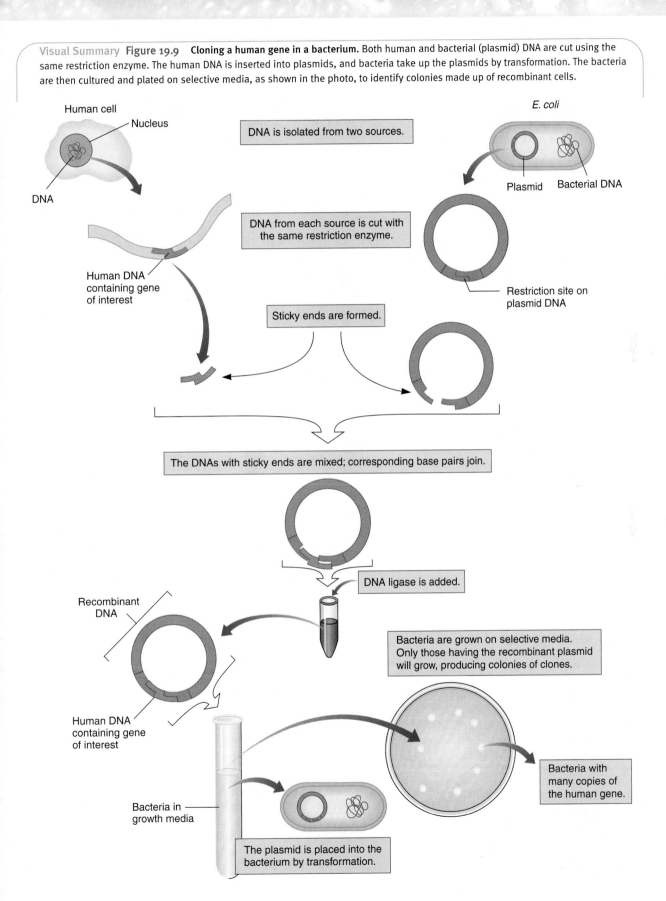

Human cell

Nucleus

DNA

E. coli

DNA is isolated from two sources.

Plasmid   Bacterial DNA

Human DNA containing gene of interest

DNA from each source is cut with the same restriction enzyme.

Restriction site on plasmid DNA

Sticky ends are formed.

The DNAs with sticky ends are mixed; corresponding base pairs join.

Recombinant DNA

DNA ligase is added.

Bacteria are grown on selective media. Only those having the recombinant plasmid will grow, producing colonies of clones.

Human DNA containing gene of interest

Bacteria in growth media

Bacteria with many copies of the human gene.

The plasmid is placed into the bacterium by transformation.

H⚙W Science
W⚙RKS Tools of Scientists

DNA Sequencing

In order to study gene organization, regulation, and function; clone and manipulate genes in the laboratory; prepare probes and primers; and carry out a variety of other procedures, scientists must first identify genes and then determine their structures. The primary structure of a gene is determined by the sequence of its nucleotide bases.

One popular method used to determine the structures of genes, a process called *DNA sequencing*, is the controlled interruption of replication. This method is usually referred to as the *dideoxy sequencing method* (DIE-dee-OX-ee). (This method was the starting point for Kary Mullis when he developed the polymerase chain reaction.) Variations of this method allow a researcher to complete single- or double-stranded DNA sequencing.

To carry out single-stranded DNA sequencing using the dideoxy method, DNA fragments are first heated and treated with chemicals, which unwind and dissociate the double-stranded molecules into complementary single strands. A preparation is made of one of these complementary DNA strands; each piece is referred to as a *template*. The template is the piece of DNA with the unknown sequence of bases, which will be determined by the sequencing process.

As shown in **Figure 19.C** (top), the template is mixed with a small amount of identical short pieces of nucleic acid called a *primer*, each piece complementary to the same specific short nucleotide base sequence on the DNA template. In our example, the primer is radioactively "labeled." This radioactive labeling is the means by which reaction products can be detected during the last step of the sequencing procedure. In addition, individual nucleotides (called deoxynucleotides and shown as dATP, dCTP, dGTP, and dTTP) and the enzyme DNA polymerase are added to the mix.

A small amount of this mixture is then added to four test tubes, each containing an *analog* of one of the nucleotides (called dideoxynucleotides and shown as ddATP, ddCTP, ddGTP, and ddTTP). An analog is something that is similar to something else. An analog nucleotide bonds to its complementary base (ddATP to thymine, and so on), but it does not allow the next nucleotides in the continuing base sequence to bond to the DNA template.

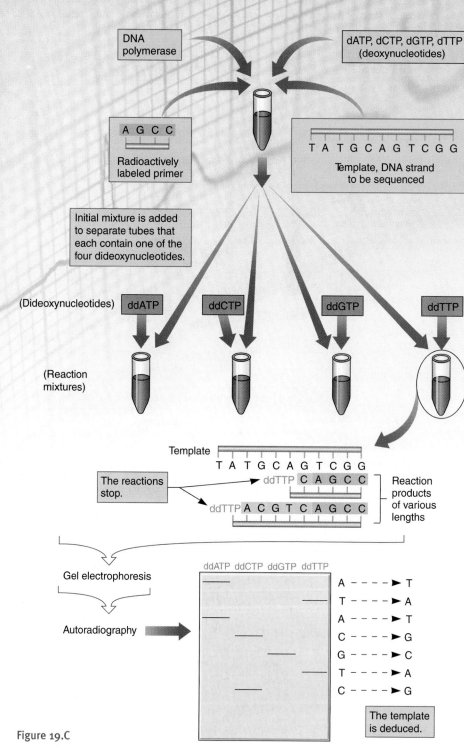

Figure 19.C

The DNA polymerase in each reaction mixture sequentially adds nucleotides to templates hybridized with primers, thus extending the primers. However, wherever the analog bonds to the DNA template instead of a deoxynucleotide, the extension of the primer stops. Since copies of the same template DNA were used in each reaction mixture, four sets of DNA fragments are produced, one set for each analog. Each set has many hybridized strands of differing lengths. The strands in each test tube, however, no matter how long or short, always end with the nucleotide analog present in that tube.

The four samples are then electrophoresed (see "Just Wondering"), and the base sequence of the new DNA is read from the autoradiogram of the four lanes as is illustrated in Figure 19.C. The template sequence is deduced from this sequence on the basis of the complementarity of the nucleotide bases of DNA. In our example autoradiogram, the topmost band is in the "A" column; therefore, the DNA strand has the com-

plementary base thymine (T) in that position. The second band down is T, so the DNA strand would have A in that position, and so forth.

DNA sequencing is automated today, using a variation of the dideoxy method. Instead of radioactively labeling the primers, scientists use different colored fluorescent dyes to label the primers. The primers added to each of the four test tubes are labeled with different colors; for example, blue-emitting primers are used in the tube in which the adenine analogs are placed, orange-emitting primers in the tube in which the thymine analogs are placed, yellow in the guanine tube, and green in the cytosine tube. After the reactions are completed, the solutions are mixed and electrophoresed together. The sequence of the colors, which is determined by measuring the wavelengths of the bands, shows the sequence of the bases. Figure 19.D shows an example using a sequence different from the one shown on the previous page.

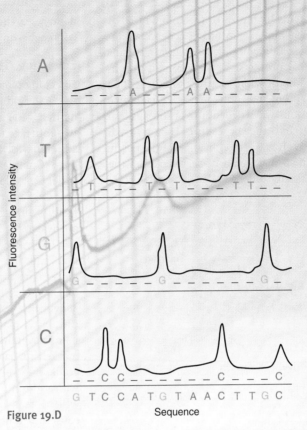

Figure 19.D

tical to one another and are considered clones because each colony grew from a single ancestral cell.

Scientists may use approaches other than the shotgun technique to clone genes, especially when more is known about the gene of interest. A process called **complementary DNA (cDNA) cloning** can be performed if the messenger RNA (mRNA) transcribed from the gene of interest is available. When a protein is synthesized by a particular tissue in the body, its mRNA is usually found in abundance there. For example, human insulin is manufactured in the pancreas; these cells are rich in insulin-specific mRNA. Scientists can extract this mRNA from pancreatic cells and manufacture molecules of DNA from mRNA using the enzyme *reverse transcriptase*. (This is the reverse of the usual situation; usually mRNA is made from DNA.) This DNA can then be inserted into bacterial or yeast cells and cloned.

Complementary DNA cloning also avoids a problem inherent in the shotgun cloning technique: the removal of introns, noncoding regions of DNA found in eukaryotes. As described in Chapter 18, eukaryotic cells cut in-

trons out of mRNA and then splice together the exons, or coding portion of the nucleotide chain, before genes can be translated into proteins. During shotgun cloning, genes having introns will not be correctly expressed in a bacterial cloning host. Additionally, bacteria cannot process mRNA the same way that eukaryotic cells can, so eukaryotic DNA put into a bacterial cell cannot make an mRNA appropriate for protein production in bacteria. In cDNA cloning, the mRNA used as the template for reverse transcriptase has no introns. Additionally, cDNA cloning avoids the eukaryotic DNA processing problem of bacteria. DNA synthesized from mRNA can readily be cloned and expressed in bacterial systems.

Gene synthesis cloning, another approach to cloning genes, can be performed if the nucleotide sequence of the gene of interest is known and the genetic engineer is able to synthesize the gene. (DNA nucleotide sequencing techniques are described in "How Science Works.") The laboratory-made gene can be inserted into bacterial or yeast cells, as described previously, and cloned. This method of cloning also avoids the intron problem.

Producing proteins by means of recombinant DNA technology is not as simple as inserting genes into microorganisms and then having these microorganisms synthesize functional proteins. There are many problems associated with the cloning processes described thus far; the intron problem is only one. Scientists solve this problem in various ways depending on the protein being synthesized and the procedures being used. Another problem involves the modification of the newly synthesized protein so that it can be secreted from the cells producing it. Many other challenges exist for ge-netic researchers in the development of new and novel approaches to meeting their goals.

Screening Clones

Screening clones to find a desired gene is necessary to select the proper clones to make large amounts of the desired gene or gene product. With shotgun cloning, the entire genome has been cloned, and consequently, screening for particular genes can be a lengthy procedure. Complementary DNA cloning results in a small variety of clones, so screening is a relatively easy process. Screening is not necessary for genes that are directly synthesized because only the synthesized genes are available for cloning.

Most screening techniques involve DNA hybridization and the use of radioactively labeled *molecular probes*. The process is illustrated in **Figure 19.10.** The master plate (petri dish with growth medium) contains the genetically engineered bacterial clones. The surface of the master plate is touched with a circular piece of sterile nitrocellulose (a thin, filter paper type material) and marked for later reference ❶. The nitrocellulose, now containing cells from the master plate, is then treated with sodium hydroxide and sodium chloride solutions, which break open the bacteria, release the DNA, and cause the doubled-stranded DNA to dissociate into single strands ❷. The filter is then baked at 80°C for two hours to affix the DNA strands to the filter ❸. A radioactively labeled probe is then added to the filter ❹.

A **probe** is a molecule that binds to a specific gene, or nucleotide sequence. Probes can be (a) molecules of mRNA purified from tissue or cells that produce the protein, (b) complementary sequences of DNA or RNA that have been synthesized using knowledge of the codons (triplet-nucleotide sequences on mRNA) for the amino acid sequence of the protein, or (c) antibodies specific for the desired protein. After the probe has been added to the filter that contains single-stranded DNA fragments or protein, the probe will bind with the complementary DNA sequences (a process called *hybridization*) or the antibodies will bind with the desired protein. Molecules of the probe that did not bind, then, are washed from the filter ❺, leaving the larger complexes of bound probe and target in the filter. These filters are then placed under x-ray film in sealed containers overnight ❻, a technique called *autoradiography*. The radioactivity exposes the x-ray film to create a photographiclike image, and only the colonies containing DNA with the hybridized probe are visualized ❼. By comparing the location of the hybridized probe on the film with the colonies on the master plate, the clones containing the searched-for gene can be isolated and the DNA collected.

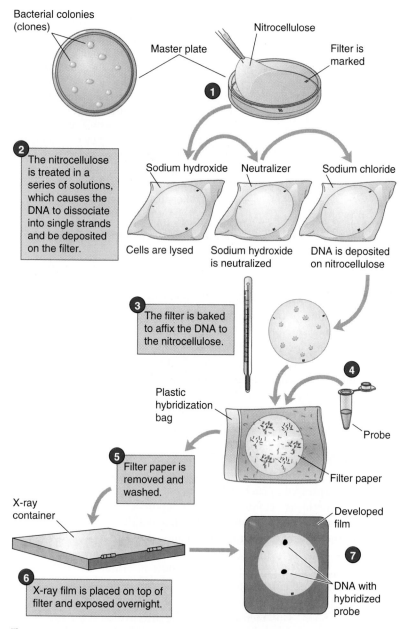

Bacterial colonies (clones)

Master plate

Nitrocellulose

Filter is marked

❶

❷ The nitrocellulose is treated in a series of solutions, which causes the DNA to dissociate into single strands and be deposited on the filter.

Sodium hydroxide — Cells are lysed

Neutralizer — Sodium hydroxide is neutralized

Sodium chloride — DNA is deposited on nitrocellulose

❸ The filter is baked to affix the DNA to the nitrocellulose.

❹

Plastic hybridization bag

Probe

❺ Filter paper is removed and washed.

Filter paper

X-ray container

Developed film

❼

❻ X-ray film is placed on top of filter and exposed overnight.

DNA with hybridized probe

Figure 19.10 **Screening cloned genes using molecular probes.**

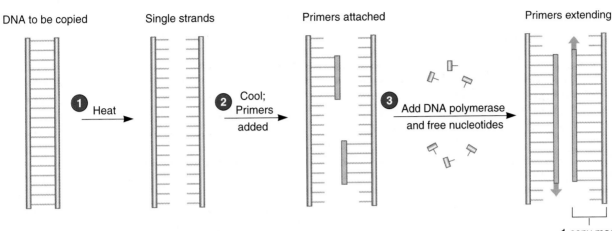

DNA to be copied   Single strands   Primers attached   Primers extending

1 Heat

2 Cool; Primers added

3 Add DNA polymerase and free nucleotides

Making Copies of a Gene: The Polymerase Chain Reaction

After a gene has been isolated, many copies of it are needed to make enough recombinant cells as an inoculum for large-scale culturing. Developed in 1983 by American scientist Kary B. Mullis, the polymerase chain reaction (PCR) is a method used to make unlimited copies of genes. This is an extremely important technique in genetic engineering today. Its applications vary widely because the DNA can be in a mixture with other molecules and can also come from anywhere, such as a cloned cell, a blood stain, or even an organism that has been dead for hundreds of years! Not only has this technique revolutionized genetic research, it has also enabled the analysis of DNA in extremely small blood stains, semen samples, or pieces of hair at crime scenes that were previously impossible to analyze. Scientists are even able to amplify small amounts of DNA found in the preserved remains of long-extinct organisms to compare their relatedness with other extinct or present-day organisms.

The polymerase chain reaction is illustrated in Figure 19.11 and proceeds the following way. First, the DNA is heated to separate the double DNA strands into single-stranded DNA **1**. A probe a few nucleotides in length (referred to here as a *primer*) is then hybridized with a portion of the DNA at the place where the "copying" is to begin **2**. The enzyme DNA polymerase and free nucleotides (adenine [A], cytosine [C], guanine [G], and thymine [T]) are added to the solution **3**. DNA polymerase, "primed" by the hybridized probe,

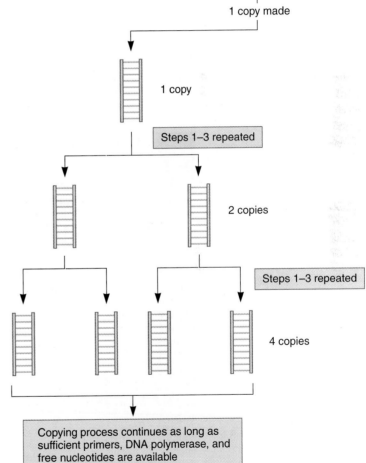

1 copy made

1 copy

Steps 1–3 repeated

2 copies

Steps 1–3 repeated

4 copies

Copying process continues as long as sufficient primers, DNA polymerase, and free nucleotides are available

Figure 19.11   **The polymerase chain reaction (PCR).**

extends the primer, adding nucleotides in sequence from the site of the primer to the end of the fragment. Each single DNA fragment is double stranded once again and is an exact replica of both the other strands and the original double strand. This process is repeated over and over again: One DNA strand is separated into two single strands; each strand is hybridized with a nucleotide primer; and DNA polymerase adds free nucleotides to complete the strands, forming two new, double-stranded molecules from one original. Two strands are then used to produce four, the four produce eight, and the numbers quickly escalate. In an afternoon, one original DNA molecule can be used to make a billion copies.

Today, PCR can be performed by machine (Figure 19.12). In fact, researchers at Lawrence Livermore National Laboratory are currently developing a miniature PCR machine. This technology would have many applications in medical diagnostics and in the military. For example, researchers think this portable device could be used by soldiers in the field to detect low levels of mi-croorganisms contaminating water or supplies, or the spores of organisms used in biological warfare.

In our example of finding, extracting, and copying a human gene for genetic engineering, PCR is used to produce as many copies of the gene as the genetic engineer needs. These genes are then inserted into bacteria or yeasts by means of cloning vectors. Populations of cells are grown on a large scale, and then the protein is extracted. As mentioned previously, technical problems are often encountered in the identification, characterization, isolation, and use of human genes to produce genetically engineered microbes for manufacturing human proteins. However, these problems can usually be overcome, as in the case of the human insulin gene. The product produced by genetic engineering is called Humulin and is used to treat more than half the new cases of diabetes in the United States. In addition to avoiding the unwanted allergic reactions sometimes caused by using insulin extracted from cattle or pig pancreas, Humulin is less expensive than animal preparations.

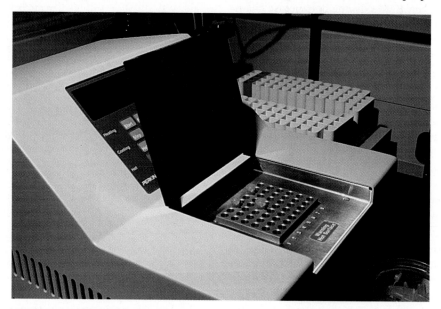

Figure 19.12 **The polymerase chain reaction (PCR) can be carried out by machine.** The PCR machine is a thermal cycler—it goes through cycles of heating and cooling to carry out this chain of reactions. The laboratory technician simply loads the DNA to be copied, primer, free nucleotides, and DNA polymerase in the pink tubes visible in the machine. A special type of DNA polymerase that is resistant to destruction by high temperature is used. The reactions will take place in a cyclic fashion as long as sufficient reactants are available.

Genetically Engineered Proteins in Use Today

Along with producing human insulin and human growth hormone, gene technology has produced proteins useful in treating a variety of human disorders or diseases. For example, the three primary types of human interferon (alpha, beta, and gamma) have been produced by genetic engineering. Interferons are proteins that interfere with the ability of a virus to invade a cell. These proteins also modulate the activity of the immune system and, in doing so, help the body stave off invasion from other disease-causing agents. Although interferons have been used to treat certain viral diseases and rheumatoid arthritis, the most successful application of genetically engineered interferon has been the use of alpha-interferon to treat hairy-cell leukemia, a rare form of cancer. An amazing 75% to 90% of new patients treated thus far have experienced remission (lessening of the symptoms) of their disease. A variety of other medically important proteins have also been genetically engineered, such as interleukin-2, which is used in certain cancer treatments and to treat particular deficiencies of the human immune system, and tissue-type plasminogen activator, a clot dissolver used in the treatment of heart attacks.

To use microbes to produce proteins for human use, scientists first find the gene in the human genome and isolate it. One method used when the researcher knows very little about the gene is shotgun cloning. This method involves cutting the DNA of the entire human genome into pieces with restriction enzymes, inserting these pieces or fragments into bacteria or yeast using cloning vectors, and allowing the organisms to reproduce, thereby making copies, or clones, of the DNA fragments. Two other methods used under different circumstances are complementary DNA cloning and gene synthesis cloning.

Most screening techniques used to find a gene of interest within a gene library involve DNA hybridization and the use of molecular probes, molecules that bind to specific genes, or nucleotide sequences.

The polymerase chain reaction is used to produce multiple copies of a gene and has wide application in molecular biotechnology.

19.6 The development of genetically engineered vaccines

A vaccine is a substance that is either injected into the body or taken orally to stimulate the immune response. Vaccines are often produced by culturing the disease-causing agent and then killing (inactivating) it or attenuating (weakening) it. However, traditional methods of producing vaccines do not always work. Certain organisms, such as the malaria parasite, are difficult to culture. Some vaccine preparations do not confer resistance well enough, whereas others produce unwanted—and sometimes dangerous—side effects.

Scientists are now applying the techniques of genetic engineering to the development of vaccines in an effort to avoid these problems. In one approach, scientists use the gene-splicing techniques described earlier in this chapter to insert one or more of a pathogen's genes into a nonpathogenic organism. (A pathogen is a disease-causing agent of infection.) Such a technique can produce, for example, a noninfective virus whose coat contains proteins of the pathogen as a result of the inserted genes being expressed. When injected into humans, these viruses do not cause disease, but they do stimulate the immune system to produce antibodies specific for the infective form of the virus.

Scientists have produced a successful preparation of hepatitis B vaccine, which helps protect against the leading cause of liver cancer. Researchers are currently working on a variety of other genetically engineered vaccines, including preparations to provide immunity against cholera, hepatitis C, and ear infections. Two new promising genetically engineered vaccines to prevent urinary tract infections are currently in clinical trials. A human immunodeficiency virus (HIV) vaccine, which consists of a viral protein called gp120, is currently in clinical trials as well.

The newest vaccine technology involves the development of DNA vaccines. DNA vaccines work by injecting the DNA coding for a particular segment of the disease-causing organism into the muscle of the person being vaccinated. The DNA becomes incorporated into the muscle cells of the recipient and causes those cells to produce the protein for which the DNA codes. The person's immune system then develops antibodies against this protein (antigen), which confers resistance against the pathogen.

Genetic engineering has a variety of applications.

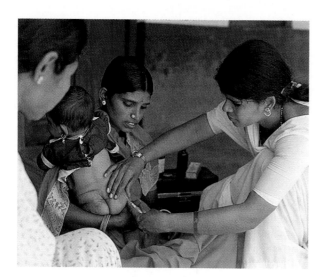

Immunization of children in India.

DNA vaccines have many advantages over other types of vaccine preparations:

1. no risk of infection by the pathogen,
2. no risk of illness from contamination of the vaccine,
3. a long-lasting immune system response, and
4. the ability to administer vaccines for a variety of diseases in a single shot.

Because these vaccines will be less costly than traditional vaccines and do not need to be refrigerated, they can be more easily used in developing countries where many of the diseases they will protect against are rampant. Researchers are currently working on DNA vaccines for a variety of diseases including malaria, rabies, epidemic (viral) diarrhea, herpes, and tuberculosis. These vaccines are new, however, and scientists are still uncertain how well they will work long term, and what future side effects might be encountered from their use.

Scientists are now using the techniques of bioengineering to develop vaccines that will provide long-lasting immunity without producing side effects or illness in the recipient. The newest technology uses the DNA of a pathogen to produce antigens within the host, thereby triggering an immune response.

19.7 Gene therapy

Recombinant DNA technology entered a new phase of application on September 14, 1990, when 4-year-old Ashanthi DeSilva became the first person to undergo **gene therapy**. Gene therapy is the treatment of a genetic disorder by the insertion of "normal" copies of a gene into the cells of a patient carrying "defective" copies of the gene. This young girl suffered from the rare genetic disorder severe combined immunodeficiency (SCID). Key immune system cells called *T cells* (see Chapter 11) were not working because they lacked the enzyme ADA (adenosine deaminase). Without these cells, Ashanthi had no defense against infection.

To treat Ashanthi, Drs. Michael Blaese, W. French Anderson, and their colleagues at the National Institutes of Health (NIH) in Bethesda, Maryland, removed some of her white blood cells, which carry the defective gene, and cultured them. To the cell culture they then added bioengineered viruses into which they had incorporated working copies of the ADA gene. When the viruses infected the blood cells, they inserted the ADA genes into the cells. The medical researchers grew these altered blood cells in the laboratory until they

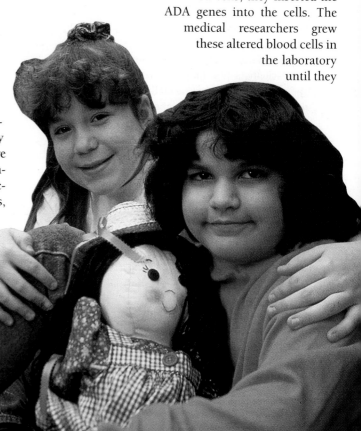

numbered in the billions and then injected them into Ashanthi's blood. Thinking that these cells could live for only a few months, Drs. Blaese and Anderson repeated the gene-insertion process quite often over a 2-year period. A few months into this process in 1991, another young ADA patient, Cynthia Cutshall, was included in this pioneering research. The girls are shown in the photo (bottom right) on page 428 in June of 1993, a few years after receiving their first gene therapy.

The gene treatment ended after 2 years, but the girls' T cells continued to express the ADA gene; their cellular and humoral immunity had been restored. The results indicated that gene transfer into long-lasting "parent" T cells was successful and that new T cells carried a working copy of the gene. Although medical researchers suggest that they still must perfect certain components of this treatment, they conclude that gene therapy can be a safe and effective addition to other treatments for some patients with SCID.

Genetic diseases most likely to be cured with gene therapy in the future are those, like SCID, which are caused by a single defective gene. In addition, genetic disorders that result in the deficiency of a cellular product are more likely to be curable than diseases that result in an overabundance of a product or a harmful substance. Scientists have learned how to turn genes on to produce a product, but have not yet determined how to turn them off so that they no longer produce an unwanted product.

The Human Genome Project and Gene Therapy

SCID is only one of more than 4000 diseases such as cystic fibrosis, sickle cell anemia, and achondroplasia (a form of dwarfism) known to result from an abnormality in a gene. The underlying idea of transferring a normal copy of a gene into an individual who has a mutant copy requires that the location of the gene is known and that the gene has been isolated. If gene transfer therapy is going to be applied to a wide range of genetic defects, scientists need a detailed map of the human chromosomes.

Launched in 1990, the **Human Genome Project (HGP)** is a worldwide effort to map the positions of all the genes and to sequence the 3 billion DNA base pairs of the human genome. The highest priority goal of the Human Genome Project is to provide a complete sequence of human DNA to the research community as a publicly available resource. This resource will be significant, not only because it will give scientists a clearer picture of the genes that cause various diseases, but also

because it will give scientists the information they need to study the relationships between the structure of genes and the proteins they produce.

Internationally, scientists are working on developing three increasingly detailed maps of the DNA in cells: (1) *a genetic linkage map,* which shows the distances between genetic markers (identified reference points such as genes for particular diseases) on the chromosomes; (2) *a physical map,* which shows the number of nucleotides between the markers; and (3) *an ultimate map,* which shows the sequence of nucleotides in a chromosome and describes its genes and the proteins they make. Work on the genetic linkage and physical maps is complete; sequencing the entire genome will take much longer.

By the end of 1998, scientists had completed a "working draft" of approximately 50% of the human genome; only 6% of the genome had been sequenced in "final draft" form. (The difference between the working and final drafts lies in the accuracy of the sequencing.) Scientists expect to complete the sequencing of 90% of the human genome as a working draft by 2001, with 33% in final draft. A complete and accurate sequencing of the entire genome is expected to be finished by the end of 2003, the fiftieth anniversary of the discovery of the double helix structure of DNA by James Watson and Francis Crick.

In 1998, U.S. scientific instrument maker Perkin-Elmer and Dr. J. Craig Venter of the Institute for Genomic Research announced that they were entering the initiative to sequence the human genome. They stated that they would use different sequencing strategies than the governmentally funded effort. The National Institutes of Health and the Department of Energy, which are sponsoring the U.S. component of the HGP, hope that cooperation between individuals working on the private and public projects will result in completion even earlier than anticipated.

Targeted Gene Replacement and Gene Therapy

Gene therapy is being helped by recent research in targeted gene replacement. This research is currently being conducted at the University of Utah by Mario Capecchi and his associates. The process of gene targeting involves changing the nucleotide sequence of a particular gene, which changes the function of the gene, and then observing the resultant changes in the anatomy, physiology, or behavior of the organism. Pooling these data with information from the Human Genome Project will give scientists great insight into the connection between the structure and function of genes.

Targeted gene replacement research is currently being conducted on mice, whose genome is surprisingly similar to the human genome. In fact, approximately 99% of the genes in mice and in humans are the same and serve the same functions. Therefore, scientists are hopeful that this research will provide data regarding how specific parts of the body (such as the brain) operate; how mutations in cells cause diseases such as cancer; how genes affect the development of cells, tissues, and organs; and, as previously mentioned, how inherited defects in the genome result in inherited disorders.

Gene therapy is the treatment of a genetic disorder by the insertion of "normal" genes into a patient's cells. Genetic diseases most likely to be cured with gene therapy in the future are those that are caused by a single defective gene or that result in the deficiency of a cellular product. However, if gene transfer therapy is going to be applied to a wide range of genetic defects, scientists need a detailed map of the human chromosomes. The Human Genome Project is a worldwide effort to accomplish this task. Researchers plan to map the positions of all the genes and to sequence the 3 billion DNA base pairs of the human genome by 2003. Data from targeted gene replacement research, which involves changing the nucleotide sequence (function) of a particular gene and then observing the resultant changes in the anatomy, physiology, or behavior of the organism, will also help researchers develop gene therapy procedures. These data will provide scientists with insight into the connection between the structure and function of genes.

19.8 Biotechnology in agriculture

Biotechnology is being used in a variety of ways in the food industry. It focuses on increasing the yields of various foods and creating products with superior qualities. The most widespread use of biotechnology in the food industry is in agriculture. Through the use of gene splicing, scientists are working to improve crops and forest trees by making them more resistant to disease, frost, and herbicides (chemicals that kill weeds). Scientists expect that plant improvement through the use of biotechnology will be an important part of increasing food production to supply the burgeoning world population, which is predicted to reach 10.7 billion by the year 2030.

Transgenic Plants

In early 1994, the U.S. Food and Drug Administration approved the first genetically engineered food: a tomato containing a gene that allows it to ripen longer on the vine yet reach the supermarket without softening. Later in 1994, many genetically altered foods were approved, such as a squash that resists viruses, cotton and soybean plants that resist certain herbicides, and a potato that produces a pesticide that kills Colorado potato beetles (see the Part 5 Bioissues). Such genetically altered plants, which carry genes from other organisms, are called **transgenic plants** (tranz-GEE-nik).

Since 1980, a plasmid of the bacterium *Agrobacterium tumefaciens,* which causes a tumorlike disease called *crown gall* in plants, has been the main vehicle used to introduce foreign genes into broadleaf plants such as tomatoes, tobacco, and soybeans. However, to use the plasmid to genetically engineer plants, scientists first had to remove *A. tumefaciens'* disease-causing genes—a technique known as disarming—while leaving its natural ability to transfer DNA intact.

After *A. tumefaciens* has been disarmed, it can be used as a vector to shuttle desired genes into plant cells. A part of its plasmid integrates into the plant DNA, carrying whichever genes the genetic engineer has inserted into the plasmid genome. (**Figure 19.13,** ❷ on the next page, diagrams the steps of this process.) New plants can be micropropagated, or grown from these transformed cells in tissue culture. The cells develop into plantlets, which can be grown in conventional ways.

Many plant species are not natural hosts for the *Agrobacterium* organisms; thus, scientists have been looking for other ways to insert genes into plants. One commonly used method shoots microscopic metal pellets coated with DNA into plant cells (see Figure 19.13, ❶). Then the cells are cultured and propagated.

Transgenic and Cloned Animals

In 1997, when cloning produced a sheep named Dolly, many people thought that cloning organisms was a new scientific breakthrough. However, plants have been cloned for hundreds of years; you may have even done it yourself. Whenever a cutting from a plant is rooted, the new plant is a clone—that is, it is an individual plant genetically identical to another plant. In this case, the rooted cutting is a clone of the parent.

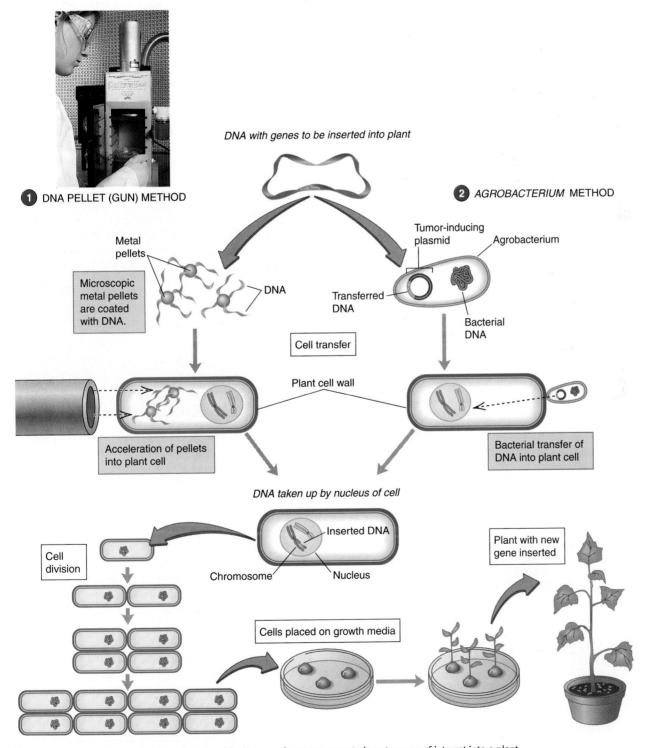

Figure 19.13 **Genetic engineering of plants.** This diagram shows two ways to insert a gene of interest into a plant.

Animals have been cloned since the 1970s. Scientists have taken embryos at early stages of development and split them in two. The split embryos were clones of one another. What was different about the cloning of Dolly was that the genetic material was taken from an *adult* cell of Dolly's 6-year-old "parent." The nucleus from a single mammary gland cell was inserted into an egg cell cleared of its hereditary material, in a process called *nuclear transfer.* The resulting embryo was implanted in the oviduct of a surrogate mother. That embryo grew to be Dolly, a sheep that at birth had nearly 7-year-old DNA.

The technique of nuclear transfer provides another way for scientists to produce transgenic animals (those containing genes from other animals) that may be useful in a variety of ways such as producing substances in their milk for treating human diseases or for organs for transplantation. Nexia Biotechnologies in Quebec, Canada, for example, is currently cloning goats that have the genetic information to produce synthetic spider silk in their milk. This material, called BioSteel, will be used to produce artificial tendons and ligaments; it has a variety of other medical applications as well.

So far, 13 transgenic species of fishes have been developed. One major goal of marine biotechnological research is to produce fishes (the focus has been Atlantic salmon) with an enhanced tolerance to cold water. Genetic engineers are experimenting with injecting certain genes of cold-tolerant species of fishes into salmon eggs, which are large and relatively easy to microinject with DNA. Cold-tolerant fishes produce proteins that act like antifreeze in their blood; scientists think that by injecting the genes that code for the production of these proteins into salmon eggs, the fishes will grow into cold-tolerant adults. Marine biologists are also experimenting with the development of disease-resistant fishes.

The beef industry may also use bioengineering in years to come to produce better and more inexpensive cuts of meat. In 1998, scientists discovered the gene

that controls "double muscling" in cattle. Double muscling refers to a trait in cattle that provides more meat per animal on the same food intake as ordinary animals. Double muscling was first reported in 1807 in a strain of cattle called Belgian Blues. Notice how heavily muscled the Belgian Blue bull is in the photo below, left. Unlike other types of cattle, whose meat gets tough when its muscles hyperdevelop, the Belgian Blue's muscles stay tender yet are relatively low in fat.

When they isolated the double muscling gene in 1998, researchers discovered that it was a mutant of a gene that makes a protein called myostatin. Myostatin normally limits muscle growth, but the mutation blocks its effects and muscles grow larger. Scientists are not pursuing the development of herds of Belgian Blues with the double muscling mutation at this time because the offspring of the cows are so large that they must be delivered by Cesarean section. Researchers are searching for a less extreme myostatin mutation and are also trying to identify another gene with a less drastic influence on muscle mass so that calves can be delivered without surgery.

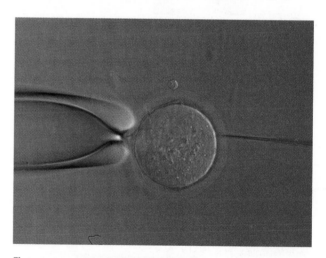

Figure 19.14 **Introducing foreign genes into the nucleus of a mouse egg.** The blunt tip of the glass micropipet supporting the egg with a slight suction can be seen on the left. The egg is in the center, and the microneedle delivering the DNA is on the right. After this process is completed, the egg will be placed in the oviduct of a surrogate mother so it can develop.

Since the 1980s, transgenic animals have been produced by the microinjection of DNA fragments into single-celled zygotes. (Figure 19.14 shows a mouse egg undergoing this process.) However, producing transgenic animals by microinjection of DNA is inefficient and cost-prohibitive. Therefore, the technique used to clone Dolly may have aspects that can provide a more viable approach. One technique being used at this time is as follows: fibroblasts (precursors of connective tissue cells) are taken from an animal fetus (Figure 19.15 ❶). The fibroblasts are placed in cell culture and allowed to grow and multiply. Foreign DNA is then mixed with the fibroblasts along with certain chemicals, and some of

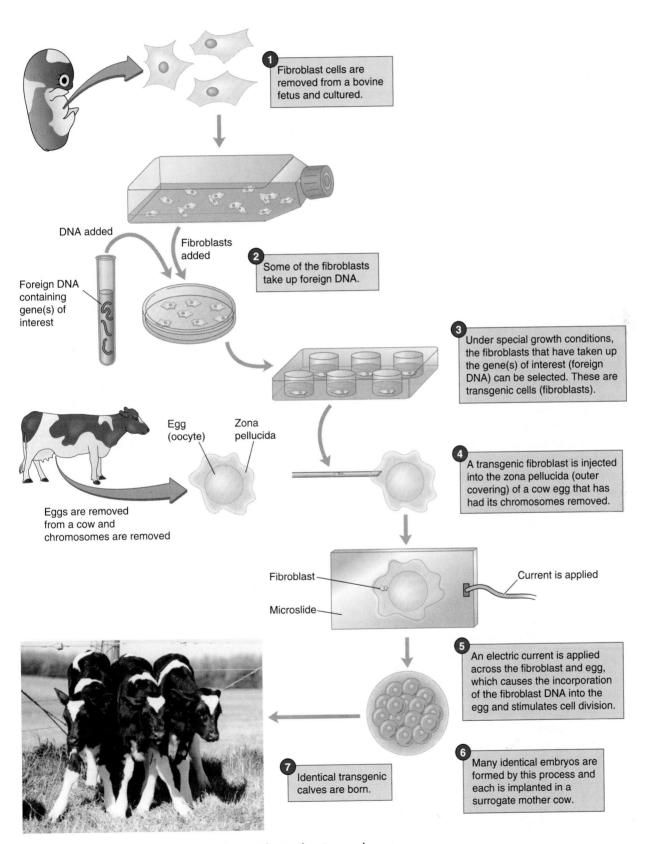

1 Fibroblast cells are removed from a bovine fetus and cultured.

DNA added

Fibroblasts added

Foreign DNA containing gene(s) of interest

2 Some of the fibroblasts take up foreign DNA.

3 Under special growth conditions, the fibroblasts that have taken up the gene(s) of interest (foreign DNA) can be selected. These are transgenic cells (fibroblasts).

Egg (oocyte)   Zona pellucida

4 A transgenic fibroblast is injected into the zona pellucida (outer covering) of a cow egg that has had its chromosomes removed.

Eggs are removed from a cow and chromosomes are removed

Fibroblast

Microslide

Current is applied

5 An electric current is applied across the fibroblast and egg, which causes the incorporation of the fibroblast DNA into the egg and stimulates cell division.

6 Many identical embryos are formed by this process and each is implanted in a surrogate mother cow.

7 Identical transgenic calves are born.

Figure 19.15   An example of a nuclear transfer procedure to clone transgenic cows.

the cells take up the foreign DNA ❷. The fibroblasts are then grown under special conditions that allow scientists to select those cells that have successfully incorporated the foreign DNA ❸. The nuclei of these transgenic fibroblasts are transferred to cow oocytes (eggs) that have had their nuclei removed. To produce each embryo, a transgenic fibroblast is injected into the zona pellucida (outer coating) of a bovine oocyte ❹. These two cells are placed on a special microslide, and an electrical current is passed across them ❺. The current fuses the cells and stimulates them to divide ❻. The embryos are implanted in surrogate cows, and after the gestation period, transgenic cows are born ❼. These cows will produce (usually in their milk) the product for which the foreign DNA codes.

The Use of Genetically Engineered Hormones in Milk Production

In the dairy industry, the focus has been the production of genetically engineered hormones that, when injected into cows, would increase their milk production. In 1993, the FDA approved genetically engineered bovine somatotropin (bST), sometimes called rbST (recombinant bovine somatotropin) or rbGH (recombinant bovine growth hormone). Recombinant bST is made by a process similar to the production of synthetic human insulin and is identical to the one naturally produced by cows. Before giving its approval for commercial use, the federal Food and Drug Administration conducted more than 120 studies on bST and concluded that milk and meat from bST-treated cows are safe to consume. In addition, pasteurization destroys approximately 90% of the bST in milk. These conclusions have been confirmed by a variety of regulatory agencies, such as the National Institutes of Health in the United States as well as those in Canada and abroad.

Biotechnology, which is being used in a variety of ways in agriculture, focuses on increasing the yields of various foods and creating products with superior qualities. The most widespread use is with plants; scientists are working to improve crops and forest trees by making them more resistant to disease, frost, and herbicides. Many genetically engineered agricultural products are on the market today. Regarding animals, research is being conducted to develop varieties of cold-tolerant and disease-resistant fishes as well as transgenic animals that would provide useful products for human consumption. Also, in the dairy industry, genetically engineered hormones have been produced that increase the milk production of cows.

19.9 Ethical guidelines for genetic engineering

There has been considerable discussion in the scientific community and among nonscientists about the potential danger of inadvertently creating undesirable or potentially dangerous organisms in the course of a recombinant DNA experiment. For example, what if someone fragmented the DNA of a cancer cell and then incorporated these DNA fragments at random into viruses that are propagated within bacterial cells? Might there not be a danger that one of the resulting bacteria or viruses could be capable of infecting humans and causing a disease—even cancer?

Even though most recombinant DNA experiments are not dangerous, such concerns are taken seriously. It was the scientists themselves who first realized the possible dangers associated with genetic engineering technology. These scientists, along with government agencies, such as the National Institutes of Health, drew up formal guidelines that govern all such experimentation in the United States. Many other countries have done the same. Many scientific organizations have ethics committees to deal with such issues also, especially as they relate to gene therapy in humans.

Scientists and governmental regulatory agencies not only monitor experiments but also study products to make sure that they do not cause increased health risks to consumers or pose environmental hazards. Genetically engineered products on the market today, especially pharmaceutical and agricultural products, have been shown to be safe and effective.

Although genetic engineering may sound rather ominous to some, it is important to remember that many of the ways that scientists manipulate genes occur naturally. These processes are occurring right now—*without* the help of scientists. In addition, scientists have been manipulating genes for centuries using classical biotechnological techniques—techniques with which most people are comfortable. Molecular biotechnology is simply more specific, quick, and versatile.

Scientists and governmental agencies regulate genetic engineering experimentation and the products that result.

Key Concepts

The use of biotechnology began about 10,000 years ago.

19.1 Biotechnology is the manipulation of organisms to yield specific characteristics, useful products, and information that would otherwise not be known.

19.1 Classical biotechnology, practiced for more than 10,000 years, uses the techniques of selection, mutation, and hybridization.

19.2 The advent of molecular biotechnology, which involves the manipulation of the genes themselves, began with the discovery of the structure of the genetic material DNA in the 1950s.

Genes can be transferred from organism to organism.

19.3 Certain bacteria can naturally transfer genetic material from one cell to another by transformation, transduction, and conjugation.

19.4 Scientists were able to manipulate these mechanisms of gene transfer in the 1950s and 1960s and, by the 1970s, discovered how to combine DNA from different species of organisms and manipulate genes from eukaryotes.

19.4 Techniques of gene manipulation are called genetic engineering or recombinant DNA technology.

Genetic engineering has a variety of applications.

19.5 One of the first applications of recombinant DNA technology was the insertion of human genes into bacteria, causing these organisms to produce human proteins for therapeutic use.

19.5 The genes that code for the desired protein must first be identified within the human genome to produce human proteins by means of recombinant DNA technology.

19.5 One method of finding a gene is shotgun cloning, which is used when researchers know very little about the gene they are trying to find. This process results in a complete gene library: a collection of copies of DNA fragments that represent the entire genome of an organism.

19.5 Shotgun cloning involves cutting the DNA of the entire genome into pieces with restriction enzymes, inserting these pieces or fragments into bacteria or yeast with plasmids or viruses, and allowing the organisms to reproduce, making copies, or clones, of the DNA fragments.

19.5 Two other methods of finding a gene, used under different circumstances, are complementary DNA cloning and gene synthesis cloning.

19.5 Most screening techniques used to find a gene of interest within a gene library involve DNA hydridization of molecular probes, molecules that bind to specific genes, or nucleotide sequences.

19.5 After a gene has been isolated, many copies of it are needed to make enough recombinant cells as an inoculum for large-scale culturing.

19.5 The polymerase chain reaction is a process used to produce multiple copies of a gene. This technique has wide application in molecular biotechnology.

19.5 Many products have been developed by genetic engineering to treat a variety of human disorders or diseases. These include human insulin, used to treat diabetes; human growth hormone, used to treat children who do not produce enough of this hormone for normal growth; and the human interferons, proteins that interfere with the ability of a virus to invade a cell.

19.6 Scientists are beginning to develop genetically engineered vaccines.

19.6 The development of DNA vaccines is a promising avenue of research to develop a variety of effective, low-cost vaccines that will be safe.

19.7 Genetic engineering has also been used to treat diseases by means of a technique called gene therapy.

19.7 Gene therapy is the treatment of a genetic disorder by the insertion of "normal" genes into the cells of a patient to replace nonfunctional or malfunctional genes.

19.7 The Human Genome Project (HGP) is a worldwide effort to sequence the DNA of the entire human genome. This project will help identify genes that cause various diseases.

19.7 Along with information from targeted gene replacement research, the HGP will help clarify the relationships between the structure of genes and the proteins they produce.

19.7 The process of gene targeting involves changing the nucleotide sequence of a particular gene, which changes the function of the gene, and then observing the resultant changes in the anatomy, physiology, or behavior of the organism.

19.8 Genetic engineering is being used in a variety of ways in the food industry. There, it focuses on increasing the yields of various foods and creating products with superior qualities.

19.8 In agriculture, scientists are improving crops and forest trees by making them more resistant to disease, frost, and herbicides.

19.8 Regarding animals, research is being conducted to develop varieties of cold-tolerant and disease-resistant fishes as well as transgenic animals that may be able to produce products useful to humans.

19.8 In the dairy industry, genetically engineered hormones have been produced that increase the milk production of cows.

19.9 Scientists and governmental agencies regulate genetic engineering experimentation and the products that result from the research.

Key Terms

biotechnology
(BYE-oh-tek-NOL-uh-jee) *412*

classical biotechnology *412*

clones *417*

cloning vector *418*

complementary DNA (cDNA)
cloning *423*

conjugation
(CON-juh-GAY-shun) *414*

gene library *417*

gene synthesis cloning *423*

gene therapy *428*

genetic engineering (recombinant
DNA technology) *416*

genome (GEE-nome) *414*

Human Genome Project (HGP) *429*

ligase (LYE-gase) *420*

plasmids (PLAZ-mids) *414*

probe *424*

restriction enzymes *417*

restriction fragments *418*

restriction sites *418*

shotgun cloning *417*

transduction (tranz-DUK-shun) *414*

transformation *414*

transgenic plants (tranz-GEE-nik) *430*

KNOWLEDGE AND COMPREHENSION QUESTIONS

1. Define the term *biotechnology*. Distinguish between biotechnology and genetic engineering. (*Comprehension*)

2. What are the differences between classical biotechnology and molecular biotechnology? Are both being practiced today? Support your answer with evidence. (*Comprehension*)

3. Name and describe the three methods of natural gene transfer among bacteria. (*Knowledge*)

4. Describe one way in which scientists insert human genes into bacteria using the natural ability of certain bacteria to transfer genetic material. (*Knowledge*)

5. Define the term *gene cloning*. Name three types of gene cloning and briefly characterize each. (*Knowledge*)

6. What is the role of restriction enzymes in genetic engineering? (*Knowledge*)

7. What is a molecular probe? What is its function in screening clones? (*Knowledge*)

8. Why is the polymerase chain reaction an important technique in genetic engineering today? (*Knowledge*)

9. List two genetically engineered proteins in use today and describe their use. (*Knowledge*)

10. How does the process of gene therapy treat disease? Is this technique in widespread use today? Support your answer with evidence. (*Comprehension*)

K E Y
Knowledge: Recalling information.
Comprehension: Showing understanding of recalled information.

e learning

Location: http://www.jbpub.com/biology

www.jbpub.com/biology

e-Learning is an on-line student review area located at this book's web site www.jbpub.com/biology. The review area provides a variety of activities designed to help you study for your class and build your learning portfolio.

Review Questions The review questions test your knowledge of the important concepts and applications in each chapter. The review provides feedback for each correct or incorrect answer. This is an excellent test preparation tool.

Figure-Labeling Exercises Sharpen your visual thinking by matching terms and labels for important illustrations.

Flash Cards Studying biology requires learning new terms. Virtual flash cards help you master the new vocabulary for each chapter.

Just Wondering As you read and study from this text, you may find that you have unanswered questions. Through this site you can ask the author, Sandy Alters, your "just wondering" questions.

Do you prefer the speed and reliability of a CD-ROM? All of the features contained on the eLearning portion of the web site are also available on the Student CD-ROM.

CRITICAL THINKING REVIEW

1. If scientists can engineer bacteria to produce human proteins, why can't they genetically engineer bacteria to synthesize gold or other precious metals? *(Application)*

2. What advantages do you think exist for a person using genetically engineered human insulin versus insulin extracted and purified from an animal source? Discuss a recipient's possible reaction to human (bacterial) insulin versus animal-derived insulin. *(Synthesis)*

3. Blood was taken from two sets of twins and restriction fragments were prepared for electrophoresis. One set is identical twins and one is fraternal. If you were to see the genetic fingerprints of these pairs of twins, how could you tell them apart? *(Application)*

4. Is bacterial conjugation a type of sexual reproduction? Using your knowledge of the characteristics of sexual reproduction (see Section 22.1) and the process of bacterial conjugation, answer this question, giving reasons for your answer. *(Synthesis)*

5. Joe had a strep throat, and his infection was not responding to the antibiotic his physician had prescribed. His physician explained to Joe that he must be harboring a resistant strain of bacteria, and that she would have to test the ability of various antibiotics to kill these bacteria. The physician swabbed Joe's throat and then swabbed the surface of a petri dish containing a special blood agar. She then placed three different antibiotic disks, spaced well apart, on the agar surface. After incubating the plates overnight, the physician saw the results shown below. Which antibiotic will work best against Joe's strep throat? Explain your answer. *(Application)*

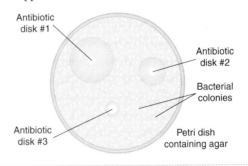

KEY
Application: Using information in a new situation.
Synthesis: Putting together information from different sources.

Visual Thinking Question

The diagram below shows the cutting and splicing of DNA from two different organisms to produce recombinant DNA. In our example, one organism is a human and one is a bacterium. Why are human genes being combined with those of a bacterium? Define the term *recombinant DNA* and explain how it is being produced as shown in this illustration.

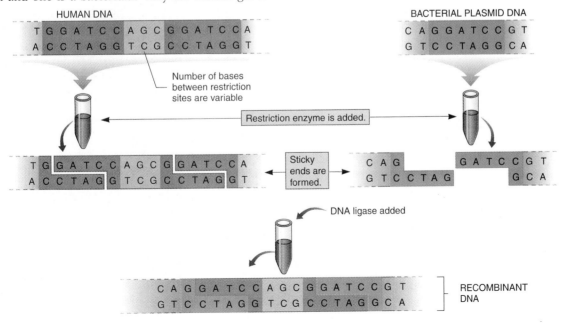

Patterns of Inheritance

The same yet different—how often have you heard that phrase? What does it really mean? Scientists would answer the last question with a single word: variation.

Variation can be seen in the group of flamingos in the photograph. Although the flamingos are all the same—with long legs and beaks and pink feathers covering their bodies—they are all different. Look closely at the same characteristic in each, such as the length of the legs or the coloration of the feathers. Their size varies somewhat—some are slightly larger than others. What differences do you see?

Clearly, each species of living things exhibits variation. Humans, for example, all have characteristics you recognize as human, but each person (except for identical twins) looks different from all others. What is the source of this variation? How are these differences distributed among populations of living things?

Organization

For centuries, humans have sought to understand inheritance.

Mendel studied inheritance using pea plants.

Other scientists built on Mendels' work.

For centuries, humans have sought to understand inheritance.

20.1 Historical views of inheritance

Today it is common knowledge that organisms inherit characteristics from their parents. During sexual reproduction, parents pass on traits to their offspring by means of genetic material within the eggs of the mother and sperm of the father. The intermingling of parental genes that takes place at fertilization, the union of the egg and sperm, is the material of variation. Organisms produced by asexual reproduction exhibit less variation for just this reason. Their genetic makeup is derived from one parent only, so offspring are genetically identical to that parent, with the exception of mutations. For example, a plant produced asexually by rooting a cutting will have the same genes as the "mother" plant, but a plant produced by cross-pollination—a type of sexual reproduction—will have certain characteristics of both its parents.

Although genetic inheritance seems obvious today, this fact was not always obvious to scientists and philosophers. Hippocrates (460–377 BC) believed that a child inherited traits from "particles" given off by all parts of the bodies of the father and mother. These particles, he suggested, travel to the sex organs. During intercourse, the father's particles merge with the mother's particles to form the child. This idea of inheritance was held by many until the mid-1800s.

Another idea widely held about inheritance was that the male and female traits blended in the offspring. Thus, according to the theory of blending inheritance, a parent with red hair and a parent with brown hair would be expected to produce children with reddish-brown hair, and a tall parent and a short parent would produce children of intermediate height. However, taken to its logical conclusion, this theory suggests that all individuals within a species would eventually look like one another as their traits continually blended together.

Other ideas regarding inheritance were formulated after the invention of a simple, handheld microscope. Anton van Leeuwenhoek (1632–1723) observed sperm for the first time with the microscope, drew pictures of them, and developed hypotheses regarding inheritance based on his observations. A widely held notion before the nineteenth century was that each sperm contained a tiny but whole human called a *homunculus*, as shown in the illustration to the right. The miniature human was supposedly implanted in the uterus during fertilization, where it grew to maturity. Because the human was fully formed, the body contained sperm that each encased another preformed individual, and so forth . . . ad infinitum. Another theory held that it was the eggs of the mother that contained minute humans.

Not until the mid-1800s with the work of Schleiden, Schwann, and Virchow (see Chapter 4) did scientists realize that new life arose from old life in the form of new cells arising from old cells. By means of the growth and division of these cells, new organisms developed from individual cells of parent organisms. Scientists of the late 1800s studied the nuclei of cells to uncover the mysteries of cell growth and division. They observed the complex process of mitosis (see Chapter 18) and wondered why cells went through this intricate process. Would it not be more efficient for a cell simply to pinch in two along its middle? By 1883, scientists knew that the complex process of cell division ensured an equal distribution of the nuclear material to two daughter cells. Not only did each daughter cell receive the same amount of nuclear material; each received a complete amount of the nuclear material.

At this same time, scientists also observed that an even more complex series of nuclear events (now known as *meiosis* [see Chapter 18]) preceded the formation of eggs and sperm. By 1885, several scientists independently concluded that this nuclear material was the physical bond that linked generations of organisms. However, scientists still did not understand how nuclear material regulated the development of fertilized eggs or how it was related to heredity and variation. Around 1900, scientists began to answer this question by piecing together research of the day with research that had long been ignored: the work of Gregor Mendel. Mendel was an Austrian monk trained in botany and mathematics at the University of Vienna.

Today, scientists know that sexual reproduction introduces variation within a species because offspring inherit characteristics from both their parents. Asexual reproduction does not introduce variation because the offspring is an exact duplicate of its parent. Historical views on inheritance varied widely, but include ideas that sperm or eggs carried preformed individuals. Not until the late 1800s did scientists realize that the nuclear material was the physical bond that linked generations of organisms.

20.2 The birth of the study of inheritance

Approximately 25 years before scientists had discovered a link between heredity and the complex processes of mitosis and meiosis, Mendel began his work with the garden pea. Mendel chose the pea plant because it was an annual plant that was small, easy to grow, and had a short generation time. Therefore, he could conduct experiments involving numerous plants and obtain results relatively quickly.

Pea plants are well suited to studies of inheritance because each pea flower contains both female parts (stigma, style, and ovules [where fertilization occurs and the fertilized egg develops]) and male parts (filaments that support anthers on which pollen is found). Both are enclosed and protected by the petals as shown in **Figure 20.1.** The petals remain closed until after pollination and fertilization take place, unlike many other species of plants. The gametes, or sex cells, produced within each flower—pollen grains (containing sperm) within the anthers and eggs within the ovules—are able to fuse and develop into seeds of new plants. (Reproduction in angiosperms such as pea plants is illustrated in Figure 28.9).

Fertilization of this sort, called *self-fertilization*, takes place naturally within individual pea flowers if they are not disturbed. As a result, the offspring of self-fertilizing garden peas are derived from one pea plant, not two. After generations of self-fertilization, some plants produce offspring consistently identical to the parent with respect to certain defined characteristics; these plants are said to be **true-breeding**.

Mendel selected a number of different true-breeding *varieties* of pea plants with which to work. Differing varieties, or strains, of an organism belong to the same species, but each has one or more distinctive characteristics that are passed from parent to offspring. To study the inheritance patterns of these characteristics, Mendel took true-breeding plants and artificially cross-fertilized them. *Cross-fertilization* occurs when the pollen (sperm) of one plant fertilizes the eggs of another plant. To do this, Mendel removed the anthers from a flower before they shed pollen and then dusted the stigma of the flower with pollen from another plant. In this way, Mendel was able to perform experimental crosses between two different true-breeding varieties of pea plants that exhibited differences regarding particular traits. The offspring, or progeny, of the cross between two different varieties of plants of the same species are called *hybrids*. (In other contexts, the word hybrid may also refer to the cross between two different species of organisms. A mule, for example, is the hybrid offspring of a horse and a donkey.)

www.jbpub.com/biology

Gregor Mendel began his study of inheritance 25 years before scientists discovered a link between heredity and the processes of mitosis and meiosis. Mendel worked with pea plants, studying the inheritance patterns of particular characteristics.

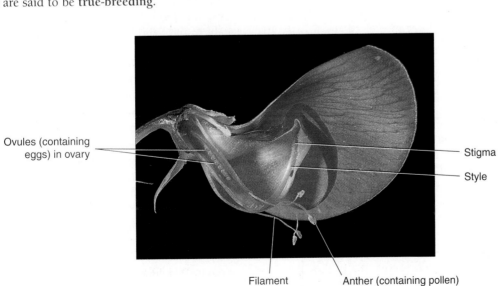

Figure 20.1 Anatomy of a pea plant.

Ovules (containing eggs) in ovary

Stigma

Style

Filament

Anther (containing pollen)

Mendel studied inheritance using pea plants.

20.3 Mendel's study of the inheritance of individual traits

Mendel first designed a set of experiments that involved crossing varieties of pea plants differing from one another in a single characteristic. Mendel chose seven different characteristics, or **traits**, to study, which are listed and shown in **Figure 20.2**. Although various other varieties of pea plants exist, Mendel chose only those varieties that differed from one another clearly and distinctly with respect to one or more of these seven traits.

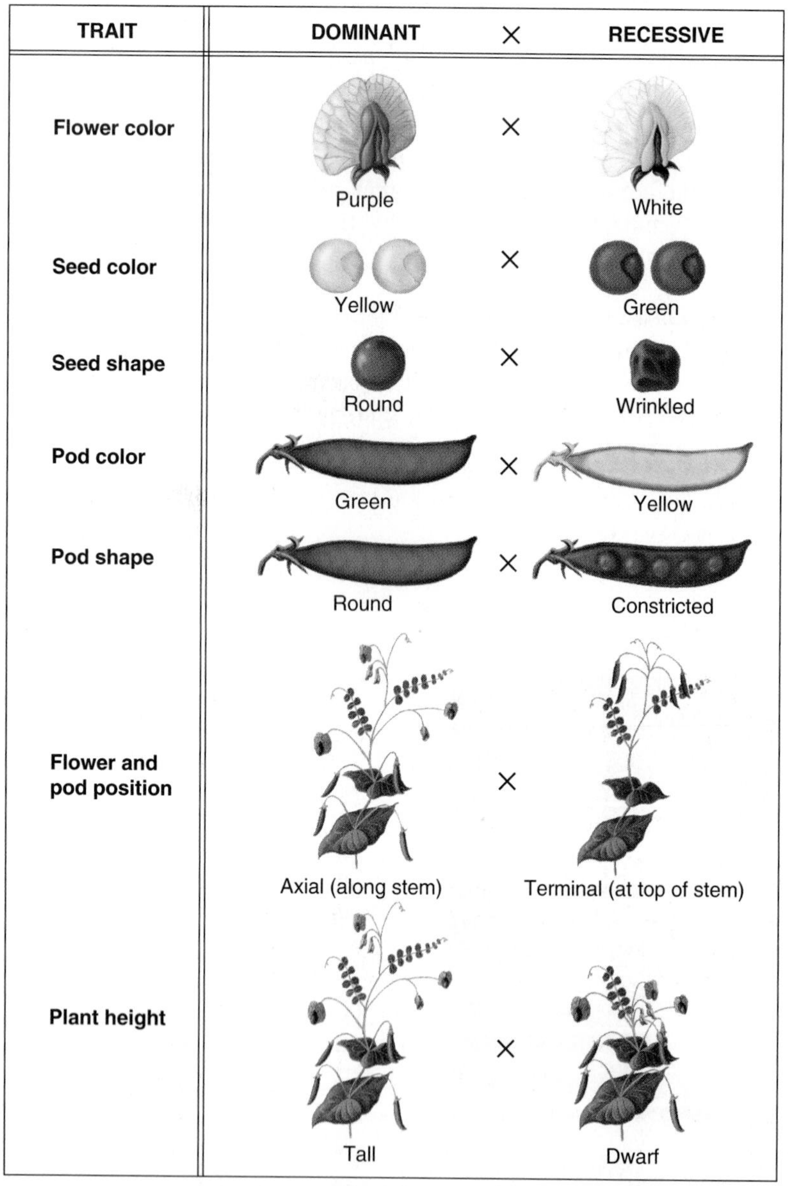

The purpose of Mendel's experiments was to observe the offspring from the crossing of each pair of plants and to look for patterns in the transmission of single traits. Although he was not the first to perform such experiments, he was the first to count and classify the peas that resulted from his crosses and compare the proportions with mathematical models. Mendel planned to use these data to try to deduce the laws by which these traits are passed from generation to generation.

First, Mendel crossed true-breeding plants, each having contrasting forms of the single traits listed in Figure 20.2 by artificially fertilizing one with the other. Mendel called these plants the **parental (P) generation** and called their hybrid offspring the **first filial (F_1) generation**. (The word *filial* is from Latin words meaning "son" and "daughter.") The F_1 offspring, or progeny, are called *monohybrids* because they are the product of two plants that differ from one another in a single trait.

When Mendel crossed two contrasting varieties, such as purple-flowered plants with white-flowered plants (P generation) as shown in **Figure 20.3**, the hybrid offspring (F_1) that he obtained were not intermediate in flower color, as the theory of blending inheritance predicted. Instead, all hybrid offspring in each case resembled *only one* of their parents. Thus, in a cross of white-flowered plants with purple-flowered plants, *all* the F_1 offspring had purple flowers. In a different cross of tall parent plants and short parent plants, *all* the F_1 offspring were tall plants.

Mendel referred to the form of a trait that was expressed in the F_1 plants as *dominating*, or **dominant**, and to the alternative form, which was not expressed in the F_1 plants, as **recessive**. He chose the word recessive because this form of the trait receded, or disappeared entirely, in the hybrids. For each of the seven contrasting pairs of traits that Mendel examined, one member of each pair was dominant; the other was recessive. Figure 20.2 shows which traits Mendel found to be dominant and which traits he found to be recessive.

Then Mendel went a step further. He allowed each F_1 plant to mature and self-fertilize. He collected and planted seeds from each plant, which produced the **second filial (F_2) generation**. He found that most of the F_2 plants exhibited the dominating form of the trait and looked like the F_1 plants, but some exhibited the recessive form (see Figure 20.3). None of the plants had blended characteristics. Therefore, Mendel was able to count the numbers of each of the two contrasting varieties of F_2 progeny and compare these results. His counts for experiments with all seven traits are shown in **Figure 20.4**.

Earlier in history, scientists had carried out hybridization experiments. However, the plants these scientists chose often produced hybrids that differed in ap-

TRAIT	DOMINANT	X	RECESSIVE
Flower color	Purple	X	White
Seed color	Yellow	X	Green
Seed shape	Round	X	Wrinkled
Pod color	Green	X	Yellow
Pod shape	Round	X	Constricted
Flower and pod position	Axial (along stem)	X	Terminal (at top of stem)
Plant height	Tall	X	Dwarf

Figure 20.2 **The seven pairs of contrasting traits in the garden pea studied by Mendel.**

pearance from both their parents because they exhibited blended traits. Also, scientists had not quantified the results of their experiments. Mendel's change in experimental design—counting the progeny—was a key component of his ability to unravel the mystery of certain inheritance patterns.

When Mendel quantified the traits he observed in the F₂ generation, he discovered that for every three plants exhibiting the dominant form of the trait, one exhibited the recessive trait. Put another way, both contrasting forms of the parental characteristics reappeared in the F₂ generation in an approximate 3:1 ratio: Three fourths of the plants exhibited the dominant form (determined in the F₁ generation), and one fourth exhibited the recessive form. Notice that the numbers in Figure 20.4 reflect a 3:1 ratio of dominant-to-recessive forms for each trait. Notice also that in his experiments, Mendel used hundreds and sometimes thousands of plants. Why do you think that such a procedure is part of a good experimental design?

Mendel went on to examine what happened when each F₂ plant was allowed to self-fertilize (see Figure 20.3). He found that the recessive one fourth were always true-breeding. For example, self-fertilizing white-flowered F₂ plants reliably produced only white-flowered offspring. By contrast, only one third of the dominant purple-flowered F₂ individuals (one fourth of the total offspring) proved true-breeding, whereas two thirds of the purple-flowered plants were not true-breeding. This last class of plants produced dominant and recessive F₃ individuals in a ratio of 3:1. The ratio, therefore, of individuals in the F₂ population was the following:

1 true-breeding dominant: 2 not true-breeding dominant: 1 true-breeding recessive

Mendel conducted experiments on pea plants to look for patterns in the transmission of single traits. When Mendel crossed two contrasting varieties, the hybrid progeny (the F₁ generation) resembled only one of their parents. Mendel called the form of a trait that was expressed dominant and the alternative form recessive. He allowed the F₁ plants to self-fertilize. He found that most of the F₂ plants exhibited the dominating form of the trait, but some exhibited the recessive form in a 3:1 ratio. Self-pollinating the F₂ generation showed that some of the F₂ plants were true-breeding and some were not.

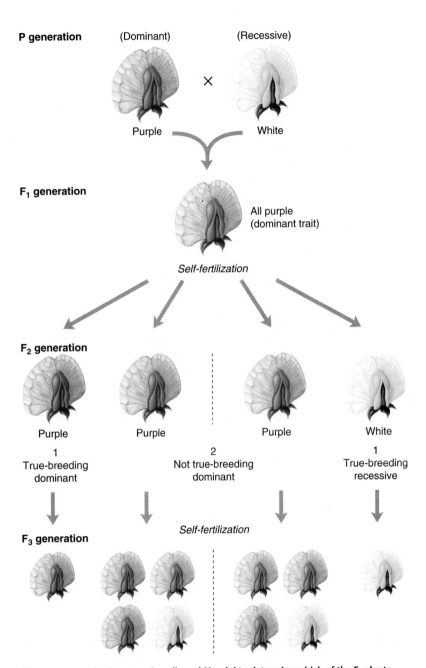

Figure 20.3 The F₃ generation allowed Mendel to determine which of the F₂ plants were true-breeding. By allowing the F₂ generation to self-pollinate, Mendel reasoned from the F₃ offspring that the F₂ generation exhibited the ratio of one true-breeding dominant, to two not-true-breeding dominant, to one true-breeding recessive.

	Flower color	Seed color	Seed shape	Pod color	Pod shape	Flower position	Plant height
Dominant trait	705	6022	5474	428	882	651	787
Recessive trait	224	2001	1850	152	299	207	277

Figure 20.4 Monohybrid crosses of 7 traits of pea plants: Results of the self-fertilization of the F₁ generation.

20.4 Conclusions Mendel drew from his experiments

From his experimental data, Mendel drew conclusions regarding the nature of heredity—conclusions that have withstood further experimentation over time. In fact, Mendel's work is historically looked upon as the birth of **genetics**, the branch of biology dealing with the principles of heredity and variation in organisms. The statements that follow regarding Mendel's conclusions are considered to be the first established principles of genetics.

First, as mentioned previously, Mendel observed that the plants exhibiting the traits he studied did not produce progeny of intermediate appearance when crossed. These observations did not uphold the theory of blending inheritance but suggested instead that traits are inherited as discrete "packets" of information that are either present or absent in a particular generation. Mendel called these discrete bits of information *factors*. These factors, hypothesized Mendel, act later in the offspring to produce the trait. Today, scientists call these factors **genes**, the units of transmission of hereditary characteristics in an organism. In addition, scientists know that a gene is a segment, or piece, of DNA occupying a particular place on a particular chromosome. (These concepts regarding the molecular nature of genes and chromosomes are explained more fully in Chapter 18.)

Second, for each pair of traits that Mendel examined, one alternative form was not expressed in the F_1 hybrids, although it reappeared in some F_2 individuals. As mentioned earlier, Mendel referred to the form of the trait that was expressed in the F_1 plants as "dominating"; today, the preferred term is dominant. He referred to the alternative form, which was not expressed in the F_1 plants, as recessive. He inferred from these observations that each individual, with respect to each trait, contains two factors. Each pair of factors may contain information for (be a code for) the same form of a trait or each member of the pair may code for an alternative form of a trait. Today, scientists call each member of a factor pair an **allele** (uh-LEEL). An allele is a particular form of a gene. Each human, for example, receives one allele for each gene from the mother's egg and one allele for each gene from the father's sperm. The 46 chromosomes of each human cell are actually 23 paired chromosomes, one member of each pair from the sperm and one from the egg.

Third, Mendel hypothesized that because (1) the two factors (alleles) that coded for a trait remained "un-contaminated," not blending with one another as the theory of blending inheritance predicted and (2) the results obtained from experiments on various traits showed similar results, then pea hybrids must form egg cells and pollen cells (gametes) in which the factors (alleles) for each trait separate from one another "in equal shares" during gamete formation. This concept is referred to as **Mendel's law of segregation**. Put in today's terms with today's understandings: Each gamete receives only one of an organism's pair of alleles. Chance determines which member of a pair of alleles becomes included in a gamete. This random segregating process takes place during the process of meiosis, or reduction division of the nuclei of cells destined to be sex cells.

Fourth, Mendel realized that plants exhibiting the dominant trait in his monohybrid crosses of pea plants, when self-fertilized, would breed true or would produce plants exhibiting either the dominant or the recessive form of the characteristic in a 3:1 ratio, respectively. He observed that plants exhibiting the recessive trait, when self-fertilized, would always breed true. These data suggested to Mendel that true-breeding plants receive *only* the dominant factors or the recessive factors from each of their parents and that non-true-breeding plants were hybrids, which received the dominant and the recessive factors in equal shares. Today, scientists call an individual having two identical alleles for a trait **homozygous** (hoe-muh-ZYE-gus) for that trait. The prefix *homo* means "the same"; the suffix *zygous* refers to the zygote, or fertilized egg. An individual having two different alleles for a trait is said to be **heterozygous** (*hetero* means "different") for that trait.

Scientists understand that Mendel's results were well defined because he was studying traits that exhibited complete dominance. This is not always the case. Incomplete dominance is a situation in which neither member of a pair of alleles exhibits dominance over the other. In fact, the "blending" of genes observed by some early investigators may have been visible expressions of incomplete dominance. Since Mendel's time, many examples of incomplete dominance have been found for various traits in both plants and animals (see Chapter 21).

Mendel suggested that traits are inherited as discrete "packets" of information. Each trait contains two factors (a pair of alleles). Mendel's first law of inheritance, the law of segregation, states that each gamete receives only one allele of each pair of alleles in an organism's genetic makeup.

20.5 Analyzing Mendel's experiments

Looking back to Mendel's experiments, you can use the information from the conclusions he drew to further analyze and understand his experiments and data. Mendel used letters to represent alleles, with the dominant allele commonly denoted by an uppercase letter and the recessive allele by the lowercase version of the same letter. In the parental generation, the cross of true-breeding (therefore homozygous) pea plants with purple flowers (the dominant trait) and true-breeding pea plants with white flowers (the recessive trait) can be represented as PP × pp as shown in **Figure 20.5.** The × denotes a cross between two plants (as used previously). Because the

purple-flowered parent can produce only P gametes and the white-flowered parent can produce only p gametes, the union of an egg and a sperm from these parents can produce only heterozygous Pp offspring in the F_1 generation. Because the P allele is dominant, all of the F_1 individuals have purple flowers. The p allele, although present, is not visibly expressed.

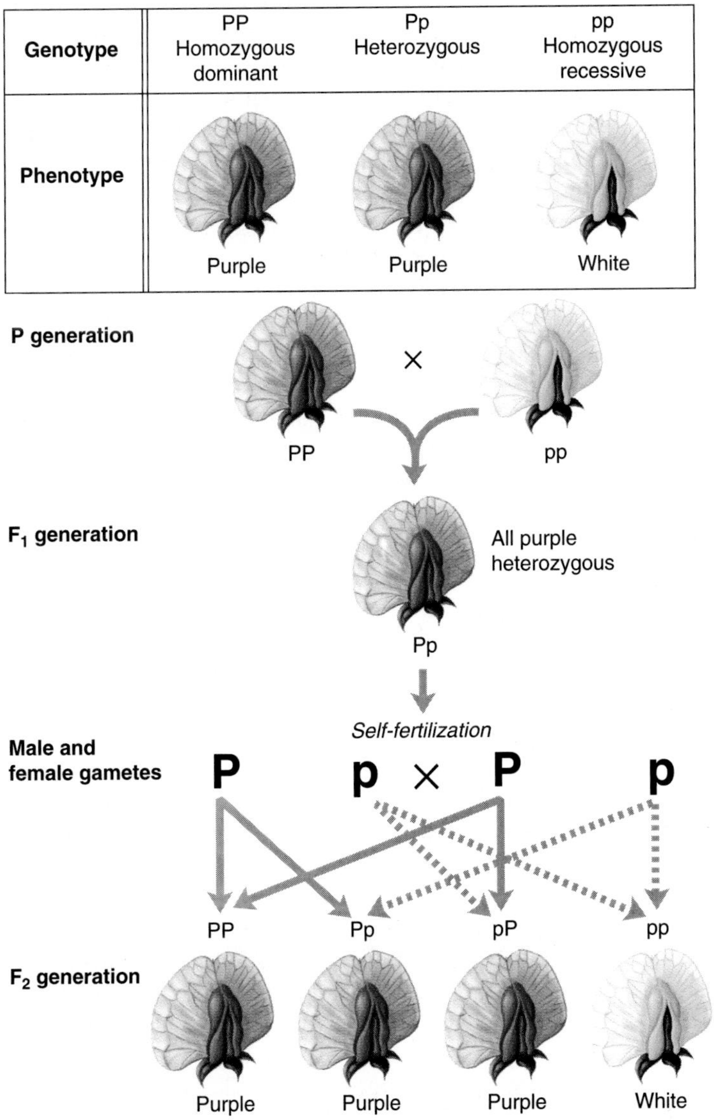

To distinguish between the presence of an allele and its expression, scientists use the term **genotype** (JEEN-uh-type) to refer to an organism's allelic makeup and the term **phenotype** (FEE-nuh-type) to refer to the expression of those genes. The phenotype—the organism's outward appearance—is the end result of the functioning of the enzymes and the proteins coded by an organism's genotype. Figure 20.5 shows the difference between these two terms at the top of the illustration.

The genotype of the F_1 generation of pea plants from the cross of true-breeding plants with purple flowers (PP) and true-breeding plants with white flowers (pp) is Pp, but the phenotype of the flowers of the hybrids is purple because P is the dominant allele. When these F_1 plants are allowed to self-fertilize, the P and p alleles segregate randomly during gamete formation. Their subsequent union at fertilization to form F_2 individuals is also random. Figure 20.5 (bottom) shows the possible combinations of the gametes formed by the F_1 plants through self-fertilization. Their random combination produces plants in an approximate ratio of 1 PP : 2 Pp : 1 pp. The phenotypes are 3 purple : 1 white.

Using a simple diagram called a **Punnett square** (PUN-et) is another way to visualize the possible combinations of genes in a cross. Named after its originator, the English geneticist Sir Reginald Punnett, the Punnett square is used to align possible female gametes with possible male gametes in an orderly way. The male gametes are shown along one side of the square; the female gametes are shown along the other side. As in **Figure 20.6,** the square is divided into smaller squares as columns are drawn vertically and rows are drawn horizontally, providing a "cell" for each possible combination of gene pair.

Whether using a Punnett square or visualizing the gametes and their recombinations as in Figure 20.5, you can see the expected ratios of the three kinds of F_2 plants: one fourth are true-breeding pp white-flowered plants, two fourths are heterozygous Pp purple-flowered plants,

Figure 20.5   Analysis of Mendel's experiments.

Figure 20.6 **Punnett square analysis of Mendel's F₁ generation.** Each smaller cell within the square contains the possible F₂ phenotypic combinations from this cross of F₁ plants. The phenotypic ratio is 3:1, but the genotypic ratio is 1:2:1. Look back at Figure 20.3 and see that the genotypic ratio matches the expected F₂ ratio of true-breeding dominant to non-true-breeding dominant to true-breeding recessive.

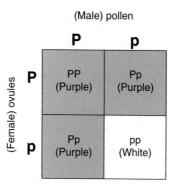

(Male) pollen

	P	**p**
P	PP (Purple)	Pp (Purple)
p	Pp (Purple)	pp (White)

(Female) ovules

and one fourth are pure-breeding PP purple-flowered individuals. The 3:1 phenotypic ratio is really a 1:2:1 genotypic ratio.

> The outward appearance of an individual (the expression of the genes) is referred to as its phenotype. The genetic makeup of an individual is referred to as its genotype. The genotype represents the alleles for given genes that are present. The use of a Punnett square is one way to visualize the genotypes of progeny in simple Mendelian crosses and to illustrate their expected ratios.

20.6 How Mendel tested his conclusions

To test his conclusions further and to distinguish between homozygous dominant and heterozygous phenotypes, Mendel devised two procedures. The first test, already mentioned, was to self-fertilize the plant having a dominant phenotype. If the plant is homozygous dominant, it breeds true: purple-flowered plants produce progeny with purple flowers, tall plants produce tall plants, and so forth. If the plant is heterozygous, on the other hand, it produces dominant and recessive offspring in a 3:1 ratio when self-fertilized.

The other procedure Mendel used is called a *testcross* and is illustrated in Figure 20.7. In this procedure, Mendel crossed the phenotypically dominant test plant whose genotype is unknown (shown in the middle of the illustration) with a known homozygous recessive plant. He predicted that if the test plant is homozygous for the dominant trait (as shown in the right side of the illustration), the progeny will all be hybrids, Pp for example, and will therefore look like the test plant. Alternatively, if the test plant is heterozygous (as shown on the left side of the illustration), then *half* the progeny will be heterozygous and look like the test plant, but half the progeny will be homozygous recessive and therefore will exhibit the recessive characteristic. Figure 20.7 uses the characteristics that Mendel actually employed in his first testcross experiments.

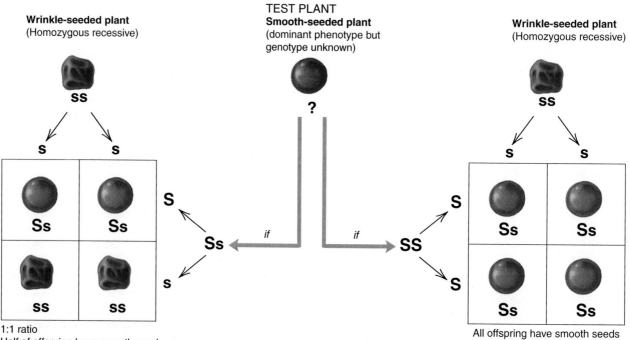

Figure 20.7
A testcross.

Wrinkle-seeded plant (Homozygous recessive) — ss

TEST PLANT
Smooth-seeded plant (dominant phenotype but genotype unknown) — ?

Wrinkle-seeded plant (Homozygous recessive) — ss

if Ss

S	S
Ss	Ss
ss	ss

1:1 ratio
Half of offspring have smooth seeds
Half of offspring have wrinkled seeds

if SS

S	S
Ss	Ss
Ss	Ss

All offspring have smooth seeds

What did Mendel's data reveal? When he performed the testcross using hybrid F_1 plants having smooth seeds (Ss) and crossed them with a variety of plant having wrinkled seeds (ss), he obtained 208 plants: 106 with smooth seeds, and 102 with wrinkled seeds—a 1:1 ratio just as he predicted. These data confirmed the primary conclusion Mendel drew from earlier work: Alternative factors (alleles) segregate from one another in the formation of gametes, coming together in the progeny in a random manner.

To determine the genotype of a phenotypically dominant test plant, it is crossed with a known homozygous recessive plant. If the test plant is homozygous dominant, the progeny will all have the dominant phenotype. If the test plant is heteozygous, half the progeny will have the dominant phenotype and half the progeny will have the recessive phenotype. This procedure is known as a testcross.

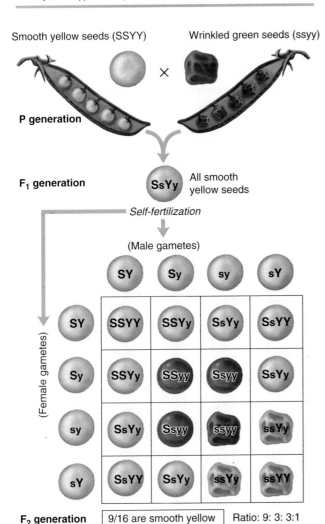

Smooth yellow seeds (SSYY)   Wrinkled green seeds (ssyy)

P generation

×

F_1 generation   SsYy   All smooth yellow seeds

Self-fertilization

(Male gametes)

SY   Sy   sy   sY

(Female gametes)

SY	SSYY	SSYy	SsYy	SsYY
Sy	SSYy	SSyy	Ssyy	SsYy
sy	SsYy	Ssyy	ssyy	ssYy
sY	SsYY	SsYy	ssYy	ssYY

F_2 generation

9/16 are smooth yellow
3/16 are smooth green
3/16 are wrinkled yellow
1/16 are wrinkled green

Ratio: 9: 3: 3:1

20.7 Mendel's study of the inheritance of pairs of traits

Expanding on his law of segregation, Mendel asked a new question. Do the pairs of factors (alleles) that determine particular traits segregate independently of factor pairs that determine other traits? In other words, does the segregation of one factor pair influence the segregation of another?

To answer his question, Mendel first developed a series of true-breeding lines of peas that differed from one another with respect to two of the seven pairs of characteristics with which he had worked in his monohybrid studies. He then crossed pairs of plants that exhibited contrasting forms of the two characteristics and that bred true. For example, he crossed plants having smooth, yellow seeds with plants having wrinkled, green seeds. (**Figure 20.8** shows this cross.) From his monohybrid studies, Mendel knew that the traits "smooth seeds" (S) and "yellow seeds" (Y) are dominant to "wrinkled seeds" (s) and "green seeds" (y). Therefore, the genotypes of the true-breeding parental (P) plants were SSYY and ssyy.

Mendel's F_1 progeny are *dihybrids*—the product of two plants that differ from one another in two traits. As in his monohybrid crosses, all the F_1 progeny had smooth, yellow seeds—the dominant phenotype. Mendel then allowed the F_1 dihybrids to self-fertilize. The seeds from these self-crosses grew into 315 plants having smooth, yellow seeds; 101 plants having wrinkled, yellow seeds; 108 plants having smooth, green seeds; and 32 plants having wrinkled, green seeds—an approximate ratio of 9:3:3:1. Other dihybrid crosses also produced offspring having the same approximate ratio.

Mendel reasoned that if the factors (alleles) for seed color and seed shape segregated into gametes independently of one another and were therefore inherited independently of one another, then the outcome for each trait would exhibit the 3:1 ratio of a monohybrid cross. Looking at Mendel's results, 315 + 108 (423) plants had smooth seeds and 101 + 32 (133) plants had wrinkled seeds. Put simply, *three times* as many plants had smooth seeds as had wrinkled seeds—a 3:1 ratio. Study Mendel's numbers for the trait "seed color." How many plants produced yellow seeds? How many produced green seeds? What is the approximate ratio of plants having yellow seeds to plants having green seeds?

Figure 20.8   A Punnett square showing the results of Mendel's self-cross of dihybrid smooth yellow-seeded plants. The approximate ratio of the four possible combinations of phenotypes is predicted to be 9:3:3:1, the ratio that Mendel found.

The contrasting factors (alleles) for the genes of seed shape and seed color assort independently from one another during gamete formation. This concept is referred to as **Mendel's law of independent assortment.** Put in today's terms with today's understandings: The distribution of alleles for one trait into the gametes does not affect the distribution of alleles for other traits.

Even though you may understand the concept of the independent assortment of alleles, you may not understand how two 3:1 ratios combine to produce a 9:3:3:1 ratio. These ratios and the relationship between these ratios are governed by the laws of *probability,* or chance. Each ratio is not simply a statement of the comparative numbers of plants Mendel found, but is also a predictive statement of events that could occur in the future under the same conditions.

Analysis of a Dihybrid Cross Using Probability Theory

Figure 20.9 (left) shows the Punnett square results of the self-cross of hybrid plants having smooth seeds. The genotype of these plants is Ss. The male gametes (pollen) S and s are shown at the top of the Punnett square, and the female gametes (eggs) S and s are shown along the left side. These four gametes can combine in four ways. What is the probability that this cross will produce a plant having wrinkled seeds?

There is only a one-in-four possibility that the ss gametes will combine because there are only four possible combinations of gametes and the alleles segregate randomly as the gametes are formed. Likewise, there is a three-in-four probability that gametes can combine in such a way that they will produce plants having smooth seeds: SS, Ss, and sS. Although Ss and sS are the same genotypes, the reversed position of the alleles represents the dominant and recessive alleles as contributed by each of two parent plants; thus, the 3:1 phenotypic ratio. The alleles for the trait seed color (shown in Figure 20.9, right), when looked at independently of seed shape, segregate in the same manner to produce plants having yellow seeds and plants having green seeds in a 3:1 ratio, respectively.

To illustrate further, you can simulate the combination of the gametes S and s of two parents by performing a simple activity. Take two pennies and tape the letter "S" on one side of each and "s" on the other. Flip the two pennies 100 times and record your results. The probability of flipping an S on one of the pennies is one out of two, or $\frac{1}{2}$. The chance of flipping an S on the other penny is also $\frac{1}{2}$. The probability that an event will occur at the same time as another independent event is simply the product of their individual probabilities. Therefore, the probability of flipping SS with the two pennies is $\frac{1}{2} \times \frac{1}{2}$, or $\frac{1}{4}$. (Did you flip SS approximately 25 times

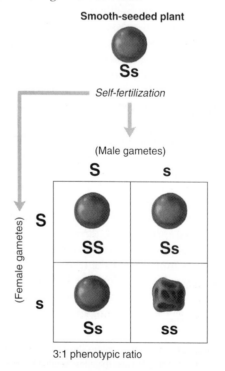

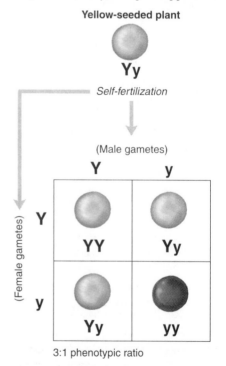

Figure 20.9 **Mendel's law of independent assortment.** As this diagram demonstrates, the contrasting alleles for seed shape (S and s) and color (Y and y) assort independently from one another during gamete formation.

out of 100?) Likewise, the probability of flipping ss is ¼, flipping Ss is ¼, and flipping sS is ¼. (Do your data match?) If the pennies were in fact gametes, ¾ would produce plants having smooth seeds and ¼ would produce plants having wrinkled seeds—a 3:1 ratio.

Now look at these traits as they occur together in a dihybrid cross. The probability that a plant with wrinkled, green seeds will appear in the F_2 generation is equal to the probability of observing a plant with wrinkled seeds (¼) times the probability of observing a plant with green seeds (¼), or 1/16. The probability that a plant with smooth, green seeds will appear in the F_2 generation is equal to the probability that the F_1 parents will produce a plant with smooth seeds (¾) times the probability that they will produce a plant with green seeds (¼), or 3/16. Can you figure out the probability that a plant with wrinkled, yellow seeds will appear in the F_2 generation or that a plant with smooth, yellow seeds will appear?

Analysis of a Dihybrid Cross Using a Punnett Square

Figure 20.8 portrays the self-cross of dihybrid F_1 plants having smooth, yellow seeds, showing how the gametes segregate to produce the F_2 plants. The eight gametes (four female gametes [eggs] and four male gametes [sperm]) of the two parents can combine in 16 different ways. What is the probability that this cross will produce a plant having wrinkled, green seeds?

There is only a 1 in 16 possibility that an sy egg will combine with an sy sperm producing ssyy offspring because there are only 16 possible combinations of gametes, assuming that the alleles segregate randomly and assort independently as the gametes are formed. Likewise, there are only 3 combinations of gametes out of 16 possible combinations that will produce plants having wrinkled, yellow seeds: ssYY, ssYy, and ssyY. Following the same reasoning and using the Punnett square, notice that there are 3 possible combinations of gametes that will form plants producing smooth, green seeds and 9 possible combinations of gametes that will form plants producing smooth, yellow seeds. So whether you calculate the expected offspring in a dihybrid cross by using a Punnett square or whether you do this analysis by using probability theory, the ratio of the F_2 progeny is predicted to be 9:3:3:1.

Mendel's second law of inheritance, the law of independent assortment, states that the distribution of alleles for one trait into the gametes does not affect the distribution of alleles for other traits.

20.8 The connection between Mendel's factors and chromosomes

During the years Mendel was experimenting with pea plants to determine the principles of heredity, other scientists were studying the structure of cells. Although Mendel published his theories of inheritance in 1866, few scientists read his work or understood its significance. In the late 1870s, scientists had described the process of nuclear division now known as mitosis, and the process of nuclear reduction-division now known as meiosis (see Chapter 18). In fact, in 1888, scientists gave the name *chromosomes* to the discrete, threadlike bodies that form as the nuclear material condenses during these processes of cell division. However, they still had no idea of the link between Mendel's *factors* of inheritance and the newly named chromosomes.

In 1900, three biologists, Carl Correns, Hugo de Vries, and Eric von Tschermak, independently worked out Mendel's principles of heredity. However, they knew nothing of Mendel's work until they searched the literature before publishing their results. The rediscovery of Mendel's work—his hypotheses supported by the independent work of others—helped scientists begin to make connections between Mendel's ideas and chromosomes.

By the late 1800s, scientists had observed the process of fertilization and knew that sexual reproduction required the union of an egg and a sperm. However, they did not know how each contributed to the development of a new individual. An American graduate student, Walter Sutton, suggested that if Mendel's hypotheses were correct, then each gamete must make equal hereditary contributions. Sutton also suggested that because sperm contain little cytoplasm, the hereditary material must reside within the nuclei of the gametes. He noted that chromosomes are in pairs and segregate during meiosis, as did Mendel's factors. In fact, the behavior of chromosomes during the meiotic process paralleled the behavior of the hereditary factors. Using this line of reasoning, Sutton suggested that Mendel's factors were located on the chromosomes. Years later, the most conclusive evidence to uphold Sutton's chromosomal theory of inheritance was provided by a single, small fly.

By the late 1800s, scientists had discovered chromosomes in the nucleus of the cell but did not yet connect their function to heredity and Mendel's work. Near the turn of the century, after three biologists independently upheld Mendel's ideas, scientists began to make links between chromosomes and inheritance.

20.9   Sex linkage

In 1910, Thomas Hunt Morgan, studying the fruit fly *Drosophila melanogaster,* detected a male fly that differed strikingly from normal flies of the same species. This fly had white eyes, as shown in the upper photo of **Figure 20.10.** Normal fruit fly eyes are red (Figure 20.10, bottom). Morgan quickly designed experiments to determine whether this new trait was inherited in a Mendelian fashion.

Morgan first crossed the white-eyed male fly to a normal female to see whether red or white eyes were dominant. All F_1 progeny had red eyes, and Morgan therefore concluded that red eye color was dominant over white. Following the experimental procedure that Mendel had established long ago, Morgan then crossed flies from the F_1 generation with each other. Eye color did indeed segregate among the F_2 progeny as predicted by Mendel's theory. Of 4252 F_2 progeny that Morgan examined, 782 had white eyes—an imperfect $3:1$ ratio but one that nevertheless provided clear evidence of segregation. Something was strange about Morgan's result, however—something totally

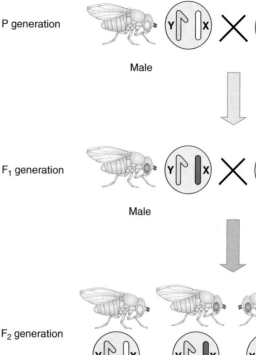

P generation

Male   Female

F_1 generation

Male   Female

F_2 generation

Males   Females

Figure 20.11   Morgan's experiment demonstrating the chromosomal basis of sex-linkage in *Drosophila melanogaster.*

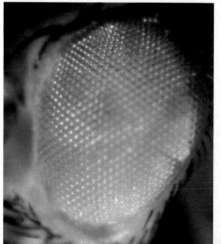

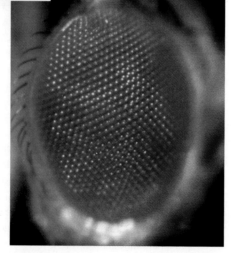

Figure 20.10   Red-eyed and white-eyed *Drosophila melanogaster.*

unpredicted by Mendel's theory: All of the white-eyed F_2 flies were males!

How could this strange result be explained? The solution to this puzzle involves gender. In *Drosophila* (as in most animals), the gender of the fly is determined by specific chromosomes called X and Y chromosomes. Female flies have four pairs of chromosomes, with one of those pairs being two X chromosomes. Male flies, on the other hand, have three pairs plus an X and Y chromosome. To explain his results, Morgan deduced that the white-eyed trait is located on the X chromosome but is absent from the Y chromosome. (Scientists now know that the Y chromosome carries relatively few functional genes.) Because the white-eye trait is recessive to the red-eye trait, Morgan's result was a natural consequence of the Mendelian segregation of alleles.

Figure 20.11 shows Morgan's experiment demonstrating the chromosomal basis of sex-linkage. In the P generation (top), white-eyed mutant male flies are crossed with normal red-eyed females. The white-eyed trait is shown on the male's X chromosome, while the Y chromosome does not contain genes for eye color. Be-

Just Wondering.... *Real Questions Students Ask*

Why do scientists study tiny organisms like fruit flies?
Why don't they choose something larger?

Drosophila melanogaster, known as a fruit fly to most people, is an excellent organism for geneticists to study for many reasons. First, fruit flies are easy to breed and their life cycles are short. A single female can lay several hundred eggs, which develop into adults within 12 days. Therefore, a geneticist can study multiple generations of flies within a short period of time. Second, fruit flies are easy to maintain and take up little room in the laboratory. Populations of flies can be kept in small containers with easily prepared media. (Thomas Hunt Morgan, one of the first to study fruit-fly genetics, housed fruit flies in half-pint bottles.) Third, fruit flies exhibit variations in certain inherited traits such as eye color and wing formation that are easy to see under a dissecting microscope. The geneticist can easily anesthetize the population, study the flies, and place them back in the bottle before they revive. Fourth, fruit flies have only four pairs of chromosomes: three pairs of autosomes (body chromosomes) and one pair of sex chromosomes. Experiments can often be simpler using organisms with few chromosomes rather than organisms with a large number of chromosomes.

Underlying the use of various organisms for genetic studies is the assumption that genetic principles are universal. Thus, Mendel and Morgan were able to articulate laws of genetics that apply to all living things, even though they studied only fruit flies and peas.

www.jbpub.com/biology

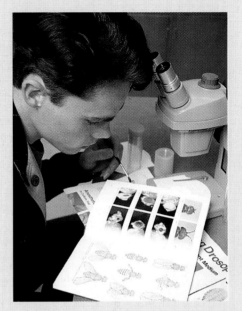

cause the male has only one trait for eye color, that trait is expressed in the phenotype, regardless of whether it is dominant or recessive. The red-eyed trait is shown on each of the female's X chromosomes.

The F_1 generation flies all exhibit red eyes. The male flies have inherited the chromosome carrying the red-eyed trait from their mothers and the Y chromosome from their fathers. Therefore, all the males will be red-eyed. The females have inherited the chromosome carrying the red-eyed trait from their mothers and the white-eyed trait from their fathers. Since the red-eyed trait is dominant to white, all females in the F_1 generation will be red-eyed as well.

In the F_2 generation, males can inherit either the red-eyed or white-eyed trait from their mothers. Therefore, half the males will have red eyes and half will have white eyes. The females can be either homozygous red-eyed or heterozygous; in either case they will be red-eyed.

Morgan's experiment is one of the most important in the history of genetics because it presented the first clear evidence upholding Sutton's theory that the factors determining Mendelian traits are located on the chromosomes. When Mendel observed the segregation of alternative traits in pea plants, he was observing the outward reflection of the meiotic segregation of homologous chromosomes. Morgan's work also revealed that certain traits may be located on the chromosomes that determine the gender of an organism. Such traits are said to be *sex-linked* or *X-linked.*

Throughout human history, an understanding of the nature of heredity was highly speculative until the work of Mendel, Sutton, and Morgan. These early geneticists took the first real steps toward solving the puzzles of inheritance and laid the foundation for one of the great scientific advancements of the twentieth century: an understanding of the nature of genetic material and how it is transmitted from generation to generation. This basic outline of heredity led to a long chain of questions and was to be modified and refined as scientists provided additional experimental data and evidence. In one line of questioning described in Chapter 18, scientists such as Beadle and Tatum probed the structure and function of genes at the molecular level. Another line of questioning regards inheritance in humans and is the topic of Chapter 21.

> From his observation of chromosomes during meiosis and his knowledge of Mendel's work, Walter Sutton hypothesized that the hereditary factors were found on the chromosomes. Thomas Hunt Morgan provided experimental evidence to uphold this hypothesis, also revealing that certain traits may be located on the chromosomes that determine the gender of an organism. Such traits are said to be *sex-linked or X-linked.*

Key Concepts

For centuries, humans have sought to understand inheritance.

20.1 Prior to the nineteenth century, some scientists and philosophers suggested that parents gave off particles from their bodies that traveled to their sex organs during reproduction and were blended in the offspring.

20.1 Some scientists and philosophers held that either sperm or eggs carried tiny preformed beings within them.

20.1 Not until the mid-1800s did scientists realize that new life arose from old life in the form of new cells arising from old cells.

20.2 In the mid-1800s, Gregor Mendel, an Austrian monk trained in botany and mathematics, studied patterns of inheritance using garden pea plants.

20.2 Using true-breeding varieties of pea plants that exhibited alternative forms of seven different traits, Mendel artificially cross-fertilized them.

Mendel studied inheritance using pea plants.

20.3 The purpose of Mendel's experiments was to observe the offspring from the crossing of each pair of plants and to look for patterns in the transmission of traits.

20.3 Mendel found that alternative traits may mask each other's presence.

20.3 From counting progeny types, Mendel learned that the alternative traits that were masked in hybrids appeared only 25% of the time when the hybrids were self-crossed (self-fertilized).

20.3 This finding regarding alternative traits, which led directly to Mendel's model of heredity, is usually referred to as the Mendelian ratio of 3:1 dominant to recessive traits.

20.4 Mendel deduced from the 3:1 ratio that traits are specified by discrete factors that do not blend in the offspring.

20.4 Today, scientists refer to Mendel's factors as genes and to alternative forms of his factors as alleles.

20.4 Because Mendel observed that traits did not blend in the offspring, Mendel hypothesized that alleles separate from one another during gamete formation; we call this concept Mendel's law of segregation.

20.5 To distinguish between the presence of an allele and its outward expression, scientists use the term phenotype for the outward appearance of an individual and genotype for the genetic makeup of an individual.

20.5 The use of a Punnett square is one way to visualize the genotypes of progeny in simple Mendelian crosses and to illustrate their expected ratios.

20.6 To determine the genotype of a phenotypically dominant test plant, it is crossed with a known homozygous recessive plant, which is a procedure known as a testcross.

20.6 If the test plant used in a testcross is homozygous dominant, the progeny will all have the dominant phenotype; if the test plant is heterozygous, half the progeny will have the dominant phenotype and half the progeny will have the recessive phenotype.

20.7 Working with pea plants that differed from one another in two characteristics, Mendel discovered that alleles of different genes assort independently during gamete formation, a concept referred to as Mendel's law of independent assortment.

Other scientists built on Mendel's work.

20.8 Near the turn of the twentieth century, after three biologists independently upheld Mendel's ideas, scientists began to make links between chromosomes and inheritance.

20.9 Thomas Hunt Morgan provided the first clear evidence that genes reside on chromosomes.

20.9 Morgan demonstrated that the segregation of the white-eye trait in *Drosophila melanogaster* was associated with the segregation of the X chromosome, the one responsible for sex determination.

Key Terms

allele (uh-LEEL) *444*
dominant *442*
first filial (F$_1$) generation
  (FIL-ee-uhl) *442*
genes *444*
genetics (juh-NET-iks) *444*
genotype (JEEN-uh-type) *445*

heterozygous
  (het-uhr-uh-ZYE-gus) *444*
homozygous
  (hoe-muh-ZYE-gus) *444*
Mendel's law of independent
  assortment *448*
Mendel's law of segregation *444*
parental (P) generation *442*

phenotype (FEE-nuh-type) *445*
Punnett square (PUN-et) *445*
recessive *442*
second filial (F$_2$) generation
  (FIL-ee-uhl) *442*
traits *442*
true-breeding *441*

KNOWLEDGE AND COMPREHENSION QUESTIONS

1. Genetically, how do offspring of sexual reproduction differ from those produced by asexual reproduction? Compare the amount of variation introduced within a species that reproduces sexually with a species that reproduces asexually. (*Comprehension*)

2. Distinguish between self-fertilization and cross-fertilization in plants. (*Comprehension*)

3. What is Mendel's law of segregation? When does this segregation occur? (*Knowledge*)

4. Assume that "L" represents the dominant trait of having long leaves, and "l" represents the recessive short-leafed trait in a plant. In the parental generation, you cross a homozygous long-leafed plant with a homozygous short-leafed plant. Draw a Punnett square illustrating this cross, and give the genotypes and phenotypes of the F_1 generation. (*Comprehension*)

5. Using a Punnett square, show the genotypes and phenotypes of the F_2 generation if the F_1 plants in question 4 are self-fertilized. (*Comprehension*)

6. In question 5, what is the probability that the F_2 plants will have short leaves? (*Comprehension*)

7. What was Mendel testing when he used a testcross? What procedure did he use, and what was the outcome? (*Knowledge*)

8. What is Mendel's law of independent assortment? (*Knowledge*)

9. In a dihybrid cross between organisms that are heterozygous for both traits, what is the probability that their offspring will exhibit the phenotype for both recessive traits? Both dominant traits? (*Comprehension*)

10. How did the work of Walter Sutton and Thomas Hunt Morgan change the way scientists viewed the role of sperm in heredity and reproduction? (*Knowledge*)

KEY
Knowledge: Recalling information.
Comprehension: Showing understanding of recalled information.

CRITICAL THINKING REVIEW

1. Was Mendel, in your opinion, able to successfully assess the phenotype of his plant crosses? Was he able to assess genotype with complete accuracy? What may affect the assessment of a genotype? (*Application*)

2. Both Gregor Mendel and Thomas Hunt Morgan chose organisms to study that were well suited to their particular investigations. Provide evidence to uphold this claim. (*Application*)

3. Can you think of a distinctive human trait that is *not* inherited? How would you test this hypothesis? (*Synthesis*)

4. Using your understanding of the process of meiosis from Chapter 19 and your understanding of Mendel's Law of Independent Assortment from this chapter, draw cells during the stages of meiosis that show how two genes on nonhomologous chromosomes assort independently. (*Synthesis*)

5. If purple-flowered pea plants were not dominant to white-flowered pea plants, and if the inheritance of these traits did "blend," what would you expect the phenotype of F_1 plants to be if produced from a cross of true-breeding purple-flowered plants and true-breeding white-flowered plants? If the F_1 plants were allowed to self-fertilize, what would you expect would be the phenotype(s) of the F_2 generation of plants? If more than one phenotype would occur, give the ratio of the phenotypes. (*Analysis*)

KEY
Application: Using information in a new situation.
Analysis: Breaking down information into component parts.
Synthesis: Putting together information from different sources.

1. Among Hereford cattle, there is a dominant allele called *polled;* the individuals that have this allele lack horns. After college, you become a cattle baron and stock your spread entirely with polled cattle. You have many cows and few bulls. You personally make sure that each cow has no horns. Among the calves that year, however, some grow horns. Angrily, you dispose of them and make certain that no horned adult has gotten into your pasture. The next year, however, more horned calves are born. What is the source of your problem? What should you do to rectify it?

2. Many animals and plants bear recessive alleles for albinism, a condition in which homozygous individuals completely lack any pigments. An albino plant lacks chlorophyll and is white. An albino person lacks melanin. If two normally pigmented persons heterozygous for the same albinism allele have children, what proportion of their children would be expected to be albino?

3. Your uncle dies and leaves you his racehorse, Dingleberry. To obtain some money from your inheritance, you decide to put the horse out to stud. In looking over the studbook, however, you discover that Dingleberry's grandfather exhibited a rare clinical disorder that leads to brittle bones. The disorder is hereditary and results from homozygosity for a recessive allele.

   If Dingleberry is heterozygous for the allele, it will not be possible to use him for stud because the genetic defect may be passed on. How would you go about determining whether Dingleberry carries this allele?

4. In *Drosophila*, the allele for dumpy wings (d) is recessive to the normal long-wing allele (D). The allele for white eye (w) is recessive to the normal red-eye allele (W). In a cross of DDWwxDdww, what proportion of the offspring are expected to be "normal" (long wing, red eye)? What proportion "dumpy, white"?

5. Your instructor presents you with a *Drosophila* named Oscar. Oscar has red eyes, the same color that normal flies possess. You add Oscar to your fly collection, which also contains Heidi and Siegfried, flies with white eyes, and Dominique and Ronald, which are from a long line of red-eyed flies. Your previous work has shown that the white-eyes trait exhibited by Heidi and Siegfried is caused by their being homozygous for a recessive allele. How would you determine whether Oscar was heterozygous for this allele?

6. In some families, children are born who exhibit recessive traits (and who therefore must be homozygous for the recessive allele specifying the trait), even though one or both of the parents do not exhibit the trait. What can account for this occurrence?

Complete the dihybrid cross shown here by first filling in the genotypes of the male gametes and the female gametes. Then complete the cross, filling in the Punnett square. What is the ratio of the phenotypes?

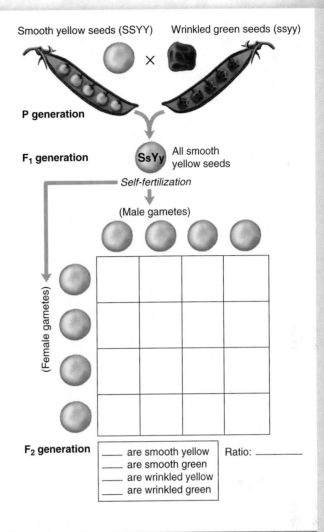

Smooth yellow seeds (SSYY)   Wrinkled green seeds (ssyy)

×

P generation

F₁ generation   SsYy   All smooth yellow seeds

Self-fertilization

(Male gametes)

(Female gametes)

F₂ generation

_____ are smooth yellow   Ratio: _____
_____ are smooth green
_____ are wrinkled yellow
_____ are wrinkled green

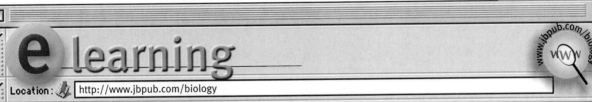

Human Genetics

It is a royal pedigree—one of interest not just to historians but to geneticists as well. Queen Victoria of Britain, shown in this portrait surrounded by members of her immediate family, would introduce into the royal line a genetic mutation that would affect not only the British royal family, but those of Spain and Russia as well.

Queen Victoria affected many of her descendants because she carried a single allele—a mutant, or changed gene that codes for a disorder known as hemophilia. Hemophilia is a hereditary condition in which the blood clots slowly or not at all. It is a recessive disorder, expressed only when an individual does not have a normal blood-clotting allele that masks the mutant's appearance. In addition to being recessive, the allele for this type of hemophilia is sex-linked—it is located on the X chromosome, one of the chromosomes that determines sex. This mutant gene probably arose in one of the sex cells from which Queen Victoria developed because the disease was not manifest in earlier generations.

CHAPTER 21

Organization

Karyotypes are used to study the inheritance of chromosomal abnormalities.

Mutations play a key role in the development of cancer.

Human traits are inherited in a variety of ways.

21.1 The inheritance of abnormal numbers of autosomes

In addition to a pair of **sex chromosomes** (two X chromosomes in females and an X and a Y chromosome in males), each human body cell also contains 22 pairs of **autosomes** (AW-tuh-somes). Unlike sex chromosomes, autosomes (or "body" chromosomes) are the same in both sexes.

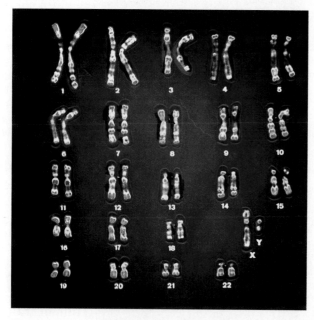

Figure 21.1 **A normal karyotype of a human male.**

The 23 pairs of human chromosomes (including the two sex chromosomes) are shown in Figure 21.1. Arranged in this manner according to size, shape, and other characteristics, the chromosome pairs make up a *karyotype*. Notice the small Y chromosome in pair 23 that is characteristic of males.

Looking at a karyotype can often help researchers see genetic disorders if they are caused by the loss of all or part of a chromosome or by the addition of extra chromosomes or chromosome fragments. Changes in single genes *cannot* be seen. The differences between alleles that code for alternative forms of a trait lie in the chemical structure of the DNA, so they are invisible in a karyotype. Permanent changes in the genetic material, whether they affect single genes, pieces of chromosomes, whole chromosomes, or entire sets of chromosomes, are called **mutations**.

Look at the karyotype in Figure 21.2a, and compare it with that in Figure 21.1. What differences do you see? Carefully examine chromosome 21. One karyotype contains an extra copy of this chromosome, a situation called *trisomy 21*. J. Langdon Down first described the developmental defect produced by trisomy 21 in 1866. For this reason, it is called *Down syndrome*. In these individuals the maturation of the skeletal system is delayed, so persons having Down syndrome generally have a short, stocky build, short hands, flattened facial features, and poor muscle tone. In addition, they are mentally retarded.

Down syndrome is only one genetic disorder caused by the inheritance of an abnormal number of autosomal or sex chromosomes. How does such a situation arise? In humans, it comes about almost exclusively as a result of errors during meiosis. Meiosis is the process of nuclear division in which the number of chromosomes in cells is halved during gamete formation (see Chapter 18).

Early in meiosis, pairs of chromosomes called *homologues* (the pairs of chromosomes shown in the karyotype) line up side by side in a process called *synapsis*. At this point in meiosis, each chromosome is made up of two sister chromatids, so each pair of homologous chromosomes has four chromatids. Gametes can gain or lose chromosomes at this time if two homologous chromosomes fail to separate, or disjoin.

Figure 21.2 **Down syndrome.** (a) Karyotype of a male with Down syndrome. (b) Down syndrome child with his mother.

Figure 21.3 shows what happens if **nondisjunction** (non-dis-JUNK-shun) occurs at either the first meiotic division or the second meiotic division. Figure 21.3a shows nondisjunction occurring during meiosis I, as shown at the top of the figure. (Compare with normal meiosis I in Figure 21.3b.) If a homologous pair of chromosomes does not separate during meiosis I, both chromosomes will appear in one sister cell and the other sister cell will receive one chromosome less than normal (no chromosomes for our example). When these cells undergo meiosis II, none of the resultant cells are normal. Two will contain an extra chromosome; two will be missing a chromosome.

If nondisjunction occurs during meiosis II (when two sister chromatids fail to separate), the cells resulting from this division will be abnormal, as seen in Figure 21.3b. One will have an extra chromosome and one will receive one chromosome less than normal (no chromosomes for our example). The cell in which sister chromatids separate normally will yield normal sister cells (right side).

The cause of nondisjunction is not known. However, as shown in the graph in Figure 21.4, the occurrence of nondisjunction of chromosome 21, which results in Down syndrome, increases with maternal age and increases sharply after age 35. (As a woman ages, so do her eggs.) In mothers younger than 20 years of age, the occurrence of Down syndrome children is only about 1 per 1700 births. In mothers 20 to 30 years old, the incidence is only slightly greater—about 1 per 1400. However, in mothers 30 to 35 years old, the incidence almost doubles to 1 per 750. In mothers older than 45 years of age, the incidence of Down syndrome babies is approximately 1 in 32 births.

The incidence of nondisjunction of other chromosomes also rises with maternal age. However, babies with other serious autosomal chromosome abnormalities are rare. Fertilized eggs with the improper number of chromosomes are almost always inviable; that is, they are unable to survive. These eggs do not begin normal development or implantation and are cast out of the body with the menstrual flow—a process called

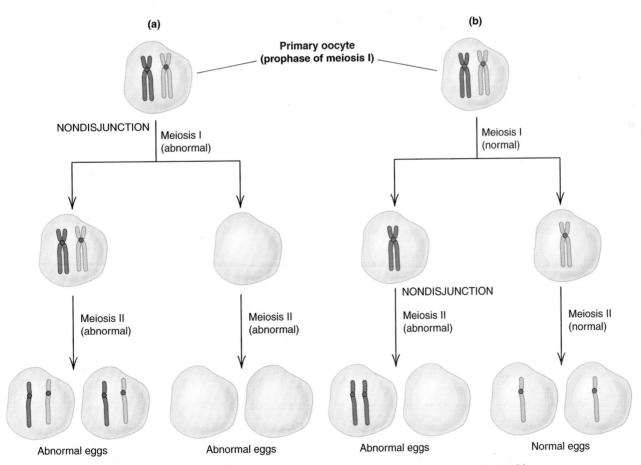

Figure 21.3 **Nondisjunction.** Nondisjunction during meiosis I (a) as compared to nondisjunction during meiosis II (b).

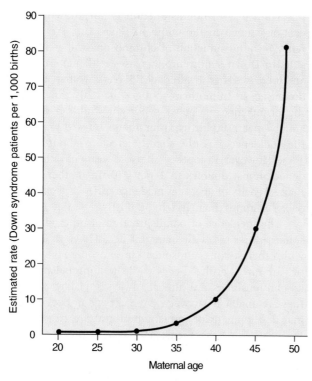

Figure 21.4 **Incidence of Down syndrome vs. maternal age.** This graph shows that there are few births of Down syndrome babies born to mothers under age 30. As maternal age rises to 40 years, 10 out of every 1000 births are of Down syndrome children (1%). As a woman approaches menopause, her chances of bearing a Down syndrome child are over 8% (80 out of 1000 births).

spontaneous abortion. This increase in the incidence of nondisjunction and its negative impact on the viability of zygotes is one reason older women often have a harder time conceiving than do younger women.

Each human cell contains 23 pairs of chromosomes. A karyotype is a picture of these chromosome pairs and shows whether a person has inherited a fewer or greater number of chromosomes than is normal or has lost or gained pieces of chromosomes.

During gamete formation, homologous chromosomes occasionally fail to separate after synapsis. This occurrence, called nondisjunction, results in gametes with abnormal numbers of chromosomes. Down syndrome is a genetic disorder produced when an individual receives three (instead of two) 21 chromosomes.

21.2 The inheritance of abnormal numbers of sex chromosomes

Nondisjunction can also occur with the sex chromosomes. However, persons inheriting an extra X chromosome or inheriting one X too few do not have the severe developmental abnormalities that persons with too many or too few autosomes do. Nevertheless, persons who inherit abnormal numbers of sex chromosomes often have abnormal physical features.

Triple X Females

When X chromosomes fail to separate in meiosis, some gametes are produced that possess both of the X chromosomes; the other gametes have no sex chromosome and are designated O. If an XX gamete joins an X gamete during fertilization, the result is an XXX (triple X) zygote. Even though triple X females usually have underdeveloped breasts and genital organs, they can often bear children. In addition, a small number of XXX people have lower-than-average intelligence. Although rare, a few individuals have been discovered to have tetra X (XXXX) and penta X (XXXXX) genotypes. Individuals having these genotypes are similar phenotypically to triple X individuals but are usually mentally retarded.

Klinefelter Syndrome

If the XX gamete joins a Y gamete, the result is quite serious. The XXY zygote develops into a sterile male, who has, in addition to male genitalia and characteristics, some female characteristics, such as breasts (Figure 21.5) and a high-pitched voice. In some cases, XXY individu-

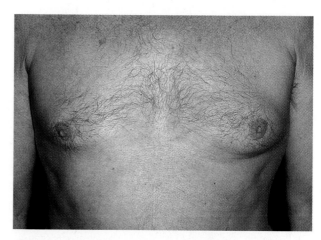

Figure 21.5 **Klinefelter syndrome.** A male with Klinefelter syndrome (XXY) exhibits some female characteristics, such as enlarged breasts (enlarged left breast shown here).

als have lower-than-average intelligence. This condition, called *Klinefelter syndrome,* occurs in about 1 out of every 600 male births.

Turner Syndrome

If an O gamete (no X) from the mother fuses with a Y gamete, the resulting OY zygote is nonviable and fails to develop further. If, on the other hand, an O gamete from either the mother or the father fuses with an X gamete to form an XO zygote, the result is a sterile female of short stature, a "webbed" neck, low-set ears, a broad chest, and immature sex organs that do not undergo puberty changes (Figure 21.6). The mental abilities of an

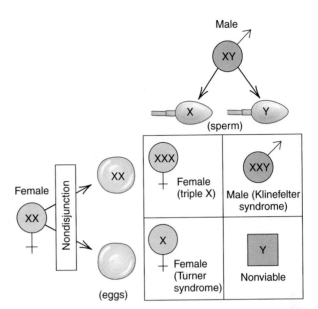

Figure 21.7   How nondisjunction can result in abnormalities in the number of sex chromosomes. Both Klinefelter syndrome and Turner syndrome result from nondisjunction of either the male or female gamete. This diagram shows how these genetic disorders occur when nondisjunction takes place in female gametes. It also shows how nondisjunction in female gametes can result in a triple X female.

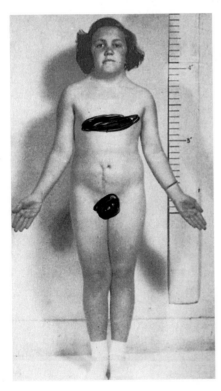

Figure 21.6   Turner syndrome. A female with Turner syndrome (XO) exhibits short stature, webbed neck, low-set ears, broad chest, and immature sex organs.

XO individual are slightly below normal. This condition, called *Turner syndrome,* occurs roughly once in every 3000 female births. The ways in which nondisjunction can result in abnormal numbers of sex chromosomes are shown in Figure 21.7.

XYY males

The Y chromosome occasionally fails to separate from its sister chromatid in meiosis II. Failure of the Y chromosome to separate leads to the formation of YY gametes. Viable XYY zygotes that develop into males are unusually tall but have normal fertility. The frequency of XYY among newborn males is about 1 per 1000. The XYY syndrome has some interesting history associated with it. In the 1960s, the frequency of XYY males in penal and mental institutions was reported to be approximately 2% (that is, 20 per 1000)—20 times higher than in the general population. This observation led to the suggestion that XYY males are inherently antisocial and violent. Further studies revealed that XYY males have a higher probability of coming into conflict with the law than XY (normal) males, but their crimes are usually nonviolent. Most XYY males, however, lead normal lives and cannot be distinguished from other males.

Persons inheriting abnormal numbers of sex chromosomes often have abnormal features and may be mentally retarded. Zygotes without an X chromosome are not viable.

21.3 Changes in chromosome structure and in single alleles

Although a person may have inherited the proper number of chromosomes, a chromosome that is structurally defective may occur if a chromosome breaks and the cell repairs the break incorrectly or does not repair the break at all. These broken and misrepaired chromosomes can then be passed on in the gametes of the parents to produce disorders in the offspring.

Chromosomes may break naturally, or breaks may be caused by outside agents, such as ionizing radiation and chemicals. **Ionizing radiation**, such as x-rays and nuclear radiation, is a form of energy known as *electromagnetic energy*. Sunlight is also a type of electromagnetic energy. Ionizing radiation, however, has a higher level of energy than ordinary light does. When ionizing radiation reaches a cell, it transfers energy to electrons in the outer shells of the atoms it encounters, raising them to a still higher energy level and ejecting them from their shells. The result is the breaking of covalent bonds and the formation of charged fragments of molecules with unpaired electrons. The charged molecular fragments, or *free radicals,* are highly reactive. They may interact with DNA, producing chromosomal breaks or changes in the nucleotide structure.

When chromosomes break and are improperly repaired by the cell, some chromosomal information may be added, lost, or moved from one location on the chromosome to another. Chromosomal rearrangement changes the way that the genetic message is organized, interpreted, and expressed.

Newly added chromosomal information can be caused by a *duplication* of a section of a chromosome. Seen in a karyotype, a chromosome having a duplication appears longer than its homologue. In Figure 21.8a you can see that one red section is shown as duplicated, which results in a longer chromosome than normal.

A *translocation* could also produce an abnormally long chromosome. Translocations are breaks in two or more nonhomologous chromosomes, followed by reattachments in new combinations. Depending on the lengths of the translocated pieces, some chromosomes could appear shorter than usual, and some longer, as shown in Figure 21.8b.

Chromosomes may also become shortened because of *deletions.* Deletions may be due to a break near a chromosome tip, with the small piece becoming lost (Figure 21.8c). Internal pieces may be lost, but this situation is rare. If a chromosomal *inversion* oc-

curs, the broken piece of chromosome reattaches to the same chromosome but in a reversed direction (Figure 21.8d).

A disorder associated with a chromosomal deletion in humans is the *cri du chat* (kree-du-shah) syndrome. Described in 1963 by a French geneticist, *cri du chat* means "cat cry" and describes the catlike cry

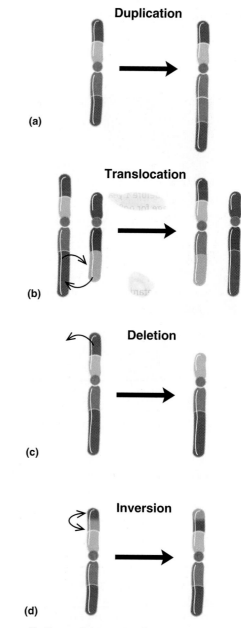

Duplication

(a)

Translocation

(b)

Deletion

(c)

Inversion

(d)

Figure 21.8 Types of chromosomal rearrangement.

made by some cri du chat babies. Other symptoms of this disorder include varying degrees of mental retardation, a round "moon face," and wide-set eyes (Table 21.1 lists the chromosomal aberrations discussed in Sections 21.1 through 21.3, along with additional trisomies.)

Sometimes changes take place in a single allele rather than entire sections of chromosomes. A change in the genetic message of a chromosome caused by alterations of molecules within the structure of the chromosomal DNA is called a **point mutation** or **gene mutation**. Through mutation, a new allele of a gene is produced. Mutations may occur spontaneously (although their occurrence is rare) or may be caused by ionizing radiation, ultraviolet radiation, or chemicals.

Ultraviolet (UV) radiation, the component of sunlight (and of the lights used in tanning beds) that leads to suntan and sunburn is much lower in energy than x-rays but still higher in energy than ordinary light. Certain molecules within the structure of chromosomes absorb UV radiation, developing chemical bonds among them that are not normally present. These unusual bonds produce a kink in the molecular structure of the DNA of chromosomes. Normally, a chromosome is able to repair itself by removing the affected molecules and synthesizing new, undamaged molecules. Sometimes, however, mistakes can take place in some part of the repair process and a change may occur in a gene.

Sometimes chemicals damage DNA directly. LSD (the hallucinogenic compound lysergic acid), marijuana (leaves of the *Cannabis sativa* plant), cyclamates (compounds widely used as non-nutritive sugar substitutes), and certain pesticides (compounds used to kill insects on crops and other plants) are a few examples of chemicals known to damage DNA. In general, chemicals alter the molecular structure of nucleotides, resulting in a mispairing of bases. As with UV radiation, mistakes can take place as the cell works to repair the DNA and changes may then occur in a gene.

Table 21.1

Summary of Chromosomal Aberrations

Condition	Characteristics	Aberration
Trisomy 13	Multiple defects, including severe mental retardation and deafness Death before 6 months of age for 90% of those who survive birth	3 13 autosomes
Trisomy 18	Facial deformities Heart defects Death before 1 year of age for 90% of those who survive birth	3 18 autosomes
Down syndrome (trisomy 21)	Developmental delay of skeletal system Mental retardation	3 21 autosomes
Trisomy 22	Similar features to Down syndrome More severe skeletal deformities	3 22 autosomes
Cri du chat	Moon face Severe mental retardation	Deletion on chromosome 5
Triple X female	Underdeveloped female characteristics	XXX XXXX XXXXX
Turner syndrome	Sterile female Webbed neck Broad chest	XO
Klinefelter syndrome	Sterile male Male and female characteristics	XXY
XYY male	Fertile male Usually quite tall	XYY

Chromosome breaks may occur naturally or may be caused by environmental agents. Mutations in individual alleles (point mutations) are usually caused by environmental factors. Three major sources of environmental damage to chromosomes are: (1) high-energy radiation, such as x-rays; (2) low-energy radiation, such as UV light; and (3) chemicals, such as LSD, marijuana, cyclamates, and certain pesticides. Structurally defective chromosomes and chromosomes with point mutations can be passed on in the gametes of the parents to produce disorders in the offspring.

21.4 The molecular biology of cancer

Mutations in the DNA of body cells play a key role in the development of cancer. Although cancer is often thought of as a single disease, more than 200 types exist. All cancers, however, have four characteristics: (1) uncontrolled cell growth, (2) loss of cell differentiation (specialization), (3) invasion of normal tissues, and (4) **metastasis** (muh-TAS-tuh-sis), or spread, to mul-

tiple sites. Uncontrolled cancer eventually causes death because the cancer cells continually increase in number while spreading to vital areas of the body, occupying the space in which normal cells would reside and carry out normal body functions.

Initiation

Cancer cells arise through a series of stepwise, progressive mutations of the DNA in normal body cells. The first stage in the development of cancer is the development of mutations in proto-oncogenes and/or tumor-suppressor genes. *Proto-oncogenes* are latent (dormant) forms of cancer-causing genes, or **oncogenes** (ON-ko-jeens), that are present in all people. (The word oncogenes is derived from the Greek word *onkos* meaning "mass" or "tumor.") Proto-oncogenes become oncogenes when some part of their DNA undergoes mutation. Oncogenes can be thought of as "on" switches in the development of cancer, signaling cells to speed up their growth and decrease their levels of differentiation.

Tumor-suppressor genes are a separate class of genes that can be thought of as "off" switches in the development of cancer, signaling cells to slow their growth and increase their levels of differentiation. The expression of tumor-suppressor genes is required for the normal functioning of a cell; they appear to allow normal cells to differentiate into mature cell types with reduced or no growth potential. These genes are inactivated when they undergo mutation, thereby allowing cancerous growth.

Mutations in proto-oncogenes and in tumor-suppressor genes may be inherited or caused by viruses, chemicals, or radiation, as shown in **Figure 21.9**. All four of these mutagenic agents are referred to as *initiators*, and the process of proto-oncogene or tumor-suppressor gene mutation is called *initiation*. Unfortunately, initiation may occur after only brief exposure to an initiator. Initiation does not directly result in cancer but in a mutated cell (precancerous cell) that may or may not look abnormal and that gives rise to other initiated cells when it divides.

Promotion

For cancer to occur, initiated cells must undergo *promotion*, the second stage in the development of cancer. Promotion is a process by which the DNA of initiated cells is damaged further, eventually stimulating these cells to grow and divide. It is a gradual process and happens over a long time as opposed to the short-term nature of initiation. Research suggests that if the pro-

Visual Summary Figure 21.9 **Stages in the development of cancer**

Normal cell

INITIATION (Stage I)

Mutation inherited or caused by viruses, chemicals or radiation (initiators)

Mutation

Gene

Precancerous cell

PROMOTION (Stage II)

Mutagenic agents (promoters)

Oncogenes activated and expressed

Benign cells (partially transformed)

(Continued promotion)

PROGRESSION (Stage III)

Malignant cells (cancer)

moter (the substance causing promotion) is withdrawn in the early stages, cancer development can be reversed. For example, if a smoker stops smoking, that person's risk of lung cancer returns eventually to that of a nonsmoker (depending on how long and how much smoking has occurred).

The same agent that caused initiation may cause promotion in the same cell or another initiator may cause it. The mode of action of promoters is unclear, but evidence suggests that they, like initiators, mutate genes or damage chromosomes. Those agents that can both initiate and promote cancer are called complete **carcinogens** (kar-SIN-uh-jens), or cancer-causing substances. Most substances linked with the development of cancer are complete carcinogens. A few, however, act only as an initiator or as a promoter. Heredity, for example, acts only as an initiator. Conversely, asbestos acts only as a promoter. Asbestos promotes cells initiated by other agents such as cigarette smoke and air pollution.

As promotion proceeds, damage to the DNA accumulates and the expression of oncogenes begins. That is, the oncogenes begin to be transcribed, producing polypeptides. These polypeptides affect specific cells in specific ways, causing cells to grow and divide when normal cells would not and causing a variety of other changes such as modifications in cells' shapes and structures, cell-to-cell interactions, membrane properties, cytoskeletal structure, protein secretion, and gene expression. The mechanisms by which a particular polypeptide interacts with a cell to produce one or more of these effects vary also. One way in which oncogene-transcribed polypeptides affect cell growth, for example, is by mimicking growth factors. This type of polypeptide binds to receptor sites on the surfaces of certain cells, activating specific enzymes within them. The activated enzymes cause these cells to grow and divide when they would normally *not* be growing and dividing.

When damage to the DNA of a cell is not drastic, most of the normal components of the cell are produced and it still responds to normal growth-inhibiting factors. Sometimes initiated cells or cells in the early stages of promotion grow and divide abnormally, forming a benign (noncancerous) tumor. **Benign tumors** (buh-NINE) are growths or masses of cells that are made up of partially transformed cells, are confined to one location, and are encapsulated, shielding them from surrounding tissues. Such tumors are not life threatening. Some benign growths, although not life threatening at the time, exhibit growth patterns that

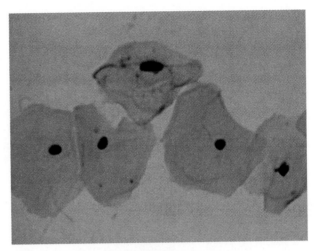

(a)

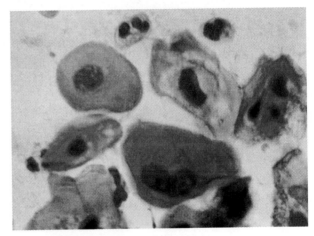

(b)

Figure 21.10   A comparison of normal and dysplastic cells.
(a) Normal cells. (b) Dysplastic cells.

are characteristic of the development of cancer cells. These cells are said to exhibit *dysplasia*. Figure 21.10a shows normal cells. These cells are all approximately the same size and have nuclei that look similar. Figure 21.10b shows dysplastic cells. These cells [stained differently from (a)] are irregular in size and the appearance of their nuclei.

Progression

As promoters continue to damage the DNA, partially transformed cells reach a point where they irreversibly become cancer cells. This point marks the beginning of the third stage in the development of cancer called *progression*. During progression, the transformed cells usually become less differentiated than benign cells and increase their rate of growth and division without

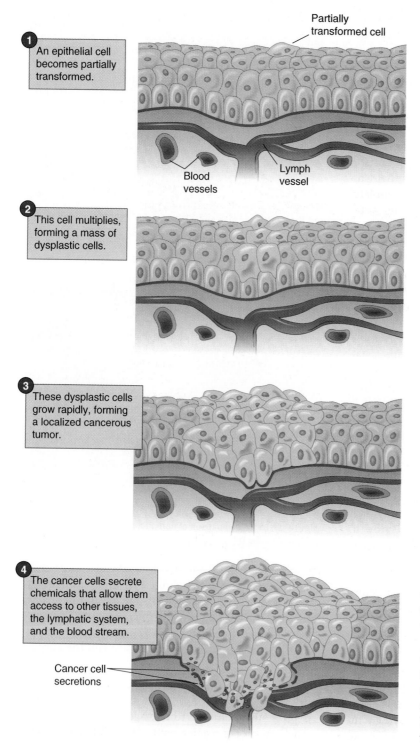

1 An epithelial cell becomes partially transformed.

Partially transformed cell

Blood vessels

Lymph vessel

2 This cell multiplies, forming a mass of dysplastic cells.

3 These dysplastic cells grow rapidly, forming a localized cancerous tumor.

4 The cancer cells secrete chemicals that allow them access to other tissues, the lymphatic system, and the blood stream.

Cancer cell secretions

Figure 21.11 How cancer cells multiply and spread.

regard to the body's needs. In addition, these cancer cells have the ability to (1) invade and kill other tissues and (2) metastasize, or move to other areas of the body. Tumors with these properties are **malignant** (muh-LIG-nunt). Interestingly, the word malignant is derived from two Latin words meaning "of an evil nature," whereas the word benign means "kind-hearted."

The spread of cancer during progression is a multistage process as is the development of the cancer cells themselves. Figure 21.11 shows these stages in the development of a carcinoma, a cancer of the epithelial tissues that tends to invade surrounding tissue. (Skin, breast, lung, and colon cancers are examples of carcinomas.) In **1** a partially transformed tumor cell is shown in the epidermis of the skin. This cell multiplies **2**, forming a mass of dysplastic cells. As the DNA of these cells is further damaged, they irreversibly become cancer cells. Their growth rate increases, and they form a cancerous tumor that is localized **3**. These cancers are referred to as *in situ*, meaning "in place," and are small, localized tumors that have not invaded the surrounding normal tissue. During the next stage of cancer progression, cancer cells invade the surrounding tissue by secreting chemicals that break down the intercellular matrix—the substances that hold cells together. Other secretions cause the cells to break apart. Cancer cells then invade the underlying dermis **4**, entering blood and lymph vessels and traveling throughout the body.

Researchers suggest that 90% of all cancers are environmentally induced; that is, they are not inherited but result from external factors. Table 21.2 lists factors that are initiators and/or promoters of cancer. Notice that many of these factors are implicated in other diseases or conditions that endanger health. Avoiding these factors may increase your chances for a longer, healthier life. Although your heredity is not a factor you can control, knowledge about your hereditary background can help you control other factors important to your health.

Table 21.2

Major Factors That Increase the Risk of Cancer

Factor	Examples of Implicated Cancers	Comments
Heredity	Retinoblastoma (childhood eye cancer) Osteosarcoma (childhood bone cancer)	Most cancers are not caused by heredity alone. Persons having family histories of certain cancers should follow physicians' recommendations.
Tumor viruses	Liver cancer Adult T cell leukemia/lymphoma Cervical cancer	Five viruses are initiators of certain cancers. (See Table 25.1.)
Tobacco use	Lung cancer Cancers of the oral cavity, esophagus, and larynx Cancers of the kidney and bladder	Cigarette smoking is responsible for approximately one third of all cancers. Nonsmokers have an increased risk of smoking-related cancers if they regularly breathe in sidestream smoke.
Alcohol consumption	Cancers of the oral cavity, esophagus, and larynx Breast cancer	The combined use of alcohol and tobacco leads to a greatly increased risk of these cancers. The mechanism of action in breast cancer is not yet known.
Industrial hazards	Lung cancer	Certain fibers, such as asbestos, chemicals such as benzene and arsenic, and wood and coal dust are prominent industrial hazards.
Ultraviolet radiation from the sun	Skin cancers	Those at greatest risk are fair-skinned persons who burn easily. However, everyone is at risk and should wear sunscreens and protective clothing when in the sun for extended periods of time. All types of UV radiation in tanning beds (UVA, UVB, & UVC) are harmful and may lead to skin cancer.
Ionizing radiation	Related to location and type of exposure	Eliminate unnecessary medical x-rays to lower cancer risk. Infants and children are particularly susceptible to the damaging effects of ionizing radiation. Check your home to detect high levels of radon gas.
Hormones (estrogen and possibly testosterone)	Breast, endometrial, ovarian, and prostate cancers	When estrogen is used with progesterone in birth control pills and hormone-replacement therapy, the risk of cancer from from this hormone is lowered and may become *less* than normal. The role of testosterone in prostate cancer is unclear.
Diet	Colon and rectum, breast (weak association), endometrium and prostate cancers (fat), stomach and esophageal cancers (nitrites)	The primary dietary factor that increases the risk of cancer is the consumption of fat, particularly polyunsaturated fats. Nitrites found in salt-cured, salt-pickled, and smoked foods also increase the risk of cancer.

One way to decrease your risk of cancer is to determine which of the previously described cancer risks applies to you and to change your behavior accordingly. Another way to decrease your cancer risk is to follow the recommendations of the American Cancer Society (ACS) regarding diagnostic tests. Your local chapter of the ACS will send these recommendations to you on request. In addition, follow the general dietary recommendations listed in **Table 21.3** and look for the danger signs of cancer listed in **Table 21.4.** If you notice any of these signs, see your doctor immediately. Otherwise, the ACS recommends a cancer-related checkup by a physician every 3 years for persons aged 20 to 39 years and annually for those 40 years of age and older. Persons at risk for particular cancers may need to see their physician more often.

Table 21.3

General Dietary Recommendations to Reduce the Risk of Cancer

1. Avoid obesity by balancing caloric intake with exercise
2. Choose foods low in fat
3. Limit consumption of meats, especially high fat meats.
4. Eat five or more servings of fruits and vegetables each day, especially green and dark yellow vegetables and those in the cabbage family
5. Eat other foods from plant sources, such as breads, cereals, grain products, soy products, rice, pasta, or beans several times each day.
6. If alcohol is consumed, limit intake to one mixed drink or glass of wine per day
7. Use moderation when consuming salt-cured, smoked, and nitrite-cured foods

Adapted from *1999 Facts & Figures,* the American Cancer Society.

Table 21.4

The Seven Warning Signs of Cancer

Change in bowel or bladder functions

A sore that does not heal

Unusual bleeding or discharge

Thickening or lump in any tissue

Indigestion (chronic) or difficulty in swallowing

Obvious change in a wart or mole

Nagging hoarseness or cough

Cancer cells arise through a series of progressive mutations in the DNA of normal body cells. These initial mutations may be inherited or may be caused by certain viruses, chemicals, or types of radiation; they activate proto-oncogenes (dormant cancer-causing genes) and/or inactivate tumor-suppressor genes.

For cancer to occur, initiated (DNA-mutated) cells must undergo promotion, a process by which the DNA is damaged further. As promotion proceeds, damage to the DNA accumulates and the expression of oncogenes begins. The polypeptides produced cause cells to grow and divide when normal cells would not, as well as a variety of other cellular changes.

As DNA is damaged more and more, partially transformed cells irreversibly become cancer cells. Cancer cells are malignant cells, meaning they can invade and kill other tissues and metastasize, or move to other areas of the body.

21.5 The use of pedigrees to study inheritance patterns

Karyotypes are useful in studying diseases or disorders caused by an abnormal number of chromosomes or the addition or deletion of chromosome pieces. However, most inherited disorders, such as the type of hemophilia discussed in the chapter opener, are caused by point mutations, which cannot be seen in a karyotype.

To study how this and other human traits are inherited, scientists study family histories—the records of relevant genetic features—to draw pedigrees. **Pedigrees** are diagrams of genetic relationships among family members over several generations. By studying which relatives exhibit a trait, it is often possible to say if the gene producing the trait is sex linked (occurs on one of the sex-determining chromosomes) or autosomal (occurs on a non-sex-determining chromosome). Pedigree analysis also helps a geneticist determine whether the trait is a dominant or a recessive characteristic. In many cases, it is also possible to infer which individuals are homozygous or heterozygous for the allele carrying the trait.

Figure 21.12 is a pedigree that shows the history of red-green color blindness in a family. (A test for color blindness and a description of this disorder are provided in Figure 21.13.) The circles in Figure 21.12 represent females, and the squares represent males. Solid-colored symbols stand for individuals who exhibit the trait being studied. In this case, those individuals would be color blind. Parents are represented by horizontal lines connecting a circle and a square. Vertical lines coming from two parents indicate their children, arranged along a horizontal line in order of their birth.

First, to determine whether a trait is sex linked or autosomal, notice whether the trait is expressed more frequently in males than in females. If so, the trait is most likely a sex-linked trait. In addition, the trait

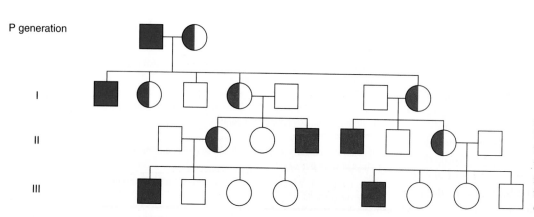

P generation

I

II

III

Figure 21.12 **A pedigree showing red-green color blindness.** From this pedigree, it can be determined that the carriers of the color blindness allele are the female in the P generation, the three females in generation I, and the first and third females in generation II. The status of the other females is unknown.

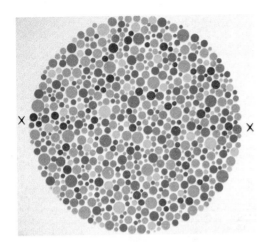

Figure 21.13   **A test for red-green color blindness.** About 8% of Caucasian males and 0.5% of Caucasian females have some loss of color vision, which is almost always caused by an X-linked recessive allele. This recessive allele causes a defect in one of three groups of color-sensitive cells in the eye. The most common form of color blindness is dichromatic vision, in which affected individuals cannot distinguish red from green or yellow from blue. The pattern tests for an inability to distinguish red from green. Those with normal vision will see a path of green dots between the points marked *X*. Those with red-green color blindness will not be able to distinguish the green dots from red and therefore will not see the green path.

would then likely be recessive because few dominant sex-linked traits are known. An autosomal trait (whether dominant or recessive) is expressed equally in both males and females. A sex-linked trait, however, is al-

ways expressed in a male, because the relatively inert Y chromosome lacks alleles that correspond to alleles on the X chromosome; one deleterious gene on a male's X chromosome results in its phenotypic expression. Is red-green color blindness a sex-linked (X-linked) trait?

Second, to determine whether a trait is dominant or recessive, notice whether each person expressing the trait has a parent who expressed the trait. In our example, if color blindness is dominant to normal color vision, then each color-blind person will have a color-blind parent. However, if the trait is recessive, a person expressing the trait can have parents who do not express the trait. Both parents may be heterozygous (may carry the trait). If each parent passes on the recessive gene to a child, the child is then homozygous recessive and the trait will be expressed. Is color blindness a dominant or a recessive characteristic?

Next, determine which individuals are carriers of the color blindness gene. By now you probably realize that color blindness is a sex-linked recessive trait. A color-blind male always contributes an X chromosome with the defective allele to his daughters; it is the only X he can contribute. His sons, however, get a Y chromosome from him and an unaffected X chromosome from their (normal) mother. A color-blind female can contribute only a chromosome with the defective allele to her children because both her X chromosomes are affected. Women who are carriers of the trait, however, may contribute either a normal or a defective X. The Punnett square in **Figure 21.14** illustrates the contributions of genes by a normal mother and a color-

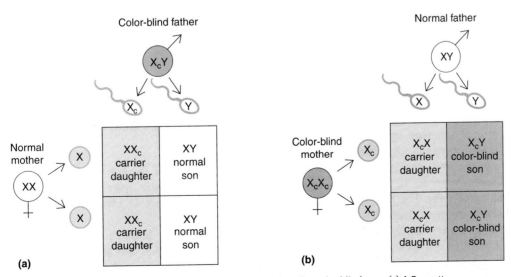

Figure 21.14   **Punnett squares depicting the outcomes for the sex-linked disorder color blindness.** (a) A Punnett square for a normal mother and color-blind father. (b) A Punnett square for a color-blind mother and normal father.

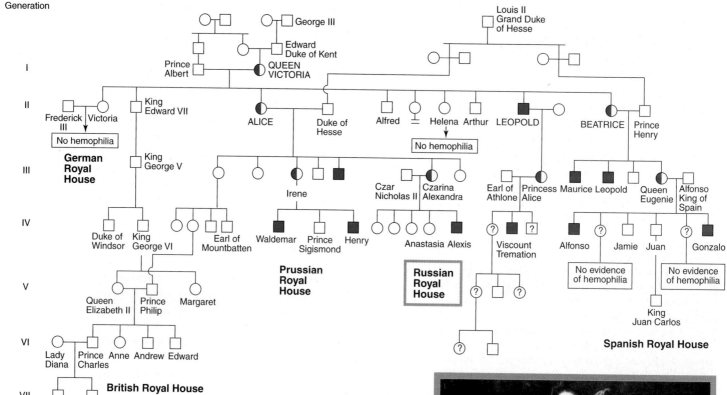

Generation

Figure 21.15   **The royal hemophilia pedigree.** (a) From Queen Victoria's daughter, Alice, the disorder was introduced into the Prussian and Russian royal houses, and from her daughter Beatrice, it was introduced into the Spanish royal house. Victoria's son Leopold, himself a victim, also transmitted the disorder in a third line of descent. (b) The last Russian royal family is shown in this photo: Czar Nicholas II, Czarina Alexandra (a carrier of the defective allele), and their five children.

blind father (Figure 21.14a) and a color-blind mother and normal father (Figure 21.14b).

Queen Victoria's family pedigree for hemophilia is shown in **Figure 21.15a**. Persons carrying a mutant allele but not having the disorder are shown by partially colored (green) symbols. Using the strategy outlined in the preceding paragraphs, analyze the pedigree to determine whether hemophilia is sex linked or autosomal and whether it is a dominant or recessive trait. Look back to the chapter opener to confirm your results.

As shown in the royal pedigree, in the six generations since Queen Victoria, 10 of her male descendants have had hemophilia. The British royal family escaped the disorder because Queen Victoria's son King Edward VII did not inherit the defective allele. Three of Victoria's nine children did receive the defective allele, however, and carried it by marriage into many of the royal families of Europe. It is still being transmitted to future generations among these family lines, except

in Russia, where the five children of Alexandra, Victoria's granddaughter, were killed soon after the Russian revolution (Figure 21.15b).

Pedigrees, diagrams of genetic relationships among families over several generations, are useful in determining whether traits are sex linked or autosomal, and dominant or recessive. Carriers of recessive genes can also be identified.

21.6 Dominant and recessive genetic disorders

Most genetic disorders are recessive; these mutant genes are able to persist in the population among carriers. Persons receiving a mutant allele from each of their parents are affected with the disease and, in the case of some diseases, may die before reaching adulthood. Likewise, persons having certain dominant disorders may die before reaching reproductive age; therefore lethal dominant disorders are less likely to persist in the population.

Huntington's disease is a dominant disorder that killed folk singer and songwriter Woody Guthrie. Guthrie

(left), who died in 1967 at the age of 55, often wrote songs about unemployment and social injustice. His most famous work is "This Land is Your Land." His son is composer and singer Arlo Guthrie (below), who wrote the 1967 hit "Alice's Restaurant."

Huntington's disease is caused by a mutant dominant allele that does not show up until individuals are older than 30 years of age—after they may have had children. This disorder causes progressive deterioration of brain cells. Other dominant genetic disorders include the following:

- *Marfan's syndrome,* which results in skeletal, eye, and cardiovascular defects
- *Polydactyly* (PAHL-ee-DAK-tih-lee), which results in extra fingers or toes
- *Achondroplasia* (ay-KON-dro-PLAY-zee-ah), which results in a form of dwarfism
- *Hypercholesterolemia* (HI-per-ko-LESS-ter-ol-EE-mee-ah), which results in high blood cholesterol levels and a higher likelihood of developing coronary artery disease

Recessive genetic disorders are often seen primarily within specific populations or races of people unless many members of the population marry and have children outside of their population or race. Among the Caucasian population, for example, the most common fatal genetic disorder is *cystic fibrosis*. Affected individuals secrete a thick mucus that clogs the airways of their lungs and the passages of their pancreas and liver. Among Caucasians, about 1 in 20 individuals has a copy of the defective gene but shows no symptoms. Approximately 1 in 1800 has two copies of the gene and therefore has the disease. These individuals inevitably die of complications that result from their disease. It was learned only recently that the cause of cystic fibrosis is a defect in the way cells regulate the transport of chloride ions across their membranes.

Sickle cell anemia is a recessive disorder most common among African Blacks and their descendants. In the United States, for example, about 9% of Blacks are heterozygous for this allele; about 0.2% are homozygous and therefore have sickle cell anemia. In some groups of people in Africa, up to 45% of the individuals are heterozygous for this allele. Heterozygous carriers of the sickle cell gene do not have the disease but can pass the gene along to their children.

Individuals afflicted with sickle cell anemia are unable to transport oxygen to their tissues properly because the molecules within red blood cells that carry oxygen—molecules of the protein hemoglobin—are defective. Red blood cells that contain large proportions of such defective molecules become sickle shaped (as shown in the top

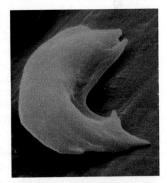

photo) and stiff; normal red blood cells (bottom photo) are disk shaped and much more flexible. As a result of their stiffness and irregular shape, the sickle-shaped red blood cells are unable to move easily through capillaries. Therefore, they tend to accumulate in blood vessels, reducing the blood supply to the organs they serve and causing pain, tissue destruction, and an early death.

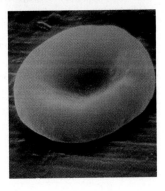

The gene for sickle cell anemia is most prevalent in the regions of Africa where malaria is prevalent. Malaria is a disease caused by microorganisms that live in a person's red blood cells. These microbes are injected into a person's bloodstream by the bite of a female *Anopheles* mosquito. A long-lasting disease, malaria affects the physical and mental development of its victims, causing damage to many body organs. Scientists have discovered that the defective hemoglobin molecules of the person with sickle cell anemia produce conditions that are unfavorable to the growth of the malaria organism, but these persons eventually die of their anemia. However, persons heterozygous for the sickle cell gene, although not afflicted with the disease, are more resistant to malaria. Heterozygotes, then, have a survival advantage with respect to malaria (and do not have sickle cell anemia although their offspring may be afflicted with the disease if they reproduce with persons also carrying the defective gene).

Tay-Sachs disease is an incurable, fatal recessive hereditary disorder. Although rare in most human populations, Tay-Sachs has a high incidence among Jews of Eastern and Central Europe and among American Jews (90% of whom are descendants of Eastern and Central Europeans). In fact, geneticists estimate that 1 in 28 individuals in these Jewish populations carries the allele for this disease and that approximately 1 in 3600 infants within this population is born with this genetic disorder. Affected children appear normal at birth but begin to show signs of mental deterioration at about 8 months of age. As the brain begins to deteriorate, affected children become blind; they usually die by the age of 5 years.

Table 21.5

Dominant and Recessive Disorders

Name	Characteristics
Dominant Disorders	
Huntington's disease	Progressive deterioration of brain cells
Marfan's syndrome	Skeletal, eye, and cardiovascular defects
Polydactyly	Extra fingers or toes
Achondroplasia	Dwarfism
Hypercholesterolemia	High blood cholesterol levels
Recessive Disorders	
Albinism	Lack of pigment in skin, hair, and eyes
Cystic fibrosis	Production of thick mucus that clogs airways
Sickle cell anemia	Defective hemoglobin in red blood cells, effect on oxygen transport
	Misshapen red blood cells, tendency to form clots in vessels
Tay-Sachs disease	Brain deterioration beginning at 8 months of age
Sex-Linked Recessive Disorders	
Color blindness	Inability to discern certain colors
Hemophilia	Improper clotting of blood
Duchenne muscular dystrophy	Degeneration of muscle
Fragile X syndrome	Easy breakage of tip of X chromosome
	Long face, squared forehead, large ears
	Mental retardation

Table 21.6

Common Dominant and Recessive Human Traits

Dominant Traits
Widow's peak
Hair on middle segment of fingers
Short fingers
Inability to straighten little finger
Brown eyes
A or B blood factor

Recessive Traits
Common baldness
Blue or gray eyes
O blood factor
Attached ear lobes

Table 21.5 lists the dominant and recessive disorders discussed in this section, as well as others. Many common human traits are also coded by either dominant or recessive genes. Table 21.6 lists some of these characteristics.

Most genetic disorders are recessive because mutant genes are able to persist in the population among carriers. Some more common recessive genetic disorders among humans are color blindness, hemophilia, cystic fibrosis, and sickle cell anemia.

Just Wondering . . . Real Questions Students Ask

How can you tell who the father of a baby is just by doing blood tests?

No test can *prove* that a particular man is the father of a child, but the paternity tests being performed today can yield a probability of paternity in excess of 99%. Paternity testing involves analyzing the blood of a child, the child's mother, and the suspected father to determine which blood antigens they possess and then using the principles of genetics to determine whether the child could have received its complement of antigens from the hypothetical parents.

There are several types of antigens that are analyzed in paternity testing. The most familiar are the antigens associated with the ABO system. The presence or absence of these antigens is governed by alleles inherited from one's parents (see "Multiple Alleles"). The alleles for each blood type are shown in Figure 21.A.

Type	Possible alleles
A	$I^A I^A, I^A i$
B	$I^B I^B, I^B i$
AB	$I^A I^B$
O	$i\ i$

Figure 21.A

Type AB Father

		I^A	I^B
Type A Mother	I^A	$I^A I^A$	$I^A I^B$
	i	$I^A i$	$I^B i$

Figure 21.B

How can you rule out a suspected father using ABO blood groups? Suppose a child's mother has type A blood with the genotype shown in the Punnett square, and the suspected father has type AB blood. The child is type O. Is the suspect the father? Examine the Punnett square in Figure 21.B. The blood types possible from an A mother (who must be heterozygous [$I^A i$] to have an O child) and an AB father are types A, B, and AB. An O child could not be born to such parents, and thus the suspect is not the father.

Paternity testing, however, is based on more than ABO blood groups. DNA analysis is now used in conjunction with antigen analysis to test paternity. It is this combination of tests that contributes to the high accuracy of the results.

21.7 Incomplete dominance, codominance, and multiple alleles

Incomplete Dominance

Alleles do not always exhibit clear-cut dominance or recessiveness. The pea plants that Gregor Mendel worked with exhibited complete dominance, and for this reason his work disputed the concept of blended inheritance (see Chapter 20). However, since Mendel's time, researchers have discovered many cases of **incomplete dominance** in which alternative alleles are *not* dominant over or recessive to other alleles governing a particular trait. Instead, heterozygotes are phenotypic "intermediates."

Incomplete dominance is found in various plant and animal traits. For example, a cross between snapdragons having red flowers and those having white flowers yields plants having pink flowers. Crossing black Andalusian chickens with white Andalusian chickens yields an intermediate hybrid that is slate blue. In humans, a curly haired Caucasian and a straight-haired Caucasian will have children with wavy hair. Persons with wavy hair will have children that are either straight haired, wavy haired, or curly haired. Figure 21.16 includes a cross between two wavy-haired people. Notice that an uppercase H denotes the "hair" allele. No lowercase letters are used because neither allele (curly or straight) is recessive. Instead, one allele is designated as *H* and the other as *H′*.

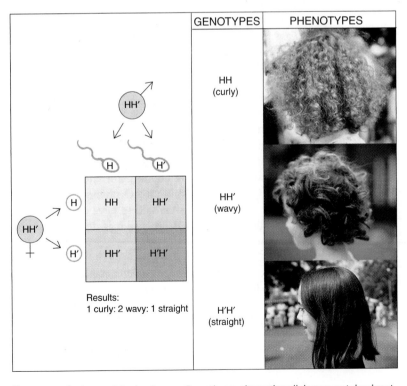

Figure 21.16 **Incomplete dominance.** Sometimes, alternative alleles are not dominant or recessive. For example, neither straight nor curly hair is dominant in Caucasians. The predicted ratio of the offspring of two wavy-haired people will be one curly-haired individual, one straight-haired individual, and two wavy-haired individuals. Wavy hair is the phenotypic intermediate.

Codominance

A slightly different situation occurs with alleles that are codominant. **Codominant** alleles are *both* dominant; both characteristics are exhibited in the phenotype. In humans the alleles that code the A, B, and AB blood types are a good example of codominance. For example, a person having allele A and allele B has blood type AB. In addition to being coded by codominant alleles, blood types are coded by more than two alleles. So far, genes that consist of only two—a pair of—alleles have been discussed. However, a gene may be represented by more than two alleles within the population. Some genes, such as the ABO blood type genes, consist of a system of alleles, or **multiple alleles**. In this system of multiple alleles, the alleles A and B exhibit codominance and the O allele is recessive.

Multiple Alleles

As mentioned, human ABO blood types are coded by multiple alleles. In fact, four phenotypes (A, B, AB, and O blood types) are determined by the presence of two out of three possible alleles (A, B, or O) in an individual. The A and B alleles each code for the production of different enzymes. These enzymes add certain sugar molecules to lipids on the surface of red blood cells. These sugars act as recognition markers for the immune system and are called *cell surface antigens*. The enzyme produced by allele A results in the addition of one type of sugar; allele B adds a different sugar. Allele O adds no sugar. Different combinations of the three possible alleles occur in different individuals, with each individual having *one pair* of alleles. Therefore, a person having an AA genotype produces only the A sugar. A person having an AB genotype produces both A and B sugars. (The A and B alleles are codominant and both are expressed.) A person having an AO phenotype will produce the A sugar (and so is said to have type A blood). The A allele is expressed because it is dominant to the O allele.

In traits exhibiting incomplete dominance, alternative forms of an allele are neither dominant nor recessive; heterozygotes are phenotypic intermediates. In traits exhibiting codominance, alternative forms of an allele are both dominant; heterozygotes exhibit both phenotypes. Some genes consist of a system of alleles, or multiple alleles, which are usually codominant to one another. The ABO system of human blood types is an example of codominant multiple alleles.

21.8 Genetic counseling

Although most genetic disorders cannot yet be cured, research scientists and physicians are continually learning a great deal about them. Exciting methods of genetic therapy are being studied. Geneticists can identify couples at risk of having children with genetic defects and can help them have healthy children. Information the genetic counselor collects helps identify or exclude individuals or their offspring from genetically based diseases or birth defects. This process of "genetic identification" is called **genetic counseling**.

By analyzing family pedigrees and applying the laws of statistical probability, for example, genetic counselors can often determine the probability that a person is a carrier of a recessive disorder. When a couple is expecting a child and both parents have a significant probability of being carriers of a serious recessive genetic disorder, the pregnancy is said to be a high-risk pregnancy. In such a pregnancy, there is a significant probability that the child will exhibit the clinical disorder. Another class of high-risk pregnancies occurs with mothers who are older than 35 years of age.

When a pregnancy is diagnosed as being high risk, many women select to undergo *amniocentesis,* a procedure that permits the prenatal diagnosis of many genetic disorders (Figure 21.17). In the fourth month of pregnancy, a sterile hypodermic needle is used to obtain a small sample of amniotic fluid from the mother. The amniotic fluid, which bathes the fetus, contains free-floating cells derived from the fetus. Once removed, these cells can be grown as tissue cultures in the laboratory. Using these tissue cultures, genetic counselors can test for many of the most common genetic disorders.

During amniocentesis, the position of the needle in relationship to the fetus is observed by means of a technique called ultrasound. The term ultrasound refers to high-frequency sound waves. Pulses of these waves are sent into the body and are reflected back in various patterns depending on the tissues or fluids the waves hit. These patterns of sound-wave reflections are then mapped to produce a picture of inner tissues as in Figure 21.18. Because ultrasound allows the position of the fetus to be determined, the person performing the amniocentesis can avoid damaging the fetus. Ultrasound also allows the fetus to be examined for the presence of major abnormalities.

www.jbpub.com/biology

Today, using ultrasound to guide them, physicians are able to sample fetal skin and certain other tissues. Cells from the outermost fetal membrane, the *chorion*, can also be sampled by a technique known as chorionic villus sampling (CVS). To perform CVS, a physician inserts a thin tube or a needle into the vagina and up into the uterus, vacuuming up fingerlike projections of the chorion called *villi*. This procedure can be performed in the second to third months of pregnancy.

The options for treating fetal abnormalities are increasing, yet are still limited. Open fetal surgery was first performed at the University of California at San Francisco in 1981. Various surgical procedures are conducted today to treat fetal urinary obstructions, certain respiratory problems, and certain growths. The use of laser scalpels and new fetal imaging technologies such as fetoscopic cameras are allowing surgeons to treat such disorders less invasively. In some cases, surgery can be performed without opening the uterus. At this time, however, most of the problems that can be diagnosed by pedigree analysis, amniocentesis, chorionic villus

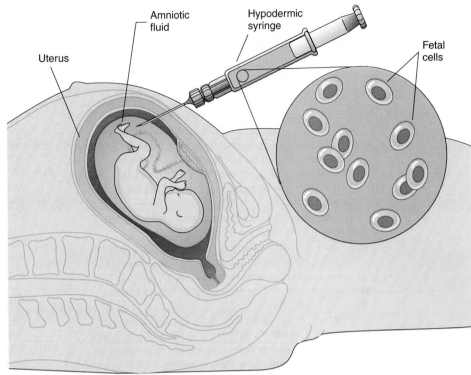

Figure 21.17 **Amniocentesis.** A needle is inserted into the amniotic cavity, and a sample of amniotic fluid, containing some free cells derived from the fetus, is drawn into a syringe. The fetal cells are then grown in tissue culture so that their karyotype and many of their metabolic functions can be examined.

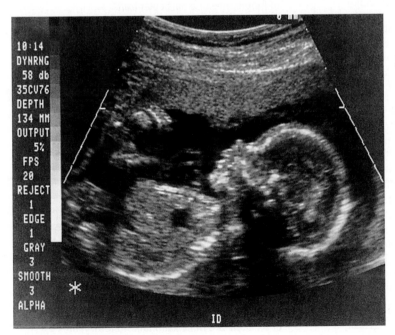

Figure 21.18 **Ultrasound of a 20-week-old fetus.**

sampling, and ultrasound techniques cannot be treated. Sometimes the only options available to a couple are to continue the pregnancy and deal with the problems after birth or to have a therapeutic abortion. Continual progress in developing gene therapies and laser technologies may, in the future, have a drastic effect on a physician's ability to treat genetic disorders before birth.

Genetic counselors collect information to identify or exclude individuals or their offspring from genetically based diseases or birth defects. Although a variety of procedures such as amniocentesis, chorionic villus sampling, and ultrasound are available to test for fetal disorders, in utero therapies are limited at this time.

Key Concepts

Karyotypes are used to study the inheritance of chromosomal abnormalities.

21.1 Human cells contain 46 chromosomes: 44 autosomes and 2 sex chromosomes; the autosomes form 22 pairs of homologous (matched) chromosomes.

21.1 Researchers often use karyotypes to study the inheritance of genetic disorders caused by the loss of all or part of a chromosome or by the addition of extra chromosomes or chromosome fragments.

21.1 A karyotype is a picture of chromosomes paired with their homologues and arranged according to size.

21.1 In humans, the inheritance of one autosome too few usually results in a nonviable zygote.

21.1 The inheritance of one autosome too many, or a trisomy, often results in a nonviable zygote with the exception of chromosomes 13, 18, 21, and 22.

21.1 Individuals with an extra copy of chromosome 21 have Down syndrome.

21.1 Down syndrome is much more frequent among pregnant mothers older than 35 years of age, and it occurs when two sister chromatids fail to separate, or disjoin, during the meiotic process.

21.2 Nondisjunction can occur with the sex chromosomes.

21.2 Persons who inherit abnormal numbers of sex chromosomes often have abnormal features and may be mentally retarded.

21.2 Examples of genetic disorders caused by abnormal numbers of sex chromosomes are poly-X females, XXY males (Klinefelter syndrome), XO females (Turner syndrome), and XYY males.

Mutations play a key role in the development of cancer.

21.3 Persons may inherit chromosomes that are structurally defective due to chromosomal breaks and misrepairs or changes taking place in the molecular structure of the chromosomal DNA.

21.3 The three major sources of environmental damage to chromosomes are (a) high-energy radiation, such as x-rays; (b) low-energy radiation, such as UV light; and (c) chemicals, such as LSD, marijuana, cyclamates, and certain pesticides.

21.4 Cancer is not a single disease but is a cluster of more than 200 diseases.

21.4 All cancers have four characteristics: (a) uncontrolled cell growth, (b) loss of cell differentiation (specialization), (c) invasion of normal tissues, and (d) metastasis, or spread, to multiple sites.

21.4 Cancer cells arise through a series of stepwise, progressive mutations, or changes, to the DNA of normal body cells.

21.4 The first set of changes (first stage) involves proto-oncogenes (dormant forms of cancer-causing genes) and/or tumor-suppressor genes (genes that can be thought of as "off" switches in the development of cancer).

21.4 These first mutations result in precancerous cells, which are termed initiated or transformed cells.

21.4 For cancer to occur, initiated cells must undergo the second stage of cancer development called promotion, a process by which the DNA of initiated cells is damaged further, eventually stimulating these cells to grow and divide.

21.4 During progression, the third stage of cancer development, the transformed cells usually become less differentiated than benign cells and increase their rate of growth and division without regard to the body's needs.

21.4 Research suggests that if the promoter (the substance causing promotion) is withdrawn in the early stages, cancer development can be reversed.

Human traits are inherited in a variety of ways.

21.5 Patterns of inheritance observed in family histories, or pedigrees, can be used to determine the mode of inheritance of a particular trait.

21.5 Pedigree analysis can often determine whether a trait is associated with a dominant or a recessive allele and whether the gene determining the trait is sex linked.

21.6 Many of the most common genetic disorders are associated with recessive alleles, the functioning of which may lead to the production of defective versions of enzymes that normally perform critical functions.

21.6 Because recessive traits are expressed only in homozygotes, the alleles are not eliminated from the human population.

21.6 Dominant alleles that lead to severe genetic disorders are less common; in some of the more frequent ones, the expression of the alleles does not occur until after the individuals have reached their reproductive years.

21.7 In traits exhibiting incomplete dominance, alternative forms of a trait are neither dominant nor recessive; heterozygotes are phenotypic intermediates.

21.7 In traits exhibiting codominance, alternative forms of an allele are both dominant; heterozygotes exhibit both phenotypes.

Key Concepts, continued

21.7 Some genes consist of a system of alleles, or multiple alleles, which are usually codominant to one another.

21.7 The ABO system of human blood types is an example of multiple alleles with two of the alleles (A and B) codominant to one another and one allele (O) recessive to the other two.

21.8 Genetic counselors collect information to identify or exclude individuals or their offspring from genetically based diseases or birth defects.

21.8 Although most genetic disorders cannot yet be cured, research scientists are making progress in developing gene therapies and other technologies that may, in the future, have an important effect on a physician's ability to treat genetic disorders before birth.

Key Terms

autosomes (AW-tuh-somes) *458*
benign tumors (buh-NINE) *465*
carcinogens (kar-SIN-uh-jens) *465*
codominant (KO-DOM-uh-nunt) *474*
genetic counseling *474*
incomplete dominance *473*

ionizing radiation (EYE-uh-nye-zing-RAY-dee-AY-shun) *462*
malignant (muh-LIG-nunt) *466*
metastasis (muh-TAS-tuh-sis) *464*
multiple alleles (uh-LEELS) *474*
mutations (myoo-TAY-shuns) *458*

nondisjunction (non-dis-JUNK-shun) *459*
oncogenes (ON-ko-jeens) *464*
pedigrees (PED-uh-greez) *468*
point mutation (gene mutation) *463*
sex chromosomes *458*

KNOWLEDGE AND COMPREHENSION QUESTIONS

1. What happens when human offspring inherit abnormal numbers of sex chromosomes? Summarize the four examples discussed in the chapter, and give the genotype of each. (*Comprehension*)

2. What are the three major sources of damage to chromosomes? (*Knowledge*)

3. Describe how heavy use of drugs could affect a woman's ova or a man's sperm. (*Comprehension*)

4. What common factors must exist for a disease to be labeled as a form of cancer? (*Knowledge*)

5. In what manner does mutation affect proto-oncogenes? In what manner does mutation affect tumor-suppressor genes? What are the clinical results? (*Knowledge*)

6. Lynn's doctor informs her that the pathology report on her annual Pap test (a screening test for cervical cancer) indicates that cells exhibiting dysplasia were seen in the smear. What does this statement mean? How would these cells look different from normal cells? (*Comprehension*)

7. What would you conclude about the inheritance pattern of each human trait in each of the following situations? (*Comprehension*)

   a. The trait is expressed more frequently in males than in females.
   b. Offspring who exhibit this trait have at least one parent who exhibits the same trait.
   c. Offspring can exhibit this trait even though the parents do not.
   d. The trait is expressed equally in both males and females.

8. Why do most genetic disorders in humans result from recessive genes? Name several examples. (*Knowledge*)

9. Distinguish between incomplete dominance and codominance. Describe the phenotype of a heterozygote in each case. (*Comprehension*)

10. If a couple want to have a child but suspect that they may be at risk for a genetic disorder, what can they do? If a pregnancy turns out to be a high risk, what options are available? (*Comprehension*)

K E Y
Knowledge: Recalling information.
Comprehension: Showing understanding of recalled information.

CRITICAL THINKING REVIEW

1. The extra chromosome 21 that is found in persons with Down syndrome is the cause of multiple developmental defects. What might this tell you about the interaction of genes on a particular chromosome? (*Synthesis*)

2. Design a family medical tree identifying known cancers in your medical history. Go back as many generations as possible and include aunts, uncles, and cousins. What cancers emerge as hereditary or potentially a health trend in your family? (*Analysis*)

3. What factors in your life and environment predispose you toward cancer (excluding heredity)? How many of these can you personally restrict or eliminate? Have you done so? (*Analysis*)

4. A woman whose blood type is AB marries a man with the same blood type. Draw a Punnett square to illustrate the possible genotypes of their children. What blood type will each genotype have? (*Application*)

5. Suppose that you and a partner were interested in having children. What information would you be able to give a genetic counselor to help him or her identify or exclude you or your offspring from genetically based diseases or birth defects? (*Analysis*)

KEY
Application: Using information in a new situation.
Analysis: Breaking down information into component parts.
Synthesis: Putting together information from different sources.

HUMAN GENETICS PROBLEMS

1. George has the same type of hemophilia as did Queen Victoria and some of her descendants. He marries his mother's sister's daughter Patricia. His maternal grandfather also had hemophilia. George and Patricia have five children: Two daughters are normal, and two sons and one daughter develop hemophilia. Draw the pedigree.

2. A couple with a newborn baby are troubled that the child does not appear to resemble either of them. Suspecting that a mix-up occurred at the hospital, they check the blood type of the infant. It is type O. Because the father is type A, and the mother is type B, they conclude that a mistake must have been made. Are they correct?

3. How many chromosomes would you expect to find in the karyotype of a person with Turner syndrome?

4. A woman is married for the second time. Her first husband was ABO blood type A, and her child by that marriage was type O. Her new husband is type B, and their child is type AB. What is the woman's ABO genotype and blood type?

5. Total color blindness is a rare hereditary disorder among humans in which no color is seen, only shades of gray. It occurs in individuals homozygous for a recessive allele and is not sex linked. A non-color-blind man whose father is totally color blind intends to marry a non-color-blind woman whose mother was totally color blind. What are the chances that they will produce offspring who are totally color blind?

6. This pedigree is of a rare trait in which children have extra fingers and toes. Which one of the following patterns of inheritance is consistent with this pedigree?
   a. Autosomal recessive
   b. Autosomal dominant
   c. Sex-linked recessive
   d. Sex-linked dominant
   e. Y-linkage

Generation

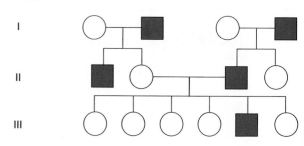

The illustration shows one way in which nondisjunction could occur during meiosis in a female.

1. Why is this process shown only in eggs and not in sperm?
2. What "goes wrong" with the meiotic process during the second meiotic division? Describe this abnormal meiosis II.
3. If each of the abnormal eggs were to join with a normal sperm, what are the possible outcomes in each case?

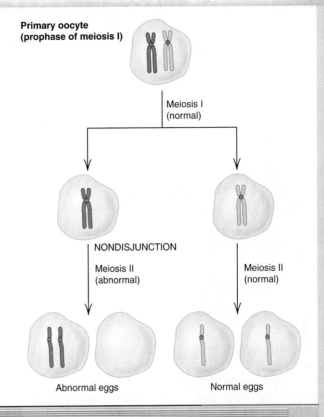

Primary oocyte
(prophase of meiosis I)

Meiosis I
(normal)

NONDISJUNCTION

Meiosis II
(abnormal)

Meiosis II
(normal)

Abnormal eggs

Normal eggs

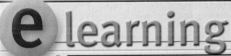

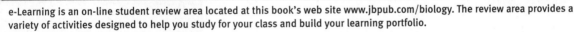

Location: http://www.jbpub.com/biology

www.jbpub.com/biology

e-Learning is an on-line student review area located at this book's web site www.jbpub.com/biology. The review area provides a variety of activities designed to help you study for your class and build your learning portfolio.

Review Questions The review questions test your knowledge of the important concepts and applications in each chapter. The review provides feedback for each correct or incorrect answer. This is an excellent test preparation tool.

Figure-Labeling Exercises Sharpen your visual thinking by matching terms and labels for important illustrations.

Flash Cards Studying biology requires learning new terms. Virtual flash cards help you master the new vocabulary for each chapter.

Just Wondering As you read and study from this text, you may find that you have unanswered questions. Through this site you can ask the author, Sandy Alters, your "just wondering" questions.

Do you prefer the speed and reliability of a CD-ROM? All of the features contained on the eLearning portion of the web site are also available on the Student CD-ROM.

Sex and Reproduction

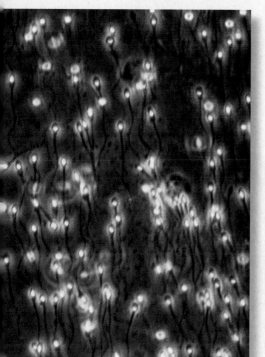

They all move toward a singular destination. Their nuclei glowing with a special fluorescent dye, these human sperm (left) are all moving in the same direction, propelled by their whiplike flagella. Only one of the hundreds of millions introduced into the vagina will be the first to encounter the female gamete—a developing egg.

The photo at right shows a human egg surrounded by sperm in the fallopian tube. Enzymes in the sperm head will dissolve the jellylike covering of the egg and pierce its membrane. Once the union of egg and sperm takes place, changes on the surface of the egg block the entry of additional sperm, ensuring that the fertilized egg will have only one complete set of hereditary material. Having more than one complete set could be fatal to a new embryo's existence. With a set of genes from each of its parents and with conditions right for survival, a single human embryo has the potential to develop into the complex network of 100 trillion cells that makes up the human body.

CHAPTER 22

Organization

Reproduction takes place both sexually and asexually.

Human reproduction is sexual and fertilization is internal.

Health Connections

Reproduction takes place both sexually and asexually.

22.1 Defining sexual and asexual reproduction

As the preceding story points out, each male sex cell, or **spermatozoon** (sperm, pl. spermatozoa), is a highly motile cell that can penetrate the membrane of the female sex cell, or *secondary oocyte* (potential egg). After penetration, a sperm joins its half of the hereditary message—the "blueprints" for the development of a new individual—with the hereditary material of the oocyte. This process, whereby a male sex cell and a female sex cell combine to form the first cell of a new individual, is called **sexual reproduction**. Although sexual reproduction usually involves the combination of sex cells, or **gametes** (GAM-eets), some sexually reproducing organisms do not produce gametes; instead, they transfer genetic material between organisms or undergo a fusion of nuclei of different mating types.

Sexual reproduction in humans is the focus of this chapter, but not all organisms reproduce sexually. Many organisms reproduce by means of **asexual reproduction**, the generation of a new individual without the union of gametes. With the exception of mutations that arise spontaneously, asexual reproduction is a cloning process—a method of exact replication of a parental organism.

Sexual and asexual reproduction both have their advantages and disadvantages with respect to survival of populations and species. A disadvantage of asexual reproduction is its outcome of genetic sameness. If environmental conditions become harmful for individuals of the cloned genetic makeup, or *genotype,* then the entire population could be at risk. An advantage of asexual reproduction is that individuals can add to a population, increasing its size quite rapidly in many cases. For example, bacteria can generate a population of billions in little more than a day from one original cell. Producing a large population from few individuals can result in excluding potential competitors and flooding the population with a particularly successful genotype.

Sexual reproduction is advantageous because it provides genetic variability among organisms of the same species. Differences among organisms within a species are the "raw materials" needed for adaptation to environmental change. When environmental conditions change, populations with genetic variability have a greater chance that some individuals will survive than do populations with little or no genetic variability. The capacity of some organisms to survive and reproduce in an environment in which others of their species cannot survive and reproduce, is in turn, the basis of natural selection. The concepts of adaptation and natural selection are described in more detail in Chapter 24.

> Sexual reproduction is a process by which a male sex cell and a female sex cell combine to form the first cell of a new individual. Asexual reproduction is a process by which new individuals are generated without the union of gametes. Sexual reproduction produces offspring that are a genetic "mix" of their parents, while asexual reproduction results in offspring that are genetic clones of each other and of their parent.

22.2 Patterns of asexual and sexual reproduction across kingdoms

Asexual reproduction occurs in all five kingdoms of life. Reproduction among bacteria is asexual; one cell divides into two, with no exchange of genetic material among cells. This process is called *binary fission* (literally, "two splitting"). Before fission, or division of the cell, the genetic material replicates and divides. Reproduction in bacteria is described in more detail in Chapter 26. Some bacteria transfer genetic material from one cell to another, but this gene transfer is not a form of reproduction because additional cells do not result. Gene transfer among bacteria is described in Section 19.3, p. 414.

Figure 22.1 Colorized electron micrograph of the bacterium *E. coli,* a normal inhabitant of the human intestine, undergoing binary fission.

Protists are another kingdom of organisms in which asexual reproduction is common. Many species of protists divide by fission, but the fission process may be binary or multiple (a splitting of the mother cell into many daughter cells). **Figure 22.1** shows a photograph of bacterium dividing by fission. Sometimes budding occurs in protists, in which an outgrowth of the mother cell develops into a new individual, which then breaks free and lives independently.

Many of the protozoans (animallike protists) reproduce sexually as well as asexually. They have no gametes but exchange genetic material through a tube connecting the cytoplasm of two individuals before fission occurs. Many of the plantlike protists (the algae) reproduce primarily by fission, but most of the brown, green, and red algae have life cycles much like plants, which include both sexual and asexual stages. Likewise, the funguslike protists (slime molds and water molds) have life cycles with sexual stages. Chapter 27 describes the reproductive patterns of protists in more detail.

Most fungi have a life cycle in which they reproduce both asexually and sexually (see Figure 27.23). Asexual reproduction in fungi usually involves budding, growing new hyphae (slender filaments) from fragments of parent hyphae, or producing *spores*. Hun-

dreds are in each spore case positioned atop individual hyphae (see photo above). Spores are reproductive bodies formed either by mitosis in a haploid parent or by meiosis in a diploid parent. (Haploid means an organism has half its normal complement of genetic material. Diploid means an organism has its full complement of genetic material.) In fungal sexual reproduction, the haploid nuclei of two genetically different mating types fuse, forming a diploid zygote. Fungi do not form gametes.

Plants undergo a life cycle that includes both asexual and sexual phases. A generalized life cycle of plants is illustrated in Figure 18.14. The sexual reproductive phase in plants' life cycles occurs when gametes form from haploid plants (gametophytes) by mitosis. These male and female gametes join during fertilization, forming diploid zygotes that grow into diploid plants (sporophytes). Asexual reproduction takes place as sporophytes form spores by meiosis. These spores germinate into the haploid gametophyte plants. Chapter 28 focuses on the reproductive diversity of plants.

In the animal kingdom, asexual reproduction is quite common among the invertebrates. Some invertebrates such as sponges, jellyfish, and sea anemones can reproduce asexually by budding (see the anemone below).

Budding anemones

The trematodes (such as liver flukes and blood flukes) can reproduce asexually when in the larval stage (an immature developmental form). The larvae simply replicate, increasing the size of the population of these flatworms. The increase in the size of the population increases the probability that some of the parasites will find hosts in which to live. Certain insects called gall midges reproduce asexually in a similar way, producing genetically identical larvae within themselves. These insects feed on fungus; increasing the size of the population increases the likelihood that some will survive to find a suitable fungal food source. Some invertebrates, such as certain flatworms and sea stars, can detach parts of their bodies, which then grow into new individuals.

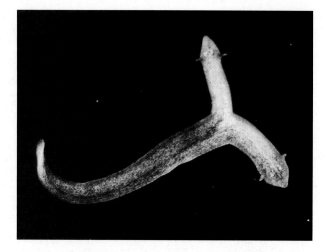

This process is called **fragmentation**. The flatworm shown above grew another head after sustaining a head injury, but the worm never completely divided.

Another type of asexual reproduction called **parthenogenesis** (PAR-theh-no-JEN-eh-sis) occurs in a variety of invertebrates such as earthworms, mites, ticks, and beetles. During this process the eggs of the female develop into adults without being fertilized. Male bees called drones are produced this way, while female bees (both workers and queen bees) develop from fertilized eggs. Parthenogenesis even occurs among vertebrates. Certain fishes, amphibians, and lizards reproduce this way. Many species of whiptail lizards reproduce exclusively by parthenogenesis but engage in behaviors that simulate mating, as shown in the photo below. The lizard on the top is simulating male behavior during mating in a closely related species of lizard that reproduces sexually.

Bacteria reproduce asexually; although certain strains of bacteria exchange genetic material at times, this process is not a form of reproduction because the population is not increased. Both asexual and sexual forms of reproduction occur in all the other kingdoms of life, but not all species in these kingdoms reproduce using both methods.

22.3 Patterns of sexual reproduction in animals

Sexual reproduction in the invertebrates is as common as asexual reproduction. Although two individuals are usually involved in sexual reproduction, a single individual, called a **hermaphrodite** (her-MAF-reh-dite), may be both male and female, thereby playing either role at any particular time. Hermaphroditism is common among invertebrates; almost all invertebrate groups have some hermaphroditic species.

Some species of invertebrates are simultaneous hermaphrodites; that is, an individual is both male and female at the same time. Barnacles, for example, are simultaneous hermaphrodites. Since they are sessile organisms (fixed to one place), they can only mate with whichever organism is beside them. Being able to function as a male or a female is certainly a reproductive advantage. (Simultaneous hermaphrodites usually do not self-fertilize.) Earthworms are another example of simultaneous hermaphrodites. **Figure 22.2** shows earthworms mating. They have aligned themselves so that they can exchange and store each other's sperm. The *clitellum* of each worm secretes a mucous covering that slips along the worms' bodies as each moves through the soil. The clitellum picks up eggs and stored sperm. These sacs then slip off each worm and remain in the soil, protecting the developing embryos.

Clitellum

Figure 22.2 Earthworms mating

A diagram of the processes of spermatogenesis and oogenesis is shown here. Fill in the missing labels. Then, describe each process, noting differences and commonalties between the two.

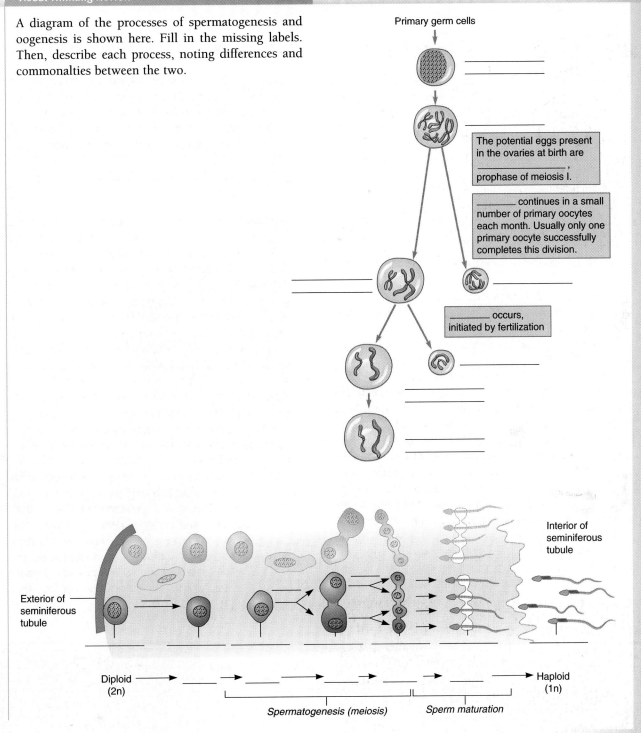

Primary germ cells

_____

_____

The potential eggs present in the ovaries at birth are _____, prophase of meiosis I.

_____ continues in a small number of primary oocytes each month. Usually only one primary oocyte successfully completes this division.

_____ occurs, initiated by fertilization

Interior of seminiferous tubule

Exterior of seminiferous tubule

Diploid (2n)

Haploid (1n)

Spermatogenesis (meiosis)

Sperm maturation

Development before Birth

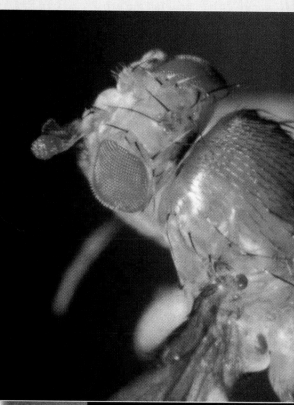

Eyes on its antennae and eyes on its wings. The fly at right gives new meaning to the phrase "having eyes in the back of your head." This mutant *Drosophila* (fruit fly) is only one of many types produced in the laboratory of Dr. Walter Gehring and his colleagues at the University of Basel in Switzerland. These mutants hold a major key to the understanding of how eyes develop. They also help us understand the developmental processes of animals in general, including ourselves. At far right is a photo of a six-week-old human embryo, whose eyes are just starting to form.

Recently, the Gehring group discovered that by turning on a gene called *eyeless* in parts of a fruit fly where it is normally turned off, they could produce flies with eyes in unusual places—their legs, antennae, and wings. Some flies grew as many as 14 eyes. The results of their research indicate that the eyeless gene may be a "master control gene"—a gene that, by itself, triggers the development of an organ or other structure.

Master control genes are also called homeotic genes, and were discovered in fruit flies in 1983. The eyeless gene is one of the latest homeotic gene discoveries. These genes produce transcription factors, which are protein switches that bind to genes or groups of genes and turn them on and off.

Prior to the eyeless discovery, scientists had found other genes that trigger the development of different cell types in the eye such as cells and pigments that detect light. However, they still know little about the process from start to finish, in which hundreds of genes are turned on and off in a cascadelike manner, with the ultimate result being the formation of a fully functional eye. With the eyeless gene, researchers should be able to learn more. Its position as one of the first genes in the sequence of gene actions that direct eye development should help researchers reveal the remaining genes in the series.

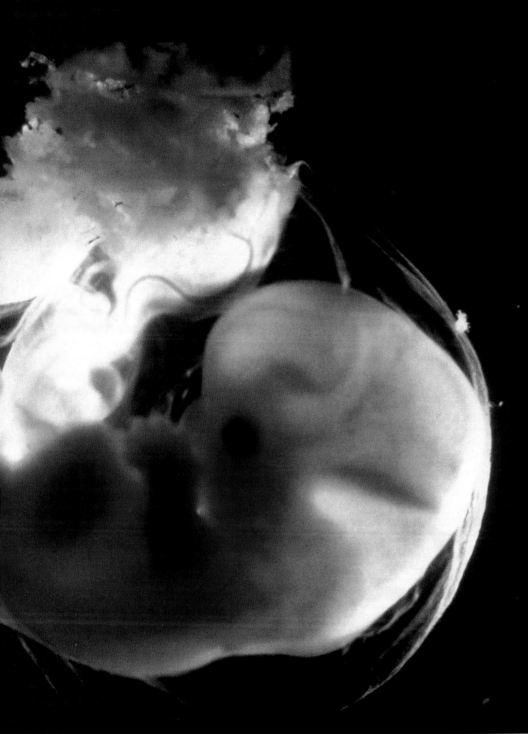

CHAPTER 23

Organization

Developmental patterns show evolutionary relationships among animals.

23.1 Patterns in embryological development

The genetic control of development appears to be much more universal than scientists once thought. One major piece of evidence that upholds this assertion is that the eyeless gene is found in organisms from flatworms to humans. It is a master control gene that crosses the boundaries of many species. This fact shows that the various types of eyes found among organisms in the animal kingdom may have evolved from a single, common ancestor millions of years ago. Therefore, what the cell biologists in the Gehring laboratory and others discover about fruit flies may teach them developmental lessons not only about flies, but also about people and a variety of other organisms as well.

Research on homeotic genes centers on one of the three processes that take place during the development of multicellular organisms: **tissue differentiation**. During tissue differentiation, groups of cells become distinguished from other groups of cells by the jobs they perform in the body. So, eye tissue becomes differentiated from skin tissue and so forth. In some invertebrate organisms, tissue differentiation occurs very early in development. However, in "higher" organisms such as mammals, the cells of very early embryos (a beginning stage of development) are capable of becoming any type of tissue. Cell differentiation takes place little by little as the embryo develops, and probably occurs first with the separation of the inner cell mass and trophoblast cells (see p. 514).

The other two processes that take place during embryological development (early development) are *cell division,* and **morphogenesis** (more-foe-JEN-uh-sis). Cell division results in the growth of a developing organism. This process (mitosis) is described in detail in Chapter 18. During morphogenesis (literally "form creation"), cells begin to move, or migrate, to shape the new individual.

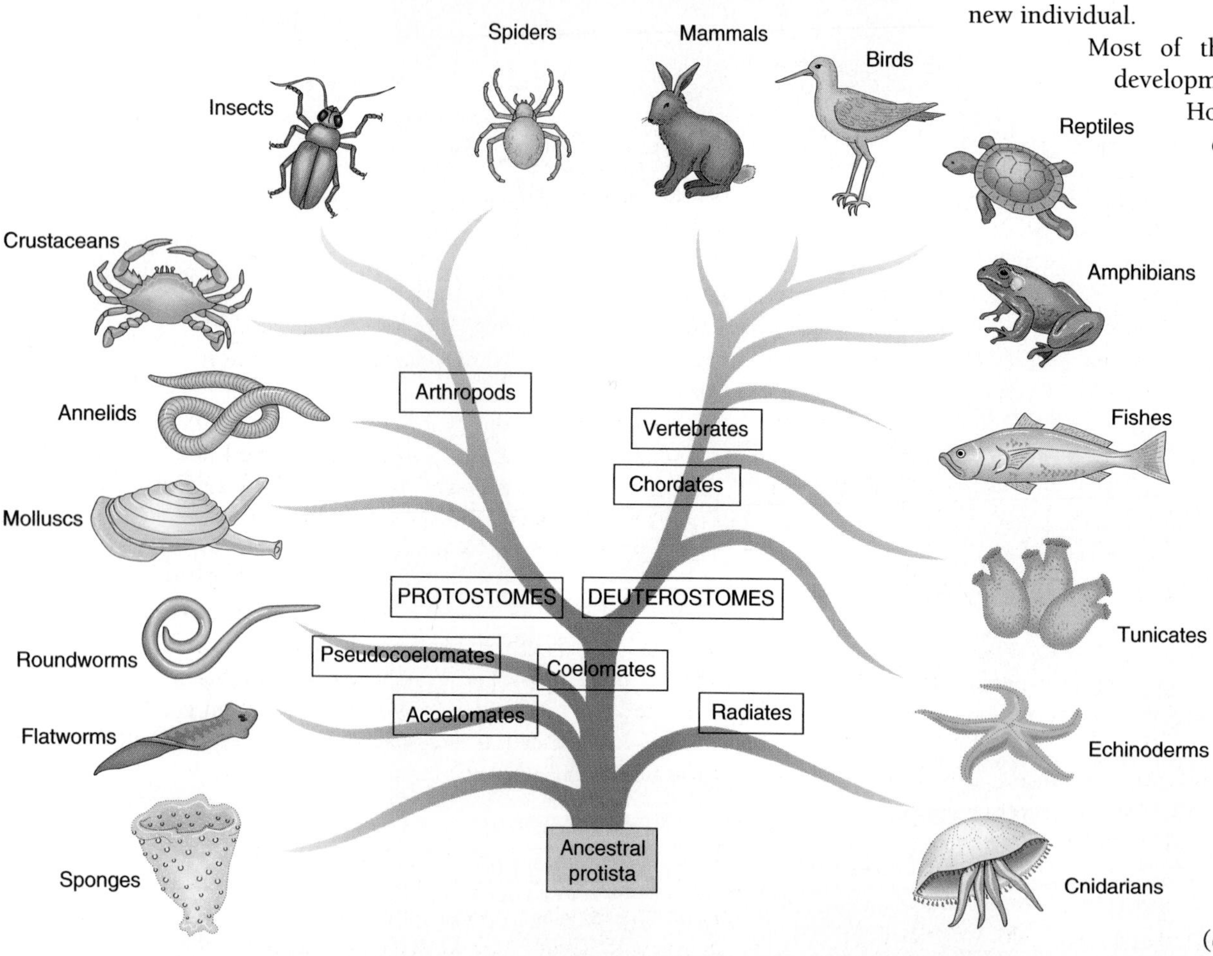

Most of this chapter describes development of the human. However, human embryological development fits into one of two distinct patterns that are present in the animal "family tree." Two main branches arise from the trunk of this tree as shown in Figure 23.1. These two branches of coelomate animals (those with a fluid-filled body cavity lined with connective tissue) represent two distinct evolutionary lines. (That is, the organisms in each line are more closely related to one another than to organisms in the other line.) One evolutionary line includes the molluscs (clams), annelids (segmented worms), and arthropods (crustaceans, insects, and spiders). These organisms are

Figure 23.1 The animal ancestral tree. All animals on the lower branches of the tree are phyla. The animals on the upper right branches are classes of phylum Chordata, subphylum Vertebrata (as shown by the boxed labels). The animals on the upper left branches of the tree are phyla of the diverse group of organisms called arthropods (as shown by the boxed label).

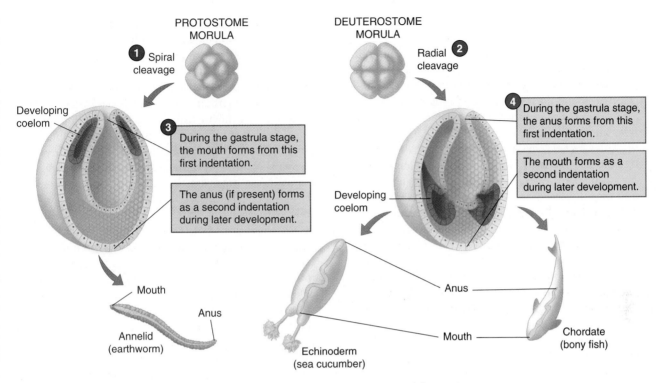

PROTOSTOME MORULA

1 Spiral cleavage

DEUTEROSTOME MORULA

Radial cleavage 2

Developing coelom

3 During the gastrula stage, the mouth forms from this first indentation.

The anus (if present) forms as a second indentation during later development.

4 During the gastrula stage, the anus forms from this first indentation.

The mouth forms as a second indentation during later development.

Developing coelom

Mouth

Anus

Annelid (earthworm)

Echinoderm (sea cucumber)

Anus

Mouth

Chordate (bony fish)

Figure 23.2   Protostomes and deuterostomes. The difference between protostomes and deuterostomes arises during embryonic development. The mouth of the protostome develops from the first indentation of the gastrula, and the anus, if present, forms as a second indentation later. The mouth of a deuterostome develops from a second indentation. The deuterostome anus develops from the first indentation.

called **protostomes** (PRO-toe-stowmz), meaning "first" (*proto*) "mouth" (*stome*). The other evolutionary line includes echinoderms (sea stars, sea cucumbers), tunicates, fishes, amphibians, reptiles, birds, and mammals. These organisms are called **deuterostomes** (DOOT-uh-row-stowmz), meaning "second mouth." (Humans are deuterostomes.) These strange names refer to events in the development of these organisms.

During early embryological development, all animals consist of a solid ball of cells usually called a **morula** (MORE-yuh-luh) from a Latin word meaning "mulberry." The morula is formed by cleavage of the fertilized egg, or zygote. (During cleavage, the fertilized egg divides into many cells without cell growth.) The protostomes and the deuterostomes show different cleavage patterns during the morula stage of development.

The cells of the protostomes divide in a way that forms a spiral pattern (Figure 23.2 ❶). The cells of the deuterostomes cleave in a radial pattern ❷. These patterns of embryological development (along with other developmental similarities) suggest close evolutionary relationships among the organisms within each group. In addition, because the protostome developmental pat-

tern occurs in all acoelomates as well, scientists think that it was the pattern of development in the ancestors of modern animals.

As development proceeds, the cells of the morula secrete fluid that fills the interior of the ball and pushes the cells to the edges. This stage is called the **blastula** (or *blastocyst* in mammals). This fluid-filled ball of cells then forms an indentation, assuming the shape of a blown-up balloon with a fist pushing in one side. This stage of development, the **gastrula** (GAS-truh-lah), gives rise to a three-layered embryo and begins the formation of the gut. In the protostomes, this first indentation becomes the mouth of the organism ❸. In the deuterostomes, the first indentation becomes the anus and a second one becomes the mouth ❹.

The protostomes (molluscs, annelids, and arthropods) differ from the deuterostomes (echinoderms and chordates) with regard to certain embryological events. The similarities in developmental patterns within each group suggest close evolutionary relationships.

23.2 Fertilization

During human development, what happens after the birth of an infant is crucial to its survival. However, what happens before birth may be even more crucial. In fact, if an infant does not develop properly before birth, normal growth after birth becomes difficult and sometimes impossible.

This chapter focuses on human life before birth, or **prenatal development**, the gradual growth and progressive changes in a developing human from conception until the fetus leaves the mother's womb. (However, the processes of development in all vertebrates are very similar to the course of human development before birth.) This time of development is also called *gestation* and lasts approximately 8½ months. (The 9-month pregnancy calculation is computed from the first day of a woman's last menstrual period, but conception usually takes place in the middle of her cycle.) The gestation period is commonly referred to as pregnancy.

The union of a male gamete (sperm) and a female gamete (egg) is called **fertilization**, or conception. In the human, fertilization usually takes place in the uterine (fallopian) tube of the female. (See Chapter 22 for a discussion of gamete formation and the events leading up to fertilization.) Therefore, development begins in the uterine tube.

Figure 23.3 shows both an egg (ovum) and a sperm; the egg is a large cell in relationship to the sperm. In fact,

ova are among the largest cells within an animal's body. Making up most of the substance of an ovum of many animals is the yolk, the nutrient material that the developing animal lives on until nutrients can be derived from the mother. In humans, the ovum contains a great deal of cytoplasm, but virtually no yolk. The developing human derives nourishment from the mother after implantation. In most egg-laying animals such as birds, for example, the young develop totally within the egg. Therefore, the yolk of these eggs must be substantial enough to maintain the organism throughout its entire course of development. The yolk of an ostrich egg, for example, is so large that it is approximately the size of a baseball.

As mentioned in Chapter 22, the secondary oocyte breaks away from the ovary during ovulation in placental mammals (which includes humans) and is swept into the uterine tube. This cell is not yet a mature ovum, or egg. Only after penetration of the oocyte by a sperm will the second meiotic division be completed, yielding the mature ovum and a second polar body. Although fertilization may sound like an easy job for sperm, only 50 to 100 sperm make it to the egg from an ejaculate that contains approximately 200 to 300 million sperm.

Many things can go wrong from the time the sperm are manufactured in the testes, stored in the epididymis, and sent out of the man's body and into the vagina of a woman. For example, up to 20% of sperm are deformed in some way. A common abnormality is having two flagella instead of one. Sperm with this abnormality do not have normal motility and cannot make the lengthy journey from the vagina, through the constricted cervix, up through the uterus, and along the uterine tube.

Although the acid environment of the female vagina is treacherous territory for sperm and they constantly swim upstream against the downward currents of female secretions, they are also aided by the female body. During the time of ovulation, a female produces strands of a special protein called *mucin*. This substance becomes a part of the cervical mucus her body produces and provides threadlike highways along which the sperm can travel. The sperm make their way up these strands of mucin to the uterus. As they move along, enzyme inhibitors located at the tips of their heads are slowly worn away, gradually uncovering enzymes that are capable of penetrating the egg's outer protective layers and its membrane. The above photo shows a sperm that has penetrated an egg's jellylike covering, the *zona pellucida*.

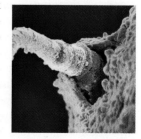

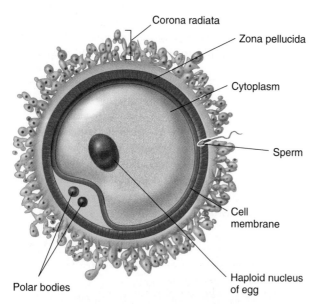

Figure 23.3 The egg. **A human egg is surrounded by a membrane and contains a haploid nucleus.** The corona radiata is a halo of cells from the ovarian follicle.

Labels on figure: Corona radiata, Zona pellucida, Cytoplasm, Sperm, Cell membrane, Haploid nucleus of egg, Polar bodies

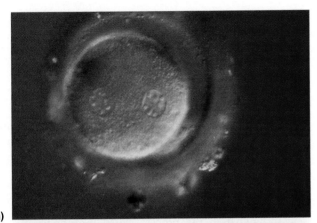

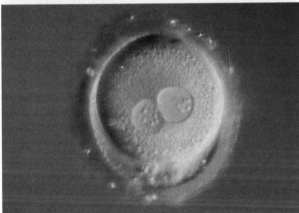

(a)

(b)

Figure 23.4 **The fusion of nuclei.** (a) The nuclei of the sperm and ovum are clearly separate. (b) The nuclei have merged to form the nucleus of the first cell of a new individual.

On entry of the cytoplasm of the oocyte, the sperm sheds its tail and its head swells. Immediate changes in the surface of the egg allow no other sperm to penetrate, and oocyte meiosis is completed. Finally, the sperm and ovum nuclei, clearly seen as separate from one another in Figure 23.4a, fuse to form the nucleus of the first cell of a new individual (Figure 23.4b). This new cell, which contains intermingling genetic material from both the mother and the father, is called a **zygote** (ZYE-goat).

With the union of the chromosomal material of both gametes, the zygote begins to divide by mitotic cell division. As division proceeds during the next 2 weeks, the developing cell mass is referred to as a **preembryo**.

The penetration of the secondary oocyte by the sperm is called fertilization. This penetration causes changes in the surface of the egg that allow no other sperm to enter and triggers the completion of meiosis. After the sperm sheds its tail, the nuclei of the egg and sperm fuse to form a zygote, the first cell of a new individual.

23.3 The first and second weeks of development: The preembryo

Within 30 hours, the one-celled zygote begins to divide rapidly: one cell into two, two into four, four into eight, and so forth, producing a cluster of cells. The photo below shows the preembryo at the eight-cell stage, floating free in the uterine tube as its cells continue to divide. (Four of the cells are hidden from view.) This process of cell division without cell growth is called *cleavage,* as mentioned previously.

Occasionally, a single, fertilized ovum splits into two cell clusters during cleavage, and each cluster continues developing on its own. Because both cell clusters arose from the same cell, they have identical genetic information and will result in the development of identical twins. Identical twins are always the same sex. Fraternal twins arise from the release of two secondary oocytes and the fertilization of both. Fraternal twins are genetically different from one another because they are two separate oocytes fertilized by two different sperm cells. Therefore, they may be different sexes.

As the cell cluster (preembryo) divides, it journeys along the remaining two-thirds of the uterine tube. This trip takes approximately 3 days. Occasionally, a preembryo gets caught in the folded inner lining of the uterine tube and implants there, creating an *ectopic pregnancy.* The possibility of this situation occurring is increased if the tubes are scarred from previous infections. Acute pelvic pain usually signals this problem, which requires immediate medical attention.

Normally, however, the preembryo reaches the uterus, or womb. Still the same size as a newly fertilized ovum, the preembryo now consists of about 16 densely clustered cells and is called a morula.

Cells of preembryo

Wall of uterine tube

The preembryo floats free in the uterus as its cells continue to divide. After 2 days in the uterus, the morula has developed into the **blastocyst** (BLAS-tuh-sist), a stage of development in which the preembryo is a hollow ball of cells as shown in Figure 23.5. The center is filled with fluid from the uterine cavity. The term blastocyst is descriptive, derived from two Greek words meaning "germ (germinal) sac."

One portion of the blastocyst contains a concentrated mass of cells destined to differentiate into the various body tissues of the new individual. It is referred to as the *inner cell mass.* Each cell in this mass has the

EARLY BLASTOCYST

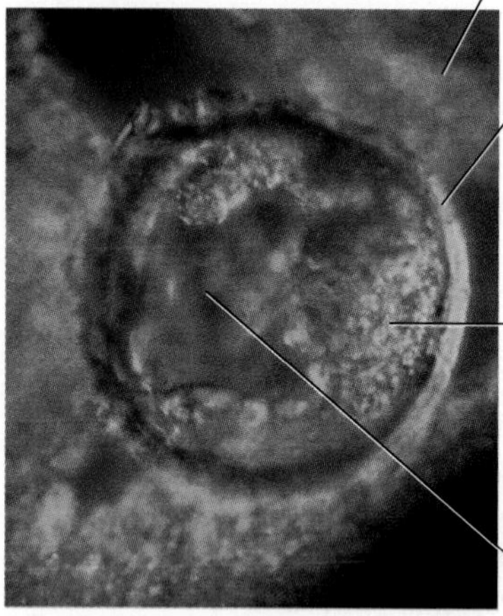

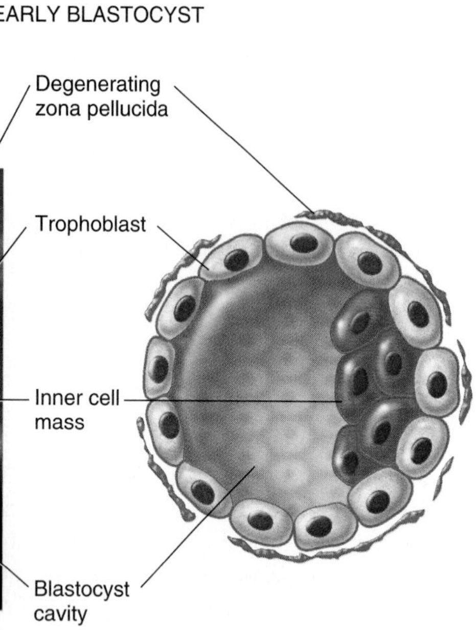

Figure 23.5 The blastocyst.

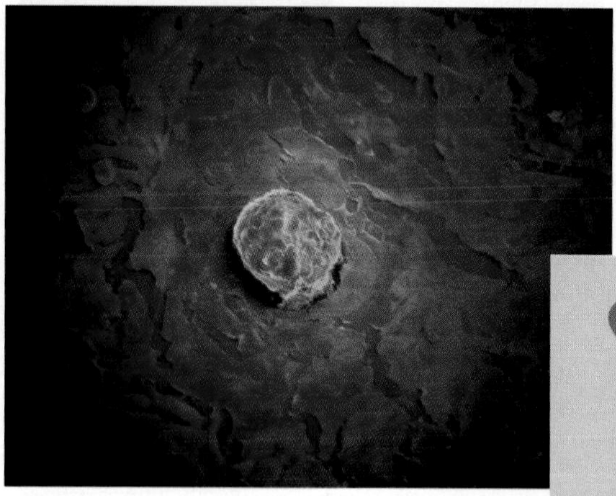

(a)

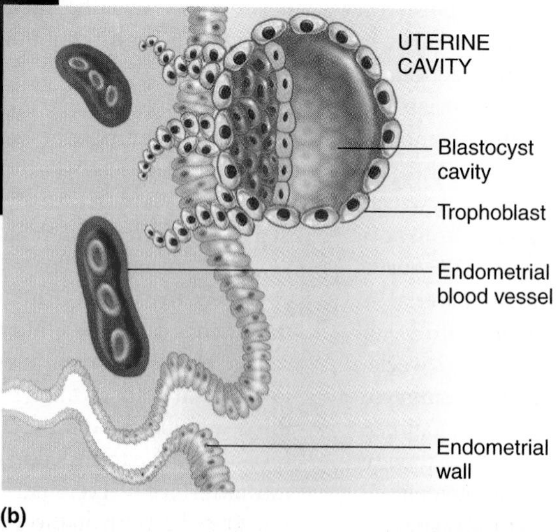

(b)

Figure 23.6 **A detailed view of implantation.** (a) About 7 to 8 days after fertilization, the blastocyst imbeds itself into the wall of the uterus. (b) Special fingerlike projections produced by the trophoblast anchor the blastocyst to the wall of the uterus.

ability to develop into a complete individual. In fact, scientists have been able to produce test-tube mice using transplanted nuclei from inner cell mass cells. The outer ring of cells, called the **trophoblast** (TRO-fuh-blast) from a Greek word meaning "nutrition," will give rise to most of the extraembryonic membranes, including much of the placenta, an organ that helps maintain the developing embryo.

The general term blastula is used to describe the saclike blastocyst of mammals and, in other animals, the stage that develops a similar fluid-filled cavity. This stage of development is significant for all vertebrates; it is the first time that cells begin to migrate to shape the new individual in the process called morphogenesis, mentioned previously. However, the major morphogenetic events occur during the third to eighth weeks.

Approximately 1 week after fertilization, the blastocyst secretes enzymes that digest a microscopic portion of the uterus. It then nestles into this site, nourished by the digested uterine cells, in a process called **implantation**. Barely visible to the naked eye [Actual length], the blastocyst most often attaches to the posterior wall of the uterus. Figure 23.6a shows a photo of an implanted blastocyst, and Figure 23.6b shows a detailed view of implantation. This diagram illustrates the fingerlike projections produced by the trophoblast, which anchor the blastocyst to the endometrial wall of the uterus. Figure 23.7 chronicles the events that take place from fertilization to implantation.

By this time, a woman is nearing the end of her menstrual cycle and the blastocyst is in danger of being swept away during menstruation. However, the blastocyst secretes a hormone called *human chorionic gonadotropin* (*HCG*), which maintains the corpus luteum in the ovary. The corpus luteum is a structure derived from the ruptured follicle from which the egg was cast out in the ovary (see Chapter 22). Sustained by HCG, the corpus luteum continues to secrete progesterone and estrogen, hormones that maintain the uterine lining and allow the development of the preembryo to proceed. After the first 3 months of development, the placenta (see p. 517) begins to secrete the estrogens and progesterone that maintain the pregnancy.

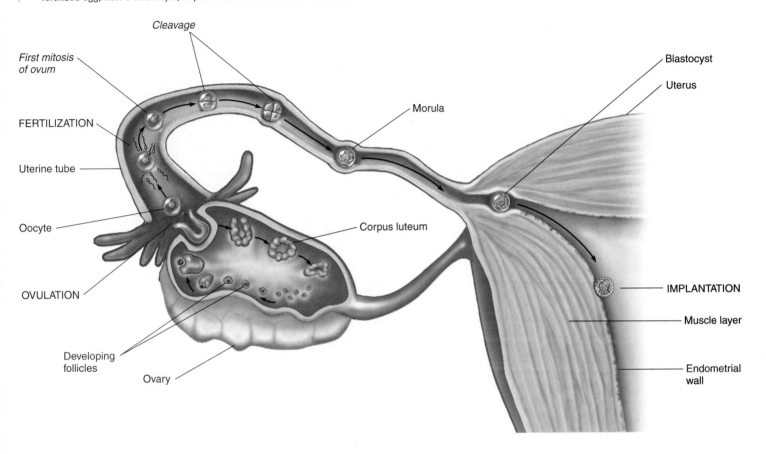

Visual Summary **Figure 23.7** **From fertilization to implantation.** The unfertilized oocyte is fertilized by a sperm in the uterine (fallopian) tube. The fertilized egg (zygote) continues to divide as it makes its way through the uterine tube to the uterus. About 7 to 8 days after fertilization, the fertilized egg, now a blastocyst, implants itself in the wall of the uterus.

Cleavage

First mitosis of ovum

FERTILIZATION

Uterine tube

Oocyte

OVULATION

Developing follicles

Ovary

Morula

Corpus luteum

Blastocyst

Uterus

IMPLANTATION

Muscle layer

Endometrial wall

During its second week of development, the preembryo completes its implantation within the uterine wall, two of its three **primary germ layers** develop, and the extraembryonic membranes begin to form. The primary germ layers are three layers of cells that develop from the inner cell mass of the blastocyst and from which all the organs and tissues of the body develop. These three layers are called the **ectoderm** ("outside skin"), **endoderm** ("inside skin"), and **mesoderm** ("middle skin"). The ectoderm forms the outer layer of skin, the nervous system, and portions of the sense organs. The endoderm gives rise to the lining of the digestive tract, the diges-

tive organs, the respiratory tract, and the lungs; the urinary bladder; and the urethra. The mesoderm differentiates into the skeleton (bones), muscles, blood, reproductive organs, connective tissue, and the innermost layer of the skin. **Figure 23.8** shows the body systems and tissues into which the germ layers develop. **Figure 23.9** diagrams the completely implanted preembryo buried within the uterine lining, showing two of the germ layers and the beginnings of the third. At the beginning of the second week, only the endoderm and the ectoderm have formed.

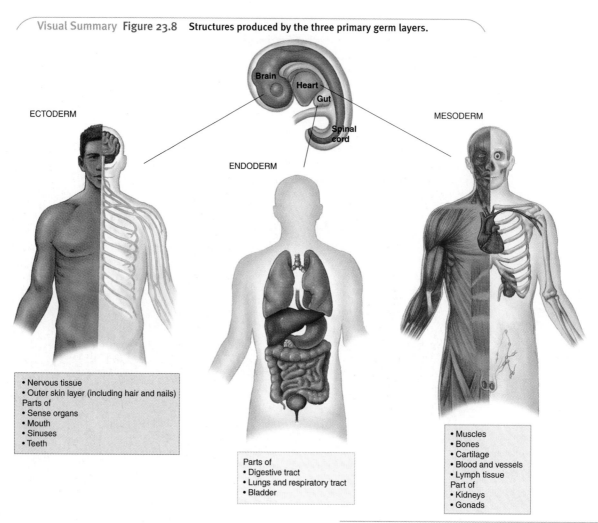

Visual Summary Figure 23.8   Structures produced by the three primary germ layers.

ECTODERM

ENDODERM

MESODERM

• Nervous tissue
• Outer skin layer (including hair and nails)
Parts of
• Sense organs
• Mouth
• Sinuses
• Teeth

Parts of
• Digestive tract
• Lungs and respiratory tract
• Bladder

• Muscles
• Bones
• Cartilage
• Blood and vessels
• Lymph tissue
Part of
• Kidneys
• Gonads

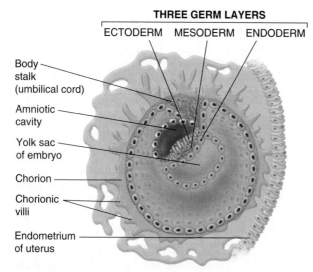

THREE GERM LAYERS

ECTODERM   MESODERM   ENDODERM

Body stalk (umbilical cord)

Amniotic cavity

Yolk sac of embryo

Chorion

Chorionic villi

Endometrium of uterus

Figure 23.9   The implanted preembryo showing the three primary germ layers during the third week of development. The endoderm and ectoderm have already formed at this stage, with the mesoderm just beginning its development.

Cleavage, or cell division of the zygote without cell growth, is the first stage in the development of humans as well as other multicellular animals. It results in the formation of a mass of cells known as a morula approximately 3 days after fertilization.

Cell migration, which begins to shape the preembryo, is the process that begins approximately 4 days after fertilization. Cell migration helps shape the developing individual in a process called morphogenesis. At this early stage, cell movement results in the formation of a hollow ball of cells known as a blastocyst in humans and in other mammals a stage more generally called the blastula.

Approximately 1 week after fertilization, the blastocyst implants in the lining of the uterus and secretes human chorionic gonadotropin (HCG). HCG acts on the corpus luteum in the ovary. The corpus luteum responds by continuing to produce estrogens and progesterone, hormones that sustain the implanted blastocyst as the placenta develops.

23.4 Early development of the extraembryonic membranes

As Figure 23.9 shows, the extraembryonic membranes have begun to form even at this early stage (third week) of development. The **extraembryonic membranes** all play some role in the life support of the preembryo, the embryo, and then the fetus. (The term fetus is used to describe the stage of development after 8 weeks.) Called extraembryonic membranes because they are not a part of the body of the embryo, these structures provide nourishment and protection. They form from the trophoblast, the ring of cells surrounding the inner cell mass. At the same time, the preembryo is forming from the inner cell mass cells. The further development of these membranes continues into the fetal period.

The **amnion** (AM-nee-on) is a thin, protective membrane that grows down around the embryo during the third and fourth weeks, fully enclosing the embryo in a membranous sac. The amniotic sac can be thought of as a shock absorber for the embryo. This thin membrane encloses a cavity that is filled with a fluid—*amniotic fluid*—in which the fetus floats and moves. This fluid also helps keep the temperature of the embryonic and fetal environment constant. The amniotic cavity is first seen at about 8 days as a slitlike space that appears between the inner cell mass and the underlying trophoblast cells that are invading the uterine wall. It appears as a better-developed structure in Figure 23.9, at the end of 2 weeks of development [Actual length].

Notice also in Figure 23.9 that a structure called the **chorion** (KORE-ee-on) is beginning to develop as tissue outside of and including the trophoblast cells that ring the developing embryo and extraembryonic tissue. The chorion is highly specialized to facilitate the transfer of nutrients, gases, and wastes between the embryo and the mother's body. It is a primary part of an organ called the **placenta**, a flat disk of tissue about the size (at birth) of a large, thick pancake that grows into the uterine wall. The placenta is made up of both chorionic and maternal tissues. Digging deeply into the uterine lining, fingerlike extensions of the chorion come into close contact with the blood-filled uterine tissues at the placenta. When the embryo reaches a stage of development in which the heart begins to beat, oxygen-poor blood filled with wastes is sent from the embryo's body through the umbilical arteries to the placenta.

The two umbilical arteries and a single vein are embedded in the connective tissue of the **umbilical cord**, the developing embryo's lifeline to the mother. This "highway" joins the circulatory system of the embryo with the placenta. At the placenta, embryonic wastes are exchanged for nutrients and oxygen through a thin layer of cells that separates the embryo's blood from the mother's blood. The embryonic blood and maternal blood do not mix. The fetal blood then travels through the umbilical vein back to the embryo. The umbilical cord develops from the body stalk, the yolk sac, and the allantois during the fourth week of development.

The **allantois** (ah-LAN-toe-us) (from a Greek word meaning "sausage") gives rise to the umbilical arteries and vein as the umbilical cord develops. Although this extraembryonic membrane is small and cannot be seen in Figure 23.9, it is shown in Figure 23.10. Appearing

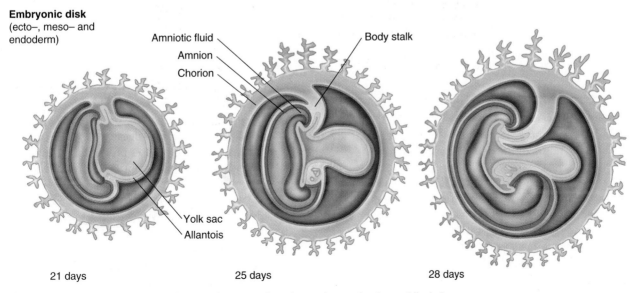

Embryonic disk
(ecto–, meso– and endoderm)

Amniotic fluid

Amnion

Chorion

Body stalk

Yolk sac

Allantois

21 days   25 days   28 days

Figure 23.10 **Extraembryonic membranes.** The extraembryonic membranes develop rapidly during a single week.

Just Wondering . . . *Real Questions Students Ask*

Can I drink a glass of wine occasionally during my pregnancy or will it hurt my baby?

Scientists have not yet determined how much alcohol, if any, a woman can drink during pregnancy and not risk damage to her fetus. Complicating the matter, researchers are unsure when the embryo/fetus is specifically most vulnerable to the effects of alcohol. So having two drinks on day 39 may not harm the embryo, for example, but having two drinks on day 40 may cause brain damage.

When a pregnant woman drinks, the alcohol she consumes crosses the placenta and intoxicates the embryo/fetus. A recent study has demonstrated that women have less of a stomach enzyme to neutralize alcohol than men do, so women become intoxicated faster because of higher levels of alcohol in their bloodstream. These high blood alcohol levels also mean that more alcohol is passed along to the embryo/fetus.

The damage that is inflicted on the embryo/fetus can be severe, or it can be more subtle. There seems to be a rough correlation between the amount of alcohol consumed and the severity of the birth defects that result. The most serious effects of drinking on the developing embryo/fetus are seen

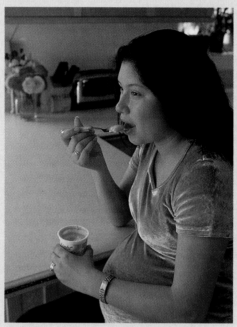

in babies born to chronic alcoholics who drank heavily during pregnancy. These effects are called *fetal alcohol syndrome* and are characterized by certain psychological, behavioral, cognitive, and physical abnormalities. These children require special education and sometimes physical therapy to help them contend with their disabilities. Women who drink less than alcoholics but drink consistently throughout their pregnancies often give birth to children with *fetal alcohol effects*. These children have learning difficulties, show poor judgment, are impulsive, are unable to learn from their mistakes, and are undisciplined. These children often have serious difficulties in school.

Experts agree that the surest way to avoid alcohol-related effects is not to drink at all during pregnancy. Some researchers believe that a woman even contemplating pregnancy should not drink. This position makes sense because many women do not know they are pregnant until weeks after conception, and unwitting alcohol consumption may have already caused damage to the embryo. Given the uncontrollable factors influencing the birth of a healthy baby (such as heredity), it makes sense to control the factors we can. Not drinking is one thing a pregnant woman can do to contribute to the health of her baby.

---

during the third week of development as a tiny, sausage-shaped pouching on the yolk sac, the allantois is initially responsible for the formation of the embryo's blood cells; it later develops into the umbilical blood vessels. Notice that the **yolk sac**, a structure established during the end of the second week, also becomes a part of the umbilical cord. Before becoming a nonfunctional part of the cord, the yolk sac produces blood for the embryo until its liver becomes functional during the sixth week of development. In addition, part of the yolk sac becomes the lining of the developing digestive tract.

During the second week of development, the pre-embryo completes implantation within the uterine wall, two of the three layers of cells develop from which organs and tissues will arise, and the extraembryonic membranes—the amnion, chorion, yolk sac, and allantois—begin to form from the trophoblast.

23.5 Development from the third to eighth weeks: The embryo

During this crucial time period, the developing individual is termed an **embryo**. Throughout the next 6 weeks, the embryo takes on a human shape by means of the ongoing process called *morphogenesis* mentioned earlier. In addition, the organs are established but need further development during the third through eighth weeks in a process called *organogenesis*. Because of the incredible array of developmental processes proceeding simultaneously during this period, the embryo is particularly sensitive to teratogens, certain agents such as alcohol that can induce malformations in the rapidly developing tissues and organs (see "Just Wondering").

The Third Week

At the end of the second week and continuing into the third week, various cell groups of the inner cell mass begin to divide, move, and differentiate, changing the two-

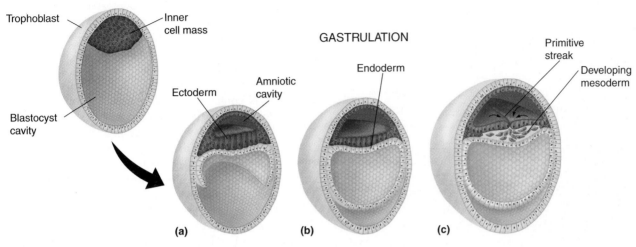

EARLY BLASTOCYST

Trophoblast

Inner cell mass

Blastocyst cavity

GASTRULATION

Ectoderm

Amniotic cavity

Endoderm

Primitive streak

Developing mesoderm

(a)   (b)   (c)

Figure 23.11   Human gastrulation. The amniotic cavity forms within the inner cell mass (a), and in its base, a layer of endoderm differentiates (b). A primitive streak develops, through which cells destined to become mesoderm migrate into the interior (c).

layered preembryo into a three-layered embryo. This process is called **gastrulation.** (The gastrula was discussed early in this chapter on p. 511.) This word is derived from a Greek word *gastros* meaning "belly." In fact, the prefix *gastr-* is found in many words denoting parts of the human body, such as gastric, referring to the stomach. This term is descriptive of the fact that at this stage in many animals, a primitive gut is formed by the invagination (infolding) of the blastula, but the intestines develop differently in humans and many animals.

At the beginning of the third week, the embryo is an elongated mass of cells barely one tenth of an inch long (2.5 mm) [Actual length] . A streak (called the *primitive streak*) runs down the midline of what will be the back side of the embryo. Cells at the streak migrate inward, producing the mesoderm that develops during the gastrulation period. Cells at the head end of the streak grow forward to form the beginnings of the **notochord** (NO-toe-kord). The notochord is a structure that forms the midline axis along which the vertebral column (backbone) develops. An embryonic notochord forms in all vertebrate animals. However, in humans and in most other vertebrates, it degenerates and disappears long before birth. Gastrulation ends with the completion of the notochord midway through the third week. By the end of gastrulation, a layer of ectoderm covers the notochord tissue. **Figure 23.11** diagrams the events of gastrulation.

In the photo of the embryo at 4½ weeks (see Figure 23.14), its head can be seen to have the beginnings of eyes, a sign that **neurulation** (NOOR-oo-LAY-shun) has

taken place. Neurulation is the development of a hollow nerve cord, which later develops into the brain, spinal cord, and related structures such as the eyes.

Neurulation begins in the third week with the folding of the ectoderm lying above the notochord, forming an indentation along the back of the embryo. This indentation is called the *neural groove.* On either side of the groove are areas of tissue called *neural folds.* In **Figure 23.12,** the neural folds of the 3-week embryo have

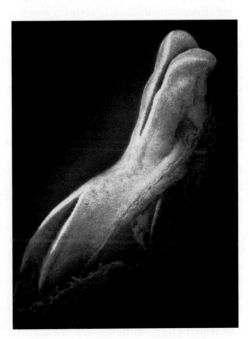

Figure 23.12   Neurulation.

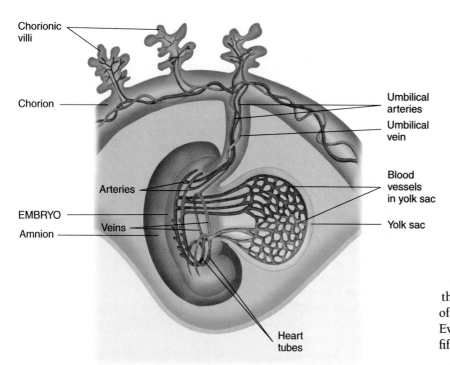

Figure 23.13   Cardiovascular system of a 3-week-old embryo.

come together at one spot and have fused. This spot is destined to be the neck region of the developing individual. The tissue above the fused region, looking somewhat like a pair of lips, will develop into the brain. The less broad area of tissue below the fused region will develop into the spinal cord. Eventually, the edges of the groove will fuse along its length, forming a neural tube, the precursor to the structures noted above. Neurulation results in an embryo called a *neurula* and signals the developmental process of *tissue differentiation* mentioned at the beginning of this chapter (p. 510).

During the third week, the embryo's heart also begins its development—as a pair of microscopic tubes. In fact, the cardiovascular system is the very first system to become functional in the embryo. At the end of the third week of development, the heart tubes have fused and have linked with blood vessels in the embryo, body stalk, chorion, and yolk sac to provide a primitive circulation of blood as shown in Figure 23.13, and the heart begins to beat. At the same time, fingerlike structures called *villi* begin to protrude from the chorion. These blood-filled projections dig into the uterine tissues of the mother, increasing the surface area over which gases, nutrients, and wastes can be exchanged.

At 8 to 10 weeks of development, doctors can take some tissue of the chorionic villi to detect whether genetic abnormalities exist in the embryo. Termed *chorionic villus sampling* (*CVS*), this procedure is performed by inserting a suction tube through the vagina, into the uterus, and to the chorionic villi. Because chorion cells and fetal cells contain identical genetic information, doctors can use these cells for genetic studies.

Colorized with a blue-gray tint, paired segments of tissue are also prominent in the embryo as shown in Figure 23.14. These chunks of mesoderm are called *somites* (from a Greek word meaning "a body"). They will give rise to most of the axial skeleton (see Chapter 16) with its associated skeletal muscles and most of the dermis of the body—tissues that underlie the epidermis of the skin. The first somites appear during the third week of development and are added as the embryo grows. Eventually, 42 to 44 pairs develop by the middle of the fifth week.

The Fourth Week

During the fourth week of development, the embryo begins to curl (see Figure 23.14). The curling occurs as folds are formed at the head and tail end of the embryo. As the embryo curls, the yolk sac becomes squeezed into a narrow stem that fuses with the body stalk connecting the embryo to the placenta (see Figure 23.10). As this fused body stalk lengthens and as its surface is tightly covered by the growing amnion membrane, it is properly called the *umbilical cord*. Blood cells continue to be produced by the yolk sac until the liver completely takes over this job in the

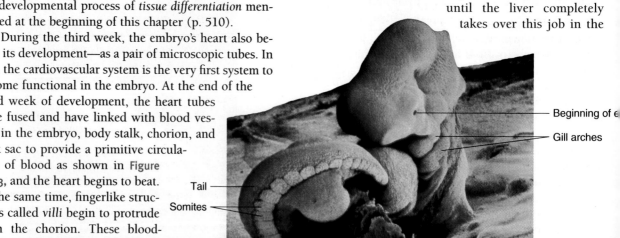

Figure 23.14   A human embryo at 4½ weeks.

sixth week. The umbilical arteries and vein, already functional, arose from the pouch of the allantois in the body stalk and have been transporting blood between the embryo and the placenta since the third week.

Four visible sets of swellings called the *branchial arches* develop on either side of the head end of the embryo during the fourth week. The word branchial is derived from a Greek word meaning "gill" and better describes similar structures that develop into gill supports in fish. In humans, these arches develop into structures such as the middle ear, eustachian tube, tonsils, thymus, and parathyroids.

The presence of gill arches and a tail (see Figure 23.14) in the human embryo indicates a relationship between humans and other vertebrates (fish and amphibians). In the nineteenth century, the German scientist Ernst Haeckel described this link as "ontogeny recapitulates phylogeny." What Haeckel meant was that the embryological development of an individual organism (ontogeny) repeats (recapitulates) the evolutionary history of its ancestors (phylogeny). This statement, however, is untrue. Embryonic stages of particular vertebrates are not a replay of the succession of its adult ancestors. Rather, the embryonic stages of a species show similarities to the embryonic stages of closely related species. Although a species' embryological development can be affected by evolutionary processes, many of the embryological similarities among closely related organisms are vestiges of their evolutionary pasts. Humans have a yolk sac, for example, as do its vertebrate relatives, the reptiles. Yolk is important in providing nutrition for a developing reptile within an egg and is much less important in the developing human nourished via the placenta. Therefore, the human yolk sac is virtually empty in comparison to that of the snake. This structure is an embryological remnant that points out the relatedness of these organisms.

The Fifth through Eighth Weeks

During the fifth week, the embryo doubles in length from 4 millimeters (3/16 inch) to about 8 millimeters (3/8 inch) | Actual length |. A nose begins to take shape as tiny pits. Although *limb buds* were first visible during the fourth week, first the arms and then the legs, at this stage of development the limbs look like microscopic flippers. The brain grows rapidly this week, resulting in an embryo with a large head in proportion to the rest of the developing body.

Seen floating within the fluid-filled amniotic sac in Figure 23.15, the 6-week-old embryo now has arm buds with distinct wrists. Fingers are just beginning to form, as are the ears. The retina of the eyes is now

darkly pigmented. Although the trunk of the body straightens out somewhat as the liver and digestive system grow, the neck area remains bent, forcing the head to rest on the chest above the red protruding heart, which now beats 150 times per minute. At the end of 6 weeks, the embryo is little more than half an inch long | Actual length |.

During the seventh week of development, eyelids begin to partially cover the eyes, and the ears develop more fully. The face begins to take on a human appearance. Each arm develops an elbow. Individual fingers can be distinguished, but they are connected with webs of skin that disappear during the eighth week. The legs, slower in their development than the arms, develop ankles and the suggestion of toes. During the eighth week—the last week as an embryo—the developing individual grows to 30 millimeters (slightly over an inch) in length. | Actual length |

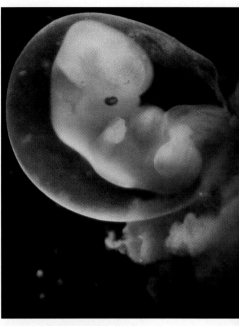

Figure 23.15 Embryo at 6 weeks.

Important processes of preembryonic and embryonic development are cleavage, cell migration, morphogenesis, and tissue differentiation. The four stages of development of the preembryo and the embryo are the morula, the blastocyst, the gastrula, and the neurula.

The embryo's heart begins development during the third week. During the fourth week, the gill arches appear, which will later develop into such structures as the middle ear and the tonsils.

From the fourth to eighth weeks of development, the embryo grows from a length of about 3 millimeters to approximately 30 millimeters (slightly longer than an inch). A primitive circulation is established within the embryo, and a maternal-fetal exchange of nutrients, gases, and wastes begins. The central nervous system and its associated structures begin to develop. The body form is established, including the appendages.

23.6 Development from the ninth to thirty-eighth weeks: The fetus

The embryonic period is primarily one of development. By the ninth week, most of the body systems are functional. The primary job of the fetal period is the refinement, maturation, and growth of these organ systems and of the body form of the developing individual.

The Third Month

The **first trimester** ends with the completion of the third month of pregnancy. During the third month, the **fetus** grows rapidly, tripling its length to approximately 85 millimeters (3½ inches). Fine hair called *lanugo,* meaning "down," appears over its body, but this downy coat is lost before a full term birth. Changes in the shape of the face cause the eyes to face forward and appear closer together, although the forehead remains prominent as can be seen in Figure 23.16. The eyes also become well developed during this month. The eyelids have continued their development and now fuse shut. They will not open until the fetus is 7 months old. An ear can be seen on the side of the head in line with the jawbone in the 12-week-old fetus in Figure 23.17.

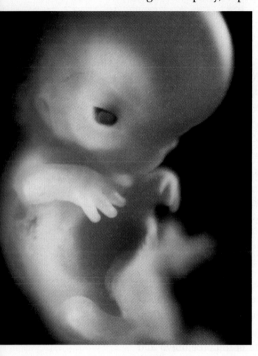

Figure 23.16   **Fetus at 11 weeks.**

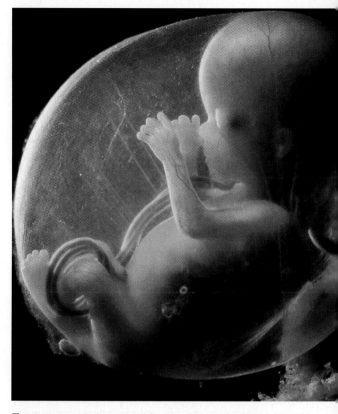

Figure 23.17   **Fetus at 12 weeks.**

During the embryonic period, male babies cannot be distinguished from female babies in their outward appearance. During early fetal development, outward differences between the sexes begin to take shape. The differentiation of male or female sex organs depends, of course, on the genetic makeup of the fetus but also on a complex interplay between hormones and tissues in the developing fetus. This interplay of factors results in the development of strikingly different reproductive systems by the end of the third month, with the penis in the male and the clitoris in the female arising from the same embryonic tissues. Likewise, the scrotum in the male is homologous to the labia majora in the female.

The Fourth through Sixth Months

During the **second trimester**, the fetus grows to about 0.6 kilogram (about 1½ pounds) and 0.3 meter (1 foot) long. The 4-month-old fetus in Figure 23.18 looks quite human. Halfway through the fourth month, the fetus can bring the hands together and suck the thumb. By 15 weeks, the sensory organs are almost completely developed, and by 16 weeks the fetus is actively turning inside the mother. Many of the bones are forming, replacing areas of cartilage. The process of bone formation, termed *ossification,* began at approximately 8 weeks of development and will continue beyond birth to the age of 18 or 19 years. By the end of the fifth month, the heartbeat of the fetus can be heard through a stethoscope.

Although the body systems have been rapidly continuing their development, the fetus is still unable to survive outside of the mother's womb. At the end of the sixth month, the head is no longer quite as large compared with the rest of the body, the eyelids have separated, and the eyelashes have formed. The fetus is capable of independent survival at the sixth month, but only with special medical intervention.

The Seventh through Ninth Months

The **third trimester** is predominantly a period of growth rather than one of development. In the seventh, eighth, and ninth months of pregnancy, the weight of the fetus doubles several times because of growth, but also because of fat that is laid down under its skin.

Figure 23.18   Fetus at 4 months.

One important developmental change, however, is that most of the major nerve tracts in the brain, as well as many new brain cells, are formed during this period. The mother's bloodstream fuels all of this growth by the nutrients it provides. Within the placenta, these nutrients pass into the fetal blood supply as shown in **Figure 23.19.** If the fetus is malnourished because the mother is malnourished, this growth can be adversely affected. The result is a severely retarded infant. Retardation resulting from fetal malnourishment is a serious problem in many underdeveloped countries where poverty is common.

By the end of the third trimester, the neurological growth of the fetus is far from complete and, in fact, continues long after birth. By this time, however, the fetus is able to exist independently.

During the third month, the fetus grows to $3\frac{1}{3}$ inches in length. Eyes become well developed, and ears begin to form. The critical stages of human development take place quite early. All the major organs of the body have been established by the end of the third month. The following 6 months are essentially a period of growth. During the fourth through sixth months, the fetus grows to 30 centimeters (12 inches) in length and begins to look human. The sensory organs become well developed, and the bones begin to form, replacing cartilage.

Figure 23.19   Placental-fetal circulation. During its development, the fetus receives all its nutrients from its mother via the umbilical vein. Wastes are removed from the fetus through the two umbilical arteries and are excreted by the mother. This is another instance (the pulmonary arteries and veins being the other) in which a vein carries oxygenated blood and the arteries carry deoxygenated blood.

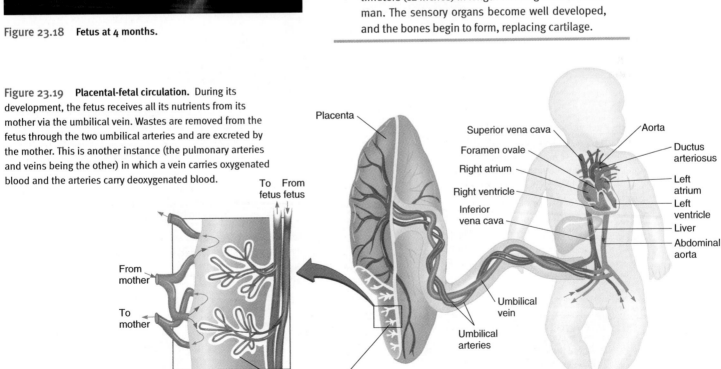

23.7 Birth

Birth takes place at the end of the third trimester, 38 weeks from conception (40 weeks from the start of the last menstrual period). Although the exact mechanism of the onset of **labor**, the sequence of events that leads to birth, is not well understood, scientists know that changing hormone levels in the developing fetus initiate this process (Figure 23.20 ❶). These hormones induce placental cells of the mother to manufacture *prostaglandins*, hormonelike substances that cause the smooth muscle of the uterine wall to contract ❷. In addition, the pressure of the fetus's head against the cervix sends nerve impulses to the mother's brain that trigger the hypothalamus to release the hormone oxytocin from her pituitary ❸ and ❹. Working together, oxytocin and prostaglandins stimulate waves of contractions in the walls of the uterus ❺, forcing the fetus downward. Initially, only a few contractions occur each hour, later in-

creasing to one every 2 to 3 minutes. Eventually, strong contractions, aided by the mother's pushing, expel the fetus through the vagina, or **birth canal**. The fetus is now a newborn.

After birth, uterine contractions continue and expel the placenta and associated membranes called the *afterbirth*. The umbilical cord is still attached to the baby, and to free the newborn, a physician or midwife ties and cuts the cord; blood clots in the cord and contraction of its muscles prevent excessive bleeding.

Birth takes place at the end of the third trimester, 38 weeks from conception. Hormones from both the fetus and the mother trigger and then sustain the birth process. After the process is initiated, the pressure of the baby's head against the cervix causes the release of oxytocin, which stimulates the uterus to contract. This positive feedback loop continues until the baby is born.

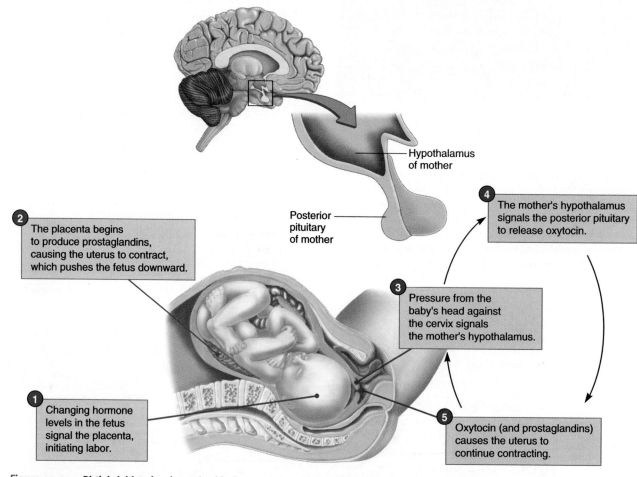

2 The placenta begins to produce prostaglandins, causing the uterus to contract, which pushes the fetus downward.

Hypothalamus of mother

Posterior pituitary of mother

4 The mother's hypothalamus signals the posterior pituitary to release oxytocin.

3 Pressure from the baby's head against the cervix signals the mother's hypothalamus.

1 Changing hormone levels in the fetus signal the placenta, initiating labor.

5 Oxytocin (and prostaglandins) causes the uterus to continue contracting.

Figure 23.20   **Birth is initiated and sustained by hormones in a positive feedback loop.** This feedback loop, in which the pressure of the fetus's head causes the release of oxytocin, continues until the baby is born.

23.8 Physiological adjustments of the newborn

At birth, the baby's lungs are not filled with air. Its first breath is therefore unusually deep. For the first time, the lungs are inflated; the baby cries. Because the baby is now obtaining oxygen from the lungs rather than from the placenta, several major changes must take place in the circulation of the blood.

Until birth, the placenta was the source of nutrients and oxygen for the fetus; in addition, it was the site for the removal of waste products from the fetal circulation. The lungs were not functional as organs of gas exchange. The fetal body had two major adaptations to limit the flow of blood to the lungs. First, a hole between the two atria called the *foramen ovale* shunted most right atrial blood directly into the left atrium, thus avoiding the right ventricle and the pulmonary circulation (see Figure 23.19). In addition, the right ventricular blood that is pumped into the pulmonary artery was mostly shunted into the aorta rather than through the lungs by the *ductus arteriosus*, a direct connection between the pulmonary artery and the aorta. At birth, the foramen ovale is closed by two flaps of heart tissue that fold together and fuse. The ductus arteriosus is shut off by contractions of muscles in its walls. Complete closure may take several months. The umbilical arteries and vein also close off.

> At birth, the circulation of the newborn changes as the lungs rather than the placenta become the organ of gas exchange.

Key Concepts

Developmental patterns show evolutionary relationships among animals.

23.1 The protostomes (molluscs, annelids, and arthropods) differ from the deuterostomes (echinoderms and chordates) with regard to certain embryological events.

23.1 The similarities in developmental patterns within the protostomes and deuterostomes suggest close evolutionary relationships.

The embryonic period is primarily one of development.

23.2 Human development before birth, or gestation, lasts approximately 8 1/2 months.

23.2 Conception, or fertilization, takes place when a sperm, or male sex cell, penetrates a secondary oocyte, or female sex cell.

23.3 The fertilized egg, or zygote, begins to divide in a process called cleavage after the genetic material of the egg and sperm unite.

23.3 Cleavage results in a ball of cells, or morula.

23.3 After 2 days in the uterus, the morula has developed into the blastocyst, a stage of development in which the embryo is a hollow ball of cells.

23.3 Blastocyst formation is significant because it is the first time that cells begin to move to shape the new individual in a process called morphogenesis.

23.3 At approximately 1 week after fertilization, the blastocyst implants in the lining of the uterine wall.

23.3 The implanted blastocyst secretes a hormone, human chorionic gonadotropin (HCG), that acts on the corpus luteum in the ovary, stimulating the body to produce estrogens and progesterone to maintain the uterine lining.

23.3 During the first 2 weeks of development, the developing mass of cells is generally referred to as a preembryo; from 3 to 8 weeks, an embryo; and from 9 to 38 weeks, a fetus.

23.4 Extraembryonic membranes (the amnion, chorion, yolk sac, and allantois) develop and help sustain the preembryo, embryo, and fetus through development.

23.5 At the end of the second week and continuing into the third week, various cell groups of the blastula begin to divide, move, and differentiate in a process called gastrulation.

23.5 During gastrulation an embryo is formed with three layers, each of which will give rise to specific tissues and organs of the developing individual.

23.5 Neurulation, the development of a hollow nerve cord that becomes the central nervous system, begins in the third week, but development of the nervous system continues even after birth.

23.5 Dramatic changes take place in the embryo from the fourth to the eighth weeks of development, such as the development of limb buds.

23.5 By the eighth week of development, the embryo is about an inch long, and most of the body systems are functional.

The fetal period is primarily one of growth.

23.6 The fetal period, from 9 to 38 weeks, is a time of refinement, maturation, and growth.

Key Concepts, continued

Hormonal changes in the fetus trigger the birth process.

23.7 Changing hormone levels in the fetus at about the thirty-eighth week trigger the onset of labor, the sequence of events that leads to birth.

23.7 Waves of contractions in the walls of the uterus force the fetus downward and out the birth canal.

23.8 At birth, the circulation of the newborn changes as the lungs rather than the placenta become the organ of gas exchange.

Key Terms

allantois (ah-LAN-toe-us) *517*
amnion (AM-nee-on) *517*
birth canal *524*
blastocyst (BLAS-tuh-sist) *513*
blastula (BLAS-chuh-luh) *511*
chorion (KORE-ee-on) *517*
deuterostomes (DOOT-uh-row-stowmz) *511*
ectoderm (EK-toe-durm) *515*
embryo (EM-bree-oh) *518*
endoderm (EN-doe-durm) *515*
extraembryonic membranes (EK-struh-EM-bree-ON-ik) *517*

fertilization *512*
fetus (FEE-tus) *522*
first trimester *522*
gastrula (GAS-truh-lah) *511*
gastrulation (GAS-truh-LAY-shun) *519*
implantation (IM-plan-TAY-shun) *514*
labor *524*
mesoderm (MEZ-oh-durm) *515*
morphogenesis (MORE-foe-JEN-uh-sis) *510*
morula (MORE-yuh-luh) *511*
neurulation (NOOR-oo-LAY-shun) *519*
notochord (NO-toe-kord) *519*

placenta (pluh-SEN-tuh) *517*
preembryo *513*
prenatal development *512*
primary germ layers *515*
protostomes (PRO-toe-stowmz) *511*
second trimester *522*
third trimester *523*
tissue differentiation *510*
trophoblast (TRO-fuh-blast) *514*
umbilical cord (um-BIL-uh-kul) *517*
yolk sac *518*
zygote (ZYE-goat) *513*

KNOWLEDGE AND COMPREHENSION QUESTIONS

1. Describe two differences between protostome and deuterostome patterns of embryological development. Which pattern do humans follow? (*Comprehension*)

2. At what point in the prenatal development of a human does the zygote exhibit a new genetic makeup different from that of either parent? (*Comprehension*)

3. What happens to a sperm and a secondary oocyte immediately after fertilization occurs? (*Knowledge*)

4. Does the development of the zygote during the first 2 weeks after fertilization involve meiotic or mitotic cell division? Explain your answer. (*Comprehension*)

5. What is the significance of morphogenesis? When does morphogenesis first begin in the prenatal development of the human? (*Comprehension*)

6. What are the extraembryonic membranes? State the function of each. (*Knowledge*)

7. Define neurulation, and explain its significance. (*Comprehension*)

8. What is a notochord, and what role does it play in humans? (*Knowledge*)

9. Summarize the changes that occur in the embryo/fetus during the first, second, and third trimesters. When do most body systems become functional? (*Comprehension*)

10. Describe the physiological adjustments that a newborn must undergo to survive. (*Knowledge*)

K E Y
Knowledge: Recalling information.
Comprehension: Showing understanding of recalled information.

CRITICAL THINKING REVIEW

1. What changes occur in a fertilized egg that prevent the penetration of a second sperm? What might be the consequences of a second sperm gaining entry? (*Synthesis*)

2. What do you think might be the consequences if the foramen ovale did not close after birth? What symptoms of this condition might be seen in the newborn? (*Analysis*)

3. Maternal nutrition is a key element in normal development of the fetus. Which fetal systems do you think are affected by maternal nutritional habits during the first 3 weeks of development? Explain your answer. (*Application*)

4. Health care providers stress that it is important that a woman preparing for pregnancy and during pregnancy avoid drinking alcohol, smoking cigarettes, and taking any drugs that were not prescribed for her. Why is this recommendation so important to prenatal development? Why should a woman preparing for pregnancy note these precautions? (*Application*)

5. Home pregnancy tests detect the presence of HCG in the urine. What is HCG and why is it present in the bloodstream of a pregnant woman? Using your understanding from Chapter 12 of how nephrons in the kidney work, explain why HCG is also found in the urine. (*Synthesis*)

KEY
Application: Using information in a new situation.
Analysis: Breaking down information into component parts.
Synthesis: Putting together information from different sources.

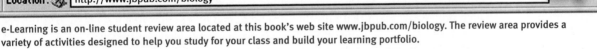

Location: http://www.jbpub.com/biology

e-Learning is an on-line student review area located at this book's web site www.jbpub.com/biology. The review area provides a variety of activities designed to help you study for your class and build your learning portfolio.

Review Questions The review questions test your knowledge of the important concepts and applications in each chapter. The review provides feedback for each correct or incorrect answer. This is an excellent test preparation tool.

Figure-Labeling Exercises Sharpen your visual thinking by matching terms and labels for important illustrations.

Flash Cards Studying biology requires learning new terms. Virtual flash cards help you master the new vocabulary for each chapter.

Just Wondering As you read and study from this text, you may find that you have unanswered questions. Through this site you can ask the author, Sandy Alters, your "just wondering" questions.

Do you prefer the speed and reliability of a CD-ROM? All of the features contained on the eLearning portion of the web site are also available on the Student CD-ROM.

Transgenic Crop Plants & Insect Resistance

I t probably arrived in North America during the early 1900s from Hungary or Italy. First noticed near Boston in 1917, it then spread west. Today it is found throughout most of the continental United States (and Canada) and poses a continual, serious threat. "It" is an insect — the European corn borer. An insect that eats corn may not seem like something important to you — and certainly nothing to worry about. However, the ability of farmers to feed the world population rests on their ability to control devastatingly destructive insects such as this one.

Transgenic corn cells growing in a laboratory petri dish. Each cell expresses an insect-resistance gene from the bacterium *Bacillus thuringiensis*.

Researchers have developed a new way to combat the corn borer and other such harmful insects: Bt crop plants. Bt plants are transgenic plants, which means that they contain genes from another species. In this case, the other species is the soil bacterium *Bacillus thuringiensis*, which produces a toxin that kills the corn borer and other damaging insects such as the cotton boll weevil and the Colorado potato beetle. Scientists have developed Bt corn, Bt cotton, and Bt potatoes, and the U.S. Department of Agriculture (USDA) has approved the use of these transgenic plants. They all have the bacterial genes that produce the toxin incorporated into their genomes, so they produce their own pesticide. Additionally, the USDA has approved about 20 other Bt crops for field testing.

Many applaud the development of genetically engineered insect-resistant crop plants (as well as the development of herbicide-resistant crop plants) as heralding a new era of the Green Revolution, the worldwide effort to increase food production for the ever-growing world population (see Chapter 33). The Green Revolution began during the 1950s. By 1966, using conventional breeding techniques, scientists had developed a high-yielding variety of rice, and by 1968, a high-yielding strain of wheat. Using these strains of rice and wheat along with chemical pesticides and fertilizers doubled world grain harvests. Since then, the world population has continued to grow, so food production must rise once again. However, the use of transgenic crop plants such as these to increase crop yields is not supported by everyone, and is now the focus of a heated debate.

Genetically engineered cotton resists attack by the boll weevil.

One major argument against the use of transgenic insect-resistant crop plants is the fear that their wide use may result in the evolution of insects resistant to Bt. A report from the Union of Concerned Scientists (a group that works to improve the environment as well as the quality of life) says that insects could begin developing resistance to Bt by 2001, because even though most of the insects that feed on Bt crop plants would die, a few would likely have a gene for resistance to the toxin. These organisms would survive and reproduce, passing the resistance trait to their offspring. Their offspring would reproduce, as would theirs and so on, until the population as a whole changed from being susceptible to the toxin produced by Bt corn (or Bt cotton or Bt potatoes), to being resistant to this toxin. This process is called natural selection (see p. 536).

Similar scenarios have occurred in the past and are occurring today with other organisms. During and after World War II, for example, the insecticide DDT was used worldwide to kill the mosquito that spreads malaria and the body louse that carries typhus. DDT kills a wide variety of arthropod pests, so was also sprayed on crop plants and livestock. By the late 1940s, however, problems related to its extensive use began to appear, and the insects it was supposed to kill were becoming resistant to its effects. Its use continued, and in 1962 biologist Rachel Carson published the book *Silent Spring* outlining the environmental problems caused by pesticides, including DDT. Carson noted that DDT was causing a dramatic decrease in the numbers of many species of birds by disrupting their reproductive processes. As a result, their offspring were often infertile or deformed. Additionally, she noted that DDT was building up in the food chain because it breaks down

---

One major argument against the use of transgenic insect-resistant crop plants is the fear that their wide use may result in the evolution of insects resistant to Bt.

---

very slowly in the soil, remaining there for decades. Animals metabolize it very slowly, so it accumulates in their bodies as they take it in. Also, as organisms higher on the food chain eat organisms lower on the food chain, toxins accumulate in the predators in a high concentration (see Figure 37.11). The use of DDT was banned in the United States in 1973, although it is still used in some parts of the world.

Supporters of insect-resistant crop plants note that, unlike pesticides sprayed on crops (such as DDT), transgenic plants can be genetically engineered to kill certain crop insects but to not harm beneficial insects, people, or wildlife. Opponents suggest that scientists do not know what wide-ranging ecological effects these plants might have (as occurred with the use of DDT).

Supporters also point out that the resistance problem is being tackled by the use of refuge areas. Refuge areas work in the following way: Insects become resistant to Bt plants because the plants produce toxin constantly. When insect populations are exposed to a deadly toxin all the time, the susceptible population is wiped out, and only resistant insects are left. Therefore, resistant insects are the only ones reproducing; their numbers increase dramatically and quickly with little competition from others of their species. To reduce the risk of this occurring, the Environmental Protection Agency (EPA) requires farmers to plant conventional (non-Bt) crops in certain areas of each crop field. These are the

Genetically engineered corn resists damage by the European corn borer.

refuge areas. Most of the insects feeding on the conventional plants in refuge areas will not be resistant to Bt. These insects breed with resistant insects feeding on the Bt plants, reintroducing the susceptibility gene into their genome.

Those cautious about this technology, such as the Union of Concerned Scientists, are urging the EPA to require larger refuge areas and adopt other approaches to reducing Bt resistance. Monsanto Company, manufacturer of Bt plants, contends that current efforts to prevent resistance seem to be working.

What effect will Bt have on beneficial insects and the rest of the environment?

What is the issue here, and what is your response to this issue?

To make an informed and thoughtful decision, it may help you to follow these steps.

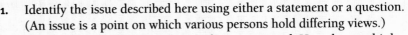

1. Identify the issue described here using either a statement or a question. (An issue is a point on which various persons hold differing views.)
2. Write your immediate reaction to the issues raised. How do you think this issue could be or should be resolved?
3. Collect new information about the issue. State the sources of that information. Explain why you believe the information to be accurate. Also determine if the information expresses a particular point of view or is biased in any other way. (Guidelines for doing efficient web searches for reliable information are available on the BioIssues website — **www.jbpub.com/biology**)
4. Determine which individuals, groups, or organizations have a stake in the issue. What does each stand to gain or lose depending on the outcome of the issue?
5. List possible outcomes (resolutions) of the issue. List the pros and cons of each outcome.
6. Which outcome do you think would be best and why? Note whether your opinion differs from or is the same as what you wrote in Step 2. Give reasons for your views and for any changes in them based on the additional information you have collected and the analysis you have done.

The Evidence of Evolution

It was a voyage of discovery. During his travels around the world, Charles Darwin (1809-1882) found some of the most exotic and interesting of creatures on an isolated archipelago in the Pactific Ocean, the Galapagos Islands. There he saw plants and animals uniquely adapted to the strange and harsh environment of the islands. Shown at right is a marine iguana eating algae off rocks in the sea. It is the only lizard on Earth that feeds in the ocean. Darwin wrote in *The Voyage of the Beagle*:

> *It is extremely common on all the islands throughout the group, and lives exclusively on the rocky sea-beaches, being never found, at least I never saw one, even ten yards in shore. . . . The lizard swims with perfect ease and quickness, by a serpentine movement of its body and flattened tail—the legs being motionless and closely collapsed on its sides.*

Other animals he observed with great interest were birds, especially the different species of *Geospiza* (finches). He wrote:

> *The most curious fact is the perfect gradation in the size of the beaks in the different species of Geospiza. Seeing this gradation and diversity of structure in one small, intimately related group of birds, one might really fancy that from an original paucity of birds in this archipelago, one species had been taken and modified for different ends.*

These observations and many others ultimately played a role in his development of a theory of evolution.

Organization

Darwin developed the theory of evolution based on his observations and other evidence.

Various types of evidence help scientists determine evolutionary relationships among organisms.

Evolutionary trees trace the lineage of organisms.

24.1 Darwin's observations

The story of Darwin and his theory begins in the early 1800s, when he was a medical student at Edinburgh. He then attended Cambridge University. In 1831, on the recommendation of one of his professors, he was selected to join a 5-year voyage (from 1831 to 1836) around the coasts of South America on H.M.S. *Beagle* (Figure 24.1). Darwin had the chance to study plants and animals on continents, islands, and seas distant from his native England. He was able to experience firsthand the remarkable diversity of living things on the Galapagos Islands off the west coast of South America. Such an opportunity clearly played an important role in the development of his thought about the nature of life on Earth.

When the *Beagle* set sail, Darwin was fully convinced that species were unchanging. (At this time in his career, Darwin's concept of species was remarkably close to the following modern definition.) A **species** (SPEE-sheez *or* SPEE-seez) is a population of organisms

Charles Darwin

that interbreeds freely in the wild and does not interbreed with other populations. In other words, species are defined by their reproductive isolation from one another. (Organisms that do not reproduce sexually are designated as species by means of their morphological and biochemical characteristics.) Darwin wrote that it was not until 2 or 3 years after his return that he began to consider seriously the possibility that species could change. Then, he began to formulate a theory integrating his observations of the trip with his understanding of geology, population biology, and the fossil record. Beginning in 1838, Darwin began to write his explanation of the diversity of life on Earth and the ways in which living things are related to one another.

During his 5 years on the ship, Darwin observed many phenomena that were of central importance to the development of his theory of evolution. While in southern South America, for example, Darwin observed fossils (preserved remains or impressions) of extinct ar-

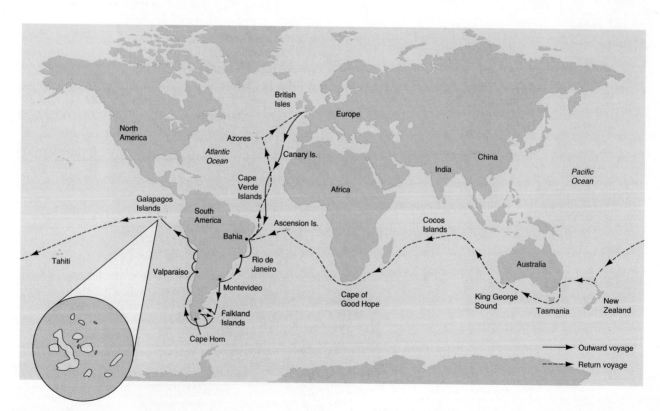

Figure 24.1 **Voyage of the H.M.S. *Beagle*.**

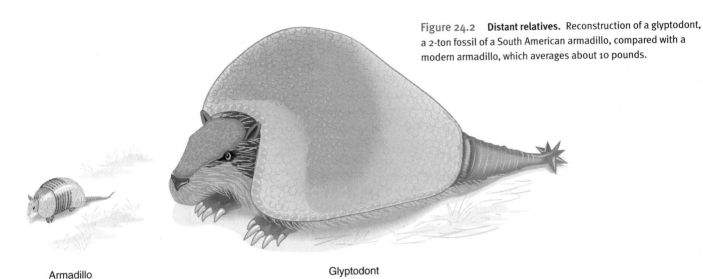

Figure 24.2 **Distant relatives.** Reconstruction of a glyptodont, a 2-ton fossil of a South American armadillo, compared with a modern armadillo, which averages about 10 pounds.

Armadillo

Glyptodont

madillos that were similar to armadillos still living in that area (Figure 24.2). He found it interesting that such similar yet distinct living and fossil organisms were found in this same small geographical area. This observation suggested to Darwin that the fossilized armadillos were related to the present-day armadillos—that they were "distant" relatives.

Another observation made by Darwin was that geographical areas having similar climates, such as Australia, South Africa, and Chile, are each populated by *different* species of plants and animals. These differences suggested to Darwin that factors other than or in addition to climate must play a role in plant and animal diversity. Otherwise, all lands having the same climate would have the same species of animals and plants. However, he noted that these organisms are of-

ten similar to one another, "shaped" by environmental similarities.

The fact that the relatively young Galapagos Islands (formed by undersea volcanoes) were home to a profusion of living organisms resembling plants and animals that lived on the nearby coast of South America also struck Darwin. Notice, for example, the similarity of the two birds in these photos. The bird on the left is a medium ground finch and is found on the Galapagos Islands. The blue-black grassquit, shown on the right, is found in grasslands along the Pacific Coast from Mexico to Chile. These observations suggested to Darwin that the Galapagos organisms were related to ancestors who long ago flew, swam, or "hitchhiked" (were transported by other organisms) to the islands from the mainland.

Darwin made observations during his 5-year voyage that suggested to him the following ideas:

1. Organisms of the past and present are related to one another.
2. Factors other than or in addition to climate play a role in the development of plant and animal diversity.
3. Members of the same species often change slightly in appearance after becoming geographically isolated from one another.
4. Organisms living in oceanic islands often resembled organisms found living on a close mainland.

Medium ground finch

Blue-black grassquit

24.3 Factors that influenced Darwin's thinking

As Darwin studied the data he collected during his voyage, he reflected on their significance in the context of what was known about geology, the breeding of domesticated animals, and population biology.

The Influence of Geology

In the late eighteenth and early nineteenth centuries, scientists studying the rock layers of the Earth noticed two things. First, scientists saw evidence that the Earth had changed over time, acted on by natural forces such as the winds, rain, heat, cold, and volcanic eruptions. Geologists began to hypothesize that the Earth was much older than the 6000 to 10,000 years many had originally thought it to be. Second, they noticed that the fossils found within the Earth's rock layers were similar to but different in many ways from living organisms—an observation Darwin himself had made on his voyage. Not only had the Earth changed, thought scientists, but evidence existed that the organisms living on Earth had changed also.

The Results of Artificial Breeding

As he pondered these ideas that the Earth and its organisms may have changed over time, Darwin reflected on the results of a process called *artificial selection*. In artificial selection, a breeder selects for desired characteristics, such as those of the chickens shown in **Figure 24.3**. At one time these chickens came from the same stock, but through artificial breeding over successive generations, their offspring have changed dramatically. In a way similar to that in which these varieties were derived, widely different species have originated in nature by means of natural selection.

Artificial selection is based on the *natural variation* all organisms exhibit. For example, looking back at p. 438 in Chapter 20, you can see that although individuals within this population of flamingos are similar, they possess characteristics that vary from individual to individual. Farmers and animal breeders, both today and in Darwin's time, take advantage of the natural variation within a population to select for characteristics they find valuable or useful. By choosing organisms that naturally exhibit a particular trait and then breeding that organism with another of the same species exhibiting the same trait, breeders are able (over successive breedings) to produce animals or plants having a desired, inherited trait. This trait will breed true in successive generations when these organisms are bred with one another.

For example, dogs have been artificially bred for centuries. Although your collie may look much different from your neighbor's terrier, both animals belong to the same species but have been artificially bred to retain traits that are characteristic of their breeds (**Figure 24.4**). Even the turkey you eat on Thanksgiving has been artificially bred for large cavities for stuffing.

The Study of Populations

Pondering his observations, Darwin began to study Thomas Malthus's *Essay on the Principles of Population*. Malthus, an economist who lived from 1766 to 1834, pointed out that populations of plants and animals (including humans) tend to increase exponentially. In an exponential progression, a population (for example) increases as its number is multiplied by a constant factor. In the exponential progression 2, 4, 8, 16, and so forth, each number is two times the preceding one. (These numbers can also be expressed as 2^1, 2^2, 2^3, and 2^4. The number, or power, to which 2 is raised is termed the *exponent,* from which the term *exponential growth* is derived.) **Figure 24.5a** shows how the numbers in a exponential progression increase quickly. Malthus suggested that although populations grow exponentially, food supplies increase only arithmetically. An arithmetic progression, in contrast, is

Figure 24.3 Artificial selection: Another clue in the natural selection puzzle. Breeders can select for traits they wish to perpetuate, resulting in different varieties of a given species. These decorative chickens were produced after many generations of selecting chickens that exhibited showy head plumage.

one in which the elements increase by a constant difference, as the progression 2, 6, 10, 14, and so forth. In this progression, each number is 4 greater than the preceding one. Figure 24.5b shows graphically how the numbers in an arithmetic progression increase.

Although Malthus suggested that populations grow at an exponential rate, he realized that factors existed to limit this astounding growth. If populations grew unchecked, organisms would cover the entire surface of the Earth within a surprisingly short time. But the world is not covered in ants, spiders, or poison ivy. Instead, populations of organisms vary in number within a certain limited range. Space and food are limiting factors of population growth; death limits infinite population growth. Malthus noted that in human populations, death was caused by famine, disease, and war. Sparked by Malthus's ideas, Darwin saw that in nature, although every organism has the potential to produce many offspring, thereby contributing to a exponential growth rate of its population, only a limited number of organisms actually survive to reproductive age. Darwin realized that factors similar to those limiting human populations must also act to limit plant and animal populations in nature.

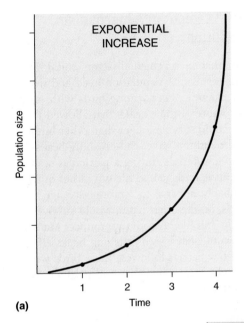

(a)

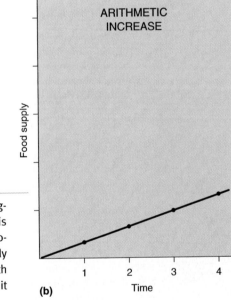

(a)

Figure 24.5 Types of mathematical progressions.

(b)

Figure 24.4 **Artificial selection leads to different breeds of the same species.** All dogs belong to the species *Canis familiaris,* but through artificial selection, breeders have been able to choose which traits each breed should retain. Notice that the collie in (a) has long, straight hair and a pointed snout in contrast to the short, curly hair and less pointed snout of the terrier in (b). How many other differences can you note?

Geological evidence suggests that the Earth is much older than geologists thought in the early 1800s, and that the Earth and organisms living on it have changed over time.

Breeders of plants and animals are able to alter the characteristics of organisms by selecting those with desired, inheritable traits and breeding them. After successive breedings, these inherited traits will consistently appear in offspring. Darwin hypothesized that a similar type of selection might take place in nature and result in changes within populations of organisms over time.

A key contribution to Darwin's thinking was Malthus's concept of exponential population growth. Real populations do not expand at this rate, and this implies that nature acts to limit population numbers.

Darwin developed the theory of evolution based on his observations and other evidence.

24.3 Natural selection: A mechanism of evolution

Darwin realized that environmental factors could influence which organisms in a population lived and which ones died. For example, berry-eating birds with variations in the structure of their beaks that allowed them to crush seeds would survive longer than other birds of their species who did not have seed-crushing beaks—if the bushes bore few berries during a particular season. The non-seed-eating birds would die out rather quickly and would probably not live to reproductive age. The seed-eating birds, on the other hand, would survive the bad berry season and likely would reproduce. Many of their progeny would have seed-crushing beaks. If another season of few berries followed, these birds would have a survival advantage over other birds that were unable to live on an alternative food source.

Darwin made associations between the process of artificial breeding and reproduction within natural populations. His ideas were expressed in his autobiography:

> I soon perceived that selection was the keystone of man's success in making useful races of animals and plants. But how selection could be applied to organisms living in a state of nature remained for some time a mystery to me.
>
> In September 1838, that is, fifteen months after I had begun my systematic enquiry, I happened to read for amusement "Malthus on Population" and being well prepared to appreciate the struggle for existence which everywhere goes on from long-continued observation of the habits of animals and plants, it at once struck me that under these circumstances favourable variations would tend to be preserved, and unfavourable ones to be destroyed. The result of this would be the formation of new species. Here then I had at last got a theory by which to work. . . .

Darwin was saying that those individuals that possess physical, behavioral, or other attributes well-suited to their environment are more likely to survive than those that possess physical, behavioral, or other attributes less suited to their environment. The survivors have the opportunity to pass on their favorable characteristics to their offspring. These characteristics are naturally occurring inheritable traits found within populations and are called **adaptations**. (Populations are individuals of a particular species inhabiting a locale or region.)

Notice that the term adaptation is used differently than in its everyday sense. Here, it refers to naturally occurring inheritable traits present in a population of organisms rather than noninheritable traits in individuals.

Adaptive traits are inherited characteristics that confer a reproductive advantage to the portion of the population possessing them. In its everyday sense, an adaptation refers to something a single individual does to change how it responds to the environment. For example, you may adapt to getting up early for an 8:00 AM class. But this is not an inherited trait in the entire population that confers a reproductive advantage!

As adaptive, or reproductively advantageous, traits are passed on from surviving individuals to their offspring, the individuals carrying these traits will increase in numbers within the population, and *the nature of the population as a whole will gradually change*. Darwin called this process, in which organisms having adaptive traits survive in greater numbers than those without such traits, **natural selection**.

Change in populations of organisms therefore occurs over time because of natural selection: The environment imposes conditions that determine the results of the selection and thus the direction of change. The driving force of change—natural selection—is often referred to as *survival of the fittest*. Again, the term fittest does not have the everyday meaning of the healthiest, strongest, or most intelligent. You may be fit if you work out at the local health club regularly. But fitness in the context of natural selection refers to reproductive fitness—the ability of an organism to survive to reproductive age in a particular environment and to produce viable offspring.

Natural selection provides a simple and direct explanation of biological diversity—why animals are different in different places. Environments differ; thus organisms are "favored" by natural selection differently in different places. The nature of a population gradually changes as more individuals are born that possess the "selected" traits. **Evolution** by means of natural selection is this process of change over time by which existing populations of organisms develop from ancestral forms through modification of their characteristics.

Natural selection is a process in which organisms having adaptive traits survive in greater numbers than those without such traits. Changes in populations of organisms therefore occur over time.

24.4 An example of natural selection: Darwin's finches

The results of evolution by natural selection can actually be seen if the process takes place relatively quickly (over a period of years to several thousand years), resulting in the existence of groups of closely related species from an original ancestral species. The results that can be seen are clusters of these closely related species found living near one another. Such clusters of species are often found on a group of islands, in a series of lakes, or in other environments that are close to but separated from one another. Organisms living in such sharply discontinuous habitats are said to be geographically isolated from one another.

The Galapagos Islands are a particularly striking example of sharply discontinuous habitats, providing a "natural laboratory" to view the results of natural selection. The islands are all relatively young in geological terms (several million years) and have never been connected with the adjacent mainland of South America or with any other area. Made up of 13 major islands (and some very tiny islands), the Galapagos are separated from one another by distances of up to 100 miles and are 600 miles from the South American mainland (see Figure 24.1). As a group, they exhibit diverse habitats. For example, the lowlands of the Galapagos are covered with thorn scrub. At higher elevations, attained only on the larger islands, there are moist, dense forests.

Formed by undersea volcanoes, the Galapagos Islands were uninhabited when they appeared above the surface of the water. The ancestors of all the organisms found on the Galapagos today reached these islands by crossing the sea by water or wind or on the bodies of other organisms. Only eight species of land birds reached the islands. One of these species was the finch, which fascinated Darwin. Hypothetically, the ancestor of Darwin's finches reached these islands earlier than any of the other birds. If so, all the types of habitats where birds occur on the mainland were unoccupied on the Galapagos—and the ancestral finches were able to take advantage of them all!

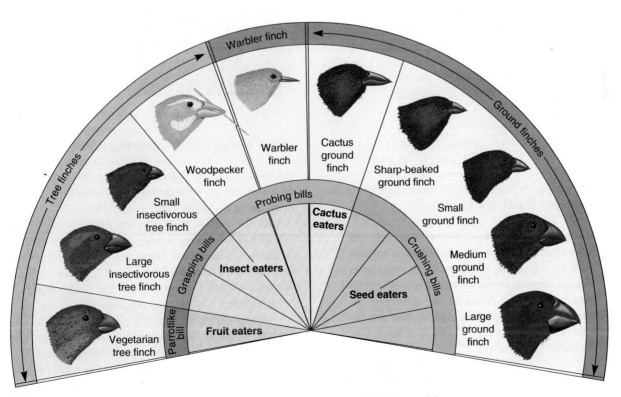

Figure 24.6 **Darwin's finches.** Ten species of Darwin's finches from Indefatigable Island, one of the Galapagos Islands, showing differences in bills and feeding habits. The bills of several of these species resemble those of different, distinct families of birds on the mainland. All of these birds are thought to have been derived from a single common ancestor.

As the finches moved into these vacant habitats, the ones best suited to each particular habitat were selected for by nature. In other words, those birds possessing naturally occurring variations in their characteristics that were beneficial to survival lived to reproduce. Their offspring also possessed these inheritable traits. Over time, the population of finches occupying each habitat changed, and the ancestral finches split into a series of diverse populations. This phenomenon, by which a population of a species changes as it is dispersed within a series of different habitats within a region, is referred to as **adaptive radiation**. Some of these populations became so changed from the others that interbreeding was no longer possible: New species of finches were formed. This process, by which new species are formed during the process of evolution, is termed **speciation**.

The evolution of Darwin's finches on the Galapagos Islands provides one of the classic examples of speciation. The descendants of the original finches that reached the Galapagos Islands now occupy many different kinds of habitats on the islands (**Figure 24.6**) and are found nowhere else in the world. Among the 13 species of Darwin's finches that inhabit the Galapagos (ten are shown in Figure 24.6), there are three main groups: ground finches, tree finches, and warbler finches. The ground finches feed on seeds of different sizes. The size of their bills is related to the size of the seeds on which the birds feed. The tree finches, as their name suggests, eat insects, buds, or fruit found in the trees. Again, the size and shape of their bills are related to their food. The most unusual member of this group is the woodpecker finch. This bird carries around a twig or a cactus spine, which it uses to probe for insects in deep crevices. It is an extraordinary example of a bird that uses a tool. And last, the warbler finches, named for their beautiful singing, search continually with their slender beaks over leaves and branches for insects.

www.jbpub.com/biology

The evolution of Darwin's finches illustrates the same kinds of processes by which species are originating continuously in all groups of organisms. Isolated populations subjected to unique combinations of selective pressures (the conditions imposed by nature) diverge biologically from one another and may ultimately become so different that they are distinct species.

24.5 The publication of Darwin's theory

Darwin drafted the overall argument for evolution by natural selection in 1842 and continued to refine it for many years. The stimulus that finally brought it into print was an essay that he received in 1858. A young English naturalist named Alfred Russel Wallace (1823–1913) sent the essay to Darwin from Malaysia; it concisely set forth the theory of evolution by means of natural selection (**Figure 24.7**). Like Darwin, Wallace had been influenced greatly in his development of this theory by reading Malthus's 1798 essay. After receiving Wallace's essay, Darwin arranged for a joint presentation of their ideas at a seminar in London. Darwin then proceeded to complete his own book, which he had been working on

Figure 24.7   Alfred Russel Wallace.

for some time, and submitted it for publication in what he considered an abbreviated version.

Darwin's book, *On the Origin of Species by Means of Natural Selection,* appeared in November 1859 and caused an immediate sensation. Some called the book "glorious" and were in complete agreement with Darwin's theories. Others criticized the book, admiring some parts and asserting that other parts were "totally false." Still others attacked Darwin on religious grounds. Many of the clergy, however, openly agreed with Darwin, saying that the theory of evolution did not deny the existence of God.

At the end of June 1860, a debate was held at a meeting of the British Association for the Advancement of Science. The debate lasted many days and attracted huge crowds of people. The debate was heated, but the outcome pleased Darwin: People read his book and gave his arguments serious consideration.

Within 20 years or so after the publication of *Origin,* the concept that species have changed over time was well accepted. However, scientists continually disagreed concerning the definition of the word "species." In fact, Darwin's own concept of species had changed since the 1830s; he eventually stopped trying to define the term, explaining that species continue to evolve, and therefore they cannot be defined.

Another hot debate within the scientific community regarded the mechanism of evolutionary change—natural selection. At that time, no one had any concept of genes or of how heredity works, and so it was impossible for Darwin to explain completely how evolution occurs. Gregor Mendel (see Chapter 20) had begun his groundbreaking work in the study of inheritance but had not yet published it. In fact, the science of genetics was not established until the beginning of the twentieth century, 40 years after the publication of Darwin's book. An understanding of the laws of inheritance and the mechanism by which inheritable traits are passed on from one generation to the next helped scientists understand the process of natural selection.

More than a century has elapsed since Charles Darwin's death in 1882. During this period, the evidence supporting his theory has grown progressively stronger. In fact, evolution is a theory that has been upheld countless times as it is tested and retested. Some consider evolution to be a fact; that is, a proposition that is so uniformly upheld (or replicated) that little doubt exists as to its accuracy. Scientists are continuously learning more, however, about the complexities of the mechanism of natural selection and the history of the evolution of life on Earth.

What is the evidence that upholds the theory of evolution? Scientists find evidence in the fossil record, using widely accepted techniques to assess the age of the rocks of which the fossils are formed or in which the fossil remains are found, while gathering a picture of the history of the Earth and its organisms. In addition, the tools of comparative anatomy help researchers understand relationships among organisms alive today. Significant scientific advances, such as those in genetics and molecular biology, have given scientists tools Darwin did not have; consequently, scientists today understand more fully than Darwin ever could about how organisms change over time.

Darwin published his argument for the theory of evolution in 1859, in a book entitled *On the Origin of Species by Means of Natural Selection*. His ideas were presented at a scholarly meeting along with those of another scientist, Alfred Russel Wallace, who had independently developed a theory of evolution. These ideas were hotly debated at that time; the mechanism of evolution—natural selection—was not well accepted or understood.

The concept of evolution by means of natural selection is established as valid within the scientific community today, and scientists are continually developing their understandings of natural selection.

24.6 The fossil record

A **fossil** is any record of a dead organism. Fossils may be nearly complete impressions of organisms or merely burrows, tracks, molecules, or other traces of their existence. Unfortunately, only a minute fraction of the organisms living at any one time are preserved as fossils.

Most fossils are preserved in sedimentary rocks, which are made up of particles of other rocks, cast off as they weather and disintegrate. Running water, such as a river or stream, picks up these pieces of rock and carries them to lakes or oceans where they are deposited as sediment, better known as mud, sand, or gravel. Over time, the sediment hardens into rock. But while some sediment is hardening, other sediment is still being deposited, creating layers of rock formed one on top of the other. Therefore, most sedimentary rock has a stratified appearance, such as that seen in this photo of the Grand Canyon.

During the formation of sedimentary rock, dead organisms are sometimes washed along with the mud or sand and eventually reach the bottom of a pond or lake. Dead marine organisms fall to the bottom of the ocean. As the sediments harden into rock, they harden around the bodies of these dead organisms. The hard parts of these organisms, such as their skeletons, may become preserved or may be broken down and replaced with other minerals.

Fossils of organisms having hard parts are the type most often found (Figure 24.8) rather than fossils formed from soft body parts, which usually decay quickly and leave no trace of their existence. Sometimes, however, soft-bodied animals are preserved in exceptionally fine-grained muds, in conditions in which the supply of oxygen was poor while the muds were being deposited,

Figure 24.8 **A fossil skeleton.** This
paleontologist is in his lab with the hind foot
of a Columbian mammoth on his workbench.
This specimen is an excellent example of a
fossil formed from the hard parts of an
organism.

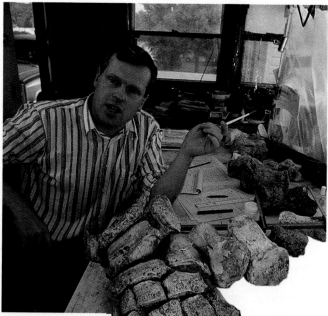

thus slowing the decomposition of the
organism. Eventually, the soft parts of
an organism decay completely, leaving behind a mold, or impression, of its body.
Molds may become filled with minerals,
such as lime or silica found in underground water,
forming casts, which resemble the original organism or
body part (Figure 24.9). In general, however, fossils of
soft-bodied organisms such as worms are rare. Even
though soft-bodied animals undoubtedly evolved before
their hard-bodied counterparts, there is comparatively
little evidence of their history in the fossil record.

Fossils provide an actual record of organisms that
once lived, an accurate understanding of where and
when they lived, and some appreciation of the environment in which they lived. Limestone that contains
corals, for example, would have been deposited when
the location was an ocean. Oak leaves found in sandstone suggest that a location was once a continent. In
this way, fossils and the rock in which they are embedded provide information about the history of an area,
which gives scientists clues to the location of the continents, ponds, lakes, and oceans and how their positions
have changed over time.

Scientists can determine the age of fossils and use
this information to establish the broad patterns of the
progression of life on Earth (Figure 24.10). One of the
ways of determining the age of particular fossils is to
compare the sequences in which they appear in different layers, or strata, of sedimentary rock. Sedimentation
deposits new layers mostly on older ones; thus, the fossils in upper layers mostly represent younger species

than the fossils in lower layers. Fossils found in
the same strata are assumed to be of the same age.
Such correlations, in fact, were known at the
time of Darwin and were used by him to formulate the theory of evolution.

Fossils, impressions of organisms that once lived,
provide a record of the past.

Figure 24.9 **A fossil "mold."** Shown here is a fossil crinoid, an
echinoderm having a small, cup-shaped body with branched arms
attached to the body by a stalk. This fossil illustrates the
remarkable preservation of soft-bodied fossils in fine-grained
sedimentary rocks.

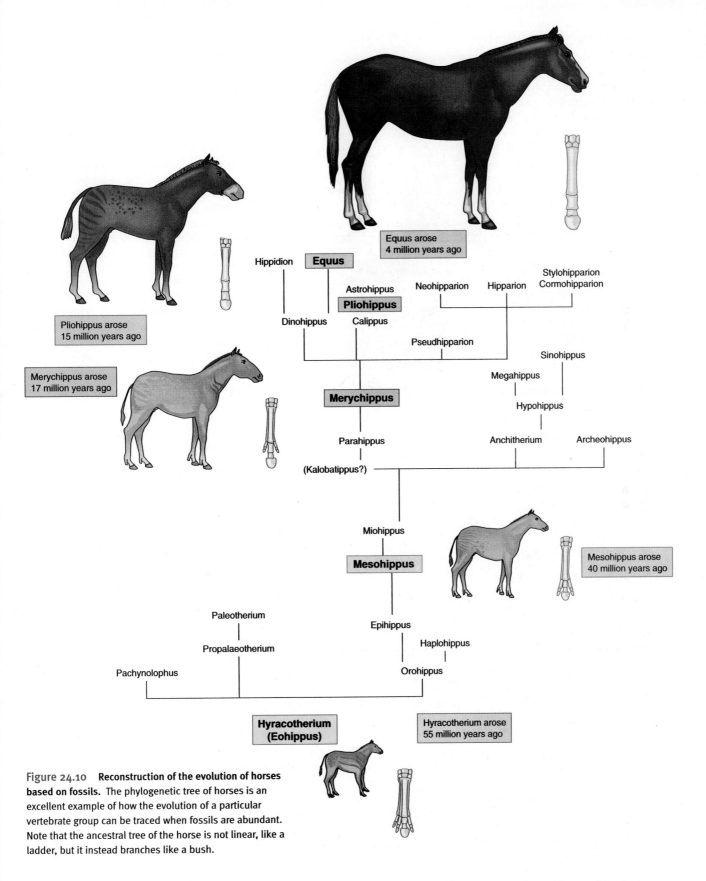

Figure 24.10 **Reconstruction of the evolution of horses based on fossils.** The phylogenetic tree of horses is an excellent example of how the evolution of a particular vertebrate group can be traced when fossils are abundant. Note that the ancestral tree of the horse is not linear, like a ladder, but it instead branches like a bush.

Just Wondering . . . *Real Questions Students Ask*

Various types of evidence help scientists determine evolutionary relationships among organisms.

How is the age of fossils determined, and how accurate is it?

There are two methods of geological dating: relative dating and absolute dating. Relative dating is briefly described on p. 540. This method of observing the layers of the Earth's crust results in a time scale with no dates but tells the age of rocks relative to one another on the basis of the unique sets of fossils embedded within them. The Earth's crust is seen as a calendar of sorts, from which scientists derived the five main divisions, or "eras," associated with the five major rock strata. Most have a number of subdivisions called periods, and some are further subdivided into epochs (see Figure 25.3).

Well before the theory of evolution was developed, fossil-bearing rocks were used to determine the stages of the history of the world. As new specimens are found, they are matched with known species. This new information is the evidence by which scientists continually revise (usually simply extending) the geologic ranges of fossil species.

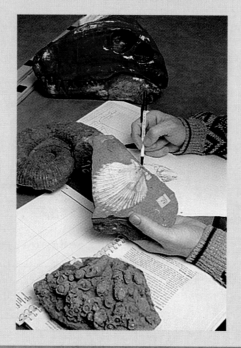

The only evidence as to the "accuracy" of this system is from the mineral and petroleum industries. Geologists in these industries use fossils to determine where to drill. If, for example, a petroleum company finds oil in a particular rock stratum in one part of the world, they look for oil in that same stratum in another location. This method is much more successful in finding oil than is random drilling.

Absolute dating, the process by which radioactive isotopes are used to determine the age of fossils and the rocks within which they are embedded, is described in detail in Section 24.7. This method (also called *radiometric dating*) does have sources of error but has yielded a great deal of consistent data. In addition, the relative ages of the rocks correlate well with the absolute dates determined by the use of isotopes. Therefore, radiometric dating is a reliable way to determine the exact ages of the rock strata.

24.7 Fossil dating

www.jbpub.com/biology

Direct methods of dating fossils first became available in the late 1940s. This process depends on naturally occurring isotopes of certain elements that are found in rock. Isotopes are atoms of an element that have the same number of protons but different numbers of neutrons in their nuclei (see Chapter 3). They therefore differ from one another in their atomic numbers. *Radioactive isotopes* are unstable; their nuclei decay, or break apart, at a steady rate, producing other isotopes and emitting energy. After decay, some radioactive isotopes may give rise to elements. Many different isotopes are used in radioactive dating. Some methods give scientists information about the age of rocks; others measure the length of time since the death of an organism.

One of the most widely used methods of dating, the carbon-14 method, estimates the relative amount of the different isotopes of carbon present in a fossil (or other organic material). Most carbon atoms have an atomic weight of 12 (6 protons and 6 neutrons); the symbol of this particular isotope of carbon is ^{12}C. A fixed proportion of the atoms in a given sample of carbon, however, consists of carbon with an atomic weight of 14 (^{14}C), an isotope that has two more neutrons than ^{12}C. Interestingly, ^{14}C is produced from ^{14}N (the chemical symbol for nitrogen-14) as these atoms are bombarded by cosmic rays—high-energy particles from space. The cosmic rays (usually protons) bump a proton from the nucleus of a nitrogen atom, leaving it with 6 protons (the atomic number of carbon) and 7 neutrons. As a result of the collision, the atom also captures a neutron and becomes an atom of ^{14}C. This newly created ^{14}C reacts with oxygen, becoming carbon dioxide, a gas commonly found in the air. Plants use this carbon during photosynthesis, incorporating it into the sugars and starches they make. By eating plants, animals incorporate ^{14}C into their bodies as well.

The carbon that is incorporated into the bodies of living organisms consists of the same fixed proportion of ^{14}C and ^{12}C that occurs in the atmosphere. After an organism dies, however, and is no longer incorporating carbon, the ^{14}C in it gradually decays back to nitrogen by emitting a beta particle. (A beta particle is an electron discharged from the nucleus when a neutron splits into a proton and an electron.) It takes 5730 years for half of the ^{14}C present in a sample to be converted by this process; this length of time is called the *half-life* of the ^{14}C isotope. By measuring the amount of ^{14}C in a fossil, scientists can estimate the proportion of the ^{14}C to all other carbon that is still present and compare that with the ratio of these isotopes as they occur in the atmosphere. In this way, scientists can then estimate the length of time over which the ^{14}C has been decaying, which is the same as the length of time since the organism died.

For fossils older than 50,000 years, the amount of ^{14}C remaining is so small that it is not possible to measure it precisely enough to provide accurate estimates of age. These fossils may be dated by using the isotope thorium-230, which has a half-life of 75,000 years and decays from uranium-238. These techniques have been most useful to scientists who study deep-sea sediments too old to be dated with ^{14}C.

With the use of radioactive dating methods, knowledge of the ages of various rocks has become more precise. The oldest rocks on Earth that have been dated include rocks from South Africa, southwestern Greenland, and Minnesota that are approximately 3.9 billion years old. Meteorites have been dated at about 4.6 billion years. Recently, rocks brought back to Earth from the moon have been dated from 3.3 to 4.6 billion years old. These pieces of evidence suggest that the Earth and the moon, most likely formed from the same processes at the same time, are about 4.6 billion years old.

What significance does the age of the Earth hold for the theory of evolution? The "accumulation" of adaptations and the development of new species usually take thousands and probably millions of years. Until the time of Darwin, most held the belief that the Earth was approximately 6000 to 10,000 years old. This time frame would not allow enough time for the process of evolution to take place. In fact, Sir Isaac Newton (1642–1727), an English physicist, calculated that it would take 50,000 years just for the Earth, after its formation, to cool to a temperature that would sustain life. (See Chapter 25 for a discussion of the formation of the Earth.) Although scientists such as William Thomson, Lord Kelvin (1824–1907), used nineteenth century techniques to date the Earth at about 100 million years, this time span still seemed too short to allow for evolution. Not until the present techniques of radioactive dating were developed could scientists begin to solve the "time problem" of evolution and accurately measure the geological age of the Earth and its fossils.

By studying the comparative amounts of certain radioactive isotopes found within fossils, scientists can determine the age of fossils and use this information to establish patterns of life's progression. Studies based on radioactive dating methods provide evidence that the Earth is 4.6 billion years old. This time is sufficient for organisms to have developed and evolved from "original" forms.

24.8 Comparative anatomy

Comparative studies of animal anatomy provide strong evidence for evolution. After Darwin proposed his theory, scientists began looking for evolutionary relationships in the anatomical structures of organisms. If derived from the same ancestor, organisms should possess similar structures, with modifications reflecting adaptations to their environments. Such relationships have been shown most clearly in vertebrate animals.

Within the subphylum Vertebrata, the classes of organisms, such as birds, mammals, and amphibians, have the same basic anatomical plan of groups of bones (as well as nerves, muscles, and other organs and systems), but these bones are put to different uses among the classes. (See Chapter 2 for a discussion of classification.) For example, the forelimbs seen in **Figure 24.11** are all constructed from the same basic array of bones, modified in one way in the wing of a bat, in another way in the leg

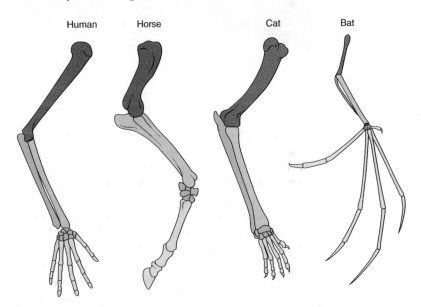

Figure 24.11 **Homology among vertebrate limbs.** Homologies among the forelimbs of four mammals show the ways that the proportions of the bones have changed in relation to the particular way of life of the organism, but that the forelimb of each animal has the same basic bone structure.

of a horse, and in yet another way in the arm of a human. The bones are said to be **homologous** in the different vertebrates—that is, of the same evolutionary origin, with the same basic anatomy but now differing in function. Although these vertebrates deviated from one another in their evolution, they all use the same bones in the same relative positions to do different jobs.

In some cases, homologous structures exist among related organisms, but they are no longer useful. These structures have diminished in size over time. Figure 24.12 shows tiny leg bones of the python that no longer serve a purpose. Such organs or structures that are present in an organism in a diminished size but are no longer useful are called *vestigial organs*.

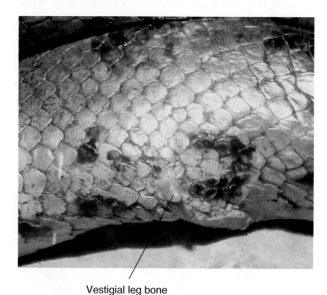

Vestigial leg bone

Figure 24.12 **A vestigial structure in a python.** Pythons possess tiny leg bones that serve no purpose in locomotion.

In contrast to homologous structures, similar structures often evolve within organisms that have developed from different ancestors. Such body parts or organs, called **analogous** structures, have a similar function but have different basic anatomies and evolutionary origins. To show that structures are indeed analogous, comparative anatomists do detailed dissections and study the embryological development of organisms. Analogous structures arise developmentally from different tissues.

The wings of birds and insects, which evolved independently but are similar in design, are analogous structures. Plants also show analogous structures. For example, three different families of flowering plants—the cacti, euphorbia, and milkweeds—have all developed thick, barrellike fleshy stems that store water as an adaptation to a desert environment. In fact, these plants look so much like one another that the casual observer might think they were all cacti. However, they evolved independently of one another in different parts of the world (southwest North America, Africa, and the Mediterranean, respectively). Figure 24.13a and b shows the striking similarity between the cacti and euphorbia.

The presence of analogous structures shows how in similar habitats, natural selection can lead to similar but

(a)

(b)

Figure 24.13 **An example of convergent evolution.** (a) North American cactus. (b) *Euphorbiaceae* found in Africa.

not identical anatomical structures. Change over time among different species of organisms having different ancestors that results in similar structures and adaptations is called **convergent evolution**.

Comparative studies of animal anatomy show that many organisms have groups of bones, nerves, muscles, and organs with the same anatomical plan but with different functions. These homologous structures provide evidence of evolutionary relatedness. The presence of analogous structures, which have similar functions but different basic anatomies and evolutionary origins, shows how natural selection can lead to similar but not identical anatomical structures when organisms live in similar habitats.

24.9 Comparative embryology

Embryologists, scientists who study the development of organisms from conception to birth, noticed as early as the nineteenth century (around the time of Darwin) that various groups of organisms, although different as adults, possessed early developmental stages that were quite similar. For example, the embryological development of vertebrate animals is similar in that all vertebrate embryos have gill arches, seen as pouches below the head. Only fish, however, actually develop gills; the pouches develop into different structures in other animals. Likewise, the embryological development of the backbone of vertebrates is similar, but some organisms develop a tail, and others such as humans do not. The similarities among embryos are shown in Figure 24.14. "Tailbones"—the fused coccyx bones at the end of the spine—are actually vestigial structures.

These similar developmental forms tell scientists that similar genes are at work during the early developmental stages of related organisms. The genes active during development have been passed on to distantly related organisms from a common ancestor. Over time, new instructions are added to the old, but both are expressed at different times and sometimes "old" instructions are deleted. Evolution acts on all stages of an organism, so there is no one "universal" stage through which all vertebrates pass, but similarities certainly exist. These similar embryos develop into adult organisms that are quite different from one another.

Comparative embryological studies show that many organisms have early developmental stages that are similar. These similar developmental forms provide evidence of evolutionary relatedness.

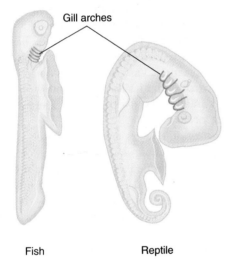

Fish   Reptile

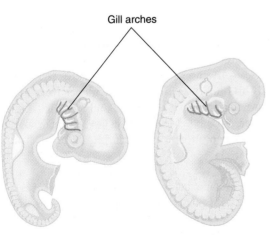

Bird   Human

Figure 24.14   **The embryos of various groups of vertebrate animals.** These embryos show the primitive features that all vertebrate animals share early in development, such as gill arches and a tail.

24.10 Molecular biology

Today, biochemical tools provide additional evidence for evolution and give scientists new insights into the evolutionary relationships among organisms. Molecular biologists study the progressive evolution of organisms by looking at their hereditary material, their DNA. According to evolutionary theory, every evolutionary change involves the formation of new alleles from the old by mutation; favorable new alleles persist because of natural selection. In this way, a series of evolutionary changes in a species involves a progressive accumulation of genetic change in its DNA. Organisms that are more distantly related will therefore have accumulated a greater number of changes in their DNA than organisms that more recently evolved from a common ancestor.

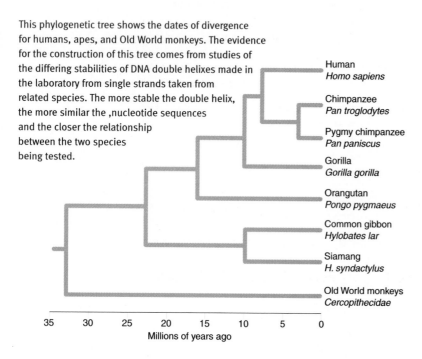

This phylogenetic tree shows the dates of divergence for humans, apes, and Old World monkeys. The evidence for the construction of this tree comes from studies of the differing stabilities of DNA double helixes made in the laboratory from single strands taken from related species. The more stable the double helix, the more similar the ,nucleotide sequences and the closer the relationship between the two species being tested.

Within the last 30 years, molecular biologists have learned to study DNA by "reading" genes, much as you read this page. They have learned to recognize the order of the nitrogenous bases—the "letters"—of the long DNA molecules. (See the "How Science Works" box in Chapter 19, p. 422.) By comparing the sequences of bases in the DNA of different groups of organisms, scientists can show the degree of relatedness among them.

> Molecular biologists study the progressive evolution of organisms by comparing the sequences of bases in their hereditary material.

24.11 Constructing evolutionary trees

As you can see, various lines of inquiry help scientists determine the evolutionary relationships among organisms. In the mid-1800s, Ernst Haeckel, an artist and a scientist, came up with the idea of pictorially representing the taxonomic groups of organisms (see Chapter 2) as trees that depicted these groups' evolutionary relationships. Scientists still use this technique today, but today's evolutionary trees are structured more like graphs and less like the actual trees of Haeckel's time.

An **evolutionary tree** depicts the pattern of relationships among major groups of organisms. Most evolutionary trees place information about the pattern of relationships among organisms on the horizontal axis and information about time on the vertical axis. However, scientists who develop evolutionary trees use a wide variety of data to construct trees. Some scientists may use only a single set of data and a particular type of data.

For example, the evolutionary tree in Figure 24.15 derives some information from molecular biology. Organisms are positioned on this tree on the basis of the differences in the nucleotide base sequence of the gene that codes for an enzyme called *cytochrome C oxidase*. The theory that underlies the development of the tree is that these differences exist because of mutations in the bases that occurred sometime in the past. The higher the number of differences among the bases, the greater the number of mutations and therefore the longer ago in time these organisms had a common ancestor (the more distantly they are related). The fewer the number of nucleotide base differences, the more recently these organisms had a common ancestor (the more closely they are related).

Evolutionary trees constructed from analyses of molecular differences are similar to those built by use of anatomical studies. For example, both approaches show whales, dolphins, and porpoises clustering together on each type of tree, as do the primates and the hoofed animals. (Note the evolutionary closeness of humans (primates) to the hoofed animals in Figure 24.15.) By the study and interpretation of the evidence from the fossil record, comparative anatomy, and genetic studies, it is often possible for scientists to estimate the rates at which evolution is occurring in different groups of organisms.

> An evolutionary tree depicts the pattern of relationships among major groups of organisms. The pattern of change seen in the molecular record supports other evidence for evolution and provides strong, direct evidence for change over time.

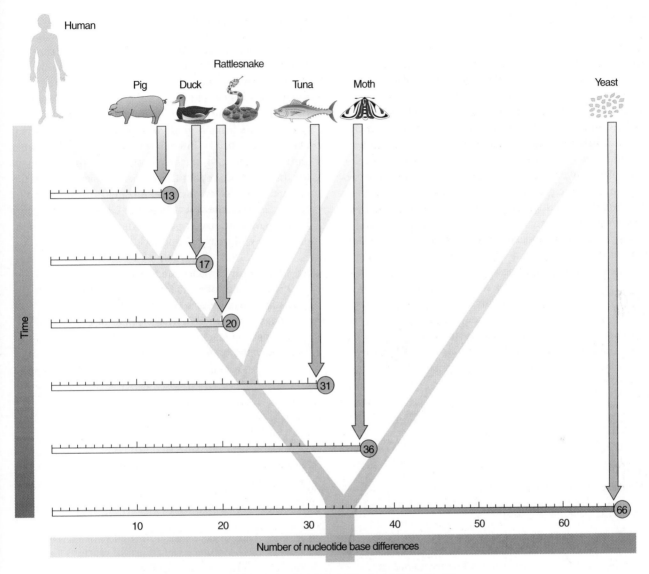

Figure 24.15 **An evolutionary tree constructed using information from molecular biology.** All of the organisms shown in this illustration have an enzyme called cytochrome C oxidase, which helps transfer electrons in the electron transfer system (see Chapter 6). The genes that code for this enzyme in these organisms are not the same, due to changes over time as organisms diverged from one another. This evolutionary tree shows the nucleotide base differences between humans and other organisms.

24.12 Tracing the mode and tempo of evolution

In 1972, a new perspective was brought to the scientific community regarding the manner and pace of evolutionary change. This theory, developed by American paleontologists Niles Eldredge and Stephen Jay Gould, is called **punctuated equilibrium**. This theory states that new species arise suddenly and rapidly as small subpopulations of a species split from the populations of which they were a part. This theory differs from the theory of **phyletic gradualism** proposed by Darwin, which asserts that new species develop slowly and gradually as an entire species changes over time. (The term phyletic gradualism is a modern phrase used to refer to these ideas of Darwin.) Table 24.1 compares the main ideas embodied in both theories.

The gradualist idea of slow change means that an entire species changes little by little over time until the species has transformed to a new species. The punctuationalist idea of rapid change means that the period of time during which speciation occurs is short with respect to the period of time of *stasis*, small variations that do not lead to speciation. The term rapid does not imply that speciation takes any particular (short) length of time. That time may vary from species to species and has been estimated to fall between 50,000 and 500,000 years—incredibly short in geological terms. The evolutionary trees shown in Figure 24.16 illustrate these ideas. In Figure 24.16a, the dots depict apparent gaps in the fossil record from a gradualist point of view: gradual change of a species but no speciation event. *Apparent gaps* are areas in which few if any intermediate forms in the fossil record are found. (Speciation is depicted in the gradualist model by one line splitting from another.) In Figure 24.16b, the dots show the same apparent gaps from a punctuationalist point of view: a speciation event occurring within a short period of time.

> The theory of punctuated equilibrium states that new species arise suddenly and rapidly as small subpopulations of a species split from the populations of which they were a part. The theory of phyletic gradualism asserts that new species develop slowly and gradually as an entire species changes over time.

Table 24.1

A Comparison of the Tenets of Phyletic Gradualism and Punctuated Equilibrium

Phyletic Gradualism	Punctuated Equilibrium
Principal Proponent: Darwin	**Principal Proponents: Eldredge and Gould**
• New species develop gradually and slowly with little evidence of stasis (no significant change).	• New species develop rapidly and then experience long periods of stasis.
• The fossil record should contain numerous transitional forms within the lineage of any one type of organism.	• The fossil record should contain few transitional forms with the maintenance of given forms for long periods of time.
• New species arise via the transformation of an ancestral population.	• New species arise as lineages are split.
• The entire ancestral form usually transforms into the new species.	• A small subpopulation of the ancestral form gives rise to the new species.
• Speciation usually involves the entire geographical range of the species (sympatry).	• The subpopulation is in an isolated area at the periphery of the range (allopatry).

From Alters BJ et al: *The American Biology Teacher*, 56 (6), 334-339, 1994.

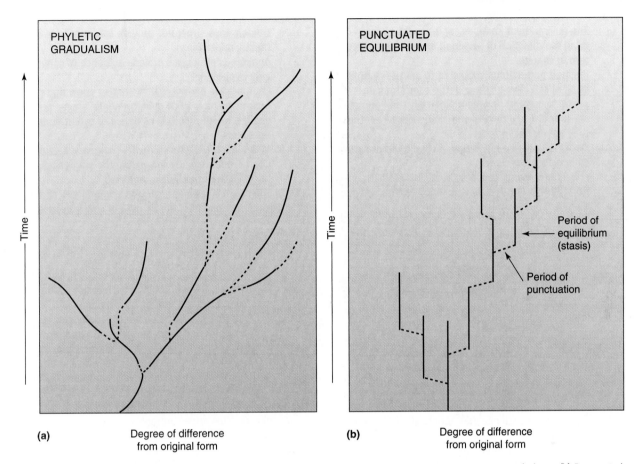

Figure 24.16 **Phyletic gradualism compared to punctuated equilibrium.** (a) Phyletic gradualism: time versus morphology . (b) Punctuated equilibrium: time versus morphology. In both representations, time progresses up the *Y* axis. Morphology, however, does *not* progress to the right along the *X* axis. This axis simply denotes morphological change, whether to the left or the right.

Key Concepts

Darwin developed the theory of evolution based on his observations and other evidence.

24.1 One of the central theories of biology is Darwin's theory of evolution, which states that living things change over time by means of natural selection.

24.1 While studying the animals and plants of oceanic islands, Darwin accumulated a wealth of evidence that organisms have changed over time.

24.2 Other factors that influenced Darwin's thinking were geological evidence that the Earth had changed over time, the changes in organisms that breeders of domesticated animals were able to attain by using artificial breeding methods, and Malthus's ideas on population dynamics.

24.3 Darwin proposed that evolution occurs as a result of natural selection.

24.3 The mechanism of natural selection works in the following manner: Some individuals have traits (adaptations) that make them better-suited to a particular environment, allowing more of them to survive to reproductive age and produce more offspring than individuals lacking these traits.

24.3 Adaptations are naturally occurring inherited traits found within populations.

24.3 Fitness is a measure of the tendency of some organisms to leave more offspring than competing members of the same population.

24.3 Genetic traits possessed by fitter individuals will appear in greater proportions among members of succeeding generations.

24.3 Traits allowing greater reproduction will increase in frequency over time.

Key Concepts, continued

24.3 The environment imposes conditions that determine the direction of selection and thus the direction of change.

24.4 Isolated populations subjected to unique combinations of selective pressures (the conditions imposed by nature) diverge biologically from one another and may ultimately become so different that they are distinct species.

24.5 Darwin published his theory in 1859 in a book titled *On the Origin of Species*.

24.5 A wealth of evidence since Darwin's time has supported his proposals that evolution occurs and that its mechanism is natural selection.

24.5 By the 1860s, natural selection was widely accepted as the correct explanation for the process of evolution, but the mechanism of natural selection was not understood.

24.5 The field of evolution did not progress much further until the 1920s because of the lack of a suitable explanation of how hereditary traits are transmitted.

Various types of evidence help scientists determine evolutionary relationships among organisms.

24.6 The fossil record, which exhibits an account of progressive change correlated with age, provides a direct line of evidence that upholds the theory of evolution.

24.7 By studying the comparative amounts of certain radioactive isotopes found within fossils, scientists can determine the age of fossils and use this information to establish patterns of life's progression.

24.8 Comparative studies of animal anatomy show that many organisms have homologous structures: groups of bones, nerves, muscles, and organs with the same anatomical plan but with different functions.

24.8 Homologous structures provide evidence of evolutionary relatedness.

24.8 Analogous structures provide evidence of convergent evolution.

24.9 Comparative embryological studies show that many organisms have early developmental stages that are similar, which provides evidence of evolutionary relatedness.

24.10 Molecular biologists study the progressive evolution of organisms by comparing the sequences of bases in their hereditary material.

24.10 The molecular record exhibits accumulated changes among organisms; the amount of change correlates with age as determined in the fossil record.

Evolutionary trees trace the lineage of organisms.

24.11 Evolutionary trees depict the patterns of relationships among major groups of organisms.

24.11 The pattern of change seen in the molecular record supports other evidence for evolution and provides strong, direct evidence for change over time.

24.12 In 1972, American paleontologists Niles Eldredge and Stephen Jay Gould developed a theory called punctuated equilibrium.

24.12 The theory of punctuated equilibrium states that new species arise suddenly and rapidly as small subpopulations of a species split from the populations of which they were a part.

24.12 Punctuated equilibrium differs from the theory of phyletic gradualism, which asserts that new species develop slowly and gradually as an entire species changes over time.

Key Terms

adaptations *536*
adaptive radiation *538*
analogous (uh-NAL-eh-gus) *544*
convergent evolution *544*
evolution (EV-uh-LOO-shun) *536*
evolutionary tree *546*

fossil *539*
homologous (hoe-MOL-eh-gus) *543*
natural selection *536*
phyletic gradualism (fye-LET-ik GRADJ-oo-uh-LIZ-um) *548*

punctuated equilibrium (PUNGK-choo-AY-ted EE-kwuh-LIB-ree-um) *548*
speciation (SPEE-shee-AY-shun) *538*
species (SPEE-sheez *or* SPEE-seez) *532*

KNOWLEDGE AND COMPREHENSION QUESTIONS

1. Summarize four of Darwin's conclusions that were inspired by his observations during his voyage. *(Comprehension)*

2. What three factors, in addition to his voyage on the H.M.S. *Beagle,* influenced Darwin's thinking on evolution? *(Knowledge)*

3. A dachshund and a Siberian husky are both dogs, but they look very different from each other. By what process did the two breeds come to look so different? What observable genetic principle is this process based on? *(Comprehension)*

4. What are adaptations? Explain their significance. *(Comprehension)*

5. Explain the phrase *survival of the fittest.* What does "fit" mean in this context? *(Comprehension)*

6. Distinguish between adaptive radiation and speciation. *(Comprehension)*

7. Explain how scientists can use the ^{14}C method to date a fossil. *(Comprehension)*

8. How old is the Earth according to radioactive dating methods? Why is the age of the Earth significant with respect to evolution? *(Comprehension)*

9. What are vestigial organs? Give an example. *(Knowledge)*

10. Explain how studies of comparative anatomy and comparative embryology support the theory of evolution. *(Comprehension)*

K E Y
Knowledge: Recalling information.
Comprehension: Showing understanding of recalled information.

CRITICAL THINKING REVIEW

1. Most species of bears are black, brown, or gray. Why are polar bears white? *(Application)*

2. If a population were composed of only a few organisms, what effects might natural selection have on its continued survival? Would a larger population have a better chance of survival? Why or why not? *(Synthesis)*

3. Humans shape their environment in ways that other organisms cannot. Are humans subject to the same pressures of natural selection as other organisms? Why or why not? *(Analysis)*

4. Organisms are often referred to as "progressing 'up' the ladder of evolution." Can evolution be accurately compared to a ladder? Can some organisms accurately be referred to as being "higher" on the ladder than others? Why or why not? *(Analysis)*

5. One problem that we are facing worldwide today is the appearance of antibiotic-resistant strains of bacteria. Diseases that were once easily treated with antibiotics are becoming difficult or impossible to treat. Using your understanding of evolutionary processes from this chapter, explain how bacteria can develop resistance to antibiotics. *(Application)*

K E Y
Application: Using information in a new situation.
Analysis: Breaking down information into component parts.
Synthesis: Putting together information from different sources.

The evolutionary tree below, constructed from information about cytochrome C oxidase, has a graph superimposed on it showing the number of nucleotide base differences among the organisms shown.

1. Which two organisms shown in the tree are the most distantly related based on their differences in cytochrome C oxidase? Explain your answer.
2. Which two organisms shown in the tree are the most closely related based on their differences in cytochrome C oxidase? Explain your answer.
3. To which organism shown on the tree are humans most closely related? Explain your answer.

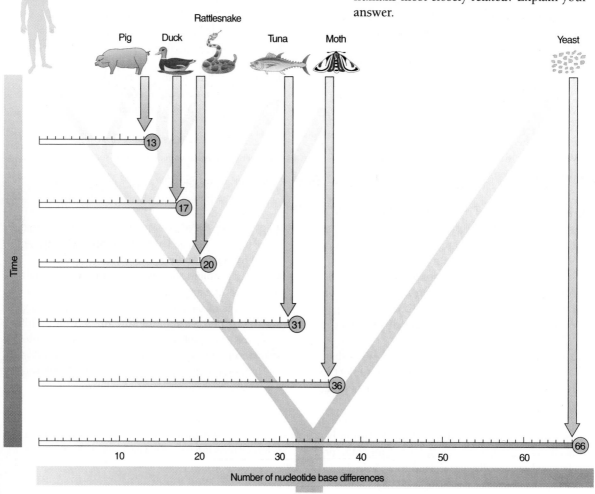

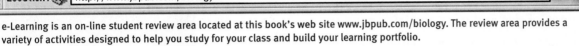

e learning

Location: http://www.jbpub.com/biology

e-Learning is an on-line student review area located at this book's web site www.jbpub.com/biology. The review area provides a variety of activities designed to help you study for your class and build your learning portfolio.

Review Questions The review questions test your knowledge of the important concepts and applications in each chapter. The review provides feedback for each correct or incorrect answer. This is an excellent test preparation tool.

Figure-Labeling Exercises Sharpen your visual thinking by matching terms and labels for important illustrations.

Flash Cards Studying biology requires learning new terms. Virtual flash cards help you master the new vocabulary for each chapter.

Just Wondering As you read and study from this text, you may find that you have unanswered questions. Through this site you can ask the author, Sandy Alters, your "just wondering" questions.

Do you prefer the speed and reliability of a CD-ROM? All of the features contained on the eLearning portion of the web site are also available on the Student CD-ROM.

The Evolution of Life

Vast areas of gas and dust particles swirl in the Orion nebula M42. Orion is a constellation, or group of stars, that can be seen in the night sky of the northern hemisphere. Surrounding some of its stars are cloudlike concentrations of gases and "stardust" called nebulae. In fact, nebulae are found in many sections of the Milky Way galaxy and are part of a cosmic life cycle—the birth, life, and death of stars.

Nebulae are "born" from the dust and gases hurled from unstable stars as they explode. Over time, this material condenses, giving birth to second- and later-generation stars. Then, in turn, these stars become unstable and explode, spewing dust and gases into space. And so the nebula life cycle continues.

The Earth (and the solar system) may have originated in the same way nebulae do. The American astronomer Edwin Hubble (1889–1953) observed that the galaxies in the universe are moving away from one another. In other words, the universe is expanding. Scientists theorize that this movement means that (approximately 15 billion years ago) *all* the material of the universe was once condensed into a single small space. It then exploded (like an unstable star) in an event called the Big Bang. The universe—as it is known today—did not exist before that time. The gases and dust from the Big Bang produced an early generation of stars. Then, over billions of years, these stars exploded and their "space debris" formed other stars and planets. The solar system probably formed in this way some 4.6 billion years ago. For almost a billion years or so, the molten Earth cooled, eventually forming a hardened, outer crust. Approximately 3.8 billion years ago, the oldest rocks on Earth formed. After that time, some 3.5 billion years ago, life began.

25

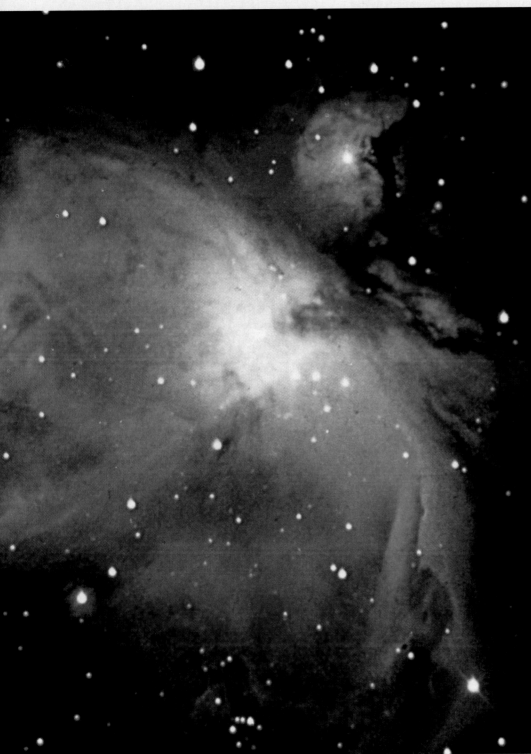

Organization

25.1 Hypotheses about the origin of simple organic molecules

Primitive Earth was an incubator for life, but scientists know very little about that incubator. In trying to understand what the Earth was like 3.8 billion years ago, scientists often use information from events that are observable today as starting points to build hypotheses. For example, scientists have determined that the dust and gases ejected from unstable stars contain mostly hydrogen as well as helium and varying amounts of other elements such as nitrogen, sodium, sulfur, and carbon. Some of these elements combine to form compounds such as hydrogen sulfide, methane, water, and ammonia. As the Earth coalesced, or came together into one mass, it may have contained some of these gases and vapors produced by the planet as it cooled. As the water vapor in this mix condensed, it could have caused millions of years of torrential rains extensive enough to form oceans. Other scientists suggest that water and gas deep within the Earth were vented to the surface by volcanoes.

Scientists agree that the environment of primitive Earth was harsh and violent, bathed in ultraviolet radiation from the sun and subjected to violent electrical storms and constant volcanic eruptions. Under these conditions, many scientists hypothesize that the elements and simple compounds of the primitive atmosphere reacted with one another. New and more complex molecules were formed, capturing the surrounding energy within their bonds. In 1953, Stanley Miller and Harold Urey, working at the University of Chicago, designed an experiment that modeled an environment that included methane (CH_4), ammonia (NH_3), water vapor (H_2O), and hydrogen gas (H_2). They wondered if organic molecules—the molecules that make up the structure of all living things—could have spontaneously formed under such conditions.

Using an apparatus similar to that in Figure 25.1, Miller and Urey filled a glass chamber with the four gases. Electrodes within the chamber shot sparks of electricity through the mixture while condensers cooled it. Miller and Urey wondered if complex molecules would form in their "atmosphere," dissolve in the water vapor, and fall into their "ocean." Within a week, they had their answer. A total of 15% of the carbon that was originally present as methane gas had been converted into other, more complex compounds of carbon.

Miller and Urey, joined later by many other scientists in laboratories around the world, performed experiment after experiment using various combinations of simple compounds that might have made up the primitive atmosphere. In addition, scientists experimented

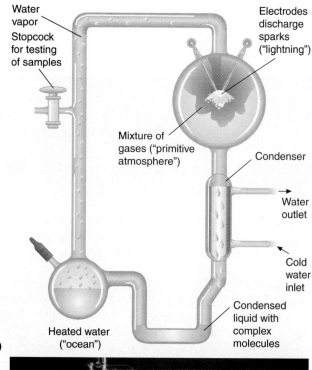

Water vapor

Stopcock for testing of samples

Electrodes discharge sparks ("lightning")

Mixture of gases ("primitive atmosphere")

Condenser

Water outlet

Cold water inlet

Condensed liquid with complex molecules

Heated water ("ocean")

(a)

(b)

Figure 25.1 The Miller-Urey experiment. (a) Miller and Urey's apparatus consisted of a closed tube connecting two chambers. The upper chamber contained a mixture of gases thought to resemble the Earth's atmosphere. Any complex molecules formed in the atmosphere chamber would be dissolved and carried in droplets to the lower "ocean" chamber, from which samples were withdrawn for analysis. (b) Dr. Stanley Miller standing in front of the apparatus in 1991.

with a variety of energy sources, such as ultraviolet light, heat, and radioactivity. The outcome was always the same: More complex organic molecules formed from simpler ones. These newly formed molecules included amino acids—the building blocks of protein. Sugars, including glucose, ribose, and deoxyribose, were also shown to form under the conditions of the primitive atmosphere. As scientists experimented fur-

How does science reconcile the differences between religious beliefs and scientific theories such as evolution?

It is not within the scientific domain to reconcile science with other disciplines such as social studies, English, or theology. What you may really be asking is how individual scientists reconcile such differences.

With respect to evolution, this theory fits easily into the belief systems of the majority of the world's religions. How the theory fits varies from religion to religion. For those scientists embracing religious beliefs in opposition to the theory of evolution, many hold their religion separate from their scientific work and do not try to reconcile the two. A small minority of scientists find the differences irreconcilable.

ther, they discovered that many of the molecules important to life—and important to the structure and function of cells—could be synthesized under certain primordial conditions. Not only were ribose and deoxyribose, components of nucleic acids, synthesized in their methane environment, but so were purine and pyrimidine bases. Fatty acids, used in membranes and storage tissues of living organisms, were also made. From this array of experiments, scientists developed the primordial soup theory, which states that life arose in the primitive seas or in smaller lakes as complex organic molecules formed from the simple molecules in the ancient atmosphere.

During the past decade, some scientists have begun to doubt the primordial soup theory as described by the experiments of Miller, Urey, and others. Using computerized reconstructions of the atmosphere, James C. G. Walker of the University of Michigan at Ann Arbor suggests that the major components of the primitive atmosphere were carbon dioxide and inert (unreactive) nitrogen gases spewed forth from volcanoes. These gases could not have formed complex organic molecules—the precursors to life—in the seas or tidal pools of the Earth's surface. Nitrogen in the form of ammonia, as used by Miller and Urey, would have been scarce.

Recently, some scientists have shifted from adherence to the primordial soup theory to thinking that life might have arisen in hydrothermal vents, spots deep in the oceans where hot gases and sulfur compounds shoot from cracks in the Earth's crust (Figure 25.2). This hypothesis was originally proposed in 1988 by Günter Wächtershäuser of the Technical University of Munich. Some scientists think that the vents could have supplied the energy and nutrients needed for life to arise and be sustained. In addition, the vents may have been a source of ammonia on the early Earth. The minerals deposited there appear to catalyze reactions that convert nitrogen into ammonia at the high temperatures and pressure of the vents. In fact, the hydrothermal vents in existence today support extensive communities of living organisms, such as tube worms, clams, and bacteria.

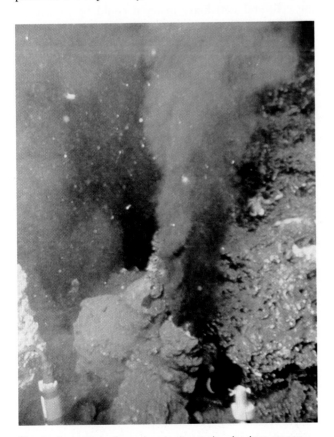

Figure 25.2 **Hydrothermal vents.** Located under deep oceans, hydrothermal vents are cracks in the Earth's crust that release hot gases and sulfur compounds.

The long-held primordial soup theory states that life arose in the sea as elements and simple compounds of the primitive atmosphere reacted with one another to form simple organic molecules such as amino acids and sugars. Recently, scientists have begun to reexamine the premises of this theory and have proposed alternative hypotheses.

25.2 Hypotheses about the origin of complex molecules

Although scientists can replicate various plausible conditions of the primitive Earth, they cannot duplicate the millions of years that passed. Even though they can observe the "birth" of simple organic molecules, they have not generated life from nonlife. However, at some point in the history of the Earth, organisms appeared. Scientists are still trying to unravel this mystery of life's beginnings. The work of many scientists over many years has led to various lines of reasoning regarding the evolution of the precursors to life and life itself.

The first step toward the evolution of life must have been the synthesis of even more complex organic molecules than the ones previously described. Scientists know that complex molecules such as polymers (for example, polysaccharides and proteins) do not spontaneously develop from a mixture of their simpler building blocks, or monomers (in this case, sugars and amino acids). Most synthesis reactions are dehydrations and depend on the removal of water and the input of energy to chemically link one molecule to another. This process, called dehydration synthesis (see Chapter 3) could have taken place if condensing agents were present with the monomers. Condensing agents are molecules that combine with water and release energy. If condensing agents were not present, heat and evaporation could also promote dehydration synthesis and the synthesis of polymers.

In the 1950s, the American scientist Sidney Fox and his co-workers developed a technique in which they used heat to produce polymers, which Fox called proteinoids, from dry mixtures of amino acids. In 1998, Günter Wächtershäuser and colleague Claudia Huber reported that they linked amino acids together into short, proteinlike chains (peptide chains) under hydrothermal vent conditions. Many scientists agree that these peptides were made under plausible prebiotic conditions, adding more support for the hydrothermal vent hypothesis of the origin of life. Some scientists do not agree, however, and think that such reactions would not be adequate to form proteins that contain many amino acids.

The formation of early proteins is significant, because proteins are not only structural molecules; many are enzymes and increase the rates of organic reactions.

Metal ions and clays (composed chiefly of minerals) could also have served as early enzymes. When combined into sequences, enzymatic reactions can be thought of as the beginning of metabolic systems. To be considered "life" as scientists define it, metabolic systems must be organized within a cellular structure, be able to "carry" information about themselves, and have the ability to pass this information on by the process of replication.

Scientists suspect that self-replicating systems of RNA molecules may have started the process of evolution; they think these molecules were formed from substances present in either a primordial soup or near hydrothermal vents. RNA molecules have the ability, as do all polynucleotides, of acting as a template, or guide, for the synthesis of a second polynucleotide based on the complementarity of its bases. For replicating RNA (or DNA) to be considered life however, it must have become enclosed by some sort of a membrane forming the first cells. Within the boundaries of a membrane, enzymes could be organized to carry on life functions. Chemicals could be selectively allowed into or kept out of the cell's interior. Wastes could be eliminated from the cell, and hereditary material could be passed on from cell to cell.

How much time passed before biochemical evolution became biological evolution? How did this evolution take place? There is no scientific consensus on these points. Scientists do agree, however, that single-celled life existed in the shallow seas of the primitive Earth.

The first step toward the evolution of life is the synthesis of highly complex organic molecules. Such complex molecules might have formed as pools of a primordial soup evaporated under the heat of the sun, promoting the linking of small molecules. Another theory holds that they may have formed at hydrothermal vents. Proteins may have served as enzymes, promoting organic reactions.

After the synthesis of complex organic molecules had taken place, self-replicating systems of RNA molecules may have begun the process of evolution. To be considered life, however, this "early" genetic material must have been organized within a cellular structure.

Multicellular organisms evolved from the first cells over millions of years.

25.3 The Archean era: Oxygen-producing cells appear

When discussing the history of life on Earth, scientists divide the time from the formation of the Earth until the present day into five major time periods or **eras** of geological time, as shown in **Figure 25.3**. The eras are subdivided into shorter time units called **periods**. In addition, the periods of the Cenozoic era are subdivided into **epochs** (EP-uks *or* EE-poks). However, each time unit (such as an era or a period) does not stand for a consistent length of time because the distinctive events that mark the beginning and end of a time unit occurred over varying amounts of time. Additionally, early geologists

Visual Summary **Figure 25.3** **Geological time.** A summary of the major biological and geological events in Earth's history. (MYA = millions of years ago.)

ERA	PERIOD	EPOCH	(MYA)	MAJOR BIOLOGICAL AND GEOLOGICAL EVENTS
Cenozoic	Quaternary	Pleistocene	1.8	
	Tertiary	Pliocene	7	First humans.
		Miocene	26	Origin of first humanlike forms.
		Oligocene	38	Monkeylike primates appear.
		Eocene	54	Origin of *Hyracotherium (Eohippus)*.
		Paleocene	65	Small mammals undergo adaptive radiation.
Mesozoic	Cretaceous		145	**Major extinction of the dinosaurs and many marine organisms.** Flowering plants appear.
	Jurassic		210	Large dinosaurs dominate the Earth. Pangea (supercontinent) forms. First birds.
	Triassic		250	Small dinosaurs appear. First mammals. Insects become more diverse.
Paleozoic	Permian		285	**Major extinction occurs; most species disappear.** Conifers appear.
	Carboniferous		360	First reptiles and arthropods. Coal deposits formed. Horsetails, ferns, and seed-bearing plants abundant. Major insect diversity arose.
	Devonian		410	"Age of the fishes." Fishes with bones and jaws appear. Amphibians appear. **Major extinction of marine invertebrates and fishes.** Insects first appear.
	Silurian		430	Notochord becomes flexible as single rod is replaced with separate pieces as seen in the Ostracoderms (armored fish without bones, jaws or teeth). Plants invade the land.
	Ordovician		505	First vertebrates appear. **Major extinction of marine species.** First fungi appear.
	Cambrian		520	**Major extinction of the trilobites.** Origin of the main invertebrate phyla.
Proterozoic			2600	Multicellular eukaryotic animals appear. First eukaryotic cells appear. Oxygen-producing bacteria present; atmosphere and oceans oxygenated.
Archean			4600	Stromatolites formed. Chemical evolution resulting in formation of first cells. First rocks formed. Earth is born.

December 31, *14 sec to midnight (2000 ya)* Birth of Christ *80 seconds to midnight (15000 ya)* End of last Glacial Age *6 PM (2 mya)* Humans appear

December 26, *(65 mya)* End of dinosaurs

December 10, *(200 mya)* Supercontinent

November 28, *(410 mya)* First land plants

November 12, *(600 mya)* Beginnings of well-known geology

November 1, *(800 mya)* Primitive higher animals

October 1, *(1000 mya)* Normal oceans and atmosphere

April 16, *(3500 mya)* First known life

February 10, *(3800 mya)* Oldest dated rocks

January 1, *(4600 mya)* Origin of Earth

defined and named these time units as they discovered them. A 1-year geological "calendar" is shown in Figure 25.3; this calendar relates the geological time scale to a single year, making it easier to understand how long each geological time unit is in relation to the others.

Scientists speculate that the first cells to evolve fed on organic materials in the environment. Because there would have been a very limited amount of suitable organic food available, a type of nutrition must have soon evolved in which cells were able to capture energy from inorganic chemicals. The evolution of a pigment system that could capture the energy from sunlight and store it in chemical bonds most likely evolved after that. Then, oxygen-producing bacteria evolved. Probably similar to present-day cyanobacteria ([SIGH-an-oh-back-TEAR-ee-ah], formerly called "blue-green algae"), these single-celled organisms became key figures in the evolution of life as it is known today. Their photosynthesis gradually added oxygen to the atmosphere and the oceans around 2 billion years ago, as evidenced by Archean and Proterozoic

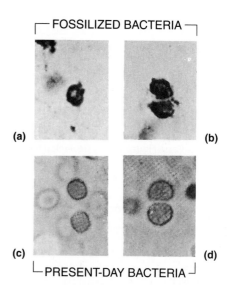

FOSSILIZED BACTERIA

(a) (b)

(c) (d)

PRESENT-DAY BACTERIA

Figure 25.5   Fossilized bacteria. The fossilized bacteria in (a) and (b) were found in stromatolites located in South Africa. These stromatolites are about 3.4 billion years old. The bacteria in (c) and (d) show living, similar bacteria. The fossilized bacteria bear a striking resemblance to the living bacteria.

sediments. Scientists have also discovered fossils of the cyanobacteria.

Although fossils of single-celled organisms are usually difficult to find, the cyanobacteria sometimes grew in "piles," creating fossilized columns of organisms and the sediments that collected around them. These ancient columns of cyanobacteria are called stromatolites (stroh-MAT-eh-lites) and date back 3.5 billion years. In Figure 25.4a, the tops of stromatolites are visible above the surface of the water in which they form. Like icebergs, much of the stromatolite is hidden under water, as shown in Figure 25.4b. Actively developing stromatolite formations can be observed even today in the warm, shallow waters of places such as the Gulf of California, western Australia, and San Salvador (Watling Island), Bahamas. The shapes and sizes of the bacteria within present-day stromatolites look very much like the bacteria found within fossil stromatolites (Figure 25.5). In the absence of competing organisms, cyanobacteria produced stromatolites abundantly in all fresh water and marine communities until about 1.6 billion years ago.

Scientists speculate that the first cells to evolve fed on organic materials in the environment. Because of limited organic material, the evolution of a pigment system that could capture the energy from sunlight and store it in chemical bonds most likely evolved relatively quickly. Then, oxygen-producing bacteria evolved, gradually adding oxygen to the atmosphere.

(a)

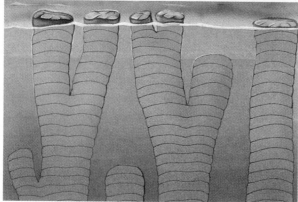

(b)

Figure 25.4   Stromatolites. (a) These stromatolites are located in Shark Bay, Western Australia. The largest structures are about 1.5 meters (approximately 4½ feet) across. These stromatolites formed about 4000 years ago. (b) This diagram shows that much of the stromatolite formation is underwater.

25.4 The Proterozoic era: The first eukaryotes appear

The beginning of Proterozoic (PRAHT-er-eh-ZOE-ik) time, which extends from 2.6 billion to 520 million years ago, is marked by the formation of a stable oxygen-containing atmosphere. The term Proterozoic is derived from two Greek words meaning "prior" (*proteros*) "life" (*zoe*) and is descriptive of the fact that, until a few decades ago, scientists did not find evidence of life during this time. Once scientists began looking for microfossils, however, they found life in both this era and the Archean era.

During the Proterozoic era, the first eukaryotic organisms appeared on Earth. A fossil stromatolite found in California dating back 1.4 to 1.2 billion years ago contains larger and more complex microfossils than the fossil cyanobacteria previously found. Scientists think these microfossils may have been eukaryotes, organisms separate from the cyanobacteria that may have lived in harmony with stromatolites or have fossilized with them. Other fossils in rocks of the same time add additional evidence to the theory that eukaryotes appeared around 1.4 billion years ago.

How did the eukaryotes arise? The most widely accepted hypothesis at this time is called the **endosymbiotic theory**. (*Symbiosis* means "living together," and *endo-* means "within.") According to the endosymbiotic theory, bacteria became attached to or were engulfed by preeukaryotic (host prokaryotic) cells. A variety of hypotheses attempt to explain how this association first occurred. One widely held hypothesis suggests that the first eukaryotic cells engulfed bacteria as food. Some of the bacterial prey escaped digestion and remained within their host. The bacteria, or endosymbionts, benefited by living within the protective environments of their hosts. The bacteria carried out chemical reactions necessary for living in an atmosphere that was increasing in its amounts of oxygen, which benefited their

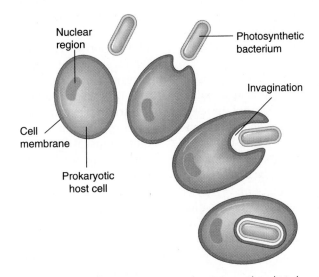

Figure 25.6 Endosymbiosis. How pre-eukaryotes are thought to have engulfed bacteria, which may have become mitochondria and chloroplasts as eukaryotic cells evolved.

host cells. The cells that engulfed the bacteria thus had a reproductive advantage over those cells not associated with bacteria and therefore flourished.

The cell membranes of the pre-eukaryotes are thought to have pouched inward during these symbiotic relationships, loosely surrounding the bacteria. The bacteria are thought to be early mitochondria. Symbiotic events similar to those postulated for the origin of mitochondria also seem to have been involved in the origin of chloroplasts, which are thought to be derived from symbiotic photosynthetic bacteria (Figure 25.6). Scientists believe the double membrane of these organelles is derived from the bacterium's plasma membrane and the surrounding cell membranes of the pre-eukaryotes that engulfed them. In addition, both mitochondria and chloroplasts are similar in size to bacteria, have their own DNA, have ribosomes similar to those of bacteria, and produce a limited portion of their own enzymes and proteins.

Scientists theorize, based on evidence in the fossil record, that the first eukaryotes, larger and more complex than the cyanobacteria, appeared around 1.4 billion years ago. The chloroplasts and mitochondria of eukaryotic cells are thought to be derived from bacteria (referred to as endosymbionts) that came to live within pre-eukaryotic cells.

Vertebrates evolved from the chordates, organisms with a stiffening rod down the back.

25.5 The Paleozoic era: The occupation of the land

The Paleozoic era (PAY-lee-oh-ZOE-ik; 520 to 250 million years ago) is marked by an abundance of easily visible, multicellular fossils. (The earliest such fossils are Proterozoic [approximately 630 million years old] and are found in rocks of southern Australia.) For many years, however, the Paleozoic fossils were the oldest known, which led scientists to name this era after this "old" (*paleos*) "life" (*zoos*).

Of the roughly 250,000 different kinds of fossils that have been identified, described, and named, only a few dozen are more than 630 million years old. In 1998, an international team of scientists discovered what appear to be worm tracks in 1.1-billion-year-old sandstone. However, many other scientists think that the sample is poorly preserved and may not be fossilized ancient worms.

The Paleozoic era is divided into six shorter time spans called periods. (These periods are shown in Figure 25.3.) The Cambrian (KAM-bree-un) period, which ended roughly 505 million years ago, is the oldest period within the Paleozoic era. It represents an important point in the evolution of life; all of the main phyla and divisions of organisms that exist today (except for the chordates and land plants) evolved by the end of the Cambrian period. Because so many new kinds of organisms appeared in such a relatively short time span, paleontologists (scientists who study fossil life) speak of a Cambrian "explosion" of living forms and often refer to geological strata older than Cambrian time as Precambrian.

The evolution of Cambrian organisms took place in the sea. Figure 25.7 shows fossils of the organisms that lived on the sea floor during the Cambrian period—quite unusual by today's standards! These fossils, together with well over 100 other species, have been found in the Burgess Shale, which are geological strata formed from fine-grained mud. During past movements of the Earth's crust, these geological strata (which were originally underwater) were uplifted and are now part of the Rocky Mountains of British Columbia, Canada.

Many kinds of multicellular organisms that thrived in the early Paleozoic era have no relatives living today, while others ultimately led to the contemporary phyla of organisms. The trilobites, for example (Figure 25.8a), appear to be derived from the same evolutionary line that gave rise to one living group of arthropods, the horseshoe crabs (Figure 25.8b).

As the Paleozoic era continued with the Ordovician period (505 to 430 million years ago), aquatic animals continued their evolution and the fungi, a kingdom of organisms that live on dead and decaying matter, invaded the land. The ancestry of fungi is not well understood. Scientists think fungi were successful because of their chitin-rich cell walls. Chitin is a stiff, hard substance (also found in the outer skeletons of insects and some other arthropods) that helps fungi remain "drought resistant." Fungi are also thought to be successful land organisms because of their roles (along with certain bacteria) as decomposers.

Also during the Ordovician period, wormlike aquatic animals similar to lancelets (Figure 25.9a) and lampreys began to evolve from ancient flatworms. Characteristic of these organisms was the stiff, internal rod that ran down the back, parallel to the central nerve. This stiffening rod is called a notochord (meaning "back cord"); these organisms are therefore called **chordates** (KOR-dates). One of the earliest known chordate fossils is shown in Figure 25.9b.

The approximately 42,500 species of chordates that exist today include fishes, amphibians, reptiles, birds,

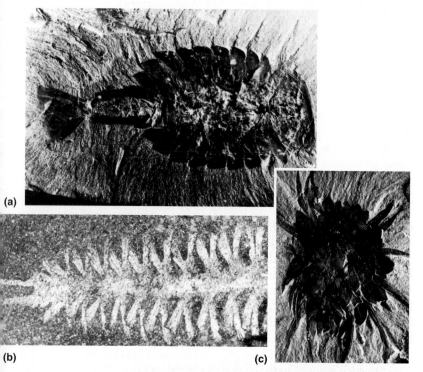

(a)

(b)

(c)

Figure 25.7 **Cambrian period fossils from the Burgess Shale, British Columbia, Canada, about 520 million years old.** (a) A fossil of *Sidneyia inexpectans*, an arthropod. (b) *Burgessochaeta setigera*, a segmented worm. The photograph shows the front end of the worm, which had a pair of tentacles. Footlike structures are attached to the worm in pairs along the sides of the body. The specimen is 16.5 millimeters (approximately ⅝ inch) long. (c) *Wiwaxia corrugata*. The body of this animal was covered with scales and also bore spines. This specimen is 30.5 millimeters (approximately 1¼ inches) across, excluding the spines. It may have been a distant relative of the molluscs.

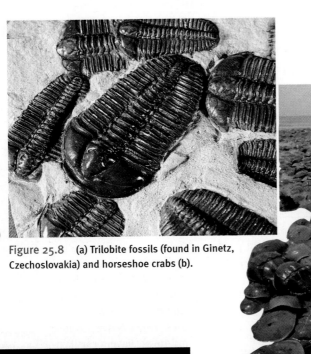

(a)

Figure 25.8   (a) Trilobite fossils (found in Ginetz, Czechoslovakia) and horseshoe crabs (b).

(b)

(a)

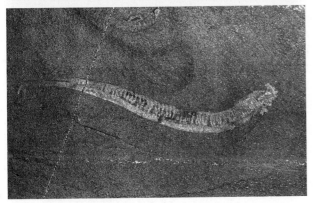

(b)

Figure 25.9   **Chordates.** (a) Two lancelets partly buried in shell gravel. Note the resemblance to *Pikaia gracilens* (b), the earliest known chordate. *Pikaia gracilens* is a small fishlike animal, and it is one of the first organisms with a notochord. This particular fossil was found in the Burgess Shale.

and mammals. (**Mammals** are warm-blooded vertebrates that have hair and whose females secrete milk from mammary glands to feed their young.) Chordates are distinguished by three principal features: (1) a single hollow *nerve cord* located along the back; (2) a rod-shaped *notochord*, which forms between the nerve cord and the developing gut (stomach and intestines); and (3) *pharyngeal (gill) arches* and *slits*, which are located at the throat (pharynx). (These three chordate features are shown in Figure 25.10 as they appear in the embryo.) All the features of chordates are evident in their embryos, even if they are not present in the adult form of the organism.

Each of the three chordate characteristics played an important role in the evolution of the chordates and the vertebrates. The dorsal nerve cord increased the animals' responsiveness to the environment. In the more

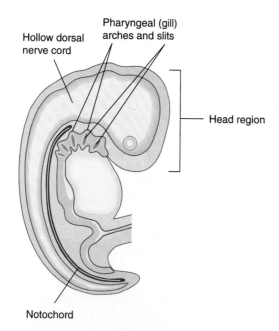

Figure 25.10   **Chordate features.**

Labels on figure: Hollow dorsal nerve cord; Pharyngeal (gill) arches and slits; Head region; Notochord

Figure 25.9a), all chordates are **vertebrates** (VER-teh-brits *or* VER-teh-braytes). Vertebrates (a subphylum of the chordates) differ from these two chordate groups in that the adult organisms have a vertebral column, or backbone, that develops around and replaces the embryological notochord. In addition, most vertebrates have a distinct head and a bony skeleton, although the living members of two classes of fishes, *Cyclostomata* (lampreys and hagfishes) (Figure 25.12) and *Chondrichthyes* (sharks, skates, and rays), have a cartilaginous skeleton.

Figure 25.12   **A lamprey.**
Lampreys attach themselves to fish by means of their suckerlike mouths, rasp a hole in the body cavity, and suck out blood and other fluids from within. Lampreys do not have a distinct jaw.

advanced vertebrates, it became differentiated into the brain and spinal cord. The notochord was the starting point for the development of an internal stabilizing framework, the backbone and skeleton, which provided support for locomotion. The bony structures that support the pharyngeal arches evolved into jaws with teeth, as shown in Figure 25.11, allowing these organisms to feed differently than their ancestors.

With the exception of two groups of relatively small marine animals, the tunicates and lancelets (see

The Devonian Period: Evolution of the Fishes

The first vertebrates to evolve were jawless fishes, members of the class *Cyclostomata*, about 500 million years ago (see Figure 25.12). Although traces of their fossils are found in Cambrian strata, most date back to the Ordovician and Silurian periods, about 505 to 410 million years ago. Jaws first developed among vertebrates that lived about 410 million years ago, toward the end of the Silurian and beginning of the Devonian periods. The

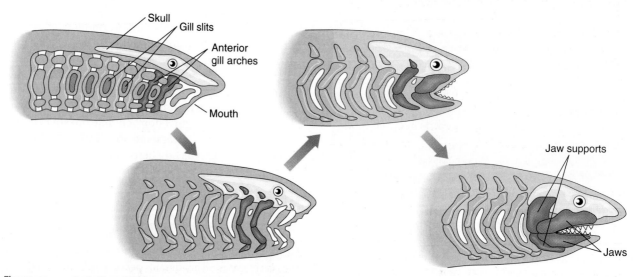

Labels on figure: Skull; Gill slits; Anterior gill arches; Mouth; Jaw supports; Jaws

Figure 25.11   **Evolution of the jaw.** Jaws evolved from the anterior gill arches of the jawless fishes.

first jawed fishes that evolved were the placoderms, ancestors of today's bony fishes (such as salmon, trout, cod, and tuna) and the cartilaginous fishes (such as sharks, skates, and rays). They were also one of the first groups of fishes to have paired fins. During the Devonian period, fishes having a bony skeleton appeared: the ray-finned fishes and the fleshy-finned fishes. The ray-finned fishes are the ancestors of the wide variety of bony fishes that exists today.

In the late Devonian period, the lungfishes and the lobe-finned fishes appear in the fossil record. Fleshy-finned fishes are ancestors to both. Only a few species of lungfishes exist today; one is shown in the photo. Lung-fishes generally live in oxygen-poor stagnant ponds and swamps. The lungs of these fishes are supplementary to their gills and allow them to surface and gulp air. Most of the Devonian period lobe-finned fishes also had primitive lungs similar to the lung-fishes.

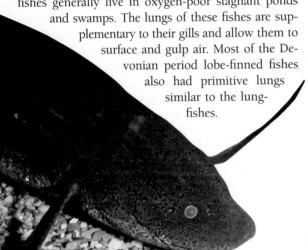

One group of the lobe-finned fishes is believed to be the ancestor of the land-living tetrapods, or four-limbed vertebrates. The first land vertebrates were amphibians, animals able to live both on land and in the water. Although a mature amphibian spends time on land, it must return to the water frequently to keep its thin skin moist. Amphibians also lay their eggs in water or in moist places and live in water during their early stages of development. The early amphibians arose in the Devonian period and had fishlike bodies, short stubby legs, and lungs (Figure 25.13). They flourished during the next period of the Paleozoic era: the swampy Carboniferous period.

The Carboniferous Period: Land Plants Arise

At the interface of the Silurian and Devonian periods, approximately 410 million years ago, the first terrestrial plants evolved. The ancestors of these plants are believed to be the green algae, because their chloroplasts are biochemically similar to those of the plants. The green algae are an extremely varied group of more than 7000 protist species, found growing in the water as well as on tree trunks or in the soil. Many of these algae are microscopic and unicellular. However, the plants that appeared during the Silurian period were multicellular and had developed mechanisms for transporting water within their bodies as well as for conserving it. Within the next 100 million years, plants became abundant and diverse on the land and eventually formed extensive forests.

The Carboniferous period (360 to 285 million years ago) is named for the great coal deposits formed during this time. Much of the land was low and swampy due to a worldwide moist, warm climate—conditions that contributed to the fossil preservation of the still-prevalent forests. These coal deposits provide a relatively complete record of the horsetail plants, ferns, and primitive seed-bearing plants of the Carboniferous period. The conifers—a group of seed-bearing plants that is represented today by pines, spruces, firs, and similar trees and shrubs—also originated then.

During the Cambrian period of the early Paleozoic era, an array of species evolved in the sea. Some of these species still exist today. The first vertebrates to evolve were the jawless fishes, about 500 million years ago. Also about that time fungi appeared, a kingdom of organisms that live on dead and decaying matter. Jawed fishes evolved toward the end of the Silurian period, about 410 million years ago. Amphibians arose from lobe-finned fishes during the Devonian period. At the interface of the Silurian and Devonian periods, approximately 410 million years ago, the first terrestrial plants evolved. The ancestors of these plants are believed to be the green algae. Within the next 100 million years, plants became abundant and diverse on the land and eventually formed extensive forests.

Figure 25.13 **Reconstruction of *Ichthyostega*, one of the early amphibians.** *Ichthyostega* had efficient limbs for crawling on land, an improved olfactory sense associated with a lengthened snout, and a relatively advanced ear structure for picking up airborne sounds. Despite these features, *Ichthyostega*, which lived about 350 million years ago, was still quite fishlike in overall appearance.

25.6 The Mesozoic era: The age of reptiles

The Mesozoic era, or "middle life" (250 to 65 million years ago), was a time of adaptive radiation of the terrestrial plants and animals that had been established during the mid-Paleozoic era but that had not been wiped out by the mass extinctions of the Permian period (see "How Science Works"). The radiation of these organisms led to the establishment of the major groups of organisms living today. The amphibians, the dominant land vertebrates for about 100 million years, were gradually replaced by the reptiles as the dominant organisms as the climate became drier.

In contrast to amphibians, reptiles have water-resistant skin and more efficient lungs. These adaptations to dry climates gave reptiles a survival advantage over the amphibians. In addition, reptiles developed a reproductive advantage—the amniotic egg. An amniotic egg has a thick shell that encloses the developing embryo and a nutrient source within a water sac (see

Chapter 31). This type of egg protects the embryo from drying out, nourishes it, and enables it to develop away from water. Birds also have amniotic eggs. During the course of the Mesozoic era, one group of reptiles evolved into mammallike reptiles, precursors to mammals; another group led to the evolution of the birds; and yet another developed into the dinosaurs. Reptiles also invaded the sea and the air, but many of these kinds of reptiles eventually died out.

The Mesozoic era is divided into three periods: the Triassic, the Jurassic, and the Cretaceous. (These names refer to geological events or formations of the eras, not to the evolution of life during this time.) During the Triassic period, small dinosaurs and primitive mammals appeared. During the Jurassic period, large dinosaurs dominated the Earth, and birds first appeared. The earliest known bird, *Protoavis* ("first bird") (Figure 25.14a), was reported in 1986 by Sankar Chaterjee of Texas Tech University. The fossil bones of this organism were found in rocks dated to be 225 million years old, very early in

(a)

(b)

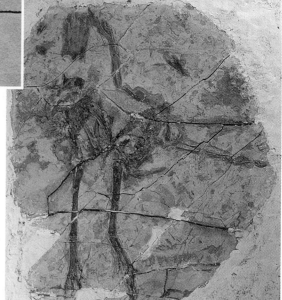

(c)

Figure 25.14 **Links to the birds.** (a) *Protoavis*, a 225 million-year-old fossil from west Texas, may be an ancestor of the birds. If so (this claim is highly disputed), it will extend the history of birds back 75 million years. (b) About the size of the crow, *Archaeopteryx* lived in the forests of central Europe 150 million years ago. The teeth and long, jointed tail are features not found in any modern birds. Discovered in 1862, *Archaeopteryx* was cited by Darwin in later editions of *On the Origin of Species* in the support of his theory of evolution. (c) *Protachaeopteryx*. This feathered dinosaur provides evidence that birds might be descended from the dinosaurs. Not all scientists agree with this hypothesis, however, and suggest that both birds and dinosaurs evolved from a common ancestor.

HOW Science WORKS Discoveries

Mass Extinctions: The Development of Hypotheses to Explain These Events

One of the most prominent features of the history of life on Earth has been periodic *mass extinctions*, the global death of whole groups of organisms. There have been five such events. Four of these events occurred during the Paleozoic era, the first occurring near the end of the Cambrian period (about 500 million years ago). At that time, most of the existing families of trilobites became extinct. Additional mass extinctions occurred about 440 million years ago and about 360 million years ago. The fourth and most drastic mass extinction happened at the close of the Permian period. It is estimated that approximately 96% of all species of marine animals living at that time died out! The fifth major extinction event occurred at the close of the Mesozoic era, 65 million years ago. This was the famous event when dinosaurs became extinct.

Major episodes of extinction produce conditions that can lead to the rapid evolution of those relatively few plants, animals, and microorganisms that survive. Habitats, or places to live, that were formerly occupied suddenly open up, providing the "raw material" for *adaptive radiation*, a phenomenon by which populations of a species change as they disperse within a series of different habitats within a region (see Chapter 24). In fact, many paleontologists suggest that extinction episodes have been key events in the pattern of evolution.

Scientists have long discussed hypotheses regarding the extinction event of the late Cretaceous period. Some propose that the climate cooled. Others have suggested that the sea level changed or that the vegetation cover of the Earth changed. These reasons, however, would not account for the change in the plankton community.

In 1980, a group of scientists headed by Luis Alvarez of the University of California, Berkeley, presented a dramatic hypothesis about the reasons for these changes. Alvarez and his associates observed that the usually rare element iridium was abundant in a thin layer in the geological strata that marked the end of the Cretaceous period. Alvarez and his colleagues proposed that if a large, iridium-rich meteorite or asteroid had struck the surface of the Earth, a dense particulate cloud would have been thrown up. This cloud would have darkened the Earth for a time, greatly slowing or temporarily halting photosynthesis and driving many kinds of organisms (particularly photosynthetic plankton alive at that time) to extinction. By disrupting and killing plant life, other organisms dependent on the plants would die out as well. Then, as its particles settled, the iridium in the cloud would have been incorporated in the layers of sedimentary rock that were being deposited at that time. Although the "meteorite hypothesis" is the most plausible one put forth to date, scientists have not reached consensus regarding this hypothesis.

Crater in Arizona formed by a much smaller meteorite than proposed by Alvarez and his group as causing an extinction event.

---

the Mesozoic era. Until this find, the oldest known fossil bird was *Archaeopteryx*, estimated to be about 150 million years old (Figure 25.14b). There is no consensus regarding whether *Archaeopteryx* arose from the same evolutionary line as *Protoavis* or independently from another dinosaur. In addition, there is no consensus among paleontologists (scientists who study fossils) whether *Protoavis* is actually a bird. Many scientists question the validity of Chaterjee's "find."

In 1998, a team of paleontologists from China, Canada, and the United States announced that they had discovered two new species of small dinosaurs covered with feathers (Figure 25.14c). This find adds more evidence to the theory that birds evolved from the dinosaurs. These dinosaur fossils are 30 million years younger than *Archaeopteryx*, however, dating back 120 million years.

Archaeopteryx had feathers, but apparently not enough of them to fly very effectively. Feathers, thought to have evolved from reptilian scales, probably helped early birds glide from tree to tree and provided them with insulation. Ultimately, feathers made possible the subsequent evolution of birds that could fly well. This ability allowed birds to inhabit unoccupied habitats and resulted in the evolution of a large and diverse class of organisms. Today, there are about 9000 different species of birds.

The Cretaceous Period: Flowering Plants Appear

During the early Cretaceous period, about 140 million years ago, flowering plants (angiosperms) began to appear, becoming the dominant form of plant life about 100 million years ago. Seed-bearing plants with fernlike leaves, similar to the living cycads, were abundant at that time. Today, with about 240,000 species of flowering plants, the angiosperms still dominate the plant kingdom, greatly outnumbering all other kinds of plants.

Until recently, all evidence suggested that insects and flowering plants *coevolved*. That is, as flowering plants changed over time, so did insects because their feeding habits were linked to the characteristics of the flowers. As the flowering plants became more diverse and plentiful, so did the insects. However, recent fossil evidence analyzed by American paleontologists Conrad Labandeira and Jack Sepkoski suggests that insects (which arose in the Devonian period) experienced a major radiation in diversity during the Carboniferous period. Then, during the mass extinctions of the Permian period, about 30% of all insect orders died out. The adaptive radiation of the insects began again in the Triassic period, *before* the diversification of the angiosperms, and it continues today. In fact, the Labandeira and Sepkoski research suggests that the diversification of the insects actually *slowed down* as angiosperm diversity speeded up.

The End of the Mesozoic Era: Mass Extinctions and Continued Change

Around 65 million years ago, as the Cretaceous period (and the Mesozoic era) ended, sudden shifts occurred in the kinds of marine organisms that existed. Some became extinct and others began to flourish. Scientists can infer these changes in the populations of marine organisms living at that time by studying marine fossils exposed in certain European geological strata. For example, many of the larger plankton (free-drifting aquatic organisms) disappeared about 65 million years ago and a fewer number of smaller ones took their place. The same rapid changes occurred in at least some non-planktonic marine animal groups, such as the bivalve molluscs (clams and their relatives). The ammonites, a large and diverse group related to octopuses but which have shells, abruptly disappeared. A major extinction occurred on land: The dinosaurs disappeared, although at a pace much slower than that of the large plankton just mentioned.

The Mesozoic era was a time of adaptive radiation of the terrestrial plants and animals, which led to the establishment of the major groups of organisms living today. Reptiles gradually replaced amphibians as the dominant organisms and in turn gave rise to mammals (about 200 million years ago) and birds (at least 150 million years ago).

25.7 The Cenozoic era: The age of mammals

The mass extinctions of the Cretaceous marked the end of the Mesozoic era and heralded in the Cenozoic era about 65 million years ago. Extending to the present, the Cenozoic era has two periods—the Tertiary and the Quaternary—which are subdivided into epochs. The epochs of the Tertiary period are listed in Figure 25.3. Notice also that most of the "year" has passed on the geological calendar. Although 65 million years may seem like an incredibly long time, in terms of the history of life, it is short.

At the beginning of the Cenozoic era during the Paleocene epoch, the small mammals that survived the extinctions of the Mesozoic era underwent adaptive radiation, quickly filling the habitats vacated by the dinosaurs. Organisms having *both* mammalian and reptilian characteristics, called transitional forms, first appear in the fossil record approximately 245 million years ago. Then, about 200 million years ago in the early Mesozoic era, the first known mammals appear. These early mammals resemble today's shrews, as seen in the photo below. They were small, fed on insects, and were probably nocturnal, or active at night. About 65 million years ago, following the extinction of the dinosaurs, mammals became abundant.

Various natural selection pressures resulted in the emergence of a number of significant anatomical and physiological changes in the mammals as they evolved from their reptilian ancestors. These changes resulted in a class of organisms that not only survived but also flourished in a wide variety of habitats. Changes occurred in the reptilian arrangement of limbs as the mammals evolved, raising them high off the ground and allowing them to walk quickly and to run. A hinge developed between the lower jaw and the skull, and the teeth became differentiated, allowing mammals to eat a wide variety of foods. The reptilian heart, having two ventricles with an incomplete separation, developed a complete wall in the mammals, preventing the mixing of oxygenated and deoxygenated blood. (See Figure 31.9.) This change was significant in that mammals had also developed *warm bloodedness:* a constant internal body temperature. Warm-blooded animals are also called *endotherms,* meaning "within" (*endo-*) "temperature" (*-therm*). Their ancestors were *cold blooded,* having internal body temperatures that followed the temperature of their environments within certain limits. Cold-blooded animals are also called *ectotherms,* meaning "outside" (*ecto-*) "temperature" (*-therm*). Endotherms have a higher rate of metabolism than ectotherms and need more oxygen for the increase in the rate of cellular respiration.

Important changes also occurred in reproduction as the mammals evolved from the reptiles. The most "reptilian" type of reproduction is seen in the monotremes, one of the three subclasses of mammals living today. Monotremes lay eggs with leathery shells and incubate

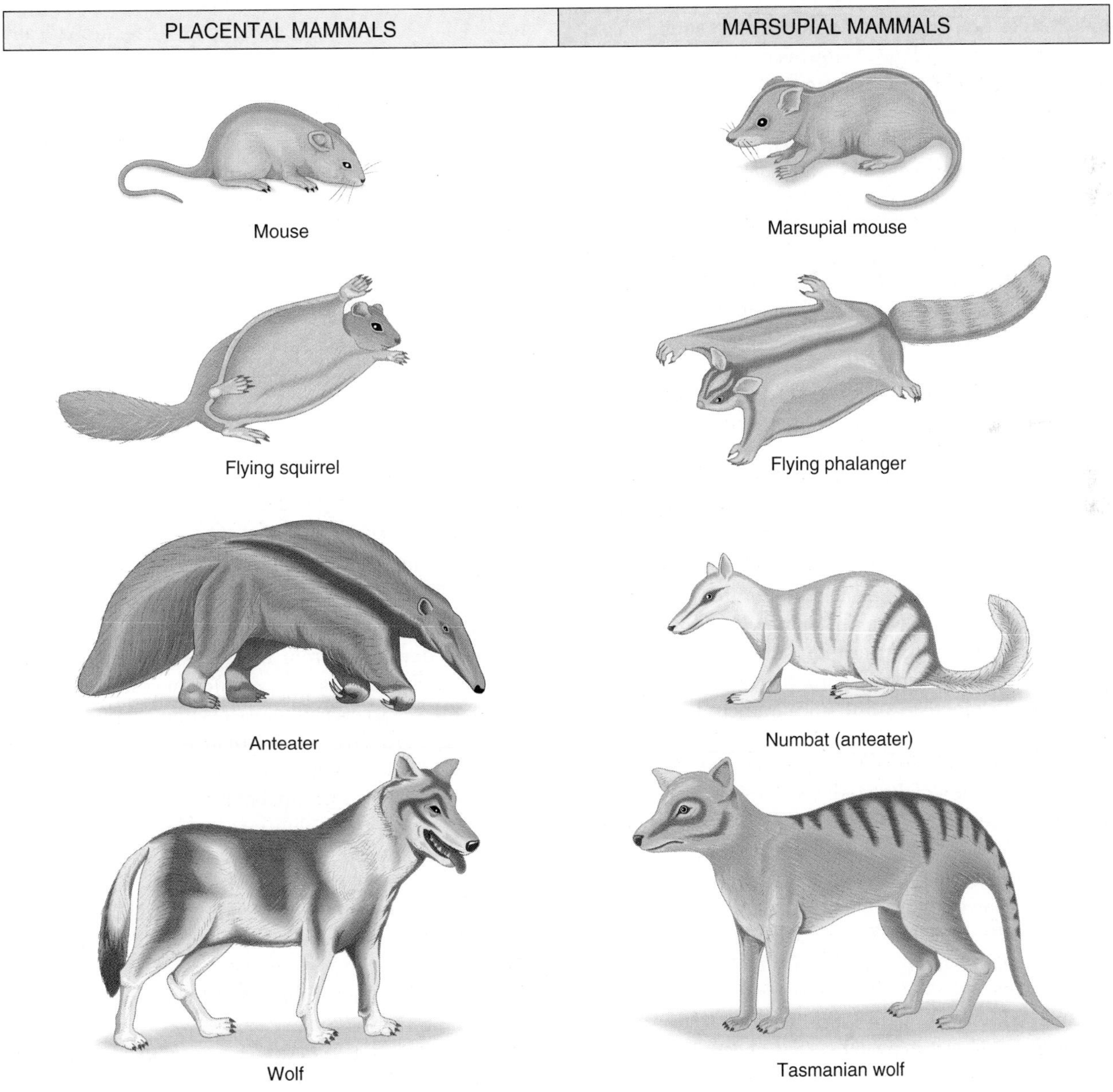

PLACENTAL MAMMALS	MARSUPIAL MAMMALS
Mouse	Marsupial mouse
Flying squirrel	Flying phalanger
Anteater	Numbat (anteater)
Wolf	Tasmanian wolf

Figure 25.15 **Convergent evolution.** This illustration demonstrates the convergent evolution of marsupials in Australia and placental mammals in the rest of the world.

Vertebrates evolved from the chordates, organisms with a stiffening rod down the back.

Primates reflect a tree-dwelling heritage.

these eggs in a nest. The underdeveloped young that hatch from these eggs feed on their mother's milk until they mature. The ability of a mother to feed her young with milk she produces in mammary glands is another characteristic that has been advantageous to the adaptive radiation of the mammals. Present-day examples of monotremes are platypuses and spiny anteaters of Australia. These organisms are thought to have arisen on Gondwana, which is today's South America, Africa, and Australia (see "How Science Works" on p. 574), and were then isolated as the continents drifted apart.

The marsupials, a second subclass of mammals, do not lay eggs but give birth to immature young. These blind, embryonic-looking creatures crawl to the mother's pouch and nurse until they are mature enough to venture out on their own. Like the monotremes, marsupials were present in Gondwana; today most marsupials, such as kangaroos, wombats, and koalas, are found in Australia. The marsupials of Australia resemble the placental mammals that are present on the other continents. Figure 25.15 (see page 569) compares individual members of these two sets of mammals. The members of each pair have similar habitats and find their food in similar ways. These characteristics, along with their strikingly similar anatomy, suggest that the evolution of these two subclasses of mammals is a product of **convergent evolution**: the development of similar structures having similar functions in different species as the result of the same kinds of selection pressures.

In placental mammals, the young develop to maturity within the mother. They are named for an organ formed during the course of their embryonic development, the placenta. The placenta is located within the walls of the uterus, or womb. Composed of both maternal and fetal tissues, the placenta is connected to the fetus by the umbilical cord. At the placenta, fetal wastes pass into the bloodstream of the mother, and oxygen and nutrients in the mother's bloodstream pass into the bloodstream of the fetus, a mechanism that allows the fetus to develop within the mother until it reaches a certain age of maturity.

Mammals evolved from the reptiles approximately 200 million years ago. About 65 million years ago, the small mammals that survived the extinctions of the Mesozoic era underwent adaptive radiation, quickly filling the habitats vacated by the dinosaurs. Changes in the structure and placement of their limbs, the structure of their heart, and reproductive strategies resulted in their ability to survive in a variety of climates and to eventually become abundant.

25.8 Characteristics of primates

There are 14 orders of placental mammals (see Chapter 2 for a discussion of classification), which include many animals familiar to you, such as dogs, cats, horses, whales, squirrels, rabbits, bats, and a variety of others. The order of placental mammals that humans belong to is the **primates**. Primates are mammals that have characteristics reflecting an arboreal, or tree-dwelling, lifestyle. Among these characteristics are hands and feet that are able to grasp objects (such as tree branches), flexible limbs, and a flexible spine. Table 25.1 lists primate characteristics. (Figure 25.16 diagrammatically represents the taxonomic relationships of mammals living today.

Table 25.1

Primate Characteristics

- Ability to spread toes and fingers apart
- Opposable thumb (thumb can touch the tip of each finger)
- Nails instead of claws
- Omnivorous diet (teeth and digestive tract adapted to eating both plant and animal food)
- A semi-erect to an erect posture
- Binocular vision (overlap of visual fields of both eyes) resulting in depth perception
- Well-developed eye-hand coordination
- Bony sockets protecting eyes
- Flattened face without a snout
- A complex brain that is large in relation to body size

Some time in the early Cenozoic era, the primates may have developed from the shrewlike mammals. As thousands of years passed, primates' ancestors probably moved into the trees, eating insects that fed on the fruits and flowers growing on tree branches. Selection pressures must have been great for these arboreal creatures. Those that survived in this habitat developed excellent depth perception, the ability to see in more than one plane as they moved from tree to tree. Depth perception is a function of stereoscopic vision, vision created by two eyes focusing on the same object but from a slightly different angle. As primates evolved from their shrewlike ancestors, the eyes moved closer together from their placement on either side of the head, allowing stereoscopic vision.

The primates also developed long limbs with flexible hands and feet adapted to grasping and swinging from branch to branch. Primates have two bones in the lower part of a limb that enable the wrists and ankles to rotate. In addition, the hands and feet of primates have digits that can be spread apart from one another, helping primates balance themselves when walking or running or enabling them to grasp objects. An opposable thumb, one that can touch the tip of each finger, helps in grasping. Most primates also developed flattened nails at the end of the digits, replacing the claws of their mammalian relatives.

The primates, an order of mammals that humans belong to, reflect an arboreal heritage. The earliest primates arose in the early Cenozoic era and lived on the ground, feeding on plants and insects. Over time, the primates moved into the trees, developing excellent depth perception and flexible, grasping hands.

25.9 Prosimians and anthropoids

The primates are divided into two suborders: the prosimians and the anthropoids (the suborder that includes humans). The **prosimians** (pro-SIM-ee-uns; meaning "before ape") are small animals such as lemurs, indris, aye-ayes, and lorises; they range in size from less than a pound to approximately 14 pounds—about the size of a cat or a small dog. Most are nocturnal and have large ears and eyes to help them see and hear at night. Prosimians have elongated snouts, reflecting their highly developed sense of smell. These characteristics are clearly seen in the prosimian pictured in Figure 25.17, in addition to its elongated rear limbs, which help it leap from tree to tree in its tropical rain forest habitat. By the end of the Eocene epoch (38 million years ago), prosimians were abundant in North America and

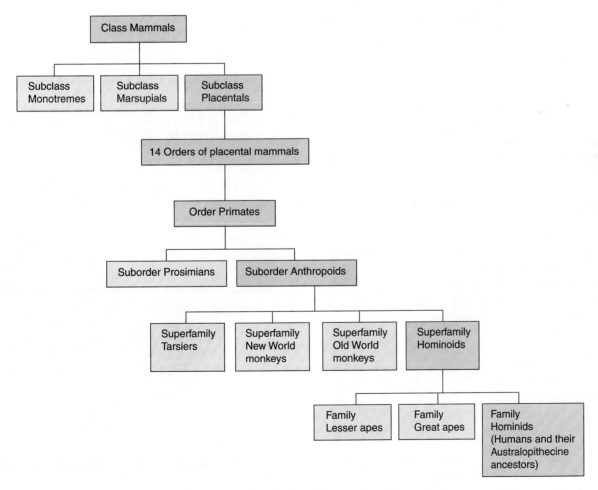

Figure 25.16 **Classification of mammals.** There are 14 orders of placental mammals, and humans belong to the primate order.

Figure 25.17   **Prosimian.** The prosimian shown in the picture is a ringtail lemur, *Lemur catta*. All living lemurs are restricted to the island of Madagascar, and all are in danger of extinction as the rain forests are destroyed.

Eurasia and were probably also present in Africa. However, their descendants now live only in the tropics of Asia, in tropical Africa, and on the island of Madagascar.

The **anthropoids** (AN-thruh-poyds; meaning "humanlike") include the monkeys, apes, gorillas, chimpanzees, and humans. These primates differ from the prosimians in the structure of their teeth, brain, skull, and limbs. The prosimians have pointed molars and horizontal lower front teeth. These horizontal teeth are used to comb the coat or to get at food. The anthropoids have more rounded molars and no horizontal lower front teeth. The brain of the prosimian is much smaller in relation to body size than the brain of an anthropoid. The face of an anthropoid is somewhat flat. In addition, the eyes of an anthropoid are closer together than those of a prosimian. Lastly, a prosimian's front limbs are short in relation to its long hind limbs, whereas both the front and hind limbs of an anthropoid are long. Compare these features in the prosimian (see Figure 25.17) and the anthropoids (Figures 25.18 and 25.19).

In addition to these structural differences, prosimians and anthropoids also exhibit behavioral differences. Anthropoids are diurnal; that is, active during the day, whereas the prosimians are nocturnal. The anthropoids have evolved color vision, probably in rela-

(a)   (b)

Figure 25.18   **Facial differences between an Old World monkey (a) and a New World monkey (b).**
(a) Hamadryas baboon. (b) Geoffroy's marmoset.

Figure 25.19 **The apes.** (a) Gorilla, *Gorilla gorilla.* (b) White-handed gibbon, *Hylobates lar.*
(c) Orangutan, *Pongo pygmaeus.* (d) Chimpanzee, *Pan troglodytes.*

tion to their diurnal existence. Also, the anthropoids live in groups in which complex social interaction occurs. In addition, they tend to care for their young for prolonged periods.

The **hominoids** (HOM-uh-noyds; also meaning "humanlike") are one of four superfamilies of the anthropoid suborder (see Figure 25.16). This superfamily includes apes and humans. The other three anthropoid superfamilies include (1) tarsiers (formerly classified with the prosimians, Figure 25.20 on page 576), (2) New World monkeys, and (3) Old World monkeys. (Both New and Old World monkeys are shown in Figure 25.18.) In

general, New World monkeys are found in South and Central America, and they have flat noses with nostrils that face outward. Old World monkeys have noses similar to humans with nostrils next to each other that point downward; these monkeys are found in Africa, southern Asia, Japan, and Indonesia.

The hominoid superfamily is divided into three families: the lesser apes, the great apes, and the hominids, or humans. A variety of characteristics put the apes in a different superfamily than that of the monkeys: Most apes are bigger than monkeys, have larger brains than monkeys, and lack tails.

H⚙W WORKS Science *Discoveries*

The Merging of Sciences: Using the Theory of Plate Tectonics to Understand Evolution

During the Mesozoic era, the continents did not exist as they do today. Instead, they were all joined in one giant continent scientists called *Pangea,* meaning "all Earth." Pangea began to break up into smaller pieces during the Mesozoic era, with this movement continuing during the following era, the Cenozoic. These changes greatly affected evolution.

After the Earth coalesced approximately 4.6 billion years ago, its surface was hot and violent for about 600 to 700 million years. At first, these conditions were too inhospitable for living things or even for biochemical evolution to take place. Scientists hypothesize that geological activity eventually resulted in the

formation of fragments of continents. Some geologists think that these land masses were formed as melted rock material rose to the surface of the Earth, pushed upward by other heavier, sinking, melted metals such as iron and nickel. Others suggest that the highly active volcanoes prevalent at that time spewed so much lava into the seas that it eventually accumulated, forming land masses.

Approximately 200 million years ago, small land masses united to form the single, large "supercontinent" Pangea. Its northern half was called *Laurasia* and consisted of the present-day North America, Europe, and Asia. Its southern half was called *Gondwana* and

was a combination of present-day South America, Africa, India, Antarctica, and Australia. Pangea remained as a supercontinent for approximately 100 million years. However, forces were at work beneath it that eventually divided it—once again—into smaller land masses.

Scientists now have some insight into the movement of land masses over time, such as those that resulted in the formation of Pangea and its subsequent division. No one knows exactly the speed of the movement long ago; the current rate of movement is 1 to 15 cm per year—incredibly slow.

Today, the outer shell of the Earth is made up of six large "plates" and several smaller ones (see Figure 25.A). According to the *plate tectonics theory,* these plates are rigid pieces of the Earth's crust, which are from 75 to 150 kilometers (approximately 50 to 100 miles) thick.

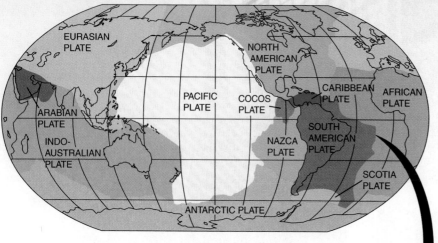

Figure 25.A **Plate tectonics theory.** Plate tectonics describes and explains the movement of the continents.

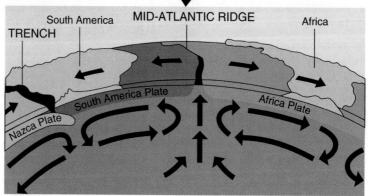

Figure 25.B **Why do the plates move?** The semiliquid rock on which the plates rest rises as it heats and sinks as it cools, causing movement.

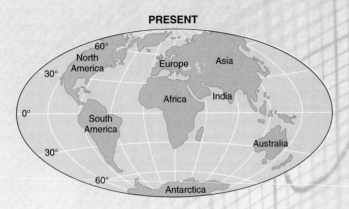

PRESENT

70 MILLION YEARS AGO

240 MILLION YEARS AGO
PANGEA

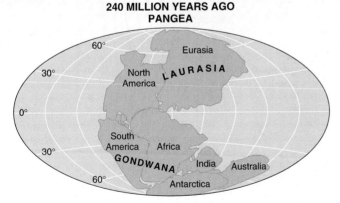

Figure 25.C **Movement of the continents.** The changes that occurred in the position of all the continents as a result of shifting plates have played a major role in the distribution of organisms seen today.

The rocks beneath these plates are less rigid than the plates they support. The surface plates move, or drift, over this underlying semiliquid rock. Scientists think that this "liquid" rock flows slowly, rising as it heats, and sinking as it cools (see Figure 25.B). At the mid-oceanic ridges—long, narrow mountain ranges under the sea—the plates move away from one another in a process called *sea floor spreading*. Molten rock wells up from deep within the Earth, pushing the plates apart and filling in the space between. At other places, the plates move toward each other. As they meet, one plate may sink below the other, or they may collide, pushing up mountain ranges in the process. The Himalayas, for example, have been thrust up to the highest elevations on Earth as a result of the grinding, prolonged collision of the Indian subcontinent with Asia. Other types of violent geological activity were also associated with the movements of the crustal plates. The earthquakes that destroyed San Francisco in 1906 and the one that occurred in the same region in 1989 resulted from two plates sliding past each other.

Figure 25.C summarizes the movements of the continents as they gradually moved apart from their positions as parts of Pangea 200 million years ago. The South Atlantic Ocean opened about 125 to 130 million years ago as Africa and South America began to move apart from one another. South America then moved slowly toward North America and, approximately 3.6 to 3.1 million years ago, became connected to it by the Isthmus of Panama. Changes occurred in the positions of all the continents, which played a major role in the distribution of organisms that are seen today.

As land masses move away from one another, populations inhabiting those lands become separated from one another. Geographically isolated, populations of the same species often undergo speciation as differing pressures of natural selection act on them. Similarly, as land masses collide, species are thrust together and must compete with one another to survive. Shorelines disappear and with them go a variety of habitats and species. And as continents move, ocean currents change, causing climatic changes and yet new selection pressures.

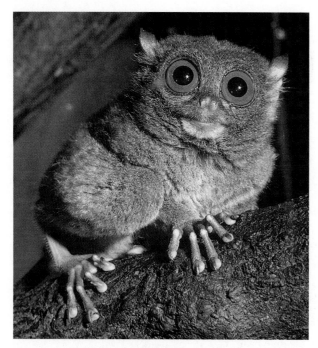

Figure 25.20 **Anthropoid.** Tarsier, *Tarsius syrichta,* tropical Asia. Note the large eyes of the tarsier, an adaptation to nocturnal living.

The *lesser apes* include the gibbons, which are the smallest hominoids and closest in size to monkeys. They are about the size of a small dog, weighing 4 to 8 kilograms (9 to 18 pounds). Like all monkeys and apes, they live in tropical rain forests, where they leap and swing from tree to tree with their long arms.

The *great apes* include the orangutans, gorillas, and chimpanzees. These apes are much larger than the gibbons (see Figure 25.19). The orangutans are about the size of humans, weighing between 50 and 100 kilograms (110 to 220 pounds). These apes exhibit sexual dimorphism; the females' weight is about half that of the males. Like the gibbons, the orangutans have long arms, but they walk—they do not swing—from branch to branch of the rain forest trees. The gorillas are the largest of the apes; they weigh about 160 kilograms, or 350 pounds. The females are smaller, ranging in weight from 165 to 240 pounds. Gorillas spend most of their time on the ground. When alarmed, male gorillas beat on their chests in a behavioral display. However, they are not the fierce animals humans think they are. In fact, gorillas are quite peaceable. The smallest of the great apes are the chimpanzees—humans' closest relatives. These animals weigh between 40 and 50 kilograms, or 90 to 110 pounds; females are slightly lighter. Like the other apes (except the gorillas), they spend their time in the trees.

The **hominids** (HOM-uh-nids), the family that includes humans of today, are the most intelligent of the

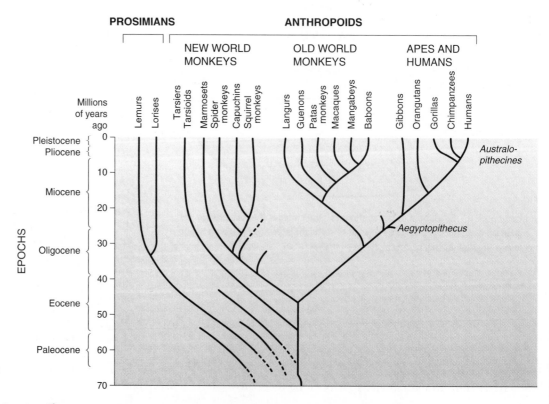

Figure 25.21 **The primate family tree.** This tree is based on DNA comparisons between living species. The dating of the branches is based on the ages of fossils.

hominoids. They are distinguished from the other families of hominoids in that they are *bipedal;* that is, they walk upright on two legs. In addition, hominids communicate by language and exhibit *culture*—a way of life that is passed on from one generation to another. Humans are the only living hominids.

Scientists have found jaw fragments that suggest that the ancestors of monkeys, apes, and humans began their evolution approximately 50 million years ago. The tarsioids may have begun their evolution even earlier—about 54 million years ago. Scientists do not know what adaptive pressures resulted in the appearance of these early anthropoid primates, nor do they know which prosimian was their ancestor.

Although the only fossil evidence of the emergence of the anthropoids consists of pieces of jaw, biochemical studies complement the knowledge gained from the fossil record. Together, these techniques tell scientists a great deal about the evolution of the anthropoids and the relationships among humans, apes, and monkeys. The evolutionary tree of the anthropoids is shown in Figure 25.21. (See Chapter 24 for a discussion of evolutionary trees.) Fossils and biochemical studies show that the New World monkeys branched from the line leading to the Old World monkeys and the hominoids about 45 million years ago in the mid-Eocene epoch. In addition, scientists have discovered many fossils in North Africa that date from 25 to 30 million years ago in the Oligocene epoch. One of the earliest of these fossils, which scientists have named *Parapithecus*, probably led to the line of Old World monkeys, splitting from the hominoids approximately 30 million years ago. The others, scientists think, are probably members of the evolutionary line that leads to the hominoids.

The most well-known of the Oligocene fossils found in North Africa is called *Aegyptopithecus*. This fossil is from the late Oligocene epoch and is thought to be the ancestor to the early Miocene hominoids of Africa. Looking at the partially restored skull in Figure 25.22, you can see that *Aegyptopithecus* had some prosimian characteristics. It had a pronounced snout, leading scientists to believe that its sense of smell was still highly specialized, much like that of the prosimians. In addition, like the prosimians, *Aegyptopithecus* lived singly rather than in social groups.

Fossil evidence is still being accumulated that will help tell the story of hominoid evolution. From mid-Miocene times on, the hominoid fossil record is quite extensive, but consists primarily of skull and teeth fragments. Investigators have calculated that the evolutionary line leading to gibbons diverged from the line leading to the other apes about 18 to 22 million years ago; the line leading to orangutans split off roughly 13 to

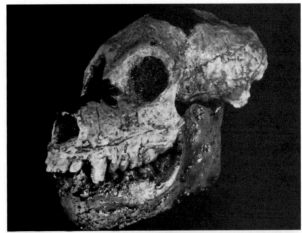

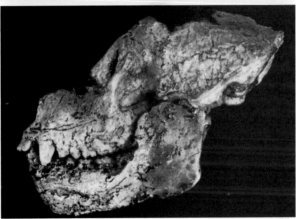

Figure 25.22 *Aegyptopithecus zeuxis.* This primate fossil dates to the late Oligocene era and is thought to be the ancestor of the hominoids that lived in Africa during the early Miocene era.

16 million years ago; the line leading to gorillas diverged 8 to 10 million years ago; and the split between hominids and chimps occurred approximately 5 to 8 million years ago. This last statement suggests that chimpanzees and gorillas are humans' closest relatives, with a common relative alive 5 to 8 million years ago.

The prosimians and anthropoids are the two suborders of primates. By the end of the Eocene epoch (38 million years ago), prosimians were abundant in North America and Eurasia and were probably also present in Africa. Fossil evidence and biochemical studies show that the anthropoid suborder, which includes humans, monkeys, and apes, first appeared about 50 million years ago. The New World monkeys branched from the evolutionary line leading to the Old World monkeys and the hominoids about 45 million years ago. The Old World monkeys split from the hominoids about 30 million years ago.

25.10 The first hominids: *Ardipithecus*

The two critical steps in the evolution of humans were the evolution of bipedalism and the enlargement of the brain. For many years, scientists have hypothesized how and when bipedalism arose. The current hypothesis is that bipedalism arose as a preadaptation because of the way humans' arboreal, apelike ancestors moved through the trees. In other words, the skeletons and muscles of human ancestors were structured in a way that allowed these hominoids to walk bipedally even though they lived in the trees. These structural adaptations developed as a part of their arboreal life and the types of locomotion they exhibited in the trees. **Figure 25.23** shows the movements of gibbons and other related brachiators—those hominoids that move through trees by hanging from the branches with their arms, "walking" themselves along. As the weather in Africa cooled somewhat and became seasonal, patches of savannah grasslands began to invade former areas of tropical forest. Current theory holds that the hominoids best able to walk efficiently on the ground survived, and bipedalism evolved.

The oldest evidence of the hominids was found in 1994 and is 4.4 million-year-old skull fragments of *Ardipithecus ramidus*. Its teeth have characteristics intermediate between the apes and *Australopithecus* (see next paragraph). It is unclear to scientists whether *A. ramidus* was fully bipedal, although researchers are considering *Ardipithecus* a hominid. Additionally, *Ardipithecus* is thought to have been a forest dweller, which may cause a modification in the hypothesis that bipedalism arose as hominoids moved to the savannah.

In 1995, hominid skeletal remains were found in Africa by an international research team led by Dr. Meave G. Leakey of the National Museums of Kenya in Nairobi. These remains were not reliably dated until 1998. It was then confirmed that the oldest human ancestor definitely known to walk erect, *Australopithecus anamensis*, was older than the previously known oldest upright-walking human ancestor *Australopithecus afarensis*. *A. anamensis* had apelike jaws, teeth, wrist

Figure 25.23 Brachiation and bipedalism. Brachiators, such as gibbons and siamangs, locomote by hanging from branches with their arms and reaching from hold to hold. Brachiation has preadapted these animals to bipedal walking, which they do on broad branches (as the siamang in the illustration is doing) or on the ground.

bones, and a small brain. Their arm and leg bones were more humanlike than apelike, which allowed them to walk upright rather than on all fours. *Anamensis* existed between 4.2 and 3.9 million years ago.

A. afarensis, which existed between 3.9 and 3.0 million years ago, was found in 1976 by Mary Leakey and an international team of scientists who found the fossil footprints shown in **Figure 25.24.** Anthropologists think that the individuals who made the footprints are human ancestors or are very closely related to human ancestors. Fossil bones discovered a few years earlier, at a site in Ethiopia about 2000 kilometers (1250 miles) north of the footprints, support this hypothesis. At 3.5 million years old, these hominid bones are somewhat younger than the footprints. These fossils were named *Australopithecus afarensis,* meaning "southern ape of Afar," by Donald Johanson and an international team of scientists who found them in the Afar region of Ethiopia. In 1974, this team also found and pieced together one of the most complete fossil skeletons of *A. afarensis,* which has since become famous (**Figure 25.25**). This "first" hominid was named Lucy after the Beatles song "Lucy in the Sky with Diamonds," which was playing on their tape machine at the time.

Although *Australopithecus* and *Ardipithecus* are hominids, they are not humans (members of the genus *Homo*). Their brains were still small in comparison to present-day human brains (about the size of the great apes), and they had long, monkeylike arms. In addition,

Figure 25.24 Fossil footprints. These fossil footprints, found in the Afar region of Ethiopia, have been preserved in volcanic ash for over 3½ million years. They were made by our nonhuman ancestors, *Australopithecus afarensis,* and give scientists important clues to our heritage. The differing sizes of the footprints may reflect our ancestors' sexual dimorphism, or size difference with gender. Also, the footprints show that these organisms walked erect on two legs instead of four, a characteristic called bipedalism. Bipedalism is only one of the many evolutionary changes that led to the appearance of modern humans.

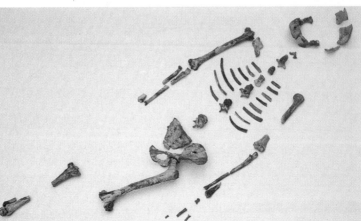

Figure 25.25 Lucy, from Ethiopia, is the most complete skeleton of *Australopithecus* discovered so far.
The reconstruction was made by a careful study of muscle attachments to the skull.

Primates reflect a tree-dwelling heritage.

their faces were apelike, as shown in the photo of a reconstruction of Lucy in Figure 25.25.

A. afarensis may have evolved into two (and possibly more) lineages, including the species *A. africanus*, *A. robustus*, and *A. boisei*. These australopithecines lived on the ground in the open savannah of eastern and southern Africa. Their diets consisted primarily of plants. At night, they probably slept in the few trees that existed in these grasslands, much like the savannah baboons do today to protect themselves from predators.

The evolutionary relationships among these australopithecines is not clear; a family tree that is widely accepted at this time is shown in **Figure 25.26.** Notice from the diagram that no australopithecines are alive today; the last ones disappeared about 1 million years ago.

The two critical steps in the evolution of humans were the evolution of bipedalism (walking on two feet) and the enlargement of the brain. The earliest hominids and the direct ancestors of humans belong to the genus *Australopithecus*. The oldest undisputed evidence of the hominids is 4.4 million years old.

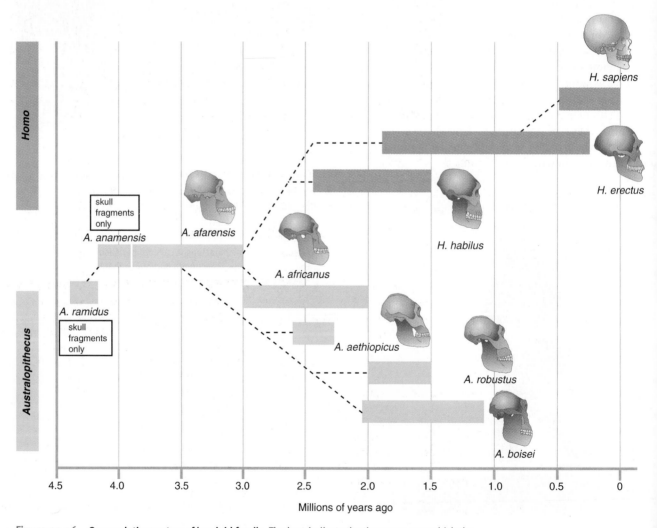

Millions of years ago

Figure 25.26   **One evolutionary tree of hominid fossils.** The bars indicate the time span over which the hominid lived.

25.11 The first humans: *Homo habilis*

Climatic changes during the Pleistocene epoch (1.8 million to 12,000 years ago) may have contributed to the eventual disappearance of the australopithecines and the survival of a new, more intelligent genus of hominids: the human (genus *Homo*). To emphasize their intelligence, these early humans were named **Homo habilis**, or "skillful human." *H. habilis* coexisted for at least 500,000 years with the smaller-brained Australopithecus. *Homo habilis* existed between 2.4 and 1.5 million years ago. The first *H. habilis* fossils were discovered in 1964 by Louis Leakey, Philip Tobias, and John Napier in the Olduvai Gorge of eastern Africa.

Judging from the structure of the hands, *H. habilis* regularly climbed trees as did their australopithecine ancestors, although the *H. habilis* people spent much of their time on the ground and walked erect on two legs. Skeletons found in 1987 indicate that the *H. habilis* people were small in stature like the australopithecines, but fossil skulls and teeth reveal that the diet of *H. habilis* was more diverse than that of the australopithecines, including meat as well as plants. The tools found with *H. habilis* were made from stones fashioned into implements for chopping, cutting, and pounding food. The use of stone tools by these early humans marks the beginning of the *Stone Age*, a time that spans approximately 2 million to 35,000 years ago.

The first human fossils date back about 2 million years and are of the extinct species *H. habilis*, meaning "skillful human." This species is considered human because by making tools they exhibited an intelligence far greater than their ancestors.

25.12 Human evolution continues: *Homo erectus*

All of the early evolution of the genus *Homo* seems to have taken place in Africa. There, fossils belonging to the second, also extinct species of *Homo*—*Homo erectus*—are widespread and abundant from 1.8 million to about 300,000 years ago. *Homo habilis*, *Australopithecus*, and *Ardipithecus* fossils have been found only in Africa, but *H. erectus* fossils have been found in Africa, Asia, and Europe.

The *H. erectus* people were about the size of modern-day humans, were fully adapted to upright walking, and had brains that were roughly twice as large as those of their ancestors. However, they still retained prominent brow ridges, rounded jaws, and large teeth. The tools of *H. erectus* were much more sophisticated than those of *H. habilis* and were used for hunting, skinning, and butchering animals. These people were hunter-gatherers, which means that they collected plants, small animals, and insects for food while occasionally hunting large mammals. Researchers have found the first evidence of the use of fire by humans at 1.4 million-year-old campsites of *H. erectus* in the Rift Valley of Kenya, Africa. Fire is characteristically associated with populations of this species from that time onward. All of these activities (tool making, hunting, using fire, and building shelters) are signs of culture, or a way of life that depends on intelligence and the ability to communicate knowledge of the culture to succeeding generations.

Inherent in the concept of culture is the concept of language. The ability to communicate enhances the ability of a species to survive, especially in harsh conditions such as an ice age, by enabling then to share survival tactics and to warn each other of danger. Anthropologists think that the development of language was probably one of the most important factors in the appearance of *Homo sapiens*. The results of recent studies indicate that language skills may have been present in primitive humans up to 300,000 years ago, earlier than the appearance of symbolic behavior such as cave paintings and symbols.

Fossils of a second extinct species of human, *H. erectus*, date back to 1.8 million years ago. These humans made sophisticated tools, built shelters, used fire, and probably communicated with language.

25.13 Recent humans: *Homo sapiens*

The earliest fossils of **Homo sapiens** (meaning "wise humans") are about 500,000 years old. They most likely evolved from the *H. erectus* species in Africa, although some scientists contend they evolved simultaneously from *H. erectus* populations in Asia, Europe, and Africa. The oldest *H. sapiens* are not considered to be anatomically modern, that is, to have the same anatomical features of today's humans. This species is therefore referred to as an early or archaic form. In general, these early *H. sapiens* had larger brains, flatter heads, more sloping foreheads, and more protruding brow ridges and faces than today's humans.

The fossil record shows gradual change of the species *H. sapiens*, with the early form evolving over a 75,000-year span to a subspecies of *H. sapiens* called **Neanderthal** (also spelled Neandertal) (nee-AN-dur-thol or nee-AN-dur-tal). This subspecies was named after the Neander Valley in Germany where their fossils were first found. The Neanderthals lived from about 230,000 to 30,000 years ago in Europe and the Middle East.

Figure 25.27 **Cave painting.** Cave paintings, almost always showing animals and sometimes hunters, were made by Cro-Magnon people, our immediate ancestors. These paintings are found primarily in Europe and were made for about 20,000 years, until 8000 to 10,000 years ago.

Compared with modern humans, the Neanderthal people were powerfully built, short, and stocky. Their skulls were massive, with protruding faces, projecting noses, and rather heavy bony ridges over the brows. Their brains were about the same size as those of modern humans, but differently shaped. The frontal lobe, in which a large proportion of "thinking" is done, was flattened in the Neanderthal like that of the earlier hominids. The Neanderthals made diverse tools, including scrapers, borers, spear heads, and hand axes. Some of these tools were used for scraping hides, which they used for clothing. They lived in hutlike structures or in caves. Neanderthals took care of their injured and sick and commonly buried their dead, often placing food and weapons and perhaps even flowers with the bodies. Such attention to the dead strongly suggests that they believed in a life after death. This is the first evidence of the kinds of thought processes—including symbolic thought—that are characteristic of modern *H. sapiens*.

Approximately 10,000 years before the Neanderthal subspecies died out, the "modern" subspecies of *H. sapiens* made their appearance. This modern subspecies (our subspecies) is called **Homo sapiens sapiens**. The early members of this subspecies are the Cro-Magnons and are named after a cave in southwestern France where scientists found some of their fossils.

The **Cro-Magnons** (KRO-MAG-nuns) had a stocky build, much like the Neanderthals, but their heads, brow ridges, teeth, jaws, and faces were much smaller than the Neanderthals and were more similar to today's humans. However, just as modern humans show variation among races, so too did the Cro-Magnons. The Cro-Magnons used sophisticated tools that were made not only from stone but also from bone, ivory, and antler—materials that were not used by earlier peoples. Hunting was an important activity for the Cro-Magnons, evidenced by the abundance of animal bones found with human bones and elaborate cave paintings of animals and hunting scenes (Figure 25.27). The paintings appear to have been part of a ritual to ensure the success of the hunt.

The subspecies of *H. sapiens* that preceded us showed a gradual development of culture and society, which was the foundation for the development of "modern" culture and society. About 10,000 years ago, the last ice age came to a close, the global climate began to warm, and various groups of *H. sapiens sapiens* began to cultivate crops and breed animals for food. Archeologists have uncovered the remains of small, ancient cities, such as those of Jericho shown in Figure 25.28, which give evidence that by 9000 years ago, humans had developed complex social structures. By 5000 years ago, the first large cities and great civilizations appeared, such as those in Egypt and Mesopotamia. The final break occurred with the hunter-gatherer way of life.

Homo sapiens likely evolved from *Homo erectus*. The oldest *H. sapiens* fossils date back 500,000 years. The early form of *H. sapiens* evolved to a subspecies called Neanderthal. The Neanderthals used more sophisticated tools than earlier human ancestors, and also exhibited symbolic thought, as evidenced by the ways in which they treated the dead. Cro-Magnons were early members of "modern" humans, or *Homo sapiens sapiens*.

Figure 25.28 **The beginnings of civilization.** This photo shows the remains of a house in Jericho, which dates to about 7000 B.C. The ruins of Jericho also contain the remnants of city walls and towers, demonstrating that by about 9000 years ago, many of our ancestors had moved away from the hunter-gatherer lifestyle into an agricultural lifestyle.

Key Concepts

Various scientific hypotheses attempt to explain the origin of life.

25.1 Approximately 13 billion years ago, scientists theorize, all the material of the universe condensed into a single small space and then exploded in an event called the "big bang," forming an early generation of stars.

25.1 Approximately 4.6 billion years ago, these stars exploded, forming the Earth and the other planets of the solar system.

25.1 Somehow, in the 300 million years between the time when the Earth cooled and the first bacteria appeared, biochemical evolution became biological evolution.

25.1 The long-held primordial soup theory states that life arose in the sea as elements and simple compounds of the primitive atmosphere reacted with one another to form simple organic molecules such as amino acids and sugars.

25.1 Recently, scientists have begun to reexamine the premises of theory and have proposed alternative hypotheses.

25.2 After the synthesis of complex organic molecules had taken place, self-replicating systems of RNA molecules may have begun the process of biological evolution.

25.2 To be considered life, this "early" genetic material must have been organized within a cellular structure.

Multicellular organisms evolved from the first cells over millions of years.

25.3 Scientists speculate that the first cells to evolve fed on organic materials in the environment.

25.3 Because of limited organic material, the evolution of a pigment system that could capture the energy from sunlight and store it in chemical bonds most likely evolved relatively quickly.

25.3 After the evolution of an energy-capturing pigment system, oxygen-producing bacteria evolved, gradually oxygenating the atmosphere.

25.4 The first unicellular eukaryotes appeared about 1.4 billion years ago during the Proterozoic era.

25.4 Eukaryotic cells likely arose as "preeukaryotic" cells became hosts to endosymbiotic prokaryotes.

Key Concepts, continued

Vertebrates evolved from the chordates, organisms with a stiffening rod down the back.

25.5 Multicellular organisms first appeared about 630 million years ago.

25.5 Many of the phyla of organisms in existence today, except the chordates and the land plants, appear to have evolved during the Cambrian period (520 to 505 million years ago) of the Paleozoic era.

25.5 Chordates and then vertebrates evolved later in the Paleozoic era, with the first vertebrates appearing about 430 million years ago.

25.5 Fishes became abundant and diverse during the Devonian period (410 to 360 million years ago).

25.5 Terrestrial plants appeared about 410 million years ago and flourished during the Carboniferous period (360 to 285 million years ago).

25.6 The amphibians, organisms that spend part of their time on land and part in the water, evolved from the fishes, appearing about 360 million years ago.

25.6 About 300 million years ago, the amphibians gave rise to the first reptiles.

25.6 A major extinction event occurred during the Permian period, wiping out many of the species living at that time.

25.6 The Mesozoic era (250 to 65 million years ago) was a time of the adaptive radiation of the organisms surviving the Permian extinction, dominated by the evolution of the reptiles.

25.6 The reptiles gave rise to the mammals (about 200 million years ago) and the birds (at least 150 million years ago).

25.7 Mammals are warm blooded (endotherms), or able to maintain a constant internal body temperature.

25.7 Other than a few organisms such as the birds, all other living animals are cold blooded (ectotherms); their body temperatures vary with the temperature of the environment.

25.7 Changes in the structure and placement of the mammalian limbs, the structure of the heart, and reproductive strategies resulted in mammals' ability to survive in a variety of climates and to eventually become abundant.

Primates reflect a tree-dwelling heritage.

25.8 Primates, one of the 14 orders of mammals, first appeared in the early Cenozoic era.

25.8 Primates have large brains in proportion to their bodies, binocular vision, and five digits, including an opposable thumb; they exhibit complex social interactions.

25.9 The primates are divided into two suborders: the prosimians (small animals such as lemurs, indris, and aye-ayes) and the anthropoids (a group that includes monkeys, apes, and humans).

25.9 The hominoids (meaning "humanlike") are one of four superfamilies of the anthropoid suborder, which includes apes and humans.

25.9 The other three anthropoid superfamilies include tarsiers, New World monkeys, and Old World monkeys.

25.9 The New World monkeys branched from the evolutionary line leading to the Old World monkeys and the hominoids about 45 million years ago.

25.9 The Old World monkeys split from the hominoids approximately 30 million years ago.

25.9 Ancestors to the apes gave rise to the gibbons, orangutans, chimpanzees, gorillas, and hominids.

25.10 The two critical steps in the evolution of humans were the evolution of bipedalism (walking on two feet) and the enlargement of the brain.

25.10 The earliest hominids belong to the genus *Ardipithecus* and are 4.4 million years old.

25.10 The direct ancestors of humans belong to the genus *Australopithecus*.

25.11 The first species of the human genus *Homo* is *H. habilis*, appearing in Africa about 2.4 million years ago.

25.11 Now extinct, *H. habilis* is considered human because they exhibited an intelligence far greater than their ancestors by making and using tools.

25.12 The second species of *Homo*, *H. erectus*, appeared approximately 1.8 million years ago.

25.12 *H. erectus* lived in Africa, Asia, and Europe.

25.12 *H. erectus* used fire, built shelters, fashioned sophisticated tools, and exhibited culture.

25.13 Early *H. sapiens* probably evolved from *H. erectus* about 500,000 years ago.

Key Terms

anthropoids (AN-thruh poyds) *572*
chordates (KOR-dates) *562*
convergent evolution (kun-VUR-junt EV-uh-LOO-shun) *570*
Cro-Magnons (KRO-MAG-nuns) *582*
endosymbiotic theory (EN-do-SIM-bye-OT-ik) *561*
epochs (EP-uks *or* EE-poks) *559*
eras (EAR-uhs *or* AIR-uhs) *559*

hominids (HOM-uh-nids) *576*
hominoids (HOM-uh-noyds) *573*
Homo erectus (HOE-moe ih REK tus) *581*
Homo habilis (HOE-moe HAB-uh-lus) *581*
Homo sapiens (HOE-moe-SAY-pee-unz) *582*
Homo sapiens sapiens *582*

mammals (MAM-uhls) *563*
Neanderthal (nee-AN-dur-thol *or* nee-AN-dur-tal) *582*
periods *559*
primates *570*
prosimians (pro-SIM-ee-uns) *571*
vertebrates (VER-tuh-bruts *or* ver-tuh-braytes) *564*

KNOWLEDGE AND COMPREHENSION QUESTIONS

1. What is the primordial soup theory, and what does it attempt to explain? Describe an alternative hypothesis that explains the same event. *(Comprehension)*

2. Explain the importance of a cell membrane to the evolution of early cells. *(Comprehension)*

3. What are cyanobacteria, and when did they first appear? Explain their evolutionary significance. *(Knowledge)*

4. When was the Cambrian period, and why was it important? *(Knowledge)*

5. Describe the major events of the Carboniferous period. What were the living conditions on Earth during this time? *(Knowledge)*

6. Summarize the events of the Cenozoic era. When did it begin? *(Comprehension)*

7. What are primates? What two characteristics have helped them to be successful? *(Knowledge)*

8. Distinguish between the prosimians and the anthropoids. Give an example of each. *(Knowledge)*

9. Which statement is true? (Or are they both false?) Explain your answer *(Comprehension)*:
   a. All hominids are hominoids.
   b. All hominoids are hominids.

10. How did *Homo sapiens sapiens* differ from earlier *Homo sapiens*? *(Knowledge)*

KEY
Knowledge: Recalling information.
Comprehension: Showing understanding of recalled information.

CRITICAL THINKING REVIEW

1. Many movies have shown early humans battling ferocious dinosaurs. Is this correct? Explain. *(Analysis)*

2. During the course of human evolution, as climate changes resulted in the disappearance of some forests, early hominids were forced to the savannah. Would that environment favor bipedalism? Why or why not? *(Synthesis)*

3. Extinctions are said to be evidence that evolution has occurred. Explain the reasoning behind this statement. *(Synthesis)*

4. At school, you overhear two of your classmates having a disagreement. One says that humans evolved from the apes. The other says that is not true. Which student is correct and why? *(Application)*

5. Mars rocks land on Earth as often as every three days. By 1999, the hypothesis that life on Earth arose on Mars and traveled to Earth during its early days was considered plausible by many NASA and university scientists. If life did arise on Mars and travel to Earth on rocks, how will that information change current hypotheses regarding the origin and evolution of life? *(Analysis)*

KEY
Application: Using information in new situation.
Analysis: Breaking down information into component parts.
Synthesis: Putting together information from different sources.

1. List the name(s) of the species of organism(s) considered human (hominid) on this evolutionary tree.
2. Are any species on this evolutionary tree considered apes or monkeys?
3. Which species is the direct ancestor of humans?
4. Which species shown on this evolutionary tree is/are living today?

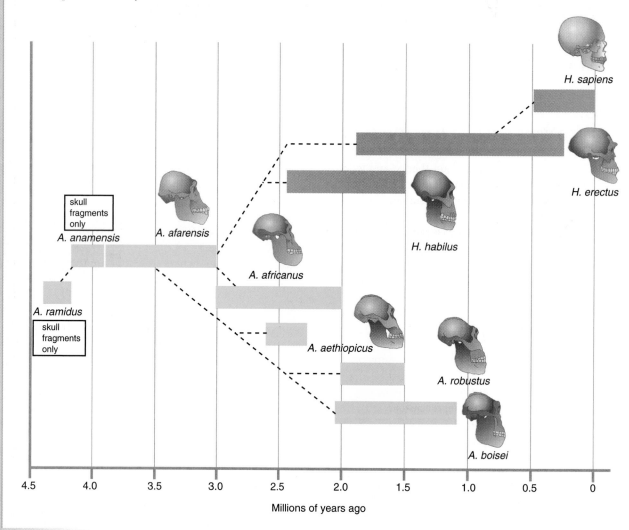

e learning

Location: http://www.jbpub.com/biology

e-Learning is an on-line student review area located at this book's web site www.jbpub.com/biology. The review area provides a variety of activities designed to help you study for your class and build your learning portfolio.

Review Questions The review questions test your knowledge of the important concepts and applications in each chapter. The review provides feedback for each correct or incorrect answer. This is an excellent test preparation tool.

Figure-Labeling Exercises Sharpen your visual thinking by matching terms and labels for important illustrations.

Flash Cards Studying biology requires learning new terms. Virtual flash cards help you master the new vocabulary for each chapter.

Just Wondering As you read and study from this text, you may find that you have unanswered questions. Through this site you can ask the author, Sandy Alters, your "just wondering" questions.

 Do you prefer the speed and reliability of a CD-ROM? All of the features contained on the eLearning portion of the web site are also available on the Student CD-ROM.

The Use of Animals in Research

The masked intruders burglarized laboratories on the Twin Cities campus, smashing computers and microscopes in the process. Did they steal top secret formulas, expensive equipment, or potent strains of bacteria? No, this attack on the University of Minnesota campus in April, 1999 had nothing to do with stealing scientific secrets or with biological warfare. It was an attack by the Animal Liberation Front (ALF) and yielded a "take" of 27 pigeons, 48 mice, 36 rats, and 5 salamanders. The goal of the ALF was to "free" the animals and find new homes for them where they would no longer be subjects in scientific experiments.

Animal rights groups such as the ALF believe that all species are equal and that no animals should be used in research experiments. Such groups insist that it is wrong to use animals for human gain. The largest animal rights organization in the world, People for the Ethical Treatment of Animals (PETA), asserts that animals should not be eaten, worn, experimented on, or used for entertainment. Many scientists and other people who disagree with animal rights activists argue that medical progress cannot be made without animal research, citing links between animal research and testing and the medical treatments humans receive. There are also some people who take a compromise position: They approve only of some kinds of research on certain animals.

Scientists use animals in research (often referred to as "whole-body" research) in studies that could not be conducted with use of tissue cultures or other research methods alone, as when they study the regulation of bodily processes or the functioning of systems. Computer modeling strategies cannot be used unless appropriate animal data are first collected. (Scientists cannot ask a computer to answer a question for which it has no data.) Scientists' choices of which animals to use are also important. If a particular line of animal research has applications for humans, vertebrates are generally used because they model human systems best. Rodents (such as rats and mice) are used in 90% of all such animal research. At the University of Minnesota, the animals taken from the locked laboratories were involved in research on neurological disorders such as Alzheimer's and Parkinson's diseases, brain tumors, and other cancers.

People who favor the use of animals in scientific and medical research agree that animals should not suffer unnecessarily. Animal rights groups often claim that there are no laws governing the treatment of animals in medical research. However, there are two governmental agencies that provide regulations scientists must follow in animal research: the U.S. Department of Agriculture (USDA) and the Department of Health and Human Services

A laboratory chimpanzee and his caretaker. There are strict regulations concerning the care of animals used in medical experiments.

Approval ratings for experiments performed on mice

- Approve
- Disapprove
- Don't Know

Purpose of Experiment

To develop a new drug to cure leukemia in children

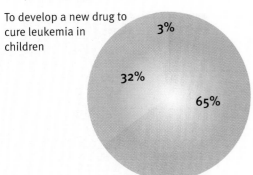

3%
32%
65%

To enable scientists to study how the sense of hearing works

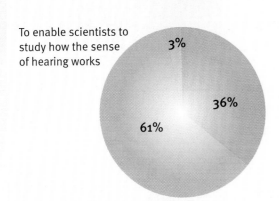

3%
36%
61%

To test whether an ingredient for use in cosmetics will be harmful to people

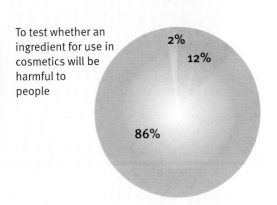

2%
12%
86%

These are the results from a poll taken in Great Britain in 1999. Approval for experimentation on mice (when mice were subjected to pain, illness or surgery) depended on the purpose of the experiment. (Source: *New Scientist,* May 22, 1999, www.newscientist.com)

(DHHS). If scientists do not follow these regulations, they become ineligible for any type of federal funding for their research.

To enforce the regulations regarding animal research, USDA representatives (veterinarians and occasionally animal care technicians) make unannounced semiannual visits to facilities that use warm-blooded animals (other than rodents) in biomedical research, teaching, or exhibits. In addition, they require an annual review of all animal activities. Protocols require that researchers document their training in animal welfare and sign written statements that any proposed research is not unnecessarily duplicative. If the proposed activity has the possibility of causing "unrelieved pain or distress," the researcher must provide a literature search to support the argument that there is no alternative approach that might have less potential for causing distress. (Only 6% of animal experimentation involves pain. The goal of such experiments is often to develop safer and more effective pain relievers or anesthetics.) The USDA also requires that any research, teaching, or exhibits involving warm-blooded animals be reviewed and approved by an Institutional Animal Care and Use Committee (IACUC) composed of a veterinarian, a chairperson, a nonscientist, and a member with no affiliation to the institution.

Many scientists state that animal research has been essential for the development of therapies for treating heart problems, cancer, and other fatal and serious diseases, as well as anesthetics used in surgery. Animal rights activists oppose these claims and contend that improved nutrition, sanitation, and other behavioral and environmental factors are responsible for the decline in deaths since 1900 from the most common infectious diseases. They hold that medicine has had little to do with increased life expectancy. Additionally, they suggest that medical research can be advanced with human studies and by means other than the use of animals.

What is the issue here, and what is your response to this issue?

www.jbpub.com/biology

To make an informed and thoughtful decision, it may help you to follow these steps.

1. Identify the issue described here using either a statement or a question. (An issue is a point on which various persons hold differing views.)
2. Write your immediate reaction to the issues raised. How do you think this issue could be or should be resolved?
3. Collect new information about the issue. State the sources of that information. Explain why you believe the information to be accurate. Also determine if the information expresses a particular point of view or is biased in any other way. (Guidelines for doing efficient web searches for reliable information are available on the BioIssues website — **www.jbpub.com/biology**)
4. Determine which individuals, groups, or organizations have a stake in the issue. What does each stand to gain or lose depending on the outcome of the issue?
5. List possible outcomes (resolutions) of the issue. List the pros and cons of each outcome.
6. Which outcome do you think would be best and why? Note whether your opinion differs from or is the same as what you wrote in Step 2. Give reasons for your views and for any changes in them based on the additional information you have collected and the analysis you have done.

Viruses and Bacteria

The statistics are astounding! A three-year survey at Rutgers University in New Jersey, published in February of 1998, found that 60% of female students are harboring a sexually transmissible virus that may be associated with cancer of the cervix. Computer color-enhanced, these virus particles (formally known as human papillomavirus, or *HPV*, are shown in the photo as red spheres speckled with green. The human cell they are destroying is yellow.

Human papillomavirus causes a sexually transmissable disease (STD) commonly known as *genital warts*. HPV comes in 60 different varieties. Infection with any one of the 60 types can lead to an outbreak of warts, although some are more likely to cause warts than others are. Ironically, the types that most often result in warts are the least dangerous. Infection with HPV types such as HPV 16 rarely cause warts or infection but are the most likely to be associated with cancer and the least likely to be detected.

Two other viral diseases transmissible by sexual contact are genital herpes and human immunodeficiency virus (HIV) infection. Unlike sexually transmissible diseases spread by bacteria, viral STDs cannot be cured. The drugs available to threat these diseases only help ease the symptoms but cannot destroy the viruses. Why are these diseases different in that regard? What makes a virus so hard to control?

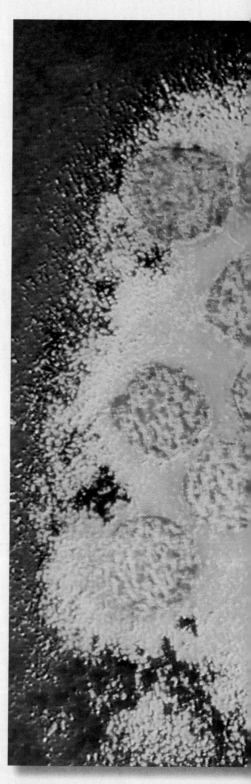

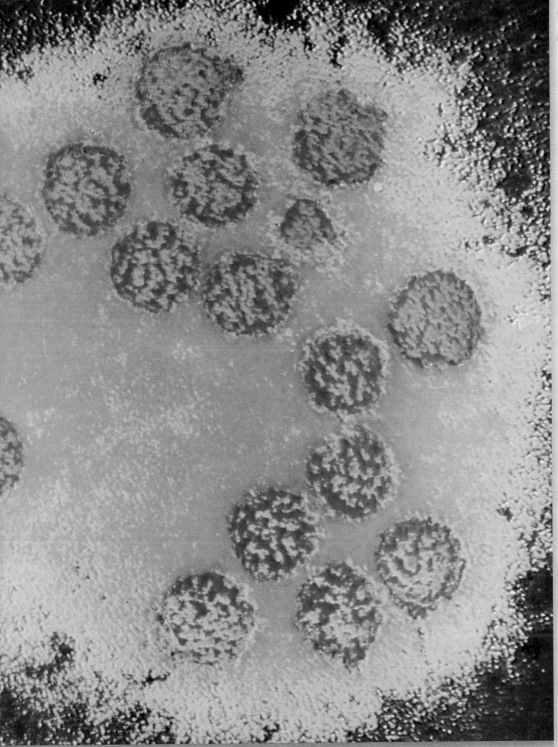

26

Organization

26.1 The discovery of viruses

Viruses are *infectious agents*, which means that they enter living organisms, causing disease. Although they invade living things and cause cells to make more viruses, the viruses themselves are not living! They do not have a cellular structure, which is the basis of all life. They are nonliving *obligate parasites*, which means that viruses cannot reproduce outside of a living system. They must exist in association with and at the expense of other organisms. Unfortunately, that "other organism" may be you!

At the end of the nineteenth century, several groups of European scientists working independently first realized that viruses existed. As they filtered fluids derived from plants with tobacco mosaic disease and cattle with hoof-and-mouth disease, the scientists discovered that the infectious agents passed right through the fine-pored filters they used, which were designed to hold back bacteria. They concluded that the infectious agents associated with these diseases were *not* bacteria—they were too small. As they studied the filtrate containing these mysterious agents, the scientists also discovered that the disease-causing agents could multiply only within living cells. These infection agents, they hypothesized, must lack some of the critical "machinery" cells use to reproduce.

For many years after their discovery, viruses were regarded as very primitive forms of life, perhaps the ancestors of bacteria. Today, scientists know that this view is incorrect: Viruses are not living organisms. The true nature of viruses became evident in the 1930s after the groundbreaking work of an American scientist, Wendell Stanley. Stanley prepared an extract of tobacco mosaic virus (TMV), purified it, and studied its chemical composition. His conclusion: TMV was a protein, and he was partially right. Scientists later discovered that TMV also contains ribonucleic acid (RNA). In the late 1930s, with the development of the electron microscope, scientists were able to see the virus that Stanley purified as shown in **Figure 26.1a**.

Because viruses are not living things, they are not included in the five kingdoms of life. Scientists have, however, devised a classification scheme for viruses that

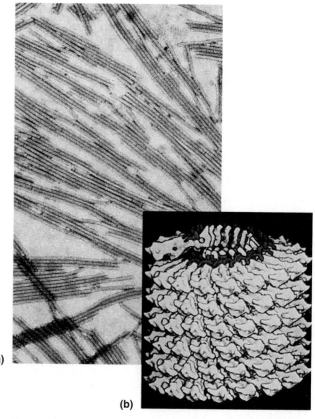

(a)

(b)

Figure 26.1 Tobacco mosaic virus. (a) An electron micrograph of purified tobacco mosaic virus. (b) Computer-generated model of a portion of tobacco mosaic virus. An entire virus consists of 2130 identical protein molecules—the yellow knobs—which form a cylindrical coat around a single strand of RNA (*red*).

is based on the host they infect. Viruses are first grouped according to whether they infect plants, animals, or bacteria. Further classification usually focuses on differences in morphology (shape and structure), type of nucleic acid, and manner of replication.

Scientists first discovered viruses around 1900. They were at first considered primitive forms of life, but scientists realized about 30 years after their discovery that viruses are acellular, and therefore nonliving.

26.2 The structure and classification of viruses

Viruses infect primarily plants, animals, and bacteria. A specific virus can infect only a certain species. Thus, you cannot be infected by a bacterial virus (bacterial viruses can only infect bacteria), nor can your dog catch your cold. Some viruses, however, can infect more than one species. For example, both monkeys and people can be infected with HIV, although the infection is usually more serious in humans.

Each virus has its own unique shape, as shown in Figure 26.2, but all contain the same basic parts: a nucleic acid **core** (either DNA or RNA) and a protein "overcoat" called a **capsid**. The structure of the TMV is shown in Figure 26.1b, and illustrates one way that a virus is put together. This virus is helical, with its single strand of RNA coiled like a spring, surrounded by a spiraling capsid of protein molecules. Many viruses have another chemical layer over the capsid called the **envelope**, which is rich in proteins, lipids, and carbohydrate molecules. Figure 26.3a is an electron micrograph of a typical enveloped virus, the causative agent of herpes. Its structure is illustrated in Figure 26.3b.

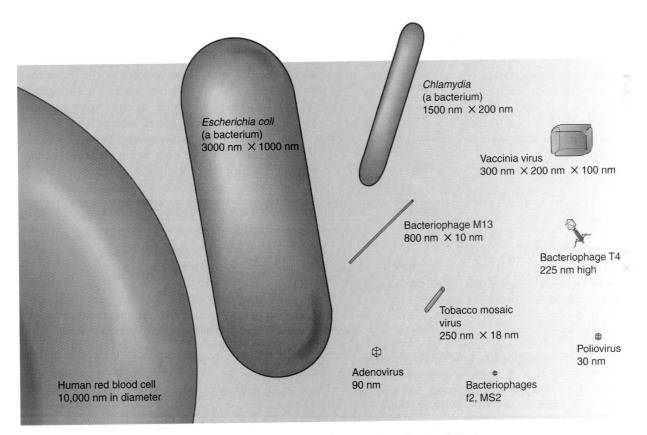

Chlamydia
(a bacterium)
1500 nm × 200 nm

Escherichia coli
(a bacterium)
3000 nm × 1000 nm

Vaccinia virus
300 nm × 200 nm × 100 nm

Bacteriophage M13
800 nm × 10 nm

Bacteriophage T4
225 nm high

Tobacco mosaic virus
250 nm × 18 nm

Poliovirus
30 nm

Adenovirus
90 nm

Bacteriophages
f2, MS2

Human red blood cell
10,000 nm in diameter

Figure 26.2 **The shapes and sizes of viruses.** The sizes of various viruses are shown here in relation to bacteria and a human red blood cell. Dimensions are given in nanometers.

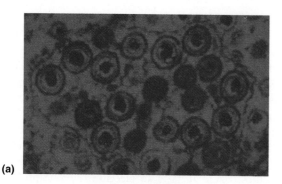

(a)

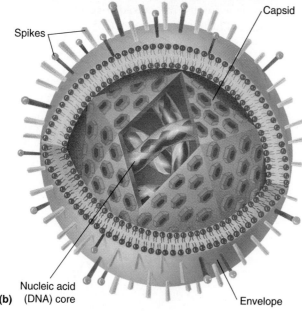

Spikes

Capsid

Nucleic acid
(b) (DNA) core

Envelope

Figure 26.3 **Structure of a typical virus.** (a) Electron micrograph of the herpesvirus. (b) Structure of the herpesvirus.

It is hard to conceptualize how small viruses are, but Figure 26.2 helps by showing the size of a few viruses relative to the size of a bacterium and a human red blood cell. Some viruses, such as the poliovirus, are as small as the width of the plasma membrane on a human cell. The largest viruses are barely visible with a light microscope. Most viruses can be seen only by using an electron microscope.

> Viruses are made up of a nucleic acid core surrounded by a protein covering called a capsid. Some viruses also have an additional covering, or envelope.

26.3 Viral replication

Viruses cannot multiply on their own. They must enter a cell and use the cell's enzymes and ribosomes to make more viruses. This process of viral multiplication within cells is called *replication*. Various patterns of viral replication exist. Some viruses enter a cell, replicate, and then cause the cell to burst, releasing new viruses. This pattern of viral replication is called the **lytic cycle** (LIT-ic). Other types of viruses enter into a long-term relationship with the cells they infect, with their nucleic acid replicating as the cells multiply. This pattern of viral replication is called the **lysogenic cycle** (lye-suh-JEN-ik).

The Lytic Cycle

The process of viral replication has been studied most extensively in bacteria because bacteria are easier to grow in the laboratory and to infect with viruses than are plant or animal cells. Many bacterial viruses (usually called *bacteriophages* (back-TEE-ree-oh-FAY-gez) or simply *phages* follow a lytic cycle pattern of replication.

As shown in Figure 26.4, a bacteriophage first attaches to a receptor site on a bacterium **1**. The virus then injects its nucleic acid into the host cell while its protein capsid is left outside the cell **2**. Next, the viral genes take over cellular processes and direct the bacterium to produce viral "parts" that will be used to assemble whole viruses **3**. After their manufacture, these strands of nucleic acid and proteins are assembled into mature viruses **4** that *lyse* (LICE), or break open, the host cell **5**. Each bacterium releases many virus particles. Each "new" virus is capable of infecting another bacterial cell.

Some animal viruses infect cells in a manner similar to bacterial virus infection, but they enter animal cells by endocytosis and must be uncoated before they can cause the cell to manufacture viruses. These cells may die as new virus particles are released in a lytic infection, or they may survive if virus particles are slowly budded from the cell by a process similar to exocytosis. This "slow budding" causes a persistent infection.

Plant cells are somewhat protected from viral infection by their rigid cell walls and protective outer, waxy cuticles. Viruses can enter plants only if they are damaged or if other organisms, such as sucking insects or fungi, assist them.

The Lysogenic Cycle

Some viruses, instead of killing host cells, integrate their genetic material with that of the host (Figure 26.4, **2a**). Then, each time the host cell reproduces **2b**, the viral

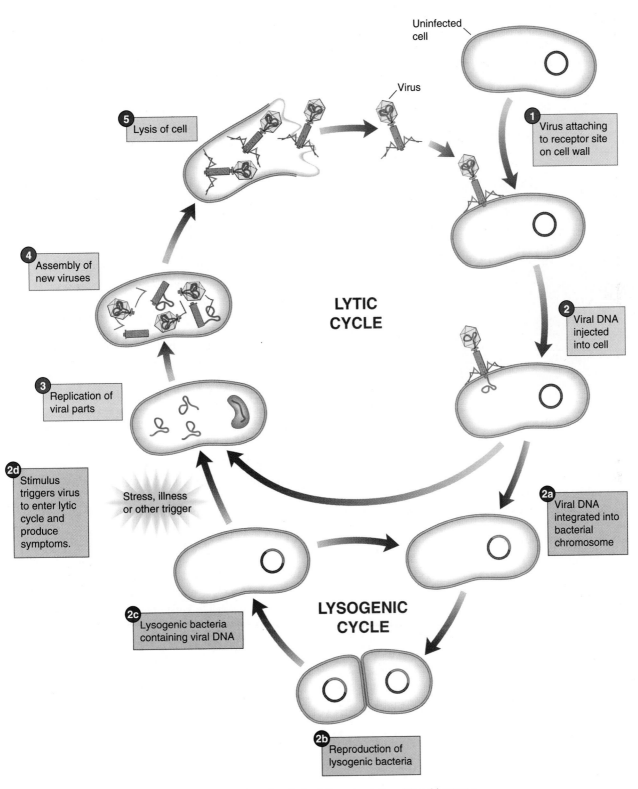

Figure 26.4 **Viral replication: The lytic and lysogenic cycles.** In the lytic cycle, viral nucleic acid enters a cell and causes it to burst, releasing new virus. In the lysogenic cycle, the virus is integrated into the host cell, and the virus's nucleic acid is replicated as the host cell multiplies.

Just Wondering . . . *Real Questions Students Ask*

How come scientists haven't found a cure for the common cold?

Although we often refer to the causative agent of colds as `"the cold virus," there are actually more than 200 viruses that cause colds. Rhinoviruses and coronaviruses top the list, but within these viral groups alone there are well over 100 cold-causing viral types. Developing a vaccine against 200 agents of infection is not practical. However, in late 1998, a California-based drug company started clinical trials on a drug developed to inactivate all types of rhinovirus, the most common cause of colds. This drug is similar to the protease inhibitors that help battle HIV infection. It inactivates a key enzyme any rhinovirus needs to replicate.

www.jbpub.com/biology

Another line of investigation to find a cold cure is to develop an antibody that would block viral receptor sites on human cells. Although more than 200 viruses cause colds, most of them use the same two receptor sites on cells of the upper respiratory tract mucosa. As yet, this cold-fighting antibody has not been developed.

Antibiotics, those "magic bullets" against bacterial diseases, do not work against viruses. Antibiotics attack bacterial structures and metabolic processes, which is the reason they cause little harm to human cells. Viruses are much different from bacteria, and like human cells, viruses are *not* affected by these infection-fighting drugs.

Your immune system helps you fight an invasion of cold viruses you have already encountered. You develop immunity against colds as you age. Each time you contract a cold, you develop immunity to the particular cold virus that infected you. Therefore, by the time you turn 60 years of age, you should be catching only one cold per year, on average, compared with the three or four you probably contracted when you were a child.

Some medications can reduce cold-related symptoms. Decongestants can lessen nasal congestion and antihistamines may help dry up a "runny" nose. (Saline nasal sprays have recently been shown to have no effect on congestion.) The only treatment studied that appears to reduce the duration of cold symptoms is zinc lozenges when taken as cold symptoms first appear. Health care providers also urge cold-sufferers to rest, get sufficient sleep, and drink plenty of fluids (avoiding caffeine and alcohol).

---

nucleic acid is replicated as if it were a part of the cell's genetic makeup ②c. In this way, the virus is passed on from cell to cell. Infection of this sort is called a *latent infection*. These integrated latent genes may not cause any change in the host for a long time. Then, triggered by an appropriate stimulus, the virus may enter a lytic cycle and produce symptoms ②d.

The herpes simplex virus causes latent infections of the skin. Herpes nucleic acid remains in nerve tissue (sensory ganglia) without damaging the host until a cold, a fever, or other factor such as ultraviolet radiation from the sun acts as a trigger, and the cycle of cell damage begins. This "damage" manifests as cold sores or fever blisters. The herpes zoster virus (the chickenpox virus) can also act in the same way. This virus may remain latent in the nerve tissue of a person who has had chickenpox, only to be triggered at a later time, causing the painful nerve disorder *shingles*.

During a lytic pattern of viral infection, a bacterial virus injects its nucleic acid into a host cell, whereas an animal virus enters by endocytosis. The viral nucleic acid directs the cell to produce "new" viral nucleic acid and protein coats. After assembly of these parts, the bacterial virus particles cause the cell to burst open, releasing them. Animal viruses often leave the cell by slow budding.

During a lysogenic pattern of viral infection, a virus integrates its genetic material with that of a host and is replicated each time the host cell reproduces.

26.4 Viruses and cancer

Five viruses are initiators of cancer: hepatitis B virus, human T-cell lymphotropic/leukemia virus, human papillomavirus, human cytomegalovirus, and the Epstein-Barr virus. These viruses are listed in **Table 26.1.**

Table 26.1

Tumor Viruses

Virus	Cancer Type
Hepatitis B virus (HBV)	Primary liver cancer
Human T cell lymphotropic/ leukemia virus (HTLV)	Leukemias and lymphomas
Cytomegalovirus (CMV)	Kaposi's sarcoma
Human papillomavirus (HPV)	Vaginal/vulval/cervical cancer
	Penile cancer
Epstein-Barr virus (EBV)	Burkitt's lymphoma
	Nasopharyngeal cancer

Hepatitis B Virus (HBV)

The hepatitis B virus is transmissible in blood or blood products (such as blood serum or plasma) and by contaminated needles or syringes. Therefore, the individuals most likely to contract this virus are health care workers (although their precautions are strict), intravenous (IV) drug users who share contaminated needles, and individuals having sexual contact with infected persons. (Hepatitis is not usually classified as a sexually transmissible disease, however, because sexual contact is not the primary route of transmission.) In addition, an infected pregnant woman can pass this virus to her developing fetus.

The hepatitis B virus causes a serious infection of the liver. Individuals with prolonged hepatitis B liver disease, especially those who develop cirrhosis, are at risk for developing liver cancer. (Cirrhosis is a condition in which liver tissue is destroyed and is replaced by scar tissue.) Liver cancer as a result of hepatitis B infection is relatively rare in the United States, but is quite common in developing countries of Africa and Asia. This cancer is usually fatal; most patients die within six months to a year after diagnosis.

Human T-cell Lymphotropic/Leukemia Virus (HTLV)

The human T-cell lymphotropic/leukemia virus causes adult T-cell leukemia/lymphoma (ATLL). This disease is rare in the United States, but is common in Japan and the Caribbean. The cancers caused by HTLV are T-cell leukemias or lymphomas.

Leukemia is a disease of the red bone marrow (the substance that produces the body's blood cells and platelets), which results in the manufacture of a greater than normal number of white blood cells that are immature, abnormal, and unable to perform their infection-fighting roles. These cells are shown in **Figure 26.5a** contrasted with normal white blood cells in Figure 26.5b.

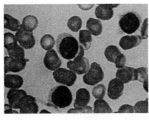 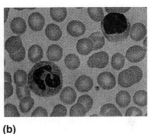

(a)   **(b)**

Figure 26.5 **A hairy cell leukemia blood smear contrasted with a normal blood smear.** (a) This micrograph of a hairy cell leukemia blood smear shows three abnormal white blood cells among red blood cells. (b) This micrograph of a normal blood smear shows one neutrophil and one monocyte, both types of white blood cells, among red blood cells.

Lymphoma is a malignant condition of the lymphoid tissue—the fluid or tissues of the lymphatic system (see Chapters 10 and 11). T lymphocytes, or T cells, are a type of white blood cell that develops in the bone marrow but matures in the thymus, an organ of the lymphatic system. T cells are integral to the immune response, your defense against disease (see Chapter 11).

ATLL spreads quickly in its victims and often results in an enlarged liver, spleen, and lymph nodes. Researchers are still uncertain about how the HTLV is transmitted, but they think possible routes may involve sexual activity, the sharing of needles and syringes among intravenous drug users, and transfusion with contaminated blood. These pathways of infection are the same as those of HBV and the human immunodeficiency virus (HIV), which causes AIDS.

Cytomegalovirus

Although HIV is a "cousin" to HTLV (it was previously known as "HTLV-III"), HIV is not a tumor virus itself. However, it causes a breakdown of immunity that leaves its victims susceptible to other infections and cancers. The most common of the "AIDS cancers" is Kaposi's sarcoma, or skin cancer characterized by flat or raised red or purplish lesions. The cytomegalovirus (CMV) is thought to be the initiator of this cancer.

Human Papillomavirus (HPV)

More than 60 types of human papillomaviruses have been isolated. Some of these viruses initiate only benign tumors, such as warts of the hands and feet. (Warts on the soles of the feet are called plantar warts.) Other human papillomaviruses cause the sexually transmissible disease described in the chapter opener commonly known as genital warts (or more formally known as *Condylomata acuminata*). At least two types of these papillomaviruses (referred to as HPV-16 and HPV-18) are strongly linked to the development of cancer, particularly cervical carcinoma. (The cervix is the tissue surrounding the opening to the uterus, or womb.) Although genital warts may appear on the cervix, within the vagina, or on the labial tissues in women, researchers think that the cervix has an area of tissue that is particularly vulnerable to viral infection. Researchers also suggest that the development of cancers of the cervix linked to HPV infection may be due to an interaction of factors, and they speculate that infection with other viruses, as well as smoking, may act as tumor initiators, with HPV types 16 or 18 functioning as promoters. The results of studies reported in late 1998 support the hypothesis that cigarette smoking may be an important factor in the development of virus-related abnormalities in cervical cells. Although past studies showed conflicting results about whether HPV is linked to prostate cancer in men, the results of recent studies show that the virus plays no role in causing this disease.

People contract the papillomavirus during sexual intercourse with an infected partner. The risk of infection with the virus rises as a person's number of sexual partners rises. This means that a woman's risk of developing cervical cancer rises as well. Also, women who become sexually active at an early age are at a greater risk than women who become sexually active at an older age. Scientists are unsure whether the tissues of a young woman are simply more vulnerable or whether women who are sexually active at an early age are more likely to have a greater number of partners than other women.

Epstein-Barr Virus (EBV)

Cancers caused by the Epstein-Barr virus are rare in the United States. Americans are most familiar with the noncancerous condition it causes: mononucleosis. The virus is found in saliva because it is shed from cells lining the nose and throat. For this reason, mononucleosis is often called the "kissing disease," but the virus can

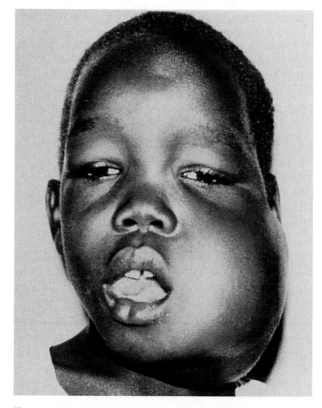

Figure 26.6 Burkitt's lymphoma. This cancer of the lymphatic system is usually manifested as a large mass in the jaw or abdomen.

also be transmitted on contaminated cups, eating utensils, and similar objects.

The cancer-causing effects of the Epstein-Barr virus are seen in people living in Africa and China. In fact, cancers of the nose and upper throat linked to this virus are one of the leading causes of cancer deaths in China. In Africa, the more common cancer in which the EBV is implicated is Burkitt's lymphoma, a cancer of the lymph system (Figure 26.6).

Five viruses are associated with the development of particular cancers. HBV causes a serious infection of the liver that can lead to liver cancer. HTLV causes adult T-cell leukemia/lymphoma. Leukemia is cancer of the red bone marrow and lymphoma is cancer of the lymphoid tissue. CMV is thought to be the initiator of Kaposi's sarcoma, a type of skin cancer common to AIDS patients. Some types of HPV are strongly linked to the development of cervical cancer and cancers of the genitals. Cancers caused by EBV are rare in the United States, but are linked to deadly cancers in Africa and China.

CRITICAL THINKING REVIEW

5. If you were to graph the growth of a population of bacteria, it would look like an S-shaped curve. Describe how a population of bacterial cells grows, showing how it would result in an S-shaped curve when graphed. (*Application*)

Visual Thinking Review

The lytic cycle of viral replication is shown in the illustration. Describe what is happening at each numbered stage of the cycle.

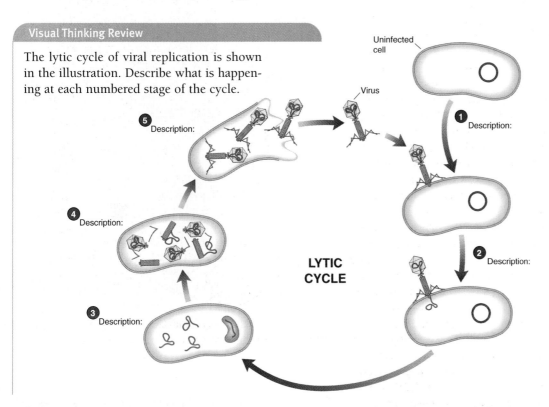

LYTIC CYCLE

Uninfected cell

Virus

1 Description:

2 Description:

3 Description:

4 Description:

5 Description:

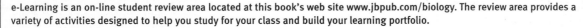

e-learning

Location: http://www.jbpub.com/biology

www.jbpub.com/biology

e-Learning is an on-line student review area located at this book's web site www.jbpub.com/biology. The review area provides a variety of activities designed to help you study for your class and build your learning portfolio.

Review Questions The review questions test your knowledge of the important concepts and applications in each chapter. The review provides feedback for each correct or incorrect answer. This is an excellent test preparation tool.

Figure-Labeling Exercises Sharpen your visual thinking by matching terms and labels for important illustrations.

Flash Cards Studying biology requires learning new terms. Virtual flash cards help you master the new vocabulary for each chapter.

Just Wondering As you read and study from this text, you may find that you have unanswered questions. Through this site you can ask the author, Sandy Alters, your "just wondering" questions.

Do you prefer the speed and reliability of a CD-ROM? All of the features contained on the eLearning portion of the web site are also available on the Student CD-ROM.

Protists and Fungi

"Please don't drink the water!" It might seem like an unnecessary warning if you were camping along this pristine-looking mountain stream. However, after an invigorating hiking trip, you could return home with nausea, cramps, bloating, and diarrhea—all symptoms of giardiasis, or "hiker's diarrhea." The culprit of this uncomfortable ailment is a single-celled organism, a protist no larger than a red blood cell: *Giardia lamblia*.

The organism *G. lamblia* is found throughout the world, including all parts of the United States and Canada. It occurs in water, including the clear water of mountain streams and the water supplies of some cities. In addition to humans, it infects at least 40 species of wild and domesticated animals. These animals can transmit *G. lamblia* to humans by contaminating water with their feces.

Flagella protrude from one end of *G. lamblia*, allowing it to move along the intestinal wall of its host. This motile form of the protist exists only while inside the body of its victim. Dormant, football-shaped cysts are expelled in the feces of a host animal. These cysts can survive for long periods of time outside their hosts—especially in the cool water of mountain streams. When ingested by other hosts, the cysts develop into their motile, feeding form.

What should you do to prevent infection by *G. lamblia* when hiking or camping? First, never drink untreated water, no matter how clean it looks. Because *G. lamblia* is resistant to usual water treatment agents, such as chlorine and iodine, you should boil the water you drink for at least a minute. Better still, *do not* drink the water—bring your own water to be safe.

Organization

Protists, although primarily single-celled, are an extremely diverse kingdom of eukaryotes.

Fungi are multicellular eukaryotic organisms that feed on dead or decaying organic material.

Protists, although primarily single-celled, are an extremely diverse kingdom of eukaryotes.

27.1 An overview of the protists

The **protists** (Kingdom Protista) are a varied group of eukaryotic organisms that live in moist or aquatic environments. Many are single-celled, although some phyla of protists include descendants that are multicellular or colonial (single cells that live together as a unit). Although protists are *not* animals, plants, or fungi, within this kingdom there are animallike, plantlike, and funguslike organisms. These groupings are useful to help organize this diverse array of organisms.

The animallike protists are called **protozoans** (PRO-tuh-ZOE-unz) and are considered animallike because they are heterotrophs: They take in and use organic matter for energy. *G. lamblia,* for example, is a flagellated protozoan (phylum Archaeprotista), see **Figure 27.1**. The plantlike protists are the **algae** (AL-jee), and include *diatoms*. These organisms are plantlike because they are eukaryotic photosynthetic autotrophs: They manufacture their own food using energy from sunlight. The funguslike protists consist of *plasmodial slime molds* and *water molds*. The plasmodial slime molds live on dead and decaying material. The water molds extend funguslike threads (hyphae) into organisms, release digestive enzymes, and absorb the "predigested" food.

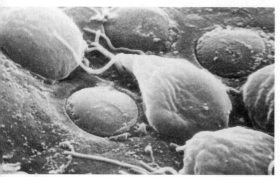

Figure 27.1 *Giardia lamblia* moving along the small intestine.

Although many protists are single celled, these organisms are incredibly different from prokaryotic single-celled organisms, the bacteria. The single-celled protists are much larger than bacteria, having approximately 1000 times the volume of bacteria, and they contain typical eukaryotic cellular organelles. This compartmentalization of eukaryotic cells by membrane-bounded organelles increases their organization to a level much more complex than that of bacteria and allows single-celled eukaryotic organisms to carry out cell functions that support their larger cell volume.

Most of the protist phyla are unicellular eukaryotes, although some phyla contain multicellular forms. Protists can be grouped as animallike (protozoans), plantlike (algae), or funguslike (plasmodial slime molds and water molds) according to their mode of nutrition.

27.2 Animallike protists: Protozoans

The animallike protists, or protozoans, obtain their food in diverse ways, although all are heterotrophic. Protozoans' food-getting characteristics are linked to the ways that they move and provide a means by which they can be grouped. Five major groups of protozoans (consisting of six phyla) are amebas and cellular slime molds, forams, flagellates, ciliates, and sporozoans.

Amebas and Cellular Slime Molds (Rhizopoda)

Amebas (uh-MEE-buhs) appear as soft, shapeless masses of cytoplasm. Within the cytoplasm lie a nucleus and other eukaryotic organelles. The ameba's cytoplasm continually flows, pushing out certain parts of the cell while retracting others. These cytoplasmic extensions are called *pseudopods* (SUE-doe-pods; from the Greek meaning "false feet") and are a means of both locomotion and food procurement. In fact, these cell extensions give the amebas their phylum name, Rhizopoda (rye-zoh-POH-dah), which means "rootlike feet." Shown in **Figure 27.2**, the pseudopods simply stream around the ameba's prey, engulfing it within a *food vacuole*. En-

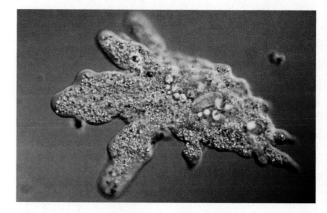

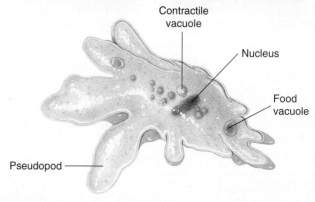

Figure 27.2 **Structure of *Amoeba proteus*.** The pseudopods of this ameba are constantly flowing and changing as it moves and takes in food.

zymes digest the contents of the vacuole, which are then absorbed into the cytoplasm to be further broken down for energy and biosynthesis. A second type of vacuole, the *contractile vacuole,* pumps excess liquid from the ameba. Amebas reproduce by binary fission: The nucleus reproduces by mitosis and then the cell splits in two.

Amebas are abundant throughout the world in fresh water and saltwater, as well as in the soil. Many species are parasites of animals, including humans, and can cause diseases such as amebic dysentery, an infection of the digestive system that produces a diarrhea containing blood and mucus. Although amebic dysentery is a disease associated with poor sanitation and is found primarily in the tropics, medical researchers estimate that about 2 million Americans are infected with the causative agent, *Entamoeba histolytica* (EN-tah-ME-bah HISS-tow-LIH-tih-cah).

New to the Rhizopoda are the **cellular slime molds.** Most of their lives, the cellular slime molds, shown in **Figure 27.3,** look and behave like amebas, moving along and capturing bacteria by means of pseudopods ❶. For this reason they have been recently reclassified with the single-celled amebas. They do, however, have characteristics that are plantlike and funguslike as well as animallike, so their classification is continually argued.

At a certain phase of their life cycle, which is often triggered by a lack of food, the individ-ual organisms move toward one another ❷ and form a large, moving mass called a *slug* ❸. (The cellular form produces a chemical attractant, acrasin, which causes the cells to aggregate.) The slug eventually stops moving and begins to rise vertically ❹, transforming into a fruiting body ❺. The tips of the fruiting bodies contain dormant cystlike forms of the amebalike cells and are called *spores.* Some of these spores fuse and undergo a type of sexual reproduction before being released ❻; others do not. Each spore becomes a new "ameba" if it falls onto a suitably moist habitat such as damp soil, decaying plant material, dung, or fallen logs ❼. The amebas begin to feed and continue the life cycle.

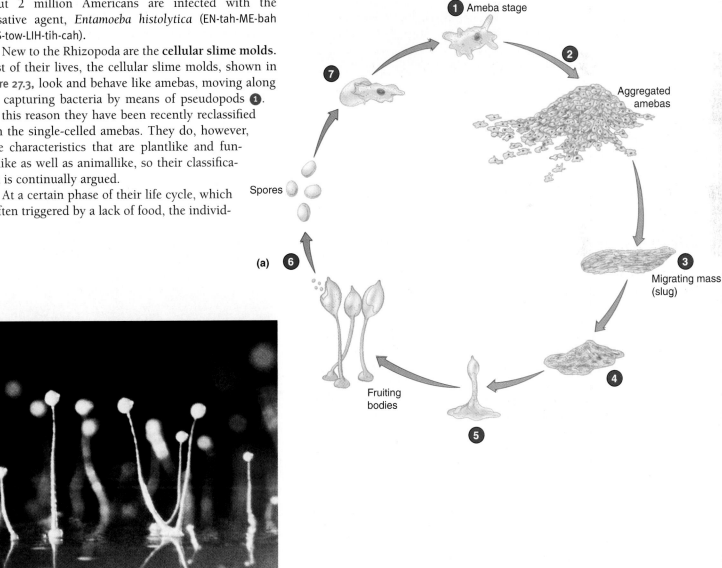

Figure 27.3 (a) Life cycle of a cellular slime mold. (b) Many fruiting bodies together.

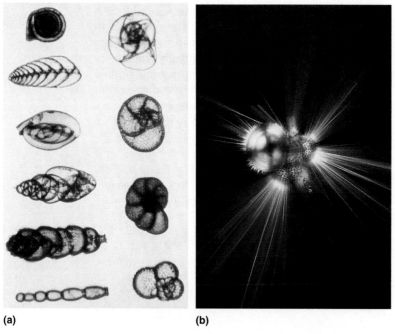

(a) **(b)**

Figure 27.4 **Foraminiferan shells.** (a) These beautiful foraminiferan shells are made out of calcium carbonate. (b) A live foraminaferan.

classified into multiple phyla based on the presence or absence of mitochondria.

All flagellates have a relatively simple cell structure. They do not have cell walls or protective outer shells as some of the amebas or ciliates do. They also have no complex internal digestive system of organelles as the ciliates do. A flagellate simply absorbs food through its cell membrane, sometimes using its flagella to ensnare food particles.

The flagellates are generally found in lakes, ponds, or moist soil where they can absorb nutrients from their surroundings. The colonial flagellate *Codosiga* shown in Figure 27.6a, for example, consists of groups of cells that are often found anchored to the bottom of a lake or pond by a cellular stalk. The flagella create currents in the water that draw food toward the cells. The flagellate *Trichonympha* (TRICK-eh-NYM-fah), shown in Figure 27.6b, lives a protected life in the gut of termites, digesting wood particles the termite eats. Many flagellates are found living within other organisms; some of these relationships are not harmful to the hosts, but other relationships are. Figure 27.6c shows the flagellate

Foraminifera (Granuloreticulosa)

Some amebas secrete shells that cover and protect their cells. The *foraminifera* (feh-RAM-eh-NIF-er-eh) (or forams) secrete beautifully sculpted shells made of calcium carbonate ($CaCO_3$) (Figure 27.4a). The name foraminifera means "hole bearers" and refers to the microscopic holes in their shells through which their pseudopods protrude. Figure 27.4b shows spinelike pseudopods protruding from this foram's shell. Food particles stick to these cellular extensions and are then absorbed into the cell.

Forams are abundant in the sea—so abundant, in fact, that their shells litter the sea floor. When studying geological strata, scientists often use the forams as indicators of geological age by noting the types of forams present in ancient rock. Interestingly, the white cliffs of Dover (Figure 27.5) are actually masses of foram shells, uplifted millions of years ago with the sea floor in an ancient geological event.

Flagellates (Discomitochondria and Archaeprotista)

Flagellates are an interesting group because they are so diverse; a few representative genera of flagellates are shown in Figure 27.6. Although they all have at least one *flagellum* (a long, whiplike organelle of motility), some members of this phylum have many flagella, and some have thousands! Recently, the flagellates have been re-

Figure 27.5 **The white cliffs of Dover.** The picturesque white cliffs are actually composed of masses of foram shells, which were uplifted from the sea floor millions of years ago.

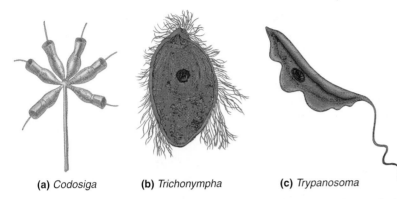

(a) *Codosiga* **(b)** *Trichonympha* **(c)** *Trypanosoma*

Figure 27.6 **The flagellates.** (a) *Codosiga,* a colonial flagellate that remains attached to its substrate. (b) *Trichonympha,* one of the flagellates that inhabits the gut of termites and woodfeeding cockroaches. *Trichonympha* ingests wood cellulose, which it finds in abundance in the digestive tracts of its hosts. (c) *Trypanosoma* causes sleeping sickness in humans. It has a single flagellum.

Trypanosoma (trih-pah-neh-SO-mah), which lives in a parasitic relationship with certain mammals, including humans, causing the disease sleeping sickness.

One group of flagellates, the **euglenoids** (yoo-GLEE-noyds), generally have chloroplasts and make their own food by photosynthesis, so some biologists classify them with the green algae. However, they have certain pigments not present in the green algae and many forms are heterotrophic, which places them with the other flagellates. Additionally, many of the photosynthetic forms become heterotrophs in the absence of light.

Each euglenoid has two flagella, a short one and a long one, that are located on the anterior end of the cell. They move by whipping the long flagellum (Figure 27.7). The euglenoids reproduce asexually by transverse fission, a process in which the parent cell divides across its short axis (see Figure 27.11a). No sexual reproduction is known among this group, which is named after its most well-known member: *Euglena.*

Euglena lives in ponds and lakes and can withstand stagnant water. It has a hard, yet flexible covering called a *pellicle* beneath its plasma membrane, with ridges spiraling around its body. These ridges can be seen clearly in Figure 27.7. Two organelles, an *eyespot* and a *photoreceptor,* help *Euglena* stay near the light. (The name *Euglena,* in fact, means "true eye.") The photoreceptor, located near the base of its longer flagellum, is shaded by the nearby eyespot. As light filters through the pigment of the eyespot, the receptor senses the direction and intensity of the light source. Information from the receptor assists the movement of *Euglena* toward the light, a behavior known as **positive phototaxis.**

One species of flagellate, *Trichomonas vaginalis* (trih-KOM-oh-nas vah-gin-NAH-lis) (Figure 27.8), causes genital infections in both women and men (despite its name). *T. vaginalis* feeds on bacteria and cell secretions in the urogenital structures it infects. In women, it infects the vagina, cervix, and vulva. However, it can cause infection only when the pH level of the vagina is elevated because it cannot survive in the vagina's normally acidic environment. (The pH of the vagina may become elevated when a woman is taking antibiotics. These drugs sometimes kill the resident vaginal bacteria that secrete acids.)

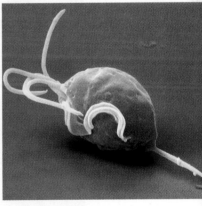

Figure 27.8 **Trichomonas vaginalis.** This flagellate has 5 flagella in the front and one flagellum in the back.

Trichomoniasis (TRIK-oh-moh-NYE-ah-sis) (infection with *Trichomonas*) in females is characterized by itching, burning, and a profuse discharge that may be bloody or frothy. This organism can survive on objects for some time; thus an infection can be contracted from a toilet seat or garments that are contaminated with the organism. Transmission is usually by sexual intercourse, however.

In males, trichomoniasis usually causes an inflammation of the urethra, but the epididymis, prostate gland, and seminal vesicles can also become infected. An infected man may have no symptoms or may experience a slight discharge from the urethra, painful urination, or increased urination. Trichomoniasis is treated with antiparasitic drugs.

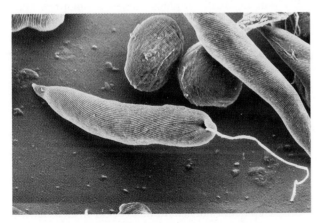

Figure 27.7 **Euglena.** Although the short flagellum of the Euglena in the center of this colorized scanning electron micrograph cannot be seen, its long flagellum, used for movement, is visible.

Ciliates (Ciliophora)

Ciliates get their name from a Latin word meaning "eyelash"—a name that is descriptive of the fact that all or parts of these cells are covered with hairlike extensions called *cilia* (**Figure 27.9**). These cilia beat in unison, moving the cell about or creating currents that move food particles toward the gullet of the cell.

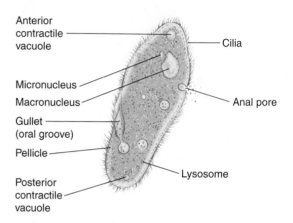

Figure 27.10   *Paramecium.*

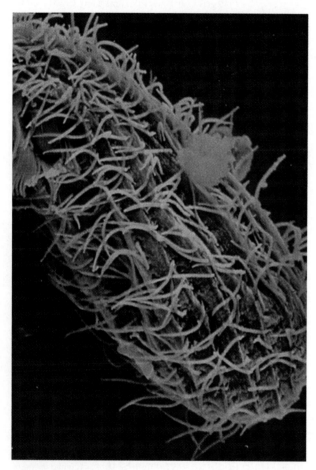

Figure 27.9   **The ciliates.** Cilia can be seen covering the paramecium shown in this photo. Note the concentration of cilia at the gullet, visible in the upper left corner.

Ciliates possess a wide array of cellular organelles that perform functions similar to the organs of multicellular organisms. An example of this interesting cellular organization is shown in the diagram of *Paramecium* in **Figure 27.10**. The paramecium is a ciliate that is classically used as one example of this group. Cilia protrude through holes in the paramecium's outer covering, or *pellicle*. The micronuclei (there may be several) func-

tion in sexual reproduction, whereas the single macronucleus controls cell metabolism and growth. The beating cilia of the paramecium sweep food into its *gullet*. From the gullet, food passes into a *food vacuole* where enzymes and hydrochloric acid aid in digestion. After absorption of the digested material is complete, the vacuole empties its waste contents into the *anal pore,* located in a special region of the pellicle. The waste then leaves the cell by a process similar to exocytosis. The contractile vacuoles expel excess water from the cell.

Paramecia reproduce asexually by *transverse fission* (**Figure 27.11a**). In addition, paramecia exchange genetic material in a process called *conjugation,* as shown in **Figure 27.11b**. Although conjugation does not produce offspring cells, it does promote genetic variability among cells that normally produce clones of identical cells when they reproduce. Genetic variability enhances the ability of the population to survive. Some algae, fungi, and bacteria also exchange genetic material in similar processes, which are also termed conjugation.

Most ciliates live in fresh water or salt water and do not infect other organisms. However, one species called *Balantidium coli* (BAL-an-TIH-dee-um KOH-lie) inhabits the intestinal tracts of pigs and rats. People who come in contact with this protozoan on farms or in slaughterhouses can become infected. The ciliates embed themselves in the lining of the intestines, producing sores and causing dysentery, similar to amebic dysentery. Occasionally, epidemics of balantidiasis occur in areas having poor sanitation.

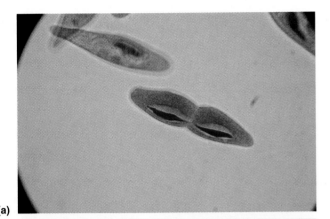

(a)

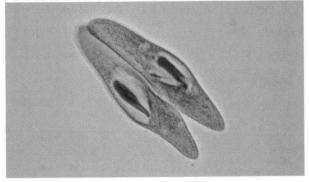

(b)

Figure 27.11 (a) Transverse fission and (b) Conjugation in *Paramecium*.

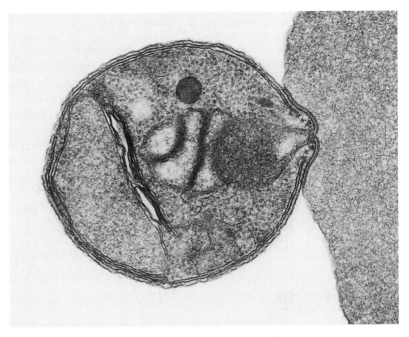

Figure 27.12 **An apicomplexan attached at its apex to a red blood cell.** Notice that many cell organelles occupy this end of the apicomplexan, which is in the center of the photo. Eventually, the entire apicomplexan will invade the red blood cell (right).

Sporozoans (Apicomplexa)

Sporozoans are nonmotile spore-forming parasites of vertebrates such as cows, chickens, and humans. Many cause serious and sometimes fatal diseases. This group of protists is most recently referred to as apicomplexans (AY-pee-com-PLEX-ans). Their old name (phylum Sporozoa) focused on their spore-forming capabilities. Their new name refers to the "apical complex," a group of cell organelles (complex) at one end of the cell (apex) specialized for penetrating host cells and tissues (Figure 27.12). (A few of the sporozoans are now classified in phyla different from the apicomplexans.)

Apicomplexans have complex life cycles that involve both asexual and sexual phases and may require several species of host. Their spores are small, infective bodies that are transmitted from host to host by various species of insects. The sporozoan used classically to represent this phylum of protists is *Plasmodium*, the causative agent of malaria. Approximately 1 million people die of this disease each year; therefore, it is considered one of the most serious diseases in the world.

The animallike protists (protozoans) are all heterotrophic. Their food-getting characteristics are linked to the ways that these organisms move and provide a means by which they can be grouped.

Amebas are protozoans that have changing shapes brought about by cytoplasmic streaming, which forms cell extensions called pseudopodia. They capture food and move by means of these "false feet." Cellular slime molds are classified with the amebas and have an ameboid stage.

Flagellates are protozoans characterized by fine, long, hairlike cellular extensions called flagella. These whiplike organelles propel them as well as help in obtaining food. Euglenoids are flagellates, most of which have chloroplasts (this is one group that is both heterotrophic and autotrophic).

Ciliates are protozoans characterized by fine, short, hairlike cellular extensions called cilia. Cilia propel these organisms and also create currents in the water that move food particles toward the cell.

Apicomplexans (sporozoans) are nonmotile protozoans that are parasites of vertebrates, including humans. They undergo complex life cycles in which they are passed from host to host.

27.3 Plantlike protists: Algae

Algae are widely distributed in the oceans and lakes of the world, floating on or near the surface of the water, with their photosynthetic parts no lower than the sun's rays can reach. They are eukaryotic organisms that contain chlorophyll and carry out photosynthesis, so they can be thought of as plantlike. However, algae lack true roots, stems, leaves, and vascular tissue (an internal water-carrying system). In addition, many contain other pigments that mask the green chlorophyll. Different pigments are associated with different groups of algae; this characteristic is used in the classification of these organisms.

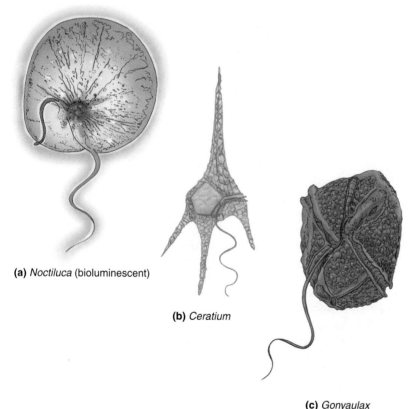

(a) *Noctiluca* (bioluminescent)

(b) *Ceratium*

(c) *Gonyaulax*

Figure 27.13 The dinoflagellates. (a) *Noctiluca* lacks the heavy cellulose armor characteristic of most dinoflagellates. (b) *Ceratium*. (c) *Gonyaulax*.

Some systems of classification place the unicellular algae in the protist kingdom and the multicellular algae in the plant kingdom. However, current thinking favors placing all the algae with the protists, broadening the scope of this kingdom. Formerly, the protist kingdom included only the protozoans and the unicellular algae. It now includes multicellular algae and the funguslike protists, each of which lacks some of the characteristics of either plants or fungi. These changes reflect the ongoing debate among scientists regarding the evolutionary relationships among organisms and how organisms should be classified to reflect these relationships. Classifying all algae as protists suggests closer evolutionary relationships among the various phyla of algae and other protists than between the algae and the plants. The green algae, thought to be the ancestors to the land plants, are sometimes still placed in the plant kingdom.

There are many phyla of unicellular algae. Two well-known types of unicellular algae are the *dinoflagellates* and the *diatoms*. There are three phyla of multicellular algae: the *brown algae,* the *green algae* (which includes many unicellular forms), and the *red algae.* These are the phyla that were formerly classified with the plants.

Dinoflagellates (Dinomastigota)

Although the **dinoflagellates** (DYE-no-FLAJ-uh-luts) have flagella, they look nothing like the flagellates. Many dinoflagellates have outer coverings of stiff cellulose plates, which give them very unusual appearances (Figure 27.13). Their flagella beat in two grooves, one encircling the cell like a belt and the other perpendicular to it. As they beat, the encircling flagellum causes the dinoflagellate to spin like a top; the perpendicular flagellum causes movement in a particular direction. In fact, the word dinoflagellate comes from Latin words meaning "whirling swimmer," and the phylum name Dinomastigota means "whirling whip."

Most dinoflagellates live in the sea and carry on photosynthesis. Their photosynthetic pigments are usually golden brown, but some are green, blue, or red. The red dinoflagellates are also called fire algae. In coastal areas, these organisms often experience population explosions or "blooms," causing the water to take on a reddish hue referred to as a red tide (Figure 27.14). Red tides destroy other living things because many species of dinoflagellates produce powerful toxins. These poisons kill fishes, birds, and marine mammals. In addition, shellfish strain these dinoflagellates from the water and store them in their bodies. Although the shellfish are not harmed, they are poisonous to humans and other animals that eat them. Many species of dinoflagellates are bioluminescent and can be seen as twinkling light in the ocean at night.

Dinoflagellates reproduce primarily by longitudinal cell division, but sexual reproduction has also been shown to occur in more than 10 genera of dinoflagellates.

Figure 27.14 A red tide at Bountiful Islands, Gulf of Carpentaria, NW Queensland, Australia.

Diatoms

Diatoms (Diatoms, DIE-uh-toms) look like microscopic aquatic pillboxes made up of top and bottom shells that fit together snugly (Figure 27.15a). These organisms reproduce asexually by separating their top from their bottom, each half then regenerating another top or bottom shell. The shells are composed of silica so the diatoms look somewhat glasslike. Their shells are so characteristically striking and intricate (Figure 27.15b) that it would be hard to confuse them with any other group of protists.

The shells of fossil diatoms often form very thick deposits on the sea floor, which are sometimes mined commercially. The resulting "diatomaceous earth" is used in water filters, as an abrasive, and to add the sparkling quality to products such as the paint used on roads and frosted fingernail polish.

An unusual characteristic of the diatoms is that they store food as oil. Some forms smell "fishy," giving unpleasant odors to the freshwater lakes they inhabit.

Brown Algae (Phaeophyta)

The **brown algae** (Phaeophyta; literally "brown plants") are the dominant algae of the rocky, northern shores of the world and are the largest protists. The types of brown algae that grow attached to rocks at the shoreline are known as rockweed (Figure 27.16a). Their puffy air bladders keep the plant afloat during high tide. One type of rockweed is also called sargasso weed and gave the Sargasso Sea its name. The Sargasso Sea is an area of ocean in the mid-Atlantic, east of Bermuda, with unusual water and current patterns that cause it to be quite calm. This calmness allows floating species of the Sargasso weed to proliferate and dominate the area.

The large brown alga with enormous leaflike structures is kelp (Figure 27.16b). These algae are an important source of food for fish and invertebrates as well as

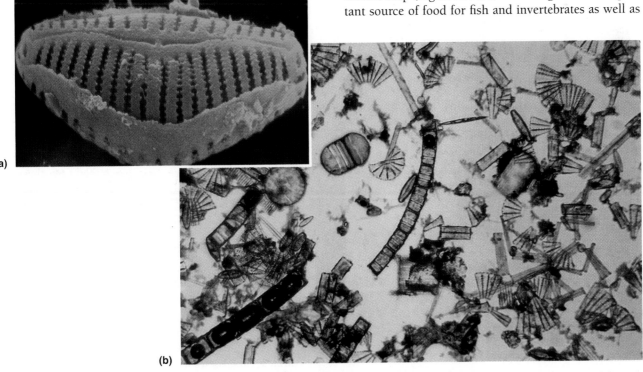

Figure 27.15 Diatoms. (a) Diatoms are composed of a top and bottom shell that fit together. (b) Several different types of diatoms.

(a)

(b)

Figure 27.16   The brown algae. (a) Rockweed. The air bladders help keep the plant afloat. (b) Giant kelp. Notice the stemlike stripe, leaflike blades, and air bladders. (Note: Brown algae does not always appear brown.)

some marine mammals and birds that live among these seaweeds. Some genera of the kelps are among the longest organisms in the world (rivaling the height of the giant sequoia trees), reaching lengths of up to 100 meters (328 feet)! These algae usually have a structure in which a rootlike portion, descriptively termed a *holdfast*, anchors the seaweed to the ocean floor or to rocks. A stemlike *stipe* carries the leaflike *blades* of the seaweed, which float on or in the water, capturing the sun's rays.

Most brown algae have life cycles that parallel the generalized life cycle of plants (see Chapter 28 and Figure 28.2b). The other multicellular algae (green algae and red algae) also have plantlike life cycles. However, the life cycles of the red algae are quite complex.

Green Algae (Chlorophyta)

The **green algae** (Chlorophyta; literally, "green plants"), are an extremely varied phylum of protists. In fact, more

than 16,000 species have been described. Of these species, most are aquatic (as are other algae) and make up a major component of phytoplankton, single-celled plantlike organisms that drift near the water surface and feed other organisms. Some green algae are semiterrestrial. Semiterrestrial algae live in moist places on land, such as on tree trunks, on snow, or in the soil. These algae are primarily unicellular microscopic forms, but some are multicellular.

Green algae show many similarities to land plants: They store food as starch, they have a similar chloroplast structure, many genera have cell walls composed of cellulose, and their chloroplasts contain chlorophylls *a* and *b*. For these reasons, scientists think the green algae were ancestors to the plant kingdom.

Well known among the unicellular green algae is the genus *Chlamydomonas* **(Figure 27.17a)**. Individuals are microscopic, green, and rounded and have two flagella at their anterior ends. They are aquatic and move rapidly in the water as a result of the beating of their flagella in opposite directions. They have eyespots that are sensitive to sunlight. The life cycle of *Chlamydomonas* is very simple. This haploid organism reproduces asexually by cell division and sexually by the functioning of some of its cells as gametes. After gametes fuse, the diploid zygote divides by meiosis, restoring the haploid state.

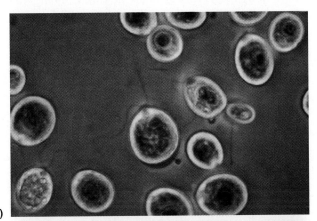

(a)

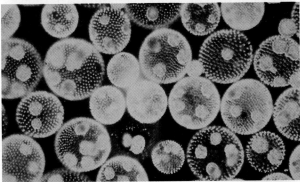

(b)

(c)

Figure 27.17 **The green algae.** (a) *Chlamydomonas*. (b) *Volvox*. (c) *Ulva*, or sea lettuce.

Some genera of green algae live together in groups. *Volvox* is one of the most familiar of these colonial green algae. Each colony is a hollow sphere made up of a single layer of individual, biflagellated cells (Figure 27.17b). The flagella of all of the cells beat in such a way as to rotate the colony in a clockwise direction as it moves through the water. In some species of *Volvox* there is a division of labor among the different types of cells, making them truly multicellular organisms.

The multicellular forms of green algae grow in either fresh water or salt water, such as the sea lettuce *Ulva* shown in Figure 27.17c. Sea lettuce is extremely plentiful in the ocean and is often found clinging to rocks or pilings. This alga consists of sheets of tissue only two cells thick.

Red Algae (Rhodophyta)

Almost all **red algae** (Rhodophyta; literally, "red plants"), are multicellular, and most of their species are marine. Their color comes from the types and amount of photosynthetic pigments present in their chloroplasts. Many species have a predominance of red pigments in addition to chlorophyll and so are red (as their name suggests). Some red algae have a predominance of other pigments so that they look green, purple, or greenish black. The red pigment, however, is especially efficient in absorbing the green, violet, and blue light that penetrates into the deepest water. (It *reflects* red light.) This enables some of the red algae to grow at greater depths than other algae and inhabit areas in which most algae cannot exist.

The red algae produce substances that make them interesting both ecologically and economically. The coralline algae, for example, deposit calcium carbonate (limestone) in their cell walls **(Figure 27.18)**. Along with coral animals, these red algae play a major role in the formation of coral reefs. Also, all red algae have gluelike substances in their cell walls: agar and carrageenan. Agar is used to make gelatin capsules, is an ingredient in dental impression material, and is used as a base for cosmetics. It is also a main component of laboratory media on which bacteria, fungi, and other organisms are often grown. Carrageenan is used mainly as a stabilizer and thickener in dairy products such as creamed soups, ice cream, puddings, and whipped cream, and as a stabilizer in paints and cosmetics. Some of the red algae are used as food in certain parts of the world, such as in Japan.

Algae are photosynthetic plantlike protists that live in aquatic or moist environments. They lack true roots, stems, and leaves. Dinoflagellates, a type of flagellated unicellular alga, are characterized by stiff outer coverings. Their flagella beat in two

(a)

(b)

Figure 27.18 The red algae. Red algae do not always appear red. Some algae show different colors depending on their daily exposure to light. (a) *Ahnfeltia plicata* growing on rocks. This type of alga is edible and is considered a delicacy in many Asian cultures. (b) *Bossiella* is a coralline red alga, which means that it secretes the hard substance calcium carbonate.

grooves, one encircling the cell like a belt and the other perpendicular to it. Diatoms are also unicellular algae and look like microscopic aquatic pillboxes. They store food as oil. Their shells are mined commercially and are used as abrasives and for other commercial applications.

There are three phyla of multicellular algae that were formerly classified with the plants: the brown algae, the green algae, and the red algae. Brown algae are large, multicellular algae found predominantly on northern, rocky shores. Some grow to enormous sizes. Green algae include both unicellular and multicellular forms. Most forms are aquatic, as are other algae, but some species live in moist places on land. The red algae play a major role in the formation of coral reefs and produce gluelike substances that make them commercially useful.

27.4 Funguslike protists

Plasmodial (Acellular) Slime Molds (Myxomycota)

The slime molds make up two unique and interesting phyla of protists. The cellular slime molds are now classified with the amebas because of their evolutionary closeness and because these slime molds have a cellular ameboid stage. They are described on p. 615. The other phyla of slime molds are the **plasmodial slime molds**, or Myxomycota (MIK-sah-my-COH-tah; literally, mucus mold). Both phyla of slime molds are called "molds" because they give rise to moldlike (funguslike) spore-bearing stalks during one stage of their life (see Figure 27.24). Both are heterotrophs and feed on bacteria, which they find in damp places rich in nutrients, such as rotting vegetation (especially rotting logs), damp soil, moist animal feces, and water.

The plasmodial slime molds have an ameboid stage to their life cycle (like the cellular slime molds), but these "amebas" are quite unusual. These bizarre organisms stream along as a **plasmodium** (plaz-MOH-dee-um)—a nonwalled, multinucleate mass of cytoplasm—which resembles a moving mass of slime (Figure 27.19). The plasmodia engulf and digest bacteria, yeasts, and other small particles of organic matter as they move along. At this stage of its life cycle, a plasmodium may reproduce asexually; the nuclei undergo mitosis simultaneously, and the entire mass grows larger.

Figure 27.19 The plasmodial slime mold. Plasmodial slime molds move about as a plasmodium, a multinucleated mass of cytoplasm.

(a)

(b)

Figure 27.20 **Spore cases of two types of plasmodial slime molds.** (a) *Lycogala.* (b) *Physarium.*

When food or moisture is in short supply, the plasmodium moves to a new area and forms spores. These spores are held in spore cases, which have a characteristic look for each genera of acellular slime mold. Fig-

ure 27.20 shows two different types of spore cases. The spores are resistant to unfavorable environmental influences and may last for years if they remain dry. Meiosis occurs in the spores. When conditions are favorable, the spore cases open and release spores that germinate into flagellated, haploid cells called *swarm cells*. The swarm cells can divide, producing more swarm cells, or can act as gametes. Gametes can fuse and form a new plasmodium by repeated mitotic divisions.

Water Molds (Oomycota)

Some scientists disagree as to whether the **water molds** should be considered protists or fungi. A primary issue is that water molds have flagellated spores, which are not characteristic of fungi. In this textbook, these organisms are classified with the funguslike protists.

If you have an aquarium and have seen white fuzz on any of your fish, you have been introduced to the water molds. They live not only in fresh water and salt water and on aquatic animals but also in moist soil and on plants. Some of the plant diseases caused by this group are late blight and downy mildew. Although this group gets the name water molds because many species thrive in moisture, they are sometimes called *egg fungi* because of the large egg cells present during their sexual reproduction. Oomycota (OH-ah-my-COH-tah), the phylum name, means egg fungus.

Slime molds are organisms that are funguslike in one phase of their life cycle and amebalike in another phase of their life cycle.

Water molds thrive in moist places and aquatic environments, parasitizing plants and animals. During sexual reproduction, they produce large egg cells.

Fungi are multicellular eukaryotic organisms that feed on dead or decaying organic material.

27.5 An overview of the fungi

Fungi are a separate kingdom of mostly multicellular eukaryotic organisms that are **saprophytic** (SAP-roe-FIT-ik); that is, they feed on dead or decaying organic material. To do this, fungi secrete enzymes onto a food source to break it down and then absorb the breakdown products. Some fungi are **parasites** and feed off living organisms in the same way (as happens in ringworm and athlete's foot).

Another example of a parasitic fungus is the pine pitch canker (*Fusarium subglutinans*) that has been attacking various species of pine trees in California since 1986. This fungus, which is transmitted by certain types of tree-inhabiting beetles, now threatens to destroy 85% of native Monterey pines within the next decade. There are only 5 stands of Monterey pines in the world, and

California is home to three of them. The fungus infects the branches first, turning the needles brown. The infection also creates a flow of amber pitch that runs down the trunk (this area is called a canker) and gives the fungus its name. Figure 27.21a shows a canker on the trunk of a Monterey pine. Figure 27.21b shows a Monterey pine that died from its infection.

Structurally, most fungi are multicellular. Unlike plants, fungi have no chloroplasts and do not produce their own food by photosynthesis. They are composed of slender filaments that may form cottony masses or that may be packed together to form complex structures, such as mushrooms.

The slender filaments of fungi are barely visible to the naked eye. Termed *hyphae* (HI-fee; sing. *hypha* [HI-fah]), they may be divided into cells by cross walls called *septa* (sing. *septum*) or may have no septa at all. The septa rarely form a complete barrier, however; thus, cytoplasm streams freely throughout either type of hypha. Because of this streaming, proteins made throughout the hyphae may be carried to their actively growing tips. As a result, the growth of fungal hyphae may be very rapid when food and water are abundant and the temperature is optimal.

A mass of hyphae is called a *mycelium* (my-SEE-lee-um; pl. *mycelia*). Part of the mycelial mass grows above the food source (substrate) and bears reproductive structures (Figure 27.22). The rest grows into the substrate. The part of the mycelium embedded in the food source secretes enzymes that digest the food. For this reason, many fungi are harmful because their mycelia grow into and decay, rot, and spoil foods (and sometimes other organic substances, such as leathers). In ad-

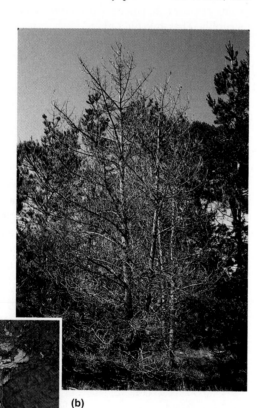

(b)

(a)

Figure 27.21 The effects of pine pitch canker. (a) A canker oozing with sap on a Monterey pine. (b) This Monterey pine has died from the effects of pine pitch canker.

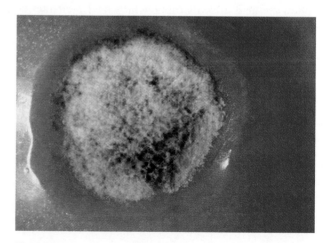

Figure 27.22 The structure of mold. A cottony mold growing on a tomato.

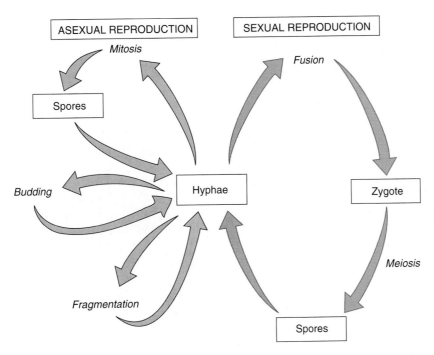

Figure 27.23 **A generalized life cycle for fungi.** Most fungi alternate between sexual and asexual reproductive stages.

dition, some fungi cause serious diseases of plants and animals, including humans.

In their roles as decomposers, fungi may seem troublesome to humans, but they are essential to the cycling of materials in ecosystems (see Chapter 35). In addition, many antibiotics (drugs that act against bacteria) are produced by fungi. Yeasts, which are single-celled fungi, are also useful in the production of foods such as bread, beer, wine, cheese, and soy sauce. These foods all depend on the biochemical activities of yeasts, in which they produce certain acids, alcohols, and gases as by-products of their metabolism of sugar. Yeasts can also cause infections in humans, such as vaginal and oral yeast infections.

Most fungi reproduce both asexually and sexually. (Figure 27.23 shows a generalized life cycle for many fungi.) During sexual reproduction, hyphae of two genetically different mating types (called + and − strains) come together, and their haploid nuclei fuse and produce a diploid zygote. This zygote germinates into diploid hyphae, which bear **spores**. Spores are reproductive bodies formed by cell division (mitosis or meiosis) in the parent organism. These spores are formed by meiosis in a diploid parent and by mitosis in a haploid

parent; the spores are always haploid. They germinate into haploid hyphae after being released and finding appropriate growth conditions. The cycle comes full circle when their hyphae fuse with those of a genetically different mating type, forming zygotes. Some fungi may reproduce asexually by budding, by growing new hyphae from fragments of parent hyphae, or by producing spores by mitosis.

There are three phyla of fungi and they include those organisms that have a distinct sexual phase of reproduction: the *zygote-forming fungi,* the *sac fungi,* and the *club fungi.* In the past, a fourth phylum of fungi (the imperfect fungi, or Deuteromycota) existed as a "catch-all" division in which organisms were placed because a sexual stage of reproduction had never been observed. However, these organisms are now classified with their relatives, either the sac fungi or the club fungi.

Fungi are eukaryotic organisms that feed on dead or decaying organic material or parasitize living organisms. Although some fungi cause disease in plants and animals, some are important in the production of food and most are ecologically important decomposers.

27.6 Zygote-forming fungi (Zygomycota)

You may have seen black mold (*Rhizopus nigricans*; rye-ZOH-pus-NYE-gri-cans) growing on bread or other food (Figure 27.24a). It is a decomposer like other molds, but it also helps produce products useful to humans. It carries out fermentations during the manufacture of steroids such as cortisone that are used to control inflammation.

Rhizopus nigricans is only one example of the **zygote-forming fungi** (phylum Zygomycota; ZYE-gah-my-COH-tah), which are generally found on decaying food and other organic material. They are characterized by their formation of sexual spores called **zygospores**, which gives this phylum its name. The life cycle of these fungi (Figure 27.24b) parallels the generalized life cycle shown in Figure 27.23 quite closely.

Zygote-forming fungi, such as black bread mold, live on decaying organic material and are characterized by their formation of sexual spores called zygospores.

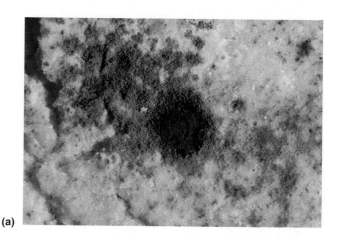

(a)

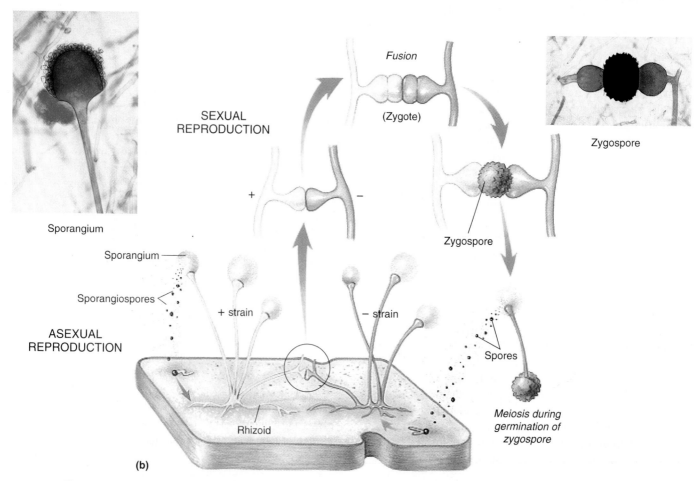

Sporangium

Sporangium

Sporangiospores

+ strain

− strain

ASEXUAL
REPRODUCTION

SEXUAL
REPRODUCTION

Fusion

(Zygote)

+ −

Zygospore

Zygospore

Spores

Meiosis during
germination of
zygospore

Rhizoid

(b)

Figure 27.24 Black bread mold. (a) Black bread mold growing on food. (b) The life cycle of black bread mold.

27.7 Sac fungi (Ascomycota)

Sac Fungi Having Sexual Stages

The **sac fungi** are the largest division of fungi, having at least 30,000 named species. They live in a wide variety of places, such as in the soil, in salt water and fresh water, on dead plants and animals, and on animal feces. They are called sac fungi because their sexual spores, which occur in groups of eight, are enclosed in saclike structures called **asci** (AS-kye; sing. ascus). Their life cycles are similar to the club fungi, which will be discussed shortly.

Among the sac fungi (phylum Ascomycota) are such familiar and economically important fungi as yeasts, cup fungi (Figure 27.25a), and truffles. This class of fungi also includes many of the most serious plant pathogens, including the causative agent of Dutch elm disease, which has killed millions of elms in North America and Europe. The spores of the mold causing this disease are carried from tree to tree by Elm and European bark beetles. The bark beetles, which carry the fungus, were transported to the United States in shipments of wood from the Netherlands in the 1920s.

Although single cells, the yeasts are also sac fungi and are one of the most economically important of the phylum because of their use in the production of various foods. Most of the reproduction of the yeasts is asexual and takes place by binary fission or by budding (the for-mation of a smaller cell from a larger one) (Figure 27.25b). Sometimes, however, whole yeast cells may fuse, forming sacs with zygotes. These cells divide by meiosis and then mitosis, forming eight spores within each sac. When the spores are released, each functions as a new cell.

One species of yeast, *Candida albicans* (CAN-dih-dah AL-bih-cans), causes superficial infections of the mucous membranes of the vagina. (Other species may cause such infections, but *C. albicans* is most often the cause.) *Candida* organisms are normally found in small numbers in the vagina, but the acid environment of the vagina and the proliferation of normal vaginal bacteria limit its growth. When the environment of the vagina changes,

Trees with Dutch elm disease.

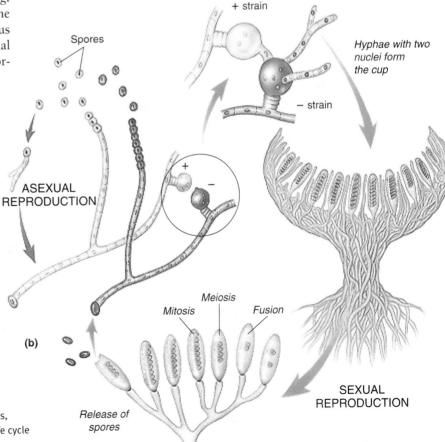

Figure 27.25 Sac fungi (phylum Ascomycota). (a) A cup fungus, *Cookeina tricholoma*, from the rain forest of Costa Rica. (b) The life cycle of a cup fungus.

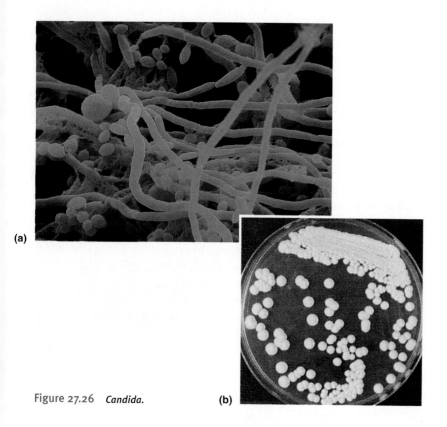

Figure 27.26 *Candida.* (b)

this organism often flourishes, producing raised gray or white patches on the vaginal walls, a scanty but thick whitish discharge, and itching. Situations that change the environment of the vagina are pregnancy and the use of birth control pills, which often change the vaginal pH, and the taking of broad-spectrum antibiotics such as tetracycline, which kills the normal vaginal bacteria. In addition, patients with AIDS or uncontrolled diabetes often develop yeast infections. Candidiasis (CAN-dih-DIE-ah-sis) can also occur in other areas of the body, such as the mouth, hands, feet, skin, and nails. Figure 27.26a is a photomicrograph that shows the elongated structures termed *pseudohyphae* that *Candida* produces as it grows.

Although candidiasis is not considered a sexually transmissible disease because it is usually contracted without sexual contact, a vaginal yeast infection can spread to a partner during sexual activity. The infection in the male is characterized by the growth of small, elevated yeast colonies on the penis, similar to those shown in Figure 27.26b. In addition, like most sexually transmissible diseases, a vaginal yeast infection can be transmitted to a newborn as it passes through the birth canal. A number of antifungal drugs are available that can be applied locally to treat infections caused by *Candida* organisms.

Sac Fungi Having No Sexual Stages

Certain sac fungi have lost the ability to reproduce sexually, or their sexual stages of reproduction have not been observed. Although these fungi do not reproduce sexually, hyphae of different mating types often fuse, providing some genetic recombination. This characteristic is important to the survival of each species, especially in the plant pathogens, because mutation and recombination produce new strains that can still infect plants bred for resistance to these fungal diseases.

The "asexual" sac fungi reproduce by spores. An enormous range of diversity occurs in the structure of the spore cases found at the tips of their hyphae. These variations represent adaptations of the fungi to dispersal of their spores. For example, spores distributed by insects usually have sticky, slimy spore cases with odors attractive to insects. Those dispersed by the wind have dry spore cases.

Among the economically important genera of asexual sac fungi are *Penicillium* and *Aspergillus*. Their hyphae and spores are shown in Figure 27.27a,b. Some species of *Penicillium* are sources of the well-known antibiotic penicillin, and other species of the genus give the characteristic flavors and aromas to cheeses such as Roquefort and Camembert. Species of *Aspergillus* are used for fermenting soy sauce and soy paste, processes in which certain bacteria and yeasts also play important roles.

Other asexual sac fungi cause diseases in humans. Some cause infections of the skin, such as athlete's foot (Figure 27.28) and other forms of ringworm. The term ringworm refers to any fungal infection of the skin, including the scalp and nails. No worms are present in such an infection. The name comes from a pattern of infection on the skin that sometimes looks like a worm has burrowed beneath the epidermis.

(a)   (b)

Figure 27.27   Sac fungi with no sexual stage of reproduction. (a) Hyphae of *Aspergillis* bearing spores. (b) Hyphae of *Penicillum* bearing spores.

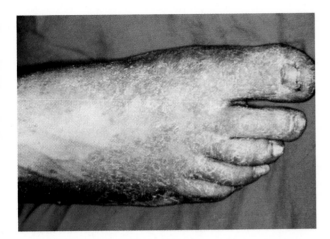

Figure 27.28 Athlete's foot, *Tinea pedis*. This itchy, painful condition is caused by members of the sac fungi, including the genus *Trichophyton*.

There is also a small group of asexual sac fungi that cause diseases of various organs, including the brain. Some of these diseases are spread by birds, bats, and contaminated soil. Such diseases are usually contracted by breathing air heavily contaminated with spores of the causative agent. Other fungal diseases occur only in persons weakened by other diseases, making them more susceptible to infection by pathogenic fungi.

Lichens

Lichens (LIE-kins) **(Figure 27.29)** are associations mostly between sac fungi and either cyanobacteria or green algae (or sometimes both), so they are now classified with the sac fungi. Most of the visible body of a lichen consists of fungus, but the photosynthetic organism lives within the fungal tissues.

Lichens provide an example of *mutualism,* a living arrangement in which both the fungus and the alga benefit. The alga provides nutrients for itself and the fungus. Specialized fungal hyphae penetrate or envelop the photosynthetic cells and transfer water and minerals to them, although this part of the mutualistic relationship has been questioned recently. Researchers now suspect that the fungus acts more like a parasite than a mutualistic partner.

Lichens are able to invade the harshest of habitats at the tops of mountains, in the farthest northern and southern latitudes, and on dry, bare rock faces in the desert. In such harsh, exposed areas, lichens are often the first colonists, breaking down the rocks and setting the stage for the growth of other plants. These amazing alga-fungus partnerships are able to dry or freeze and then recover quickly and resume their normal metabolic activities. The growth of lichens may be extremely slow in harsh environments—so slow, in fact, that many small lichens appear to be thousands of years old and are among the oldest living things on Earth.

The sac fungi live in both aquatic and terrestrial environments and are characterized by sexual spores borne in saclike structures. Familiar examples of this group are cup fungi and yeasts. Some of the sac fungi have no sexual stage of reproduction and reproduce asexually by spores. Some of these organisms are sources of antibiotics or are important in food production, but many cause disease in humans.

Lichens are associations mostly between sac fungi and either cyanobacteria or green algae (or sometimes both) and are now classified with the sac fungi. These organisms are able to live in harsh environments.

(a)

Figure 27.29 Three types of lichens. (a) Crustose (encrusting) lichens growing on a rock. (b) A fruticose (shrubby) lichen. Fruticose lichens predominate in deserts because they are more efficient in capturing water from moist air than either of the other two types. (c) A foliose (leafy) lichen growing on the bark of a tree.

(b)

(c)

27.8 Club fungi (Basidiomycota)

Club fungi include the mushrooms, toadstools, puffballs, jelly fungi, and shelf fungi (**Figure 27.30**). Some species of club fungi are commonly cultivated; for example, the button mushroom—commonly served in salad bars—is grown in over 70 countries, producing a crop with a value of over $15 billion. Other kinds of club fungi are represented by the rusts and smuts, which are devastating plant pathogens. Wheat rust, for example, causes huge economic losses to wheat farmers.

The club fungi are so-named because they have club-shaped structures from which unenclosed spores are produced. These structures are called **basidia** (buh-SID-ee-uh; sing. basidium) and give the phylum name of Basidiomycota. In mushrooms, basidia line the gills found under the cap. (**Figure 27.31** shows these gills and the life cycle of the mushroom.) In rusts and smuts, basidia arise from hyphae at the surface of the plant.

As shown in Figure 27.31, a mushroom is formed when pairs of hyphae (often of two different mating strains) fuse and intermingle their cytoplasm and haploid nuclei. The fused hyphae develop into the mushroom, including the gills on the underside of its cap. The basidia develop as single cells on the free edges of the gills. Within the cells that develop into basidia, the nuclei present from the fusion of hyphae now fuse to form zygotes. Each zygote undergoes meiosis, producing four haploid spores in each basidia. The zygotes within some basidia undergo mitosis, producing basidia having eight haploid spores. Turgor pressure (water buildup within the basidia) bursts the basidia and hurls

(a) **(b)**

Figure 27.30 Club fungi (phylum Basidiomycota). (a) *Cyathus*, bird's nest fungi. (b) *Gymnosporangium*, apple-cedar rust.

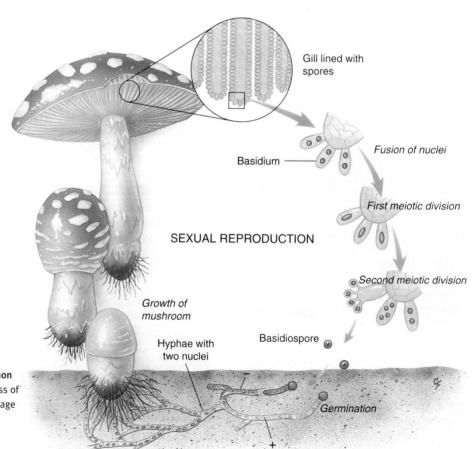

Gill lined with spores

Basidium

Fusion of nuclei

First meiotic division

SEXUAL REPRODUCTION

Second meiotic division

Growth of mushroom

Basidiospore

Hyphae with two nuclei

Germination

Figure 27.31 Life cycle of a common club fungus—a mushroom. This class of fungi rarely undergoes an asexual stage of reproduction.

I love to eat mushrooms and often see them growing in my yard or when I go hiking in the woods. How do I tell which are safe to eat?

You should never eat mushrooms unless you are absolutely sure they are safe. For most people, that means eating only the ones they buy at the grocery store.

The mushrooms most often sold in grocery stores are *Agaricus bisporus,* a fungus closely related to the common field mushroom. Other popular edible mushrooms are shiitake mushrooms, *Lentinus edodes.* Together, these two species make up about 86% of the world crop of mushrooms.

Within other genera of mushrooms such as *Amanita,* for example, there are both poisonous and edible species. For an untrained person, these closely related organisms are difficult to tell apart; even mushrooms from different genera can be difficult to identify with certainty. Because the stakes are so high (one bite of a highly poisonous *Amanita* could kill you), it makes no sense to take a chance eating wild mushrooms. In addition, certain mushrooms produce powerful hallucinogenic drugs; ingesting these drugs can be extremely dangerous. Therefore, unless you are a trained mushroom expert, stick with the grocery store varieties!

the spores from the mushroom—at an average rate of 40 million per hour! These spores germinate to produce hyphae, which fuse to begin a sexual cycle of reproduction once again. Asexual reproduction is rare in this class of fungi.

The club fungi, named because they have club-shaped structures from which unenclosed spores are produced, include the mushrooms, puffballs, and shelf fungi.

Key Concepts

Protists, although primarily single-celled, are an extremely diverse kingdom of eukaryotes.

27.1 The protists (Kingdom Protista) are a varied group of eukaryotic organisms.

27.1 Many protists are single-celled, although some phyla of protists include multicellular or colonial forms.

27.1 Within the protist kingdom are animallike cells, which take in organic matter for nutrition; plantlike cells, which manufacture their own food; and funguslike cells, which live on dead or decaying organic matter.

27.2 Animallike protists, the heterotrophs, are usually referred to as protozoans.

27.2 The protozoans are classified according to the means by which they move and feed.

27.2 The amebas move and eat by means of cell extensions, or pseudopods.

27.2 The flagellates move by means of long, whiplike cellular extensions, their flagella.

27.2 The ciliates move and sweep food toward themselves by means of short, hairlike cellular processes, their cilia.

27.2 The sporozoans (apicomplexans) are nonmotile parasites of vertebrate animals, carried from one host to another by insects.

27.3 The algae (plantlike protists) contain chlorophyll, carry out photosynthesis, but lack true roots, stems, leaves, and vascular tissue.

27.3 The dinoflagellates and the diatoms are unicellular organisms, whereas brown algae, green algae, and red algae are multicellular.

27.3 The green algae have certain unicellular species.

27.4 The slime molds have unique life cycles that have funguslike spore-bearing stages and "slimy" amebalike stages.

27.4 The water molds grow in fresh water and salt water, on aquatic animals, in moist soil, and on plants.

27.4 Water molds produce large egg cells during sexual reproduction.

Key Concepts, continued

Fungi are multicellular eukaryotic organisms that feed on dead or decaying organic material.

27.5 Fungi are important decomposers that are essential to the cycling of materials in ecosystems, are important in the production of certain foods, and also cause certain diseases.

27.5 Most fungi are composed of slender microscopic filaments, or hyphae, that form cottony masses or compact plantlike structures.

27.6 The zygote-forming fungi (such as black bread mold) are characterized by their formation of sexual spores called zygospores.

27.7 The sac fungi (such as cup fungi and yeasts) have sexual spores that occur in groups of eight and are enclosed in saclike structures called asci.

27.7 Many fungi that have no sexual stages are related to the sac fungi and are now classified with them.

27.7 The "asexual" sac fungi include organisms that produce antibiotics and are used in the production of certain foods but that also cause certain diseases in humans.

27.7 Lichens are associations of sac fungi with green algae and/or cyanobacteria.

27.7 Lichens are able to exist in harsh environments and live for long periods of time.

27.8 The club fungi (such as mushrooms, puffballs, and shelf fungi) are so-named because they have club-shaped structures (basidia) from which unenclosed spores are produced.

Key Terms

algae (AL-jee) *614*
amebas (uh-MEE-buhs) *614*
asci (AS-kye) *629*
basidia (buh-SID-ee-uh) *632*
brown algae *621*
cellular slime molds *615*
ciliates (SIL-ee-uts *or* SIL-ee-ates) *618*
club fungi *632*
diatoms (DIE-uh-toms) *621*
dinoflagellates (dye-no-FLAJ-uh-luts *or* DYE-no-FLAJ-uh-lates) *620*

euglenoids (yoo-GLEE-noyds) *617*
flagellates (FLAJ-uh-luts *or* FLAJ-uh-lates) *616*
green algae *622*
lichens (LIE-kins) *631*
parasites *626*
plasmodial (acellular) **slime molds** (plaz-MOE-dee-uhl) *624*
plasmodium *624*
positive phototaxis (foe-toe-TAK-sis) *617*
protists *614*

protozoans (PRO-tuh-ZOE-unz) *614*
red algae *623*
sac fungi *629*
saprophytic (SAP-roe-FIT-ik) *626*
spores *627*
sporozoans (SPOR-uh-ZOH-uns) *619*
water molds *625*
zygospores (ZYE-go-SPORZ) *628*
zygote-forming fungi *628*

KNOWLEDGE AND COMPREHENSION QUESTIONS

1. Explain how the following organisms are related: protists, protozoa, algae, diatoms, and slime molds. (*Comprehension*)

2. Describe an ameba. How does it move and take in food? (*Knowledge*)

3. What do flagellates and ciliates have in common? How do they differ? (*Comprehension*)

4. What are sporozoans (apicomplexans)? Briefly summarize a sporozoan's life cycle. (*Knowledge*)

5. What do dinoflagellates and diatoms have in common? Describe each one. (*Comprehension*)

6. Identify the phylum name of each of the following: diatoms, brown algae, and green algae. (*Knowledge*)

7. Fill in the blanks: The slime molds are protists that are _____ like in one phase of their life cycle, and _____ like in another phase of their life cycle. (*Knowledge*)

8. Fungi are both helpful and harmful to humans. Give some examples of each. (*Knowledge*)

9. Match each term with the most appropriate statement:
   Club fungi
   Zygote-forming fungi
   Sac fungi
   Water molds
   a. Their sexual spores are borne in saclike structures.
   b. Their sexual spores are produced by basidia.

c. Their sexual spores are called zygospores.

d. During sexual reproduction, they produce large egg cells. (*Knowledge*)

10. What are lichens, and where are they found? (*Knowledge*)

KEY
Knowledge: Recalling information.
Comprehension: Showing understanding of recalled information.

CRITICAL THINKING REVIEW

1. Some single-celled protists like *Volvox* form spherical colonies of hundreds of cells. Within the colony, some of the cells become specialized for sexual reproduction. Would such a colony be considered a multicellular organism or a colonial organism? Why? (*Synthesis*)

2. If the protist *Euglena* is "plantlike" because it photosynthesizes in the presence of light, and it is "animallike" because it ingests food in the absence of light, in which protist group does it belong? What does this quandary say about the usefulness of such an approach to classifying protists? (*Analysis*)

3. Many freshwater lakes, when polluted by high-phosphate detergents, become overgrown with mats of photosynthetic green algae. If the pollution continues, all other life in the lake soon dies. Why? Why can you not avoid this lethal result by simply poisoning the algae? (*Synthesis*)

4. Certain antibiotics kill certain types of bacteria. Many of these antibiotics are produced naturally by fungi. Penicillin, for example, is produced by the mold *Penicillium*. What would be the adaptive advantage for molds to produce antibiotics? (*Synthesis*)

5. Although the protists are an extremely diverse kingdom of organisms, they all are aquatic or live in moist environments. What shared characteristics of this diverse group necessitate such an environment? (*Application*)

KEY
Analysis: Breaking down information into component parts.
Application: Using information in a new situation.
Synthesis: Putting together information from different sources.

 learning

e-Learning is an on-line student review area located at this book's web site www.jbpub.com/biology. The review area provides a variety of activities designed to help you study for your class and build your learning portfolio.

Review Questions The review questions test your knowledge of the important concepts and applications in each chapter. The review provides feedback for each correct or incorrect answer. This is an excellent test preparation tool.

Figure-Labeling Exercises Sharpen your visual thinking by matching terms and labels for important illustrations.

Flash Cards Studying biology requires learning new terms. Virtual flash cards help you master the new vocabulary for each chapter.

Just Wondering As you read and study from this text, you may find that you have unanswered questions. Through this site you can ask the author, Sandy Alters, your "just wondering" questions.

 Do you prefer the speed and reliability of a CD-ROM? All of the features contained on the eLearning portion of the web site are also available on the Student CD-ROM.

Plants: Reproductive Patterns and Diversity

Although this does not look like a family photo of two generations of individuals—it is! The green, leaflike structures at the base are plants of one generation, and the stalks are another generation. Both are the same plant—the hairy-cap moss *Polytrichum*.

The green plants are the **gametophyte** generation. Gametophytes do as their name suggests: they produce gametes, or sex cells. These male and female gametes fuse during fertilization to form zygotes.

The generation of plants that develops from the zygotes, the stalklike plants called **sporophytes**, looks very different from its gametophyte parents. The sporophyte generation is named because it produces spores. The gametophyte generation is in turn produced from these spores, resulting in a cycling of two very different generations of the same plant.

CHAPTER 28

Organization

Plants are multicellular, photosynthetic organisms that typically live on land.

Plants have two different multicellular generations that alternate within their life cycles.

In nonvascular plants, one generation grows out of the tissues of the other.

In vascular plants, one generation of plants lives separately from the other.

Plant hormones regulate plant growth.

28.1 Characteristics of plants

Plants are multicellular, eukaryotic, photosynthetic autotrophs (producers). That is, they produce their own food (which most plants store as starch) by using energy from the sun and carbon dioxide from the atmosphere (a process described in Chapter 6). However, not all organisms fitting this description are plants. The multicellular algae, at one time classified as plants and now classified by most scientists as protists, also fit this description. However, there are other characteristics that distinguish plants from these algae.

Unlike multicellular algae, which are aquatic, plants are adapted primarily for *life on land*. The few species of plants that live in the water evolved on land and have adapted to aquatic evironments. Additionally, unlike multicellular algae, plants develop from embryos, which are multicellular structures enclosed in maternal tissue. **Figure 28.1** is a micrograph of a longitudinal section of a plant embryo. A third characteristic that differentiates the multicellular algae from the plants is their type of chlorophyll, the pigment that absorbs energy from the sun during the process of photosynthesis. Plants have the pigments chlorophyll *a* and chlorophyll *b*. The red and brown algae do not have chlorophyll *b* and so are thought not to have evolved from the same ancestors as the plants. The green algae do have chlorophylls *a* and *b* but, because they have many single-celled forms, are also classified as protists. The green algae are thought by most scientists to be the evolutionary predecessors of the land plants.

Food for embryo
(endosperm)   Embryo

Figure 28.1 Longitudinal section of a plant embryo surrounded by stored food and maternal tissue.

Table 28.1

The twelve phyla of plants

Group	Phylum	Examples
Nonvascular plants	Bryophyta	Mosses
	Hepatophyta	Liverworts
	Anthocerophyta	Hornworts
Seedless vascular plants	Psilophyta	Whisk ferns
	Lycophyta	Club mosses
	Sphenophyta	Horsetails
	Filicinophyta	Ferns
Vascular plants with naked seeds (Gymnosperms)	Coniferophyta	Conifers
	Cycadophyta	Cycads
	Ginkgophyta	Ginkgos
	Gnetophyta	Gnetae
Vascular plants with protected seeds (Angiosperms)	Anthophyta	Flowering plants
	Class Monocotyledons	Grasses, irises
	Class Dicotyledons	Flowering trees, shrubs, roses

All plants can be placed into one of two groups: the nonvascular plants and the vascular plants. Most plants (approximately 95% of all living plant species) are **vascular plants** (VAS-kyuh-ler), those having specialized tissues to transport fluids. There are three major groups of vascular plants: seedless vascular plants (such as ferns), vascular plants with naked seeds (such as pine trees), and vascular plants with protected seeds (such as flowering plants). Plants lacking these specialized transport tissues (such as mosses and liverworts) are called **nonvascular plants**.

The plant kingdom has 12 phyla. **Table 28.1** lists these phyla, including examples. It shows which phyla make up the previously described groupings of plants.

Plants are multicellular, eukaryotic photosynthetic autotrophs, which are primarily adapted for life on land. They develop from embryos and have the pigments chlorophyll *a* and chlorophyll *b*. There are three phyla of nonvascular plants (those having no specialized tissues to transport fluids) and nine phyla of vascular plants (those having specialized tissues to transport fluids).

Plants have two different multicellular generations that alternate within their life cycles.

28.2 The general pattern of reproduction in plants

The series of events that takes place from one stage during the life span of an organism, through a reproductive phase, and until a stage similar to the original is reached in the next generation is called a **life cycle**. For example, a bacterium's life cycle can be described as beginning when it (along with another daughter cell) is formed by fission from a single mother cell. This life cycle is complete when the bacterium itself splits into two new cells. This is an example of an asexual life cycle.

Animals, plants, some protists, and some fungi have sexual life cycles, which are characterized by the alternation of meiosis and fertilization. For example, your life cycle could be described as all of the events of growth and development that began at fertilization and continued after your birth through the time of your sexual maturation. This is the time you began producing sex cells by the process of meiosis. The cycle would begin again, or have come "full circle," when you have a child—the next generation.

Figure 28.2a diagrams this type of sexual life cycle, which is common to most animals. Notice in the diagram that during the animal life cycle, the multicellular organism has a double set of hereditary material; it is diploid. (One set was contributed by a father and one set was contributed by a mother.) This diploid, multicellular phase is the dominant phase of the animal life cycle (as it is in vascular plants). The gametes represent the haploid phase of the life cycle; they are individual, haploid cells.

In some protists and many fungi, the haploid phase of the life cycle is multicellular—the phase that is dominant and noticeable. Gametes are produced from these organisms by mitosis rather than by meiosis. Only the zygotes are diploid (see Chapter 27). This phase may go unnoticed to the casual observer.

Plants and some species of algae have life cycles that differ from the two sexual life cycles already described. They have both a multicellular haploid phase and a multicellular diploid phase. Because both phases of the life cycle are multicellular, this type of life cycle is called **alternation of generations** (Figure 28.2b).

The alternating generations of plants are called the sporophyte (spore-plant) generation and the gametophyte (gamete-plant) generation. The **gametophyte (haploid) generation** usually dominates the life cycles of the nonvascular plants, whereas the **sporophyte (diploid) generation** dominates the life cycles of the vascular plants.

As mentioned previously, **gametophytes** form gametes, or sex cells (eggs and sperm). Because they are

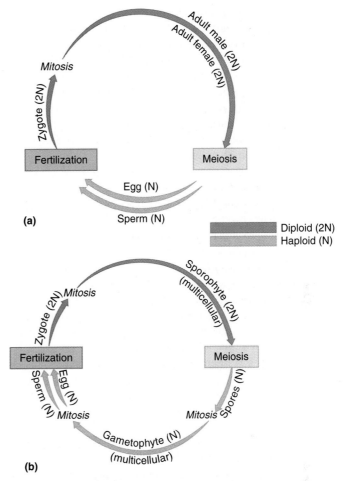

Figure 28.2 **Sexual life cycles.** (a) The animal sexual life cycle. The organism that results from the union of haploid gametes is diploid. (b) A generalized plant life cycle showing the alternation of generations. The sporophyte generation (diploid) alternates with the gametophyte generation (haploid).

haploid (HAP-loyd) organisms, having *half* the usual number of chromosomes for that species, they form gametes (which are haploid) by the process of mitosis. These gametes fuse during fertilization. The plants produced from fertilized eggs are **diploid** (DIP-loyd), having a full complement of genetic material. These diploid plants are the **sporophytes**, and they use the process of meiosis to produce haploid spores. When dispersed, spores grow into gametophyte plants.

Each species of nonvascular and vascular plant has its own variation of a life cycle, but each follows the general pattern of a dominant gametophyte generation if a nonvascular plant, and a dominant sporophyte generation if a vascular plant. The dominance of the sporophyte generation in the vascular plants reflects an adaptation of the vascular plants for life on land as they evolved from earlier, nonvascular forms. These adaptations include the

In nonvascular plants, one generation grows out of the tissues of the other.

development of spores with protective walls able to tolerate dry conditions, efficient water and food-conducting systems, and gametophyte generations that became protected by and nutritionally dependent on the sporophyte generation (Figure 28.2b).

Plants produce gametes and spores in specialized structures. In the nonvascular plants and several phyla of vascular plants, spores are borne in spore cases called **sporangia** (spoh-RAN-jee-uh; sing. **sporangium**). (This terminology is used with the fungi as well.) Eggs are formed in female reproductive structures called **archegonia** (AR-kih-GO-nee-uh; sing. **archegonium**). The archegonium is flask-shaped and contains a single egg. Sperm are produced in structures called **antheridia** (AN-thuh-RID-ee-uh; sing. **antheridium**).

In the more specialized vascular plants (except in a few species of seed-bearing plants), the gamete-producing structures became smaller during the course of evolution as the sporophyte generation began to dominate the life cycle. Eggs and sperm differentiate from a small number of haploid cells within the sporophyte. These haploid cells are actually the gametophyte generation, rather than specialized structures of a gametophyte plant.

Animal life cycles are characterized by a dominant, multicellular diploid phase and a unicellular haploid phase. The multicellular diploid organisms produce haploid gametes by meiosis. These gametes fuse during fertilization, forming the first cell of a new diploid multicellular organism.

In contrast, plant life cycles are marked by an alternation of generations of diploid sporophytes with haploid gametophytes. As a result of meiosis, sporophytes produce spores, which grow into gametophytes. Gametophytes produce gametes as a result of mitosis. These gametes fuse during fertilization and grow into sporophytes. In the nonvascular plants, the gametophyte generation is dominant. In the vascular plants, the sporophyte generation is dominant.

28.3 Patterns of reproduction in nonvascular plants

The nonvascular plants comprise three phyla of small, low-growing plants that are commonly found in moist places. The three phyla of nonvascular plants are mosses (such as the hairy-cap moss shown on p. 637), liverworts (see Figure 28.5), and hornworts (see Figure 28.6). In all three, the gametophyte is the dominant generation. The sporophyte generation grows out of the gametophyte and is dependent on it for nutrition.

The Mosses (Bryophyta)

The largest phylum of nonvascular plants and probably the one most familiar to you is the mosses. The gametophytes of most mosses have small, simple leaflike structures often arranged in a spiral around stemlike structures. Two examples of mosses are shown in the photos on p. 637 and in **Figure 28.3a**. Their "stems" and

(a)

(b)

Figure 28.3 *Sphagnum,* peat mosses. (a) The glistening black, round objects are the spore cases, which contain spores. The spore cases of *Sphagnum* have a lid that blows off explosively, releasing the spores. (b) Peat bog in Scotland. Cut sections have been laid out to dry.

"leaves" are different from the stems and leaves of vascular plants; these differences are described in Chapter 29. In both photos you can see the green, leafy gametophyte generation. The sporophyte generation rises above. In the *Sphagnum*, the glistening black, round objects are the spore cases (sporangia), held above the gametophye generation by the short stalks of the sporophyte. In *Polytricum*, the sporophyte stalks are much taller and bear capsulelike sporangia.

Many small, carpetlike plants are mistakenly called mosses. For example, *Spanish moss* is actually a flowering plant, a relative of the pineapple. Even eliminating these impostors, however, there are still some 10,000 species of "true" mosses, and they are found almost everywhere on Earth.

One kind of moss that is most important economically is *Sphagnum*. This moss grows in boggy places (low-lying, wet, spongy ground), forming dense and deep masses that are often dried and sold as peat moss. (A peat bog is pictured in Figure 28.3b.) Peat moss is used in gardening as mulch; it is layered around trees or plants to protect the roots from temperature fluctuations and to retain moisture, control weeds, and enrich the soil. It functions well as mulch and as a soil additive because its tissues have special water storage cells. These cells allow the peat to absorb and retain up to 90% of its dry weight in water, whether or not the moss is alive.

The hairy-cap moss (*Polytrichum*) has a life cycle that is very typical of the mosses. **Figure 28.4** shows its life cycle. Out of the parent gametophyte, a slender sporophyte stalk grows ❶. This stalk is initially green, but its chlorophyll disintegrates as it matures, leaving the stalk yellow or brown. At this stage, the sporophyte stalk derives its nourishment from the gametophyte. Each sporophyte stalk bears a sporangium (spore capsule) near its tip. Haploid spores are produced by meiosis within this capsule. When the top of the sporangium pops off, the spores are freed ❷. Under the proper conditions, these spores germinate into threadlike fila-

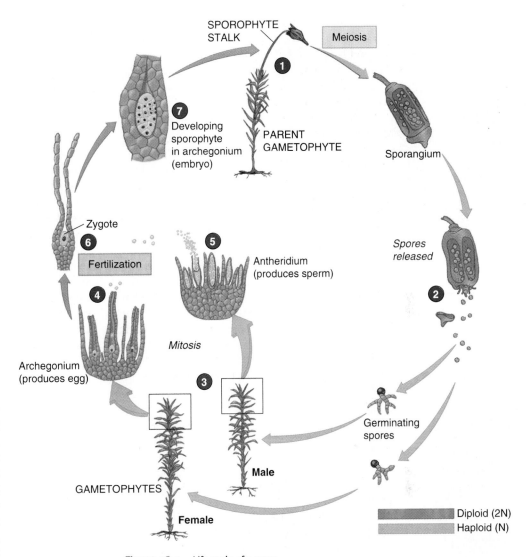

Figure 28.4   Life cycle of a moss.

ments; the characteristic leafy gametophytes arise from buds that form on these filaments ❸. The flask-shaped archegonia ❹, each with one egg, are found among the top "leaves" of the female gametophyte plants. Antheridia ❺ are found in a similar place in the male gametophytes. Each antheridium produces many sperm. The flagellated sperm swim through drops of water from rain, dew, or other sources into the neck of the archegonium and then to the egg. After fertilization takes place, the zygote ❻ develops into a young sporophyte embryo within the archegonium ❼. The cycle begins anew when the embryo then grows out of the archegonium and differentiates into a slender sporophyte stalk.

Figure 28.5 Liverwort, *Marchantia*.

Liverworts

Liverworts were given their name in medieval times when people believed that plants resembling particular body parts were good for treating diseases of those organs. Some liverworts are shaped like a liver and were thought to be useful in treating liver ailments. The ending *-wort* simply means "herb."

A well-known example of a liverwort is *Marchantia*, shown in **Figure 28.5**. It has green, leafy gametophytes that grow close to the ground. Antheridia and archegonia develop within the umbrellalike portion of the stalks that grow up from the gametophyte. The sporophytes develop encased within these tissues. Their sporangia are the dark spheres under the "umbrella" structures and are somewhat difficult to see in the photo. Spores are freed from these structures.

Hornworts

The *hornwort* is so named because it has elongated sporophytes that protrude like horns from the surface of the creeping gametophytes (**Figure 28.6**). These tall sporophytes are each made up of a foot and a long, cylindrical sporangium. The life cycle of the hornwort parallels quite closely the life cycles of the mosses and liverworts.

Figure 28.6 Hornwort, *Antheros*.

The life cycles of all three phyla of nonvascular plants—the mosses, liverworts, and hornworts—are somewhat uniform, all having two distinct phases. The sporophyte generation grows out of or is embedded in the tissues of the gametophyte generation (the dominant generation) and depends on the gametophyte generation for nutrition.

28.4 Seedless vascular plants

As listed in **Table 28.1**, the members of four phyla of vascular plants do not form seeds. **Seeds** are structures from which new sporophyte plants grow; they protect the embryonic plant from drying out or being eaten when it is at its most vulnerable stage. Seeds also contain stored food for the new plant. The seedless plants overcome these problems in interesting ways. **Figure 28.7** diagrams the life cycle of a fern—a familiar member of the **seedless vascular plants**. Its life cycle is representative of this group.

When a fern plant is mature ❶, it produces spores by meiosis. Each cluster of sporangia ❷ looks like a dot on the underside of the fern leaf, or *frond*. (These clusters of sporangia are also called *sori* [sing. *sorus*] in the fern.) Because it produces sporangia and spores, the fern is the sporophyte generation—the dominating form in the life cycle of the seedless vascular plants. After its spores are dispersed ❸, those that settle in a moist environment will germinate into haploid plants that look very *unlike* ferns. Each plant is a small, ground-hugging, heart-shaped gametophyte called a *prothallus* (pl. *prothalli*), which is anchored to the ground by filaments of cells called *rhizoids* ❹. Their antheridia and archegonia ❺ are protected somewhat by being located on the underside of the plant. Sperm ❻, when released, swim through moisture collected on the underside of the gametophyte to the archegonia. Each fertilized egg (zygote) starts to grow within the protection of the archegonium and develops into an embryo ❼. After this initial protected phase of growth, the fern sporophyte is able to grow on its own and becomes much larger than the gametophyte ❽.

Many of the seedless vascular plants that lived about 270 million years ago were converted long ago to a fuel used today—coal. Club mosses that grew on trees, horsetails, ferns, and tree ferns made up great swamp forests during the Carboniferous period (see Chapter 25). Areas of New York State, Pennsylvania, and West Virginia, for example, were lying near the equator at that time. Dead plants did not completely decay in the stagnant, swampy waters, and they accumulated. These swamps were later covered by ocean waters. Marine sediments piled on top of the plant remains. Pressure and heat acted on the layers of dead plant material beneath the ocean floor and converted the remains to coal. When you burn fossil fuels such as coal, you are burning a resource that was formed under special conditions that have not been repeated in the last 270 million years. Coal is therefore called a *nonrenewable resource*. Burning coal adds carbon dioxide to the atmosphere that had been locked in this fossil fuel for millions of years. This gas adds to the amount of atmospheric carbon dioxide and contributes to the problems of global warming (see Chapter 37).

The life cycles of seedless vascular plants are similar to one another. The sporophyte (diploid) generation is dominant and lives separately from the gametophyte. It produces spores, which germinate to form the gametophyte plants. The gametophytes produce motile sperm that need water to swim to the eggs. The fertilized eggs produce young sporophytes that grow protected within gametophyte tissues but eventually become free living.

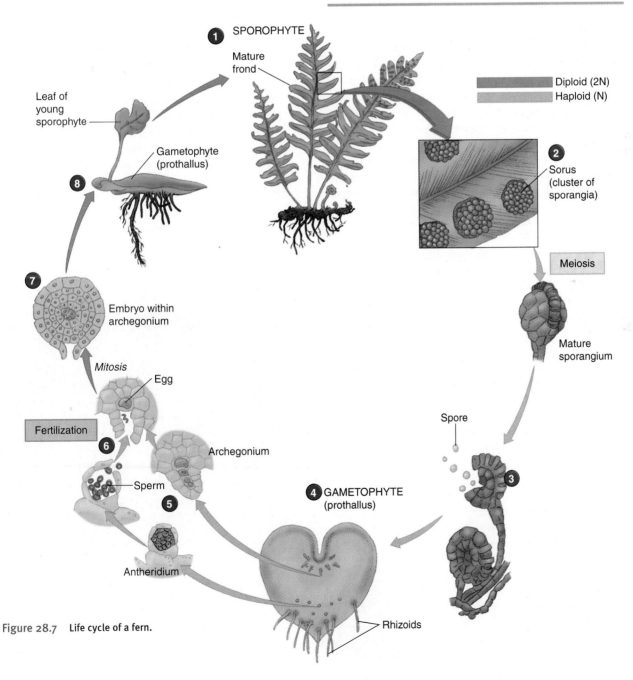

Figure 28.7   Life cycle of a fern.

28.5 Vascular plants with naked seeds

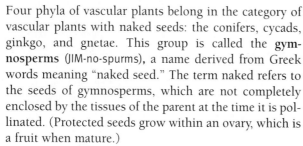

Four phyla of vascular plants belong in the category of vascular plants with naked seeds: the conifers, cycads, ginkgo, and gnetae. This group is called the **gymnosperms** (JIM-no-spurms), a name derived from Greek words meaning "naked seed." The term naked refers to the seeds of gymnosperms, which are not completely enclosed by the tissues of the parent at the time it is pollinated. (Protected seeds grow within an ovary, which is a fruit when mature.)

The most familiar phylum of gymnosperm is the conifers, or cone-bearing trees. The conifers include pines, spruces, firs, redwoods, and cedars. (In Chapter 37 there is a discussion of a parasitic fungus that has been attacking various species of pine trees and now threatens to destroy 85% of native Monterey pines within the next decade.) Figure 28.8 diagrams the life cycle of a pine.

The pine tree is the sporophyte (diploid) generation. Pines and most other conifers bear both male cones and female cones ❶ on the same tree. You can tell them apart because the female cones are the larger of the two and often grow on the lower branches of the tree, with the male cones growing higher up. In some pines, the male cones are borne on the ends of the branches while the female cones grow further in toward the trunk.

Both types of cones produce spores. In the male cones, the microspore mother cells ❷ undergo meiosis to form pollen ❸. Each pollen grain is a gametophyte that arose from a haploid spore. The male gametophytes produce sperm, which remain located in the pollen grains.

In the female cones, the megaspore mother cells ❷ undergo meiosis to form the female gametophyes containing egg cells ❸. Both the male and female gametophytes are so small that they are not visible to the naked eye. Their lives are also comparatively short.

Each of the multicellular female gametophytes produces two to three eggs. The eggs develop within protective structures called ovules within the scales of the female pinecones. In the spring, the male cones release their pollen, which is blown about by the wind ❹. (Those allergic to pine pollen know this fact quite well—although allergies to pine pollen are rare.) As some of this pollen passes by female cones, a sticky fluid produced by the now open female cones traps it there. As this fluid evaporates, the pollen is drawn further into the cone. When the pollen comes into contact with the outer portion of the ovule, it germinates and forms a pollen tube that slowly makes its way into the ovule—to the egg ❹.

After 15 months the tube reaches its destination and discharges its sperm. Fertilization takes place, producing a zygote ❺. The development of the zygote into an embryo takes place within the ovule, which matures into a seed ❻. Eventually the seed falls from the cone ❼ and germinates, and the embryo resumes growing and becomes a new pine tree ❽.

In pine trees, a representative conifer and an example of the gymnosperms, both male cones and female cones produce spores, which undergo meiosis to form male and female gametophytes. The male gametophytes (pollen) produce sperm, which remain in the pollen grains. Each of the female gametophytes produces eggs, which develop in ovules within the scales of the female pinecones. In the spring, the male cones release their pollen into the wind. When pollen comes into contact with the ovule, it germinates and forms a pollen tube that reaches the egg and discharges sperm. After fertilization, the zygote develops into an embryo within the ovule, which matures into a seed.

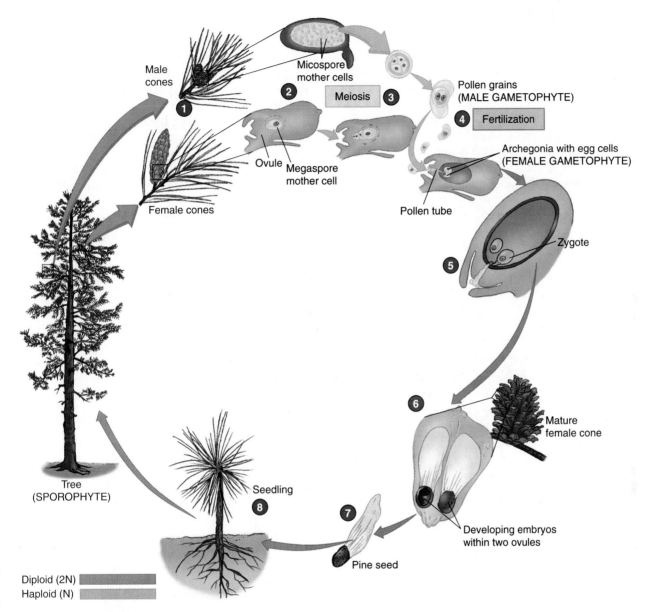

Figure 28.8 Life cycle of a pine.

Diploid (2N)
Haploid (N)

Labels in figure:
Male cones
Micospore mother cells
Meiosis
Pollen grains (MALE GAMETOPHYTE)
Fertilization
Ovule
Megaspore mother cell
Archegonia with egg cells (FEMALE GAMETOPHYTE)
Female cones
Pollen tube
Zygote
Tree (SPOROPHYTE)
Seedling
Pine seed
Mature female cone
Developing embryos within two ovules

28.6 Vascular plants with protected seeds

Flowers are the organs of sexual reproduction in the vascular plants with protected seeds—the **angiosperms** (AN-jee-oh-spurms). These plants bear seeds in a fruit; therefore, the seeds are referred to as "protected." The flower is structured to promote sexual reproduction—the union of gametes that ultimately develops into an embryo within the seed.

Angiosperms are the dominant plants in the world today. There are over 250,000 species of angiosperms, and many of these plants provide products that are crit-ically important to the survival of humans as well as some products that make life easier or more pleasant. Angiosperms provide such diverse products as

- rice, wheat, and corn;
- cotton;
- medicines such as codeine and the breast cancer drug taxol (see the Just Wondering box in Chapter 29);
- lumber;
- rubber;
- coffee; and
- oils for perfumes.

Angiosperms provide food, shelter, and nesting sites for other members of the animal kingdom as well.

Developmentally, angiosperm flower parts are actually modified leaves. The outermost whorl, or ring, of modified leaves are the sepals (Figure 28.9). Sepals enclose and protect the growing flower bud. In flowers such as roses, the sepals remain small, green, and somewhat leaflike. In other flowers, such as tulips, the sepals become colored and look like the petals, the next whorl of flower parts.

Flower petals are frequently prominent and colorful, and attract pollinating animals, especially insects

(see Section 28.7). Although the sepals and petals are the dominant outward features of flowers, they are not the organs of sexual reproduction. Many flowers either do not have sepals or petals, or have inconspicuous sepals or petals—particularly flowers such as grasses that are pollinated by the wind. The sex organs of the flower are the innermost modified leaves, the male stamens, and, at the center of the flower, the female pistil.

The plants on which flowers are produced (Figure 28.9 ❶) are sporophytes (therefore they produce spores). Each stamen consists of an anther ❷, a com-

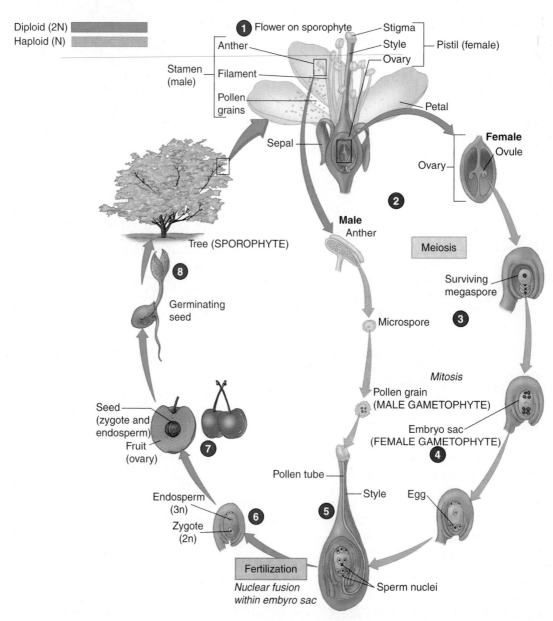

Figure 28.9 **Angiosperm life cycle.** The embryo sac is fertilized twice by sperm in the process of double fertilization. The embryo and endosperm that result develop inside a seed, which forms inside a fruit.

partmentalized structure in which haploid microspores ❸ are produced by meiosis from diploid cells. Haploid pollen grains ❹ are produced from these microspores by mitosis. Each pollen grain is a male gametophyte enclosed within a protective outer covering. This gametophyte produces sperm by mitosis, which remain enclosed within the pollen grain. A long, thin filament bears and supports the anther, exposing the anther (and therefore the pollen) to wind or pollinating animals.

The pistil consists of three parts. At its tip is a sticky surface called the stigma to which pollen grains can adhere ❶. At the base of the pistil is the ovary, a chamber that completely encloses and protects the ovules ❷. Within each ovule, a single mother cell (megaspore) develops and then divides meiotically, producing four cells ❸. One of these cells develops into a female gametophyte. When mature, the female gametophyte is called an embryo sac ❹. Within this sac are typically eight cells, one of which is the egg. The ovary will become a seed when its eggs are fertilized by the male gamete.

The style is a narrow stalk arising from the top of the ovary that bears the stigma. The style may be either long or short to facilitate the best exposure of the stigma to the method of pollination that characterizes the plant. Pollination in the flowering plants is the transfer of pollen from the anther to the stigma.

After a pollen grain lands on the stigma, it produces a long pollen tube that grows from the pollen grain down the style and penetrates the ovary, entering an ovule ❺. One of the two haploid nuclei (sperm) in the pollen tube fertilizes the haploid egg nucleus in the ovule, producing a diploid zygote ❻. The zygote will become a new plant embryo. The other sperm nucleus fuses with two other nuclei of the embryo sac, producing a triploid endosperm nucleus. The endosperm nucleus develops into tissue that will feed the embryo as it grows into a plant. The ovule becomes the seed within which the embryo develops, and the ovary ripens into a fruit ❼. The fruit is a food that attracts animals, which play a role in seed dispersal. The seed will eventually germinate, and the embryo will grow into a new plant ❽.

Sexual reproduction in the angiosperms produces seeds by the union of male gametes released from pollen grains and female gametes within ovules. Germinating pollen tubes convey sperm to eggs. The seeds, formed within the ovary after the union of sperm and eggs, protect and nourish the young sporophytes as they begin to develop into new plants.

28.7 Mechanisms of pollination

Pollination takes place in many ways, such as by insects, animals, and the wind. Insects and other animals often visit the flowers of angiosperms for a liquid food called nectar, which is rich in sugars, amino acids, and other substances.

In certain angiosperms and in most gymnosperms, pollen is blown about by the wind and reaches the stigmas passively. However, compared with insects or other animals, wind does not carry pollen far or precisely. Therefore, plants pollinated by the wind usually produce large quantities of pollen and grow close together as shown in Figure 28.10.

Figure 28.10 Ponderosa pine shedding pollen.

In some angiosperms, the pollen does not reach other individuals at all. Instead, it is shed directly onto the stigma of the same flower, sometimes in bud. This process is termed self-pollination.

For plants to be effectively pollinated by animals, a particular insect or other animal must visit many plants of the same species. Flowers have evolved various colors and forms that attract certain pollinators, thereby promoting effective pollination. Yellow flowers, for example, are particularly attractive to bees, whereas red flowers attract birds but not insects (Figure 28.11). Insects in turn have evolved a number of special traits that enable them to obtain food efficiently from the flowers of the plants they visit. For example, the copper butterfly has a long, coiled, tonguelike organ that it uses to extract nectar from flowers. Hummingbirds, such as the one pictured in Figure 28.11b, have long curved beaks for the same purpose.

(a)

Figure 28.11 Different modes of pollination. (a) Pollination by a bumblebee. As this bumblebee, *Bombus,* collects **(b)** nectar from the flame azalea, the stigma contacts its back and picks up any pollen that the bee might have acquired there during a visit to another flower. (b) Pollination by a hummingbird. A long-tailed hermit hummingbird extracting nectar from the flowers of *Helinconia imbricata* in the forests of Costa Rica. There are pollen grains on the tip of the bird's beak.

Animals, insects, or the wind can pollinate plants.

28.8 Fruits and their significance in sexual reproduction

Parallel to the evolution of the angiosperms' flowers and nearly as spectacular has been the evolution of their fruits. Fruits have evolved a diverse array of shapes, textures, and tastes and exhibit many differing modes of dispersal.

Fruits that have fleshy coverings—often black, bright blue, or red—are normally dispersed by birds and other vertebrates.

Just as red flowers attract birds, the red fruits signal an abundant food supply. By feeding on these fruits, birds and other animals carry seeds from place to place and thus transfer the plants from one suitable habitat to another **(Figure 28.12a)**. Other fruits, such as snakeroot, beggar ticks, and burdock (Figure 28.12b), have evolved hooked spines and are often spread from place to place because they stick to the fur of mammals or the clothes of humans. Others, such as the coconut and those that occur on or near beaches, are regularly spread by water (Figure 28.12c). Some fruits have wings and are blown about by the wind. The dandelion is a familiar example (Figure 28.12d).

Seeds are dispersed by sticking to or being eaten by animals, by being blown by the wind, or by floating to new environments across bodies of water.

(a)

(b)

(c)

(d)

Figure 28.12 How seeds are dispersed. (a) Bird dispersing seeds. This cedar waxwing feeds berries to the waiting young in the nest. Birds that eat fruits digest them rapidly so that much of the seed is left intact. What is excreted from the bird can grow into a mature plant. (b) Dog covered with spiny hedge parsley seeds. (c) The seeds of a coconut, *Cocos nucifera.* One of the most useful plants for humans in the tropics, coconuts have become established even on the most distant islands by drifting in the waves. (d) The seeds of a dandelion, *Pyropappus caroliniana,* are dispersed by the wind. The "parachutes" disperse the fruits of dandelions widely in the wind, much to the gardener's despair.

28.9 Seed formation and germination

The long chain of events between fertilization and maturity is development. During development, cells become progressively more specialized, or differentiated. Development in seed plants results first in the production of an embryo, which remains dormant within a seed until the seed germinates, or sprouts.

Seed **germination** depends on a variety of environmental factors, especially the presence of water. However, the availability of oxygen (for aerobic respiration in the germinating seed), a suitable temperature, and sometimes the presence of light are also necessary. The first step in germination of a seed occurs when it imbibes, or takes up, water. Once this has taken place, metabolism within the embryo resumes.

Germination and early seedling growth require the mobilization of food storage reserves within the seed. A major portion of almost every seed consists of food reserves. Angiosperms fall into two groups regarding the placement of stored food in their seeds: the monocots and the dicots. In **dicots**, most of the stored food is in the cotyledons, or seed leaves. In **monocots**, most of the food is stored in extraembryonic tissue called endosperm (**Figure 28.13**). (Although dicots store some food in endosperm, the amount of endosperm varies among species of plants.) The single cotyledon of the monocots, which is embedded within the endosperm, absorbs food from the endosperm and shuttles it to the embryo.

Monocots and dicots also differ from one another in a number of features. Monocots usually have parallel veins (fluid-carrying tissues) in their leaves, and their flower parts are often in threes. Among the monocots are the lilies, grasses, cattails, orchids, and irises (**Figure 28.14a–c**). Dicots usually have netlike veins in their leaves, and their flower parts are in fours or fives. The dicots include the great majority of familiar plants: almost all kinds of flowering trees and shrubs and most garden plants, such as snapdragons, chrysanthemums, roses, and sunflowers (Figure 28.14d–f).

Usually, the first portion of the embryo to emerge from the germinating seed is the young root, or radicle, which anchors the seed and absorbs water and minerals from the soil. Then, the shoot of the young seedling elongates and emerges from the ground.

As the shoot emerges from the soil, the first true leaves are protected by a straight sheath or a curved stem. Multiplication of the cells in the tips of the stem and roots, along with their elongation and differentiation, initiates and continues the growth of the young seedling.

The embryo of a plant remains dormant within the seed until the seed germinates. Germination begins when the seed takes up water and begins to sprout.

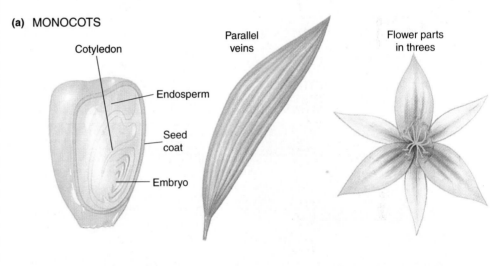

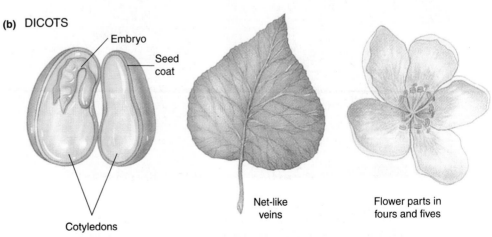

Figure 28.13 **Monocots and dicots.** (a) In monocots, most of the food for the embryo is stored in endosperm. Other characteristics of monocots are the parallel veins in their leaves and the occurrence of their flower parts in threes. (b) Dicots store their food in cotyledons, or seed leaves. They have netlike veins in their leaves, and their flower parts occur in fours or fives.

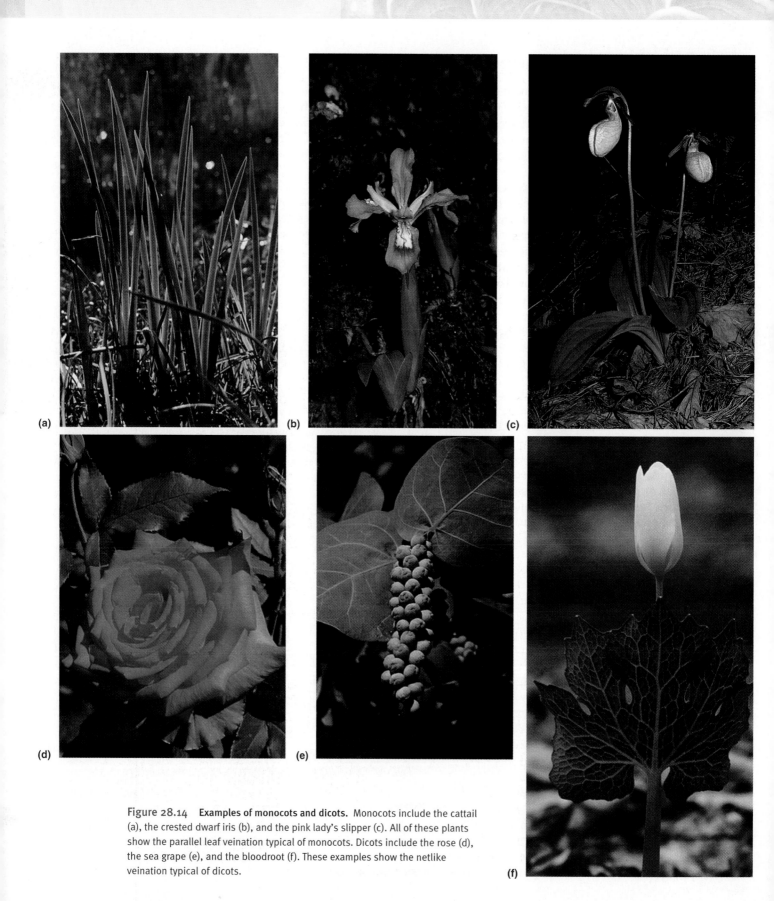

(a)

(b)

(c)

(d)

(e)

Figure 28.14 Examples of monocots and dicots. Monocots include the cattail (a), the crested dwarf iris (b), and the pink lady's slipper (c). All of these plants show the parallel leaf veination typical of monocots. Dicots include the rose (d), the sea grape (e), and the bloodroot (f). These examples show the netlike veination typical of dicots.

(f)

28.10 Types of vegetative propagation in plants

Vegetative propagation is an asexual reproductive process in which a new plant develops from a portion of a parent plant. Some plants, such as irises and grasses, produce new plants along underground stems called rhizomes (RYE-zomz). Other plants, such as strawberries, have horizontal stems that grow above the ground called runners or stolons (STOW-lunz). These plants produce new roots and shoots at nodes (places where one or more leaves are attached) along these stems. New plants can also arise vegetatively from specialized underground storage stems called tubers. A white potato, for example, is a tuber and can grow a new plant from each of its eyes (which are nodes containing lateral buds).

In some plants, new shoots can arise from roots and grow up through the surface of the soil. For example, a group of aspen trees often consists of a single individual that has given rise to a colony of genetically identical trees by producing new shoots from its horizontal roots (Figure 28.15).

In a few species of plants, even the leaves are reproductive. The plant in **Figure 28.16** is commonly called the maternity plant because small plants arise in the notches along the margins of the leaf. When mature, they drop to the soil and take root. Gardeners commonly propagate African violets from leaf cuttings and many other plants from stem cuttings.

A major breakthrough in the asexual propagation of plants has been the development of **cell culture techniques** (Figure 28.17). Using these techniques, scientists are able to remove individual cells from a parent plant and grow these cells into new individuals. These techniques are successful, however, only because individual plant cells (unlike animal cells) have the inherent capability to direct the growth and development of a new plant.

Using cell culture techniques, botanists are able to produce virtually unlimited numbers of genetically identical offspring. These techniques have been particularly useful in propagating plants that are slow to multiply on their own, such as coconut palms and redwoods, and in cultivating varieties of individual plants with special characteristics, such as large flowers. Award-winning varieties of orchids, for example, are often produced in this way. Cell culture is also used for the commercial production of tremendous numbers of plants in a short period of time—such as

Figure 28.15 Clones of aspen. The contrast between the golden quaking aspens *(Populus tremuloides)* and the dark green Engelmann spruces *(Picea engelmannii)* evident in this autumn scene near Durango, Colorado, makes it possible to see their clones, large colonies produced by individuals' roots spreading underground and sending up new shoots periodically.

Figure 28.16 *Kalanchoe daigremontiana.* The small plants growing in the notches along the leaf margins will drop to the soil and take root. For this reason, *Kalanchoe daigremontiana* is called the "maternity plant."

Plant hormones regulate plant growth.

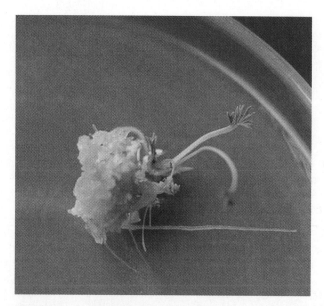

Figure 28.17 **A plant growing in a cell culture.** The small shoots in the petri dish are growing from a mass of undifferentiated callus tissue and are genetic clones of carrots. Researchers use clones and tissue cultures for a variety of purposes such as finding heartier and more insect-resistant varieties of plants.

the chrysanthemums you may buy at the grocery store. The timber industry uses cell culture techniques in developing rapidly growing conifers such as the Douglas fir.

The use of cell culture techniques in agriculture will revolutionize the way in which farmers produce some crops. In the future, farmers will be able to grow certain crop plants that have been mass produced by cell culture and are genetically superior in some way, including having disease resistance, a high yield per plant, or a desirable taste. However, variations (mutations) arise frequently in cell culture and limit the usefulness of cell culture techniques in the mass production of many crops.

The use of genetic engineering techniques is another way in which scientists may revolutionize worldwide agriculture. Scientists are able to insert genes that express favorable characteristics into the chromosomes of culture cells, producing plants with desirable characteristics. Genetic engineering and its applications are described in Chapter 19.

Many plants can reproduce vegetatively by growing new plants from their roots, stem, or leaves. Scientists, using cell culture techniques, can also propagate many plants from a single cell of the parent. Use of these techniques, along with the genetic engineering of plants, will revolutionize the way that farmers produce some crops.

28.11 Plant hormones

As a plant grows, it is influenced by environmental factors such as the amount of water and light it receives. However, a plant's growth, differentiation, maturation, flowering, and many other activities are also regulated by chemicals called **hormones**. Hormones are chemical substances produced in small, often minute quantities in one part of an organism and then transported to another part of the organism where they bring about physiological responses.

There are at least five major kinds of hormones in plants. (1) *Auxins* (AWK-sinz), (2) *gibberellins* (JIB-eh-REL-enz), and (3) *cytokinins* (SIGH-toe-KYE-ninz) promote and regulate growth. Cytokinins stimulate cell division, and auxins and gibberellins promote growth through cell elongation (Figure 28.18). The differences in concentration of auxin from one side of the stem or root to another, for example, control the bending of plants toward

Figure 28.18 **Gibberellins.** The California poppy on the left is the control (untreated) plant. The California poppy on the right was treated with gibberellins.

or away from light (phototropism) (Figure 28.19) and toward or away from gravity (gravitropism). In contrast, (4) *abscisic* (ab-SIS-ick) *acid* is a growth inhibitor that induces and maintains dormancy or otherwise opposes the three growth-promoting hormones. (5) *Ethylene* is released by plants as a gas and effects the ripening of fruit and leaf drop in nearby plants.

Other chemicals, such as special plant photoreceptor pigments called *phytochromes,* also affect a plant's response to its environment. Phytochromes change form in response to the length of the day and thereby stimulate or inhibit flowering. The study of plant hormones and other regulatory chemicals, especially how they produce their effects, is an active and important field of research today.

I heard that one reason we need to save the tropical rain forests is so that we won't lose plants that could cure diseases. How do plants cure diseases? And how do scientists find these plants?

Some plants produce chemicals that affect animals. These chemicals can be poisons, such as the blowgun poison curare, for example, or the breast cancer drug taxol. You may have used a medicine derived from plants yourself, because approximately 25% of all prescription drugs dispensed each year in North America were originally derived from flowering plants and ferns alone! (Many are now synthesized in

the laboratory.) Interestingly, the science of botany (the study of plants) was considered a branch of medicine until the early to mid-1800s.

The only way that scientists can discover whether plants produce biologically active chemicals is to test them in the laboratory. Testing all the plants in the rain forest (or in any area) for their medicinal benefits is, of course, impossible. Therefore, researchers

have devised various methods to hunt for therapeutic plants.

One method drug hunters use is to study the therapeutic plant uses of those people indigenous to a particular area. They often choose cultures that have populated a particular area for many generations and pass their "folk medicine" knowledge from generation to generation. Scientists also choose cultures living in areas exhibiting diversity in plant life. Researchers then isolate and study the biologically active chemicals they extract from these plants and determine whether they are active against the type of condition for which they are used by folk healers. Many useful drugs have been developed this way, such as reserpine, which is derived from the Indian snakeroot (*Rauwolfia serpentina*) and is used to treat hypertension, and two cancer drugs, which are both extracted from rosy periwinkle (*Catharanthus roseus*).

Another approach researchers use is to study plants that are closely related to those shown to have medicinal value. Evolutionarily, plants that produce chemicals that help them survive are more reproductively fit than other plants in their species. These plants would flourish and carry this adaptive trait from generation to generation. As species diverge from one another, traits conferring a survival

advantage tend to be retained in a population. Therefore, it makes evolutionary sense that closely related plants might have similar biochemical traits.

Yet another approach is to study plants in their natural surroundings and to choose those plants that remain untouched by predators such as insects. These plants are likely candidates to produce chemicals that affect animals; some may produce chemicals that have desirable and useful druglike characteristics.

In their laboratories, scientists run automated tests to screen the chemicals they extract from plants. Plant extracts can be quickly screened to see whether they are active against certain types of cancer cells or whether they affect enzymes active in particular diseases. After initial screenings, drug companies often take over further analyses to decide whether the drug holds promise as an addition to their pharmaceutical line.

It takes many years for a plant with suspected healing powers to reach your local pharmacy as a prescription drug. Many never make it. However, the drugs that prove to be useful are extremely important additions to our arsenal against disease. Thus, saving the rain forest, where about two thirds of the world's plant species are found, not only will preserve biological diversity but also will preserve plants that one day may cure many of our most devastating illnesses.

Plant growth is influenced by hormones as well as by environmental factors. Hormones are chemical substances produced in one part of an organism and then transported to another part of the organism, where they bring about physiological responses. Plants have at least five major kinds of hormones.

Figure 28.19 **Phototropism.** Plants grow toward the light because auxin, a plant hormone that controls plant growth, is redistributed in a plant that is darker on one side, resulting in a higher concentration of this hormone on the darker side of the stem. Thus, the side of the plant away from the light will grow faster, causing the plant to bend toward the light. The stems of this blood sorrel are oriented toward the window, which lets in plenty of sunlight.

Key Concepts

Plants are multicellular, photosynthetic organisms that typically live on land.

28.1 Plants are multicellular, eukaryotic, photosynthetic autotrophs that are primarily adapted for life on land.

28.1 Plants contain chlorophyll *a* and chlorophyll *b*, and develop from embryos.

28.1 Plants are both nonvascular (lacking specialized transport tissues for fluids) and vascular (having specialized transport tissues for fluids).

Plants have two different multicellular generations that alternate within their life cycles.

28.2 Animals, plants, some protists, and some fungi have sexual life cycles that are characterized by the alternation of the processes of meiosis and fertilization.

28.2 Plants and some species of algae have life cycles that have both a multicellular haploid phase and a multicellular diploid phase. Because both phases of the life cycle are multicellular, this type of life cycle is called alternation of generations.

28.2 In plant life cycles, multicellular haploid plants produce gametes by mitosis and are gametophytes.

28.2 The gametes produced by haploid plants fuse during fertilization and grow into the multicellular, diploid spore-producing plants, or sporophytes.

28.2 Sporophytes produce haploid spores by meiosis, which grow into gametophytes.

28.2 In general, the gametophyte generation often dominates the life cycles of the nonvascular plants, whereas the sporophyte generation dominates the life cycles of vascular plants, a difference that reflects an adaptation of the vascular plants for life on land.

In nonvascular plants, one generation grows out of the tissues of the other.

28.3 The nonvascular plants include three phyla: the mosses, liverworts, and hornworts.

28.3 The life cycles of most nonvascular plants have two distinct phases, with the gametophyte phase dominating.

28.3 In nonvascular plants, the sporophyte plants live on and derive nutrients from the gametophyte plants.

In vascular plants, one generation of plants lives separately from the other.

28.4 The sporophyte generation of the seedless vascular plants (such as ferns) produce spores, which germinate to form gametophyte plants.

28.4 Gametophytes of the seedless vascular plants produce motile sperm that need water to swim to the eggs.

28.4 The fertilized eggs of seedless vascular plants produce young sporophytes that grow protected within gametophyte tissues but eventually become free living.

28.5 In pine trees (an example of plants with naked seeds), both male cones and female cones of the sporophyte generation produce spores, which undergo meiosis to form male and female gametophytes.

28.5 The male gametophytes (pollen) of pine trees produce sperm, which remain in the pollen grains.

28.5 The female gametophytes of pine trees produce eggs, which develop in ovules within the scales of the female pinecones.

28.5 In the spring, the male pinecones release their pollen grains into the wind, which germinate when they come into contact with ovules.

28.5 After a sperm reaches an egg in the pine cone ovule and fertilization takes place, the zygote develops into an embryo, which matures into a seed.

28.6 The organs of sexual reproduction in the flowering plants, or angiosperms, are in the flower.

28.6 In angiosperms, the male gamete-producing cells are pollen grains and the female gamete-producing cells are in the ovules in the ovary.

28.6 Angiosperm seeds, formed within the ovary after the union of sperm and eggs, protect and nourish the young sporophytes as they begin to develop into new plants.

28.6 The seeds of angiosperms remain within the ovary, which develops into a fruit.

28.7 Insects (particularly bees), birds and other animals, and the wind transfer pollen.

28.8 Seeds are dispersed by sticking to or being eaten by animals, by being blown by the wind, or by floating to new environments across bodies of water.

28.9 Seeds germinate only when they receive water and appropriate environmental cues.

28.9 Developing embryos in seeds use the food reserves stored in the cotyledons and in the endosperm during initial development.

28.10 Plants can reproduce asexually by vegetative propagation; a new plant grows from a portion of another plant.

Plant hormones regulate plant growth.

28.11 Plant growth is influenced by at least five different kinds of chemical messengers, or hormones.

Key Terms

alternation of generations *639*
angiosperms (AN-jee-oh-spurms) *645*
antheridia (singular, antheridium)
    (an-thuh-RID-ee-uh /
    AN-thuh-RID-ee-um) *640*
archegonia (singular, archegonium)
    (ar-kih-GO-nee-uh /
    AR-kih-GO-nee-um) *640*
cell culture techniques *651*
dicots *649*

diploid (DIP-loyd) *639*
gametophyte *639*
gametophyte (gamete-plant) genera-
    tion (guh-MEE-toe-fite) *639*
germination *649*
gymnosperms (JIM-no-spurms) *644*
haploid (HAP-loyd) *639*
hormones (HORE-mones) *652*
life cycle *639*
monocots *649*

nonvascular plants
    (non-VAS-kyuh-ler) *638*
seedless vascular plants *642*
seeds *642*
sporangia (spoh-RAN-jee-uh) *640*
sporophyte *639*
sporophyte (spore-plant) generation
    (SPOR-uh-fite) *639*
vascular plants (VAS-kyuh-ler) *638*
vegetative propagation *651*

KNOWLEDGE AND COMPREHENSION QUESTIONS

1. What are seeds? Explain their function(s). (*Comprehension*)

2. Summarize the life cycle of seedless vascular plants. Give an example of this type of plant. (*Comprehension*)

3. Match each of the following with the most appropriate term (*Knowledge*):
   a. Seedless vascular plants
   b. Nonvascular plants
   c. Vascular plants with naked seeds
   d. Vascular plants with protected seeds
   (1) Gymnosperms
   (2) Angiosperms
   (3) Bryophytes
   (4) Ferns

4. Draw a generalized diagram of a flower. Label the following: sepals, petals, stamens, pistil, anther, pollen grains, stigma, ovary, ovules, and style. (*Knowledge*)

5. Summarize the commonalities of sexual reproduction in the angiosperms and the gymnosperms. (*Comprehension*)

6. Summarize the significance of fruits in sexual reproduction. (*Comprehension*)

7. Place the following events in the correct sequence (*Knowledge*):
   a. The radicle emerges from the seed.
   b. Fertilization occurs.
   c. The shoot emerges from the ground.
   d. An embryo develops.
   e. The seed germinates.

8. What do monocots and dicots have in common? How do they differ? (*Comprehension*)

9. What is vegetative propagation? Give an example. (*Knowledge*)

10. What are hormones? Summarize the function(s) of the five major kinds of plant hormones. (*Comprehension*)

KEY
Knowledge: Recalling information.
Comprehension: Showing understanding of recalled information.

CRITICAL THINKING REVIEW

1. There are fruits and vegetables for sale in produce sections of grocery stores. Pick one fruit and one vegetable, such as an apple and a tomato. Describe the characteristics of each and explain why both are really fruits. (*Analysis*)

2. You probably know what a dandelion looks like when it goes to seed—a delicate white sphere at the tip of a slender stalk. Blowing on these dandelion "flowers" is something almost all children do. However, the "real" dandelion flowers are a familiar yellow, growing unwanted in many lawns. If the white spheres are not flowers, what are they? What role do they play in the reproduction of the dandelion plant? (*Application*)

3. In vascular plants, the sporophyte generation is dominant, yet in many vascular plants such as the conifers and the angiosperms, spores are not a dominating structure. When are spores formed in the life cycle of the conifers and the angiosperms? Into what structures do they develop? Do you think they are easily visible to the casual observer? (*Analysis*)

4. In the description of the life cycles of the mosses and the ferns in this chapter, it is noted that water is necessary for the sexual reproduction of both. However, water is not needed for the sexual reproduction of the conifers and the angiosperms. Explain the differences in the life cycles of these plants that account for this. (*Analysis*)

5. In some forests, conifers such as firs, spruces, and pines dominate over angiosperms, or the flowering trees and plants. Many factors determine the distributions of these plants. Discuss how their differing types of seed dispersal might affect where they grow. (*Application*)

KEY

Application: Using information in a new situation.
Analysis: Breaking down information into component parts.

 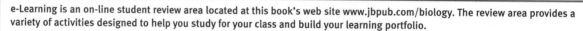

Location: http://www.jbpub.com/biology

e-Learning is an on-line student review area located at this book's web site www.jbpub.com/biology. The review area provides a variety of activities designed to help you study for your class and build your learning portfolio.

Review Questions The review questions test your knowledge of the important concepts and applications in each chapter. The review provides feedback for each correct or incorrect answer. This is an excellent test preparation tool.

Figure-Labeling Exercises Sharpen your visual thinking by matching terms and labels for important illustrations.

Flash Cards Studying biology requires learning new terms. Virtual flash cards help you master the new vocabulary for each chapter.

Just Wondering As you read and study from this text, you may find that you have unanswered questions. Through this site you can ask the author, Sandy Alters, your "just wondering" questions.

 Do you prefer the speed and reliability of a CD-ROM? All of the features contained on the eLearning portion of the web site are also available on the Student CD-ROM.

1. In each of the life cycles shown, label the sporo-phyte(s) and gametophytes.
2. For all stages labeled indicate whether they are haploid (1N) or diploid (2N). Indicate where meiosis and fertilization occur in each cycle.
3. What generation is dominant in each cycle?

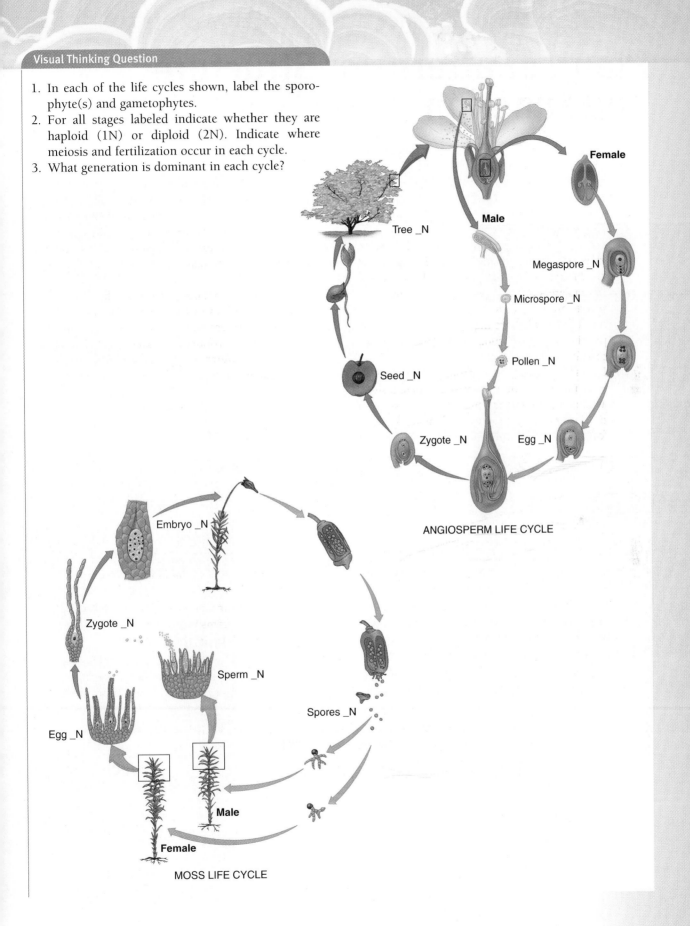

ANGIOSPERM LIFE CYCLE

MOSS LIFE CYCLE

Plants: Patterns of Structure and Function

Although the photo below looks like the discarded shell of a huge turtle, it is actually a plant. In fact, one nickname for this plant is the turtleback. The turtleback's "shell" is really a modified stem that stores water for the plant. This helps the turtleback survive the dry season in its tropical environment. During the rainy season, a system of green branches grows out its top, manufacturing food and bearing leaves, flowers, and fruits. Through the years these branches die, covering the turtleback with an intertwining mass.

Turtlebacks commonly grow in Africa. The native people call this plant *Hottentot's bread* because its fleshy storage organ can be eaten and has a texture similar to a turnip. This organ can weigh up to 318 kilograms (700 pounds).

Another example of a stem that is modified as a giant water storage organ is found in the tall cacti (shown at right) that are characteristic of the American Southwest.

You will often notice in your study of plants and animals that structures give you clues about function. In this case, the enlarged, fleshy stems of these two plants that live on opposite sides of the Earth from each other perform the same function, water storage.

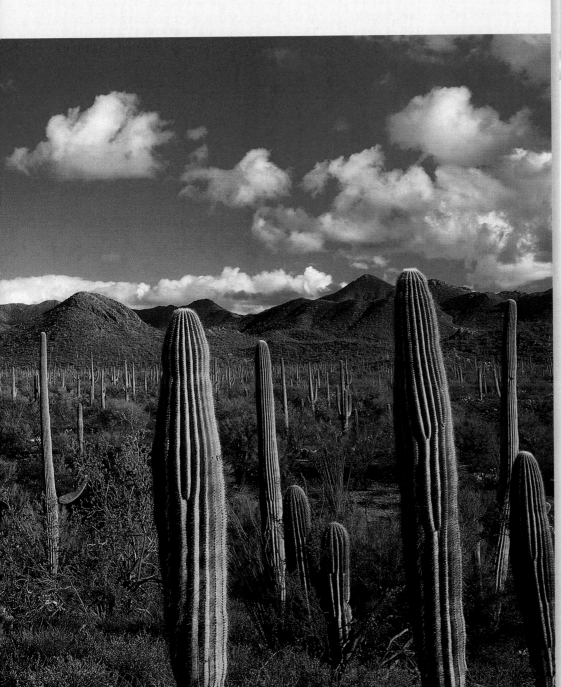

Organization

29.1 The organization of vascular plants

For any plant to grow as tall as the turtleback, it must have a means to transport substances to and from its various parts. Most plants (approximately 95% of all living species) have a system of specialized tissues called vascular tissue that transports water and nutrients. Plants with these tissues are called **vascular plants** (see Chapter 28 for discussion of their reproductive cycles). Despite their reproductive diversity, however, vascular plants all have the same basic architecture.

A vascular plant is organized along a vertical axis as shown in Figure 29.1. The part below the ground is called the **root**. The part above the ground is called the **shoot**. The root penetrates the soil and absorbs water and various ions crucial for plant nutrition. It also anchors the plant. The shoot consists of a **stem** and **leaves**. The stem serves as a framework for the positioning of the leaves, where most photosynthesis takes place. The arrangement, size, and other characteristics of the leaves are critically important in the plant's production of food. Flowers and ultimately fruits and seeds are also formed on the shoot of the flowering plants.

> Vascular plants are those having a system of tissues that transports water and nutrients. Vascular plants have the same basic architecture: They are made up of underground roots and aboveground shoots. A shoot consists of a stem and leaves.

29.2 Overview of the tissues of vascular plants

The organs of a vascular plant—the leaves, roots, and stem—are made up of different mixtures of tissues, just as your legs are composed of different tissues such as bone and muscle. A tissue is a group of cells that work together to carry out a specialized function. Vascular plants have three types of differentiated tissues, groups of cells having specific structures to perform specific functions: *vascular tissue, ground tissue,* and *dermal tissue* (see Figure 29.1).

The word vascular comes from a Latin word meaning vessel. Thus, **vascular tissue** forms the "circulatory system" of a plant, conducting water and dissolved inorganic nutrients (see Table 29.1, p. 663) up the plant and the products of photosynthesis throughout the plant. **Ground tissue** stores the carbohydrates the plant produces. It is the tissue in which the vascular tissue is embedded and it forms the substance of the plant. **Dermal tissue** covers the plant, protecting it—like skin—from water loss and injury to its internal structures. Plants also contain **meristematic tissue** (MER-uh-stuh-MAT-ik), or growth tissue, an undifferentiated tissue in which cell division occurs. This tissue, often just called *meristem,* is considered undifferentiated because the cells it produces will eventually become one of the other three cell types.

> Vascular plants have three types of differentiated tissues: vascular tissue, ground tissue, and dermal tissue. Meristematic tissue is undifferentiated growth tissue.

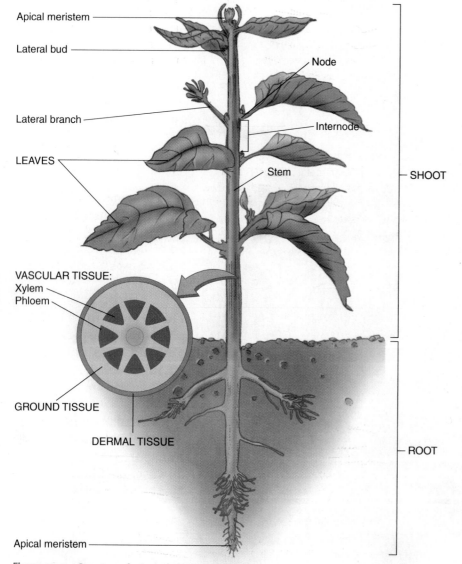

Apical meristem

Lateral bud

Lateral branch

LEAVES

Node

Internode

Stem

SHOOT

VASCULAR TISSUE:
Xylem
Phloem

GROUND TISSUE

DERMAL TISSUE

ROOT

Apical meristem

Figure 29.1 **Structure of a typical plant.**

29.3 Vascular tissue: Fluid movement

There are two principal vascular tissues: **xylem** (ZY-lem), which conducts water and dissolved inorganic nutrients, and **phloem** (FLO-em), which conducts carbohydrates the plant uses as food with other needed substances. Each vascular tissue is made up of different, specialized conducting cells.

Figure 29.2a and b shows two types of xylem cells. Both types of cells conduct water and dissolved inorganic nutrients only after they die and lose their cytoplasm, becoming hollow and thick walled. Stacked end to end, they form pipelines that extend throughout the plant. The cells pictured in Figure 29.2a are *tracheids* (TRAY-kee-idz), which are stacks of cells with tapering ends that have connections between them called pits. A layer of cell wall material covers the pits, but water moves through them nevertheless. Figure 29.2b shows *vessel elements*, which are stacks of cells that have connections between them called perforations. Because these openings are not covered by tissues, water moves through perforations unimpeded. Along the lengths of both cell types are pits, which allow water to move laterally to cells surrounding the xylem pipelines. They work much like soaking hoses used in gardens.

Figure 29.2c, shows two types of phloem cells: a *sieve-tube member* and *companion cell*. The conducting cells (sieve-tube members) are alive and contain cytoplasm but are not typical cells. They lack nuclei, for example. The ends of these elongated cells have pores that allow the easy passage of sugar-filled water from where it is produced—in the leaves, for example—to where it is used rapidly—at the reproductive structures, for example. Companion cells, which contain all of the organelles commonly found in plant cells (including nuclei), secrete substances into and remove substances from the sieve-tube members. These substances include sugars produced during photosynthesis.

There are two principal vascular tissues: xylem, which conducts water and dissolved inorganic nutrients, and phloem, which conducts carbohydrates the plant uses as food with other needed substances. Two types of xylem cells are tracheids and vessel elements. Two types of phloem cells are sieve tubes and companion cells.

29.4 Ground tissue: Food storage

Ground tissue contains *parenchyma* (pah-RENG-keh-mah) cells, which function in photosynthesis and storage. These thin-walled cells contain large vacuoles and may also be packed with chloroplasts. They are the most common of all plant cell types and form masses in leaves, stems, and roots. Fleshy storage roots, such as carrots or sweet potatoes, contain a predominance of storage parenchyma. The flesh of most fruits is also made up of these cells.

Sclerenchyma (skli-RENG-keh-mah) cells, also found in the ground tissue, are hollow cells with strong walls. These cells help support and strengthen the ground tissue. The gritty texture of a pear is due to schlerenchyma cells that are dispersed among the softer parenchyma cells. Both types are in the photos at right.

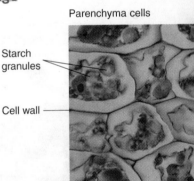
Parenchyma cells

Starch granules

Cell wall

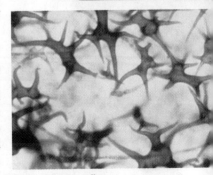
Sclerenchyma cells.

Ground tissue contains parenchyma cells, which function in photosynthesis and storage, and sclerenchyma cells, which help support and strengthen the ground tissue.

(a)

Perforation plate

Pits

Pores

(b)

Companion cell

Nucleus

Sieve-tube member

(c)

Figure 29.2 Vascular tissues of a plant. (a) Tracheids and (b) vessel elements are two types of xylem cells. (c) Sieve tube members and companion cells are two types of phloem cells.

29.5 Dermal tissue: Protection

Dermal tissue covers the outside of a plant, with the exception of woody shrubs and trees that have protective bark in its place. Bark is made up of other tissue types (see p. 668). Epidermal cells are the most abundant type of cell found in the dermal tissue. These cells are often covered with a thick, waxy layer called a cuticle that protects the plant and provides an effective barrier against water loss (Figure 29.3a).

Other types of cells found in the dermis are guard cells, which surround openings in the leaves through which gases and water vapor enter and leave (Figure 29.3b), and trichomes (TRY-kohmz; Figure 29.3c), which are outgrowths of the epidermis (much like the hairs on your body) that have various functions. For example, on "air plants," trichomes help provide a large surface area for the absorption of water and inorganic nutrients. On leaves or fruits, fuzzlike trichomes reflect sunlight, which helps control water loss. In some desert plants, white multicellular trichomes reflect enough sunlight to reduce temperatures in internal tissues. Some trichomes defend a plant against insects or larger animals with their sharpness or by secreting chemicals.

> Dermal tissue covers the outside of a plant. Various types of cells can be found in dermal tissue serving various functions, such as protection, gas exchange, and prevention of water loss.

29.6 Meristematic tissue: Growth

Plants contain *meristems,* areas of undifferentiated cells that are centers of plant growth. Every time one of these cells divides, one of the two resulting cells remains in the meristem. In this way, meristem cells remain "forever young," capable of repeated cell division. The other cell goes on to differentiate into one of the three kinds of plant tissue, ultimately becoming part of the plant body.

Plants can grow only in relationship to where their meristematic tissue is located. Growth tissue located at the tips of the roots and the tips of the shoots are called apical meristems. (Apex means "tip.") The location of this type of tissue is pointed out in Figure 29.1. Apical meristems allow plants to grow taller and their roots to grow deeper into the ground. This type of plant growth, which occurs mainly at the tips of the roots and shoots, is called **primary growth**.

Some plants not only grow taller but grow thicker as well. This type of growth is called **secondary growth** and occurs in all woody trees and shrubs such as pines, oaks, and rhododendrons. These plants have a cylinder of meristematic tissue along the length of their stems and branches, which is called lateral meristem.

Herbaceous (nonwoody) plants, such as tulips, have only apical meristems and therefore only primary growth. However, their stems do grow somewhat thicker as they develop. One reason for this growth is that plant development involves both cell division *and* cell growth; as cells enlarge, plant parts enlarge. In addition, some cell division occurs in the cells differentiating from meristematic tissue.

> Meristem, an undifferentiated type of tissue, produces new plant cells during growth. Apical meristem produces growth in length, while lateral meristem (present only in woody plants) produces growth in width.

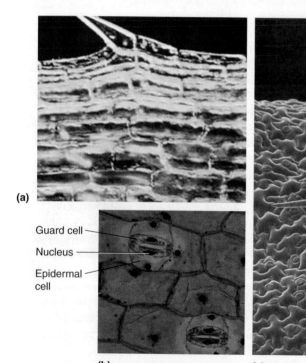

(a)

Guard cell
Nucleus
Epidermal cell

(b)

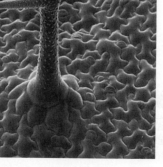

(c)

Figure 29.3 Dermal tissue.
(a) Epidermal cells. (b) Guard cells.
(c) Trichomes.

29.7 Roots

Organs of Vascular Plants

Roots, stems, and leaves are the organs of vascular plants. Together, the stems and leaves make up the shoot. The roots and shoots of all vascular plants share the same basic architecture, but there are differences. This chapter discusses the structure of the plants that dominate the plant world today: the flowering plants, or **angiosperms**. The principal differences between the structures of the roots, stems, and leaves in the seedless vascular plants and those with naked seeds lie primarily in the relative distribution of the vascular and ground tissue systems.

Roots

The function of a root system is to anchor a plant in the soil and to absorb water and inorganic nutrients. During the process of photosynthesis, plants use the water they absorb along with carbon dioxide they capture from the air to produce carbohydrates (see Chapter 6). Primarily, water absorbed at the roots replaces the water released by the plant into the air in a process called *transpiration* (see p. 669). Leaves are the principal organs of transpiration.

The essential inorganic nutrients that plants use are listed in **Table 29.1**. Why do plants need to absorb inorganic nutrients? The answer is simple. Although plants manufacture carbohydrates during photosynthesis, these sugars are not the only substances that plants need in order to live. Plants need nucleic acids, proteins, fats, and vitamins. These substances are formed from the carbohydrates plants manufacture and from the in-

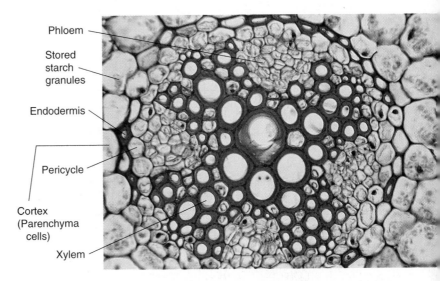

Figure 29.4   Dicot root. This typical dicot root from a buttercup has a central column of xylem with radiating arms. Phloem tissue lies between the xylem arms (red).

Labels on figure:
- Phloem
- Stored starch granules
- Endodermis
- Pericycle
- Cortex (Parenchyma cells)
- Xylem

organic nutrients that plants take in from the soil and concentrate. Many plants, therefore, are an important source of inorganic nutrients in the human diet. Broccoli and cabbage, for example, are excellent sources of calcium. Bananas provide potassium.

Structurally, the roots of dicots (plants having net-like veins in their leaves; see Chapter 28) have a central column of xylem with radiating arms. Between these arms are strands of phloem **(Figure 29.4)**. Ringing this column of vascular tissue (often called the vascular cylinder) and forming its outer boundary is another cylinder of cells called the pericycle. This tissue is made up of parenchyma cells able to undergo cell division to produce branch roots (roots that arise from other older roots). Surrounding the pericycle is a mass of parenchyma called the *cortex*. These cells store food for the growth and metabolism of the root cells. The innermost layer of the cortex is called the endodermis, which consists of specialized cells that regulate the flow of water between the vascular tissues and the outer portion of the root. The outer layer of the root is the epidermis, which absorbs water and inorganic nutrients. These cells have extensions called *root hairs* that provide the epidermal cells with a larger surface area over which absorption can take place—like the function of microvilli in the intestinal walls of animals.

Monocot roots (plants having parallel veins in their leaves; see Chapter 28) are similar to dicot roots with one important exception: Monocot roots often have centrally located parenchyma (storage) tissue called

www.jbpub.com/biology

Table 29.1

Inorganic nutrients important to plants

Macronutrients*	Micronutrients*
Hydrogen	Chlorine
Carbon	Iron
Oxygen	Boron
Nitrogen	Manganese
Potassium	Zinc
Calcium	Copper
Magnesium	Molybdenum
Phosphorus	
Sulfur	

*Macronutrients are present in greater than 10,000 ppm (parts per million); nutrients of lesser concentrations are considered *micro*nutrients. Parts per million equals units of an element by weight per million units of oven-dried plant material.

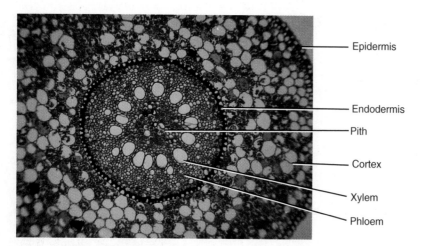

Figure 29.5 Monocot root.

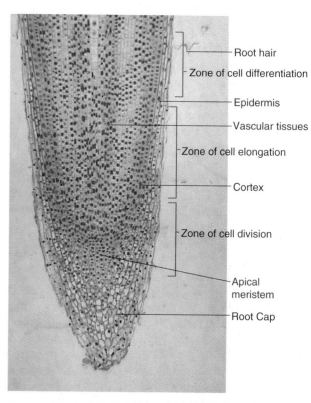

Figure 29.6 Longitudinal section of an onion root tip.

pith. The xylem and phloem are arranged in rings around the pith (**Figure 29.5**).

The end of a root is tipped with apical meristem as is shown in **Figure 29.6**. These growth cells divide and produce cells inwardly, toward the body of the plant, and outwardly. Outward cell division results in the formation of a thimblelike mass of relatively unorganized cells, the root cap, which covers and protects the root's apical meristem as it grows through the soil. Just behind its tip, the root cells elongate. Velvety root hairs, the tiny projections from the epidermis, form above this area of elongation. Cells also differentiate in this portion of the root to produce specialized cell types.

If you have ever pulled a plant up by its roots, you probably noticed that the roots of one type of plant may look different from those of another type of plant. These differences in appearance are linked to differences in function. Many dicots, such as dandelions, for example, have a single, large root called a *taproot* (see **Figure 29.7a**). Even a small section of

Figure 29.7 Types of roots. (a) Taproots in a dandelion, *Taraxacum officinale*. (b) Prop roots in corn, *Zea mays*, are adventitious—they arise from stem tissue and take over the function of the main root. (c) Fibrous roots in a daffodil.

(a)

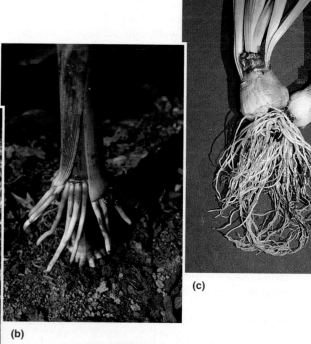

(b)

(c)

If all plants are producers, then why do Venus fly traps consume flies?

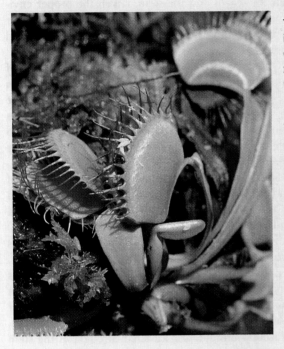

The Venus flytrap (*Dionaea muscipula*) is one of a few carnivorous (animal-eating) plants. Carnivorous plants often grow in soils depleted of nutrients, especially nitrogen. Digesting insects is the way in which these plants obtain nutrients they are unable to get from the soil. Like other plants, however, carnivorous plants make their own carbohydrates during the process of photosynthesis, and thus are producers.

All carnivorous plants must first trap their prey and then digest it to absorb nutrients. The Venus flytrap ensnares prey by means of its touch response, which works as follows. On the inner sides of each pair of cusplike modified leaves of the plant are sensitive hairs. Attracted by nectar on the leaves' surfaces, insects touch the hairs as they walk onto the leaves. If two hairs are touched in succession or if one hair is touched twice (the plant's way of distinguishing between a living and a nonliving stimulus) the leaves close on the insect, imprisoning it. The leaf closure results from biochemical changes that occur within the epidermal cells of the leaves when the sensitive hairs are stimulated. These biochemical changes result in the expansion of the outer epidermal cells of each leaf, whereas the inner epidermal cells do not change. As the leaves change shape because of the cellular changes, they come together.

The Venus flytrap produces digestive enzymes, storing them in vacuoles within the cells of the leaves. As the insect becomes enclosed by the leaves, it gets pressed against the inner surfaces of the leaves. This stimulus results in the discharge of the digestive enzymes by the vacuoles onto the trapped prey. As the organism is digested, the nutrients are absorbed by the leaf cells. Seems like quite a gory story for the plant world, doesn't it?

these taproots can regenerate a new plant, which is one reason why dandelions are so difficult to eliminate from lawns and gardens. Taproots grow deep into the soil, firmly anchoring the plant. Some taproots, such as carrots and radishes, are fleshy because they are modified for food storage. The plant draws on these food reserves when it flowers or produces fruit; thus, taproot crops are harvested before that time. Plants with taproots also have extensive secondary root systems, which are much smaller in diameter than the taproot and so are lost and therefore not seen when the taproot is pulled up. The secondary root system is the major water and nutrient absorption organ of the plant.

The taproot that develops in monocots, on the other hand, often dies during the early growth of the plant, and new roots develop from the lower part of the stem. These are called *adventitious roots*, (AD-ven-TISH-us) and they develop from an aboveground structure. Often, adventitious roots help anchor a plant, such as "prop" roots in corn (Figure 29.7b). Certain dicots, such as ivy plants, also develop adventitious roots. The adventitious roots of ivy plants help them cling to walls.

Have you ever pulled up a clump of grass and looked at its roots? Grass has *fibrous roots*, a type of root system that has no predominant root. Most monocots have fibrous roots also, such as the daffodil shown in Figure 29.7c. They work well to anchor the plant in the ground and absorb nutrients efficiently because of their large surface area. They also help prevent soil erosion by holding soil particles together.

Roots, the part of a plant usually found below ground, absorb water and minerals as well as anchor the plant. Most dicot roots have a central column of xylem with radiating arms and strands of phloem between these arms. Surrounding this vascular tissue is a layer of cells called the pericycle that is capable of cell division. Parenchyma (storage) cells of the cortex surround the pericycle. The entire root is covered with a protective epidermis. Monocot roots are structured similarly but often have an additional storage tissue called pith, which is located in the center of the root; they also contain more xylem elements.

29.8 Stems

Plant shoots grow aboveground and are made up of stems and leaves. As with roots, the growing end of a shoot is tipped with apical meristematic tissue. Young leaves cluster around the apical meristem, unfolding and growing as the stem itself elongates.

Looking back at Figure 29.1, you can see that the leaves form on the stem at locations called **nodes**. The portions of the stem between the nodes are called

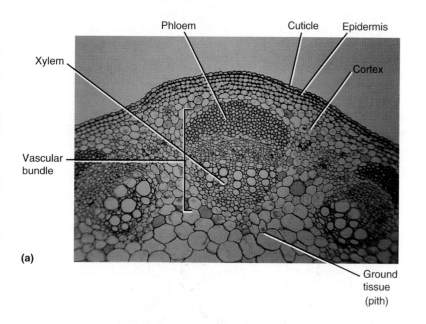

(a)

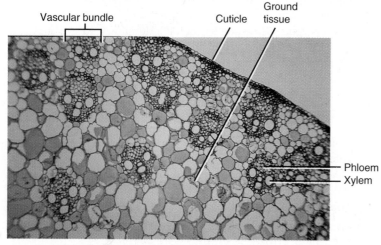

(b)

Figure 29.8 Stems. (a) A dicot stem from the common sunflower, in which the vascular bundles are arranged around the outside of the stem. (b) A monocot stem from corn, showing the scattered vascular bundles characteristic of monocots.

the **internodes**. As the leaves grow, tiny undeveloped side shoots called **lateral buds** develop at the angles between the leaves and the stem. Given the proper environmental conditions, these buds, which contain their own embryonic leaves, may elongate and form lateral branches.

One purpose of stems is to support the parts of plants that carry out photosynthesis. This process takes place primarily in the leaves, which are arranged on the stem so that light will fall on them. In addition, stems conduct water and inorganic nutrients from the roots to all plant parts and bring the products of photosynthesis to where they are needed or stored.

The stem "transportation system" is made up of strands of xylem and phloem tissue that are positioned next to each other, forming cylinders of tissue called vascular bundles (often referred to as the veins of leaves) **(Figure 29.8)**. The xylem tissue characteristically forms the part of each bundle closer to the interior of the stem. The phloem lies closer to the epidermis. In herbaceous dicots (those with soft stems rather than woody, treelike stems), the vascular bundles are arranged in a ring near the periphery of the stem (Figure 29.8a). In monocots, the vascular bundles are scattered throughout the stem (Figure 29.8b).

Because of the arrangement of their vascular bundles, dicot stems have a mass of ground tissue in the center of the stem called the pith and a ring of ground tissue between the epidermis and the vascular bundles called cortex. Monocot stems also have ground tissue, but because it surrounds scattered vascular bundles, ground tissue does not form areas of pith or cortex. The epidermis of both monocot and herbaceous dicot stems is covered with a protective waxy coating called the cuticle.

Dicots with woody stems or trunks (such as flowering trees) and gymnosperms (such as pine trees) have lateral meristems called *cambia*. One type of cambium in woody stems is called vascular cambium. As the name suggests, this growth tissue lies between the vascular tissue—the xylem and phloem—connecting the bundles to form a ring **(Figure 29.9)**. As the cambial cells divide during secondary growth, one of the resulting daughter cells remains as a cambial cell, and the other differentiates into either a xylem or a phloem cell. This new xylem and phloem is called secondary xylem and secondary phloem **(Figure 29.10)**.

The wood of trees is actually accumulated secondary xylem. The wood of dicot trees (such as cherry, hickory, oak, and walnut) is commonly referred to as

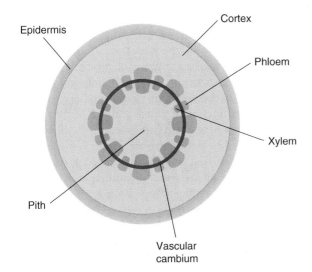

Figure 29.9 **The vascular cambium in a stem.** The vascular cambium (*red*) is the growth tissue in woody stems and lies between the xylem and the phloem.

hardwood, whereas the wood of conifers (such as fir, cedar, pine, and spruce) is called softwood. However, these names are not accurate descriptions of each group. Each contains trees having woods of varying hardness. The hardness of wood relates to its density, which depends on its proportion of wall substance to the space bounded by the cell wall. The denser a wood, the more wall substance it has in relation to the space it bounds and the stronger it is. When used for building, denser woods are harder to nail and machine, but they generally shrink and swell less than less dense, softer woods. Denser woods are also better fuel woods.

When growth conditions are favorable, as in the spring and early summer in most temperate regions, the cambium divides most actively, producing large, relatively thin-walled cells. During the rest of the year, the cambium divides more slowly, producing small, thick-walled cells. This pattern of growth results in the formation of rings in the wood. These rings are called an-

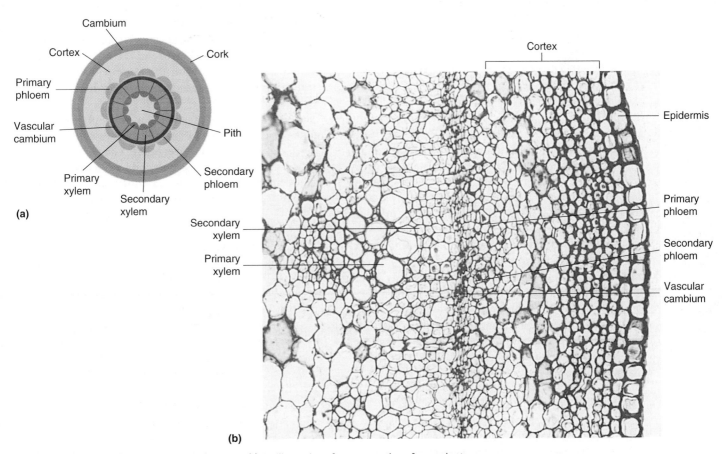

Figure 29.10 **Secondary growth in woody stems.** (a) An illustration of a cross-section of a woody stem. This older stem has an outer layer of cork and underlying cork cambium. Its epidermal cells have died. (b) A photomicrograph of a cross-section of a woody stem. This younger stem has an epidermis as its outer layer. The cork and cork cambium have not yet formed. Secondary growth can be observed on both stems.

Figure 29.11   Annual rings in the cross-section of a tree trunk.

www.jbpub.com/biology

nual rings and can be used to calculate the age of a tree (Figure 29.11).

Other lateral meristem tissue called cork cambium lies just under the epidermis. The outermost cells of this growth tissue produce densely packed cork cells. Cork cells have thick cell walls that contain fatty substances, making these cells waterproof and resistant to decay. When mature, the cork cells lose their cytoplasm and become hardened in a manner similar to the epidermal cells on your skin. The innermost cells of the cork cambium produce a dense layer of parenchyma cells. The cork and the cork cambium make up the outer protective covering of the plant called the bark.

Stems support the photosynthetic structures—the leaves—and transport water and dissolved substances throughout plants. Herbaceous dicot stems are characterized by an inner cylinder of ground tissue called pith surrounded by a ring of vascular bundles. Encircling the ring of vascular bundles is additional ground tissue called cortex. Monocot stems are characterized by scattered vascular bundles embedded in ground tissue.

Woody stems have lateral meristem tissue that provides for secondary growth, which increases the diameter of the stem. Other lateral meristem tissue produces a dense layer of cells called bark that protects the plant.

29.9 Leaves

Leaves, outgrowths of the shoot apex, are the light-capturing photosynthetic organs of most plants. Some exceptions exist; a major exception is found in most cacti, in which the stems are green and have largely taken over the function of photosynthesis for the plants.

Most leaves have a flattened portion, the **blade**, and a slender stalk, the **petiole** (Figure 29.12). Veins, consisting of both xylem and phloem, run through the leaves. In monocots, veins are usually parallel, and in dicots, they are usually netlike (Figure 29.13). Many conifers (vascular plants with naked seeds) have needlelike leaves suited for growth under dry and cold conditions. These modified leaves have thick, waxy coverings beneath, which are compactly arranged, thick-walled cells.

Microscopically, a cross-section of a typical leaf looks somewhat like a sandwich: parenchyma cells in the middle, bounded by epidermis. The vascular bundles, or veins, run through the parenchyma. The leaf parenchyma is appropriately called the *mesophyll*, or "middle leaf" (Figure 29.14).

The mesophyll of most dicot leaves is divided into two layers: the palisade layer and the spongy layer. The palisade layer lies beneath the upper epidermis of the leaf and consists of one or more layers of loosely packed columnlike cells. These cells contain most of the chloroplasts of the leaf, the organelles in which photosynthesis takes place. The spongy layer lies beneath the palisade layer and is made up of irregularly shaped cells. These cells are loosely packed as well and have many air spaces between them. These spaces are connected, directly or indirectly, with openings to the outside called **stomata**

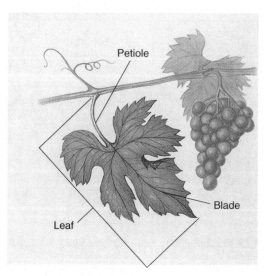

Figure 29.12   Structure of a leaf.

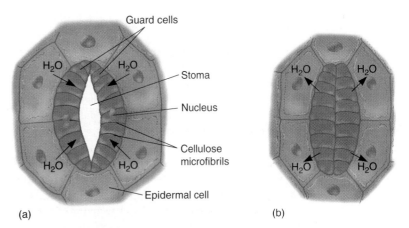

Figure 29.15 **Guard cells.** (a) When solute pressure is high within the guard cells, water moves into the two guard cells and they elongate and bow outward, opening the stomata. (b) When solute pressure is low, the guard cells lose water and close the stomata.

Figure 29.13 **Dicot versus monocot leaves.** (a) The leaves of dicots have net veination; those of monocots, like this palm (b) have parallel veination.

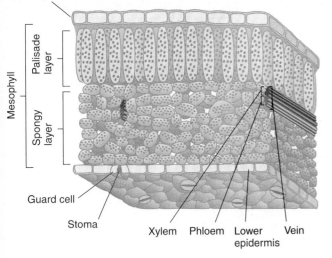

Figure 29.14 **Internal structure of a leaf.** This diagram shows the internal structure of a typical leaf.

(STOW-muh-tuh; sing. **stoma**). Each stoma is bracketed by two **guard cells** that regulate its opening and closing.

The stomata open and close because of changes in the turgor pressure of their guard cells. Opening occurs when solutes (dissolved substances) are actively accumulated in the guard cells. This accumulation of solutes results in the movement of water into the guard cells by osmosis, creating water pressure commonly referred to as *turgor pressure*. As **Figures 29.15** and 29.3 show, guard cells are long cells that are held together at their ends like sausages. Figure 29.15 illustrates that guard cells are ringed by microfibrils of cellulose, much like hoops encircle a barrel. (These microfibrils are invisible to the naked eye.) When the guard cells fill with water, they cannot increase in diameter because of the restricting cellulose microfibrils. Therefore, they increase in length and bow out. This change in the shape of the guard cells opens the stomata.

When photosynthesis is taking place, water enters the guard cells (opening the stomata) because the guard cells actively transport potassium ions to their interior. (The water follows the potassium ions by osmosis because of the osmotic gradient the ions create.) The oxygen produced by photosynthesis diffuses into the atmosphere through the stomata, whereas the carbon dioxide needed for photosynthesis to take place diffuses in. In addition, water leaves the leaf through these openings in the form of water vapor, which is a process called **transpiration**.

Leaves, the light-capturing organs of most vascular plants, are made up of parenchyma cells bounded by epidermis. The parenchyma cells contain chloroplasts, the organelles of photosynthesis. Openings in the leaves allow for the movement of gases and water vapor between the interior of the leaf and the environment.

29.10 Transpirational pull

Did you ever wonder how trees manage to move water to their uppermost leaves? This topic has been studied extensively over the past century. The theory that explains this movement is called the *adhesion-cohesion-tension theory,* or simply *transpirational pull.* **Figure 29.16** shows how transpirational pull moves water from the roots to the leaves of a plant.

The word "tension" as used here means a "pull" on the water molecules within the plant as transpiration takes place, that is, as water evaporates from the leaves. This is how the pull occurs: Water evaporates from air spaces within the leaf and moves out of the leaf via the stomata ❶. Then, water within the mesophyll cells next to these spaces moves by osmosis into the spaces. As water moves out of these cells, the concentration of wa-

ter within them decreases. Water therefore moves from adjacent cells having a higher concentration of water into those cells having a lower concentration of water. A "domino" effect is created, with water moving from cell to cell until the xylem is reached ❷, resulting in a "pull" on the water in the xylem. Notice the close placement of the xylem, the mesophyll cells, and the air spaces in the cross-sectional view of the leaf in Figure 29.16.

The column of water in the xylem of the plant "holds together" because of two other forces: cohesion and adhesion. Cohesion means that water molecules tend to stick together. (See Chapter 3 for a more complete discussion of the properties of water molecules.) Therefore, as molecules at the top of the xylem are pulled up, this force is transmitted all the way down the xylem to the root, from one water molecule that sticks to the next and so forth ❸. Additionally, the water mol-

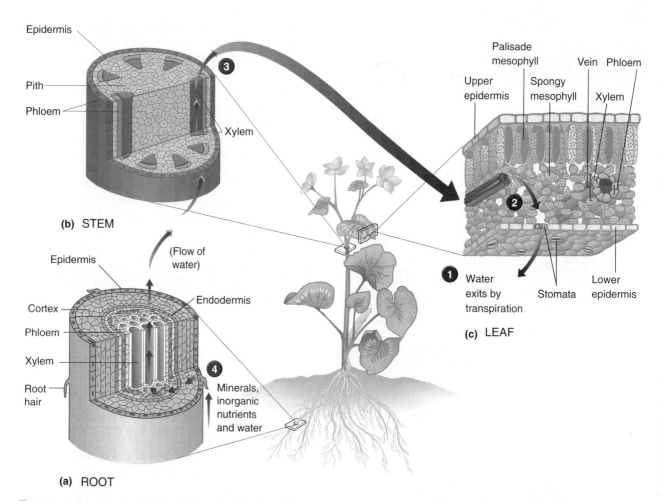

Figure 29.16   Water movement through a plant (transpiration). Water is pulled up through the plant by a combination of forces. (a) Minerals and other inorganic nutrients enter root hairs by active transport. Water enters root hairs by osmosis. (b) These substances travel up the plant through the xylem. (c) Water vapor passes out of the plant through the stomata.

ecules adhere to the walls of the very narrow tracheids and vessel elements of the xylem, a property known as adhesion. These two forces link water molecules together and to the sides of the xylem with weak chemical bonds called hydrogen bonds.

Where does the water come from that is in the xylem of plants? Most of the water absorbed by a plant comes in through its root hairs ❹. Inorganic nutrients also pass into the cells of the root hairs by means of special cellular "pumps." In this way, root cells maintain a higher concentration of dissolved inorganic nutrients than the concentration of inorganic nutrients in the water of the soil. Therefore, water tends to steadily move into the root hair cells from an area of higher concentration (in the soil) to an area of lower concentration (in the root hairs) by osmosis, developing a force called root pressure. Once inside the roots, the water and dissolved inorganic nutrients pass inward to the conducting elements of the xylem.

Water rises in a vascular plant because transpiration from its leaves produces a force that pulls up on the entire water column all the way down to the roots. The forces of cohesion and adhesion work to maintain an unbroken column of water.

29.11 Mass flow

Fluid is also transported in plants by the phloem. As mentioned previously, these tissues transport the products of photosynthesis—sugars—dissolved in water. This sugary solution is commonly called sap. Have you ever seen the collection of sap from maple trees? If so, you realize that plants can rapidly transport large volumes of fluid very quickly. Phloem sap contains 10% to 25% sucrose in addition to inorganic nutrients, amino acids, and plant hormones and may travel as fast as 1 meter (about 1 yard) per hour!

The forces of diffusion and osmosis alone cannot account for this rapid movement of phloem sap. Instead, a pressure-flow, or mass-flow, system performs this function. The mass-flow system is shown in **Figure 29.17** and works in the following way. Sucrose is produced at a source, such as a photosynthesizing leaf, and is actively transported into sieve-tube members by companion cells. As the concentration of sucrose increases in the phloem, water follows by osmosis. In the roots below or at some other sink where sucrose is used, companion cells actively transport sucrose out of the phloem. Water again follows by osmosis. The high hydrostatic (water) pressure in the phloem near the source and the low pressure near the sink cause the rapid flow of the sap. After the sap reaches its destination in the plant, the water can be recycled by moving back to the source through the xylem.

Sucrose from the source is transported into sieve-tube members by companion cells, and water follows by osmosis. At a sink—a place where sucrose is used—sucrose is actively transported from the phloem, and water again follows by osmosis. The high water pressure in the phloem near the source and the low pressure near the sink cause this flow of sucrose, which can be very rapid.

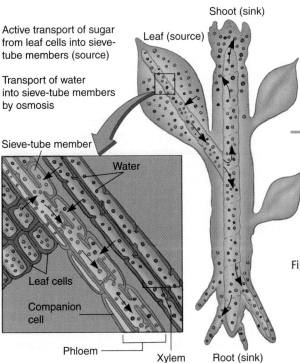

Active transport of sugar from leaf cells into sieve-tube members (source)

Transport of water into sieve-tube members by osmosis

Sieve-tube member

Water

Leaf cells

Companion cell

Phloem

Xylem

Leaf (source)

Shoot (sink)

Root (sink)

Figure 29.17 **Mass flow.**

Nonvascular plants have less
sophisticated transport systems than do
vascular plants.

29.12 The organization of nonvascular plants

The nonvascular plants comprise three phyla: the "true" mosses, the liverworts, and the hornworts (see Chapter 28). These plants are not organized in the same way as the vascular plants. None have true roots, stems, or leaves. They range from forms that look like filamentous algae to forms that look somewhat like certain vascular plants.

Some nonvascular plants have distinct stems and leaflike structures, such as most mosses (see Chapter 28), whereas other nonvascular plants do not. The stems and leaflike structures of nonvascular plants look outwardly different from the stems and leaves of vascular plants and are anatomically similar but simpler. These organs have an outer layer of epidermis made up of protective cells and growth cells. The cortex is made of parenchyma cells, much like those found in vascular plants. In addition, these stems and leaflike structures have a central area of water-conducting tissue.

Nonvascular plants have no roots, but some have slender, usually colorless projections called **rhizoids** (RYE-zoyds) that anchor these simple plants to their substrate (see photo at right). Unlike roots, however, rhizoids consist of only a few cells and do not play a major role in the absorption of water or inorganic nutrients. These substances often enter a nonvascular plant directly through its stems or leaves.

In general, the nonvascular plants have no specialized vascular tissues. The sporophytes of many moss species and the gametophytes of some moss species do have a central strand of somewhat specialized water-conducting tissue in their stems, and food-conducting tissue has been identified in a few genera. Even when such tissues are present, however, their structures are much less complex than those found in the vascular plants. As a consequence of having less sophisticated transport systems than the vascular plants do, the bryophytes (mosses) do not grow very tall. Some look as though they are creeping over their substrate, or food source.

The mosses, liverworts, and hornworts are primarily low-growing plants. Some have stems and leaflike structures, but these structures are anatomically much simpler than those of vascular plants. Although most bryophytes (mosses) have no specialized vascular tissues, a few species have somewhat specialized water-conducting tissues. Food-conducting tissues are rare.

Key Concepts

Vascular plants are organized along a vertical axis.

29.1 The body of a vascular plant has two parts: an underground root and an aboveground shoot.

Vascular plants have specialized tissues.

29.2 Vascular plants are made up of four types of tissue: vascular tissue, ground tissue, dermal tissue, and meristematic tissue.

29.2 Vascular tissue conducts water and dissolved inorganic nutrients up the plant; ground tissue stores the carbohydrates the plant produces and forms the substance of the plant; dermal tissue covers the plant; and meristematic tissue provides for growth.

29.3 Vascular tissue is of two types: xylem and phloem.

29.3 Xylem conducts water and dissolved inorganic nutrients from the roots through the stem and to the leaves.

29.3 Phloem conducts water and dissolved sugars (sap).

29.4 Ground tissue contains parenchyma cells, which function in photosynthesis and storage, and scle-

renchyma cells, which help support and strengthen the ground tissue.

29.5 Dermal tissue covers the outside of a plant.

29.5 Various types of cells can be found in dermal tissue serving various functions, such as protection, gas exchange, and prevention of water loss.

29.6 Plants undergo primary growth (growth in length) by cell elongation and apical meristems, zones of active cell division at the ends of the roots and the shoots.

29.6 Secondary growth (growth in width) in both stems and roots takes place in woody trees and shrubs by means of lateral meristems along the length of their stems and branches.

Roots, stems, and leaves are the organs of vascular plants.

29.7 The root system anchors a plant in the ground and absorbs water and inorganic nutrients from the soil.

29.7 Most dicot roots have a central column of xylem with radiating arms and strands of phloem between these arms.

Key Concepts, continued

29.7 Surrounding the vascular tissue in dicot roots is a layer of cells called the pericycle, which is capable of cell division, surrounded by parenchyma (storage) cells of the cortex.

29.7 The entire dicot root is covered with a protective epidermis.

29.7 Monocot roots often have an additional storage tissue called pith, which is located in the center of the root.

29.7 In monocot roots, the xylem and phloem are arranged in rings around the pith.

29.8 Stems support the leaves, conduct water and inorganic nutrients from the roots to all plant parts, and bring the products of photosynthesis to where they are needed or stored.

29.8 Together, the xylem and phloem are called vascular bundles.

29.8 Herbaceous dicot stems are characterized by an inner cylinder of ground tissue called pith surrounded by a ring of vascular bundles.

29.8 Encircling the ring of vascular bundles in herbaceous dicot stems is additional ground tissue called cortex.

29.8 Monocot stems are characterized by scattered vascular bundles embedded in ground tissue.

29.8 Dicots with woody stems and gymnosperms have lateral meristem tissue, or cambium.

29.8 The vascular cambium lies between the xylem and phloem; its dividing cells form xylem toward the interior (secondary xylem) and phloem (secondary phloem) toward the exterior.

29.8 Wood is accumulated secondary xylem; it often displays rings because it exhibits different rates of growth during different seasons.

29.9 Leaves, the photosynthetic organs of most vascular plants, are made up of specialized ground tissue cells, or parenchyma, bounded by epidermis.

29.9 Vascular bundles run through the parenchyma, and parenchymal cells contain chloroplasts, the organelles in which photosynthesis takes place.

29.9 Openings in the epidermis of the leaf (stomata) allow the carbon dioxide needed for photosynthesis to enter and the oxygen produced by photosynthesis to escape.

29.9 Water vapor evaporates from a plant through its stomata.

Fluids move in vascular plants by transpirational pull and mass flow.

29.10 Water flows through plants in a continuous column, driven mainly by the evaporation of water vapor from the stomata.

29.10 The cohesion of water molecules and their adhesion to the walls of the narrow xylem through which they pass are important factors in maintaining the flow of water to the tops of plants.

29.11 The process of mass flow moves sucrose from where it is produced in a plant to where it is used.

29.11 During mass flow, sucrose is actively transported into phloem cells, where it is produced, and is actively transported out of the phloem cells, where it is used.

29.11 These active transport processes produce a sugar gradient and a water pressure gradient, which cause the movement of sugar and water.

Nonvascular plants have less sophisticated transport systems than do vascular plants.

29.12 The organization and transport systems of nonvascular plants are much less complex than those of the vascular plants.

Key Terms

angiosperms
  (AN-jee-oh-spurms) *663*
blade *668*
dermal tissue *660*
guard cells *669*
ground tissue *660*
internodes *666*
lateral buds *666*
leaves *660*

meristematic tissue
  (MER-uh-stuh-MAT-ik) *660*
nodes *666*
petiole (PET-ee-ole) *668*
phloem (FLO-em) *661*
primary growth *662*
rhizoids (RYE-zoyds) *672*
root *660*
secondary growth *662*

shoot *660*
stem *660*
stomata (singular, stoma)
  (STOW-muh-tuh/STOW-muh) *668*
transpiration *669*
vascular plants *660*
vascular tissue *660*
xylem (ZY-lem) *661*

KNOWLEDGE AND COMPREHENSION QUESTIONS

1. Match each type of tissue to its function:
   a. Stores food manufactured by the plant
   b. Protects the plant
   c. Produces new plant cells during growth
   d. Conducts water, inorganic nutrients, carbohydrates, and other substances throughout the plant
      (1) Meristem
      (2) Dermal tissue
      (3) Ground tissue
      (4) Vascular tissue
   (*Knowledge*)

2. What do xylem and phloem have in common? How do they differ? (*Comprehension*)

3. Identify and give a function for each of the following: (a) parenchyma cells, (b) epidermal cells, (c) stomata, (d) root cap. (*Knowledge*)

4. Distinguish between primary and secondary growth. (*Knowledge*)

5. Draw two diagrams, one showing the root of a "typical" monocot, the other the root of a "typical" dicot. Label the pericycle, xylem, cortex, endodermis, epidermis, and pith. (*Knowledge*)

6. What type of root system would you expect to find in a dandelion, an ivy plant, and a clump of grass? What are the advantages of each type of root? (*Comprehension*)

7. How can annual rings help you estimate the age of a tree? What type of tissue is involved? In which type of growth does this result? (*Knowledge*)

8. Describe the process by which water rises in a vascular plant. What forces are involved? (*Comprehension*)

9. Diagram the movement of phloem sap in a mass-flow system. Label the source and sink, and show the direction of flow. (*Knowledge*)

10. In general, how does the appearance of nonvascular plants differ from that of vascular plants? Why? (*Comprehension*)

KEY
Knowledge: Recalling information.
Comprehension: Showing understanding of recalled information.

CRITICAL THINKING REVIEW

1. In tropical climates, many tall plants shut their stomata during the hot days and open them at night. If their stomata are closed during the day, why doesn't the water within the plant fall down the stem? (*Application*)

2. The roots of many plants have permanent mutualistic associations with fungi. What might be the advantage of this association to the plant? (*Synthesis*)

3. Compare and contrast the structure and function of the "circulatory system" of a vascular plant with the circulatory system of a human. List and compare at least three major differences and two similarities. (*Synthesis*)

4. When you were in middle/junior high school, you carved your initials about 4½ feet up the trunk of a tree in the woods near your house. It's 10 years later, you've moved, and that tree has grown 10 feet taller. If you stopped to "visit" the tree, how high up the trunk would you look to find your initials, and why? (*Application*)

5. If you look at the food pyramid in Chapter 8 on p. 172, you can see that a large proportion of the pyramid—that is, a large proportion of a healthy diet—consists of "plant foods." Using your knowledge of plant structure and physiology from this chapter, describe two benefits of eating "plant food." (*Application*)

KEY
Application: Using information in a new situation.
Synthesis: Putting together information from different sources.

Label the structures of a typical plant (shown below) by filling in the blanks.

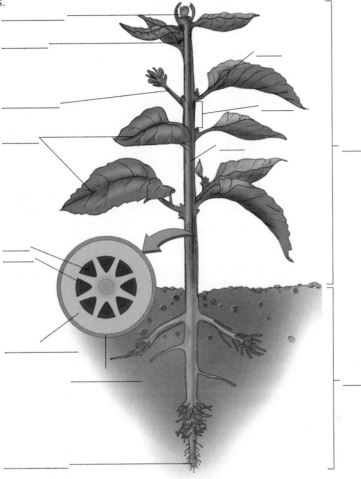

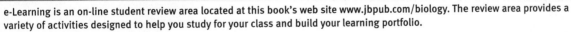

e-learning

Location: http://www.jbpub.com/biology

www.jbpub.com/biology

e-Learning is an on-line student review area located at this book's web site www.jbpub.com/biology. The review area provides a variety of activities designed to help you study for your class and build your learning portfolio.

Review Questions The review questions test your knowledge of the important concepts and applications in each chapter. The review provides feedback for each correct or incorrect answer. This is an excellent test preparation tool.

Figure-Labeling Exercises Sharpen your visual thinking by matching terms and labels for important illustrations.

Flash Cards Studying biology requires learning new terms. Virtual flash cards help you master the new vocabulary for each chapter.

Just Wondering As you read and study from this text, you may find that you have unanswered questions. Through this site you can ask the author, Sandy Alters, your "just wondering" questions.

Do you prefer the speed and reliability of a CD-ROM? All of the features contained on the eLearning portion of the web site are also available on the Student CD-ROM.

Invertebrates

Hermit crabs are masters of self-defense. Like other crabs and lobsters, they have two pinching claws, but unlike other crabs, they can retreat into their portable shelter (usually the leftover shell of a snail that has died). The hermit crab carries its shell wherever it goes, holding on to the shell with two specialized hooks at the end of its abdomen.

Several species of hermit crab go one step further. This coral reef dwelling hermit crab at right employs the services of another invertebrate, the sea anemone. The hermit crab encourages the growth of anemones on its shell by allowing anemones to settle there, and even picks up anemones and places them on the shell. The anemones (relatives of jellyfish) have stinging tentacles that dissuade predators from attacking the hermit crab, and also help the hermit crab blend in with the background environment. The anemones, in turn, benefit from being moved to many different locations, enhancing their feeding opportunities on the coral reef. This is an example of symbiosis between two invertebrates.

Later in this chapter you will learn about the patterns of structure and function that invertebrate animals have in common despite their incredible diversity. But first you will look at the "bigger picture" of the animal kingdom to discover its unifying themes.

Organization

Animals are multicellular, heterotrophic eukaryotes.

Sponges are asymmetrical and have no coelom.

Hydra and jellyfish are radially symmetrical and have no coelom.

Flatworms are bilaterally symmetrical and have no coelom.

Roundworms are bilaterally symmetrical and have a false coelom.

Molluscs, annelids, arthropods, and echinoderms are bilaterally symmetrical and have a coelom.

Animals are multicellular, heterotrophic eukaryotes.

30.1 Characteristics of vertebrate and invertebrate animals

How are animals different from the other kingdoms of living things? Like the plants, fungi, and protists, animals are eukaryotic organisms, having a distinct nucleus and a cellular structure different from the prokaryotic structure of bacteria (see Chapter 26). Also like the plants, most fungi, and some protists, animals are multicellular. No single-celled animals exist. Only their gametes are single celled, but these cells are not independently living organisms. As soon as fertilization takes place, the development of a new multicellular individual begins. (See Figure 28.2a, which diagrams the sexual life cycle of animals.)

Animals are heterotrophs, unable to make their own food. Therefore, animals must eat plants, other organisms, or organic matter for food. Some simple animals, such as the sponges, take organic matter directly into their cells. Most animals digest food within a body cavity. The resulting molecules are then taken into the body cells to be broken down further by the chemical reactions of cellular respiration (see Chapter 6). The end product of cellular respiration is energy, which is used to drive the activities of life, including growth, maintenance, reproduction, and response to the external environment. As part of this response, most animals are capable of movement to capture food or to protect themselves from injury.

The cells of animals are organized into tissues, which are groups of cells combined into structural and functional units (see Chapter 7). In most animals, the tissues are organized into organs, complex structures made up of two or more kinds of tissues. Organs that work together to perform a function are organ systems.

Animals are extraordinarily diverse in their forms and how they function. This diverse kingdom is often divided informally into two subgroups: the invertebrates

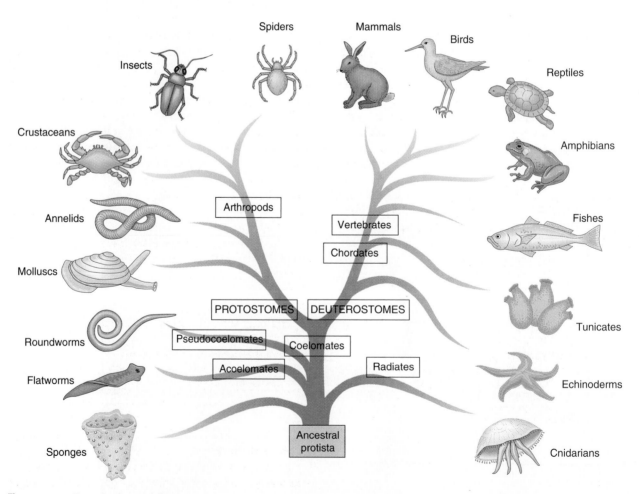

Figure 30.1 The animal ancestral tree.

and the vertebrates. The invertebrates are animals without a backbone, a series of bones that surrounds and protects a dorsal (back) nerve tube and that is used as a lever system in movement. In some invertebrates, the dorsal nerve, as well as the backbone, does not exist. Examples of invertebrates are spiders, sponges, jellyfish, snails, and worms. The vertebrates (a group that includes you) have both a backbone and a dorsal nerve tube. Examples of vertebrates are bears, fish, dogs, cats, birds, and frogs. Interestingly, although the vertebrates are usually larger and more commonly known, the invertebrates make up more than 95% of all animal species.

Most biologists agree that the animals arose from protist ancestors (see Chapter 25). The evolutionary relationships among the animals and their ancestors are shown in Figure 30.1. Each animal on the lower branches of the tree represents a present-day phylum in the animal kingdom. The animals on the upper right branches—the mammals, birds, reptiles, amphibians, and fishes—are all classes of animals in the subphylum Vertebrata (phylum Chordata). Tunicates (sea squirts) are not vertebrates but are a phylum of chordate marine animals that have a notochord instead of a bony vertebral column. In addition, the tunicates lack a brain, an organ common to all vertebrates. The animals on the upper left branches in Figure 30.1—the crustaceans, insects, and spiders—all represent phyla of the diverse group of organisms called arthropods. Arthropods are animals that have a hard outer shell, jointed legs, and a segmented body.

Figure 30.1 shows that the sponges, cnidarians (jellyfish, corals), and flatworms are the present-day phyla of animals most closely related to the evolutionary ancestors of the animals. Likewise, animals higher on the tree represent present-day phyla and classes of organisms that are less closely related to the ancestors of the animals. Each branch of this phylogenetic tree represents an evolutionary pathway that diverged from an ancestral pathway.

Animals are a diverse group of eukaryotic, multicellular, heterotrophic organisms. This diverse group is often informally divided into two subgroups: the invertebrates and the vertebrates. An invertebrate is an animal without a backbone; a vertebrate is an animal with a backbone that surrounds and protects a dorsal nerve cord. Most zoologists agree that the animals evolved from protist ancestors.

30.2 Sponges (Porifera)

All animals (except the sponges) exhibit either radial symmetry or bilateral symmetry (see Chapter 7, p. 133). Radial symmetry means that the body parts emerge from a central axis. Bilateral symmetry means that the right side is a mirror image of the left side. Sponges are asymmetrical (without symmetry or regularity of form).

Sponges (phylum Porifera [poe-RIF-er-ah]) are aquatic organisms; most species live in the ocean rather than in fresh water. These sessile creatures are considered simpler than other animals in their organization because they have no tissues (although biologists debate this; see the next paragraph), no organs, and no coelom, or body cavity (see Chapter 7, p. 134). A coelom is a fluid-filled enclosure within an organism that is lined with connective tissue. The asymmetrical bodies of sponges consist of little more than masses of cells embedded in a gelatinous material, or matrix, as shown in Figure 30.2a.

The body of a sponge is shaped like a sac or vase. Thin, vase-shaped sponges are shown in the photograph in Figure 30.2b and in the accompanying illustration. The body wall is covered on the outside by a layer of flattened cells called the *epithelial wall*. (Some call it epitheliumlike because this epithelium is different from that of most other animals. Yet others contend that sponges have no tissues because each specialized cell functions on its own.) Lining the inside cavity of the sponge are specialized, flagellated cells called *collar cells*. The matrix makes up the substance of the sponge, sandwiched between the outer epithelial layer and the inner layer of collar cells. Within the matrix are ameboid-type cells, needlelike crystals of calcium carbonate or silica, and tough protein fibers. Pores, channellike openings that span the matrix, are dispersed throughout the sponge. These pores are integral to the movement of water, dissolved substances, and particulate matter to the interior of the sponge. They give the sponges their phylum name Porifera, which means "pore bearers."

As they beat, the flagella of the collar cells create a current of water that flows from the outside of the sponge, through pores in the matrix, to the internal cavity of the sponge, and then out again through the large opening at the top of the sponge. The circulation of water in this way brings the nutrients in the water to the collar cells.

Sponges reproduce both sexually and asexually. Sponges frequently reproduce asexually by fragmentation; groups of cells become separated from the body of the sponge and develop into new individuals. In addition, sponges may develop branches that grow over the

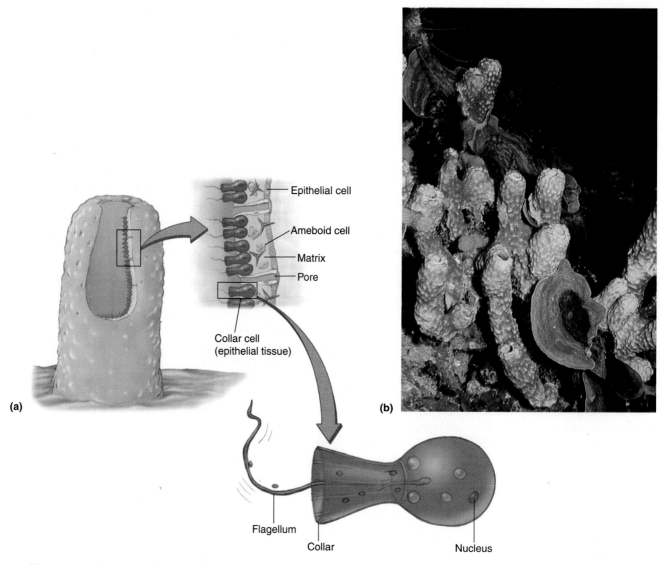

Figure 30.2   **The sponge (phylum Porifera).** (a) Diagram of a sponge, with detail of a collar cell. (b) Yellow tube sponges.

rocks on the sea floor, much like a plant develops underground runners. Colonies of sponges grow along these branches.

During sexual reproduction, most species of sponges produce both female sex cells (eggs) and male sex cells (sperm), which arise from cells in the matrix. Both types of sex cells are produced within the same organism. Such individuals are called **hermaphrodites** (hur-MAF-rah-dytes), after the Greek male god Hermes and the female goddess Aphrodite. The sperm are re-

leased into the cavity of the sponge and are carried out of the sponge with water currents and into neighboring sponges through their pores. Fertilization occurs in the gelatinous matrix where the eggs are held. There, the fertilized eggs develop into flagellated, free-swimming larvae that are released into the sponge's cavity. After the larvae leave the interior of the sponge, they settle on rocks and develop into adults.

Sponges (phylum Porifera) are aquatic, asymmetrical, acoelomate organisms shaped somewhat like a vase. They are considered the simplest of animals because they have no tissues or organs.

30.3 Hydra and jellyfish (Cnidaria)

Animals other than the sponges have a definite shape and symmetry. Only two phyla are classified as radially symmetrical: the Cnidaria (neye-DARE-ee-uh), which includes jellyfish, hydra, sea anemones, and corals, and Ctenophora (teh-NOF-ah-rah), a minor phylum that includes the comb jellies. Two examples of Cnidarians are shown in Figure 30.3, and a comb jelly is shown in Figure 30.4.

There are two basic body plans exhibited by the cnidarians: **polyps** (POL-ups) and **medusae** (meh-DOO-see). The polyp form is exhibited by the jellyfish in Figure 30.3a, and the medusa form is exhibited by the coral in Figure 30.3b. Both forms are illustrated in Figure 30.5.

Figure 30.4   A comb jelly. Note the comblike plates and two tentacles.

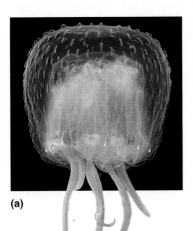

(a)

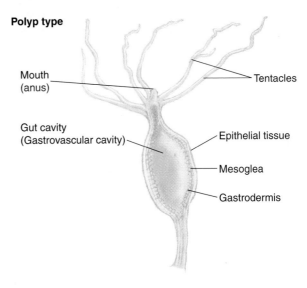

Polyp type

Mouth (anus)

Gut cavity (Gastrovascular cavity)

Tentacles

Epithelial tissue

Mesoglea

Gastrodermis

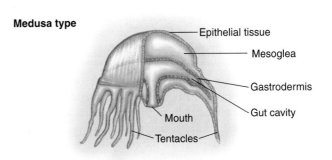

Medusa type

Epithelial tissue

Mesoglea

Gastrodermis

Gut cavity

Mouth

Tentacles

(b)

Figure 30.3   Representatives of two classes of cnidarians (phylum Cnidaria). (a) Jellyfish. (b) Orange cup coral.

Figure 30.5   Body forms of cnidarians: The polyp and the medusa.

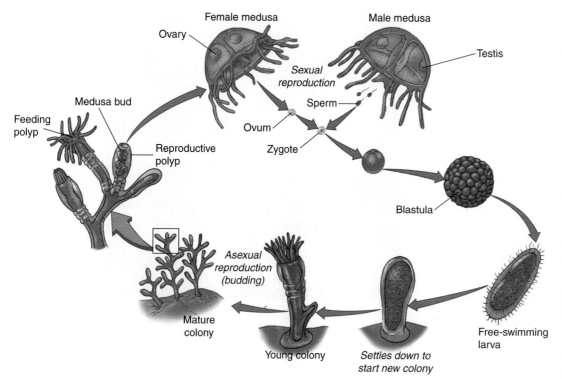

Figure 30.6   The life cycle of *Obelia*, a marine cnidarian. This life cycle demonstrates the alternation of medusa and polyp forms found in cnidarians.

Polyps are aquatic, cylindrical animals with a mouth at one end that is ringed with tentacles. (The name polyp actually means "many feet.") Polyps such as the sea anemones and corals live attached to rocks. Like the corals, many polyps build up a hard outer shell, an internal skeleton, or both. Some polyps are free floating, such as the freshwater *Hydra*.

In contrast, most medusae are free floating and are often umbrella shaped. Commonly known as jellyfish, medusae have a thick, gelatinous interior. The mouth of a medusa is usually located on the underside of its umbrella shape, with its tentacles hanging down around the umbrella's edge.

Structurally, epithelial tissue covers the outside of cnidarians; an inner tissue layer, the gastrodermis, lines the gut cavity (see Figure 30.5). The mesoglea (literally, "middle glue") lies between. This layer is quite thick within medusae and gives them their jellylike appearance. A network of nerve cells extends through cnidarians, but they have no brainlike controlling center.

Some cnidarians such as the hydra, sea anemones, and corals occur only as polyps. Simple polyps such as hydra usually reproduce asexually by budding (see p. 398). However, some of their tissue is organized into primitive ovaries that produce eggs and testes that produce sperm, as do the more complex polyps such as sea anemones and corals. (Some species of these organisms *never* reproduce by budding.) Some hydra, like sponges, are hermaphrodites. Others exist as separate sexes. Eggs remain attached to the hydra but exposed to the water. Sperm are discharged from the testes and swim to the egg. After fertilization, developing hydras grow while attached to the parent.

Some cnidarians exist only as medusae. Medusae reproduce sexually. Ovaries hang from the underside of female medusae, and testes hang from the males. Eggs and sperm are shed into the water, where fertilization takes place. The fertilized egg develops into a larva that never settles down to become a sessile polyp but develops directly into a medusa. However, in most species, medusae have a life cycle in which the larvae develop into polyps. Some of these polyps produce medusae. This type of alternating life cycle is shown in **Figure 30.6.**

The tentacles of a cnidarian help it capture prey—other animals such as small fishes, shrimp, and aquatic worms. The tentacles bear stinging cells called *cnidocytes,* which give the phylum its name. You can see these cells as tiny dots in the tentacles of the yellow cup coral shown in Figure 30.3b. If you have ever been stung by a jellyfish, you know how powerful its sting can be. These stinging cells work much like harpoons. Powered by water pressure, threadlike stingers are jettisoned out of the cells, spearing and immobilizing the prey. The tentacles then draw the prey back to the mouth.

Cnidarians have two layers of tissues and a nerve net that coordinates cell activities. They exist either as polyps (corals and sea anemones), which are cylindrically shaped animals that anchor to rocks, or as medusae (jellyfish), which are free-floating, umbrella-shaped animals. In some cnidarians, these two forms alternate during the life cycle of the organism.

Flatworms are bilaterally symmetrical and have no coelom.

30.4 Flatworms (Platyhelminthes)

The remaining phyla of animals are all bilaterally symmetrical or (in the case of the echinoderms) are thought to have evolved from bilaterally symmetrical forms. In addition, they all develop embryologically from three layers of tissue: an inner layer or **endoderm**, an outer layer or **ectoderm**, and a middle layer or **mesoderm**. (See p. 515 in Chapter 23 and also Figure 23.8, which shows these body layers in the human.) Although cnidarians have a middle layer between their two tissue layers, it consists of a jellylike material with only widely dispersed cells. It is not considered a tissue layer.

The bilaterally symmetrical animals with the simplest body plan are the flatworms (phylum Platyhelminthes [PLAT-eh-hel-MIN-these]). This phylum name comes from Greek words meaning "flat" (*platys*) "worm" (*helminthos*) and describes their flattened ribbon or leaflike shapes. Although simple in structure, the flatworms have organs and some organ systems, but because they have no coelom, the organs are embedded within the body tissues (see Figure 7.4). Flatworms are also the simplest animals to have a distinct head, a characteristic common to many of the bilaterally symmetrical animals.

There are three classes of flatworms: the turbellarians, the flukes, and the tapeworms. Examples of the turbellarians and the flukes are shown in Figure 30.7. A photo of a tapeworm head is shown in the "Just Wondering" box. The turbellarians are free living and

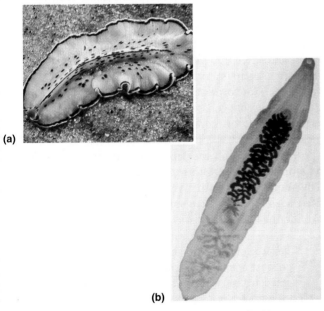

(a)

(b)

Figure 30.7 **Flatworms (phylum Platyhelminthes).** (a) A marine, free-living turbellarian. (b) The human liver fluke, *Clonorchis sinensis.*

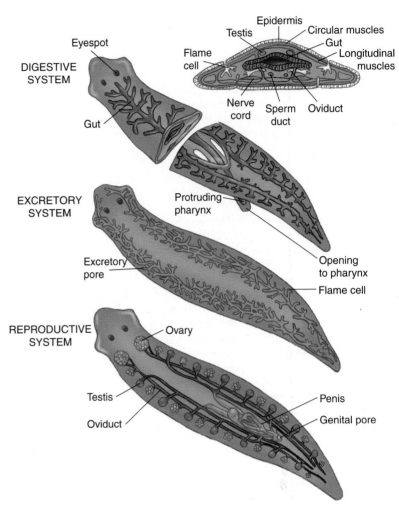

Figure 30.8 **Flatworm anatomy.** The organism shown is *Dugesia,* the familiar freshwater flatworm used in many biology laboratories.

found in fresh water, salt water, or damp soil. The flukes and tapeworms are parasites and live on or in other animals, deriving nutrition from their hosts.

Free-living flatworms move from place to place, feeding on a variety of small animals and bits of organic debris. They move by means of ciliated epithelial cells that are concentrated on their ventral surfaces. In fact, the name turbellarian comes from a Latin word meaning "to bustle or stir" and refers to the water turbulence created by their movement. Sensory pits or tentacles along the sides of their heads detect food, chemicals, and movements of the fluid in which they are moving. They also have eyespots on their heads, which contain light-sensitive cells that enable the worms to distinguish light from dark. These organs are part of the flatworm nervous system and connect to a ladderlike paired nerve cord that extends down the length of the animal (see Figure 14.1). Tiny swellings

*J*ust **Wonder***ing* . . . *Real Questions Students Ask*

How do people get tapeworms? Are they common in the United States?

The most common tapeworm that affects humans is the beef tapeworm *(Taenia saginata)*. The life cycle of this tapeworm begins when cattle (the secondary hosts) graze in areas in which tapeworm eggs contaminate the soil. After the eggs are eaten, they hatch within the intestine, and the larvae burrow through the intestinal wall until they reach muscle. Here, the larvae encyst—they become encapsulated and quiescent.

The person who eats raw or undercooked infected beef ingests the cysts along with the meat. Digestive enzymes break down the capsule surrounding the larvae, and they attach themselves to the intestinal wall with grasping hooks of their head (shown in the photo). Feeding on the digested food of the host, the tapeworms mature within a few weeks. As adults, they consist of long chains of segments called *proglottids*. Each proglottid has both male and female sex organs and can produce up to 100,000 eggs, which are shed with proglottids in the feces. If the feces contaminate areas where cattle graze and they are in-

gested, the life cycle begins once again.

Parasites such as tapeworms are more prevalent in countries having poor sanitation systems (or no sanitation systems at all) than in countries with effective sewage treatment. In the United States, water is treated and purified in sewage treatment plants before it is returned to rivers, streams, or the ocean. In addition, beef is inspected before it can be sold. However, infection is possible. To protect yourself, do not eat beef that is raw or undercooked.

Two other tapeworms, pork tapeworms and fish tapeworms, can also infect humans. Their life cycles are similar to that of the beef tapeworm; humans are their primary hosts. Therefore, avoid eating undercooked pork, raw fish (sushi), or undercooked fish.

Infection with tapeworms is serious. The organisms can live for years within the body

and absorb nutrients essential for proper nutrition. Developing malnutrition due to a tapeworm infestation is common. In addition, long tapeworms can block the movement of materials through the intestine. Tapeworm infections can be treated with certain medications, but treatment is difficult if the worms invade tissues beyond the intestines, as sometimes happens. Following the precautions mentioned here will help you avoid sharing your body with these unpleasant creatures!

---

at the cephalic, or head, end of the organism are considered a primitive brain.

The flatworm has a digestive system consisting of a digestive sac, or gut, open only at one end. (Flatworm anatomy is shown in Figure 30.8. Also refer to Figure 8.1.) Muscular contractions in the upper end of the gut of flatworms cause a strong sucking force by which the flatworms ingest their food and tear it into small bits. The cells making up the gut wall engulf these particles; most digestion takes place within these cells. Wastes from within the cells diffuse into the digestive tract and are expelled through the mouth. In addition, the flatworm excretes excess water and some wastes by means of a primitive excretory system: a network of fine tubules that runs along the length of the worm. Specialized bulblike cells, or flame cells, are located along these tubules (see Figure 30.8 and Figure 12.1). As cilia within them beat (looking like a flickering flame), they move the water and wastes into the tubules and out excretory pores.

Reproduction in flatworms is much more complicated than in sponges or cnidarians. Although most flatworms are hermaphroditic, a characteristic exhibited by the sponges and cnidarians, the organs of reproduction

are better developed. When flatworms mate, each partner deposits sperm in a copulatory sac of the other. The sperm travel along special tubes to reach the egg. In free-living flatworms, the fertilized eggs are laid in cocoons and hatch into miniature adults. In some parasitic flatworms, there is a complex succession of distinct larval forms. Flatworms are also capable of asexual reproduction. In some genera, when a single individual is divided into two or more parts, each part can regenerate an entirely new flatworm.

The flukes and the tapeworms are two classes of flatworms that live within the bodies of other animals. The adult form of both classes parasitizes humans. The life cycles of these organisms are discussed in the "Just Wondering" box.

The acoelomates, typified by the flatworms, are the most primitive bilaterally symmetrical animals. Although simple in structure, the flatworms have organs and some organ systems, including a primitive brain. Because flatworms do not have a coelom, their organs are embedded within the body tissues.

Roundworms are bilaterally symmetrical and have a false coelom.

30.5 Roundworms (Nematoda)

Seven phyla have a pseudocoelomate ("false" coelom) body plan—a body cavity with no connective tissue lining. (This type of a body plan is shown in Figure 7.4.) Only one of the seven pseudocoelomate phyla, the roundworms, includes a large number of species.

Roundworms are classified in the phylum Nematoda, a word that comes from a Greek word meaning "thread."

These cylindrical worms are diverse in size, but some of them are so small and slender that they look like fine threads. In fact, they can be so microscopically small that a spadeful of fertile soil may contain millions of these worms, which are shown in the photo to the left. Although most are similar in form, they range in length from about 0.2 millimeters (about the width of a human hair) to about 6 millimeters (the diameter of the head of a tack). Some species are abundant in fresh water or salt water.

Many members of this phylum are parasites of vertebrates. About 50 species of roundworms parasitize humans, causing problems such as blockage of the lym-

phatic vessels or intestines and infections of the muscles or lungs. The people shown in the photo to the right have elephantiasis, a condition caused by roundworms that live in the lymphatic passages and block the flow of lymph. As a result, fluids cannot drain, and swelling occurs. The swelling is noticeable in this photo in the lower legs, ankles, and feet. Roundworms are transmitted to humans as larvae by the bite of an infected mosquito.

Roundworms also parasitize invertebrates and plants. For this reason, some nematodes are being investigated as agents of biological control of insects and other agricultural pests.

Roundworms are covered by a flexible, tough, transparent multilayered tissue (cuticle) that is shed as they grow (see Figure 16.1). A layer of muscle lies beneath this epidermal layer and extends lengthwise,

which is pointed out in the illustration of a roundworm in Figure 30.9. These longitudinal muscles pull against both the cuticle and the firm, fluid-filled pseudocoelom, similar to how your muscles pull against your bones. All this effort gets them nowhere in water, but they can move in mud or soil, which provide surfaces against which the worm's body pushes. Whether or not the movements result in locomotion, they push on the fluid-filled pseudocoelom and aid in the distribution of food and oxygen throughout the worm.

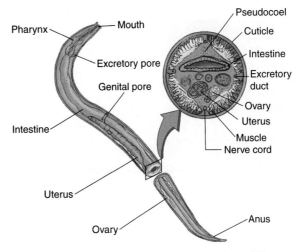

Figure 30.9 **Anatomy of _Ascaris_, a parasitic roundworm of humans.**

The roundworm digestive system has two openings—a mouth and anus. Most roundworms have raised, hairlike sensory organs near their mouths. The mouth itself often has piercing organs, or stylets. Food passes through the mouth as a result of the sucking action of a muscular pharynx. After passing through these organs, food continues through the digestive tract, where it is broken down and then absorbed. The roundworms that parasitize animals take in digested food of the host; the cells lining the digestive system simply absorb these nutrients.

The roundworms also contain primitive excretory and nervous systems. The nervous system consists of a ring of tissue surrounding the pharynx and a solid dorsal and a ventral nerve cord (unlike the hollow nerve tube of a chordate [see p. 698]). The excretory system consists of two lateral canals that unite near the anterior end to form a single tube ending in an excretory pore.

The pseudocoelomates are typified by the roundworms. Muscles attached to their thick, outer cuticle push against the fluid-filled pseudocoelom, resulting in a whiplike movement. Many of the roundworms are parasites of invertebrates, vertebrates (including humans), and plants.

Molluscs, annelids, arthropods, and echinoderms are bilaterally symmetrical and have a coelom.

30.6 Clams, snails, and octopuses (Mollusca)

The rest of the invertebrate animals have a "true" coelom. The next three groups of coelomates—the molluscs (clams, snails, and octopuses), the annelids (earthworms and leeches), and the arthropods (lobsters, insects, and spiders)—are all protostomes. That is, they show a spiral pattern of cleavage in early development (see Figure 23.2). The acoelomate animals already discussed are not protostomes, but exhibit similar developmental patterns to the protostomes. These similarities suggest an evolutionary closeness among these groups.

The molluscs are a large phylum of invertebrate animals having a muscular foot and a soft body covered by a mantle; they are usually covered with a hard shell. The phylum name Mollusca comes from a

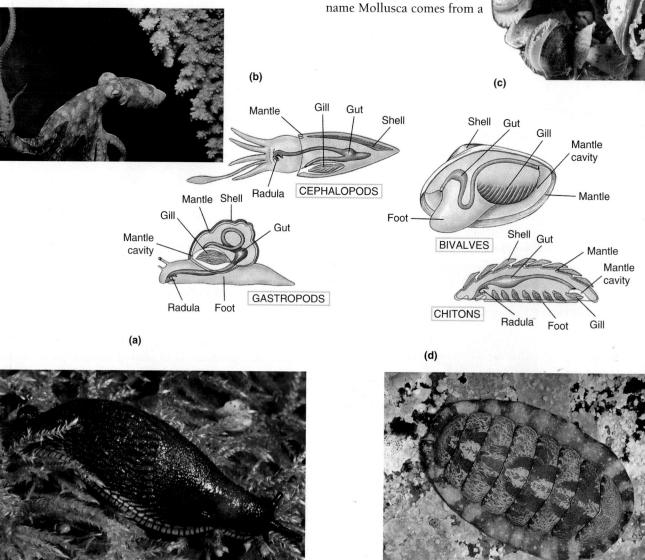

Figure 30.10 Body plans among the molluscs. The name of each group describes its prominent features. (a) *Gastropods*. The example shown in the photo is a slug. (b) *Cephalopods*. The example shown in the photo is an octopus. (c) *Bivalves*. The example shown in the photo is a mussel. (d) A *chiton* (shown in both illustration and photo).

Latin word meaning "soft bodied." The shelled molluscs include the snails, clams, scallops, and oysters. Unshelled molluscs are represented by the octopuses, squids, and slugs.

Molluscs are widespread and often abundant in marine and freshwater environments, and some, such as certain snails and the slugs, live on land. They range in size from being near microscopic to having the huge proportions of giant squid. Large giant squid measure approximately 21 meters long (almost 70 feet) and weigh approximately 250 kilograms (550 pounds). The largest giant squid ever documented, however, is over 140 feet long!

Molluscs exhibit four body plans: cephalopod, bivalve, gastropod, and chiton, which are shown in **Figure 30.10.** The name of each group describes its prominent features. Although each group (with the exception of the bivalves) has a head end, the cephalopods (literally, "head-foot") have the most well-differentiated head and the most well-developed nervous system. Along with a large head, cephalopods have long, armlike tentacles that surround a mouth and a pair of large eyes. Examples of cephalopods are octopus and squid.

The bivalves ("two-shelled") gastropods have the least well-developed nervous system of these groups. Bivalves are usually sedentary and may have a muscular foot with which they bury themselves. Examples of bivalves are clams, mussels, oysters, and scallops.

Gastropods (from Greek words meaning "stomach foot") have eyes and feelers on a distinct head. Most have a shell that is spiral or cone-shaped, and a muscular foot on its underside for locomotion. Examples of gastropods are snails, slugs, and limpets.

Chitons are marine organisms that adhere to rocks. Their name comes from a Greek word meaning "tunic," which refers to their shell of eight overlapping plates.

All molluscs have a *visceral mass,* or group of organs, consisting of the digestive, excretory, and reproductive organs. The visceral mass is covered with a soft epithelium called the mantle, which arises from the dorsal body wall and encloses a cavity between itself and the visceral mass. This cavity is *not* the coelom; the coelom surrounds the heart only. The mollusc's gills, the organs of respiration, lie within the mantle cavity. Gills are a system of filamentous projections of the mantle tissue that is rich in blood vessels. These projections greatly increase the surface area available for gas exchange. In land-dwelling molluscs, a network of blood vessels within the mantle cavity serves as a primitive lung.

Molluscs exhibit both open circulatory systems (in clams, for example) and closed circulatory systems (in squid, for example). Both types of circulatory system have a heart to pump blood. In a **closed circulatory system**, blood is enclosed within vessels as it travels throughout the body of the organism. In an **open circulatory system**, blood flows in vessels leading to and from the heart but through irregular channels called *blood sinuses* in many parts of the body. Sections 10.2 and 10.3, and Figure 10.1 describe and show the differences between open and closed circulatory systems in more detail.

Most coelomate animals have a closed circulatory system (with the notable exception of the insects); their blood vessels are intimately associated with the excretory organs, making the direct exchange of materials between these two systems possible. Molluscs were one of the earliest evolutionary lines to develop an efficient excretory system. Wastes are removed from the mollusc by tubular structures called *nephridia* (see Chapter 12, p. 244). In molluscs with open circulatory systems, wastes move from the coelom into the nephridia and are discharged into the mantle cavity. From there, they are expelled by the continuous pumping of the gills. In molluscs and other animals with closed circulatory systems, such as annelids, some molluscs, and the vertebrates, the coiled tubule of a nephridium is surrounded by a network of capillaries. Wastes move from the circulatory system to the nephridium (referred to as nephron tubules in vertebrates) for removal from the body. All coelomates (except for arthropods and chordates) have basically similar excretory systems.

Molluscs, widespread in marine and freshwater environments, exhibit four body plans as represented by snails, clams, squid, and chitons. These animals use gills for respiration in water; terrestrial species have adaptations for breathing on land. They exhibit both open and closed circulatory systems and well-developed excretory systems.

30.7 Earthworms and leeches (Annelida)

The other two major groups of protostomes, the annelids and the arthropods, have segmented bodies, whereas the molluscs do not. Segmentation underlies the organization of all of the more complex animals: the annelids (earthworms and leeches), arthropods (lobsters, insects, and spiders), echinoderms (sea urchins and sea stars), and chordates (tunicates, fishes, amphibians, reptiles, birds, and mammals). In some adult arthropods the segments are fused; their segmentation is apparent only in their embryological development. So, too, embryological development reveals segmentation in the chordates. Segmentation in this phylum is exhibited only by the vertebrates in the repeating units of their backbones and as muscle blocks, such as the abdominal muscles of humans. Because segmentation in animals is different among phyla, scientists think it arose independently in more than one line of evolution.

The annelids (AN-ul-idz; phylum Annelida) are worms characterized by a soft, elongated body composed of a series of ringlike segments. In fact, the word annelid means "tiny rings." Annelids are abundant in the soil and in both marine and freshwater environments throughout the world. Internally, their segments are divided from one another by partitions called *septa*. Circulatory, excretory, and neural structures are repeated in each segment.

There are three classes of annelids: marine worms, freshwater and terrestrial worms (such as earthworms), and leeches. The photo shows a shiny bristleworm, an example of a marine worm. Exhibiting unusual forms and sometimes iridescent colors, marine worms live in burrows, under rocks, inside shells, and in tubes of hardened mucus they manufacture. Leeches occur mostly in fresh water, although a few are marine and some tropical leeches are found in terrestrial habitats. Most leeches are predators or scavengers, and some suck blood from mammals, including humans. The best-known leech is the medicinal leech, which was used for centuries to remove what was thought to be excess blood responsible for certain illnesses. The medicinal leech is now used as

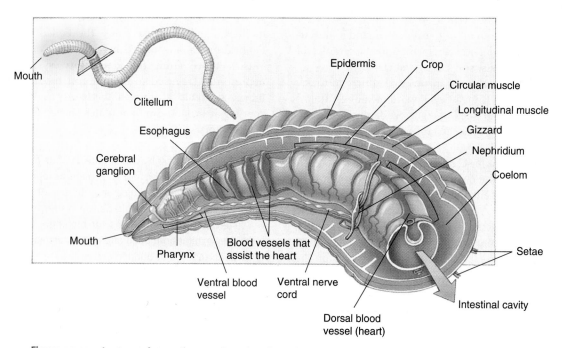

Figure 30.11 **Anatomy of an earthworm.** Note the tube within a tube body plan.

a source of anticoagulant in research that focuses on blood clotting. This animal is also used by some physicians to remove excess blood after surgery and to help restore circulation to severed body parts (such as fingers) after reattachment.

The earthworm exhibits the generalized body plan of this phylum: a tube within a tube (Figure 30.11). The digestive tract, a straight tube running from mouth to anus, is suspended within the coelom. An earthworm sucks in organic material by contracting its strong pharynx. It grinds this material in its muscular gizzard, aided by the presence of soil particles it takes in with its food.

The anterior segments of an earthworm contain a well-developed *cerebral ganglion,* or brain, and a few muscular blood vessels that act like hearts, pumping the blood through the closed circulatory system. Sensory organs are also concentrated near the anterior end of the worm. Some of these organs are sensitive to light, and elaborate eyes with lenses and retinas have evolved in certain members of the phylum. Separate nerve centers, or ganglia, are located in each segment and are connected by nerve cords. Each segment also contains both circular and longitudinal muscles, which annelids use to crawl, burrow, and swim. *Setae,* or bristles, help anchor the worms during locomotion or when they are in their burrows.

Reproduction differs among the annelid classes. In the marine worms, the sexes are usually separate, and fertilization is often external, occurring in the water and away from both parents. The earthworms and leeches, on the other hand, are hermaphroditic. When they mate, their anterior ends point in opposite directions and their ventral surfaces touch (see Figure 22.2). The *clitellum,* a thickened band on an earthworm's body, secretes a mucus that holds the worms together as they exchange sperm. Ultimately, the worms release the fertilized eggs into cocoons also formed by mucous secretions of the clitellum.

The annelids are characterized by serial segmentation and a tube within a tube body plan. The body is composed of numerous similar segments, each with its own circulatory, excretory, muscular, and neural structures.

30.8 Crabs, insects, and spiders (Arthropods)

Crabs and lobsters (phylum Crustacea); insects, centipedes, and millipedes (phylum Mandibulata); and spiders, horseshoe crabs, mites, and ticks (phylum Chelicerata; KEH-lis-er-AH-tah) are representatives of the diverse group of organisms called arthropods (ARE-thrapods). The name *arthropod* comes from the two Greek words *"arthros"* (jointed) and *"podes"* (feet) and describes the characteristic jointed appendages of all arthropods. The nature of the appendages differs greatly in different subgroups; appendages may take the form of antennae, mouthparts of various kinds, or legs.

The arthropods have a rigid external skeleton, or **exoskeleton,** which varies greatly in toughness and thickness among arthropods. Some arthropods have a tough exoskeleton, like the South American scarab beetle shown in Figure 30.12a. Others have a fragile exoskeleton, like the green darner dragonfly shown in Figure 30.12b. The exoskeleton provides places for muscle attachment, protects the animal from predators and injury, and most important, protects arthropods from water loss. As an individual outgrows its exoskeleton, that

(a)

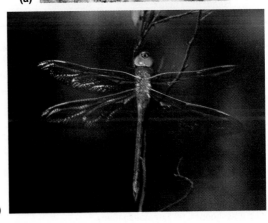

(b)

Figure 30.12 **Exoskeletons.**

exoskeleton splits open and is shed. A new soft exoskeleton lies underneath, which subsequently hardens. The animal then grows into its new "shell."

All arthropods can be placed into one of two groups: those with jaws and those without jaws. The crustacean and insect phyla have jaws, or *mandibles*. These jaws are formed by the modification of one of the pairs of anterior appendages (but *not* the first pair). The appendages nearest the anterior end are sensory antennae. The mandibles and antennae of the bulldog ant are shown in **Figure 30.13a**).

The remaining arthropods (phylum Chelicerata), which include the spiders, horseshoe crabs, mites, and ticks, lack mandibles. Their mouthparts usually take the

form of fangs (also called *chelicerae* [keh-LIS-er-ee], meaning claw), which evolved from the appendages nearest the anterior end of the animal. The fangs of the jumping spider are shown in Figure 30.13b. This spider uses these fangs to catch its prey.

An important structure of many arthropods such as bees, flies, moths, and grasshoppers is the *compound eye*. The compound eyes of the robberfly are shown in the photo above left. Compound eyes are composed of many independent visual units, each containing a lens. *Simple eyes,* composed of a single visual unit having one lens, are found in many arthropods also having compound eyes and function in distinguishing light and

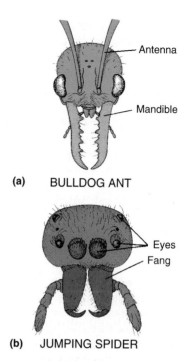

(a)   BULLDOG ANT

(b)   JUMPING SPIDER

Figure 30.13   Arthropod mouthparts.

darkness. In some flying insects, such as locusts and dragonflies, simple eyes function as horizon detectors and help stabilize the insects during flight.

In the course of arthropod evolution, the coelom has become greatly reduced, consisting only of the cavities that house the reproductive organs and some glands. **Figure 30.14** illustrates the major structural features of a grasshopper as a representative of the arthro-

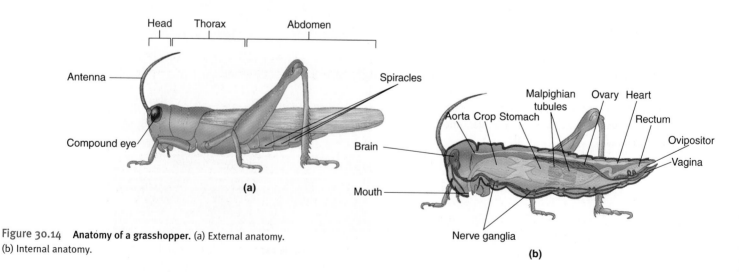

(a)

(b)

Figure 30.14   **Anatomy of a grasshopper.** (a) External anatomy.
(b) Internal anatomy.

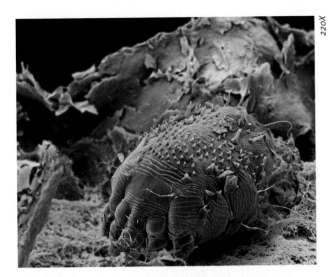

220X

Figure 30.15 A colorized scanning electron micrograph of a mite on human skin.

pods. Like the annelids, the arthropods have a tubular gut that extends from the mouth to the anus. The circulatory system of arthropods is open; their blood flows through cavities between the organs. One longitudinal dorsal vessel functions as a heart, helping move the blood along (see Figure 10.1).

Most aquatic arthropods breathe by means of gills. Their feathery looking structure provides a large surface area over which gas exchange takes place between the surrounding water and the animal's blood. The respiratory systems of terrestrial arthropods generally have internal surfaces over which gas exchange takes place. The respiratory systems of insects, for example, consist of small, branched air ducts called *tracheae* (see Chapter 9, p. 180). These tracheae, which ultimately branch into very small *tracheoles,* are a series of tubes that transmit oxygen throughout the body. The tracheoles are in direct contact with the individual cells, and oxygen diffuses from them to other cells directly across the cell membranes. Air passes into the trachea by way of specialized openings called *spiracles,* which in most insects can be closed and opened by valves.

Although there are various kinds of excretory systems in different groups of arthropods, a unique excretory system evolved in terrestrial arthropods in relation to their open circulatory system. The principal structural element is the *Malpighian tubules* (mal-PIG-ee-en), which are slender projections from the digestive tract (see Figure 12.1 and p. 244). Fluid is passed through the walls of the Malpighian tubules to and from the blood in which the tubules are bathed. The nitrogenous wastes in it are separated out as a solid (precipitated) and then emptied into the hindgut (the posterior part of the digestive tract) and

eliminated. Most of the water and salts in the fluid is reabsorbed by the hindgut, thus conserving water.

Two arthropods, *Sarcoptes scabiei* (sar-KOP-teez SKAY-be-ee; Figure 30.15) and *Phthirus pubis* (THIR-us PEW-bus; Figure 30.16), cause two common contagious parasitic skin infestations of humans: scabies (mites) and pubic lice (commonly known as *crabs*). Both of these organisms are transmitted by close physical contact and are often spread by sexual contact. Although they do not cause serious disease, both mites (organisms closely related to spiders) and lice (organisms closely related to fleas) can be vectors, or carriers, of other diseases. Both respond promptly to treatment with antiparasitic medications applied to the skin.

40X

Figure 30.16 A colorized scanning electron micrograph of an adult and a juvenile louse hanging from pubic hair.

During an infestation of scabies, female itch mites bore into the top layers of their host's skin to lay their eggs. Burrows are formed that look like fine, wavy, dark lines on the surface of the skin. The eggs hatch in a few days, increasing the level of infestation. Itching becomes intense, but scratching abrades the skin surface, which creates an environment in which a secondary bacterial infection can take hold. The common sites of infestation on the body are the base of the fingers, the wrists, the armpits, the skin around nipples, and the skin around the belt line.

The pubic louse is the most common louse to infect humans. It lives in the hairs of the anal and genital area, and the females lay their eggs there, attaching the eggs to the hairs. Infestation causes intense itching.

Arthropods are a diverse phylum of organisms having jointed appendages and rigid exoskeletons.

30.9 Sea urchins and sea stars (Echinodermata)

Sea urchins and sea stars are representatives of the phylum Echinodermata (ih-KEYE-neh-der-MA-tah). The term *echinoderm* means "spine skin," an appropriate name for many members of this phylum. Examples of this phylum are sea lilies, sea stars, brittle stars, sea urchins, sand dollars, and sea cucumbers. Figure 30.17a shows sea stars off the island of Belize. Figure 30.17b shows a brittle star and a sea urchin. The sea urchin is attached to the left side of the red vase sponge, and a brittle star is on top of the sponge. Figure 30.17c shows a sand dollar. These marine animals live on the sea floor, with the exception of a few swimming sea cucumbers.

The echinoderms are different from the other invertebrates in that they are deuterostomes, not protostomes. Deuterostomes cleave in a radial pattern during early embryological development (see Figure 23.2). The embryological differences that distance echinoderms evolutionarily from the other invertebrates connect them with the chordates. It is thought that the echinoderms and the chordates, along with two smaller phyla not mentioned here, evolved from a common ancestor.

Echinoderms are bilaterally symmetrical as larvae but radially symmetrical as adults. They are closely related to and grouped with the bilaterally symmetrical animals, however. Adult echinoderms have a five-part body plan corresponding to the arms of a sea star or the design on the "shell" of a sand dollar. Five *radial canals,* the positions of which are determined early in the development of the embryo, extend into each of the five parts of the body as shown in the illustration of a sea star in Figure 30.18. This water vascular system is used for locomotion and is unique to echinoderms.

As adults, these animals have no head or brain. Their nervous systems consist of central *nerve rings* from which branches arise (see Figure 14.1). The animals are capable of complex response patterns, but there is no centralization of function.

There is no well-organized circulatory system in echinoderms. Food from the digestive tract is distributed to all the cells of the body in the fluid that lies within the coelom. In many echinoderms, respiration takes place by means of skin gills, which are small, fingerlike projections that occur near the spines. Waste removal also takes place through these skin gills. The digestive system is simple, consisting of a mouth, gut, and anus.

(a)

Figure 30.17
Representatives of the
phylum Echinodermata. **(b)**

(c)

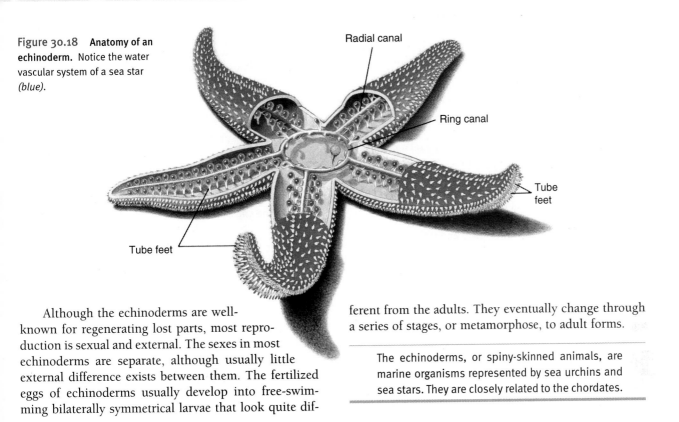

Figure 30.18 Anatomy of an echinoderm. Notice the water vascular system of a sea star (blue).

Radial canal

Ring canal

Tube feet

Tube feet

Although the echinoderms are well-known for regenerating lost parts, most reproduction is sexual and external. The sexes in most echinoderms are separate, although usually little external difference exists between them. The fertilized eggs of echinoderms usually develop into free-swimming bilaterally symmetrical larvae that look quite different from the adults. They eventually change through a series of stages, or metamorphose, to adult forms.

The echinoderms, or spiny-skinned animals, are marine organisms represented by sea urchins and sea stars. They are closely related to the chordates.

Key Concepts

Animals are multicellular, heterotrophic eukaryotes.

30.1 Animals are a diverse group of multicellular, eukaryotic, heterotrophic organisms.

Sponges are asymmetrical and have no coelom.

30.2 Sponges (phylum Porifera) are acoelomate, asymmetrical, aquatic animals.

30.2 Sponges are considered simple animals because they have no tissues or organs.

Hydra and jellyfish are radially symmetrical and have no coelom.

30.3 The hydra and jellyfish (phylum Cnidaria) are acoelomate, radially symmetrical, aquatic animals.

30.3 Cnidarians have two layers of tissues and exist either as cylindrically shaped polyps (corals and sea anemones) or medusae (jellyfish).

30.3 In some cnidarians, these two forms alternate during the life cycle of the organism.

Flatworms are bilaterally symmetrical and have no coelom.

30.4 All animals other than the sponges and the cnidarians are bilaterally symmetrical or, in the case of the echinoderms, evolved from bilaterally symmetrical ancestors.

30.4 All animals other than the sponges and the cnidarians develop embryologically from three tissue layers: endoderm, ectoderm, and mesoderm.

30.4 The flatworms (phylum Platyhelminthes) are ribbonlike worms that live in the soil or water or within other organisms.

30.4 Having no coelom, flatworms are considered to have the simplest body plan of the bilaterally symmetrical animals.

30.4 The flatworms have organs and some organ systems, including a primitive brain.

Roundworms are bilaterally symmetrical and have a false coelom.

30.5 The roundworms (phylum Nematoda) are pseudocoelomate, cylindrical worms that live in the soil, in water, or within other organisms.

Key Concepts, continued

30.5 Roundworms have simple digestive, excretory, and nervous systems.

Molluscs, annelids, arthropods, and echinoderms are bilaterally symmetrical and have a coelom.

30.6 The molluscs (phylum Mollusca) are a large phylum of coelomates that exhibit four body plans: cephalopod, gastropods, bivalve, and chiton.

30.6 All molluscs have a visceral mass, or group of organs, consisting of the digestive, excretory, and reproductive organs.

30.6 Molluscs exhibit both closed and open circulatory systems.

30.7 The annelids (phylum Annelida) are worms characterized by a soft, elongated body composed of a series of ringlike segments, each with its own circulatory, excretory, and neural structures.

30.7 Annelids have a tube within a tube body plan.

30.8 The arthropods are an extremely diverse group and include organisms such as lobsters, insects, and spiders.

30.8 Arthropods are characterized by a rigid external skeleton and jointed appendages.

30.9 The phylum Echinodermata, represented by sea urchins and sea stars, are spiny-skinned marine animals that live on the sea floor.

30.9 Echinoderms are the only deuterostome invertebrates, a characteristic that shows their relatedness to the chordates.

30.9 Although adult echinoderms are radially symmetrical, their larvae are bilaterally symmetrical.

Key Terms

closed circulatory system *687*
ectoderm (EK-toe-durm) *683*
endoderm (EN-doe-durm) *683*
exoskeleton (EK-so-SKEL-uh-tun) *689*

hermaphrodites (hur-MAF-rah-dytes) *680*
medusae (meh-DOO-see *or* meh-DOO-zee) *681*

mesoderm (MEZ-oh-durm) *683*
open circulatory system *687*
polyps (POL-ups) *681*

KNOWLEDGE AND COMPREHENSION QUESTIONS

1. What do you and jellyfish have in common? What is an important taxonomic difference between the two of you? (*Comprehension*)

2. "Animals are a diverse group of eukaryotic, multicellular, heterotrophic organisms." Explain this statement. (*Comprehension*)

3. Which organisms are considered the "simplest animals"? Why? (*Knowledge*)

4. What two basic body plans are shown by cnidarians? Summarize the differences between these two plans. (*Comprehension*)

5. Which organisms are the most primitive bilaterally symmetrical animals? Briefly describe these animals' structure. (*Comprehension*)

6. People can become very ill if infested by roundworms. Describe these organisms and why they can be a health risk to humans. (*Comprehension*)

7. Summarize the characteristics of molluscs. (*Comprehension*)

8. Briefly summarize the characteristics of arthropods. (*Comprehension*)

9. In casual conversation, you refer to a spider on the wall as an insect. A friend who has studied biology informs you that spiders are not insects. Explain what she means. (*Comprehension*)

10. How do echinoderms differ from all other invertebrates? What is the significance of this fact? (*Comprehension*)

KEY
Knowledge: Recalling information.
Comprehension: Showing understanding of recalled information.

learning portfolio

CRITICAL THINKING REVIEW

1. Some invertebrates, such as most flatworms, are hermaphroditic. Wouldn't this characteristic result in a decrease in genetic diversity in populations of these worms? (*Analysis*)

2. One theme that is evident in living things is that they often have structures that are shaped in ways that result in large surface areas. For example, the human small intestine has villi and microvilli that result in an enormous surface area over which the absorption of nutrients takes place (see Chapter 8). Name one structure in an invertebrate animal that results in a large surface area. Describe the structure and how a large surface area assists its function in this animal. (*Application*)

3. Echinoderms are grouped with the bilaterally symmetrical organisms, yet they exhibit radial symmetry. Why are they considered bilaterally symmetrical? Support the view that they evolved from a bilaterally symmetrical ancestor. (*Synthesis*)

4. Insects are a highly diverse phylum of animals. Name one characteristic that distinguishes them from all other arthropods? State a hypothesis as to why you think insects are so abundant on Earth. Give a rationale for your hypothesis. (*Analysis*)

5. What are the major differences between earthworms and flatworms? What advantage does segmentation give earthworms over the flatworms, if any? (*Application*)

K E Y
Application: Using information in a new situation.
Analysis: Breaking down information into component parts.
Synthesis: Putting together information from different sources.

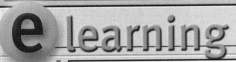 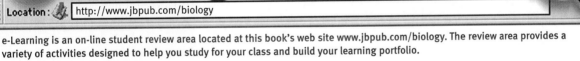

Location: http://www.jbpub.com/biology

e-Learning is an on-line student review area located at this book's web site www.jbpub.com/biology. The review area provides a variety of activities designed to help you study for your class and build your learning portfolio.

Review Questions The review questions test your knowledge of the important concepts and applications in each chapter. The review provides feedback for each correct or incorrect answer. This is an excellent test preparation tool.

Figure-Labeling Exercises Sharpen your visual thinking by matching terms and labels for important illustrations.

Flash Cards Studying biology requires learning new terms. Virtual flash cards help you master the new vocabulary for each chapter.

Just Wondering As you read and study from this text, you may find that you have unanswered questions. Through this site you can ask the author, Sandy Alters, your "just wondering" questions.

 Do you prefer the speed and reliability of a CD-ROM? All of the features contained on the eLearning portion of the web site are also available on the Student CD-ROM.

Chordates and Vertebrates

For some people, the mere mention of bats conjures up images of blood-sucking vampires. In fact, the majority of bats would be totally uninterested in your neck. This bat has just fed on the pollen and nectar of the saguaro cactus flowers. Saguaros are the giant cactuses that are a prominent feature of the Sonoran Desert in the American Southwest (and are shown in full on page 658). Much like bees and some birds, many bats are pollinators. As they feed from flowers, pollen sticks to hairs on their bodies, and can be transferred to other plants visited later. Sperm in the pollen fertilize eggs within the plant—this results in the development of fruit and seeds.

Interestingly, bats are mammals—the same class of animals to which humans belong. Bats are more closely related to you than they are to birds! As this chapter progresses, you will come to understand those similarities as you explore the patterns of structure, function, and reproduction of the chordates (including the vertebrates).

Organization

Chordates have a cartilagenous rod-shaped notochord that forms during development.

In vertebrates, the notochord develops into the vertebral column.

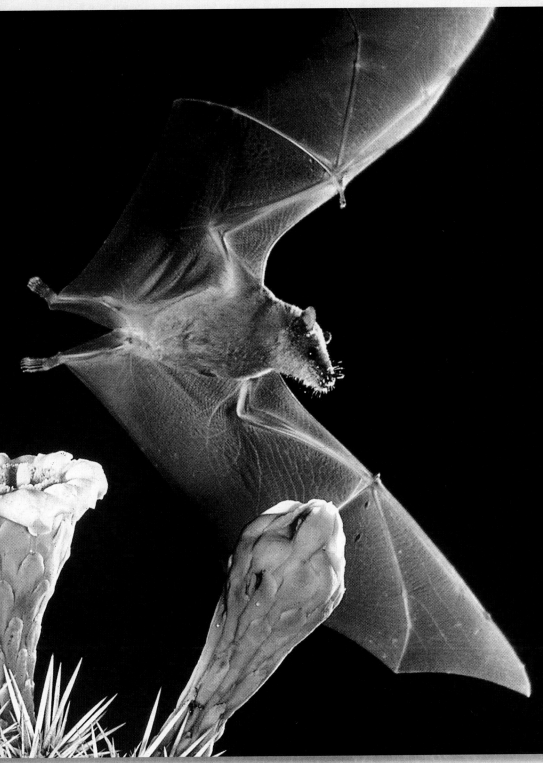

Chordates have a cartilagenous
rod-shaped notochord that forms
during development.

31.1 Characteristics of chordates (Chordata)

The **chordates** are a phylum of animals that include three subphyla: tunicates (TOO-nih-kits), lancelets (LANS-lets), and vertebrates (VER-tuh-bruts). Chordates are characterized by three principal features (see Chapter 23): (1) a single, hollow **nerve cord** located along the back, (2) a rod-shaped **notochord**, which forms between the nerve cord and the gut (stomach and intestines) during development, and (3) **pharyngeal (gill) arches**, which are located at the throat (pharynx) at some stage of life. These three features are present in the embryos of all chordates (see Figure 25.9). In addition, lancelets exhibit these characteristics as adults, and tunicates exhibit them as larvae. In vertebrates, the nerve cord differentiates into a brain and spinal cord. The notochord serves as a core around which the vertebral column develops, encasing the nerve cord and protecting it. The pharyngeal arches develop into the gill structures of the fishes and into ear, jaw, and throat structures of the terrestrial vertebrates. The presence of the gill arches in all vertebrate embryos provides a clue to the aquatic ancestry of the subphylum Vertebrata.

Along with having these three traits, chordates have many other characteristics in common. All have a true coelom and bilateral symmetry. The embryos of chordates exhibit segmentation. **Figure 31.1** shows segments of tissue called *somites* in the human embryo, which develop into the skeletal muscles. Most chordates have an internal skeleton to which their muscles are attached and work against, providing movement. (Larval tunicates and adult lancelets do not have an internal skeleton; their muscles are attached to the notochord.) Finally, chordates have a tail that extends beyond the anus, at least during embryonic development. Nearly all other animals have a terminal anus.

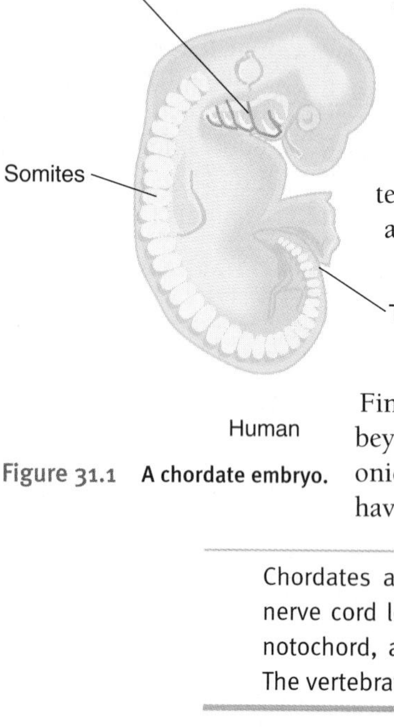

Gill arches

Somites

Tail

Human

Figure 31.1 **A chordate embryo.**

Chordates are characterized by a single, hollow nerve cord located along the back, a rod-shaped notochord, and gill arches located at the throat. The vertebrates are a subphylum of the chordates.

31.2 Tunicates (Urochordata)

The **tunicates** comprise a group of about 2500 species of marine animals, most of which look like living sacs attached to the floor of the ocean (**Figure 31.2a**). As shown in Figure 31.2b, the tunicates are not much more than a large pharynx covered with a protective *tunic*. The tunic is a tough outer "skin" composed mainly of cellulose, a substance found in the cell walls of plants and algae but rarely found in animals. Some colonial tunicates live in masses on the ocean floor and have a common sac and a common opening to the outside. Colonial tunicates reproduce asexually by budding. Individual tunicates are hermaphrodites, with each organism having both male and female sex organs.

The pharynx of a tunicate is lined with numerous cilia. As these cilia beat, they draw a stream of water through the incurrent siphon into the pharynx, which is lined with a sticky mucus. Food particles are trapped within the pharynx, and the filtered water flows out of the animal through the excurrent siphon. Because 90% of all species of tunicate have this structure and forcefully squirt water out their excurrent siphons when disturbed, they are also called *sea squirts*.

As adults, tunicates lack a notochord and a nerve cord. The gill slits (which develop from the gill arches) are the only clue that adult tunicates are chordates. Only the larvae, which look like tadpoles (Figure 31.2c), have a notochord and nerve cord. The notochord is in the tail, and the nerve cord runs dorsal to the notochord almost the entire length of the body. The subphylum name Urochordata (YUR-eh-core-DA-tah) comes from the placement of the larval notochord and literally means "tail chordate." The larvae remain free swimming for no more than a few days. Then they settle to the bottom and become sessile, attaching themselves to a suitable substrate by means of a sucker. As they mature, they adjust to a filter-feeding existence. Some tunicates that live in the world's warmer regions retain the ability to swim and never develop into a sessile form.

Adult tunicates are saclike, sessile, marine filter feeders. Their only chordate characteristic is their gill slits. Their larvae, however, have a notochord and nerve cord in the tail of a tadpolelike body.

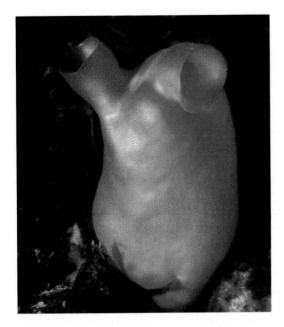

(a)

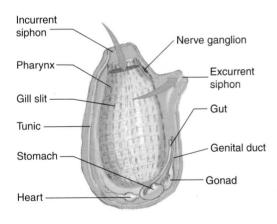

Incurrent siphon
Nerve ganglion
Pharynx
Excurrent siphon
Gill slit
Gut
Tunic
Stomach
Genital duct
Heart
Gonad

(b)

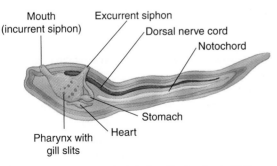

Mouth (incurrent siphon)
Excurrent siphon
Dorsal nerve cord
Notochord
Stomach
Pharynx with gill slits
Heart

(c)

Figure 31.2 Tunicates (subphylum Urochordata). (a) The sea perch *Halocynthia auranthium*. (b) The structure of a tunicate. (c) The structure of a larval tunicate.

31.3 Lancelets (Cephalochordata)

The **lancelets** are tiny, scaleless, fishlike marine chordates that are just a few centimeters long and pointed at both ends as shown in **Figures 31.3** and 25.8. They look very much like tiny surgical blades called lancets, from which they get their name. You may have had a few drops of blood taken in the doctor's office from a "fingerstick" done with a lancet.

Lancelets have a segmented appearance because of blocks of muscle tissue that are easily seen through their thin, unpigmented skin. Although they have pigmented light receptors, lancelets have no real head, eyes, nose, or ears. However, unlike the tunicates, the lancelet's notochord runs the entire length of its dorsal nerve cord, so these organisms are called *cephalochordates*, or *head chordates*. The lancelet retains its notochord throughout its lifespan.

The 23 species of lancelets live in the shallow waters of oceans all over the world. They spend most of their time partly buried in the sandy or muddy bottom with only their anterior ends protruding, feeding on plankton (floating microscopic plants and animals). In a manner similar to the tunicates, lancelets filter plankton from the water. Cilia line the anterior end of the alimentary canal, and these beating cilia create an incoming current of water. The filtered water exits at an excurrent siphon. An oral hood projects beyond the mouth, or incurrent siphon, and bears sensory tentacles. The sexes are separate, but no obvious external differences exist between them.

Oral hood with tentacles
Gill slits in pharynx
Atrium
Excurrent siphon
Gut
Notochord
Dorsal nerve cord
Anus
Tail

Figure 31.3 Lancelets (subphylum Cephalochordata). The structure of a lancelet.

> The lancelets are scaleless, fishlike marine chordates. The adult forms exhibit chordate characteristics.

In vertebrates, the notochord develops into the vertebral column.

31.4 Characteristics of vertebrates (Vertebrata)

The **vertebrates** differ from the other chordates in that most have a vertebral column in place of a notochord. A vertebral column, or backbone, is a stack of bones, each with a hole in its center, that forms a cylinder surrounding and protecting the dorsal nerve cord. (Each bone in the column is a vertebra.) One class of vertebrates that does not have a vertebral column is the jawless fishes. Present-day jawless fishes have a notochord, but their ancestors had a bony skeleton and a vertebral column. Another class, the cartilaginous fishes, have a skeleton and vertebral column composed of cartilage (a tough yet elastic type of connective tissue) rather than bone.

Although not all present-day vertebrates have a bony vertebral column, they all have a distinct head with a skull that encases the brain. (For this reason, some current classification systems use the term "craniates" [chordates with brains] instead of the term "vertebrates.") Vertebrates also have a closed circulatory system (the blood flows within vessels) and a heart to pump the blood. Most vertebrates also have a liver, kidneys, and endocrine glands. (Endocrine glands are ductless glands that secrete hormones, which play a critical role in controlling the functions of the vertebrate body.)

The human body plan is representative of the vertebrate body plan and is described extensively in Part Three. Nevertheless, vertebrates are a diverse group, consisting of animals adapted to life in the sea, on land, and in the air. There are eight classes of living vertebrates: Four classes are fishes and four classes are land-dwelling *tetrapods* (four-footed animals). Only three classes of fish are described in detail in this chapter. The fourth class is the lungfish, which only live in African freshwater lakes that regularly dry up. When this occurs, the lungfish burrow into the mud, leaving the mouth exposed to gulp air.

Vertebrates comprise a subphylum of chordates characterized by a vertebral column surrounding a dorsal nerve cord or by descent from ancestors with these features.

31.5 Jawless fishes (Cyclostomata)

Other than the lamprey eels and hagfishes, the major groups of the **jawless fishes** (Cyclostomata, meaning "round mouth") have been extinct for hundreds of millions of years. Only about 20 to 30 species of each of these two groups are alive today. Both groups are long, tubelike aquatic animals that usually live in the sea or in brackish (somewhat salty) water where the fresh water of a river meets the ocean. In addition to their lack of jaws, they have no paired fins to help them swim, and they have no scales. They do have a notochord, however, and portions of a cartilaginous skeleton that are remnants from their extinct ancestors.

Lampreys parasitize other fishes. In fact, they are the only parasitic vertebrates. They have a round mouth that functions like a suction cup (see Figure 25.11), which they attach to their prey—the bony fishes. When a lamprey attaches to a fish, it uses its spine-covered tongue like a grater, rasping a hole through the skin of the fish and then sucking out its body fluids. Sometimes, lampreys can be so abundant that they are a serious threat to commercial fisheries, preying on salmon, trout, and other commercially valuable fishes. Entering the Great Lakes from the sea, they have become important pests there; millions of dollars are spent annually on their control. Although the hagfishes are similar in size and shape to lampreys, they are not parasitic. They are scavengers, often feeding on the insides of dead or dying fishes or large invertebrates.

To reproduce, jawless fishes *spawn,* as do most fishes, amphibians, and shellfish. During spawning, the males and females deposit eggs and sperm directly into the water. Fertilization that takes place outside the body of the female is called **external fertilization.** Lampreys swim upstream to spawn as salmon do and create nests in which they deposit eggs and sperm. The fertilized eggs develop into larvae that feed on plankton. Over a period of years, the larvae mature and metamorphose (change) into parasitic adults. In contrast, the eggs of hagfishes do not develop into larvae. Completely formed hagfishes hatch directly from fertilized eggs.

Lampreys and hagfishes are tubular, scaleless, jawless organisms that lack paired fins and live in the sea or in brackish water. Lampreys parasitize bony fishes and pose a serious threat to some commercial fisheries. Hagfishes are scavengers.

Most fishes, amphibians, and shellfish have an external type of fertilization. Eggs are fertilized by sperm outside the body of the organism.

31.6 Cartilaginous fishes (Chondrichthyes)

The Chondrichthyes (kon-DRIK-thee-eeze; *chondri* meaning "cartilage," *ichthyes* meaning "fishes") include the sharks, skates, and rays. Examples are shown in Figure 31.4. Hundreds of extinct species of **cartilaginous fishes** are known from the fossil record, but less than 800 species exist today.

The skin of sharks (as well as the skates and rays) is covered with small, pointed, toothlike scales called *denticles,* which give the skin a sandpaper texture. Sharks have streamlined bodies and two pairs of fins: pectoral fins just behind the gills and pelvic fins just in front of the anal region. The dorsal (back) fins provide stability, and motions of the other fins (including

the asymmetrical tail fin), as well as sinuous motions of the whole body, give the shark lift and propel it through the water.

Many sharks are predators and eat large fishes and marine mammals. Their sharp, triangular teeth saw and rip off pieces of flesh as the shark thrashes its head from side to side. Some sharks feed on plankton rather than prey on other animals. These sharks swim with their mouths open, and the plankton are strained from the water by specialized denticles on the inner surfaces of the gill arches. Skates and rays are generally smaller than sharks and have flattened bodies with enlarged pectoral fins (see Figure 31.4) that undulate when these fishes move. Their tails, which are not a principal means of locomotion as in sharks, are thin and whiplike. They are sometimes armed with poisonous spines that are used as defense mechanisms rather than for predation. These animals have a mouth on their underside and feed mainly on invertebrates on or near the ocean floor.

The sensory systems of sharks, skates, and rays are quite sophisticated and diverse. Sharks have a lateral line system, as do all fishes and amphibians. A **lateral line system** is a complex system of mechanoreceptors that lies in a single row along the sides of the body and in patterns on the head. These receptors can detect mechanical stimuli such as sound, pressure, and movement. Sharks often detect prey by means of their lateral line systems, but, using their electroreceptors, they can also detect electrical fields emitted by other fishes. Sharks have these receptors on their heads, and rays have them on their pectoral fins. Scientists think these fishes may use their electroreceptors for navigation as well. Chemoreception is another important sense in the Chondrichthyes. In fact, sharks have been described as "swimming noses" because of their acute sense of smell. Vision is also important to the feeding behavior of sharks; they have well-developed mechanisms for vision at low-light intensities. (These senses are also described in Chapter 15, p. 312.)

Another important characteristic of fishes is their ability to regulate their buoyancy, or ability to float at various depths in the water. Bony fishes have gas-filled *swim bladders* that regulate their buoyancy. Cartilaginous fishes do not have swim bladders but can adjust the size and oil content of their livers. Because oil is less dense than water and the liver has a high oil content, adjusting the oil content and size of the liver can regulate buoyancy to a certain degree. Most species of sharks, however, swim continually to keep from sinking and to keep water flowing over their gills.

(a)

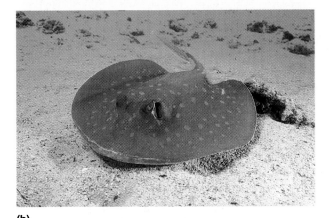

(b)

Figure 31.4 **Cartilaginous fishes (class Chondrichthyes).** (a) Caribbean reef shark. (b) Bluespotted stingray. (Skates look very much like rays.)

Aquatic organisms show various adaptations regarding **osmoregulation**, the control of water movement into and out of their bodies. Marine organisms tend to lose water to their surroundings because their body fluids usually have a solute concentration (concentration of dissolved substances) lower than that of seawater. The cartilaginous fishes maintain solute concentrations close to that of seawater, however, because they change potentially toxic nitrogen-containing wastes (such as ammonia) into a less toxic compound called urea and retain it in their bodies rather than excreting it **(Figure 31.5)**. (In fact, shark meat must be soaked in fresh water before it is eaten to remove most of the urea.)

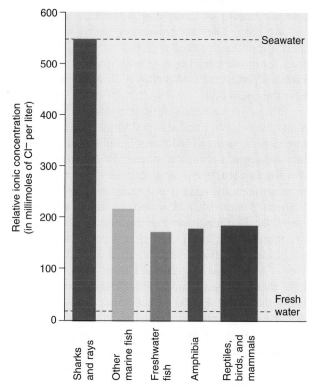

Figure 31.5 **Ion concentrations for different classes of vertebrates.** Ion concentrations in the bodies of different classes of vertebrates are roughly similar, with the exceptions of sharks and rays. Sharks and rays keep their ion concentrations close to that of seawater by adding urea to the bloodstream.

With similar solute concentrations both inside and outside the fish, water movement remains relatively equal in both directions.

Although most aquatic animals reproduce by means of external fertilization, the cartilaginous fishes have developed a method of internal fertilization. These fishes have *pelvic claspers,* which are rodlike projections between the pelvic fins of the male fish. During copula-

tion, or coupling of the male and female animals, the male inserts his clasper into the female's **cloaca** (kloe-AY-kuh). The cloaca is the terminal part of the gut into which ducts from the kidney and reproductive systems open. The fish either swim side-by-side, or the male wraps around the female or holds onto her with his jaws. He then ejaculates into a groove in the clasper, which directs the sperm into the female's cloaca. The sperm then swim up the female's reproductive tract.

In addition to having a mechanism of internal fertilization, the cartilaginous fishes are ovoviviparous. **Ovoviviparous** (OH-vo-vye-VIP-uh-rus) organisms retain fertilized eggs within their oviducts until the young hatch (*ovum* meaning "egg," *vivus* meaning "alive," *pario* meaning "to bring forth"). However, although the young are born alive, they do not receive nutrition from the mother but from the yolk of the egg. The mother acts like an internal "nest" for the eggs until the young hatch. Certain fishes, some reptiles, and many insects are ovoviviparous.

In contrast, **oviparous** (oh-VIP-uh-rus) organisms lay their eggs, and the young hatch from the eggs outside of the mother. Many skates and rays, for example, release eggs within protective egg cases. Birds and most reptiles, along with some of the cartilaginous fishes, are oviparous.

Viviparous (vye-VIP-uh-rus) organisms such as mammals bear live young as do ovoviviparous organisms, but the developing embryos primarily derive nourishment from the mother and not the egg. A few sharks, such as hammerhead sharks, blue sharks, and lemon sharks, are viviparous fishes.

The cartilaginous fishes include the sharks, skates, and rays. The skin is covered with small, pointed, toothlike scales that result in a sandpaper texture. Many sharks are predators of large fishes and marine mammals; skates and rays feed on invertebrates on the ocean floor.

All fishes and amphibians have a complex system of mechanoreceptors called a lateral line system. These receptors, which lie in a row on the sides of the body and on the head, detect mechanical stimuli such as sound, pressure, and movement.

Most cartilaginous fishes fertilize their eggs internally and are ovoviviparous, retaining fertilized eggs within their oviducts until the young hatch.

31.7 Bony fishes (Osteichthyes)

The vast majority of known species of fishes belong to the class Osteichthyes (OS-tee-IK-thee-eeze; *oste* meaning "bone"). **Bony fishes** live in both salt water and fresh water, with many species spending a portion of their lives in fresh water and another portion in the sea.

The Osteichthyes get their name from their bony internal skeletons. However, the skin of Osteichthyes is also covered with thin, overlapping bony scales. Sometimes these scales have spiny edges, which provide some protection for the animal. Bony fishes also have a protective flap that extends posteriorly from the head and protects the gills. This flap is called the **operculum (Figure 31.6)**. Along with protecting the gills, the movement of the operculum enhances the flow of water over the gills, bringing more oxygen in contact with the gas-exchanging surfaces and allowing a fish to breathe while stationary. Unlike most of the cartilaginous fishes, bony fishes can remain motionless at various depths because of the ability to more finely regulate buoyancy with a swim bladder.

The swim bladder of most bony fishes, the *ray-finned fishes* (the largest subclass of bony fishes), evolved from lunglike sacs of their ancestors. These sacs aided in respiration. In the ray-finned fishes, gas exchange between the swim bladder and the blood regulates the density of a fish and allows it to remain suspended in water without sinking to the bottom. The ray-finned fishes are the most familiar fishes and include perch, cod, trout, tuna, herring, and salmon.

The cartilaginous fishes and the bony fishes, like all chordates, have a closed circulatory system consisting of blood vessels and a pump (see Section 10.3). In the cartilaginous fishes and the bony fishes, the pump is a tubelike heart with four chambers, one right after the other (see Figure 31.9a), but these four chambers really function as a two-chambered heart. The first two chambers (*sinus venosus* and *atrium*) collect blood from the organs. The second two (*ventricle* and *conus arteriosus*) pump blood to the gills. From the gills, the blood moves to the rest of the body. Because of the great resistance in the narrow passageways of the capillaries at the gills, the movement of blood to the rest of the body is sluggish.

Although the cartilaginous fishes maintain a solute concentration of their body fluids near that of seawater, bony fishes do not. Only some bony fishes live in the sea; many groups live in fresh water. Marine and freshwater bony fishes have opposite situations regarding osmoregulation. The body fluids of marine fishes are *hypoosmotic* with respect to seawater. That is, their body fluids contain a lower concentration of dissolved substances than seawater does (see Figure 31.5). Conse-

quently, water tends to leave these fishes (at the gill epithelium) by osmosis. To regulate water balance, marine fishes drink seawater and then excrete the salt by means of active transport at the gill epithelium. The fish kidney is not able to get rid of excess salt in the urine because it has no loop of Henle (see Chapter 12). This part of the kidney nephron works to concentrate urine and enable an organism to produce urine with a high solute concentration. Only birds and mammals have loops of Henle.

Operculum

In contrast to marine fishes, the body fluids of freshwater fishes are *hyperosmotic* with respect to their freshwater environment. That is, their body fluids contain a higher concentration of dissolved substances than fresh water does. Consequently, water tends to enter freshwater fishes by osmosis. To regulate water balance, freshwater fishes do not drink water and excrete large amounts of very dilute urine. They reclaim some of the ions they lose in their urine by the uptake of sodium and chlorine ions by the gills and in the food they eat.

Figure 31.6 An example of a ray-finned bony fish (class Osteichthyes), the smallmouth bass.

Most Osteichthyes are oviparous, fertilizing their eggs externally. These eggs are often food for other marine organisms, however, and must survive other risks such as drying out. Most fishes (as well as other organisms that externally fertilize their eggs) lay large numbers of eggs, with some surviving these dangers. Many species of fishes build nests for their eggs and watch over them, which also enhances the chances of survival.

Bony fishes live in both salt water and fresh water and get their name from their bony internal skeletons. By means of gas exchange between their swim bladders and their blood, bony fishes can regulate their buoyancy better than the cartilaginous fishes do. Cartilaginous fishes regulate buoyancy by controlling the amount of oil in their livers.

Because they tend to lose water by osmosis, marine fishes drink sea water and excrete salt at the gill epithelium. Because they tend to take on water, freshwater fishes do not drink water and excrete large amounts of very dilute urine.

In vertebrates, the notochord develops into the vertebral column.

31.8 Amphibians (Amphibia)

The word amphibian means "two lives" (*amphi* meaning "both," *bios* meaning "life") and refers to both the aquatic and terrestrial existence of this class of animals. **Amphibians** (unlike reptiles, birds, and mammals) depend on water during their early stages of development. Many amphibians live in moist places such as swamps and in tropical areas even when they are mature, which lessens the constant loss of water through their thin skin. The two most familiar orders of amphibians are those that have tails—the salamanders, mudpuppies, and newts (Figure 31.7a)—and those that do not have tails—the frogs and toads (Figure 31.7b and c).

Most frogs and toads fertilize their eggs externally. The male grasps the female and sheds sperm over the eggs as they are expelled from the female as shown in the photo on p. 485. Most salamanders use internal fertilization but are still oviparous (lay their eggs). Because amphibian eggs have no shells or membranes to keep them from drying out, amphibians lay their eggs directly in water or in moist places. The photo above right shows the "nests" of foam that tropical tree frogs create in tree branches to incubate their eggs. When the tadpoles develop, they drop from the tree branches into the water. Some amphibians protect their eggs by incubating them in their mouths, on their backs, or even in their stomachs! A few amphibian species are ovoviviparous and incubate the eggs within their reproductive organs until they hatch, and a few are viviparous.

The young of frogs and toads undergo *metamorphosis* (MET-uh-MORE-feh-sis), or change, during development from a larval to an adult form. The larvae are immature forms that do not look like the adult. The larvae of frogs and toads are tadpoles, which usually live in the water and have internal gills and a lateral line system like that of fishes. They feed on minute algae. These fishlike forms develop into carnivorous adults having legs and lungs; their gills and lateral line system disappear. The lungs of the adults are inefficient, however. Much of the gas exchange takes place across the skin and on the surfaces of the mouth (see Section 9.2). The skin of amphibians must therefore remain damp to allow gases to diffuse in and out.

The adults of certain salamanders (such as mudpuppies) live permanently in the water and retain gills and other larval features as adults. Other salamanders are terrestrial but return to water to breed. They, like frogs and toads, usually live in moist places such as under stones or logs or among the leaves of certain tropical plants.

Along with having lungs rather than gills (in most cases), amphibians have a pattern of blood circulation different from that of the fishes. After the blood is pumped through the fine network of capillaries in the amphibian lungs, it does not flow directly to the body as it does in fishes. Instead, it

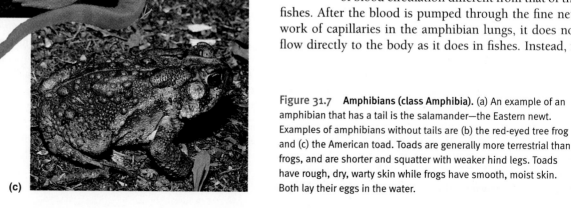

(a)

(b)

(c)

Figure 31.7 Amphibians (class Amphibia). (a) An example of an amphibian that has a tail is the salamander—the Eastern newt. Examples of amphibians without tails are (b) the red-eyed tree frog and (c) the American toad. Toads are generally more terrestrial than frogs, and are shorter and squatter with weaker hind legs. Toads have rough, dry, warty skin while frogs have smooth, moist skin. Both lay their eggs in the water.

returns to the heart. It is then pumped out to the body at a much higher pressure than if it were not returned to the heart. However, the oxygenated blood that returns to the heart from the lungs mixes with the deoxygenated blood returning to the heart from the rest of the body. Consequently, the heart pumps out a mixture of oxygenated and deoxygenated blood rather than fully oxygenated blood. Figure 31.9b, shows this pathway of blood flow. The blood from the lungs and the blood from the body enter the right and left atria of the heart, respectively. The blood in both these chambers flows into the single ventricle of the three-chambered amphibian heart and is pumped through two large vessels to both the lungs and the body.

Amphibians live both aquatic and terrestrial existences. Because they fertilize their eggs externally and their eggs have no shells or membranes, amphibians lay their eggs in water or in moist places. The young of frogs and toads change from a larval form to an adult form during their development.

Amphibians have a three-chambered heart rather than the two-chambered heart of fishes. However, oxygenated and deoxygenated blood mix in the heart.

31.9 Reptiles (Reptilia)

The three major orders of **reptiles** are the crocodiles and alligators, the turtles and tortoises, and the lizards and snakes. Representatives of the orders are shown in Figure 31.8. Reptiles have dry skins covered with scales that help retard water loss. As a result, reptiles can live in a wider variety of environments on land than amphibians can, but the crocodiles, alligators, and turtles are aquatic organisms.

The hearts of reptiles differ from amphibian hearts in that a partition called a *septum* subdivides the ventricle, the pumping chamber of the heart. The septum reduces the mixing of oxygenated and deoxygenated blood in the heart. In most crocodiles, the separation is complete. Figure 31.9c, shows that the reptilian heart closely resembles the four-chambered heart of birds and mammals shown in Figure 31.9d. It is still considered to be a three-chambered heart, however.

One of the most critical adaptations of reptiles to life on land is the evolution of the shelled **amniotic egg** (AM-nee-OT-ik). Amniotic eggs are also character-

(a)

(c)

(b)

(d)

Figure 31.8   Reptiles (class Reptilia).
(a) American alligator. (b) Sonora mountain kingsnake.
(c) Collared lizard. (d) Eastern box turtle.

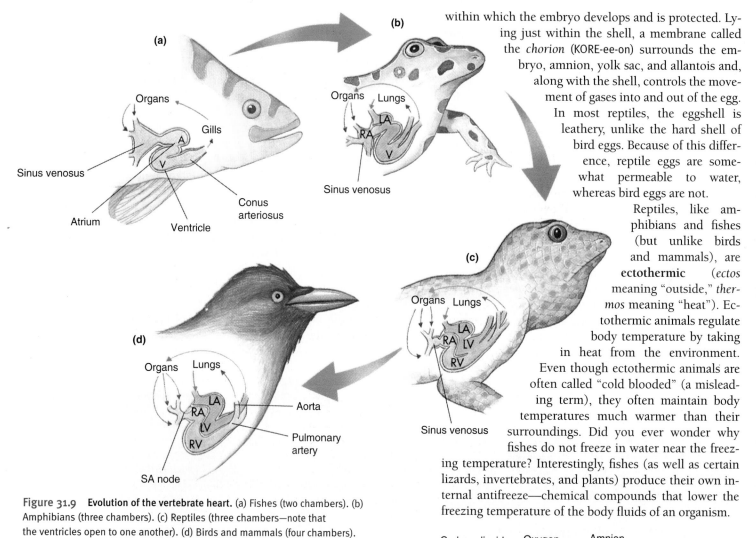

Figure 31.9 **Evolution of the vertebrate heart.** (a) Fishes (two chambers). (b) Amphibians (three chambers). (c) Reptiles (three chambers—note that the ventricles open to one another). (d) Birds and mammals (four chambers).

within which the embryo develops and is protected. Lying just within the shell, a membrane called the *chorion* (KORE-ee-on) surrounds the embryo, amnion, yolk sac, and allantois and, along with the shell, controls the movement of gases into and out of the egg. In most reptiles, the eggshell is leathery, unlike the hard shell of bird eggs. Because of this difference, reptile eggs are somewhat permeable to water, whereas bird eggs are not.

Reptiles, like amphibians and fishes (but unlike birds and mammals), are **ectothermic** (*ectos* meaning "outside," *thermos* meaning "heat"). Ectothermic animals regulate body temperature by taking in heat from the environment. Even though ectothermic animals are often called "cold blooded" (a misleading term), they often maintain body temperatures much warmer than their surroundings. Did you ever wonder why fishes do not freeze in water near the freezing temperature? Interestingly, fishes (as well as certain lizards, invertebrates, and plants) produce their own internal antifreeze—chemical compounds that lower the freezing temperature of the body fluids of an organism.

istic of birds and egg-laying mammals (monotremes). The amniotic egg protects the embryo from drying out, nourishes it, and enables it to develop outside of water.

The amniotic egg is illustrated in Figure 31.10. It contains a yolk and albumin (egg white). The yolk is the primary food supply for the embryo, and the albumin provides additional nutrients and water. The embryo's nitrogenous wastes are excreted into the allantois (ah-LAN-toe-us), a sac that grows out of the embryonic gut. Blood vessels grow out of the embryo through the sac surrounding the yolk and through the allantois to the egg's surface, where gas exchange takes place. The amnion (AM-nee-on) surrounds the developing embryo, enclosing a liquid-filled space

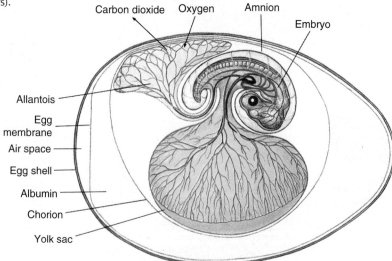

Figure 31.10 **The amniotic egg.** The amniotic egg is an important adaptation that allows reptiles (and birds and monotremes) to live in a wide variety of terrestrial habitats.

Just Wondering . . . Real Questions Students Ask

I know that chameleons are "famous" for their ability to change color. How and why do they do that?

There are over 100 species of chameleons — some more colorful than others. The one shown here (a *Chamaeleo pardalis* male) is a particularly colorful chameleon and one that is easily tamed, so you might have seen one like this in a pet store. In contrast, many species of chameleon are much less colorful, and many are very difficult to keep and breed.

The common perception is that chameleons change their color to protect themselves from predators. However, the chameleon's camouflage primarily consists of its body shape and behavior; the ability to change color plays only a small role in its blending with the environment. Most chameleons have a body shape that is flattened vertically, making them look somewhat like a leaf. Additionally, chameleons walk in a faltering or hesitating manner, looking a bit like the "leaf" is being blown by the wind. Many chameleons are shades of green or brown, which adds to the look of a live or dead leaf. Some species are camouflaged by looking like moss, and yet others like wood chips on the forest floor.

If color change is not the chameleon's primary method of camouflage, then why does color change occur . . . and how does it occur? Chameleon color change is primarily triggered by a combination of both internal and external cues. Internal factors that affect a chameleon's color are its physical health, its level of sexual maturity, and its hormonal response to the presence of other chameleons (male or female) of the same species. For example, young chameleons are usually gray, beige, or dark brown; sexually mature chameleons usually have more intense colors. Additionally, in males, color changes are often behavioral displays and part of aggressive, submissive, or sexual behaviors (see Sections 32.9 – 32.11).

External conditions such as light intensity and the temperature of the environment also affect a chameleon's color. Chameleons are lizards and therefore ectothermic. When a chameleon is cold, certain cells called iridocytes (see description below) allow more light to enter the skin. The animal flattens out and its colors darken, resulting in greater absorption of heat.

Chameleon skin is structured much like human skin (and like that of all vertebrates), with dead, hardened cells forming the topmost layer of the epidermis and living cells lying just below. Beneath the epidermis is the dermis, a layer of dense, connective tissue (see Figure 16.3). Three main types of skin cells are involved in chameleon color change: guanocytes, iridocytes, and chromatophores.

The living layer of the epidermis contains guanocytes. These cells contain non-pigmented crystals that diffract (bend) light rays. They appear yellow when light is reflected up through them by the iridocytes beneath, and blue when light is not reflected up. The iridocytes lie between the epidermis and the dermis. These cells contain reflecting plates stacked like decks of cards. The plates scatter the light, and depending on their positions, may direct light back up through the epidermis (and the guanocytes) or down into the dermis where the chromatophores are located.

The chromatophores are pigmented cells with fingerlike cell processes that extend upward toward the epidermis. Depending on the species of chameleon, these cells may contain red, brown, yellow, orange, or blue-violet pigments. Chromatophores are under neuro-hormonal control. For example, if a chameleon is threatened by another animal, the chromatophores are stimulated to send pigment to the upward-reaching cell processes. This results in a coloration of the skin since pigment is "pumped" into upper cell layers.

Chameleons are not the only animals that can change color. For example, certain fish change color (such as the freshwater cichlid shown in Figure 33.10) as a part of their behavioral repertoire. Certain frogs can also darken, camouflaging themselves when threatened. Cephalopods (octopus and squid) change colors as well when mating or hunting, or when threatened. The mechanisms of color change are similar among these animals as among others capable of this "feat." It's just that chameleons are often so strikingly colorful and their color changes so dramatic, that they have become the animals most noticed for this ability.

Along with such physiological adaptations, ectothermic animals protect themselves against the cold in behavioral ways. For example, frogs help protect themselves against freezing by spending the winter buried in the soil or in the mud at the bottom of ponds. Ectothermic animals also protect themselves from high heat by burrowing under rocks or remaining in shady, somewhat cooler areas. Desert tortoises, for example, construct shallow burrows to stay in during the summer and deeper burrows for hibernation in winter. Reptiles often bask in the sun, which raises their body temperature and their metabolic rate. When cold-blooded animals are cold, the metabolic rate slows down and they are unable to hunt for food or move about very quickly.

Reptiles have a dry skin covered with scales that help retard water loss. Another critical adaptation to their life on land (although some reptiles are aquatic) is the development of the amniotic egg, which protects the embryo from drying out, nourishes it, and enables it to develop outside of water. Amniotic eggs are also characteristic of birds and egg-laying mammals.

31.10 Birds (Aves)

The gold and blue macaw and the herons shown in **Figure 31.11** represent only two of the approximately 9000 species of birds living today. In birds, the wings are homologous to the forearms of other vertebrates. That is, they are derived from the same evolutionary origin and have the same basic anatomy but differ in function. Birds have reptilianlike scales on their legs and lay amniotic eggs as reptiles do. They have hard, horny extensions of the mouth called beaks that tear, chisel, or crush their food. They also have digestive organs called gizzards, often filled with grit, that grind food. Beaks are not limited to birds. Many reptiles have beaks, including turtles, and so do some fishes, including the parrot fish, which uses its beak to rip away fragments of coral reef. Birds, however, are the only animals that have feathers.

Feathers are flexible, light, waterproof epidermal structures. Several types of feathers form the body covering of birds, including contour feathers and down feathers. Contour feathers are flat (except for a fluffy, downy portion at the base) and are held together by tiny barbules as shown in **Figure 31.12**). These feathers provide a streamlined surface for flight, and some are modified to reduce drag on the wings or act like individual propeller blades. In addition, birds can alter the area and shape of their wings by altering the positions of their feathers. Feathers also provide birds with waterproof coats and play an important role in insulating birds against temperature changes.

Along with feathers, birds have light, hollow bones adapted to flight. Birds also have highly efficient lungs that supply the large amounts of oxygen necessary to sustain muscle contraction during prolonged flight. Unlike fishes, amphibians, and reptiles (except for the crocodiles), birds and mammals have hearts that act as a double pump. The sides of the heart are completely separated with a septum (see Figure 31.9). The right side of the heart pumps blood to the lungs. The left side of the heart pumps blood to the body. Oxygenated blood and deoxygenated blood do not mix. As you compare the pathway of blood in the hearts of fishes, amphibians, reptiles, and birds in Figure 31.9, note that the four-chambered heart of birds and mammals evolved from only two of the chambers of the fish heart.

Birds and mammals are **endothermic** (*endo* meaning "within"); they regulate body temperature internally. The evolution of the four-chambered heart with separate pathways to the lungs and the body is thought to have been important in the evolution of endothermy in the birds and mammals. More efficient circulation is necessary to support the great increase in metabolic rate that is required to generate body heat internally. In addition, blood is the carrier of heat in the body, and an efficient circulatory system is required to distribute heat evenly throughout the body. Although endothermic animals are sometimes called "warm blooded," their body temperatures may be cooler than that of the surroundings (such as when you are in 100° F heat). Usually,

(a)

(b)

Figure 31.11 Birds (class Aves). The birds are a large and successful group of about 9000 species, more than any other class of vertebrates except the bony fishes. (a) Great blue herons in a courtship display. (b) Gold and blue macaw.

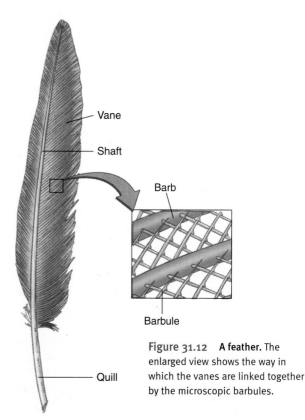

Vane

Shaft

Barb

Barbule

Quill

Figure 31.12 A feather. The enlarged view shows the way in which the vanes are linked together by the microscopic barbules.

however, because endotherms maintain high internal body temperatures (37° C [98.6° F] in humans, for example), their internal temperatures are higher than that of the environment.

Endotherms maintain a constant high body temperature by adjusting heat production to equal heat loss from their bodies under various environmental conditions. The high metabolic rate of endotherms and the energy released during these chemical reactions produce much of this body heat. Increasing the action of skeletal muscles increases the metabolic rate and the amount of heat produced. Shivering in the cold, for example, is an action that produces body heat. But because endotherms usually live in environments cooler than body temperature, restricting heat loss is usually the concern. Animals that live in extremely cold temperatures, such as arctic birds and polar bears, are well-insulated with either feathers or hair that traps air. Raising or lowering the feathers or hair adjusts the insulating capacity. Getting "goosebumps" when you are cold is your body's reaction to raise your hairs and increase your insulation. Although humans no longer have substantial body hair, this mechanism is important in reducing heat loss in other endothermic vertebrates.

Birds are winged vertebrates that are covered with feathers and have scales on their legs. They have beaks, which are horny extensions of the mouth. Their light, hollow bones are an adaptation to flight. They have a four-chambered heart as mammals do. Unlike the reptiles, amphibians, and fishes, birds regulate body temperature internally.

31.11 Mammals (Mammalia)

There are about 4500 species of living **mammals**, including humans. Mammals are endothermic vertebrates that have hair and whose females secrete milk from mammary glands to feed their young. Mammals, like birds and most crocodiles, have a four-chambered heart with circulation to the lungs and separate circulation to the body. The locomotion of mammals is advanced over that of the reptiles, which in turn is advanced over that of the amphibians. The legs of mammals are positioned much farther under the body than those of reptiles and are suspended from limb girdles, which permit greater leg mobility.

Specialized teeth have evolved in mammals, different from the teeth of fishes, amphibians, and reptiles, which are essentially the same size and shape. In mammals, evolutionary specialization has resulted in incisors, which are chisellike teeth used for cutting; canines, which are used for gripping and tearing; and molars, which are used for crushing and breaking (see Figure 8.3a). There are three subclasses of mammals: monotremes (MON-oh-treems), marsupials (mar-SOO-pee-uhlz), and placental (pluh-SENT-uhl) mammals.

The only **monotremes** that exist are the duckbilled platypus (**Figure 31.13a**) and two genera of spiny anteaters (Figure 31.13b). Monotremes lay eggs with leathery shells similar to those of reptiles. The platypus generally lays one egg and incubates it in a nest. The spiny anteater generally lays two eggs and incubates them in a pouch. When the young hatch, they feed on milk produced by specialized sweat glands of the mother.

(a)

(b)

Figure 31.13   **Monotremes (class Mammalia).** (a) Duckbilled platypus. (b) Spiny anteater (*Echidna*) from Australia.

Marsupials are mammals in which the young are born early in their development and are retained in a pouch. After birth, the embryos crawl to the pouch and nurse there until they mature. The kangaroo and koala are familiar examples of marsupials (Figure 31.14).

In **placental mammals**, the young develop to maturity within the mother. They are named for the first organ to form during the course of their embryonic development, the placenta. (Human development as an example of the development of a placental mammal is described in detail in Chapter 23.)

Placental mammals are extraordinarily diverse, as is exhibited in just the three examples shown in Figure 31.15. There are 14 orders of placental mammals (see Figure 25.16), of which one is the primates, the order that includes monkeys, apes, and humans. The bats in the opening photograph (as well as all bats) belong to the order Chiroptera (keye-ROP-ter-ah), a name that means the forelimbs are modified to form wings (*chiro* meaning "hand," *pteron* meaning "wing"). So, although humans and bats are quite different, they have remarkable similarities in their circulatory, thermoregulatory, osmoregulatory, and reproductive patterns.

Mammals are endothermic vertebrates that have hair and whose females secrete milk from mammary glands to feed their young. There are three subclasses of mammals: monotremes, marsupials, and placentals.

Figure 31.14 Marsupials (class Mammalia). (a) Eastern gray female kangaroo with a young kangaroo in her pouch. (b) Koala with young.

Figure 31.15 Placental mammals (class Mammalia). (a) Bottle-nosed dolphin. (b) Grey timber wolf. (c) Greater horseshoe bat in flight.

Key Concepts

Chordates have a cartilagenous rod-shaped notochord that forms during development.

31.1 The chordates are characterized by three main features: (1) a single, hollow nerve cord located along the back; (2) a rod-shaped notochord, which forms between the nerve cord and the gut (stomach and intestines) during development; and (3) pharyngeal (gill) arches, which are located at the throat (pharynx).

31.1 There are three subphyla of chordates: the tunicates (subphylum Urochordata), the lancelets (subphylum Cephalochordata), and the vertebrates (subphylum Vertebrata).

31.2 The tunicates are sessile, saclike marine organisms that filter food from the surrounding water.

31.2 Although adult tunicates have gill slits, only the larvae have a notochord and nerve cord.

31.3 The lancelets are tiny, scaleless, fishlike marine chordates that are just a few centimeters long.

In vertebrates, the notochord develops into the vertebral column.

31.4 Vertebrates are a subphylum of chordates characterized by a vertebral column surrounding a dorsal nerve cord.

31.4 Vertebrates have a distinct head with a skull that encases the brain, a closed circulatory system, and a heart to pump the blood.

31.4 Most vertebrates also have a liver, kidneys, and endocrine glands.

31.4 In spite of many similarities, the vertebrates are an extremely diverse subphylum of organisms.

31.4 Four classes of vertebrates are fishes; four are tetrapods—animals with four limbs.

31.5 The jawless fishes—lampreys and hagfishes—are long, tubelike animals that live in the sea or brackish water.

31.5 The jawless fishes lack paired fins and scales.

31.5 Lampreys are parasites of bony fishes, and hagfishes are scavengers.

31.6 The cartilaginous fishes—sharks, skates, and rays—are covered with toothlike scales and have sophisticated and diverse sensory systems.

31.6 Cartilaginous fishes fertilize their eggs internally, and most are ovoviviparous, retaining fertilized eggs within their oviducts until the young hatch.

31.7 The vast majority of fishes are the bony fishes.

31.7 Along with having a bony internal skeleton, the bony fishes have thin, bony, platelike scales.

31.7 Both the bony fishes and the cartilaginous fishes have two-chambered hearts that pump blood to the gills; from there, the blood moves sluggishly around the body.

31.8 The amphibians live both in water and on land.

31.8 Because amphibian eggs have no shells or membranes to keep them from drying out, amphibians lay their eggs directly in water or moist places.

31.8 The young of frogs and toads undergo change from larval to adult forms during development.

31.8 The adults of certain salamanders live permanently in the water and retain gills and other larval features as adults.

31.8 Although amphibians have a three-chambered heart, oxygenated and deoxygenated blood mix in the heart.

31.9 Reptiles are better adapted to life on land than the amphibians because of their dry, scaly skin that retards water loss and their shelled (amniotic) egg.

31.9 The amniotic egg retains a watery environment within the egg while protecting and nourishing the developing embryo.

31.9 Fishes, amphibians, and reptiles are ectothermic; that is, they regulate body temperature by taking in heat from the environment.

31.9 Ectothermic animals protect themselves from the cold and high heat in behavioral ways.

31.10 Birds are winged vertebrates that are covered with feathers and are adapted to flight.

31.10 Birds lay amniotic eggs like the reptiles but have a four-chambered heart like the mammals.

31.10 Birds, like mammals and unlike reptiles, amphibians, and fishes, are endothermic; that is, they regulate body temperature internally.

31.11 Mammals are endothermic vertebrates that have hair and whose females secrete milk from mammary glands to feed their young.

Key Terms

amniotic egg (am-nee-OT-ik) *705*
amphibians (am-FIB-ee-uns) *704*
bony fishes *703*
cartilaginous fishes
 (kar-tuh-LAJ-uh-nus) *701*
chordates *698*
cloaca (kloe-AY-kuh) *702*
ectothermic (EK-toe-THUR-mik) *706*
endothermic (EN-doe-THUR-mik) *708*
external fertilization *700*
jawless fishes *700*
lancelets (LANS-lets) *699*

lateral line system *701*
mammals *709*
marsupials (mar-SOO-pee-uhls) *710*
monotremes (MON-o-treems) *709*
nerve cord *698*
notochord (NO-toe-kord) *698*
operculum (oh-PUR-kyuh-lum) *703*
osmoregulation
 (OZ-mo-reg-you-LAY-shun *or*
 OS-mo-reg-you-LAY-shun) *702*
oviparous (oh-VIP-uh-rus) *702*

ovoviviparous
 (OH-vo-vye-VIP-uh-rus) *702*
pharyngeal (gill) arches (fuh-RIN-jee-
 uhl *or* FAR-in-JEE-uhl) *698*
placental mammals
 (pluh-SENT-uhl) *710*
reptiles *705*
tunicates (TOO-nih-kits *or*
 TOO-nih-kates) *698*
vertebrates (VER-tuh-bruts *or*
 VER-tuh-braytes) *700*
viviparous (vye-VIP-uh-rus) *702*

KNOWLEDGE AND COMPREHENSION QUESTIONS

1. List three distinguishing characteristics of chordates. *(Knowledge)*

2. Match each term to the most appropriate description: *(Knowledge)*
 Descriptions
 a. Scaleless, fishlike marine chordates that retain a notochord throughout life
 b. Tubular, scaleless organisms without paired fins; most members of this class have been extinct for millions of years
 c. Saclike, sessile, marine filter feeders
 Terms
 (1) Jawless fishes
 (2) Tunicates
 (3) Lancelets

3. What is a lateral line system? Which animals have one, and why is it important? *(Comprehension)*

4. What is osmoregulation, and what is its significance? *(Comprehension)*

5. Distinguish among ovoviviparous, oviparous, and viviparous. *(Comprehension)*

6. What does the term *amphibian* mean? *(Knowledge)*

7. Distinguish between ectothermic and endothermic. Give an example of an ectotherm and an endotherm. *(Comprehension)*

8. Describe two important adaptations of reptiles to life on land. *(Comprehension)*

9. What are feathers, and what functions do they perform for birds? *(Knowledge)*

10. What characteristics differentiate mammals from other vertebrates? *(Knowledge)*

KEY
Knowledge: Recalling information.
Comprehension: Showing understanding of recalled information.

CRITICAL THINKING REVIEW

1. How are the characteristics of the three subphyla of chordates related to the way of life of each group? *(Application)*

2. What limits the ability of amphibians to occupy the full range of terrestrial habitats and allows other terrestrial vertebrates to live in them successfully? *(Analysis)*

3. Most species of snakes inhabit tropical and subtropical countries. Based on information from this chapter relating to the physiology of reptiles, why do you think this is so? Why don't snakes live in the arctic? *(Application)*

4. In what way is the heart of a frog more efficient than the heart of a fish? *(Analysis)*

5. Compare and contrast the structure of the heart in fishes, amphibians, reptiles, and birds. Which is the most efficient and why? *(Analysis)*

KEY
Application: Using information in a new situation.
Analysis: Breaking down information into component parts.

Compare the extraembyronic structures in the amniotic egg shown below to the extraembryonic structures of the developing human. (1) List the structures that are common to both and note their functions. (2) Describe the relative importance of the yolk sac to each. (3) Describe the differences in oxygen and carbon dioxide exchange in each.

Developing human

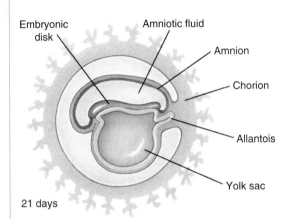

21 days

Amniotic egg

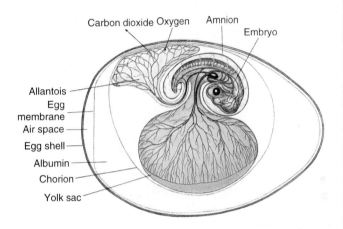

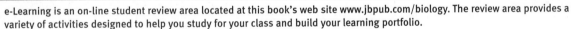

e-Learning is an on-line student review area located at this book's web site www.jbpub.com/biology. The review area provides a variety of activities designed to help you study for your class and build your learning portfolio.

Review Questions The review questions test your knowledge of the important concepts and applications in each chapter. The review provides feedback for each correct or incorrect answer. This is an excellent test preparation tool.

Figure-Labeling Exercises Sharpen your visual thinking by matching terms and labels for important illustrations.

Flash Cards Studying biology requires learning new terms. Virtual flash cards help you master the new vocabulary for each chapter.

Just Wondering As you read and study from this text, you may find that you have unanswered questions. Through this site you can ask the author, Sandy Alters, your "just wondering" questions.

 Do you prefer the speed and reliability of a CD-ROM? All of the features contained on the eLearning portion of the web site are also available on the Student CD-ROM.

Tropical Rain Forest DEFORESTATION

Did you know that . . .

- Tropical rain forests are located at or near the equator.

- The largest rain forests are in Brazil (South America), Zaire (Africa), and Indonesia (islands found near the Indian Ocean).

- The Amazon rain forest in South America is the largest of all and is about 2/3 the size of the United States.

- Four square miles of rain forest contain up to 1500 species of flowering plants, 750 species of trees, 125 species of mammals, 400 species of birds, 100 species of reptiles, 60 species of amphibians, and 150 species of butterflies.

- Every day, an area of rain forest about the size of New York City is destroyed.

- Every year, about 1000 rain forest species become extinct due to rain forest destruction.

Starbucks Coffee has begun an initiative to sell coffee brewed from beans grown underneath the rain forest canopy. This effort promotes conservation of the rain forest and provides employment for local people.

The tropical rain forest is home to nearly half the world's species. As the habitats of species are lost, species die out and biological diversity declines. This decline erodes the biological richness of the Earth and eliminates potential sources of medicines and natural genotypes for selective breeding or genetic engineering. For example, many rain forest species have yielded important drugs to treat various cancers, heart ailments, and high blood pressure—yet only 1% of rain forest species have been examined for their medicinal properties. Many species are already lost forever.

Species extinction is only one effect of the destruction of tropical rain forests. Another effect is the changing of the rain forest ecosystem—a canopy of giant trees, with lush vegetation below, which provides a habitat for many animal species. When the land is cleared, such forests become grasslands that have a totally different ecology from rain forest ecology. (Part Seven defines ecological terms and gives background for the ecological concepts mentioned here.)

Changes in the ecology of rain forests may lead to a third effect: regional and global changes in the carbon and water cycles. Carbon, which is normally taken from

the atmosphere and incorporated into carbon compounds during photosynthesis within rain forest plants, remains in the atmosphere when these plants no longer exist. This buildup of carbon dioxide may add to the greenhouse effect. During this process, the blanket of carbon dioxide in the atmosphere allows heat to enter the Earth's atmosphere but prevents it from leaving. As a result, the worldwide temperature may increase, a situation called global warming. Warmer temperatures are predicted to cause changes in weather patterns, water supplies, ocean levels, growing conditions, and the northward spread of tropical diseases.

With deforestation, water cycles change as well. Rainwater no longer falls on the rich growth of leaves of rain forest trees and other plants, eventually evaporating and returning to the atmosphere. Instead, all the rainwater reaches the ground, increasing the runoff into rivers and streams. In some areas where the land and rivers cannot hold the water, periodic flooding becomes a problem.

So, what are the issues of rain forest deforestation if deforestation clearly seems to be a serious ecological problem? One issue has to do with the survival of indigenous (native) peoples. In the Amazon, for example, families known as smallholders cut and burn the rain forest so they can grow food. This process is known as slash-and-burn agriculture. To raise crops, smallholders cut and-burn four or five acres of forest. The resulting ash fertilizes the soils, which are poor; the nutrients are held in the vegetation. When the soil is depleted of nutrients from the ash after a few years, they slash and burn another four or five acres. Many indigenous peoples use sustainable farming methods; that is, they reuse land that has been allowed to regrow for 10 to 50 years. However, slash-and-burn agriculture accounts for about 25% of the yearly Amazonian deforestation and the fires release carbon dioxide into the atmosphere. But indigenous peoples need to grow food to live.

Cattle ranching is another practice that results in rain forest destruction. In fact, it has been implicated as the main cause of irreversible deforestation in Central and South America. Government officials in these areas sold large tracts of rain forest to beef producers inexpensively to encourage economic growth in their region. To be useful for cattle ranching, portions of rain forest land are converted into pastures using slash-and-burn techniques. However, this process quickly uses up the nutrients in the soil, and when the soil can no longer support the cattle, ranches are moved to other tracts of rain forest land where the process begins again. The worldwide demand for inexpensive beef supports cattle ranching in tropical areas. The United States, a highly competitive beef market, is the largest consumer of Central and South American beef.

An example of slash-and-burn farming techniques. Land prepared and used in such a way is quickly depleted of its nutrients—usually within 5 years.

A variety of other practices have interests that conflict with rain forest conservation and add to deforestation. Tropical hardwoods are cut to supply markets for these woods. Timber cutting is not matched by replanting. Certain minerals, such as tin ore, are mined in some tropical forests. To mine for ore, companies cut down forested areas and often contaminate streams with the chemicals they use in ore extraction. They also use the trees to produce charcoal, a fuel used in many mining and refining operations. Environmentalist groups also charge that the exploration for and the development of oil resources has destroyed vast tracts of rain forest and rivers.

What is the issue here, and what is your response to this issue?

To make an informed and thoughtful decision, it may help you to follow these steps.

1. Identify the issue described here using either a statement or a question. (An issue is a point on which various persons hold differing views.)
2. Write your immediate reaction to the issues raised. How do you think this issue could be or should be resolved?
3. Collect new information about the issue. State the sources of that information. Explain why you believe the information to be accurate. Also determine if the information expresses a particular point of view or is biased in any other way. (Guidelines for doing efficient web searches for reliable information are available on the BioIssues website — **www.jbpub.com/biology**)
4. Determine which individuals, groups, or organizations have a stake in the issue. What does each stand to gain or lose depending on the outcome of the issue?
5. List possible outcomes (resolutions) of the issue. List the pros and cons of each outcome.
6. Which outcome do you think would be best and why? Note whether your opinion differs from or is the same as what you wrote in Step 2. Give reasons for your views and for any changes in them based on the additional information you have collected and the analysis you have done.

Behavior

These "kissing" fish are not as affectionate as they seem. In fact, their behavior is one of aggression. Engaged in a territorial "war," each wants to win the rights to its own "space" within the water.

Their fight begins as they encounter one another at a territorial boundary. First, each tries to intimidate the other, puffing up its body with air (using accessory breathing organs atop the gill chambers)—to appear as big and threatening as possible. Then they swim side by side, pushing currents of water past one another with their fins. These currents provide clues to each contender regarding the size of the opponent. If neither retreats, the tug-of-war begins, mouths locked in battle. Soon, one fish emerges the victor. The loser signals submission, and the "tournament" is over.

Organization

32.1 The history of the study of animal behavior

Animals (including humans) continually exhibit a wide range of behaviors. **Behaviors** include the patterns of movement, sounds (vocalizations), and body positions (postures) exhibited by an animal. In addition, behaviors include any type of change in an animal, such as a change in coloration or the releasing of a scent that can trigger certain behaviors in another animal.

Some behaviors are simple, automatic responses to environmental stimuli. A bacterium "behaves" when it moves toward higher concentrations of sugar. You behave when you slam on the brakes to avoid a car accident. Other types of behaviors, such as the eating behavior of the sea otter, are quite complex. In the photo below, a sea otter is shown having dinner while swimming on its back. It is using the rock as a hard surface against which to hit the clam and break it open. Often a sea otter will keep a favorite rock for a long time, suggesting that it understands the manner in which it will use the rock. The sea otter may learn this pattern of eating behavior from others while young, but the capacity to use tools and consciously foresee their future use certainly depends on inherited abilities. Complex behaviors such as these are limited to multicellular animals with a neural network that can sense stimuli, process these stimuli in a central nervous system (brain), and send out appropriate motor impulses.

Before the late 1950s, the study of animal behavior was dominated by **ethology** (ee-THOL-uh-jee). Using a physiological perspective, ethologists observed and interpreted the behavior of animals in their natural environments. They broke down behavior patterns into recognizable units, named these patterns, and categorized them. They often focused on the activity of the nervous system and also sought to understand the connection between an animal's behavior and its genetic makeup. They were particularly interested in the types of motor movements and responses to stimuli that typically occurred within closely related species as they attempted to uncover the biological significance of animal behavior—the importance of a behavior for a particular species in its natural environment.

One of the most famous ethologists is the Austrian scientist Konrad Z. Lorenz (see photo on p. 724), referred to by some as the father of modern ethology. In 1973, Lorenz, Dutch ethologist Nikolaas Tinbergen, and Austrian zoologist Karl von Frisch won the Nobel Prize for their contributions to the study of animal behavior. Lorenz based his work on the premise that animal behaviors are evolutionary adaptations, as are physical adaptations. He referred to behaviors as being part of an animal's "equipment for survival." This concept is still a fundamental premise in the study of animal behavior. Today, behaviors that enhance the ability of members of a population to live to reproductive age and that tend to occur at an increased frequency in successive generations are called **adaptive behaviors**.

Another group of animal behavior researchers prominent before the late 1950s was the *behaviorists*. These psychologists focused on behaviors themselves, studying them in the laboratory *without* focusing on the mental (cognitive) events that took place during the behaviors. In the late 1950s and the early 1960s, the fields of ethology and behavioral psychology merged to form the discipline of **animal behavior**. This approach includes many features derived from both the behaviorists and the ethologists.

The field of animal behavior expanded in the late 1970s as the science of *behavioral ecology* emerged. This biological discipline has an evolutionary focus. The central assumption of behavioral ecology is that an animal will behave in ways that will benefit it in the short term and that will maximize its Darwinian fitness, or ability to achieve reproductive success.

Today, the field of animal behavior is composed of researchers from a variety of disciplines: psychology, sociobiology (see p. 730), neurobiology, and behavioral ecology. Researchers from these fields study animal behavior from distinct perspectives, using different methodologies and basing their work on underlying assumptions and philosophies of their fields. Many biologists suggest that behavioral ecology is the dominant approach to the study of animal behavior today. Other scientists disagree. Whatever the case, a main difference between the work of animal behaviorists today and the ethologists of the past is explanation. Ethologists described, characterized, and cataloged behavior; animal behaviorists today pose hypotheses and test their predictions in an effort to determine *how* behavioral mechanisms work and *why* they evolved.

"How" and "why" questions are two types of causal questions. "How" questions refer to the *immediate cause*—the mechanisms underlying the particular behavior. "Why" questions refer to the *ultimate cause*. Behavioral ecologists, for example, frame "why" questions in evolutionary terms, asking why natural selection favored a particular behavior and not another.

Before the late 1950s, ethology dominated the study of animal behavior. Ethologists study the behaviors of animals in their natural environments, focusing on the patterns of movement, sounds, and body positions exhibited by animals. They often focused on the activity of the nervous system and also sought to understand the connection between an animal's behavior and its genetic make-up. One important concept rising out of ethology was that behaviors, like physical traits, can be adaptive. Today, researchers from a variety of disciplines study animal behavior. Behavioral ecology, a prominent biological approach to the study of animal behavior, has an evolutionary focus.

32.2 The link between genetics and behavior

All behaviors depend on nerve impulses, hormones, and other physiological mechanisms such as sensory receptors. Therefore, genes play a role in the development of behaviors because they direct the development of the nervous system. In addition, automatic responses depend on specific nerve pathways within the central nervous system of an organism. These pathways are neural programs and are genetically determined.

Margaret Bastock of Oxford University conducted classic experiments in the 1950s to show that certain behavioral traits are under the control of single genes. She found that male fruit flies (*Drosophila melanogaster*) having a mutant sex-linked gene termed "yellow" displayed a courtship pattern with less wing vibration than that of males having the "normal" gene. This altered courtship pattern resulted in the mutant males being less successful than normal males in mating normal females.

Genetically determined neural programs are part of the nervous system at the time of birth or develop at an appropriate point in maturation, resulting in **innate behaviors**. These instinctive or inborn behaviors are performed in a reasonably complete form the first time they are exhibited. A human newborn, for example, will turn to suckle when touched on the cheek near the mouth. Innate behaviors are important to the survival of an animal because they help it stay alive in certain situations and provide adaptive advantages that contribute to its fitness, or ability to achieve reproductive success.

Ethologists such as Lorenz and animal behaviorists of today would argue about the role of environment and learning on innate behaviors. **Learning** is an alteration in behavior based on experience. Today, the rigid distinction between innate and learned behavior that was once held by ethologists no longer exists. Evidence suggests that both genetic factors *and* the experience of the individual influence many aspects of behavior. Innate behaviors are now thought to be those that occur without *obvious* environmental influence.

The development of the nervous system and certain automatic nervous responses that are "preprogrammed" within the nervous system are directed by genes and occur without obvious environmental influence.

Just Wondering . . . *Real Questions Students Ask*

Why do people yawn?

Your question could be broadened to ask why most animals yawn—there is nothing particularly human about yawning. Most carnivores yawn. Few herbivores seem to yawn, although there are exceptions—hippos have enormous yawns. Frogs yawn. Even fish appear to yawn. Whatever is going on, it must be a basic behavior, like eating or sleeping, that is exhibited in many animals.

It used to be hypothesized that a yawn was a silent scream for oxygen that usually occurs when people are tired or bored—a type of deep breath to increase oxygen in the blood or to get rid of excess carbon dioxide. Not so. When a group of freshman psychology students at the University of Maryland inhaled air

containing different mixtures of oxygen and carbon dioxide and counted their yawns, only breathing rates went up or down to compensate for changing levels of oxygen and carbon dioxide. Yawning rates did not change.

Some researchers think that yawning is the body's way of promoting arousal in situations where you have to stay awake. Most humans yawn when stimulation is lacking. When a team of yawn counters observed people engaged in various activities, they found data to support this "lack of stimulation" idea. For example, they found that people riding subway cars yawned far more often when the cars were empty. The arousal hypothesis helps explain why people driving late at night on the high-

way yawn a lot and why very few people yawn when they are actually in bed—they don't need to stimulate themselves with a yawn because it's okay to go to sleep.

Whether this hypothesis is plausible or not, one thing is certain: Yawning is highly contagious. Seeing another person yawn releases a powerful urge for you to yawn yourself. This trait seems to be restricted to humans; no other animal responds in this way. Although yawning is not a hotbed of research, there is still much to learn about this behavior, because yawning seems associated with many diseases in ways not yet understood—brain lesions and epilepsy often lead to excessive yawning, but schizophrenics yawn very little.

32.3 Coordination and orientation behaviors

To survive, animals must respond to the environment. To do this, they must coordinate their movements in ways that result in effective responses. Such behaviors are called *coordination behaviors*. In addition, they must orient their movements in relation to external stimuli. These types of behaviors are termed *orientation behaviors*. Certain types of responses are characteristic of a species and thus are considered innate. A reflex is the simplest type of innate reaction to a stimulus and is an example of a coordination behavior involving various muscles. Kineses and taxes are simple types of orientation behaviors.

Reflexes

A **reflex** is an automatic response to nerve stimulation. The knee jerk is one of the simplest types of reflexes in the human body. If the tendon just below the kneecap is struck lightly, the lower leg automatically "kicks" or extends. In complex organisms, reflexes play a role in survival, such as when you jerk your hand away from a hot stove before you consciously realize that your hand hurts. In animals with extremely simple nervous systems such as the cnidarians (hydra, jellyfish, sea anemones, and coral, see Figure 14.1), most behaviors are the result of reflexes, although some simple learning can take place. In these animals, a stimulus is detected by sensory

neurons, and the impulse is passed on to other neurons in the animal's nerve net, eventually reaching the body muscles and causing them to contract. There is no associative activity in which other neurons can influence the outcome, no control of complex actions, and little coordination. The nerve net of the cnidarians possesses only the barest essentials of nervous reaction (see Habituation, p. 724).

Kineses

A **kinesis** (pl. kineses) is the change in the speed of the random movements of an animal with respect to changes in certain environmental stimuli. Put simply, movement slows down in an environment favorable to the animal's survival and speeds up in an unfavorable one. Have you ever picked up a rotting log or a clump of damp leaves in a wooded area? The pillbugs that you may have seen living under the leaves or the log stay in this favorable environment because of their low levels of activity and then scurry when the log is rolled over.

Taxes

Kineses are nondirected types of movements. In other words, an animal is not attracted to a favorable environment; it tends to "blunder" there as it moves about quickly and then stays there as it moves about slowly. A **taxis** (pl. taxes), however, is a directed movement toward or away from a stimulus, such as light, chemicals, or heat. Animals having preprogrammed taxes also have

Figure 32.1

Figure 32.1 **Phototaxis and gravitaxis in fish.** The normal fish in the left column orient to both light and gravity. The fish in the right column, which have had their gravity-detecting organ removed, orient to light only and have difficulty staying upright. The lower left diagram points out that orientation to gravity overrides orientation to light.

NORMAL

GRAVITY-DETECTING ORGAN REMOVED

Light

Light

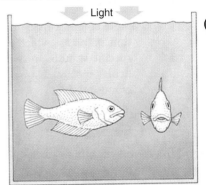

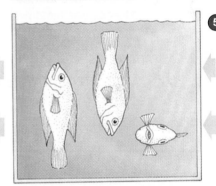

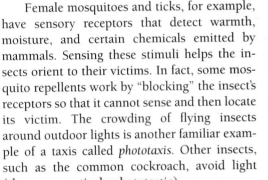

receptors that sense the particular stimuli to which the animal can orient.

Female mosquitoes and ticks, for example, have sensory receptors that detect warmth, moisture, and certain chemicals emitted by mammals. Sensing these stimuli helps the insects orient to their victims. In fact, some mosquito repellents work by "blocking" the insect's receptors so that it cannot sense and then locate its victim. The crowding of flying insects around outdoor lights is another familiar example of a taxis called *phototaxis*. Other insects, such as the common cockroach, avoid light (they are negatively phototactic).

Fish swim upright by orienting their ventral side to gravity and their dorsal side to light as shown in Figure 32.1 ❶. In ❷ the fish change position when the position of the light is changed. However, when the light is moved to the bottom of the tank ❸, the fish are unable to orient properly to both stimuli; the orientation to gravity is the stronger response. If the gravity-detecting organ in the inner ear is removed, the fish orient only to light as shown in ❹, ❺, and ❻. Certain species of fishes such as trout and salmon also automatically orient against a current and therefore face and swim upstream as shown in Figure 32.2.

Even organisms without a nervous system such as protozoans and bacteria exhibit taxes. These organisms have cellular organelles or inclusions that act as receptors or that react to specific environmental stimuli.

Coordination behaviors such as reflexes and orientation behaviors such as kineses and taxes are simple types of innate behaviors.

Figure 32.2 **Taxis in a salmon.** Salmon leaping up a small waterfall in Alaska while hungry grizzly bear watches.

32.4 Fixed action patterns

More complex than other innate behaviors are **fixed action patterns**, which are behavior patterns elicited by specific stimuli, apparently innate and specific to a particular species. The actions follow an unchanging order of muscular movements, such as a mother cardinal popping an insect into the mouth of her young. This sequence is a recognizable "unit" of behavior. When a releaser triggers the behavior, the sequence of activity begins and is carried through to completion. Fixed action patterns are often seen in body maintenance behaviors (such as a cat washing its face), courtship behavior, nest building, and attainment of food. Some researchers suggest that the term fixed action pattern should be replaced with *modal action pattern* because these behavior patterns are not identical when looked at carefully. Such terminology would indicate, more generally, a type or mode of behavior pattern.

The classic example used to illustrate a fixed (modal) action pattern behavior is the retrieval by the graylag goose of an egg that rolls out of its nest. As shown in Figure 32.3, if a goose notices that an egg has been knocked out of the nest, it will extend its neck toward the egg, reach up, and roll the egg back into the nest with its bill. Because this behavior seems so logical, it is tempting to believe that the goose saw the problem and figured out what to do. The entire behavior, however, is totally instinctive.

During experimentation, ethologists have discovered that any rounded object, regardless of size or color, acts as a releaser and triggers the response. Beer bottles, for example, are an effective releaser. Additionally, as a fixed action pattern behavior, the goose completes the action even if the egg rolls away from its retrieving bill. Instead of stopping in the middle of the behavior, the goose will continually repeat the entire behavior until the egg is brought back to the nest.

Figure 32.3   **Fixed action pattern in a graylag goose.**

Fixed action patterns—sequences of innate behaviors in which the actions follow an unchanging order of muscular movement—are more complex than other innate behaviors.

Some animals can change their behaviors based on experience.

32.5 Learned behaviors and survival

Innate behaviors are certainly important to the survival of an animal. An animal with protective coloration, for example, does not have time to learn to "freeze" when it detects a predator; its survival depends on its instinct to do so. For example, the photo below shows a small tropical frog that has markings that resemble a bird dropping. The effectiveness of this protective coloration depends on the frog remaining completely still. This stillness is an instinctive behavior on which the animal depends for survival.

Many social behaviors, such as certain mating or food-attainment behaviors, also depend on each individual's performance of instinctive behaviors. In addition, organisms with simple nervous systems, such as the cnidarians (jellyfish, corals, and hydra), have an extremely limited capacity for learning and must rely primarily on hard-wired innate behavior patterns for their survival.

Although important in many respects to the survival and fitness of animals, innate behaviors can become a liability if environmental conditions change and an animal's behavior cannot change to adapt to new conditions. For this reason, behavior patterns that can change in response to experience have adaptive advantages over the set programs of instinct. In most animals, only some behaviors are innate and permanent; many behaviors can be changed or modified by an individual's experiences during the process of learning. And, as mentioned previously, the line between learned behaviors and innate behaviors is blurring as scientists learn more about the nature of animal behavior.

Learned behaviors can help an animal become better suited to a particular environment or set of conditions, and these behaviors can be grouped into five categories: imprinting, habituation, classical conditioning, trial-and-error learning, and insight. As you read about these types of learning, notice that the first four categories regard learning as automatic and machinelike. Learning that takes place in a stimulus/response fashion, possibly reinforced by some type of reward that may or may not be readily apparent, reflects a school of thought called **behaviorism**. Behaviorism, which flourished from 1900 to 1960, is the theoretical basis for many of the rote, drill-and-practice type approaches to learning in schools during that time.

More recently, another school of thought regarding how we learn has emerged. **Cognitivism**, the prominent view in psychology today, suggests that individuals acquire and then store information in memory. Learning takes place as new information builds and merges with the old, leading to new types of behavior. Cognitivism is the theoretical basis for the constructivist approach to learning, which has gained wide acceptance in education.

Neuropsychological experiments show that we learn *both* ways. Each approach to learning uses different mechanisms and neural systems in the brain. Recent studies have confirmed that the mammalian brain (which includes your brain) has both systems.

Behaviors based on experience are learned behaviors. Learning helps an animal change its behavior to adapt to changes in environmental conditions.

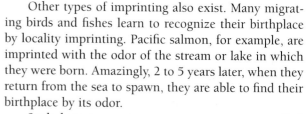

Some animals can change their behaviors based on experience.

32.6 Types of learning

Imprinting

Innate behavior and learning interact very closely in a time-dependent form of learning known as **imprinting**. Imprinting is a rapid and irreversible type of learning that takes place during an early developmental stage of some animals. Various types of imprinting exist. One type is object imprinting and has been observed in birds such as ducks, geese, and chickens. During a short time early in the bird's life (optimally 13 to 16 hours after hatching in mallard ducks, for example), the young animal forms a learned attachment to a moving object. Usually this object is its mother. However, animals can imprint on various objects regardless of size or color, such as balloons, clocks, and people. From an evolutionary viewpoint, parent-offspring imprinting enhances reproductive fitness by allowing parents and offspring to "recognize" one another, enabling parents to care for their offspring. Researchers have found that imprinting tends to occur in species that have a social organization in which attachment to parents, to the family group, or to a member of the opposite gender is important.

Konrad Lorenz performed classical experiments regarding object imprinting during the 1930s. In one of his most famous experiments, Lorenz divided a clutch of graylag goose eggs in half. He left one half of the eggs with the mother goose and put the other half in an incubator. The half that hatched with their mother displayed normal behavior, following her as she moved. As adults, these geese also exhibited normal behavior and mated with other graylag geese. The other half of the clutch, however, hatched in the incubator and then spent time with Lorenz. These goslings did not behave in the same manner as their siblings; they followed Lorenz around as if he were their mother (photo to left). After their initial time with him, Lorenz introduced the geese to their mother, but they still preferred to follow Lorenz. As adults, these geese tried to "court" adult humans!

Other types of imprinting also exist. Many migrating birds and fishes learn to recognize their birthplace by locality imprinting. Pacific salmon, for example, are imprinted with the odor of the stream or lake in which they were born. Amazingly, 2 to 5 years later, when they return from the sea to spawn, they are able to find their birthplace by its odor.

Such long-range, two-way movements are called *migrations*. The physiological and behavioral mechanisms that interact to help migrating animals navigate have interested and puzzled biologists and naturalists for centuries. Migration is discussed more fully on p. 728.

Habituation

Habituation is the ability of animals to "get used to" certain types of stimuli. People get used to stimuli all the time. A single shotgun blast would startle most people, but this response would fade at the end of a day in a firing range. Not only would you get used to the noise level (and wear ear protection), but you would perceive the gunshots as nonthreatening. Likewise, you quickly get used to the feel of your clothing after dressing in the morning. Animals also stop responding to stimuli that they learn are neither harmful nor helpful. Learning to ignore unimportant stimuli is a critical ability in an animal confronting a barrage of stimuli in a complex environment and can help an animal conserve its energy. You may have observed that pigeons living in the city are undisturbed by usual city noises, while those living in quiet settings fly away when they hear noise. In other examples, hydra stop contracting if they are disturbed too often by water currents. Sea anemones stop withdrawing if they are touched repeatedly. Young black-headed gull chicks stop crouching as familiar-shaped birds fly overhead. As Figure 32.4 shows, at first a chick will crouch at any object that flies overhead ❶. Gradually, the chick stops crouching at the familiar objects that fly overhead—the chick has gotten used to them ❷. However, the chick will crouch when an unfamiliar object poses a potential threat.

Classical Conditioning

Classical conditioning is a form of learning in which an animal is taught to associate a new stimulus with a natural stimulus that normally evokes a response in the animal. Repeatedly presenting an animal with the new stimulus in association with the natural stimulus can cause the animal's brain to form an association between the two stimuli. Eventually, the animal will respond to the new stimulus alone; the new stimulus will act as a

Figure 32.4   Habituation in gull chicks.

1 EARLY
Chick crouches at any object

2 LATER
Chick habituates to familiar objects

Chick continues to crouch at unfamiliar objects

substitute for the natural stimulus. The connection between the new stimulus and the natural stimulus is the result of a learning process, but must be reinforced periodically with the presence of the natural stimulus or the animal will stop responding to the substitute.

In his famous study of classical conditioning, the Russian psychologist Ivan Pavlov worked with dogs to condition them to salivate in response to a stimulus normally unrelated to salivation. His discovery of conditioning, however, was accidental and occurred as he investigated the physiology of the digestive system. In 1904, he was awarded the Nobel Prize for his work on digestion. As part of his research, Pavlov routinely collected saliva from his laboratory dogs. He stimulated saliva flow by placing meat powder in the dog's mouth. He soon noticed that the dog began salivating as he approached it with the meat powder. He then paired other, unrelated stimuli with the meat. For example, he shined a light on the dog at the same time that meat powder was blown into its mouth. As expected, the dog salivated. After repeated trials, the dog eventually salivated in response to the light alone. The dog had learned to associate the unrelated light stimulus with the meat stimulus (Figure 32.5). Pavlov also used a tuning fork and a bell in his experiments as the conditioning stimuli.

The natural stimulus-response connection in an animal is an inborn reflex, or an *unconditioned response*. Such innate responses are important to animals; many of them are protective, such as blinking, sneezing, vomiting, and coughing. Even the knee-jerk reflex mentioned earlier is part of the mechanism by which humans (and other vertebrates) maintain their posture. The work of researchers such as Pavlov contributed to the understanding that animal behavior depends on innate neural circuitry but that these neural programs can most often be modified and directed by the processes of learning.

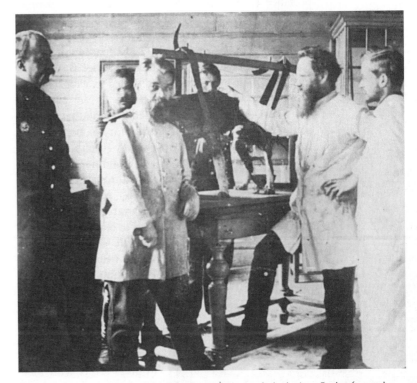

Figure 32.5   **Classical conditioning.** The Russian psychologist Ivan Pavlov (second from right) in his laboratory. In the background, a dog is suspended in a harness that Pavlov used to condition salivation in response to light.

Trial-and-Error Learning: Operant Conditioning

More complex than imprinting or habituation, **operant conditioning** (OP-uh-runt) is a form of learning in which an animal associates something that it does with a reward or punishment. (The word *operant* means "having the power to produce an effect" and is a form of the verb "to operate.") In operant conditioning, an animal must make the proper association between its response (such as pressing a lever) and a reward (the appearance of a food pellet) before it receives this reinforcing stimulus. It may also learn to avoid a behavior when the stimulus is negative. Animal trainers use the techniques of operant conditioning.

The American psychologist B.F. Skinner studied such conditioning in rats by placing each in a specially designed box (today called a *Skinner box*) fitted with levers and other experimental devices. A rat is shown in a Skinner box in **Figure 32.6**. Once inside, the rat would explore the box feverishly, running this way and that. Occasionally, it would accidentally press a lever, and a pellet of food would appear. At first, a rat would ignore the lever and continue to move about, but soon it learned to press the lever to obtain food.

This sort of trial-and-error learning is of major importance to most vertebrates in nature. For example, when a toad gobbles a bumblebee, it gets a violent sting on its tongue and spits out the bee. The toad learns quickly not to eat a bumblebee. The toad does not use reasoning to determine that the bumblebee is not good to eat—it merely becomes conditioned by experience to avoid a response that causes it pain.

Behavioral psychologists used to think that animals could be conditioned in the laboratory to perform *any* learnable behavior in response to *any* stimulus by operant conditioning. However, through experimentation, researchers have discovered that animals tend to learn only in ways that are compatible with their neural programs. Put simply, operant conditioning works only for stimuli and responses that have meaning for animals in nature. Rats can be conditioned, for example, to press levers with their paws to obtain food because in nature they obtain food with their paws.

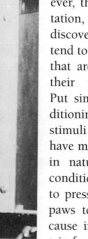

Figure 32.6 Operant conditioning.

They cannot, however, be conditioned to obtain food by jumping, an unnatural food-attaining behavior for a rat.

Insight

Best developed in primates such as chimpanzees and humans, **insight**, or **reasoning**, is the most complex form of learning. An animal capable of insight can recognize a problem and solve it mentally before ever trying out a solution. Therefore, the animal is able to perform a correct or appropriate behavior the first time it tries, without having been exposed to the specific situation.

German psychologist Wolfgang Kohler was the first to describe learning by insight, performing extensive experiments on chimpanzees in the 1920s. Kohler showed that an animal must perceive relationships and manipulate concepts in its mind to solve a problem on the first try. In his classical experiments, he placed chimpanzees in a room with a few crates, poles that could be joined together, and a banana hung high above the grasp of the animals. The chimpanzees were able to stack the crates and join the poles appropriately to retrieve the banana (**Figure 32.7a**). Unlike chimpanzees, most animals are unable to use insight to solve problems. Your dog or cat, for example, must use trial and error to "try out" solutions to problems. Your dog has no insight into the problem that having its leash wrapped around a tree is keeping it from reaching its food (Figure 32.7b). The dog can merely continue walking in various directions and ultimately free itself by chance. This experience, however, may help the dog perform the appropriate behavior in the future, having learned by trial and error.

Imprinting is a rapid and irreversible type of learning that takes place during an early developmental stage of some animals. Imprinting helps animals recognize kin or recognize their "home" or birthplace. Habituation is the ability of animals to get used to certain types of stimuli. Classical conditioning is a form of learning in which an animal is taught to associate a new stimulus with a natural stimulus that normally evokes a response in the animal. Some animals can learn by trial and error; by associating a specific behavior with a reward or punishment, they can learn to apply (or avoid) certain behaviors to affect outcomes in particular situations. Insight or reasoning is the most complex form of learning. An animal capable of insight can recognize a problem and solve it mentally before ever trying out a solution.

(a)

(b)

Figure 32.7 **Insight is the most complex form of learning.** (a) A chimpanzee is able to see a problem and develop a solution, even before taking any action. To reach the bananas, the chimpanzee sees a solution in stacking the boxes and fitting the poles together. (b) A dog, on the other hand, lacks insight and cannot develop a solution to a problem. A dog will eventually learn by trial and error to go around the tree, unwinding his leash to reach the food.

Animals exhibit regularly repeated behaviors.

32.7 Circadian rhythms and biological clocks

Organisms (including animals, plants, protists, fungi, and even bacteria) have internal clocks that regulate many of their activities. These 24-hour cycles of physiological activity and behavior are called **circadian rhythms** (sir-KAY-dee-un). Circadian rhythms are regulated by organisms' **biological clocks**. These internal clocks often interact with environmental cues, which help keep the internal clock timed to the outside world. Recently, scientists have gained some insight into exactly what a biological clock is and how it governs rhythmic patterns such as sleep and wake cycles, feeding patterns, photosynthetic activity, and animal movements such as migration.

The basis of circadian rhythms is genetic; that is, genes regulate these cyclical patterns of behavior and physiology. Scientists have recently discovered ways in which these regulatory genes are turned on and off.

In 1997, studies in mice demonstrated that a protein known as CLOCK regulates circadian rhythms in mammals. Then in 1998, scientists gained a glimpse into how this regulation occurs. In both fruit flies and mice, CLOCK binds with another protein called BMAL1. The CLOCK-BMAL1 complex binds to a specific sequence of DNA, activating a gene that drives the transcription of CLOCK proteins. The CLOCK proteins accumulate over the course of the day. A certain level of CLOCK proteins is a trigger that turns the gene off. Gene inactivation (after a lag) is, in turn, a trigger for CLOCK-BMAL1 to reactivate the gene.

This type of regulatory mechanism is known as a *negative feedback loop*. Scientists suspect that this negative feedback loop is only one factor in a variety of yet unknown factors that govern circadian rhythms. Additionally, the same biological clock genes in fruit flies, mice, and people appear to govern their circadian rhythms, suggesting that they all depend on the same molecular mechanisms to keep time.

Organisms have internal clocks, or biological clocks, that regulate 24-hour cycles of physiological activity and behavior called circadian rhythms. Circadian rhythms are genetic mechanisms governed by negative feedback loops and other, as yet unknown, factors. Environmental cues often help keep the internal clock timed to the outside world.

32.8 Migration

Migrations are movements of animals from one region to another with the change of seasons. In many animals, migrations occur once a year and result from interactions among various environmental factors (such as day length) with animals' physiological and hormonal changes. Ducks, geese, and shorebirds for example, migrate down flyways from Canada across the United States each fall and return each spring.

Perhaps the longest migration is that of the American Golden Plover, a shorebird that flies from Arctic breeding grounds to wintering areas in southeastern South America, a distance of approximately 13,000 kilometers (about 8000 miles). The migration route of this bird is shown in **Figure 32.8**. (Shorebirds include the plovers, gulls, terns, and sandpipers. They are found not only along shorelines as their name suggests, but some are also found inland, on arctic tundra, and at sea.)

Monarch butterflies (thought to be the only migrating insects) travel long distances also, from the eastern United States to Mexico and back, a journey of more than 3000 kilometers (approximately 2000 miles). The monarchs that make the trip to Mexico are the result of breedings in late August. The shorter days and colder temperatures of late August cause the emerging monarchs to postpone reproductive maturity. These butterflies will live for eight or nine months (much longer than monarchs that develop earlier in the summer will) and can endure the flight to Mexico. In spring, the eight- or nine-month-old monarchs reach sexual maturity and begin migrating back to the eastern United States. They mate all along the migratory route. Unlike their journey south the previous fall, they do not complete the return trip, but their offspring (or the next generation of offspring) do. The round-trip, therefore, takes from two to five generations of butterflies to complete.

Animals that migrate short distances, such as deer that live in mountainous regions, migrate downhill in response to snow cover and a shortage of food. Long distance migrants are triggered by their biological clocks, which produce annual physiological and behavioral rhythms just like biological clocks that produce daily rhythms. Their internal clocks (or calendars in this case) are programmed in such a way that the animals perform the behavior for a set amount of time—whatever time is necessary to complete their journeys.

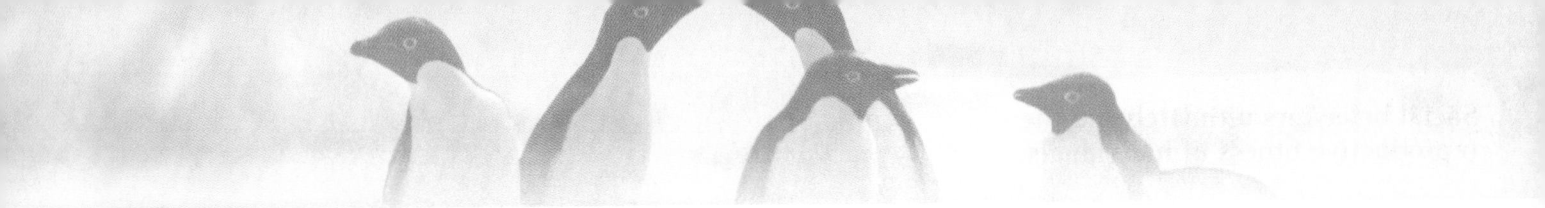

Migrating animals also have the ability to navigate in order to reach their destination and return.

Migrating organisms orient themselves in relation to an environmental cue, such as the sun, the stars, or the Earth's magnetic field. Day migrants such as some birds, ants, and bees use the sun's position to chart a course. Night migrants, a group that includes birds such as the indigo bunting, use the stars to chart a course. The biological clocks of all these animals interact with information from the environment to help migrant animals find their way.

> Migrations, the movements of animals from one region to another with the change of seasons, occur once a year in many animals. Migratory behavior results from interactions among various environmental factors (such as day length) with animals' physiological and hormonal changes.

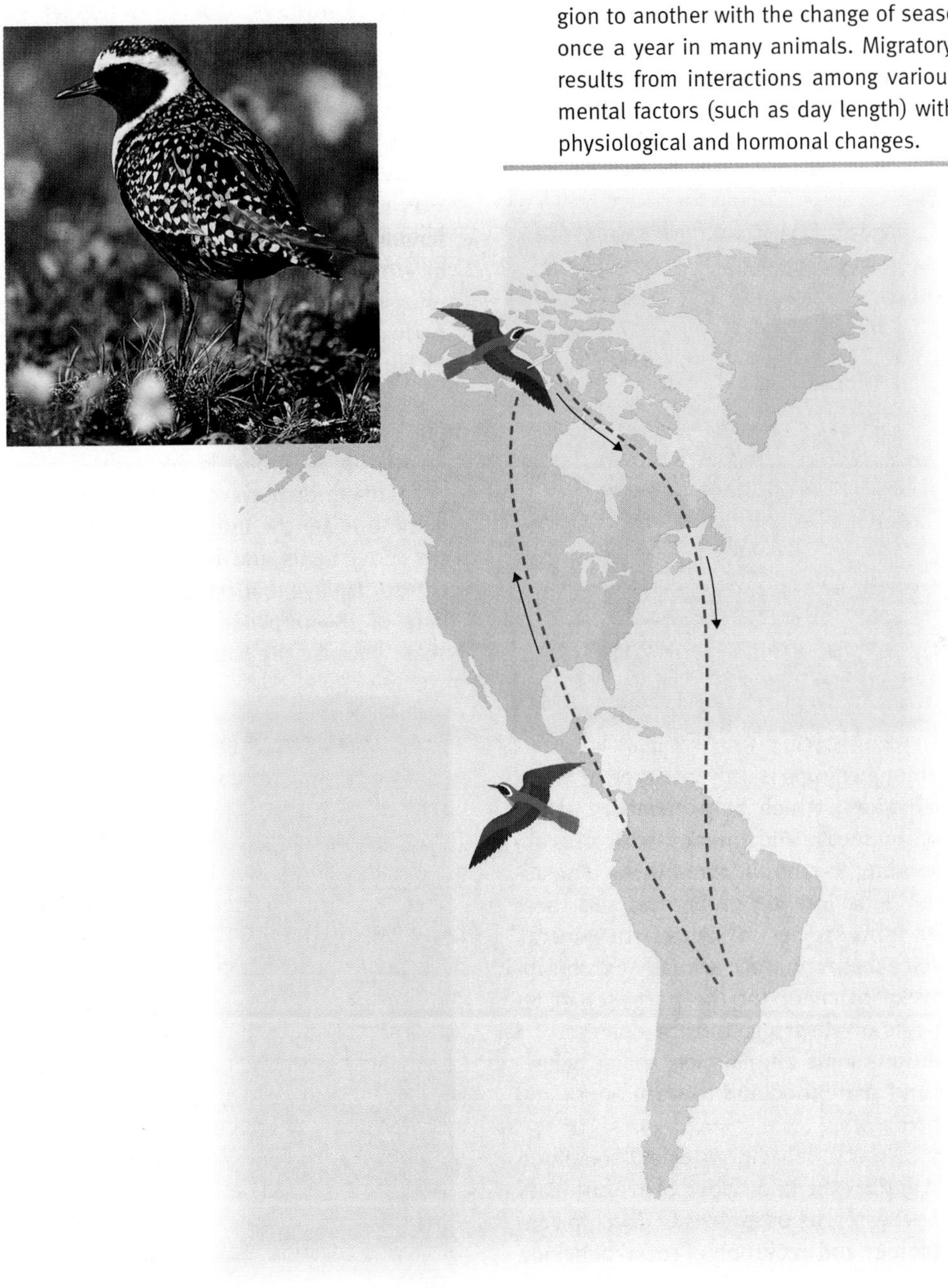

Figure 32.8   The American golden plover and its migration route.

32.9 Communication via social behaviors

If you've ever observed chimpanzees at the zoo or in a nature preserve, you've probably seen the grooming behavior exhibited in the photo. Grooming in chimps consists of the cleaning of dirt, debris, and sometimes parasitic insects such as ticks from the body. Occasionally chimps groom themselves, but social grooming (one chimp grooming another) is more common. Not only does social grooming help keep chimps clean and relatively free of parasites, it strengthens the social bond between them.

Grooming among chimps is only one type of an array of **social behaviors**, which help members of the same species communicate and interact with one another, each responding to stimuli from others. The advantages of social behaviors are numerous, and these advantages differ from species to species. In general, however, all species that reproduce sexually exhibit interactive patterns of behavior for the purposes of reproduction, the care of offspring, and the defense of a territory. In addition, some animals use social behaviors to hunt for and share food and to warn about and defend against predators.

The biology of social behavior is called **sociobiology**. This science applies the knowledge of evolutionary biology to social behavior. Its purpose is to develop general laws of the biology and evolution of social behavior.

www.jbpub.com/biology

> Social behaviors are interactive patterns of animal behavior that help members of the same species communicate and interact with one another. The biology of social behavior is sociobiology.

32.10 Competitive behaviors

When two or more individuals strive to obtain the same needed resource, such as food, water, nesting sites, or mates, they are exhibiting **competitive behavior**. This type of behavior occurs when resources are scarce.

Threat Displays

How do animals compete? Many first engage in **threat** or **intimidation displays**, a form of aggressive behavior. The purpose of these displays is to do as the name suggests: scare other animals away or cause them to "back down" before fighting takes place. Intimidation displays vary widely among species but usually involve such behaviors as showing fangs or claws, making noises such as growls or roars, changing body color to one that is a releaser of aggression in an opponent, and making the body appear larger by standing upright, making the fur or hair stand on end, or inflating a throat sac. The common toad in the photo below is exhibiting a threat display; it inflates its body and sways from side to side in an attempt to appear larger.

Threat displays are important social signals and communicate the intent to fight. Interestingly, some of the movements and body postures (body language) of threat displays that repel competitors also attract members of the opposite gender. Biologically, this makes sense because the competition may be over a mate.

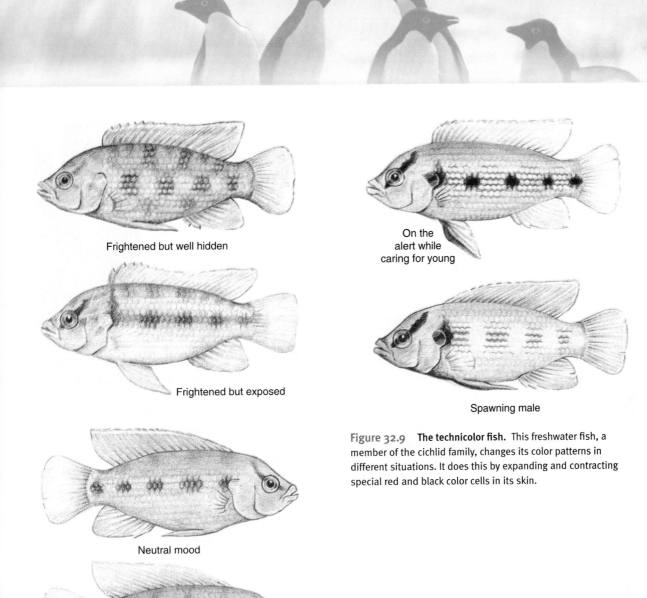

Frightened but well hidden

Frightened but exposed

Neutral mood

Mild territorial aggression

On the alert while caring for young

Spawning male

Figure 32.9 **The technicolor fish.** This freshwater fish, a member of the cichlid family, changes its color patterns in different situations. It does this by expanding and contracting special red and black color cells in its skin.

Submissive Behavior

Animals use other behaviors to avoid fighting. One type of behavior is called **submissive behavior** and is usually a behavior opposite to a threat display. The behavior might include making the body appear smaller, "putting away weapons" of fangs or claws, turning a vulnerable part of the body to an opponent, or removing body colors that are releasers of aggression. As **Figure 32.9** shows, an animal may, in fact, possess an array of color "signals." If an animal is losing a fight, it might also display submissive behaviors to stop the fight. Contrary to popular belief, animals of the same species rarely fight to the death. One reason is that most animals do not have the means to do so—they do not have sufficiently dan-gerous fangs, claws, or horns, for example. Species with dangerous weapons usually have defenses against those weapons too, such as a strong hide, long hair, or a thick layer of body fat. Scientists hypothesize that aggression within a species is meant to chase off rather than kill the rival.

Territorial Behavior

A territory is an area that an animal marks off as its own, defending it against the same-gender members of its species. These behaviors are called **territorial behaviors.** Members of the opposite gender are sometimes allowed into the territory, often for mating purposes.

Territoriality is very common in all classes of vertebrates and is even found in some invertebrates such as crickets, wasps, and praying mantises. Some animals such as dragonflies mark their territories by conspicuously patrolling their borders, resting and then moving from prominent landmarks. As an animal demonstrates the borders of its territory (Figure 32.10), it may exhibit specific movements or body postures as signals. Some animals, such as song birds, monkeys, apes, frogs, and lizards, mark their territory by producing sounds that announce their ownership. Animals with a well-developed sense of smell, such as wolves, hippopotamuses, rhinoceroses, some rodents, and even domestic dogs, mark their territories with substances that have an odor such as urine. Many species, in fact, possess special scent glands that secrete substances just for this purpose. The odor marking helps keep out members of the same gender and helps owners of large territories orient themselves to its borders.

Territorial behavior has several adaptively important consequences. In many mammalian societies, territoriality often has an important influence on the establishment of new populations, with the dominant male defeating the lesser males and leaving them to found their own populations. Individuals surviving in the surrounding marginal areas repopulate any vacant territories.

Although territoriality evolved as a result of the advantages to individuals, this system has an effect on the resources available to various members within a population. When resources such as nesting sites and food are limited, each member of a population is in danger of the resources being spread too thin, so no member gets an adequate amount of what it needs. In addition, breeding pairs of animals will not reproduce unless they have adequate resources; territorial behavior ensures that at least some members of the population will have their own space, food, and nesting sites so that they will survive and reproduce. Territorial behaviors also enhance reproductive capabilities by placing the peak time of competition and aggressive behavior at the time of the marking of the territory, *before* the time of reproduction and raising of young. The white-tailed buck in the photo is exhibiting territorial behavior by marking vegetation with a scent gland near his eye. The higher the marking, the more impressive it is to other male deer, suggesting a larger animal.

Individuals within the same species compete with each other for various resources. Competition among animals includes a repertoire of behaviors such as threat displays, submissive behavior, and territorial behavior. Such competitive behaviors evolved as a result of the advantages to individuals, but they have an effect on the resources available to various members within a population.

Figure 32.10 Territory of a male Demoiselle dragonfly. The male dragonfly sits on a perch above the egg deposit site. He periodically flies around his territory (red arrows) or circles within the territory.

32.11 Reproductive behaviors

Sexual reproduction—the union of a male sex cell (sperm) and a female sex cell (egg) to form the first cell of a new individual (zygote)—takes place in most animals. Even some animals that can reproduce asexually, such as sponges that can bud off bits of themselves that grow into new individuals, reproduce sexually at certain times.

Sexual reproduction is biologically significant because it provides genetic variability among organisms of the same species. Put simply, the combining of genes from two parents produces offspring that are similar but different from their parents. In addition, because the genes of each parent become assorted independently during the meiotic process that produces sex cells (see Chapters 18 and 22), the offspring are different not only from their parents but from one another. Differences among organisms within a species are the "raw materials" needed for adaptation to environmental change. The ability of some organisms to survive and reproduce in an environment in which others of their species cannot is, in turn, the basis of natural selection.

Genetic variability does not exist (with the exception of mutation) in organisms that reproduce asexually; these organisms are clones of one parent. That is, organisms produced by means of asexual reproduction develop from one or a few cells of a single parent and are therefore genetically identical to that parent. (Chapter 22, pp. 482–484 describes asexual reproduction in more detail.) Because organisms within a population of clones do not vary from one another, the entire population can be at risk if environmental conditions change.

For reproduction to take place, animals of most species must communicate and cooperate with one another. Individuals within many species of animals form pair bonds for the purposes of reproduction and parenting. An important outcome of these processes is the retention in the population of the genes responsible for the various facets of social behavior. For natural selection to favor a particular gene, the advantages of the particular social trait it encodes must outweigh the disadvantages—for the individual or for the individual and its close relatives (*kin*).

Behaviors that promote successful sexual reproduction, then, are highly adaptive behaviors. Organisms that reproduce sexually must have patterns of male-female interactions that lead to fertilization—the penetration of an egg by a sperm. Fertilization can be preceded by copulation, the joining of male and female reproductive organs. However, some organisms such as many fish and frogs do not copulate; instead the male releases sperm over the eggs after they have been deposited by the female. The term **mating** refers to male-female behaviors that result in fertilization, regardless of whether copulation occurs. The term **courtship** refers to the behavior patterns that lead to mating.

For courtship to take place, males and females must first find each other. (The opening vignette in Chapter 15 discusses how male and female elephants use infrasound to find each other for mating.) In some species, males and females live together in pairs or groups, but many animal species do not. Most often the job of attracting a mate goes to the male of the species. Notable exceptions, however, are often found within insect populations.

Males often attract a mate by marking a territory and defending it against intrusion by other males. As mentioned previously, aggressive behaviors against other males are often the same behaviors that attract a female. In addition, males can attract females by displaying body colors or markings like the peacock shown in the photo. Many birds sing songs to attract a mate, and certain frogs and many insects use other sounds or mating calls for this purpose. Many male mammals and insects produce odors that are attractive to females. In some species, the males congregate and perform various dances or songs as a group to attract their mates (Figure 32.11).

After the formation of a male-female pair, a **courtship ritual**, unique to each species,

Figure 32.11   Courtship dance of the sage grouse. At one point in this series of behaviors, males display their fanned-out tail feathers to a female.

> Social behaviors ultimately aid the reproductive fitness of individuals.

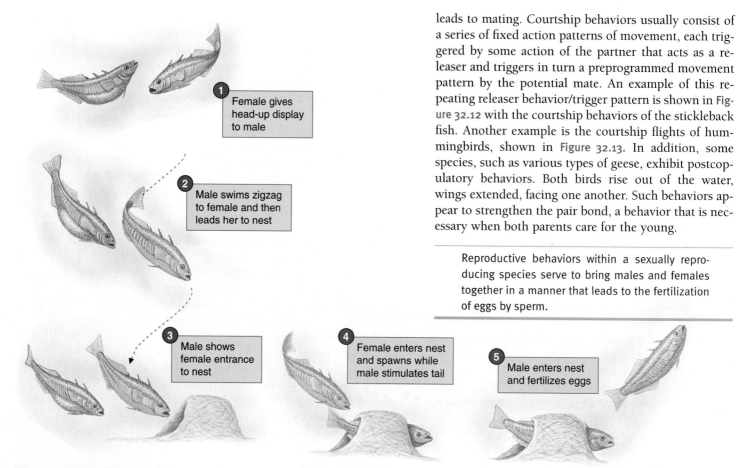

Figure 32.12 Stickleback courtship. Each set of movements in the courtship ritual is a fixed action pattern that triggers the next set of movements by the partner.

1 Female gives head-up display to male

2 Male swims zigzag to female and then leads her to nest

3 Male shows female entrance to nest

4 Female enters nest and spawns while male stimulates tail

5 Male enters nest and fertilizes eggs

leads to mating. Courtship behaviors usually consist of a series of fixed action patterns of movement, each triggered by some action of the partner that acts as a releaser and triggers in turn a preprogrammed movement pattern by the potential mate. An example of this repeating releaser behavior/trigger pattern is shown in Figure 32.12 with the courtship behaviors of the stickleback fish. Another example is the courtship flights of hummingbirds, shown in Figure 32.13. In addition, some species, such as various types of geese, exhibit postcopulatory behaviors. Both birds rise out of the water, wings extended, facing one another. Such behaviors appear to strengthen the pair bond, a behavior that is necessary when both parents care for the young.

> Reproductive behaviors within a sexually reproducing species serve to bring males and females together in a manner that leads to the fertilization of eggs by sperm.

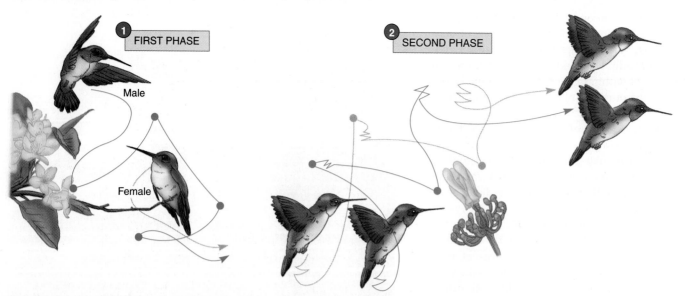

1 FIRST PHASE

Male

Female

2 SECOND PHASE

Figure 32.13 Courtship flight in hummingbirds. In the first phase of the courtship flight, the male flies around the female in an undulating pattern. In the second phase, both partners fly around each other and may mirror their partner's movements.

32.12 Parenting behaviors and altruism

For the offspring to survive, grow, and eventually reproduce, parents must either make preparations for the care of their young or care for them themselves. Parenting behaviors are a type of **altruistic behavior**. Such behaviors benefit one at the cost of another. In animal behavior, this term is different from the everyday use of the word. When we speak of altruistic behaviors of humans, we imply that the person performing the altruistic act was knowingly and willingly risking or giving up something for another. For example, good Samaritans who risk their lives to save another person are performing an altruistic act. With animals other than humans, however, altruism does not carry this meaning. In addition, its focus is evolutionary: An altruistic act is one that increases the individual fitness of the recipient while decreasing the individual fitness of the donor. Natural selection favors genes that promote altruistic behavior among closely related individuals (kin) such as parents and offspring.

Parenting behaviors are found in almost all animal groups, but the participation of the parents varies. In many species of birds and fishes, both parents care for the young, and a division of labor usually exists, with each parent carrying out a specific role. Because female mammals produce milk to nourish the young, most often the mother assumes the parenting role in this class of animals. In only a few species, such as certain fishes, sea horses, and birds, does the male alone take care of the offspring.

To prepare for their young, animals build protective structures such as nests or cocoons. Usually a food supply is gathered, and the eggs are deposited in areas that are protected and that may also lie near a ready food source. Marsupials such as kangaroos—mammals in which the young are born developmentally early—keep their young safe and nourish them as they complete their development by storing them in a pouch, conveniently situated at the mammary glands.

Some species exhibit quite interesting and unique parenting behaviors. Very early in his studies of behavior, Nikolaas Tinbergen noted that certain birds remove the shell fragments of broken eggs from their nests. The adaptive advantage of this parenting behavior, Tinbergen discovered, was that the odor from shell fragments attracted predators and their prompt removal helped protect the other eggs in the nest.

> Parenting behaviors lead to the reproductive fitness of offspring by helping individuals survive and grow to reproductive age. These acts are altruistic; they increase the individual fitness of the recipient while decreasing the individual fitness of the donor.

32.13 Group behaviors

In addition to pair-bond relationships, the individuals of many species form temporary or permanent associations with many other members of their species, forming social groups. Animals within social groups work together for common purposes. Individuals within a group, for example, have more protection against predators than do single individuals or pairs. They can warn each other of danger by using calls or chemical signals or cluster together when a predator appears, forcing the predator to either separate individuals from the group or fight the entire group. In addition, some animals such as musk oxen and bison place their young within the center of the group, forming a circle around them much like the early American settlers did with their covered wagons.

Many animals living within social groups search for food together and, scientists have observed, are more successful as group hunters than as single individuals. Food-gathering strategies become group strategies; often animals such as those that prey on fish will drive the prey to one location, encircle them, and feast. The eastern white pelicans shown in **Figure 32.14** hunt for fish in this way.

In social groups, animals can also ration the "work load" so that each animal does not have to perform an

(a)

(b)

Figure 32.14 Hunting behaviors of pelicans. Eastern white pelicans hunt in groups (a) and then circle their prey (b).

array of daily tasks but may focus on one task (Figure 32.15). Many species of insects, particularly bees and ants, are organized into social groups having a division of labor. These social groups are called **insect societies**.

Figure 32.15 **Ant society.** This worker ant, a member of the genus *Polyergus*, is responsible for capturing slaves for the nest. The slave, a member of the genus *Formica*, is in its pupa stage. When it emerges as an adult, it will be a nonreproductive worker slave.

Insect Societies

In insect societies, individuals are organized into highly integrated groups in which each member of the society performs one special task or a series of tasks that contributes to the survival of the group. The role an individual plays depends on its body structure, which in turn depends on its gender or age. In some animal societies, the same animal performs different roles during the course of its life. In addition, a common feature among social insects is a queen that outlives other members of the society. The role of the queen is to produce offspring and to promote cooperation among members of the group.

Figure 32.16 **A bee colony.** The queen bee has a red spot on her back painted on by a researcher.

A honeybee colony, for example, is made up of three different types of bees called *castes:* the queen, the workers, and the drones. The queen is the focus of the colony; she lays the eggs from which all the other bees develop (Figure 32.16). The drones are the male bees that fertilize the eggs of the queen. They develop from unfertilized eggs and so have no male parent themselves—and only one set of chromosomes. The workers are females that develop from fertilized eggs. Genetically, a queen and a worker are no different; any fertilized egg can develop into either a queen or a worker depending on how the egg is housed and fed.

In one hive, there may be up to 50,000 workers, 5000 drones, and, of course, 1 queen. The workers perform all the tasks of the hive except for mating and egg laying. In the winter months, the average length of the worker bee's life is 6 months. In the summer, workers live only 38 days! During the first 20 days of a worker's life, she performs various duties within the hive sequentially as she matures: feeding larvae, producing wax, building the honeycomb, passing out food, and guarding the hive. Although a worker may go on short play flights around the hive during this time, she does not venture very far. It is only after the first 20 days of her life that the worker takes long flights away from the hive to forage for food.

An interesting means of communication among the workers has evolved that enables them to communicate the location of flowers at which they can collect pollen and nectar. This communication is a form of body language called the *waggle dance,* a behavior discovered by Austrian zoologist Karl von Frisch (who, with Dutch ethologist Nikolaas Tinbergen, won the Nobel Prize in 1973 for their contributions to the study of animals). The manner in which a worker shakes, or waggles, her abdomen communicates the distance to the flowers. The angle at which the worker dances indicates the direction her co-workers must fly (Figure 32.17).

The drones play almost no role in the hive other than to mate with the queen. They help keep the larvae warm by congregating near them in the morning while the temperature is still cool. Drones also help distribute food. At the time of mating, many drones follow the queen out of the hive, following her pheromone trail (see next paragraph), and mate in flight. After mating, the drones die. Drones that do not mate have no better fate, because the workers pinch them, sting them, and then throw them out of the hive as the winter approaches.

The queen bee controls the worker bees with a chemical she produces known as *queen substance.* This substance is a **pheromone** (FARE-uh-moan), a chemical produced by one individual that alters the physiology or

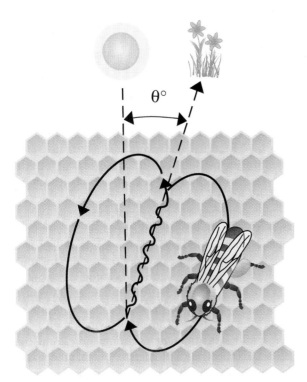

Figure 32.17 **The waggle dance.** This dance is performed by a forager bee who, on spotting a food supply (a flower), goes back to the hive and communicates to the other bees through dance the exact location of the flower. The dance tells how far away the flower is and also the angle of the flower to the sun. The part of the dance that yields both kinds of information is the straight run, *(zigzag line between two circles)*. The forager shakes its abdomen back and forth and buzzes with its wings while moving in the straight run. Then it makes a semicircle and returns to the starting point of the straight run. After completing this run, it circles in the opposite direction to land again at the beginning of the straight run and so on. How fast the bee performs the straight run determines how far away the flower is—a distance of 500 meters equals a straight run duration of about 1 second. The direction of the flower is indicated by the angle that the straight run deviates from the vertical. This angle equals the angle of the flower to the sun.

behavior of other individuals of the same species. The bees ingest the queen substance by licking the queen and then pass the substance around from one to the other as they pass food. Queen substance renders the worker bees sterile and also inhibits the workers from making queen cells, compartments in which fertilized eggs can develop into queens. If the hive becomes too congested, however, the queen may begin producing less queen substance, resulting in the removal of the inhibition to produce queen cells. Workers make a half dozen or more new queen cells in which replacement queens begin to develop. The old queen and a swarm of

females and male drones leave to establish a new hive. The first new queen to emerge may kill the other candidate queens and assume rule or may create another swarm and leave to establish another hive.

The lifestyles of social insects can be so unusual as to be bizarre, none more so than the little reddish ants—the leafcutters (Figure 32.18). Leafcutters live in the tropics, organized into colonies of up to several million individuals. These ants are farmers, growing crops of fungi beneath the ground. Their moundlike nests look like tiny underground cities covering more than 100 square yards, with hundreds of entrances and chambers as deep as 16 feet underground. Long lines of leafcutters march daily from the mound to a tree or bush, hack its leaves into small pieces, and carry the pieces back to the mound. Each ant finds it way by following a trail of secretions left by those who came before it. Although these ants are nearly blind, they follow the scent by holding their antennae close to the ground. At the underground site, worker ants chew the leaf fragments into a mulch that they spread like a carpet in underground chambers. They wet the leaf mulch with their saliva and fertilize it with their feces. Soon a luxuriant lawn of fungi is growing, which serves as the sole food for all the ants, no matter what their age. Nurse ants even feed this fungus to the larvae, carrying them around to browse on choice spots.

The complex caste system of bees and ants apparently evolved a long time ago. Ant fossils 80 million years old exhibit three castes (males, queens, and

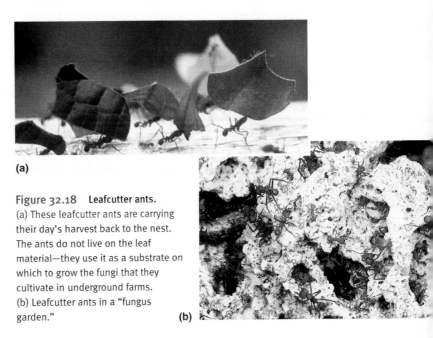

Figure 32.18 **Leafcutter ants.**
(a) These leafcutter ants are carrying their day's harvest back to the nest. The ants do not live on the leaf material—they use it as a substrate on which to grow the fungi that they cultivate in underground farms.
(b) Leafcutter ants in a "fungus garden."

workers), indicating that their complex social system had already evolved.

Rank Order in Vertebrate Groups

A social hierarchy, or **rank order**, exists in groups of fishes, reptiles, birds, and mammals. In chickens, the rank order is often called a **peck order** because these animals establish the order by pecking at one another. In many species of animals, the rank order is linear, with the highest-ranking individual dominant over all the others and the lowest-ranking individual submissive to all others. Some rank orders show certain complexities within the ranking, such as a rank order in which animal "A," for example, may be dominant over animal "B" and "B" dominant over "C," but "C" may be dominant over "A." In some rankings, the sex of the animal may play no role. In other rankings, males and females are ranked separately, and in still other vertebrate groups, the female takes on the rank of the male. Such rankings help reduce aggression and fighting within social groups, focusing aggressive behavior on the time that the ranking is developed. This behavior is only one of a repertoire of behaviors, including territoriality and submissive behaviors, that helps contribute to the stability of the social relationships among animals and within groups.

> Many animals not only interact with other members of their species but form social groups and aid one another in various ways. Many species of insects, particularly bees and ants, form highly organized social groups exhibiting a division of labor.

32.14 Human behavior

Humans are unique animals. Because of this uniqueness, many scientists disagree with applying the principles of sociobiology to human behavior. Many scientists question the scientific validity of sociobiological research on human activity. Others question the social and political implications of human sociobiology, proposing that such endeavors might lead to yet another form of racism, sexism, or other type of group stereotyping. In addition, the study of human sociobiology has inherent limitations: Scientists cannot experimentally manipulate the genes or vary the environments of humans for scientific studies. It is therefore extremely difficult to learn whether observed behaviors are a product of heredity or a person's environment.

This "nature or nurture" dichotomy continually raises questions that scientists have a difficult time researching. Interesting observations have been made, however, in the study of identical twins raised by different families in different environments. From these studies and others, most biologists hold that much of the behavioral *capacity* of humans is genetically determined, but that the neural circuits specified by genes can be shaped and molded—within certain limits—by learning.

> Many scientists disagree with applying the principles of sociobiology to human behavior. Most biologists hold that much of the behavioral capacity of humans is genetically determined, but that behavior can be modified, within limits, by learning.

Key Concepts

Researchers from various disciplines study animal behavior.

32.1 The study of animal behavior encompasses patterns of change in animals, including movement, sounds, and body positions.

32.1 Ethologists, the prominent animal behavior researchers before the late 1950s, study the behavior of animals in their natural environments, observing and interpreting their behavior in the context of physiology while attempting to uncover the biological significance of the behavior.

32.1 Today, the field of animal behavior is composed of researchers from a variety of disciplines: psychology, sociobiology, neurobiology, and behavioral ecology.

32.2 Some patterns of behavior are inborn, or innate, resulting from neural pathways developed before birth.

32.2 Such instinctive behaviors are important to animals, providing them with fixed patterns of survival, such as those used to elude predators, to mate, and to care for young.

Innate behaviors are performed correctly the first time they are attempted.

32.3 Certain animals slow down in suitable environments and speed up in unsuitable ones by means of kineses.

32.3 Some animals move toward or away from certain stimuli by means of taxes.

32.3 Both kineses and taxes are simple types of innate orientation behaviors.

32.3 During innate coordination behaviors called reflexes, animals automatically respond to nerve stimulation.

32.4 Fixed action patterns are more complex innate behaviors in which organisms perform sequences of movements.

Key Concepts, continued

Some animals can change their behaviors based on experience.

32.5 Behaviors are changed within certain limits by means of various types of learning: imprinting, habituation, trial and error, and conditioning.

32.6 Certain animals learn to recognize their kin or birthplace by means of imprinting, a rapid and irreversible type of learning that takes place during an early developmental stage.

32.6 Habituation helps animals adapt to certain types of stimuli.

32.6 During classical conditioning, animals learn to associate a new stimulus with a natural stimulus that normally evokes a response in the animal.

32.6 Some animals learn to associate a behavior with a reward or punishment by means of trial-and-error learning.

32.6 An animal capable of insight can recognize a problem and solve it mentally by using experience, thereby performing an appropriate behavior the first time it tries.

Animals exhibit regularly repeated behaviors.

32.7 Organisms have internal clocks, or biological clocks, that regulate 24-hour cycles of physiological activity and behavior called circadian rhythms.

32.7 Circadian rhythms are genetic mechanisms governed by negative feedback loops and other, as yet unknown, factors.

32.8 Migratory behavior results from interactions among various environmental factors (such as day length) with animals' cyclical physiological and hormonal changes (circadian rhythms).

Social behaviors ultimately aid the reproductive fitness of individuals.

32.9 Social behaviors help members of the same species communicate and interact with one another.

32.10 Animals of the same species naturally compete for resources they need to survive.

32.10 Within animals' repertoire of behaviors are aggressive behaviors that threaten and intimidate rivals, and submissive behaviors—signals of "backing down."

32.10 Animals also use territorial behavior to limit aggression within their species.

32.10 A territory is an area that an animal marks off as its own, defending it against same-gender members of its species.

32.11 Reproductive behaviors include patterns of male-female interactions that lead to successful fertilization—the penetration of an egg by a sperm.

32.12 Parenting behaviors include preparations for the care of young and their direct care to aid their survival and ultimate reproduction.

32.12 Parenting behaviors are an example of altruistic behaviors: those that are advantageous to the recipient, increasing its individual fitness, while disadvantageous to the donor, decreasing its individual fitness.

32.13 Animals within social groups work together for a variety of common purposes.

32.13 Some species of insects, particularly various species of bees and ants, are organized into social groups having a division of labor.

32.13 Behaviors relating to rank order (the social hierarchy) help contribute to the stability of the social relationships among animals and within groups.

32.14 Some scientists question the scientific validity of applying sociobiological principles to human behavior.

Key Terms

adaptive behaviors *718*
animal behavior *718*
altruistic behavior
    (AL-troo-ISS-tik) *735*
behaviorism *723*
behaviors *718*
biological clocks *728*
circadian rhythms
    (sir-KAY-dee-un) *728*
classical conditioning *724*
cognitivism (KOG-nih-tiv-izm) *723*
competitive behavior *730*
courtship ritual *733*
courtship *733*

ethology (ee-THOL-uh-jee) *718*
fixed action patterns *722*
habituation
    (huh-BICH-yoo-AY-shun) *724*
imprinting *724*
innate behaviors *719*
insect societies *736*
insight (reasoning) *728*
kinesis (kih-NEE-sis) *720*
learning *719*
mating *733*
migrations *728*
operant conditioning
    (OP-uh-runt) *726*

peck order *738*
pheromone (FARE-uh-moan) *736*
rank order *738*
reflex *720*
social behaviors *730*
sociobiology (SO-see-oh-bye-OL-uh-gee
    or SO-shee-oh-bye-OL-uh-gee) *730*
submissive behavior *731*
taxis (TAK-sis) *720*
territorial behaviors *731*
threat displays/intimidation
    displays *730*

KNOWLEDGE AND COMPREHENSION QUESTIONS

1. Explain this statement: Behaviors are part of an animal's equipment for survival. (*Comprehension*)

2. Explain the importance of innate and learned behaviors. How do they complement each other? (*Comprehension*)

3. Identify the specific type of behavior involved in each of the following situations: (*Comprehension*)
   a. Fireflies are attracted to the flashing luminescence of other fireflies.
   b. While studying your biology text, you hear a screen door slam. It startles you at first, but then you barely notice it when you realize the wind is occasionally opening and closing it.
   c. You praise your puppy every time it sits down when you say "sit." Soon you have trained it to sit on command.

4. Distinguish between operant conditioning and classical conditioning. (*Comprehension*)

5. What is insight? Explain its significance. (*Knowledge*)

6. Identify the specific type of behavior involved in each of the following situations: (*Comprehension*)

   a. Every year, purple martins fly to Brazil and return to the United States around April.
   b. When a bright light flashes near your eyes, you automatically blink.
   c. Your cat frequently grooms itself by licking its fur and rubbing its paws over its face.

7. What is competitive behavior? (*Knowledge*)

8. Identify the type of behavior shown in each of the following situations: (*Comprehension*)
   a. The hair on a dog's back stands on end as the dog growls at an approaching stranger.
   b. When you scold the dog for growling, it rolls over on its back and exposes its belly.
   c. A robin builds its nest.
   d. A male peacock extends its tail feathers into a colorful fan.
   e. A cat chases another cat away from its food bowl.

9. Summarize the social structure of a honeybee colony. (*Comprehension*)

10. Explain the term *rank order*. What is its significance, and in what animal(s) does it appear? (*Comprehension*)

KEY
Knowledge: Recalling information.
Comprehension: Showing understanding of recalled information.

CRITICAL THINKING REVIEW

1. Very few mammals exhibit the complex societies seen among bees and ants—but a few do. Naked mole rats of the Middle East, for example, maintain large colonies with queens and special worker castes, organized remarkably like the colonies of bees. Why do you think such societies are so much more rare among mammals? (*Synthesis*)

2. Pacific salmon are born at the headwaters of rivers, then swim downstream hundreds of miles to the sea where they spend their adult lives. Years later, when it is time to spawn (that is, to lay and fertilize the eggs that will be the next generation), the adults swim up the same rivers to the precise location where they were born. Using information from this chapter, propose a hypothesis regarding how the fish know which way to go when they come to a fork in the river. (*Application*)

3. Do you think the principles of sociobiology can be applied to human behavior? Explain your answer using information from this chapter and from other sources. Cite your sources. (*Evaluation*)

4. Researchers at the University of Chicago published a study in 1998 in which they contend that human pheromones exist. In the study, women between the ages of 20 to 35 wore a pad in their armpit for at least 8 hours during an early stage of their menstrual cycles. The researchers then placed these pads under the noses of 20 women in the same age group. The menstrual cycles of the second group of women shortened after this treatment. Why might these results be evidence of the existence of human pheromones? (Be sure to define "pheromone" in your answer.) (*Application*)

5. Why are genes that encode social behaviors retained in the gene pool (the total of all the genes of the breeding individuals in a population at a particular time)? (*Analysis*)

KEY
Application: Using information in a new situation.
Analysis: Breaking down information into component parts.
Synthesis: Putting together information from different sources.
Evaluation: Making informed decisions.

Fish respond to both gravity and light. Explain the positioning of the fish in each tank.

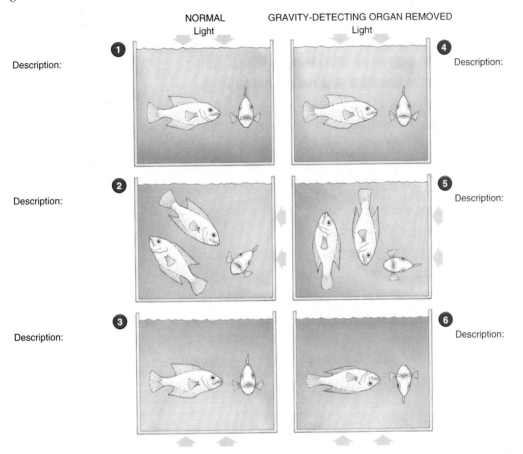

NORMAL
Light

GRAVITY-DETECTING ORGAN REMOVED
Light

1 Description:

2 Description:

3 Description:

4 Description:

5 Description:

6 Description:

Population Ecology

Looking like homes for oversized mud wasps, these cliff-swallow nests hang from the face of a rock outcropping. Any rough, vertical surface will serve as a nesting site for these birds. In fact, cliff swallows have recently discovered that the sides of bridges work well for this purpose . . . and come complete with protective overhangs! As a result, cliff swallows, once found mainly in the western part of the United States, can now be found inhabiting the prairie, nesting on the sides of bridges that span the major prairie rivers.

Cliff swallows that inhabit these nests are a population of organisms. A **population** consists of the individuals of a given species that occur together at one place and at one time. This flexible definition allows the use of this term in many contexts, such as the world's human population, the population of protozoans in the gut of an individual termite, or the population of blood-sucking bugs living in the feathers of a cliff swallow.

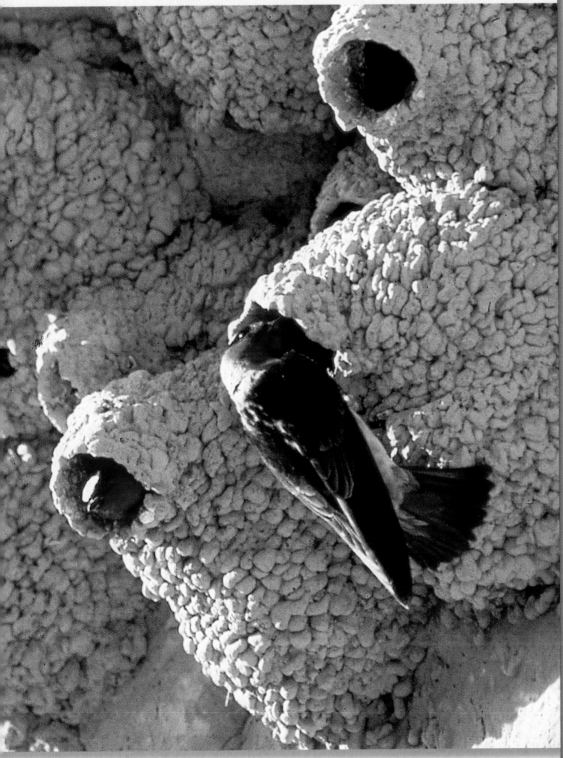

CHAPTER 33

Organization

Populations generally grow exponentially, then level off.

In 2000, the global human population of about 6 billion is growing at a rate of 1.25%.

33.1 Population growth

Population ecologists study how **populations** grow and interact. The science of ecology considers more than populations of organisms, however; it includes the study of interactions between organisms and the environment. Underlying questions to this study are "What factors determine the distribution patterns of organisms?" and "What factors control their numbers in the locations in which they are distributed?"

Scientists usually classify the study of ecological interactions into four levels: populations, communities, ecosystems, and the biosphere. This chapter discusses the first level of this hierarchy: populations.

Most populations will grow rapidly if optimal conditions for growth and reproduction of its individuals exist. Why, then, is the Earth not completely covered in bacteria, cockroaches, or houseflies? Why do some populations change from season to season or year to year?

Population Size and Growth Rate

To answer these questions, you need to first understand how the size of a population is determined. The size of a population at any given time is the result of additions to the population from births and from **immigration**, the movement of organisms into a population, and deletions from the population from deaths and **emigration**, movement of organisms out of a population. Put simply:

(Births + immigrants) − (Deaths + emigrants) = Population change.

These statistics (births and deaths) are often expressed as a *rate*: numbers of individuals per thousand per year. For example, the population of the United States at the beginning of 1996 was approximately 264 million people. During 1996, the following occurred:

- 3,891,494 live births: The birth rate was 3,891,494 per 264,000,000 people, or 14.7 births per 1000.
- 2,314,690 deaths: The death rate was 2,314,690 per 264,000,000 people, or 8.8 deaths per 1000.
- 915,900 (legal) immigrations: The immigration rate was 915,900 per 264,000,000 people, or 3.5 legal immigrants per 1000.
- 222,000 emigrations*: The emigration rate was 222,000 per 264,000,000 people, or 0.8 emigrants per 1000.

www.jbpub.com/biology

*The collection of statistics on emigration from the United States was discontinued in 1957; no direct measure of emigration has been available since then. The U.S. Bureau of the Census currently uses an annual emigration figure of 222,000.

The population change in the United States in 1996 can be calculated as follows:

(14.7 births/1000/year + 3.5 legal immigrants/1000/year) − (8.8 deaths/1000/year + 0.8 emigrants/1000/year) = 8.6 people/1000/year.

This figure can also be expressed as a population change of 0.86%—an increase of slightly less than 1%.

In natural populations of plants and animals, immigration and emigration are often minimal. Therefore, a determination of the **growth rate** of a population does not include these two factors. Growth rate (r) is determined by subtracting the death rate (d) from the birth rate (b):

$$r = b - d$$

Using the figures from our previous example:

$r = 14.7$ births/1000/year − 8.8 deaths/1000/year

$r = 5.9$ people/1000/year or 0.0059 (slightly more than ½ percent)

To figure out the number of individuals added to a population of a specific size (N) in a given time *without* regard to immigration and emigration, r is multiplied by N:

$$\text{Population growth} = rN$$

Therefore, the population growth in the United States in 1996 solely from births and deaths was $0.0059 \times 264,000,000$ people = 1,557,600 people.

Exponential Growth

Even though the rate of increase in a population may stay the same, the actual size of the population (the number of individuals) grows. This sort of growth pattern is similar to the growth pattern of money in the bank as interest is earned and compounded. If you put $1000 in the bank at 4% per year, the first year you will earn $40. The second year you will earn 4% interest on $1040, or $41.60. Although your interest rate has stayed the same, the amount of money you earn grows as your money grows. The third year you will earn 4% on $1081.60 or $43.26, raising your savings to $1124.86. (Actually, in a bank, your earnings would grow even more quickly because interest would be posted and compounded more often than once a year.)

Figure 33.1 illustrates this principle with a population of bacteria in which each individual divides into two every half hour. The rate of increase remains constant, but the actual increase in the number of cells accelerates rapidly as the size of the population grows. This type of mathematical progression found in the growth pattern of bacteria is termed **exponential**

Figure 33.1
Exponential growth
in a population of
bacteria.

Time (hours)	Number of bacteria
10	1,048,576
$9\frac{1}{2}$	524,288
9	262,144
$8\frac{1}{2}$	131,072
8	65,536
$7\frac{1}{2}$	32,768
7	16,384
$6\frac{1}{2}$	8,192
6	4,096
$5\frac{1}{2}$	2,048
5	1,024
$4\frac{1}{2}$	512
4	256
$3\frac{1}{2}$	128
3	64
$2\frac{1}{2}$	32
2	16
$1\frac{1}{2}$	8
1	4
$\frac{1}{2}$	2
0	1

Growth curve

Number of bacteria (×100,000) vs Time (hours)

growth. (For example, two cells split to form 2^2 or 4, 4 becomes 2^3 or 8, and so on. The number, or power, to which 2 is raised is called an exponent.) Exponential growth refers to the rapid growth in numbers of a population of any species of organism, even though most organisms reproduce sexually and one organism does not split into two "new" organisms.

A period of exponential growth can occur only as long as the conditions for growth are ideal. In nature, exponential growth often takes place when an organism begins to grow in a new location having abundant resources. Such a situation occurred when the prickly pear cactus was introduced into Australia from Latin America. The species flourished, overrunning the ranges. In fact, the cactus became so abundant that cattle were unable to graze (Figure 33.2a). Scientists regulated the population by introducing a cactus-eating moth to the area. The larvae of the moth fed on the pads of the cactus and rapidly destroyed the plants. Within relatively

(a) **(b)**

Figure 33.2 A cactus takes over Australia. After an initial period in which prickly pear cacti, introduced from Latin America, choked many of the pastures of Australia with their rampant growth, they were controlled by the introduction of a cactus-feeding moth from the areas where the cacti were native. (a) An infestation of prickly pear cacti in scrub in Queensland, Australia, in October 1926. (b) The same view in October 1929, after the introduction of the cactus-feeding moth.

few years, the moth had reduced the population; the prickly pear cactus became rare in many regions where it was formerly abundant (Figure 33.2b).

Carrying Capacity

No matter how rapidly a population may grow under ideal conditions, however, it cannot grow at an exponential rate indefinitely. As a population grows, each individual takes up space, uses resources such as food and water, and produces wastes. Eventually, shortages of important growth factors will limit the size of the population. In some populations such as bacteria, a buildup of poisonous wastes may also limit population growth. Ultimately, a population stabilizes at a certain size, called the **carrying capacity** of the particular place where it lives. The carrying capacity is the number of individuals within a population that can be supported within a particular environment for an indefinite period. A population actually rises and falls in numbers at the level of the carrying capacity, but tends to be maintained at an average number of individuals (Figure 33.3). The exponential growth of a population and its subsequent stabi-

lization at the level of the carrying capacity is represented by an S-shaped **sigmoid growth curve** (after the Greek letter sigma).

Population Size and Ability to Survive

The size of a population has a direct bearing on its ability to survive. Very small populations are less able to survive than large populations and are more likely to become extinct. Random events or natural disturbances can wipe out a small population, whereas a large population—simply due to its larger numbers and wider geographical distribution—is more likely to have survivors. Inbreeding—reproduction between closely related individuals—is also a negative factor in the survival of small populations. Inbreeding tends to produce many homozygous offspring (see photo below and Chapter 20), which results in the expression of many recessive deleterious traits that are usually masked by dominant genes. In addition, inbreeding reduces the level of variability in the gene pool (the genes of all breeding individuals) of the population, detracting from the population's ability to adjust to changing conditions. Loss of genetic diversity therefore increases the probability of extinction of that species.

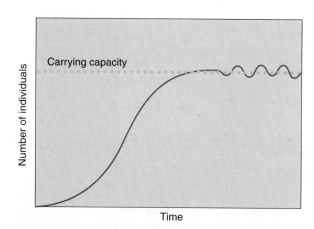

Figure 33.3   **The sigmoid growth curve.**

> The size of a population is the result of additions to the base population due to births and immigrations, and deletions from the population due to deaths and emigrations.
>
> The determination of the growth rate of a population does not include the factors of immigration and emigration. It is determined by subtracting the death rate from the birth rate.
>
> Under ideal conditions, populations grow at an exponential rate and show some stability in size at the carrying capacity of that place for that species.

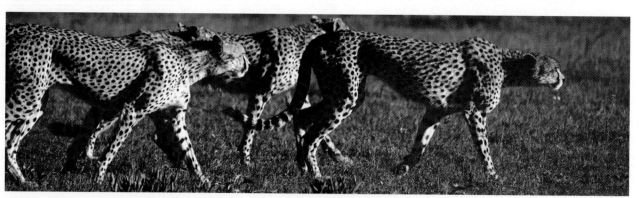

Due to hunting and loss of habitat, the world's cheetah population has been greatly reduced. The remaining members of the population are very inbred.

33.2 Population density and dispersion

In addition to a population's size, its **density**—the number of organisms per unit of area—influences its survival. For example, if the individuals of a population

(a) Uniform

(b) Random

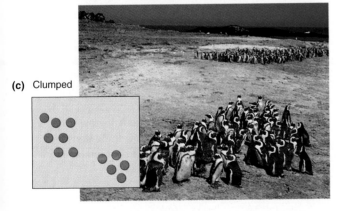

(c) Clumped

Figure 33.4 **Distribution patterns in populations.** (a) Creosote bushes in Death Valley, California. (b) Pine trees in Tahoe National Forest, California. (c) Molting African (Jackass) penguins in Namibia, South Africa.

are spaced far from one another, they may rarely come into contact. Sexually reproducing animals cannot produce offspring if they do not mate. Therefore, the future of such a population may be limited even if the absolute numbers of individuals over a wide area are relatively high.

A factor related to population density is **dispersion**, the way in which the individuals of a population are arranged. In nature, organisms within a population may be distributed in one of three different patterns: uniform, random, and clumped (Figure 33.4). Each of these patterns reflects the interactions between a given population and its environment, including the other species that are present.

Uniform, or evenly spaced, distributions are rare in nature and generally are indications of competition or interference. For example, populations of plants exhibiting allelopathy (AL-eh-LOP-eh-thee), the secretion of toxic chemicals that harm other plants, often show a uniform distribution. The creosote bush, often the dominant vegetation covering wide areas of deserts of Mexico and the southwestern United States, grows well spaced and evenly dispersed. This uniform pattern of distribution is probably due to chemicals secreted by the bush that retard the establishment of other individuals near established ones.

Random distributions occur if individuals within a population do not influence each other's growth and if environmental conditions are uniform—that is, if the resources necessary for growth are distributed equally throughout the area. Random distributions are often seen in plants as the result of certain types of seed dispersal, such as scattering by the wind. Because environmental conditions are rarely "purely" uniform, most ecologic patterns are clumped to some degree.

Clumped distributions are by far the most frequent in nature. Organisms that show a clumped distribution are close to one another but far from others within the population. Clumping occurs as a result of the interactions among animals, plants, microorganisms, and unevenly distributed resources in an environment. Organisms are found grouped in areas of the environment that have resources they need. Furthermore, animals often congregate for a variety of other reasons, such as for hunting, mating, and caring for their young.

> In nature, organisms within a population may be dispersed (arranged) in one of three different patterns: uniform, random, and clumped. Each pattern reflects the interactions between a population and its environment.

33.3 Regulation of population size

As a population grows and its density increases, competition among organisms for resources such as food, shelter, light, and mating sites increases and poisonous waste products accumulate. Factors that result from the growth of a population regulate its subsequent growth and are "nature's way" of keeping the population size of every species in check. Such factors increase in effectiveness as population density increases and are appropriately termed **density-dependent limiting factors**. Other factors such as the weather, availability of soil nutrients, and physical disruptions of an area (such as volcanoes or earthquakes) can also limit the growth of a population. Because these factors operate regardless of the density of a population, they are called **density-independent limiting factors**.

Density-Independent Limiting Factors

A variety of environmental conditions can limit populations. For example, freak snowstorms in the Rocky Mountains of Colorado in the summer can kill butterfly populations there. The size of insect populations that feed on pollen and flower tissues varies seasonally with the blooming of flowering plants. Humans, too, can affect the sizes of populations. Poachers have killed so many African elephants for their ivory, for example, that the species may become extinct.

Density-Dependent Limiting Factors

Individuals within a species and individuals of differing species compete for the same limited resources. (Chapter 34 discusses interspecific competition in detail and points out that competition among different species of organisms is often greatest between those that obtain their food in similar ways.) In fact, if two species are competing with one another for the same limited resource in a specific location, the species able to use that resource most efficiently will eventually eliminate the other species in that location. This concept is called the principle of **competitive exclusion**. In fact, competition among organisms of the same and differing species was described by Charles Darwin as resulting in natural selection and survival of the fittest or most well-adapted organisms. Competition, therefore, not only limits the

sizes of populations but is also one of the driving forces of evolutionary change. (However, the fittest members of populations are often the ones that successfully avoid competition.)

Predation is another factor that limits the size of populations and works most effectively as the density of a population increases. Predators are organisms of one species that kill and eat organisms of another—their prey. Predators include animals that feed on plants (such as cows grazing on grass) and plants that feed on insects (such as the Venus flytrap; see the Just Wondering box in Chapter 29). The intricate interactions between predators and prey are an essential factor in the maintenance of diverse species living in the same area. By controlling the levels of some species, the predators make the continued existence of other species in that same community possible. In other words, by keeping the numbers of individuals of some of the competing species low, the predators prevent or greatly reduce competitive exclusion. In fact, a given predator may very often feed on two or more different kinds of plants or animals, switching from one to the other as their relative abundance changes. Similarly, a given prey species may be a primary source of food for an increasing number of predator species as it becomes more abundant, a factor that will limit the size of its population automatically.

Parasitism also limits the size of populations by weakening or killing host organisms. Parasites live on or in larger species of organisms and derive nourishment from them. As a population increases in density, parasites such as bacteria, viruses, and a variety of invertebrates can more easily move from one organism to another, infecting an increasing proportion of a population. Once again, this limiting factor to population size acts in negative feedback fashion, becoming more effective as the density of the population increases.

Factors that regulate the growth of a population and that operate regardless of its density (such as weather conditions) are density-independent limiting factors. Factors that result from the growth of a population and regulate its subsequent growth are density-dependent limiting factors. Three density-dependent limiting factors are competition, predation, and parasitism.

33.4 Mortality and survivorship

A population's growth rate depends not only on the availability of needed resources and on the ability of its individuals to survive and compete effectively for those resources, but also on the ages of the organisms in it. Interestingly, when a population lives in a constant environment for a few generations, its **age distribution**—the proportion of individuals in the different age categories—becomes stable. This distribution, however, differs greatly from species to species and even to some extent within a given species from place to place.

Scientists express the mortality characteristics of a population by means of a survivorship curve. **Mortality** is the death rate. **Survivorship** refers to the proportion of an original population that survives to a certain age. The curve is developed by graphing the number of individuals within a population that survive through various stages of the life span. Samples of survivorship curves that represent certain types of populations are shown in **Figure 33.5.**

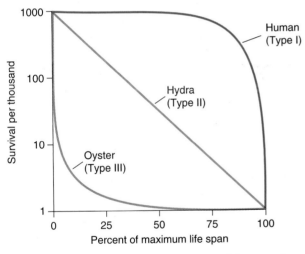

Figure 33.5 Survivorship curves. The shapes of the respective curves are determined by the percentages of individuals in populations that are alive at different ages.

In the hydra, individuals are equally likely to die at any age, as indicated by the straight survivorship curve (type II) shown in Figure 33.5. This type of survivorship curve is characteristic of organisms that reproduce asexually, such as hydra, bacteria, and asexually reproducing protists.

Oysters, on the other hand, produce vast numbers of offspring, but few of these offspring live to reproduce. The death rate of organisms that survive and reach reproductive age is extremely low (type III survivorship curve). This type of survivorship curve is characteristic of organisms producing offspring that must survive on their own and therefore die in large numbers when young because of predation or their inability to acquire the resources they need.

The survivorship curve for humans and other large vertebrates is much different from that for hydra and oysters. Humans, for example, produce few offspring but protect and nurture them; therefore, most humans (except in areas of great poverty, hunger, and disease) survive past their reproductive years (type I survivorship curve).

Many animal and protist populations have survivorship curves that lie somewhere between those characteristic of type II and type III. Many plant populations, with high mortality at the seed and seedling stages, have survivorship curves close to type III. Humans have probably approached type I more and more closely through the years, with the life span being extended because of better health care and new medical technology.

When a population lives in a constant environment for a few generations, the proportion of individuals in various age categories becomes stable. Characteristic survivorship curves occur among species and are closely linked in animal populations to parental care for offspring.

33.5 Demography

Demography is the statistical study of human populations. The term comes from two Greek words: *demos,* "the people" (the same root in the word *democracy*), and *graphos,* "to write." Demography therefore means the description of peoples and the characteristics of populations. Demographers predict the ways in which the sizes of populations will change in the future, taking into account the age distribution of the population and its changing size through time.

A population whose size remains the same through time is called a **stable population**. In such a population, births plus immigration exactly balance deaths plus emigration. In addition, the number of females of each age group within the population is similar. If this were not the case, the population would not remain stable. For example, if there were many more females entering their reproductive years than older females leaving the population, the population would grow.

The age distribution of males and females in human populations of Kenya, the United States, and Austria in 1997, and the projected populations in 2025, are shown as population pyramids in Figure 33.6. A **population pyramid** is a bar graph that shows the composition of a population by age and gender. Males are conventionally enumerated to the left of the vertical age axis and females to the right.

By using population pyramids, scientists can predict the future size of a population as shown in Figure 33.6b. First, the number of females in each age group is multiplied by the average number of female babies that women in that age group bear. These numbers are added for each age group to see whether the new number will exceed, equal, or be less than the number of females in the population being studied. By such means, the future growth trends of the human population as a whole and of individual countries and regions can be determined.

The population pyramids show the differences in the pattern of a rapidly growing population (Kenya [sub-Saharan Africa]), a slowly growing population (the United States), and a country experiencing negative growth (Austria [central Europe]). The 1997 population pyramid of Kenya is characteristic of developing countries (those that have not yet become industrialized) such as countries in Africa, Asia, and Latin America. Each of these countries has a population pyramid like that of Kenya, with a broad base reflecting the large numbers of individuals yet to enter their reproductive years. (Demographers consider the reproductive years to be ages 15–44.) In Kenya, for example, about 45% of the population is younger than 15 years of age. These children will reach reproductive age in the near future. The fertility rate (number of children a woman will have in her lifetime) is dropping in Kenya and other developing countries due to rising illness from HIV/AIDS (in sub-Saharan countries) and increased use of contraception. However, it is still high at 4.85. (However, the AIDS/HIV epidemic in sub-Saharan Africa affects infant and adult mortality more than it does the fertility rate.) The base of the population pyramid for Kenya will remain wide in 2025 as shown in Figure 33.6b due to relatively high fertility rates and large numbers of women in their reproductive years, in spite of deaths from HIV/AIDS.

In the United States, birth rates are higher than death rates at present, producing a growth rate of approximately 0.7%. The high birth rate is not due to couples having large families—the fertility rate is approximately 2.0, about the replacement rate. Rather, the growth rate in the United States is due to the large size of the "baby boom" generation that is just passing the peak of its reproductive years. (The baby boom generation can be seen as the bulge in the U.S. population pyramid.) Individuals in this age group were born within the 20 years or so after World War II. The large number of women in this group causes the births to still outnumber the deaths. Notice in the population pyramid for the United States for 2025, that the pyramid will become more "squared off" as the baby boomers move past their reproductive years.

Austria and the United States are both experiencing a decline in fertility and mortality. However, Austria's population does not include as high a percentage of women in their childbearing years as the United States, so deaths are outnumbering births. Additionally, Austria's fertility rate is lower than that of the United States. It is approximately 1.47, well under the replacement rate. For these reasons, the population pyramid has a narrow base, which continues to narrow by the year 2025.

Developing countries—those that have not yet become industrialized—have proportionately young populations. Although fertility is declining and mortality is rising in many of these countries, they are experiencing rapid growth due to still-high fertility rates and large numbers of women entering their reproductive years. Developed countries have populations with similar proportions of their populations in each age group and are growing very slowly or not at all.

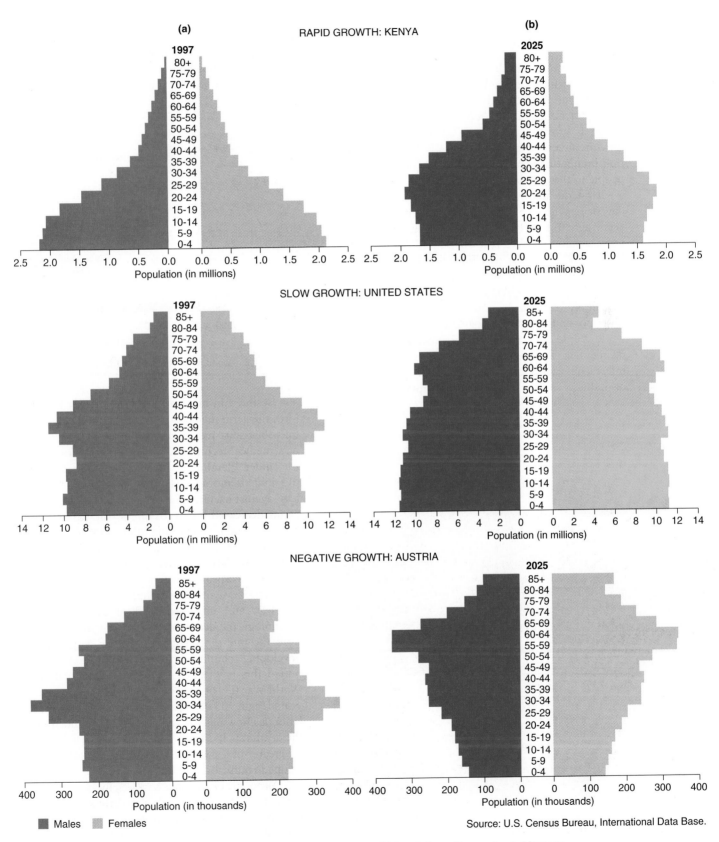

(a) RAPID GROWTH: KENYA **(b)**

SLOW GROWTH: UNITED STATES

NEGATIVE GROWTH: AUSTRIA

■ Males   ■ Females

Source: U.S. Census Bureau, International Data Base.

Figure 33.6 **Three patterns of population change.** (a) Population patterns in 1997. (b) Population patterns estimated for 2025.

In 2000, the global human population of about 6 billion is growing at a rate of 1.25%.

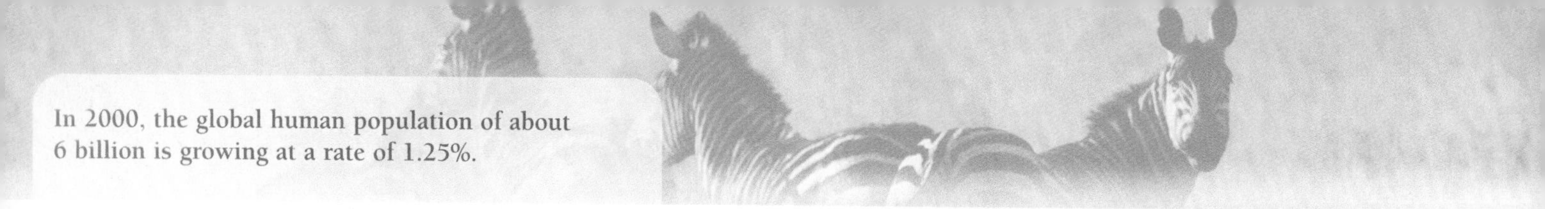

33.6 The human population explosion

www.jbpub.com/biology

Although some countries have populations that are no longer growing, such as Denmark, Germany, Hungary, and Italy, and some countries such as Austria are declining in numbers, the population of the world as a whole is growing at the rate of 1.25% a year. This growth rate may sound low, but with the world population numbering about 6 billion, it adds over 76 *million* people to the population each year.

How did the human population reach its present-day size? With the development of agriculture 11,000 years ago **(Figure 33.7)**, human populations began to grow steadily. Villages and towns were first organized about 5000 years ago, and human effects on the environment began to intensify. In these centers of civilization, however, the specialization of professions such as metallurgy (the science and technology of metals) became possible and technology advanced. By 1660, the world population totaled approximately 500 million people. The Renaissance in Europe, with its renewed interest in science, ultimately led to the establishment of industry in the seventeenth century and to the Industrial Revolution of the late eighteenth and early nineteenth centuries. **Figure 33.8** shows the slow growth of the human population until it began an exponential increase around the beginning of the nineteenth century.

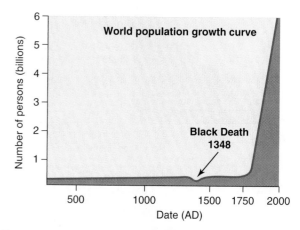

Figure 33.8 World population growth through history.

Figure 33.7 The development of agriculture was a key step in the growth of human populations. By producing abundant supplies of food, agriculture made the growth of cities and the future development of human culture possible.

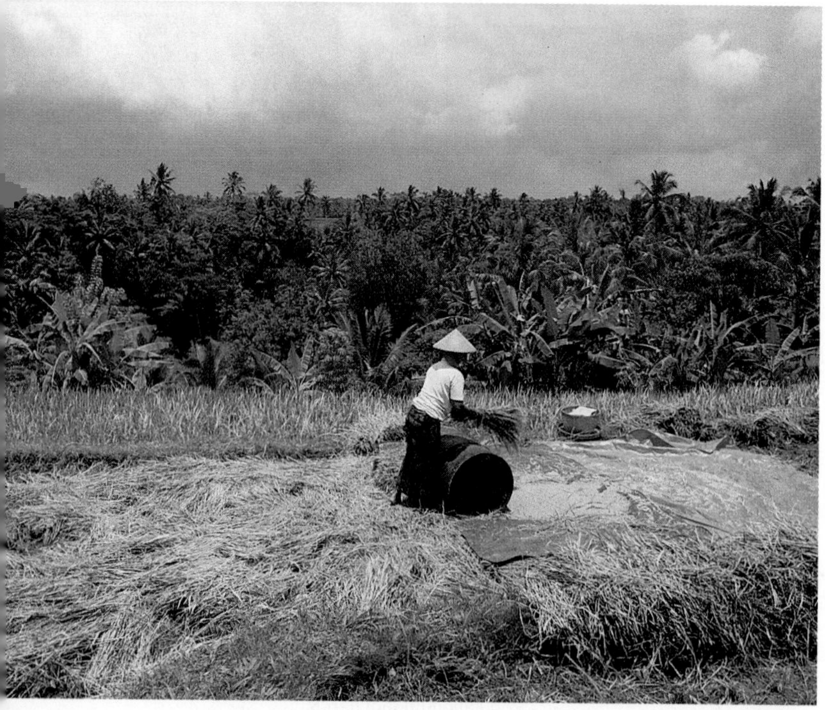

By the mid-nineteenth century, Louis Pasteur put forth the germ theory of disease, the understanding that microbes cause infection. With this understanding came new medical technology and discoveries. In the 1920s, Alexander Fleming discovered penicillin and opened the door to antibiotic therapy—medicine's "magic bullets" against bacterial infection. These medical advancements decreased the death rate by increasing the number of individuals surviving infection.

The advent of the Industrial Revolution also heralded new farming and transportation technology, which helped provide better nutrition for many people, especially those in industrialized countries. With better nutrition and increased medical understanding and technology, the death rate fell steadily and dramatically from the mid-nineteenth century on **(Figure 33.9a)**. In developing countries, international foreign aid imported this new technology along with food aid after World War II. The mortality rate plunged in a matter of years (Figure 33.9b). The natural increase in the population is shown as the space between the birth rate and the death rate in Figure 33.9. Notice how the increase is far greater in less developed countries than in more developed countries.

Although birth rates have declined in developing countries, they are still high. Presently, about 80% of the people in the world are living in less developed countries; about 60% are living in countries that are at least partly tropical or subtropical and about 20% are living in

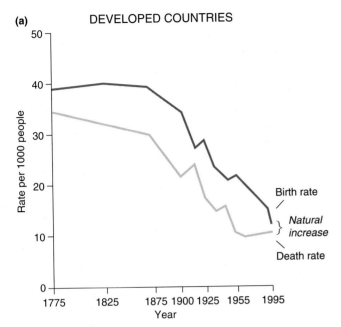

DEVELOPED COUNTRIES

(a)

Birth rate

} Natural increase

Death rate

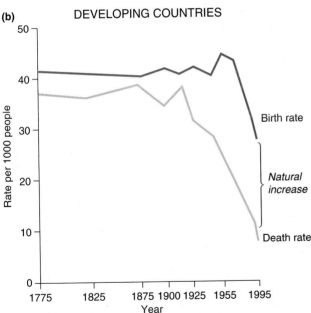

DEVELOPING COUNTRIES

(b)

Birth rate

} Natural increase

Death rate

Figure 33.9 **Population growth from 1775 to 1995.** (a) In developed countries, the Industrial Revolution caused mortality (death) to drop. Birth rates also began to drop at the turn of the century. (b) In developing countries, mortality began to decline after World War II, but birth rates remained high. Taken together, these changes caused human population growth rates to soar, especially in developing countries.

China. The remaining 20% are located in the more developed countries of Europe, the Commonwealth of Independent States (formerly the Soviet Union), Japan, the United States, Canada, Australia, and New Zealand (Fig-

ure 33.10). **Figure 33.11** graphically depicts the share of the world population living in less developed countries from 1970 through today, with projections through 2020.

Of the estimated 2.8 billion people living in the tropics in the late 1980s, the World Bank estimated that about 1.2 billion people were living in poverty. (The World Bank is an international bank that provides loans and technical assistance for economic development projects in developing member countries.) These people cannot reasonably expect to be able to consistently provide adequate food for themselves and their children. Even though some experts estimate that enough food is produced in the world to provide an adequate diet for everyone in it (see the Just Wondering box), the distribution is so unequal that large numbers of people live in hunger. The United Nations International Children's Emergency Fund (UNICEF) estimates that in the developing world today, about 6 million children younger than 5 years of age—about 16,500 per day—die each year, mainly of malnutrition and the complications associated with it.

The size of human populations, like those of other organisms, is or will be controlled by the environment. Early in its history, human populations were regulated by both density-dependent and density-independent limiting factors, including food supply, disease, and predators; there was also ample room on Earth for migration to new areas to relieve overcrowding in specific regions. In the past century, however, humans have been able to expand the carrying capacity of the Earth because of their ability to develop technological innovations. Gradually, changes in technology have given humans more control over their food supply and enabled them to develop superior weapons to ward off predators as well as the means to cure diseases. Improvements in transportation and housing have increased the efficiency of migration. At the same time, improvements in shelter and storage capabilities have made humans less vulnerable to climatic uncertainties.

As a result of the ability to manipulate these factors, the human population has been able to grow explosively to its present level of about 6 billion people. Both the current human population level and the projected rate of growth have potential consequences for the future that are extremely grave. Pressures are placed on the land, water, forests, and other natural resources. Although industrialization raises the standard of living by increasing the availability of goods and services, it adds to air and water pollution. In the developed countries, nations of consumers have developed into "throwaway" societies, adding billions of tons of solid waste to landfills every year.

In 2000, the global human population of about 6 billion is growing at a rate of 1.25%.

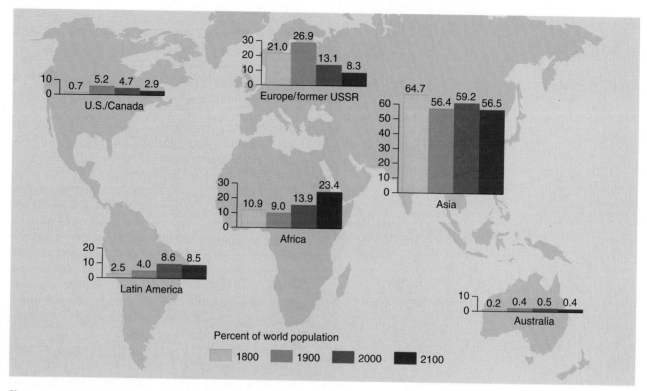

Figure 33.10 **World population distribution by region, 1880 to 2100.** If current trends continue, Asia will have 57% of the total world population in 2100, Africa nearly 25%, and Europe's share will drop to less than 10%.

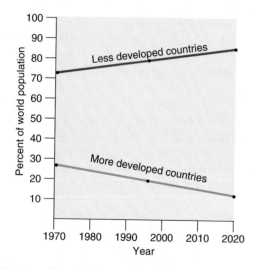

Figure 33.11 **Share of the world population.**

The most effective means of dealing with the population explosion has been the support of governments to encourage small families, the establishment of family planning clinics, improvement in education, and socioeconomic development. The developed countries of Western Europe and North America, Japan, and Australia have very low birth rates at this time. China, Indonesia, Thailand, South Korea, Hong Kong, and Singapore have had considerable success with lowering their birth rates. Countries such as Mexico and India have had some success in reducing their birth rates but are still striving toward this goal. The countries in sub-Sahara Africa have the highest birth rate of any countries in the world. Family planning programs are now being implemented across the continent.

The world population rose sharply and dramatically after the Industrial Revolution because of new technology in agriculture, transportation, industry, and medicine. Although fertility rates and birth rates have dropped in recent years, the world population has reached 6 billion people and continues to grow at the rate of 1.25%. Growth rates in developing countries are higher than growth rates in developed countries. Due to this historical trend, approximately 80% of the world's people live in developing countries, while only 20% live in developed countries.

Just Wondering . . . Real Questions Students Ask

On the news I often see photos of starving children. Why are their stomachs bloated? Also, is the world running out of enough food to feed everybody?

Starving children often have swollen bellies, a condition known as kwashiorkor (KWASH-ee-ORE-kore). This disease is a sign of protein deficiency in the child. Normally, a certain level of proteins is present in the blood. These blood proteins affect the movement of water into and out of the blood. When a child does not take in enough protein, blood proteins become diminished. Fluid leaks out of the blood and into the belly and legs, building up there.

Starvation and malnutrition are issues of grave concern as the world faces such astonishing growth. There are no easy answers, and scientists themselves are polarized on these issues. The only thing both sides can agree on is that the demand for food will grow enormously as the population swells.

One group within the scientific community contends that new technologies, if managed properly, will allow humans to expand the world food production and feed all of its people. Recent research assessing the climate and soil conditions in 93 developing countries suggests that three times as much land as is currently cultivated could be put to agricultural use. In areas without additional land to farm, the number of crops grown each year on land presently in use could be increased—an agricultural practice called *multicropping*. In addition, the use of high-yield crop varieties, fertilizer, and irrigation in areas where they are not currently employed could increase crop yields.

An opposing view is held by many scientists who suggest that intensifying agricultural practices as suggested will cause serious ecological damage to our world, such as extensive deforestation, loss of species diversity, erosion of the soil, and the pollution of aquifers, streams, and rivers from pesticides and fertilizers. In addition, they assert that our natural resources will be unable to support this future demand, suggesting that crop yields would have to rise by 112% to feed all the people of the developing (not yet industrialized) nations in 2050. To also raise the standards of their presently inadequate diets would translate into each acre of land increasing its yield more than six times. Many scientists think that these goals are impossible to achieve unless new technologies are developed.

Unfortunately, in countries torn apart by war, food production is only one of the many serious problems facing its citizens. War-torn countries and those with inadequate natural resources and technologies will have to rely on food aid, placing even more pressure on the resources available in the rest of the world. There are no easy answers to your question, and feeding the people of the world will continue to be a critical issue. Possibly researchers in biotechnology, agriculture, and related areas will make contributions that will help solve this problem for future generations. However, if we solve our immediate problems of food production but the world population continues to grow at the same rate as it does today, what will happen beyond 2050? Will we reach a final limit?

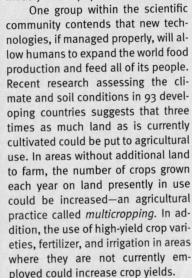

www.jbpub.com/biology

Key Concepts

Populations generally grow exponentially, then level off.

33.1 Populations consist of the individuals of a given species that occur together at one place and at one time.

33.1 The rate of growth of any population is the difference between the birth rate and the death rate.

33.1 The actual change in a population may also be affected by emigration from the population and immigration into it, but these factors are only considered in human populations.

33.1 Most populations exhibit a sigmoid growth curve, which implies a relatively slow growth, a rapid increase, and then a leveling off when the carrying capacity of the species' environment is reached.

33.2 Individuals in a population may be dispersed in a uniform, clumped, or random manner.

33.2 Clumped dispersion patterns are the most frequent, with uniform distributions being rare (most showing some degree of clumping).

33.3 Each population grows in size until it eventually reaches the limits of its environment to support it; resources are always limiting.

Key Concepts, continued

33.3 Some of the limits to the growth of a population are related to the density of that population, but others are not.

33.3 Factors that regulate the growth of a population and that operate regardless of its density (such as weather conditions) are density-independent limiting factors.

33.3 Factors that result from the growth of a population and regulate its subsequent growth are density-dependent limiting factors.

33.3 Three density-dependent limiting factors are competition, predation, and parasitism.

33.4 Survivorship curves are used to describe the characteristics of growth in different kinds of populations.

33.4 Type I populations are those in which a large proportion of the individuals approach their physiologically determined limits of age.

33.4 Type II populations have a constant mortality throughout their lives.

33.4 Type III populations have very high mortality in their early stages of growth, but an individual surviving beyond that point is likely to live a very long time.

In 2000, the global human population of about 6 billion is growing at a rate of 1.25%.

33.5 Developing countries—those that have not yet become industrialized—have proportionately young populations.

33.5 Although fertility is declining and mortality is rising in many developing countries, they are experiencing rapid growth due to still-high fertility rates and large numbers of women entering their reproductive years.

33.5 Developed countries have populations with similar proportions of their populations in each age group and are growing very slowly or not at all.

33.6 The world population rose sharply and dramatically after the Industrial Revolution because of new technology in agriculture, transportation, industry, and medicine.

33.6 Today, the global population is about 6 billion people and is growing at a rate of 1.25%.

33.6 Growth rates in developing countries are higher than growth rates in developed countries.

33.6 Approximately 80% of the world's people live in developing countries, while only 20% live in developed countries.

Key Terms

age distribution *749*

carrying capacity *746*

competitive exclusion *748*

demography (dih-MOG-ruh-fee) *750*

density *747*

density-dependent limiting factors *748*

density-independent limiting factors *748*

dispersion *747*

emigration (EM-uh-GRAY-shun) *744*

exponential growth (ek-spo-NEN-shul) *744*

growth rate *744*

immigration (IM-uh-GRAY-shun) *744*

mortality *749*

parasitism (PARE-uh-suh-tiz-um) *748*

population *744*

population ecologists (ih-KOL-uh-jists or ee-KOL-uh-jists) *744*

population pyramid *750*

predation *748*

sigmoid growth curve *746*

stable population *750*

survivorship *749*

KNOWLEDGE AND COMPREHENSION QUESTIONS

1. What is ecology? State one question that underlies ecological inquiry. (*Knowledge*)

2. In the early twentieth century, the United States experienced a large influx of immigrants from many European nations. Does immigration affect the size or growth rate of the population in the United States? (*Comprehension*)

3. How is the work of population ecologists similar to that of demographers? How does it differ? (*Comprehension*)

4. Which is more likely to survive, a small population or a large one? Why? (*Comprehension*)

5. Distinguish between density and dispersion. How does each affect a population's chances for survival? (*Comprehension*)

6. Identify the three patterns of population dispersion found in nature. Into which pattern do human populations fall? (*Knowledge*)

KNOWLEDGE AND COMPREHENSION QUESTIONS

7. Distinguish between density-dependent and density-independent limiting factors. Give an example of each. (*Comprehension*)

8. What do predation and parasitism have in common? How do they differ? (*Comprehension*)

9. Draw type I, type II, and type III survivorship curves. Summarize the types of organisms that are characteristic of each curve, and give an example of each. (*Comprehension*)

10. Compare the typical population pyramids of a developing country and an industrialized country. (*Comprehension*)

K E Y

Knowledge: Recalling information.
Comprehension: Showing understanding of recalled information.

CRITICAL THINKING REVIEW

1. Both the current human population level and the projected rate of growth worldwide have potential consequences for the future that are extremely grave. Explain why. (*Analysis*)

2. Suppose that you were given the political power to deal with the world's population explosion. What steps would you take? (*Evaluation*)

3. Draw a hypothetical population pyramid of a stable population. Describe your pyramid and explain how it represents a stable population. (*Application*)

4. What conclusions can you draw about a population that has a low proportion of its members under reproductive age? (*Analysis*)

5. If the most successful parasite does not kill its host, then why are parasites considered to be density-dependent limiting factors? (*Application*)

K E Y

Application: Using information in a new situation.
Analysis: Breaking down information into component parts.
Evaluation: Making informed decisions.

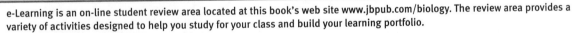

e learning

Location: http://www.jbpub.com/biology

e-Learning is an on-line student review area located at this book's web site www.jbpub.com/biology. The review area provides a variety of activities designed to help you study for your class and build your learning portfolio.

Review Questions The review questions test your knowledge of the important concepts and applications in each chapter. The review provides feedback for each correct or incorrect answer. This is an excellent test preparation tool.

Figure-Labeling Exercises Sharpen your visual thinking by matching terms and labels for important illustrations.

Flash Cards Studying biology requires learning new terms. Virtual flash cards help you master the new vocabulary for each chapter.

Just Wondering As you read and study from this text, you may find that you have unanswered questions. Through this site you can ask the author, Sandy Alters, your "just wondering" questions.

 Do you prefer the speed and reliability of a CD-ROM? All of the features contained on the eLearning portion of the web site are also available on the Student CD-ROM.

Interactions within Communities

This beautiful coral reef is home to a variey of organisms.
Each organism contributes to the array of colors evident here.
The reef provides a shallow-water environment favorable to
many organisms. Nutrients are abundant as are surfaces for
attachment and hiding places into which animals can burrow.
Although a number of species characteristic of coral reef
populations are visible in the photograph, more than 3000
species may coexist in a large reef. Its populations interact in a
variety of ways. The coral reef organisms compete with one
another and with other organisms for space and food. Sponges
growing on particular coral surfaces will eventually bore into the
coral, destroying it. Schools of fish swim within and around the
reef, eating algae and plankton that drift in the surrounding
water. Together, the interacting populations of a coral reef form
a vibrant, colorful marine community.

Organization

34.1 Ecosystems and communities

Scientists have long known that the nonliving or **abiotic factors** (AYE-bye-OT-ik) within the environment—such as air, water, and even rocks—affect an organism's survival, as do the living or **biotic factors** (by-OT-ik)—such as surrounding plants, animals, and microorganisms. All the biotic and abiotic factors together within a certain area are called an **ecosystem** (EH-koe-SIS-tem). Therefore, the rock is as important to the makeup of the ecosystem as the barnacles.

Within an ecosystem, each living thing has a home, an actual area in which it resides. This space, including the factors within it, is an organism's **habitat**. Organisms not only reside in their habitats, they interact with the biotic and abiotic factors within it and use them to survive. Each organism also plays a special role within an ecosystem; this role is called a **niche** (nich). The term niche refers to the organism's use of the biotic and abiotic resources in its environment. Thus, it may be described with reference to space, food, temperature, appropriate conditions for mating, and requirements for moisture, for example. A full portrait of an organism's niche also includes the organism's behavior and the ways in which this behavior changes at different seasons and at different times of the day. These concepts are important to the understanding of the concept of communities.

The interactions among organisms within ecosystems are varied. Individuals of the same species make up a **population** of organisms. (The individuals within a population interact with one another in ways described in Chapters 32 and 33.) Populations also interact with one another, forming communities. A **community** is a grouping of populations of different species living together in a particular area at a particular time. Thus, an ecosystem can be thought of as a community of organisms, along with the abiotic factors with which the community interacts.

The magnificent redwood forest that extends along the coast of central and northern California and into the southwestern corner of Oregon is an example of a community. Within it, the most obvious organisms are redwood trees. The huge trunk of a redwood can be seen in **Figure 34.1**. Populations of other organisms live in the filtered light beneath the towering redwoods, such as the rhododendrons visible in the photo, as well as sword ferns, ground beetles, and deer. The coexistence of these various populations is made possible in part because of the special conditions that are created by the redwoods: shade, water (dripping from the branches), and relatively cool temperatures. For this reason and because the redwoods visually dominate the area, this distinctive group of populations is known as the redwood community.

Scientists study the interactions among organisms and between organisms and their environments in the laboratory and in nature. This specialized field of biology is called **ecology** (eh-KOL-uh-gee); the scientists who work within this field are ecologists. Ecologists study the physical and biological variables governing the distribution and growth of living things. Ecologists also study the theoretical bases of these interactions. Some ecologists use computers to develop mathematical models of ecological systems. The knowledge gained by ecologists is essential to the basic understanding of the world and provides a foundation for finding solutions to the many environmental problems created by humans.

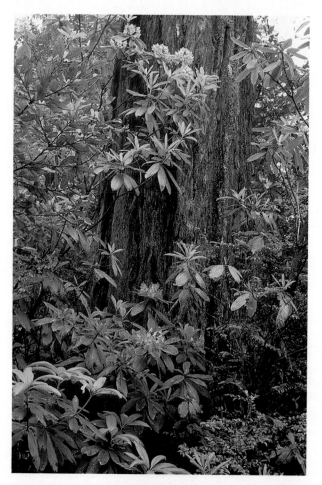

Figure 34.1 The redwood community.

Individuals of a species comprise a population; many populations make up a community. These populations of living things interact with one another in many complex ways.

34.2 Competition

Competition is a situation in which organisms that live near one another strive to obtain the same limited resources. When organisms use the same limited resource for survival, they must compete for that resource. Complex animals such as vertebrates compete by using innate behaviors such as threat displays and territorial behavior (see Chapter 32). Lower forms of animals and plants do not exhibit complex behaviors; they compete with one another simply by their adaptive fitness—by reproducing more and better-adapted offspring that crowd out opponents.

Competition among different species of organisms is often greatest between organisms that obtain their food in similar ways. Thus, green plants compete mainly with other green plants, meat-eating animals with other meat-eating animals, and so forth. Competition within a genus or between individuals of the same species occurs as well.

Competition among organisms has been observed by scientists for a long time. More than 50 years ago, the Russian scientist G.F. Gause formulated the principle of **competitive exclusion**, which was based on his experimental work. This principle states that if two species are competing with one another for the same limited resource in a specific location, the species able to use that resource most efficiently will eventually eliminate the other species in that location. A modification of this principle is the concept of *coexistence with niche subdivision*. In this situation, two species still exist in the same geographical area, and one or both species occupies a more restricted niche as the result of competition with the other. This concept is illustrated in the next section.

The Study of Competition in the Laboratory

Scientists sometimes study competition between species in the laboratory so that they can control the environmental conditions. John Harper and his colleagues at the University College of North Wales, Australia, for example, performed competition experiments with two species of clover: white clover and strawberry clover. Each species was sown with the other at one of two densities: 36 or 64 plants per square foot. Various plots of the two species were planted, using all the possible combinations of the two densities of plants. One plot contained 36 white clover plants and 36 strawberry clover plants per square foot. Another contained 64 white and 36 strawberry per square foot and so forth. The white clover initially formed a dense canopy of leaves in each experimental plot. However, the slower-growing strawberry clover, whose leaf stalks are taller, eventually produced leaves that grew above the white clover leaves. In competing more effectively for light, the strawberry clover overcame the white clover, causing it to die out. The outcome was the same, regardless of the initial densities at which the seeds of the plants were sown.

The Study of Competition in Nature

As well as competing for sunlight, plants compete for soil nutrients. The roots of one species, for example, may outcompete another species by using up minerals in the soil essential to both species. In addition, one species may secrete poisonous substances that depress the growth of other species. Sage plants, for example, inhibit the establishment of other plant species nearby, producing bare zones around populations of these plants. In the aerial photograph of sage plants taken in the mountains above Santa Barbara, California (Figure 34.2), you can see these bare zones around the colonies of plants.

Figure 34.2   Bare zones around colonies of sage plants.

Acorn barnacles demonstrate another interesting view of competition in nature. Highly adapted to their environment, acorn barnacles are typically found in the intertidal zone of rocky shores—the narrow strip of land exposed during low tide and covered during high tide. When submerged, an acorn barnacle feeds by extending appendages from the hole in its shell. Spread out, these appendages act like a net, sweeping the water and collecting food that it then brings into its shell and eats. When exposed to the air, an acorn barnacle pulls in its feeding appendages and shuts down, actually using much less oxygen than when underwater. Interestingly, barnacles of the genus *Balanus* have been kept out of the water as long as 6 weeks without detectable ill effects. However, a relative of *Balanus* organisms, barnacles of the genus *Chthamalus* (pronounced with the first two letters silent), have been kept out water for 3 years, being submerged only 1 or 2 days a month—and they survived!

Although both organisms have adaptations that make them well suited to the intertidal environment, their differences play an important role in determining where each genus lives. Of the two, *Chthamalus* barnacles live in shallower water, where they are often exposed to air as the tide rolls in and out. *Balanus* barnacles live deeper in the intertidal zone and are covered by water most of the time. Figure 34.3 is a photograph of both these organisms growing together on a rock. *Balanus* are the larger organisms of the two genera.

Figure 34.3 *Chthamalus* and *Balanus* barnacles growing together on a rock.

In studying these two genera of barnacles, J.H. Connell of the University of California, Santa Barbara, found that in this deeper zone, *Balanus* barnacles could always outcompete *Chthamalus* barnacles. *Balanus* organisms would crowd *Chthamalus* barnacles off the rocks, replacing them even where they had begun to grow. When Connell removed *Balanus* barnacles from the area, however, *Chthamalus* organisms were easily able to occupy the deeper zone, indicating that no physiological or other general obstacles prevented it from becoming established there. *Balanus* barnacles, however, must use the resources of the deeper zone more efficiently than *Chthamalus* organisms do, even though *Chthamalus* barnacles are able to survive there in the absence of its competitor. In contrast, *Balanus* barnacles cannot survive in the shallow water where *Chthamalus* organisms normally occur. *Balanus* barnacles evidently do not have the special physiological and morphological adaptations that allow *Chthamalus* barnacles to occupy this zone.

Along with illustrating the principle of competitive exclusion, these experiments with *Balanus* and *Chthamalus* barnacles illustrate that the role an organism plays in an ecosystem—its niche—can vary depending on the biotic and abiotic factors in the ecosystem. In this example, the niche occupied by *Chthamalus* barnacles is its **realized niche**—the role it *actually plays* in the ecosystem (Figure 34.4, pink solid line). It is distinguished from its **fundamental niche**—the niche that it *might* occupy if competitors were not present. Thus, the fundamental niche of the barnacle *Chthamalus* (pink dotted line) in Connell's experiments included the fundamental niche of

Balanus barnacles (purple dotted line), but its realized niche was much narrower because *Chthamalus* organisms were outcompeted by *Balanus* organisms. However, the realized and fundamental niches of *Balanus* barnacles are the same (purple solid and dotted lines).

Gause's principle of competitive exclusion can be restated in terms of niches as follows: No two species can occupy exactly the same niche indefinitely. Certainly, species can and do coexist while competing for the same resources. Nevertheless, Gause's theory predicts that when two species do coexist on a long-term basis, one or more features of their niches will always differ; otherwise, the extinction of one species will inevitably result. The factors that are important in defining a niche are often difficult to determine, however; thus, Gause's theory can sometimes be difficult to apply or investigate.

Organisms of different species compete with one another when they need the same limited resource for survival. This competition is greatest among organisms that obtain their food in similar ways.

In a competitive situation and under controlled laboratory conditions, the species able to use a particular resource more efficiently will be the species to survive.

The role an organism plays in an ecosystem can vary depending on the biotic and abiotic factors in the ecosystem.

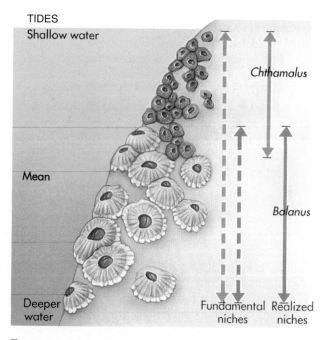

Figure 34.4 Competition can limit niche use.

34.3 Predation

Predation is a relationship in which an organism of one species (a **predator**) kills and eats an organism of another (the **prey**). (Animals that kill and eat members of their own species are cannibals.) Predation includes one kind of animal capturing and eating another, an animal feeding on plants, and even a plant, such as the Venus flytrap shown on p. 665 in Chapter 29, capturing and eating insects. In a broad sense, parasitism is also considered a form of predation. *Parasitism* is a close association between two organisms in which the parasite (predator) is much smaller than the prey but feeds on the prey, harming it and benefiting the parasite.

How do predator populations affect prey populations? When experimental populations are set up under very simple conditions in the laboratory, the predator often exterminates its prey and then becomes extinct itself because it has nothing to eat. This fact was illustrated nicely in experiments performed by Gause. In his experiments, Gause used populations of the two protozoans shown in **Figure 34.5**: *Didinium* and its prey *Paramecium*. As shown in Figure 34.5a, when *Didinium* protozoans are introduced into a growing population of *Paramecium* protozoans, the population of paramecia instantly begins to decline and quickly dies out. The *Didinium* population lives on for a short while, then dies out itself.

If refuges are provided for the prey, however, its population can be driven to low levels but can recover. In another experiment, Gause provided sediment in the bottom of the test tubes in which he was growing *Didinium* and *Paramecium* protozoans. Interestingly, as *Didinium* began to prey on *Paramecium,* only those organisms in the clear fluid of the test tubes were killed. Those in the sediment were not eaten. Eventually, *Didinium* protozoans died from lack of food; meanwhile, the *Paramecium* prey multiplied (Figure 34.5b) and overtook the culture!

In another series of experiments, Gause discovered that when he introduced new prey at successive intervals (Figure 34.5c), the decline and rise in the numbers of the predator-prey populations followed a cyclical pattern. As the number of prey increased, the number of predators increased. As the numbers of the prey were

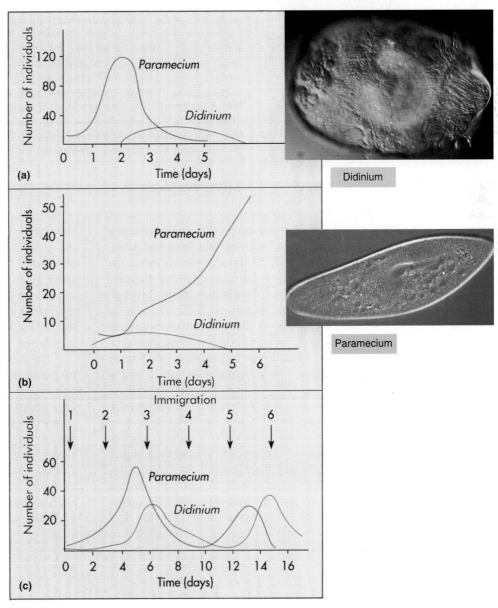

Figure 34.5 Outcome of Gause's experiments with *Paramecium* and *Didinium*.

Just Wondering . . . *Real Questions Students Ask*

Why were killer bees given that name? Is their sting so serious that you could die of it?

www.jbpub.com/biology

African honeybees were dubbed "killer bees" by the news media, not scientists. The sting of a so-called killer bee will not kill a person (unless he or she is allergic to bee stings); one bee dispenses no more venom than any other bee. However, there are aspects of the behavior of African honeybees that make them potentially life threatening.

When a bee stings a victim, glands attached to the stinger release alarm pheromone. Pheromones are chemicals produced by one individual that alter the physiology or behavior of other individuals of the same species. African honeybees release more alarm pheromone when they sting than do other bees, and they are more sensitive to this chemical trigger than are other bees. Instead of a dozen bees pursuing a victim for 100 yards as a reaction to alarm pheromone, an entire colony of "killer bees" may pursue a victim for a mile. (A small swarm may contain 30,000 bees.) Luckily, however, people can outrun African honeybees. The problem arises if the victim is an animal that is tied down and cannot flee or a human who falls while fleeing, allowing the bees to catch up and inflict numerous stings. One death occurred in 1986 when a University of Miami graduate student on a field trip in Costa Rica stepped in a crack inside a cave, which disturbed a killer bee colony. The student caught his foot in the crack and was unable to run. He died of 8000 bee stings.

Anita Collins, a U.S. Department of Agriculture research leader specializing in honeybees, thinks that African honeybees are a minimal threat to the public. She is more concerned that African bees will upset the pattern of crop pollination established by the European honeybees now in the United States. Steps are currently being undertaken to inject European honeybee queens with semen from African honeybees to produce a hybrid that will be as high a honey producer as the African variety and, it is hoped, that will be less sensitive to alarm pheromone. If scientists are successful in their breeding goals, interactions within the bee community and between the bee community and human populations may become more productive and less threatening!

lowered by predation, the large predator population did not have enough food to eat; some died and the predator population declined. As this decline occurred, the prey recovered (aided by the addition of new organisms) and again became abundant, starting the cycle once again.

At one time, scientists thought that predator-prey populations always cycled in this manner. However, they have come to realize that in nature, conditions for survival are complex and do not always lead to such a cycling of populations. From experiments such as those described, scientists know that predators cannot survive when the prey population is low. Immigration of prey (movement of new prey into the community) may be necessary to sustain the predator population. From other experiments, scientists learned that changes in predator-prey populations also depend on how prey are dispersed in an area and the manner in which the predator searches for the prey. Factors other than the relationship between a single predator population and a single prey population also influence the survival and abundance of both predator and prey. For example, adverse weather conditions may result in the death of the predator and/or prey species; the prey may be eaten by more than one predator; or fluctuations may occur in the food source of the prey, limiting the survival of this population.

The intricate interactions between predators and prey often affect the populations of other organisms in a community. By controlling the levels of some species, for example, predators help species survive that may compete with their prey. In other words, predators sometimes prevent or greatly reduce competitive exclusion by limiting the population of one of the competing species. Such interactions among organisms involving predator-prey relationships are key factors in determining the balance among populations of organisms in natural communities.

Plant-Herbivore Coevolution

Plants, animals, protists, fungi, and bacteria that live together in communities have changed and adjusted to one another continually over millions of years. Such interactions, which involve the long-term, mutual evolutionary adjustment of the characteristics of the members of biological communities in relation to one another, are examples of *coevolution.*

Plants and plant-eating predators called *herbivores* are a group of organisms that change and adjust to one another over time. Natural selection favors plants that have developed some means of protection against herbivores. In the dynamic equation of coevolution, however, natural selection also favors adaptations that enable animals to prey on plants in spite of their protective mechanisms.

To avoid being eaten, for example, some plants have developed hard parts that are difficult to eat or are unpalatable. In fact, certain grasses defend themselves by incorporating silica (a component of glass) in their structure. If enough silica is present, the plants may simply be too tough to eat. Some groups of herbivores, however, have developed strong, grinding teeth and powerful jaws. In addition, herbivores such as cattle have developed adaptations of their digestive systems. One such adaptation allows them to store the grass they have eaten in a digestive pouch called a *rumen.* Bacteria that live in the rumen attack the grass chemically, aiding in the digestive process. This stored food is then regurgitated and rechewed at a later time, providing a better breakdown of the cell walls within the grass.

Some plants have developed chemical defenses against herbivores. The best-known plant groups with toxic effects are the poison ivy, poison oak, and poison sumac plants. All contain the contact poison urushiol (oo-ROO-she-all). Castor bean seeds are also toxic to a wide variety of animals, producing a protein that attaches to ribosomes, blocking protein synthesis. Other plants produce toxins that inhibit the growth of bacteria, fungi, and roundworms. Still others produce chemicals having odors that act as a warning or as a repellent to a predator. Today, using the techniques of genetic engineering, scientists have been able to grow plants that chemically repel certain predators, thereby reducing the need for artificial pesticides **(Figure 34.6).**

However, associated with each family or other group of plants naturally protected by a particular kind

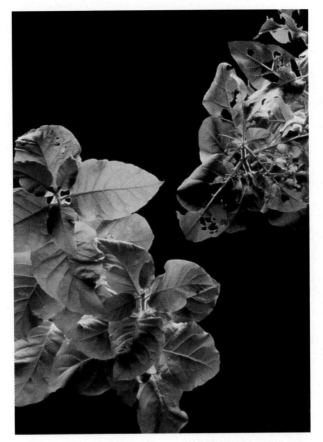

Figure 34.6 **Defense through genetic engineering.** Recently, scientists have developed plants that, through genetic engineering, chemically repel predators. The tobacco plant on the right is a nonengineered plant and shows the effects of insect predation. The tobacco plant on the left, however, has been engineered to produce an insect toxin that deters insects and protects the plant from insects.

of chemical compound are certain groups of herbivores that are adapted to feed on these plants, often as their exclusive food source. For example, the larvae of cabbage butterflies feed almost exclusively on plants of the mustard and caper families, which are characterized by the presence of protective chemicals—the mustard oils. Although these plants are protected against most potential herbivores, the cabbage butterfly caterpillars have developed the ability to break down the mustard oils, rendering them harmless. In a similar example of co-

Figure 34.7 **Monarch butterflies make themselves poisonous.** All stages of the life cycle of the monarch butterfly are protected from predators by the poisonous chemicals that occur in the milkweeds and dogbanes on which they feed as larvae. Both caterpillars and adult butterflies advertise their poisonous nature with warning coloration.

evolution, the larvae of monarch butterflies are able to harmlessly feed on the toxic plants of the milkweed and dogbane families (Figure 34.7).

Protective Coloration

Some groups of animals that feed on toxic plants receive an extra benefit—one of great ecological importance. When the caterpillars of monarch butterflies feed on plants of the milkweed family, for example, they do not break down the chemicals that protect these plants from most herbivores. Instead, they store them in fat within their bodies. As a result, the caterpillars and all developmental stages of the monarch butterfly are protected against predators by this "plant" poison. A bird that eats a monarch butterfly quickly regurgitates it. Although this is no help to the eaten insect, the bird will soon learn not to eat another butterfly with the bright orange and black pattern that characterizes the adult monarch. Such conspicuous coloration, which "advertises" an insect's toxicity, is called *warning coloration*. Warning coloration is characteristic of animals that have effective defense systems, such as poisons, stings, or bites. Other examples of animals that exhibit warning coloration are shown in Figure 34.8.

During the course of their evolution, many unprotected species have come to resemble distasteful ones that exhibit warning coloration. Provided that the unprotected animals are present in low numbers relative to the species they resemble, predators will avoid them as

(a)

(b)

Figure 34.8 **Warning coloration.** The coloring of all these animals is meant to warn other animals to stay away. (a) The red and black African grasshopper feeds on highly poisonous *Euphorbia* plants. (b) This tropical frog is so poisonous that Indians in western Columbia use frog venom to poison their blow darts.

well. If the unprotected animals are too numerous, of course, predators that have not yet learned to avoid individuals with a particular set of characteristics will eat many of them. Such a pattern of resemblance is called *Batesian mimicry*, after the British naturalist H.W. Bates, who first described this concept in the 1860s. Many of the best-known examples of Batesian mimicry occur among butterflies and moths such as the mimicry of the poisonous monarch butterfly by the viceroy. Both are shown in Figure 34.9.

Another kind of mimicry, *Müllerian mimicry*, was named for the German biologist Fritz Müller, a contemporary of Bates. Interestingly, in Müllerian mimicry, the protective colorations of different animal species come to resemble one another as in Batesian mimicry. However, unlike Batesian mimicry, the organism and its mimic *do* possess similar defenses.

(a)

(b)

Figure 34.9 **Mimicry.** The viceroy butterfly (a) is a Batesian mimic of the poisonous monarch (b).

Some organisms are colored so as to blend in with the surroundings—a protective coloration called *camouflage*. Both cabbage caterpillars (Figure 34.10) and cabbage butterflies have evolved a green coloration,

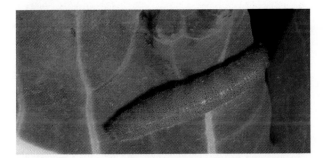

Figure 34.10 **Insect herbivores are well suited to their hosts.** The green caterpillars of the cabbage butterfly are camouflaged on the leaves of cabbage and other plants on which they feed. These caterpillars are able to break down the toxic mustard oils that prevent most insects from eating cabbage.

allowing them to hide while feeding. Insects such as these that eat plants lacking specific chemical defenses are seldom brightly colored. If they do feed on toxic plants, they break down the toxin into harmless chemicals rather than storing the toxin in their bodies. Instead, their coloration helps them become "invisible" to predators. **Figure 34.11a and b** show examples of other animals who are camouflaged from their predators.

(b)

(a)

Figure 34.11 **Two striking examples of camouflage.** (a) A leaf insect displaying effective camouflage on a plant in Malaysia. (b) A spotted scorpionfish in the waters off Fiji.

The effect of predator populations on prey populations is difficult to predict because their complex interactions depend on their interactions with other organisms in the community, movement of new organisms into and out of the community, and the abiotic factors that influence their survival.

Over time, plants have developed various morphological and chemical adaptations that help protect them against plant eaters, or herbivores. In turn, however, herbivores have changed and adjusted to the plants and their adaptations. Such interactions, involving long-term mutual evolutionary adjustment, characterize coevolution.

Some organisms within communities exhibit conspicuous coloration, or warning coloration, that advertises their ability to poison, sting, or bite. Organisms without specific defenses are often colored, or camouflaged, so as to blend in with their surroundings.

34.4 Commensalism

In **commensalism**, an organism of one species benefits from its interactions with another, whereas the other species neither benefits nor is harmed. Many examples of the "one-sided" relationship of commensalism exist in nature. Often, the individuals deriving benefit are physically attached to the other species in the relationship. As shown in **Figure 34.12** for example, plants called *epiphytes* (EP-ih-fits) grow on the branches of other plants. The epiphytes derive their nourishment from the air and the rain—not from the plants to which they attach for support. Similarly, various marine animals such as barnacles grow on other, often actively moving sea animals (**Figure 34.13**). These "hitchhikers" gain more protection from predation than if they were fixed in one place, and they continually reach new sources of food. They do not, however, harm the organisms to which they are attached.

Figure 34.12 **Commensalism: Epiphytes and trees.**

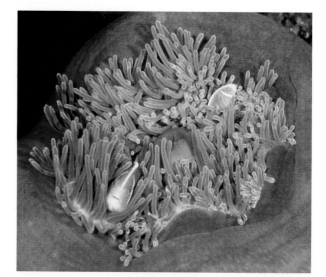

Figure 34.14 **Commensalism: Clownfish and sea anemones.** Two clownfish peer out from the tentacles of a large red sea anemone off the coast of Australia.

Possibly one of the best-known examples of commensalism involves the relationship between certain small tropical fish (the clownfish) and sea anemones (marine animals that have stinging tentacles). The fish have developed an adaptation that allows them to live among the deadly tentacles of the anemones (**Figure 34.14**). These tentacles quickly paralyze other species of fishes, protecting the clownfish against predators.

Commensalism is a relationship in which one organism benefits while the other neither benefits nor is harmed. Often, the individuals deriving benefit are physically attached to the other species in the relationship.

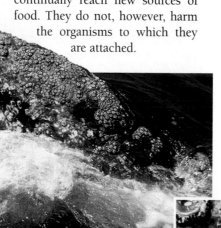

(a)

Figure 34.13 **Commensalism: Barnacles and whales.** (a) This breaching gray whale displays the barnacles on its skin. (b) A close-up of a gray whale's skin reveals hitchhikers—lice and barnacles. The lice are actually parasites, whereas the barnacles cause no harm to the whale.

(b)

34.5 Mutualism

Mutualism is a relationship in which two species live together in close association, both benefiting from the relationship. A particularly striking example of mutualism involves one genus of stinging ants and a Latin American plant of the genus *Acacia.* The modified leaves of acacia plants appear as paired, hollowed thorns. These thorns provide a home for the ants, protecting them and their larvae. In addition, the plants produce nectar that the ants eat. In turn, the ants attack any herbivore that lands on the branches or leaves of an acacia and clear away vegetation that comes in contact with their host shrub, increasing the plant's ability to survive.

Figure 34.15   Mutualism: Red billed oxpeckers on impala in South Africa.

Many other interesting examples of mutualism exist in nature. Certain birds, for example, spend most of their time clinging to grazing animals (such as cattle), picking insects from their hides **(Figure 34.15).** In fact, the birds carry out their entire life cycles in close association with the cattle. The birds are provided with food, and the cattle benefit by having their parasites removed. In another similar mutualistic relationship, ants use the tiny insect aphids, or greenflies, as a provider of food. The aphids suck fluids from the phloem of plants, extracting a certain amount of sucrose and other nutrients. However, many of these nutrients are not absorbed within the digestive tract of the aphid. A substantial portion runs out (somewhat altered) through the anus. The ants use this nutritional excrement as a food source and, in turn, actually carry the aphids to new plants so that they can continue eating!

> Mutualism is a relationship in which two species live together in close association, both benefiting from the relationship.

34.6 Parasitism

Parasitism is a relationship in which an organism of one species (the parasite) lives in or on another (the host). The parasite benefits from this relationship, whereas the host is usually harmed. Parasitism is sometimes considered a form of predation. However, unlike a true predator, the successful parasite does not kill its host.

Parasites include viruses, many bacteria, fungi, and an array of invertebrates. A different species of organism, usually larger than the parasite itself, is "home" for a parasite. During the intimate relationship between parasite and host, the parasite derives nourishment and the host is harmed.

Many instances of parasitism are well known. Intestinal hookworms, for example, are parasites **(Figure 34.16).** A person is infected when walking barefoot in soil containing hookworm larvae. These larvae are able to penetrate the skin, entering the bloodstream. The blood carries the larvae to the lungs. From there, they are able to migrate up the windpipe to the esophagus. The larvae are then swallowed and reach the intestines. After growing into adult worms, they attach to the inner lining of the intestines. They remain attached there, feeding on the blood of the host.

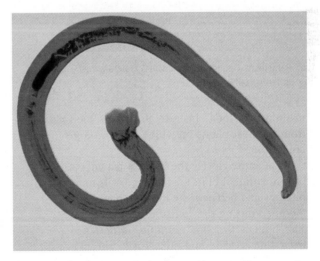

Figure 34.16   Parasitism: Hookworms and humans. These parasites of humans live in the intestine, feeding on the blood of the host.

Some parasites do not live within an organism as hookworms do, but attach to the outer surface of a plant or an animal. The attachment may be fleeting, as with the bite of a mosquito, or may take place over a longer period, such as the burrowing of mites.

Many fungi and some flowering plants are parasitic. The dodder plant, for example, has lost its chlorophyll and leaves in the course of its evolution and is unable to

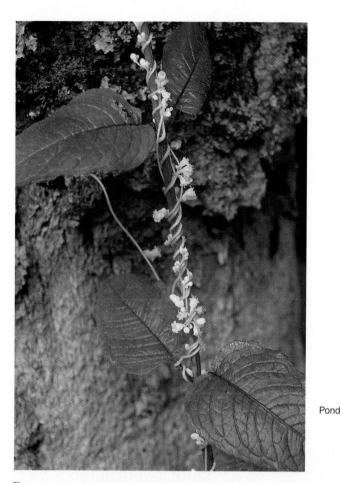

Figure 34.17 **Parasitism.** Dodder plant entwined around a host plant, deriving nutrients from it.

www.jbpub.com/biology

manufacture food. Instead, it obtains food from the host plants on which it grows (Figure 34.17).

The more closely the life of a parasite is linked with that of its host, the more its morphology and behavior are likely to have been modified during the course of its evolution. The human flea, for example, is flattened from side to side and slips easily through hair. The ancestors of this species of flea were brightly colored, large, winged insects. The structural and behavioral modifications of the human flea have come about in relation to a parasitic way of life.

Parasitism is a relationship in which an organism of one species (the parasite) benefits from living in or on another (the host), which is usually harmed.

34.7 Succession

Communities change over time. Even when the climate of a given area remains stable year after year, the composition of the species making up a community, as well as the interactions within the community, show a dynamic process of change known as **succession**. During the process of succession, a sequence of communities replaces one another. This process is familiar to anyone who has seen a vacant lot or cleared woods slowly become occupied with plants and animals, or has seen a pond become filled with vegetation (Figure 34.18).

Primary succession takes place in areas not previously supporting organisms. Primary succession occurs in lakes formed from the retreat of glaciers, for example, or on volcanic islands that may rise above the sea. In the latter case, succession may begin as lichens take hold on bare rock. Lichens are made up of an alga and a fungus

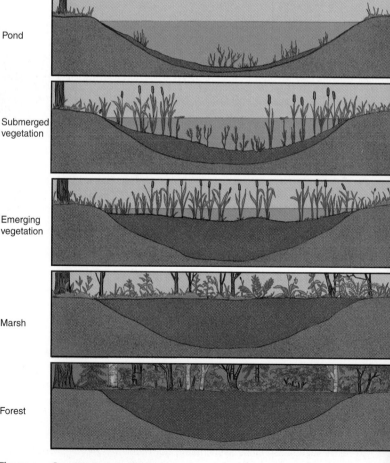

Pond

Submerged vegetation

Emerging vegetation

Marsh

Forest

Figure 34.18 **Succession of a pond.**

living together in a symbiotic relationship (see. p. 631, Chapter 27). As lichens grow, they produce acids that can break down rock, forming small pockets of soil. When enough soil has accumulated, mosses may begin to grow. This first community, a pioneer community, consists of plants that are able to grow under harsh conditions. The pioneer community paves the way for the growth and development of vegetation native to that climate. Over many thousands of years or even longer, the rocks may be completely broken down, and vegetation may cover a once-rocky area. As plants take hold, the area becomes able to sustain other forms of life. As the plant community changes, so too do the other living things. Eventually, the mix of plants and animals becomes somewhat stable, forming what is termed a **climax community**. However, with an increasing realization that (1) climates may change, (2) the process of succession is often very slow, and (3) the nature of a region's vegetation is determined to a great extent by human activities, ecologists do not consider the concept of a climax community as useful as they once did.

Succession occurs not only within terrestrial communities but in aquatic communities as well. A lake poor in nutrients, for example, may gradually become rich in nutrients as organic materials accumulate (see Figure 34.18). Plants growing along the edges of the lake, such as cattails and rushes, and those growing submerged, such as pondweeds, may contribute to the formation of a rich organic soil as they die and are decomposed by bacteria. As this process of soil formation continues, the pond may become filled in with terrestrial vegetation. Eventually, the area where the pond once stood may become an indistinguishable part of the surrounding vegetation.

Secondary succession occurs in areas that have been disturbed and that were originally occupied by organisms. Humans are often responsible for initiating secondary succession throughout portions of the world that they inhabit. Abandoned farm fields, for example, undergo secondary succession as they revert to forest. Secondary succession may also take place after natural disasters such as a forest fire or a volcanic eruption producing ash (rather than lava flows) **Figure 34.19** shows secondary succession on Mt. St. Helens.

Communities change over time by means of the dynamic process of succession. During this process, a sequence of communities replaces one another.

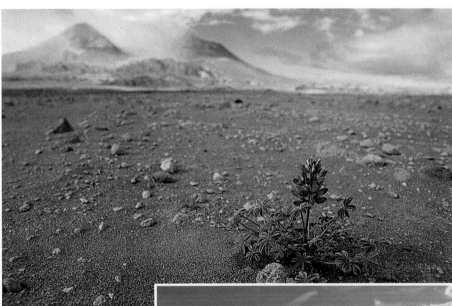

(a)

(b)

Figure 34.19 Secondary succession. Mount St. Helens in the state of Washington erupted violently on May 18, 1980. The lateral blast devastated more than 600 square kilometers of forest and recreation lands within 15 minutes. (a) shows an area called Pumice Plain (with crater in the background) 4 years after the blast. (b) Nineteen years after the blast (in 1999), succession is underway at the same spot in Pumice Plain.

Key Concepts

Populations of different species living together compose a community.

34.1 All of the living (biotic) factors and nonliving (abiotic) factors within a certain area are an ecosystem.

34.1 Within an ecosystem, each living thing has a habitat, an area in which it lives that is characterized by its physical (abiotic) properties.

34.1 Each organism plays a special role within an ecosystem, which is its niche.

34.1 Of the biotic factors within an ecosystem, groups of organisms of the same species are populations.

34.1 Various populations of organisms living together in a particular area at a particular time make up a community.

34.1 An ecosystem can be thought of as a community of organisms, along with the abiotic factors with which the community interacts.

Each organism plays a special role within a community.

34.2 Competition involves organisms striving to obtain the same needed resource.

34.3 During predation, one species (the prey) becomes a resource, being killed and eaten by another species (the predator).

34.4 Commensalism is a one-sided relationship: One species in a relationship between organisms of different species benefits, whereas the other species neither benefits nor is harmed.

34.5 In mutualism, both species benefit.

34.6 In a parasitic relationship, one species (the parasite) benefits, but the other (the host) is harmed.

34.6 Interactions that involve the long-term, mutual evolutionary adjustment of the characteristics of the members of biological communities in relation to one another are forms of coevolution.

Communities change over time by succession.

34.7 Communities, like species, change over time.

34.7 This dynamic process of change, during which a sequence of communities replaces one another, is succession.

34.7 Primary succession takes place in areas that are originally bare, such as rocks or open water.

34.7 Secondary succession takes place in areas where the communities of organisms that existed initially have been disturbed.

Key Terms

abiotic factors (AYE-bye-OT-ik) *760*
biotic factors (bye-OT-ik) *760*
climax community *771*
commensalism *768*
community *760*
competition *761*
competitive exclusion *761*
ecology (eh-KOL-uh-gee *or* ee-KOL-uh-gee) *760*

ecosystem (EH-koe-SIS-tem *or* EE-koe-SIS-tem) *760*
fundamental niche *762*
habitat *760*
mutualism *769*
niche (nich) *760*
parasitism *769*
population *760*

predation *763*
predator *763*
prey *763*
primary succession *770*
realized niche *762*
secondary succession *771*
succession *770*

KNOWLEDGE AND COMPREHENSION QUESTIONS

1. Distinguish among the terms population, community, and ecosystem. (*Comprehension*)

2. Fill in the blanks: Within an ecosystem, each organism has a(n) _____, an area in which it resides. Each organism plays a role within the ecosystem, which is its _____. (*Knowledge*)

3. Interactions within communities can be grouped into five categories. Name and describe each category. (*Knowledge*)

4. State the principle of competitive exclusion in your own words. (*Comprehension*)

5. Explain the terms *realized niche* and *fundamental niche*. (*Comprehension*)

6. Summarize the pattern of the predator-prey relationship as shown in Gause's experiments. (*Comprehension*)

7. What other factors, in addition to those revealed in Gause's work, can affect the balance between predators and prey? (*Comprehension*)

8. What is coevolution? Give an example of co-evolution involving a herbivore. *(Knowledge)*

9. What type of adaptation is shown in each of the following: *(Comprehension)*
   a. The bright yellow and black stripes of a bee's body.
   b. The unobtrusive color of some lizards that makes them difficult to see against surrounding rocks.

10. What is primary succession? Summarize the process. *(Comprehension)*

K E Y
Knowledge: Recalling information.
Comprehension: Showing understanding of recalled information.

CRITICAL THINKING REVIEW

1. Give a hypothetical example in which a seemingly commensal relationship could become mutualistic. *(Application)*

2. Recently, two researchers at the University of Florida at Gainesville conducted "taste tests" with birds. They fed the birds the bodies of both Viceroy and Monarch butterflies, and discovered that the birds found both to be unpalatable. How do these findings challenge the idea that these butterflies are Batesian mimics? What does this research suggest with respect to mimicry? *(Application)*

3. Would the most successful predator feed on the very young of a population, the adults of a population, or the very old? Give a rationale for your answer. *(Synthesis)*

4. What factors determine the type of organisms that grow in a climax community? *(Synthesis)*

5. Herbivory is a situation in which grazing animals feed on plants. Under what conditions would herbivory be considered predation? Under what conditions would herbivory be considered parasitism? *(Application)*

K E Y
Application: Using information in a new situation.
Synthesis: Putting together information from different sources.

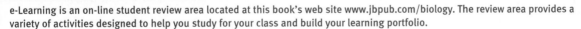

e**learning**

Location: http://www.jbpub.com/biology www.jbpub.com/biology

e-Learning is an on-line student review area located at this book's web site www.jbpub.com/biology. The review area provides a variety of activities designed to help you study for your class and build your learning portfolio.

Review Questions The review questions test your knowledge of the important concepts and applications in each chapter. The review provides feedback for each correct or incorrect answer. This is an excellent test preparation tool.

Figure-Labeling Exercises Sharpen your visual thinking by matching terms and labels for important illustrations.

Flash Cards Studying biology requires learning new terms. Virtual flash cards help you master the new vocabulary for each chapter.

Just Wondering As you read and study from this text, you may find that you have unanswered questions. Through this site you can ask the author, Sandy Alters, your "just wondering" questions.

Do you prefer the speed and reliability of a CD-ROM? All of the features contained on the eLearning portion of the web site are also available on the Student CD-ROM.

Ecosystems

A wriggling beetle may not be your idea of a gourmet meal, but to this collared lizard it is a five-star dinner. Beetles and other small animals, including smaller lizards, are on the daily menu of the collared lizard. This inhabitant of dry, rocky regions in western America was named for the black bands encircling its neck. Pictured here is just one link in the food chain that describes a series of organisms each feeding on one another. Minutes before this photograph was taken, it was the beetle that was feeding on the fresh seedlings of desert plants just starting to emerge after a spring rain. Unknown to the lizard, a hawk soars overhead, waiting for the right moment to swoop down and pick up the foot-long lizard in its talons. The hawk will add yet another link to this chain of relationships among the plants and animals of this ecosystem.

Organization

An ecosystem is a community of organisms and their environment.

35.1 Populations, communities, and ecosystems

Nutrients cycle and energy flows through ecosystems.

35.2 Food chains and webs
35.3 Food pyramids

All elements essential to life are cycled through the atmosphere or the soil.

35.4 The water cycle
35.5 The carbon cycle
35.6 The nitrogen cycle
35.7 The phosphorus cycle

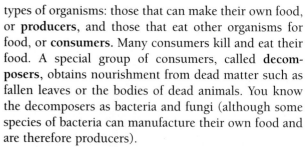

35.1 Populations, communities, and ecosystems

Individuals of a species, such as bullfrogs and mice, are each part of an individual population of organisms. Together, interacting populations are communities (see Chapters 33 and 34). The living organisms in a community interact not only with each other but with the nonliving substances in their environment, such as the soil, water, and air, to form an ecological system, or ecosystem.

An **ecosystem** is a community consisting of plants, animals, and microorganisms that interact with one another and with their environments and that are interdependent on one another for survival. The living, or biotic, components of an ecosystem are made up of two types of organisms: those that can make their own food, or **producers**, and those that eat other organisms for food, or **consumers**. Many consumers kill and eat their food. A special group of consumers, called **decomposers**, obtains nourishment from dead matter such as fallen leaves or the bodies of dead animals. You know the decomposers as bacteria and fungi (although some species of bacteria can manufacture their own food and are therefore producers).

Figure 35.1 is a diagram of an ecosystem. Notice that the abiotic, or nonliving, components of the environment (such as decomposed material in the soil) contribute substances needed for the ecosystem to function. In addition, notice that the exchanges of nutrients and other chemical substances among the organisms within the ecosystem form a cyclical pattern. Energy, however,

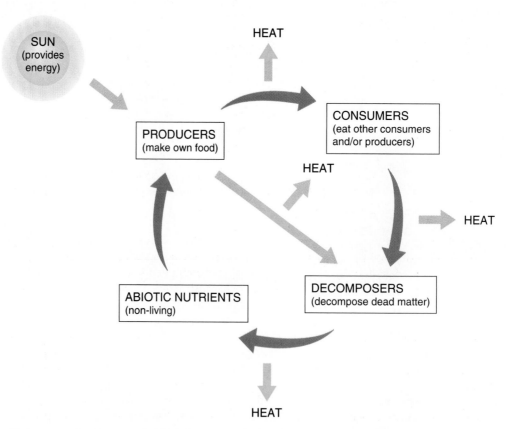

Figure 35.1 An ecosystem. Nutrients (represented by green) flow within an ecosystem in a cyclic pattern, whereas energy (represented by orange) flows *through* an ecosystem.

Figure 35.2 **Two distinct ecosystems.** In the coastal ranges of California, the boundary between the evergreen shrub ecosystem known as chaparral and the grassland ecosystem is often sharp, as shown in this photograph taken along the western edge of the Santa Clara Valley near Morgan Hill. These two ecosystems also include characteristic sets of nonliving factors.

eventually flows *through* the ecosystem, first captured from the sun by the producers, then used by herbivores that eat the producers, and ultimately used by all the consumers living in the ecosystem. However, much of the sun's energy originally captured by producers is lost to the environment as heat. Therefore, it is said to *flow through* the ecosystem rather than cycle. Thus, ecosystems are systems in which there is a regulated transfer of energy and an orderly, controlled cycling of nutrients. The individual organisms and populations of organisms in an ecosystem act as parts of an integrated whole, adjust over time to their roles in the ecosystem, and relate to one another in complex ways that are only partly understood.

Where does one ecosystem begin and another end? Some ecosystems have clearly recognizable boundaries, such as that of a pond or a puddle or those found within the coastal ranges of California (Figure 35.2). Sometimes humans produce artificial ecosystems with human-made boundaries, such as the glass walls of an aquarium or the fencing surrounding a cultivated field. The boundaries of many natural ecosystems blend with one another, sometimes almost imperceptibly. Ecosystems also change over time and with climate changes, slowly becoming modified into new ecosystems whose characteristics differ increasingly from those that preceded them.

An ecosystem is made up of communities of organisms living within a defined area and the nonliving environmental factors with which they interact. The organisms of an ecosystem—the producers, consumers, and decomposers—each play a specific role within it, contributing to the flow of energy and the cycling of nutrients.

Nutrients cycle and energy flows through ecosystems.

What happens if one species in a food chain dies off? Do all the other species in the chain die too?

A food chain is part of an intricate food web, and the organisms that make up the chain enter into a variety of relationships (in addition to predator-prey relationships) with other organisms within an ecosystem. As species die off, new species move in and other species move out of the ecosystem as relationships among organisms change. Species that have the greatest impact on ecosystems are *keystone species*.

Keystone species are those that influence the structure and function of an ecosystem to a much greater degree than would be expected simply from the size of their populations. When a keystone is removed from an ecosystem (and therefore from a food chain and web), changes are certain to take place, and those changes depend on the role of the keystone species. For example, in the forests of Peru (located on the Pacific coast of South America), only a dozen or so species of fig and palm trees support an entire community of fruit-eating birds and mammals during the time of year when fruits are least available (see photo). Loss of these few tree species would probably result in the loss of most of the fruit-eating animal species even if hundreds of other tree species remained. Conversely, if a keystone predator species dies, the populations of the species it kept in check through predation may soar. These prey may now overtake their competitors as predators of other species, or they may become prey for species that move into the area. Thus, the loss of a keystone species can have a great impact on a food chain and the ecosystem of which it is an integral part.

Scientists have also discovered that the number of species lost is critical to ecosystems. As more species are lost, the more the functioning of an ecosystem is degraded. In other words, diversity contributes to the health of an ecosystem—it survives better.

Therefore, the effects of the loss of a single species in a food chain vary, and they depend on whether the species was a keystone species and on the other adjustments made among species as relationships change in response to the loss. Because of the importance of keystone species, however, scientists are anxious to identify them, because their extinction *could* result in the extinction of many other species in food chains, food webs, communities, and ecosystems.

Macaw eating fruit of Mauritia Palm in Amazon rainforest, Peru.

35.2   Food chains and webs

The energy that flows through an ecosystem comes from the sun. Green plants, the primary producers of terrestrial ecosystems, are able to capture some of the sun's radiant energy that falls on their leaves and convert it to chemical energy during the process of photosynthesis (see Chapter 6). Producers, then, are the key to life on Earth, because no other organisms can capture this energy for use in living systems. **Primary consumers**, or herbivores, feed directly on the green plants, incorporating some of this energy into molecules that make up their bodies and using the rest to perform the activities of life. **Secondary consumers** are meat eaters, or carnivores, that feed in turn on the herbivores. The chain continues, with one living thing feeding on another, passing energy along that was once captured from the sun.

The refuse or waste material of an ecosystem is known as **detritus** (dih-TRITE-us). Organisms that are decomposers break down the organic materials of detritus into inorganic nutrients that can be reused by plants (Figure 35.3). The decomposers use some of the energy still held in the tissues of once-living things, but they are the last link in this transfer of energy among organisms.

All the feeding levels previously described and additional levels such as tertiary consumers are represented in any fairly complicated ecosystem. These feeding levels are called **trophic levels** (TROW-fik), from the Greek word *trophos*, which means "feeder." (In fact, this

Greek word is the root for the words **heterotroph**, or "other feeder"—another word for consumer, and for **autotroph**, or "self-feeder"—another word for producer.) Organisms from each of these levels, feeding on one another, make up a series of organisms called a **food chain**.

An example of a food chain can be seen in a pond ecosystem in which water fleas (primary consumers) feed on green algae (producers). Sunfish (secondary consumers) eat the water fleas, but in turn are eaten by green herons (tertiary consumers) (Figure 35.4). The length and complexity of food chains vary greatly.

In reality, it is rare for any species of organism to feed on only one other species. Organisms feed on many different species and types of organisms and are, in turn, food for two or more other kinds. Shown in a diagram such as Figure 35.5, these relationships appear as a series of branching and overlapping lines rather than as one straight line. The organisms in an ecosystem that have such interconnected and interwoven feeding relationships make up a **food web**. Figure 35.5 shows a food web in a salt marsh ecosystem.

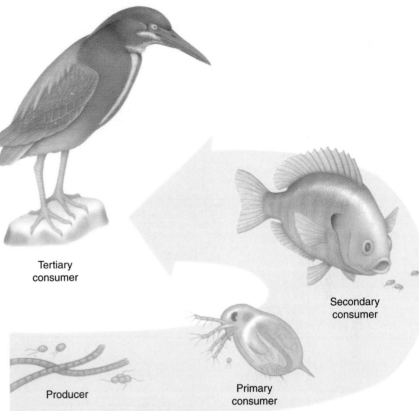

Tertiary consumer

Secondary consumer

Producer

Primary consumer

Figure 35.4   A typical pond food chain. A food chain is a series of organisms from each trophic level that feed on one another.

Figure 35.3   A decomposer doing its job. The shelf fungi (phylum Basidiomycota) growing on this hardwood stump are decomposing it, converting the organic materials contained within it to nutrients that can be reused by plants.

The producers of an ecosystem capture energy from the sun and convert it to chemical energy usable by themselves and consumers. Consumers feed on the producers and other consumers in the ecosystem, passing energy along that was once captured from the sun. Decomposers break down the organic molecules of dead organisms, serving as the last link in the flow of energy through an ecosystem and contributing to the recycling of nutrients within the environment.

A linear relationship among organisms that feed on one another is a food chain. Food chains that interweave are food webs.

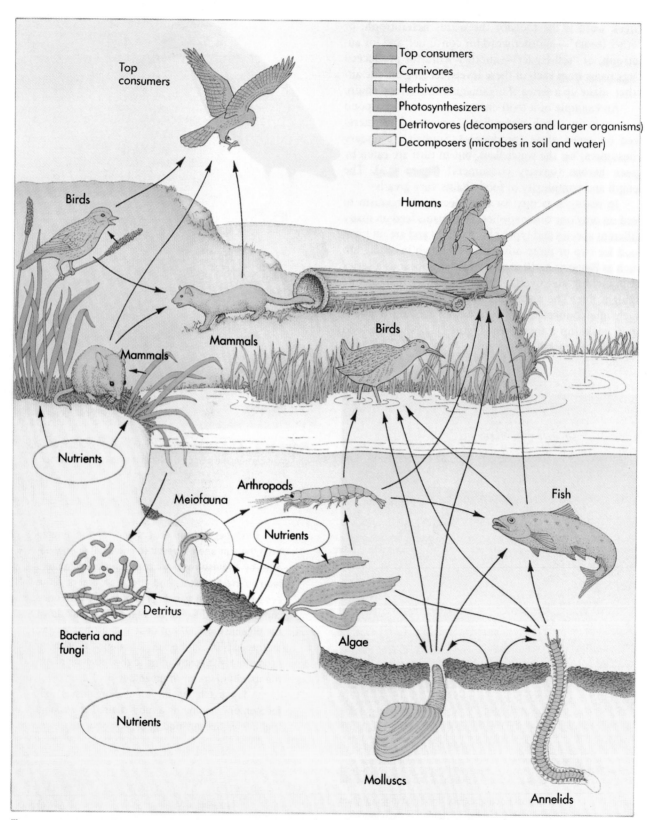

Figure 35.5 **The food web in a salt marsh.** A food web is a group of interwoven food chains within an ecosystem. In this diagram, each color represents a trophic level. Each trophic level feeds on, or gains energy from, the layer below. (The term *meiofauna* refers to a group of animals that live in the spaces between grains of sand.)

35.3 Food pyramids

In any ecosystem, the number of organisms (or their total mass [biomass]) and the amount of energy making up each successive trophic level are often less than the level that precedes it. Lamont Cole of Cornell University illustrated this concept in his study of the energy flow in a freshwater ecosystem in Cayuga Lake, New York. He calculated that approximately 150 calories of each 1000 calories of energy "fixed" by producers during photosynthesis was transferred into the bodies of small heterotrophs that feed on these plants and bacteria (Figure 35.6). Smelt, which are tiny fish, eat the het-

erotrophs; these secondary consumers obtain about 30 calories of each original 1000. If humans eat the smelt, they gain about 6 calories from each 1000 calories that originally entered the system. If trout eat the smelt and humans eat the trout, humans gain only about 1.2 calories from each original 1000.

These types of calculations show that, on average, only 10% of plants' accumulated energy is actually converted into the bodies of the organisms that consume them. What happens to the rest? A certain amount of the energy that is ingested by organisms goes toward heat production. (It is actually "lost" as heat, consistent with the second law of thermodynamics, which states that no transformation of energy is 100% efficient.) A great deal of energy is used for digestion and work, and usually 40% or less goes toward growth and reproduction. An invertebrate, for example, typically uses about a quarter of this 40% for growth. In other words, about 10% of the food that an invertebrate eats is turned into new body tissue. This figure varies from approximately 5% for carnivores to nearly 20% for herbivores, but 10% is an average value for the amount of energy (or organic matter) that organisms incorporate into their bodies from the energy available in the previous trophic level.

If shown diagrammatically (Figure 35.7), the relationships already described between trophic levels appear as pyramids. Diagrams that depict the total weight of organisms supported at each trophic level in an ecosystem are referred to as *pyramids of biomass* (Figure 35.7a). Those that depict the energy flow through an ecosystem are called *pyramids of energy* (Figure 35.7b). *Pyramids of number* depict, as the name suggests, the total number of organisms at each feeding level (Figure 35.7c). Generalized pyramids incorporating these three concepts are referred to as food pyramids (Figure 35.8).

Occasionally, pyramids of biomass and/or pyramids of number can be inverted. For example, in an ocean community, the photosynthesizing organisms (usually bacteria and protists) reproduce so rapidly that a small but constant biomass can feed a much larger biomass of herbivores. An inverted pyramid of biomass would depict this relationship. In another example, within a forest community, a single tree can support many herbivorous insects, resulting in an inverted pyramid of numbers.

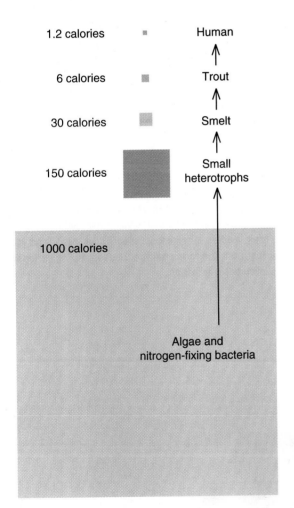

Figure 35.6 The flow of energy in Lake Cayuga. The experiments in Lake Cayuga demonstrated that the number of organisms and the amount of energy making up each successive trophic level is smaller than the preceding level.

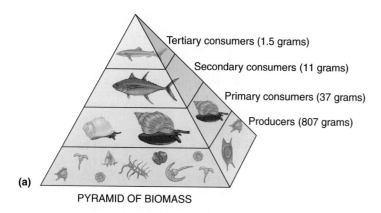

Tertiary consumers (1.5 grams)

Secondary consumers (11 grams)

Primary consumers (37 grams)

Producers (807 grams)

(a)

PYRAMID OF BIOMASS

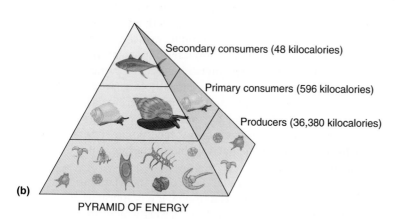

Secondary consumers (48 kilocalories)

Primary consumers (596 kilocalories)

Producers (36,380 kilocalories)

(b)

PYRAMID OF ENERGY

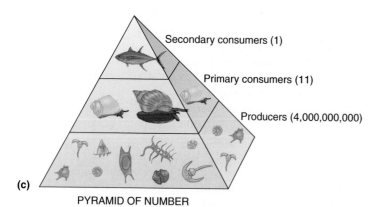

Secondary consumers (1)

Primary consumers (11)

Producers (4,000,000,000)

(c)

PYRAMID OF NUMBER

Figure 35.7 Pyramids of (a) biomass, (b) energy, and (c) number for an aquatic ecosystem.

The relationships among organisms at the various trophic levels of a food pyramid make it clear that herbivores have more food available to them than do carnivores. In other words, the lower a population eats in the food chain, the higher the number of individuals that can be fed. Such considerations are important as humans work to maximize the food available for a hungry and increasingly overcrowded world.

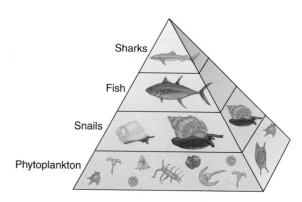

Sharks

Fish

Snails

Phytoplankton

Figure 35.8 **Food pyramid.** All food pyramids are basically the same—each successive trophic level is smaller than the preceding level.

A plant captures some of the sun's energy that falls on its green leaves. The successive members of a food chain process about 10% of the energy available in the organisms on which they feed into their own bodies. The rest is released as heat (and is lost from the food chain) or is used during various metabolic activities.

35.4 The water cycle

Although energy flows through ecosystems and most is lost at each successive level in food pyramids, the matter making up the organisms at each level is not lost. All of these substances are recycled and are used only temporarily by living things. Hydrogen, carbon, nitrogen, and oxygen—the principal elements that make up all living things—are primarily held in the atmosphere in molecules of water, carbon dioxide, nitrogen gas, and oxygen gas. Other recycled substances necessary for life such as phosphorus, potassium, sulfur, magnesium, calcium, sodium, iron, and cobalt are held in rocks and, after weathering, enter the soil. The atmosphere and rocks are therefore referred to as the *reservoirs* of inorganic substances that cycle within ecosystems.

The cycling of materials in ecosystems is usually described as beginning at the reservoirs. Living things incorporate substances into their bodies from their reservoirs or from other living things, passing these materials along the food chain. Ultimately these substances, with the help of decomposers, move from the living world back to the nonliving world, becoming part of the soil or the atmosphere once again.

Heated by the sun, water evaporates into the atmosphere from the surfaces of oceans, lakes, and streams **(Figure 35.9)**. In terrestrial (land-based) ecosystems, as much as 90% of the water that reaches the atmosphere comes from plants as they release water vapor into the air during the process of transpiration (see Chapter 29). Because oceans cover three fourths of the Earth's surface, these bodies contribute most of the water to the atmosphere worldwide.

Atmospheric water condenses in clouds and eventually falls back to the Earth as precipitation. Most of it falls directly into the oceans. Some falls onto the land, flowing into surface bodies of water or trickling through layers of soil and rock to form subsurface bodies of fresh water called *groundwater*. Plants take up water as it trickles through the soil, almost in a continuous stream. Crop plants, for example, use about 1000 kilograms of water just to produce 1 kilogram of biomass. Animals

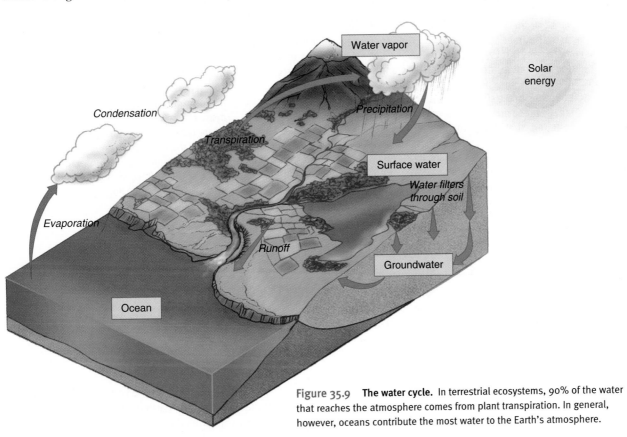

Figure 35.9 **The water cycle.** In terrestrial ecosystems, 90% of the water that reaches the atmosphere comes from plant transpiration. In general, however, oceans contribute the most water to the Earth's atmosphere.

H W Science
WORKS Discoveries

Elements essential to life are cycled through the atmosphere or the soil.

The Hubbard Brook Experiment: What Have We Learned about the Cycling of Nutrients?

In 1963, scientists at the Yale School of Forestry and Environmental Studies began an interesting experiment designed to show the effects of the cycling of nutrients (chemicals) within ecosystems. (Their work is still ongoing today and is called *long-term ecological research*.) Their studies have yielded much of the information that is now known about nutrient cycles, and their ingenious experimental design has provided the basis for the development of the experimental methods used in the study of other ecosystems.

www.jbpub.com/biology

Hubbard Brook is the central stream located in a temperate deciduous forest in New Hampshire. For measurement of the flow of water and nutrients within the Hubbard Brook ecosystem, concrete weirs with V-shaped notches were built across six tributary streams that were selected for study. All of the water that flowed out of the six valleys (the areas surrounding the six streams) had to pass through the notches. The precipitation that fell in the six streams was measured, and the amounts of nutrients present in the water flowing in the six streams were also determined. By these methods, the scientists demonstrated that the undisturbed forests in this area were very efficient in retaining nutrients. The small amounts of nutrients that fell from the atmosphere with the rain and snow were approximately equal to the amounts of nutrients that ran out of the six valleys. For example, there was only a small net loss of calcium—about 0.3% of the total

calcium in the system per year—and small net gains of nitrogen and potassium.

Then the researchers disturbed the ecosystem. In 1965, the investigators felled all of the trees and shrubs in one of the six valleys and prevented their regrowth by spraying the area with herbicides. The effects of these activities were dramatic. The amount of water running out of the valley increased by 40%, indicating that water normally taken up by the trees and shrubs—and evaporated into the atmosphere from their leaves—was now running off. The amounts of nutrients running out of the system also increased. The loss of calcium was 10 times higher than it had been previously. The change in the status of nitrogen was partic-

ularly striking. The undisturbed ecosystem in this valley had been accumulating nitrogen at a rate of 2 kilograms per hectare per year, but the cut-down ecosystem lost it at a rate of about 120 kilograms per hectare per year! The nitrate level of the water rapidly increased to a level exceeding that judged safe for human consumption, and the stream that drained the area generated massive blooms of cyanobacteria and algae.

The Hubbard Brook experiment demonstrated that nutrient cycling depends, among other things, on the vegetation present in the ecosystem. The fertility of the deforested valley decreased rapidly, and at the same time the danger of flooding greatly increased. The Hubbard Brook experiment is particularly instructive in the 2000s because large areas of tropical rain forest are being and have been destroyed to make way for cropland. Many of the insights gleaned from the Hubbard Brook experiment can help scientists understand some of the consequences of rain forest destruction that may not have been readily apparent otherwise.

obtain water directly from surface water or from the plants or other animals they eat. In the United States, groundwater provides about a quarter of the water used by humans for all purposes and provides about half of the population with drinking water.

About 2% of the groundwater in the United States is polluted, and the situation is worsening. Pesticides are one source, being carried to *aquifers,* underground reservoirs in which the groundwater lies within porous rock as rain washes the chemicals from the surfaces of leaves and the topsoil. Chemical wastes, stored in surface pits, ponds, and lagoons, are another key source of

groundwater pollution. Scientists have no technology that will remove pollutants from underground aquifers.

Within ecosystems, matter cycles from its reservoir in the environment to the bodies of living organisms and back to the environment.

Water in the atmosphere condenses in clouds and falls back to the Earth as precipitation. Plants take up water from the soil, and animals obtain water from surface water or from the plants or other animals they eat. Water returns to the atmosphere through the evaporation of surface water and transpiration by plants.

35.5 The carbon cycle

The carbon cycle is based on carbon dioxide (CO_2), which makes up about 0.03% of the atmosphere and is found dissolved in the oceans (Figure 35.10). Terrestrial as well as marine producers use CO_2—along with energy from the sun—to build carbon compounds such as glucose during the process of photosynthesis (see Chapter 6). The producers and the consumers that eat them break down these carbon compounds during cellular respiration and use the energy locked in their chemical bonds to carry on the metabolic processes of life. Consumers use some of the carbon atoms and compounds from the food they eat to produce needed substances. However, most of this carbon is waste and is released to the atmosphere (or to the oceans) as CO_2.

Intimately linked to the cycling of carbon is the cycling of oxygen (O_2), as shown in Figure 35.11. As plants use CO_2 during the process of photosynthesis, they produce O_2 as a by-product. This O_2 is released into the atmosphere and becomes available to organisms for the process of cellular respiration.

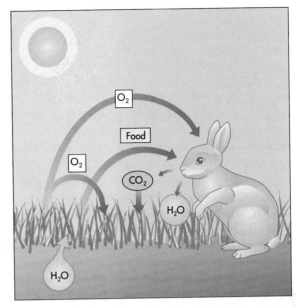

Figure 35.11 **How carbon dioxide is linked to oxygen cycling.** Plants use carbon dioxide in photosynthesis and give off oxygen. This oxygen is available to animals for the process of cellular respiration. Animals breathe out carbon dioxide; some carbon dioxide is liberated from the decomposition of dead organisms.

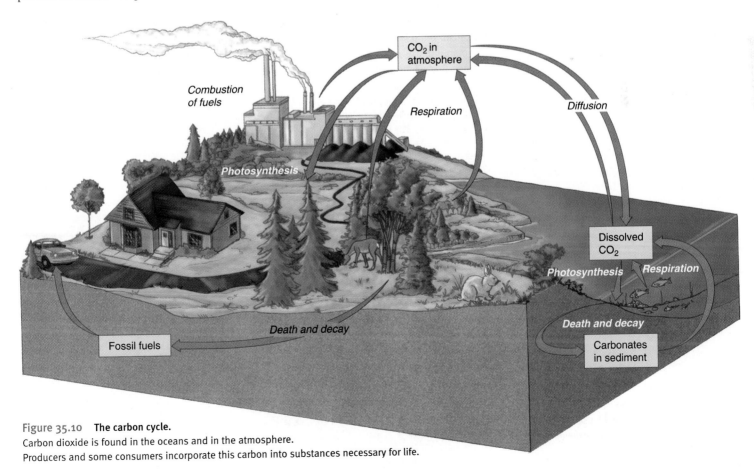

Figure 35.10 **The carbon cycle.**
Carbon dioxide is found in the oceans and in the atmosphere.
Producers and some consumers incorporate this carbon into substances necessary for life.

Some aquatic organisms such as molluscs use the CO_2 dissolved in water and combine it with calcium to form their calcium carbonate ($CaCO_3$) shells. When these organisms die, their shells collect on the sea floor. Years of exposure to water slowly dissolves the $CaCO_3$, releasing the CO_2 and making its carbon once again available to aquatic producers to use in the process of photosynthesis.

When organisms die, decomposers break down the carbon compounds making up their bodies. Some carbon-containing compounds, such as the cellulose found in the cell walls of plants, are more resistant to breakdown than others, but certain bacteria, fungi, and protozoans are able to accomplish this feat. Some cellulose, however, accumulates as undecomposed organic matter. Over time and with heat and pressure, this undecomposed matter results in the formation of fossil fuels such as oil and coal. When these fuels are burned, as when wood is burned, the CO_2 is returned to the atmosphere. The release of this carbon as CO_2, a process that is proceeding rapidly as a result of human activities, may change global climates (see Chapter 37).

> Carbon is held in the atmosphere in the form of carbon dioxide. This gas is taken in by photosynthetic organisms and is used to build the carbon compounds that plants manufacture during the process of photosynthesis. Both producers and consumers use these compounds as an energy source, metabolizing them during cellular respiration and releasing carbon dioxide to the atmosphere.

35.6 The nitrogen cycle

Although nitrogen gas (N_2) makes up 78% of the Earth's atmosphere (**Figure 35.12**), only a minute amount is incorporated into chemical compounds in the soil, oceans, and bodies of organisms. However, this N_2 is an essential part of the proteins within living things. Relatively few kinds of organisms—only a few genera of bacteria—can convert N_2 into a form that can be used for biological processes, playing a crucial role in the cycling of nitrogen. These bacteria are called *nitrogen-fixing bacteria*, and they convert N_2 to ammonia (NH_3). Living things depend on this process of nitrogen fixation. Without it, they would ultimately be unable to continue to synthesize proteins, nucleic acids, and other necessary nitrogen-containing compounds.

Certain of the nitrogen-fixing bacteria are free living in the soil. Others form mutualistic relationships (see Chapter 34) with plants by living within swellings, or nodules, of plant roots. Some of these plants are legumes (LEG-yoomz *or* lih-GYOOMZ)—plants such as soybeans, alfalfa, and clover. Plants having mutualistic associations with nitrogen-fixing bacteria can grow in soils having such low amounts of available nitrogen that they are unsuitable for most other plants. Growth of a leguminous crop can enrich the nitrate level of poor soil enough to benefit the next year's nonleguminous crop. This is the basis for crop rotation in which, for example,

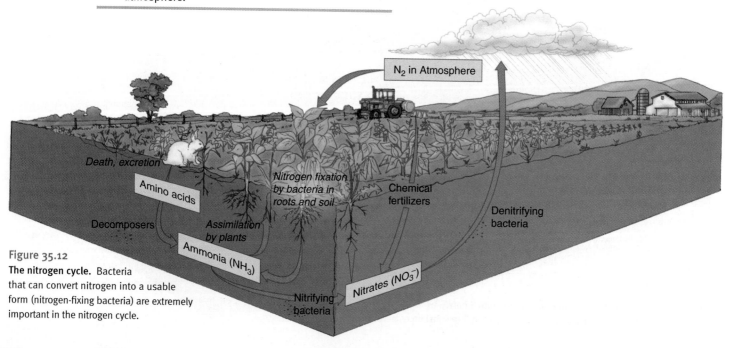

Figure 35.12

The nitrogen cycle. Bacteria that can convert nitrogen into a usable form (nitrogen-fixing bacteria) are extremely important in the nitrogen cycle.

Figure 35.13   The use of nitrogen-fixing bacteria in agriculture. The nitrogen-fixing bacteria *Anabaena azollae* (a) lives in the spaces between the leaves of the floating water fern *Azolla* (b), which is deliberately introduced into the rice paddies of the warmer parts of Asia. By means of this mutualistic relationship, the *Anabaena* thrives and not only fixes nitrogen for the water fern but also enriches the nitrate level of the rice paddy. Rice, here cultivated in Bali (c), is the major food for well over one fourth of the human race.

a field may be planted with soybeans (a legume) and corn (a nonlegume) in alternating years.

The roots of a few other kinds of plants form associations with nitrogen-fixing bacteria of the group actinomycetes. Some of the plants involved are alders and mountain lilac. In addition, nitrogen-fixing bacteria of the genus *Anabaena* contribute large amounts of nitrogen to the rice paddies of China and Southeast Asia (Figure 35.13).

Other bacteria—certain decomposers (Figure 35.1)—play another key role in the nitrogen cycle, producing NH_3 (ammonia) from amino acids that make up the proteins and wastes of dead organisms. Still other bacteria convert NH_3 to nitrates (NO_3^-). NO_3^- is also produced by lightning, which causes N_2 to react with O_2 in the atmosphere. Humans add NO_3^- to the soil by spreading chemical fertilizers. Plants are able to use the nitrogen within molecules of NH_3 and NO_3^- to build their own proteins, nucleic acids, and vitamins. The nitrogen cycle comes full circle as nitrogen is continuously returned to the environment by bacteria that break down NO_3^-, liberating N_2 to the atmosphere.

Although nitrogen gas constitutes about 78% of the Earth's atmosphere, it becomes available to organisms only through the metabolic activities of a few genera of bacteria, some of which are free living and others of which live mutualistically on the roots of legumes and some other plants. This bound nitrogen is released back to the atmosphere by other microorganisms capable of breaking down certain nitrogen compounds.

35.7 The phosphorus cycle

The reservoir of the nutrients phosphorus, potassium, sulfur, magnesium, calcium, sodium, iron, and cobalt is in rocks and minerals rather than in the atmosphere.

Phosphorus, more than any of the other required plant nutrients except nitrogen, is apt to be so scarce that it limits plant growth. In the soil, phosphorus is found as relatively insoluble phosphate (PO_4^{-3}) compounds, present only in certain kinds of rocks (Figure 35.14). For this reason, phosphates exist in the soil only in small amounts. Therefore, humans add millions of tons of phosphates to agricultural lands every year. Plants take up the phosphates from the soil, and animals obtain the phosphorus they need by eating plants or other plant-eating animals. When these organisms die, decomposers release the phosphorus incorporated in their tissues, making it again available for plant use.

Humans unnecessarily alter the phosphorus cycle by adding up to four times as much phosphate as a crop requires each year. Much of this phosphate runs off the land or is eroded into rivers, streams, and the oceans. In addition, phosphates in sewage and waste water from homes and industry find their way into surface water. Rivers and streams become *eutrophic,* or "well fed," caus-

Figure 35.15 **A guano coast.** Marine animals incorporate phosphates from runoff into their bodies. The seabirds that feast on these marine animals deposit feces rich in phosphorus on certain coasts.

ing algae and other aquatic plants to overgrow. As they die, the decomposers feed on them, using up much of the oxygen in the water and choking out other forms of life.

Rivers and streams carry phosphates to the oceans. Some of these phosphates become incorporated into the bodies of fishes and other marine animals. The seabirds

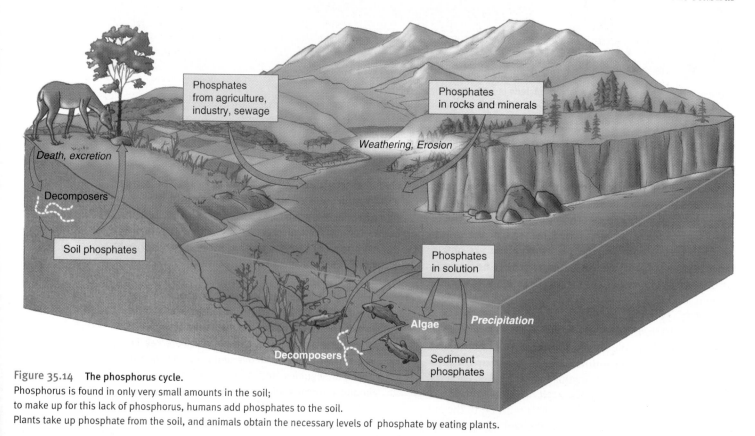

Figure 35.14 **The phosphorus cycle.**
Phosphorus is found in only very small amounts in the soil; to make up for this lack of phosphorus, humans add phosphates to the soil.
Plants take up phosphate from the soil, and animals obtain the necessary levels of phosphate by eating plants.

that eat these animals deposit enormous amounts of guano (feces) rich in phosphorus along certain coasts (Figure 35.15). These deposits have traditionally been used for fertilizer. Phosphates not incorporated into the bodies of animals precipitate out of the water and become part of the bottom sediment. These phosphates become available again only if the sea floor rises up during eras of climatic change, such as along the Pacific coast of North and South America.

Phosphorus is held in the soil in the form of relatively insoluble phosphate compounds, found in most soils in small amounts. Therefore, humans add phosphate compounds to the soil for plants to take up, and animals obtain the phosphates they need by eating plants or plant-eating animals. When organisms die, decomposers release phosphorus back to the soil.

Key Concepts

An ecosystem is a community of organisms and their environment.

35.1 Ecosystems are communities of organisms and the nonliving factors of their environments through which energy flows and within which nutrients cycle.

Nutrients cycle and energy flows through ecosystems.

35.2 Through photosynthesis, plants growing under favorable circumstances capture and lock up some of the sun's energy that falls on their green parts.

35.2 Plants may be eaten by herbivores (primary consumers), which in turn may be eaten by secondary consumers (carnivores).

35.2 Decomposers break down the remains of all organisms.

35.2 This sequence of organisms, one feeding on another, constitutes a food chain.

35.2 Each link in a food chain is a trophic, or feeding, level.

35.3 Each trophic level includes organisms that can transfer about 10% of the energy that exists at each level to the next level.

35.3 From one level to the next, 90% of the energy in a food chain is lost as heat or is used for various metabolic activities.

Elements essential to life are cycled through the atmosphere or the soil.

35.4 Carbon dioxide, nitrogen gas, oxygen gas, and water are the atmospheric reservoirs of the carbon, nitrogen, oxygen, and hydrogen used in biological processes; all of the other elements that organisms incorporate into their bodies come from the Earth's rocks.

35.4 Water in the atmosphere condenses in clouds and falls to the Earth as precipitation.

35.4 Plants take up water from the soil, and animals obtain water from surface water or from the plants or other animals they eat.

35.4 Water returns to the atmosphere through the evaporation of surface water and transpiration by plants.

35.5 Carbon is held in the atmosphere in the form of carbon dioxide.

35.5 Carbon dioxide is taken in by photosynthetic organisms and is used to build the carbon compounds that plants manufacture during the process of photosynthesis.

35.5 Both producers and consumers use the carbon compounds plants manufacture as an energy source, metabolizing them during cellular respiration and releasing carbon dioxide to the atmosphere.

35.6 Atmospheric nitrogen is converted to ammonia by several genera of symbiotic and free-living bacteria.

35.6 The ammonia, in turn, is assimilated into amino groups in proteins of cells or is converted to nitrites and then to nitrates by other bacteria.

35.6 Nitrates are incorporated into the bodies of plants and are converted back into ammonium ions, which are used in the manufacture of many kinds of molecules in the bodies of living organisms.

35.6 The breakdown of nitrogen-containing molecules either converts them to recyclable forms or results in the release of atmospheric nitrogen.

35.7 Phosphorus is a key component of many biological molecules; it weathers out of soils and is transported to the world's oceans, where it tends to be lost.

35.7 Phosphorus is relatively scarce in rocks; this scarcity often limits or excludes the growth of certain kinds of plants.

35.7 Humans add phosphate compounds to the soil for plants to take up, and animals obtain the phosphates they need by eating plants or plant-eating animals.

35.7 When organisms die, decomposers release phosphorus back to the soil.

Key Terms

autotroph *779*
consumers *776*
decomposers
   (DEE-kum-POE-zurs) *776*
detritus (dih-TRITE-us) *778*

ecosystem (EE-ko-SIS-tem
   or EH-ko-SIS-tem) *776*
food chain *779*
food web *779*
heterotroph (HET-uhr-uh-TROFE) *779*

primary consumers *778*
producers *776*
secondary consumers *778*
trophic levels (TROW-fik) *778*

KNOWLEDGE AND COMPREHENSION QUESTIONS

1. What is an ecosystem? Briefly describe three of the living groups it contains with respect to mode of nutrition. (*Knowledge*)

2. Draw a diagram that illustrates the relationships between the biotic and abiotic components of a generalized ecosystem. (*Comprehension*)

3. Fill in the blanks: _____ feed directly on green plants. _____ feed on herbivores. Other organisms known as _____ live on the refuse or waste material of an ecosystem, called _____. (*Knowledge*)

4. Distinguish among trophic level, food chain, and food web. (*Comprehension*)

5. Summarize what happens to the sun's energy as it travels through a food chain. (*Comprehension*)

6. Explain how chemicals are cycled within an ecosystem. (*Comprehension*)

7. Describe how water is cycled within ecosystems. (*Comprehension*)

8. What is groundwater, and how can it become polluted? Explain why this is a serious problem. (*Comprehension*)

9. Summarize the carbon cycle. (*Comprehension*)

10. Why can crop rotation make soil more fertile? To what chemical cycle does this relate? (*Comprehension*)

KEY
Knowledge: Recalling information.
Comprehension: Showing understanding of recalled information.

CRITICAL THINKING REVIEW

1. Imagine you had a hollow glass sphere the size of a basketball and you wanted to create within it a self-sustaining stable ecosystem that needed only sunlight and moderate temperature to persist indefinitely. What would you put in this ecosphere? (*Analysis*)

2. Most terrestrial food chains have only three or rarely four links. Why do you think this is so? Why not forest food chains with 10 links? (*Application*)

3. Extensive cutting and burning of tropical rain forests often result in a drastic and permanent lowering of rainfall in the cleared area (an effect noted by Alexander von Humboldt more than 100 years ago). Why? (*Application*)

4. Many farmers fertilize their crops heavily with phosphates, believing that this will improve

the soil. Explain how this practice can affect an ecosystem. (*Analysis*)

5. Compare the following two nature preserves. One has 200 deer living on eroded land with sparse vegetation and few other animals. The other has only 20 deer but the land has abundant vegetation. A variety of other species of animals inhabit the area. In which ecosystem are the deer more likely to survive? Why? Would you consider one ecosystem more "healthy" than the other? Why? (*Synthesis*)

KEY
Application: Using information in a new situation.
Analysis: Breaking down information into component parts.
Synthesis: Putting together information from different sources.

learning portfolio

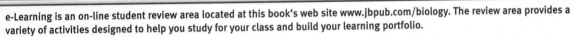

e-learning

Location: http://www.jbpub.com/biology

www.jbpub.com/biology
WWW

e-Learning is an on-line student review area located at this book's web site www.jbpub.com/biology. The review area provides a variety of activities designed to help you study for your class and build your learning portfolio.

Review Questions The review questions test your knowledge of the important concepts and applications in each chapter. The review provides feedback for each correct or incorrect answer. This is an excellent test preparation tool.

Figure-Labeling Exercises Sharpen your visual thinking by matching terms and labels for important illustrations.

Flash Cards Studying biology requires learning new terms. Virtual flash cards help you master the new vocabulary for each chapter.

Just Wondering As you read and study from this text, you may find that you have unanswered questions. Through this site you can ask the author, Sandy Alters, your "just wondering" questions.

Do you prefer the speed and reliability of a CD-ROM? All of the features contained on the eLearning portion of the web site are also available on the Student CD-ROM.

Biomes and Life Zones of the World

The jaguar is an awesomely powerful yet graceful animal. Its home is the tropical rainforests of Central and South America. Equally awe-inspiring is the realization that many organisms, mostly tiny insects and remote plants, are still unidentified and unnamed creatures of this threatened biome.

Providing a home for about half of the planet's 5–10 million plant and animal species, rain forests contain a wealth of agricultural and medicinal resources. Treatments for leukemia, Hodgkin's disease, heart ailments, hypertension, and arthritis have been derived from rain-forest plants. However, fewer than 1% of rain forest species have been tested for their medicinal properties, and scientists estimate that there are at least 30,000 plant species in the rain forest which have yet to be discovered. Unfortunately, some of these plants may be driven to extinction before scientists are able to fully examine them. At least 50 million acres of rain forest are destroyed per year by natives and outsiders alike seeking to take advantage of the wealth provided by this bountiful biome.

A biome is a major terrestrial community characterized by a distinct climate and inhabited by a particular species of plants and animals. You will learn about the rain forests and other biomes on Earth in this chapter.

Organization

Particular biomes occur within specific climatic regions.

Biomes cover wide geographic areas.

Aquatic life zones cover the majority of the Earth's surface.

Particular biomes occur within specific climatic regions.

36.1 The sun and its effects on climate

Biomes are ecosystems of plants and animals that occur over wide areas of land within specific climatic regions and are easily recognized by their overall appearance. Each biome is similar in its structure and appearance wherever it occurs on Earth and differs significantly from other biomes. Biomes are sometimes named by the climax vegetation (stable plant communities) of the region (see Chapter 34), such as the tropical rain forest.

The characteristics of biomes are a direct result of their temperature and rainfall patterns. These patterns result from the interaction of the features of the Earth itself (such as the presence of mountains and valleys) with two physical factors:

1. The amounts of *heat from the sun* that reach different parts of the Earth and the seasonal variations in that heat
2. *Global atmospheric circulation* and the resulting patterns of oceanic circulation

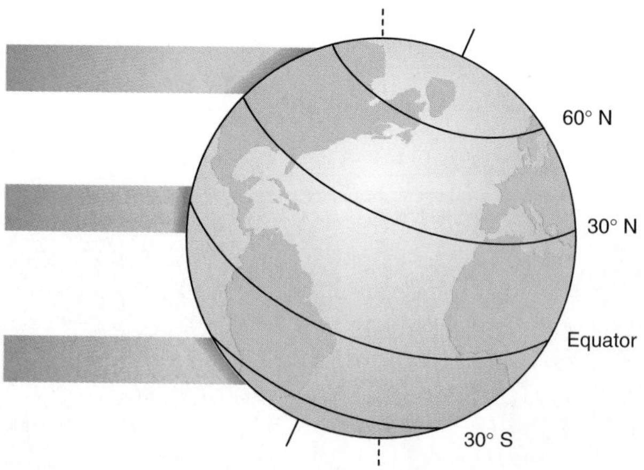

Figure 36.1 The angle of solar energy striking the Earth affects climate.

Together these factors determine the local climate, including the amounts and distribution of precipitation.

Because the Earth is a sphere, some parts receive more energy from the sun than others do. The tropics (near the equator) are warmer than the temperate regions further north or south. As is shown in **Figure 36.1**, the sun's rays arrive almost perpendicular to regions near the equator. They fall on a smaller surface of the Earth than they do farther north or south, so their warmth is concentrated in a smaller area. Therefore, the greater the latitude (distance north or south of the equator, measured in degrees), the colder the climate.

The northern and southern hemispheres also experience a change of seasons as well as the gradations in temperature that vary with latitude. As is shown in **Figure 36.2**, seasons occur because the Earth is tilted on its axis and, as it takes its year-long journey around the sun, the northern and southern hemispheres receive unequal amounts of sunlight at various times. One of the poles is closer to the sun than the other except during the spring and autumn equinoxes.

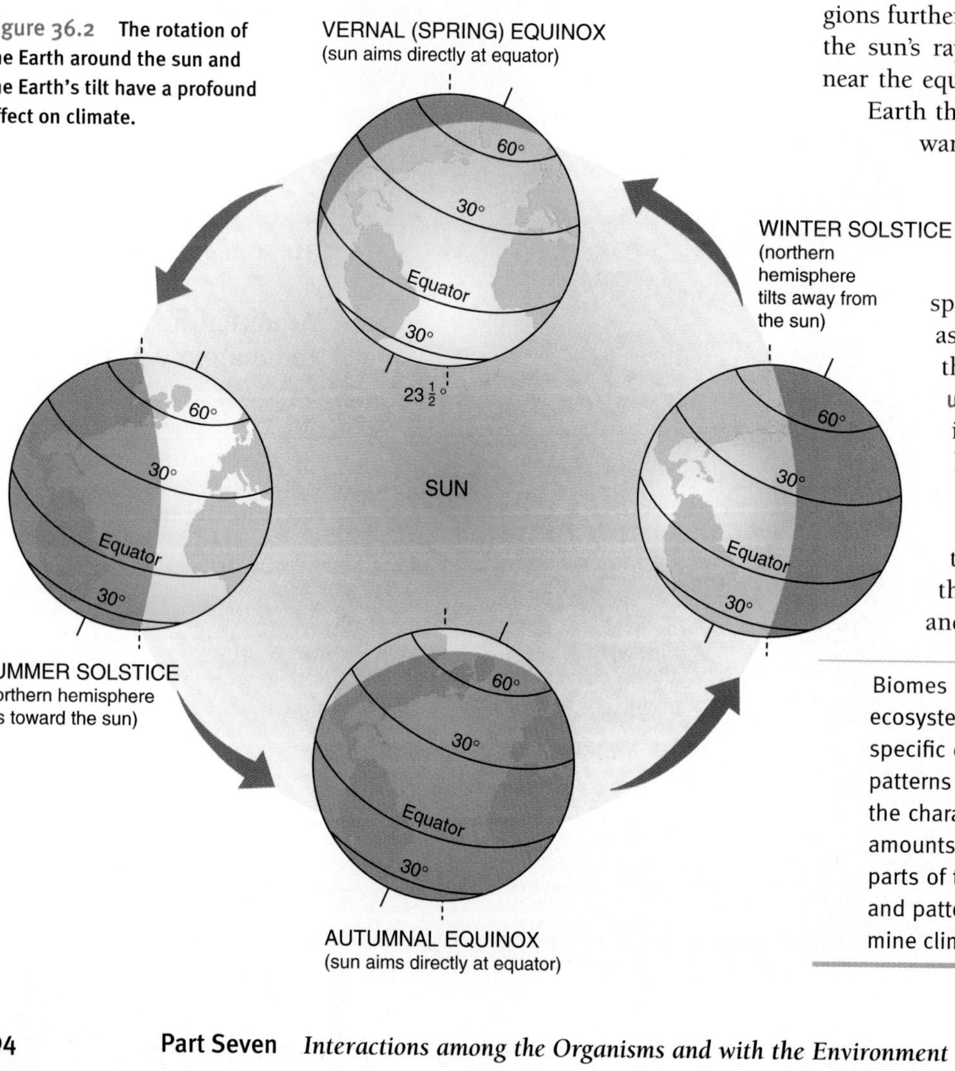

Biomes are the largest recognizable terrestrial ecosystems and occur over wide areas and within specific climatic regions. Temperature and rainfall patterns (climate) are the primary determinants of the characteristics of these huge ecosystems. The amounts of heat from the sun that reach different parts of the Earth, global atmospheric circulation, and patterns of oceanic circulation together determine climate.

Figure 36.2 The rotation of the Earth around the sun and the Earth's tilt have a profound effect on climate.

VERNAL (SPRING) EQUINOX
(sun aims directly at equator)

WINTER SOLSTICE
(northern hemisphere tilts away from the sun)

SUN

SUMMER SOLSTICE
(northern hemisphere tilts toward the sun)

AUTUMNAL EQUINOX
(sun aims directly at equator)

36.2 Atmospheric circulation and its effects on climate

Near the equator, warm air rises and flows toward the poles. If the Earth were a stationary sphere, this air would reach the poles, cool, fall to the ground, and flow back toward the equator. The Earth is not stationary, however. As it spins, its rotary movement breaks the air into six "coils" of rising and falling air that surround the Earth, which are shown in Figure 36.3a. With increasing latitude, each air mass is cooler than the one before, but warmer than the next. The air in each mass rises at the region of its lowest latitude where it is warmest, moves toward the poles, sinks to the ground at the region of its highest latitude where it is coldest, and flows back toward its lowest latitude. As it warms, it rises again, completing the cycle. However, the rotation of the Earth also affects these coils of air. As the Earth rotates, it deflects the winds from the vertical paths illustrated in Figure 36.3a to the patterns shown in Figure 36.3b. Additionally, because the surface of the Earth near the equator moves faster than that at the poles, it causes the coils of air at the equator to blow from east to west, creating the Northern and Southern trade winds. It also causes the coils of air in the temperate zones north and south of the equator to blow from west to east, creating the westerlies.

These moving masses of air also have an effect on precipitation. The moisture-holding capacity of air increases when it is warmed and decreases when it is cooled. Therefore, precipitation is relatively high near 60 degrees north and south latitude (see Figure 36.3a) and at the equator, as shown in Figure 36.4, where it is rising and being cooled. Conversely, precipitation is generally low near 30 degrees north and south latitude, where air is falling and being warmed (see Figure 36.3a).

Partly as a result of these factors, all the great deserts of the world lie near 30 degrees north or 30 degrees south latitude, and some of the great temperate forests are near 60 degrees north or south latitude (see Figure 36.3a). Tropical rain forests lie near the equator. However, other factors come into play. Some major deserts are formed in the interiors of the large continents because these areas have limited precipitation due to their distance from the sea, the ultimate source of most precipitation. Other deserts sometimes occur because mountain ranges intercept the moisture-laden winds from the sea.

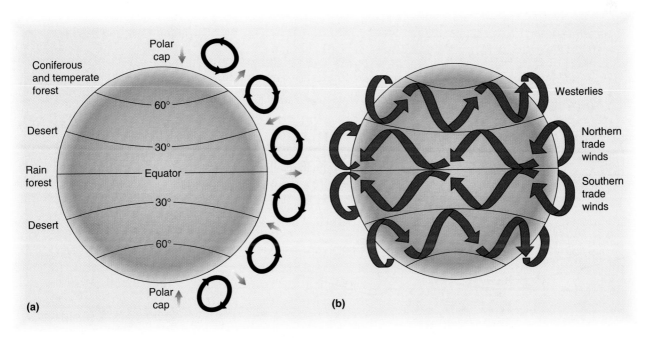

Figure 36.3 **The Earth's atmospheric circulation influences climate.** (a) Air surrounding the rotating Earth is broken into six coils of rising and falling air. (b) The prevailing wind patterns around the Earth are set up by the six moving air masses and the rotation of the Earth.

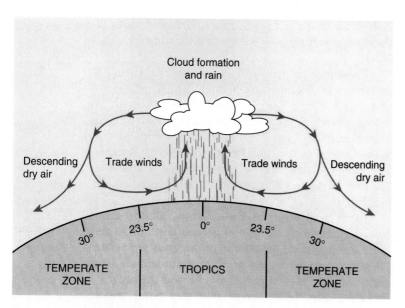

Figure 36.4   **As hot or warm air rises and is cooled, it can hold less moisture.**

As shown in Figure 36.5, as the air travels up a
mountain (its windward side), it is cooled, and precipitation forms. As the air descends the other side of the
mountain (its leeward side), it is warmed; its moisture-
holding capacity increases. For these reasons, the windward sides of mountains are much wetter than the leeward sides, and the vegetation is often very different.
This phenomenon is the rain shadow effect. Seattle,

Washington, for example, lies on the windward side
of the Cascade Mountain range in the northwest-
ern United States. This city receives 99 centimeters
(39 inches) of rainfall per year. Yakima, Washington,
slightly south from Seattle on the leeward side of this
mountain range, receives only 20 centimeters (8 inches)
of rain per year.

The patterns of atmospheric circulation influence
patterns of circulation in the ocean, modified by the location of the land masses around and against which the
ocean currents must flow. Oceanic circulation is dominated by huge surface gyrals (Figure 36.6), which move
around the subtropical zones of high pressure between
approximately 30 degrees north and 30 degrees south

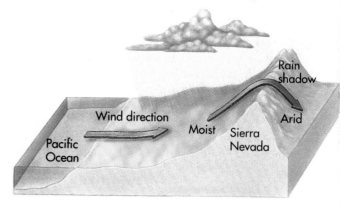

Figure 36.5   **The rain shadow effect.**

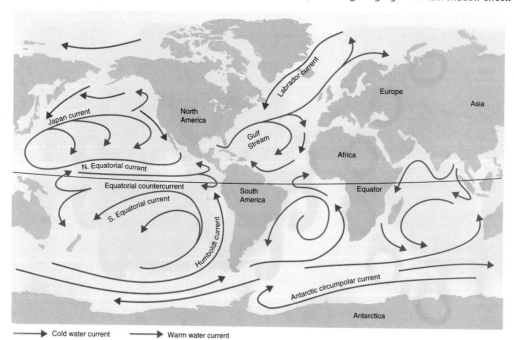

Figure 36.6   **Ocean circulation.** The circulation in the oceans moves in great surface spiral patterns, or gyrals,
and affects the climate on adjacent lands.

Just Wondering . . . *Real Questions Students Ask*

What is the difference between El Niño and La Niña?

El Niño and La Niña are both oceanic phenomena. The story of El Niño starts with the Humbolt current, which sweeps up the west coast of South America from the south (see Figure 36.6). Simultaneously, southern trade winds push the warm, nutrient-poor surface water away from the coast, resulting in an upwelling of the Humbolt water. An upwelling brings up cool, nutrient-rich waters from the depths, replacing the surface water as it is blown away.

Once every 3 to 7 years, the southern trade winds diminish due to changes in atmospheric pressures in certain parts of the world. Therefore, the upwelling of the Humbolt current diminishes, and warm water flows south, down the coast to southern Peru and northern Chile. This warm water flow has been named "El Niño" (meaning "the little boy") by local fisherman because the warm current occurs around the Christmas season. Their reference is to the Christ child. Fishermen are familiar with this event because the warm, nutrient-poor water causes massive fish kills. These events can be devastating to South American fisheries, especially in Chile and Peru.

Conversely, La Niña (the little girl) refers to colder than average sea surface temperatures in the central or eastern equatorial Pacific region. Many scientists call this a *cold event* rather than La Niña. A cold event is a more widespread cooling of the eastern tropical Pacific than is the warming of the area off the cost of South America during El Niño.

El Niño and La Niña affect the weather near and far from the eastern tropical Pacific. Both affect the winds that blow over these waters, changing the weather those winds carry. Rain patterns shift and temperatures shift.

Scientists are still working to understand these unusual weather patterns and cannot predict them with accuracy. Nevertheless, scientists have used the knowledge they have gained to recognize such events in their early stages. Further study may hold the key to understanding the complex interplay of atmospheric and water currents around the globe and both the positive and negative effects of El Niño and La Niña on weather patterns.

www.jbpub.com/biology

---

latitude. These gyrals move clockwise in the northern hemisphere and counterclockwise in the southern hemisphere. They profoundly affect life not only in the oceans but also on coastal lands because they redistribute heat. For example, the Gulf Stream in the North Atlantic swings away from North America near Cape Hatteras, North Carolina, and reaches Europe near the southern British Isles. Because of the Gulf Stream, western Europe is much warmer and thus more temperate than eastern North America at similar latitudes. As a general principle, the western sides of continents in the temperate zones of the northern hemisphere are warmer than their eastern sides; the opposite is true in the southern hemisphere.

The climate of a region is determined primarily by its latitude and wind patterns. These factors, interacting with surface features of the Earth such as mountains and distance from the ocean, result in the particular rainfall patterns. The temperature, rainfall, and altitude of an area, in turn, provide conditions that result in the growth of vegetation characteristic of that area.

36.3 The classification of biomes

Biomes are often classified in seven categories: (1) tropical rain forests, (2) savannas, (3) deserts, (4) temperate grasslands, (5) temperate deciduous forests, (6) taiga, and (7) tundra. (The distribution of these biomes around the world is shown in **Figure 36.7.**) This list is arranged by distance from the equator, but the biomes do not encircle the Earth in neat bands. Their distribution is greatly affected by the climatic effects caused by the presence of mountains, the irregular outlines of the continents, and the temperature of the surrounding sea. In addition, the climate (and vegetation) changes with the elevation of land similar to changes with increasing latitude. Thus, as shown in **Figure 36.8,** tundralike vegetation occurs near the top of a mountain in the tropics, as well as near the North and South Poles.

The distribution of the seven biomes of the world is correlated to climate, which is affected by latitude, the presence of mountains, the temperature of the surrounding sea, and the elevation of the land.

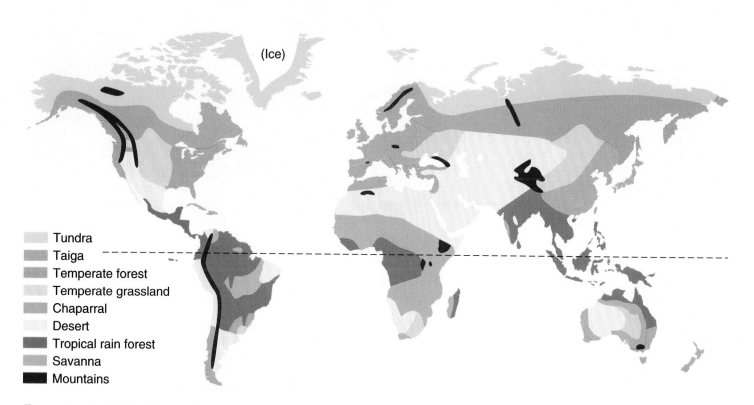

(Ice)

Tundra
Taiga
Temperate forest
Temperate grassland
Chaparral
Desert
Tropical rain forest
Savanna
Mountains

Figure 36.7 **Distribution of biomes.** Many mountain ranges and chaparrals are also shown on this map. Chaparrals are temperate shrublands that border grasslands and deserts in certain parts of the world.

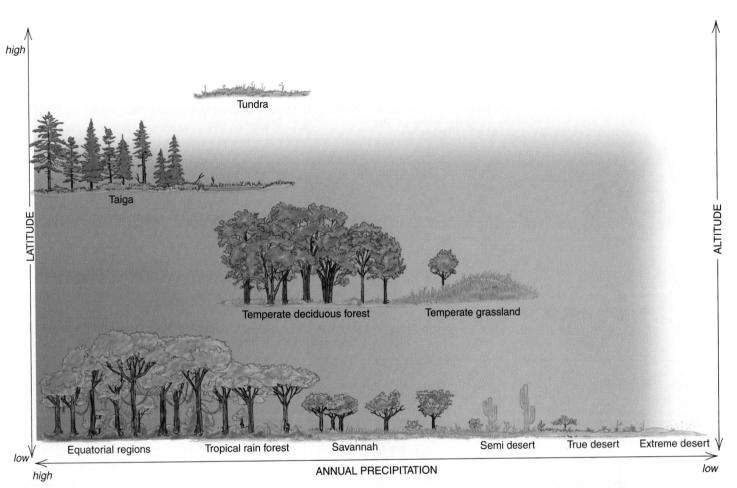

Figure 36.8 **Elevation and biomes.** Biomes that normally occur far north and far south of the equator at sea level occur also in the tropics (and at other latitudes) but at high mountain elevations. Thus on a tall mountain in southern Mexico or Guatemala, you might see a sequence of biomes like the one illustrated here.

36.4 Tropical rain forests

Tropical rain forests occur in Central America; in parts of South America, particularly in and around the Amazon Basin; in Africa, particularly in central and west Africa; and in southeast Asia. As the name suggests, the tropical rain forests occur in regions of high temperature and rainfall, generally 200 to 450 centimeters (80 to 175 inches) per year, with little difference in its distribution from season to season. The temperature averages 25° C (77° F). As a comparison, Houston, Texas—one of the hottest and wettest cities in the United States—receives an average of 121 centimeters (48 inches) of rainfall per year; the average temperature is 20.5° C (69° F). New Orleans, Louisiana—another hot, wet city—receives an average of 145 centimeters (57 inches) of rainfall per year; the average temperature is 20° C (68° F). These cities almost seem dry and cool in comparison to the tropical forest!

You may have the idea that the tropical rain forest (shown in the opener photo) is thick with lush vegetation and creeping vines, creating a network too dense to penetrate without a machete. These forests *are* thick

Figure 36.9 **The colorful diversity of the tropical rainforest.** (a) An orchid from Peru. (b) A chestnut-mandibled toucan from Costa Rica. (c) A flag-legged insect from Costa Rica, with leaflike extensions of its hind legs.

and lush—but not at the forest floor. Little can grow on the ground far beneath the canopy of trees whose branches and leaves form an overlapping roof to the forest. In fact, only 2% of the light shining on the forest canopy reaches its floor! Plants that do grow there have large, dark green leaves adapted to conducting photosynthesis at low light levels. Other types of vegetation have interesting adaptations that enable them to compete successfully with the large trees for sunlight. Vines, for example, have their roots anchored in the soil but they climb up the trees with their long stems. They reach the canopy where leaves grow to capture sunlight. Other interesting plants are the epiphytes, or "air plants." *Epiphytes,* such as those shown in Figure 34.12, grow on the trees or other plants for support, but draw their nourishment mostly from rain water. Some epiphytes catch moisture with modified leaves or flower parts; others have roots that hang free in the air and absorb water (with its dissolved minerals) from the rain.

The giant trees of the tropical forest support a rich and diverse community of animals on their branches. Figure 36.9 provides examples of just a few of these interesting organisms. The roots of the trees are interesting also, spreading out from thickened trunks into a thin layer of soil, often no more than a few centimeters deep. These roots transfer the nutrients from fallen leaves and other organic debris quickly and efficiently back to the trees after bacteria and fungi break down the debris. Very few nutrients remain in the soil. Therefore, when humans cut down and then burn these trees to clear the land for agriculture, they are "burning away" the nutrients held in the trees, as well as breaking down organic matter to carbon dioxide. The small amount of ash that is left provides few nutrients for the crops farmers try to grow. In two to three years of farming, these few remaining nutrients are depleted from the soil, and the land becomes barren.

The tropical rain forest, a biome that occurs in regions of high temperature and rainfall, is characterized by tall trees that support a variety of plant and animal life on their branches.

36.5 Savannas

Not all areas near the equator are wet; some areas experience prolonged dry seasons or have a lower annual rainfall than the tropical forests do (generally, about 90 to 150 centimeters per year, or 35 to 60 inches). The heat, periodic dryness, and poor soils cannot support a forest, but have led to the evolution of **savannas**: open grasslands with scattered shrubs and trees, as shown in Figure 36.10. These areas, situated between the tropical rainforests and deserts, cover much of central and southern Africa, western India, northern Australia, large areas of northern and east-central South America, and some of Malaysia.

The vegetation of the savanna supports large grazing herbivores such as buffalo, wildebeests, and zebra. These animals, in turn, are food for carnivores such as lions. The savanna also supports a large number of plant-eating invertebrates, such as mites, grasshoppers, ants, beetles, and termites. The termites, in fact, are one of the most important soil organisms in the savanna.

Figure 36.10 **Savannas: dry, tropical grasslands.** Marabou storks, elephants, and zebras visit the water on an African savanna in Tsavo Park, Kenya.

The termites of the savanna break down dried twigs, leaves, and grass to usable nutrients, thereby improving the nutrient status of the soil in and around their huge, complex mounds. Additionally, these mounds (Figure 36.11) provide passageways for rainwater to deeply penetrate the ground rather than just running off or evaporating from the surface. The mounds are built of termite excrement or of soil held together with saliva, and they have air cooling holes through which rain may pass. The mound is connected to underground chambers and passageways, and water can trickle through the mound and into the ground below.

Like tropical forests, savannas are found near the equator but in areas having less annual rainfall. This climate supports grasslands with only scattered trees and shrubs.

Figure 36.11   Termite mound in Australia.

36.6 Deserts

Deserts are biomes that have 25 centimeters (10 inches) or less of precipitation annually. For this reason, the vegetation in deserts is characteristically sparse. The higher the annual rainfall a desert has, however, the greater the amount of vegetation it will be able to support.

Ecologists classify deserts based on their annual rainfall: Semideserts receive about 25 centimeters (10 inches) per year (Phoenix, Arizona, and San Diego, California, for example); true deserts receive less than 12 centimeters (4.7 inches) per year (Las Vegas, Nevada); and extreme deserts average below 7 centimeters (2.8 inches) per year (Namib Desert in southwestern Africa). The photos of these three types of deserts in **Figure 36.12** show the differences in their patterns of vegetation.

Major deserts occur around 20 to 30 degrees north and south latitude, where the warm air that rose from the equator falls. As previously mentioned, the air at the equator rises, cooling and releasing its moisture, which falls on the tropical forests. The dry air then falls over desert regions, resulting in little precipitation. Deserts also occur in the interiors of continents far from the moist sea air, especially in Africa (the Sahara Desert), Eurasia (Europe and Asia), and Australia. Some deserts, such as the Baja region of California, are near the ocean yet are dry; the winds blow from the north, carrying little moisture because they are cool. High pressure areas off the West Coast of the United States also deflect storms moving down from the north. In addition, some deserts form on the leeward side of mountain ranges, such as in the Great Basin of Nevada and Utah in the United States.

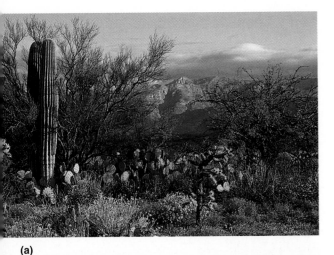

(a)

(b)

(c)

Figure 36.12 **Deserts.** (a) The Sonoran Desert in Arizona, a semidesert. (b) Death Valley, California, a true desert. (c) Namib Desert, Namibia (western Africa), an extreme desert.

Because desert vegetation is sparse and the skies are usually clear, deserts radiate heat rapidly at night. This situation results in substantial daily changes in temperature, sometimes more than 30 degrees Centigrade (approximately 55 degrees Fahrenheit) between day and night. Although both hot deserts (the Sahara, for example) and cool deserts (the Great Basin of North America) exist, summer daytime temperatures in all deserts are extremely high, frequently exceeding 40° C (104° F). In fact, temperatures of 58° C (136.4° F) have been recorded both in Libya and in San Luis Potosi, Mexico—the highest that have been recorded on Earth.

Plants have developed a wide variety of adaptations in this difficult environment. Annual plants are often abundant in deserts and simply bypass the unfavorable dry season in the form of seeds. After sufficient rainfall, many germinate and grow rapidly, sometimes forming spectacular natural displays. Characteristic of deserts, of course, are the many species of succulent plants, those with tissues adapted to store water, such as cacti (see Figure 36.12a). The trees and shrubs that live in deserts often have deep roots that reach sources of water far below the surface of the ground. The woody plants that grow in deserts may be either deciduous, losing their leaves during the hot, dry seasons of the year, or evergreen, with hard, reduced leaves. The creosote bush of the deserts of North and South America is an example of an evergreen desert shrub. Near the coasts in areas where there are cold waters offshore, deserts may be foggy, and the water that the plants obtain from the fog may allow them to grow quite luxuriantly.

Desert animals, too, have fascinating adaptations that enable them to cope with the limited water of the deserts. Many limit their activity to a relatively short period of the year when water is available or even plentiful. They resemble annual plants in this respect. Many desert vertebrates live in deep, cool, and sometimes moist burrows. Organisms that are active for much of the year emerge from their burrows only at night, when temperatures are relatively cool. Other organisms, such as camels, can drink large quantities of water when it is available and store it for use when water is unavailable. A few animals simply migrate to or through the desert and exploit food that may be abundant seasonally. When the food disappears, the animals move on to more favorable areas.

Deserts occur around 20 to 30 degrees north and south latitude and in other areas that have 25 centimeters (10 inches) or less of precipitation annually. Desert life is somewhat sparse but exhibits fascinating adaptations to life in a dry environment.

36.7 Temperate grasslands

Temperate grasslands have various names in different parts of the world: the prairies of North America, steppes (STEPS) of Russia, pusztas (PUZ-taz) of Hungary, veld (VELT) of South Africa, and pampas (PAM-pahz) of South America. All temperate grasslands have 25 to 75 centimeters (10 to 30 inches) of rainfall annually, much less than that of savannas but more than that of deserts. Temperate grasslands also occur at higher latitudes than savannas but are often found bordering deserts, as savannas do.

Temperate grasslands are characterized by large quantities of perennial grasses; the rainfall is insufficient to support forests or shrublands. Grasslands are often populated by burrowing rodents, such as prairie dogs and other small mammals, and herds of grazing mammals, such as the North American bison (Figure 36.13). The grazing of herbivores contributes to the maintenance of this biome by preventing woody vegetation from becoming established.

The soil in temperate grasslands is rich; in fact, much of the temperate grasslands are farmed for this reason. Grasslands are often highly productive when they are converted to agriculture, and many of the rich agricultural lands in the United States and southern Canada were originally occupied by prairies.

Temperate grasslands experience a greater amount of rainfall than deserts but a lesser amount than savannas. They occur at higher latitudes than savannas but, like savannas, are characterized by perennial grasses and herds of grazing mammals.

Figure 36.13 **A herd of bison on temperate grasslands.**

36.8 Temperate deciduous forests

The climate in areas of the northern hemisphere such as the eastern United States and Canada and an extensive region in Eurasia supports the growth of trees that lose their

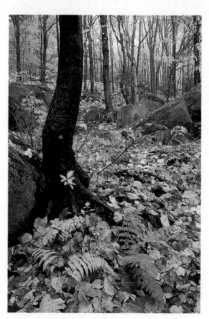

leaves during the autumn. Such trees are called *deciduous* (dih-SIJ-you-us) from a Latin word meaning "to fall," dropping their leaves and remaining dormant throughout the winter. These vast areas of trees are therefore **temperate deciduous forests** (Figure 36.14) and thrive in climates where summers are warm, winters are cold, and the precipitation is moderate, generally from 75 to 150 centimeters (30 to 60 inches) annually. Precipitation is well distributed throughout the year, but water is generally unavailable during the winter because it is frozen.

The temperate forest differs from the tropical forest in that more vegetation grows near the forest floor than in the tropical forest. Although the temperate forest does have an upper canopy of dominant trees such as beech, oak, birch, and maple, there is a lower tree canopy and a layer of shrubs beneath. On the ground, herbs, ferns, and mosses abound (see Figure 36.14). In addition, animal life in the temperate forest is abundant on the ground as well as in the trees; in the tropical forest, animal life is primarily arboreal (in the trees), with the

Figure 36.14 **Temperate deciduous forests.** This photograph was taken in the fall at Baxter State Park, Maine.

exception of large mammals, a few bird species, and soil invertebrates.

In areas having less than 75 centimeters (30 inches) of precipitation annually, temperate deciduous forests are replaced by grassland, as in the prairies of North America and the steppes of Eurasia. Where conditions are more limiting—restricted, for example, by intense cold—these forests may be replaced by coniferous forests.

> Temperate deciduous forests occur in areas having warm summers, cold winters, and moderate amounts of precipitation. The trees of this forest lose their leaves and remain dormant throughout the winter.

36.9 Taiga

The northern coniferous forest is called **taiga** (TIE-gah). The cone-bearing trees of this forest are primarily spruce, hemlock, and fir and extend across vast areas of Eurasia and North America (Figure 36.15). The taiga is characterized by long, cold winters with little precipitation. Most of the precipitation falls in the summers. Because of the latitude where taiga occurs, the days are short in winter (as little as 6 hours of daylight) and correspondingly long in summer (as little as 6 hours of darkness). The light, warmth, and rainfall of the summer allows plants to grow rapidly, and crops often attain a large size in a surprisingly short time.

The trees of the taiga occur in dense stands of one or a few species of cone-bearing trees. Alders, a common species, harbor nitrogen-fixing bacteria in nodules on their roots; for this reason, they are able to colonize the infertile soils of the taiga. Marshes, lakes, and ponds also characterize the taiga; they are often fringed by willows or birches. Many large mammals can also be found there, such as the caribou (Figure 36.16). Other herbivores, including moose, elk, and deer, are stalked by carnivores such as wolves, bear, lynx, and wolverines.

To the south, taiga grades into temperate forests or grasslands, depending on the amount of precipitation. Coniferous forests also occur in the mountains to the south, but these are often richer and more diverse in species than those of the taiga. Northward, the taiga gives way to open tundra.

> The taiga, or northern coniferous forest, consists of evergreen, cone-bearing trees. The climate of this biome is characterized by long, cold winters with little precipitation.

Figure 36.15 **Taiga: great evergreen forests of the north.** This photograph was taken in Alaska.

Figure 36.16 **Caribou in the taiga in winter.**

36.10 Tundra

Farthest north in Eurasia, North America, and their associated islands—between the taiga and the permanent ice—is the open, often boggy community known as the **tundra** (Figure 36.17). Dotted with lakes and streams, this enormous biome encircles the top of the world, covering one fifth of the Earth's land surface. (A well-developed tundra does not occur in the Antarctic because there is no land at the right latitude.) The tundra is amazingly uniform in appearance, dominated by scattered patches of grasses and sedges (grasslike plants), heathers, and lichens. Some small trees do grow, but they are primarily confined to the margins of streams and lakes.

Annual precipitation in the tundra is very low, similar to desertlike precipitation of less than 25 centimeters (10 inches) annually. In addition, the precipitation that falls remains unavailable to plants for most of the year because it freezes. During the brief Arctic summers, some of the ice melts. The permafrost, or permanent ice found about a meter down from the surface, never melts, however, and is impenetrable to both water and roots. When the surface ice melts in the summer, it has nowhere to go and forms puddles on the land. In contrast, the alpine tundra found at high elevations in temperate or tropical regions does not have this layer of permafrost.

The tundra teems with life during its short summers. As in the taiga, perennial herbs grow rapidly then, along with various grasses and sedges. Large grazing mammals, including musk oxen, caribou, and reindeer, migrate from the taiga. Many species of birds and waterfowl nest in the tundra in the summer and then return to warmer climates for the winter. Populations of lemmings, small rodents that breed throughout the year beneath the snow, rise rapidly and then crash on a 3- to 4-year cycle, influencing the populations of the carnivores that prey on them, such as snowy owls and arctic foxes.

Figure 36.17 **Tundra.** Mount McKinley National Park in Alaska.

The tundra encircles the top of the world. This biome is characterized by desertlike levels of precipitation, extremely long and cold winters, and short, warmer summers.

36.11 Freshwater life zones

Only 2% of the Earth is covered by fresh water, found standing in lakes and ponds or moving in rivers and streams. Freshwater ecosystems lie near and are intertwined with terrestrial ecosystems. For example, some organisms such as amphibians may move from one ecosystem to another. In addition, organic and inorganic material continuously enters bodies of fresh water from terrestrial communities. Often, the wet, spongy land of marshes and swamps provides habitats intermediate between the two.

Ponds and lakes have three life zones, or regions, in which organisms live: the shore zone, the open-water zone, and the deep zone, which are shown in **Figure 36.18**. The **shore zone** is the shallow water near edges of a lake or pond in which plants with roots, such as cattails and water lilies, may grow. Consumers such as frogs, snails, dragonflies, and tiny shrimplike organisms live among these producers. The **open-water zone** is the main body of water through which light penetrates. Floating and drifting algae and plantlike organisms, or phytoplankton, grow here. Floating "animals" such as protozoans (zooplankton) feed on the phytoplankton in this aquatic food chain. They, in turn, are eaten by small fish, which are eaten by larger fish. The **deep zone**, the water into which light does not penetrate, is devoid of producers. This dark zone is inhabited mainly by decomposers and other organisms such as clams that feed on the organic material that filters down to them. Ponds differ from lakes in that they are smaller and shallower. Therefore, light usually reaches to the bottom of all levels of a pond; it has no deep zone.

Rivers and streams differ from ponds and lakes primarily in that their water flows rather than remains stationary. The nature of this ecosystem is therefore different from that of a pond or lake. One difference is that the level of dissolved oxygen is usually much higher in a river or stream than in a standing body of water because moving water mixes with the air as it churns and bubbles along. A high level of dissolved oxygen allows an abundance of fish and invertebrates to survive. In addition, only a few types of producers inhabit rivers and streams: various species of algae that grow on rocks and a few types of rooted plants such as water moss.

A river or stream is characterized as an open ecosystem. That is, it derives most of its organic material from sources other than itself. Detritus (debris or

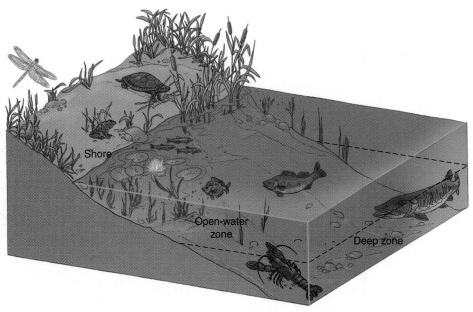

Figure 36.18 The three life zones of a lake.

decomposing material) flows from upstream or enters from the land. Leaves and woody material drop into the stream from vegetation bordering its banks. Rain water washes organic material from overhanging leaves. In addition, water seeps into a river or stream from below the surface of adjoining land, carrying with it organic materials and, in some cases, fertilizers and other chemicals. These nutrients feed the producers and small consumers. As commonly occurs in pond and lake food chains, large fish feed on smaller fish that feed on tiny invertebrates. The river/stream ecosystem is largely heterotrophic and is strongly tied to terrestrial ecosystems that surround it.

As the water from a stream or river flows into a lake or the sea, the velocity of the water decreases. As the water slows, sediment carried along by faster water now sinks to the bottom. These deposits form fan-shaped areas called **deltas**. Accumulated sediment breaks the path of the water into many small channels that course through the delta. Occasionally, the path of the water is actually stopped as sediment accumulates. For this reason, a delta created by a large river, such as the Mississippi River delta, consists of a great deal of swampy, marshy land.

Freshwater ecosystems lie near and are intertwined with terrestrial ecosystems. From them, organic and inorganic material continuously enters freshwater ecosystems.

36.12 Estuaries: Life between rivers and oceans

As rivers and streams flow into the sea, an environment called an **estuary** (ESS-choo-ER-ee) is created where fresh water joins salt water. In the shallow water of the estuary, rooted grasses often grow. Other producers of the estuary are various types of algae and phytoplankton. Consumers are primarily molluscs, crustaceans, fish, and various zooplankton. All organisms inhabiting estuaries, however, have adaptations that allow them to survive in an area of moving water and changing salinity. (*Salinity* refers to the concentration of dissolved salts in the water.) For example, many estuarine organisms are bottom dwellers and attach themselves to bottom material or burrow in the mud. Each species of organism is found living in a region of salinity optimal for its survival.

Oysters are one of the most important bottom dwellers in estuaries, providing a habitat for many other species of organisms. These molluscs either bury themselves in the mud, forming oyster beds, or cement themselves in clusters to the partially buried shells of dead oysters, forming oyster reefs. Many other invertebrates, such as sponges and barnacles, attach themselves to the oysters and feed on plankton. Other species, such as crabs, snails, and worms, live on, beneath, and between the oysters, feeding on the oysters themselves or on detritus (dead, decaying material) trapped in the oyster reef. One researcher, in fact, has documented more then 300 species of organisms living in association with a single oyster reef!

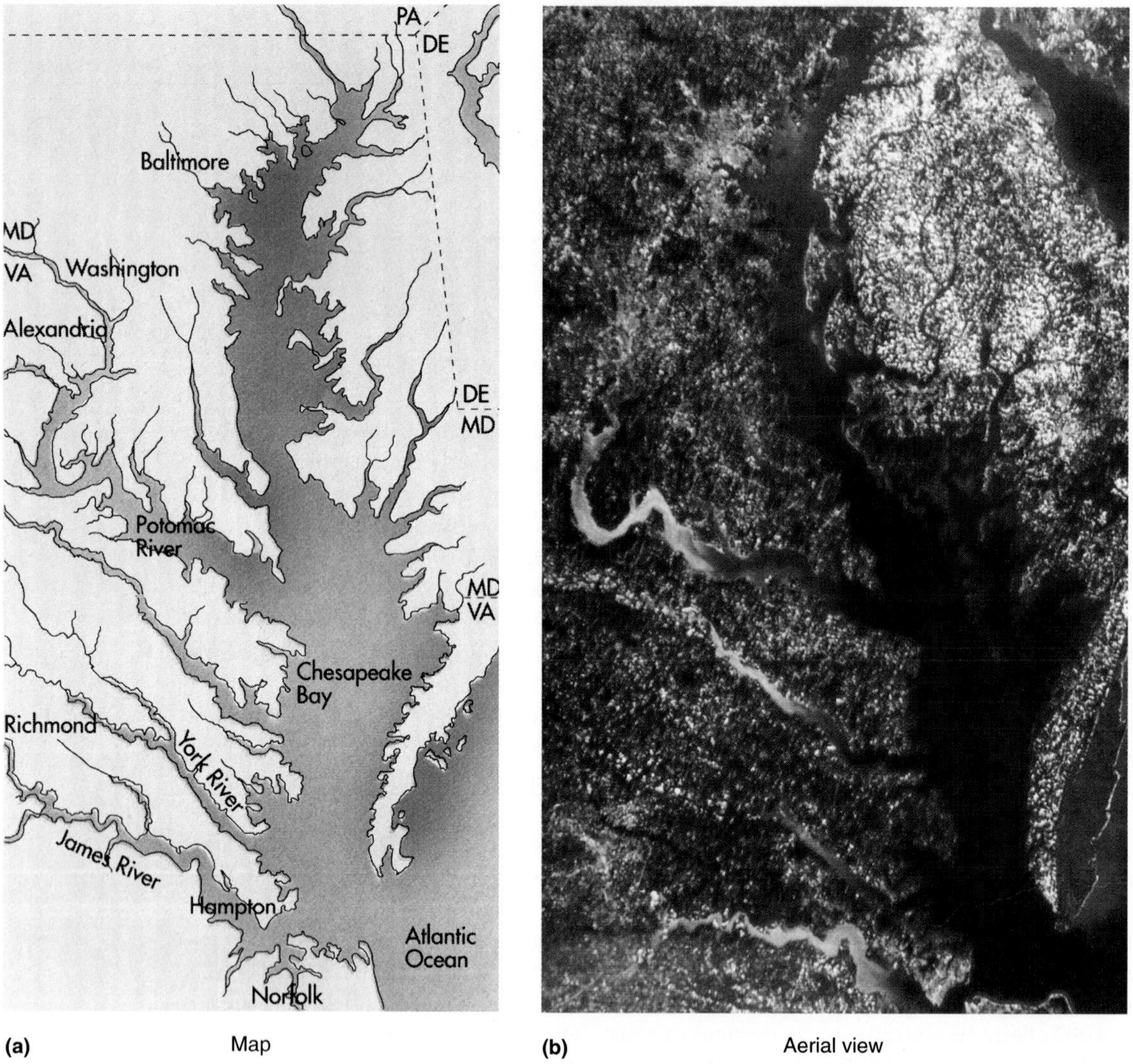

(a)   Map

(b)   Aerial view

Figure 36.19 **Chesapeake Bay.** Chesapeake Bay has more than 11,300 kilometers (7020 miles) of shoreline and drains more than 166,000 square kilometers (64,100 square miles) in one of the most densely populated and heavily industrialized areas in North America. The body of open water is about 320 kilometers (199 miles) long and, at some points, nearly 50 kilometers (31 miles) wide. (a) The large metropolitan areas (noted on the map) and shipping facilities make the bay one of the busiest natural harbors anywhere. (b) Uncontrolled erosion from certain agricultural practices, pesticides, and increases in nutrients block the light needed for photosynthesis and upset the delicate ecological balance on which the productivity of the bay depends. This material can be seen in some of the rivers that empty into the bay in this aerial view. The states that border the bay are cooperating, with the assistance of the Environmental Protection Agency, to try to bring the bay back to its former productivity.

The motile organisms of the estuary are primarily crustaceans such as crabs, lobsters, and shrimp and various species of fish. Fish exhibit interesting reproductive adaptations to the varying salinities of the estuary. Some species, such as the striped bass, spawn upstream from the estuary where the salinity of the water is low. The larvae and young fish move downstream through increasing concentrations of salt as they develop, moving into the ocean in adulthood. In a similar manner, shad spawn upstream in fresh water, and the young spend their first summer in the estuary before swimming to the sea.

Nutrients are more abundant in estuaries than in the open ocean because estuaries are close to terrestrial ecosystems and derive much of their nutrients from them as rivers and streams do. Unfortunately, estuaries are also easily polluted from these sources. In Chesapeake Bay (Figure 36.19a), for example, complex systems of rivers enter the Atlantic Ocean, forming one of the most biologically productive bodies of water in the world. In the 1960s, the bay yielded an annual average of about 275,000 kilograms (600,000 pounds) of fish. As the human population in this area increased, however, along with oil transport and commercial shipping, pollution also increased. More than 290 oil spills were reported in the bay in 1983 alone. In addition, uncontrolled erosion from certain agricultural practices increased the level of nutrients (primarily nitrogen and phosphorus) in the water. These nutrients resulted in the growth of algae, which clouds the water and blocks the sunlight needed by the Bay grasses. (This material can be seen in some of the rivers that empty into the bay in the aerial view in Figure 36.19b). The grasses then die, sink, and decompose, a process that uses the dissolved oxygen in the water. Organisms living in the water can no longer survive when oxygen is depleted, so they either leave the area or they die. This situation caused a 90% decrease in the yield of fish from the Chesapeake Bay area in the 1980s.

In the late 1990s, the primary problem of the Chesapeake Bay area continued to be the abundance of nutrients in the water. However, due to the implementation of agricultural "best management" practices, these levels steadily declined during the 1990s. Maryland and Virginia, the states that border the bay, along with the Environmental Protection Agency, are working to bring the Chesapeake Bay estuary back to its former productivity.

Estuaries are places where the fresh water of rivers and streams meets the salt water of oceans. These ecosystems are made up of plentiful communities of organisms exhibiting behaviors and growth characteristics adapted to changing salinity.

36.13 Ocean life zones

Although only 2% of the Earth is covered by fresh water, nearly three quarters of the Earth's surface is covered by ocean. These seas have an average depth of 4 kilometers (approximately 2 1/2 miles), and they are, for the most part, cold and dark. The concentration of oxygen, as well as the availability of light and food, is a factor that limits life in the ocean. Although cold water is able to "hold" more oxygen than warm water, the warmer sea water near the surface of the ocean mixes with the oxygen in the atmosphere. Therefore, oxygen is present in its highest concentrations in the upper 200 meters (650 feet) or so of the sea. Light is most abundant in the top 100 meters (325 feet).

The marine environment provides a variety of habitats, but it can be divided into three major life zones:

1. The *intertidal zone,* the area between the highest tides and the lowest tides
2. The *neritic zone* (nih-RIT-ik), the area of shallow waters along the coasts of the continents, which extends from the low tide mark to waters down to 200 meters deep
3. The *open-sea zone,* comprising the remainder of the ocean

The Intertidal Zone

The wind-swept shoreline is a harsh place for organisms to live. As the tide rolls in and out, environmental conditions change from hour to hour: wet to dry, sun protected to sun parched, and wave battered to calm. Nevertheless, life abounds in the intertidal zone, exhibiting interesting adaptations and characteristics necessary for survival.

Figure 36.20 shows the rocky shore at low tide at the Pacific Grove Monterey Bay in California. The exposed rocks are teeming with life, but these organisms vary along a continuum from the driest areas (those least often covered with water) to the wettest areas (those most often covered with water). Highest up on the rocks, in areas that the high tide sometimes does not reach, grow certain lichens and algae. Somewhat lower on the rocks grow barnacles. Then oysters, blue mussels, and limpets (molluscs that have conical shells) take over, followed by brown algae and red algae. "Forests" of large brown algae, or kelp, take over in areas that are exposed for only short periods of time. All of these organisms have adaptations such as hard shells or gelatinous coverings that keep them from drying out and are either anchored within the sand or stick to the surfaces of rocks so that they will not be washed away.

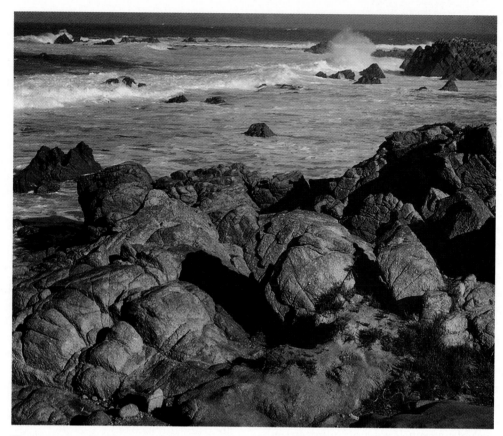

Figure 36.20   A rocky shore in Pacific Grove in Monterey Bay, California.

In contrast to the rocky shore, the sandy shore (Figure 36.21) may look as though no life is present. However, life is plentiful beneath the sand and mud. Because organisms have no large surfaces to which they can attach, they are adapted to burrowing under the sand during low tide. Copepods (KO-peh-pods), tiny "micro" crustaceans, are predominant organisms. In addition, worms, crabs, and molluscs such as clams burrow to safety when the tides roll out. In areas near the low tide mark, sea anemones, sea urchins, and sea stars make their home.

The intertidal zone has plentiful light and is home to a variety of producers. Along with the algae, phytoplankton float in the water and are used for food by the zooplankton and many other consumers. In addition, the heterotrophs of the intertidal zone have the waves to thank for bringing fresh organic material to them and for washing away their wastes.

The Neritic Zone

Surrounding the continents of the world is a shelf of land that extends out from the intertidal zone usually 50 to 100 kilometers (30 to 60 miles), sloping to a depth

Figure 36.21   Sandy beach and sand dune at Cape Cod National Seashore, Massachusetts.

of about 200 meters (approximately 650 feet) beneath the sea. This margin of land is called the **continental shelf**. The waters lying above it make up the neritic zone, which is derived from a Greek root referring to the sea. Because light reaches the waters of most of this zone, it supports an abundant array of plant and therefore animal life.

One outstanding community in the neritic zone is the **coral reef**. (A photgraph of a coral reef is shown in the opener for Chapter 34, p. 758.) The term reef refers to a mass of rocks in the ocean lying at or near the surface of the water. Coral reefs are built by marine animals, or corals (phylum Cnidaria), that secrete calcium carbonate, a hard, shell-like substance. With the help of algae that reside in their bodies, the coral build on already shallow portions of the continental shelf or on submerged volcanoes in the ocean. Complex and fascinating ecosystems, coral reefs provide habitats for a variety of invertebrates and fishes. Along with the tropical rain forests, coral reefs are the most highly productive ecosystems in terms of biomass (see Chapter 35).

Figure 36.22 Zones of the ocean.

The Open-Sea Zone

Beyond the continental shelf lies the great expanse of the open ocean. This open-sea zone is often referred to as the **pelagic zone**, a term derived from another Greek word meaning "ocean" **(Figure 36.22)**. Within this huge ecosystem exist many diverse forms of life, some with which you are familiar, such as the floating plankton, squid, and various species of fishes **(Figure 36.23a)**. But other forms of life are unfamiliar—even bizarre—such as the flashlight fish shown in Figure 36.23b.

Organisms live in the vast expanse of the ocean in relationship to available light and food. Temperature, salinity, and water pressure also play roles in creating the various habitats of the ocean. Light is available to organisms from the water's surface only to an approximate depth of 200 meters (650 feet). In fact, this area of the open ocean is called the **photic zone** for this reason. (Photic means "light.") Phytoplankton (small plantlike organisms that float or drift in the water) thrive in this well-lighted layer of the ocean (especially within the better-lighted upper 100 meters [325 feet]), drifting freely with the ocean currents and serving as the base of oceanic food webs. Zooplankton (small animallike organisms that float or drift in the water) float with the

(a)

(b)

Figure 36.23 **The open ocean.** (a) A squid. (b) A flashlight fish with an illuminating organ under the eye.

phytoplankton and are first-order consumers in many photic food webs. Other typical heterotrophs of this zone are most air-breathing mammals such as whales, porpoises, dolphins, seals, and sea lions and fishes such as herring, tuna, and sharks. These fishes and mammals of the sea, which are called *nekton* (NEK-ton), feed on the plankton and on one another. Together, the organisms that make up the plankton and the nekton provide all of the food for those that live below.

Little light penetrates the ocean from 200 to 1000 meters (650 to 3250 feet). Therefore, no photosynthetic organisms live in this zone known as the **mesopelagic zone** (MEZ-oh-puh-LAJ-ik), or middle ocean. At this depth, temperatures remain somewhat steady throughout the year but are cooler than the water above; water pressure increases steadily with depth. Under these conditions, many bizarre organisms have evolved, such as those that exhibit bioluminescence (Figure 36.23a), which they use to communicate with one another or to attract prey. The organisms common to middle-ocean life are fish with descriptive names such as swordfish, lanternfish, and hatchetfish; certain sharks and whales; and cephalopods such as octopi and squid.

Peculiar creatures also live in the ocean depths of 1000 or more meters—a mile or so (and more) beneath the surface. This region of the ocean is called the **abyssal zone** (uh-BISS-uhl), meaning "bottomless," because it is so deep that it seems to be bottomless. The water at this tremendous depth contains high concentrations of salt, is under immense pressure from the water above, and is very cold. The organisms living at this level cannot make trips into the photic zone to capture food as some mesopelagic organisms do, but must feed on material that settles from above.

Organisms collectively referred to as *benthos* (BEN-thos), or bottom dwellers, live on the ocean floor. Sea cucumbers and sea urchins crawl around eating detritus. Various species of clams and worms burrow in the mud, feeding on a similar array of decaying organic material. Bacteria are also rather common in the deeper layers of the sea, playing important roles as decomposers as they do on land and in freshwater habitats. Such organisms (sea cucumbers, sea urchins, clams, worms, and bacteria) are found in other ocean zones as well; however, different species inhabit different ocean zones.

Of the three main life zones in the ocean, the intertidal zone, the neritic zone, and the photic area of the open-sea zone support the most life because of the presence of light and the availability of oxygen. Organisms that live in waters beneath the photic zone must feed on organic material and detritus that falls from above and must be adapted to increased water pressure and salinity.

Key Concepts

Particlular biomes occur within specific climatic regions.

36.1 Biomes are the largest recognizable terrestrial ecosystems, which occur over wide areas and within specific climatic regions.

36.1 The characteristics of biomes are a direct result of temperature and rainfall patterns.

36.2 The climate of a region is determined primarily by its latitude and wind patterns.

36.2 These factors interact with the surface features of the Earth, resulting in particular rainfall patterns.

36.2 The temperature, rainfall, and altitude of an area provide conditions that result in the growth of vegetation characteristic of that area.

Biomes cover wide geographic areas.

36.3 Seven categories of biomes, arranged by distance from the equator, are (1) tropical rain forests,

(2) savannas, (3) deserts, (4) temperate grasslands, (5) temperate deciduous forests, (6) taiga, and (7) tundra.

36.4 Tropical rain forests occur near the equator, receive an enormous amount of rain year round, and are characterized by the growth of tall trees and lush vegetation.

36.5 Savannas also lie near the equator but experience less rain than tropical rain forests and sometimes have prolonged dry spells.

36.5 Savannas are characterized by open grasslands with scattered trees and shrubs.

36.6 Deserts are extremely dry biomes.

36.6 Hot deserts are hot year round, whereas cool deserts are hot only in the summer.

36.6 Deserts are of great biological interest due to the extreme behavioral, morphological, and physiological adaptations of the plants and animals that live there.

Key Concepts, continued

36.7 Temperate grasslands receive less rainfall than savannas but more than deserts.

36.7 The soil in temperate grasslands is rich, so they are well suited to agriculture.

36.8 Temperate deciduous forests receive moderate precipitation that is well distributed throughout the year.

36.8 The climate of temperate deciduous forests differs from tropical forests in that they receive less rainfall, are found at a higher and cooler latitude, and experience cold winters.

36.8 The trees in temperate deciduous forests lose their leaves and remain dormant throughout the winter.

36.9 The taiga is the coniferous forest of the north.

36.9 The taiga consists primarily of cone-bearing evergreen trees, which are able to survive long, cold winters and low levels of precipitation.

36.10 Even farther north than the taiga is the tundra, which covers about 20% of the Earth's land surface and consists largely of open grassland, often boggy in summer, which lies over a layer of permafrost.

Aquatic life zones cover the majority of the Earth's surface.

36.11 Freshwater ecosystems make up only about 2% of the Earth's surface; most of them are ponds and lakes.

36.11 Ponds and lakes have a shore zone and an open-water zone, while lakes also have a deep zone.

36.11 Rivers and streams differ from ponds and lakes because they contain moving water that mixes with the air to provide high levels of oxygen for its fish and invertebrate inhabitants.

36.12 Estuaries are places where fresh water meets salt water as rivers empty into the ocean.

36.12 Estuaries receive nutrients from the surrounding land and usually support a large number of organisms.

36.13 The marine environment consists of three major life zones: the intertidal zone, between the highest tides and the lowest tides; the neritic zone, the area of shallow water that lies over the continental shelf; and the open-sea zone.

36.13 The ocean supports the most life in areas that have light and sufficient quantities of dissolved oxygen.

Key Terms

abyssal zone (uh-BISS-uhl) *812*
biomes (BYE-omes) *794*
continental shelf *811*
coral reef (KOR-ul) *811*
deep zone *807*
deltas *807*
desert *802*

estuary (ESS-choo-ER-ee) *808*
open-water zone *807*
mesopelagic zone
   (MEZ-oh-puh-LAJ-ik) *812*
pelagic zone (puh-LAJ-ik) *811*
photic zone (FOE-tik) *811*
savanna *801*

shore zone *807*
temperate deciduous forest *804*
temperate grasslands *804*
taiga *805*
tropical rain forest *800*
tundra *806*

KNOWLEDGE AND COMPREHENSION QUESTIONS

1. What is a biome? *(Knowledge)*
2. January is a winter month in the United States but a summer month in Australia. Why? *(Comprehension)*
3. What factors determine the climate of a region? *(Knowledge)*
4. Identify the terrestrial biome associated with each of the following *(Comprehension)*:
   a. Animals in these areas live mostly in trees.
   b. Dominant trees such as oak and beech allow enough light so that herbs, ferns, and mosses grow on the ground.
   c. Because of permafrost, when surface ice melts, the ground cannot absorb the water.
   d. Although located near the equator, these regions are mostly open grasslands.
   e. These coniferous forests receive most of their precipitation during the summer.
   f. Herds of North American bison once roamed these areas, grazing on large expanses of perennial grasses.

   g. These regions radiate heat rapidly at night, causing wide temperature differences between day and night.
5. What factors limit rainfall in a desert? *(Knowledge)*
6. Distinguish between taiga and tundra. *(Comprehension)*
7. Describe the three life zones found in fresh water. What types of organisms live in each? *(Knowledge)*
8. What is an estuary? Summarize how its inhabitants are adapted to its conditions. *(Comprehension)*
9. Why are estuaries and freshwater regions so vulnerable to human pollution? *(Comprehension)*
10. Identify and briefly describe the three major life zones of the ocean. *(Knowledge)*

KEY
Knowledge: Recalling information.
Comprehension: Showing understanding of recalled information.

CRITICAL THINKING REVIEW

1. Much of the world's most productive agriculture is carried out on soil of temperate grasslands. Agriculture in the tropics is far less productive. Why do you think temperate grassland soil is so much richer than soil in a tropical rain forest? *(Application)*
2. There is a thin strip of desert along the west coast of South America (mid-country). Immediately to the east of this desert land lie the Andes mountains. In the United States, the Mojave desert lies east of the Sierra Madre mountains. Explain why these deserts do not occur on the same side of the mountain ranges to which they are adjacent. *(Application)*
3. Analyze the graph below, which gives climate information for a particular city in North

America. In which biome is this city located? Explain the reasons for your answer. *(Analysis)*
4. In which biome do you live? List the characteristics of your biome and describe how they match the characteristics of the biome you chose. *(Synthesis)*
5. If a river or stream were dammed to form a pond or lake, what changes would you expect to see in the plant and animal life in that aquatic environment? *(Application)*

KEY
Application: Using information in a new situation.
Analysis: Breaking down information into component parts.
Synthesis: Putting together information from different sources.

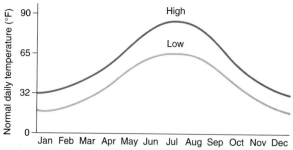

Using the illustration below to guide your thinking, answer the following questions:

1. In many areas of the United States, it is cold in the winter and warm in the summer. Why is this so?

2. In the summer in the United States a person standing in the sun for a half hour generally has greater exposure to the ultraviolet rays from the sun than in the winter. Why?

3. When the Earth is in the same position as the "Earth" farthest to the left, which season would it be in the southern hemisphere?

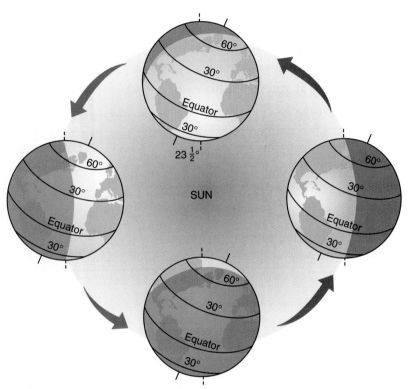

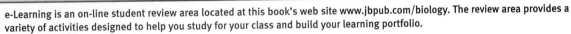

e learning

www.jbpub.com/biology

Location: http://www.jbpub.com/biology

e-Learning is an on-line student review area located at this book's web site www.jbpub.com/biology. The review area provides a variety of activities designed to help you study for your class and build your learning portfolio.

Review Questions The review questions test your knowledge of the important concepts and applications in each chapter. The review provides feedback for each correct or incorrect answer. This is an excellent test preparation tool.

Figure-Labeling Exercises Sharpen your visual thinking by matching terms and labels for important illustrations.

Flash Cards Studying biology requires learning new terms. Virtual flash cards help you master the new vocabulary for each chapter.

Just Wondering As you read and study from this text, you may find that you have unanswered questions. Through this site you can ask the author, Sandy Alters, your "just wondering" questions.

Do you prefer the speed and reliability of a CD-ROM? All of the features contained on the eLearning portion of the web site are also available on the Student CD-ROM.

The Biosphere: Today and Tomorrow

Just imagine that you could hold the world in your hands and make a difference in its future. Well, imagining isn't even necessary. All people do hold the world in their hands in a figurative way. Everyone's actions have a direct impact on the Earth and on the quality of life that you and others will experience for generations to come.

However, you must ask yourself some crucial questions. How are you affecting the Earth? What environmental problems do people face today? And what can you do to deal with those problems to be sure that the quality of life on Earth will be enhanced for yourself and your children? The answers to these questions are the focus of this chapter and will help you understand more about how you shape this fragile planet on which you live.

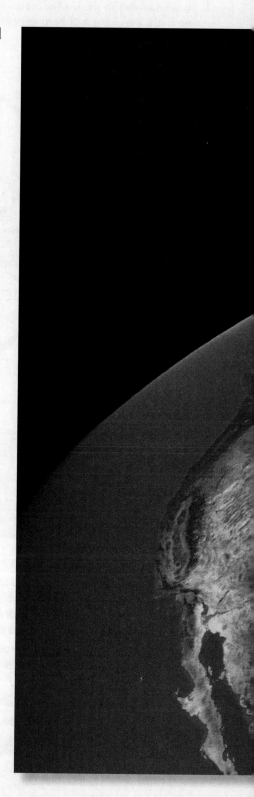

Organization

37.1 The biosphere

Life on Earth is confined to a region called the **biosphere**, the global ecosystem in which all other ecosystems exist. The biosphere extends from approximately 9000 meters (30,000 feet) above sea level to about 11,000 meters (36,000 feet) below sea level. You can think of it as extending from the tops of the highest mountains (such as Mount Everest or some of the Himalayas) to the depths of the deepest oceans (such as the Mariana Trench of the Pacific Ocean)—the part of our Earth in which the land, air, and water come together to help sustain life (Figure 37.1).

The biosphere is often spoken of as the **environment**. This general term refers to everything around you—not only the land, air, and water but other living things as well. You can speak, for example, of the environment of an ant, a water lily, or all the peoples of the world. Your particular environment can change during the day from a home environment, to a classroom environment, and then to an office environment. The environment can include a great deal—or very little—of the total biosphere and its living things.

> The biosphere is the part of the Earth where biological activity occurs. Sometimes the terms biosphere and environment are used interchangeably.

Figure 37.1 **The biosphere.** The Mariana Trench in the western Pacific is the deepest part of the ocean. It is 11,022 meters below sea level. The deepest dive was to 10,920 meters, and bottom fish and crustaceans were found at that level. Mount Everest in the Eastern Himalayas is the highest mountain in the world and rises to 8,848 meters. Spiders and insects have been found as high as 6,100 meters.

37.2 Nonrenewable resources

The land gives humans many things: fossil fuels, wood, food, and minerals. Water, too, is an important natural resource. Some of these resources—fossil fuels and minerals—are finite or **nonrenewable resources**; they are formed at a rate much slower than their consumption. Coal and oil are examples of nonrenewable resources.

Fuel Resources

Many of the fuels used to heat homes and run cars are **fossil fuels**. These substances—coal, oil, and natural gas—are formed over time (acted on by heat and pressure) from the undecomposed carbon compounds of organisms that died millions of years ago. Once these fuels are used up, they are gone forever. (Currently, 75% of the world's energy is derived from coal, oil, and natural gas.) Additionally, burning fossil fuels releases the stored carbon to the atmosphere. This carbon (in the form of carbon dioxide) is referred to as a greenhouse gas, which can result in global warming. Greenhouse gases and global warming are discussed in Section 37.4. Burning coal also releases sulfates and nitrates into the atmosphere, compounds that can result in acid precipitation and air pollution. These topics are discussed in Section 37.8.

Nuclear power, once considered a viable alternative energy source, contributes no carbon, sulfates, or nitrates to the atmosphere. Although nuclear power is dependent on uranium, a finite but abundant natural resource, comparatively small amounts of uranium are used to produce electricity. For example, 2.2 pounds of U-235 can yield as much energy as 2,200 tons of coal. The problems of nuclear power lie primarily with the cost of building nuclear power plants, the difficulty of disposing of highly radioactive wastes, and the public fear regarding safety. Because of these problems, nuclear power is minimally used and contributes only about 6% of the world's energy.

Today's nuclear power plants rely on fission (splitting) reactions to produce energy. The power produced by nuclear power plants is produced when neutrons having a particular level of energy strike the nuclei of uranium isotopes (U-235). These reactions produce neutrons capable of splitting more uranium nuclei, so chain reactions result. Such reactions produce a tremendous amount of energy, which is released as heat. In light-water reactors (the most prevalent type in the United States), the energy heats water surrounding the reactor core where the U-235 fuel rods are located. The chain reactions in the reactor core are carefully con-trolled to regulate the heat produced and prevent a meltdown, such as happened at the Chernobyl nuclear power plant in the Ukraine.

On April 26, 1986, the Chernobyl plant's No. 4 reactor exploded while an experiment was being conducted. A series of miscalculations resulted in a neutron buildup in one area of the core, and the nuclear reaction suddenly went out of control. The resulting power surge shattered the fuel rods, causing explosions that blew the lid off the reactor. The containment structure was unable to withstand the pressure and radioactivity entered the environment. Thirty-one persons were killed and 500 others were hospitalized. Scientists are continuing to study the long-term effects of this disaster.

Fission reactors also produce nuclear waste, which must be buried or contained so that its radioactivity does not contaminate the environment or harm humans. The waste disposal problem would be virtually eliminated if nuclear fusion reactors were developed to replace today's nuclear fission reactors. In the process of nuclear fusion, hydrogen isotopes are combined (fused) at extreme temperatures, forming helium and releasing energy. However, this method of energy production is extremely costly, involving extremely high reaction temperatures (100 million to 1 billion degrees Celsius). Only after decades of use, would nuclear fusion as a source of electricity become cost efficient. Even if current attempts to develop fusion reactors are successful, this energy source probably will not be commercially available before the middle of the twenty-first century.

Bioenergy refers to the use of living plants to produce energy. The most obvious type of bioenergy, the burning of wood, was first used by our ancestors more than 1 million years ago. In fact, until the Industrial Revolution of the 1800s, wood, not coal, supplied most of the world's energy. Today, wood supplies 12% of the world's energy, primarily in Latin America, Asia, India, and Africa. Unfortunately, however, the world is experiencing a fuel wood crisis—demand is exceeding the supply. The reasons for this crisis are complex but include the high world population and therefore high demand for wood, the degradation of woodlands without proper reforestation (replanting) techniques, and the cutting and burning of huge areas of tropical rain forests (see Section 37.4).

Mineral Resources

Minerals are inorganic substances that occur naturally within the Earth's crust. Zinc, lead, copper, aluminum, and iron are the minerals that humans use in large quantities. However, other minerals are also mined, such as gold, silver, and mercury. These minerals are

www.jbpub.com/biology

present in the Earth in fixed amounts; once they are used, they are gone. Each American consumes 40,000 pounds of new minerals every year.

The American culture has been called the throwaway society because Americans use, and then dispose of, a large amount of and variety of consumables. The average American uses six times more zinc, lead, and iron than does a person living outside the United States. In addition, the average American uses 11 times more copper and 14 times more aluminum. The amount of garbage Americans generate per person is still one of the highest rates of any industrialized nation in the world—almost one ton per person every year (see Figure 37.2).

Researchers agree that increased use of plastics and technology such as microelectronics will lower the demand for certain minerals. Plastics, however, are derived from petroleum products, so their manufacture increases consumption of fossil fuels.

Recycling is another way to reduce the consumption of minerals. Aluminum soda cans are useful items to recycle. These cans are produced from the claylike mineral bauxite, an aluminum ore that must be processed to produce aluminum metal. Producing cans from recycled aluminum uses 95% less energy than producing them from bauxite, saves mineral resources, and saves waste disposal costs and problems.

Plastics, on the other hand, are not easily recyclable and are not "truly" recyclable. That is, plastics cannot be refashioned into the product from which they were claimed; plastic milk jugs cannot be made into plastic milk jugs but must be made into park benches or parking lot curbs. Grocery store plastic bags are the closest truly recycled plastic products, being made from used plastic bags and other additives. Therefore, purchasing goods in recyclable containers such as paper, glass, and aluminum, and recycling those containers is a wise choice. Scrap metal and old automobiles are also recyclable, as are automobile tires and batteries.

Reuse is another way to help avert a mineral shortage, and it avoids the consumption of energy necessary to produce recycled items. Obviously, some items cannot be reused, but some can be put to the same or a different use: A plastic bread bag can be used to carry your lunch to work, glass juice bottles can be used as containers for a variety of liquids or other items, and paper printed on one side only can be cut up and used as notepaper. Many of the items we use regularly can be put to creative reuse.

Nonrenewable resources are finite; they are formed at a rate much slower than their consumption. These resources must be conserved or they will be depleted in years to come. Energy conservation will help preserve fossil fuels and proper reforestation techniques will help forests keep up with the demand for wood. Recycling and reusing paper, glass, and various metals will help preserve mineral resources as well as other natural resources.

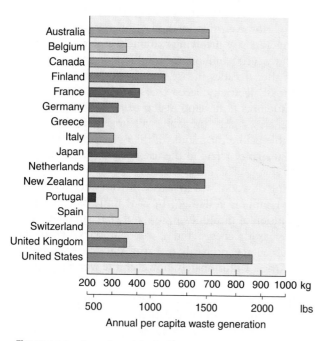

Figure 37.2 **Annual municipal waste generated per person in selected developed nations.**

37.3 Renewable resources

Environmental scientists suggest that instead of depending on nonrenewable fossil fuels for energy—fuels that are finite—people should work toward a *sustainable society* that uses nonfinite and renewable sources of fuel (see the "How Science Works" box). In a sustainable society, the needs of the society are satisfied without compromising the ability of future generations to survive and without diminishing the available natural resources.

Renewable resources are those produced by natural systems that replace themselves quickly enough to keep pace with consumption. The primary renewable and nonfinite sources of power that are available are the sun, the wind, moving water, geothermal energy, and bioenergy.

Solar power is the use of the sun for heating or to produce electricity. Solar cells can collect energy from the sun, convert it to electricity, and store it in large batteries for later use (Figure 37.3a). Currently under development is a solar power satellite system, which would collect solar energy in space, convert it to electricity, transmit it to Earth, and then convert it to a form that could be delivered to the existing electric power distribution system. Solar power is also being used to power automobiles. In March, 1999, Royal Dutch Shell opened its first stations in Germany and the Netherlands to charge electric cars with solar power.

Passive solar architecture is another way to use the energy from the sun. Homes and businesses can be built so that windows are located where the rays of the winter sun, low in the sky, enter the building, heating it. In the summer, an overhang protects the windows from the sun, which is high in the summer sky.

Water has been used as an energy source for decades and is currently supplying approximately 25% of the world's electricity. Figure 37.3b shows the Hoover Dam, a typical hydroelectric plant. Turbines located at the base of the dam rotate by the force of the falling water. The turbines drive generators, which produce electricity.

A new form of hydropower called *wave power* is now being tested. Wave power uses the vertical motion of sea waves to produce electricity. In addition, *tidal power*, the use of the movement of the tides of the oceans, is being used St. Malo, France, and in the Bay of Fundy, Canada. Figure 37.3c shows this facility. The problems with producing electricity from tidal and wave sources are the high cost and the unsteady rate at which the electricity is produced. Therefore, neither method appears useful to fulfill the total energy demands of an area. In late 1998, a

(a)

(b)

(c)

(d)

Figure 37.3 **Renewable energy sources.** (a) An array of solar panels at a solar power station near Sacramento, California. (b) Hoover Dam in Arizona. (c) Tidal power plant in the Bay of Fundy. (d) Modern windmills in southern California.

H**O**W Science
W**O**RKS *Applications*

Can Technology Save Us from Ourselves?

As the human population grows, the amount of resources available to sustain our population shrinks. In addition, humans create waste and pollute the environment. One result of these increases in population, pollution, and wastes and the decrease in natural resources is that global ecological stability becomes difficult to maintain. Are we on the brink of disaster? Some think the picture may not be as gloomy as it looks. Humans have made astonishing technological advances in a variety of areas while their numbers were "exploding." Can technology be an answer to our problems? Can technology help avert an ecological crisis resulting from the human population explosion and all its ramifications?

In recent years, advances in gene technology have allowed scientists to genetically alter bacteria, producing new strains that can break down hazardous pollutants such as oil from coastal oil spills. Other bacteria have been "gene spliced" to consume potentially harmful medical refuse such as waste blood. Although solutions such as these are important in that they alleviate serious ecological damage that would otherwise occur, they respond to immediate problems only. To ensure a bright ecological future for all organisms living on the Earth, scientists must develop structured strategies that allow for economic development and ensure sustainability of the environment. To approach sustainability, we must meet human needs without continuing to degrade our environment and diminish our resource base, while concurrently using technologies that help our human ecosystem imitate biological ecosystems.

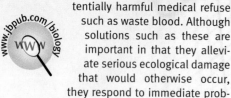

www.jbpub.com/biology

In biological ecosystems, plants produce their own food using carbon dioxide from the atmosphere, nutrients from the soil, and energy from the sun. This food allows plants to grow but also provides food for plant-eating organisms, the herbivores. Herbivore populations grow and, in turn, feed meat-eating organisms, the carnivores. Wastes from these organisms and their remains cycle through yet another group of organisms, the decomposers (fungi and bacteria), serving eventually as nutrients for plants.

Learning from this process of cycling by-products and degraded materials in natural systems, individuals and organizations are finding ways to reduce waste products or reclaim them. For example, one industrial manufacturing leader recovers hydrochloric acid, a hazardous waste product of their manufacturing process, which was formerly discarded. Now this company either reuses the acid or sells it to others who use it. Another company, which disposed of its used (and hazardous) alumina catalysts in the past, now sells them to cement makers for use in their manufacturing processes. And fly ash, previously a problematic waste product for energy companies, is being used to build roads.

Similarly, in agriculture, with the help of new technologies and the resurgence of ancient methods, wastes are being minimized or reclaimed, providing more ecologically sound agricultural systems. One new irrigation method, called *trickle irrigation,* involves carrying water to plants by tubing and delivering it through nozzles rather than spraying it into the air to fall on the plants. In standard irrigation systems, nearly two thirds of the water sprayed into the air is lost by evaporation; trickle irrigation decreases this amount dramatically. In addition, as water evaporates through standard methods, the salts dissolved in the water become concentrated. With trickle irrigation, salt concentrations increase only slightly and the soil suffers less salt buildup. An additional technology also helps trickle irrigation systems trap runoff water and reuse it, thereby further reducing water use and reducing the pollution of nearby lakes and streams by herbicides and pesticides in the runoff.

Another agricultural management system is multiple cropping. This ancient method, in which several crops are planted together, results in a decrease in the growth of crop pests such as weeds, insects, and fungi because the multiple-crop environment is more diverse than that of a single-crop environment. For example, if a particular fungus normally attacks one type of crop, it can spread less easily if other crop plants are growing between the fungus and its host. This type of diverse environment also allows predators from one crop to inhibit pests from another crop. Thus, even the use of pesticides and herbicides, sources of serious groundwater pollution, can be minimized or eliminated. In one multiple cropping system in the tropics, family gardens are created by collecting soil from unprofitable swamp land and building raised gardens on which several crops are grown. While the garden is established in such a way as to recycle nutrients, it also makes previously undervalued swamp forests more valuable, decreasing the possibility that these forests will be destroyed.

In our households, many Americans have focused on reclaiming and recycling materials to prevent the by-products of our daily lives from being discarded as waste products. Aluminum recycling not only maintains the use of the aluminum involved but also conserves nearly 20 times the energy it would cost to manufacture the same amount of aluminum from ore. Similarly, recycling plastic bottles and jugs into other plastic products such as polyester fiber and synthetic lumber substantially reduces the cost of fuel to create the products from "scratch" and conserves natural resources.

A sustainable society is indeed a future possibility if we can integrate the ecosystem concept into our farming, industrial, business, and home activities, recycling by-products and spent products repeatedly through our human ecosystem. There are two major challenges to actualizing such a system, however. The first challenge is developing more technologies that can be based on the use of by-products and reused products. The second challenge is more difficult and is where the analogy of biological ecosystems to human-centered ecosystems breaks down. The source of energy for natural ecosystems is almost entirely from solar energy, whereas the source of energy for human-centered ecosystems arises mostly from fossil fuels such as coal and oil—fuels that will someday be depleted. Until this problem is rectified, the maintenance of a sustainable human ecosystem will not be possible. And even then—as we approach the "biological model" of a human ecosystem—can the Earth support a continually growing human population? Although recycling and reuse technologies, as well as the use of resources that are infinite such as solar power, may help increase our carrying capacity, we cannot raise the carrying capacity of the Earth past the levels of the amount of food we can produce and the technologies we can provide amid the space that is necessary for humans to enjoy civilized, healthy lives. ◌

British company announced plans to build a tidal power plant off the coast of England.

The wind has been used for centuries as a source of energy. Today, windmills are being used in developing countries to pump water to livestock and to irrigate the land. In developed countries, however, windmills are being used to generate electricity. This "new breed" of windmill has rigid blades fashioned from lightweight materials and is shown in Figure 37.3d. At present, the United States, Denmark, and China are the three countries that lead in the use of **wind power** to generate electricity. In fact, some scientists predict that the United States will be generating 10% to 20% of its electricity from wind power by the year 2030.

Geothermal energy refers to the use of heat deep within the Earth. In some places, reservoirs of hot water or steam exist that can be extracted from the Earth by drilling procedures much like those used to tap into the Earth's oil and natural gas reserves. Alternatively, dry, hot rock can be drilled and water flushed through it. After the water is heated within the Earth, it can be used directly for heating purposes or as part of a process to produce electricity. Currently, this technology is being used and further developed in the United States, Russia, England, Italy, New Zealand, Japan, China, Indonesia, Kenya, Mexico, and the Philippines. The largest plant in the world is in the United States, near Geyserville, California (north of San Francisco).

In developing countries such as China, India, and Africa, the use of *biogas machines* is helping to ease the shortage of fuel wood already described. These stoves use microorganisms to decompose animal manure, harvest waste, wood waste, and even human sewage in a closed container. This process yields a methane-rich gas that can be used to fuel stoves, light lamps, and produce electricity. The substances left over after combustion can be used as fertilizer.

A newer form of bioenergy is the use of plants such as corn and sugar cane to produce carbohydrates that are fermented, producing liquid fuels such as ethanol. This technology, however, produces a product that appears to be an expensive alternative to fossil fuels. In addition, many cars do not function well on gasoline with ethanol additives. Critics say that food crops should be used to feed people, not produce energy (see "Just Wondering," Chapter 33).

Renewable resources are those produced by natural systems that replace themselves quickly enough to keep pace with consumption. The primary renewable and nonfinite sources of power that are available are the sun, the wind, moving water, geothermal energy, and bioenergy.

37.4 Tropical rain forests

The forests of the world are being depleted faster than they are being replanted. The most severe crisis is that of **tropical rain forest deforestation**. Tropical rain forests (see Chapter 36) are located in Central and South America, tropical Asia, and central Africa, forming a belt around the equatorial "waist" of the Earth. Although this belt of forest covers only 2% of the Earth's surface, it is home to more than *half* the world's species of plants, animals, and insects. These organisms contribute 25% of medicines, along with fuel wood, rubber, charcoal, oils, and nuts. In addition, the tropical rain forests play an important role in the world climate.

Population and poverty are both high in rain forest countries. People with few resources move from towns and cities to the rain forest and cut the trees for sale as lumber or burn them to clear a patch of land to grow crops and raise cattle to sustain themselves and their families (Figure 37.4). Commercial ranchers also cut and burn the forests to make way for pastureland to feed beef-producing cattle. However, the soil of the rain forests is poor, with few nutrients, and does not support crops. Before it is cut, the forest sustains itself because of mutualistic relationships between the trees and microorganisms that quickly decompose dead and dying material on the forest floor. These "processed" nutrients are quickly reabsorbed by the tree roots. Few

Figure 37.4 **Deforestation: a desperate effort to stave off poverty results in an environmental disaster.** These farmers live near the Andasibe reserve in Madagascar, an island where the per capita income is less than $250 a year. Clearing the rain forest allows these impoverished farmers to plant rice and graze cattle. But the environmental price may be too steep. In Madagascar alone, 80% of the rain forest has been destroyed.

Figure 37.5 **A rain forest reserve.** In 1979, the World Wildlife Fund established a research program—the Biological Dynamics of Forest Fragments Project. The program isolates "islands" of rain forests out of the larger forest during conversion to pastureland. Scientists can study these islands to determine what their optimal size and shape should be if they are to provide a suitable habitat for various species.

nutrients stay in the soil; most of the nutrients are in the vegetation. Cutting these trees down and burning them releases the nutrients from the trees. Crops grow poorly on this land after it is stripped. After a year or two, crops will not grow at all. The people move on, cutting yet another portion of the forest.

Commercial logging also takes its toll on the tropical rain forests. Many of the trees are cut to supply fuel wood, paper, wood panels such as plywood, and charcoal and to supply furniture manufacturers with ma-

hogany and other woods demanded by consumers around the world. At this time, the tropical forest is being slashed and burned at a devastating rate. By 1950, two thirds of the forests of Central America had been cleared. Madagascar, an island off the southwestern coast of Africa, had lost about half its rain forest by 1950, and lost half again by 1990. The United Nations Food and Agriculture Organization estimates that 100,000 square kilometers (62,000 square miles) of rain forest are now being lost each year worldwide—an area about the size of New England. At the present rate, scientists estimate, nearly all the tropical rain forests will be gone—including their rich diversity of animal life—within the next 50 years.

In 1999, the Environment Ministry of Brazil announced that the rate at which its rain forests are being destroyed jumped by 27% in 1998. These forests, which are the world's largest, were being cleared at the rate of about 6,350 soccer fields per day. These figures did not include damage done by massive fires near the Brazilian/Venezuelan border that year. In response, Brazil's Environment Ministry suspended all new permits for clearing land in the Amazon River Basin. However, at the same time, the Brazilian Office of Land Reform was settling landless people in the middle of the jungle, which would result in their clearing the land and counteracting policies to save the Amazonian rain forest.

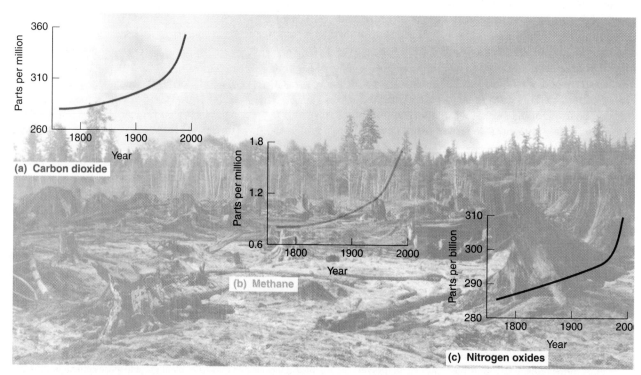

Figure 37.6 Global greenhouse gas concentrations over the last 200 years: (a) carbon dioxide, (b) methane, (c) nitrogen oxides.

Research programs are currently being conducted to explore the concept of conserving forest "fragments" as reserves that will provide suitable habitats for species (Figure 37.5). Research results in 1998 showed that forest fragments with rounded corners, rather than with square corners (as shown in the photo), were more suitable for growing shade-loving trees. At corners, young trees are more exposed to sun, wind, and competition from invading species. To avoid the problem, reserve designers are considering planting sun-tolerant species of trees to round out the edges.

In addition to losing a rich natural resource, many scientists agree that the burning of the tropical forests is adding tremendous quantities of carbon dioxide (CO_2) to the air. (Burning fossil fuels also releases CO_2.) CO_2 acts like the glass in a greenhouse (or the windows in your car), allowing heat to enter the Earth's atmosphere but preventing it from leaving. This blocking of outward heat radiation from the atmosphere by high CO_2 levels is called the **greenhouse effect**. Rising levels of atmospheric CO_2 may therefore result in a worldwide temperature increase, a situation called **global warming**. Scientists are currently debating how greenhouse gases (which also include nitrous oxide and methane) will affect (and are affecting) global temperatures and climate. Figure 37.6 shows atmospheric concentrations of three greenhouse gases. Warmer temperatures are predicted to cause changes in weather patterns, water supplies, ocean levels, growing conditions, and the northward spread of tropical diseases.

A study conducted at the University of Massachusetts in 1999 concluded that the 1990s was the warmest decade of the millennium. This study adds to the growing body of evidence that global warming is occurring. Other studies show that portions of the Greenland ice sheet are melting, a further suggestion of global warming.

In an effort to combat this climate change, an international team of scientists seeded a 21 square mile patch of ocean in the Antarctic with iron in February, 1999. Iron is a phytoplankton nutrient and resulted in the increased growth of this microscopic floating plant life and therefore an increased absorption of carbon dioxide from the air. Scientists are experimenting with methods such as these in an effort to reverse the probable global warming trend.

> The world's forests are diminishing, resulting in the loss not only of this resource but of the habitats of thousands of species of organisms. In addition, the burning of the forests adds tremendous quantities of carbon dioxide to the air, a situation that may result (and probably is resulting) in a worldwide temperature increase, or global warming.

37.5 Species richness and diversity

Although scientists have classified approximately 1.7 million species of organisms, they estimate that approximately 40 million exist. However, many of these species are becoming extinct—dying out—and will never be seen again. An estimated 1000 extinctions are taking place each year, a number that translates into more than two species per day. Although such mass extinctions have occurred in the past, as with the dying out of the dinosaurs, these former extinctions were caused by climatic and geophysical factors. Much of today's species' extinction is caused by human activity.

One way in which humans destroy species is by destroying their habitats. Widespread habitat destruction is taking place as the tropical rain forests are cut down. This one factor alone will cause the extinction of one third of the world's species.

Humans are also destroying coral reefs at an alarming rate. Because the base of the coral reef food chain is algae, the reef ecosystem depends on sunlight for its existence. In areas where soil erosion muddies the water, the reef dies from insufficient sunlight. In some cases, the water may become polluted with fertilizer runoff or with sewage, which causes the algae to overgrow, smothering the corals. In some cases, humans use dynamite to kill and harvest fish, a practice that obviously destroys the coral reef. In addition, coral is often harvested to sell to tourists.

Although habitat destruction and pollution are the most serious threats to the existence of certain species, exploitation of commercially valuable species threatens them as well. The Convention on International Trade in Endangered Species (CITES) regulates trade in live wildlife and products. However, illegal trade still takes place because of smuggling and the inability of law-enforcement inspectors to check all shipments of goods. Certain types of alligator, crocodile, sea turtle, snake, and lizard trade are illegal and endanger the existence of various species. Buying other products such as coral and ivory endangers the existence of coral reefs and African elephants. Even certain plants such as cacti and succulents should be purchased only if cultivated in greenhouses; they should not be taken from the wild. Organizations such as the World Wildlife Fund are invaluable sources of information regarding which products consumers should avoid so that they can help stop trade that threatens certain species.

The extinction of many species leads to a reduction in **biological diversity**, a loss in numbers of different kinds of species. Each species is not only the result of millions of years of evolution and can never be repro-

Figure 37.7   **A germ plasm bank at the United States Department of Agriculture.** Researchers are in a cold storage room that contains seeds from around the world.

duced once it is gone, but also is an intricate part of the interwoven relationships among organisms within the ecosystems of the world. The extinction of just one species can affect ecosystems in many unforeseen ways as well as affecting sources of food, medicine, and other substances.

In 1998, the results of two studies on prairie chicken populations in the United States and in butterfly populations in Australia suggested that inbreeding of small populations plays a role in the extinction of a species. Inbreeding is reproduction among members of a species that are genetically closely related. In these studies, a lack of genetically different mates in butterflies and prairie chickens led to changes in birth and survival rates, making it less likely that an endangered population would survive. These results emphasize the importance of genetic diversity in breeding programs designed to bolster populations of species whose numbers are in decline.

Recently, for example, the genetic diversity in crop plants has declined as scientists have used selective breeding techniques to produce plants with specific characteristics, such as resistance to particular diseases or pests, hardiness, and other characteristics considered important. However, when only a few species or varieties of a species are cultivated or survive, the genetic diversity of the organism declines. Populations of species that have little diversity are more vulnerable to being wiped out by new diseases or climatic changes.

In the 1960s, the United Nations' Food and Agriculture Organization (FAO) made recommendations that have led to the establishment of "banks" that store plant seeds and genetic material, or germ plasm (Figure 37.7). Zoos around the world are making similar efforts to preserve genetic diversity among animals by establishing and carrying out sophisticated breeding programs to increase genetic diversity among endangered species (Figure 37.8).

> Serious threats to the survival of species are habitat destruction, habitat pollution, and the illegal trade of endangered species.

Figure 37.8   **Endangered species: The Asian (Indian) elephant.** This Asian elephant is receiving a tooth and mouth examination at the National Zoo in Washington, DC.

37.6 Solid waste

The average American produces about 19 pounds of garbage and trash—referred to as **solid waste**—per week. If that does not sound like very much, multiply that number by 52, and you will realize that each person produces approximately a *half ton* of solid waste per year. In late 1999, the population of the United States was approximately 273 million—a population that produced approximately 135 million tons of waste per year, not including the waste produced by schools, stores, and manufacturing.

What happens to your trash when you put it out for collection? The burial site for your throwaways is the **sanitary landfill**, an enormous depression in the ground where trash and garbage is dumped, compacted, and then covered with dirt (Figure 37.9). In 1983, the Federal Resource Conservation and Recovery Act forced the closing of all *open dumps* or required them to be converted to landfill sites. At an open dump, solid waste is heaped on the ground, periodically burned, and left uncovered. Landfills are considered superior to dumps because landfill wastes are covered, reducing the number of flying insects and rodents that are attracted to the site and reducing the odor produced by open, rotting organic material. In addition, wastes are not burned at landfills, decreasing the problem of air pollution. Further, when the capacity of a landfill site is reached, it may be used as a building site or recreational area. Examples of landfill reuse are Mount Trashmore recreational complex in Evanston, Illinois, and Mile High Stadium in Denver, Colorado.

Problems do exist with landfills, however. First of all, space is running out. In just one year, the population of New York City alone produces enough trash to cover more than 700 acres of land 10 feet deep. Another problem with landfills is that liquid waste can trickle down through a landfill, reaching and contaminating groundwater below. Liquids leaching from landfills can also pollute nearby streams, lakes, and wells (therefore, it is unwise to put batteries, paint solvents, drain cleaners, and pesticides in with the trash). In addition, as the organic material compacted in landfills is decomposed in the absence of oxygen, methane gas is produced. This highly explosive gas rises from landfills and can seep into buildings constructed on or near reclaimed sites.

Recycling paper, glass, aluminum cans, and even clothing is one way to reduce solid waste and landfill

Figure 37.9   A sanitary landfill in California.

problems. An important part of recycling is buying recycled goods so that a market is maintained for them. Additional recycling behaviors are purchasing products in recyclable containers, avoiding the purchase of "over-packaged" products, reusing products when feasible, and composting yard waste. Composting involves piling (and periodically turning to aerate) grass clippings, wood shavings, and similar yard wastes. Bacteria degrade these substances, and they can be used to fertilize flower beds and vegetable gardens. Composting helps eliminate the second largest waste by volume in landfills, which is yard waste.

In the future, careful planning will be necessary to ensure that landfill sites are located away from streams, lakes, and wells and that they have proper drainage and venting systems. Most likely, landfills will be only a portion of a solid waste disposal system that incorporates recycling, safe incineration, and waste-to-energy reclamation.

Solid waste is disposed of in sanitary landfills, depressions in the ground where trash and garbage is dumped, compacted, and covered with dirt. Solid waste management could be improved with increased and better-organized recycling efforts, safe incineration, better-engineered landfill sites, and waste-to-energy reclamation.

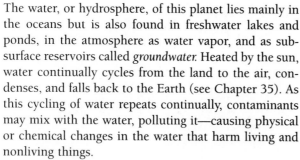

37.7 Surface water and ground water

www.jbpub.com/biology

The water, or hydrosphere, of this planet lies mainly in the oceans but is also found in freshwater lakes and ponds, in the atmosphere as water vapor, and as subsurface reservoirs called *groundwater.* Heated by the sun, water continually cycles from the land to the air, condenses, and falls back to the Earth (see Chapter 35). As this cycling of water repeats continually, contaminants may mix with the water, polluting it—causing physical or chemical changes in the water that harm living and nonliving things.

Surface water can be polluted by factories, power plants, and sewage treatment plants that dump waste chemicals, heated water, or human sewage into a lake, stream, or river. These sources of pollutants are called *point sources* because they enter the water at one or a few distinct places. Other types of pollutants may enter surface water at a variety of places and are called *nonpoint sources.* Examples of nonpoint sources of pollution are (1) sediments in land runoff caused by erosion from poor agricultural practices (a major type of water pollution); (2) metals and acids draining from mines; (3) poisons leaching from hazardous waste dumps (Figure 37.10); and (4) pesticides, herbicides, and fertilizers washing into surface waters after a rain. These pollutants affect both aquatic organisms and terrestrial organisms that drink the water. The manner in which a pollutant affects living things depends on its type: nutrient, infectious agent, toxin (poison), sediment, or thermal pollutant.

Organic nutrients are sometimes discharged into rivers or streams by sewage treatment plants, paper mills, and meat-packing plants. These "organics" are food for bacteria. If high amounts of organic nutrients are available to bacteria, their populations will grow exponentially. As they grow and reproduce, they use oxygen that fish need. Therefore, as the bacterial populations rise, the only organisms that survive are those that can live on little oxygen. So-called "trash fish" such as carp can outsurvive other species such as trout and bass, but if oxygen levels become extremely low, all the fish die, survived only by various worms and insects.

The accumulation of *inorganic nutrients* in a lake is called **eutrophication,** meaning "good feeding." Certain inorganics such as nitrogen and phosphorus, which come from croplands or laundry detergents, stimulate plant growth. Although heavy plant growth makes swimming, fishing, or boating difficult, it does not cause most of its problems until the autumn, when the plants die (in many regions of the United States). At that time, bacteria decompose the dead plant material, and prob-

Figure 37.10 A toxic chemical dump in northern New Jersey. Toxic chemical dumps are serious threats to groundwater and surface water. Pollution occurs when the drums rust through and release their contents, which then enter the surface water and may eventually trickle down to the groundwater.

lems similar to those of organic nutrient pollution arise. In addition, the decomposed materials begin to fill the bottom of the lake. Eventually, the lake may become transformed into a marsh and then into a terrestrial community by an accelerated process of succession (see Chapter 34). *Sediment* that flows into lakes from erosion of the land caused by certain agricultural practices, mining, and road construction also fills in lakes and hastens natural succession.

An example of eutrophication that began leading to the successional "death" of a lake occurred in Lake Washington near Seattle. The problem stemmed from local communities dumping their wastes into the lake. In 1968, the dumping was halted, and Lake Washington is now fully recovered.

Surface waters are rarely polluted with *infectious agents,* or disease-causing microbes, in the United States. However, in areas such as Africa, Asia, and Latin America, waterborne diseases are common. Surface waters become polluted from untreated human wastes and from animal wastes, causing diseases such as hepatitis, polio, amebic dysentery, and cholera.

An array of *toxic substances* pollutes surface waters worldwide. Toxic substances include both organic compounds such as PCBs (polychlorinated biphenyls) and phenols and inorganic substances such as metals, acids, and salts. These toxic, or poisonous, substances come from a diverse array of sources such as industrial discharge, mining, air pollution, soil erosion, old lead

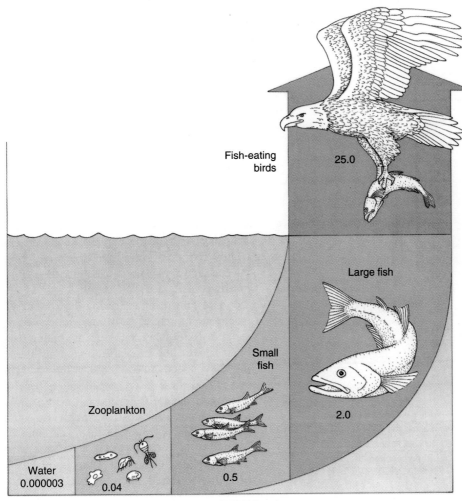

Fish-eating birds 25.0

Large fish 2.0

Small fish 0.5

Zooplankton 0.04

Water 0.000003

Amount of DDT (ppm)
ppm = parts per million

Figure 37.11 Biological magnification. The amount of DDT in an organism increases as you go up the food chain. This illustration depicts an 8 million-fold amplification of the DDT.

pipes, and many natural sources. The effects on humans from drinking these substances in water range from numbness, deafness, vision problems, and digestive problems to the development of cancers.

Most toxic pollutants do not degrade, or break down, and are therefore present in bottom sediments of surface waters for decades or more. In fact, some organisms accumulate certain chemicals (often deadly ones) within their bodies, a process called **biological concentration**. Oysters, for example, accumulate heavy metals such as mercury; thus, these organisms might be highly toxic when living in waters with relatively low concentrations of this metal. Also, as organisms higher on the food chain eat organisms lower on the food chain, toxins accumulate in the predators in a high concentration, a concept called **biological magnification**. As discussed in Chapter 35, the relationships between trophic levels of food chains are pyramidal, so a smaller number of organisms eat a larger number of organisms as you

progress "up" the chain. (**Figure 37.11** illustrates this concept.) Scientists estimate that the concentration of a toxin in polluted water may be magnified from 75,000 to 150,000 times in humans who consume tainted fish.

The electric power industry and various other industries such as steel mills, refineries, and paper mills use river water for cooling purposes, discharging the subsequently heated water back into the river (**Figure 37.12**). Small levels of **thermal pollution** do not cause serious problems in aquatic ecosystems, but sudden, large temperature changes kill heat-intolerant plants and animals. For example, ecosystems that have slowly adjusted to artificially heated waters are damaged if the heat source is shut down, as when a power plant closes.

The surface water and rainwater that trickle through the soil to underground reservoirs is groundwater. This water can be contaminated with some of the same substances as surface water. The primary groundwater contaminants are toxic chemicals that seep into

Preventing contamination is essential to ensuring a safe water supply.

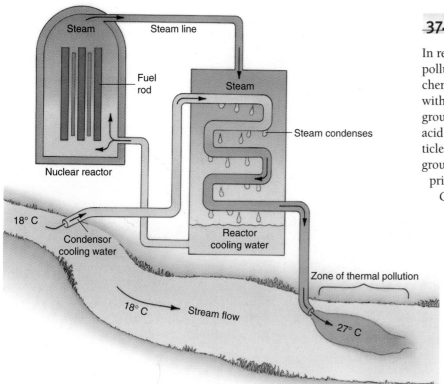

Figure 37.12 **Thermal pollution.** Water from streams or rivers is used to cool nuclear power plants and to condense steam to water. As it cools the reactor, the water is heated, forming steam, which is then cooled again in the condenser. The cool stream water used in the condenser is discharged into the stream after it has absorbed heat from the steam.

37.8 Precipitation

In recent years, the water in the atmosphere has become polluted with sulfur dioxide and nitrogen dioxide, two chemical compounds that form acids when combined with water. As this water vapor condenses and falls to the ground, it is commonly referred to as **acid rain**, although acid precipitation also falls as snow or as dry "micro" particles, mixing with water when it reaches surfaces on the ground. "Normal" rain has a pH of approximately 5.7, primarily because of dissolved carbon dioxide (see Chapter 3 for a discussion of pH). The pH of acid rain is lower than 5.7 and usually falls between 3.5 and 5.5. However, rainfall samples taken in the eastern United States have measured as low as 1.5—a pH lower than that of lemon juice and approaching that of battery acid.

Acid rain results in many devastating effects on the environment. As it mixes with surface water, it acidifies lakes and streams, killing fish and other aquatic life. It seeps into groundwater, causing heavy metals to leach out of the soil. The result is that these heavy metals enter the groundwater and surface water, posing health problems for humans as well as fish. Acid rain also eats away stone buildings, monuments, and metal and painted surfaces (Figure 37.13).

The effect of acid rain falling on plant life has been hypothesized but undocumented until re-

the ground from hazardous waste dump sites and chemicals such as pesticides used in agriculture. Although groundwater does get filtered and "cleansed" of some substances as it trickles through the soil, toxic chemicals consist of molecules too small to be filtered in this manner. At this time, prevention of contamination is the cheapest and most feasible way to end groundwater contamination. Pumping contaminated water from underground sources to the surface, purifying it, and returning it to the ground is extremely costly.

Surface water can be polluted by waste chemicals, heated water, or human sewage put into the water by factories, power plants, and sewage treatment plants or by silt or chemicals that can leach into the water from surrounding sites. These substances all affect the aquatic ecosystem in different ways. Additionally, surface water contaminants and contaminants in precipitation can trickle through the soil, polluting underground reservoirs called groundwater.

Figure 37.13 **Acid rain damage.** This statue has been eroded by acid precipitation.

Figure 37.14 **Effects of acid rain.** These balsam fir trees in North Carolina have been killed by acid rain. The chemicals in the acid rain, which are carried on prevailing winds, come from as far away as the Midwest.

cently. Botanists realized that acid rain not only has leached many of the minerals essential to plant growth from the soil but also has liberated toxic minerals such as aluminum. Recent research suggests that acid rain also kills or damages microorganisms living in symbiotic associations with forest trees, helping them extract water and needed minerals from the soil. Without these organisms, the trees die (Figure 37.14).

Although sulfur and nitrous oxides are produced naturally during volcanic eruptions and forest fires, humans produce more than half these chemicals from the burning of coal by electricity-generating plants, industrial boilers, and large smelters that obtain metals from ores. In addition, nitrogen oxides are emitted by cars and trucks. The situation is complicated because countries pollute both their own air and the air of other countries. Emissions produced in the Midwest and eastern portions of the United States affect not only those areas, for example, but are carried by the wind into Canada. Emissions produced in England move into the Scandinavian countries.

Solving the problem of acid rain and curtailing these emissions is not easy or inexpensive. One technique used to reduce sulfur dioxide is to put *scrubbers* on coal-burning power plants. This technology can remove up to 95% of the sulfur dioxide emissions produced. Unfortunately, the United States lags behind most sulfur-emitting countries of the world in implementing this technology. A solution to this problem depends on international agreements to reduce emissions, to employ energy-conservation measures, and to increase the use of public transportation.

Sulfur dioxide and nitrogen dioxide, gases emitted primarily by coal-fired power plants, combine with water in the atmosphere to produce acid rain, a precipitation that harms both living and nonliving things.

The burning of fossil fuels affects the atmosphere.

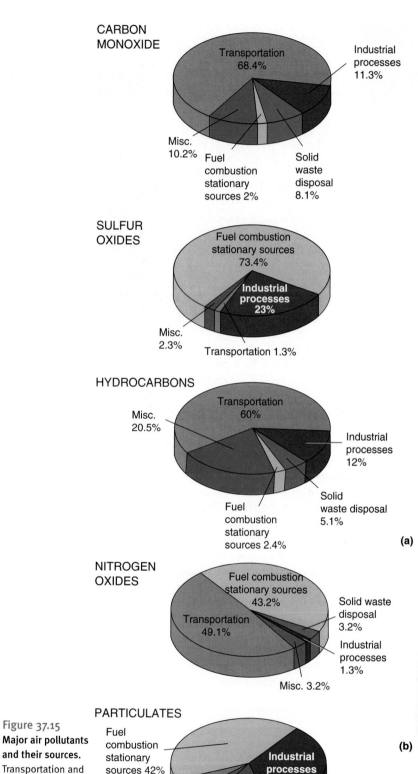

CARBON MONOXIDE

- Transportation 68.4%
- Industrial processes 11.3%
- Misc. 10.2%
- Fuel combustion stationary sources 2%
- Solid waste disposal 8.1%

SULFUR OXIDES

- Fuel combustion stationary sources 73.4%
- Industrial processes 23%
- Misc. 2.3%
- Transportation 1.3%

HYDROCARBONS

- Misc. 20.5%
- Transportation 60%
- Industrial processes 12%
- Solid waste disposal 5.1%
- Fuel combustion stationary sources 2.4%

NITROGEN OXIDES

- Fuel combustion stationary sources 43.2%
- Solid waste disposal 3.2%
- Transportation 49.1%
- Industrial processes 1.3%
- Misc. 3.2%

PARTICULATES

- Fuel combustion stationary sources 42%
- Industrial processes 34.8%
- Transportation 5.5%
- Solid waste disposal 4.5%
- Misc. 13.2%

Figure 37.15
Major air pollutants and their sources. Transportation and fuel combustion at stationary sources are the main contributors to pollution.

37.9 The air

The Earth's atmosphere actually extends much higher than the portion within the biosphere, a part of the atmosphere more technically called the *troposphere.* The troposphere extends approximately 11 kilometers (36,000 feet) into the atmosphere but slopes downward toward the poles and upward toward the equator. The word troposphere literally means "turning over," a name extremely descriptive of the atmosphere. As the Earth is heated by the sun, air rises from its surface, cooling as it ascends. Cooler air then falls, resulting in a constant turnover of the air, aided by the prevailing winds.

About 99% of the clouds, dust, and other substances in the atmosphere are located in the troposphere. Some of these "other substances" are nitrogen and sulfur oxide pollutants. Other major air pollutants are carbon monoxide, hydrocarbons, and tiny particles, or particulates. As **Figure 37.15** shows, sulfur oxides and particulates are produced primarily by the burning of

(a)

(b)

Figure 37.16 **Air pollution.** (a) "Gray air" in New York City is caused by fossil fuel pollutants from power plants. (b) "Brown air" in Los Angeles. Sunlight reacts with the chemicals spewed from automobiles to create ozone, which in the upper atmosphere protects you from the sun's harmful rays but in the lower atmosphere is poisonous.

coal in electricity-generating plants. Carbon monoxide and hydrocarbons are emitted primarily by cars, buses, and trucks; all these sources spew out nitrogen oxides.

The type of air pollution in a city depends not only on which of the pollutants are in the air but also on the climate of the city. *Gray-air cities* (Figure 37.16a), such as New York, Philadelphia, and Pittsburgh, are located in relatively cold but moist climates and have an abundance of sulfur oxides and particulates in the air. The haze, or **smog**, that can be seen in the air is the result of the burning of fossil fuels in power plants, other industries, and homes. *Brown-air cities* (Figure 37.16b), such as Los Angeles, Denver, and Albuquerque, have an abundance of hydrocarbons and nitrogen oxides in the air. In these sun-drenched cities, the hydrocarbons and nitrogen oxides undergo photochemical reactions that produce "new" pollutants called *secondary pollutants*. The principal secondary pollutant formed is **ozone** (O_3), a chemical that is extremely irritating to the eyes and the upper respiratory tract. Smog caused by pollutants reacting in the presence of sunlight is called **photochemical smog**. Figure 37.17 shows the average maximum levels of ground-level ozone in 12 cities.

The solutions to cutting down on smog are the same solutions to problems of acid rain formation and fossil fuel depletion: the use of coal-fired power plant scrubbers, energy conservation, and the recycling of resources leading to a reduction in energy use for manufacturing.

The primary type of air pollution in the first 11 kilometers of the atmosphere consists of smoke and fog, or smog. Smog is caused by sulfur oxides, nitrogen oxides, hydrocarbons, and particulates in the air as a result of the combustion of fossil fuels.

37.10 The ozone layer

Ironically, humans are producing ozone in the troposphere that is polluting the environment, but destroying it in the *stratosphere* where it is needed. The stratosphere is the layer of the atmosphere directly above the biosphere. It contains a layer of ozone that is formed when sunlight reacts with oxygen. Although ozone is harmful in air that is breathed, it is helpful in the stratosphere, acting as a shield against the sun's powerful ultraviolet (UV) rays. Excess exposure to UV rays can cause serious burns, cataracts (an opacity of the lens of the eye), and skin cancers and can harm or kill bacteria and plants. Scientists have measured "holes" in the ozone layer where it is thinnest, over the polar ice caps (Figure 37.18). These holes are huge—larger than the United States and Canada combined. Additionally, the ozone layer is being reduced over much of the globe.

The chemical chlorofluorocarbon (CFC) is the main culprit. For example, freon, which is a refrigerant used in air conditioners, is a CFC. In addition, CFCs are used in the manufacture of styrofoam and foam insulation. CFCs damage the ozone layer because they contain chlorine. At ground level, CFCs are stable, but when they reach the stratosphere, ultraviolet rays break them down, liberating

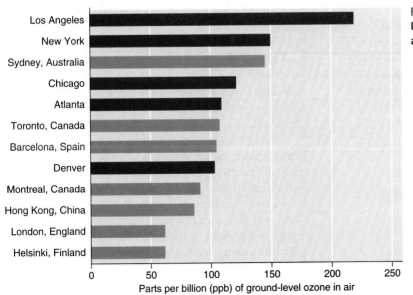

Figure 37.17 Ground-level ozone in 12 cities around the world.

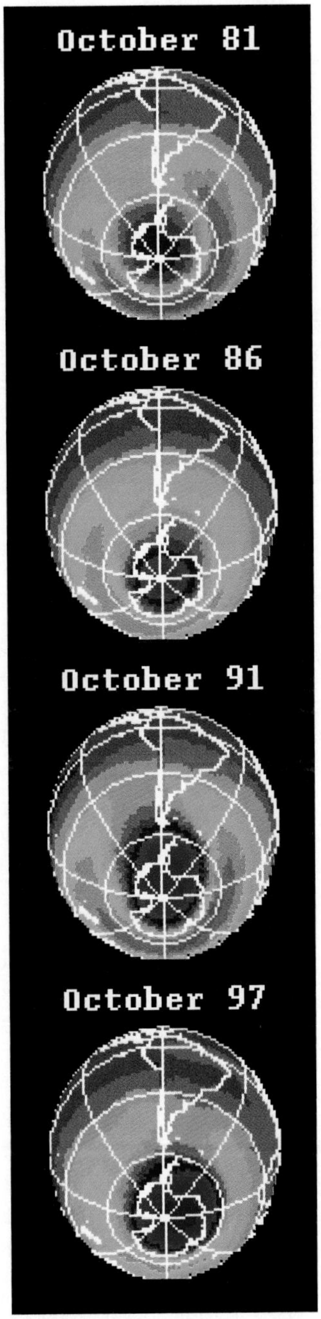

Los Angeles
New York
Sydney, Australia
Chicago
Atlanta
Toronto, Canada
Barcelona, Spain
Denver
Montreal, Canada
Hong Kong, China
London, England
Helsinki, Finland

0   50   100   150   200   250
Parts per billion (ppb) of ground-level ozone in air

Figure 37.18 Depletion of the ozone layer over the South Pole. The colors denote the thickness of the ozone layer. In order of dercreasing thickness (i.e., thickest to thinnest) the colors are orange → green → blue → purple.

October 81
October 86
October 91
October 97

Just Wondering . . . *Real Questions Students Ask*

What direct effect does the depletion of the ozone layer have on our weather? What can be done to prevent further depletion?

Your question shows confusion between the ozone-hole problem and the greenhouse effect. These situations are not the same, and you are not alone in your confusion. The results of a study published in 1997 in the *Bulletin of the American Meterological Society* shows that many students confuse the two.

Chlorofluorocarbons (CFCs) are the major cause of the ozone hole, the depletion of the stratospheric "blanket" of ozone that protects the Earth from the sun's damaging ultraviolet (UV) rays. However, CFCs are only a minor cause of the greenhouse effect, the blocking of outward heat radiation from the Earth. Its major cause is increasing levels of carbon dioxide in the atmosphere from the burning of wood and fossil fuels.

The thinning of the ozone layer in the stratosphere will not affect the climate. It affects the amount of ultraviolet radiation that reaches the surface of the Earth. UV rays are damaging to the skin and may result in increased numbers of skin cancers, for example. Increased levels of UV radiation will also penetrate bodies of water and may affect photosynthesis and therefore the growth of phytoplankton and algae.

The greenhouse effect, on the other hand, is expected to affect climate worldwide. (Climate differs from weather. Weather is a day-to-day occurrence. Climate refers to atmospheric conditions over decades.) Scientists are currently debating climate effects that may already be taking place, and refer to the long-term climate change caused by the greenhouse effect as global warming. With a long-term rise in temperature come other climate changes as well, such as patterns of more destructive storms and flooding. Additionally, the polar ice caps will begin to melt and the seas will rise.

Strategies to deal with the problems of the ozone hole and global warming are discussed in Section 37.10.

---

chlorine gas. This gas reacts with ozone, producing chlorine oxide and oxygen. Chlorine oxide then breaks down, releasing molecules of chlorine gas that continue to destroy ozone. In fact, one molecule of chlorine gas can destroy up to 10,000 ozone molecules.

In 1987, 37 nations signed the Montreal Protocol on Substances that Deplete the Ozone Layer. They agreed to limit their release of CFCs and to cut CFC emissions in half by the year 2000, with some exceptions given to developing countries. The protocol was strengthened in 1990 and again in 1992. In the United States, the Clean Air Act of 1990 and its subsequent amendments enforces the Montreal Protocol.

www.jbpub.com/biology

Ozone in the stratosphere acts as a shield against the sun's ultraviolet (UV) rays. Chlorofluorocarbons (CFCs) damage the ozone layer because they contain chlorine, which reacts with ozone, breaking it down. In 1987, 37 nations signed the Montreal Protocol, agreeing to limit their release of CFCs and to cut CFC emissions in half by the year 2000.

37.11 Overpopulation and environmental problems

Some scientists hold that the enormous world population is the key to the problems of pollution and diminishing natural resources. Others state that technology is the culprit. Most, however, would agree that neither factor alone is the cause of these complex problems. Many factors interact to affect population size and the "health" of the environment.

The size of the human population influences the environment because it puts a demand on resources. However, the social, economic, and technological development of a country affects the demand its population places on resources. A person living in a rural area of Kenya, for example, does not use the same amount and kind of environmental resources as a person living in a large city in the United States. The lifestyle and per capita consumption of a population make a big difference on the impact of that population on the environment.

Using resources and disposing of wastes in certain ways pollute the environment. The pollution of one aspect of the environment (such as the air) can lead to pollution of other aspects of the environment (such as the water through acid rain). Pollution, in turn, can limit population size by increasing the death rate. However, living with pollution can lead to attitudinal changes among members of a population, which may result in the development of laws to better manage the use of resources and to curb the pollution resulting from their use.

The impact of the human population on the environment is extremely complex and has multiple causes and effects. Many factors, such as population size, per capita consumption, technology, and politics, interact in complex ways, resulting not only in the problems faced today but in solutions for the future. Through research, environmental scientists will help everyone understand how each person can live on this planet without harming it or endangering everyone's future existence.

Many factors interact to affect population size and the "health" of the environment. Some factors that modify the effects of a population on the environment include lifestyle, per capita consumption of natural resources, the manner of resource use, and the effects of resource use.

Key Concepts

The biosphere is the part of the Earth that supports life.

37.1 The biosphere is the interface of the land, air, and water, extending from approximately 9000 meters (30,000 feet) above sea level to 11,000 meters (36,000 feet) below.

The land provides a variety of resources.

37.2 Many of the natural resources of the Earth are finite and nonrenewable; that is, they cannot be replaced.

37.2 Coal, oil, and natural gas are fossil fuels, nonrenewable resources formed over time from the remains of organisms that lived long ago.

37.3 Alternative renewable energy sources are solar, wind, water, geothermal power, and bioenergy.

37.4 Approximately 100,000 square kilometers (62,000 square miles) of rain forest are being lost each year because of slashing and burning for agriculture and logging, and to create pastures for cattle.

37.4 Scientists estimate that at the present rate, nearly all the tropical rain forest will be gone within the next 50 years.

37.4 The atmospheric rise in carbon dioxide levels, which is caused by the burning of the tropical forests and the combustion of fossil fuels, is acting as a barrier against the escape of heat from the surface of the Earth.

37.4 This greenhouse effect could affect global temperatures, rainfall patterns, and agricultural lands.

37.5 Species are dying out at a rate of 1000 extinctions per year.

37.5 A major cause of species extinction is habitat loss, which occurs when the tropical rain forests are cut or other habitats are destroyed.

37.5 The extinction of many species leads to a reduction in biological diversity, a loss of species richness.

37.5 The loss of species also affects ecosystems and diminishes future sources of food, medicine, and other substances.

37.6 The United States produces approximately 135 million tons of solid waste (garbage and trash) per year, not including the waste produced by schools, stores, and manufacturing.

37.6 Solutions to this problem lie in recycling, reusing, lowering consumption, composting yard waste, and implementing landfill technology that incorporates recycling, safe incineration, and energy reclamation.

Preventing contamination is essential to ensuring a safe water supply.

37.7 Surface water is contaminated by factories, power plants, and sewage treatment plants that dump waste chemicals, heated water, and human sewage into lakes, streams, and rivers.

37.7 Other sources of surface water pollution are sediment runoff during erosion and the leaching of chemicals from mines, hazardous waste dumps, and croplands.

37.7 Chemical pollutants affect aquatic life in different ways and can become concentrated in their bodies.

37.7 Underground reservoirs of water can be contaminated by the same sources as surface water when the contaminants trickle through the soil; however, the primary groundwater contaminants are toxic chemicals.

37.7 At this time, prevention of contamination is the cheapest and most feasible way to end groundwater contamination.

37.8 The atmosphere has become polluted with sulfur and nitrogen oxides, primarily from the combustion of fossil fuels in automobiles and electricity-generating plants.

Key Concepts, continued

37.8 As sulfur and nitrogen oxides mix with water vapor in the atmosphere, they form acid rain, which harms both living and nonliving things.

The burning of fossil fuels affects the atmosphere.

37.9 Along with sulfur and nitrogen oxides, fuel combustion in cars, at electricity-generating plants, and in various industrial processes produces carbon monoxide, hydrocarbons, and particulates.

37.9 As the nitrogen oxides and hydrocarbons react with sunlight, they form photochemical smog, an upper respiratory irritant that is a health hazard.

37.10 Ozone in the stratosphere acts as a shield against the sun's ultraviolet (UV) rays.

37.10 Chlorofluorocarbons (CFCs) damage the ozone layer because they contain chlorine, which reacts with ozone, breaking it down.

37.10 In 1987, 37 nations signed the Montreal Protocol, agreeing to limit their release of CFCs and to cut CFC emissions in half by the year 2000.

37.11 Environmental problems are the result of many interacting factors.

37.11 Some factors that modify the effects of a population on the environment include lifestyle, per capita consumption of natural resources, the manner of resource use, and the effects of resource use.

Key Terms

acid rain *830*
bioenergy (BYE-oh-EN-ur-jee) *819*
biological concentration *829*
biological diversity *825*
biological magnification *829*
biosphere (BYE-oh-sfear) *818*
environment *818*
eutrophication
    (yoo-trowf-uh-KAY-shun) *828*

fossil fuels *819*
geothermal energy
    (JEE-oh-THUR-mul) *823*
global warming *825*
greenhouse effect *825*
nonrenewable resources *819*
nuclear power (NOO-klee-ur) *819*
ozone (O₃) (OH-zone) *833*
photochemical smog *833*

renewable resources *821*
sanitary landfill *827*
smog *833*
solar power *821*
solid waste *827*
thermal pollution *829*
tropical rain forest deforestation *823*
wind power *823*

KNOWLEDGE AND COMPREHENSION QUESTIONS

1. Many scientists think people should work toward creating a "sustainable society." What does this mean? (*Comprehension*)

2. Name several energy sources that could be used to help create a sustainable society. Explain your answers. (*Comprehension*)

3. What can be done to help prevent a mineral shortage in the future? Why are preventive measures important? (*Comprehension*)

4. What is global warming? Why is it dangerous? (*Knowledge*)

5. Define biological diversity. How are humans affecting the biological diversity of the world's species? Why is this serious? (*Comprehension*)

6. How can the amount of solid waste thrown away be reduced and waste management improved? (*Comprehension*)

7. Distinguish between point and nonpoint sources of surface water pollution. Give an example of each. (*Knowledge*)

8. Describe the different effects of pollution by organic nutrients, inorganic nutrients, infectious agents, and toxic substances in surface water. (*Knowledge*)

9. How is acid rain produced? What are its effects? (*Knowledge*)

10. Ozone is a dangerous pollutant in air that is breathed. Why should people be concerned that the ozone layer in the stratosphere is being depleted? (*Knowledge*)

KEY
Knowledge: Recalling information.
Comprehension: Showing understanding of recalled information.

CRITICAL THINKING REVIEW

1. Suppose the President of the United States asked you to put together a plan of action for addressing environmental problems. What steps would you recommend? *(Evaluation)*

2. Imagine that nothing was done to address the environmental problems you identified in your answer to question #1. Outline a scenario of the likely consequences—what would such a world be like in 20 years? *(Evaluation)*

3. Why is the destruction of tropical rain forests a serious problem for everyone—not just the people living in tropical countries? *(Application)*

4. Measure how much solid waste you (or your family) produce each week. Sort your solid waste into categories of yard waste, trash, and garbage (food waste). How much (by weight) of each do you produce weekly? Monthly? Yearly? What proportion of the total is each?

What steps could you take to reduce the amount of solid waste you contribute to the landfill? *(Evaluation)*

5. Contact local environmental agencies to determine the "health" of the land, air, and water of your city or town. Do you live in a gray air city, a brown air city, or neither? What is the level and type of air pollution in your region, city, or town? What is the status of the surface water? Do you have acid rain in your area? How much solid waste is produced in your region and where is it disposed? Develop as complete an environmental picture of your area as possible from available sources. *(Application)*

K E Y
Application: Using information in a new situation.
Evaluation: Making informed decisions.

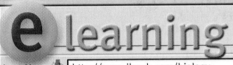

Location: http://www.jbpub.com/biology

e-Learning is an on-line student review area located at this book's web site www.jbpub.com/biology. The review area provides a variety of activities designed to help you study for your class and build your learning portfolio.

Review Questions The review questions test your knowledge of the important concepts and applications in each chapter. The review provides feedback for each correct or incorrect answer. This is an excellent test preparation tool.

Figure-Labeling Exercises Sharpen your visual thinking by matching terms and labels for important illustrations.

Flash Cards Studying biology requires learning new terms. Virtual flash cards help you master the new vocabulary for each chapter.

Just Wondering As you read and study from this text, you may find that you have unanswered questions. Through this site you can ask the author, Sandy Alters, your "just wondering" questions.

 Do you prefer the speed and reliability of a CD-ROM? All of the features contained on the eLearning portion of the web site are also available on the Student CD-ROM.

Chapter 3

Knowledge and Comprehension Questions

1. a. See the illustration in Section 3.1, bottom left. Your atom would be hydrogen and would have one proton, one electron, and from zero to two neutrons.

   b. The atom would still have one proton and one electron, but it would have a different number of neutrons from the atom in question 1.a., within the range mentioned.

2. Three factors that influence how an atom interacts with other atoms are: (1) the tendency of electrons to occur in pairs, (2) the tendency of atoms to balance positive and negative charges, and (3) the tendency of the outer shell (energy level) of electrons to be full. The third point is known as the octet rule because an atom with an unfilled outer shell tends to interact with other atoms to fill this outer shell. For many atoms, eight (an octet of) electrons fill the outer shell.

3. The atoms in nitrogen gas are bound together by covalent bonds, as are the hydrogen and oxygen atoms of water molecules. Both involve the sharing of electrons between molecules. However, nitrogen gas molecules have triple bonds, which are stronger bonds than the single covalent bonds in water.

4. Water is unusual because it is the only common molecule on Earth that exists as a liquid at the Earth's surface. This liquid enables other molecules dissolved or suspended in it to move and interact. A second important characteristic is its ability to form hydrogen bonds with itself and other molecules because it is a polar molecule. Water molecules are strongly attracted to ions and other polar molecules, giving water another important trait: It is an excellent solvent. Chemical interactions readily take place in water because so many molecules are water soluble.

5. Organic molecules tend to be large, covalently bonded, and carbon based. Inorganic molecules tend to be small, they do not usually contain carbon, and they interact by means of ionic bonding. If you wanted to study living things—such as humans—you would primarily learn about organic molecules.

6. This environment is slightly basic. Like all basic environments, the intestinal environment would have a relatively low level of free H^+ ions.

7. Water is a polar molecule. It quickly and easily forms hydrogen bonds, which, although not very strong, allow water to react with many other molecules. Carbon lends itself to being the basis of living material because of its ability to form four bonds with other atoms and therefore interact with other molecules in myriad ways.

8. Monosaccharides, disaccharides, and polysaccharides are all carbohydrates. Most organisms use carbohydrates as a primary fuel. Monosaccharides are among the least complex carbohydrates. Many organisms link monosaccharides to form disaccharides that are less readily broken down as they are transported within the organism. To store the energy from carbohydrates, organisms convert monosaccharides and disaccharides into polysaccharides, long insoluble polymers of sugars.

9. Plants and animals must store energy as insoluble polysaccharide compounds, and glucose is a very soluble sugar in either blood or water. Plants generally store sugars as starch, whereas animals store it as glycogen.

10. The three classes of macromolecules taken in as food energy by humans are carbohydrates, fats, and proteins. All three are made up of smaller units that, when disassembled, release energy. Because these molecules were all put together by dehydration synthesis, breaking them apart (digesting them) entails hydrolysis, an opposite process.

Critical Thinking Review

1. Oil is a nonpolar molecule, which means that it does not form hydrogen bonds with water and therefore does not dissolve. Answers will vary as to why an oil spill should be cleaned up, but one reason based on its nonpolarity is that the oil will form a film on the top of the water, affecting the amount of light entering the water. Lowered light levels would affect aquatic plant life. In addition, organisms that come to the surface of the water would get coated with this film of oil, which would probably affect their ability to survive.

2. The macromolecules present in this food are lipids, carbohydrates, and proteins. This food could be a part of a heart-healthy diet because it supplies only 17% of its calories from fat, well under the 25% to 30% recommendation. In addition, the amount of saturated fats as compared with the total fat is low—30%.

3. In living systems, proteins are used as enzymes and as structural components. Because DNA codes for the sequence of amino acids in a protein, an error in DNA can result in an error in the sequence of amino acids in a protein. This sequence is called the primary structure of a protein, and it affects the formation of the secondary and tertiary levels of structure. If a protein does not have the correct sequence of amino acids and is not structurally configured correctly, it cannot perform its job.

4. When each of these molecules is broken down into its component parts in the body, much more water is lost during hydrolysis in carbohydrates and proteins than in long chains of fatty acids, which are simply hydrolyzed from the glycerol backbone. Put simply, proteins and carbohydrates contain more water than do fats (lipids). (Additionally, fats are less polar than carbohydrates and proteins and therefore attract less water. A gram of dry glycogen binds about two grams of water!)

5. The chemical structure of a vitamin is the same whether it is made in the laboratory or derived from "natural" sources such as plants. They are no better. (However, nutrition scientists know that some vitamins and minerals in pills may not be absorbed as effectively by the body as are those same nutrients eaten in food.)

Visual Thinking Question

1. Reaction 1: dehydration synthesis
   Reaction 2: hydrolysis
2. Reaction 1: glucose, fructose
   Reaction 2: serine, valine, tyrosine, cysteine
3. Reaction 1: sucrose
   Reaction 2: polypeptide
4. Reaction 1: Two monosaccharides are being joined in a dehydration synthesis reaction, producing a disaccharide. Reaction 2: A polypeptide is being broken down by hydrolysis, producing amino acids.

Chapter 4

Knowledge and Comprehension Questions

1. The basic principles of the cell theory are: (a) All living things are made up of one or more cells; (b) the smallest living unit of structure and function of all organisms is the cell; and (c) all cells arise from preexisting cells.
2. The human body is composed of millions of cells that interact and work together, allowing the organism to survive. Cells are specialized for different tasks, yet all cells take in nutrients, maintain homeostasis, react to stimuli, and are composed of smaller units called organelles.
3. Large, complex cells that are active often have many nuclei. A single nucleus could not control the activities of such a large cell.
4. The cell would have produced the protein along ribosomes located on the rough endoplasmic reticula. The protein would then enter the inner space of the ER, and eventually be encased in a vesicle at the smooth ER. This vesicle would travel to the Golgi complex and fuse with its membrane. Within the Golgi complex it might be modified. The "finished" protein would be encased in a vesicle that would "bud" from the Golgi, pass to the plasma membrane, and leave the cell via exocytosis.
5. If the lysosomes stopped working, your cells would soon fill up with old cell parts and foreign substances such as bacteria. Your cells would probably die as these substances accumulated.
6. See Table 4.1.
7. See Table 4.2 for a summary of the differences. For example, the cell with chloroplasts would be the plant, the cell with mitochondria but no chloroplasts would be the animal, and the one with neither mitochondria nor chloroplasts would be the bacterium.

8. No. The ancient bacteria that were endosymbionts to other cells and developed into chloroplasts or mitochondria appeared millions of years ago. Today, mitochondria and chloroplasts are functioning organelles that are part of the organism itself, as are other cell parts.
9. Both diffusion and facilitated diffusion occur along a concentration gradient and do not require energy. However, unlike diffusion, facilitated diffusion involves specific transport proteins that assist the movement of molecules across the membrane.
10. See Figure 4.25. White blood cells use endocytosis to engulf bacteria and other foreign substances.

Critical Thinking Review

1. Nutrients will enter the cell in different ways, depending on the nutrient. Water (because it is a small molecule) and the lipid-soluble vitamins (A, D, and E) are the only nutrients that can pass directly through the plasma membrane. The other nutrients, fatty acids and glycerol (lipids), amino acids (proteins), sugars (carbohydrates), and minerals must move through the membrane by means of carrier molecules or endocytosis.
2. a. Facilitated diffusion (passive).
   b. Diffusion (passive).
   c. Active transport (active).
3. The solute is the instant coffee crystals, the solvent is water, and the solution is the liquid coffee.
4. Blood is made up of a solution of water with various dissolved substances and cells; the fluid portion of the blood is isotonic with respect to the red and white blood cell contents. Plain water is hypotonic to blood and to blood cells. The dehydrated patient would run the risk of the blood cells absorbing water until they eventually lyse if water were put into the blood, causing the fluid portion of the blood to become hypotonic to the blood cells.
5. Cilia in the respiratory tract of humans sweep invading bacteria and particles up the trachea and away from the lungs. If ciliary effectiveness is reduced because of smoking, invading organisms or particles could pass down the windpipe and enter the lungs, causing damage to the delicate lung tissue that might result in disease.

Visual Thinking Question
See Figures 4.5 and 4.6.

Chapter 5

Knowledge and Comprehension Questions

1. Free energy of activation destabilizes existing chemical bonds in the substrates of a reaction, allowing the reaction to proceed.

2. The reactions involved in building complex molecules, such as blood cells, from simpler substrates would be anabolic reactions. Energy is used to build the chemical bonds, and the product contains more energy than the substrates. The process of breaking down monosaccharides, such as glucose, is a catabolic reaction, which releases energy.

3. The first law of thermodynamics states that energy cannot be created or destroyed; it can only be changed from one form to another. The second law states that disorder in the universe constantly increases and energy is continually lost (as heat) to this disorder.

4. Entropy is the energy lost to disorder. Your friend is restating the second law of thermodynamics, which holds that disorder in the universe is constantly increasing.

5. Metabolic pathways are the chains of reactions within your body that move, store, and free energy. By means of metabolic pathways, people obtain energy from food, repair damaged tissues, and in general, avoid increasing entropy.

6. A catalyst increases the rate and often the probability of a chemical reaction while remaining unchanged itself. It may reduce the free energy of activation needed by placing stress on the bonds of a substrate and bringing substrates together.

7. The cells of your body require such a high number of enzymes because there is a high degree of specificity in enzymatic reactions; only specific enzymes catalyze specific reactions.

8. Both activators and inhibitors bind to specific sites on enzymes and affect enzymatic activity. However, activators stimulate catalysis whereas inhibitors restrict it. Inhibitors play a key role in negative feedback loops. As end products are produced in metabolic pathways, they may serve as inhibitors by binding to the first enzyme of the metabolic pathway and causing enzymatic action to stop at that point, thus shutting down the pathway.

9. Cofactors are special nonprotein molecules that help enzymes catalyze chemical reactions; one example is a zinc ion that helps digestive enzymes break down proteins in food. Coenzymes are cofactors that are nonprotein organic molecules. Many vitamins are synthesized into coenzymes in your body.

10. ATP is a high-energy compound that can be broken down easily to release energy for use by cells. It is called "energy currency" because cells can save energy released in exergonic reactions by storing it as ATP; cells can also spend ATP to provide necessary energy for endergonic reactions.

Critical Thinking Review

1. The speed of the ATP recycling process allows the cell to fuel a large variety of cell processes as quickly as possible. The process also has endergonic and exergonic aspects so that it can make full use of the cell's nutrient supply as it varies.

2. Most enzymes function best within a narrow range of pH. Water and ions enter the cells of the plants from their environment, possibly changing the internal environment of the plants. If the plants' internal environment becomes too acidic, enzymes could be affected, causing chemical reactions to stop, which ultimately would affect the metabolic pathways of organisms.

3. Enzymes are proteins. The directions for making proteins are encoded in the DNA of an organism. A feedback system might trigger particular enzymes to begin transcribing those portions of the DNA that code for the particular enzymes needed.

4. The energy that is released when ATP is broken down to ADP and inorganic phosphate is used to do cell work or synthesize molecules. Some of this energy may be lost as heat. It is no longer available to be stored once again in ATP. "New" energy stored in the chemical bonds of food is used to keep the ADP–ATP cycle functioning in living systems.

5. The Earth constantly receives energy from the sun, which replaces energy "lost" as heat. This situation does not contradict the first law of thermodynamics because energy "lost" as heat is simply unavailable to do work, and the "new" energy is simply energy that has reached the Earth and become available to do work. Photosynthetic organisms capture this energy and convert it to chemical energy, storing it in the bonds of the carbohydrates they manufacture. When photosynthesizers are eaten by other organisms, this energy becomes available to them as these carbohydrates are broken down. The photosynthetic organisms are key to capturing the energy from the sun for all the organisms that eat plants or eat the organisms that eat plants.

Visual Thinking Question

See Figure 5.8.

Chapter 6

Knowledge and Comprehension Questions

1. Autotrophs are organisms that produce their own food by photosynthesis or by using chemical energy rather than light energy. Heterotrophs are organisms that cannot produce their own food; they depend on the autotrophs and the sun for survival. Humans are heterotrophs.

2. The visible light in sunlight consists of many wavelengths. The prism separated the various wavelengths so that each color became visible as the light shone on the floor.

3. Photosynthesis involves two series of reactions: (a) synthesis of ATP and NADPH and (b) the use of ATP and NADPH to produce sucrose and other sugars. The reactions that produce ATP and NADPH are called light-dependent reactions because they occur only in the presence of light.

During the light-independent reactions (which do not need light to drive the reactions), ATP and NADPH are used to produce glucose from carbon dioxide.

4. A photocenter is an array of pigment molecules within the thylakoid membranes of a chloroplast. It channels photon energy to chlorophyll *a*, a special molecule of chlorophyll that participates in photosynthesis. Two kinds of chlorophyll *a* act as reaction centers in the two photosystems of photosynthesis, absorbing and then ejecting energized electrons that provide the energy (originally from the sun) for the reactions of photosynthesis.

5. Such a dust cloud would interfere with photosynthesis by blocking the light from the sun. All life on Earth ultimately depends on the process of carbon fixation (photosynthesis) powered by the sun's light, and on the byproduct, oxygen. Eventually all organisms would die if photosynthesis did not take place.

6. Cellular respiration and fermentation are both metabolic processes that break down nutrient molecules (glucose) to yield ATP (useable energy), although fermentation yields less ATP per molecule of glucose. All organisms that use oxygen break down nutrient molecules by means of aerobic cellular respiration. Organisms that do not "breathe" oxygen make ATP by either anaerobic cellular respiration or fermentation. Fermentation consists of glycolysis, an initial series of reactions of cellular respiration, and one or two additional reactions.

7. Your body metabolizes foods by breaking down complex molecules into simple ones. The digestive system breaks down the proteins (hamburger meat) into amino acids, the carbohydrates (bun and fries) into simple sugars, and the lipids (within hamburger meat and fried potatoes) into fatty acids and glycerol. The carbohydrates (now glucose) are further broken down to release energy within the cells by means of glycolysis, the Krebs cycle, and the electron transport chain. The amino acids, fatty acids, and glycerol enter the Krebs cycle without undergoing glycolysis.

8. Three main events of glycolysis are that (1) glucose is converted to pyruvate, (2) ADP + P_i is converted to ATP, and (3) NAD^+ is converted to NADH.

9. See Figure 6.16.

10. Oxygen is the electron acceptor (thus it is reduced) for both the electrons and protons of the electron transport chain. Water is formed as a result of this reaction. It is an important end product of cellular respiration; 1 H_2O molecule is produced for each NAD or $FADH_2$ molecule that enters the electron transport chain.

Critical Thinking Review

1. The extra plant mass came from the process of carbon fixation (photosynthesis), in which carbon dioxide and energy from the sun were used to manufacture sucrose and other carbohydrates for the plant. This plant had made more organic material than it needed as fuel, and it was used to build new plant parts for growth, as well as to provide starch for storage. In addition, as the plant grows, more water is taken up by the plant and held in its tissues, which would be reflected in its weight.

2. The roots do not manufacture ATP via photosynthesis but via cellular respiration. Sucrose would be transported to the roots from the photosynthetic parts of the plant, where it would be broken down during respiration, yielding ATP.

3. The absorption spectrum for chlorophyll *a* depicted in Figure 6.4 shows that chlorophyll *a* absorbs many violet and red wavelengths. These are the wavelengths effective for photosynthesis because they are the wavelengths that are absorbed. The others are reflected. Other wavelengths effective for photosynthesis are absorbed by other pigments; chlorophyll *b* absorbs many of the violet and red wavelengths chlorophyll *a* does not. Although it is not shown in the figure, the carotenoids absorb wavelengths in the green and blue range (as well as violet), increasing the wavelengths effective for photosynthesis.

4. True. The breakdown of glucose during cellular respiration is exergonic in that energy is released from the glucose being cleaved. The energy is then used endergonically to produce molecules of ATP.

5. Answers may vary, but this chapter shows how the mitochondria are the place in which a more complete breakdown of food molecules occurs than in the cytoplasm. It seems likely that a symbiotic relationship between certain cells with bacteria could result millions of years later in eukaryotic cells having bacterialike organelles. The structure of the mitochondria is discussed more fully in this chapter in terms of its function—the inner membrane (possibly the original bacterial membrane) is the site of biochemical activity as it is in bacteria. The outer membrane (which is relatively inactive in cellular respiration) could be derived from the plasma membrane of the cell, and not be bacterial in origin.

Visual Thinking Question

See Figure 6.10. The events that occur outside the mitochondrion are those of glycolysis. The events of the Krebs cycle and the electron transport chain occur within the mitochondrion.

Chapter 7

Knowledge and Comprehension Questions

1. (Examples will vary.) A cell is the smallest structure in an organism that is capable of performing all the functions necessary for life. An example is a cardiac muscle cell. The cells of the body are organized into tissues (groups of similar cells that work together to perform a function, such as

cardiac muscle tissue). Several different tissues group together to form a structural and functional unit—an organ (the heart). An organ system is a group of organs that works together to carry out the body's principal activities (such as the circulatory system).

2. (Examples will vary.) The four types of tissue in the body are: (1) epithelial (lining of the body cavities), (2) connective (tendons), (3) muscle (skeletal muscles such as leg muscles), and (4) nervous (the brain).

3. (Examples will vary.) See Summary Figure 7.6. Squamous (flattened) cells are found in the air sacs of the lungs, cuboidal cells line ducts in glands, and columnar cells line much of the digestive tract.

4. Lymphocytes, erythrocytes, and fibroblasts are all cells found in connective tissue, but they have different functions. In addition, lymphocytes and erythrocytes are found in the blood, but fibroblasts are found in other types of connective tissue. Lymphocytes defend the body against infection; erythrocytes pick up and deliver gases in the blood; and fibroblasts produce fibers that are found in various types of connective tissue.

5. Oxygen delivery by means of red blood cells is necessary for cells to use the aerobic process of cellular respiration. This process is the primary means of ATP (hence, energy) production in our bodies.

6. Yes, fat cells have their uses. In addition to storing fuel, fat tissue helps shape and pad the body and insulate it against heat loss.

7. Skeletal muscle; cardiac muscle.

8. A coelomate (example: human) has a fluid-filled body cavity lined with connective tissue. Organs are located within this body cavity. A pseudocoelomate (example: roundworm) has a fluid-filled cavity that houses the organs, but that is not lined by connective tissue. An acoelomate (example: flatworm) has no body cavity and its organs are embedded within the other tissues of the body.

9. See Table 7.1.

10. Homeostasis is the maintenance of a stable environment inside your body despite varying conditions in the external environment. The molecules, cells, tissues, organs, and organ systems in your body must all work together to maintain this internal equilibrium.

Critical Thinking Review

1. Like most living tissues, bone is constantly renewing itself. Blood vessels within the bone bring calcium and other substances necessary to heal a break to the area. In addition, the bone cells are able to lay down new matrix to help repair a break.

2. This is part of the body's inflammation response to injury. Mast cells produce histamine, which dilates blood vessels and increases blood flow to the area. This makes the area turn red and feel warm.

3. The connective tissue matrix may provide a nutrient pool for connective tissue cells, allow for waste removal, and provide an environment enhancing cell mobility. Proteins secreted into the matrix may provide a fibrous, stronger structure to matrix, such as that found in bone and cartilage.

4. (Answers may vary, for the chapter gave only the basics on which to base an answer.) Nervous tissue is specialized for conducting electrical impulses. The brain is specialized, in addition, for interpreting electrical impulses. This suggests that neurons are connected to one another in particular pathways, unlike other tissues such as skin. Therefore, a reasonable suggestion might be that nervous tissue in the brain and spinal cord is incapable of regeneration based on properties of its specialization regarding carrying messages and, more importantly, interpretation (in the brain). Other tissues do not have such a specialization. These pathways, once severed, may be too complex to "hook up" to one another again.

5. The coelom acts as a transportation system because it is fluid-filled. Substances move about the body dissolved in its fluids. The blood is the major transport mechanism of the body, but substances move among all the body fluids by passive and active transport.

Chapter 8

Knowledge and Comprehension Questions

1. The six classes of nutrients are carbohydrates, fats, proteins, vitamins, minerals, and water. The first three are organic compounds that your body uses as a source of energy; they are also used as building blocks for growth and repair and to produce other important substances. Vitamins, minerals, and water help body processes take place. Some minerals are also part of body structures.

2. Three types of enzymes help to digest the energy nutrients. Proteases break down proteins into peptides and amino acids. Amylases break down carbohydrates (starches and glycogen) to sugars. Lipases break down the triglycerides in lipids to fatty acids and glycerol.

3. The process of digestion begins in the mouth, as the saliva lubricates food and the salivary amylase breaks down starches into maltose.

4. You would bite off the mouthful with your incisors, your canines would help you tear the food away, and your premolars and molars would allow you to crush and grind the food thoroughly.

5. Peristalsis is a series of involuntary muscular contractions. The esophagus transports the food from mouth to stomach, but no digestive processes occur here (except for any continuing digestion by salivary amylase mixed with the food).

6. The hydrochloric acid in the stomach converts pepsinogen to the active form of pepsin, softens connective tissue in

food, denatures large proteins, and kills bacteria. The stomach wall is protected by thick mucus.

7. The duodenum is the first part of the small intestine; digestion is completed there. Starch and glycogen are broken down to disaccharides and then to monosaccharides. Proteins and polypeptides are broken down to shorter peptides and then to amino acids. Triglycerides are digested to fatty acids and glycerol.

8. These are hormones that help regulate digestion. Gastrin controls the release of hydrochloric acid in the stomach. Secretin stimulates the release of sodium bicarbonate to neutralize acid in the chyme and increases bile secretion in the liver. Cholecystokinin stimulates the gallbladder to release bile into the small intestine and stimulates the pancreas to release digestive enzymes.

9. Increasing the internal surface area of the small intestine helps absorb more nutrients from the same amount of food. Three features accomplish this: inner folds, projections (villi) on the folds, and additional projections on the projections (microvilli).

10. Glucose is broken down by glycolysis, the Krebs cycle, and the electron transport chain. Amino acids and triglycerides are converted to substances that can be metabolized by two of these pathways. Fatty acids are metabolized by another pathway.

Critical Thinking Review

1. Meat and animal products provide protein to the diet and some fat. Because we Americans eat too much fat, eating a vegetarian diet would most likely lower fat intake, a positive dietary consequence. To replace the protein, a vegetarian could eat legumes (dried peas, beans, peanuts, or soy-based food), grains, nuts, and seeds. To take in the essential amino acids, legumes should be combined with any grain, nut, or seed supplemented with small amounts of milk, cheese, yogurt, or eggs. Persons who are not strict vegetarians might also consider eating small amounts of fish or poultry.

2. The shape of the teeth on the skull would indicate the type of food eaten by the animal. Herbivores (plant-eaters) have flat, grinding teeth that look like human molars.

3. The liver and pancreas are not organs in which the digestive products are altered and pass through and so are called accessory organs. They are not a stop in the digestive tract itself.

4. Sodium bicarbonate neutralizes the acid in chyme. Thus, it can help reduce the uncomfortable symptoms of acid indigestion.

5. Answers will vary. See Section 8.14 and the "How Science Works" box. Comparisons should be made to percentage intake of fats, proteins, and carbohydrates, or to numbers of servings of categories of foods recommended by the Food Pyramid.

Chapter 9

Knowledge and Comprehension Questions

1. Organisms without special organs of respiration exchange respiratory gases over cell and skin surfaces. These organisms are either single-celled, multicellular with a large surface-to-volume ratio, or large with thin, moist skins supplied with abundant blood vessels.

2. Respiration is the uptake of oxygen and the release of carbon dioxide by the body. Cellular respiration is the process at the cellular level that uses the oxygen you breathe in and produces the carbon dioxide you breathe out. Internal respiration is the exchange of oxygen and carbon dioxide between the blood and tissue fluid; the exchange of the two gases between the blood and alveoli is external respiration.

3. Your vocal cords produced the sound for speech as air rushed by and made them vibrate. Your lungs served as a power supply and volume control for your voice, and your lips and tongue formed the sounds into words.

4. As the volume of the thoracic cavity increases, the air pressure within it decreases. Outside air is pulled in, equalizing the pressures inside and outside the thoracic cavity. As the volume of the thoracic cavity decreases, the air pressure inside it increases and forces air out, equalizing the pressures.

5. At the alveoli, carbon dioxide diffuses out of the blood into the alveoli because it moves down the pressure gradient (its pressure in the blood is greater than its pressure within the alveoli). Meanwhile, oxygen diffuses from the alveoli into the blood because its pressure within the lungs is greater than in the blood.

6. Hemoglobin is known as the oxygen-carrying molecule, but it also carries almost 25% of circulating carbon dioxide in the bloodstream.

7. The hemoglobin in the deoxygenated blood of your veins absorbs oxygen in the air as soon as it leaves the body. Thus, wounds appear to bleed bright red blood even when a vein is involved.

8. At the capillaries, oxygen moves down a pressure gradient to diffuse from the blood into the tissue fluid. Meanwhile, carbon dioxide moves down its own pressure gradient and diffuses from the tissue fluid into the blood.

9. You apply the Heimlich maneuver when a victim cannot cough or talk and is unable to breathe.

10. At high altitudes, the pressure of the oxygen molecules in the air is lower than at sea level. This means that the pressure gradient is lower at the alveoli and less oxygen diffuses into the blood. This can cause you to feel dizzy and short of breath.

Critical Thinking Review

1. During hibernation, the metabolism of the frog slows, reducing the amount of oxygen the frog uses. Buried in the mud, the frog breathes through its thin skin, which has an

abundant supply of blood vessels just beneath the skin surface. Air is available in the mud, surrounding the microscopic spaces between particles of dirt and sand.

2. The sound of the hiccup is produced as the sudden inhaled air of the spasm causes the glottis at the back of the throat to close suddenly. The opening of the vocal cords and quick closing produce a loud gasp.

3. The circulatory system provides a transport system that distributes gases throughout the body. Without it, you would not survive because it would take too long for oxygen to diffuse from the lungs to the rest of the body.

4. One of the primary symptoms of respiratory diseases is that patients tire easily. This is because their decreased lung capacity does not allow for sufficient reoxygenation of blood and produces an overall effect of less oxygen reaching the brain and other organ systems.

5. Water and blood flow over the gill surface in opposite directions, which is called countercurrent flow. As the water gives up its oxygen, its partial pressure of oxygen drops steadily. As blood picks up oxygen, its partial pressure of oxygen rises steadily. Since the blood is moving in a direction opposite to the water, the partial pressure of oxygen in the water never drops below that in the blood. If the flow of these two fluids were in the same direction, diffusion would soon stop as the partial pressure of oxygen in the blood rose while it fell in the water.

Visual Thinking Question

See Figures 9.4 and 9.7d.
1. Arteries, arterioles, and eventually capillaries at the aveoli.
2. carbon dioxide
3. pulmonary vein
4. the heart (left side), to be pumped to the body

Chapter 10

Knowledge and Comprehension Questions

1.

Open system	Closed system
Circulating fluid is not enclosed in vessels; may have open-ended vessels	Circulating fluid is enclosed in vessels
Circulating fluid bathes internal organs directly	Circulating fluid does not bathe internal organs directly
May have a heart to pump the circulating fluid	Has a heart or specialized vessels to pump the circulating fluid
Hemolymph (blood mixed with tissue fluid) is the circulating fluid	Blood is the circulating fluid

2. The human circulatory system has four main functions: (a) nutrient and waste transport, (b) oxygen and carbon dioxide transport, (c) temperature maintenance, and (d) hormone circulation.

3. The blood carries sugars, amino acids, and fatty acids to the liver, where some of the molecules are converted to glucose and released into the bloodstream. Excess energy molecules are stored in the liver for later use. Essential amino acids and vitamins pass through the liver into the bloodstream. The cells release their metabolic waste products into the blood, which carries them to the kidneys.

4. As blood flows through tissue, from capillaries to venules, it loses gases (primarily oxygen) and nutrients to the tissues. The blood in the venules is deoxygenated and carries waste products from the tissues.

5. When people are scared (or cold), the walls of the arterioles can contract and the blood flow decreases, routing more blood to other areas of the body. This can cause light-skinned people to "turn pale."

6. The walls of arteries have more elastic tissue and smooth muscle than those of veins; this helps arteries accommodate the pulses and high pressures of blood pumped from the heart. The lumen of veins is larger because blood flows through them at lower pressures; a larger diameter offers less resistance. The walls of capillaries are only one cell thick, which permits the exchange of substances between the cells and the blood.

7. The heart has left and right sides, which both serve as pumps yet are not directly connected with one another. Oxygenated blood from the lungs enters and is pumped from the left side, while deoxygenated systemic blood passes through the right side into the lungs. These two types of blood must be separated—hence the two-pump system.

8. Valves are necessary in heart function to prevent the backflow of blood.

9. The SA node is the pacemaker of the heart; this cluster of cells initiates the excitatory impulse that causes the atria to contract and the impulse to be passed along to other cardiac cells. The AV node conducts the impulse from the SA node to other cardiac cells, initiating the contraction of the ventricles.

10. See Visual Summary Figure 10.8. Plasma is the fluid in which various substances, such as nutrients and ions, are dissolved. Plasma carries these dissolved substances and the formed elements throughout the body and provides the fluid in which chemical reactions and the movement of substances takes place.

Critical Thinking Review

1. The student should list family members who have (or have had) heart disease. The student should also list lifestyle factors that may contribute to heart disease such as whether they consume a high-fat diet, if they are overweight, if they smoke cigarettes, and if they exercise regularly.

2. Obviously, one cannot change one's genetics. The other factors are behavioral and can be changed. Answers will vary depending on the answer to question #1.
3. The left ventricle of the heart is the "workhorse" portion of the heart. It has the thickest musculature, pumping blood out to the body. The right ventricle has a less ominous job, pumping blood to the nearby lungs. The atria simply squeeze blood to the ventricles. Therefore, replacing the pumping action of the left ventricle is the most important job in heart-replacement technology.
4. The first heart sound occurs at the start of ventricular contraction (systole), so should be shown on your diagram just after the "R" peak. The second heart sound occurs at the end of ventricular contraction, and should coincide with the end of ventricular systole (part-way through the T wave).
5. Lymph nodes filter the lymph as it passes through. When a cancer spreads, some cells break away from the original cancer and travel to other parts of the body in the blood and lymph. If cancer cells are traveling from the original site of the cancer (the breast), some of these cells may become trapped by the lymph nodes nearest to that site (nodes in the arm pits).

Chapter 11

Knowledge and Comprehension Questions

1. See Table 11.1 and descriptions in Section 11.3.
2. In this case, the immune response (specific resistance) is not involved. Nonspecific resistance includes cases where foreign matter is trapped by mucous membranes and expelled from the body, probably via tears or eye watering.
3. The transplanted cell would be identified as "non-self"; therefore, it would act as an antigen and probably produce an immune response in the new environment.
4. The five principal types of T cells are: (a) helper T cells (which initiate the immune response), (b) cytotoxic T cells (which break apart infected and foreign cells), (c) inducer T cells (which oversee the development of T cells in the thymus), (d) suppressor T cells (which limit the immune response), and (e) memory T cells (which respond quickly and vigorously if the same foreign antigen reappears).
5. During the cell-mediated immune response, cytotoxic T cells recognize and destroy infected body cells in addition to destroying transplanted and cancer cells. Helper T cells initiate the response, activating cytotoxic T cells, macrophages, inducer T cells, and finally, suppressor T cells.
6. During the antibody-mediated response, B cells recognize foreign antigens and, if activated by helper T cells, produce antibodies. The antibodies bind to the antigens and mark them for destruction.

7. Antibodies have highly specific binding sites that fit only one specific antigen with the exact amino acid composition necessary for complete binding.
8. Injection with antigens causes an antibody (B cell) response, with the production of memory cells. A booster shot induces these memory cells to differentiate into antibody-producing cells and form still more memory cells.
9. The human immunodeficiency virus (HIV), which is responsible for AIDS, is dangerous because it destroys helper T cells, thus destroying the immune system's ability to mount a defense against any infection.
10. An allergic reaction occurs when the immune system mounts a defense against a harmless antigen. When class E antibodies are involved and bind to the antigens, they cause a strong inflammatory response that can dilate blood vessels and lead to symptoms ranging from uncomfortable to life-threatening.

Critical Thinking Review

1. The phagocytes serve to sound the alarm for an immune response if the body is being invaded by foreign microbes. They offer a quick response by engulfing the invader, stimulating the maturation of large numbers of monocytes into macrophages, and inducing an inflammation or fever to resist microbial growth.
2. Memory T cells were developed by your immune system as part of your immune response. These cells give you an accelerated response during subsequent exposures to measles, promptly defending you against infection.
3. This is a case of passive immunity, in which an organism is protected not by its own cell-mediated immune response, but by short-term protection from another source (in this case, the maternal immune system). This protection is short-term because these cells are not the baby's own. Eventually they die; only the baby's immune cells are propagated by the baby's body.
4. Answers will vary, but students should compare their activity level to the data given.
5. Antihistamines suppress the action of histamines, which are released during an allergic reaction. Histamines cause a variety of responses, such as the dilation of blood vessels. With a respiratory allergy, for example, these responses could make you feel "stuffed up." An antihistamine would block the production of histamine, diminishing its effects.

Visual Thinking Question

1. Nonspecific defenses are a "first line" of defense because they include a wide variety of defenses and work against any foreign invader. On first exposure to a pathogen, the body's specific resistance takes time to mount a defense, and the nonspecific defenses help give the body that time.

2. Helper T cells are key to stimulating both the cell-mediated immune response and the antibody-mediated immune response. Helper T cells are so important to specific immune system function that the progression of disease in an HIV-infected individual coincides with progressive depletion of T cells.

3. Specific immunity is triggered when the macrophages engulf and display antigen. Interaction with the Helper T cells stimulates both "arms" of the immune system.

4. Nonspecific resistance acts against any invader. Cell-mediated immunity recognizes and destroys infected body cells. Antibody-mediated immunity kills bacteria and viruses outside of body cells.

Chapter 12

Knowledge and Comprehension Questions

1. Excretion is the removal of metabolic wastes and excess water from the body. Without excretion, your body would soon become poisoned by the build-up of metabolic waste products. The process also helps maintain the balance of water and ions that is necessary for life.

2. Organisms with no excretory organs rid themselves of wastes by means of diffusion, osmosis, and active transport.

3. The primary metabolic waste products of humans are carbon dioxide, water, salts, and nitrogen-containing molecules.

4. Urea, urine, and uric acid are all forms of metabolic waste products (nitrogenous wastes). Urea is the primary excretion product from the deamination of amino acids and is a compound formed from ammonia and carbon dioxide. Uric acid forms from the breakdown of the nucleic acids you eat and from the metabolism of your own nucleic acids and ATP. Urine is the excretion product consisting of water, urea, creatinine, uric acid, and other substances.

5. See Figure 12.4.

6. Kidneys help conserve water by reabsorbing it from the filtrate. By changing the concentrations of salts and urea in the filtrate, the kidneys create an osmotic gradient that results in water moving from the filtrate into the surrounding tissue fluid.

7. Urine can reveal the presence of certain substances in the body because the kidney detoxifies the blood. When these substances, such as drugs, are removed from the blood, they become part of the filtrate.

8. Urine is formed when the blood that flows through the kidneys is filtered, removing most of its water and all but its largest molecules and cells. The kidneys then selectively reabsorb certain substances.

9. Homeostasis is the maintenance of constant physiological conditions in the body. ADH and aldosterone are two hormones that help regulate the water and ion balance necessary for homeostasis. ADH regulates that amount of water reabsorbed at the collecting ducts; aldosterone promotes the retention of sodium (and therefore water), while promoting the excretion of potassium.

10. Two problems with kidney function are kidney stones and renal failure. Kidney stones are crystals of certain salts that can block urine flow. Renal failure is a reduction in the rate of filtration of blood in the glomerulus.

Critical Thinking Review

1. Carbon dioxide is a waste product of cellular respiration and is used by the liver to form urea. Other excess carbon dioxide is excreted by the lungs.

2. Neither. You need to conserve water. Alcohol acts as a diuretic; drinking the beer will cause the body to excrete water. Salt will be reabsorbed to its threshold; remember, the kidney's main job is regulating salt and water balance. Drinking salt water will cause the body to excrete water with the excess salt that is taken in, because water follows salt osmotically at the collecting ducts.

3. The body rids itself of harmful drugs by tubular secretion. These drugs are processed by the liver, and their breakdown products are secreted into the filtrate within the kidney tubules, becoming part of the urine.

4. Persons with diabetes mellitus have a lack or partial lack of insulin or the inability of tissues to respond to insulin, which leads to increased levels of glucose in the blood. (Insulin allows blood to enter the cells and leave the blood.) In the kidneys, glucose is filtered out of the blood and returned to only a threshold level; the excess glucose is excreted in the urine. Additionally, high levels of glucose in the blood result in water moving out of the body's tissues by osmosis. The kidneys remove this excess water through increased urine production.

5. The mitochondria in the region of contractile vacuoles generate the ATP necessary for the contraction of the vacuole.

Chapter 13

Knowledge and Comprehension Questions

1. Neurons and hormones are both forms of communication that integrate and coordinate body functions. They differ in that neurons transmit rapid signals that report information or initiate quick responses in specific tissues. Hormones are chemical messengers that trigger widespread prolonged responses, often in a variety of tissues.

2. Resting potential refers to the difference in electrical charge along the membrane of the resting neuron. The action of sodium-potassium transmembrane pumps and ion-specific membrane channels separates positive and negative ions along the inside and outside of the membrane, which creates the resting potential. The difference in electrical charges is the basis for the transmission of nerve impulses.

3. A nerve impulse travels because nearby transmembrane proteins respond to the electrical changes that accompany depolarization of the nerve cell membrane. The adjacent section of membrane depolarizes, followed by another section, leading to a wave of depolarization.

4. Saltatory conduction occurs along myelinated neurons when impulses jump to unmyelinated areas and skip over myelinated portions. Impulses travel much more quickly by this method than by continuous waves of depolarization that occur along continuous portions of unmyelinated membrane.

5. The decision to move your finger is made in the brain. A nerve impulse passes along neuronal pathways that end at the muscles in your finger. At the neuromuscular junction, acetylcholine is secreted from the presynaptic vesicles of the neuron. This neurotransmitter passes across the synapse and binds to receptors in the muscle cell membrane, opening sodium channels. A depolarization of the muscle cell membrane follows, which ultimately triggers muscle contraction.

6. The decision to stop moving your finger is made in the brain and would result in inhibitory neurotransmitters being released to postsynaptic neurons that make up the pathway of neurons to your muscle. The ability of the postsynaptic membrane to depolarize would be reduced, and the muscle would no longer be stimulated to contract.

7. Neurons in the CNS receive both excitatory impulses and inhibitory impulses. They "sum up" (integrate) these opposing impulses and respond accordingly.

8. All drugs, whether stimulatory or inhibitory, act by interfering with the normal activity of neurotransmitters.

9. Although a few alcoholic drinks or sleeping pills (barbiturates) are not themselves dangerous, the combination may be deadly. Barbiturates and alcohol enhance the sedative activities of each other, leading to a potentially dangerous combination.

10. These stimulants block the breakdown and recycling of neurotransmitters, causing a depletion of the former. At this point, the drug abuser needs consistently higher doses of stimulant (tolerance has developed).

Critical Thinking Review

1. The nerve impulse is not a chemical reaction because there are no changes in molecules in which one substance is changed to another. It consists of the movement of charged particles called ions, along the length of the neuron membrane. Therefore, the nerve impulse is actually an electrical-chemical event.

2. Psychoactive drugs, by affecting neurotransmitter transmission in certain parts of the brain, can produce emotional and physical states that are, at least temporarily, extremely exciting or pleasant to the drug abuser. This effect often outweighs the person's intellectual considerations of the drug's potential for harm if abused. Drug addiction may oc-

cur quickly, and withdrawal symptoms make it very difficult to stop the drug use without professional help.

3. Drug tolerance develops over time, not only causing an addict to require increased drug doses for the desired effect but also increasing the drug's side effects and swifter degradation in the body. Obtaining these higher doses may become financially unmanageable and lead to criminal behavior or cause side effects that are antisocial and harder for an addict to hide. Heavy drug use also may interfere with judgment and assessment of situations and interactions between people.

4. Marijuana has a variety of health-related effects as described in Section 13.11 (Hallucinogens). If this drug were legalized and its use increased, incidence of these health-related effects would rise as well.

5. Intensity of a sensation such as pain varies from occurrence to occurrence because (1) the number of neurons carrying the message differs, and (2) the brain interprets these impulses modified by a person's emotional and psychological response to the particular situation.

Visual Thinking Question

See Visual Summary Figure 13.11.

Chapter 14

Knowledge and Comprehension Questions

1. The simplest type of invertebrate nervous system is the nerve net, a system of interconnecting nerve cells with no central controlling area.

2. A ganglion is an aggregation of nervous tissue. It is not really a brain because it is not capable of conscious thought or emotion, but simply processes incoming and outgoing impulses.

3. The vertebrate central nervous system (brain and spinal cord) is the site of information processing. The peripheral nervous system (nerves) shuttles messages to and from the central nervous system.

4. The somatic nervous system consists of motor neurons that control voluntary responses (for example, moving your leg to take a step). The autonomic nervous system consists of motor neurons that control involuntary activities (for example, digesting a meal).

5. Integration would occur in the spinal cord for such a simple reflex.

6. Integration would occur, ultimately, in the cerebrum, but the sensory input travels to the cerebrum through the limbic system.

7. The brainstem (midbrain, pons, medulla) contains tracts of nerve fibers that carry messages to and from the spinal cord. Nuclei located there control important body reflexes.

8. Sensory receptors change stimuli into nerve impulses. The

nerve fibers of sensory neurons carry this information to the central nervous system, where interneurons interpret them and direct a response (in this case, turning your head). The axons of motor neurons conduct these impulses to the appropriate muscles.

9. The way you perceive stimuli depends on which receptors are stimulated. You registered this stimulus as a sound because it stimulated your auditory nerve.

10. These systems act in opposition to each other to maintain homeostasis and help you respond to environmental changes. The sympathetic nervous system generally mobilizes the body for greater activity (faster heart rate, increased respiration), whereas the parasympathetic nervous system stimulates normal body functions such as digestion.

Critical Thinking Review

1. The areas of the hands and face are associated with the primary senses (touch, taste, sight) and, as such, require more of the cortex area to interpret and react to increased input from sensory and motor neurons.

2. Infants and toddlers are quickly developing nervous system connections and expanding motor skills. The myelin sheath of many nerve fibers, particularly in the cerebral white matter and motor neurons, is critical for nerve signal transmission and normal development.

3. The limbic systems of both a lion and a human respond to similar stimuli (pain, sexual drive, hunger, thirst) and would be somewhat similar, whereas their associative cortexes would be quite dissimilar in function, because primates (such as humans) have a larger portion of the cerebrum devoted to associative activities than do nonprimate mammals (such as lions).

4. As the spinal cord extends down the back from the brain, many sensory and nerve tracts emerge from it along its length. The higher the point of an injury on the spinal cord that severs it from the brain, the greater the number of sensory and motor nerve tracts that are no longer connected to the brain because of the injury.

5. The left side of the grandmother's brain was affected because she experienced numbness on the right side of her body. (Most of the tracts of neurons cross over one another within the medulla.) Most likely the speech center in the frontal lobe was affected as well as the primary sensory area on the left parietal lobe.

Visual Thinking Question

1. See Figure 14.12.

2. The incoming sensory message would enter the spinal cord about the level of the hand via sensory neurons. These neurons would synapse directly in the spinal cord with a motor neuron, which would carry messages to the muscles of the hand and arm to pull away from the stove.

3. The spinal cord is the location of the integration.

4. The brain plays no role in the integration.

Chapter 15

Knowledge and Comprehension Questions

1. For you to sense a stimulus, it must be of sufficient magnitude to open ion channels within the membrane of the receptor cell. This depolarizes the membrane, creating a generator potential that leads to an action potential (nerve impulse) in the sensory neurons with which the receptor synapses. Nerve fibers conduct the impulse to the central nervous system.

2. A roller coaster ride would most probably include an extreme rise in proprioception stimuli as the person's spatial orientation changes quickly.

3. Satiation is a sensation involving the body's information about the internal environment, particularly those receptors in the digestive system.

4. Olfactory receptors detect different smells because of specific binding of airborne gases with the receptor chemicals located in the cilia of the nasal epithelium.

5. On its way to the olfactory area of the cerebral cortex, the nerve impulse travels through the limbic system, the area of the brain that is responsible for many of your drives and emotions. Thus, certain odors become linked in your memory with emotions and events.

6. Taste buds are taste receptors concentrated on the tongue. They detect chemicals in food and register an overall taste that consists of different combinations of sweet, salty, sour, and bitter. The sense of taste interacts with the sense of smell to produce a taste sensation.

7. See Figure 15.9.

8. Rod cells, which can detect low light levels, are the primary cell type stimulated in a dark room, but in a well-lit setting, the cone cells (which sense color) would be equally activated.

9. Your outer ear funnels sound waves toward the eardrum, which changes these waves into mechanical energy. This energy is then transmitted to the bones of the middle ear, which increases its force. Receptor cells in the inner ear change the mechanical energy into nerve impulses.

10. You can detect movements of your head because of the otoliths embedded in the jellylike layer that covers the cilia in your inner ear. When you move your head, the otoliths slide and bend the underlying cilia. This generates signals to the brain, which interprets the type and degree of movement.

Critical Thinking Review

1. The birds oriented in some way to the direction of migration, but that orientation was superseded by an orientation

to a magnet. Because the birds were placed in a cage lined with carbon paper, they would be unable to see movements of the sun or moon for navigational purposes. Possibly these birds have a sensory system in which they can detect and orient to the Earth's magnetic field, which is their means of navigation. However, placing a magnet near them would provide a stronger stimulus than that of the Earth's magnetic field, so the birds would orient to it.

2. The inability to feel pain is a disadvantage to a person's well-being. The reason we avoid many injuries is that we detect potentially injurious stresses we place on our bodies because of the pain it causes, and then we stop the activity. For example, a child who jumped from a staircase and slightly twisted an ankle would not continue to walk on it and would probably not jump from that height again. However, the child who felt no pain would continue to walk and jump, injuring the ankle further, and might not be deterred from this behavior in the future.

3. The drops affected the set of smooth muscles that cause the iris to open, enlarging the pupil.

4. Answers will vary.

5. During a middle ear infection, fluid builds up in the middle ear, surrounding the malleus, incus, and stapes; pressing on the eardrum, filling the eustachian tube. Hearing can be affected because the tympanic membrane cannot vibrate as usual, and the bones of the middle ear cannot move as usual.

Visual Thinking Question

See Figure 15.15 and accompanying text.

Chapter 16

Knowledge and Comprehension Questions

1. Your movement resulted from the contraction of your skeletal muscles, which are anchored to bones. The muscles use bones like levers to direct force against an object. When you raised your hand, your skin stretched to accommodate the change in position.

2. The role of skin as a sensory surface is of great use to small children. As we age, we tend to rely less heavily on such input and more on visual perception.

3. Bone is a type of connective tissue consisting of living cells that secrete collagen fibers into the surrounding matrix. The bones of the skeletal system support the body and permit movement by serving as points of attachment and acting as levers against which muscles can pull. They also protect delicate internal structures, store important minerals, and produce red and white blood cells.

4. Bone marrow transplants entail the removal and transfer of red bone marrow from spongy bone in the end of long bones.

5. Spongy bone makes up most of the tissue of the smaller bones of the axial skeleton.

6. Excess calcium taken in by the body and not needed at the moment in cellular activities is stored in bone tissue until a later time.

7. A "slipped disk" refers to one of the intervertebral disks of fibrocartilage that separate the vertebrae from each other (except in the sacrum and coccyx). These disks act as shock absorbers, provide the means of attachment between vertebra, and allow the vertebral column to move.

8. These are all types of synovial (freely movable) joints. They differ in the type of movement they allow. A hinge joint allows movement in one plane only, a ball-and-socket joint allows rotation, and a pivot joint permits side-to-side movement.

9. Tendons connect muscle to bone. Ligaments connect muscles together.

10. Actin is a protein that makes up the thin myofilaments of muscle; the protein myosin makes up the thick myofilaments. Myofilaments are the microfilaments of muscle cells. Muscle cells contract when the actin and myosin filaments slide past each other. Changes in the shape of the ends of the myosin molecules (located between adjacent actin filaments) cause the myosin molecule to move along the actin, causing the myofilament to contract.

Critical Thinking Review

1. Answers may vary. People generally exhibit less elasticity, wrinkling, and stretching or sagging skin with age as well as lower touch sensation. Bones often become brittle, causing more fractures and breaking more easily. Joints often stiffen, and osteoarthritis is common.

2. Bone functions as one type of connective tissue that supports the skin, giving the body some of its shape and strength, as well as providing a place of attachment for muscles. Certain minerals, such as calcium, give hardness to bone but the bone also provides a storage place for them. In addition, the bone marrow is located within spaces of bones. Red bone marrow, located within spongy bone, produces most of the body's blood cells. Artificial bones can support the skin, helping to impart shape to the body. Artificial joints can allow for movement where bones meet bones. However, artificial bones cannot be storehouses for minerals, nor can they provide bone marrow.

3. The trunk of the elephant has a hydrostatic skeleton. Muscles are attached to the flexible walls of the elephant's trunk. When muscles pull on one part of the trunk, they compress the fluid, causing it to push out another part.

4. Human skin provides a relatively dry environment with little food for bacteria (the top layer of keratinized skin is dead, hardened cells). However, when the skin is broken,

an environment opens that has fluid and tissue on which bacteria can reproduce in great numbers. (Additionally, the normal microbiota colonize the skin and therefore, disease-producing bacteria have no place to live. However, when the skin is broken, pathogens can invade.)

5. Answers will vary, but moderate amounts of weight-bearing exercise help maintain bone density. Excessive strenuous exercising, irregular or sudden impacts to the limbs, and applying heavy loads to the limbs can damage the cartilage in joints.

Visual Thinking Question

See Visual Summary Figure 16.7.

Chapter 17

Knowledge and Comprehension Questions

1. Hormones are chemical messages secreted by cells that affect other cells. They are an important method the body uses to integrate the functioning of various tissues, organs, and organ systems.

2. The nervous system sends electrical messages to glands and muscles, regulating glandular secretion and muscular contraction. Endocrine hormones carry chemical messages to virtually any type of cell in the body.

3. Peptide hormones are not lipid soluble and so do not pass through lipid bilayers of cell membranes. The binding of peptide hormones to a cell membrane receptor triggers the production of a second messenger within the cell. Cyclic AMP is this second messenger, and it stimulates the activity of enzymes within the cell that cause the cell to alter its functioning.

4. A feedback loop controls hormone production by initially stimulating a gland to produce the hormone. After the hormone has exerted its effect on the target cell, the body feeds back information to the endocrine gland. In a positive feedback loop, the feedback causes the gland to produce more hormone; in a negative feedback loop, it causes the gland to slow down or stop hormone production.

5. ADH levels would be monitored and controlled by the hypothalamus. This structure produces "releasing hormones" that trigger ADH production and release from the posterior pituitary.

6. Tropic hormones stimulate other endocrine glands. The four tropic hormones are: (1) follicle-stimulating hormone (FSH). In women, FSH triggers the maturation of eggs in the ovaries and stimulates the secretion of estrogens. In men, it triggers the production of sperm. (2) Luteinizing hormone (LH). In women, LH stimulates the release of an egg from the ovary and fosters the development of progesterone. In men, it stimulates the production of testosterone. (3) Adrenocorticotropic hormone (ACTH). ACTH stimu-lates the adrenal cortex to produce steroid hormones. (4) Thyroid-stimulating hormone (TSH). TSH triggers the thyroid gland to produce the thyroid hormones.

7. Parathyroid hormone (PTH) and calcitonin (CT) work antagonistically to maintain appropriate calcium levels. If the level becomes too low, PTH stimulates osteoclasts to liberate calcium from the bones and stimulates the kidneys and intestines to reabsorb more calcium. When levels grow too high, more CT is secreted, which inhibits the release of calcium from bones and speeds up its absorption.

8. Over a prolonged period of stress, the body reacts in three stages: (1) alarm reaction (quickened metabolism triggered by adrenaline and noradrenaline); (2) resistance (glucose production and rise in blood pressure); hormones involved are ACTH, GH, TSH, mineralocorticoids and glucocorticoids; and (3) exhaustion (loss of potassium and glucose and organs become weak and may stop functioning).

9. The pancreatic islets of Langerhans secrete two hormones that act antagonistically to one another to regulate glucose levels. Glucagon raises the glucose level by stimulating the liver to convert glycogen and other nutrients into glucose, whereas insulin decreases glucose levels in the blood by helping cells transport it across their membranes.

10. Neurosecretory cells are specialized nerve cells that make and secrete hormones. The endocrine glands made up of neurosecretory cells are the hypothalamus, the posterior pituitary, and the adrenal medulla.

Critical Thinking Review

1. People living far from the ocean faced little chance of receiving their iodine from seafood, a primary source of iodine. Until this century, iodized salt was not freely available. This lack of dietary iodine caused goiter.

2. Chronic stress leads to long-term stimulation by the autonomic nervous system, which heightens metabolism, raises blood pressure, and speeds up internal chemical reactions. Over time, this can physically stress the body and lead to health problems.

3. No. Alcohol suppresses the release of ADH (antidiuretic hormone), which means that it encourages more water to leave your body in your urine. If you are already hot and thirsty, this will only dehydrate you further.

4. If andro causes estrogen levels to rise in the male body, then a male might experience the emergence of some female secondary sexual characteristics such as the development of (excess) breast tissue.

5. Most youngsters of middle-school age undergo puberty. During this time, the sex hormones cause the development of the primary and secondary sexual characteristics. In boys, these include a deepening of the voice and the growth of facial hair. In girls, these include the onset of menstruation and development of the breasts. Because of

the hormonal changes that occur during puberty, adolescents undergo a wide variety of both physiological and psychological changes, which are often accompanied by behavioral and mood "swings."

Visual Thinking Question

See Figure 17.7 and Section 17.5.

Chapter 18

Knowledge and Comprehension Questions

1. The chemical instructions that determine our specific personal characteristics are located within chromosomes. The DNA within our chromosomes directs the millions of complex chemical reactions that govern our growth and development.

2. The two types of nucleic acid are DNA (deoxyribonucleic acid) and RNA (ribonucleic acid). Both consist of nucleotides that are made up of three molecular parts: a sugar, a phosphate group, and a base. The sugar in RNA is ribose, whereas the sugar in DNA is deoxyribose. DNA contains the bases adenine, thymine, cytosine, and guanine; RNA contains uracil instead of thymine.

3. Double helix refers to the DNA molecule, which is shaped like a double-stranded helical ladder. The bases of each strand together form rungs of uniform length, and alternating sugar-phosphate units form the ladder uprights.

4. Chargaff's experiments showed that the proportion of bases varied in the DNA of different types of organisms. This suggested that DNA has the ability to be used as a molecular code, because its base composition varies as its code varies from organism to organism.

5. The type of RNA found in ribosomes is ribosomal RNA (rRNA). Transfer RNA (tRNA) is in the cytoplasm; during polypeptide synthesis, tRNA molecules transport amino acids to the ribosomes and position each amino acid at the correct place on the elongating polypeptide chain. Messenger RNA (mRNA) brings information from the DNA in the nucleus to the ribosomes in the cytoplasm to direct which polypeptide is assembled.

6. Sexual reproduction, whereby a new human is formed by the fusion of egg and sperm meiotically produced, leads to tremendous variation. The alternative, mitosis, produces identical daughter cells and is responsible for human growth during the life cycle.

7. The correct sequence is c (prophase), a (metaphase), d (anaphase), b (telophase). This is the process of mitosis.

8. The correct sequence is c (prophase I), d (metaphase I), a (anaphase I), b (telophase I), e (telophase II). This is the process of meiosis I.

9. Mitosis produces two daughter cells from one parent cell; the daughter cells are identical with each other and with the parent. Meiosis produces four sex cells, each of which contains half the amount of the parent cell's hereditary material.

10. Genetic recombination during meiosis generates variability in the hereditary material of the offspring. It is the principal factor that has made the evolution of eukaryotic organisms possible.

Critical Thinking Review

1. There are many answers, and student responses may vary. Possible answers include the following: a nonviable DNA strand is produced; an error causes the chemical message to vary and results in genetic mutation of some sort; and the change does not directly affect instruction because of repetition of the instruction elsewhere on the strand.

2. There may be only one gene coding for a particular enzyme, and many enzymes are individually essential to the viability of an organism. A "backup" system of repetition may ensure enzyme production despite a single mutation or alteration in genetic code.

3. The codons in mRNA after transcription from this DNA strand would be: CUA AUG UCU AAU GUG GAU CGA UAG. Using Table 18.3 as noted in the instructions reveals the following sequence of amino acids from this sequence of codons: Leu-Met-Ser-Asn-Val-Val-Arg-Termination.

4. A deletion of the first guanine in the sequence would probably cause the greatest change in the polypeptide because all the codons would change, thereby changing all the amino acids. A deletion of the last guanine would cause a change in only two codons, but would leave the last codon with only two bases.

5. This egg cell could have undergone a normal meiosis I, which separates homologous chromosomes. In meiosis II, one of these cells probably underwent a normal division, yielding the 2 haploid cells. However, the other cell could not have undergone normal meiosis II. The cell must have split, but the sister chromatids did not, so that one cell received no chromosomes, while the other cell received the still-joined sister chromatids.

Visual Thinking Question

Plants have an alternation of generations and animals do not. That is, in plants, there is a haploid organism (the gametophyte) that produces gametes by mitosis. In animals, no haploid organisms exist (in general). Only the gametes are haploid. Diploid organisms (adult animals) produce gametes by meiosis. In plants, meiosis is the process that forms spores from the diploid sporophyte. Spores grow into haploid gametophyes. In both plants and animals, mitosis produces cells for growth of the diploid organism. However, in plants, mitosis also produces cells for growth of the gametophyte (haploid) organism and for the production of gametes from the gametophyte. Although in the first examples mitosis produces cells in diploid organisms, and in the second examples it produces cells in haploid organisms, the mitotic process is still the same.

Chapter 19

Knowledge and Comprehension Questions

1. Biotechnology is the manipulation of organisms to yield organisms with specific characteristics, organisms that produce particular products, and information about an organism or tissue that would otherwise not be known. Genetic engineering refers to techniques of molecular biotechnology (as opposed to classical biotechnology) in which genes are manipulated, not just organisms.

2. Classical biotechnology employs the techniques of selection, mutation, and hybridization, manipulating the genetics of organisms at the organism level. Molecular biotechnology employs the techniques of genetic engineering, or recombinant DNA technology, in which genes of organisms are manipulated. Yes, both are being practiced today. Dog breeders, for example, still practice classical biotechnology in breeding programs, whereas molecular biologists continually break new ground in the applications of molecular biotechnology. (Evidence will vary among students.)

3. Three methods of natural gene transfer among bacteria are transformation, transduction, and conjugation. In transformation, free pieces of DNA move from a donor cell to a recipient cell. During transduction, DNA from a donor cell is transferred to a recipient cell by a virus. During conjugation, a donor and a recipient cell make contact, and the DNA from the donor is injected into the recipient cell.

4. After identifying and isolating a desired human gene, scientists can insert the gene or genes into bacteria using either plasmids or viruses. Plasmids can be extracted from bacteria or yeasts, induced to incorporate genes (restriction fragments) into their genomes, and reinserted into the cell, much like a cell would take up DNA fragments during the natural process of transformation. Scientists can also incorporate restriction fragments into viral genomes and use the viruses to infect bacteria similar to the natural process of transduction.

5. Gene cloning is the process of making copies of genes. In shotgun cloning, no one gene is targeted for cloning—all the genes of a genome are cloned from restriction fragments. In complementary DNA cloning, genes are "manufactured" by using the messenger RNA of the gene and the

enzyme reverse transcriptase. Genes can also be synthesized in the laboratory when the nucleotide sequence of the gene is known; this process is called gene synthesis cloning.

6. Restriction enzymes play an important role in genetic engineering because they are tools that scientists use to cut long, intact DNA strands into fragments that contain one or a few genes. These small pieces are more easily studied than long DNA strands.

7. Molecular probes are molecules that bind to specific genes or nucleotide sequences. Probes help scientists identify searched-for genes.

8. The polymerase chain reaction (PCR) makes many copies of a gene or segment of DNA. This is necessary so that: (a) there are sufficient copies of a gene available for large-scale culturing or (b) extremely small amounts of DNA can be amplified for analysis.

9. Answers may vary, but two genetically engineered proteins in use today are human insulin, used to treat diabetes, and human growth hormone, used to treat growth disorders in children.

10. Gene therapy is the treatment of a genetic disorder by the insertion of "normal" genes into a patient's cells. This technique is not in widespread use today; only a few persons have undergone gene therapy, with the first case occurring in the fall of 1990.

Critical Thinking Review

1. Scientists must insert human genes into the bacterial genome so that bacteria can express those genes, producing human protein. Gold and other precious metals are not produced by living things and thus are not coded for by DNA. Although it sounds like a great idea to have bacteria produce gold, it won't work.

2. Humans produce antibodies against foreign proteins. Although genetically engineered insulin is synthesized by bacteria, it is still human (not foreign) protein because it is coded for by human DNA. Animal-derived insulin, on the other hand, is coded for and synthesized by an animal other than a human. The human body is likely to react by producing antibodies against such protein.

3. The identical twins would have identical genetic fingerprints, while the fraternal twins would have some restriction fragments in common, but not all.

4. Although conjugation involves DNA from one bacterium being transferred to another bacterium and is an important method of genetic recombination in bacteria, a new cell is not produced from two "mating" cells during this process. Therefore, conjugation is not reproduction.

5. Antibiotic #1 will work the best because it shows the largest zone of inhibition.

Visual Thinking Question

Human genes are being combined with those of a bacterium to incorporate the "code" for a human product into the bacterial genome. The DNA resulting from combining genes from different organisms is called recombinant DNA. In this way, the bacterium, when grown in culture, will produce this product. The production of recombinant DNA is explained in the caption of Figure 19.8.

Chapter 20

Knowledge and Comprehension Questions

1. Offspring of sexual reproduction inherit characteristics from both parents; those of asexual reproduction are exact duplicates (with the exception of mutations) of one parent. Sexual reproduction introduces variation within a species; asexual reproduction does not.

2. Self-fertilization occurs when the male gametes of one plant fertilize the female gametes of the same organism. Cross-fertilization occurs when the male gametes of one plant fertilize the female gametes of another.

3. Mendel's law of segregation states that each gamete receives only one of an organism's pair of alleles. Random segregation of the pair members occurs during meiosis.

4. All of the F_1 offspring would have an Ll genotype and the phenotype of long leaves.

	L	L
l	Ll	Ll
l	Ll	Ll

5. The F_2 genotype would be 1 LL:2 Ll:1 ll and the phenotypes would be 3 long-leaved and 1 short-leaved.

	L	l
L	LL	Ll
l	Ll	ll

6. The probability is 1/4.

7. Mendel used a testcross to determine whether a phenotypically dominant plant was homozygous or heterozygous for the dominant trait. He crossed the plant with a homozygous recessive plant. When the test plant was homozygous, the offspring were hybrids and phenotypically dominant. When it was heterozygous, half the progeny were heterozygous and looked like the test plant, and half were homozygous recessive and resembled the recessive parent.

8. Mendel's law of independent assortment states that the distribution of alleles for one trait into the gametes does not affect the distribution of alleles for other traits, unless they are on the same chromosome.

9. Both recessive traits: 1/16. Both dominant traits: 9/16.

10. Sutton and Morgan provided a framework that explained the equal genetic role of both egg and sperm in the hereditary material of offspring. Sutton suggested the presence of hereditary material within the nuclei of the gametes, whereas Morgan gave the first clear evidence that genes reside on chromosomes.

Critical Thinking Review

1. Because phenotype assessment involves the outward appearance of an individual, it is relatively easy and accurate. However, genotype determination is far more complex, especially in the era in which Mendel lived. Genotype may be complicated by cases of incomplete dominance, more than one gene coding of the same trait, genetic linkage, and other complex relationships between individual genes or chromosomes.

2. Pea plants were well suited to Mendel's investigations because they are small and easy to grow. Additionally, they contain male and female parts that are enclosed by the flower, so that Mendel could easily experiment with self-fertilization in these plants and develop true-breeding varieties. Also, pollen is easy to extract from a pea plant and use to perform cross-fertilization experiments. Another factor is that pea plants have varieties of the same species with distinctive characteristics Mendel could study. Morgan studied sex linkage in *Drosophila*, which are also small and easy to grow. These flies exhibit sex-linked characteristics that are easy to see and to study under the microscope.

3. Answers will vary regarding distinctive human traits that are not inherited. All physical traits are inherited (other than those caused by environmental influences in utero or later in life). Students may mention traits that are behavioral and suggest tracing these traits in their own families for evidence of inheritance. (Scientists are still probing this question; thus, student results are merely speculative.)

4. The drawing should follow the pattern of division in Figure 18.22, with one gene on one of the larger chromosomes and another gene on one of the smaller chromosomes. The movement of one of these genes into a particular cell during the meiotic process is not dependent on the movement of the other.

5. The phenotype of these F_1 plants would be a light purple, a blending of the purple and white colors. When these plants were self-fertilized, they would yield plants in the following ratio: 1 purple: 2 light purple: 1 white.

Genetics Problems

1. Somewhere in your herd, you have cows and bulls that are not homozygous for the dominant gene "polled." Because you have many cows and probably only one or some small number of bulls, it would make sense to concentrate on the bulls. If you have only homozygous "polled" bulls, you could never produce a horned offspring regardless of the genotype of the mother. The most efficient thing to do would be to keep track of the matings and the phenotype of

the offspring resulting from these matings and prevent any bull found to produce horned offspring from mating again.

2. Albinism, a, is a recessive gene. If heterozygotes mated, you would have the following:

	A	a
A	AA	Aa
a	Aa	aa

One fourth would be expected to be albinos.

3. The best thing to do would be to mate Dingleberry to several dames homozygous for the recessive gene that causes the brittle bones. Half of the offspring would be expected to have brittle bones if Dingleberry were a heterozygous carrier of the disease gene. Although you could never be 100% certain Dingleberry was not a carrier, you could reduce the probability to a reasonable level.

4. Your mating of DDWw and Ddww individuals would look like the following:

	Dw	Dw	dw	dw
DW	DDWw	DDWw	DdW	DdWw
Dw	DDww	DDww	Ddww	Ddww
DW	DDWw	DDWw	DdWw	DdWw
Dw	DDww	DDww	Ddww	Ddww

Long-wing, red-eyed individuals would result from 8 of the possible 16 combinations, and dumpy, white-eyed individuals would never be produced.

5. Breed Oscar to Heidi. If half of the offspring are white eyed, Oscar is a heterozygote.

6. Both parents carry at least one of the recessive genes. Because it is recessive, the trait is not manifested until they produce an offspring who is homozygous.

Visual Thinking Question

See Figure 20.8.

Chapter 21

Knowledge and Comprehension Questions

1. People who inherit abnormal numbers of sex chromosomes often have abnormal features and may be mentally retarded. Examples are: (a) triple X females (XXX zygote), underdeveloped females who may have lower than average intelligence, (b) Klinefelter syndrome (XXY zygote), sterile males with some female characteristics, (c) Turner syndrome (XO zygote), sterile females with immature sex organs, and (d) XYY males, fertile males of normal appearance.

2. The three major sources of damage to chromosomes are high-energy radiation (such as x-rays), low-energy radia-

tion (such as UV light), and chemicals (such as certain legal and illegal drugs).

3. Heavy use of drugs could affect ova and sperm by damaging the DNA. In general, chemicals add or delete molecules from the structure of DNA. This can result in chromosomal breaks or changes as the cell works to repair the damage.

4. All cancers exhibit uncontrolled cell growth, loss of cell differentiation, invasion of normal tissues, and metastasis to other sites.

5. Mutation of the DNA of proto-oncogenes causes them to become active oncogenes. These act as the "on" switches in the development of cancer. Mutation of the DNA of tumor-suppressor cells causes them to be inactivated and they cannot stop the development of cancerous tissue. As a result of these mutations, cancer will develop in a patient.

6. Dysplastic cells are benign cells that exhibit growth patterns that are characteristic of the development of cancer cells. These cells are irregular in size and in the appearance of their nuclei.

7. a. X-linked, b. dominant, c. recessive, d. autosomal

8. Most human genetic disorders are recessive because those genes are able to persist in the population among carriers; people carrying lethal dominant disorders are more likely to die before reproducing. Recessive disorders include cystic fibrosis, sickle cell anemia, and Tay-Sachs disease.

9. In incomplete dominance, alternative forms of an allele are neither dominant nor recessive; heterozygotes are phenotypic intermediates. In codominance, alternative forms of an allele are both dominant; thus heterozygotes exhibit both phenotypes.

10. Couples who suspect that they may be at risk for genetic disorders can undergo genetic counseling to determine the probability of this risk. When a pregnancy is diagnosed as high risk, a woman can undergo amniocentesis (analysis of a sample of amniotic fluid) to test for many common genetic disorders.

Critical Thinking Review

1. The fact that a genetic abnormality such as trisomy 21 can exist shows that there are often groups of genes located near each other on a chromosome that may be transferred or altered as a group and cause multiple alterations that occur as a discrete group.

2. Answers will vary.

3. Answers will vary.

4. Their children's genotypes would be 1:2:1 (AA [type A]: AB [type AB]: BB [type B]).

	A	B
A	AA	AB
B	AB	BB

5. Answers will vary, but information should be about the known genetics of the individual's family tree.

Human Genetics Problems

1.

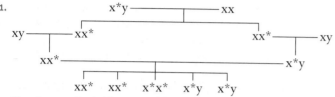

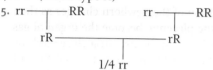

2. No. A type O child is possible.
3. 45 (44 autosomes, one X)
4. $I^A I^O$ (blood type A)
5.

rr ⊤ RR   rr ⊤ RR
   rR ————— rR
        1/4 rr

6. e. Y-linkage

Visual Thinking Question

1. Sperm are produced continuously and do not age. Primary oocytes, however, are all produced prenatally, so they age as a woman ages. As primary oocytes age, errors in the meiotic process are more likely to occur.
2. During meiosis II, the sister chromatids in one of the cells produced during meiosis I do not separate. Therefore, the product of meiosis II is one cell with both sister chromatids, and one cell with no chromatids.
3. If each of these cells were to combine with a sperm, one zygote would have an extra chromosome and one would have one less chromosome than the normal complement.

Chapter 22

Knowledge and Comprehension Questions

1. Sexual reproduction, as in human reproduction, is the process in which a male and a female sex cell combine to form the first cell of a new individual. Asexual reproduction, as in growing a new plant from a cutting, is the generation of a new individual without the union of gametes.
2. A spermatozoon has a head that contains the hereditary material. Located at its leading tip is an acrosome that contains enzymes helping the sperm penetrate an egg's membrane. The sperm also has a flagellum that propels it and mitochondria that produce the ATP from which sperm derives the energy to power the flagellum.
3. These are accessory glands that add fluid to the sperm to produce semen. The seminal vesicles supply a fluid containing fructose, which serves as a source of energy for the sperm. The prostate gland adds an alkaline fluid that neutralizes the acidity of any urine in the urethra and the acid-

ity of the female vagina. The bulbourethral glands also contribute an alkaline fluid.

4. In spermatogenesis, which occurs continually in a sexually mature male, four fully functional (haploid) sperm are produced from each diploid spermatogonium. In oogenesis, only one fully functional (haploid) ovum is produced from each diploid oogonium. The development of oogonia to primary oocytes is completed before birth. Each month, after puberty, one ovum is produced.
5. The reproductive cycle of females occurs roughly every 28 days. The primary oocyte matures, is released from the ovary during ovulation, and journeys through the uterine tube to the uterus. The endometrial lining of the uterus has thickened to prepare for implantation; if fertilization does not occur, it sloughs off during menstruation. The hormones FSH, LH, estrogen, and progesterone orchestrate these events.
6. Once the date of ovulation is determined, a couple must have intercourse within 24 hours to fertilize the egg. The egg is viable only up to 1 day after ovulation.
7. The only methods of birth control that may prevent the transmission of sexually transmissible diseases are abstinence, male latex condoms, and possibly female condoms. Nonoxynol–9, a spermacide, may inactivate some viruses and may reduce the risk of gonorrhea and chlamydia transmission.
8. Birth control pills contain estrogen and progesterone, which shut down the production of FSH and LH. By maintaining high levels of estrogen and progesterone, the pills cause the body to act as if ovulation has already occurred; the ovarian follicles do not mature, and ovulation does not occur.
9. The most effective birth control methods are vasectomies and tubal ligation. Birth control pills are very effective. Condoms and diaphragms are effective when used correctly, but mistakes are common. Least reliable are the rhythm method and withdrawal, which have high failure rates.
10. The only completely effective protection against sexually transmissible diseases is abstinence. There is decreased risk if latex condoms are used correctly along with the spermicide nonoxynol-9.

Critical Thinking Review

1. At birth, females have all of the oocytes that they will ever produce. As a woman ages, so do her oocytes, and the odds of a harmful mutation increase appreciably after age 35. Because a man's gametes are (at the most) a few days old, this risk of mutation due to aging of the gametes does not exist in men. Therefore, women older than 35 years of age are more likely to bear children with genetic abnormalities than are younger women (in general), but this is not the case with men.

2. When a cervical cap or diaphragm is used, there is still fluid-to-fluid contact between sexual partners during intercourse. This contact permits the transfer of sexually transmissible diseases. Diaphragms and cervical caps could be protective in some cases because they cover the cervix, and the spermicide nonoxynol-9 is used with them.

3. Answers will vary.

4. Answers will vary. Students will answer the question based on self-knowledge of their behavior. If the student thinks that he or she will use a birth control method unreliably, then the actual data should carry more weight. If he or she is meticulous with use, the theoretical data might carry more weight in decision making.

5. Sea stars can reproduce asexually. Any piece that contained part of the central ring of the sea star has the ability to grow into a new organism.

Visual Thinking Question
See Figures 22.5 and 22.8.

Chapter 23

Knowledge and Comprehension Questions

1. Two differences between protostome and deuterostome development are that: (1) the mouth of the protostome develops from the first indentation of the gastrula, but the mouth of a deuterostome develops from a second indentation, and (2) early protostome cleavage is spiral, while early deuterostome cleavage is radial. Humans are deuterostomes.

2. At the instant of union and fertilization between a male and female gamete, the zygote formed has a unique genetic composition.

3. After fertilization, the oocyte's surface changes so that no other sperm can penetrate, and oocyte meiosis is completed. The sperm sheds its tail, and the sperm and egg nuclei fuse.

4. This period consists of mitotic cell division, the type of cell division involved in growth and duplication of cells.

5. Morphogenesis is significant because it is the first time that cells begin to migrate to shape the new individual. The major morphogenetic events occur during the third to eighth weeks in humans.

6. The extraembryonic membranes play a role in the life support of the preembryo/embryo/fetus. They are the: (a) amnion (cushions the embryo in amniotic fluid and keeps the temperature constant), (b) chorion (facilitates the exchange of nutrients, gases, and wastes between the embryo and mother), (c) yolk sac (produces blood for the embryo before its liver is functional and becomes the lining of the digestive tract), and (d) allantois (responsible for formation of the embryo's blood cells and vessels).

7. Neurulation is the development of a hollow nerve cord. During neurulation, cells lying over the notochord curl upward to form a tube that will develop into the central nervous system.

8. The notochord is a structure that forms the midline axis of an embryo. In humans and in most other vertebrates, it develops into the vertebral column.

9. The first trimester is primarily a time of development; by the ninth week, most body systems are functional. During the second trimester, the fetus grows, ossification is well underway, and the fetus has a heartbeat. The third trimester is primarily a period of growth; by birth, the fetus is able to exist on its own.

10. At birth, the circulation of the newborn changes as the lungs, rather than the placenta, become the organ of gas exchange. The hole between the two atria close, and the ductus arteriosis (the direct connection between the pulmonary artery and the aorta) is shut off.

Critical Thinking Review

1. Cell surface changes occur on the egg that prevent entry of a second sperm. If a second sperm gained entry, the zygote would not be viable because it would have 1-1/2 times the normal amount of hereditary material.

2. If the foramen ovale did not close, some of the deoxygenated blood from the right side of the heart would flow directly to the left side of the heart without going to the lungs to be oxygenated. This condition results in blood with a deep red color rather than the bright red of oxygenated blood. Such babies are termed "blue babies," because the blood has a bluish cast as seen through the skin. This abnormality must be repaired by surgery.

3. Along with general systemic development, neural development is a key aspect of the first 3 weeks of fetal development, and the fate of the central nervous system is dependent on good maternal nutrition.

4. A woman preparing for pregnancy should behave as though she were pregnant regarding these behaviors, because she might be pregnant without realizing it. As mentioned in the answer to question #3, neural development is key in the first 3 weeks of development. In general, development proceeds at a fast pace during this time. Alcohol, drugs, and cigarette smoke could be extremely harmful.

5. HCG is human chorionic gonadotropin, a hormone that maintains the corpus luteum so that it will continue to secrete progesterone and estrogen. Progesterone is needed for the maintenance of the uterine lining and the continued attachment of the fetus to this lining. This substance would be filtered from the blood in the kidney during filtration and become part of the urine. Although some may be reabsorbed at the nephron, reabsorption would take place only to a threshold level.

Chapter 24

Knowledge and Comprehension Questions

1. Darwin's observations led him to conclude that: (a) organisms of the past and present are related, (b) factors other than or in addition to climate are involved in the development of plant and animal diversity, (c) members of the same species often change slightly in appearance after becoming geographically isolated from each other, and (d) organisms living on oceanic islands often resembled those living on a nearby mainland.

2. Three other factors that influenced Darwin were: (a) evidence of geological layers and fossils that suggested the Earth was much older than traditionally thought, (b) the observed changes in species that occur with artificial selection, and (c) Malthus's writings on geometric population growth.

3. The two dogs look so different today because of artificial selection; people who bred them selected for desired characteristics, so that over time the breeds changed. Artificial selection is based on the principle that all organisms within a sexually reproducing species exhibit variation.

4. Adaptations are naturally occurring inheritable traits found within populations that confer a reproductive advantage on organisms that possess them. As adaptive (advantageous) traits are passed on from surviving individuals to their offspring, the individuals carrying those traits will increase, and the nature of the general population will gradually change.

5. Survival of the fittest refers to the idea that natural selection tends to favor those organisms that are most fit to survive to reproductive age in a particular environment and to produce offspring.

6. In adaptive radiation, the population of a species changes as it disperses into different habitats. Eventually, some of these populations change, with the result that interbreeding is no longer possible. This is speciation, the formation of new species through the process of evolution.

7. The carbon-14 method can help establish a date by estimating the relative amounts of the different isotopes of carbon present in a fossil. The half-life of carbon-14 is 5730 years (the amount of time for half of the ^{14}C to decay into nitrogen). Scientists can estimate the length of time that the carbon-14 has been decaying, which means the time that has elapsed since the organism died.

8. According to radioactive dating, the Earth is 4.6 billion years old. This is significant because it allows sufficient time for evolution to occur.

9. Vestigial organs are structures that are present in an organism of today but are no longer useful. The tiny leg bones of a python are an example.

10. Comparative studies of anatomy show that many organisms have groups of bones, nerves, muscles, and organs with the same anatomical plan but with different functions. These homologous structures imply evolutionary relatedness. Studies in comparative embryology show that many organisms have early developmental stages that are similar. This again implies evolutionary relatedness.

Critical Thinking Review

1. Polar bears live in the Arctic where there is snow year round. Those bears that are white survive in greater numbers than those that have discolorations, spots, or are more darkly colored. The white color serves as a camouflage. Over the years, natural selection "weeds out" nonwhite bears.

2. A population composed of only a few organisms has limited genetic variation and mating possibilities. Should conditions for survival change and should some of these organisms die due to their lack of characteristics that could accommodate that change, the population would become smaller over time and could eventually die out. Larger populations generally have wider genetic variation and are better able to withstand changes that affect the survival of the population.

3. All organisms are subject to the pressures of natural selection—those pressures differ from species to species and depend on the environment in which they live. Although humans shape their environments, they are still subject to environmental pressures that affect the reproductive fitness of the species. Additionally, not all humans shape their environments, such as those living in third world countries who experience environmental pressures such as famine.

4. No, evolution cannot be considered a ladder that organisms can climb, because there is no goal to reach. Each surviving species is successful simply because it continues to survive in the face of environmental pressures. Additionally, survival under differing environmental pressures does not imply that one species is "better" than another.

5. When a population of a particular species of bacterium is exposed to an antibiotic to which that species is susceptible, most bacteria are killed. However, some bacteria may have a chance mutation that confers resistance to that antibiotic. Even if one bacterium were to survive, if given the proper growth conditions, that bacterium could reproduce asexually, resulting in a new population of resistant bacteria.

Visual Thinking Question

1. Humans and yeast.

2. The duck and the rattlesnake diverged most closely in time from a common relative. They have the fewest number of differences in their base sequences of cytochrome C oxidase than any other two organisms.

3. Of the organisms shown on the tree, the pig is most closely related to the human because it has the fewest number of

differences from humans in its base sequences of cytochrome C oxidase.

Chapter 25

Knowledge and Comprehension Questions

1. The primordial soup theory attempts to explain how life began on Earth. It states that life arose in the sea as elements and simple compounds in the atmosphere interacted to form simple organic molecules. One alternative to this theory is the hypothesis that life might have arisen in hydrothermal vents in the oceans.

2. The formation of some sort of membrane would have allowed chemical reactions to occur in a closed environment. Enzymes could be organized to carry on life functions, chemicals could be selectively absorbed, wastes could be eliminated, and hereditary material could be passed on to future cells.

3. Cyanobacteria are single-celled organisms that are similar to the first oxygen-producing bacteria. These evolved during the Archean era and played an important evolutionary role by gradually oxygenating the atmosphere and oceans.

4. The Cambrian period was the oldest period in the Paleozoic era (roughly 590 to 500 million years ago). All of the main phyla and divisions of organisms that exist today (except chordates and land plants) evolved by the end of the Cambrian period.

5. During the Carboniferous period, reptiles developed from amphibians and anthropods moved from the sea onto land. Insects, one arthropod group, evolved water-conserving characteristics such as a cuticle. Fungi evolved during the late Carboniferous period. Much of the land was low and swampy with extensive forests; the worldwide climate was warm and moist.

6. The Cenozoic era, which began about 65 million years ago, saw the rapid evolution and growth of mammals, including primates.

7. Primates are the order of mammals with characteristics reflecting an arboreal lifestyle. They have developed two especially helpful characteristics: depth perception resulting from stereoscopic vision and flexible, grasping hands with opposable thumbs.

8. These are both suborders of the primates. Prosimians (lower primates such as lemurs) are small animals, usually nocturnal, with a well-developed sense of smell. Anthropoids (higher primates such as apes and humans) have larger brains, flatter faces, eyes that are closer together, and relatively long front and hind limbs. Anthropoids are also diurnal and possess color vision.

9. Statement (a) is correct; statement (b) is false. All hominids are a family within the hominoid superfamily.

10. More modern *Homo sapiens* had smaller heads, brow ridges, teeth, jaws, and faces than earlier species did. They also fashioned sophisticated tools from stone and from other substances such as bone and ivory. Over time, groups of *Home sapiens sapiens* moved away from the hunter-gatherer lifestyle to develop agriculture, complex social structures, and civilization.

Critical Thinking Review

1. This portrayal is inaccurate. Dinosaurs died out at the end of the Cretaceous period (about 65 million years ago), long before the first primates appeared.

2. An arboreal (tree) environment favors the ability to use both the hands and feet while moving from branch to branch and tree to tree. However, the savannah is rather open grassland with few bushes. Bipedalism would be favored because organisms could move more quickly to escape from predators, who could easily see the hominids without tree cover. Meanwhile, the hands would be freed to carry things and possibly perform tasks that would aid survival.

3. Scientists have discovered that after mass extinctions, many species proliferate that were not prevalent before, and there is often an adaptive radiation of species. Many more species exist today, for example, than immediately after the last mass extinction.

4. Humans did not evolve from the apes. Both apes and humans evolved from a common ancestor.

5. Current hypotheses regarding the origin of life on Earth all begin with the premise that the life arose on this planet and did not come from another planet. Therefore, all these hypotheses take into account the environment of the Earth at the time and how molecules might evolve into organisms. If life arose on Mars and were transported to Earth, that information would change that initial premise of all the major hypotheses currently held.

Visual Thinking Question

1. *H. habilis, H. erectus, H. sapiens*
2. No.
3. *A. afarensis.*
4. modern *H. sapiens*

Chapter 26

Knowledge and Comprehension Questions

1. Viruses are infectious agents that lack a cellular structure, so they are not living. They are nonliving obligate parasites that must exist in association with and at the expense of other organisms.

2. See Figure 26.3.

3. Both cycles are patterns of viral replication. In the lytic cycle, viruses enter a cell, replicate, and cause the cell to burst and release new viruses. In the lysogenic cycle, viruses enter into a long-term relationship with the host cells, with their nucleic acid replicating as the cells multiply.

4. Bacteria perform many vital functions, such as decomposing organic materials and recycling inorganic compounds. They were also largely responsible for creating the properties of the atmosphere and soil that are present today.

5. No. Sexually transmissible diseases are not caused by poor personal hygiene or socioeconomic class. They are communicable illnesses transmitted between persons during sexual contact.

6. The extreme simplicity of bacterial design—small size, lack of discrete organelles, lack of a membrane-bound nucleus—seems to support the theory that they evolved before eukaryotes.

7. Gummas are tumorlike masses exhibited during the tertiary stage of syphilitic infection. These lesions are not communicable, but may develop in the cardiovascular or central nervous systems and cause paralysis or death.

8. In men, gonorrheal infection, if untreated, may spread to other urogenital structures from the urethra. Infection of the epididymis or ductus deferens may result in scar tissue, causing male sterility. Chronic gonorrheal infections in women may lead to pelvic inflammatory disease (PID), and if uterine tubes are involved, scar tissue may develop, causing sterility.

9. These diseases may be acquired without the knowledge of the person or partner. Occasionally, patients do not have extreme or even normal symptoms of infection and may unwittingly pass on these diseases.

10. The only completely effective protection against sexually transmissible diseases is abstinence. There is decreased risk if latex condoms are used correctly along with the spermicide nonoxynol-9.

Critical Thinking Review

1. Apparently, it takes 10 days for the antibiotic your physician prescribed to kill the bacteria causing your infection, or inhibit their growth so that your body can mount its defenses and destroy these disease-causing cells. Some cells may still be living and able to reproduce after only 4 days of the antibiotic. These cells could begin to reproduce; when they enter the exponential phase of growth, they would trigger a strep infection once again.

2. A cold sore results from the lysogenic cycle of cell damage caused by the herpes simplex virus. The virus remains latent in nervous tissues until something triggers it, such as a cold. This virus could be transmitted to a partner's genitals during oral sex if the person was "shedding" virus at the time.

3. Answers will vary.

4. Viruses are composed of nucleic acid (sometimes with a protein coat), but it is the nucleic acid that is infective. Prions have only protein and no nucleic acid, the type of self-replicating molecule found in viruses and in all living things.

5. Bacteria grow exponentially; that is, one divides into two, two into four and so forth. Therefore, as a population of cells begins to multiply at first, the numbers are low, such as when 5 cells double to 10 cells (the bottom of the "S"). As the numbers increase, a doubling of the population results in enormous growth, such as when 1,000 cells double to 2,000 cells (the curve of the "S;" see Figure 33.3). Eventually, the size of the population levels off when food becomes limited and wastes accumulate (the top curve of the "S").

Visual Thinking Question

See Figure 26.4.

Chapter 27

Knowledge and Comprehension Questions

1. Protists are single-celled eukaryotic organisms; this kingdom contains organisms that are animallike (protozoa including cellular slime molds), plantlike (algae and diatoms), and funguslike (acellular slime molds).

2. An ameba is a protozoan that has a changing bloblike shape. It moves and obtains food by means of pseudopods (cytoplasmic extensions). Its pseudopods engulf organic matter, which the ameba then digests.

3. Both flagellates and ciliates are protozoans, and both have hairlike cellular processes that help them move and obtain food. However, flagellates have flagella (long hairlike processes), whereas ciliates have cilia (short hairlike extensions). Ciliates have a more complex internal organization than flagellates do.

4. Sporozoans are nonmotile protozoans that parasitize animals. They undergo complex life cycles in which they are passed from host to host by means of a vector.

5. Dinoflagellates and diatoms are both phyla of plantlike protists. Dinoflagellates have stiff outer coverings, and their flagella beat in two grooves. The diatoms look like microscopic pillboxes made up of top and bottom shells.

6. diatoms, *Diatoms*; brown algae, *Phyaeophyta*; green algae, *Chlorophyta*.

7. Fungus; ameba.

8. All fungi could be described as helpful, because as decomposers they are crucial to the cycling of materials in the environment. More specifically, helpful fungi include those that produce antibiotic drugs. Yeasts are fungi that are crucial in the production of bread, cheese, and beer. Other

fungi, however, cause diseases in plants and animals, such as Dutch elm disease, wheat rust, and yeast infections.

9. Club fungi (b); zygote-forming fungi (c); sac fungi (a); water molds (d).

10. Lichens are associations between fungi and photosynthetic partners. Lichens are able to tolerate harsh living conditions; thus, they are found in a wide range of habitats (such as deserts, extreme northern and southern latitudes, and mountaintops).

Critical Thinking Review

1. It is difficult to give a definitive answer to this question. As groups of cells within colonial organisms specialize, taking on particular jobs, many would suggest that they should be considered multicellular organisms. However, in this case, only one group of cells is specialized; if the cells could exist independently, the group might still be considered as a colonial organism.

2. Euglena is difficult to place using this frame of reference. Clearly, classifying protists on the basis of such criteria is difficult and, in some cases, may be impossible. Protists are a varied group; classifying these organisms is the substance of debates among scientists who still ponder the evolutionary relationship among members of this group. Using a simplistic approach such as "plantlike" and "animallike" may not appropriately reflect evolutionary closeness or distance among organisms, and it may be an arbitrary way to group organisms.

3. Eventually the exponential growth of the algae will slow as nutrients are used and the surface of the water becomes covered. Some algae will then begin to die. As the algae die, bacteria and other decomposers feed on it. As they do, they use oxygen dissolved in the water. As the oxygen level decreases, other organisms dependent on the oxygen begin to die also, providing yet more food for decomposers. Poisoning the algae would have a devastating effect on the lake, because a great deal of food would become available to decomposers. Unless oxygen is returned to the system, as happens in a rapidly flowing stream splashing over rocks, the lake will die.

4. Both fungi and bacteria are decomposers and often compete for space and nutrients. A mold that could inhibit bacterial competitors from growing near it would have a survival and reproductive advantage over molds that could not suppress the growth of bacterial competitors.

5. Wet or moist environments are necessary for food-getting and food absorption in all protists. Some protists are motile and can move toward nutrients by chemotaxis. Some sessile protists have cilia that create currents in the water to channel food to the organism. In others, nutrients flow past the organism carried by water currents. Food absorption takes place across cell surfaces in protists, sometimes

helped by means of specialized organelles. Additionally, most protists have little protection against drying out and could therefore not live in a dry environment.

Chapter 28

Knowledge and Comprehension Questions

1. Seeds are structures from which sporophyte plants grow. They protect the embryonic sporophyte plant from drying out or being eaten, and they contain stored food for the plant.

2. In seedless vascular plants, the sporophyte generation is dominant and lives separately from the gametophyte. The gametophytes produce motile sperm that swim to the eggs. The fertilized eggs produce sporophytes that at first grow within gametophyte tissues but eventually become free-living. A fern is an example of this type of plant.

3. Gymnosperms (c), angiosperms (d), bryophytes (b), ferns (a).

4. See Figure 28.9.

5. Sexual reproduction in gymnosperms and angiosperms produces seeds by the union of male and female gametes. Germinating pollen tubes convey the sperm to the eggs. The seeds protect and nourish the young sporophytes as they develop into new plants, the dominant generation. The sporophytes produce the male and female gametophytes.

6. Fruits have evolved a range of shapes, textures, and tastes that help disperse their seeds effectively. Fruits with brightly colored flesh are often eaten by animals and thus dispersed to different habitats. Other fruits have evolved hooked spines that stick to animals' fur; others have wings that are blown by the wind.

7. b, d, e, a, c.

8. Monocots and dicots are both angiosperms. They differ in the placement of the stored food in their seeds. In dicots, most of the stored food is in the cotyledons (seed leaves), whereas in monocots, most of it is stored in endosperm. In addition, monocots usually have parallel veins in their leaves, and dicots have netlike veins. Monocots also have flower parts in threes, whereas dicots have flower parts in fours and fives.

9. Vegetative propagation is an asexual reproductive process in which a new plant develops from a portion of a parent plant. An example is the white potato, a tuber which can grow new plants from its "eyes."

10. Hormones are chemical substances produced in one part of an organism and then transported to another part of the organism, where they bring about physiological responses. The five kinds of plant hormones are auxins, gibberellins, cytokinins (promote and regulate growth), abscisic acid (inhibits growth), and ethylene (affects the ripening of fruit).

Critical Thinking Review

1. Fruits are really angiosperm ovaries containing seeds ("protected seeds"). The seeds usually have a colorful, fleshy protective covering. These characteristics can easily be seen in fruits such as apples, peaches, and pears, but it can also be seen in vegetables such as tomatoes, peppers, and eggplants.

2. The reproductive advantage of the dandelion flower is the adaptation of dispersal of its seeds by the wind. (Actually the dandelion "flower" is not a flower at all but is many fruits.)

3. In the life cycle of both conifers and angiosperms, spores are formed in the diploid adult sporophyte generation by meiosis, in the cones of conifers and in the flowers of angiosperms. These spores would not be visible to the casual observer because they are enclosed within the cones and flowers and develop into gametes (gametophyte generation).

4. In mosses and ferns, sperm develop in gametophyte plants as flagellated cells. In these low-growing, small plants, flagellated sperm swim easily to the eggs in dew or rainwater on the gametophyte. In larger and higher-growing conifers and angiosperms, the male gametes are pollen grains that travel to the structures housing female gametes (eggs), blown by the wind or, in the case of angiopserms, more commonly transported by insects. The pollen then germinates a tube that makes its way to the egg.

5. In conifers, the cones (which contain the seeds) fall from the trees. Unless animals move these cones, the seeds would not be widely dispersed from the area in which the trees were growing. In angiosperms, animals are often involved in the dispersal of seeds. Sometimes seeds get attached to animals and are dispersed in that way, or the seeds are eaten with fruits, and are dispersed when the animal defecates. See Figure 28.12.

Visual Thinking Question

1-2. Refer to Figures 28.4 and 28.9. Haploid stages are those that follow meiosis, before fertilization occurs; diploid stages are those that follow fertilization before meiotic cell division occurs.

3. The sporophyte generation is dominant in the angiosperm life cycle; the gametophyte generation is dominant in the moss life cycle.

Chapter 29

Knowledge and Comprehension Questions

1. (1) c, (2) b, (3) a, (4) d.
2. Xylem and phloem are both types of vascular tissue, but they have different functions and contain different types of conducting cells. Xylem conducts water and dissolved inorganic nutrients, whereas phloem conducts carbohydrates and other substances.

3. a. Cells in ground tissue that function in photosynthesis and storage.
   b. The most abundant type of cell in dermal tissue, often covered with a thick waxy layer that protects the plant and retards water loss.
   c. Openings that connect air spaces inside dicot leaves with the exterior; they are opened and closed by the change in shape of surrounding guard cells. Water vapor and gases move into and out of the leaves through these openings.
   d. Thimblelike mass of unorganized cells that covers and protects a root's apical meristem as it grows through soil.

4. Primary growth occurs mainly at the tips of roots and shoots, making plants taller; secondary growth occurs in the lateral meristem and makes plants larger in diameter.

5. See Figures 29.5 and 29.6.

6. A dandelion has a taproot, a single major root that can grow deep as a firm anchor. Some taproots also store food for the plant. An ivy plant has adventitious roots that develop from the lower part of the stem and help anchor the plant. Grass has fibrous roots, a root system without one major root; this type of system anchors the plant and absorbs nutrients over a large surface area.

7. Annual rings provide a clue to a tree's age by illustrating the annual pattern of rapid and slow growth in the cambium (lateral meristem tissue). This results in secondary growth.

8. Water rises beyond the point at which it would be supported by air pressure because evaporation from the plant's leaves (transpiration) produces a force that pulls upward on the entire column of water. The forces of adhesion and cohesion maintain an unbroken column of water.

9. See Figure 29.18.

10. In general, nonvascular plants grow closer to the ground than vascular plants because they lack the sophisticated transport systems of vascular plants.

Critical Thinking Review

1. Air pressure still pushes on water in the ground, pushing water up the thin xylem tubes of the plant. In addition, the water molecules adhere to the walls of the xylem tubes and stick to one another (cohesion), which helps maintain the column.

2. The fungi, because they are at the roots, probably function in making inorganic nutrients in the soil available to the plants.

3. Answers may vary. Here are some examples.
   Differences:

"Circulatory System" of Plants	Circulatory System of Animals
Has no circulating pump.	Has a heart—a circulating pump
Circulation is not a continuous loop; water moves from the roots to the leaves.	Circulation is a continuous loop throughout the body.
Fluid is water with dissolved inorganic nutrients	Fluid is blood

Similarities: In both, the "circulating" fluid is enclosed in vessels and is essential to the survival of the organism. Death results if too much of the fluid is lost.

4. You would look at the same spot—4 ½ feet up the tree. The tree grows taller from the apical meristem, which would have been above the initials.

5. Answers may vary. Examples are: Plants contain fibrous material, which is beneficial in the diet. Dietary fiber helps substances move through the large intestine and gives the muscles of the large intestine bulk against which to push. Plants also contain inorganic nutrients, which supply the human body with minerals it needs for proper nutrition.

Visual Thinking Question
See Figure 29.1.

Chapter 30

Knowledge and Comprehension Questions

1. Both you and jellyfish are animals. An important difference is that jellyfish are invertebrates (they lack a backbone), whereas you are a vertebrate (with a backbone and dorsal nerve cord).

2. Eukaryotic organisms have a distinct nucleus and a cellular structure different from that of bacteria. Animals are also multicellular (having more than one cell), and as heterotrophs, they are unable to make their own food. Therefore, they must eat other organic matter for food. Beyond these basic similarities, however, animals are a diverse group.

3. Sponges are called the simplest animals because they lack tissues and organs.

4. Cnidarians exist as either polyps (cylindrically shaped animals that anchor to rocks) or as medusae (free-floating, umbrella-shaped animals).

5. The acoelomates are the most primitive bilaterally symmetrical animals. They have organs and some organ systems, including a primitive brain. Their organs are embedded within body tissues.

6. Roundworms can be dangerous if they parasitize humans, absorbing digested food from their hosts. Roundworms are cylindrical pseudocoelomates encased in a flexible outer cuticle; they have primitive excretory and nervous systems. They move by using muscles attached to the cuticle that push against the pseudocoelom.

7. Molluscs are coelomates with a muscular foot and soft body, usually covered by a shell. They use gills for respiration in water, whereas terrestrial species have adaptations to breathing on land. They show both open and closed circulatory systems and well-developed excretory systems.

8. Arthropods are a diverse phylum of organisms having jointed appendages and rigid exoskeletons.

9. Insects belong to a group of arthropods that have jaws (mandibles). Spiders belong to a different group of arthropods that lack mandibles.

10. Echinoderms are deuterostomes, whereas the other invertebrates are protostomes. These embryological differences make echinoderms more similar to the chordates, which implies that echinoderms and chordates evolved from a common ancestor.

Critical Thinking Review

1. Generally, no. Most hermaphroditic invertebrates such as flatworms and earthworms still reproduce sexually, although each organism has both male and female reproductive glands. Sponges produce both sperm and eggs, but these gametes float through the water and fertilization generally takes place between gametes of different organisms.

2. Answers may vary. Example: Malpighian tubules in terrestrial arthropods such as grasshoppers are slender projections from the digestive tract (see Summary Figure 12.1). These projections, which function in waste removal, provide a large surface area over which fluid is passed to and from the blood in which the tubules are bathed.

3. Echinoderms have bilaterally symmetrical larval forms and are deuterostomes: They cleave in a radial pattern during early embryological development (see Figure 23.2). These characteristics connect them evolutionarily with the chordates and suggest that they evolved from a common ancestor to the (bilaterally symmetrical) chordates.

4. Insects are the only arthropods (actually the only invertebrates) that can fly. Hypotheses will vary as to why insects are so abundant on Earth, but should center on the idea that insect species have adapted to a wide variety of environments, both aquatic and terrestrial.

5. Flatworms have a much simpler body plan than earthworms do and have no coelom. Earthworms are coelomates with a tube within a tube body plan. The structural units of the earthworm are repeated in its segments, giving organi-

zation to its long body. Advantages of having a segmented body are more apparent in organisms such as the arthropods, however, than in the annelids. In the arthropods, for example, the segments have become specialized for different functions.

Chapter 31

Knowledge and Comprehension Questions

1. Nerve cord; notochord; pharyngeal (gill) slits.
2. a (3), b (1), c (2).
3. A lateral line system (found in all fishes and amphibians) is a system of mechanoreceptors that detects sound, pressure, and movement.
4. Osmoregulation is the control of water movement into and out of organisms' bodies. Maintaining the proper balance of water and solutes is vital to life. Terrestrial organisms must conserve water, whereas aquatic organisms must have mechanisms to maintain an appropriate water balance living in either a freshwater or saltwater environment. Freshwater fishes tend to take on water, whereas marine fishes tend to lose water.
5. Ovoviviparous organisms (some fish, reptiles; many insects) retain fertilized eggs within their oviducts until the young hatch; the young receive nourishment from the egg. Oviparous animals (birds) lay eggs, and their young hatch outside the mother. Viviparous organisms (humans) bear their young alive, and the young are nourished by the mother, not the egg.
6. Amphibian means "two lives." This refers to the fact that amphibians live both aquatic and terrestrial existences.
7. Ectothermic animals (fish, amphibians, reptiles) regulate body temperature by taking in heat from the environment. Endothermic animals (birds, mammals) regulate their body temperatures internally.
8. Reptiles have dry skins covered with scales, retarding water loss. They also lay amniotic eggs, which contain nutrients and water for the embryos and protect the embryos from drying out.
9. Feathers are flexible, light, waterproof epidermal structures; birds are the only animals that have them. Feathers are important to flight; they also give birds waterproof coverings and insulate them against temperature changes.
10. Mammals are endothermic vertebrates that have hair and whose females secrete milk from mammary glands to feed their young.

Critical Thinking Review

1. The three subphyla of chordates are tunicates, lancelets, and vertebrates. The tunicates are marine saclike forms with no notochord in the adult. The structure of these bloblike creatures fits their lifestyle as sessile filter feeders. Their structure allows them to draw in organisms and permits filtered water to exit. The lancelets are tiny, scaleless fishlike marine chordates. Their knifelike shape allows them to bury the posterior parts of their bodies in the sand while their anterior ends stick out and feed on plankton. Vertebrates each have a vertebral column, distinct head, and a closed circulatory system; they are much more complex organisms than the other two subphyla of chordates. These characteristics allow for the range of lifestyles of the vertebrates, from those that swim in the oceans, fly through the air, walk on four legs on land, or walk upright. The vertebral column provides support and acts as a lever system for movement.
2. The skin of amphibians limits their ability to remain on land for long periods of time because they dry out unless their skin is kept moist. Likewise, other terrestrial organisms are able to live successfully on land because of their skins (and their osmoregulatory mechanisms), all of which are designed to conserve water.
3. Reptiles are ectothermic and regulate their body temperature by taking in heat from the environment. In the arctic, snakes would not be able to maintain a body temperature high enough to survive.
4. The heart of a frog is more efficient than the heart of a fish because blood returns to the heart after picking up oxygen at the lungs, and is then pumped out to the body. In fish, the heart pumps blood to the gills but it continues on to the rest of the body without another push.
5. The fish heart has four chambers in a row. The first two receive blood from the body, and the second two send it to the lungs. The amphibian heart has three chambers that work as noted in the previous answer, but the ventricles are open to one another and oxygenated and deoxygenated bloods mix. The reptile heart has (generally) an incomplete septum between the ventricles that reduces the amount of mixing of the bloods, and the vertebrate heart (birds) has four separate chambers. The birds have the most efficient heart of this group because the heart acts as a double pump, pumping blood to the lungs and then to the body, without a mixing of bloods.

Visual Thinking Question

1. *allantois:* fluid-filled sac that receives wastes in amniotic egg and gives rise to the umbilical arteries and vein as the umbilical cord develops in the human
   amnion: protective watery sac
   chorion: controls the movement of gases into and out of the amniotic egg and between the embryo and the mother in humans
   yolk: food for the embryo
   embryo: developing organism
2. In humans, the yolk sac contains little yolk because the developing human derives nourishment from the mother after

implantation. In most egg-laying animals such as birds, for example, the young develop totally within the egg. The yolk of these eggs is substantial enough to maintain the organism throughout its entire course of development.

3. In the amniotic egg, gas exchange takes place through the chorion. In the developing human, gas exchange takes place at the mother's lungs as soon as the umbilical cord develops.

Chapter 32

Knowledge and Comprehension Questions

1. This statement expresses a fundamental idea in the study of animal behavior—that animal behaviors are evolutionary adaptations that make an organism more fit to survive in and adjust to its environment.

2. Innate behaviors protect an animal from environmental hazards without it having learned to do so; many social behaviors also depend on innate behaviors. Organisms with simple nervous systems rely primarily on innate behaviors for survival. Learned behaviors, however, allow an animal to adjust its behavior on the basis of its experiences, helping it to adapt better to its environment or to environmental changes.

3. Type of behavior:
   a. Taxis (phototaxis).
   b. Habituation.
   c. Trial-and-error conditioning (operant conditioning).

4. Operant conditioning is a form of learning in which an animal associates something that it does with a reward or punishment. In classical conditioning, an animal learns to associate a new stimulus with a natural stimulus that normally evokes a particular response. Eventually, the animal will respond to the new stimulus alone.

5. Insight (reasoning) is the most complex form of learning in which an animal recognizes a problem and solves it mentally before trying out the solution. Insight allows an animal to perform a correct behavior the first time it tries; it helps the animal adjust and respond to new situations and perhaps determine better ways to deal with its environment.

6. a. Locality imprinting (migration).
   b. Reflex.
   c. Fixed action pattern.

7. Competitive behavior results when two or more individuals are striving to obtain the same resource (such as food, territories, or mates).

8. a. Threat display.
   b. Submissive behavior.
   c. Parenting behavior.
   d. Courtship ritual (reproductive behavior).
   e. Territorial behavior.

9. A honeybee colony consists of three castes: (a) the queen, which lays the eggs, (b) the drones, male bees that fertilize

the queen's eggs, and (c) the workers, female bees that develop from fertilized eggs and do everything around the hive except lay eggs and fertilize them. (This includes feeding larvae, guarding the hive, and foraging for food.)

10. A rank order is the social hierarchy that appears in groups of fishes, reptiles, birds, and mammals. This hierarchy helps reduce aggression and fighting within social groups, focusing these behaviors to a short time when the rank order develops.

Critical Thinking Review

1. Answers will vary. However, the evolutionary basis of group social behaviors is that the advantages of a particular social trait must outweigh the disadvantages for the individual and its kin. Individuals within colonies of bees, for example, are closely related. Mammals do not produce large numbers of offspring as occurs in bee and ant colonies; large groups of mammals are generally not closely related. Research has shown, however, that naked mole rats live within colonies and are closely related—an unusual situation for mammals.

2. Answers will vary, but hypotheses should focus on imprinting and other means of animal migration mentioned in the chapter. Research shows, however, that Pacific salmon become imprinted with the smells of their native stream. Because they can discriminate between the waters of two rivers coming together at a fork, they know which way to swim.

3. Answers will vary; human behavior appears to result from a combination of hereditary and environmental factors.

4. A pheromone is a chemical produced by one individual that alters the physiology or behavior of other individuals of the same species. Since the chemicals on the pads of the first group of women altered the physiology of the second group of women, these chemicals fit the definition of a pheromone.

5. The genes that encode social behaviors are retained in the gene pool because they increase the reproductive fitness of the species.

Visual Thinking Question

See Figure 32.1 and accompanying narrative.

Chapter 33

Knowledge and Comprehension Questions

1. Ecology is the study of interactions between organisms and the environment. Underlying questions to this study are "What factors determine the distribution patterns of organisms?" and "What factors control their numbers in the locations in which they are distributed?"

2. The population size is affected by immigration. The growth rate does not include factors of emigration and immigration.

3. Demographers study human populations, describing peoples and the characteristics of populations. They predict the ways in which the sizes of populations will alter in the future, taking into account the age distribution of the population and its changing size through time. Population ecologists study how populations of any species grow and interact, investigating the factors that determine the distribution patterns of organisms and the factors that control their numbers in the locations in which they are distributed. Although both study populations, demographers are concerned only with human populations.

4. A large population is more likely to survive because it is less vulnerable to random events and natural disasters and because its size reflects greater genetic diversity than does a small population.

5. Density refers to the number of organisms per unit of area; dispersion refers to the way in which the individuals of a population are arranged. If individuals are too far apart, they may not be able to reproduce. If population becomes too dense, however, factors can arise that limit the population's size (disease, predation, starvation).

6. The three patterns are (a) uniform (evenly spaced), (b) random, and (c) clumped (organisms are grouped in areas of the environment that have the necessary resources). Human populations are distributed in the clumped pattern.

7. Density-dependent limiting factors (such as competition) come into play when a population size is large, whereas density-independent factors (such as weather) operate regardless of population size.

8. Both predation and parasitism are density-dependent limiting factors. Predators are organisms that kill and eat organisms of another species; parasites live on or in other species and derive nourishment from them but do not necessarily kill them.

9. See Figure 33.5. Type I is typical of organisms that tend to survive past their reproductive years (such as humans). Type II characterizes organisms that reproduce asexually and are equally likely to die at any age (such as the hydra). Type III is characteristic of organisms that produce offspring that must survive on their own and therefore die in large numbers when young (oysters).

10. Developing countries tend to have a population pyramid with a broad base, reflecting a rapidly growing population with large numbers of individuals entering their reproductive years. Industrialized countries tend to have populations with similar proportions of their populations in each age group and that are growing slowly or not at all.

Critical Thinking Review

1. People are already putting pressure on our natural resources (land, water, forests, atmosphere). Increasing industrialization leads to rising levels of pollution and solid waste. Even though some experts suggest that there is enough food produced to feed the world's population, many people live—and die—in poverty and hunger.

2. Answers will vary.

3. In a population pyramid of a stable population, births should equal deaths and the number of females of each age group within the population should be similar. Therefore, the pyramid would not be a pyramid at all, but would be "squared off" similar to Figure 33.6b, United States, only with a wider "top."

4. Such a population probably does not include a high percentage of women in their childbearing years, and deaths are probably outnumbering births. The fertility rate is probably lower than the replacement rate. The population pyramid would have a narrow base; this population would be one of negative growth.

5. Parasites can (and do) weaken their hosts and make them less fit to compete for space and food. As the density of a population increases, competition for these resources increases, and the effects of parasites increase. The reproductive fitness of parasitized organisms would therefore diminish as density of the population increased.

Chapter 34

Knowledge and Comprehension Questions

1. A population consists of individuals of the same species. Populations that interact with one another form a community. An ecosystem is a community of organisms along with the abiotic factors with which the community interacts.

2. habitat; niche.

3. Interactions can be grouped into: (a) competition (organisms of different species living near each other strive to obtain the same limited resources); (b) predation (organisms of one species kill and eat organisms of another species); (c) commensalism (an organism of one species benefits from its interaction with another while the other species neither benefits nor is harmed); (d) mutualism (two species live together in close association, both benefiting from the relationship); and (e) parasitism (an organism of one species [the parasite] lives in or on another [the host]. The parasite benefits from this relationship while the host is usually harmed.)

4. The principle of competitive exclusion states that if two species are competing for the same limited resource in a location, the species that can use the resource most efficiently will eventually eliminate the other species in that area.

5. An organism's realized niche is the role it actually plays in the ecosystem; its fundamental niche is the role it might play if competitors were not present.

6. Gause's experiments showed that predator-prey relationships follow a cyclical pattern: As the number of prey increases, the number of predators increases. When the prey decrease be-

cause of predation, the predator population also declines until an upsurge in the prey population starts the cycle again.

7. The balance between predator and prey populations is complex because interactions depend on many factors: interactions with other organisms, movement of new organisms into or out of the community, and abiotic factors in the ecosystem.

8. Coevolution is the long-term, mutual evolutionary adjustment of characteristics of members of biological communities in relation to one another. Cattle are one example: Their digestive systems are adapted (grinding teeth, strong jaws, a rumen) to digesting silica in certain grasses.

9. a. Warning coloration.
   b. Camouflage.

10. Primary succession takes place in areas that were originally bare. It could begin, for instance, with lichens growing on bare rock, producing acids that break down the rock to form soil. This gives rise to a pioneer community, which in turn leads to more vegetation, eventually making the area able to support other forms of life. At some point the mix of plants and animals becomes somewhat stable, forming a climax community.

Critical Thinking Review

1. Answers will vary. However, students must show how a one-sided relationship (commensalism) between two organisms of different species really benefits both rather than only one.

2. A Batesian mimic is an unprotected species that has come to resemble a distasteful species that exhibits warning coloration. Since the birds found both the monarch butterfly (the poisonous and distasteful "protected" species) and the viceroy (the supposedly unprotected species) distasteful, perhaps these butterflies are Mullerian mimics in which both species possess a similar defense.

3. The most successful predator would feed on those organisms past reproductive age. In this way, the predator's food source continues to reproduce, ensuring food for the future.

4. The factors are: climate (including annual precipitation), types of available habitats, types of predators, and level and type of human activity.

5. When an herbivore eats plants but does not destroy them, allowing them to continue to grow, the relationship would be considered parasitism. If, however, the herbivore destroys the plants and they are unable to continue growing, the relationship would be considered predation.

Chapter 35

Knowledge and Comprehension Questions

1. An ecosystem is a community of plants, animals, and microorganisms that interact with one another and their environment and that are interdependent. The living (biotic) groups it contains include producers (organisms that can make their own food), consumers (organisms that eat other organisms for food), and decomposers (consumers that eat dead matter).

2. See Figure 35.1.

3. Primary consumers; secondary consumers; decomposers; detritus.

4. A trophic level is a feeding level within a food chain. A food chain is a linear relationship among organisms that feed one on another. Food webs are food chains that interweave with one another.

5. A green plant captures some of the sun's energy. The successive members of a food chain, in turn, incorporate into their own bodies about 10% of the energy in the organisms on which they feed. The rest is released as heat or is used to power metabolic activities.

6. Chemicals are stored in the atmosphere and in rocks (reservoirs). These substances enter the soil, are incorporated into the bodies of living organisms, and are passed along the food chain. Ultimately, the chemicals, with help from decomposers, return to the nonliving reservoirs.

7. Water in the atmosphere falls to the Earth as precipitation. Plants take up water from the soil, and animals obtain water from drinking it or from eating plants or other animals. Water returns to the atmosphere through the evaporation of surface water and transpiration by plants.

8. Groundwater includes subsurface bodies of fresh water. It can be polluted by pesticides that are washed from plants and topsoil by the rain and by chemical wastes that are dumped in surface water. It is serious because at the present time there is no technology to remove these pollutants from groundwater; already about 2% of groundwater in the United States is polluted, and the situation is getting worse.

9. Carbon is held in the atmosphere as carbon dioxide. Photosynthetic organisms take in this gas and use it to build carbon compounds. Both producers and consumers use carbon compounds as an energy source, metabolizing them and releasing carbon dioxide to the atmosphere again.

10. Crop rotation can improve soil by alternating a nonleguminous crop (which consumes nitrogen from the soil) with a leguminous crop (which can enrich the nitrate level of depleted soil by forming mutualistic associations with nitrogen-fixing bacteria). This relates to the nitrogen cycle.

Critical Thinking Review

1. Student answers will vary. However, the ecosphere must include producers to serve as food for the consumers. Decomposers are necessary to break down dead material, making the breakdown products available for re-use. The photosynthesizing organisms (producers) will generate oxygen for the nonphotosynthesizing organisms (con-

sumers). Both producers and consumers will generate carbon dioxide through cell respiration, which is needed by the producers. A moderate amount of water must be present in a terrestrial environment to be continually cycled throughout the system. The specific organisms would, of course, need to be carefully chosen for their ability to survive in such a system. An aquatic self-sustaining closed system might consist of snails and elodea plants.

2. Producers ultimately support all the other organisms in a food chain. Organisms highest in the food chain have the least food available to them; the total number of organisms able to be fed at each succeeding level of the food chain diminishes because energy is lost as heat and is used for work at each level. Only 10% of the energy from food is incorporated into the bodies of the organisms at each level. Therefore, there is a limit to the number of links in the chain and how far removed an organism can be from producers.

3. Plants transpire, or release water vapor into the atmosphere. This water vapor is a critical part of the water cycle. As the rain forest is cut down, the plants are lost and less water is available in the atmosphere to return to the ground as precipitation.

4. Adding more phosphates than crops need causes the extra phosphates to wash off or erode into water. Bodies of water can become eutrophic; algae and other aquatic plants overgrow and decomposers feed on them, using much of the water's oxygen and choking out other organisms.

5. The deer are more likely to survive in the ecosystem with abundant vegetation, assuming that the other animals in the nature preserve are not predators of the deer. These deer have abundant food and will reproduce, even though their numbers are low. In the other situation, although the population is higher, food is scarce. This environmental pressure will undoubtedly increase competition for food, and the mortality rate will likely be high. The land is eroded, so growth of new vegetation is unlikely. Things will probably only get worse in this more "unhealthy" ecosystem.

Chapter 36

Knowledge and Comprehension Questions

1. A biome is an ecosystem of plants and animals that occurs over wide areas within specific climatic regions.

2. Australia is in the Southern Hemisphere, whereas the United States is in the Northern Hemisphere. Thus, when the Earth tilts so that the United States is tilted away from the sun, people in the United States experience winter but it is summer in Australia (which is tilted toward the sun). When the Earth tilts so that the United States is tilted toward the sun, it is summer in the United States but winter in Australia.

3. The climate of an area is determined by its latitude, wind patterns, and surface features of the Earth.

4. a. Tropical rain forests
   b. Temperate deciduous forests
   c. Arctic tundra
   d. Savannas
   e. Taiga
   f. Temperate grasslands
   g. Desert

5. Desert rainfall is limited by latitude because the air, which has released its moisture over the rain forests, is dry over desert latitudes. Deserts can also occur in the interior of continents, far from the ocean, and on the leeward side of mountains that block rainfall.

6. The taiga is the northern coniferous forest with long, cold winters and little precipitation. The tundra lies farther north and is open, often boggy, with desertlike levels of precipitation.

7. The three freshwater life zones are: (a) shore zone (shallow water in which plants with roots may grow; some consumers live here), (b) open-water zone (the main body of water through which light penetrates; floating plants, microscopic floating animals, and fish live here), and (c) deep zone (light does not penetrate, so there are no producers; inhabited mainly by decomposers or organisms that feed on organic material that filters down).

8. An estuary is an environment where fresh water and salt water meet. All organisms living here are adapted to living in an area of moving water and changing salt concentrations. Some are bottom dwellers that attach to bottom material or burrow into the mud; others spawn in less salty water.

9. Estuaries are ecosystems where freshwater rivers and streams meet the ocean. They are vulnerable to pollutants that flow from upstream, as well as to pesticides that leach from the land and soil.

10. The three major life zones of the ocean are: (a) intertidal (area between high and low tides), (b) neritic (shallow waters along the coasts of the continents), and (c) open sea (the rest of the ocean).

Critical Thinking Review

1. Tropical rain forests make poor farmland because most nutrients are found in the vegetation, not the soil. Root systems are poor and shallow; the soil itself is thin, often only a few centimeters deep. Cutting down and burning the trees remove the nutrients and break down the organic matter to carbon dioxide. In 2 to 3 years, the land becomes barren. Conversely, in grasslands, root systems are extensive and deep. Much of the nutrients are held in the root systems and in the soil rather than in the above ground vegetation.

2. The windward sides of mountains are much wetter than the leeward sides. In fact, the leeward sides can have desertlike conditions as cited in this question. In the Northern Hemisphere, winds blow from the west over the Sierra Madre Mountains, so that desert conditions exist on the eastern side of this mountain range. However, in the Southern Hemisphere, winds blow from the east, so those desertlike conditions exist on the western side of the Andes.

3. This city is located in temperate deciduous forest, in which the annual rainfall is from 30 to 60 inches, and summers are warm and winters are cold. Based on the temperatures shown and the above information, this city is probably in the northeastern United States. (Just so you won't wonder, this city is Hartford, Connecticut.)

4. Answers will vary.

5. Answers will vary, but the changes in plant and animal life should follow from the differences between a flowing river and a pond. For example, the water in a pond may not be aerated as much as in a stream, so with less oxygen, the numbers and types of fish might diminish.

Visual Thinking Question

1. In the winter, the Northern Hemisphere is tilted away from the sun (see the Earth on the right), and in the summer, the Northern Hemisphere is tilted toward the sun (see the Earth on the left).

2. When the Northern Hemisphere is tilted toward the sun in the summer, its rays cover a smaller area (see Figure 36.1). Therefore, in the summer a person will have greater exposure to UV rays because they are more concentrated than in the winter, when the Northern Hemisphere is tilted away from the sun and its rays cover a larger area.

3. Winter.

Chapter 37

Knowledge and Comprehension Questions

1. A sustainable society uses nonfinite, renewable sources of energy. Society's needs are satisfied without compromising future generations and natural resources.

2. People could use solar, wind, and water power, as well as the Earth's geothermal energy.

3. People can reuse and recycle paper, glass, and metal products to help preserve mineral resources. This is important because mineral supplies are finite; once used up, they are gone forever.

4. Global warming refers to a worldwide increase in temperature. Even a small increase in temperature could melt ice at the poles, raising the sea level and flooding one fifth of the world's land area. Higher temperatures also affect rain patterns and agriculture.

5. Biological diversity refers to the richness of species. Humans are reducing the world's biological diversity through habitat destruction, pollution, exploitation of commercially valuable species, and selective breeding. Reducing diversity makes species more vulnerable to being wiped out by disease or environmental changes.

6. Humans can improve solid waste management by increasing and improving recycling, using safe incineration, engineering better landfill sites, and practicing waste-to-energy reclamation.

7. Point sources are sources of pollutants that enter the water at one or a few distinct places (industrial waste dumped by a specific factory). Nonpoint sources refer to pollutants that enter water at a variety of places (metals and acids draining from mines).

8. Organic nutrients are food for bacteria; as the bacteria multiply, they use up oxygen in the water, killing the fish. Inorganic nutrients can stimulate plant growth in the water. Bacteria decompose the plants after they die; again, when the bacteria multiply, they consume the water's oxygen. The decomposed materials also begin to fill in the body of water, eventually turning it into a terrestrial community. Pollution with infectious agents can cause serious diseases. Toxic substances in water can poison the organisms that take in this water; sometimes organisms accumulate these substances in their bodies.

9. Acid rain results from gases such as sulfur dioxide and nitrogen dioxide that combine with water in the atmosphere. When this acidic rain falls, it kills aquatic life, pollutes groundwater, damages plants, and eats away at stone, metal, and painted surfaces.

10. Ozone in the stratosphere helps shield you from the sun's ultraviolet rays. Depletion of this layer can damage bacteria and plants and cause burns and skin cancers.

Critical Thinking Review

1. Answers will vary.

2. Answers will vary.

3. Tropical rain forests are home to more than half the world's species of plants, animals, and insects from which people get many medicines as well as wood, charcoal, oils, and nuts. Burning the forests adds carbon dioxide to the air, which may raise temperatures worldwide. Destroying the forests also destroys the habitat for many species and leads to extinctions.

4. Answers will vary.

5. Answers will vary.

Appendix B   Classification of Organisms

The purpose of the classification scheme presented here is for student reference. It is not intended to include all of the phyla of organisms currently identified by taxonomists, but it includes all of the phyla described and referred to in *Biology: Understanding Life*, Third Edition. Because this textbook describes humans in its extensive physiology section and includes a chapter on chordate and human evolution, the classification of the chordates is listed in more detail than other phyla.

The classification scheme below is based on: Margulis, L. and Schwartz, K.V. (1998). *Five Kingdoms: An Illustrated Guide to the Phyla of Life on Earth,* Third Edition [New York: W.H. Freeman and Company], with the exception of the classification of the Urochordata, Cephalochordata, and Vertebrata (Craniata) as subphyla of the Phylum Chordata, rather than as individual phyla. Additionally, the classification scheme below uses the widely accepted subkingdom name Archaebacteria rather than Archaea, as well as the familiar kingdom designation Protista rather than Protoctista.

Kingdom Bacteria

Prokaryotic, single-celled organisms that have neither membrane-bounded nuclei nor membrane-bounded organelles.

> Subkingdom Archaebacteria: Methanogens (methane-producing bacteria), halophiles (bacteria that live in salt marshes), thermoacidophiles (bacteria that live in hot springs and deep-sea vents). A separate kingdom or alternatively, a separate domain, has been proposed for this subkingdom.
> Subkingdom Eubacteria: All other bacteria, including nitrogen-fixing bacteria and cyanobacteria.

Kingdom Protista

A varied group of eukaryotic organisms. Many are single celled, although some phyla include multicellular or colonial forms. Some forms are photosynthetic.

> Phylum Rhizopoda: Amebas and cellular slime molds
> Phylum Granuloreticulosa: Foraminifera
> Phylum Discomitochondria: Flagellates
> Phylum Archaeprotista: Flagellates
> Phylum Ciliophora: Ciliates
> Phylum Apicomplexa: Sporozoans
> Phylum Dinomastigota: Dinoflagellates
> Phylum Diatoms: Diatoms
> Phylum Phaeophyta: Brown algae
> Phylum Chlorophyta: Green algae
> Phylum Rhodophyta: Red algae
> Phylum Myxomycota: Plasmodial (acellular) slime molds
> Phylum Oomycota: Water molds, white rusts, downy mildews

Kingdom Fungi

Mostly multicellular eukaryotic organisms that lack undulipodia at all life stages, and that are saprophytes, feeding on dead or decaying organic material. Some fungi are parasitic, feeding on living organisms.

> Phylum Zygomycota: Zygote-forming fungi (black bread mold)
> Phylum Ascomycota: Sac fungi (yeasts, cup fungi, *Penicillium,* lichens)
> Phylum Basidiomycota: Club fungi (mushrooms, puffballs, shelf fungi)

Kingdom Plantae

Multicellular, eukaryotic organisms that evolved on land and that perform photosynthesis, producing their own food by using energy from the sun and carbon dioxide from the atmosphere.

Phylum Bryophyta: Mosses
Phylum Hepatophyta: Liverworts
Phylum Anthocerophyta: Hornworts
Phylum Psilophyta: Whisk ferns
Phylum Lycophyta: Club mosses
Phylum Sphenophyta: Horsetails
Phylum Filicinophyta: Ferns
Phylum Coniferophyta: Conifers
Phylum Cycadophyta: Cycads
Phylum Ginkgophyta: Ginkgos
Phylum Gnetophyta: Gnetae
Phylum Anthophyta: Flowering plants (angiosperms)
   Class Monocotyledons: Grasses, irises
   Class Dicotyledons: Flowering trees, shrubs, roses

Kingdom Animalia

Multicellular, eukaryotic organisms that have a blastula embryo, and are heterotrophic, eating other organisms for food.

Phylum Porifera: Sponges
Phylum Cnidaria: Hydra, jellyfish, corals, sea anemones
   Class Hydrozoa: Hydra
   Class Scyphozoa: Jellyfish
   Class Anthozoa: Corals, sea anemones
Phylum Ctenophora: Comb jellies, sea walnuts
Phylum Platyhelminthes
   Class Turbellaria: Free-living flatworms
   Class Trematoda: Flukes
   Class Cestoda: Tapeworms
Phylum Nematoda: Roundworms
Phylum Mollusca
   Class Polyplacophora: Chitons
   Class Gastropoda: Snails and slugs
   Class Bivalvia: Bivalves
   Class Cephalopoda: Octopuses, squids, nautilus
Phylum Annelida
   Class Polychaeta: Marine worms
   Class Oligochaeta: Earthworms and fresh water worms
   Class Hirundinea: Leeches

Arthropods
Phylum Chelicerata
   Class Arachnida: Spiders, mites, ticks
   Class Merostomata: Horseshoe crabs
   Class Pycnogonida: Sea spiders
Phylum Crustacea: Lobsters, crayfish, shrimps, crabs
Phylum Mandibulata
   Class Myriapoda: Centipedes, millipedes
   Class Insecta: Insects

Phylum Echinodermata
    Class Crinoidea: Sea lilies
    Class Asteroidea: Sea stars
    Class Ophiuroidea: Brittle stars
    Class Echinoidea: Sea urchins and sand dollars
    Class Holothuroidea: Sea cucumbers
Phylum Chordata: Chordates
Subphylum Urochordata: Tunicates (sea squirts)
Subphylum Cephalochordata: Lancelets
Subphylum Vertebrata (Craniata): Vertebrates
    Class Cyclostomata: Jawless fishes (lamprey, eels, hagfishes)
    Class Chondrichthyes: Cartilaginous fishes (sharks, skates, rays)
    Class Osteichthyes: Bony fishes (perch, cod, and trout)
    Class Amphibia: Salamanders, frogs, toads
    Class Reptilia: Reptiles (lizards, snakes, turtles, crocodiles)
    Class Aves: Birds
    Class Mammalia: Mammals
      Subclass Prototheria: Egg-laying mammals (duck-billed platypus, spiny anteater)
      Subclass Metatheria: Pouched mammals or marsupials (oppossums, kangaroos, wombats)
      Subclass Eutheria: Placental mammals
        Order Edentata: Anteaters, armadillos, sloths
        Order Lagomorpha: Rabbits, hares, pikas
        Order Rodentia: Squirrels, rats, woodchucks
        Order Insectivora: Hedgehogs, tenrecs, moles, shrews
        Order Carnivora: Dogs, wolves, cats, bears, weasels
        Order Pinnipeda: Sea lions, seals, walruses
        Order Cetacea: Whales, dolphins, porpoises
        Order Proboscidea: Elephants
        Order Perissodactyla: Horses, asses, zebras, tapirs, rhinoceroses
        Order Artidactyla: Swine, camels, deer, hippopotamuses, antelopes, cattle, sheep, goats
        Order Tubulidentia: Aardvarks
        Order Scandentia: Tree shrews
        Order Chiroptera: Bats
        Order Primates: Lemurs, monkeys, humans
          Suborder Strepsirhini: Prosimians (lemurs, aye-ayes)
          Suborder Haplorhini: Anthropoids
            Superfamily Tarsioidea: Tarsiers
            Superfamily Ceboidea: New World monkeys
            Superfamily Cercopithecoidea: Old World monkeys
            Superfamily Hominoidea
              Family Hylobatidae: Gibbons, orangutans, chimpanzees
              Family Pongidae: Gorillas
              Family Hominidae: Humans

Appendix C   Periodic Table of the Elements

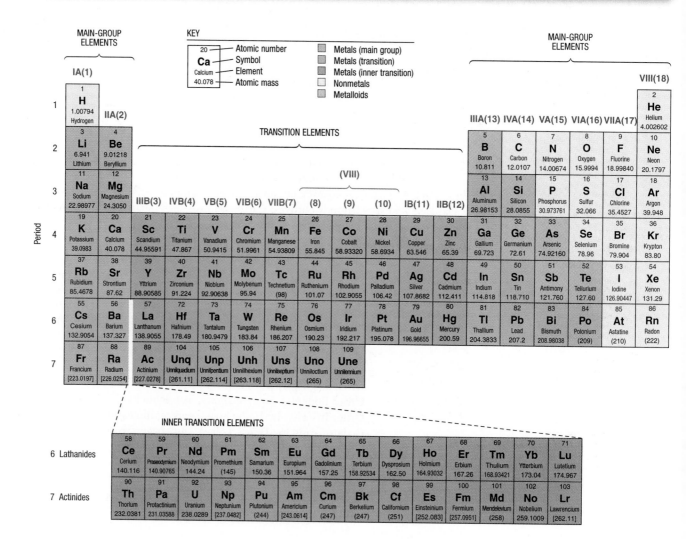

Appendix D   Units of Measurement

Metric Units	Metric Equivalents	Symbols	U.S. Equivalents
Measures of length			
1 kilometer	= 1000 meters	km	0.62137 mile
1 meter	= 10 decimeters or 100 centimeters	m	39.37 inches
1 decimeter	= 10 centimeters	dm	3.937 inches
1 centimeter	= 10 millimeters	cm	0.3937 inch
1 millimeter	= 1000 micrometers	mm	
1 micrometer	= 1/1000 millimeter or 1000 namometers	μ	
1 nanometer	= 10 angstroms or 1000 picometers	nm	No U.S. equivalent
1 angstrom	= 1/10,000,000 millimeter	Å	
1 picometer	= 1/1,000,000,000 millimeter	pm	
Measures of volume			
1 cubic meter	= 1000 cubic decimeters	m³	1.308 cubic yards
1 cubic decimeter	= 1000 cubic centimeters	dm³	0.03531 cubic foot
1 cubic centimeter	= 1000 cubic millimeters or 1 milliliter	cm³(cc)	0.06102 cubic inch
Measures of capacity			
1 kiloliter	= 1000 liters	kl	264.18 gallons
1 liter	= 10 deciliters	L	1.0567 quarts
1 deciliter	= 100 milliliters	dl	0.4227 cup
1 milliliter	= volume of 1 gram of water at standard temperature and pressure	ml	0.3381 ounce
Measures of mass			
1 kilogram	= 1000 grams	kg	2.2046 pounds
1 gram	= 100 centigrams or 1000 milligrams	g	0.0353 ounce
1 centigram	= 10 milligrams	cg	0.1543 grain
1 milligram	= 1/1000 gram	mg	
1 microgram	= 1/1,000,000 gram	μg	
1 nanogram	= 1/1,000,000,000 gram	ng	
1 picogram	= 1/1,000,000,000,000 gram	pg	

Note that a micrometer was formerly called a micron (μ), and a nanometer was formerly called a millimicron (mμ).

Comparative Temperature Scales

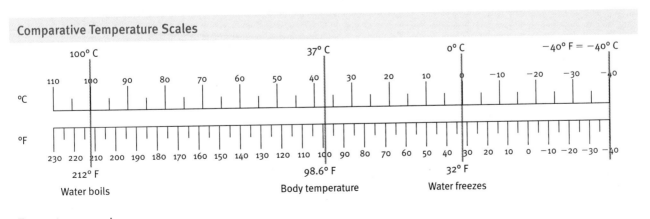

Temperature conversions:
Fahrenheit (° F) to Celsius (° C) = ⁵/₉ (° F − 32)
Celsius (° C) to Fahrenheit (° F) = ⁹/₅ (° C + 32)

Appendix E A Scientific Journal Article Examined

ELSEVIER

Biological Conservation 85 (1998) 63–68

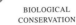

BIOLOGICAL
CONSERVATION

Experimental measurement of nesting substrate preference in Caspian terns, *Sterna caspia*, and the successful colonisation of human constructed islands

James S. Quinn *, Jane Sirdevan

Biology Department, McMaster University, Hamilton, Ontario, Canada, L8S 4K1

Received 7 June 1997; accepted 30 September 1997

Abstract

Caspian terns, *Sterna caspia*, recently bred in Hamilton Harbour, at the western end of Lake Ontario, on private property that is likely to be developed in the next decade. To reduce this land-use conflict and to promote the current level of biodiversity of colonial nesters in the area, artificial islands were built in the winter of 1995-1996 with different areas designated for a variety of nesting waterbirds including Caspian terns. In 1994, prior to island construction, we tested three substrate types for tern nesting preferences so that an appropriate substrate could be placed on the Caspian tern designated portion of the new islands. We found a preference for sand over pea-gravel and crushed stone, and indirect evidence for a preference favouring the experimental substrates over the pre-existing substrate of hard-packed ground. Based on these results, the small area of the island designed for Caspian tern nesting was surfaced with sand and was subsequently colonised successfully. The colony established and reproduced successfully on the designated site in 1996 and grew in numbers of nesting pairs in 1997. © 1998 Elsevier Science Ltd. All rights reserved

Keywords: Habitat preference; Restoration; Colonial nesting waterbirds

1. Introduction

Caspian terns, *Sterna caspia*, have probably never been abundant in North America, but they have nested in the Great Lakes since as early as 1896 (Blokpoel and Scharf, 1991). In 1986, a small colony of 48 nests was established in Hamilton Harbour (43° 16′N, 79° 46′W; Dobos et al.; 1988) at the western end of Lake Ontario. Between 1990 and 1994, the main colony was located on the site of the experiment reported here and numbers in the harbour increased from 184 to 331 pairs, representing over 16% of Lake Ontario's Caspian terns (Moore et al., 1995). This immigration to Hamilton Harbour and subsequent growth of the colony, particularly the increase to 134 pairs in 1987 (Dobos et al., 1988), coincided with the decline and desertion of a Caspian tern colony site at the Eastern Headland, a man-made area extending into Lake Ontario from the Toronto waterfront (Morris et al., 1992). The success of this thriving new colony in Hamilton Harbour is encouraging given the questionable status of the species, ranging from rare to endangered (proposed) in Ontario and the States surrounding the Great Lakes (Blokpoel and Scharf, 1991).

Development plans for an area of the Hamilton Harbour shoreline included the Caspian tern main colony site. During the winter of 1995–1996 three islands were constructed to provide nesting habitat for the six species of colonial nesters currently in the area, a major component of the remedial action plan for the rehabilitation of Hamilton Harbour (Quinn et al., 1996). The motivation for island construction was to maintain current levels of diversity of colonial nesters in the harbour and to reduce land-use conflict with the current property owners. We tested substrate preferences of Caspian terns on an experimental site on the mainland where the colony had been previously to facilitate the establishment of a colony on one of the new islands.

Little is known about the nesting substrate preferences of Caspian terns. Caspian terns generally nest in dense colonies situated in open and largely unvegetated areas (Peck and James, 1983). Although descriptions

* Corresponding Author. Tel.: (905) 525-9140 ext. 23194; fax: (905) 522-6066; email: quinn@mcmaster.ca

0006-3207/98/$19.00 © 1998 Elsevier Science Ltd. All rights reserved
PII: S0006-3207(97)00142-0

Marginal annotations:

1 The name and logo of the publisher of the journal

2 The journal title

3 The journal title, volume number, year, and page numbers of the article

4 TITLE OF THE PAPER. The title should be descriptive of the study and should have a clear focus. This title states the independent variable (IV—[nest] substrate) and the dependent variable (DV—nesting site preference [on the experimental grid]).

5 AUTHORS OF THE STUDY. The author to whom queries should be addressed is noted by an asterisk, with contact information stated in a footnote at the bottom of the research page.

6 Institutional affiliation of the authors

7 This paper was received by the editor of the journal on June 7, 1997. It then was sent to peer reviewers (other scientists in this field). Peer reviewers give their opinion as to a paper's acceptability for publication. This paper was accepted for publication on September 30, 1997.

8 ABSTRACT. An abstract is a concise summary of the research paper. Notice that it includes the statement of the problem, procedures, principal results, and conclusions. The hypothesis is also often stated in the abstract. The abstract gives readers an overview of the paper so they know whether it contains information they want or need. Additionally, it helps readers understand the paper by creating a context for the specific sections of the paper that follow.

9 KEYWORDS. Keyword computer searches using scientific literature databases will yield titles, abstracts, and sometimes full texts of research papers related to the keywords. (Many such databases are available at university libraries.)

10 INTRODUCTION. In this section, the investigators present the rationale (statements of their reasons) for their study and its foundation in the literature. The specific references cited here are listed at the end of the paper so that other researchers can find and read these papers. This section typically concludes with the purpose of the study and research hypotheses.

11 The purpose of this paragraph is to provide the rationale for studying Caspian terns (i.e., why study conservation issues related to this species).

12 This paragraph further explains the rationale for the study. The investigators discuss plans to develop the area in which a large colony of terns is nesting and the consequent construction of three islands to provide new places for the birds to nest. At the end of this paragraph, the investigators state that they tested nesting materials for use on these three islands' nesting sites.

13 This paragraph gives background about tern preferences in nesting materials (substrates) and sites (habitats). The authors also mention the problems ring-billed gulls create with Caspian tern nesting.

(14) Statement of the "problem"; importance of the study

(15) PURPOSE OF THE STUDY. The investigators do not overtly state a hypothesis, but it is implied in the title and background given here: If the terns are provided with nesting substrates of gravel, sand, and stone, then the terns will favor one of these materials as a nesting substrate. The IV is the nesting substrate, and the DV is tern preference.

(16) METHODS. In the methods section, investigators describe the materials and procedures they used to conduct the study. Enough detail is included so that another researcher could repeat (replicate) the study.

(17) This section describes the selection of the study site and the design of the study.

Photo A Photograph of the distribution of substrates in the experimental grid. There are nine cells within the frame (see Figure 1).

(18) Here, the investigators cite references that further describe methods not described in detail in the paper.

(19) The investigators describe how they determined offspring (chick) survival. They wanted to determine whether chick survival would vary with substrate type. This was measured as it may provide insight as to why Caspian terns may prefer particular nesting substrates.

---

of nesting habitat are available (Peck and James, 1983; Quinn et al., 1996), experimental studies have not been reported and descriptive studies may not take habitat availability into account. For example, at sites where ring-billed gulls, *Larus delawarensis*, nest, only areas not used by gulls are available to terns as terns begin nesting after ring-billed gulls have become well established (Quinn et al., 1998). The later nesting terns are unable to take over areas that have become occupied by ring-billed gulls, a problem encountered also by common terns, *Sterna hirundo* (Morris et al., 1992). Ring-billed gulls began nesting as a colony of 17 pairs in Hamilton Harbour in 1978 (Dobos et al., 1998). Since then the colony has grown to about 40000 pairs when last counted in 1990 (Moore et al., 1995).

(14) The Hamilton Harbour Caspian tern colony is one of only five on the lower Great Lakes. In 1994, this colony was located on a site that is slated for development in the next 4 to 9 years. Caspian terns are sensitive to human disturbance and, unlike ring-billed gulls, will not nest in close proximity to human activity. The focus of **(15)** this study was to evaluate Caspian tern nesting preferences when given the opportunity to nest on one of three commercially available substrates, or on the hard-packed ground found on the colony site. The choice of materials for the experiment was based on examinations of Caspian tern nesting substrates in the Great Lakes and the Gulf of Mexico (J. Quinn, pers. obs.). Results of this experiment were used to determine substrate type placed on a designated area of recently built wildlife islands in Hamilton Harbour. The implementation resulted in the successful establishment of a colony of Caspian terns on the designated site. This colony has nested successfully for two seasons (1996 and 1997).

(16) 2. Methods

(17) Our study site was Pier 26 in Hamilton Harbour, at the west end of Lake Ontario. Caspian terns nested at a main site in the midst of a large colony of ring-billed gulls, estimated at about 40 000 pairs in 1990 (Moore et al., 1995), and at a separate sub-colony site about 100m to the North-East. On 8 April 1994, prior to the arrival of Caspian terns in the area, we established a 9 × 9 m frame subdivided into 3 × 3 m cells within the area of the 1993 Caspian tern colony site. Nest markers from the 1993 nesting season had been left in place and the number of clutches that had been initiated in each experimental cell was recorded (Fig. 1). We removed the old stakes and then placed the substrates within the grid, in a 3 × 3 Latin Square design, so that each substrate type was represented once in each row and once in each column, thus controlling for any effects due to position. The substrates tested were: (1) construction grade sand with a few small stones; (2) crushed stone

	West	Middle	East	Total
North	Gravel n=5(5)	Sand n=17(7)	Stone n=12(3)	34
Middle	Sand n=18(10)	Stone n=9(9)	Gravel n=7(9)	34
South	Stone n=6(3)	Gravel n=3(9)	Sand n=11(4)	20
Total	29	29	30	88

Fig. 1. Distribution of substrates in the experimental grid. Each grid cell was 3x3m. Numbers of clutches initiated in each cell is given for 1994 and (1993)—prior to establishment of the grid. Row and column totals are for 1994 data.

See Photo A.

with sharp edges, approximately 1 cm in diameter; and (3) 'pea' gravel with rounded edges, approximately 1 cm in diameter. During the period of ring-billed gull nest building, prior to the arrival of terns, a field assistant visited the Caspian tern nesting area five times per day from 5 to 26 April 1994 and destroyed all ring-billed gull nests as they were started.

Upon tern clutch initiation we marked nests initially with numbered tongue depressors (wooden spatulas) placed about 15 cm from the edge of the nest scrape. Tongue depressors were replaced with 20 cm tall numbered wooden stakes on 17 May. Prior to 2 June we placed nest covers (Quinn, 1984) over most nests with **(18)** eggs, particularly those near the periphery of the colony, to protect them from the ring-billed gull depredation. We typically checked nests every second day until 2 June, and every sixth day thereafter, weather permitting (Fig. 2). We placed chick shelters (Burness and Morris, 1992) near where eggs had or were about to hatch to provide shelter from harsh weather or predators during parental absences.

To examine whether offspring survival varied with substrate types we followed the fates of eggs and chicks which were banded by age 7 days post-hatching. Caspian tern nestlings generally remain in the vicinity of the **(19)** nest until fledging. However, to restrict their movements when we checked the colony, we fenced the grid with 30 cm tall 6mm² mesh hardware cloth on 8 June, before chicks became mobile, and used hatching and survival to at least age 24 days post-hatching as measures of breeding success.

On several occasions we observed a ring-billed gull that flew into the colony and removed an egg from an unattended clutch. Each time, the bird flew into a particular

NOTE: A clutch is a nest of eggs or newly hatched chicks. When chicks acquire the feathers necessary for flight, they are said to be fledged—they become fledglings.

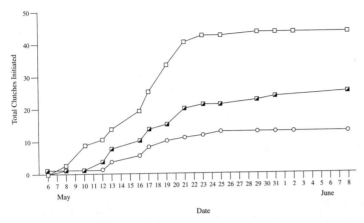

Fig. 2. Cumulative frequency of clutches initiated on three subset types. Clutches from the three replicates of each substrate type were pooled. Data are presented for each nest census date.

part of the colony just outside the grid fence and ate the egg. Yolk stains on the gull's plumage and behavioural idiosyncrasies made it individually recognisable. Our observations from a blind located 13m south of the edge of the colony suggest that the same individual took at least 30 eggs. On 15 June, this bird was shot under Canadian Wildlife Service permit and egg loss virtually stopped.

During the winter of 1995–1996 three wildlife islands measuring about 100 by 20m were constructed under the direction of the Hamilton Harbour Fish and Wildlife Habitat Restoration Project, part of the Hamilton Harbour Remedial Action Plan (Quinn et al., 1996). We covered the tern designated areas with heavy gauge plastic sheeting anchored with rocks prior to the beginning of gull nesting (to prevent nesting by gulls) and removed sheeting shortly after the arrival of the terns. Nesting activities were monitored in 1996 and 1997 through twice-weekly visits to the islands which ceased when chicks became mobile.

We tested the spatial and temporal patterns of clutch initiations and reproductive parameters using the likelihood L^2 statistic (Hays, 1981). We employed a Spearman Rank Correlation (Siegal, 1988) on numbers of clutches initiated within the specific area of each of the cells between 1993 (prior to establishment of the experimental site; nest markers were left in place over winter) and 1994 (after establishment of the experimental site) to test for the possibility that pairs of terns established nests in the same location between years. To show that the frequencies of clutches initiated in the cells between 1993 and 1994 we used a multi-sample χ^2 test with 2 degrees of freedom.

3. Results

See Photo B.

Between 6 May and 8 June 1994, 88 clutches were initiated on the experimental substrates. Although the available area of the Caspian tern colony was larger than that of the experimental grid (approximately 280 m^2 vs 81 m^2), 20 of the first 21 clutches were initiated on the experimental substrates suggesting that Caspian terns preferred the experimental substrates over the natural hard-packed ground at the colony site. Additionally, despite smaller numbers of nests initiated in the main colony site in 1994 (n=177) compared with 1993 (n=242), the number initiated on the experimental substrates in 1994 exceeded that on the same area in 1993 by about a third (Fig. 1). Patterns of nest placements were independent of nest positions in 1993 as the number of nests positioned within the location of each cell of the experimental grid in 1993 and 1994 were not correlated (Spearman Rank Correlation, n_1=59, n_2=88, r_s=0·15, ns; Fig. 1). Furthermore, the nest frequencies in the cells (grouped by 1994 substrate-type) changed significantly between 1993 and 1994 (multi-sample χ^2=9·07, df=2, p <0·025).

The number and timing of clutches initiated on each substrate were analysed using pooled values of the three replicates. Row position had an effect on the number of clutches initiated, with significantly fewer clutches initiated on the South row than on the middle or North rows of the experimental substrates (χ^2=4·76, df=1, p <0·050). Because substrate preference patterns were constant for each row, replicates were pooled for analysis.

Clutch initiations were not at random with regard to substrate type (L^2=16.8, p <0·0005). We found that

---

Photo B Caspian terns nesting in the grid. The eggs or newly hatched nestlings are not visible in this photograph. The fence prevents them from leaving the area during nest checks by the researchers or during other disturbances. Caspian terns are nesting outside of the grid area also, but 20 of the first 21 eggs laid were laid on the grid.

NOTE: The article uses spelling and style conventions typical of Canadian and British publications. This includes a raised decimal point, as seen on this page.

After the preliminary nesting studies were completed, the Hamilton Harbour Remedial Action Plan's (HHRAP) Fish and Wildlife Habitat Restoration Project constructed the islands. The investigators were involved in the design of the topography, substrates, and vegetation of the islands.

RESULTS. The results section of the paper present the data collected. The data are summarized in narrative form and often have accompanying charts, graphs, and/or tables. Statistical analyses of the data are presented; "raw" data are usually not presented unless it is manageable and useful, as in this study (i.e., Figure 1).

First, the investigators describe the results of their preliminary testing of the three substrate types for tern nesting preferences.

66   *J.S. Quinn, J. Sirdevan/Biological Conservation 85 (1998) 63-68*

Table 1

Reproductive parameters for caspian tern pairs nesting on experimental substrates in 1994

Substrate type	Sand	Gravel	Stone
Clutches (n)	46	15	27
Clutch size (mean±1SD)	2·17±0·74	2·00±0·63	2·07±0·87
Clutches hatching ≥1 egg	28	9	10
Hatching success[a]	0·61	0·60	0·37
Clutches fledging ≥1 Chick	21	6	7
Fledging Success[b]	0.75	0.67	0·70
Chicks fledged/pair[c] (mean ± 1 SD)	0·61±0·74	0·67±0·90	0·44±0·80

a Number of clutches hatching one or more eggs divided by total number of clutches initiated.

b Number of clutches fledging one or more chicks divided by total number of clutches hatching one or more eggs. Chicks were considered fledged if they survived to age 24 days.

c Number of clutches where one or more chicks survived to age 24 days divided by total number of clutches initiated.

more clutches were initiated on sand than on the other substrates (Fig. 2; $L^2=13\cdot3$, $p <0.0005$). There were no significant differences in the tendency to nest on stone compared with gravel ($L^2=3\cdot5$, ns). Nests initiated on sand were generally edged with a circle of pea gravel and/or crushed stone along the edge of the nest scrape. Although more pairs laid on sand, the timing clutch initiations did not differ among substrate types. The median clutch initiation date, 17 May, was the same for all three substrates.

Reproductive parameters were similar for pairs nesting on all three substrates (Table 1). Hatching success was defined as the proportion of successful clutches in which one or more eggs hatched, whereas fledging success was determined as the proportion of clutches where one or more eggs hatched and at least one chick fledged (survived to at least 24 days). Values for hatching and fledging success were highest on sand, but the differences were not statistically significant ($L^2=2\cdot1$ and 0·38, $p <0·05$). Egg failures were categorised as: (1) failed to hatch; (2) cracked or broken; (3) disappeared. The proportion of eggs which disappeared or were cracked or

broken was the greatest in the South row of the experimental grid (Table 2). Chick loss prior to fledging was classified as death or disappearance. The most common source of chick loss was death (chicks found dead; sand 77%; stone 67%; gravel 71%). While we tried to minimise the effect of our nest chicks, we cannot dismiss the possibility that out infrequent nest visits affected these results. Notably, the activities of a family of red foxes (Vulpes vulpes) observed on Pier 26 in the area of the colony may have affected chick survival, however we have no direct data from 1994 to address this possibility. In 1995, fox activity was observed to cause colony desertion at night for periods of up to 42 min (J. Sirdevan, per. obs.).

During the winter of 1995 to 1996 three wildlife islands were constructed in Hamilton Harbour (Quinn et al., 1996). The design of the islands utilised the results reported here for the construction of one 200m² knoll surfaced with sand and a small amount of 1 cm diameter pea gravel for lining of nest rims. Most of the area was covered with plastic sheeting (held down with stones) to discourage nesting by ring-billed gulls until the terns arrived. Despite intentions to use decoys to attract Caspian terns to the site in 1996 (Quinn et al., 1996) they had already begun to lay eggs on the exposed sand that remained uncovered before we were able to remove the plastic and install decoys. After uncovering the sand covered habitat terns nested over the entire sandy knoll. In total there were 226 clutches (86% of clutches initiated in Hamilton Harbour) initiated on the designated Caspian tern site in 1996. About 37 clutches (14%) were initiated on a raft (H. Blokpoel, pers. comm.) which was positioned in a confined disposal facility near the original Caspian tern colony (Lampman et al., 1996). Again in 1997 we covered most of the sandy Caspian tern habitat with plastic. The first Caspian tern nests were found on 6 May, 7 days after the plastic was removed. There were 319 clutches initiated (80% of clutched initiated in Hamilton Harbour) on the Caspian tern site with a median initiation date of 21 May. A small proportion of pairs nested on the CDF tern raft (estimated

Table 2

Classification of egg failures on experimental substrates during the 1994 breeding season

Substrate	Row[a]	Total eggs	Proportion of eggs		
			Disappeared	Cracked/broken	Failed to hatch
Sand	N	37	0.19	0.05	0.05
	M	37	0.51	0	0.03
	S	26	0.69	0.08	0
Gravel	N	10	0	0.10	0
	M	14	0.50	0	0
	S	6	0.83	0.17	0
Stone	N	28	0.39	0	0
	M	16	0.75	0.13	0
	S	12	0.92	0.08	0

a Row indicates position in experimental grid (N, north; M, middle; S, south)

N=15, 4%; H. Blokpoel, pers. Comm.). Sixty-three late-nesting or re-nesting pairs initiated clutches (16%) on another site on the wildlife islands with a median initiation date of 10 June. The secondary colony, on an island slightly to the south, was located on a site that had been targeting common terns. The area was open and covered with pea gravel (1 cm diameter).

4. Discussion   `See Photo C.`

Caspian terns demonstrated a statistical preference for nesting on sand over the other substrates. A preference for a particular substrate may be due to a variety of parameters. Substrates differed in colour, size and shape of grains. Construction grade sand is light brown in colour while crushed stone and pea gravel are shades of grey. Caspian tern eggs are light tan with dark brown spots. While they match each of the substrates quite well, they are relatively easy to see, even on sand. Chaniot (1970) noted changes in the frequency of tan-coloured relative to darker and lighter phase downy young Caspian terns in San Francisco Bay since 1943 and attributed this frequency change to natural selection favouring the colour morph which best matched the nesting substrate colour. Nesting substrate preference could reflect coloration of eggs or chicks. Nevertheless, the scrapes in sand were usually edged with crushed stone or pea gravel gathered by the birds from nearby cells with other substrates. This lining of nest edges made the nests more visible.

Sand may be marginally easier for terns to make scrapes. The nest depression is formed by resting the breast on the substrate while kicking substrate out behind. Additionally, sand, being more dense and darker in colour may carry thermal advantages.

The higher density of terns nesting on sand did not lead to diminished reproductive success as both hatching and fledging success values were comparable or slightly better than the alternative substrates. In contrast, Richards and Morris (1984) found that common terns nesting on their preferred experimental substrate, which included structures providing shelter from predators, realised significantly improved fledging success. Larger sample sizes or different environmental conditions (e.g. absence of terrestrial predators) might be required to demonstrate differences in reproductive success if they exist. Fledging success reported here was low compared with previously reported fledging success rates for Hamilton Harbour Caspian terns (Ewins et al., 1994). Our sampling techniques differed from Ewins et al. (1994). Additionally, the presence of foxes in 1994 may have resulted in temporary desertions of the colony leaving chicks exposed to ambient conditions, particularly at night when foxes are active (Southern et al., 1995). Indeed, subsequent observations in 1995 revealed

that Caspian tern adults left the colony for periods of greater than 45 min when the foxes were active in the vicinity (J. Sirdevan, pers. obs.). Parental abandonment during periods of nocturnal disturbances by foxes caused ring-billed gull chick mortality due to prolonged exposure to low ambient temperatures (Southern et al., 1985). All Caspian terns nesting on the Hamilton Harbour mainland sites in 1995 eventually abandoned their eggs, probably due to the activity of six or more foxes observed on the site. Foxes remained active in the area in 1996 and 1997 and no nesting of Caspian terns was observed on the mainland at Hamilton Harbour.

The first eggs in 1996 were laid on the edge of the sandy knoll designed for Caspian terns on sand beside the plastic sheeting. Apparently the terns were attracted to the habitat, as no decoys or other efforts had been used to entice them. The colony grew from that nucleus of nests. The 200 m² Caspian tern area was the only part of the human-made island chain with a sand substrate and represented a very small area of the island chain (Quinn et al., 1996). It was one of the only two raised areas (about 1m higher than surrounding substrate) on the islands. The other, not included in the earlier design plan (Quinn et al., 1996), was a ridge placed on the middle island to encourage herring gull nesting. While it is possible that topography played a role, it should be noted that the first nests were at the edge of the plastic at the same elevation as the rest of the island. The substantial increase in the number of pairs nesting on the site in 1997 is a very good sign suggesting that the site will continue to be used. The Caspian tern nesting activity observed on the middle island in 1997 began after hatching had begun on the crowded main Caspian tern site. It is likely that the high density and the feeding activities at nests with chicks made the main colony site less attractive to the late nesters. The site on middle island was one of two surfaced with a smaller pea gravel and intended for common terns. While not the most preferred of substrates, smaller-sized pea gravel was shown to be acceptable to Caspian terns in our experiment.

This study shows a statistical preference for nesting on sand by Caspian terns. Additionally, we found that terns nesting at greater densities on sand were not disadvantaged reproductively compared with those nesting on other substrates. We concluded that construction grade sand is a suitable substrate for attracting nesting Caspian terns and recommended its use, with a small addition of pea gravel for nest lining, in a habitat creation project. The sandy raised knoll on the Northern end of the Northern-most wildlife island in Hamilton Harbour (Quinn et al., 1996) has attracted most of the Caspian terns nesting in Hamilton Harbour and this colony has grown in numbers from 1996 to 1997. The application of the results of our experiment to this habitat creation project has been a great success.

`See Photo D.`

Photo C The three types of substrates tested for nesting.

Photo D A new colony of Caspian terns on a human-made wildlife island. This site was designed for the terns based on the results of the experiment. The covers are chick shelters, which provide protection from weather or predators and provide a place into which the nestlings tend to settle when researchers perform nest checks.

DISCUSSION. The focus in this section is the interpretation of the data; that is, the investigators make sense of the data. They compare findings with other research, propose explanations, note discrepancies they may have found, and draw conclusions. Scientists compare their results with their hypotheses to determine if the data support or do not support the hypotheses. Sometimes researchers then make suggestions for improvements in their procedures if they are to continue study in this area and make recommendations for further study.

The investigators discuss possible explanations for the terns' preference for nesting on sand.

The investigators compare their finding with those of other researchers.

The investigators draw conclusions from their study.

ACKNOWLEDGEMENTS. In this section, the authors of the paper (the investigators) give credit to all who helped them conduct their study or reviewed their paper before submission to the journal. Sources of funding for the study are usually noted here as well. They are noted in this paper.

REFERENCES. The reference section lists all the books, papers, and journal articles that are listed in the paper. They are listed alphabetically by author last name. Authors of scientific papers give references whenever they cite prior work or make statements based on prior work. In this way, the reader can access the original source of that information.

Acknowledgements

The authors thank John Hall and the Hamilton Harbour Remedial Action Plan team for getting us involved in this research and for incorporating our findings and ideas into their plans. They gratefully acknowledge the reliable assistance of Cynthia Pikerik, who prevented nesting by ring-billed gulls on the study site. Robert Dawson, Sarah Hopkin, Andrea Kirkwood, Vanessa Lougheed, Brent Murray, Joe Minor, Salmon Cuso, Cynthia Pekaric, Jennifer Startek, and Carole Yauk worked hard to help construct the substrate experimental grid for peanuts (and beer). Dedicated field assistance was provided by Carrie Rongits in 1994, Cynthia Pekaric and Angelo Nicassio in 1996 and Cheryl Fink and Cindy Anderson in 1997. The Hamilton Harbour Commissioners kindly provided access to their property and permission to construct the substrate grid. The authors also thank Chip Weseloh (CWS) for providing the services of Cynthia Peraric and the Canadian Wildlife Service for permission to carry on our studies of the colonial nesters. Finally, they are pleased to acknowledge the McMaster Eco-Research Project (MERP) for logistical and financial support. Funding for MERP was provided courtesy of the Tri-council Green Plan. Ralph D. Morris and two anonymous reviewers provided valuable comments on earlier versions of this paper.

References

Blokpoel, H., Scharf, W.C., 1991. Status and conservation of seabirds nesting in the Great Lakes of North America. ICBP Technical Publication 11, pp. 17-41.

Burness, G.P., Morris, R.D., 1992. Shelters decrease gull predation on chicks at a common tern colony. Journal of Field Ornithology 63, 186-189.

Chaniot, G.E. Jr., 1970. Notes on color variation in downy Caspian terns. Condor 72, 460-465.

Dobos, R.Z., Struger, J., Blokpoel, H., Weseloh, D.V., 1988. The status of colonial waterbirds nesting at Hamilton Harbour, Lake Ontario, 1959-1987. Ontario Birds 6, 51-60.

Ewins, P.J., Weseloh, D.V., Norstrom, R.J., Legierse, K., Auman, H.J., Ludwig, J.P., 1994. Caspian terns on the Great Lakes: organochlorine contamination, reproduction, diet and population changes. Canadian Wildlife Service, Occasional paper No. 85.

Hays, W.L., 1981. Statistics. 3rd ed. Holt, Rinehart, and Winston, New York.

Lampman, K.P., Taylor, M.E., Blokpoel, H., 1996. Caspian terns, Sterna caspia, breed successfully on a nesting raft. Colonial Waterbirds 19, 135-138.

Moore, D.J., Blokpoel, H., Lampman, K.P., Weseloh, D.V., 1995. Status, ecology, and management of colonial waterbirds nesting in Hamilton Harbour, Lake Ontario, 1988-1994. Technical Report Series No. 213, Canadian Wildlife Service, Ontario Region.

Morris, R.D., Blokpoel, H., Tessier, G.D., 1992. Management efforts for the conservation of common tern Sterna hirundo colonies in the Great Lakes: two case histories. Biological Conservation 60, 7-14.

Peck, G.K., James, R.D., 1983. Breeding Birds of Ontario: Nidology and Distribution, Vol. 1: Nonpasserines. Royal Ontario Museum Life Science, Toronto.

Quinn, J.S., 1984. Egg predation reduced by nest covers during researcher activities in a Caspian tern colony. Colonial Waterbirds 7, 419-151.

Quinn, J.S., Morris, R.D., Blokpoel, H., Weseloh, D.V., Ewins, P.J., 1996. Design and management of bird nesting habitat: tactics for conserving colonial waterbird biodiversity on artificial islands in Hamilton Harbour Ontario. Canadian Journal of Fish and Aquatic Sciences 53 (Suppl.), 44-56.

Richards, M.R., Morris, R.D., 1984. An experimental study of nest site selection in common terns. Journal of Field Ornithology 55, 457-466.

Siegal, S., 1988. Nonparametric Statistics for Behavioural Sciences, 2nd ed. McGraw-Hill, New York.

Southern, W.E., Patton, S.R., Southern, L.K., Hanners, L., 1985. Effects of nine years of fox predation on two species of breeding gulls. Auk 102, 827-833.

Glossary

A

abiotic factors (AYE-bye-OT-ik) nonliving factors within the environment, such as air, water, and rocks.

abyssal zone (uh-BISS-uhl) the deepest region of the ocean. The abyssal zone includes ocean depths of 1000 or more meters.

acid any substance that dissociates to form H^+ ions when it is dissolved in water.

acid rain precipitation having a low pH (usually between 3.5 and 5.5) produced when sulfur dioxide and nitrogen dioxide pollutants combine with water in the atmosphere.

acquired immunodeficiency syndrome (AIDS) the final stage of HIV infection in which the helper T cell count falls below 200 cells/mm³ and a person exhibits certain serious diseases typical of this end stage.

actin (AK-tin) a protein that makes up the thin myofilaments in a muscle fiber.

action potential the rapid change in a membrane's electrical potential caused by the depolarization of a neuron to a certain threshold.

active immunity a kind of immunity in which a vaccination or the contraction of a disease causes the body to build up antibodies against that particular disease.

active sites the grooved or furrowed locations on the surface of an enzyme where reactions are catalyzed.

active transport the movement of a solute across a membrane against the concentration gradient with the expenditure of energy. This process requires the use of a transport protein specific to the molecule(s) being transported.

adaptations naturally occurring inheritable traits present in a population of organisms that confer reproductive advantages on organisms that possess them.

adaptive behaviors behaviors that enhance the ability of members of a population to live to reproductive age and that tend to occur at an increased frequency in successive generations.

adaptive radiation the phenomenon by which a population of a species changes as it is dispersed within a series of different habitats within a region.

adenosine diphosphate (ADP) (uh-DEN-o-seen dye-FOS-fate) the molecule that remains after ATP has been used to drive an endergonic reaction.

adenosine triphosphate (ATP) (uh-DEN-o-seen try-FOS-fate) the primary molecule used by cells to capture energy and later release it during chemical reactions.

adrenal cortex (uh-DREE-nul KOR-tecks) the outer, yellowish portion of each adrenal gland that secretes a group of hormones known as corticosteroids in response to the hormone ATCH, or adrenocorticotropic hormone.

adrenal glands (uh-DREE-nul) two triangular glands located above the kidneys; each adrenal gland consists of two parts: the adrenal cortex and the adrenal medulla.

adrenal medulla (uh-DREE-nul muh-DULL-uh) the inner, reddish portion of each adrenal gland. The adrenal medulla is surrounded by the cortex and secretes the hormones adrenaline and noradrenaline.

aerobic cellular respiration the chemical process by which cells break down fuel molecules using oxygen, producing carbon dioxide, and releasing energy.

age distribution the proportion of individuals in the different age categories of a population.

agranulocytes (ay-GRAN-yuh-low-sites) one of the two major groups of leukocytes; they have neither cytoplasmic granules nor lobed nuclei.

aldosterone (al-DOS-ter-own) a hormone that regulates the level of sodium ions and potassium ions in the blood; it promotes the retention of sodium and the excretion of potassium.

algae (AL-jee) plantlike protists that are photosynthetic autotrophs. Algae contain chlorophyll and manufacture their own food using energy from sunlight.

allantois (ah-LAN-toe-us) an extraembryonic membrane that gives rise to the umbilical arteries and vein as the umbilical cord develops.

allele (uh-LEEL) each member of a factor pair containing information for an alternative form of a trait that occupies corresponding positions on paired chromosomes.

allergic reaction a specific immune response that results from the immune system mounting a major defense against a harmless antigen.

alternation of generations a type of life cycle that has both a multicellular haploid phase and a multicellular diploid phase.

altruistic behavior (AL-troo-ISS-tik) a kind of behavior that benefits one at the cost of another; parenting behaviors are a type of altruistic behavior.

alveoli (al-VEE-uh-lye) microscopic air sacs in the lungs where oxygen enters the blood and carbon dioxide leaves.

amebas (uh-MEE-buhs) protozoans that have changing shapes brought about by cytoplasmic streaming, which forms cell extensions called pseudopodia.

amnion (AM-nee-on) a thin, protective membrane that grows down around the embryo during the third and fourth week of prenatal development. Fluid fills the cavity between the amnion and the embryo or fetus.

amniotic egg (am-nee-OT-ik) an egg, characteristically produced by birds, reptiles, and monotremes, that protects the embryo from drying out, nourishes it, and enables it to develop outside of water.

amphibians (am-FIB-ee-uns) a class of vertebrates capable of living on land and in the water. Amphibians depend on water during their early stages of development.

amylases (AM-uh-lace-es) digestive enzymes that break down starches and glycogen to sugars.

anabolic reactions (AN-uh-BOL-ick) chemical reactions that use energy to build complex molecules from simpler molecules.

analogous (uh-NAL-eh-gus) of differing evolutionary origins and basic anatomy, now similar in function.

anaphase (ANN-uh-faze) the third phase of mitosis, meiosis I, and meiosis II. During anaphase, sister chromatids separate and move to opposite poles of the cell.

angiosperms (AN-jee-oh-spurms) vascular plants with protected seeds and flowers that act as their organs of sexual reproduction.

animal behavior a scientific discipline that was formed when the fields of ethology and behavioral psychology merged. The field of animal behavior applies both the physiological perspective of the ethologist and the psychological perspective of the behaviorist to animal study. Today, the field of animal behavior is composed of researchers from a variety of disciplines.

antennal gland *(green gland)* paired excretory glands of crustaceans (lobsters) that consist of a sac, a long coiled tube, and a bladder.

antheridia (singular, antheridium) (an-thuh-RID-ee-uh / AN-thuh-RID-ee-um) specialized structures in which sperm are produced in nonvascular plants and several divisions of vascular plants.

anthropoids (AN-thruh-poyds) a suborder of primates that includes the monkeys, apes, gorillas, chimpanzees, and humans; they differ from prosimians in the structure of the teeth, brain, skull, and limbs.

antibodies proteins produced by plasma cells (B lymphocytes) that recognize foreign antigens and prevent them from causing disease.

antibody-mediated immune response one of the two branches of the specific immune response; it is initiated by helper T cells that have been activated by interleukin-1 and the presence of antigens, and results in B cells producing antibodies. The antibodies bind to the antigens they encounter and mark them for destruction.

anticodon a portion of a tRNA molecule with a sequence of three base pairs complementary to a specific mRNA codon.

antidiuretic hormone (ADH) (AN-ti-dye-yoo-RET-ik) the hormone that regulates the rate at which water is lost or retained by the body; it is secreted by the pituitary gland at the base of the brain.

antigens (ANT-ih-jens) foreign molecules that induce the formation of antibodies, which specifically bind to the foreign substance and mark it for destruction.

anus the opening of the rectum for the elimination of feces.

aorta (ay-ORT-uh) the largest artery in the body. It receives blood from the left ventricle and has many vessels branching from it that carry blood throughout the body (with the exception of the lungs).

appendicular skeleton (AP-en-DIK-you-lur) the portion of the human skeleton, consisting of 126 bones, that forms the bones of the appendages (arms and legs) and the bones that help attach the appendages to the axial skeleton.

aqueous humor (AYK-wee-us HYOO-mur) the watery fluid that fills the chamber behind the cornea.

archegonia (singular, archegonium) (AR-kih-GO-nee-uh / AR-kih-GO-nee-um) specialized structures in which eggs are produced in nonvascular plants and several phyla of vascular plants.

arteries (ART-uh-rees) blood vessels that carry blood away from the heart.

arterioles (are-TEER-ee-oles) small arteries that lead from arteries to capillaries.

arteriosclerosis (ar-TEER-ee-oh-skluh-ROW-sis) a thickening and hardening of the walls of arteries.

articulation (ar-TIK-you-LAY-shun) another term for a joint.

asci (AS-kye) saclike structures that enclose the sexual spores of sac fungi.

asexual reproduction the generation of a new individual without the union of gametes.

atherosclerosis (ATH-uh-ROW-skluh-ROW-sis) a disease in which the inner walls of the arteries accumulate fat deposits, narrowing the passageways.

atoms submicroscopic particles that make up all matter.

atrioventricular (AV) bundle a strand of impulse-conducting muscle located in the septum of the heart that conducts heartbeat impulses from the AV node to the ventricles of the heart.

atrioventricular (AV) node (AY-tree-oh-ven-TRIK-yuh-lur) a group of specialized cardiac muscle cells located in the base of the atria that receives the impulses initiated by the sinoatrial node and conducts them to the atrioventricular bundle in the heart septum.

auditory canal (AWD-uh-tore-ee) a 1-inch long canal that receives sound waves funneled from the pinna and carries them directly to the eardrum.

autonomic nervous system (awe-tuh-NOM-ik) the branch of the peripheral nervous system consisting of motor neurons that control the involuntary and automatic responses of the glands and the nonskeletal muscles of the body.

autosomes (AW-tuh-somes) chromosomes that carry the majority of an individual's genetic information but do not determine gender.

autotroph (AW-tuh-TROFE) a self-feeder; an organism that produces its own food by photosynthesis or chemosynthesis.

axial skeleton (AK-see-uhl) the central column of the skeleton, from which the appendages (arms and legs) of the appendicular skeleton hang. The axial skeleton consists of 80 bones, including the skull, vertebral column, and rib cage.

axon (AK-son) a single, long cell extension of a neuron that often makes distant connections. It may give out branches, which usually split off at right angles, and usually has a myelin covering, a type of insulation.

B

B cells or B lymphocytes (LIM-foe-sites) white blood cells that develop and mature in the bone marrow. As they mature, B cells develop the ability to identify bacterial antigens and respond by developing into plasma cells, which are specialized to produce antibodies. B cells provide the antibody-mediated immune response.

ball-and-socket joint a kind of joint that allows rotation, or the movement of a bone around an axis.

basal nuclei (BAY-suhl) groups of nerve cell bodies located at the base of the cerebrum that control large, subconscious movements of the skeletal muscles.

base any substance that combines with H^+ ions when it is dissolved in water.

basidia (buh-SID-ee-uh) in club fungi, club-shaped structures from which unenclosed sexual spores are produced.

basophils (BAY-soh-filz) one of three kinds of granulocytes; basophils contain granules that rupture and release chemicals that enhance the body's response to injury or infection; they also play a role in causing allergic responses.

behaviorism a school of thought that suggests that learning takes place in a stimulus/response fashion, possibly reinforced by some type of reward that may or may not be readily apparent.

behaviors the patterns of movement, sounds (vocalizations), and body positions (postures) exhibited by an animal. Behaviors also include any type of change in an animal, such as a change in coloration or the releasing of a scent, that can trigger certain behaviors in another animal.

benign tumors (buh-NINE) growths or masses of cells that are made up of partially transformed cells, are confined to one location, and are encapsulated, shielding them from surrounding tissues.

bilateral symmetry having two sides in which the right side is a mirror image of the left side.

bile a collection of molecules secreted by the liver that help in the digestion of lipids.

bile pigments substances produced by the liver from the breakdown of old, worn-out red blood cells; bile pigments enter the small intestine with the bile and are the cause of the characteristic color of the feces.

binary fission (BYE-nuh-ree FIZH-un) a type of asexual reproduction in which one cell divides into two with no exchange of genetic material among cells; bacteria reproduce in this way.

binomial nomenclature (bye-NO-mee-uhl NO-men-clay-chur) literally, "two-name naming"; the system of determining the scientific name of an organism using its genus and species classifications. Thus, an organism's genus becomes its first name, and its species becomes its last name.

bioenergy (BYE-oh-EN-ur-jee) the use of living plants to produce energy; the burning of wood to produce energy is an example of bioenergy.

biological clocks the governing mechanisms of circadian rhythms

biological concentration a process by which some organisms accumulate within their bodies certain harmful or deadly chemicals that are present in their environments or in the food they eat.

biological diversity the richness and variety of species on Earth.

biological magnification a process by which toxins accumulate in organisms in high concentrations when they consume tainted organisms lower on the food chain. This effect increases with progression up the food chain.

biomes (BYE-omes) ecosystems of plants and animals that occur over wide areas of land within specific climatic regions easily recognized by their overall appearance.

biosphere (BYE-oh-sfear) the global ecosystem of life on Earth that extends from the tops of the tallest mountains to the depths of the deepest seas; the part of the Earth where biological activity exists.

biotechnology (BYE-oh-tek-NOL-uh-jee) the use of scientific and engineering principles to manipulate organisms.

biotic factors (bye-OT-ik) living factors within the environment, such as plants, animals, and microorganisms.

birth canal the vagina, through which the fetus passes during birth.

birth control pills (oral contraceptives) drugs that contain estrogen and progesterone (or progesterone alone), which shut down the production of the pituitary hormones FSH and LH, preventing the maturation of the secondary oocyte and ovulation.

blade the flattened portion of a leaf.

blastocyst (BLAS-tuh-sist) a stage of development in which the preembryo is a hollow ball of cells; its center is filled with fluid.

blastula (BLAS-chuh-luh) the general term used to describe the saclike blastocyst of mammals and, in other animals, the embryonic stage that develops a similar fluid-filled cavity.

bone a type of connective tissue consisting of widely separated bone cells embedded in a matrix of collagen fibers and mineral salts; it forms the vertebrate skeleton.

bony fishes a class of vertebrates whose members have skeletons made of bone.

brain one of the two main parts of the central nervous system; a complicated maze of interconnected neurons linked to the hormone-producing glands, the muscles, and other tissues. The brain consists of four main parts: the cerebrum, the cerebellum, the diencephalon, and the brainstem.

brainstem the part of the brain consisting of the midbrain, pons, and medulla that brings messages to and from the spinal cord and controls important body reflexes such as the rhythm of the heartbeat and rate of breathing.

breathing the movement of air into and out of the lungs.

brown algae a phylum of multicellular algae that was formerly classified with the plants; brown algae are found predominantly on northern, rocky shores, and can grow to enormous sizes.

bulbourethral glands (BULL-bo-yoo-REE-thrul) a set of tiny glands lying beneath the prostate that secrete an alkaline fluid into the semen.

C

Calvin cycle another name for the light-independent reactions that take place during photosynthesis.

capillaries (KAP-uh-LARE-ees) fine latticeworks of microscopic blood vessels that permeate tissues.

capsid (KAP-sid) a protein "overcoat" that covers the nucleic acid core of a virus.

carbohydrates molecules that contain carbon, hydrogen, and oxygen, with the concentration of hydrogen and oxygen atoms in a 2:1 ratio.

carbon fixation a process by which organisms use the ATP and NADPH produced by light-dependent photosynthetic reactions to build organic molecules from atmospheric carbon dioxide.

carboxypeptidase (kar-BOK-see-PEP-ti-dace) an enzyme produced by the pancreas, which, together with trypsin and chymotrypsin, completes the digestion of proteins in the small intestine.

carcinogens (kar-SIN-uh-jens) cancer-causing substances.

cardiovascular system (KAR-dee-oh-VAS-kyuh-lur) the heart and blood vessels, the "plumbing" of the circulatory system.

carotenoids (kuh-ROT-uh-noids) pigments that absorb photons of green, blue, and violet wavelengths and reflect red, yellow, and orange; they are second only to the chlorophylls in importance in photosynthesis.

carrying capacity the number of individuals within a population that can be supported within a particular environment for an indefinite period.

cartilaginous fishes (kar-tuh-LAJ-uh-nus) a class of vertebrates whose members have skeletons made of cartilage rather than bone and that includes the sharks, skates, and rays.

catabolic reactions (CAT-uh-BOL-ick) chemical reactions that release energy by breaking down complex molecules into simpler molecules.

cell a microscopic mass of protoplasm; the unit of structure of all living things.

cell body the region of the neuron around the nucleus. The main organelles of the cytoplasm are in this portion of the cell.

cell culture techniques methods that enable scientists to remove cells from a parent plant and grow them into new plants.

cell cycle the time from the generation of a new cell until it reproduces. The cell cycle includes interphase, nuclear division, and cytoplasmic division.

cell theory a statement regarding the nature of living things, which holds that all living things are made up of cells; that the smallest living unit of structure and function of all organisms is the cell; and that all cells arise from preexisting cells.

cell wall a rigid structure that surrounds some cells' plasma membranes.

cell-mediated immune response one of the two branches of the specific immune response; it is initiated by helper T cells that have been activated by interleukin-1 and the presence of antigens, and results in cytotoxic T cells recognizing and destroying body cells.

cellular slime molds protozoans (classified with the amebas) that look and behave like amebas during much of their life cycle but can group together and form dormant, cystlike spores when food is scarce. These spores then revert to the amebalike form when conditions are favorable again.

central nervous system one of the two main parts of the vertebrate nervous system; the site of information processing within the nervous system. The central nervous system is made up of the brain and spinal cord.

cerebellum (SER-uh-BELL-um) the part of the brain located below the occipital lobes of the cerebrum that coordinates subconscious movements of the skeletal muscles.

cerebral cortex (suh-REE-brul KOR-tecks) a thin layer of tissue, called gray matter, that forms the outer surface of the cerebrum, and within which most of the activity of the cerebrum occurs, including higher cognitive processes such as learning and memory.

cerebrospinal fluid (suh-REE-brow-SPY-nul) a liquid filtrate of blood, formed by specialized structures in the ventricles of the brain, that acts as a shock absorber, cushioning the brain and spinal cord.

cerebrum (suh-REE-brum) the largest and most dominant part of the human brain, which is divided into two halves, or hemispheres, connected by the corpus callosum. The cerebrum governs motor, sensory, and association activity.

cervix (SUR-viks) the narrower, bottom part of the uterus that opens into the vagina.

chemical bonds forces that hold atoms together.

chemiosmosis (KEM-ee-oz-MOH-sis) the synthesis of ATP using the potential energy of a hydrogen ion gradient across a membrane to phosphorylate ADP. Chemiosmosis fuels most ATP synthesis in cells.

chemoautotrophs (KEE-mo-AWE-toe-trofes) organisms that make their own food by deriving energy from inorganic molecules.

chiasmata (keye-AZ-muh-tuh) in meiosis, the point of crossing over where parts of chromosomes have been exchanged during synapsis; under a light microscope, a chiasma appears as an X-shaped structure.

chlorophyll *a* the name given to special molecules of chlorophyll that absorb wavelengths of light in the 680- to 700-nanometer range, in the far-red portion of the light spectrum.

chlorophylls (KLOR-uh-fils) pigments that absorb photons of violet-blue and red wavelengths and reflect green and yellow; chlorophylls are the primary light gatherer in all plants and algae and in almost all photosynthetic bacteria.

chloroplasts (KLOR-oh-plasts) energy-capturing organelles that are found in the cells of plants and algae; these eukaryotic cell organelles manufacture carbohydrates during photosynthesis using energy from the sun and carbon dioxide.

cholecystokinin (CCK) (KOL-uh-SIS-tuh-KINE-un) one of the hormones that control digestion in the small intestine.

chordates (KOR-dates) animals having a hollow dorsal nerve cord, a dorsal rod-shaped notochord that forms during development, and gill arches at some stage of life; includes fishes, amphibians, reptiles, birds, mammals, and humans.

chorion (KORE-ee-on) an extraembryonic membrane that facilitates the transfer of nutrients, gases, and wastes between the embryo and the mother's body. It is a primary part of the placenta.

choroid (KOR-oyd) a thin, dark-brown membrane that lines the sclera of the eye and contains blood vessels that nourish the retina, and a dark pigment that absorbs light rays so they will not be reflected within the eyeball.

chromatin (KRO-muh-tin) the complex of deoxyribonucleic acid (DNA) and protein that makes up the chromosomes of eukaryotes.

chromosomes (KROW-muh-somes) shortened, thickened structures consisting of DNA coiled tightly around proteins.

chronic bronchitis (bron-KYE-tis) an inflammation of the bronchi and bronchioles that lasts for at least 3 months each year for 2 consecutive years with no accompanying disease as a cause; one of the disorders commonly included in chronic obstructive pulmonary disease.

chronic obstructive pulmonary disease (COPD) a term used to refer to disorders that block the airways and impair breathing.

chymotrypsin (kye-moe-TRIP-sin) an enzyme produced by the pancreas, which, together with trypsin and carboxypeptidase, completes the digestion of proteins in the small intestine.

cilia (SILL-ee-uh) hairlike extensions that often cover the surface of some eukaryotic cells and allow them to move or to move substances across their surfaces.

ciliary muscle (SILL-ee-err-ee) a tiny circular muscle that slightly changes the shape of the eye's lens by contracting or relaxing.

ciliates (SIL-ee-uts *or* SIL-ee-ates) protozoans whose members are characterized by fine, short, hairlike extensions called cilia.

circadian rhythms (sir-KAY-dee-un) 24-hour cycles of physiological activity and behavior

circulatory system a transport system that uses a fluid to move substances such as nutrients, wastes, and gases throughout an organism.

class a taxonomic subcategory of phyla; as organisms appear to be increasingly related to each other, they are more and more narrowly classified within taxonomic subcategories.

classical biotechnology the use of scientific and engineering principles to manipulate organisms, relying on the traditional techniques of selection, mutation, and hybridization.

classical conditioning a form of learning in which an animal is taught to associate a new stimulus with a natural stimulus that normally evokes a response in the animal.

climax community a community in which the mix of plants and animals becomes stable; the last stage of succession.

clitoris (KLIT-uh-ris) a small mass of erectile and nervous tissue in the female genitalia that responds to sexual stimulation.

cloaca (kloe-AY-kuh) the terminal part of the gut of most vertebrates into which ducts from the kidney and reproductive systems open.

clones copies.

cloning vector a plasmid, virus, or other piece of self-replicating DNA that can carry an inserted DNA fragment into a bacterial, yeast, or human cell.

closed circulatory system a transport system in which blood is enclosed within vessels as it travels throughout an organism.

club fungi one of three phyla of fungi. Club fungi have club-shaped structures from which unenclosed spores are produced. Mushrooms, puffballs, and shelf fungi are all types of club fungi.

cochlea (KOCK-lee-uh) a winding, snail-shaped tube that contains the organ of hearing and forms a portion of the inner ear.

codominant (KO-DOM-uh-nunt) a term referring to traits in which the alternative forms of an allele are both dominant, and both characteristics are exhibited in the phenotype.

codons sequences of three nucleotide bases in transcribed mRNA that code for specific amino acids, which are the building blocks of polypeptides.

coelom (SEE-lum) a body cavity found within most bilaterally symmetrical organisms and the echinoderms that is a fluid-filled enclosure lined with connective tissue. In humans, the coelom is divided into two parts: the thoracic cavity and the abdominal cavity.

coenzyme (KO-ENN-zyme) cofactor that is an organic molecule.

cofactors nonprotein molecules that bind to enzymes and help them catalyze chemical reactions.

cognitivism (KOG-nih-tiv-izm) a school of thought that suggests that individuals acquire and then store information in memory; learning takes place as new information builds and merges with the old, leading to changes in behavior.

colon the large intestine; its function is to absorb sodium and water, and to eliminate wastes.

commensalism a relationship in which an organism of one species benefits from its interactions with another, whereas the other species neither benefits nor is harmed.

community a grouping of populations of different species living together in a particular area at a particular time.

competition a situation in which organisms that live near one another strive to obtain the same limited resources.

competitive behavior a type of behavior that occurs when two or more animals strive to obtain the same needed resource, such as food, water, nesting sites, or mates.

competitive exclusion a principle that states that if two species are competing with one another for the same limited resource in a specific location, the species able to use that resource most efficiently will eventually eliminate the other species in that location.

complement proteins that circulate in the blood and that are part of both nonspecific and specific immunity. When triggered by an antigen, these proteins poke holes in the cell membranes of pathogens.

complementary DNA (cDNA) cloning the process of synthesizing DNA from mRNA and then making copies of this DNA using yeast or bacterial cells.

compounds molecules or ionically bonded substances made up of atoms of different elements.

condom (KON-dum) a sheath for the penis, constructed of thin rubber or other natural or synthetic materials, that is designed to prevent sperm from reaching the uterine tubes. The latex condom helps reduce the risk of contracting sexually transmissible diseases.

cones light receptors located within the retina at the back of the eye that function in bright light and detect color.

conjugation (CON-juh-GAY-shun) one of the ways in which bacteria transfer genetic material. Conjugation occurs as a donor and a recipient bacterium make contact, and the DNA from the donor is transferred to the recipient cell.

connective tissue groups of similar cells that provide a framework for the body, join its tissues, help defend it against foreign invaders, and act as a storage site for specific substances.

consumers organisms that cannot make their own food and must feed on producers or other consumers.

continental shelf the margin of land that extends out from the intertidal zone usually 50 to 100 kilometers (30 to 60 miles) and slopes to a depth of about 200 meters (approximately 650 feet) beneath the sea.

contractile vacuoles cellular organelles found in protozoans and sponges that collect wastes and excess water, squeezing them out of the cell when the vacuole fills.

control the standard against which experimental effects may be checked to establish their validity.

controlled experiment a method of testing a hypothesis in which a factor is changed and other factors that respond due to this change are observed. All other factors are held constant.

convergent evolution (kun-VUR-junt EV-uh-LOO-shun) the development of similar structures having similar functions in different species as the result of the same kinds of selection pressures.

coral reef (KOR-ul) a community of organisms that live on masses of rocks in the ocean lying at or near the surface of the water. Coral reefs are partially made up of marine animals called corals (phylum Cnidaria) that secrete calcium carbonate, a hard shell-like substance. Other marine organisms also grow on the rocks and make up part of the coral reef.

core viral nucleic acid.

cornea (KOR-nee-uh) the rounded, transparent portion of the eye's outer layer that permits light to enter the eye.

corpus callosum (KOR-pus kuh-LOW-sum) a single, thick bundle of nerve fibers that connects the two sides of the cerebrum in humans and primates.

corticosteroids (KORT-ik-oh-STARE-oydz) a group of hormones, secreted by the adrenal cortex in response to ACTH, that act on the nuclei of target cells, causing the cell's hereditary material to produce certain proteins.

courtship or **courtship ritual** patterns of behavior, usually consisting of a series of fixed action patterns of movement, designed to lead to mating. Courtship rituals are unique to each species and occur after the formation of a male-female pair.

covalent (ko-VAY-lent) a type of chemical bond that is caused by the electrical attraction created by atoms sharing electrons.

cranial nerves any of the 12 pairs of nerves that enter the brain through holes in the skull.

creatinine (kree-AT-uh-neen) a nitrogenous waste found in small amounts in the urine that is formed from a nitrogen-containing molecule in muscle cells called creatine.

Cro-Magnons (KRO-MAG-nuns) early members of *H. sapiens sapiens* whose anatomical features were similar to modern humans; they made sophisticated tools, hunted, and created elaborate cave paintings.

crossing over a complex series of events during meiosis in which homologous nonsister chromatids cross over one another. These crossed-over pieces break away from the chromatids to which they are attached and reattach to the nonsister chromatid.

cytokinesis (sye-toe-kuh-NEE-sis) the physical division of the cytoplasm of a cell into two daughter cells.

cytoplasm (SYE-toe-PLAZ-um) a viscous fluid within a cell that contains all cell organelles except the nucleus.

D

decomposers (DEE-kum-POE-zurs) organisms such as bacteria and fungi that obtain their energy by breaking down organic material in dead organisms and contribute to the recycling of nutrients to the environment.

deep zone the area of pond or lake water into which light does not penetrate.

dehydration synthesis the process by which monomers are put together to form polymers.

deltas fan-shaped areas of accumulated sediment deposited by a stream or river as it enters an open body of water such as a lake or the ocean.

demography (dih-MOG-ruh-fee) the statistical study of human populations.

dendrite a cell extension of a neuron that is usually much shorter than an axon, has no myelin covering (insulation), and is specialized to receive impulses from sensory cells or from axons of other neurons. Most neurons contain multiple dendrites.

density the number of organisms or individuals in a population per unit of area.

density-dependent limiting factors environmental factors that result from the growth of a population but act to limit its subsequent growth. Density-dependent limiting factors increase in effectiveness as population density increases.

density-independent limiting factors environmental factors that operate to limit a population's growth, regardless of its density.

deoxyribonucleic acid (DNA) (de-OK-see-RYE-boh-new-KLAY-ick) the hereditary material; DNA controls all cell activities and determines all of the characteristics of organisms.

dependent variable the factor that varies in response to changes in the independent variable during a controlled experiment.

depressant (dih-PRESS-unt) drugs that slow down the activity of the central nervous system.

dermal tissue the outer protective covering of virtually all plants and animals.

desert a biome that occurs around 20 to 30 degrees north and south latitude and in other areas that have 25 centimeters (10 inches) or less of precipitation annually. Desert life is somewhat sparse, but exhibits adaptations to life in a dry environment.

detritus (dih-TRITE-us) the refuse or waste material of an ecosystem.

deuterostomes (DOOT-uh-row-stowmz) one of two distinct evolutionary lines of coelomates that includes the echinoderms and chordates. Their embryological development is characterized by the anus developing from the first indentation of the gastrula and a radial pattern of cleavage.

dialysis (dye-AL-uh-sis) a method of treating renal failure in which blood is filtered through a machine called an artificial kidney or by using the peritoneum.

diaphragm (DYE-uh-fram) 1. a sheet of muscle that forms the horizontal partition between the thoracic cavity and the abdominal cavity. 2. a rubber dome that is inserted immediately before intercourse to cover the cervix and prevent the entry of sperm into the uterine tubes.

diastole (dye-AS-tl-ee) the time of relaxation of a heart chamber, when it is filling.

diatoms (DIE-uh-toms) unicellular algae that look like microscopic pillboxes because they are made up of top and bottom shells that fit together.

dicots one of two groups of angiosperms that differ in the placement of stored food in their seeds; dicots store food in their cotyledons, or seed leaves.

diencephalon (DYE-un-SEF-uh-lon) the part of the brain consisting of the thalamus and hypothalamus.

diffusion the net movement of molecules from a region of higher concentration to a region of lower concentration, eventually resulting in a uniform distribution of the molecules.

digestion a process in which food particles are broken down into small molecules that can be absorbed by the body.

digestive system a series of organs specialized for breaking down food and ridding the body of waste.

dinoflagellates (dye-no-FLAJ-uh-luts or DYE-no-FLAJ-uh-lates) a type of flagellated unicellular algae that is characterized by stiff outer coverings. Their flagella beat in two grooves, one encircling the cell like a belt and the other perpendicular to it.

diploid (DIP-loyd) having a full complement of genetic material for that species.

disaccharidases (dye-SACK-uh-rye-DAYS-is) enzymes, produced by specialized epithelial cells of the small intestine, that break down the disaccharides maltose, sucrose, and lactose to the monosaccharides glucose, fructose, and galactose.

disaccharide (dye-SACK-uh-ride) two monosaccharides linked together.

dispersion the way in which the individuals of a population are arranged within their environment.

dominant in an organism carrying a pair of contrasting alleles for a particular trait, the form of the trait (the allele) that will be expressed.

drug addiction a compulsive urge to continue using a psychoactive drug, physical and/or psychological dependence on the drug, and a tendency to increase the dosage of the drug.

drug tolerance a decrease in the effects of the same dosage of a drug in a person who takes the drug over time.

ductus deferens (DUCK-tus DEF-uh-renz) paired tubes that ascend from the epididymides into the pelvic cavity, looping over the side of the urinary bladder and eventually joining at the urethra.

duodenum (DOO-oh-DEE-num or doo-ODD-un-um) the initial short segment of the small intestine that is actively involved in digestion and absorption of nutrients.

E

ecology (eh-KOL-uh-gee or ee-KOL-uh-gee) the study of the interactions among organisms and between organisms and their environments.

ecosystem (EH-koe-SIS-tem or EE-koe-SIS-tem) all the biotic and abiotic factors within a certain area. A community consisting of plants, animals, and microorganisms that interact with one another and with their environments and are interdependent on one another for survival.

ectoderm (EK-toe-durm) the outer layer of cells formed during the early development of the embryos of all bilaterally symmetrical animals; the primary germ layer that gives rise to the outer layer of skin, the nervous system, and portions of the sense organs.

ectothermic (EK-toe-THUR-mik) a term referring to animals such as reptiles, amphibians, and fishes that regulate body temperature by taking in heat from the environment and by their behavior. Their internal body temperature fluctuates.

electrocardiogram (ECG) (ih-LEK-trow-KARD-ee-uh-GRAM) a recording of the electrical impulses that pass throughout the heart as it contracts and relaxes.

electron transport chain a group of electron carriers located on the inner mitochondrial membrane and on the thylakoid membranes that, during redox reactions, release energy used to make ATP during photosynthesis and cellular respiration.

electrons particles that surround the core of an atom. Electrons carry a negative electrical charge.

elements pure substances that are made up of a single kind of atom and cannot be separated into different substances by ordinary chemical methods.

elimination a process whereby unabsorbed digestive wastes leave the body during defecation.

embryo (EM-bree-oh) the early stage of development in humans, from the third to eighth weeks.

emigration (EM-uh-GRAY-shun) the movement of organisms out of a population.

emphysema (em-fi-SEE-muh) a chronic obstructive pulmonary disease in which mucus plugs various bronchioles, trapping air within alveoli and often causing them to rupture. In this condition, the lungs lose their elasticity and the ability to recoil during exhalation, and instead stay filled with air.

endergonic reaction (ENN-der-GON-ick) a chemical reaction in which the products contain more energy than the substrates; energy must be supplied for this kind of reaction to proceed.

endocrine glands (ENN-doe-krin) ductless glands that secrete hormones and spill them directly into the bloodstream.

endocrine system the collective term for the 10 different endocrine glands of the human body, which secrete over 30 different hormones.

endocytosis (ENN-doe-sye-TOE-sis) a process in which cells engulf large molecules or particles and bring these substances into the cell packaged within vesicles.

endoderm (EN-doe-durm) the inner layer of cells formed during the early development of the embryos of all bilaterally symmetrical animals; the primary germ layer that gives rise to the digestive tract lining, the digestive organs, the respiratory tract, the lungs, the urinary bladder, and the urethra.

endometrium (en-do-MEE-tree-um) the inner lining of the uterus. This lining has two layers: one functional, transient layer that is in contact with the uterine cavity, and an underlying permanent layer. The functional layer is shed each month an embryo is not present in the uterus.

endoplasmic reticulum (EN-doe-PLAZ-mik ri-TIK-yuh-lum) an extensive system of interconnected membranes that forms flattened channels and tubelike canals within the cytoplasm of a cell.

endoskeleton a rigid support system within an animal body.

endosymbiotic theory (EN-do-SIM-bye-OT-ik) the most widely accepted theory regarding how eukaryotes arose. According to this theory, bacteria became attached to or were engulfed by host prokaryotic cells. Mitochondria are thought to have originated from aerobic bacteria and chloroplasts from photosynthetic bacteria.

endothermic (EN-doe-THUR-mik) a term referring to organisms such as birds and mammals that maintain a stable internal body temperature by means of internal regulatory mechanisms.

entropy (ENN-truh-pee) the energy lost to disorder; entropy is a measure of the disorder of a system.

envelope a chemical layer over the capsid of many viruses that is rich in proteins, lipids, and carbohydrate molecules.

environment a general term for the biosphere, encompassing the land, air, water, and every living thing on Earth.

enzymes (ENN-zymes) biological catalysts that reduce the amount of free energy of activation needed for a chemical reaction to take place, thus speeding up the reaction.

eosinophils (EE-oh-SIN-oh-filz) one of three kinds of granulocytes; eosinophils are believed to be involved in allergic reactions and also act against certain parasitic worms.

epididymis (ep-ih-DID-uh-mis) a long, coiled tube that sits on the back side of the testes and in which sperm mature.

epiglottis (ep-ih-GLOT-iss) a flap of tissue that folds back over the opening to the larynx, thus preventing food or liquids from entering the airway.

epithelial tissue (ep-uh-THEE-lee-uhl) groups of similar cells that cover body surfaces and line body cavities.

epochs (EP-uks *or* EE-poks) subdivisions of geological periods.

eras (EAR-uhs *or* AIR-uhs) geological time periods. Scientists divide the time from the formation of the Earth until the present day into five eras.

erythrocytes (ih-RITH-row-sites) or red blood cells, one type of formed element of the blood that resembles flat disks with central depressions and that are packed with the oxygen-carrying molecule hemoglobin.

esophagus (ih-SOF-uh-gus) the food tube that connects the pharynx to the stomach.

essential amino acids the eight amino acids that humans cannot manufacture and therefore must obtain from proteins in the food they eat.

estrogens (ES-truh-jens) various hormones that develop and maintain the female reproductive structures, such as the ovarian follicles, the lining of the uterus, and the mammary glands.

estuary (ESS-choo-ER-ee) a place where the fresh water of rivers and streams meets the salt water of oceans.

ethology (ee-THOL-uh-jee) the study of animal behavior in the natural environment. Ethology examines the biological basis of the patterns of movement, sounds, and body positions of animals.

euglenoids (yoo-GLEE-noyds) flagellate protists that have chloroplasts and make their own food by photosynthesis.

eukaryotes (you-KARE-ee-oats) cells that each have organelles and a membrane-bounded nucleus.

eustachian tube (yoo-STAY-kee-un *or* yoo-STAY-shun) a structure that connects the middle ear with the nasopharynx (the upper throat); it equalizes air pressure on both sides of the eardrum when the outside air pressure is not the same as the pressure in the middle ear.

eutrophication (yoo-TROWF-uh-KAY-shun) the accumulation of inorganic nutrients in a lake, which stimulate plant growth. When the plants die, the decomposition process takes oxygen from the water, leading to the death of other organisms.

evolution (ev-oh-LOO-shun) a scientific theory of organismal change over time originally developed by Charles Darwin; it embodies the ideas that species alive today are descendants of species living long ago, and that species have changed and diverged from one another over billions of years; the process of change over time by which existing populations of organisms develop from ancestral forms through modification of their characteristics.

evolutionary tree a representation of the pattern of relationships among major groups of organisms.

excretion (ex-SKREE-shun) a process whereby metabolic wastes and excess water and salt are removed from the blood and passed out of the body during urination.

exergonic reaction (EK-sur-GON-ick) a chemical reaction in which energy is released and, therefore, the products contain less energy than the substrate.

exocrine glands (EK-so-krin) glands whose secretions reach their destinations by means of ducts.

exocytosis (EK-so-sye-TOE-sis) the reverse of endocytosis; the discharge of material from a cell by packaging it in a vesicle and moving the vesicle to the cell surface.

exoskeleton (EK-so-SKEL-uh-tun) a rigid support system that covers an animal body.

expiration expelling air from the lungs.

exponential growth (ek-spo-NEN-shul) a type of growth pattern in which a population (for example) increases as its number is multiplied by a constant factor.

external fertilization a process in which a female sheds eggs and the male deposits sperm on them after they leave her body; the union of a male gamete (sperm) and a female gamete (egg) outside the body of the female.

external genitals (JEN-uh-tuls) sexual organs, such as the penis, that are located on the outside of the body.

external respiration the exchange of carbon dioxide and oxygen gases at the alveoli in the lungs.

extraembryonic membranes (EK-struh-EM-bree-ON-ik) structures that form from the trophoblast and provide nourishment and protection for the preembryo, the embryo, and then the fetus.

F

facilitated diffusion a type of transport process in which molecules move across the cell membrane by means of a carrier protein, but down the concentration gradient without an input of energy by the cell.

fact proposition that is so uniformly upheld that little doubt exists as to its accuracy.

family a taxonomic subcategory of order.

fats large molecules made up of carbon, hydrogen, and oxygen, with a hydrogen-to-oxygen ratio higher than 2:1; one type of lipid.

feedback loops mechanisms by which information regarding the status of a physiological situation or system is fed back to the system so that appropriate adjustments can be made.

female condom a birth control device that is shaped somewhat like a male condom, but has a ring at each end. One ring fits over the cervix and the other hangs outside the vagina.

fermentation the chemical process by which cells extract energy from glucose without using oxygen.

fertility awareness (natural family planning) a form of birth control that involves avoiding sexual relations for three days preceding and one day following ovulation.

fertilization the union of a male gamete (sperm) and a female gamete (egg).

fetus (FEE-tus) the term used to describe the stage of prenatal development after 8 weeks and until birth.

first filial (F_1) generation (FIL-ee-uhl) the hybrid offspring of the parental (P) generation.

first law of thermodynamics a law stating that energy cannot be created or destroyed; it can only be changed from one form or state to another.

first trimester the first 3 months of pregnancy.

fixed action patterns sequences of innate behaviors in which the actions follow an unchanging order of muscular movements.

flagella (fluh-JELL-uh) whiplike extensions from some cells that allow them to move.

flagellates (FLAJ-uh-lates *or* FLAJ-uh-luts) protozoans that are characterized by fine, long, hairlike cellular extensions called flagella.

fluid mosaic model the most well-accepted theory regarding the nature of the cell membrane; it describes the fluid nature of a lipid bilayer studded with a mosaic of proteins.

food chain a series of organisms from each trophic level that feed on one another.

food web the interwoven and interconnected feeding relationships of an ecosystem; interwoven food chains.

formed elements the solid portion of blood that is suspended in blood plasma, composed principally of erythrocytes, leukocytes, and platelets.

fossil preserved remains or impression of a dead organism.

fossil fuels substances such as coal, oil, and natural gas that are formed over time (acted on by heat and pressure) from the undecomposed carbon compounds of organisms that died millions of years ago.

fovea (FOE-vee-uh) a spot on the retina that has the highest concentration of cones; the lens focuses images on this spot, resulting in sharp vision.

fragmentation a form of asexual reproduction in which an organism detaches parts of its body, which then grow into new individuals.

frontal lobe one of the four sections of the cerebral cortex of each hemisphere of the brain; the frontal lobe integrates motor activity and certain aspects of speech.

functional group a group of atoms with definite chemical properties that is attached to the carbon-based core of an organic molecule.

fundamental niche the role that an organism might play in an ecosystem if competitors were not present.

G

gallbladder a sac attached to the underside of the liver, where bile is stored and concentrated.

gametes (GAM-eets) sex cells; the female gamete is the egg, and the male gamete is the sperm.

gametophyte (guh-MEE-toe-fite) plant that produces haploid gametes by mitosis; also known as the gamete-plant generation.

gametophyte (gamete-plant) generation the haploid phase of a plant life cycle. This phase tends to dominate the life cycles of nonvascular plants.

gastric glands glands dotting the inner surface of the stomach that secrete a gastric juice of hydrochloric acid and pepsinogen.

gastrin a digestive hormone of the stomach that controls the production of gastric juice.

gastrula (GAS-truh-lah) three-layer embryo.

gastrulation (GAS-truh-LAY-shun) during prenatal development, the process by which various cell groups of the inner cell mass migrate, divide, and differentiate resulting in a three-layer embryo.

gene a unit of heredity formed of a sequence of nucleotides that codes for the amino acid sequence of a polypeptide.

gene library a collection of clones of DNA fragments, which together represents the entire genome of an organism.

gene synthesis cloning the process of synthesizing DNA in the laboratory based on knowledge of its nucleotide sequence and then making copies of this DNA using yeast or bacterial cells.

gene therapy the treatment of a genetic disorder by the insertion of "normal" genes into the cells of a patient.

genetic counseling the process in which a geneticist discusses a couple's genetic history early in a pregnancy or before conception to determine if their offspring may be at risk for a variety of genetic disorders.

genetic engineering (recombinant DNA technology) techniques of molecular biology that involve the manipulation of genes.

genetics (juh-NET-iks) the branch of biology dealing with the principles of heredity and variation in organisms.

genital herpes (JEN-ih-tul HUR-peez) a sexually transmitted disease, caused most often by the herpes simplex virus type 2 (HSV-2), that produces blisterlike sores on the genitals.

genital warts a sexually transmissible disease caused by the human papillomavirus (HPV). The warts are soft, pink, flat, or raised growths that appear singly or in clusters on the external genitals and rectum.

genome (GEE-nome) an organism's total complement of genetic material.

genotype (JEEN-uh-type) an organism's allelic (genetic) makeup.

genus (GEE-nus) a taxonomic subcategory of family.

geothermal energy (JEE-oh-THUR-mul) the use of heat from deep within the Earth for heating purposes or as part of a process to produce electricity.

germination the sprouting of a seed, which begins when it receives water and has appropriate environmental conditions.

gills evaginations of the body surface that form a respiratory surface in large, multicellular aquatic animals and that are confined to one part of the body as in bony fishes or are distributed throughout an organism as in aquatic worms.

global warming a worldwide temperature increase that could result from the greenhouse effect.

glottis (GLOT-iss) the space between the vocal cords; the opening to the larynx and trachea.

glycogen (GLYE-ko-jen) highly branched polysaccharides that are the storage form of sugar in animals.

glycolysis (glye-KOL-uh-sis) the first of the three series of chemical reactions of cellular aerobic respiration, in which glucose is broken down to pyruvate.

Golgi apparatus (GOL-gee) an organelle that collects, modifies, and packages molecules that are made at different locations within the cell and prepares them for transport.

gonadotropins (GON-ah-duh-TROP-inz) 2 of the 4 tropic hormones; gonadotropins affect the male and female sex organs.

gonads (GO-nads) the male and female reproductive organs that produce the sex cells, or gametes.

gonorrhea (GON-uh-REE-uh) a sexually transmissible disease, caused by the bacterium *Neisseria gonorrhoeae,* that causes a primary infection and inflammation of the urethra (in men and women) and the vagina and cervix (in women).

gradient differences in concentrations, pressures, and electrical charges that result in the net movement of molecules in a particular direction.

grana (GRA-nuh) stacks of thylakoid membranes within chloroplasts.

granulocytes (GRAN-yuh-low-sites) one of the two major groups of leukocytes distinguished by their cytoplasmic granules and lobed nuclei.

green algae a phylum of multicellular algae; green algae include both unicellular and multicellular forms. Most forms of green algae are aquatic, but some species live in moist places on land.

greenhouse effect the blocking of outward heat radiation from the Earth by carbon dioxide in the atmosphere.

ground tissue stores the carbohydrates the plant produces. It forms the substance of the plant and is the tissue in which vascular tissue is embedded.

growth rate the number of individuals added to a population during a given time. The growth rate of a population is determined by subtracting its death rate from its birth rate.

guard cells a pair of cells that brackets a stoma and regulates its opening and closing.

gymnosperms (JIM-no-spurms) vascular plants with naked seeds. Four phyla of plants fall into this category: the conifers, cycads, ginkgos, and gnetophytes.

H

habitat a place where an organism lives or grows.

habituation (huh-BICH-yoo-AY-shun) the ability of animals to "get used to" certain types of stimuli that they perceive as nonthreatening.

hallucinogens (huh-LOO-suh-nuh-jens) psychedelic drugs that cause sensory perceptions that have no external stimuli; they cause a person to see, hear, smell, or feel things that do not exist.

haploid (HAP-loyd) having half the usual number of chromosomes for that species.

Haversian canal (huh-VUR-shun) in long bones, a narrow channel that runs parallel to the length of the bone and contains blood vessels and nerves. Some Haversian canals run crosswise to the length of the bone to connect the lengthwise canals.

heart the muscular pump of the circulatory system.

hemoglobin (Hb) (HEE-muh-glow-bin) an oxygen-carrying molecule within red blood cells that also carries some carbon dioxide.

hermaphrodites (hur-MAF-rah-dytes) an animal or plant having both male and female reproductive organs.

heterotroph (HET-uhr-uh-trofe) an organism that cannot produce its own food; a consumer in an ecosystem or a food chain.

heterozygous (het-uhr-uh-ZYE-gus) a term referring to an individual who has two different alleles for a trait.

hinge joint a kind of joint that allows movement in one plane only.

homeostasis (HOE-mee-oh-STAY-sis) the maintenance of a stable internal environment despite what may be a very different external environment.

hominids (HOM-uh-nids) the family of hominoids consisting of humans; the only living hominid is *Homo sapiens sapiens.*

hominoids (HOM-uh-noyds) one of four superfamilies of the anthropoid suborder. This superfamily includes apes and humans.

Homo erectus (HOE-moe ih-REK-tus) a second extinct species of hominids whose fossil record dates back 1.8 million years; members of this species were fully adapted to upright walking, made sophisticated tools, and probably communicated with language.

Homo habilis (HOE-moe HAB-uh-lus) an extinct species of hominids whose fossil record dates back about 2 million years; members of this species are thought to have been the first humans and have been given a name that emphasizes their intelligence (*Homo habilis* means "skillful human").

Homo sapiens (HOE-moe SAY-pee-unz) an early or archaic form of modern humans (*Homo sapiens sapiens*) that are extinct. Their fossil record dates back about 500,000 years.

Homo sapiens sapiens modern humans; the subspecies of hominids who made their appearance 10,000 years before the Neanderthal subspecies died out, and whose early members are called Cro-Magnons.

homologous (hoe-MOL-eh-gus) of the same evolutionary origin and basic anatomy, now differing in function.

homologous chromosomes (homologues) (hoe-MOL-uh-gus KRO-muh-somes/HOME-uh-logs) pairs of chromosomes that all organisms produced by sexual reproduction receive. Half of these chromosomes are from one parent organism and half from the other. Homologous chromosomes each contain genes that code for the same inherited trait.

homozygous (hoe-muh-ZYE-gus) a term referring to an individual who has two identical alleles for a trait.

hormone (HORE-mone) a chemical messenger secreted and sent by a gland to other cells of the body.

Human Genome Project (HGP) a worldwide scientific project to decipher the DNA code of all 46 human chromosomes.

human immunodeficiency virus, or HIV (IM-yoo-no-de-FISH-un-see VYE-rus) a deadly virus that weakens the ability of the immune system to mount a defense against infection because it attacks and destroys helper T cells.

hydrogen bonds weak electrical attractions between the H of an NH group or an OH group, and other oxygen or nitrogen atoms.

hydrolysis (hi-DROL-uh-sis) the process by which polymers are disassembled into monomers.

hydrophilic (HI-droe-FIL-ick) a term referring to polar molecules that form hydrogen bonds with water.

hydrophobic (hi-dro-FO-bick) a term referring to nonpolar molecules that cannot form hydrogen bonds with water.

hydroskeleton a support system that uses a fluid under pressure as a scaffolding for an organism or body part.

hypertonic (HI-per-TAWN-ick) refers to a solution with a solute concentration higher than that of another fluid.

hypothalamus (HYE-poe-THAL-uh-muss) a mass of gray matter located beneath the thalamus that regulates vital body functions such as respiration and the heartbeat and directs the hormone secretions of the pituitary gland.

hypothesis (hi-POTH-uh-sis) a tentative explanation or prediction that guides scientific inquiry.

hypotonic (HI-poe-TAWN-ick) refers to a solution with a solute concentration lower than that of another fluid.

I

ileum (ILL-ee-um) the third and final part of the small intestine, which is highly specialized for absorption of nutrients. The ileum follows the jejunum.

immigration (IM-uh-GRAY-shun) the movement of organisms into a population.

immune system the body structures that perform specific defense responses. The organs of the immune system are the red bone marrow and the thymus gland (in which disease-fighting cells [lymphocytes] arise and/or develop) and the spleen and lymph nodes (in which mature lymphocytes accumulate and function).

immunity (ih-MYOON-ih-tee) protection from disease, particularly infectious disease. The two types of immunity are nonspecific and specific.

implantation (IM-plan-TAY-shun) the embedding of the developing blastocyst into the posterior wall of the uterus approximately 1 week after fertilization.

imprinting a rapid and irreversible type of stimulus/response learning that takes place during an early developmental stage of some animals.

incomplete dominance a term referring to traits in which alternative alleles are neither dominant over nor recessive to other alleles governing a particular trait; heterozygotes for incomplete dominance traits are phenotypic "intermediates."

incus (ING-kus) one of the three bones of the middle ear that amplify sound vibrations and carry them from the outer ear to the inner ear. The incus is also known as the anvil.

independent variable the factor that is manipulated during a controlled experiment.

inferior vena cava a large vein that collects blood from the lower body and returns it to the right atrium of the heart.

innate behaviors behaviors resulting from genetically determined neural programs that are part of the nervous system at the time of birth or develop at an appropriate point in maturation.

insect societies social groups that are formed by many species of insects, particularly bees and ants, and are characterized by a division of labor.

insight (reasoning) the capability of recognizing a problem and solving it mentally before ever trying out a solution.

inspiration taking air into the lungs.

integument (in-TEG-you-ment) in vertebrates, the skin, hair, and nails.

integumentary system (in-TEG-you-MEN-tuh-ree) another name for the vertebrate body's integument: the skin, hair, and nails.

intercostal muscles (inn-ter-KOS-tul) literally, "between-the-rib muscles"; muscles that extend from rib to rib and assist the diaphragm in the breathing process.

internal fertilization a process in which the male deposits sperm in or near the female reproductive tract and fertilization (the union of a male and a female gamete) takes place within the body of the female.

internal respiration the exchange of oxygen and carbon dioxide between the blood and the tissue fluid.

interneurons nerve cells located within the brain or spinal cord that integrate incoming information with outgoing information.

internodes the portions of a plant's stem that lie between the nodes.

interphase (IN-tur-faze) the portion of the cell cycle preceding cell division in which the cell grows and carries out life functions. During this time the cell also doubles in size and produces an exact copy of its heredity material, DNA, as it prepares for cell division.

intervertebral disks (IN-tur-VER-tuh-brul) disks of fibrocartilage, positioned between all vertebrae except the sacrum and coccyx, that act as shock absorbers, provide the means of attachment between one vertebra and the next, and permit movement of the vertebral column.

ionic (eye-ON-ick) a type of chemical bond between atoms that is caused by the attraction of oppositely charged particles formed by the gain or loss of electrons.

ionization (EYE-uh-ni-ZAY-shun) the spontaneous formation of charged particles (ions) very typical of water molecules, caused by the breaking of the molecules' covalent bonds.

ionizing radiation (EYE-uh-nye-zing RAY-dee-AY-shun) a form of electromagnetic energy that can cause chromosomes to break or can cause changes in the nucleotide structure of DNA. X-rays and nuclear radiation are kinds of ionizing radiation.

ions (EYE-onz) charged particles formed by atoms that have either acquired or lost electrons and have therefore developed a negative or positive charge.

iris (EYE-rus) a diaphragm lying between the cornea and lens that controls the amount of light entering the eye.

islets of Langerhans (EYE-lets of LANG-ur-HANZ) the separate types of cells within the exocrine cells of the pancreas that produce the hormones insulin and glucagon.

isotonic (EYE-so-TAWN-ick) refers to solutions having equal solute concentrations to one another.

isotopes (EYE-suh-topes) atoms that have the same number of protons but different numbers of neutrons.

J

jawless fishes tubular, scaleless, jawless vertebrates that live in the sea or in brackish water. They have no paired fins.

jejunum (ji-JOO-num) the second portion of the small intestine, which is highly specialized for absorption of nutrients and extends from the duodenum to the ileum.

joint a place within the skeletal system where bones, or bones and cartilage, come together; also known as an articulation.

K

kidney stones crystals of certain salts that can develop in the kidney and block urine flow.

kidneys along with the lungs, the primary organs of excretion. Kidneys excrete the ions of salts and the nitrogenous wastes urea, creatinine, and uric acid, along with small amounts of other waste products.

kinesis (kih-NEE-sis) the change in the speed of the random, nondirected movements of an animal with respect to changes in certain environmental stimuli.

kinetic energy (kuh-NET-ick) energy actively doing work; the energy of motion. (Energy is the ability to do work.)

kingdoms broad categories in which taxonomists group all living things; the taxonomic system used in this book recognizes five kingdoms of life: *Bacteria, Protista, Plantae, Fungi,* and *Animalia.*

Krebs cycle the second series of chemical reactions of aerobic cellular respiration, in which pyruvate, the end product of glycolysis, is oxidized to carbon dioxide. Also called the *citric acid cycle.*

L

labia majora (LAY-bee-uh muh-JORE-uh) two longitudinal folds of skin that run posteriorly from the mons in the exterior genitals of the female.

labia minora (LAY-bee-uh mu-NORE-uh) folds of skin covered by the labia majora in the exterior genitals of the female.

labor the sequence of events that leads to birth.

lancelets (LANS-lets) a subphylum of the chordates. Tiny, scaleless, fishlike marine organisms that are just a few centimeters long and pointed at both ends. Their adult forms exhibit chordate characteristics.

larynx (LAIR-inks) the voice box, which is located at the beginning of the trachea.

lateral buds tiny undeveloped side shoots that develop at the angles between a plant's leaves and its stem.

lateral line system a complex system of mechanoreceptors possessed by all fishes and amphibians that detects mechanical stimuli such as sound, pressure, and movement.

laws of thermodynamics two laws that govern all of the changes in energy that take place in the universe.

learning an alteration in behavior based on experience.

leaves the parts of a plant's structure where most photosynthesis takes place.

left atrium (AY-tree-um) the upper left chamber of the heart, into which oxygenated blood enters from the lungs.

left ventricle (VEN-truh-kul) a chamber in the heart into which blood flows from the left atrium. The left ventricle then pumps blood through the aorta into the arteries.

lens a structure in the eye lying just behind the aqueous humor that plays a major role in focusing the light that enters the eye onto the retina at the back of the eye.

leukocytes (LOO-ko-sites) or white blood cells, one type of formed element of the blood that is larger than red blood cells and is essentially colorless. There are several kinds of leukocytes, including macrophages and lymphocytes, but all of them function in defending the body against invading microorganisms and foreign substances.

lichens (LIE-kins) associations between sac fungi and either cyanobacteria or green algae.

life cycle the progression of stages an organism passes through from its conception until it conceives another similar organism.

ligaments (LIG-uh-munts) bundles or strips of dense connective tissue that hold bones to bones.

ligase (LYE-gase) a sealing enzyme that helps re-form the bonds between pieces of DNA that have been cut with the same restriction enzymes.

light-dependent reactions the reactions of photosynthesis that produce ATP and NADPH, which can occur only in the presence of light.

light-independent reactions the reactions of photosynthesis that use ATP and NADPH to provide energy for the formation of sucrose from carbon dioxide; to occur, these reactions do not require light.

limbic system a network of neurons, which together with the hypothalamus forms a ringlike border around the top of the brainstem; the operations of the limbic system are responsible for many deep-seated drives and emotions of vertebrates, including pain, anger, sexual drive, hunger, thirst, and pleasure.

lipases (LYE-pays-es) digestive enzymes that break down the triglycerides in lipids to fatty acids and glycerol.

lipids composite molecules made up of glycerol and fatty acids (in the case of oils and fats) or carbon rings (in the case of steroids).

liver a large, complex organ weighing over 3 pounds, lying just under the diaphragm, that performs more than 500 functions in the body, including the secretion of bile, which aids in the digestion of lipids.

lungs invaginations of the body surface, which are confined to one part of the body and form a respiratory surface in large, multicellular land animals.

lymph (LIMF) the name given to tissue fluid in the vessels of the lymphatic system.

lymph nodes (LIMF NODES) small, ovoid, spongy structures located in various places in the body along the routes of the lymphatic vessels, which filter lymph as it passes through them.

lymphatic system (lim-FAT-ik) the body's one-way, passive circulatory system, which collects and returns to the blood tissue fluid that does not return directly from the body's tissues.

lymphocytes (LIM-foe-sites) white blood cells that are formed in the bone marrow, circulate in the blood and lymph, and reside in lymph tissue such as the lymph nodes, spleen, and thymus. Lymphocytes recognize and react to substances that are foreign to the body, sometimes producing a protective immunity to disease.

lysogenic cycle (lye-suh-JEN-ik) a pattern of viral replication in which a virus integrates its genetic material with that of a host and is replicated each time the host cell replicates.

lysosomes (LYE-so-somes) membrane-bounded organelles that are essentially bags of many different digestive enzymes. Lysosomes break down old cell parts or materials brought into the cell from the environment and are extremely important to the health of a cell.

lytic cycle (LIT-ik) a pattern of viral replication in which a virus enters a cell, replicates, and then causes the cell to burst, releasing new viruses.

M

macrophages (MAK-row-FAY-djus) enlarged, amebalike cells that entrap microorganisms and particles of foreign matter by phagocytosis.

malignant (muh-LIG-nunt) a term describing cancerous tumors that have the ability to invade and kill other tissues and move to other areas of the body.

malleus (MAL-ee-us) a very small bone, connected to the internal side of the eardrum, that works with two other small bones, the incus and stapes, to amplify sound vibrations and carry them from the outer ear to the inner ear. The malleus is also known as the hammer.

Malpighian tubules excretory organs of insects and spiders that consist of blind-ended tubes lying in the blood-filled body cavity and emptying into the intestine.

mammals (MAMM-uhls) warm-blooded vertebrates that have hair and whose females secrete milk from mammary glands to feed their young.

mammary glands milk-producing glands that lie over the chest muscles in female mammals.

marsupials (mar-SOO-pee-uhls) a subclass of mammals that gives birth to immature young that are carried in a pouch.

mating male-female behaviors that result in fertilization, regardless of whether copulation occurs.

medulla (mih-DULL-uh) the lowest portion of the brainstem, continuous with the spinal cord below. The medulla is the site of neuron tracts, which cross over one another, delivering sensory information from the right side of the body to the left side of the brain and vice versa.

medusae (meh-DOO-see *or* meh-DOO-zee) free-floating and often umbrella-shaped aquatic animals with the mouth usually located on the underside of the umbrella shape and tentacles hanging down around the umbrella's edge.

meiosis (my-OH-sis) a type of cell division by means of which the sex organs of a mature animal produce gametes. During meiosis, one parent cell produces four sex cells; each gamete produced during this process contains half the number of chromosomes of the original parent cell. Consists of meiosis I and meiosis II.

meiosis I and meiosis II (my-OH-sis) the two-staged process of nuclear division in which the number of chromosomes in cells is halved during gamete formation. Both meiosis I and meiosis II can be further divided into four stages: prophase, metaphase, anaphase, and telophase.

Mendel's law of independent assortment the concept that the distribution of alleles for one trait into the gametes does not affect the distribution of alleles for other traits.

Mendel's law of segregation the concept that each gamete receives only one of an organism's pair of alleles. Chance determines which member of a pair of alleles becomes included in a gamete.

meninges (muh-NIN-jeez) the three layers of protective membranes covering both the brain and the spinal cord.

menopause (MEN-uh-pawz) the permanent cessation of menstrual activity in a woman, usually between the ages of 50 and 55; the end of the menses.

menstruation (MEN-stroo-AY-shun) the monthly sloughing of the blood-enriched lining of the uterus when pregnancy does not occur; the lining degenerates and causes a flow of blood, tissue, and mucus from the uterus out through the vagina.

meristematic tissue (MER-uh-stuh-MAT-ik) an undifferentiated type of tissue in a vascular plant in which cell division occurs during growth.

mesoderm (MEZ-oh-durm) the middle layer of cells formed during the early development of the embryos of all bilaterally symmetrical animals; the primary germ layer that gives rise to the skeleton, muscles, blood, reproductive organs, connective tissue, and the innermost layer of the skin.

mesopelagic zone (MEZ-oh-puh-LAJ-ik) a term referring to the middle ocean zone; little light penetrates this ocean zone from 200 to 1000 meters (650 to 3250 feet).

messenger RNA (mRNA) a type of RNA that brings information from the DNA within the nucleus to the ribosomes in the cytoplasm and directs polypeptide synthesis.

metaphase (MET-uh-faze) the second phase of mitosis, meiosis I, and meiosis II. Metaphase begins when the chromosomes align themselves equidistantly from the two poles of the cell.

metastasis (muh-TAS-tuh-sis) one of the characteristics of cancer cells: the ability to spread to multiple sites throughout the body.

midbrain the top part of the brainstem; it contains nerve tracts that connect the upper and lower parts of the brain, and nuclei that act as reflex centers.

middle ear the portion of the ear that contains three bones that amplify the force of sound vibrations as they conduct them from the tympanic membrane to the oval window of the inner ear.

migrations long-range, two-way movements by animals, often occurring yearly with the change of seasons.

mitochondria (MITE-oh-KON-dree-uh) oval, sausage-shaped, or threadlike cellular organelles approximately the size of bacteria that have their own DNA; these eukaryotic cell organelles break down fuel molecules, releasing energy for cell work.

mitosis (my-TOE-sis *or* mih-TOE-sis) a process of cell division that produces two identical cells from an original parent cell.

mixed nerves nerves that are made up of both sensory (incoming or afferent) and motor (outgoing or efferent) nerve fibers.

molecule two or more atoms held together by shared electrons.

monocots one of two groups of angiosperms that differ in the placement of stored food in their seeds; monocots store most of their extra food in extraembryonic tissue called endosperm.

monocytes (MON-oh-sites) a group of agranulocytes that circulates as the granulocytes do; monocytes are attracted to the sites of injury or infection, where they mature into macrophages and engulf any bacteria or dead cells that neutrophils may have left behind.

monosaccharides (MON-o-SACK-uh-rides) simple sugars.

monotremes (MON-oh-treems) a subclass of mammals that lays eggs with leathery shells similar to those of reptiles.

mons pubis (monz PYOO-bis) the mound of fatty tissue that lies over the place of attachment of the two pubic bones in females.

morphogenesis (MORE-foe-JEN-uh-sis) the early stage of development in a vertebrate when cells begin to move, or migrate, thus shaping the new individual.

mortality the death rate of a population.

morula (MORE-yuh-luh) a stage of development in which the preembryo consists of about 16 densely clustered cells and is still the same size as a newly fertilized ovum.

motor neurons nerve cells that are specialized to transmit information from the central nervous system to the muscles and glands.

multiple alleles (uh-LEELS) a system of more than two alleles that govern certain traits.

muscle tissue groups of similar cells that are capable of contraction.

mutations (myoo-TAY-shuns) permanent changes in the genetic material that alter the original expression of a gene or genes. Mutations can affect single genes, pieces of chromosomes, whole chromosomes, or entire sets of chromosomes.

mutualism a relationship in which two species live together in close association, both benefiting from the relationship.

myelin sheath (MY-eh-len) the fatty wrapping surrounding some axons and long dendrites of sensory neurons created by multiple layers of many Schwann cell membranes.

myosin (MY-uh-sin) a protein that makes up the thick myofilaments in muscle fiber.

N

nasal cavities two hollow areas, located above the oral cavity and behind the nose, that are bordered by projections of bone covered with moist epithelial tissue.

natural selection the process in which organisms with adaptive traits survive in greater numbers than organisms without such traits.

Neanderthal (nee-AN-dur-thol *or* nee-AN-dur-tal) a subspecies of *Homo sapiens* named after the Neander Valley in Germany, where their fossils were first found. Neanderthals were short and powerfully built, with large brains; they made diverse tools, took care of the sick and injured, and buried their dead.

negative feedback loop a feedback loop in which the response of the regulating mechanism is opposite with respect to the output. Most of the body's regulatory mechanisms work by means of negative feedback loops.

nephridia (neh-FRID-ee-uh) excretory tubes found in the platyhelminthes (flatworms) and annelids (earthworms).

nephrons (NEF-rons) the microscopic filtering systems of the vertebrate kidneys in which urine is formed.

nerve cord a single, hollow cord along the back that carries sensory and motor impulses and that is a principal feature of chordates; in vertebrates, the nerve cord differentiates into a brain and spinal cord.

nerve impulses the electrochemical message signals sent by nerves.

nerve net a system of interconnecting nerve cells with no central controlling area, or brain.

nervous system the quick message system of the body, the nervous system in all complex animals is made up of nerve cells that transmit electrical signals throughout the body.

nervous tissue groups of similar cells that are specialized to conduct electrical impulses, and other supporting cells.

neuromuscular junction (NER-oh-MUS-kyuh-lur) a synapse between a neuron and a skeletal muscle cell.

neurons (NER-ons) nerve cells.

neurotransmitters (NER-oh-TRANS-mit-urs) chemicals released when nerve impulses reach the axon tip of a nerve cell; neurotransmitters then cross the synaptic cleft to combine with receptor molecules on the target cell.

neurulation (NOOR-oo-LAY-shun) the development of a hollow nerve cord.

neutrons (NOO-trons) particles found at the core of an atom. Neutrons carry no electrical charge.

neutrophils (NOO-truh-filz) one of three kinds of granulocytes; neutrophils migrate to the site of an injury and engulf microorganisms and other foreign particles.

niche (nich) the role each organism plays within an ecosystem.

nitrogenous wastes (nye-TROJ-uh-nus) nitrogen-containing molecules that are produced as waste products from the body's breakdown of proteins and nucleic acids.

noble gases atoms of elements that have equal numbers of protons and electrons and have full outer-electron energy levels and thus do not react readily with other elements.

nodes locations on the stem of a plant where leaves form.

nodes of Ranvier (ron-VYAY) uninsulated spots on the myelin sheath between two Schwann cells.

nondisjunction (non-dis-JUNK-shun) the failure of homologous chromosomes to separate after synapsis during meiosis, resulting in gametes with abnormal numbers of chromosomes.

nongonococcal urethritis (NGU) a disease caused by bacteria other than the gonorrhea bacterium (such as *Chlamydia trachomatis*) that has gonorrhealike symptoms.

nonrenewable resources those formed at a rate much slower than their consumption; they are thus finite in supply. Fossil fuels and minerals are nonrenewable resources.

nonspecific immunity a set of defenses the body has to keep out any foreign invader; these defenses include the skin and mucous membranes.

nonvascular plants (non-VAS-kyuh-ler) plants that lack specialized transport tissues.

notochord (NO-toe-kord) a rod-shaped structure that forms between the nerve cord and the gut (stomach and intestines) during the development of all chordates. The notochord forms the midline axis along which the vertebral column (backbone) develops in all vertebrate animals.

nuclear power (NOO-klee-ur) an energy source that derives its power from the splitting apart of the nuclei of large atoms (nuclear fission) or from the combining of the nuclei of certain small atoms (nuclear fusion). All nuclear reactors in use today use nuclear fission reactions to produce energy.

nucleic acid (noo-KLAY-ick) a long polymer of repeating subunits called nucleotides; the two types of nucleic acid within cells are deoxyribonucleic acid (DNA) and ribonucleic acid (RNA); nucleic acids store information about the structure of proteins.

nucleolus (noo-KLEE-oh-lus) a darkly staining region within the nucleus of a cell that contains a special area of DNA that directs the synthesis of ribosomal ribonucleic acid, or rRNA.

nucleotide (NOO-klee-o-tide) a single unit of nucleic acid, consisting of a five-carbon sugar bonded to a phosphate group and a nitrogen-containing base.

nucleus (NOO-klee-us) the control center of the cell, which is made up of an outer, double membrane that encloses the cell's chromosomes and one or more nucleoli.

nutrients carbohydrates, lipids, proteins, vitamins, minerals, and water.

O

occipital lobe (ock-SIP-uh-tul) one of the four sections of the cerebral cortex in each hemisphere of the brain; the occipital lobe receives and interprets nerve impulses regarding vision.

octet rule (OCK-tet ROOL) one of three factors that influence whether an atom will interact with other atoms; the octet rule states that an atom with an unfilled outer shell has a tendency to interact with another atom or atoms in ways that will complete this outer shell.

olfactory receptors (ole-FAK-tuh-ree) neurons whose cell bodies are embedded in the nasal epithelium; they detect smells when different airborne chemicals bind with receptor chemicals in their ciliated dendrite endings.

oncogenes (ON-ko-jeens) cancer-causing genes.

one-gene-one-enzyme theory a scientific explanation that states that the production of a given enzyme is under the control of a specific gene. If the gene mutates, the enzyme will not be synthesized properly or will not be made at all. Therefore, the reaction it catalyzes will not take place, and the product of the reaction will not be produced.

open circulatory system a transport system in which blood flows in vessels leading to and from the heart but through irregular channels called blood sinuses in many parts of an organism. In an open circulatory system, blood mixed with tissue fluid bathes the organs and tissues directly.

open-water zone the main body of pond or lake water through which light penetrates.

operant conditioning (OP-uh-runt) a form of learning in which an animal associates something that it does with a reward or punishment.

operculum (oh-PUR-kyuh-lum) a flap in a bony fish that extends posteriorly from the head over the gills, protects the gills, and enhances water flow over the gills.

opiates (OH-pee-uts) compounds derived from the poppy plant that act as narcotic analgesics, which stop or reduce pain without causing a person to lose consciousness.

optic nerve the nerve that carries impulses from the retina to the brain.

order a taxonomic subcategory of class.

organ two or more tissues grouped together to form a structural and functional unit.

organ of Corti (KORT-ee) the organ of hearing; the collective term for the hair cells, the supporting cells of the basilar membrane, and the overhanging tectorial membrane.

organ system a group of organs that function together to carry out the principal activities of an organism.

organic compounds the carbon-containing molecules that make up living things.

organisms living things. Organisms can be either multicellular or unicellular.

osmoregulation (OZ-mo-reg-you-lay-shun or OS-mo-reg-you-LAY-shun) the control of water movement within an organism.

osmosis (os-MOE-sis) a special form of diffusion in which water molecules move from an area of higher concentration to an area of lower concentration across a differentially permeable membrane.

osmotic pressure the pressure that water exerts on a cell as it diffuses into the cell.

oval window the entrance to the cochlea of the inner ear. The stirrup fits into this membrane-covered opening.

ovaries (OH-vah-reez) the female gonads, in which secondary oocytes develop.

oviparous (oh-VIP-uh-rus) a term that describes an organism that lays eggs in which the embryo develops after egg laying.

ovoviviparous (OH-vo-vye-VIP-uh-rus) a term that describes an organism that retains fertilized eggs within the oviducts until the young hatch.

ovulation (OV-yuh-LAY-shun) the monthly process by which a secondary oocyte (potential egg) is expelled from the ovary.

oxidation (OK-si-day-shun) a reaction that involves an atom losing an electron.

ozone (O_3) (OH-zone) a chemical air pollutant formed as a secondary pollutant when hydrocarbons and nitrogen oxides undergo photochemical reactions; it is extremely irritating to the eyes and the respiratory tract. In the stratosphere, ozone protects the Earth from UV rays from the sun.

P

pancreas (PANG-kree-us *or* PAN-kree-us) a long gland that lies beneath the stomach and is surrounded on one side by the curve of the duodenum; it secretes a number of digestive enzymes and the hormones insulin and glucagon.

pancreatic amylase (pang-kree-AT-ick *or* pan-kree-AT-ick AM-uh-lace) a digestive enzyme secreted by the pancreas that works within the small intestine to break down starch and glycogen to maltose.

pancreatic lipase (pang-kree-AT-ick LYE-pace *or* LIP-ace) a digestive enzyme secreted by the pancreas that works in the small intestine to break down triglycerides to fatty acids and glycerol.

parasites organisms that feed on other, living organisms.

parasitism (PARE-uh-suh-tiz-um) an interaction in which an organism of one species (the parasite) lives in or on another (the host).

parasympathetic nervous system (PARE-uh-SIM-puh-THET-ik) a subdivision of the autonomic nervous system that generally stimulates the activities of normal internal body functions and inhibits alarm responses; acts in opposition to the sympathetic nervous system.

parathyroid glands (PARE-uh-THIGH-royd) four small glands embedded in the posterior side of the thyroid that produce parathyroid hormone.

parental (P) generation the members of a cross between pure-breeding organisms, giving rise to offspring called the F_1.

parietal lobe (puh-RYE-uh-tul) one of the four sections of the cerebral cortex of each hemisphere of the brain; the parietal lobe integrates sensory activities and stores information required for speech.

parthenogenesis (PAR-theh-no-JEN-eh-sis) a form of asexual reproduction in a variety of invertebrates and some vertebrates in which eggs of the female develop into adults without being fertilized.

passive immunity a kind of specific immunity conferred on an individual when he or she receives antibodies from another source. An example of passive immunity is a baby receiving antibodies during breastfeeding.

passive transport molecular movement down a gradient but across a cell membrane; three types of passive transport are diffusion, osmosis, and facilitated diffusion.

peck order in chickens, another term for the rank order chickens establish by pecking at one another.

pectoral girdle (PEK-tuh-rul) the part of the appendicular skeleton that is made up of two pairs of bones: the clavicles, or collarbones, and the scapulae, or shoulder blades. The pectoral girdle is also known as the shoulder girdle.

pedigrees (PED-uh-greez) diagrams of genetic relationships among family members over several generations.

pelagic zone (puh-LAJ-ik) the open ocean.

pelvic girdle the part of the appendicular skeleton that is made up of the two bones called coxal bones, pelvic bones, or hip bones. The pelvic girdle is also known as the hip girdle.

penis (PEE-nis) a cylindrical organ that transfers sperm from the male reproductive tract to the female reproductive tract.

pepsin an enzyme of the stomach that digests only proteins, breaking them down into short peptides.

peptidases (PEP-teh-DACE-is) enzymes produced by cells in the intestinal epithelium that work as a team with trypsin, chymotrypsin, and carboxypeptidase to break down polypeptides into shorter chains and then to amino acids.

peptide hormones one of the two main classes of endocrine hormones. Peptide hormones are made of amino acids and are unable to pass through cell membranes; instead, they bind to receptor molecules embedded in the membranes of target cells.

peptides short chains of amino acids.

periods shorter time units into which eras are subdivided.

peripheral nervous system (puh-RIF-uh-rul) one of the two main parts of the vertebrate nervous system; it consists of an information highway of nerves that bring messages to and from the brain and spinal cord.

peristalsis (PEAR-ih-STALL-sis) successive waves of contractions of the esophagus and small intestine that move food along these parts of the digestive system.

petiole (PET-ee-ole) the slender stalk of a leaf.

pH scale a succession of values (scale) based on the slight degree of spontaneous ionization of water that indicates the concentration of H^+ ions in a solution.

phagocytosis (FAG-oh-sye-TOE-sis) a type of endocytosis in which a cell ingests an organism or some other fragment of organic matter.

pharyngeal (gill) arches (fuh-RIN-jee-uhl *or* FAR-in-JEE-uhl) a principal feature of the embryos of all chordates. Pharyngeal arches develop into the gill structures of fish, and the ear, jaw, nose, and throat structures of terrestrial vertebrates.

pharynx (FAIR-inks) the upper part of the throat that extends from behind the nasal cavities to the openings of the esophagus and larynx.

phenotype (FEE-nuh-type) the outward appearance or expression of an organism's genes.

pheromone (FARE-uh-moan) a chemical produced by one individual that alters the physiology or behavior of other individuals of the same species.

phloem (FLO-em) a type of vascular tissue that conducts carbohydrates the plant uses as food, along with other needed substances.

phosphorylation (FOS-for-ih-LAY-shun) the bonding of a phosphate group to a molecule; the conversion of a molecule into a phosphorus compound.

photic zone (FOE-tik) an area of the open ocean where light is available to organisms from the water's surface to an approximate depth of 200 meters (650 feet).

photoautotrophs (FOTE-oh-AWE-toe-trofes) organisms that make their own food by photosynthesis, using the energy of the sun.

photochemical smog a type of air pollution caused by reactions between hydrocarbons and nitrogen oxides taking place in the presence of sunlight.

photosynthesis (foe-toe-SIN-thuh-sis) the process whereby energy from the sun is captured by organisms and used to produce carbohydrates and other organic molecules.

photosystem I the first photosynthetic pathway to evolve; in it, energy is transferred to a molecule of chlorophyll *a* called P700, which then acts as the photosystem's reaction center. Photosystem I, along with hydrogen ions, reduces NADP$^+$ to NADPH.

photosystem II a photosynthetic pathway in which energy is transferred to a molecule of chlorophyll *a* called P680, which then acts as the photosystem's reaction center. Energized electrons ejected from photosystem II are passed down an electron transport chain, triggering events of ATP production.

phyla (FYE-luh) subcategories of kingdoms into which taxonomists place organisms having similar evolutionary histories.

phyletic gradualism (fye-LET-ik GRADJ-oo-uh-LIZ-um) a theory that states that new species develop slowly and gradually as an entire species changes over time.

pigments molecules that absorb some visible wavelengths of light and transmit or reflect others.

pineal gland (PIN-ee-uhl) a gland resembling a tiny pine cone embedded deep within the brain; though its exact function remains a mystery, it is believed to be the site of the control center that regulates the body's daily rhythms. It may also stimulate the onset of puberty.

pinna (PIN-uh) the ear flap located on the outside of the head.

pinocytosis (PIE-no-sye-TOE-sis) a type of endocytosis in which a cell ingests liquid material containing dissolved molecules.

pituitary (puh-TOO-ih-tare-ee) a small but powerful gland hanging from the underside of the brain that is controlled by the hypothalamus and secretes 9 major hormones.

pivot joint a kind of joint that allows side-to-side movement.

placenta (pluh-SEN-tuh) a flat disk of tissue that grows into the uterine wall, made up of both chorionic and maternal tissues, through which the embryo and fetus are supplied with food, water, and oxygen and through which wastes are removed.

placental mammals (pluh-SENT-uhl) a subclass of mammals that nourishes their developing young within the body of the mother.

plaques (PLAKS) masses of cholesterol and other lipids that can build up within the walls of large and medium-sized arteries in a disease called atherosclerosis; the accumulation of such masses impairs the arteries' proper functioning.

plasma (PLAZ-muh) a straw-colored fluid that forms the liquid portion of blood. Plasma is made up of water and dissolved substances such as nutrients, hormones, respiratory gases, wastes, salts, ions, and proteins.

plasma membrane a thin, flexible lipid bilayer that encloses the contents of a cell.

plasmids (PLAZ-mids) extrachromosomal pieces of DNA that replicate independently of the main chromosome in bacteria.

plasmodial (acellular) slime molds (plaz-MOE-dee-uhl) a phylum of funguslike protists that have an ameboid stage in their life cycle; they spend much of their time as a nonwalled, multinucleated mass of cytoplasm called a plasmodium, but can form spores when food or moisture is in short supply.

plasmodium a nonwalled, multinucleate mass of cytoplasm that resembles a moving mass of slime. Plasmodial slime molds spend much of their life cycle in this form.

platelets cell fragments present in blood that play an important role in blood clotting.

point mutation (gene mutation) a change in the genetic message of a chromosome caused by alterations of molecules within the structure of the chromosomal DNA.

polar molecules molecules (two or more atoms sharing electrons) that have opposite partial charges at either end because electrons are shared unequally.

polyps (POL-ups) aquatic, cylindrical animals with a mouth at one end that is ringed with tentacles.

polysaccharides (POL-ee-SACK-uh-rides) long, insoluble polymers composed of sugar.

pons (ponz) a part of the brainstem that consists of bands of nerve fibers that act as bridges, connecting various parts of the brain to one another; the pons also brings messages to and from the spinal cord.

population a group that consists of the individuals of the same species that occur together at one place and at one time.

population ecologists (ih-KOL-uh-jists *or* ee-KOL-uh-jists) scientists who study how populations grow and interact.

population pyramid a bar graph that shows the composition of a population by age and sex.

positive feedback loop a feedback loop in which the response of the regulating mechanism is intensified with respect to the output.

positive phototaxis (foe-toe-TAK-sis) an innate movement of an organism toward a source of light.

potential energy energy not actively doing work but having the capacity to do so.

predation a relationship in which an organism of one species kills and eats an organism of another.

predator an organism of one species that kills and eats organisms of another.

preembryo a term referring to the developing cell mass formed by the zygote as it begins to divide by mitotic cell division during the first 2 weeks of development.

prenatal development life before birth; the gradual growth and progressive changes in a developing human from conception until the time the fetus leaves the mother's womb.

prey an organism killed and eaten by a predator.

primary bronchi (BRON-keye) the two airways branching immediately from the trachea into each lung.

primary consumers organisms that feed directly on green plants.

primary germ layers three layers of cells that develop from the inner cell mass of the blastocyst and from which all the organs and tissues of the body develop.

primary growth occurs mainly at the tips of the roots and shoots. During primary growth, the plant grows taller and its roots grow deeper into the ground.

primary succession succession that takes place in areas not previously supporting organisms.

primates placental mammals that have characteristics reflecting an arboreal, or tree-dwelling, lifestyle.

probe a molecule that binds to a specific gene or nucleotide sequence.

producers organisms that can make their own food.

products substances obtained as the result of chemical reaction; substrates that have undergone a chemical change.

prokaryotes (pro-KARE-ee-oats) unicellular organisms having no membrane-bounded organelles or nucleus.

prophase (PRO-faze) the first stage of mitosis, meiosis I, and meiosis II. Prophase begins when the chromosomes have condensed; they become visible under a light microscope.

proprioceptors (PRO-pree-oh-SEP-turz) receptors located within skeletal muscles, tendons, and the inner ear that give the body information about the position of its parts relative to each other and to the pull of gravity.

prosimians (pro-SIM-ee-uns) a suborder of primates that includes lemurs, indris, aye-ayes, and lorises; they are small animals, mostly nocturnal, with large ears and eyes, and elongated snouts and rear limbs.

prostate a chestnut-sized gland in males that adds an alkaline fluid to the semen.

proteases (PRO-tee-ACE-es) digestive enzymes that break down proteins to smaller polypeptides, and polypeptides to amino acids.

proteins long, complex chains of amino acids linked end to end.

protists members of the Kingdom Protista, which includes unicellular eukaroytes as well as some multicellular forms.

protons particles found at the core of an atom. Protons carry a positive electrical charge.

protostomes (PRO-toe-stowmz) one of two distinct evolutionary lines of coelomates that includes the molluscs, annelids, and arthropods. Their embryological development is characterized by the mouth developing from the first indentation of the gastrula and a spiral pattern of cleavage.

protozoans (PRO-tuh-ZOE-unz) animallike protists that are heterotrophs: They take in and use organic matter for energy.

pulmonary artery extends from the right ventricle of the heart; it branches into two smaller pulmonary arteries that carry deoxygenated blood to the lungs.

pulmonary circulation (PUHL-muh-NARE-ee sur-kyuh-LAY-shun) the part of the human circulatory system that carries blood to and from the lungs.

pulmonary veins (PUHL-muh-NARE-ee VAYNZ) large blood vessels that carry oxygenated blood from the lungs to the left atrium of the heart.

punctuated equilibrium (PUNGK-choo-AY-ted EE-kwuh-LIB-ree-um) a theory that states that new species arise suddenly and rapidly as small subpopulations of a species split from the populations of which they were a part.

Punnett square (PUN-et) a simple diagram that provides a way to visualize the possible combinations of genes in a cross and that illustrates their expected ratios. Male gametes are shown along one side of the square and female gametes are shown along the other. Each possible combination of gametes is shown in cells within the square.

pupil the opening in the center of the iris through which light passes.

purines (PYOOR-eens) double-ring compounds that are components of nucleotides. Both DNA and RNA contain the purines adenine and guanine.

pyrimidines (pih-RIM-uh-deens *or* pye-RIM-uh deens) single-ring compounds that are components of nucleotides. Both DNA and RNA contain the pyrimidine cytosine; in addition, DNA contains the pyrimidine thymine, and RNA contains the pyrimidine uracil.

Q

qualitative data descriptive results not based on numerical measurements.

quantitative data results based on numerical measurements.

R

radial symmetry having parts that emerge, or radiate, from a central point, much like spokes on a wheel.

rank order a social hierarchy that exists in many groups of fishes, reptiles, birds, and mammals.

reactants substances entering into a chemical reaction.

realized niche the role an organism actually plays in the ecosystem.

recessive in an organism carrying a pair of contrasting alleles for a particular trait, the form of the trait (the allele) that recedes or disappears entirely.

rectum the lower portion of the colon, which terminates at the anus.

red algae a phylum of multicellular algae; red algae play an important role in the formation of coral reefs and produce gluelike substances that make them commercially useful.

redox reaction a term used to describe the occurrence of oxidation and reduction reactions, which always happen together.

reduction a reaction that involves an atom gaining an electron.

reflex an automatic response to nerve stimulation.

reflex arcs pathways that integrate responses to certain types of stimuli within the spinal cord.

regulatory sites in eukaryotic cells, specific nucleotide sequences that control the transcription of genes.

renal failure a disease that occurs when the filtration of the blood at the glomerulus either slows or stops; in acute renal failure, filtration stops suddenly, while in chronic renal failure, the filtration of blood at the glomerulus slows gradually.

renewable resources those produced by natural systems that replace themselves quickly enough to keep pace with consumption; wind power and solar power are renewable resources.

repeated trials performing experiments more than one time to increase the reliability of the results.

reptiles a class of vertebrates characterized by dry skin covered with scales that help retard water loss.

respiration (respire) a term that, when used alone, refers to the uptake of oxygen and the release of carbon dioxide by the whole body.

respiratory bronchioles (BRON-kee-olz) the last division of bronchi. Bronchioles have thousands of tiny air passageways whose walls have clusters of tiny pouches, or alveoli.

respiratory system collectively, all the organs of respiration in an organism.

resting potential an electrical potential difference, or electrical charge, along the membrane of a resting neuron.

restriction enzymes proteins that recognize certain nucleotide (base) sequences in a DNA strand and break the bonds between the nucleotides at those points.

restriction fragments the pieces of DNA that have been cut from larger pieces of DNA by restriction enzymes.

restriction site a particular nucleotide (base) sequence in DNA that is acted upon by a specific restriction enzyme.

retina (RET-uh-nuh) tissue at the back of the eye that is composed of rod and cone cells and that is sensitive to light.

rhizoids (RYE-zoyds) in nonvascular plants, slender, rootlike projections that anchor the plant to its substrate; unlike roots, however, they consist of only a few cells and do not play a major role in the absorption of water or minerals.

ribonucleic acid (RNA) (RYE-boh-noo-KLAY-ick) one of two types of nucleic acid found in cells; the nucleotides of RNA contain the 5-carbon sugar ribose.

ribosomal ribonucleic acid (rRNA) (RYE-buh-SO-mull RYE-boh-new-KLAY-ick) a type of ribonucleic acid that is manufactured by the DNA within a eukaryotic cell's nucleolus; rRNA molecules are structural components of ribosomes.

ribosomes (RYE-buh-somes) minute, round structures found on the endoplasmic reticulum of eukaryotic cells or free in the cytoplasm of both prokaryotes and eukaryotes; ribosomes are the places where proteins are manufactured.

rods sensory cells in the retina of the eye that are stimulated by low levels of light and detect only white light.

root the part of a vascular plant that exists below the ground. The root penetrates the soil and absorbs water and various ions that are crucial for plant nutrition. It also anchors the plant.

rough ER a kind of endoplasmic reticulum that is covered with ribosomes and hence resembles long sheets of sandpaper; in tandem with its ribosomes, rough ER manufactures and transports proteins designed to leave the cell.

round window a membrane-covered hole at the wider end of the cochlea in the inner ear.

S

sac fungi one of the three phyla of fungi. Sac fungi live in both aquatic and terrestrial environments and are characterized by sexual spores borne in saclike structures. Familiar examples of this group are cup fungi and yeasts. Certain sac fungi have no sexual stages.

saccule (SAK-yool) the smaller of two membranous sacs within the vestibule of the inner ear, containing both ciliated and nonciliated cells, that functions in the maintenance of bodily equilibrium and coordination.

salivary amylase (SAL-i-VER-ee AM-uh-lace) a digestive enzyme found in saliva that breaks down starch into molecules of the disaccharide maltose.

salivary glands (SAL-ih-VER-ee GLANDZ) the paired glands of the mouth that secrete saliva.

sanitary landfill an enormous depression in the ground where trash and garbage are dumped, compacted, and then covered with dirt.

saprophytes organisms that live and feed on dead organisms.

saprophytic (SAP-roe-FIT-ik) a term describing an organism that feeds on dead or decaying organic matter; most fungi are saprophytic.

sarcomere (SAR-koe-mere) the repeating bands of actin and myosin myofilaments that appear between two Z lines in a muscle fiber; the contractile unit of muscles.

savanna a biome found near the equator but in areas having less annual rainfall than tropical rain forests. This climate supports grasslands with only scattered trees and shrubs.

Schwann cells special neuroglial cells that are wrapped around many of the long cell processes of sensory and motor neurons of the peripheral nervous system.

scientific laws descriptions of patterns of regularity with respect to natural phenomena.

scientific theory hypotheses or groups of related hypotheses that consistently account for (explain) existing data and consistently predict new data.

sclera (SKLEAR-uh) the tough outer layer of connective tissue that covers and protects the eye.

scrotum (SKRO-tum) a sac of skin, located outside the lower pelvic area of the male, which houses the testicles, or testes.

second filial (F_2) generation (FIL-ee-uhl) the offspring of the F_1 generation.

second law of thermodynamics a law stating that disorder in the universe constantly increases. Energy spontaneously converts to less organized forms.

second trimester the second 3-month period of pregnancy.

secondary bronchi (BRON-keye) outbranchings of each of the two primary bronchi. Three secondary bronchi serve the three right lobes of the lungs, whereas two secondary bronchi serve the two left lobes of the lungs.

secondary consumers meat eaters that feed on the primary consumers (herbivores).

secondary growth occurs at the stem. During secondary growth, a plant grows wider (thicker).

secondary succession succession that takes place in areas that have been disturbed and that were originally occupied by organisms.

secretin (sih-KRE-tin) one of the hormones that control digestion in the small intestine.

sedative-hypnotics (SED-uh-tiv-hip-NOT-iks) central nervous system depressants that induce sleep (sedatives) and reduce anxiety (hypnotics).

seedless vascular plants plants that reproduce by means of spores rather than seeds; a familiar member of this group is the fern.

seeds structures from which new sporophyte plants grow; seeds protect the embryonic plant from drying out or being eaten when it is at its most vulnerable stage.

semicircular canals three looped fluid-filled canals in the inner ear that detect the direction of the body's movement.

seminal vesicles (SEM-un-uhl VES-uh-kuls) two accessory glands that secrete a thick, clear fluid, primarily composed of fructose, that forms a part of the semen.

seminiferous tubules (SEM-uh-NIF-uhr-us TOO-byools) tightly coiled tubes within each testis, where sperm cells develop.

sensory neurons nerve cells that are specialized to change specific stimuli into nerve impulses and transmit this information to the central nervous system.

septum tissue that separates the two sides of the heart.

sex chromosomes chromosomes that determine the gender of an individual as well as certain other characteristics.

sexual reproduction a process whereby a male sex cell and female sex cell combine to form the first cell of a new individual.

shell (energy level) a term for the volume of space around an atom's nucleus where an electron is most likely to be found.

shoot the part of a vascular plant that exists above the ground; the stem, leaves, and other plant structures are formed at the shoot.

shore zone the shallow water near the edges of a lake or pond in which plants with roots, such as cattails and water lilies, may grow.

shotgun cloning the process of synthesizing DNA by cutting the DNA of an entire genome into pieces, isolating and purifying the genomic DNA, and then making copies of this DNA using yeast or bacterial cells.

sigmoid growth curve the exponential growth of a population and its subsequent stabilization at the level of its environment's carrying capacity.

sinoatrial (SA) node (sye-no-AY-tree-uhl NODE) a small cluster of specialized cardiac muscle cells that are embedded in the upper wall of the right atrium and automatically and rhythmically send out impulses that initiate each heartbeat.

skeletal system another term for the skeleton.

skeleton the collective term for a vertebrate's internal bony scaffolding. The skeleton provides support for the body, helps the body move (along with the muscles), and protects delicate internal structures.

skull the framework of the head, which in humans consists of 8 cranial and 14 facial bones.

small intestine the tubelike portion of the digestive tract that begins at the pyloric sphincter and ends at its T-shaped junction with the large intestine.

smog a type of air pollution that results from the burning of fossil fuels.

smooth ER a kind of endoplasmic reticulum that does not have ribosomes attached to its surface and hence manufactures no proteins; instead, smooth ER helps produce carbohydrates and lipids.

social behaviors activities of animals that help members of the same species communicate and interact with one another, each responding to stimuli from others.

sociobiology (SO-see-oh-bye-OL-uh-gee or SO-shee-oh-bye-OL-uh-gee) the biology of social behavior. This science applies the knowledge of evolutionary biology to the study of animal behavior.

sodium-potassium pump membrane proteins that transport sodium ions out of the cell and potassium ions into the cell.

solar power the use of the sun for heating or electricity.

solid waste garbage and trash.

soluble (SOL-you-ble) able to be dissolved in water.

somatic nervous system (soe-MAT-ik) the branch of the peripheral nervous system consisting of motor neurons that send messages to the skeletal muscles and control voluntary responses.

speciation (SPEE-shee-AY-shun) the process by which new species are formed during the process of evolution.

species (SPEE-shees or SPEE-sees) a population of organisms that interbreed freely in their natural settings and do not interbreed with other populations. The most narrow and specific taxonomic classification; a taxonomic subcategory of family.

specific immunity a set of cellular and molecular defenses that protect the body from each particular microbe that may invade it.

spermatozoon (sperm) (spur-MAT-uh-ZOO-un) a male sex cell.

spinal cord one of the two main parts of the central nervous system; the spinal cord runs down the neck and the back. It receives information from the body, relays this information to the brain, and sends information from the brain to the body. Some simple reflexes are also integrated there.

spinal nerves nerves by which the spinal cord receives and transmits information; spinal nerves include sensory and motor neurons.

spiracles small openings for breathing; in tracheal systems, spiracles are muscular openings that can open and close, directed by nerves and triggered by carbon dioxide levels.

sporangia (spoh-RAN-jee-uh) spore cases.

spores reproductive bodies formed by cell division (mitosis or meiosis) in the parent organism. Spores are formed by meiosis in a diploid parent and mitosis in a haploid parent; the spores are always haploid.

sporophyte (SPOR-ih-fite) a plant that produces haploid spores by meiosis; also known as the spore-plant generation.

sporophyte (spore-plant) generation the diploid phase of a plant life cycle. This generation tends to dominate the life cycles of vascular plants.

sporozoans (SPOR-uh-ZOH-uns) protozoans that are nonmotile, spore-forming parasites of vertebrates, including humans. Also called apicomplexans.

stable population a population whose size remains the same through time.

stapes (STAY-peez) one of the three bones of the middle ear that amplify sound vibrations and carry them from the outer ear to the inner ear. The stapes is also known as the *stirrup*.

starches polysaccharides that are the storage form of sugar in plants.

stem a part of a plant's structure that serves as a framework for the positioning of its leaves.

steroid hormones one of the two main classes of endocrine hormones. Steroid hormones are made of cholesterol and are able to pass through cell membranes.

stimulants drugs that enhance the activity of two neurotransmitters, norepinephrine and dopamine, resulting in an alerting and euphoric effect.

stomach a muscular sac in which food is collected and partially digested by hydrochloric acids and proteases.

stomata (singular, stoma) (STOW-muh-tuh/STOW-muh) openings in the epidermis of a leaf bordered by guard cells.

stretch receptors sensory neuron endings that are stimulated to fire when tension is placed on them.

stroma (STROH-muh) within a chloroplast, a fluid that surrounds the thylakoids and contains the enzymes of the light-independent reactions.

submissive behavior behavior that animals often exhibit in response to a threat display in order to avoid fighting. Submissive behavior may include making the body appear smaller or withdrawing fangs or claws.

substrates substances on which enzymes act.

succession a process of change during which a sequence of communities replaces one another in an orderly and predictable way.

superior vena cava (VEE-nuh KAY-vuh *or* VAY-nuh KAH-vuh) a large vein that drains blood from the upper body and returns it to the right atrium of the heart.

survivorship the proportion of an original population that lives to a certain age.

suture (SOO-chur) a type of immovable joint.

sympathetic nervous system (SIM-puh-THET-ik) a subdivision of the autonomic nervous system that generally mobilizes the body for greater activity; acts in opposition to the parasympathetic nervous system.

synapse (SIN-aps) a place where a neuron communicates with another neuron or effector cell.

synapsis (sih-NAP-sis) a process during prophase I of meiosis in which homologous chromosomes line up side-by-side, initiating the process of crossing over.

synaptic cleft (sin-AP-tik KLEFT) the space or gap between two adjacent neurons, which the nerve impulse must cross.

synovial joints (suh-NO-vee-uhl) a freely movable joint in which a fluid-filled space exists between the articulating bones.

syphilis (SIF-uh-lis) a sexually transmitted disease, caused by the bacterium *Treponema pallidum*, that produces three stages of infection, from localized to widespread.

systemic circulation the pathway of blood vessels to the body regions and organs other than the lungs.

systole (SIS-tl-ee) the time of contraction of a heart chamber, in which blood is forced out of the chamber.

T

T cells or T lymphocytes white blood cells that develop in the bone marrow, but migrate to the thymus to mature. As they mature, T cells develop the ability to identify body cells that have been infected with fungi, some viruses, and certain bacteria. They also recognize foreign body cells. T cells provide the cell-mediated immune response.

taiga a biome that consists of evergreen, cone-bearing trees. The climate of this biome is characterized by long, cold winters with little precipitation. It is also called the northern coniferous forest.

taste buds microscopic receptors embedded within the papillae of the tongue that work with olfactory receptors to produce taste sensations; humans have taste buds specialized to respond to four different kinds of sensations: salty, sweet, sour, and bitter.

taxis (TAK-sis) a directed movement by an animal toward or away from a stimulus, such as light, chemicals, or heat.

taxonomy (tack-SAHN-uh-me) a method of classification of the diverse array of species based on their common ancestry.

telophase (TEL-uh-faze) the last phase of mitosis, meiosis I, and meiosis II. During telophase, the mitotic apparatus assembled during prophase is disassembled, the nuclear envelope is reestablished, and the normal use of the genes present in the chromosomes is reinitiated.

temperate deciduous forest a biome that occurs in areas having warm summers, cold winters, and moderate amounts of precipitation. The trees of this forest lose their leaves and remain dormant throughout the winter.

temperate grasslands a biome that experiences a greater amount of rainfall than deserts but a lesser amount than savannas. They occur at higher latitudes than savannas but, like savannas, are characterized by perennial grasses and herds of grazing mammals.

temporal lobe one of the four sections of the cerebral cortex of each hemisphere of the brain; the temporal lobe receives and interprets nerve impulses regarding hearing and smell.

tendons cords of dense connective tissue that attach muscles to bones.

territorial behaviors behaviors an animal may exhibit that involve marking off an area as its own and defending it against the same-gender members of its species.

testes (TES-teez) the male gonads, where sperm production occurs.

thalamus (THAL-uh-muss) paired oval masses of gray matter at the base of the cerebellum, lying close to the basal ganglia, that receive sensory stimuli, interpret some of these stimuli, and send the remaining sensory messages to appropriate locations in the cerebrum.

thermal pollution a change in water temperature that occurs when industries release heated water back into rivers after using the water for cooling purposes.

third trimester the third 3-month period of pregnancy.

thoracic cavity (thu-RASS-ick) the chest cavity in humans, located within the trunk of the body and extending above the diaphragm and below the neck.

threat displays/intimidation displays a form of aggressive behavior that animals may exhibit during a competitive situation. The purpose of these displays is to scare other animals away or cause them to "back down" before fighting takes place.

threshold potential The level at which a stimulus results in an action potential along the nerve cell membrane.

thylakoids (THIGH-luh-koidz) flattened, saclike membranes within chloroplasts. These membranes contain the enzymes necessary for the light-dependent reactions.

tidal volume the amount of air inspired or expired with each breath.

tissue differentiation a prenatal developmental process in which groups of cells become distinguished from other groups of cells by the jobs they will perform in the body.

tissues groups of similar cells that work together to perform a function.

trachea (TRAY-kee-uh) the windpipe; the air passageway that runs down the neck in front of the esophagus and that brings air to the lungs.

tracheal system a branching network of microscopic air tubes found throughout the bodies of most spiders and insects.

traits distinguishing features or characteristics.

transcription (trans-KRIP-shun) the first step in the process of polypeptide synthesis and gene expression, in which a gene is copied into a strand of messenger RNA.

transduction (tranz-DUK-shun) one of the ways in which bacteria transfer genetic material. Occurs as DNA from a donor bacterium is transferred to a recipient bacterium by a virus.

transfer RNA (tRNA) a type of ribonucleic acid found in the cytoplasm that transports amino acids, used to build polypeptides, to the ribosomes; tRNA molecules also position each amino acid at the correct place on the elongating polypeptide chain.

transformation one of the ways in which bacteria transfer genetic material. Transformation occurs as DNA from a lysed donor cell is taken up from the surrounding medium by a competent cell.

transgenic plants (tranz-GEE-nik) plants that are genetically altered using the techniques of genetic engineering.

translation (trans-LAY-shun) the second step of gene expression, in which the mRNA, using its copied DNA code, directs the synthesis of a polypeptide.

transpiration the process by which water vapor leaves through the stomata of a leaf.

triglyceride (try-GLISS-er-ide) a fat molecule in which three fatty acids are attached at each of the three carbon atoms of a glycerol molecule.

trophic levels (TROW-fik) the different feeding levels within an ecosystem.

trophoblast (TRO-fuh-blast) the outer ring of cells of the blastocyst that will give rise to most of the extraembryonic membranes as the embryo develops.

tropic hormones (TROW-pik) 4 of the 7 hormones produced by the anterior pituitary gland; the tropic hormones turn on, or stimulate, other endocrine glands.

tropical rain forest a biome that occurs in regions of high temperature and rainfall, characterized by tall trees that support a variety of plant and animal life on their branches.

tropical rain forest deforestation the loss of the plants and animals of the tropical rain forest (a biome that occurs in regions of high temperature and rainfall) through cut-and-burn agriculture and logging.

true-breeding a term referring to plants that produce offspring consistently identical to the parent with respect to certain defined characteristics after generations of self-fertilization.

trypsin (TRIP-sin) an enzyme produced by the pancreas, which, together with chymotrypsin and carboxypeptidase, completes the digestion of proteins in the small intestine.

tubal ligation (TOO-bul lye-GAY-shun) in females, an operation that involves the removal of a section of each of the two uterine tubes, through which the oocyte travels to the uterus, in order to bring about sterilization.

tundra a biome that encircles the top of the world. The tundra is characterized by desertlike levels of precipitation, extremely long and cold winters, and short, warmer summers.

tunicates (TOO-nih-kits *or* TOO-nih-KATES) a subphylum of the chordates. A group of about 2500 species of marine animals, most of which look like living sacs attached to the floor of the ocean. The only chordate characteristic adult tunicates exhibit is their gill slits; however, larvae have a notochord and nerve cord.

tympanic membrane (tim-PAN-ik) a thin piece of fibrous connective tissue that is stretched over the opening to the middle ear; the tympanic membrane is also known as the eardrum.

U

umbilical cord (um-BIL-uh-kul) the developing embryo's lifeline to the mother; it joins the circulatory system of the embryo with the placenta.

urea (yoo-REE-uh) the primary excretion product from the deamination of amino acids.

ureter (YER-ih-ter) the tube that takes urine from the kidney to the urinary bladder.

urethra (yoo-REE-thruh) a muscular tube that brings urine from the urinary bladder to the outside; in men, the urethra also carries semen to the outside of the body during ejaculation.

uric acid (YOO-rik) a nitrogenous waste found in small amounts in the urine that is formed from the breakdown of nucleic acids found in the cells of food and from the metabolic turnover of nucleic acids and ATP.

urinary bladder a hollow muscular organ in the pelvic cavity that acts as a storage pouch for urine.

urinary system a set of interconnected organs that not only remove wastes, excess water, and excess ions from the blood but store this fluid until it can be expelled from the body.

urine an excretion product composed of water and the substances excreted by the kidneys.

uterine tube (YOO-tur-in) the passageway from an ovary to the uterus; commonly called a fallopian tube.

uterus (YOO-tur-us) the female organ in which a fertilized egg can develop; the womb.

utricle (YOO-trih-kul) the larger of two membranous sacs within the vestibule, containing both ciliated and nonciliated cells, that functions in the maintenance of bodily equilibrium and coordination.

V

vaccination (VAK-suh-NAY-shun) a procedure that involves the injection (or oral administration, in some cases) of a weakened or killed microbe, or laboratory-made antigen proteins into a person or animal in order to confer resistance to a disease-causing pathogen.

vacuole (VACK-yoo-ole) membrane-bounded storage sacs within eukaryotic cells that hold such substances as water, food, and wastes; most often found within plant cells.

vagina (vuh-JINE-uh) an organ in the body of a female whose muscular, tubelike passageway to the exterior has three functions: It accepts the penis during intercourse; it is the lower portion of the birth canal; and it provides an exit for the menstrual flow.

vascular plants (VAS-kyuh-ler) plants that have systems of specialized tissues that transport water and nutrients; they are made up of underground roots and aboveground shoots.

vascular tissue forms the "circulatory system" of a vascular plant. Vascular tissue conducts water and dissolved inorganic nutrients up the plant and carries the products of photosynthesis throughout the plant.

vasectomy (va-SEK-tuh-mee) in males, an operation that involves the removal of a portion of the ductus (vas) deferens, the tube through which sperm travel to the penis, in order to bring about sterilization.

vegetative propagation an asexual reproductive process in which a new plant develops from a portion of a parent plant.

veins (VAYNS) blood vessels that carry blood to the heart.

venules (VAYN-yooles) small veins that connect capillaries to larger veins.

vertebral column (VER-tuh-brul) the collection of 26 vertebrae, stacked one on top of the other along the midline of the back, that acts as a strong, flexible rod and supports the head in a human skeleton. The vertebral column is also known as the backbone.

vertebrates (VER-tuh-bruts *or* ver-tuh-braytes) a subphylum of chordates; vertebrates have a bony (vertebral) column surrounding a dorsal nerve cord.

vestibule (VES-tuh-byool) a structure within the inner ear that detects the effects of gravity on the body.

villi (VIL-eye) fine, fingerlike projections of the epithelium of the small intestine.

viruses (VYE-russ-es) nonliving infectious agents that enter living organisms and cause disease.

viviparous (vye-VIP-uh-rus) a term that describes an organism that gives birth to live offspring.

vocal cords two pieces of elastic tissue, covered with a mucous membrane, that are stretched across the upper end of the larynx and are involved in the production of sound.

vulva (VUL-vuh) the collective term for the external genitals of a female.

water molds funguslike protists that predigest and absorb food as fungi do. Water molds thrive in moist places and aquatic environments and parasitize plants and animals. Also called egg fungi because they produce large egg cells during sexual reproduction.

W

wind power the use of the wind to generate electricity, usually by means of windmills.

X

xylem (ZY-lem) a type of vascular tissue in plants that conducts water and dissolved inorganic nutrients.

Y

yolk sac in humans, a membranous sac that produces blood for the embryo until its liver becomes functional. In addition, part of the yolk sac becomes the lining of the developing digestive tract.

Z

zygospores (ZYE-go-SPORZ) sexual spores formed by the zygote-forming fungi.

zygote (ZYE-goat) a new cell that contains intermingling genetic material from both the sperm and egg cells; the fertilized ovum.

zygote-forming fungi a phylum of fungi. Zygote-forming fungi have a distinct sexual phase of reproduction that is characterized by the formation of sexual spores called zygospores.

Index

Note: Page numbers followed by *f* refer to figures; those followed by *t* refer to tables.

1 ATOP
12 of 20

Credits

Title page Stephen Dalton/Photo Researchers, Inc. / page iv ©PhotoDisc / page vi (middle) ©PhotoDisc; (bottom) ABC-TV (courtesy Kobal) / page vii (top) Courtesy M. Gillott, from *Electron Microscopy* 2/e by John J. Bozzola and Lonnie D. Russell, Jones and Bartlett Publishers, Inc.; (bottom) Courtesy, Santa Cruz Beach Boardwalk / page viii ©PhotoDisc / page ix (left) Nina Lampen/Photo Researchers, Inc.; (right) Dr. Gopal Murti/Science Photo Library/Photo Researchers, Inc. / page x Jeffrey L. Rotman/Peter Arnold, Inc. / page xi (left) Columbia Pictures (courtesy Kobal); (right) Lee Simon/Science Photo Library/Photo Researchers, Inc. / page xii (top) Ted Thai/Time Magazine; (bottom) Jean-Michel Labat/Ardea London / page xiii (left) Courtesy of Winfield Sterling; (right) Lennart Nilsson/Albert Bonniers Publishing Company, *A Child Is Born*, Dell Publishing Co. / page xiv ©PhotoDisc / page xv Tim Flach/Tony Stone Images / page xvi (top left) Sterling Zumbrunn (courtesy Conservation International); (bottom left) Gregory G. Dimijian/Photo Researchers, Inc.; (right) M.P.L. Fogden/Bruce Coleman, Inc. / page xvii Kennan Ward/The Stock Market. All banner images ©PhotoDisc.

Chapter 1

Page 1 John Warden/Tony Stone Images / page 2 NASA / page 3 (both) Stewart Halperin / page 4 (all) Stewart Halperin / page 5 Stewart Halperin / page 6 John Mitchell/Photo Researchers, Inc. / page 7 The Granger Collection, New York / page 8 Tozzer Library/Harvard University / page 9 (left) Courtesy Tom Cirrito, Washington University, St. Louis; (right) ©PhotoDisc. Part 1 banner image ©PhotoDisc.

BioIssues, Part 1

Page 14 (top) David Phillips/Photo Researchers, Inc.; (bottom) ©PhotoDisc / page 15 ©PhotoDisc. All background images ©PhotoDisc. Part 1 banner image ©PhotoDisc.

Chapter 2

Page 16 Gamma Liaison / page 17 Dr. Jeremy Burgess/Science Photo Library/Photo Researchers, Inc. / page 18 Fred LeBlanc / page 19 (top left) Manfred Kage/Peter Arnold, Inc.; (bottom left) Kim Taylor/Bruce Coleman, Inc.; (right) Stephen Dalton/NHPA/Photo Researchers, Inc. / 2.1 (biosphere) NASA/Galileo Imaging Team, (ecosystem, organism, DNA) ©PhotoDisc, (community) Joe McDonald/Visuals Unlimited, (population) Will Troyer/Visuals Unlimited, (tissue) Biophoto Associates/Photo Researchers, Inc., (cell) Richard Gross, Biological Photography / page 21 (top left) Will Troyer/Visuals Unlimited; (bottom left) John D. Cunningham/Visuals Unlimited; (right) ©PhotoDisc / page 22 ©PhotoDisc / 2.2 (all) Richard Gross/Biological Photography / 2.3a R. Arndt/Visuals Unlimited, 2.3b Richard Gross/Biological Photography, 2.3c Terry Hazen/Visuals Unlimited; page 24 (feather) Stewart Halperin; 2.4a ©PhotoDisc, 2.4b Tom McHugh/Photo Researchers, Inc. / 2.5 O. Louis Mazzatenta/NGS Image Collection; page 25 (feather) Stewart Halperin / page 26 (left) From *Electron Microscopy* 2/e by John J. Bozzola and Lonnie D. Russell. Reprinted with permission from Jones and Bartlett Publishers, Inc.; (right) K.G. Murti/Visuals Unlimited / 2.6 (plantae, fungi) Richard Gross/Biological Photography, (animalia) ©PhotoDisc, (protista) K.G. Murti/Visuals Unlimited, (bacteria) From *Electron Microscopy* 2/e by John J. Bozzola and Lonnie D. Russell. Reprinted with permission from Jones and Bartlett Publishers, Inc. / page 28 (left) ©PhotoDisc; (top right) Stewart Halperin; (middle right) Roger Angel; (bottom right) G. and C. Merker/Visuals Unlimited / page 29 ©PhotoDisc.

Chapter 3

Page 32 Kim Taylor/Bruce Coleman, Inc. / page 33 Paolo Fioratti / 3.A SIU/Visuals Unlimited / 3.4 Nathan Cohen/Visuals Unlimited; page 39 (right) Hermann Eisenbeiss/Photo Researchers, Inc. / page 40 Joe Sohm/Tony Stone Images / page 42 Robert Brons/BPS/Tony Stone Images / 3.11 From *Electron Microscopy* 2/e by John J. Bozzola and Lonnie D. Russell. Reprinted with permission from Jones and Bartlett Publishers, Inc.; 3.12a J.D. Litvay/Visuals Unlimited.

BioIssues, Part 2

Page 56 (left) ABC-TV (courtesy Kobal); (right) Patrick L. Pfister/Stock, Boston / page 57 Alfred Pasieka, Science Photo Library/Photo Researchers, Inc. All background images ©PhotoDisc. Part 2 banner image courtesy M. Gillott, from *Electron Microscopy* 2/e by John J. Bozzola and Lonnie D. Russell, Jones and Bartlett Publishers, Inc.

Chapter 4

Page 59 David M. Phillips/Visuals Unlimited / 4.1a M. Abbey/Visuals Unlimited, 4.1b Richard Gross/Biological Photography, 4.1c Science Pictures Ltd/Science Photo Library/Photo Researchers, Inc. / 4.2 Linda Sims/Visuals Unlimited / 4.4 David M. Phillips/Visuals Unlimited / 4.A Leonard Lessin/Peter Arnold, Inc.; 4.B Bruce Iverson; 4.C Robert Brons/Biological Photo Service; 4.D Lester Lefkowitz/The Stock Market / 4.E Courtesy M. Brown, Jr. and 4.F courtesy S. Schmitt from *Electron Microscopy* 2/e by John J. Bozzola and Lonnie D. Russell, Jones and Bartlett Publishers, Inc. / 4.7 J. David Robertson / 4.10a From Bozzola and Russell *Electron Microscopy* 2/e / 4.11 From Bozzola and Russell *Electron Microscopy* 2/e / 4.14 From Bozzola and Russell *Electron Microscopy* 2/e / 4.15 From Bozzola and Russell *Electron Microscopy* 2/e; 4.16 Ed Reschke / page 73 (top) Kevin & Betty Collins/Visuals Unlimited; 4.17 From Bozzola and Russell *Electron Microscopy* 2/e / 4.18 From Bozzola and Russell *Electron Microscopy* 2/e / 4.19 Courtesy M. Gillott, from Bozzola and Russell *Electron Microscopy* 2/e. / 4.20 Courtesy W. Dougherty, from Bozzola and Russell *Electron Microscopy* 2/e. / page 77 Courtesy B. Crandall-Statler, from Bozzola and Russell *Electron Microscopy* 2/e. / 4.21 Barbara Cousins / page 82 Charles L. Sanders/Biological Photo Service / page 83 Dr. Birgit H. Satir.

Chapter 5

Page 89 Courtesy, Santa Cruz Beach Boardwalk / page 91 Nick Bergkessel/Photo Researchers, Inc. / page 92 NASA / 5.5 (both) Dr. Thomas A. Steitz, Yale University / page 96 AP/Wide World Photos / page 98 P.A. Hinchliffe/Bruce Coleman, Inc.

Chapter 6

Page 103 Dr. E.R. Degginger / page 104 Ernest A. Janes/Bruce Coleman, Inc. / page 105 L.L. Rue III/Earth Scenes / page 107 ©PhotoDisc / 6.5 (left) Richard Gross/Biological Photography, (right) Courtesy M. Gillott, from *Electron Microscopy* 2/e by John J. Bozzola and Lonnie D. Russell, Jones and Bartlett Publishers, Inc. / page 109 Elizabeth Gentt, Visuals Unlimited / 6.18a Richard Gross/Biological Photography, 6.18b Stewart Halperin / page 124 Stewart Halperin.

Part 3, BioIssues

Page 128 (both) ©PhotoDisc / page 129 (table) ©PhotoDisc; (flowers) Courtesy of Tom Clothier. Part 3 banner image (outside edge) courtesy M. Doran and S. Pelok, from Bozzola and Russell *Electron Microscopy* 2/e; (other images) ©PhotoDisc.

Chapter 7

Page 131 Herb Snitzer/Stock, Boston / page 132 (top) ©PhotoDisc; (bottom) M.I. Walker/Science Source/Photo Researchers, Inc. / page 133 (top) Sherman Thomson/Visuals Unlimited, (bottom) 133 Rod Barbee/Visuals Unlimited / 7.6 (1-3) Ed Reschke, (4) Biophoto Associates/Science Source/Photo Researchers, Inc., (5-6) Ed Reschke, (athlete) ©PhotoDisc / 7.7 Emma Shelton; page 137 (right) Brian Eyden/Science Photo Library/Photo Researchers, Inc. / page 139 (top left) J.V. Small and F. Rinnerthaler; 7.8 *St. Louis Globe-Democrat* / 7.9 (10-19) Ed Reschke, (7) Andrew Syred/Science Photo Library/Photo Researchers, Inc., (8) Manfred Kage/Peter Arnold, Inc., (9) Nina Lampen/Photo Researchers, Inc., (athlete) ©PhotoDisc / 7.10 Lennart Nilsson, *Behold Man* (Little, Brown and Co.) / page 143 Miki Koren, courtesy of Sigma Chemical Co. / 7.12 (20) M.I. Walker/Science Source/Photo Researchers, Inc., (21-22) Ed Reschke, (athlete) ©PhotoDisc / 7.13 Digital Stock / page 146 Stewart Halperin / 7.15 Bob Daemmrich/Stock, Boston.

Chapter 8
Page 154 News Office/Woods Hole Oceanographic Institution / page 155 Keren Su/Tony Stone Images / 8.2 Nadine Sokol / page 162 (skulls) Glenn Oliver/Visuals Unlimited; 8.3a Nadine Sokol, 8.3b G. David Brown / page 164 (top) Stewart Halperin; (botttom) NASA / 8.5 G. David Brown; 8.6 (photo) Ed Reschke, (art) Bill Ober / 8.9 (photo) Michael Webb/Visuals Unlimited / 8.11 (photos) Stewart Halperin.

Chapter 9
Page 177 Frank S. Balthis / page 180 Norm Thomas/Photo Researchers, Inc. / 9.5 (all) AT&T Archives; 9.6 Ellen Dirkson/Visuals Unlimited / 9.7 (top) Lennart Nilsson, Behold Man (Little, Brown and Co.), (bottom) Art Siegel, University of Pennsylvania / page 187 David M. Grossman/Photo Researchers, Inc. / page 188 Oliver Meckes/Photo Researchers, Inc. / page 192 (both) Stewart Halperin / page 193 Stewart Halperin.

Chapter 10
Page 197 SIU, School of Medicine/Peter Arnold, Inc. / pages 198, 199 Jane Shaw/Bruce Coleman, Inc. / 10.4a Cabisco/Visuals Unlimited, 10.4b Ed Reschke, 10.4c (top) John D. Cunningham/Visuals Unlimited, (bottom) Cabisco/Visuals Unlimited / page 205 Bill Ober / page 207 Biophoto Associates/Science Source/Photo Researchers, Inc. / page 210 Andrew Syred/Science Photo Library/Photo Researchers, Inc. / 10.9 CNRI/Science Photo Library/Photo Researchers, Inc. / 10.11 (both) Harry Ransom Humanities Research Center / page 215 AP/Wide World Photos / page 216 Stewart Halperin.

Chapter 11
Page 221 Lennart Nilsson / 11.1 Mike Abbey/Visuals Unlimited / page 223 Virginia Weinland/Photo Researchers, Inc. / 11.6 Prof. S.H.E. Kaufman & Dr. J.R. Golecki/SPL/Photo Researchers, Inc. / 11.7 (both) Dr. A. Liepins/SPL/Photo Researchers, Inc. / 11.10a CNRI/Science Photo Library/Photo Researchers, Inc., 11.10b Dr. Gopal Murti/Science Photo Library/Photo Researchers, Inc. / 11.13 Ken Eward/Biografx/Photo Researchers, Inc. / 11.15 (left) Ed Reschke, (right) David Scharf/Peter Arnold, Inc.; page 235 (top right) Stewart Halperin.

Chapter 12
Page 240 Stewart Halperin / page 241 Michael Sewell/Peter Arnold, Inc. / page 242 Andrew Syred/Science Photo Library/Photo Researchers, Inc. / 12.1 (left) David M. Phillips/Visuals Unlimited, (right) A. Kerstitch/Visuals Unlimited / page 244 ©PhotoDisc / page 246 SIU/Photo Researchers, Inc. / 12.4a Cabisco/Visuals Unlimited / 12.5 Courtesy, Kenjiro Kimura, M.D., PhD. / page 250 Ted Horowitz/The Stock Market / page 251 (left) Stewart Halperin; (right) Tom McHugh/Photo Researchers, Inc. / 12.7 (detail) Barbara Cousins; page 254 (left) Stephen J. Krasemann/Photo Researchers, Inc.; (right) Michael Heron/The Stock Market.

Chapter 13
Page 259 George Tiedemann/Sports Illustrated / 13.1 after Aidley, D.S. (1989) The Physiology of Excitable Cells, 3/e, Cambridge: Cambridge University Press. / page 262 ©PhotoDisc / 13.4 (left) David M. Phillips/Visuals Unlimited, (right) Don W. Fawcett/Visuals Unlimited / page 268 Brian Bailey/Tony Stone Images / 13.11 Dr. John Heuser / 13.12 Science VU/E.R. Lewis / page 272 ©PhotoDisc / 13.13 Dr. Michael J. Kuhar, NIDA Addiction Research Center / 13.14 Courtesy, Brookhaven National Laboratory / page 276 ©PhotoDisc.

Chapter 14
Page 282 William Gage / page 283 Lunagrafix, Inc./Photo Researchers, Inc. / page 284 Jeffrey L. Rotman/Peter Arnold, Inc. / 14.3 Nadine Sokol / page 288 Marcus Raichle, M.D. / 14.5 (photo) Dr. C. Chumbley/SPL/Photo Researchers, Inc., (art) Karen Waldo / page 295 Courtesy Ronald Reagan Library / 14.10 Nadine Sokol / 14.11 Barbara Cousins / page 300 Richard H. Gross

Chapter 15
Page 307 Nicholas Parfitt/Tony Stone Images / page 308 ©PhotoDisc / 15.3 Barbara Cousins; page 312 Heather Angel/Biofotos / page 313 (left, both) Heather Angel/Biofotos; 15.4 Joe Mcdonald/Bruce Coleman, Inc. / 15.6 Kelvin Aitken/Peter Arnold, Inc. / 15.7 Raychel Ciemma; page 315 (right) Dr. Diana Wheeler / 15.8a Marsha J. Dohrmann, 15.8b Christine Oleksyk / page 317 (left) Tom E. Adams/Peter Arnold, Inc.; (right) Cathlyn Melloan/Tony Stone Images / page 318 (top) Richard Walters/Visuals Unlimited; 15.9 Marsha J. Dohrmann / 15.11 Scott Mittman / 15.12 A.L. Blum/Visuals Unlimited / 15.13 Marsha J. Dohrmann / page 323 SEMs by Robert S. Preston, courtesy of Dr. J.E. Hawkins, Kresge Hearing Research Institute / 15.16 (athletes) ©PhotoDisc.

Chapter 16
Page 331 Belinda Wright/DRK Photo / page 332 Richard Gross / page 336 Oliver Meckers/Photo Researchers, Inc. / 16.3 (art) Ronald J. Ervin, (photo) Ed Reschke / 16.4 Laurie O'Keefe/John Daugherty / 16.6 Nadine Sokol / page 343 Zigy Kaluzny/Tony Stone Images / 16.9 VideoSurgery/Photo Researchers, Inc. / page 346 ©PhotoDisc / 16.12a Barbara Cousins, 16.12b Don W. Fawcett/Visuals Unlimited / 16.14 Dr. Michael Klein/Peter Arnold, Inc.

Chapter 17
Page 357 Larry Williams/The Stock Market / page 358 ©PhotoDisc / page 359 Michael Lustbader/Photo Researchers, Inc. / page 365 (left) Bettina Cirone/Photo Researchers, Inc.; (right) American Journal of Medicine. 20 (1956). 133. / 17.8 Martin Rotker/Phototake / page 368 (left) Dr. P. Marazzi/Science Photo Library/Photo Researchers, Inc.; (right) Stephen J. Krasemann/Photo Researchers, Inc. / 17.12a Ed Reschke.

Part 4, BioIssues
Page 378 (lower left) Peter Menzel/Stock, Boston / page 379 Columbia Pictures (courtesy Kobal). Part 4 banner image ©PhotoDisc.

Chapter 18
Page 380 ©PhotoDisc / page 381 Art Wolfe/Tony Stone Images / page 382 (left) CNRI/Science Photo Library/Photo Researchers, Inc.; (right) ©PhotoDisc / 18.1 (top) Ada L Olins/Biological Photo Service, (bottom) Biofoto Associates/Science Source/Photo Researchers, Inc. / 18.4 Cold Springs Harbor Laboratory Archives/Photo Researchers, Inc. / 18.5b Cold Springs Harbor Laboratory Archives/Photo Researchers, Inc. / page 389 Courtesy, Cellmark Diagnostics, Inc., Germantown, Maryland / 18.A Lee Simon/Science Photo Library/Photo Researchers, Inc. / 18.12a E. Kifelva - D. Fawcett/Visuals Unlimited / page 398 Richard Gross/Biological Photography / 18.16 (all photos) M. Abbey/Photo Researchers, Inc. / 18.17 Dr. A.S. Bajer / 18.18 B.A. Palevitz and E.H. Newcomb/BPS/Tom Stack and Associates / 18.20 James Kezer/University of Oregon.

Chapter 19
Page 411 Miki Koren, courtesy of Sigma Chemical Co. / page 412 ©PhotoDisc / 19.2 John Durham/Photo Researchers, Inc. / page 414 ©PhotoDisc / 19.4 H. Potter - D. Dressler/Visuals Unlimited / 19.5 David P. Allison, Oak Ridge National Laboratory/Biological Photo Service / 19.6 Visuals Unlimited / 19.12 Wedgworth/Custom Medical Stock Photo / page 428 (top) S. Nagandra/Photo Researchers, Inc.; (bottom) Ted Thai/Time Magazine / page 430 ©PhotoDisc / 19.13 (1) Matt Meadows/Peter Arnold, Inc. / page 432 (bottom) Courtesy, Shiloh Creek Farm, Fresno, Ohio; 19.14 M. Baret/Science Source/Photo Researchers, Inc. / 19.15 Reprinted with permission from SCIENCE Magazine. Volume 280, May 22, 1998, Figure 2, page 1257 ©1999 American Association for the Advancement of Science. Photograph courtesy Dr. James M. Robl, University of Massachusetts.

Chapter 20
Page 439 Wes Thompson/The Stock Market / page 440 E. Rohne Rudder / 20.1 Jean-Michel Labat/Ardea London / 20.10 (both) Carolina Biological Supply Company/Phototake / page 451 Stewart Halperin.

Chapter 21
Page 457 The Royal Collection ©HM Queen Elizabeth II / 21.1 CNRI/Science Photo Library/Photo Researchers, Inc. / 21.2a Courtesy, Colorado Genetics Laboratory, University of Colorado Health Science Center, Denver, 21.2b J. Cancalosi/DRK Photo / 21.5 Dr. P. Marazzi/Science Photo Library/Photo Researchers, Inc. / 21.6 Earl Plunkett; from Valentine, 1986 / 21.10 (both) Cabisco/Visuals Unlimited / 21.13 Adam Hart-Davis / 21.15b Corbis/Bettman / page 471 (left) AP/Wide World Photos; (center) Corbis/Henry Diltz; (right) Stan Flegler/Visuals Unlimited / 21.16 (all photos) Stewart Halperin / 21.18 Med. Illus. SBHS/Tony Stone Images.

Chapter 22
Page 480 Lennart Nilsson, *Behold Man*, Little, Brown and Company / page 481 Yorgos Nikas/Tony Stone Images / 22.1a CNRI/SPL/Photo Researchers, Inc., 22.1b Richard Gross / page 483 (left) Carolina Biological Supply/Phototake; (right) Stuart Westmorland/Photo Researchers, Inc. / 484 (top left) Tom Adams/Visuals Unlimited; (bottom left) David Crews; 22.2 David Dennis/Tom Stack and Associates / page 485 (top) Gregory G. Dimijian/Photo Researchers, Inc.; (bottom) Ed Reschke / page 489 Lennart Nilsson, *Behold Man*, Little, Brown and Company / 22.10 Ed Reschke / 22.13a Scott Camazine & Sue Trainor/Photo Researchers, Inc., 22.13b-d Stewart Halperin; 22.14 Laura J. Edwards / page 500 (top) Stewart Halperin; (middle, bottom) Saturn Stills/Science Photo Library Photo Researchers, Inc. / page 502 Oliver Meckes/Photo Researchers, Inc.

Chapter 23
Page 508 Dr. Walter J. Gehring, Biozentrum University of Basel, Switzerland / page 509 Nestle/Petit Format/Photo Researchers, Inc. / 23.4a-b, page 513 (lower right) Lennart Nilsson/Albert Bonniers Publishing Company, *A Child Is Born*, Dell Publishing Co. / 23.5, 23.6a Lennart Nilsson *A Child Is Born* / page 518 David Young-Wolff/PhotoEdit / 23.12 Lennart Nilsson *A Child Is Born* / 23.14 Lennart Nilsson *A Child Is Born* / 23.15 Lennart Nilsson *A Child Is Born* / 23.16, 23.17 Lennart Nilsson *A Child Is Born* / 23.18 Lennart Nilsson *A Child Is Born*.

BioIssues, Part 5
Page 528 (petri dish) Courtesy of David J. Tenenbaum; (boll weevil) Courtesy of Winfield Sterling / page 529 (corn borer) Scott Camazine/Photo Researchers, Inc.; (flower) ©PhotoDisc. All background images ©PhotoDisc. Part 5 banner images ©PhotoDisc.

Chapter 24
Page 531 Tui De Roy/Minden Pictures / page 532 Science VU/Visuals Unlimited / 533 (left) Frank B. Gill/VIREO, (right) John S. Dunning/VIREO / 24.3 (both) Eric Sander / 24.4a Ralph A. Reinhold/Animals Animals, 24.4b Gerald Lacz/Animals Animals / 24.7 The National Portrait Gallery, London / page 539 Kathy Naylor / 24.8 Phil Degginger/Bruce Coleman, Inc.; 24.9 Richard Gross/Biological Photography / page 542 Stewart Halperin / 24.12 Don and Pat Valenti/Tom Stack & Associates; 24.13a Arthur Gloor/Earth Scenes, 24.13b George H.H. Huey/Earth Scenes / page 546 adapted from Sibley, C.G. and J.E. Ahlquist, 1984. The Phylogeny of Primites as Indicated by DNA-DNA Hydridization. *J. Mol. Evol.*, 20, 22-15.

Chapter 25
Page 555 NASA / 25.1b Dr. Stanley Miller / 25.2 Dudley Foster/Woods Hole Oceanographic Institute; (top) ©PhotoDisc / 25.4a M.R. Walter, MacQuarie University, Australia; 25.5 Andrew W. Knoll, Botanical Museum of Harvard University / 25.7a David Bruton, 25.7b Simon Conway Morris, University of Cambridge, 25.7c Simon Conway Morris, University of Cambridge / 25.8a Richard Gross/Biological Photography, 25.8b Frans Lanting/Minden Pictures / 25.9a Heather Angel Biofotos, 25.9b Courtesy of Smithsonian Institution / 25.12 Steve Martin/Tom Stack and Associates / page 565 (left) Gary Milburn/Tom Stack and Associates; 25.13 Bill Ober / 25.14a Barbara Lating/Black Star, 25.14b John D. Cunningham/Visuals Unlimited, 25.14c O. Louis Mazzatenta, National Geographic Society / page 567 Dell R. Foutz/Visuals Unlimited / page 567 Stouffer Productions Ltd/Animals, Animals / 25.17 Sidney Bahri/Photo Researchers, Inc.; 25.18a Stuart Westmorland/Tony Stone Images, 25.18b S.R. Maglione/Photo Researchers, Inc. / 25.19a Russell A. Mittermeier, 25.19b J.C. Carton/Bruce Coleman, Inc., 25.19c Anup Shah/DRK Photo, 25.19d K & K Ammann/Bruce Coleman, Inc. / 25.20 Tom McHugh/Photo Researchers, Inc. / 25.22 (both) Alan E. Mann / 25.24 John Reader; 25.25 (left) The Living World, St. Louis Zoo, St. Louis, MO, (right) John Reader / 25.26 after S.J. Gould (ed.) *The Book of Life: An Illustrated History of the Evolution of Life on Earth*. W.W. Norton (1993). / 25.27 Douglas Waugh/Peter Arnold, Inc. / 25.28 Roy King/Superstock.

BioIssues, Part 6
Page 588 (mouse) ©PhotoDisc; (monkey) A. Ramey/Stock, Boston. All background images ©PhotoDisc. Part 6 banner image ©PhotoDisc.

Chapter 26
Page 591 Science Source/Photo Researchers, Inc. / 26.1a K.G. Murti, Visuals Unlimited , 26.1b K. Namba and D.L.D Caspar / 26.3a Visuals Unlimited, 26.3b Carlyn Iverson / page 596 Stewart Halperin / 26.8 Biophoto Associates/Science Source/Photo Researchers, Inc. / 26.9a Biophoto Associates/Science Source/Photo Researchers, Inc. / 26.9b Science Source/Photo Researchers, Inc. / 26.11 Elizabeth Gentt/Visuals Unlimited / 26.12 Runk/Schoenberger/Grant Heilman/Photography, Inc.; 26.13 Jo Handelsman and Steven A. Vicen / page 605 Dr. E. Walker/SPL/Photo Researchers, Inc. / 26.14 Photo Researchers/CDC / 26.15 Ken Greer/Visuals Unlimited; 26.16a Biophoto Associates/Photo Researchers, Inc. / 26.16b Visuals Unlimited.

Chapter 27
Page 613 Craig Tuttle/The Stock Market /27.1 Stanley Erlandson; 27.2 M. (photo) Abbey/Visuals Unlimited, (art) Carlyn Iversen / 27.3b Higuchi Bioscience Laboratory / 27.4a Richard Gross/Biological Photography, 27.4b Manfred Kage/Peter Arnold, Inc.; 27.5 Diana Laulaien-Schein / 27.7 Biophoto Associates/Photo Researchers, Inc.; 27.8 Oliver Meckes/Photo Researchers, Inc. / 27.9 K.G. Murti/Visuals Unlimited / 27.11a-b Richard Gross/Biological Photography; 27.12 Hisashi Fujioka & Masamichi Aikawa (1999) *The Malaria Parasite and its Life Cycle.* / 27.14 Bill Bachman/Photo Researchers, Inc.; 27.15a David M. Phillips/Visuals Unlimited, 27.15b Richard Gross/Biological Photography / 27.16a Philip Sze/Visuals Unlimited, 27.16b Gregory Ochocki/Photo Reseachers, Inc. / 27.17a Biophoto Associates/Photo Researchers, Inc., 27.17b-c John D. Cunningham/Visuals Unlimited / 27.18a William C. Jorgensen/Visuals Unlimited, 27.18b Gary K. Robinson/Visuals Unlimited; 27.19 E.S. Ross / 27.20a-b Richard Gross/Biological Photography / 27.21a-b N.H. (Dan) Cheatham/DRK Photo; 27.22 Dwight Kuhn / 27.24a Stewart Halperin, 27.24b (both photos) Ed Reschke, (art) Carlyn Iversen / 27.25a Visuals Unlimited, 27.25b Carlyn Iversen; page 629 (top right) Gordon Langsbury/Bruce Coleman Ltd. / 27.26a Manfred Kage/Peter Arnold, Inc.; 27.27a David M. Phillips/Visuals Unlimited, 27.27b Dennis Kunkel/Phototake / 27.28 Ken Greer, Visuals Unlimited; 27.29a-c Richard Gross/Biological Photography / 27.30a-b Richard Gross/Biological Photography; 27.31 Carlyn Iversen / page 633 Stewart Halperin.

Chapter 28
Page 637 John D. Cunningham/Visuals Unlimited / 28.1 Courtesy James D. Mauseth / 28.3a Stewart Halperin, 28.3b Heather Angel/Biofotos / 28.5 Kirtley-Perkins/Visuals Unlimited; 28.6 Ken Davis/Tom Stack and Associates / page 644 (both) ©PhotoDisc / 28.10 Heather Angel/Biofotos; 28.11a Whit Bronough, 28.11b Michael and Patricia Fogden / 28.12a Ron Spomer/Visuals Unlimited, 28.12b D. Cavagnaro/Visuals Unlimited, 28.12c-d Stewart Halperin / 28.13a-b Carlyn Iversen / 28.14a Richard Gross/Biological Photography, 28.14f Whit Bronough / 28.15 Stewart Halperin / 28.16 Jack M. Bostrack/Visuals Unlimited / 28.17 E. Webber/Visuals Unlimited; 28.18 R. Lyons/Visuals Unlimited / page 653 Stewart Halperin; 28.19 John D. Cunningham/Visuals Unlimited.

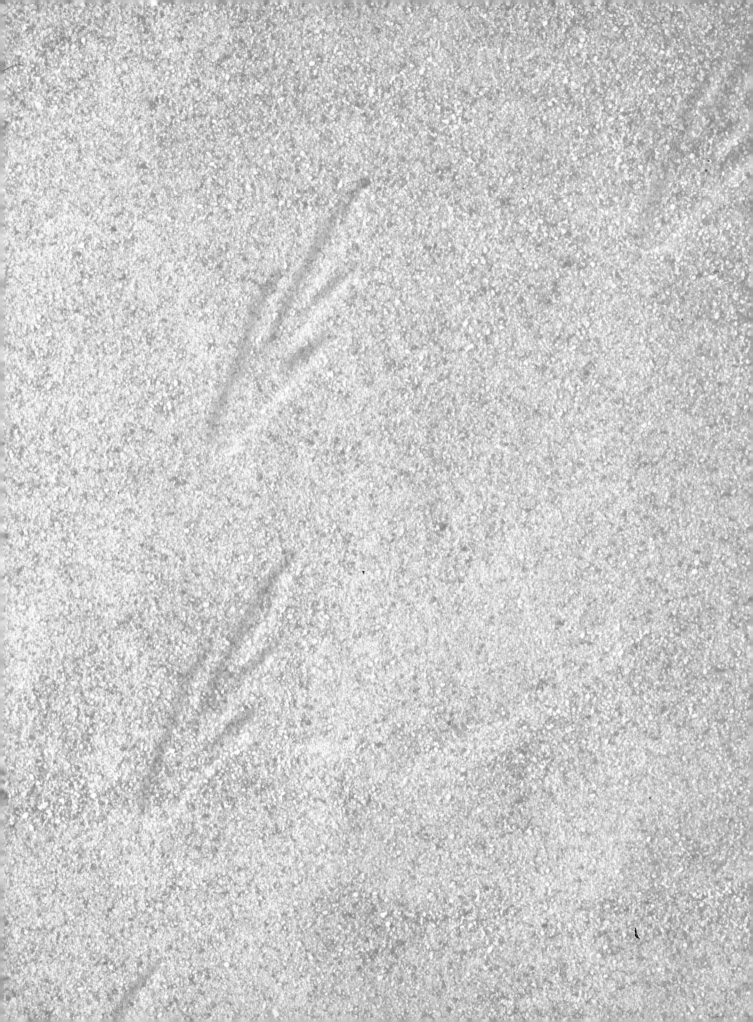